T0363919

Cancer Genomics

From Bench to Personalized Medicine

ELSEVIER *science & technology books*

Companion Web Site:

http://store.elsevier.com/product.jsp?&isbn=9780123969675

Cancer Genomics: From Bench to Personalized Medicine

Graham Dellaire, Jason N. Berman and Robert J. Arceci, Editors

Resources available:

- All figures from the book available as both Power Point slides and .jpeg files
- Links to web sites carefully chosen to supplement the content of the textbook
- Contact the editors with questions and/or suggestions

Cancer Genomics

From Bench to Personalized Medicine

Edited by

Graham Dellaire, Ph.D.

Departments of Pathology and Biochemistry &
Molecular Biology Dalhousie University, Halifax, Canada

Jason. N. Berman, M.D., FRCPC, FAAP

IWK Health Centre, Departments of Pediatrics,
Pathology and Microbiology and Immunology
Dalhousie University, Halifax, Canada

Robert J. Arceci, M.D., Ph.D.

The Ron Matricaria Institute of Molecular Medicine at
Phoenix Childrens Hospital, Department of Child Health,
University of Arizona, College of Medicine,
Phoenix, USA

AMSTERDAM • BOSTON • HEIDELBERG • LONDON
NEW YORK • OXFORD • PARIS • SAN DIEGO
SAN FRANCISCO • SINGAPORE • SYDNEY • TOKYO
Academic Press is an imprint of Elsevier

Academic Press is an imprint of Elsevier
32 Jamestown Road, London NW1 7BY, UK
225 Wyman Street, Waltham, MA 02451, USA
525 B Street, Suite 1800, San Diego, CA 92101-4495, USA

British Library Cataloguing-in-Publication Data
A catalogue record for this book is available from the British Library

Library of Congress Cataloging-in-Publication Data
A catalog record for this book is available from the Library of Congress

ISBN : 978-0-12-396967-5

For information on all Academic Press publications
visit our website at www.store.elsevier.com

Typeset by MPS Limited, Chennai, India
www.adi-mps.com

Contents

List of Contributors

Kenneth C. Anderson Department of Medical Oncology, Jerome Lipper Multiple Myeloma Center, Dana-Farber Cancer Institute and Department of Medicine, Harvard Medical School, Boston, MA, USA

Samuel A. Aparicio British Columbia Cancer Agency, Vancouver, BC, Canada; Department of Pathology and Laboratory Medicine, University of British Columbia, Vancouver, BC, Canada; Molecular Oncology, British Columbia Cancer Research Centre, Vancouver, BC, Canada

Robert J. Arceci Children's Center for Cancer and Blood Disorders, Hematology/Oncology and The Ron Matricaria Institute of Molecular Medicine, Phoenix Children's Hospital, Department of Child Health, University of Arizona, College of Medicine, Phoenix, AZ, USA

Alejandro Berlin Departments of Radiation Oncology and Medical Biophysics, University of Toronto, Toronto, ON, Canada; Ontario Cancer Institute and Princess Margaret Cancer Center (University Health Network), Toronto, ON, Canada

Jason N. Berman Departments of Pediatrics, Microbiology and Immunology, and Pathology, Dalhousie University and IWK Health Centre, Halifax, NS, Canada

Maria Grazia Borrello Molecular Mechanisms Unit, Department of Experimental Oncology and Molecular Medicine, Fondazione IRCCS Istituto Nazionale dei Tumori, Milan, Italy

Robert G. Bristow Ontario Cancer Institute and Princess Margaret Cancer Center (University Health Network), Toronto, ON, Canada; Departments of Radiation Oncology and Medical Biophysics, University of Toronto, Toronto, ON, Canada

Moamen Bydoun Department of Pathology, Dalhousie University, Halifax, NS, Canada

George A. Calin Department of Experimental Therapeutics and Leukemia, The University of Texas MD Anderson Cancer Center, Houston, TX, USA

Priscilla P.L. Chiu Department of Surgery, The Hospital for Sick Children, Toronto, ON, Canada

Hye-Jung E. Chun British Columbia Cancer Agency, Vancouver, BC, Canada; Canada's Michael Smith Genome Sciences Centre, Vancouver, BC, Canada

Francesca Cottini Department of Medical Oncology, Jerome Lipper Multiple Myeloma Center, Dana-Farber Cancer Institute and Department of Medicine, Harvard Medical School, Boston, MA, USA

Janet E. Dancey Ontario Institute for Cancer Research, Toronto, ON; NCIC Clinical Trials Group, Queen's University, Kingston, ON, Canada

Graham Dellaire Departments of Pathology and Biochemistry & Molecular Biology, Dalhousie University, Halifax, NS, Canada

Amal M. EL-Naggar Department of Molecular Oncology, BC Cancer Research Center, Vancouver, BC; Department of Pathology and Laboratory Medicine, University of British Columbia, Vancouver, BC, Canada

Conrad V. Fernandez Departments of Pediatrics and Bioethics, Head, Division of Pediatric Hematology/Oncology, Dalhousie University and IWK Health Centre, Halifax, NS, Canada

Michael Fraser Ontario Cancer Institute and Princess Margaret Cancer Center (University Health Network), Toronto, ON, Canada

Daniel Gaston Department of Pathology, Dalhousie University, Halifax, NS, Canada

Carman Giacomantonio Departments of Pathology and Surgery, Dalhousie University, Halifax, NS, Canada

Angela Greco Molecular Mechanisms Unit, Department of Experimental Oncology and Molecular Medicine, Fondazione IRCCS Istituto Nazionale dei Tumori, Milan, Italy

Erin Hedlund Computational Biology Department, St Jude Children's Research Hospital Memphis, TN, USA

Matthew Herder Health Law Institute, Faculties of Medicine and Law, Dalhousie University, Halifax, NS, Canada

Peter Hohenstein The Roslin Institute, University of Edinburgh, Easter Bush Campus, Midlothian, UK

Meredith S. Irwin Department of Paediatrics, Hospital for Sick Children and University of Toronto, Toronto, ON, Canada

Steven J.M. Jones Canada's Michael Smith Genome Sciences Centre, British Columbia Cancer Agency, Vancouver, BC, Canada

Katayoon Kasaian Canada's Michael Smith Genome Sciences Centre, British Columbia Cancer Agency, Vancouver, BC, Canada

Javed Khan Oncogenomics Section, Pediatric Oncology Branch, Advanced Technology Center, National Cancer Institute, Bethesda, MD, USA

Jaswinder Khattra British Columbia Cancer Agency, Vancouver, BC, Canada; Department of Pathology and Laboratory Medicine, University of British Columbia, Vancouver, BC, Canada; Molecular Oncology, British Columbia Cancer Research Centre, Vancouver, BC, Canada

Martin Krzywinski British Columbia Cancer Agency, Vancouver, BC, Canada; Canada's Michael Smith Genome Sciences Centre, Vancouver, BC, Canada

Cécile Le Page Centre hospitalier de l'Université de Montréal, Institut du Cancer de Montréal, Montreal, QC, Canada

Gabriel Leprivier Department of Molecular Oncology, BC Cancer Research Center, Vancouver, BC; Department of Pathology and Laboratory Medicine, University of British Columbia, Vancouver, BC, Canada

Yvonne Y. Li Canada's Michael Smith Genome Sciences Centre, British Columbia Cancer Agency, Vancouver, BC, Canada

Hui Ling Department of Experimental Therapeutics and Leukemia, The University of Texas MD Anderson Cancer Center, Houston, TX, USA

Anthony M. Magliocco H. Lee Moffitt Cancer Center and Research Institute, Tampa, FL, USA

David Malkin Division of Hematology/Oncology, The Hospital for Sick Children and Department of Pediatrics, University of Toronto, Toronto, ON, Canada

Branavan Manoranjan McMaster Stem Cell and Cancer Research Institute, McMaster University, Hamilton, ON, Canada; Michael G. DeGroote School of Medicine, McMaster University, Hamilton, ON, Canada

Paola Marcato Department of Pathology, Dalhousie University, Halifax, NS, Canada

Marco A. Marra British Columbia Cancer Agency, Vancouver, BC, Canada; Canada's Michael Smith Genome Sciences Centre, Vancouver, BC, Canada; Department of Medical Genetics, University of British Columbia, Vancouver, BC, Canada

Anne-Marie Mes-Masson Université de Montréal, Department of Medicine, Montreal, QC, Canada; Centre hospitalier de l'Université de Montréal, Institut du Cancer de Montréal, Montreal, QC, Canada

Soheil Meshinchi Clinical Research Division, Fred Hutchinson Cancer Research Center, Department of Pediatrics, University of Washington School of Medicine, Seattle, WA, USA

Claudia Miranda Molecular Mechanisms Unit, Department of Experimental Oncology and Molecular Medicine, Fondazione IRCCS Istituto Nazionale dei Tumori, Milan, Italy

Stephen B. Montgomery Departments of Pathology and Genetics, Stanford University, Stanford, CA, USA

Daniel A. Morgenstern Department of Paediatrics, Hospital for Sick Children and University of Toronto, Toronto, ON, Canada

Ryan D. Morin Department of Molecular Biology and Biochemistry, Simon Fraser University, Burnaby, BC, Canada

Veronique Ouellet Centre de Recherche du Centre hospitalier de l'Université de Montréal, Montreal, QC, Canada

Louise Parker Departments of Medicine and Pediatrics, Dalhousie University, Halifax, NS, Canada

Matthew Parker Computational Biology Department, St Jude Children's Research Hospital Memphis, TN, USA

Elizabeth J. Perlman Ann & Robert H. Lurie Children's Hospital of Chicago, Northwestern University Feinberg School of Medicine, Chicago, IL, USA

Marco A. Pierotti Scientific Directorate, Fondazione IRCCS Istituto Nazionale dei Tumori, Milan, Italy

Shahrad Rod Rassekh Child and Family Research Institute, British Columbia's Children's Hospital, Vancouver, BC, Canada; Department of Pediatrics, Division of Pediatric Hematology/Oncology/BMT, University of British Columbia, Vancouver, BC, Canada

Colin J.D. Ross Child and Family Research Institute, British Columbia's Children's Hospital, Vancouver, BC, Canada; Department of Pediatrics, Division of

Translational Therapeutics, University of British Columbia, Vancouver, BC, Canada; Centre for Molecular Medicine and Therapeutics, University of British Columbia, Vancouver, BC, Canada

Fred Saad Department of Surgery, Centre hospitalier de l'Université de Montréal, Montreal, QC, Canada; Montreal Cancer Institute, Montreal, QC, Canada

Sheila K. Singh McMaster Stem Cell and Cancer Research Institute, McMaster University, Hamilton, ON, Canada; Michael G. DeGroote School of Medicine, McMaster University, Hamilton, ON, Canada; Surgery, Faculty of Health Sciences, McMaster University, Hamilton, ON, Canada

Poul H. Sorensen Department of Molecular Oncology, BC Cancer Research Center, Vancouver, BC; Department of Pathology and Laboratory Medicine, University of British Columbia, Vancouver, BC, Canada

Giovanni Tonon Division of Molecular Oncology, Functional Genomics of Cancer Unit, San Raffaele Scientific Institute, Milan, Italy

Chitra Venugopal McMaster Stem Cell and Cancer Research Institute, McMaster University, Hamilton, ON, Canada

Parvez Vora McMaster Stem Cell and Cancer Research Institute, McMaster University, Hamilton, ON, Canada

Jun S. Wei Oncogenomics Section, Pediatric Oncology Branch, Advanced Technology Center, National Cancer Institute, Bethesda, MD, USA

Zhaolin Xu Department of Pathology, Dalhousie University and Queen Elizabeth II Health Sciences Centre, Halifax, NS, Canada

Jinghui Zhang Computational Biology Department, St Jude Children's Research Hospital Memphis, TN, USA

Shile Zhang Oncogenomics Section, Pediatric Oncology Branch, Advanced Technology Center, National Cancer Institute, Bethesda, MD, USA; Program in Bioinformatics, Boston University, Boston, MA, USA

Part 1

Introduction

Cancer Genomics

Historical Perspective and Current Challenges of Cancer Genomics

Graham Dellaire[1]
Robert J. Arceci[2]

[1]Departments of Pathology and Biochemistry & Molecular Biology, Dalhousie University, Halifax, NS, Canada
[2]Children's Center for Cancer and Blood Disorders, Hematology/Oncology and The Ron Matricaria Institute of Molecular Medicine, Phoenix Children's Hospital, Department of Child Health, University of Arizona, College of Medicine, Phoenix, AZ, USA

Contents

Cancer Genomics. DOI: http://dx.doi.org/10.1016/B978-0-12-396967-5.00001-3

Key Concepts

- Advances in nucleic acid sequencing technology have had a profound impact on the field of cancer genomics and have enabled the interrogation of the genetic basis of cancer at the single nucleotide level
- Cancer genomics has provided a detailed view of the complexity of the cancer genome, including the extraordinary ability to sustain and thrive on alterations of DNA
- Translating cancer genomics into clinically real time and actionable personalized medicine is beginning to be tested, although significant advances in the speed of providing molecular data, analysis, utilizing combination targeted agents and understanding clinical response or no response will require new generations of technology and bioinformatic tools

A HISTORICAL PERSPECTIVE ON THE DEVELOPMENT OF CANCER GENOMICS

Several hundred years BC, Hippocrates is attributed with providing us with the term "carcinoma" and thus "cancer", originating possibly from the image of finger-like extensions (veins) from a tumorous (main body of the tumor) breast lesion that shared resemblance to the shape of a crab. Around 400 years later, the Roman physician Celsus translated the Greek (*karkinos*) into the Latin word for crab, which led to the term "cancer" [1,2]. A relative late-comer to this nomenclature narrative, in 168 BC, Galen introduced the terms "*oncos*", meaning "swelling" to describe tumors, leading to the term defining the field of oncology [3,4].

The next 2000 years witnessed several key events that helped to refine further the still ongoing main areas of cancer investigation and treatment. Maimonides in AD 1190 appears to have been the first to document surgically removing tumors [3]. The recognition of cancer clustering in distinct populations was introduced in 1713 with Razmazzini's observation of the low cervical but high breast cancer incidence in nuns [2]. The observation that environmental and occupational exposures can be associated with increased incidence of specific cancers also became evident. In the first half of the 1800s, Recmier appears to have reignited the flare for nomenclature by writing about "metastasis" in 1829 to describe the movement of some cancers to different parts of the body [3]. Muller's notes on the cellular origin of cancer also in 1838, and Paget's subsequent "seed and soil" hypothesis over 50 years later in 1889, established the cognitive paradigm for the cell biological basis for cancer and the concept of microenvironmental niches [2,3]. The first half of the 20th century was ushered in by a set of remarkable observations in cancer biology that were made before the discovery of DNA. These included the theories of Rous regarding the potential viral origin of some cancers in 1910, derived from his work on avian sarcomas, and the concept of the somatic mutation theory of cancer by Boveri in 1914 that stemmed from his work on polyspermic development in invertebrates [5,6].

While not necessarily providing a deeper understanding of the mechanisms of carcinogenesis, the first half of the 20th century in many ways broke open the gates of cancer treatment. After being commissioned by the US government to understand the physiological consequences of nitrogen mustard gas used in warfare, Louis Goodman and Alfred Gilman recognized the key bone marrow toxicity of this agent and subsequently introduced its intravenous use for the treatment of lymphoid malignancies in 1946 [7]. Soon afterward in 1948, the antimetabolite aminopterin was used to treat several children with acute lymphoblastic leukemia by Farber and colleagues, a treatment built on the work of the chemist Subbarao [8]. A decade later, in 1958, Hertz, Li and colleagues reported the first cure of a metastatic tumor, namely a gestational-related choriocarcinoma, with another antimetabolite, methotrexate [9].

However, despite such encouraging forays into treating patients, few cures were achievable with surgery, radiation therapy and chemotherapy. In this regard, the extraordinary efforts of Ms Mary Lasker following her husband's death from cancer should not go unmentioned. Through her efforts and the Citizens Committee for the Conquest of Cancer through the 1960s, they challenged government, physicians and scientists to push forth with a "War on Cancer" [10–12]. And this was in spite of a significant number of naysayers who had concluded in various publications that we knew enough to cure cancer and all that was needed was to translate the knowledge that was available at the time. Such a lack of vision was thankfully thwarted by those who propitiously concluded that only through scientific discovery and its ongoing application would improvements in cancer outcomes occur. In 1971, the US National Cancer Act was passed by Congress and then President Nixon signed it into law within 2 weeks, an astonishingly rapid accomplishment on the part of government and one that should inform current, often stalled efforts [10–13]. The consequences of the above investment, along with other efforts across the globe [14], led to an infusion of intellectual engagement and financial support for conquering cancer. The results included the establishment of clinical trial groups, comprehensive cancer centers, an explosion of new anticancer agents from the lab and from nature, and the beginning of work focused on the biological understanding of cancer. This latter work built of course on the model and profound implications of the seminal discovery of the structure of the DNA double helix by Watson and Crick in 1953 [15], a discovery that would earn them the Nobel Prize in Physiology or Medicine in 1962, an accolade they shared with colleague Maurice Wilkins. In many ways, the journey to our present age of high-throughput and genome-wide discoveries in cancer biology began with this fundamental description of the fabric of life and, as such, this discovery makes a suitable origin point from which to chronicle the key events in cancer genomics in the last 60 years (Figure 1.1).

With the structure of the molecule of heredity in hand, the latter half of the 20th century saw a number of major contributions to our understanding of the biochemical and genetic underpinnings of cancer. These contributions included the identification of the "Philadelphia chromosome" as a genetic marker of chronic myelogenous leukemia (CML) by Nowell and Hungerford in 1960, and the subsequent identification of chromosomes 9 and 22 as the translocation partners underlying this anomalous chromosome by Rowley in 1973 [16,17]; the identification of the first cellular proto-oncogene, *SRC*, by Varmus and Bishop in the 1976 [18] leading to the realization that cellular genes could become deregulated resulting in tumorigenesis; and the identification of the p53 protein in 1979 as the primary molecular target underlying transformation by the DNA tumor virus, simian virus 40 (SV40) [18–20]. Empowered by the molecular tools developed for the study of tumor viruses, scientists studying cancer made many more seminal discoveries in the 1980s and early 1990s, including the identification of several tumor suppressor genes including *retinoblastoma* (*RB*) [21,22], the gene encoding p53 (*TP53*) [18–20,23,24] and *adenomatous polyposis coli* (*APC*) gene [25–27]. All these discoveries were foreshadowed by Knudson's "two hit" hypothesis of tumorigenesis and his pioneering epidemiological studies of retinoblastoma in 1971 [28], which laid the conceptual framework for how loss of heterozygosity (LOH) of a tumor suppressor gene contributes to cancer development. It was during the 1980s that non-genetic mechanisms of oncogene regulation were first identified. One such epigenetic mechanism of gene regulation was the loss of cytosine nucleotide methylation in CpG doublets, known as hypomethylation, which was first demonstrated by Feinberg and Vogelstein in 1983 and later shown to regulate the expression of oncogenes such as *HRAS* [29]. Soon after, in 1986, it would be demonstrated by Baylin et al. that increased CpG methylation, termed hypermethylation, also occurred in cancer cells by studying the methylation pattern of the *calcitonin* gene [30], and, in 1989, Horsthemke and colleagues demonstrated hypermethylation of the *RB* tumor suppressor gene in retinoblastoma [31].

Through the 1990s and into the twenty-first century further groundbreaking work would identify key genes underlying hereditary susceptibilities to breast cancer (e.g. *BRCA1*) [32] and colon cancer (e.g. *MSH2*) [33,34], the identification of the telomerase gene and demonstration of its role in subverting senescence [35–38], and the discovery of cancer stem cells in leukemias and solid tumors [39–41].

The sum total of an organism's genetic information was first referred to as a genome by Hans Winkler in 1920, who used the term *genom* to describe the haploid chromosomal set of an organism [42]. The term "genomics" would later be coined in 1986 by Thomas H. Roderick, a geneticist at the Jackson Laboratory in Bar Harbor, Maine [42] and refers to the study of an organism's entire complement of genetic information. The history of genomics and the application of genomics to cancer biology is a story of technological advances and, arguably, DNA sequencing technologies have had the greatest impact and form the foundation of cancer genomics. These advances began in 1977 with the development of chemical sequencing of DNA by Maxam and Gilbert and the dideoxy nucleotide method of sequencing by Sanger and Coulson (reviewed in [43]). The Sanger technique in particular would go on to dominate DNA sequencing for nearly three decades before the development of massively parallel approaches to sequencing such as pyrosequencing (e.g. 454 sequencing), reverse-termination (e.g. Illumina sequencing), sequencing by ligation (e.g. massively parallel signature sequencing (MPSS)), polony sequencing, and single molecule sequencing (reviewed in [44]). Another transformative technology was the development of DNA microarrays in 1995 [45]. Although DNA microarrays have been largely supplanted by next-generation sequencing and approaches such as RNA-sequencing (RNA-seq) developed in 2008 (reviewed in [46]), they were, and remain, a relatively cost-effective platform for the profiling of gene expression and genotyping. In addition, the development of DNA microarrays led to important advances in our understanding of copy number variation in cancer by dramatically increasing the resolution by which chromosomal changes could be observed when compared to predecessor technologies such as comparative genome hybridization (CGH) [47,48]. The emergence of these novel technologies was leveraged to provide novel insights into the driver mechanisms by which normal cells are transformed into cancer.

In the late 1990s, the Cancer Genome Anatomy Project (CGAP) [49] was initiated on the shoulders of The Human Genome Project that had begun in the early 1990s and completed in 2003, leading to the identification of a "standard" set of approximately 25 000 human genes [50]. In 2006, the National Cancer Institute (NCI) launched The Cancer Genome Atlas (TCGA), first as a pilot and then as a full project in 2009, which was rapidly

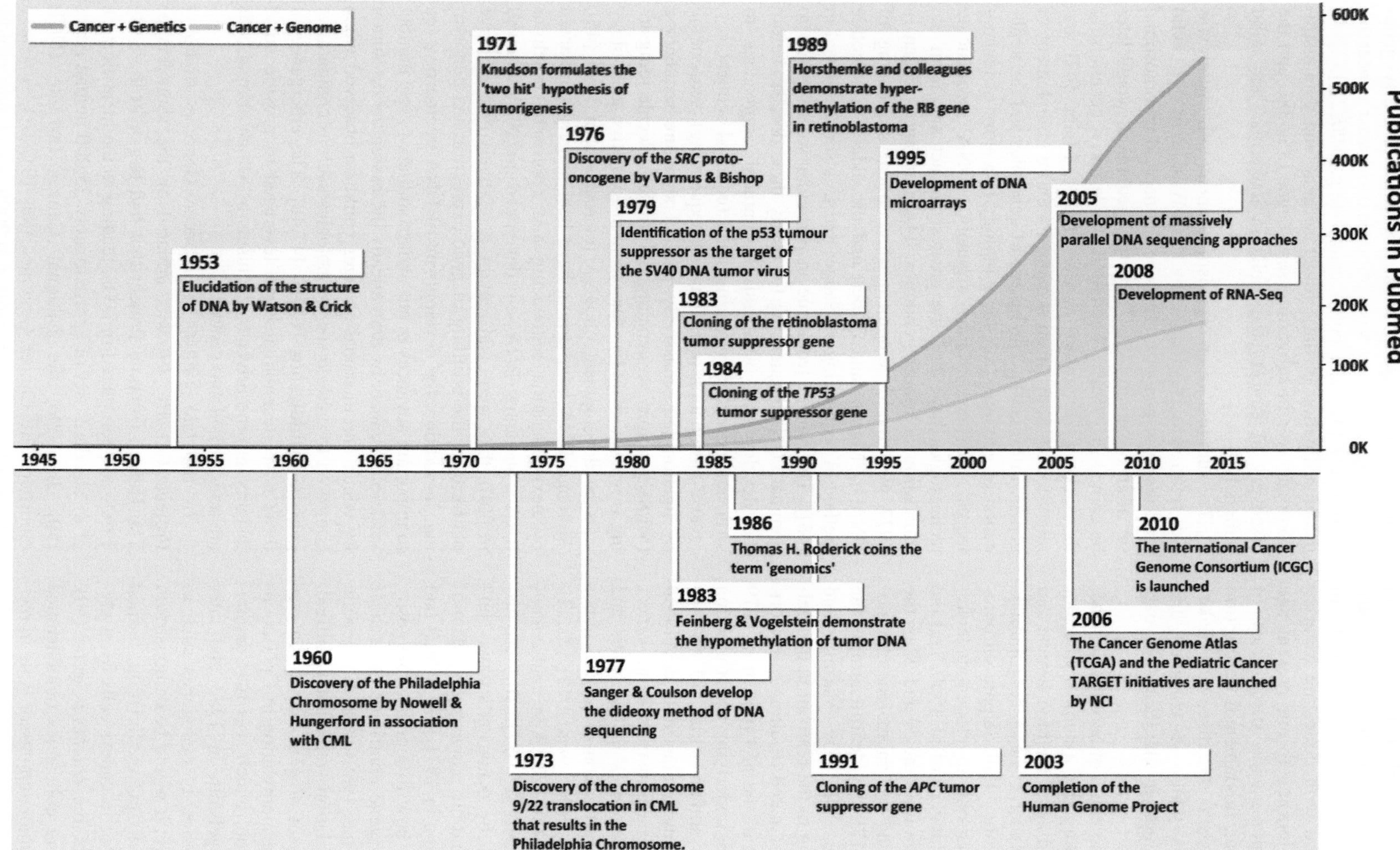

FIGURE 1.1 Historical milestones in cancer genomics. Key milestones in the field of cancer genomics are depicted starting with the elucidation of the structure of DNA by Watson and Crick in 1953. These milestones are depicted over a line graph of the total number of publications listed in the Pubmed database of the National Center for Biotechnology Information (NCBI) with the keywords "Cancer + (Genetics or Gene)" (in blue), or "Cancer + (Genomics or Genome)" (in green) from 1945 to 2013.

followed by the formation of the International Cancer Genome Consortium (ICGC) in 2010 that set a goal of characterizing 25 000 cancer genomes in 50 different cancer types [51]. More recent efforts of NCI have led to the creation of the Cancer Genome Characterization Initiative (CGCI), the Cancer Target Discovery and Development (CTD^2) Program, and a pediatric cancer version of TCGA that links clinical treatment and outcomes with detailed genomic analysis called the Therapeutically Applicable Research to Generate Effective Treatments (TARGET) program (http://ocg.cancer.gov/).

In the wake of the development of massively parallel sequencing technologies and the launch of large-scale cancer genome projects such as TCGA and the ICGC, there have been a number of important observations in the genetic basis of cancer development and progression that may have never been uncovered without high-throughput and integrated genomics approaches (reviewed by Garraway and Lander [52]). These observations include the discovery of distinctive chromosome shattering events termed "chromothripsis", characterized first in chronic lymphocytic leukemia (CLL) but highly prevalent in bone cancer and occurring in up to 3% of all cancers [53]; the discovery of chromosomal chains referred to as "chromoplexy" in prostate cancer genomes [54,55]; the identification of a novel mechanism of highly localized hypermutation in a single chromosomal region termed "kataegis" to describe the shower of mutations associated with DNA deaminases-induced breaks in breast cancer genomes [56–58]; and the demonstration of a high degree of previously unrecognized tumor heterogeneity between both cancers of the same tissue and among the clonal populations of cells within a single individual's tumor [57,59–61]. The latter discovery may be the most profound in that it surely darkens the prospect of successful personalized therapy if strategies are not developed to tackle tumor heterogeneity.

CURRENT AND FUTURE CHALLENGES

Building on such discoveries, the 21st century has seen and will continue to see the exploitation of cancer genomic and transcriptomic sequencing efforts for the development of mutation directed, targeted therapies, such as imatinib mesylate, a tyrosine kinase inhibitor initially developed to inhibit the growth of leukemia cells carrying the BCR/Abl fusion gene product encoded by the Philadelphia chromosome in chronic myelogenous leukemia [62]. Subsequently, a plethora of other targeted agents have been developed that would make Paul Ehrlich, the originator of the concept of "Magic Bullets" to treat cancer, a proud progenitor [63]. In addition, there has been an expansion in tumor selective immunologically based therapies using monoclonal antibodies, such as those targeting the HER2 receptor in breast cancer and the epidermal growth factor receptor (EGFR) in lung and other solid tumors [64], built on the work of Kohler and Milstein in the 1970s [65], Cooley, Steinman [66,67] and others (reviewed by [66]) [67,68], who collectively built the scientific foundation from which the promise of immune cell-based therapies and cancer vaccines has been fostered [69].

Sixty years since the discovery of the structure of DNA, all the progress we have described in our understanding of cancer biology now seems like it is just a prelude to the current era of cancer genomics and personalized medicine. The extraordinarily rapid advances in nucleic acid and protein sequencing have led us to the brink of generating more data than the human brain can usefully manage. This has in turn led to the development of increasingly sophisticated analytical tools. The added complexities of the role of epigenetic regulation, compensatory cell signaling pathways and the profoundly complex universe of the non-coding "Dark Matter" portion of the genome, have both informed and humbled most investigators and clinicians [52].

The purpose of this book is to provide a snapshot of where cancer "omics" and our understanding of the genetic basis of cancer stand today. But we should also reflect on what the key challenges are in translating cancer genomics into meaningful outcomes for cancer patients. While predicting the future is usually futile, there are several observations that seem on solid ground in terms of next steps to applying "omics" to the effective treatment of patients with cancer as well as providing for earlier detection and predisposition. Although DNA, RNA and protein sequencing are becoming more rapid and accurate, the ability to provide complete sequence coverage, integration for pathway analysis and functional assessment is not yet a reality in clinically actionable time frames for most patients. There are limited numbers of gene mutation and expression-based diagnostics that can certainly be turned around in short order. However, while such information can frequently provide information that can be used to offer novel treatments to patients, the contextual basis to understand treatment responses and, importantly, lack of response, is lacking with such limited approaches.

The "real time" ability to provide sufficiently detailed information for informed molecularly guided therapy still requires the development of more rapid, less expensive and integrated sequencing, functional and analytical assessment of a patient's cancer (including the clonal complexity), as well as their own germline and its associated idiosyncrasies of drug metabolism and sensitivity [70–72]. This further emphasizes the need for physicians to be trained in not only the art of medicine, but also for the ability to act on integrated "omic" information and interact with multidisciplinary teams of not just other medical specialties, but those skilled in the technical and bioinformatic aspects of disease analysis. In addition, the ability to understand contextual responses is likely to require new approaches to computer learning that can be applied to the cancer problem [73]. The promise of *in silico* testing of new treatments that has a high level of clinical predictability may eventually pre-empt the need for larger scale testing in humans, an expensive and often risky business at best.

Our goals (and abilities) to (1) cure everyone, (2) not hurt them while doing it, and (3) do so without spending a disproportionate amount of money, should emerge as guiding principles from these beginning steps into applying genomics to cancer diagnosis and treatment. There is no lack of creativity and passion on the part of those committed to these goals. However, there are real and potential obstacles that represent "real enemies of success", as Cyrano de Bergerac's last words make evident. There is the "Falsehood" that our knowledge of cancer is sufficient to cure this disease, which we must overcome by recognizing with humility that there is still much to be learned; there is "Compromise" that is made in choosing treatments for cancer, which must be balanced by striving for better therapies and better outcomes; there is "Prejudice" that limits our ability to think outside of our own boxes; and there is "Treachery", whether intended or not, clothed in the form of overregulation, underfinancing, and overbearing risk aversion. We should not lapse into thinking that treatments have killed more people than cancer.

While not nearly a completed story of our understanding of cancer, this text and the incredible work of investigators and the profound generosity of patients, who contribute to a goal beyond themselves by taking part in clinical trials, will hopefully continue to drive us forward towards the goal of eradicating the burden of cancer. In this regard, cancer genomics represents both the latest toolkit for the characterization of malignancies and the next step in the evolution of our understanding of the mechanisms of cancer development and progression which, ultimately, will be used to develop safer and more effective cancer therapies.

GLOSSARY

Chromoplexy From the Greek *plexy*, meaning weave or braid, this term refers to large chains of rearrangements of chromosomes than can affect multiple chromosomes.

Chromothripsis From the Greek *thripsis*, meaning "shattering", this term refers to chromosome shattering with subsequent multiple rearrangements.

DNA methylation Addition of a methyl group to the cytosine base of DNA. The methylation status of a regulatory DNA sequence can silence or promote the expression of genes. DNA hypomethylation and hypermethylation refer to reduced or increased DNA methylation, respectively.

DNA microarrays High-throughput gene expression quantification technology based on DNA probe-target hybridization (of RNA or DNA) and subsequent fluorescence detection. Also referred to as gene chips.

Epigenetic A term used to describe the regulation of gene expression by mechanisms that modify DNA but do not change the sequence of the gene. These mechanisms include the methylation of DNA as well as the post-translational modification of histones by a host of events including phosphorylation, methylation, acetylation and ubiquitination. Although epigenetic changes are dynamic they are also heritable and can persist during cell division and be transmitted to offspring.

Kataegis A Greek word meaning "thunder" or "thunderstorm", this term refers to patterns of localized hypermutation.

Massively parallel sequencing A term used to refer collectively to high-throughput DNA sequencing approaches that employ miniaturized and highly parallel platforms to sequence millions of short sequence reads of usually 50–400 nucleotides. These techniques are also referred to as next-generation sequencing (NGS) or second-generation sequencing approaches.

Philadelphia chromosome A chromosomal abnormality created by the translocation of human chromosomes 9 and 22 (i.e. t(9;22)(q34;q11)) that is associated with chronic myelogenous leukemia (CML). The translocation fuses the BCR and ABL kinase genes resulting in the expression of an oncogenic fusion protein that can be targeted by the tyrosine kinase inhibitor imatinib mesylate.

RNA-sequencing (RNA-seq) High-throughput (or next-generation) sequencing of a sample's cDNA to characterize its transcriptome.

ABBREVIATIONS

CGAP Cancer Genome Anatomy Project
CGCI Cancer Genome Characterization Initiative
CGH Comparative genomic hybridization
CLL Chronic lymphocytic leukemia
CML Chronic myelogenous leukemia
CTD^2 Cancer Target Discovery and Development program
ICGC International Cancer Genome Consortium
LOH Loss of heterozygosity
MPSS Massively parallel signature sequencing
TCGA The Cancer Genome Atlas
TARGET Therapeutically Applicable Research to Generate Effective Treatments

REFERENCES

[1] Castiglioni A. A History of medicine. New York: Alfred A. Knope; 1941.

[2] American Cancer Society. The History of Cancer. Available from: <http://www.cancer.org/acs/groups/cid/documents/webcontent/002048-pdf.pdf>; 2012 [accessed June 2013].

[3] Morton LTM. A chronology of medicine and related sciences. Aldershot, England: Scholar Press; 1997.

[4] Gurunluoglu R, Gurunluoglu A. Paul of Aegina: landmark in surgical progress. World J Surg 2003;27(1):18–25.

[5] Manchester KL. Theodor Boveri and the origin of malignant tumours. Trends Cell Biol 1995;5:384–7.

[6] Rous P. A sarcoma of the fowl transmissible by an agent separable from the tumor cells. J Exp Med 1911;13:397–411.

[7] Goodman LS, Wintrobe MM, Dameshek W, Goodman MJ, Gilman A, McLennan MT. Nitrogen mustard therapy; use of methyl-bis (beta-chloroethyl) amine hydrochloride and tris (beta-chloroethyl) amine hydrochloride for Hodgkin's disease, lymphosarcoma, leukemia and certain allied and miscellaneous disorders. J Am Med Assoc 1946;132:126–32.

[8] Farber S, Diamond LK. Temporary remissions in acute leukemia in children produced by folic acid antagonist, 4-aminopteroyl-glutamic acid. N Engl J Med 1948;238:787–93.

[9] Hertz R, Bergenstal DM, Lipsett MB, Price EB, Hilbish TF. Chemotherapy of choriocarcinoma and related trophoblastic tumors in women. J Am Med Assoc 1958;168:845–54.

[10] DeVita Jr VT, Chu E. A history of cancer chemotherapy. Cancer Res 2008;68:8643–53.

[11] Chabner BA, Roberts Jr TG. Timeline: chemotherapy and the war on cancer. Nat Rev Cancer 2005;5:65–72.

[12] Gaudilliere JP. Essay review: cancer and science: the hundred years war. J Hist Biol 1998;31:279–88.

[13] Cairns J. The evolution of cancer research. Cancer Cells 1989;1:1–8.

[14] Pinell P, Brossat S. The birth of cancer policies in France. Sociol Health Illn 1988;10:579–607.

[15] Watson JD, Crick FH. Molecular structure of nucleic acids; a structure for deoxyribose nucleic acid. Nature 1953;171:737–8.

[16] Nowell PC, Hungerford DA. Chromosome studies on normal and leukemic human leukocytes. J Natl Cancer Inst 1960;25:85–109.

[17] Rowley JD. Letter: A new consistent chromosomal abnormality in chronic myelogenous leukaemia identified by quinacrine fluorescence and Giemsa staining. Nature 1973;243:290–3.

[18] Stehelin D, Varmus HE, Bishop JM, Vogt PK. DNA related to the transforming gene(s) of avian sarcoma viruses is present in normal avian DNA. Nature 1976;260:170–3.

[19] Linzer DI, Maltzman W, Levine AJ. The SV40 A gene product is required for the production of a 54,000 MW cellular tumor antigen. Virology 1979;98:308–18.

[20] Lane DP, Crawford LV. T antigen is bound to a host protein in SV40-transformed cells. Nature 1979;278:261–3.

[21] Cavenee WK, Dryja TP, Phillips RA, Benedict WF, Godbout R, Gallie BL, et al. Expression of recessive alleles by chromosomal mechanisms in retinoblastoma. Nature 1983;305:779–84.

[22] Friend SH, Bernards R, Rogelj S, Weinberg RA, Rapaport JM, Albert DM, et al. A human DNA segment with properties of the gene that predisposes to retinoblastoma and osteosarcoma. Nature 1986;323:643–6.

[23] Wolf D, Rotter V. Inactivation of p53 gene expression by an insertion of Moloney murine leukemia virus-like DNA sequences. Mol Cell Biol 1984;4:1402–10.

[24] Rotter V, Wolf D, Pravtcheva D, Ruddle FH. Chromosomal assignment of the murine gene encoding the transformation-related protein p53. Mol Cell Biol 1984;4:383–5.

[25] Groden J, Thliveris A, Samowitz W, Carlson M, Gelbert L, Albertsen H, et al. Identification and characterization of the familial adenomatous polyposis coli gene. Cell 1991;66:589–600.

[26] Kinzler KW, Nilbert MC, Su LK, Vogelstein B, Bryan TM, Levy DB, et al. Identification of FAP locus genes from chromosome 5q21. Science 1991;253:661–5.

[27] Nishisho I, Nakamura Y, Miyoshi Y, Miki Y, Ando H, Horii A, et al. Mutations of chromosome 5q21 genes in FAP and colorectal cancer patients. Science 1991;253:665–9.

[28] Knudson Jr AG. Mutation and cancer: statistical study of retinoblastoma. Proc Natl Acad Sci USA 1971;68:820–3.

[29] Feinberg AP, Vogelstein B. Hypomethylation distinguishes genes of some human cancers from their normal counterparts. Nature 1983;301:89–92.

[30] Baylin SB, Hoppener JW, de Bustros A, Steenbergh PH, Lips CJ, Nelkin BD. DNA methylation patterns of the calcitonin gene in human lung cancers and lymphomas. Cancer Res 1986;46:2917–22.

[31] Greger V, Passarge E, Hopping W, Messmer E, Horsthemke B. Epigenetic changes may contribute to the formation and spontaneous regression of retinoblastoma. Hum Genet 1989;83:155–8.

[32] Hall JM, Lee MK, Newman B, Morrow JE, Anderson LA, Huey B, et al. Linkage of early-onset familial breast cancer to chromosome 17q21. Science 1990;250: 1684–9.

[33] Leach FS, Nicolaides NC, Papadopoulos N, Liu B, Jen J, Parsons R, et al. Mutations of a mutS homolog in hereditary nonpolyposis colorectal cancer. Cell 1993;75:1215–25.

[34] Fishel R, Lescoe MK, Rao MR, Copeland NG, Jenkins NA, Garber J, et al. The human mutator gene homolog MSH2 and its association with hereditary nonpolyposis colon cancer. Cell 1993;75:1027–38.

[35] Nakamura TM, Morin GB, Chapman KB, Weinrich SL, Andrews WH, Lingner J, et al. Telomerase catalytic subunit homologs from fission yeast and human. Science 1997;277:955–9.

[36] Meyerson M, Counter CM, Eaton EN, Ellisen LW, Steiner P, Caddle SD, et al. hEST2, the putative human telomerase catalytic subunit gene, is up-regulated in tumor cells and during immortalization. Cell 1997;90:785–95.

[37] Bodnar AG, Ouellette M, Frolkis M, Holt SE, Chiu CP, Morin GB, et al. Extension of life-span by introduction of telomerase into normal human cells. Science 1998;279:349–52.

[38] Vaziri H, Benchimol S. Reconstitution of telomerase activity in normal human cells leads to elongation of telomeres and extended replicative life span. Curr Biol 1998;8:279–82.

[39] Al-Hajj M, Wicha MS, Benito-Hernandez A, Morrison SJ, Clarke MF. Prospective identification of tumorigenic breast cancer cells. Proc Natl Acad Sci USA 2003;100:3983–8.

[40] Bonnet D, Dick JE. Human acute myeloid leukemia is organized as a hierarchy that originates from a primitive hematopoietic cell. Nat Med 1997;3:730–7.

[41] Bhatia M, Wang JC, Kapp U, Bonnet D, Dick JE. Purification of primitive human hematopoietic cells capable of repopulating immune-deficient mice. Proc Natl Acad Sci USA 1997;94:5320–5.

[42] Yadav SP. The wholeness in suffix -omics, -omes, and the word om. J Biomol Tech 2007;18:277.

[43] Hutchison III CA. DNA sequencing: bench to bedside and beyond. Nucleic Acids Res 2007;35:6227–37.

[44] Pettersson E, Lundeberg J, Ahmadian A. Generations of sequencing technologies. Genomics 2009;93:105–11.

[45] Schena M, Shalon D, Davis RW, Brown PO. Quantitative monitoring of gene expression patterns with a complementary DNA microarray. Science 1995;270:467–70.

[46] Wang Z, Gerstein M, Snyder M. RNA-Seq: a revolutionary tool for transcriptomics. Nat Rev Genet 2009;10:57–63.

[47] Pollack JR, Perou CM, Alizadeh AA, Eisen MB, Pergamenschikov A, Williams CF, et al. Genome-wide analysis of DNA copy-number changes using cDNA microarrays. Nat Genet 1999;23:41–6.

[48] Pinkel D, Segraves R, Sudar D, Clark S, Poole I, Kowbel D, et al. High resolution analysis of DNA copy number variation using comparative genomic hybridization to microarrays. Nat Genet 1998;20:207–11.

[49] Strausberg RL. The Cancer Genome Anatomy Project: new resources for reading the molecular signatures of cancer. J Pathol 2001;195:31–40.

[50] Timeline of genomics. Genomics Proteomics Bioinformatics 2004;2:132–42.

[51] Hudson TJ, Anderson W, Artez A, Barker AD, Bell C, Bernabe RR, et al. International network of cancer genome projects. Nature 2010;464:993–8.

[52] Garraway LA, Lander ES. Lessons from the cancer genome. Cell 2013;153:17–37.

[53] Stephens PJ, Greenman CD, Fu B, Yang F, Bignell GR, Mudie LJ, et al. Massive genomic rearrangement acquired in a single catastrophic event during cancer development. Cell 2011;144:27–40.

[54] Berger MF, Lawrence MS, Demichelis F, Drier Y, Cibulskis K, Sivachenko AY, et al. The genomic complexity of primary human prostate cancer. Nature 2011;470:214–20.

[55] Baca SC, Prandi D, Lawrence MS, Mosquera JM, Romanel A, Drier Y, et al. Punctuated evolution of prostate cancer genomes. Cell 2013;153:666–77.

[56] Taylor BJ, Nik-Zainal S, Wu YL, Stebbings LA, Raine K, Campbell PJ, et al. DNA deaminases induce break-associated mutation showers with implication of APOBEC3B and 3A in breast cancer kataegis. Elife 2013;2:e00534.

[57] Nik-Zainal S, Alexandrov LB, Wedge DC, Van Loo P, Greenman CD, Raine K, et al. Mutational processes molding the genomes of 21 breast cancers. Cell 2012;149:979–93.

[58] Lada AG, Dhar A, Boissy RJ, Hirano M, Rubel AA, Rogozin IB, et al. AID/APOBEC cytosine deaminase induces genome-wide kataegis. Biol Direct 2012;7:47 [discussion].

[59] Gerlinger M, Rowan AJ, Horswell S, Larkin J, Endesfelder D, Gronroos E, et al. Intratumor heterogeneity and branched evolution revealed by multiregion sequencing. N Engl J Med 2012;366:883–92.

[60] Ding L, Ley TJ, Larson DE, Miller CA, Koboldt DC, Welch JS, et al. Clonal evolution in relapsed acute myeloid leukaemia revealed by whole-genome sequencing. Nature 2012;481:506–10.

[61] Shah SP, Morin RD, Khattra J, Prentice L, Pugh T, Burleigh A, et al. Mutational evolution in a lobular breast tumour profiled at single nucleotide resolution. Nature 2009;461:809–13.

[62] Goldman JM, Melo JV. Chronic myeloid leukemia – advances in biology and new approaches to treatment. N Engl J Med 2003;349:1451–64.

[63] Strebhardt K, Ullrich A. Paul Ehrlich's magic bullet concept: 100 years of progress. Nat Rev Cancer 2008;8:473–80.

[64] Dienstmann R, Markman B, Tabernero J. Application of monoclonal antibodies as cancer therapy in solid tumors. Curr Clin Pharmacol 2012;7:137–45.

[65] Kohler G, Milstein C. Continuous cultures of fused cells secreting antibody of predefined specificity. Nature 1975;256:495–7.

[66] Lugosi L. Theoretical and methodological aspects of BCG vaccine from the discovery of calmette and guerin to molecular biology. A review. Tuber Lung Dis 1992;73:252–61.

[67] Steinman RM, Cohn ZA. Identification of a novel cell type in peripheral lymphoid organs of mice. II. Functional properties in vitro. J Exp Med 1974;139:380–97.

[68] Steinman RM, Cohn ZA. Identification of a novel cell type in peripheral lymphoid organs of mice. I. Morphology, quantitation, tissue distribution. J Exp Med 1973;137:1142–62.

[69] Zigler M, Shir A, Levitzki A. Targeted cancer immunotherapy. Curr Opin Pharmacol 2013.

[70] Horwitz RI, Cullen MR, Abell J, Christian JB. Medicine. (De)personalized medicine. Science 2013;339:1155–6.

[71] Sandmann T, Boutros M. Screens, maps & networks: from genome sequences to personalized medicine. Curr Opin Genet Dev 2012;22:36–44.

[72] Dammann M, Weber F. Personalized medicine: caught between hope, hype and the real world. Clinics (Sao Paulo) 2012;67 (Suppl. 1):91–7.

[73] Welch BM, Kawamoto K. Clinical decision support for genetically guided personalized medicine: a systematic review. J Am Med Inform Assoc 2013;20:388–400.

Part 2

Genomics Technologies, Concepts and Resources

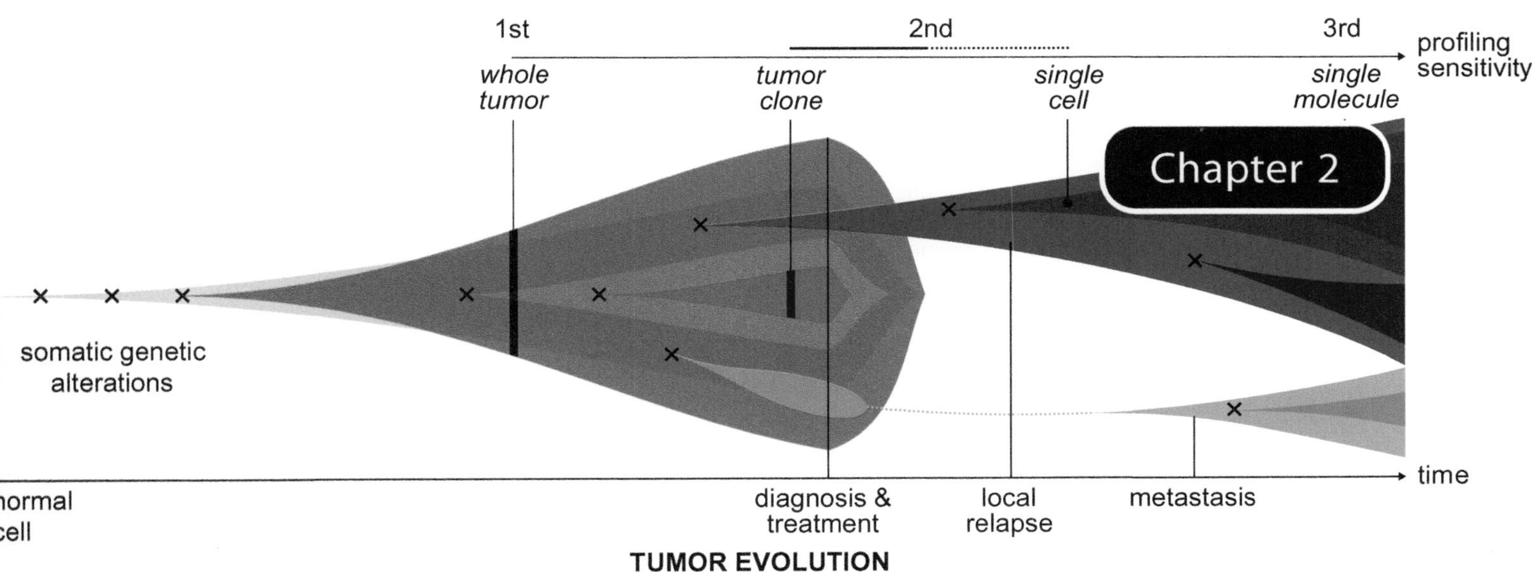

Cancer Genomics

Second-Generation Sequencing for Cancer Genome Analysis

Hye-Jung E. Chun[1,2]
Jaswinder Khattra[2,3,4]
Martin Krzywinski[1,2]
Samuel A. Aparicio[2,3,4]
Marco A. Marra[1,2,5]

[1]*Canada's Michael Smith Genome Sciences Centre, Vancouver, BC, Canada*

[2]*British Columbia Cancer Agency, Vancouver, BC, Canada*

[3]*Department of Pathology and Laboratory Medicine, University of British Columbia, Vancouver, BC, Canada*

[4]*Molecular Oncology, British Columbia Cancer Research Centre, Vancouver, BC, Canada*

[5]*Department of Medical Genetics, University of British Columbia, Vancouver, BC, Canada*

Contents

Cancer Genomics. DOI: http://dx.doi.org/10.1016/B978-0-12-396967-5.00002-5

Key Concepts

- Cancer is a disease driven by mutations, which can be comprehensively profiled using second-generation sequencing technologies. Understanding the heterogeneous genetics of tumors and subclonal population structure are important for data analysis and interpretation
- First-generation Sanger sequencing technology that relies on chain termination rather than amplification, enabled discoveries of well-known oncogenes and tumor suppressor genes
- Second-generation sequencing technologies use *in vitro* amplification of DNA templates and innovative cycle sequencing chemistries and methods, which drastically increased sequencing capacity. These technologies escalated the scale and scope of cancer genomics research to an unprecedented level
- Third-generation sequencing technologies bypass *in vitro* clonal amplification of the DNA templates and may be used to perform single molecule, single strand and single cell sequencing
- Unbiased, genome-wide profiling of cancer genomes and transcriptomes enabled discoveries of mutations in genes that were previously not implicated in tumorigenesis or those with less well-understood roles in cancer. Novel oncogenic phenomena such as frequent chromosomal rearrangements in epithelial cancer types and spatial clusters of substitution mutations were observed through analysis of sequencing data. Detection of nucleic acids from microbes in tumor sequencing data also highlighted the linkage between infection and oncogenesis
- Sequencing technologies are critical in enhancing cancer genome research which, in turn, offers much promise to molecular evidence-based cancer diagnosis, patient stratification and cancer therapeutics

INTRODUCTION

In 1866, Gregor Mendel reported on his 8-year-long plant hybridization experiment that studied behaviors of seven distinct observable traits of garden peas [1]. With this historic work, Mendel established the fundamental foundation of genetics by demonstrating that genetic information of an organism was inherited by its offspring following particular rules of heredity, and in discrete and separable entities (termed *genes*) that were responsible for observable traits of an organism. Major milestones in genetics were made following Mendel's discovery, among which include the identification of the biochemical identity of genes [2], the elucidation of the DNA structure [3] and the discovery of the genetic code [4]. More recently, it has become feasible to determine the complete genetic information of an organism, referred to as the *genome*, which has spawned the field of *genomics*.

The first steps towards the current era of genomics began in the late 1970s, when Frederick Sanger pioneered the development of a DNA sequencing technology that determined the linear nucleotide sequence of DNA in a genome. Propelled by automated Sanger sequencing, much effort ensued to decipher genetic information at a genome-wide scale, including the monumental achievement of the reference human genome sequencing at the turn of the 21st century [5,6]. The reference genome sequence served as the common comparator for aligning re-sequenced individual human genomes and annotating single nucleotide and structural variants in DNA sequences to infer their phenotypic consequences. Insights gained by sequencing genomes have greatly impacted all fields of biomedical research and, in particular, cancer biology.

Cancer predisposition and progression are associated with DNA mutations. Hence, mutation discovery through sequencing can reveal cancer properties, including the molecular and genetic basis of cancer. Since the discovery of the first cancer-specific mutation in *HRAS* in the T24 bladder cancer cell line was made using a sequencing approach [7], sequencing technologies have been instrumental in identifying oncogenes that promote cancer development and tumor suppressor genes whose inactivation leads to oncogenesis. However, the relatively low sequencing efficiency of slab gel-based and capillary-based sequencing approaches restricted the scope of cancer characterization studies to small numbers of samples and to sub-genome scale analyses by evaluating expressed sequence tags (ESTs), serial analysis of gene expression (SAGE) and products of single or multiplex polymerase chain reaction (PCR) and the like.

The arrival of second-generation sequencing technologies has elevated the scale and scope of genomics research to unprecedented and ever-increasing levels. The adoption of these technologies has led to a fundamental shift away from serial Sanger sequencing towards massively parallelized sequencing-by-synthesis approaches (reviewed in [8]). These approaches are relatively scalable, cheap and accessible, and are amenable to automation. While the new technologies have enhanced all types of sequencing analysis approaches, it is the access to entire genome sequences – and many of them – that has become particularly important in comprehensive mutation discovery in cancers. Multiple cancer genomes sequenced at high coverage revealed molecular heterogeneity even within the same bulk tumor, and provided evidence of subclonal populations whose structure changed as disease progressed [9]. Other studies showed that environmental influences, such as exposure to ultraviolet rays, can shape the mutation spectrum, and therefore the molecular heterogeneity in a cancer genome [10]. With continued advancement of sequencing and microfluidics technologies, cancer characterization may be achieved at the single cell level [11]. The impact of cancer genome sequencing projects is already being felt, and has facilitated insights into cancer development, progression and treatment resistance that are fundamentally changing the way researchers think about cancer biology.

CANCER CHARACTERIZATION USING SEQUENCING APPROACHES

Cancer Is a Genetic Disease

More than a century ago, Theodor Boveri made prescient observations of abnormal chromosomes in malignant cells during mitosis [12]. The discovery of proto-oncogenes in the late 1970s provided evidence that cancers were caused by genetic aberrations that transformed normal cells to malignant ones [13]. Knudson's two hit hypothesis introduced the concept of mutation accumulation in cancer from statistical analysis of 48 unilateral (one eye affected) and bilateral (both eyes affected) retinoblastomas whose frequency distribution could be explained by a series of two *hits* (i.e. mutations) that occurred at approximately equal rates: an inherited (i.e. germline) mutation followed by a somatic one in familial cases, or two somatic mutations in sporadic cases [14]. The two hit hypothesis provided a framework for understanding how a germline mutation could increase predisposition to cancer requiring fewer somatic mutations to reach malignancy. Loeb and colleagues [15] advanced the concept of a mutator phenotype that accelerated the accumulation of nucleotide variants that could serve as substrates for Darwinian selection, favoring the growth of certain mutated cells above the growth of others. The concept of accumulated mutations causing malignant progression was subsequently advanced using colorectal cancer as a model system [16]. Furthermore, as predicted by Boveri, the discovery of the Philadelphia chromosome in chronic myeloid leukemia (CML) showed the oncogenic effect of a specific chromosomal rearrangement [17].

Presently, cancers are understood to develop from the accumulation of multiple genetic and epigenetic alterations, predominantly of somatic origin and are therefore not heritable (i.e. present in the cancer cells, but not in normal cells from the same person). Somatic alterations in cancers include substitution mutations, copy number alterations, insertions and deletions (collectively referred to as *indels*), structural rearrangements such as gene fusions, altered gene expression and the presence of nucleic acids of viral and other microbial origin; all of which can be detected using sequencing approaches. The multigenic nature of cancers lends well to comprehensive interrogation at single nucleotide resolution enabled by the use of second-generation sequencing technologies. These technologies allow putative pathologic lesions in the DNA and RNA of an individual's tumor to be mapped comprehensively and economically, offering an opportunity to decipher the functional basis of malignant phenotypes and aiding our understanding of biomolecular networks in cancer.

Cancer Properties Are Amenable to Second-Generation Sequence Analysis

Cancers are heterogeneous entities. At the sample level, a bulk tumor can contain an admixture of cancer cells, non-cancer normal cells, necrotic cells, immune and stromal cells representing the tumor microenvironment. Also, cancer stem cells may be present, which possess self-renewal capacity and may contribute to therapeutic resistance and relapsed disease [18]. At the molecular level, cancer genomes have nucleotide sequences and/or chromosomal structures that are different not only from their normal tissue, but also differ within the same tumor mass due to multiple subclonal populations of cancer cells (Figure 2.1).

Cancers are also clonal in origin [19]. The term *clone* generally refers to a mass of cells with a common cell of origin. To define the clonal structure of a cell population, a stable biomolecular feature is required. One such feature derived from the second-generation sequencing is the frequency of a DNA sequence variant as measured by its

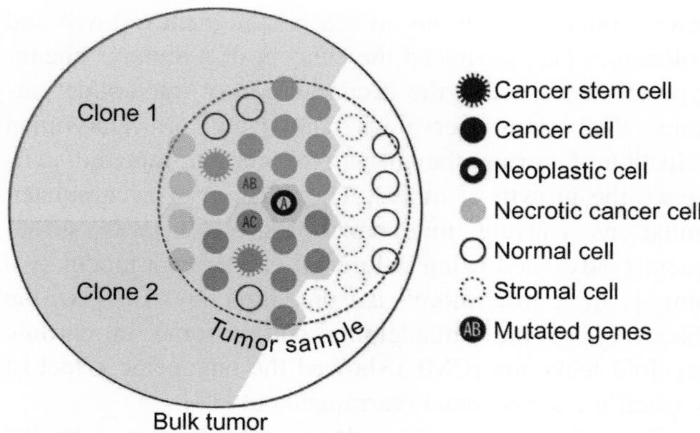

FIGURE 2.1 Tumor heterogeneity at multiple levels. A tumor biopsy sample from a bulk tumor mass typically contains subpopulations of various cell types: cancer cells, non-cancer normal cells, and stromal cells including fibroblasts, immune cells and endothelial cells from blood vessels. Within a single biopsy sample, tumor subclonal populations with different mutational spectra have differential selective advantage as they arise over time.

relative abundance in sequence reads. This mutant allele frequency can be a function of tumor cellularity and the size of a subclonal population harboring that variant. Another molecular feature routinely identified using second-generation sequencing is the chromosomal rearrangement events. In a study by Campbell and colleagues [20], structural rearrangements in pancreatic cancer genomes were analyzed to infer clonal structures and phylogenetic relationships among metastases. Heritable genetic and epigenetic variations in cells in a bulk tumor confer unequal traits such as cellular growth that can be either positively or negatively selected during cancer evolution (reviewed in [21]). This selection pressure can result in growth or purge of subclonal populations as a consequence of stresses such as the limited availability of nutrients or oxygen to the tumor, the immune response or anticancer drug treatment. The ability to track changes in the mutational spectrum of tumors as they respond to such stresses can provide insight to the mechanistic role of mutations in shaping tumor progression.

Second-generation sequencing technologies have been instrumental in profiling somatic mutations and uncovering molecular heterogeneity in cancer genomes at single nucleotide resolution. For example, Shah and colleagues [22] reported a continuous distribution of somatic non-synonymous mutations across 104 breast cancer samples with the triple-negative subtype, irrespective of tumor cellularity or copy number variation. Other studies revealed yet another kind of heterogeneity, that of intratumoral spatial heterogeneity, and showed distinct mutational spectra in different regions of the same tumor tissue [23]. These studies suggested that the current paradigm of using single tumor biopsy samples for cancer genome characterization is flawed, allowing only a partial representation of the true mutational landscape in cancers.

Impact of Tumor Heterogeneity on Clinical Diagnosis and Treatments

Traditional cancer classification has relied on gross assessments based on anatomic and morphological features. Although semi-quantitative nuclear and histological grading allows some quantitative measurements of tumor heterogeneity, this clinical standard is often limited to assigning the tumor to one grade even if co-occurring pockets in the same tissue section exhibit another grade [24]. On the other hand, apparently identical histopathological features between patients' individual tumors can show differential progression and therapeutic response [25].

Tumor heterogeneity has become an important clinical feature for diagnosis and therapeutic decision making. Several studies demonstrated the effect of histological and molecular heterogeneity on cancer progression and on different sensitivity levels to anticancer treatments [26]. In the case of breast cancers, heterogeneity of hormone receptor status predicted differential response to the estrogen receptor antagonist, Tamoxifen™, which became one of the first evidence-based molecular targeted therapies [27].

Tumor heterogeneity is also a confounding factor in the identification of molecular profiles based on single tumor biopsy samples. Single biopsy samples are subject to random sampling error, which can reduce the accuracy of prognostic gene expression signatures, as shown in renal carcinoma [23], or can obscure important differences between primary and metastatic tumors [28]. Assessing molecular signatures from subclonal populations, and charting the three-dimensional architecture and spatial relationship among them, can be informative in addition to taking multiple biopsy samples. An overlay of clonal genotype and functional imaging assays can also enhance our understanding of genotype variants that confer malignant potential. Sampling error is ever more of an issue in the analysis of the clinical relevance of micrometastases and rare circulating tumor cells, requiring much more investigation to determine the utility of such analyses for patient treatment and follow up. Finally, cancers of an unknown primary origin may also benefit from genome analysis by allowing the identification of the most likely cell of origin based on gene expression, miRNA or epigenetic profiling [29–31].

ADVANCES IN SEQUENCING TECHNOLOGIES

The DNA sequencing technology in the late 1970s revolutionized the way in which scientists could understand biology by sequencing genes. Commonly referred to as

Sanger or dideoxy sequencing, the chain-termination method developed by Sanger's group [32] became the dominant sequencing platform for the next three decades.

The second wave of sequencing technologies came in 2005 with the arrival of so-called *next-generation* sequencing approaches. These technologies generally shared an emphasis on massively parallel capacity with dramatically increased sensitivity and cost-effectiveness, allowing re-sequencing of individual genomes, and thus comprehensively characterizing whole genomes at an unprecedented scale. A rapid development of these next-generation DNA sequencing technologies resulted in various sequencing platforms that have different DNA template preparation and amplification, sequencing and detection strategies (reviewed in [8]). In this chapter, we use the term second-generation sequencing technologies for the commercially available next-generation sequencing technologies that rely on clonal amplification of single DNA molecules. The upcoming sequencing technologies that bypass DNA template amplification, thus enabling single molecule sequencing, are referred to as third-generation and are discussed later in this chapter.

First- and Second-Generation Sequencing Technologies

First-generation sequencing technology included the Sanger sequencing method described in 1977 [32]. The Sanger method had the advantage of reduced quantities of toxic chemicals and radioactive isotopes, which made it the preferred sequencing platform among other methods. The Sanger method is based on DNA polymerase-dependent synthesis of a DNA template using natural 2′-deoxynucleotides (dNTPs) and termination of synthesis of the template by incorporating 2′,3′-dideoxynucleotides (ddNTPs). The competitive incorporation of dNTPs or ddNTPs to a growing chain of oligonucleotides results in stochastic termination of the oligonucleotide sequence and generates DNA strands with varying lengths. These strands are then separated according to their lengths using polyacrylamide gel electrophoresis, and the chain-terminating ddNTP moiety is revealed. Many technological innovations were followed to increase the sequencing throughput and efficiency, including the development of capillary-based polymer gel electrophoresis that allowed faster DNA fragment separation at higher resolution compared to the slab gel-based electrophoresis [33]. The automated high-throughput Sanger sequencing became the gold standard for accurate DNA sequencing [34]. Also, the long read length (approximately 800 bp) [8] is useful for *de novo* sequencing and assemblies. However, the method requires a laborious and lengthy process of *in vivo* amplification of DNA templates in bacterial hosts, which results in some loss of DNA fragments during the cloning process [35]. Also, the expense associated with Sanger sequencing limited the application of the method to large population-based DNA sequencing experiments, such as those typically now done in cancer genomics studies. Furthermore, since the individual nucleotide identity is revealed by the fluorescence trace peak that is generated from multiple DNA molecules, the technology is limited to detection of rare variant alleles or multiple variant alleles in a heterogeneous population such as the communities of cells comprising tumors (reviewed in [34]).

In 2005, Roche/454 Life Sciences developed the first commercially available second-generation sequencing technology [36]. A major innovation of 454 sequencing was the application of emulsion PCR for *in vitro* DNA template amplification and the development of pyrosequencing in picoliter-scale reactors that achieved parallelized sequencing capacity [37]. For emulsion PCR, sheared genomic DNA fragments are ligated to a common PCR adapter, and individual single-stranded fragments are then captured using streptavidin beads. An individual fragment with a bead is then incorporated into an oil emulsion where single droplets containing theoretically one DNA fragment and all necessary PCR reagents are suspended in oil. Thus, the emulsion droplet serves as an individual PCR reactor and produces millions of clonal copies of the template adhering to the streptavidin bead. After amplification, the emulsion droplets are broken open, and each streptavidin bead with amplified DNA sequences is sequenced in a picoliter-sized fiberoptic well using pyrosequencing. Pyrosequencing is a sequencing-by-synthesis approach that relies on the use of pyrophosphate (PP_i) released after a complementary ddNTP incorporation and light emission caused by chemiluminescent ATP sulfurylase and luciferase [38]. The light emission and the pattern of specific ddNTP incorporation events detected at each sequencing cycle determine the template sequence. Since a single type of ddNTP (corresponding to either A, G, T or C) is introduced at a time, the sequencing is asynchronous, resulting in varying lengths of reads depending on the sequence composition of the templates and the order of ddNTP introduced at each sequencing cycle. A major drawback of the 454 sequencing method is the relatively inaccurate detection of homopolymers (i.e. a series of consecutive bases of the same nucleotide, e.g. AA or AAAAA) due to incorrect estimation of the number of PP_i released using signal intensities. Thus, the majority of sequencing errors from the 454 technology result from insertions and deletions. On the other hand, a key advantage of 454 sequencing is the long read lengths that are up to 1000 bp, which is particularly useful for *de novo* sequencing, such as in ancient DNA sequencing and metagenomics (reviewed in [39]).

The Illumina sequencing method circumvents *in vitro* cloning amplification of DNA templates by amplifying DNA templates that are tethered onto a solid glass surface known as a *flow cell* using a process called "bridge" PCR [40]. The process is facilitated by the unique design of the flow cells which are densely populated by forward and reverse PCR primer adapters. Upon introduction of an adapter-ligated DNA template molecule onto the flow cell surface, the molecule arches over and hybridizes to the complementary adapter, creating a *bridge* that serves as the substrate for amplification. Millions of templates are amplified over iterative cycles, each generating over a thousand clonal copies from a single DNA molecule. Like the 454 system, the Illumina method also uses a sequencing-by-synthesis approach, but a key difference is the inclusion of modified nucleotides with reversible terminators. This innovation allows the addition of only one base per sequencing cycle, thus enhancing the accuracy of sequencing through homopolymer runs. At each sequencing cycle, a mixture of four nucleotides labeled with chemically cleavable fluorescent dyes is added. The reversible terminators have a chemically cleavable moiety at the 3′ OH position, which allows for controlled incorporation of one nucleotide per sequence cycle [41]. The identity of the incorporated nucleotide is revealed by fluorescence. While this sequencing method results in accurate sequencing of homopolymer repeats, the reliable read length is relatively short due to both fluorescence decay and dephasing over longer cycles, which results from incomplete cleavage of fluorescence tags or reverse terminating moieties within template clusters. This tends to yield substitution errors and a high proportion of such errors occur in the base after guanine [42]. Short read lengths can limit the ability to align (or *map*) uniquely the sequence reads to the reference genome, especially in repeat regions of the genome. Obtaining such short reads from each end of the DNA fragments in *paired-end sequencing*, as opposed to sequencing only one end of template fragments, significantly improves the accuracy of read mapping and thus the assessment of redundant coverage of the genome [43]. Paired-end sequencing is particularly useful for identifying genomic rearrangement events, and does not depend on high sequence coverage to detect such events provided that they are properties of a dominant clone within the tumor sample.

Another second-generation sequencing technology is the supported oligonucleotide ligation and detection (SOLiD™) system [44]. Similar to the 454 sequencing method, single-stranded DNA templates are amplified *in vitro* using emulsion PCR. The uniqueness of the SOLiD system is the sequencing-by-ligation approach, in which DNA ligase is used instead of DNA polymerase to add short oligonucleotides [45]. After the emulsion PCR stage, *polonies* (amplified polymerase colonies) are transferred to a glass slide for sequencing. At each sequencing cycle, fluorescently labeled octamers of degenerate bases are ligated and then chemically cleaved between the fifth and sixth base from the 3′ end of the nucleotide sequence. The cycle of octamer ligation, fluorescence detection and chemical cleavage of the octamer to remove fluorescent tags is repeated over multiple iterations. In each subsequent iteration, ligation of octamers is offset by one base. This strategy is designed to increase the base calling accuracy by the double interrogation of each base of the templates. The sequence identity is deduced from a di-base color encoding system based on the double interrogations of each base by the resultant colors associated with a nucleotide at a specific position in the octamer.

Ion Torrent sequencing technology also utilizes the emulsion PCR amplification method and the sequencing-by-synthesis approach. However, unlike other second-generation sequencing technologies, it does not depend on the use of light to detect signals [46]. Instead, it uses a semiconductor chip known as an ion-sensitive field effect transistor (ISFET), to measure the change in pH as hydrogen ions are released by DNA polymerase during DNA replication [47]. pH changes are converted into voltage changes that are detected for sequence readout. The number of bases incorporated is proportional to the magnitude of voltage change. Parallelized sequencing can be achieved by having multiple ISFETs on a chip. Similar to sequences generated using 454 sequencing, Ion Torrent generates reads with frequent insertion and deletion sequencing errors (approximately 1.5 indels per 100 bp), the majority of which are associated with homopolymer repeats [48].

Although short reads generated using second-generation sequencing technologies (ranging between 100 and 200 bp, with the exception of 454 sequencing) are usually not suitable for *de novo* sequencing, short reads generated at high volume from re-sequenced genomes are highly economical and useful for variant discovery by aligning them to the reference genome to identify genetic alterations. Highly redundant coverage of a genome enables accurate detection of such alterations including rare ones. Thus, re-sequencing of individual tumor genomes for mutation discovery has been the most common application of the second-generation sequencing technologies in cancer genomics. The various sequencing technologies, their optimal use and the caveats associated with each technique are summarized in Table 2.1.

Progress in Cancer Genomics Research Using Sequencing Approaches

Prior to the emergence of second-generation sequencing technologies, many cancer genomics studies were carried out using targeted approaches that focused on profiling a

TABLE 2.1 Summary of First-, Second- and Third-Generation Sequencing Technologies

Generation	Name	Technology	Optimal Use	Pros	Cons	Example Ref.
1st	Sanger	Sequencing-by-synthesis; *in vivo* DNA amplification; stochastic chain termination; DNA fragment separation using capillary-based gel electorophoresis; base readout from fluorescent signals	*De novo* sequencing (e.g. reference genomes) and assemblies	Accuracy (*gold standard* for sequencing); long read length (<1000 bp)	Expensive; lengthy and laborious library construction processes; difficult to detect rare or multiple variant alleles in a sample	[5,6]
2nd	Roche/454	*In vitro* DNA amplification by emulsion PCR; sequencing-by-synthesis using pyrosequencing; base readout based on light intensity	*De novo* sequencing (e.g. ancient DNA sequencing and metagenomics)	Long read length (<1000 bp)	Inaccurate sequencing of homopolymers	[36]
	Illumina	*In vitro* DNA amplification by bridge PCR; sequencing-by-synthesis using reversible terminators; base readout from fluorescent signals	Re-sequencing of individual genomes for mutation discovery including rare or multiple variants	Enhanced accuracy of sequencing homopolymers; currently the most widely used platform	Short read length (≤100 bp)	[64]
	SOLiD	Sequencing-by-ligation; *in vitro* DNA amplification by emulsion PCR; double interrogation of each base in DNA templates; base identity determined using a di-base encoding system	Re-sequencing of individual genomes for mutation discovery including rare or multiple variants	Inherent sequencing error checks by double interrogations of bases	Short read length (≤50 bp)	[44]
	Ion Torrent	Sequencing-by-synthesis; *in vitro* DNA amplification by emulsion PCR; use of a semiconductor chip (ISFET) as a sequencing platform; base readout using voltage changes converted from pH changes	Re-sequencing of individual genomes for mutation discovery including rare or multiple variants	Short run time	Frequent indel errors associated with homopolymers; short read length (~100 bp)	[46]
3rd	HeliScope	Sequencing-by-synthesis of a single DNA molecule using reversible terminators; base readout from fluorescent signals	Single molecule sequencing; direct RNA sequencing	Single molecule sequencing	High error rate from incomplete inhibition of base incorporation	[95]
	Pacific Biosciences	Single molecule real-time sequencing; continuous base readout from fluorescent signals detected in a zero-mode waveguide chamber; multiple re-sequencing using circular DNA templates	Single molecule sequencing; direct methylated DNA sequencing; full-length transcriptome sequencing	Long read length (~1000 bp)	High error rate	[99]
	Oxford Nanopore Technologies	Single molecule sequencing using electrical nanopore structures; base readout based on voltage changes	Single molecule sequencing; direct RNA, methylated DNA and protein sequencing	Short run time; potential for a portable sequencing device	High error rate	[101] (review)

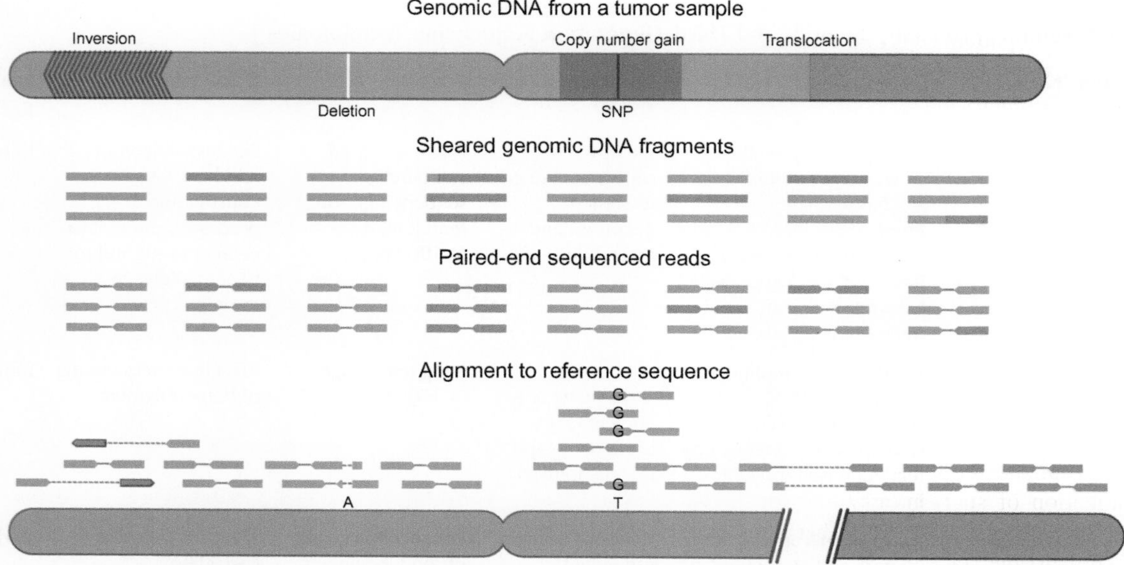

FIGURE 2.2 Profiling molecular alterations using sequencing approaches. Second-generation sequencing is an effective method to profile simultaneously multiple modalities of genetic alterations in cancer genomes. Such alterations include SNVs, chromosomal copy number alterations and structural rearrangements. Sheared DNA fragments are sequenced as *reads*, which are then aligned against the reference human genome sequences. Mismatches in alignments, relatively different depths of sequence coverage and unusual alignment patterns (e.g. gaps in the middle of a read, reversed alignment orientation in read pairs from paired-end sequencing) are analyzed to identify putative genetic alterations.

small number of targeted genes across large sample cohorts [49] or profiling protein-coding regions of the genome from small sample cohorts using a PCR-based targeted sequencing approach [50]. Other genomics projects such as the Cancer Genome Anatomy Project, the Human Cancer Genome Project and the Cancer Genome Project used Sanger sequencing to sequence partial transcripts (known as *tags*) to catalog gene expression in tumors (reviewed in [51]). These studies used methods such as expressed sequence tags (ESTs), serial analysis of gene expression (SAGE), cap-analysis of gene expression (CAGE) [52] and massively parallel signature sequencing (MPSS) [51]. Tag-based methodologies effectively quantified gene expression levels and identified transcription start sites, but only partial transcript coverage could be achieved, making the data less than optimal for discovering novel and alternatively spliced transcripts. Using these methods, more than 1% of protein-coding genes were found to have recurrent somatic mutations that were causally implicated with tumorigenesis (reviewed in [53]). These early technologies provided important glimpses into how entire genome sequences would be crucial to inform fully on cancer biology.

The second-generation sequencing technologies enabled genome-wide molecular profiling at a single nucleotide resolution, and enabled simultaneous interrogation of multiple modalities of molecular alterations that include single nucleotide variations (SNVs), indels, copy number variations (CNVs) and structural rearrangements in chromosomal DNA such as inversions and translocations (Figure 2.2).

With affordable sequencing technology, it was possible to sequence cancer and matched normal genomes to identify somatic mutations in cancer. In the seminal study in which the first cancer genome was sequenced [54], Ley and colleagues identified somatic mutations by comparing the genomic DNA sequence of cytogenetically normal acute myeloid leukemia (AML) to the genomic DNA sequence generated from a matched normal sample. Since this work, burgeoning numbers of genomes from various cancer types have been sequenced. Typically, approximately 30-fold redundant sequence coverage (denoted as *30×*) was used for whole genome sequencing using second-generation sequencing technologies [55]. However, especially for detecting SNVs, 30× in sequence coverage is often inadequate, and 60× or more in sequence coverage was shown to be necessary for accurately determining the genotype of at least 95% of the genome [56]. Even higher sequence coverage would be recommended for sequencing tumor samples with low enrichment of tumor cells or for aneuploid cancer genomes. Identifying rare mutations requires even higher sequence coverage as demonstrated by Shah and colleagues, in which approximately 20 000× in sequence coverage enabled the identification of rare mutations representing clonal subpopulations [22].

With highly accessible routine re-sequencing of cancer genomes, cancer genomics studies have begun to characterize genomes of almost all major cancer types using thousands of patient samples. Such large-scale studies include The Cancer Genome Atlas (http://cancergenome.nih.gov), International Cancer Genome Consortium (http://icgc.org), and Molecular Taxonomy of Breast Cancer International Consortium [57].

In contrast to relatively straightforward sequence data generation, data analysis and management has become a significant bottleneck. Analysis tools need to be developed for accurately distinguishing cancer-specific alterations from false positives. Cancer genomics studies are increasingly complex, involving whole genome, transcriptome and epigenome profiles. Hence, particularly needed is the development of computational tools for integrative data analysis and statistical models for testing the significance of combinatorial molecular alterations in cancer, and visualization of such interactions for effective communication. Some of the biggest obstacles in data interpretation lie in the lack of analysis transparency in published literature, the difficulty in accessing the appropriate high-performance computing infrastructure and the lack of *best practices* established for data analyses, similar to those established in the 1000 Genomes project (reviewed in [58]). These informatic challenges must be met to translate the progress made using second-generation sequencing technologies in cancer research into actionable information in cancer clinics.

Applications of Second-Generation Sequencing Technologies

In addition to accurate sequencing of whole genomes, second-generation sequencing technologies have wide applications for targeted re-sequencing of subgenomic regions, and transcriptome and epigenome profiling (reviewed in [59]).

Selective capturing methods followed by second-generation sequencing increase the sequence depth across targeted regions of interest. In exome sequencing (also known as exome-seq or targeted exome-capture), the exonic regions of the genome are captured using probe hybridization, and then amplified and sequenced [60]. Highly redundant nucleotide interrogation makes exome-seq ideal for mutation discovery in the coding regions of the genome. But, this approach is limited to detecting mutations in predetermined captured regions only (e.g. exons), and is subject to variability in hybridization arising from inconsistent capture efficiency across targeted regions and non-specific probe hybridization.

Currently, whole transcriptome profiling (RNA-seq) is achieved by sequencing cDNAs that are reverse transcribed from RNA transcripts. Typically, protein-coding transcripts are profiled using, for example, poly A-tail capturing methods to enrich for mRNA before sequencing. Depletion of rRNA in total RNA has also been used as an enrichment method [61]. Rare transcripts and non-coding RNAs such as miRNA or small interfering RNAs have been profiled using capturing methods such as targeted probe hybridization [62] or fragment size selection [63]. Since the sequence coverage directly correlates with transcript abundance, RNA-seq is a sensitive approach to measure gene expression, and to detect rare transcripts and quantify the expression levels of novel and alternative transcript isoforms [64]. Also, combined with the genomic DNA information, RNA-seq can reveal allele-specific gene expression, RNA-editing events, and somatic mutations that are expressed and thus likely have functional consequences [65].

Second-generation sequencing methods have been applied to characterize chromatin structure, epigenetic alterations and interactions between proteins and DNA molecules. In the bisulfite-seq method (also known as methyl-seq), sequencing bisulfite-treated DNA, which contains uracil that was converted from unmethylated cytosine while methylated cytosine remained unaltered, allows detection of allele-specific DNA methylation events. A modification of this method, the reduced representation of bisulfite sequencing (RRBS), focuses on the DNA methylation events in CpG islands using specific restriction enzymes that cut around the CpG islands (reviewed in [66]). Chromatin immunoprecipitation techniques followed by second-generation sequencing (ChIP-seq) are used to profile genomic sites that directly bind to regulatory proteins such as post-translationally modified histones, transcription factors or components of the transcription machinery (reviewed in [67]). In DNase-seq, DNase I-digested DNA is sequenced to profile transcriptionally active open chromatin regions that are prone to DNase-I-induced cleavage [68]. Similarly, the FAIRE-seq method (formaldehyde-assisted isolation of regulatory elements [69]) utilizes differences in cross-linking efficiency between nucleosomes and sequence-specific regulatory elements to profile open chromatin regions in a genome. These alternatives to standard DNA sequencing are summarized in Table 2.2.

DISCOVERIES USING SECOND-GENERATION SEQUENCING TECHNOLOGIES

Novel Properties of Cancer

The sequencing of cancer genomes revealed novel mechanisms that had not been previously implicated in cancer development. Notably, an emerging role for the reprogramming of energy metabolism in oncogenesis was elucidated in different cancer types. The whole genome sequencing of AML [70] revealed recurrent somatic mutations in *IDH1*

TABLE 2.2 Some Applications of Second-Generation Sequencing Technologies

Method	Description and Purpose	Example Ref.
Exome-seq	Coding regions of a genome are captured using probe hybridization and then sequenced to achieve highly redundant sequence coverage across the captured regions; used to discover mutations in the predetermined captured regions	[60]
RNA-seq	RNA transcripts are enriched using a capturing method (e.g. enrichment of mRNA using a poly A-tail capturing method) to be reverse transcribed to cDNA and then sequenced; used for whole transcriptome profiling to measure gene expression and detect RNA-edits, alternative transcript isoforms, and rare and novel transcripts	[64]
Bisulfite-seq	Bisulfite-treated DNA is sequenced to detect methylated cytosines across a genome	[66] (review)
RRBS	Targeted bisulfite-sequencing using specific restriction enzymes that enrich for CG-rich regions in a genome to profile DNA methylation events in subgenomic regions	[66] (review)
ChIP-seq	Chromatin immunoprecipitation followed by sequencing to profile genomic sites that directly bind to regulatory proteins, transcription factors or post-translationally modified histones	[67] (review)
DNase-seq/ FAIRE-seq	DNase I-digested DNA (DNase-seq) or formaldehyde-fixed DNA (FAIRE-seq) is sequenced to profile nucleosome-depleted regions in a genome, which is indicative of active transcription	[68,69]

and *IDH2* genes, encoding enzymes involved with metabolizing isocitrate to α-ketoglutarate, which were previously found in glioblastoma multiforme (GBM) patients using targeted PCR-based exon sequencing [71].

The role of aberrant epigenetic modification in tumorigenesis has also been revealed. Recurrent mutations in *DNMT3A* (DNA methyltransferase) were found in AML patients [72]. Recurrent gain-of-function mutations in *EZH2*, a catalytic subunit of the polycomb PRC2 complex that methylates H3K27, were observed in lymphomas [29,73], whereas recurrent loss-of-function mutations in *EZH2* were found in myeloid disorders [74]. Other histone modifying genes such as *MLL2* (an H3K4 methyltransferase) and *MEF2B* (a calcium-dependent transcription factor that recruits multiple histone modifiers) were also frequently mutated in various lymphoma types [75]. Genes involved in chromatin remodeling such as *PBRM1* and *ARID1A* were recurrently mutated in various cancer types such as renal carcinoma and specific subtypes of ovarian cancer [76,77]. Together, these novel findings show significant contribution of aberrant histone modification and chromatin remodeling processes in cancer pathogenesis.

The ability to detect structural rearrangements and chimeric transcripts was especially enhanced by whole genome and transcriptome sequencing. Using a whole genome sequencing approach, chromosomal rearrangements, once thought to be rare in epithelial tumors, were actually found to be frequent in many types of epithelial tumors including breast, lung cancers, melanoma (reviewed in [78]), prostate cancer [79], and notably in colorectal cancer subjected to an extreme chromosome shattering event, named *chromothripsis* [80].

Mutation discovery using whole genome sequencing revealed a novel phenomenon of spatial clusters of somatic mutations. Named *kataegis* in the study of 21 breast cancer genomes, this phenomenon refers to clusters of somatic mutations found in genomic regions whose lengths range from a few hundred bases to more than 10 Mb [81]. The majority of mutations found in these regions had a distinct pattern of C > T transition mutations in Tp<u>C</u>pX trinucleotides, and the mutation clusters were frequently found to co-localize with chromosomal rearrangements. While the mechanism of kataegis is currently unknown, this phenomenon was observed in various cancer types including head and neck, colorectal, prostate carcinomas, melanoma, multiple myeloma and CML [82].

Second-generation sequencing technologies are sensitive tools for detecting DNA of foreign origins and can be an effective method to identify the presence of bacteria and viruses in tumor samples that may have a causal effect on oncogenesis. Well-known oncogenic associations of various infectious agents include hepatitis B and C virus in hepatocellular carcinoma, human papilloma virus in cervical cancer, Epstein-Barr virus in Burkitt lymphoma and *Helicobacter pylori* in gastric cancer. Recent studies that used whole genome and transcriptome sequencing approaches found significant enrichment of *fusobacteria*, invasive proinflammatory anaerobic bacteria, in colorectal cancers compared to their matched normal tissues [83,84], suggesting a novel linkage between *fusobacteria* and colorectal cancer development.

Novel Biomarkers and Therapeutic Targets

Novel mutations identified from whole genome characterization studies may be used as biomarkers and as therapeutic and diagnostic targets. For example, *FOXL2* mutation, which occurred in 97% of patients with the granulosa-cell tumor subtype of malignant ovarian sex cord-stromal tumor [85], is a highly specific marker to diagnose this particular subtype of ovarian cancer. The findings of recurrent mutations in the *EZH2* gene mentioned above has led to the development of novel EZH2 inhibitors as potential therapeutic agents for lymphomas and other malignancies [86].

High sensitivity detection of rare DNA molecules using second-generation sequencing technologies has allowed detection of rare circulating tumor cells from a patient's blood, serving as a less invasive *liquid biopsy* for cancer diagnostics or for monitoring tumor burden. Dawson and colleagues measured the sensitivity in detecting tumor-specific somatic mutations in circulating cell-free DNA in breast cancer patients using digital PCR assay or deep amplicon sequencing, and compared it to other blood-based biomarkers [87]. They showed that sequencing-based detection of biomarkers from cell-free DNA was the most sensitive (found in 97% of the patients) compared to circulating tumor cell counts (87%) and immunoassay of a known cancer antigen (78%). In a study by Leary and colleagues, tumor-specific structural rearrangement events were identified from whole genome sequencing [88]. The rearranged DNA sequences were then used as specific probes for targeted re-sequencing of cell-free DNA with tumor-specific rearrangements in plasma samples. The study further demonstrated the feasibility of quantitatively measuring the proportion of tumor DNA and showed the potential usage of this type of biomarker to monitor tumor progression after surgery or chemotherapy treatments.

Development of Personalized Medicine

Identifying cancer-driving mutations in an individual patient's cancer genome can help a therapeutic decision-making process. Earlier efforts in cancer research particularly focused on studying protein kinases (reviewed in [53]). Therapies targeting these proteins were actively pursued in clinical trials, such as a BRAF inhibitor for melanoma and crizotinib, an ALK inhibitor for non-small cell lung cancer (reviewed in [89]). The mutational status of cancer genes can be used to stratify patients who are likely to respond better to therapeutic agents. For example, lung cancer patients with *EGFR* mutation had significantly longer progression-free survival after being treated with an EGFR inhibitor, gefitinib, compared to patients without *EGFR* mutations [90]. Similarly, *BRCA1* mutation significantly increases tumor sensitivity to a PARP1 inhibitor in *BRCA1*-associated cancers, demonstrating a successful synthetic lethal approach of targeting cancers containing DNA repair pathway defects with the agent that causes DNA breaks [89].

With second-generation sequencing technologies, individual cancer genome sequencing became feasible. The feasibility of clinically actionable cancer genome characterization and genomics-guided therapy was first demonstrated by Jones and colleagues [91]. In this study, genome and transcriptome sequences were generated from a rare oral adenocarcinoma that recurred after standard therapy. Mutations and affected pathways were identified, and this information was used to infer appropriate treatment selection for the patient and suggested the potential mechanism of therapeutic resistance in recurrent metastatic tumors. In a study by Welch and colleagues, accurate diagnosis of leukemia was achieved using a whole genome sequencing approach [92]. In their study, an unusual breakpoint of the *PML–RAR*α gene fusion that eluded detection using the standard FISH-based detection method was identified from whole genome sequencing, and this information led to accurate diagnosis and appropriate treatment for the patient.

Despite the clear potential for genomics-based personalized medicine, there are challenges [89]. Completion of clinical-grade analyses in clinically relevant timelines is one of the critical challenges in making genomics-guided medicine into reality. Also, somatic mutations identified by whole genome sequencing still need functional validation to understand their true biological relevance in cancer development, which prompts for developing high-throughput functional validation systems such as cell culture-based screening and transgenic and xenograft animal models. Furthermore, the presence of evolving subclonal populations should be considered to alleviate potential shortcomings in controlling cancer progression following targeted-therapeutic treatments. To understand the true level of molecular heterogeneity in cancer, the complete assessment at the single cell level may become necessary and feasible as the sequencing technology advances to single cell and single molecule sequencing.

FUTURE OF CANCER GENOME SEQUENCING

The latest development in sequencing technology is to bypass clonal DNA amplification steps that can introduce polymerase errors and directly sequence individual DNA molecules (reviewed in [93]). This new generation of sequencing instrumentation – the *third generation* – offers real-time, strand-specific sequencing of individual

DNA molecules, enhancing the resolution of sequencing interrogation yet again.

Third-Generation Sequencing Technologies

Third-generation sequencing technologies have been developed for research and clinical DNA sequencing. Helicos Bioscience developed the first commercially available single molecule sequencing instrument, called the HeliScope. The Helicos sequencing method uses the sequencing-by-synthesis approach to sequence a single DNA molecule using a highly sensitive light detector and fluorescently tagged dNTPs. These dNTPs are 3′-unblocked reversible terminators (known as the Virtual Terminator [94]) that allow one base to be incorporated to the growing chain of nucleotides at each sequencing cycle. DNA fragments are poly-A tailed and captured onto a solid surface with poly-T oligomers. At each cycle, a known type of dNTP (i.e. A, C, G or T) is introduced to the templates. The base incorporation results in fluorescence emission for signal detection and the inhibitory moiety and the fluorescent label are chemically cleaved off for the next sequence cycle. Because each labeled dNTP does not have a terminating moiety, incomplete inhibition of a base incorporation can lead to inaccurate sequencing, especially for homopolymer repeats. The Helicos sequencer was used to sequence a human genome [95], and also performed the first direct transcriptome sequencing without intermediate cDNA template generation [96].

Pacific Biosciences, also referred to as PacBio, developed single molecule real-time (SMRT) sequencing technology. This technology captures sequencing information using movies as opposed to static images that require halting the synthesis process to achieve base readout. PacBio technology uses a *zero-mode waveguide* (ZMW) [97] detection chamber that allows the observation of the incorporation of a single phospholinked fluorescent nucleotide by a DNA polymerase affixed inside the chamber. The fluorescent tag is cleaved off at the time of nucleotide incorporation, and the corresponding fluorescence color reveals the base identity. Parallelized sequencing is achieved by having arrays of ZMW detectors with a parallel confocal imaging system. PacBio sequencing makes innovative use of circular templates, which are called SMRTbell™ templates that consist of double-stranded DNA template flanked by single-stranded universal hairpin loops [98]. Sequencing circular DNA templates multiple times results in increased accuracy in the consensus sequence generated from multiple re-sequencing [99]. The circular template sequencing also allows for simultaneous interrogation of both sense and anti-sense strands of double-stranded DNA molecules. Another application of this technology is profiling DNA methylation based on detection of differential kinetics generated from incorporating modified nucleotide by DNA polymerase [100].

Oxford Nanopore Technologies® developed electrical nanopore structures coupled with proprietary enzymes for single molecule sequencing (reviewed in [101]). As each base of a single DNA strand passes through the nanopore, a voltage change occurs, which is characteristic of a nucleotide type. This voltage change is translated into a base call. Oxford Nanopore Technologies claims to have developed a portable sequencer that is as small as a USB key. This system, called MinION, has a sequence capacity of several hundred megabases. Its bench-top version, the GridION 2k node, is expected to achieve a substantially higher sequence throughput of tens of gigabases per 24-hour run, and is claimed to be scalable to an even higher sequencing throughput. The nanopore technology can also be applied to epigenome profiling such as DNA methylation [102], single protein molecule sequencing [103] and RNA sequencing without intermediate cDNA conversion [104].

Relatively high raw error rate is a current limitation in implementing these third-generation sequencing technologies. These technologies are sensitive but are vulnerable to contaminants that are carried over from nucleic acid preparations and to damaged DNA. Nevertheless, they can achieve sequence reads in the kilobase range, approximately 100-fold longer than sequences generated using current second-generation sequencing devices. The combination of shorter, but more accurate second-generation DNA sequence reads with error-prone, but longer third-generation data can achieve the milestone of contiguous sequencing of a single bacterial chromosome, and also improve sequencing of repeat regions and complex structural rearrangements. Such approaches have been implemented in the A Hybrid Assembler (AHA) scaffolding algorithm [105] and the PacBio correct Read (PBcR) algorithm [106]. Also, having long read lengths enables sequencing full-length transcripts for accurate profiling of alternate transcript splicing. Another promising opportunity is to resolve haplotype phase by virtue of single strand sequencing using these single molecule sequencing technologies. A summary of third-generation sequencing technologies is given in Table 2.1.

Clinical Application of Second-Generation Sequencing

The clinical implementation of sequencing devices is being actively pursued, especially for routine *en masse* testing for familial cancer genes [107]. Second-generation sequencing can replace a more laborious gene-by-gene sequencing approach. Similar testing can also help guide patients to appropriate clinical trial arms, by determining mutations specifically relevant to a targeted therapy. But first, routine tumor and matched normal clinical specimen collection and storage protocols will need to incorporate steps suited for stabilizing nucleic acids. Clinical formalin-fixed, paraffin-embedded (FFPE) tissue samples that are highly variable in nucleic acid integrity will require qualification for the assays based on second-generation sequencing. Nevertheless, whole genome and transcriptome sequencing of paired tumor and normal tissue is steadily finding clinical application [108]. In 2012, a commercial Clinical Laboratory Improvement Amendments (CLIA)-certified and College of American Pathologists (CAP)-accredited individual genome sequencing service was introduced by Illumina, with a clinically relevant turnaround time of 2 weeks.

For clinical sequencing, targeted sequencing by multiplex or long-range PCR and hybridization-based capture can be effective and economical. Protocol automation for a targeted sequencing approach can be implemented using devices engineered for submicroliter reaction chemistries such as Fluidigm's Access Array™ System and RainDance™ Technologies' ThunderStorm™ System. Given the large sequence output capacity per instrument run, barcoded-libraries from different patient samples are pooled into a single run for efficient sequencing throughput [109]. However, irrespective of the scale of a clinical second-generation sequencing assay, protocols routinely involve the repeated use of common oligonucleotide primers or adaptors and PCR for library construction, which requires stringent cross-contamination control measures [110]. Given the sequencing platform-specific nature of base errors and quality scores [48], there is a clear need for rigorous standardization in the implementation of clinical second-generation sequencing.

The development of educational tools in genomic medicine for healthcare providers is in its infancy. The role of genetic counselors will need to evolve for distilling cancer data at a genomic scale into common parlance for patients. It has been more than six years since the first individual human genomes were sequenced [111], but the ethical issues surrounding the privacy of personal whole genome sequence data generated for research and clinical use are still the subject of ongoing debates with policies being drafted in some jurisdictions (2012 Presidential Commission for the Study of Bioethical Issues, http://bioethics.gov/cms/node/764). How to address the right of a patient to know about his/her genetic health risks, or how to communicate changes in interpreted results from personal genomic data as continued improvements are made in genotype–phenotype knowledge bases [112] are just a few of many prickly questions facing bioethicists [113].

Single Cell Genotyping of Bulk Tumors

Not all malignant cells in a tumor have the same mutational spectrum. Cell-to-cell or clonal subgroup interactions in tumors are unknown and may span the entire gamut typically found in biological ecosystems including competition, mutualism, predation and parasitism. These evolutionary processes continually shape molecular landscapes of tumors and impact clinical outcomes (reviewed in [114]). Open questions in tumor biology include the extent of competition and cooperation among tumor subclones through time and space, genetic dependency between allelic variants, identity of functionally neutral variants and genetic drift, three-dimensional spatial patterns of subclones in relation to cell lineage, patterns of clonal population flux (gradualism versus punctuated evolution-like chromothripsis-type events), and selective sweeps during therapy. In current experimental paradigms, often using tumor tissues from a single biopsy, molecular signals are averaged across a population of cells and so may not reflect the state of individual cells. Such data can be dominated by outliers from a small fraction of cells, which is one of the consequences of intratumoral heterogeneity and results in loss of statistical power in cancer sequencing studies. The stochastic nature of transcription in a cell, with intermittent bursts of transcriptional activity, renders quantitative PCR and RNA-seq data across bulk tumors a poor surrogate for per cell measurements [115]. Multiple biopsies from an individual tumor at multiple time points and spatial locations can help address these questions but, ultimately, single cell measurements of intratumoral heterogeneity will be helpful in advancing our understanding of a group of cells acting in concert to create a malignant tissue phenotype.

A cell-by-cell experimentation strategy requires sensitive assays and measurements from a sufficient number of cells for robust multivariate analysis (reviewed in [116]). Miniaturization and automation are key requirements, and can be met by using nanoliter reaction chambers in microfluidic devices made by multilayer soft lithography [117]. Current challenges in single cell sequencing protocols include developing DNA polymerases for single molecule templates, reducing allele dropout, addressing whole genome amplification bias, minimizing cross-contamination and developing rapid and efficient microfluidic cell isolation methods and computational approaches for single cell data analysis.

A few groups have attempted genome-wide molecular interrogation of single cells. Navin and colleagues [118] evaluated the clonality of 20 primary breast tumors in three-dimensional space by dissecting the tumors into arbitrary sectors, then flow-sorting subpopulations by DNA content to perform CGH for copy number analysis using whole genome-amplified DNA. The same group applied second-generation sequencing technology [119] to analyze partial genomes of 100 single cells from two primary breast cancer samples and a matched metastatic liver carcinoma sample [120]. They identified clonal subpopulations with distinct chromosomal copy number profiles and showed that tumors evolved through punctuated clonal expansion instead of gradual progression. Powell and colleagues [121] profiled single circulating tumor cells in breast cancers for transcriptional heterogeneity and found these cells to have profiles distinct from commonly used cancer cell lines. The parallel biomolecular profiling of hundreds to thousands of single cells from individual tumors is among key pursuits in dissecting the genetic basis of cancer.

CONCLUSION

Nobel laureate Sydney Brenner once remarked that "progress in science depends on new techniques, new discoveries and new ideas, probably in that order" [122]. Cancer genomics is no exception. Second-generation sequencing technologies enabled affordable genome-wide interrogation of nearly every base. Such sensitivity in detection allowed discoveries of novel mutations that implicated a novel role of energy metabolism and regulation of the epigenome in cancer. Also, for the first time, extensive levels of molecular heterogeneity – including intratumoral heterogeneity – were revealed by profiling mutations from subclonal populations, and changes in mutational landscape could be assessed as cancer evolved. These findings explained the sobering reality of why cancers have been very difficult to eradicate using current therapeutic approaches.

Together, these discoveries have generated new ideas for effective clinical applications including sequencing-aided cancer detection and monitoring and genomics-aided cancer diagnostics and treatments, opening the era of personalized medicine. With ongoing efforts to improve sequencing technologies and data analysis methodologies, the distance between *bench to bedside* in cancer care will continue to shorten.

GLOSSARY

Allele One of variant forms of a gene that can exist at the same locus (location on the chromosome).

Allele dropout Failure to amplify one of the two alleles in a diploid genome.

Allele frequency The proportion of the allele of interest observed in a population; a measure of the incidence of gene variants; allele frequency is calculated by dividing the number of copies of the allele of interest by the number of copies of all alleles at the same locus in a population.

Alternative splicing A process by which different mRNA transcripts are produced from the same precursor transcript through alternate ways to remove introns and connect exons.

Amplification bias Bias introduced from preferential amplification of nucleotide molecules due to extreme base composition (e.g. GC-rich), poly A slippage etc., resulting in a different coverage distribution in the amplified sample compared to the original unamplified sample.

Aneuploidy A type of chromosome mutation in which the numbers of certain chromosomes are different from the normal chromosome numbers.

Biomarker Biological marker; also called molecular marker. A characteristic that can be objectively measured as a sign of a normal, abnormal or pathogenic process. A biomarker can also be used to indicate pharmacologic responses to a therapy.

Bridge PCR A PCR process in which the DNA molecule is clonally amplified from an adapter-ligated DNA template molecule that arches over and hybridizes to complementary adapter on a flow cell, serving as the substrate for amplification.

Chimeric transcript A fusion gene transcript that arise from chromosomal rearrangement.

Chromatin A complex of DNA and proteins that forms chromosomes. Two chromatin forms exist; euchromatin is the less condensed form that is often involved in active transcription. Heterochromatin is the condensed form that is typically not involved with transcription.

Chromatin remodeling A process facilitated by specific protein complexes (known as chromatin remodeling complexes), which changes chromatin structure by restructuring, mobilizing or ejecting nucleosomes to regulate access of various regulatory factors to the DNA.

Clinical laboratory improvement amendments (CLIA) certification Regulatory standards set up by the federal government of the USA to ensure accuracy, reliability and timeliness of all clinical laboratory testing on humans.

College of american pathologists (CAP) accreditation program Internationally recognized laboratory accreditation program from the College of American Pathologists. The accreditation is based on rigorous inspection of the quality of pathology and laboratory services.

Copy number variation (CNV) DNA segments that are present in a various number of copies in a genome.

DNA methylation An epigenetic modification that results in the addition of a methyl group (-CH_3) to C5 position of cytosine pyrimidine ring. DNA methylation can directly or indirectly repress transcription of a gene.

DNA sequencing A process of identifying the precise order of nucleotides in DNA molecules.

Epigenetic alterations Changes in cellular behavior or the activity state of chromatins without altering DNA nucleotide sequence. Epigenetic alterations include DNA methylation, chromatin remodeling and histone modification.

Exome-seq A methodology to capture selectively and sequence the complete coding regions of a genome. Also known as targeted exome-capture.

Flow cell Solid glass surface onto which DNA templates are tethered for clonal amplification and sequencing using second-generation sequencing technologies such as Illumina sequencing method.

Genome The complete set of genetic information in an organism.

Genomics Study of structure, evolution, function and mapping of genetic information conducted at the genome level.

Genotype The genetic constitution of an organism; the specific set of alleles inherited at a particular locus. The expression of the genotype constitutes an observable trait of an organism.

Germline mutation A mutation that occurs in the genome of germline cells.

Haplotype phase Determining the arrangements of alleles that are co-located on the same chromosome using correlated single nucleotide polymorphisms (i.e. haplotype).

Histology Study of the microscopic structure of tissues.

Histone Small chromosomal proteins that form a structural unit, around which DNA is wrapped in a nucleosome.

Histone modification A process in which chemical groups are added to or removed from histones to alter gene expression.

Hybridization A process of complementary DNA strands binding together. Typically, a probe is used to detect a specific DNA molecule by hybridization of a probe and the target DNA strands.

Kinase Enzyme that covalently attaches phosphate groups to substrate molecules, typically proteins.

Long-range PCR A type of polymerase chain reaction technology that allows amplification of long DNA fragments (typically greater than 5000 bases).

Locus A location on a chromosome.

mRNA Messenger RNA; a class of RNA molecules that are transcribed from the DNA of genes, and from which proteins are translated using ribosomes.

miRNA MicroRNA; small non-coding RNAs of 20–22 nucleotides in length. miRNAs are involved with transcriptional and post-transcriptional regulation of gene expression by imperfect pairing with target mRNAs.

Metagenomics A field of genomics that studies the collective microbial genomes obtained from an environmental sample such as soil or human gut.

Metastatic tumor Tumor that spreads from the original place of malignancy to another location in a body.

Multiplex PCR A type of polymerase chain reaction technology that allows simultaneous amplification of multiple target DNA fragments.

Mutation A change in a gene or a chromosome that differs from the wild type.

Non-synonymous mutation A single nucleotide substitution mutation that results in an amino acid change.

Nucleosome A basic unit of eukaryotic chromosomal structure that consists of 146 bp of DNA that wraps around eight histone proteins (an octamer consisting of duplicates of H2A, H2B, H3, H4). Nucleosomes are linearly arranged like "beads-on-a-string".

Oligomer A molecule that consists of a few monomers.

Physical coverage A number of DNA fragments that cover a particular nucleotide in the reconstructed sequence such as the reference genome.

Polymorphism The existence of more than one variant, or more than one allele for a phenotypic trait in a population.

Primary tumor The original tumor.

Probe A nucleotide sequence that can be used to identify a specific DNA molecule whose sequence is complementary to the probe sequence.

rRNA Ribosomal RNA; a class of RNA molecules that form part of the ribosome, the protein-synthesizing organelle in a cell.

Read Read-out of the nucleotide sequence of a DNA fragment during sequencing.

Sequence coverage A number of sequenced reads that cover a particular nucleotide in the reconstructed sequence such as the reference genome.

Sequencing library A population of nucleic acid fragments that are flanked by sequencing adapters.

Single nucleotide variation (SNV) Change of a single nucleotide compared to the reference genome.

Somatic mutation A mutation that occurs in the genome of cells that are not germline cells in an individual.

Synthetic lethality A genetic interaction in which a combination of mutations in more than one gene leads to cell death, while a mutation in one of these genes does not lead to cell death.

Transcriptome The complete set of RNA molecules expressed by a genome.

Tumorigenesis A process by which tumor is formed.

ACRONYMS AND ABBREVIATIONS

AML Acute myeloid leukemia

APL Acute promyelocytic leukemia

CAP College of American Pathologists (laboratory accreditation program)

CGH Comparative genomic hybridization (arrays)

cDNA Complementary DNA

CLIA Clinical Laboratory Improvement Amendments (certification)

CNV Copy number variation

CML Chronic myeloid leukemia

CpG Cytosine-phosphodiester-guanine dinucleotides

DLBCL Diffuse large B-cell lymphoma

DNA Deoxyribonucleic acid

EST Expressed sequence tag

FFPE Formalin-fixed, paraffin-embedded (tissue)

FISH Fluorescence *in situ* hybridization

FL Follicular lymphoma

GBM Glioblastoma multiforme

Indels Insertions and deletions

miRNA MicroRNA

PCR Polymerase chain reaction

RNA Ribonucleic acid

SAGE Serial analysis of gene expression

SNV Single nucleotide variation

REFERENCES

[1] Mendel G. Versuche ber ptlanzenhybriden. Verh Natuturforsch Ver Brunn 1866;4:3.

[2] Avery OT, Macleod CM, McCarty M. Studies on the chemical nature of the substance inducing transformation of pneumococcal types: induction of transformation by a desoxyribonucleic acid fraction isolated from pneumococcus type III. Oswald Theodore Avery (1877–1955). Clin Orthop Relat Res 2000;379(Suppl)S3–8.

[3] Watson JD, Crick FH. Molecular structure of nucleic acids; a structure for deoxyribose nucleic acid. Nature 1953;171:737–8.

[4] Crick FH, Barnett L, Brenner S, Watts-Tobin RJ. General nature of the genetic code for proteins. Nature 1961;192:1227–32.

[5] Lander ES, Linton LM, Birren B, Nusbaum C, Zody MC, Baldwin J, et al. Initial sequencing and analysis of the human genome. Nature 2001;409:860–921.

[6] Venter JC, Adams MD, Myers EW, Li PW, Mural RJ, Sutton GG, et al. The sequence of the human genome. Science 2001;291:1304–51.

[7] Reddy EP, Reynolds RK, Santos E, Barbacid M. A point mutation is responsible for the acquisition of transforming properties by the T24 human bladder carcinoma oncogene. Nature 1982;300:149–52.
[8] Metzker ML. Sequencing technologies – the next generation. Nat Rev Genet 2010;11:31–46.
[9] Ding L, Ley TJ, Larson DE, Miller CA, Koboldt DC, Welch JS, et al. Clonal evolution in relapsed acute myeloid leukaemia revealed by whole-genome sequencing. Nature 2012;481:506–10.
[10] Pleasance ED, Cheetham RK, Stephens PJ, McBride DJ, Humphray SJ, Greenman CD, et al. A comprehensive catalogue of somatic mutations from a human cancer genome. Nature 2010;463:191–6.
[11] Wang J, Ramakrishnan R, Tang Z, Fan W, Kluge A, Dowlati A, et al. Quantifying EGFR alterations in the lung cancer genome with nanofluidic digital PCR arrays. Clin Chem 2010;56:623–32.
[12] Boveri T. Concerning the origin of malignant tumours by Theodor Boveri. Translated and annotated by Henry Harris. J Cell Sci 2008;121(Suppl. 1):1–84.
[13] Stehelin D, Varmus HE, Bishop JM, Vogt PK. DNA related to the transforming gene(s) of avian sarcoma viruses is present in normal avian DNA. Nature 1976;260:170–3.
[14] Knudson Jr AG. Mutation and cancer: statistical study of retinoblastoma. Proc Natl Acad Sci USA 1971;68:820–3.
[15] Loeb LA, Springgate CF, Battula N. Errors in DNA replication as a basis of malignant changes. Cancer Res 1974;34:2311–21.
[16] Fearon ER, Vogelstein B. A genetic model for colorectal tumorigenesis. Cell 1990;61:759–67.
[17] Rowley JD. Letter: a new consistent chromosomal abnormality in chronic myelogenous leukaemia identified by quinacrine fluorescence and Giemsa staining. Nature 1973;243:290–3.
[18] Al-Hajj M, Wicha MS, Benito-Hernandez A, Morrison SJ, Clarke MF. Prospective identification of tumorigenic breast cancer cells. Proc Natl Acad Sci USA 2003;100:3983–8.
[19] Nowell PC. The clonal evolution of tumor cell populations. Science 1976;194:23–8.
[20] Campbell PJ, Yachida S, Mudie LJ, Stephens PJ, Pleasance ED, Stebbings LA, et al. The patterns and dynamics of genomic instability in metastatic pancreatic cancer. Nature 2010;467:1109–13.
[21] Yates LR, Campbell PJ. Evolution of the cancer genome. Nat Rev Genet 2012;13:795–806.
[22] Shah SP, Roth A, Goya R, Oloumi A, Ha G, Zhao Y, et al. The clonal and mutational evolution spectrum of primary triple-negative breast cancers. Nature 2012;486:395–9.
[23] Gerlinger M, Rowan AJ, Horswell S, Larkin J, Endesfelder D, Gronroos E, et al. Intratumor heterogeneity and branched evolution revealed by multiregion sequencing. N Engl J Med 2012;366:883–92.
[24] Komaki K, Sano N, Tangoku A. Problems in histological grading of malignancy and its clinical significance in patients with operable breast cancer. Breast Cancer 2006;13:249–53.
[25] Baldus CD, Mrozek K, Marcucci G, Bloomfield CD. Clinical outcome of *de novo* acute myeloid leukaemia patients with normal cytogenetics is affected by molecular genetic alterations: a concise review. Br J Haematol 2007;137:387–400.
[26] Rouzier R, Perou CM, Symmans WF, Ibrahim N, Cristofanilli M, Anderson K, et al. Breast cancer molecular subtypes respond differently to preoperative chemotherapy. Clin Cancer Res 2005;11:5678–85.
[27] Fisher B, Redmond C, Brown A, Wickerham DL, Wolmark N, Allegra J, et al. Influence of tumor estrogen and progesterone receptor levels on the response to tamoxifen and chemotherapy in primary breast cancer. J Clin Oncol 1983;1: 227–41.
[28] Yachida S, Jones S, Bozic I, Antal T, Leary R, Fu B, et al. Distant metastasis occurs late during the genetic evolution of pancreatic cancer. Nature 2010;467:1114–7.
[29] Morin RD, Johnson NA, Severson TM, Mungall AJ, An J, Goya R, et al. Somatic mutations altering EZH2 (Tyr641) in follicular and diffuse large B-cell lymphomas of germinal-center origin. Nat Genet 2010;42:181–5.
[30] Rosenfeld N, Aharonov R, Meiri E, Rosenwald S, Spector Y, Zepeniuk M, et al. MicroRNAs accurately identify cancer tissue origin. Nat Biotechnol 2008;26:462–9.
[31] Widschwendter M, Fiegl H, Egle D, Mueller-Holzner E, Spizzo G, Marth C, et al. Epigenetic stem cell signature in cancer. Nat Genet 2007;39:157–8.
[32] Sanger F, Nicklen S, Coulson AR. DNA sequencing with chain-terminating inhibitors. Proc Natl Acad Sci USA 1977;74: 5463–7.
[33] Cohen AS, Najarian DR, Paulus A, Guttman A, Smith JA, Karger BL. Rapid separation and purification of oligonucleotides by high-performance capillary gel electrophoresis. Proc Natl Acad Sci USA 1988;85:9660–3.
[34] Strausberg RL, Levy S, Rogers YH. Emerging DNA sequencing technologies for human genomic medicine. Drug Discov Today 2008;13:569–77.
[35] Osoegawa K, Woon PY, Zhao B, Frengen E, Tateno M, Catanese JJ, et al. An improved approach for construction of bacterial artificial chromosome libraries. Genomics 1998;52:1–8.
[36] Margulies M, Egholm M, Altman WE, Attiya S, Bader JS, Bemben LA, et al. Genome sequencing in microfabricated high-density picolitre reactors. Nature 2005;437:376–80.
[37] Ghadessy FJ, Ong JL, Holliger P. Directed evolution of polymerase function by compartmentalized self-replication. Proc Natl Acad Sci USA 2001;98:4552–7.
[38] Nyren P, Pettersson B, Uhlen M. Solid phase DNA minisequencing by an enzymatic luminometric inorganic pyrophosphate detection assay. Anal Biochem 1993;208:171–5.
[39] Rothberg JM, Leamon JH. The development and impact of 454 sequencing. Nat Biotechnol 2008;26:1117–24.
[40] Adessi C, Matton G, Ayala G, Turcatti G, Mermod JJ, Mayer P, et al. Solid phase DNA amplification: characterisation of primer attachment and amplification mechanisms. Nucleic Acids Res 2000;28:e87.
[41] Turcatti G, Romieu A, Fedurco M, Tairi AP. A new class of cleavable fluorescent nucleotides: synthesis and optimization as reversible terminators for DNA sequencing by synthesis. Nucleic Acids Res 2008;36:e25.
[42] Dohm JC, Lottaz C, Borodina T, Himmelbauer H. Substantial biases in ultra-short read data sets from high-throughput DNA sequencing. Nucleic Acids Res 2008;36:e105.
[43] Pop M, Salzberg SL. Bioinformatics challenges of new sequencing technology. Trends Genet 2008;24:142–9.
[44] Shendure J, Porreca GJ, Reppas NB, Lin X, McCutcheon JP, Rosenbaum AM, et al. Accurate multiplex polony sequencing of an evolved bacterial genome. Science 2005;309:1728–32.
[45] Housby JN, Southern EM. Fidelity of DNA ligation: a novel experimental approach based on the polymerisation of libraries of oligonucleotides. Nucleic Acids Res 1998;26:4259–66.

[46] Rothberg JM, Hinz W, Rearick TM, Schultz J, Mileski W, Davey M, et al. An integrated semiconductor device enabling non-optical genome sequencing. Nature 2011;475:348–52.
[47] Sakurai T, Husimi Y. Real-time monitoring of DNA polymerase reactions by a micro ISFET pH sensor. Anal Chem 1992;64:1996–7.
[48] Loman NJ, Misra RV, Dallman TJ, Constantinidou C, Gharbia SE, Wain J, et al. Performance comparison of benchtop high-throughput sequencing platforms. Nat Biotechnol 2012;30:434–9.
[49] Greenman C, Stephens P, Smith R, Dalgliesh GL, Hunter C, Bignell G, et al. Patterns of somatic mutation in human cancer genomes. Nature 2007;446:153–8.
[50] Sjoblom T, Jones S, Wood LD, Parsons DW, Lin J, Barber TD, et al. The consensus coding sequences of human breast and colorectal cancers. Science 2006;314:268–74.
[51] Strausberg RL, Simpson AJ, Wooster R. Sequence-based cancer genomics: progress, lessons and opportunities. Nat Rev Genet 2003;4:409–18.
[52] Carninci P, Kasukawa T, Katayama S, Gough J, Frith MC, Maeda N, et al. The transcriptional landscape of the mammalian genome. Science 2005;309:1559–63.
[53] Futreal PA, Coin L, Marshall M, Down T, Hubbard T, Wooster R, et al. A census of human cancer genes. Nat Rev Cancer 2004;4:177–83.
[54] Ley TJ, Mardis ER, Ding L, Fulton B, McLellan MD, Chen K, et al. DNA sequencing of a cytogenetically normal acute myeloid leukaemia genome. Nature 2008;456:66–72.
[55] Bentley DR, Balasubramanian S, Swerdlow HP, Smith GP, Milton J, Brown CG, et al. Accurate whole human genome sequencing using reversible terminator chemistry. Nature 2008;456:53–9.
[56] Ajay SS, Parker SC, Abaan HO, Fajardo KV, Margulies EH. Accurate and comprehensive sequencing of personal genomes. Genome Res 2011;21:1498–505.
[57] Curtis C, Shah SP, Chin SF, Turashvili G, Rueda OM, Dunning MJ, et al. The genomic and transcriptomic architecture of 2,000 breast tumours reveals novel subgroups. Nature 2012;486:346–52.
[58] Nekrutenko A, Taylor J. Next-generation sequencing data interpretation: enhancing reproducibility and accessibility. Nat Rev Genet 2012;13:667–72.
[59] Shendure J, Lieberman Aiden E. The expanding scope of DNA sequencing. Nat Biotechnol 2012;30:1084–94.
[60] Ng SB, Turner EH, Robertson PD, Flygare SD SD, Bigham AW, Lee C, et al. Targeted capture and massively parallel sequencing of 12 human exomes. Nature 2009;461:272–6.
[61] Ingolia NT, Ghaemmaghami S, Newman JR, Weissman JS. Genome-wide analysis in vivo of translation with nucleotide resolution using ribosome profiling. Science 2009;324:218–23.
[62] Mercer TR, Gerhardt DJ, Dinger ME, Crawford J, Trapnell C, Jeddeloh JA, et al. Targeted RNA sequencing reveals the deep complexity of the human transcriptome. Nat Biotechnol 2011;30:99–104.
[63] Morin RD, O'Connor MD, Griffith M, Kuchenbauer F, Delaney A, Prabhu AL, et al. Application of massively parallel sequencing to microRNA profiling and discovery in human embryonic stem cells. Genome Res 2008;18:610–21.
[64] Mortazavi A, Williams BA, McCue K, Schaeffer L, Wold B. Mapping and quantifying mammalian transcriptomes by RNA-Seq. Nat Methods 2008;5:621–8.
[65] Shah SP, Morin RD, Khattra J, Prentice L, Pugh T, Burleigh A, et al. Mutational evolution in a lobular breast tumour profiled at single nucleotide resolution. Nature 2009;461:809–13.
[66] Laird PW. Principles and challenges of genome-wide DNA methylation analysis. Nat Rev Genet 2010;11:191–203.
[67] Park PJ. ChIP-seq: advantages and challenges of a maturing technology. Nat Rev Genet 2009;10:669–80.
[68] Boyle AP, Davis S, Shulha HP, Meltzer P, Margulies EH, Weng Z, et al. High-resolution mapping and characterization of open chromatin across the genome. Cell 2008;132:311–22.
[69] Giresi PG, Lieb JD. Isolation of active regulatory elements from eukaryotic chromatin using FAIRE (Formaldehyde Assisted Isolation of Regulatory Elements). Methods 2009;48:233–9.
[70] Mardis ER, Ding L, Dooling DJ, Larson DE, McLellan MD, Chen K, et al. Recurring mutations found by sequencing an acute myeloid leukemia genome. N Engl J Med 2009;361:1058–66.
[71] Parsons DW, Jones S, Zhang X, Lin JC, Leary RJ, Angenendt P, et al. An integrated genomic analysis of human glioblastoma multiforme. Science 2008;321:1807–12.
[72] Ley TJ, Ding L, Walter MJ, McLellan MD, Lamprecht T, Larson DE, et al. DNMT3A mutations in acute myeloid leukemia. N Engl J Med 2010;363:2424–33.
[73] Yap DB, Chu J, Berg T, Schapira M, Cheng SW, Moradian A, et al. Somatic mutations at EZH2 Y641 act dominantly through a mechanism of selectively altered PRC2 catalytic activity, to increase H3K27 trimethylation. Blood 2011;117:2451–9.
[74] Ernst T, Chase AJ, Score J, Hidalgo-Curtis CE, Bryant C, Jones AV, et al. Inactivating mutations of the histone methyltransferase gene EZH2 in myeloid disorders. Nat Genet 2010;42:722–6.
[75] Morin RD, Mendez-Lago M, Mungall AJ, Goya R, Mungall KL, Corbett RD, et al. Frequent mutation of histone-modifying genes in non-Hodgkin lymphoma. Nature 2011;476:298–303.
[76] Varela I, Tarpey P, Raine K, Huang D, Ong CK, Stephens P, et al. Exome sequencing identifies frequent mutation of the SWI/SNF complex gene PBRM1 in renal carcinoma. Nature 2011;469:539–42.
[77] Jones S, Wang TL, Shih I, Mao TL, Nakayama K, Roden R, et al. Frequent mutations of chromatin remodeling gene ARID1A in ovarian clear cell carcinoma. Science 2010;330:228–31.
[78] Meyerson M, Gabriel S, Getz G. Advances in understanding cancer genomes through second-generation sequencing. Nat Rev Genet 2010;11:685–96.
[79] Berger MF, Lawrence MS, Demichelis F, Drier Y, Cibulskis K, Sivachenko AY, et al. The genomic complexity of primary human prostate cancer. Nature 2011;470:214–20.
[80] Kloosterman WP, Hoogstraat M, Paling O, Tavakoli-Yaraki M, Renkens I, Vermaat JS, et al. Chromothripsis is a common mechanism driving genomic rearrangements in primary and metastatic colorectal cancer. Genome Biol 2011;12:R103.
[81] Nik-Zainal S, Alexandrov LB, Wedge DC, van Loo P, Greenman CE, Raine K, et al. Mutational processes molding the genomes of 21 breast cancers. Cell 2012;149:979–93.
[82] Drier Y, Lawrence MS, Carter SL, Stewart C, Gabriel SB, Lander ES, et al. Somatic rearrangements across cancer reveal classes of samples with distinct patterns of DNA breakage and rearrangement-induced hypermutability. Genome Res 2013;23:228–35.
[83] Kostic AD, Gevers D, Pedamallu CS, Michaud M, Duke F, Earl AM, et al. Genomic analysis identifies association of *Fusobacterium* with colorectal carcinoma. Genome Res 2012;22:292–8.
[84] Castellarin M, Warren RL, Freeman JD, Dreolini L, Krzywinski M, Strauss J, et al. *Fusobacterium nucleatum* infection is prevalent in human colorectal carcinoma. Genome Res 2012;22:299–306.

[85] Shah SP, Kobel M, Senz J, Morin RD, Clarke BA, Wiegand KC, et al. Mutation of FOXL2 in granulosa-cell tumors of the ovary. N Engl J Med 2009;360:2719–29.

[86] McCabe MT, Ott HM, Ganji G, Korenchuk S, Thompson C, van Aller GS, et al. EZH2 inhibition as a therapeutic strategy for lymphoma with EZH2-activating mutations. Nature 2012;492:108–12.

[87] Dawson SJ, Tsui DW, Murtaza M, Biggs H, Rueda OM, Chin SF, et al. Analysis of circulating tumor DNA to monitor metastatic breast cancer. N Engl J Med 2013;368:1199–209.

[88] Leary RJ, Kinde I, Diehl F, Schmidt K, Clouser C, Duncan C, et al. Development of personalized tumor biomarkers using massively parallel sequencing. Sci Transl Med 2010;2:20ra14.

[89] Chin L, Andersen JN, Futreal PA. Cancer genomics: from discovery science to personalized medicine. Nat Med 2011;17:297–303.

[90] Mok TS, Wu YL, Thongprasert S, Yang CH, Chu DT, Saijo N, et al. Gefitinib or carboplatin–paclitaxel in pulmonary adenocarcinoma. N Engl J Med 2009;361:947–57.

[91] Jones SJ, Laskin J, Li YY, Griffith OL, An J, Bilenky M, et al. Evolution of an adenocarcinoma in response to selection by targeted kinase inhibitors. Genome Biol 2010;11:R82.

[92] Welch JS, Westervelt P, Ding L, Larson DE, Klco JM, Kulkarni S, et al. Use of whole-genome sequencing to diagnose a cryptic fusion oncogene. J Am Med Assoc 2011;305:1577–84.

[93] Niedringhaus TP, Milanova D, Kerby MB, Snyder MP, Barron AE. Landscape of next-generation sequencing technologies. Anal Chem 2011;83:4327–41.

[94] Bowers J, Mitchell J, Beer E, Buzby PR, Causey M, Efcavitch JW, et al. Virtual terminator nucleotides for next-generation DNA sequencing. Nat Methods 2009;6:593–5.

[95] Pushkarev D, Neff NF, Quake SR. Single-molecule sequencing of an individual human genome. Nat Biotechnol 2009;27:847–50.

[96] Ozsolak F, Platt AR, Jones DR, Reifenberger JG, Sass LE, McInerney P, et al. Direct RNA sequencing. Nature 2009;461:814–8.

[97] Levene MJ, Korlach J, Turner SW, Foquet M, Craighead HG, Webb WW. Zero-mode waveguides for single-molecule analysis at high concentrations. Science 2003;299:682–6.

[98] Travers KJ, Chin CS, Rank DR, Eid JS, Turner SW. A flexible and efficient template format for circular consensus sequencing and SNP detection. Nucleic Acids Res 2010;38:e159.

[99] Eid J, Fehr A, Gray J, Luong K, Lyle J, Otto G, et al. Real-time DNA sequencing from single polymerase molecules. Science 2009;323:133–8.

[100] Flusberg BA, Webster DR, Lee JH, Travers KJ, Olivares EC, Clark TA, et al. Direct detection of DNA methylation during single-molecule, real-time sequencing. Nat Methods 2010;7:461–5.

[101] Branton D, Deamer DW, Marziali A, Bayley H, Benner SA, Butler T, et al. The potential and challenges of nanopore sequencing. Nat Biotechnol 2008;26:1146–53.

[102] Clarke J, Wu HC, Jayasinghe L, Patel A, Reid S, Bayley H. Continuous base identification for single-molecule nanopore DNA sequencing. Nat Nanotechnol 2009;4:265–70.

[103] Wei R, Gatterdam V, Wieneke R, Tampe R, Rant U. Stochastic sensing of proteins with receptor-modified solid-state nanopores. Nat Nanotechnol 2012;7:257–63.

[104] Kasianowicz JJ, Brandin E, Branton D, Deamer DW. Characterization of individual polynucleotide molecules using a membrane channel. Proc Natl Acad Sci USA 1996;93:13770–3.

[105] Bashir A, Klammer AA, Robins WP, Chin CS, Webster D, Paxinos E, et al. A hybrid approach for the automated finishing of bacterial genomes. Nat Biotechnol 2012;30:701–7.

[106] Koren S, Schatz MC, Walenz BP, Martin J, Howard JT, Ganapathy G, et al. Hybrid error correction and *de novo* assembly of single-molecule sequencing reads. Nat Biotechnol 2012;30:693–700.

[107] Walsh T, Lee MK, Casadei S, Thornton AM, Stray SM, Pennil C, et al. Detection of inherited mutations for breast and ovarian cancer using genomic capture and massively parallel sequencing. Proc Natl Acad Sci USA 2010;107:12629–33.

[108] Roychowdhury S, Iyer MK, Robinson DR, Lonigro RJ, Wu YM, Cao X, et al. Personalized oncology through integrative high-throughput sequencing: a pilot study. Sci Transl Med 2011;3:111–21.

[109] Kircher M, Sawyer S, Meyer M. Double indexing overcomes inaccuracies in multiplex sequencing on the Illumina platform. Nucleic Acids Res 2012;40:e3.

[110] Champlot S, Berthelot C, Pruvost M, Bennett EA, Grange T, Geigl EM. An efficient multistrategy DNA decontamination procedure of PCR reagents for hypersensitive PCR applications. PLoS One 2010;5:e13042.

[111] Levy S, Sutton G, Ng PC, FeuK L, Halpern AL, Walenz BP, et al. The diploid genome sequence of an individual human. PLoS Biol 2007;5:e254.

[112] The 1000 Genomes Project Consortium. A map of human genome variation from population-scale sequencing. Nature 2010;467:1061–73.

[113] Thorogood A, Knoppers BM, Dondorp WJ, de Wert GM. Whole-genome sequencing and the physician. Clin Genet 2012;81:511–3.

[114] Aparicio S, Caldas C. The implications of clonal genome evolution for cancer medicine. N Engl J Med 2013;368:842–51.

[115] Elowitz MB, Levine AJ, Siggia ED, Swain PS. Stochastic gene expression in a single cell. Science 2002;297:1183–6.

[116] Lecault V, White AK, Singhal A, Hansen CL. Microfluidic single cell analysis: from promise to practice. Curr Opin Chem Biol 2012;16:381–90.

[117] Leung K, Zahn H, Leaver T, Konwar KM, Hanson NW, Page AP, et al. A programmable droplet-based microfluidic device applied to multiparameter analysis of single microbes and microbial communities. Proc Natl Acad Sci USA 2012;109:7665–70.

[118] Navin N, Krasnitz A, Rodgers L, Cook K, Meth J, Kendall J, et al. Inferring tumor progression from genomic heterogeneity. Genome Res 2010;20:68–80.

[119] Baslan T, Kendall J, Rodgers L, Cox H, Riggs M, Stepansky A, et al. Genome-wide copy number analysis of single cells. Nat Protoc 2012;7:1024–41.

[120] Navin N, Kendall J, Troge J, Andrews P, Rodgers L, McIndoo J, et al. Tumour evolution inferred by single-cell sequencing. Nature 2011;472:90–4.

[121] Powell AA, Talasaz AH, Zhang H, Coram MA, Reddy A, Deng G, et al. Single cell profiling of circulating tumor cells: transcriptional heterogeneity and diversity from breast cancer cell lines. PLoS One 2012;7:e33788.

[122] Robertson M. Biology in the 1980s, plus or minus a decade. Nature 1980;285:358–9.

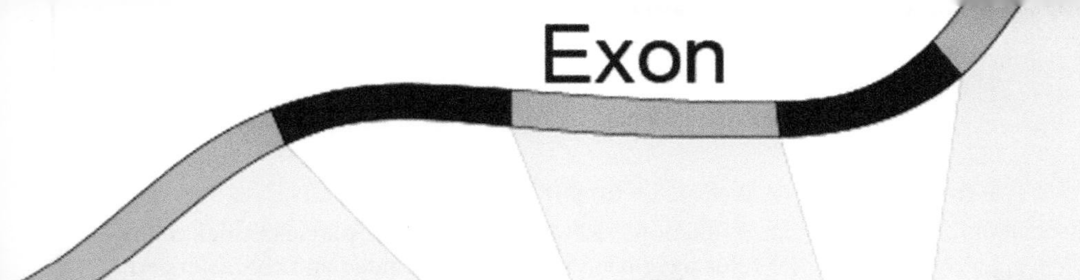

Chapter 3

mRNA

Short reads

Cancer Genomics

Cancer Transcriptome Sequencing and Analysis

Ryan D. Morin[1]
Stephen B. Montgomery[2]

[1]Department of Molecular Biology and Biochemistry, Simon Fraser University, Burnaby, BC, Canada

[2]Departments of Pathology and Genetics, Stanford University, Stanford, CA, USA

Contents

Cancer Genomics. DOI: http://dx.doi.org/10.1016/B978-0-12-396967-5.00003-7

Key Concepts

- Transcriptome analysis provides a powerful window into pathological processes in cancer, including differential gene expression, allele-specific expression, mutation, and the presence of fusion transcripts
- Generation of cancer transcriptomes requires consideration of sample origins, whether fixed, fresh, liquid or solid. Abundant tissue heterogeneity can introduce noise that masks detection of pathological differences
- Processing RNA-seq requires a suitable set of software and statistical considerations to compensate for short, spliced reads and biases inherent in the technology
- Differential expression analyses require consideration of sample preparation and sequencing biases. Processing RNA-seq data requires different statistical considerations than those applied for microarrays
- Transcriptomes generated with RNA-seq can harbor variants which inform the activity of somatic variants
- eQTL analyses can implicate the cellular function and pathways influenced by cancer-associated genetic variants
- Transcriptomes can harbor sequences not typically present in the human genome such as those deriving from oncogenic pathogens

INTRODUCTION

The abundance of functional units of the cell (proteins and non-coding RNAs) results from the transcription of genes into complementary RNA molecules. Collectively, these RNA molecules make up what is referred to as the transcriptome, which includes numerous species of RNA and post-transcriptionally modified RNA. The RNA species that is subsequently translated into proteins, namely messenger RNA (mRNA), is the most commonly studied molecule type as it is the intermediate molecule between the stable genetic components of the cell (genomic DNA) and the functional proteins that perform cellular functions (the central dogma of molecular biology). Beyond mRNA, other functional RNA molecules that play essential cellular roles have been isolated, quantitated and characterized. The most famous non-coding RNA (besides rRNA, the RNA components of the ribosome) are microRNAs, which are important post-transcriptional regulators of gene expression. A less well-understood class of potentially functional non-coding large RNA molecules, which were discovered with the aid of global mRNA sequencing approaches, are collectively referred to as long intergenic non-coding RNAs (lincRNAs) [1].

The complex and dynamic character of the transcriptome provides the opportunity to identify molecular mechanisms of pathogenesis and biology as well as biomarkers of disease progression and response to treatment. In many diseases, the cells comprising pathological tissues or organs have one or more defects reflected in some element of their transcriptome, which can range in scale from a single nucleotide replacement, gain, loss or alteration of an exon, to the altered dosage of an entire gene or dysregulation of any number of genes. Some of these alterations may result from changes to the DNA sequence and others may not. For example, mutations that affect regulatory regions can alter the transcriptional level of genes and can also impact their splicing patterns and act in *cis*. Alternatively, epigenetic changes such as DNA methylation or histone protein modification, two processes that can be deregulated in cancer, can also result in perturbations to the transcriptome. These may result in either the inappropriate silencing or deregulated expression of genes and can further be restricted to a single allele. Like regulatory mutations, epigenetic changes affecting a single allele can result in allele-specific expression (ASE), in which the amount of mRNA transcribed from each of the two copies of a gene in a cell is quantitatively different.

Methods for studying individual transcripts and their variability in disease have been developed and have evolved from low-throughput assays capable of interrogating the abundance of only single mRNA species (including Northern blots and quantitative-PCR) to current global approaches collectively referred to as transcriptomics (including microarrays and RNA-sequencing (RNA-seq)). Among these methods, the recent invention of RNA-seq has provided new potential to detect many different types of transcriptomic perturbations; it has provided the opportunity to survey previously unknown expressed elements and extended our view of the transcriptome from the simple measurement of dosage (or expression) to provide a more detailed view of the variation in transcript abundance, alternative splicing and allelic-expression. These features provide a new opportunity to understand cancer biology and this chapter will discuss methods for analyzing each feature using

RNA-seq and more general considerations for study design of cancer transcriptome experiments.

CANCER TRANSCRIPTOME SEQUENCING

Over the last decade, systematic characterization of gene expression (or transcriptome profiling) of cancer samples has been largely performed with either microarray technologies or Sanger sequencing methods. Microarrays, which are based on synthesis of oligonucleotide probes and subsequent fluorescence through target hybridization, provide high-throughput and low-cost measurement of mRNA abundance (gene expression). They have been used to survey every major type of cancer. For example, ArrayExpress [2], a public microarray repository, currently lists nearly 400 000 assays with the keyword "cancer". Despite the large impact that microarrays have had on the study of transcriptomes, they possess several limitations: first, the individual genes to be queried need to be selected in advance, such that appropriate probes can be designed; second, unless specially designed [3], the selected gene sequences convey only limited information about splicing or allelic origin; and third, each individual measurement itself is non-discrete (analog) in nature. Complementing microarrays, conventional sequencing has also been directly applied to characterize transcriptomes (cancer or otherwise). Historically, for cost and time savings, random sampling of cDNA libraries using Sanger capillary sequencing allowed the production of catalogs of expressed sequence tags (ESTs), which comprise sequence reads from the ends of cDNA clones. Though these facilitated improved annotation of the exon structure of the human genome and enabled the identification of some alternative splicing, the output of such assays was not sufficiently quantitative to facilitate differential expression analysis. Serial analysis of gene expression (SAGE) transformed the application of sequencing to transcriptome profiling; in SAGE, short cDNA tags (between 10 and 25 bp per tag) are concatenated and sequenced. The discrete counts of SAGE tags that could be unambiguously mapped to genes provided a method for quantifying transcriptional output [4]. Furthermore, SAGE studies facilitated the *de novo* identification of genes and pathways that have led to biomarker discovery of cancer gene targets [5]. Despite the potential of this approach, the cost of sequencing SAGE libraries deeply enough to provide sufficient statistical power remained restrictively expensive especially when compared to microarrays. Now, with the routine availability of high-throughput DNA sequencing, millions of sequencing fragments of lengths between 36 and 262 bp and higher can be routinely profiled [6] using either SAGE-like [7] or global sequencing strategies [8]. These approaches have provided unprecedented and unbiased resolution of cancer transcriptomes that has and will continue to impact greatly our understanding the etiology and evolution of cancer.

Experimental Designs

RNA-sequencing (RNA-seq) libraries are principally produced by fragmentation or randomly primed amplification of cDNA molecules, followed by the subsequent addition of universal sequencing adaptors. In preparing libraries, there are several choices in experimental design and some major technical caveats, which are summarized in Table 3.1. The original RNA-seq procedure yielded fragments that were uniformly derived from both strands of the cDNA, thus representing the parent mRNA strand and its complement. Kits for producing strand-specific RNA-seq libraries are now becoming available. Though currently more costly, depending on the specific needs of a researcher, these strand-specific libraries may be

TABLE 3.1 RNA-seq Library Protocols

Library Design	Pros	Cons
Ribo-depletion	Removes ribosomal RNA from sample and provides access to mRNA and non-coding RNA	Pre-mRNA retained providing reads for junctions
Poly-A selection	Removes all non-polyadenylated RNA. Enriches for mRNA	Limited quantification of non-coding RNA
Treatment with double-stranded nuclease	Reduces expression level of highly abundant transcripts	Not a well-utilized protocol; biases/impact on global expression not characterized
MicroRNA capture	Provides access to microRNA	MicroRNA recovered depends on capture technique
mRNA capture	Provides selective representation of specific transcripts and potentially unknown fusion partners	Potentially biases quantification; reduces representation of untargeted genes, which may be of interest
Multiplexed using genetic barcoding	Many samples processed together	Some read pairs might be erroneously swapped if run cluster density is too high

appealing. Beyond choosing an appropriate library construction kit, one must also decide on whether to enrich or deplete a "total" RNA sample for particular classes of RNA. Since the total RNA pool will include ribosomal RNA, both pre-mRNA and mRNA and a litany of non-coding RNA, various protocols have been developed with the aim of removing or enriching for these specific RNA species. In various cell types, the bulk of RNA is usually composed of ribosomal RNA (rRNA) genes and very highly transcribed mRNAs such as hemoglobins in whole blood, immunoglobulins in mature B cells and insulin in β-cells of the pancreas. If such transcripts are not removed prior to library construction, oversampling these transcripts consumes the bulk of reads and thus increases the cost of sufficiently sampling libraries such that other less abundant mRNAs are well represented in the data. As a result, most protocols focus on removal of rRNAs by either selection of polyadenylated RNA (termed a "poly-A" library) or by selective depletion of rRNAs using various capture and separation techniques often employing commercially available kits such as RiboMinus or Ribo-Zero (Epicentre). Furthermore, denaturing and re-annealing double-stranded sequencing libraries in the presence of double-stranded nucleases can also be used as a means to "normalize" libraries by removing highly abundant RNA species based on the fact that more abundant and homogeneous RNA molecules will anneal more rapidly, and thus succumb to enzymatic digestion [9].

Owing to the diversity of approaches for library construction, comparison of gene expression features between libraries requires monitoring and a detailed understanding of the relative technical properties, limitations and biases arising from the different methodological approaches. For instance, poly-A libraries do not facilitate accurate quantification of all non-coding RNA species when compared to ribo-depletion libraries, and even between ribo-depletion protocols there exist differences between the sizes of non-coding RNA that are retained [10]. Specific library protocols have also been designed to target selectively microRNAs (miRNAs), a class of RNAs of which some have been specifically identified to be oncogenic or tumor suppressive on their own [11,12]. Because of the size of mature miRNAs (18–23 nt) and their lack of polyadenylation, a separate strategy (discussed in detail elsewhere [13]) is needed for profiling this RNA species.

Another approach that can be used to provide suitable coverage of genes of interest involves selective capture/enrichment (rather than depletion) of cDNA fragments corresponding to those genes. This is comparable, in both principle and implementation, to approaches used to sequence the exome instead of the entire genome. One such method involves a set of biotinylated RNA baits that comprise sequences corresponding to the exons of interest. These baits are combined with the RNA-seq library and, after allowing hybridization of library fragments complementary to the baits, these complexes are retrieved using streptavidin beads. The resulting library has reduced complexity and represents only the fragments corresponding to the baits. Despite the removal of much of the RNA species during the preparation of such a library, the read coverage of retrieved transcripts retains their gene expression information. However, it is important to note that this approach is able to retrieve fragments from oncogenic fusion transcripts even if one of the gene partners was not among those targeted by the baits [14]. In addition to capture approaches, economies in sequencing mean that multiple samples can typically be indexed (e.g. by genetic "barcoding") and pooled in a single sequencing reaction. Currently, a single lane of an Illumina HiSeq 2500 will generate 750 million paired-end reads; according to the "Standards, Guidelines and Best Practices for RNA-seq" as published by the ENCODE Consortium and depending on application, this would provide capacity for adequate transcriptome coverage for 6 to 24 samples [15].

As sequencing technologies and protocols are evolving rapidly, it is important to measure accurately the influence of such changes on the quantification of gene expression. To enable monitoring of how such technical variability impacts expression values derived from RNA-seq, many laboratories have included control samples or "spike-ins" in their sequencing pipelines or libraries. For example, the PhiX bacteriophage genome can be included to monitor minimally sequencing performance and resultant base qualities. However, PhiX alone does not provide information regarding the dynamic range or sequence length biases of the sampled transcripts from RNA-seq. To address this, an approach for assessing quality and supporting data integration over time using ERRC (External RNA Control Consortium) RNA spike-ins has been described [16]. This method requires pre-loading a mix of 92 control RNAs that have been standardized across a concentration range spanning 20 orders of magnitude and supports tracking of not only biases in quantification but diverse mapping, orientation and sequencing biases. Unfortunately, the use of such spike-ins has not been universally adopted and their design does not easily support investigation of biases that may influence more complex transcribed loci (such as those that are extensively alternatively spliced in higher eukaryotes). More recent designs, however, include multiple mixes in which the differences between concentrations are known. This design can aid differential expression studies between cancer and normal samples by providing a standard curve across the quantification range that describes an experiment's power and variability to detect known differences in transcript abundance.

In addition to the choice of library construction protocol or the addition of control samples or spike-ins, the abundance of starting RNA is another factor that can limit experimental design. Common sequencing protocols such as Illumina's TruSeq RNA-sequencing kit recommends 1 μg of RNA per sample but, depending on the RNA quality, quality libraries can be generated from 100 ng of RNA or less. More recently, much effort has been placed on extending the limits of library preparation to RNA abundance low enough to allow profiling of transcriptomes from single cells [17,18]. The importance of such approaches to cancer will undoubtedly enable the dissection of cell-to-cell variation within tumors, improved targeting of cancer stem cells and circulating tumor cells (CTCs), monitoring of tumor progression (or evolution) and evaluation of therapeutic responses [19]. Indeed, by combining advances in cDNA synthesis and amplification, Ramskold et al. recently developed a single cell RNA-sequencing method and applied it to the characterization of CTCs (NG2 + putative melanoma cells). Their study indicated that the gene expression profiles of isolated cells clustered based on their identity, demonstrating that single cell expression profiling could provide a new means of identifying cancer-specific biomarkers, particularly those associated with disease burden or recurrence, in addition to furthering our understanding of the biology of tumor metastasis [20].

Archival Samples

As many variables can influence RNA quality, the source and treatment of a cancer sample prior to analysis is an important factor in designing RNA-seq experiments. Currently, the bulk of cancer samples available in tissue banks are formalin-fixed, paraffin-embedded (FFPE), a process that enhances the preservation of tissue and facilitates standard histopathology. However, this process has important impacts on RNA quality. Von Ahlfen et al. systematically explored the impact of storage time, conditions, fixation time and specimen time on the quality of RNA [21]. They reported that improved RNA quality involves reducing time between surgical removal and fixation, limiting the size of tissue subject to FFPE (to provide efficient inactivation of RNases), extracting RNA from samples within 12 months and ensuring that cross-links are reversed without furthering RNA degradation. Furthermore, for increasingly degraded RNA, cDNA synthesis and library construction approaches that enrich for mRNA using poly-A tails artificially enrich for sequences proximal to the 3′ end of mRNAs. Though this limits utility for some analyses, depending on desired application, this enrichment can be desirable as 3′ end sequencing of FFPE samples has been reported to give quantitative measurements of gene expression and has been able to pinpoint novel oncogenic pathways [22].

MicroRNAs may be more stable than larger RNA molecules such as mRNAs. Multiple studies have assessed the degree to which miRNA can be recovered from FFPE archival samples and have shown that paired and archived renal cell carcinoma and normal tissues appropriately cluster ($n = 3$) [23] and, in lung cancer, miRNA profiles between fresh frozen and FFPE samples were found to be highly correlated ($r = 0.868$) [24].

Handling Tissue Heterogeneity

For various reasons, it can be difficult to obtain a sufficiently pure (and large) population of malignant cells, and even appropriate non-malignant tissues, to perform gene expression profiling experiments such as RNA-seq. Cancer tissues can have complex histological organization with connective tissues, tumor-infiltrating immune cells and stromal tissues often mixed with cancer cells. In gene expression profiling of normal and cancer tissues, one of the major challenges in experimental design is obtaining homogeneous cell populations. This heterogeneity can confound analysis and reduce power to detect mutations and differential expression. For solid tumors, laser microdissection can aid in selecting small cell populations. In liquid tumors, known cell-surface markers can assist in accessing a pure population when combined with fluorescence-activated cell sorting. However, suitable and equally pure normal populations must also be identified. The magnitude of the challenge is largely specific to individual cancers. For instance, in chronic lymphocytic leukemia (CLL), it can be difficult to distinguish normal B cells from CLL cells from a single affected individual and, as a result, gene expression analyses have subsequently relied on comparisons to normal B cells collected from unrelated healthy individuals [25]. The chief disadvantage of this approach is that effects that are due to treatment and/or are modified by an individual's genetic background are not well captured. Similarly, the isolation of cancer and normal cell types in solid cancers can also be challenging where the cell type from which a given cancer type derives is not always known or may be equally difficult to obtain. For instance, in the four major types of lung cancer – adenocarcinoma, squamous carcinoma, small cell carcinoma and large cell carcinoma – each likely arises from different lung cell types and each of those cell types represents only a fraction of the overall diversity of normal lung cells. The latter challenge is best illustrated in small cell carcinoma where the most likely cell from which these tumors originate is a rare neuroendocrine cell that is present in less than one cell in 1000 in normal tissue [26,27]. Furthermore, even if the cell type is known, many cells are difficult to isolate. For instance, the isolation of β-cells from the pancreas requires significant effort [28,29] and, when making associated comparisons in insulinomas, issues regarding the isolation of pure cell populations are likely to be exacerbated. Despite these challenges, in carefully designed experiments, sample heterogeneity can have advantages in expression profiling studies as novel pathways can indicate the presence of independent molecular subtypes, implicate cells of origin [30] or identify unknown primary sites [31].

CANCER TRANSCRIPTOME ANALYSIS

The application of gene expression profiling in cancer has significantly aided the identification of genes involved in pathogenesis [32,33] and progression [34,35], has enabled the identification of prognostic expression signatures, and has provided a method for subgrouping heterogeneous cancer types [30,36]. Most studies have relied on gene expression microarrays to survey broadly the expression level of many genes. However, with recent advances in RNA sequencing, microarrays are no longer the gold standard for gene expression analysis. RNA-seq approaches provide higher resolution transcriptome analyses but additional and considerable new technical challenges. Specifically, expression profiling by RNA-seq offers the ability not only to survey gene expression levels, but also to access novel gene structure, alternative splicing, polyadenylation sites, gene fusions and allele-specific expression. In fact, in genetic studies of gene expression using RNA-seq, genetically correlated variability in many of these RNA features has already been observed [37,38]. Each RNA species (or change in RNA expression or structure) has the potential to inhibit or promote tumorigenesis and/or treatment response in the context of a given cancer type and, as a result, there is great interest in the development of tools which can support RNA-seq analysis integrating extra level of complexity [39]. This section will highlight how such data are analyzed to provide insights into a richer spectrum of functional variation found in cellular RNAs.

RNA-Sequencing Data and Analysis Pipelines

The general pipeline for most RNA-seq experiments includes generation of standard FASTQ-format files and sequence alignments using the latest assembly of the human genome (potentially supplemented with known splice-junctions). However, in manipulating RNA-seq data, various sources of bias need to be considered and, in many cases, procedures must be applied to compensate for these biases. One source of bias occurs due to cell line history and technical artifacts, while other biases occur during the read-mapping stage. For example, when dealing with sequence data in FASTQ files, one must carefully consider a number of parameters including the sequence diversity of reads, adaptor contamination, base qualities per cycle, nucleotide composition per cycle and the percentage of uncalled bases. In regard to alignment data, parameters that should be assessed include percentage of transcriptome mapped, percentage of reads mapped as collinear pairs, chromosomal distribution of reads, 5′ to 3′ coverage bias and coverage of the quantified transcriptome (number of exons quantified and variance in depth across a gene). When considering genotype information, one should check sample concordance and integrity by assaying the correlation of heterozygous single nucleotide polymorphisms (SNPs) between DNA and RNA, as well as allelic imbalance (in other words, the number of RNA-seq reads representing each allele) at a given genomic position of interest. The importance of the latter statistic is that it provides insight into accidental sample swaps or mixtures of two or more individuals, which can happen due to operator error, equipment contamination or over-clustering prior to Illumina sequencing. In such cases, one would observe more discordant variants, in the case of a swapped sample, or more significant patterns of allele-specific expression than expected from a single individual (due to more alleles being at {1/4} or {3/4} ratios due to combinations of heterozygotes and homozygotes).

Mapping RNA-seq data to the genome is itself more challenging than mapping DNA reads as, for most mapping algorithms to function adequately, reads must be mapped across one or more splice junctions. In essence, this is equivalent to mapping DNA sequences from an individual with a high prevalence of genomic deletions. One solution (or work-around) to this problem is to supplement the reference genome with sequences derived from all exon–exon junctions in the gene annotations. A preferred approach involves direct alignment of reads to the genome with aligners that are "aware" of splicing and able efficiently to align reads deriving from two or more exons. There are a growing number of splicing-aware mapping tools, which are specifically designed for handling RNA-seq data. A few examples of more commonly cited RNA-seq mapping tools are GEM [40], STAR [41], RUM [42], GSNAP [43], MapSplice [44] and TopHat [45]. Effort to understand the robustness of these tools has involved assessing their efficacy to map across simulated junctions in the presence of genetic variation [42]. In addition, metrics such as reference allele mapping bias, run-time and memory utilization can also inform the relative efficacy of different mappers.

Once RNA-seq reads are mapped, various approaches have been developed to infer transcript abundances from the alignments. In general, these methods aim to disentangle read diversity across splice junctions, to deconvolute and quantify transcript-specific mapping intervals, and often utilize paired-end read mapping data to inform the overall length and identity of the sequenced transcript fragment.

Methods aimed at providing these estimates include FluxCapacitor [46], iReckon [47], miso [48] and Cufflinks [49] but many more are available. Active competition, initiated through the RNA-seq Genome Annotation Assessment Project 3 (RGASP3), has been designed to evaluate these methods. Many of these tools support the use of known transcript annotation to aid in informing transcript annotation and quantities and this is a major parameter that affects the inference of transcript quantities. However, *in lieu* of guide annotation, approaches like Velvet [50] and TransABySS [51] aim to identify *de novo* transcripts; the advantage of applying unguided approaches in the analysis of cancer transcriptomes can be the identification of fusion genes (see below), alternative polyadenylation sites and improved transcript quantifications. These and other software tools described in this chapter for RNA-seq analysis are summarized in Table 3.2.

Assessing Differential Expression

Identifying genes that are differentially expressed is the central aim of the majority of gene expression profiling assays. However, compared to microarrays, there are separate statistical parameters that must be considered when using RNA-seq data. The model that best fits the distribution of RNA-seq reads across replicates was initially considered to be a Poisson distribution [59,60]. However, extra-Poisson variation or overdispersion has indicated that a negative binomial distribution improves the data fit [56]. Efforts to handle directly sources of variability, which may yield patterns of overdispersion, have resulted in two-parameter models. One such model includes the observation that sequencing reads are not equally distributed across genes, and thus models read depth and sequencing bias simultaneously as independent parameters [61,62]. In addition to this source of variability, differences in read depth and gene length can also artificially create differences between samples. For instance, differences in total sampling depth in certain libraries can make samples appear more divergent when raw counts are assessed and when comparisons are made across genes. In other words, when comparing the relative expression levels between genes, the number of reads that map to a gene is not only a product of its expression level but of its size. To address this, it is common to transform

TABLE 3.2 RNA-Seq Software

Primary Goal	Tool Name	Notes
Fusion transcript identification	DeFuse [52]	Annotation-guided; separate genome and transcript alignment followed by targeted breakpoint resolution
	TopHat-Fusion [53]	Not annotation-guided; genome alignment followed by split alignment for break/fusion point resolution
	Trans-ABySS [51]	Not annotation-guided; *de novo* assembly-based; meta-assembly with different k-mers; contig-to-genome alignment for fusion annotation
	FusionFinder [54]	Annotation-guided; transcriptome alignment with split alignment for break/fusion point resolution
Spliced alignment/ mapping	TopHat [45]	Read mapping using Bowtie (based on Burrows-Wheeler Tranform) followed by candidate exon identification; split read mapping to resolve spliced reads linking exons
	GEM [40]	Read mapping using a filtration-based approach to approximate string matching
	STAR [41]	Fast, splice junction-aware read mapping; uses sequential mappable seed search in uncompressed suffix arrays followed by a seed clustering and stitching procedure
	RUM [42]	An RNA-seq mapping pipeline which integrates the alignment tools Blat and Bowtie to recover unmapped reads
	GSNAP [43]	Variant aware aligner
	MapSplice [44]	Splicing aware aligner
Gene/isoform expression summary	Cufflinks [55]	Assembles transcripts and estimates their abundance
Differential expression analysis	Cuffdiff2 [49]	Identifies different expression of isoforms; explicitly leverages replicate libraries
	EdgeR [56]	R package; based on overdispersed Poisson model; also useful for Tag-seq data
	DESeq [57]	R package; based on negative binomial distribution and local regression
	BaySeq [58]	R package; empirical Bayesian approach; based on negative binomial distribution

read count data to obtain a gene's RPKM (reads per kilobase per million reads mapped) or FPKM (fragments per kilobase per million reads mapped) values, where the RPKM value does not account for double-counting paired-end reads and FPKM does. Despite this well used correction, a few very highly expressed genes that are unique to a sample can still significantly influence these estimates; as the total number of reads is typically fixed, a highly expressed gene will "absorb" many reads overall, thereby reducing the sequencing reads distributed to the rest of the genes and exaggerating genome-wide differences in their expression levels. A simple statistical approach to accounting for these biases is to use highly expressed genes as model covariates while being conscious of removing important biological variability. Such approaches are also well complemented by statistical methods which account for hidden covariates such as the PEER framework, which corrects hidden factors in expression data [63], and the Hidden Covariates with Prior (HCP) framework, a method that captures unwanted technical variability in gene expression estimates by modeling the influence of known and hidden confounding factors in a Bayesian framework (Mostafavi et al., unpublished observations).

There is also considerable diversity of methods that can support testing differential expression from RNA-seq data including BaySeq [58], DESeq [57], Cuffdiff [49] and EdgeR [56] (summarized in Table 3.2). Understanding the model parameters and their limitations in the context of the observed data, particularly the conditions under which the observed data can violate model assumptions, is an important context of using any of these tools [64]. Additionally, the importance of replicates in RNA-seq experiments is as important as increasing depth to be able confidently to estimate these models' parameters [65,66], and a minimum of three replicates per sample has been suggested [67]. Complementing replicates, the availability of spike-ins for differential expression studies provides a promising future direction for identifying technical variability influencing differential expression [16,68].

Quantifying Allele-Specific Expression

A specific advantage of RNA-sequencing is that sequenced transcript reads, like genomic reads of DNA sequence, provide coverage of heterozygous sites within transcripts. When there is significant coverage of RNA-sequencing reads over a heterozygous site in an individual transcript, the null hypothesis is that the ratio of observed alleles will be balanced. Deviation from this expectation can indicate allele-specific expression (ASE) due to genetic or epigenetic effects. An advantage of ASE is that it provides an internal (within individual) control for detecting mutations or polymorphisms that directly impact transcription levels. In our own work, we have demonstrated that ASE can mark both rare and common regulatory variant effects [69] as well as expression modifiers of protein coding variation [70,71] and the effects of loss of function variation [72]. Application of RNA-seq to a variety of cancers has provided evidence for ASE. For example, a study of squamous cell carcinomas and matched normal tissue [73] revealed hundreds of genes with significant ASE between the tumors and matched normal samples. Some ASE was attributed to large copy number variants (CNV) but others were found to result from heterozygous point mutations within the coding region or untranslated region (UTR) of the affected gene. These authors also provided evidence for epigenetic reprogramming of imprinted genes including *H19* and *IGF2*, supporting the notion that genomic imprinting is reset during tumorigenesis and that changes to epigenetic regulation can be detected in *cis* using RNA-seq. Similar types of patterns have further been reported in triple-negative breast cancers where patterns of monoallelic expression as acquired from RNA-seq have been used to identify somatic deletions [74] and the proportion of expressed somatic mutations [75], and in colorectal cancer where ASE of the *TGFBR1* gene has informed an ninefold increase in risk [76].

There are several challenges with ASE calling from RNA-seq data including sampling variance and overdispersion at different read depths, the influence of mapping errors, allelic alternative splicing, genotyping uncertainty and limited ability to test insertion–deletion polymorphism (a class of polymorphism enriched in loss-of-function variation). Currently, very few approaches exist to aid in testing these errors and detecting ASE. Among the available methods, an approach by Sun et al. models reads using a beta-binomial distribution at individual sites to account for overdispersion [77]. Another approach employs a Bayesian hierarchical model to integrate information across genomic sites, and between replicates and technologies to estimate allelic imbalance of genes [78]. To address further mapping errors, one can also empirically assess, for any given site, the bias in reads mapping to discrete start sites covering the variant site using all samples (i.e. to determine if the non-reference allele is as distributed across all possible cycle positions as the reference allele). Furthermore, one can assess the bias in base errors, particularly in the non-reference allele-containing reads where mismatches (in addition to the non-reference allele) could be differentially detrimental when mapping. Similarly, one can also assess allele-specific alternative splicing by comparing the allele-specific distribution of insert sizes for paired-end reads as previously reported [37]. Overall, methods for detecting ASE from RNA-seq in normal and cancer samples are promising but methods for accurately assessing the presence of ASE are under active development and are expected to reach maturity in the near future.

Interpreting Cancer Genome-Wide Association Studies

Cancer is a genetic disease that results from acquired mutations as well as, in some cases, inherited variants that can predispose to certain cancer types. In recent years, the common variants underlying the predisposition to cancer have been studied through genome-wide association studies (GWAS) where genetic variants are sought that stratify affected and control individuals. However, despite the early promise of this approach, it has been observed that, for complex traits, the effects of common genetic variants are small [79,80]. This is in contrast to the strong effect of rare predisposing mutations, for example mutations in *BRCA1* that greatly increase the risk of breast and ovarian cancer, which are typically restricted to families or certain populations. Furthermore, associated variants themselves typically only yield the target genomic intervals rather than the target genes themselves. Thus, it has been suggested by David Goldstein, director of the Center for Human Genome Variation at Duke University in Durham, North Carolina, that to improve our understanding a paradigm shift is required where "the field will transition from doing primarily association work to figuring out what implicated variants do biologically" [81]. Further, Elaine Mardis, Director of Technology Development at the Genome Center of Washington University School of Medicine in St Louis, Missouri, predicts that for this shift to happen the development of "human cell experimental systems that allow researchers to examine the impact of the mutations or other alterations of the genome on cellular pathways" [82] will be essential. Genetic studies of gene expression are transformative in this regard. Specifically, through expression microarrays and RNA-sequencing, it is possible to obtain quantitative measurements of most, if not all, of the transcriptional activity in a cell. When compared across individuals and correlated to their genetic data, genetic variants can be associated with changes in transcriptional activity as expression quantitative trait loci (eQTL) using statistical methods like linear regression, or non-parametric tests like Spearman correlation. Increasingly, GWAS in cancer utilize genetic studies of gene expression to aid in determining what associated variants do biologically by connecting these variants to gene activity [83–85]. Although this approach is promising, the intersection of trait and expression-associated variation has to be carefully considered. For instance, we and others have previously shown that eQTLs can vary widely between populations and tissues [86–92]. This specificity requires that the pathology and epidemiology of a given cancer must be considered; for example, it may make little sense to form hypotheses based on eQTL data derived from B cells when the pathological tissue is the liver. More recently, we have developed new approaches to detecting the relationship between cancer-associated genetic variants and regulatory variation by developing an ASE-based approach which we validated by identifying the heterogeneity between risk and protective alleles for follicular lymphoma [93]. This approach identifies genes which have an excess of ASE effects in individuals that are heterozygous for both the risk and protective alleles – the rationale being that these individuals should exhibit more ASE for causal genes than individuals homozygous for either allele and, unlike eQTL, the effect can be due to single or multiple independent gene variants.

Detecting Fusion Transcripts

Acquired genomic rearrangements resulting from double-strand breaks can reposition important genes such as oncogenes under the control of regulatory sequences belonging to genes that are often highly expressed, and thus deregulate their expression in that cell. In some cases, these rearrangements also fuse portions of the transcribed region of each gene, resulting in a fusion transcript, which may also yield a protein with a biologically distinct function. Beyond measuring the total transcriptional output of a gene, it was demonstrated that appropriate analysis of RNA-seq data from tumors could reveal the presence of such fusion events. Unfortunately, beyond the somatically acquired fusion events present in certain tumors, early analysis of RNA-seq data for this purpose revealed numerous examples of adjacent (or proximal) genes that exhibited significant evidence of chimerism across many of the tested samples. These chimeric transcripts arose from natural co-transcription across adjacent loci and subsequent splicing of the two genes together into a larger transcript. Such natural events are not commonly annotated and, as such, are difficult to filter automatically and thus can add a significant amount of noise to the analysis. Fortunately, with sufficient numbers of samples, these can be recognized and accounted for based on their consistent recurrence across samples, particularly if present in controls [94].

Various open-source toolkits for the robust identification of fusion transcripts from RNA-seq have been described, for example deFuse [52], TopHat-Fusion [53] and FusionFinder [54] (summarized in Table 3.2). There are many common features employed by such alignment-based methods. First, the reads are mapped to the genome or annotated transcripts and candidate fusions are identified from clusters of discordantly mapped read pairs. Further evidence can be gleaned directly from the sequences of reads that cross the point at which the two genes are fused (or spliced together). These methods also typically consider the overall (concordant) density of reads aligned to each gene involved in candidate fusion events. By collectively considering various independent levels of support, these tools are able to combat the number of false positive fusion predictions. Nonetheless, specificity of fusion detection remains a considerable problem.

A separate strategy for identifying fusion transcripts and other structural alterations begins by "assembling" the individual reads into overlapping sets of reads with shared sequences. Most assembly algorithms used for short read sequence data proceed by first producing a de Bruijn graph from the sub-sequences of all reads of length k (k-mers). Using various approaches to traverse the graph, *de novo* assembly algorithms ultimately generate contiguous stretches of consensus sequences termed "contigs" that derive from the concatenation of the k-mer sequences traversed in the graph. Many de Bruijn assemblers have been designed to handle next-generation sequencing data [50,95,96], but these are more commonly applied to *de novo* genome sequencing rather than RNA-seq applications. *De novo* assembly of RNA-seq data has nonetheless been demonstrated to yield contigs that represent portions of mRNAs which can offer benefits for identifying alternative splicing in normal cells (Figure 3.1) [97]. Notably, at different values of k, assemblies produce optimal contig lengths for transcripts at differing levels of expression. In particular, transcripts with low expression levels in a given library are more readily assembled using a lower value of k with the converse being true for abundant transcripts. To combat this issue the Trans-ABySS method involves assembly at a variety of values of k and the subsequent pooling of contigs while removing shorter contigs that are redundant sub-sequences of larger contigs. For identifying fusion transcripts, contigs from such pooled assemblies are then aligned to the reference genome to reveal situations in which disparate regions of the genome appear to be spliced together [98]. This approach has been successfully used in several studies to identify novel fusion transcripts in human cancers [98–100]. The computational time and space required to perform a *de novo* assembly for

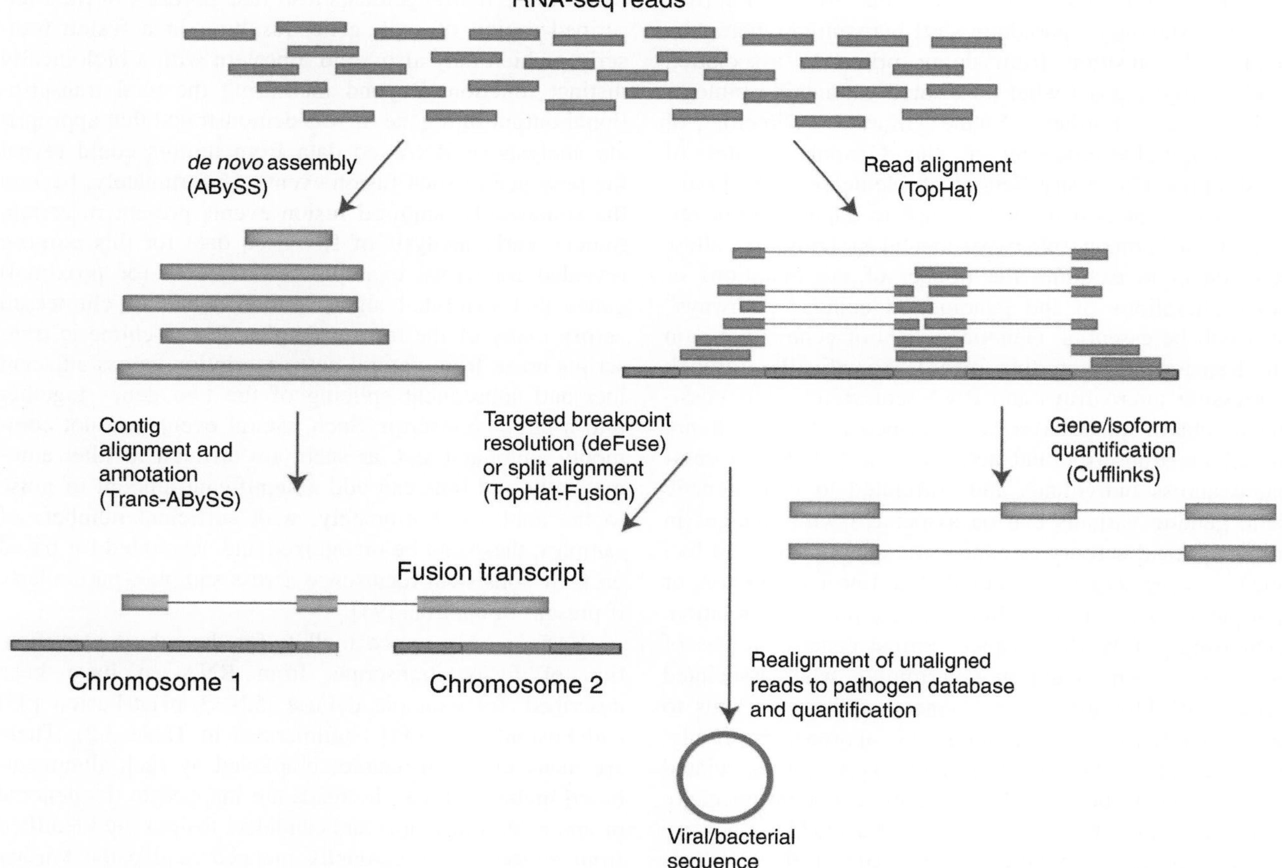

FIGURE 3.1 Example pipelines for analyzing RNA-seq data to somatic structural variant fusion genes. Reads are either built into transcripts using *de novo* assembly approaches or mapped to the genome and/or known gene models. When mapped to the genome, aligned reads inform on the location of components of a transcript (exons), which will often span introns and also may reveal deletions, inform insertions or fusion genes by mapping across multiple chromosomes. Realignment of unaligned reads can also reveal the presence of non-human sequences such as those deriving from pathogens.

mammalian genomes and transcriptomes is orders of magnitude greater than the time required to map an equal number of reads to a reference and it is still unclear whether (or in which situations) either approach may be preferable. It should be noted that neither exome sequencing nor whole genome sequencing (WGS) is a suitable method for detecting fusion transcripts. With appropriate analysis of whole genome sequencing one can indeed detect genomic breakpoints but the presence of a fusion transcript resulting from the rearrangement can only be inferred and not proven without confirming at the mRNA level. As such, even in large studies in which entire tumor genomes are sequenced, RNA-seq continues to be used as an adjunct to identify fusion transcripts, and there are numerous examples in which it has proven useful for this such as prostate cancers [94,101], leukemias [100], lymphomas [98], and sarcomas [102].

Identifying Somatic Single Nucleotide Variants

Another substantial benefit offered by RNA-seq analysis that is not possible with SAGE and array-based methods is the potential to detect nucleotide-level discrepancies between transcribed sequences and the reference genome, generally referred to as single nucleotide variants (SNVs). It is important to note that the SNVs that can be identified in RNA-seq data will be limited to those in the exons of genes with expression levels (and thus coverage) sufficient to facilitate accurate identification. As such, intergenic SNVs and those affecting introns (including splice sites) are largely invisible by this approach. As is true for whole genome sequencing experiments, the bulk of these variants correspond to common SNPs that are found in 1% or more of the population and are unlikely to be pathogenic but also, to a lesser extent, distinct inherited non-pathogenic variants unique to the individual. To complicate things further, a large number of these events have also been attributed to post-transcriptional RNA editing [103] but, owing to some issues that impede accurate alignment, a substantial proportion have been demonstrated to be artifacts [104,105]. Furthermore, even if RNA editing may impact the coding sequence of certain genes [106], its relevance in cancer remains speculative. These caveats are mentioned here mainly as a cautionary example of yet another layer of complexity of which one must be aware when analyzing these types of data.

A relatively small number of SNVs present in the transcriptome of a tumor represent events potentially relevant to cancer progression, namely the somatic point mutations. Beyond the potential artifacts inherent in these data types, a major hurdle in identifying somatic mutations from RNA-seq data is the ability reliably to distinguish these from the numerous rare variants present in the germline, which are typically found in excess of somatic SNVs. To parallel the paradigm used in WGS or exome sequencing, the ideal experimental design to enable recognition of somatic variants would involve paired sequencing of a matched population of normal cells from the same individual. Though it would be virtually impossible to ensure the same genes were expressed in both the tumor and normal samples, a sample representing the normal cell of origin from which the tumor derives would be the best approximation. Exome (or genome) sequence of genomic DNA from the same individual can also prove useful for recognizing rare variants present in the germline, but this control, because it sequences only the genomic DNA, would not allow recognition of pervasive RNA edits and systematic artifacts present in the tumor sample. The use of unpaired libraries from separate individuals has been demonstrated to facilitate recognition of these sources of variation [107]. However, this approach is still limited by potential differences between the repertoires of genes expressed in the tumor sample and the unmatched controls.

Despite the caveats and numerous potential limitations to the utility of RNA-seq for discovering point mutations, there are multiple examples of the success of this approach as well as benefits of this over strictly WGS or exome-based methods. The natural restriction of RNA-seq to represent genes that are actively transcribed can be considered a limitation but also a benefit considering that oncogenes, upon which the tumor relies for survival, should be actively transcribed in tumor cells. Oncogenic mutations also typically have focal mutation patterns, often affecting a restricted set of key amino acids. This is in contrast to tumor suppressor genes which can be inactivated by various modes and thus have diffuse mutation patterns and can also be affected by deletions and truncating mutations which can lead to nonsense-mediated decay of message. Oncogenic mutation "hot spots" have been discovered by applying RNA-seq to primary non-Hodgkin lymphomas [108] and various ovarian cancers [109,110]. Each of the mutations described in the cited studies has since been confirmed to produce dominant-acting or neomorphic changes to the functions of the affected proteins [111–113]. Another feature of RNA-seq data is that sequence coverage is proportional to transcript abundance. Beyond using transcript abundance to determine the expression level of genes, this enhanced coverage of highly expressed transcripts can also allow one to detect mutations present only in a fraction of the tumor cells. In summary, because a key goal in cancer genomics is to identify mutations that may be targeted by extant or novel therapeutic strategies, this restriction of RNA-seq to identifying potentially dominant acting events can be considered a benefit of the approach.

IDENTIFYING PATHOGEN PRESENCE IN CANCER SAMPLES

There are a variety of established relationships between the presence of a viral infection and the development of certain cancers in the affected tissue. Understanding such relationships is paramount as, in some cases, prevention or eradication of such infections might impede or even eliminate certain cancers before they develop. For example, the relationship between human papillomavirus (HPV) infection and cervical carcinoma is well established. Using a *de novo* assembly approach, an early RNA-seq study demonstrated the potential utility of RNA-seq for detecting the genomic integration site of the HPV genome in the HeLa cell line [114], suggesting that RNA-seq can be used for this purpose. Another well-studied example of virally induced cancers is chronic hepatitis B virus (HBV) infection, which strongly predisposes to the liver cancer known as hepatocellular carcinoma (HCC) [115]. Integration of portions of the HBV genome into the genome of the host cell is thought to contribute directly to the transformation of these cells. Recent exploration of HCC tumor genomes and transcriptomes found a likely mechanism underlying this transformation. Specifically, the integration sites of the HBV genome, including the viral enhancer sequence, often lies upstream of cancer-related genes such as *TERT*, *MLL4* and *CCNE1* [116]. Though the integration sites were identified from genome sequences in the cited study, RNA-seq data from matched samples enabled confirmation of some of these sites. Given the involvement of viral integration in either deregulating or disrupting the expression of host genes, general methods for detecting viral integration sites from RNA-seq and other data types are desirable. ViralFusionSeq, a tool that was designed specifically to leverage RNA-seq data for this purpose, was recently released [117].

A few examples of the association between chronic bacterial infection and certain cancers exist but, in many cases, these associations remain speculative. One such link that has been firmly established is the infection of *Heliobactor pylori*, the bacteria responsible for gastric ulcers. The presence of *H. pylori* infection induces the development of various types of gastric cancers including adenocarcinomas and MALT lymphomas [118,119]. It is thought that the inflammation induced by persistent infection contributes to oncogenesis [119]. It has also been demonstrated, in some cases, that eradicating the *H. pylori* infection, which reverses the resultant inflammation, can induce regression or slow the progression of precancerous lesions to cancers [120].

Based on this observation, it is plausible that other undiscovered links between bacterial infection and cancer exist, especially among cancers deriving from tissues susceptible to bacterial infection. A recent metagenomics study employing RNA-seq demonstrated such a link between another bacterial species, *Fusobacterium nucleatum*, and colorectal carcinoma (CRC) [121]. The basic strategy involved preparation of RNA-seq libraries from 11 bulk tumor samples and matched adjacent normal samples, which (as they are epithelial in nature) both contain a plethora of sequences deriving from intestinal microbiota. Unlike human mRNAs, transcripts originating from pathogens may not necessarily be polyadenylated, particularly those from prokaryotes. To improve chances of capturing non-polyadenylated mRNAs, the libraries were prepared by depleting human ribosomal RNA rather than by selecting for polyadenylated RNA. As the vast majority of sequences in the experiment were expected to derive from human cells, the reads were first mapped to the human genome and transcriptome and all high quality alignments were removed.

A common procedure for identifying evidence for pathogens from RNA-seq involves pre-filtering the data for reads corresponding to known human sequences and mapping the resulting data to a compendium of viral and bacterial sequences. A good example is the procedure used in the aforementioned metagenomics study of CRC. Here, all reads that could be recognized as deriving from human cells were computationally removed by aligning the remaining data to various human sequence databases (Figure 3.2). The remaining sequences were then aligned using a more sensitive aligner to a curated database containing genomic and mRNA sequences from all known bacteria and viruses. Read pairs that aligned with the expected orientation and unambiguously to a single species were used to determine the species present in the individual samples. Finally, as most microbial genomes were present in both samples from the same patient, those over-represented in the tumor were identified. Only a single species, namely *Fusobacterium nucleatum*, was found consistently over-represented in the tumor samples with more than twofold higher levels in nine of the eleven patients. This illustrated that, although present in normal microbiota, *Fusobacterium* were significantly more abundant in samples from CRC tumors and illustrates the potential for RNA-seq to uncover novel unexpected relationships between prokaryotes and human cancers.

FUTURE PROSPECTS

The analysis of cancer transcriptomes provides increasing opportunities to identify the origins and evolution of cancer and aid in predicting progression and informing treatment. In fact, high profile examples of how such data can aid in treating individual cancers have already appeared; one such case involving the identification of overexpression of *FLT3* gene in an individual with acute myeloid leukemia

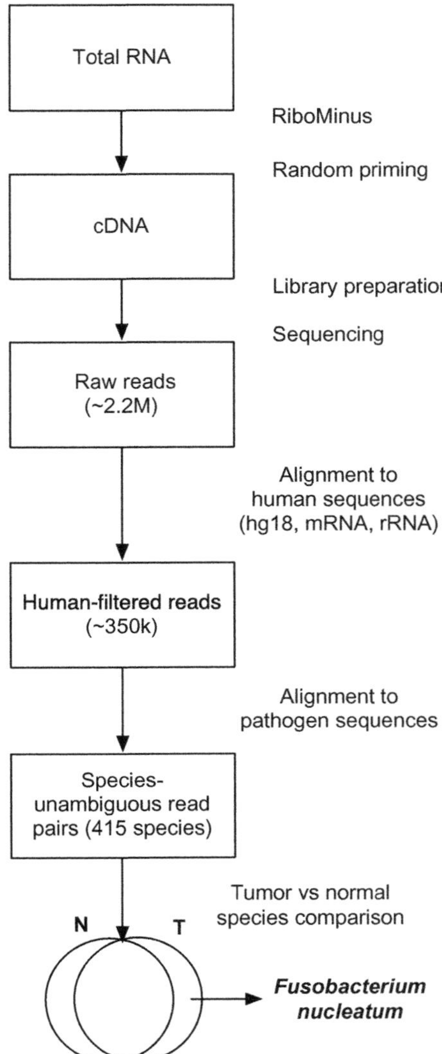

FIGURE 3.2 Approach used to identify *Fusobacterium* transcripts in colorectal cancer samples. RNA-seq libraries were created from total RNA extracted from whole tumor samples. A random priming strategy was used to ensure non-polyadenylated transcripts were captured. Each library was sequenced to a depth of approximately 2.2 million reads. Sequences were aligned to various sources of human sequence including the reference genome (hg18) and transcript annotations. The filtered reads not matching human sequences were then aligned to a compendium of viral and bacterial genomes and unambiguosly mapped read pairs allowed identification of sequences derived from *Fusobacterium nucleatum*.

allowed the successful repurposing of FLT3 inhibitor used in kidney cancer. Complementing personal approaches, analysis of genomes and transcriptomes of thousands of breast cancers has also aided in subtyping cancer and identifying groups with distinct clinical outcomes [122]. In combination with these advances, single cell approaches are rapidly improving the understanding of cancer evolution, tissue specificity and metastasis and higher-throughput sequencing is allowing increased surveillance of the metagenomes of individuals and cancer. As advances in gene expression methods and technologies continue to increase rapidly, it can be expected that these new areas of cancer research will soon be complemented by definitive identification and quantification of all RNA species.

ACKNOWLEDGMENTS

The masthead image was created by Rodrigo Goya, BC Genome Sciences Centre and is reproduced here with permission. Special thanks to Julien Sage for comments on cancer tissue heterogeneity. RDM is supported by a British Columbia Cancer Foundation Investigator Establishment Award. SBM is supported by the Edward Mallinckrodt, Jr Foundation.

GLOSSARY

Allele-specific expression (ASE) Gene expression distinguished from a particular allele. In diploid organisms, this is the relative expression of homologous genes. A gene that exhibits allele-specific expression is considered to have imbalanced expression between alleles.

Allelic imbalance See allele-specific expression.

Bayesian framework A statistical framework based on Bayes' theorem.

Bayesian hierarchical model A probabilistic graphical model or network that depicts the relationships between random variables and their conditional dependencies.

Beta-binomial distribution An extension of the binomial probability distribution where the prior distribution is modeled as a beta-distribution. The beta-binomial distribution is used to handle overdispersion in the binomial distribution (where the variance is more than expected).

Copy number variation (CNV) A class of genomic structural variation characterized by gain or loss of one or more copies of a section of DNA.

Circulating tumor cells (CTCs) Cancer cells present in the bloodstream.

de Bruijn graph A graph representation of the overlap between short sequences. Nodes represent individual sequences of length *k* and edges represent the overlap between two sequences.

Differential expression A significant difference in the abundance of total mRNA from a gene or a specific isoform between two samples or biological states.

Exome The exonic portion of the genome (approximately 1.5–2% of the human genome).

Expression quantitative trait loci (eQTL) A genomic loci where genetic variation correlates with expression level.

Extra-Poisson variation A Poisson-like distribution where the variance is proportional rather than equal to the mean.

External RNA Control Consortium (ERCC) A consortium with the aims to develop industry-wide standard controls for gene expression assays [123].

Formalin-fixed paraffin-embedded samples A process used for archiving tissues and conducting histopathological diagnosis of diseases. Formalin-fixation prevents decay by creating covalent chemical bonds between proteins in tissue. Paraffin-embedding involves dehydrating tissues and then infiltration with wax. This process provides sample stability for multiple years.

Fusion gene Two genes brought in collinear order by translocation, inversion or deletion that are transcribed and spliced together to form a functional gene.

Genome-wide association studies (GWAS) An examination of multiple genetics variants to detect those correlated with a trait.

Library construction The process for building a sequencing library. Follows extraction of RNA or DNA and precedes sequencing. In many cases, this involves fragmentation of the sample and ligation of sequencing-specific primers.

Mapping (RNA-seq) The computational approach to assigning sequencing reads to their origin within a genome. Mapping strategies for RNA need to be aware that sequencing reads may contain individual-specific variants and also cross splice-junctions.

Metagenomics Genomic approaches which investigate environmental samples typically containing communities of organisms such as bacteria.

Non-coding RNA RNA that does not encode a protein but has a separate biochemical function. This RNA includes multiple species such as transfer RNA, ribosomal RNA, microRNA, small nuclear RNA and linc-RNA.

Microarrays High-throughput gene expression quantification technology based on probe-target hybridization and subsequent luminescence.

Overdispersion A property of a statistical distribution where more variance is observed than predicted.

PEER framework The probabilistic estimation of expression residuals (PEER) framework assesses global hidden factors which influence expression variability. Such factors may be hidden biological or technical variables which influence the expression of multiple genes.

Poisson distribution A statistical distribution where the variance is equal to the mean and which assesses the probability that a number of events occur in a fixed interval.

RNA-sequencing (RNA-seq) High-throughput (or next-generation) sequencing of a sample's cDNA to characterize its transcriptome.

RNA spike-ins A positive control that is introduced into a sequencing reaction or library. Spike-ins may be at known quantities to assess the dynamic range of quantification in an experiment.

Serial analysis of gene expression (SAGE) A technique based on sequencing short concatenated cDNA tags to identify and quantify transcripts. cDNA tags are created through application of special restriction enzymes with cut sites several nucleotides from the recognition site.

Single nucleotide polymorphism (SNP) A single base DNA variation common in the population.

Single nucleotide variant (SNV) A single base DNA variation that may or may not be common (includes mutations).

Transcript quantification Computational methods for assessing the abundance of transcripts. These methods usually involve deconvoluting the transcripts present for a given gene and then resolving their individual quantities.

Transcriptome The repertoire of expressed RNA molecules within a sample.

Variant detection Computational methods for detecting genetic or somatic variants.

ABBREVIATIONS

ASE Allele-specific expression
cDNA Complementary DNA
CNV Copy number variant
CRC Colorectal carcinoma
CTCs Circulating tumor cells
eQTL Expression quantitative trait loci
EST Expressed sequence tag
FFPE Formaldehyde-fixed, paraffin-embedded
FPKM Fragments per kilobase per million reads mapped
GWAS Genome-wide association studies
HCC Hepatocellular carcinoma
HBV Hepatitis B virus
HPV Human papillomavirus
MALT Mucosa-associated lymphoid tissue
miRNA MicroRNA
mRNA Messenger RNA
RPKM Reads per kilobase per million reads mapped
rRNA Ribosomal RNA
SAGE Sequential analysis of gene expression
SNP Single nucleotide polymorphism
SNV Single nucleotide variant
UTR Untranslated region
WGS Whole genome sequencing

REFERENCES

[1] Guttman M, Amit I, Garber M, French C, Lin MF, Feldser D, et al. Chromatin signature reveals over a thousand highly conserved large non-coding RNAs in mammals. Nature 2009;458:223–7.

[2] Parkinson H, Sarkans U, Kolesnikov N, Abeygunawardena N, Burdett T, Dylag M, et al. ArrayExpress update – an archive of microarray and high-throughput sequencing-based functional genomics experiments. Nucleic Acids Res 2011;39:D1002–4.

[3] Griffith M, Tang MJ, Griffith OL, Morin RD, Chan SY, Asano JK, et al. ALEXA: a microarray design platform for alternative expression analysis. Nat Methods 2008;5:118.

[4] Velculescu VE, Zhang L, Vogelstein B, Kinzler KW. Serial analysis of gene expression. Science 1995;270:484–7.

[5] Liang P, Pardee AB. Analysing differential gene expression in cancer. Nat Rev Cancer 2003;3:869–76.

[6] Wang Z, Gerstein M, Snyder M. RNA-Seq: a revolutionary tool for transcriptomics. Nat Rev Genet 2009;10:57–63.

[7] Morrissy AS, Morin RD, Delaney A, Zeng T, McDonald H, Jones SJM, et al. Next-generation tag sequencing for cancer gene expression profiling. Genome Res 2009;19:1825–35.

[8] Mortazavi A, Williams BA, McCue K, Schaeffer L, Wold B. Mapping and quantifying mammalian transcriptomes by RNA-Seq. Nat Methods 2008;5:621–8.

[9] Yi H, Cho YJ, Won S, Lee JE, Jin Yu H, Kim S, et al. Duplex-specific nuclease efficiently removes rRNA for prokaryotic RNA-seq. Nucleic Acids Res 2011;39:e140.

[10] Huang R, Jaritz M, Guenzl P, Vlatkovic I, Sommer A, Tamir IM, et al. An RNA-Seq strategy to detect the complete coding and non-coding transcriptome including full-length imprinted macro ncRNAs. PloS One 2011;6:e27288.

[11] Calin GA, Dumitru CD, Shimizu M, Bichi R, Zupo S, Noch E, et al. Frequent deletions and down-regulation of micro-RNA genes miR15 and miR16 at 13q14 in chronic lymphocytic leukemia. Proc Natl Acad Sci USA 2002;99:15524–9.
[12] Sassen S, Miska EA, Caldas C. MicroRNA: implications for cancer. Virchows Arch 2008;452:1–10.
[13] Morin RD, Zhao Y, Prabhu AL, Dhalla N, McDonald H, Pandoh P, et al. Preparation and analysis of microRNA libraries using the Illumina massively parallel sequencing technology. Methods Mol Biol 2010;650:173–99.
[14] Levin JZ, Berger MF, Adiconis X, Rogov P, Melnikov A, Fennell T, et al. Targeted next-generation sequencing of a cancer transcriptome enhances detection of sequence variants and novel fusion transcripts. Genome Biol 2009;10:R115.
[15] Ren S, Peng Z, Mao JH, Yu Y, Yin C, Gao X, et al. RNA-seq analysis of prostate cancer in the Chinese population identifies recurrent gene fusions, cancer-associated long noncoding RNAs and aberrant alternative splicings. Cell Res 2012;22:806–21.
[16] Jiang L, Schlesinger F, Davis CA, Zhang Y, Li R, Salit M, et al. Synthetic spike-in standards for RNA-seq experiments. Genome Res 2011;21:1543–51.
[17] Tang F, Lao K, Surani MA. Development and applications of single-cell transcriptome analysis. Nat Methods 2011;8(Suppl. 4):S6–11.
[18] Tang F, Barbacioru C, Wang Y, Nordman E, Lee C, Xu N, et al. mRNA-Seq whole-transcriptome analysis of a single cell. Nat Methods 2009;6:377–82.
[19] Wang D, Bodovitz S. Single cell analysis: the new frontier in 'omics'. Trends Biotechnol 2010;28:281–90.
[20] Ramskold D, Luo S, Wang YC, Li R, Deng Q, Faridani OR, et al. Full-length mRNA-seq from single-cell levels of RNA and individual circulating tumor cells. Nat Biotechnol 2012;30:777–82.
[21] von Ahlfen S, Missel A, Bendrat K, Schlumpberger M. Determinants of RNA quality from FFPE samples. PloS One 2007;2:e1261.
[22] Beck AH, Weng Z, Witten DM, Zhu S, Foley JW, Lacroute P, et al. 3′-end sequencing for expression quantification (3SEQ) from archival tumor samples. PloS One 2010;5(1):e8768.
[23] Weng L, Wu X, Gao H, Mu B, Li X, Wang JH, et al. MicroRNA profiling of clear cell renal cell carcinoma by whole-genome small RNA deep sequencing of paired frozen and formalin-fixed, paraffin-embedded tissue specimens. J Pathol 2010;222:41–51.
[24] Kolbert CP, Feddersen RM, Rakhshan F, Grill DE, Simon G, Middha S, et al. Multi-platform analysis of microRNA expression measurements in RNA from fresh frozen and FFPE tissues. PloS One 2013;8:e52517.
[25] Klein U, Tu Y, Stolovitzky GA, Mattioli M, Cattoretti G, Husson H, et al. Gene expression profiling of B cell chronic lymphocytic leukemia reveals a homogeneous phenotype related to memory B cells. J Exp Med 2001;194:1625–38.
[26] Sutherland KD, Proost N, Brouns I, Adriaensen D, Song JY, Berns A. Cell of origin of small cell lung cancer: inactivation of Trp53 and Rb1 in distinct cell types of adult mouse lung. Cancer Cell 2011;19:754–64.
[27] Park KS, Liang MC, Raiser DM, Zamponi R, Roach RR, Curtis SJ, et al. Characterization of the cell of origin for small cell lung cancer. Cell Cycle 2011;10:2806–15.
[28] Sturm D, Marselli L, Ehehalt F, Richter D, Distler M, Kersting S, et al. Improved protocol for laser microdissection of human pancreatic islets from surgical specimens. J Visual Exp 2013;:71.
[29] Kutlu B, Burdick D, Baxter D, Rasschaert J, Flamez D, Eizirik DL, et al. Detailed transcriptome atlas of the pancreatic beta cell. BMC Med Genomics 2009;2:3.
[30] Alizadeh AA, Eisen MB, Davis RE, Ma C, Lossos IS, Rosenwald A, et al. Distinct types of diffuse large B-cell lymphoma identified by gene expression profiling. Nature 2000;403:503–11.
[31] Pavlidis N, Pentheroudakis G. Cancer of unknown primary site. Lancet 2012;379:1428–35.
[32] Whitfield ML, George LK, Grant GD, Perou CM. Common markers of proliferation. Nat Rev Cancer 2006;6:99–106.
[33] Goodarzi H, Elemento O, Tavazoie S. Revealing global regulatory perturbations across human cancers. Molec Cell 2009;36:900–11.
[34] Sotiriou C, Piccart MJ. Taking gene-expression profiling to the clinic: when will molecular signatures become relevant to patient care? Nat Rev Cancer 2007;7:545–53.
[35] Reis-Filho JS, Pusztai L. Gene expression profiling in breast cancer: classification, prognostication, and prediction. Lancet 2011;378: 1812–23.
[36] Perou CM. Molecular stratification of triple-negative breast cancers. Oncologist 2011;16(Suppl. 1):61–70.
[37] Montgomery SB, Sammeth M, Gutierrez-Arcelus M, Lach RP, Ingle C, Nisbett J, et al. Transcriptome genetics using second generation sequencing in a Caucasian population. Nature 2010;464:773–7.
[38] Pickrell JK, Marioni JC, Pai AA, Degner JF, Engelhardt BE, Nkadori E, et al. Understanding mechanisms underlying human gene expression variation with RNA sequencing. Nature 2010;464:768–72.
[39] Garber M, Grabherr MG, Guttman M, Trapnell C. Computational methods for transcriptome annotation and quantification using RNA-seq. Nat Methods 2011;8:469–77.
[40] Marco-Sola S, Sammeth M, Guigo R, Ribeca P. The GEM mapper: fast, accurate and versatile alignment by filtration. Nat Methods 2012;9:1185–8.
[41] Dobin A, Davis CA, Schlesinger F, Drenkow J, Zaleski C, Jha S, et al. STAR: ultrafast universal RNA-seq aligner. Bioinformatics 2013;29:15–21.
[42] Grant GR, Farkas MH, Pizarro AD, Lahens NF, Schug J, Brunk BP, et al. Comparative analysis of RNA-Seq alignment algorithms and the RNA-Seq unified mapper (RUM). Bioinformatics 2011;27:2518–28.
[43] Wu TD, Nacu S. Fast and SNP-tolerant detection of complex variants and splicing in short reads. Bioinformatics 2010;26:873–81.
[44] Wang K, Singh D, Zeng Z, Coleman SJ, Huang Y, Savich GL, et al. MapSplice: accurate mapping of RNA-seq reads for splice junction discovery. Nucleic Acids Res 2010;38:e178.
[45] Trapnell C, Pachter L, Salzberg SL. TopHat: discovering splice junctions with RNA-Seq. Bioinformatics 2009;25:1105–11.
[46] Griebel T, Zacher B, Ribeca P, Raineri E, Lacroix V, Guigo R, et al. Modelling and simulating generic RNA-Seq experiments with the flux simulator. Nucleic Acids Res 2012;40:10073–83.
[47] Mezlini AM, Smith EJ, Fiume M, Buske O, Savich GL, Shah S, et al. iReckon: Simultaneous isoform discovery and abundance estimation from RNA-seq data. Genome Res 2013;: Jan 28
[48] Katz Y, Wang ET, Airoldi EM, Burge CB. Analysis and design of RNA sequencing experiments for identifying isoform regulation. Nat Methods 2010;7:1009–15.
[49] Trapnell C, Roberts A, Goff L, Pertea G, Kim D, Kelley DR, et al. Differential gene and transcript expression analysis of RNA-seq experiments with TopHat and Cufflinks. Nat Protocols 2012;7:562–78.

[50] Zerbino DR, Birney E. Velvet: algorithms for de novo short read assembly using de Bruijn graphs. Genome Res 2008;18:821–9.

[51] Robertson G, Schein J, Chiu R, Corbett R, Field M, Jackman SD, et al. De novo assembly and analysis of RNA-seq data. Nat Methods 2010;7:909–12.

[52] McPherson A, Hormozdiari F, Zayed A, Giuliany R, Ha G, Sun MGF, et al. deFuse: an algorithm for gene fusion discovery in tumor RNA-Seq data. PLoS Comput Biol 2011;7:e1001138.

[53] Kim D, Salzberg SL. TopHat-Fusion: an algorithm for discovery of novel fusion transcripts. Genome Biol 2011;12:R72.

[54] Francis RW, Thompson-Wicking K, Carter KW, Anderson D, Kees UR, Beesley AH. FusionFinder: a software tool to identify expressed gene fusion candidates from RNA-Seq data. PloS One 2012;7:e39987.

[55] Trapnell C, Williams BA, Pertea G, Mortazavi A, Kwan G, van Baren MJ, et al. Transcript assembly and quantification by RNA-Seq reveals unannotated transcripts and isoform switching during cell differentiation. Nat Biotechnol 2010;28:511–5.

[56] Robinson MD, McCarthy DJ, Smyth GK. EdgeR: a Bioconductor package for differential expression analysis of digital gene expression data. Bioinformatics 2010;26:139–40.

[57] Anders S, Huber W. Differential expression analysis for sequence count data. Genome Biol 2010;11:R106.

[58] Hardcastle TJ, Kelly KA. BaySeq: empirical Bayesian methods for identifying differential expression in sequence count data. BMC Bioinformatics 2010;11:422.

[59] Jiang H, Wong WH. Statistical inferences for isoform expression in RNA-Seq. Bioinformatics 2009;25:1026–32.

[60] Marioni JC, Mason CE, Mane SM, Stephens M, Gilad Y. RNA-seq: an assessment of technical reproducibility and comparison with gene expression arrays. Genome Res 2008;18:1509–17.

[61] Srivastava S, Chen L. A two-parameter generalized Poisson model to improve the analysis of RNA-seq data. Nucleic Acids Res 2010;38:e170.

[62] Roberts A, Trapnell C, Donaghey J, Rinn JL, Pachter L. Improving RNA-Seq expression estimates by correcting for fragment bias. Genome Biol 2011;12:R22.

[63] Stegle O, Parts L, Durbin R, Winn JA. Bayesian framework to account for complex non-genetic factors in gene expression levels greatly increases power in eQTL studies. PLoS Computat Biol 2010;6:e1000770.

[64] Bullard JH, Purdom E, Hansen KD, Dudoit S. Evaluation of statistical methods for normalization and differential expression in mRNA-Seq experiments. BMC Bioinformatics 2010;11:94.

[65] Tarazona S, Garcia-Alcalde F, Dopazo J, Ferrer A, Conesa A. Differential expression in RNA-seq: a matter of depth. Genome Res 2011;21:2213–23.

[66] Glaus P, Honkela A, Rattray M. Identifying differentially expressed transcripts from RNA-seq data with biological variation. Bioinformatics 2012;28(13):1721–8.

[67] Auer PL, Doerge RW. Statistical design and analysis of RNA sequencing data. Genetics 2010;185:405–16.

[68] Devonshire AS, Elaswarapu R, Foy CA. Evaluation of external RNA controls for the standardisation of gene expression biomarker measurements. BMC Genomics 2010;11:662.

[69] Pastinen T. Genome-wide allele-specific analysis: insights into regulatory variation. Nat Rev Genet 2010;11:533–8.

[70] Montgomery SB, Lappalainen T, Gutierrez-Arcelus M, Dermitzakis ET. Rare and common regulatory variation in population-scale sequenced human genomes. PLoS Ggenet 2011;7:e1002144.

[71] Lappalainen T, Montgomery SB, Nica AC, Dermitzakis ET. Epistatic selection between coding and regulatory variation in human evolution and disease. Am J Hum Genet 2011;89:459–63.

[72] MacArthur DG, Balasubramanian S, Frankish A, Huang N, Morris J, Walter K, et al. A systematic survey of loss-of-function variants in human protein-coding genes. Science 2012;335:823–8.

[73] Tuch BB, Laborde RR, Xu X, Gu J, Chung CB, Monighetti CK, et al. Tumor transcriptome sequencing reveals allelic expression imbalances associated with copy number alterations. PloS One 2010;5:e9317.

[74] Ha G, Roth A, Lai D, Bashashati A, Ding J, Goya R, et al. Integrative analysis of genome-wide loss of heterozygosity and monoallelic expression at nucleotide resolution reveals disrupted pathways in triple-negative breast cancer. Genome Res 2012;22:1995–2007.

[75] Shah SP, Roth A, Goya R, Oloumi A, Ha G, Zhao Y, et al. The clonal and mutational evolution spectrum of primary triple-negative breast cancers. Nature 2012;486:395–9.

[76] Valle L, Serena-Acedo T, Liyanarachchi S, Hampel H, Comeras I, Li Z, et al. Germline allele-specific expression of TGFBR1 confers an increased risk of colorectal cancer. Science 2008;321:1361–5.

[77] Sun W. A statistical framework for eQTL mapping using RNA-seq data. Biometrics 2012;68:1–11.

[78] Skelly DA, Johansson M, Madeoy J, Wakefield J, Akey JM. A powerful and flexible statistical framework for testing hypotheses of allele-specific gene expression from RNA-seq data. Genome Res 2011;21:1728–37.

[79] Collins F. Has the revolution arrived? Nature 2010;464:674–5.

[80] Manolio TA, Collins FS, Cox NJ, Goldstein DB, Hindorff LA, Hunter DJ, et al. Finding the missing heritability of complex diseases. Nature 2009;461:747–53.

[81] Baker M. Functional genomics: the changes that count. Nature 2012;482(257):9–62.

[82] Mardis ER. The $1,000 genome, the $100,000 analysis? Genome Med 2010;2:84.

[83] Crawford NP, Qian X, Ziogas A, Papageorge AG, Boersma BJ, Walker RC, et al. Rrp1b, a new candidate susceptibility gene for breast cancer progression and metastasis. PLoS Genet 2007;3:e214.

[84] Spurdle AB, Thompson DJ, Ahmed S, Ferguson K, Healey CS, O'Mara T, et al. Genome-wide association study identifies a common variant associated with risk of endometrial cancer. Nat Genet 2011;43:451–4.

[85] Montgomery SB, Dermitzakis ET. From expression QTLs to personalized transcriptomics. Nat Rev Genet 2011;12:277–82.

[86] Nica AC, Parts L, Glass D, Nisbet J, Barrett A, Sekowska M, et al. The architecture of gene regulatory variation across multiple human tissues: the MuTHER study. PLoS Genet 2011;7:e1002003.

[87] Dimas AS, Deutsch S, Stranger BE, Montgomery SB, Borel C, Attar-Cohen H, et al. Common regulatory variation impacts gene expression in a cell type-dependent manner. Science 2009;325:1246–50.

[88] Stranger BE, Montgomery SB, Dimas AS, Parts L, Stegle O, Ingle CE, et al. Patterns of *cis* regulatory variation in diverse human populations. PLoS Genet 2012;8:e1002639.

[89] Stranger BE, Nica AC, Forrest MS, Dimas A, Bird CP, Beazley C, et al. Population genomics of human gene expression. Nat Genet 2007;39:1217–24.

[90] Emilsson V, Thorleifsson G, Zhang B, Leonardson AS, Zink F, Zhu J, et al. Genetics of gene expression and its effect on disease. Nature 2008;452:423–8.

[91] Heinzen EL, Ge D, Cronin KD, Maia JM, Shianna KV, Gabriel WN, et al. Tissue-specific genetic control of splicing: implications for the study of complex traits. PLoS Biol 2008;6:e1.

[92] Spielman RS, Bastone LA, Burdick JT, Morley M, Ewens WJ, Cheung VG. Common genetic variants account for differences in gene expression among ethnic groups. Nat Genet 2007;39: 226–31.

[93] Conde L, Bracci PM, Richardson R, Montgomery SB, Skibola CF. Integrating GWAS and expression data for functional characterization of disease-associated SNPs: an application to follicular lymphoma. Am J Hum Genet 2013;92(1):126–130.

[94] Maher CA, Kumar-Sinha C, Cao X, Kalyana-Sundaram S, Han B, Jing X, et al. Transcriptome sequencing to detect gene fusions in cancer. Nature 2009;458:97–101.

[95] Warren RL, Sutton GG, Jones SJ, Holt RA. Assembling millions of short DNA sequences using SSAKE. Bioinformatics 2007;23:500–1.

[96] Butler J, MacCallum I, Kleber M, Shlyakhter IA, Belmonte MK, Lander ES, et al. ALLPATHS: de novo assembly of whole-genome shotgun microreads. Genome Res 2008;18:810–20.

[97] Birol I, Jackman SD, Nielsen CB, Qian JQ, Varhol R, Stazyk G, et al. *De novo* transcriptome assembly with ABySS. Bioinformatics 2009;25:2872–7.

[98] Scott DW, Mungall KL, Ben-Neriah S, Rogic S, Morin RD, Slack GW, et al. TBL1XR1/TP63: a novel recurrent gene fusion in B-cell non-Hodgkin lymphoma. Blood 2012;119:4949–52.

[99] Morin RD, Mendez-Lago M, Mungall AJ, Goya R, Mungall KL, Corbett RD, et al. Frequent mutation of histone-modifying genes in non-Hodgkin lymphoma. Nature 2011;476:298–303.

[100] Roberts KG, Morin RD, Zhang J, Hirst M, Zhao Y, Su X, et al. Genetic alterations activating kinase and cytokine receptor signaling in high-risk acute lymphoblastic leukemia. Cancer Cell 2012;22:153–66.

[101] Maher CA, Palanisamy N, Brenner JC, Cao X, Kalyana-Sundaram S, Luo S, et al. Chimeric transcript discovery by paired-end transcriptome sequencing. Proc Natl Acad Sci USA 2009;106:12353–8.

[102] Lee CH, Ou WB, Marino-Enriquez A, Zhu M, Mayeda M, Wang Y, et al. 14-3-3 fusion oncogenes in high-grade endometrial stromal sarcoma. Proc Natl Acad Sci USA 2012;109:929–34.

[103] Peng Z, Cheng Y, Tan BC, Kang L, Tian Z, Zhu Y, et al. Comprehensive analysis of RNA-Seq data reveals extensive RNA editing in a human transcriptome. Nat Biotechnol 2012;30:253–60.

[104] Schrider DR, Gout JF, Hahn MW. Very few RNA and DNA sequence differences in the human transcriptome. PloS One 2011;6:e25842.

[105] Pickrell JK, Gilad Y, Pritchard JK. Comment on "Widespread RNA and DNA sequence differences in the human transcriptome". Science 2012;335:1302.

[106] Shah SP, Morin RD, Khattra J, Prentice L, Pugh T, Burleigh A, et al. Mutational evolution in a lobular breast tumour profiled at single nucleotide resolution. Nature 2009;461:809–13.

[107] Picardi E, Gallo A, Galeano F, Tomaselli S, Pesole G. A novel computational strategy to identify A-to-I RNA editing sites by RNA-Seq data: *de novo* detection in human spinal cord tissue. PloS One 2012;7:e44184.

[108] Morin RD, Johnson NA, Severson TM, Mungall AJ, An J, Goya R, et al. Somatic mutations altering EZH2 (Tyr641) in follicular and diffuse large B-cell lymphomas of germinal-center origin. Nat Genet 2010;42:181–5.

[109] Shah SP, Kobel M, Senz J, Morin RD, Clarke BA, Wiegand KC, et al. Mutation of FOXL2 in granulosa-cell tumors of the ovary. N Engl J Med 2009;360:2719–29.

[110] Heravi-Moussavi A, Anglesio MS, Cheng SWG, Senz J, Yang W, Prentice L, et al. Recurrent somatic DICER1 mutations in nonepithelial ovarian cancers. N Engl J Med 2012;366:234–42.

[111] Yap DB, Chu J, Berg T, Schapira M, Cheng SWG, Moradian A, et al. Somatic mutations at EZH2 Y641 act dominantly through a mechanism of selectively altered PRC2 catalytic activity, to increase H3K27 trimethylation. Blood 2011;117:2451–9.

[112] Cheng JC, Klausen C, Leung PC. Overexpression of wild-type but not C134W mMutant FOXL2 enhances GnRH-induced cell apoptosis by increasing GnRH receptor expression in human granulosa cell tumors. PloS One 2013;8:e55099.

[113] Anglesio M, Wang Y, Yang W, Senz J, Wan A, Heravi-Moussavi A, et al. Cancer-associated somatic DICER1 hotspot mutations cause defective miRNA processing and reverse-strand expression bias to predominantly mature 3 p strands through loss of 5 p strand cleavage. J Pathol 2013;229:400–9.

[114] Morin R, Bainbridge M, Fejes A, Hirst M, Krzywinski M, Pugh TJ, et al. Profiling the HeLa S3 transcriptome using randomly primed cDNA and massively parallel short-read sequencing. BioTechniques 2008;45:81–94.

[115] Fallot G, Neuveut C, Buendia MA. Diverse roles of hepatitis B virus in liver cancer. Curr Opin Virol 2012;2:467–73.

[116] Sung WK, Zheng H, Li S, Chen R, Liu X, Li Y, et al. Genome-wide survey of recurrent HBV integration in hepatocellular carcinoma. Nat Genet 2012;44:765–9.

[117] Li JW, Wan R, Yu CS, Co NN, Wong N, Chan TF. ViralFusionSeq: accurately discover viral integration events and reconstruct fusion transcripts at single-base resolution. Bioinformatics 2013;29:649–51.

[118] Stathis A, Chini C, Bertoni F, Proserpio I, Capella C, Mazzucchelli L, et al. Long-term outcome following *Helicobacter pylori* eradication in a retrospective study of 105 patients with localized gastric marginal zone B-cell lymphoma of MALT type. Ann Oncol 2009;20:1086–93.

[119] Uemura N, Okamoto S, Yamamoto S, Matsumura N, Yamaguchi S, Yamakido M, et al. *Helicobacter pylori* infection and the development of gastric cancer. N Engl J Med 2001;345:784–9.

[120] Leung WK, Lin SR, Ching JY, To KF, Ng EK, Chan FK, et al. Factors predicting progression of gastric intestinal metaplasia: results of a randomised trial on *Helicobacter pylori* eradication. Gut 2004;53:1244–9.

[121] Castellarin M, Warren RL, Freeman JD, Dreolini L, Krzywinski M, Strauss J, et al. *Fusobacterium nucleatum* infection is prevalent in human colorectal carcinoma. Genome Res 2012;22:299–306.

[122] Curtis C, Shah SP, Chin SF, Turashvili G, Rueda OM, Dunning MJ, et al. The genomic and transcriptomic architecture of 2,000 breast tumours reveals novel subgroups. Nature 2012;486:346–52.

[123] Baker SC, Bauer SR, Beyer RP, Brenton JD, Bromley B, Burrill J, et al. The External RNA Controls Consortium: a progress report. Nat Methods 2005;2:731–4.

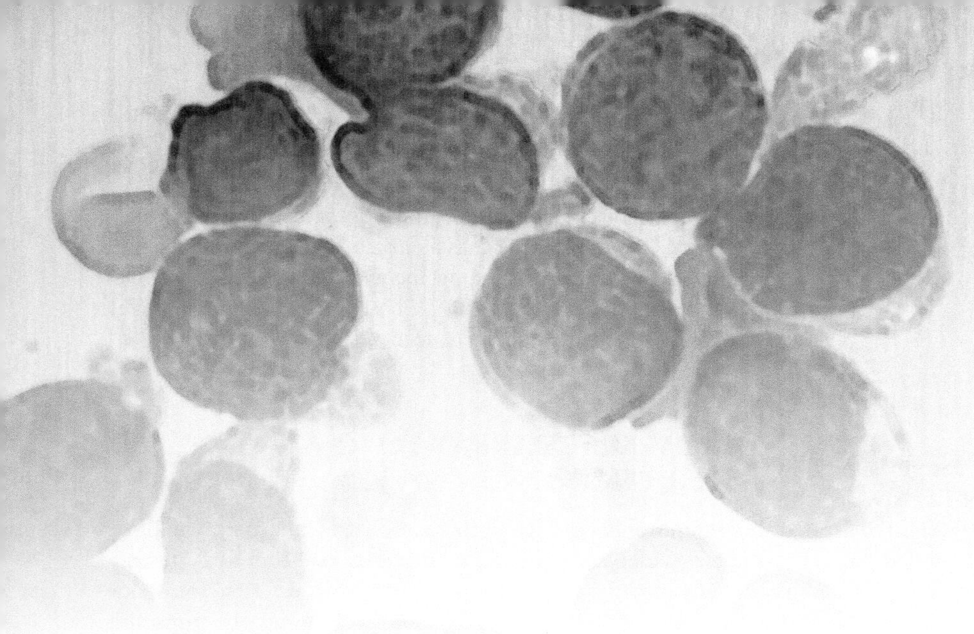

Chapter 4

Cancer Genomics

The Significance of Transcriptome Sequencing in Personalized Cancer Medicine

Shile Zhang[1,2]
Jun S. Wei[1]
Javed Khan[1]

[1]Oncogenomics Section, Pediatric Oncology Branch, Advanced Technology Center, National Cancer Institute, Bethesda, MD, USA
[2]Program in Bioinformatics, Boston University, Boston, MA, USA

Contents

Cancer Genomics. DOI: http://dx.doi.org/10.1016/B978-0-12-396967-5.00004-9

WHY IS RNA IMPORTANT?

The central dogma of molecular biology describes the flow of genetic information from genomic DNA to RNA to protein within a biological system. In early 2003, the Human Genome Project estimated 23 000 protein-coding genes representing approximately 1.2% of the human genome [1,2]. Even though the existence of non-coding RNAs such as transfer RNAs (tRNAs), ribosomal RNAs (rRNAs), small nucleolar RNAs (snoRNAs) and microRNAs was noted at the time, not until a decade later did the Encyclopedia of DNA Elements (ENCODE) project report in 2012 that 62% of the human genome is reproducibly represented in RNA molecules (>200 nucleotides) based on a conservative threshold [3]. Although many of the functions of these transcripts are not currently clear, the transcriptome is an essential component in human biology. Considering the additional contributions from alternative splicing, variable 3′ and 5′ UTRs, RNA editing and secondary/tertiary structure varieties, the diversity of RNA molecules could certainly multiply exponentially compared to the number of genomic loci.

Accordingly, RNA has inched its way into the spotlight to improve our understanding of cancer biology and for development of novel biomarkers and targeted therapies for personalized cancer medicine. Gene expression profiling has been used for diagnosis and predicting outcome in cancer patients since microarray technology first became popular [4,5]. However, the emergence of transcriptome-sequencing technology has provided us with unprecedented opportunities to profile RNA molecules in single nucleotide resolution without prior sequence knowledge. For the past 5 years, thousands of tumor transcriptomes have been deciphered. By mining the massive data generated, we are one step closer to understanding the functions of various transcripts and to developing RNA-based biomarkers and novel therapies for personalized cancer medicine.

FROM MICROARRAY TO RNA SEQUENCING

High-throughput hybridization-based microarrays have been widely used in cancer research as a tool to systematically profile gene expressions since their first introduction in 1995 [6]. This technology has greatly impacted our understanding of human diseases as well as clinical practice, such as tumor classification, treatment response and outcome prediction [7]. Despite the wide utilization of microarrays in genomic research, this technology has its limitations. First, due to the high background noise caused by cross-hybridization, it is difficult to detect single nucleotide variations or structural alterations such as balanced translocation. Second, prior knowledge of the targeted sequences is required for designing probes on microarrays. Third, it is technically challenging to detect every mutation in a given tumor sample using microarray-based strategies. On the contrary, next-generation sequencing (massively parallel sequencing or deep sequencing) does not have the limitations of microarrays and directly identifies millions or even billions of nucleic acid species in parallel in one sequencing reaction, including low-expressed transcripts. Shortly after the introduction of next-generation sequencing technology, many studies were carried out to compare the two methods. In their review of both technologies, Wang et al. [8] listed major advantages of RNA-seq over the microarray technology (Table 4.1). Compared to the hybridization-based microarray, RNA-seq produces a very low background noise and has higher dynamic range for gene

TABLE 4.1 Comparison between Microarray and RNA-Seq Technology

	Microarray	RNA-Seq
Background noise	High	Low
Resolution	From several to 100 bp	Single nucleotide
Dynamic range	Low	High
Isoform detection	Limited	Yes
Allelic expression	Limited	Yes
Required amount of RNA	High	Low
Rely on genomic sequence and annotation	Yes	No

Adapted from [8]

expression measurements to detect low-abundance transcripts and small differences in expression. The single nucleotide resolution of RNA-seq enables small variance analysis, which is important to identify somatic mutations and small indels in tumor tissues. In addition, the single nucleotide sequence readout helps to determine accurately transcription boundaries, splicing boundaries, allele-specific expression and post-transcriptional RNA editing. Moreover, by combining the analysis of transcriptome variation in gene expression, splicing and allele-specific expression of each tumor sample, the studies of expression quantitative trait loci (eQTLs) can elucidate the genetic mechanisms underlying tumorigenesis and progression.

Since RNA-seq analysis relies primarily on the human genome sequence and not on prior knowledge of expressed transcripts, this has significantly enhanced the discovery of novel transcripts such as non-coding or fusion transcripts. Although the human reference genome and RNA sequences are available, it is technically overwhelming to design and utilize microarrays to detect all possible variations existing in tumor RNA sequences, such as single nucleotide variation, structure and splicing variation, abnormal fusion events and viral integration. Additionally, tumor tissues may have different cellular pathways producing novel transcripts that are not detectable in normal tissues. Thus, RNA-seq is capable of systematically surveying the transcriptome without genome sequence bias or annotation.

WORKFLOW FOR RNA-Seq

There are three major sequential steps to generate RNA-seq data from any given sample, patient or cell line (Figure 4.1A). The first step is sample collection. Since normal tissue can provide a tissue-specific baseline for both transcriptional and post-transcriptional regulation, it is advisable to collect adjacent or distal tissue samples for tumor–normal comparison [9,10].

The next step is RNA isolation and the subsequent quality evaluation. RNA can be extracted from fresh, frozen or formalin-fixed tissues. Due to its chemical nature, RNA is much more labile than its nucleic acid counterpart DNA. In general, fresh or snap-frozen tissues will yield high-quality RNA, whereas RNA isolated from formalin-fixed, paraffin-embedded (FFPE) tissues will be fragmented and modified by chemical reactions during fixation. Therefore, RNA from fresh and snap-frozen tissue has been the first choice of molecular biological experiments. Nevertheless, FFPE samples collected during surgery or biopsies are valuable and a more abundant resource for cancer research. A recent study showed that with optimized library chemistry and data analysis, RNA-seq experiments on FFPE tissue samples validate biomarkers

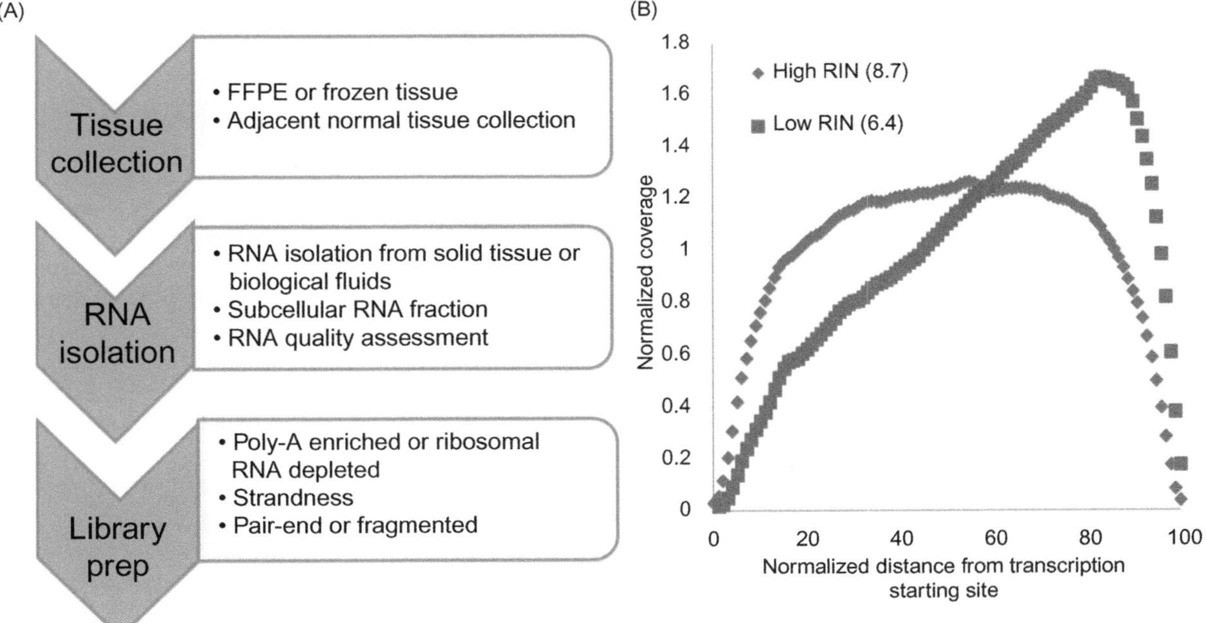

FIGURE 4.1 (A) Three major sequential steps to generate RNA-seq data from any tissue or cell line. (B) Gene coverage of poly-A enriched RNA-seq from samples with high and low RNA integrity number (RIN). For sample with an RIN of 8.7, a homogeneous coverage between transcription starting site to 3′ UTR is observed. However, for sample with a low RIN of 6.4, coverage is higher towards the 3′ UTR end.

measured by reverse transcriptase-polymerase chain reaction (RT-PCR) as well as possess sufficient precision and sensitivity for biomarker discovery [11]. Besides, total RNA can be isolated from either solid tissues or RNA-containing exosomes in biological fluids such as plasma, saliva, urine and amniotic fluid. The subcellular fraction of RNA can also be examined by nuclear/cytosol fractionation. To ensure a successful sequencing experiment, RNA qualities need to be assessed by gel electrophoresis before proceeding to the next step. Figure 4.1B shows two poly-T primed RNA-seq libraries generated from two samples with different RNA integrity numbers (RIN) measured by an Agilent bioanalyzer. Due to the 3′ bias during poly-T primed library construction, 5′ genes were largely missing for the sample with a low RIN. However, a sample with a high RIN has a homogeneous read coverage from 5′ to 3′ genes.

The third step is the actual library preparation. Polyadenylated (poly-A) enrichment and ribosomal RNA (rRNA) depletion are currently the two most popular library preparation methods. A poly-A-enriched library is made by priming with poly-dT primers. This method highly enriches for mature mRNA with poly-A tails. However, the downside of this approach is the loss of non-polyadenylated RNA which may also carry out important functions in biological systems [12]. Moreover, poly-A RNA enrichment may result in 3′ bias (see Figure 4.1B), which can be more prominent in samples with RNA degradation. An alternative is the method of rRNA depletion aiming to exclude ribosomal (rRNAs) and transfer RNAs (tRNAs), which comprise the majority of total RNA in a cell. rRNAs depletion allows the capturing of long, non-coding RNAs (lncRNAs) lacking poly-A tails. This RNA species has emerged as an important player in tumorigenesis [13,14]. Similar to the poly-A enrichment method, low RNA quality can result in poor efficiency in ribosomal RNA removal, causing a large portion of rRNA to be retained in the libraries. In addition, pre-mRNA, intermediate product and partially degraded RNA can also be detected in the rRNA-depleted libraries. These molecules can provide additional information or noise in the RNA-seq data.

Furthermore, strandness information of RNA can be extracted using strand-specific RNA library preparation methods which are particularly useful for novel transcript annotation. It enables accurate strand-specific alignment of RNA-seq reads to the genome, so that adjacent or overlapping genes transcribed on the opposite strand can be distinguished and quantified. Strand-specific RNA-seq usually requires a more complicated experimental procedure and customized downstream computational analysis [15].

In addition to different library construction strategies, current sequencing technology limitations, such as short reads, have also an important impact on the information derived from next-generation sequencing experiments. However, paired-end RNA sequencing, allowing both ends of cDNA fragments to be sequenced, overcomes some of these limitations. Paired-end sequencing also affords other significant advantages over fragment libraries, including increasing the sensitivity to detect junctions resulting from splicing or fusion events [16]. These junction reads are crucial evidence for transcript structure, as much as increasing the alignment/assembly accuracy.

The three most common RNA-seq platforms are Roche's 454, Illumina's HiSeq and Life Technology's SOLiD system. All of these platforms generate short sequencing reads (100−1000 bp). Liu et al. [17] compared the performance of these three platforms in terms of sequencing mechanism, components and cost, and application, providing scientists with a useful guideline for choosing the most suitable platform for their research project. More recently sequencers with lower throughput but faster turnaround time have been introduced for clinical applications. Life Technology's Ion Torrent Proton and Illumina's MiSeq shorten the run time from a couple of weeks to a day or even hours for an RNA-seq experiment [17,18]. Moreover, direct RNA sequencing technologies, such as Helicoscope developed by Helicos Biosciences, eliminate the error-prone cDNA synthesis step [19]. Third-generation sequencing led by PacBio and Oxford Nanopore allows single-molecule sequencing without PCR amplification steps. With a length of more than 1000 bp, these third-generation sequencing technologies have the potential to generate long reads spanning full-length RNA molecules [20].

RNA-Seq DATA ANALYSIS IN CANCER GENOME RESEARCH

The major goals of RNA-seq analysis are to elucidate the structure and sequence of transcripts and to profile transcriptome expression. Due to pre-mRNA splicing, there are gaps when mapping RNA reads to the genome. Thus, the analysis of RNA-seq is more complex than that of genomic DNA sequencing. Methods to analyze RNA-seq are in their infancy and still being developed. In this section, we focus on examples of gene expression analysis and post-transcriptional regulation analysis to give an overview of the RNA-seq data analysis workflow.

Gene Expression Analysis

Gene expression profiling using RNA-seq is based on counting the copy numbers of each transcript (Figure 4.2A). Hundreds of millions of reads are generated by the sequencer in a format that stores both nucleotide sequence and its corresponding quality score. In the

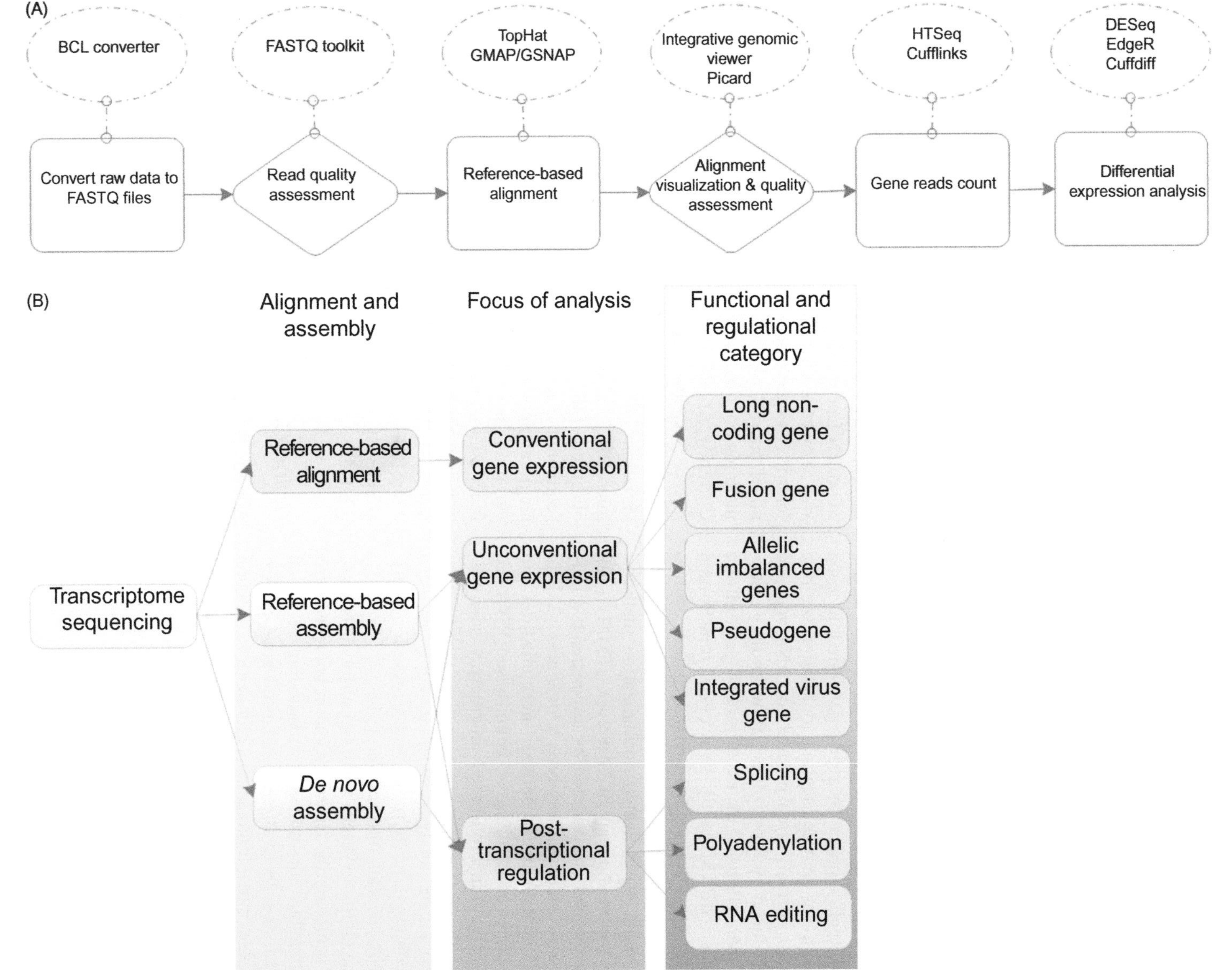

FIGURE 4.2 (A) Conventional RNA-seq gene expression analysis pipeline for profiling annotated protein-coding genes and non-coding genes. (B) RNA-seq specific analysis on unconventional gene expression and post-transcriptional regulation. Reference-based or *de novo* assemblies are required to reconstruct the transcriptome with SNVs, trans-splicing, novel splicing/polyadenylation sites, unannotated gene loci and virus integration.

initial step, reads of PCR-duplicates, ribosomal RNA and of low quality as well as adapter sequences will be filtered out to ensure the accuracy of the downstream analysis. Secondly, reads are aligned to genome sequences with a splice-aware aligner such as GMAP/GSNAP [21,22] or TopHat [23]. Sequence alignment/map (SAM) [24] and its BGZF compression (BAM) are generic formats to store the sequence alignment results. At this point, several RNA-seq quality control metrics can be generated, such as pairing percentage, mapping percentage, mapping score distribution and gene coverage distribution. If the quality assessments indicate a successful RNA-seq experiment from library preparation to sequence alignment, then gene expression can be quantified with tools such as HTSeq [25] and Cufflinks [26] followed by differential expression analysis with software packages such as Cuffdiff (a module in Cufflinks [26]), DESeq [27] and EdgeR [28]. Due to the differences in the total number of reads generated in each sequencing run per sample, scaling to library size is a widely applied normalization method for RNA-seq and was first introduced by Mortazavi et al. when they calculated reads per kilobase of exon model per million mapped reads (RPKM) [29]. However, transformed RPKM values mask the raw read counts, which contain important information that affects differential expression calls. Thus, more sophisticated normalization methods, such as the trimmed mean of M-values (TMM) [30], were later developed and adopted by downstream differential expression statistical analysis procedures [27]. In addition, due to differences between normalization and differential expression calling on count-based RNA-seq analysis and hybridization-based microarray analysis, some of the downstream examinations, such as gene ontology enrichment analysis, need to apply adjusted statistical methods in order to take the expression distribution of RNA-seq data into account [31].

Due to the ability to detect novel transcripts, RNA-seq experiments can study the transcriptome in unprecedented detail including but not limited to non-coding or fusion gene detection, as well as allele-imbalanced gene expression, viral gene integration and pseudogene expression (see Figure 4.2B). The following section discusses strategies to apply RNA-seq data to systematically study these areas that were challenging in the microarray era.

Non-Coding RNAs

Long non-coding RNAs (lncRNAs; > 200 nt), including tRNAs, rRNAs, small nuclear RNAs and small nucleolar RNAs, were once thought "housekeeping" RNAs. They are emerging as a new category of non-coding RNAs that carry important functions in cancer [32,33]. Many of the lncRNAs have not been explored or even annotated, thus RNA-seq is an ideal tool to capture systematically transcribed lncRNAs. For example, one study discovered 121 novel prostate cancer-associated non-coding RNAs and found one of them, PCAT-1, to be implicated in prostate cancer progression [34]. This report showed that lncRNAs express in a tissue- and disease-specific manner. Hence, systematically profiling lncRNA is important to understand disease biology and to develop lncRNA-based biomarkers.

Similarly, RNA-seq can be used to profile small non-coding RNA, such as microRNAs (miRNAs) in cancers. miRNAs are short (20–24 nucleotides) non-coding RNAs that modulate protein translation and mRNA stability in a tissue-specific manner and regulate proliferation, apoptosis, differentiation, growth, migration and metabolism in cancers [35]. Therefore, microRNAs can be used for early detection, diagnosis, prognosis and therapies [36]. Compared to standard RNA-seq, microRNA sequencing captures small RNAs enriched by size selection. In parallel to mRNA sequencing, microRNA sequencing has advantages in identifying novel microRNA and detecting microRNA variants over methods of hybridization-based arrays or RT-PCR. Currently, 1600 human microRNAs are annotated in the microRNA database, miRBase, with many of them identified by microRNA sequencing [37]. Moreover, variation in sequences and lengths of the reference microRNA sequences, called "isomiR", have been observed [38]. IsomiRs are the result of imprecise sequential cleavage by exoribonucleases, RNA editing, single nucleotide polymorphisms and somatic nucleotide variation in cells [39]. Although much of the isomiR functional significance is still under investigation, variation of isomiRs has been implicated to affect microRNA target efficiency and half-life [40]. Consequently, microRNA sequencing is a preferred method to study effectively various isomers over microarrays and RT-PCRs.

Fusion Genes

Abnormal fusion genes caused by chromosomal rearrangements often play key roles in tumor initiation and progression. Historically, highly recurrent fusion gene events were identified in cancers with cytogenetic techniques [41,42]. Due to the poor resolution of these cytogenetic techniques, identification of the fusion partners and their fusion products is not a trivial task. RNA-seq can be used to systematically discover genome-wide fusion events with single nucleotide resolution on fusion break points, which can be used in diagnosis or targeted clinical intervention for cancers [43–47]. In one study, investigators performed RNA-seq in high-risk acute lymphoblastic leukemia that had a Ph-like gene expression signature but lacked the BCR–ABL fusion gene and found fusion events of genes involving *ABL1*, *JAK2*, *PDGFRB*, *CRLF2* and *EPOR*. They reported that several of these genomic events result in transformation that was attenuated with tyrosine kinase inhibitors, thus indicating that these patients may be treated with targeted therapy in future trials [46]. Therefore, RNA-seq is an effective method to examine thoroughly genome-wide gene fusion including novel events with low frequency levels in tumors. In addition to fusion gene events caused by genomic rearrangement, RNA-seq can also detect trans-splicing and read-through transcripts, which were recently found to be important in cancer development [48,49]. Such events are difficult to detect by traditional cytogenetic techniques or DNA sequencing.

Viral Gene Integration

Similar to fusion gene detection, virus–human fusion transcripts also can be detected by RNA-seq experiments [50]. Viruses are important pathogenic factors in initiation and progression of certain types of cancers, such as hepatocellular carcinoma and Hodgkin lymphoma. In a recent study, hepatitis B virus (HBV) integration was observed more frequently in hepatocellular carcinoma than in adjacent liver tissues, indicating a potential mechanism of the tumorigenic role for HBV in this tumor type [51]. Apart from identifying viral gene integrating events, RNA-seq analysis also enables quantification of downstream gene expression changes due to viral integration, providing biological insights during the viral integration event.

Allelic Imbalance Expression

Allelic imbalanced (AI) gene expression of the diploid human genome has been demonstrated to contribute to tumorigenesis and progressions. For example, germline allele-specific expression of *TGFBR1* was shown to occur significantly more often in colorectal cancer patients and potentially to contribute to tumor development [52]. In addition, AI expressions in *BRCA1* and *BRCA2* genes were associated with an increased breast cancer risk [53]. Because of its single nucleotide resolution, RNA-seq can detect AI expression by separately quantifying each allele of the same gene. Together with revealing AI events, RNA-seq can be used to identify the molecular mechanisms leading to AI expressions, such as copy number alterations [54,55] or single nucleotide variances [56].

Pseudogenes

The other possible gene analysis for RNA-seq data is to identify pseudogenes. Pseudogenes are discovered to have tumor-specific expression and are subject to structural or single nucleotide somatic variations in the tumor genome. Furthermore, pseudogenes have been shown to be functionally involved in cancers via various biological mechanisms [57]. Many pseudogenes serve as "competing endogenous RNA" (ceRNA) along with their parental genes sharing the same miRNA-, siRNA- and translational machinery binding sites [58]. A recent RNA-seq study examined genome-wide pseudogene expression in multiple human cancer types and found over 2000 lineage- and cancer-specific expressed pseudogenes [59]. The finding suggests that the functional importance of pseudogenes is likely underestimated. Although detailed biochemical experiments are required to characterize their biological functions, RNA-seq is nevertheless a valuable tool to study pseudogenes systematically.

Post-Transcriptional Regulation Analysis

The structure and expression level of a transcript are not only controlled by transcription-regulating program, but also by post-transcriptional modifications including splicing, polyadenylation and RNA-editing (see Figure 4.2B). Due to the single nucleotide resolution of RNA-seq reads, these post-transcription mechanisms can be thoroughly explored using RNA-seq data. Hence the following section focuses on recent advances in post-transcriptional mechanisms using RNA-seq data.

Alternative Splicing

Alternative splicing is the most prevalent post-transcriptional RNA processing event, affecting 95% of all human multiexon genes [60]. Spliceosomes catalyze the splicing process by recognizing splice sites located at exon–intron boundaries. In some cancers, genomic mutations disrupt important splice sites that result in aberrant splicing, which can cause insertion, deletion or frame shift in the amino acid sequence. Moreover, splicing factors that bind to specific pre-mRNA binding motifs regulate splicing patterns in a tissue- and disease-specific manner [61–64]. Changes of splicing factors are shown to be important contributors during normal tissue development such as epithelial–mesenchymal transition [65], as well as tumorigenesis, such as tumor cell motility [66], metastasis [67,68], metabolism [69] and proliferation [70]. RNA-seq allows examination of the splicing program at three different levels. First, splicing boundaries can be elucidated by the RNA-seq reads at single nucleotide resolution. Also splice variant expression can be quantified by counting the reads aligning to exons specific to the particular variant. Second, genomic mutations at particular splice sites and splicing factor binding motifs can result in aberrant splicing in tumors. RNA-seq read sequences can be analyzed for such potential mutations. Third, the expression level of splicing machinery genes, as well as splicing factor genes can be quantified to study splicing control in tumors.

Alternative Cleavage and Polyadenylation

Alternative cleavage and polyadenylation (APA) is another mechanism of post-transcriptional control of eukaryotic mRNA. An estimated 69.1% of human genes carry multiple poly-A sites and APA controls the usage of poly-A sites in a tissue-specific manner [71]. Oncogenes in cancer cells are reported to undergo APA resulting in isoforms with shorter 3′ UTRs which, in turn, affect mRNA stability and translation efficiency, producing 10 times more protein than their longer 3′ UTR alternatives [72]. Similar to splicing regulation, APA is controlled by both *trans*- and *cis*-regulators. RNA-seq experiments can be used to quantify APA at a genome-wide scale, as well as to explore *trans*- and *cis*- regulatory elements.

RNA Editing

Canonical RNA editing in human cells turns an adenosine (A) into an inosine (I) through site-selective hydrolytic deamination and non-selective hyperediting clusters that mostly occur in non-coding *Alu* repeats in pre-mRNAs [73,74]. Together with germline and somatic mutations, selective RNA editing may result in protein coding changes. Non-selective RNA editing can affect gene expression by modulating promoter-binding sites, microRNA-binding seeds, splicing factor-binding motifs and 3′ UTR polyadenylation signals [73,74]. Three enzymes, ADAR1, ADAR2 and ADAR3, catalyze A-to-I RNA editing in human cells. It has been reported that RNA-editing events and RNA-editing enzymes are perturbed in various cancers [74]. For instance, a high rate of A-to-I editing in the gene *AZIN1* was observed in approximately 50% of primary hepatocellular carcinoma; the increase of the editing rate in tumor compared to adjacent normal tissue indicated its contributions to cancer initiation and progression [75]. Nevertheless, RNA-editing events are still largely unexplored in cancer and RNA-seq is evidently a valuable tool to study this transcriptome-wide event. However, an obvious challenge is the calling of RNA-editing events, which is prone to produce false positives for multiple reasons. First, variants can be introduced by reverse transcription converting RNA to cDNA. Second, sequencing errors can be introduced systematically to the reads. Third, mapping errors exist due to: (1) inaccuracy of the human reference genome; and (2) repetitive regions, splicing junction sites and paralog genes. Therefore, an adaption of the mapping algorithm and accurate modeling of potentially occurring errors are essential to avoid false positives in RNA-editing analysis.

Collectively, RNA-seq allows the systematic analyses of gene expressions and post-transcriptional regulations. For the first time, complex transcriptomes can be examined with a single RNA-seq experiment. However, accurately and comprehensively mining millions of short read sequences is a challenging computational task. As discussed previously, analytical workflow needs to be carefully evaluated to study the gene expression and post-transcriptional regulations.

Computational Approaches for RNA-Seq Analysis

To perform gene expression and post-transcriptional regulation analysis with adequate precision, the bioinformatic approach needs to be customized to achieve the goal of the analysis (see Figure 4.2B). Other than reads alignment to the reference human genome, reference-based transcriptome assembly is commonly used to facilitate the reconstruction of less well annotated gene loci, isoforms with novel splice or polyadenylation sites and fusion genes with sequencing reads spanning two distant genes. Cufflinks [26] and Scripture [76] are the two most commonly used reference-based transcriptome assemblers.

Alternatively, *de novo* assembly of sequencing reads without alignment to a reference genome or transcriptome can be used to identify novel transcripts in an RNA sample. The cancer transcriptome is expected to deviate from the reference genome or transcriptome due to somatic single nucleotide variants, trans-splicing, occurrence of novel splice and polyadenylation sites and viral integration. *De novo* transcriptome assembly has the advantage of accurately reconstructing the transcriptome with genomic/transcriptomic modification events occurring during tumor initiation and progression. However, there are some drawbacks to this approach as summarized by Martin et al. [77]: (1) higher coverage is required to reconstruct full-length transcripts; (2) *de novo* assembly algorithms are sensitive to sequence errors; (3) *de novo* assemblers require a large amount of computational resources. Thus, careful evaluation of the coverage and removal of low-quality sequencing reads prior to *de novo* assembly are highly recommended.

INCORPORATING TRANSCRIPTOME SEQUENCING ANALYSIS TO IDENTIFY GENOMIC DRIVER MUTATIONS

Next-generation sequencing technology has been used extensively to detect cancer somatic variations and indels. One earlier RNA-seq study identified a missense point mutation in *FOXL2* to be highly recurrent in adult-type granulosa-cell tumors [78]. However, compared to genomic DNA sequencing studies, there are much fewer reports of somatic mutations discovered solely by RNA-seq due to several reasons. First, normal RNA samples are under different transcription/post-transcription control, resulting in differently expressed genes/isoforms. Without matched normal RNA samples, it is extremely challenging to distinguish somatic variants from germline rare variants or RNA-editing events. Second, the coverage of RNA-seq reads is uneven across different genes, corresponding to their abundance in an RNA sample. Some transcript genes with somatic mutations might be completely absent due to mechanisms such as nonsense-mediated decay. Third, mutations introduced in cDNA synthesis of reverse transcription result in sequence artifacts [79]. Finally, as RNA-seq reads are mapped to the genome with intronic gaps, sequences adjacent to the exon/intron boundaries are prone to mapping errors, thus hindering the sensitivity and specificity of variant calling.

However, RNA-seq can be used to complement DNA sequencing to distinguish a "driver" mutation event from "passenger" mutations (Figure 4.3). DNA sequencing can identify hundreds if not thousands of genomic variants in a single cancer genome; therefore distinguishing a "driver" genomic variant that causes a selective growth or survival advantage to the tumor is a challenging task. Researchers can systematically narrow down driver mutation candidates by examining transcriptome expression and evaluating if the genomic variation is expressed in the RNAs [80]. Cancer driver genes can promote tumorigenesis or development by gain or loss of their normal functions. The genomic mechanisms can be copy number changes, structural mutations or sequence mutations, which all impact on the expression of the affected genes.

A region with copy number variation usually contains multiple genes in tumors. In this case, modulating gene expressions of potential drivers and downstream targets with key biological functions can help to identify the "driver gene" that offers a growth advantage to tumor cells [81]. Likewise, genes with single nucleotide variants (SNVs) can potentially be gain- or loss-of-function resulting in tumor growth advantages. If a mutated tumor suppressor gene gains growth advantage to the tumor, then the expression of the tumor suppressor should be decreased or lost due to nonsense-mediated decay triggered by aberrant mRNA transcripts. Vice versa, a driver oncogene mutated allele is expected to be activated and expressed to fit the gain-of-function hypothesis. In some rare cases, a typical tumor suppressor gene can also exert dominant-negative effects or gain oncogenic activities [82]. In this case, the allele frequencies of wild-type/mutated transcripts need to be carefully examined to evaluate the phenotypic outcomes. Moreover, the functional consequences of SNVs in tumor DNA can be further examined by evaluating molecular interactions of mutated genes with other genes in the same network. In one study, Ng et al. incorporated gene expression and pathway information to predict if a mutation event was neutral, gain- or loss-of-function in a given tumor sample [83]. If upstream genes of this mutant gene had high activity inferred by gene expression, while downstream genes of the mutant genes had low activity, then the mutant gene was inferred to be loss-of-function. On the contrary, if the upstream genes had low activity and the downstream genes had high activity, then the mutated gene was potentially gain-of-function. To take it together, RNA-seq can first verify if some variants called from DNA sequencing analysis are expressed. Along with expression analysis and pathway analysis, RNA-seq data can infer the functional effects of any potential genomic variance, thus distinguishing "driver" from "passenger" genomic events. Based on this hypothesis, a pilot study of personalized cancer treatment combined DNA and RNA sequencing analyses to generate an overview of an individual's genetic aberrations [84]. Derived from integrated sequencing results, therapeutic interventions for potential pathways were recommended along with approved drug candidates.

TRANSCRIPTOME SEQUENCING: FROM BENCH TO BEDSIDE

Transcriptome profiling is widely used in cancer research with potential clinical value in diagnosis and prognosis in many cancers [85]. Traditional microarray-based expression profiling technologies only detect expressions of annotated genes, which is a small portion of transcribed RNA molecules [1]. Moreover, genome-wide association study (GWAS) demonstrates that only 12% of disease-associated SNPs are located in the protein-coding regions, while the rest are in the intronic and intergenic regions [86]. Transcriptome sequencing allows unbiased capture of both known and novel RNA molecules. Likewise, single nucleotide resolution of sequencing technology provides detailed and unbiased insights into the structure and abundance of each individual transcript, through gene expression and post-transcriptional regulation analysis. These underexplored spectra of transcriptomic analysis can help our understanding of tumor biology and provide opportunities for developing new biomarkers and even novel targeted therapies [87].

From the bench side, great efforts have been concentrated on identification of cancer genes through sequencing in large-scale projects, such as the International Cancer Genome Consortium (ICGC) [88], The Cancer Genome Atlas (TCGA) [89] and Therapeutically Applicable Research to Generate Effective Treatments (TARGET) [90]. These collaborative projects attempt to characterize the recurrent key genomic and epigenetic alterations in various cancer types by integration of

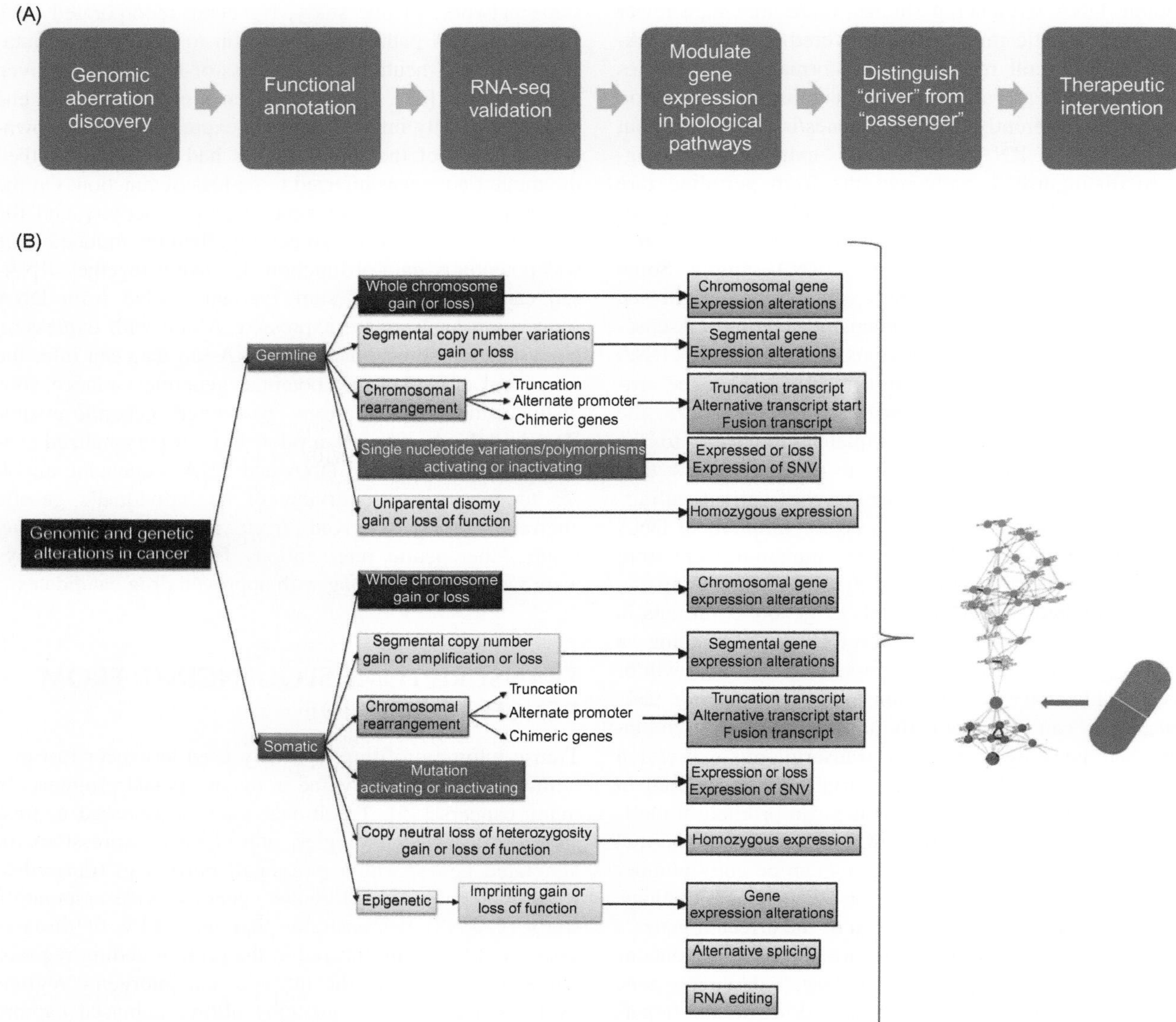

FIGURE 4.3 (A) Consequences of genomic alterations can be investigated by RNA-seq to elucidate alterations in pathway, distinguish driver from passenger mutations or direct therapy. (B) RNA-seq can be used for elucidating tumor biology and directing therapy. Tractable genomic alterations can be either germline or somatic. Alterations may be as a result of copy number alterations such as whole chromosomal gains (losses usually incompatible with prolonged life) or segmental changes. Constitutional chromosomal rearrangements have been observed in patients with cancer. These rearrangements can result in truncation of a protein, expression of a gene under the control of an alternative promoter or the production of a chimeric or novel fusion protein not normally found in nature. Single nucleotide variations (SNV) or polymorphisms (SNP) are associated or causal to an increased predisposition to cancer. These variations may be in the promoter regions leading to overexpression or suppression or may be within the protein-coding regions leading to cancer syndromes. Other constitutional alterations that predispose to cancer include paternal segmental isodisomy or, in a cancer, copy neutral loss of heterozygosity where there is no net loss of DNA but uniparental disomy with only one parental chromosome or region present in two copies. Epigenetic alterations such as silencing of genes by methylation is an increasingly important mechanism of oncogenesis. All these DNA changes result in changes in the transcript. Additionally, RNA-seq may reveal tractable changes due to alternative splicing or RNA editing. These changes can be linked to pathways or drugs for therapeutic intervention. Adapted from Pizzo and Poplack, Principles and Practices of Pediatric Oncology.

multidimensional analyses. Various databases such as COSMIC [91] and cBioPortal [92] have been used to deposit the rich datasets coming off these large cancer projects, providing detailed genomic and epigenetic information of each tumor sample for biological and clinical researchers.

Current research using transcriptome sequencing technology to guide therapies is underway. Integration of

DNA copy number aberrations, single nucleotide variants, chromosomal rearrangements identified by sequencing cancer DNA and RNA have led to discovery of oncogenic driver events of cancers, thus guiding therapeutic treatments (Figure 4.4). In a clinical sequencing study, combined DNA and RNA sequencing analysis identified a patient with *NRAS* mutation and amplification and overexpression of *CDK8*, resulting in the use of targeted therapies using CDK inhibitors and MEK/PI3K inhibitors respectively [84]. Table 4.2 provides a list of molecular biomarkers used in clinics for guiding diagnosis and therapeutic decisions. RNA-seq can assess these molecular biomarkers in a single experiment to detect potential rearrangements, point mutations and gene expression aberrations. With more genetic and epigenetic traits identified in various tumor types, as well as on-going personalized therapy clinical trials to target mutated genes and perturbed signaling pathways, we anticipate a substantial growth of this list and the applications of RNA-seq-based clinical assays in the near future.

Moreover, novel findings discovered in clinical sequencing programs can facilitate understanding of tumor biology. A good example is the novel *NAB2–STAT6* fusion in solitary fibrous tumor, which was first identified from a patient enrolled in a clinical sequencing study. Subsequently, the fusion events were tested to be 100% recurrent in another 51 solitary fibrous tumors. Functional study of the *NAB2–STAT6* fusion gene demonstrated its driver activity in the solitary fibrous tumors by converting a transcription repressor *NAB2* into a transcription activator of mitogenic pathways during tumor progression [93].

Another application of RNA-seq in the field of pharmacogenomics is to identify the mechanisms of sensitivity or resistance to a cancer therapy [94]. Recent sequencing studies have demonstrated intratumor and intertumor heterogeneity in the genomic landscapes of tumors during progression and metastasis [95]. Thus, understanding tumor temporal and spatial molecular dynamics is crucial for effective diagnosis and treatment. Furthermore, RNA-seq can be used to measure genetic and epigenetic traits that affect drug transport, drug metabolism and potential adverse drug/treatment effects [85,94].

Despite the tremendous potential of applying RNA-seq to cancer medicine, there are challenges translating RNA-seq findings into clinics. To start with, one single RNA-seq experiment generates hundreds of millions of short sequence reads. As a result, a proper computing infrastructure is required for data transfer, storage, computing and incorporation into databases and medical records. Moreover, there is greater challenge for interpreting RNA-seq data of millions of sequence reads. First, there are a limited number of tools available for data analysis to yield useful clinical information. Many of the bioinformatics tools are still in their nascent stages, with major modifications and updates coming out frequently. In addition, there is not yet agreement on "best practice" sequencing data analysis. Different workflow and different computation methods will result in heterogeneous interpretations of the same datasets [96]. As discussed in previous sections, determined by the ultimate goals of each specific analysis, some of the initial steps such as read mapping and transcript reconstruction need to be carefully examined and filtered to achieve satisfactory sensitivity and specificity. Depending on genetic background, origin and type of tumor, even with the same analysis pipelines, sensitivity and specificity of the data interpretation could differ dramatically. In addition, RNA samples of the same patient from different tumor sites or at different time points can be sequenced multiple times. Integrating DNA, RNA, protein, methylation, metabolism and interaction data is often required to interpret the tumor biology correctly, thus resulting in another layer of

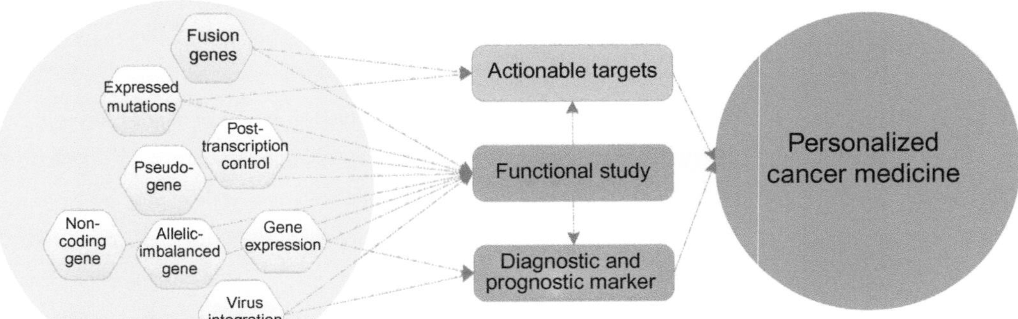

FIGURE 4.4 Studies on complex transcriptome and functional validations contribute to diagnostic and prognostic markers and actionable therapeutic targets for personalized cancer medicine.

TABLE 4.2 A List of Molecular Biomarkers Used in Clinic for Diagnosis and to Guide Prognostic and Therapeutic Decisions

Diagnostic		
Acute leukemia	PML-RARA	WHO 2008 classification of leukemia
	BCR-ABL1	
	CBFB-MYH11	
	ETV6-RUNX1	
	MLL-rearranged	
	TCF3-PBX1	
	RBM15-MKL1	
Myeloproliferative disease	JAK2	Mutations confirm diagnosis of clonal myeloproliferative disorders
Sarcomas	SS18-SSX1/SSX2	Synovial sarcoma
	PAX3/PAX7-FOXO1A	Alveolar rhabdomyosarcoma
	EWSR-FLI1	Ewing's sarcoma
	EWSR1-ERG	
	EWSR1-NR4A3	Extraskeletal myxoid chondrosarcoma
	TAF15-NR4A3	
	EWSR1-ATF1	Clear cell sarcoma (and angiomatoid fibrous histiocytoma)
	EWSR1-CREB1	
	ASPSCR1-TFE3	Alveolar soft-part sarcoma (and renal cell carcinoma)
	FUS-DDIT3	Myxoid liposarcoma
	FUS-CREB3L2	Low-grade fibromyxoid sarcoma
	JAZF1-SUZ12	Endometrial stromal sarcoma
	ETV6-NTRK3	Congenital fibrosarcoma (and secretory breast carcinoma)
Predictive		
Non-small cell lung cancer	EGFR	Mutations predict response to TKI
	ALK	Rearrangements predict response to ALK-inhibitors
Gastrointestinal stromal tumor	KIT and PDGFRA	Mutations predict response to c-KIT/PDGFRA inhibitors
Metastatic colorectal cancer	KRAS	Mutations predict lack of response to anti-EGFR antibodies
Melanoma	BRAF	Mutations predict response to specific BRAF inhibitors
Breast cancer	HER2	Amplifications predict response to anti-HER2 antibodies
Prognostic		
Chronic lymphocytic leukemia	TP53	Mutations are indicative of poor outcome
	IGHV	Lack of mutations is indicative of poor outcome
Acute myeloid leukemia	FLT3-ITD	Mutations are indicative of poor outcome
Metastatic colorectal cancer	BRAF	Mutations are indicative of poor outcome
Breast cancer	OncotypeDx	Risk stratification (21-gene expression signature)
	Mammaprint	Risk stratification (70-gene expression signature)
	IHC4	Risk stratification (4-protein IHC expression)

(*Continued*)

TABLE 4.2 (Continued)

Disease monitoring		
Chronic myeloid leukemia	BCR-ABL1	Minimal residual disease detection
Acute promyelocytic leukemia	PML-RARA	Minimal residual disease detection
Acute lymphoblastic leukemia	IGHV-TCR	Minimal residual disease detection

Adapted from Gonzalez de Castro D, Clarke PA, Al-Lazikani B, Workman P. Personalized cancer medicine: molecular diagnostics, predictive biomarkers, and drug resistance. Clin Pharmacol Ther 2013;93:252–9. ALK: anaplastic lymphoma kinase; TKI: tyrosine kinase inhibitors

complexity. Therefore, a robust and comprehensive bioinformatic pipeline needs to be worked out to standardize this technique in clinical applications. Moreover, to interpret the clinical significance of data needs multidisciplinary contributions from pathologists, pharmacists, clinicians, biologists, genetic counselors and bioinformaticians. To understand fully each of the analytical steps and work cooperatively is challenging for healthcare providers without multidisciplinary training.

In summary, although in its infancy, RNA-seq is a promising technique to generate rich information from patient samples, which can be combined with other genetic information to strategize personalized treatment for cancer patients. With decreasing cost and an acceptance in the medical field, RNA-seq technology will be routinely used in clinics as a clinical test for personalized therapy in the near future.

GLOSSARY

Allele-specific expression When two alleles of a gene are expressed at different rates.

Dark matter transcripts RNA molecules that are not associated with annotated genes.

***De novo* transcriptome assembly** Transcriptome sequencing reads assembly without reference genome guidance.

Fusion gene A chimerical gene produced by abnormal fusion of two separate genes usually as a result of chromosomal deletion or translocation. A fusion gene can be oncogenic in a cancer as a "driver" event.

Genomic "driver" event Genomic alterations that cause a selective growth or survival advantage to tumor cells.

Genomic "passenger" event Genomic alterations that have no effect on the selective growth or survival advantage of tumor cells.

Transcriptome The collection of RNA molecules that are expressed in a cell or tissue. This collection includes mRNA, rRNA, tRNA and various types of non-coding RNAs.

ABBREVIATIONS

APA Alternative cleavage and polyadenylation
ceRNA Competing endogenous RNA
COSMIC Catalogue of somatic mutations in cancer
ENCODE Encyclopedia of DNA elements
eQTL Expression quantitative trait loci
GWAS Genome-wide association study
ICGC International Cancer Genome Consortium
lncRNA Long non-coding RNA
pre-mRNA Precursor messenger RNA
qPCR Quantitative polymerase chain reaction
RIN RNA integrity number
RNA-seq RNA sequencing
rRNA Ribosomal RNA
snoRNA Small nucleolar RNA
SNV Single nucleotide variant
TCGA The Cancer Genome Atlas
TARGET Therapeutically Applicable Research to Generate Effective Treatments
tRNA Transfer RNA

REFERENCES

[1] International Human Genome Sequencing Consortium. Finishing the euchromatic sequence of the human genome. Nature 2004;431:931–45.

[2] Lander ES, Linton LM, Birren B, Nusbaum C, Zody MC, Baldwin J, et al. Initial sequencing and analysis of the human genome. Nature 2001;409:860–921.

[3] The ENCODE Project Consortium. An integrated encyclopedia of DNA elements in the human genome. Nature 2012;489: 57–74.

[4] van't Veer LJ, Bernards R. Enabling personalized cancer medicine through analysis of gene-expression patterns. Nature 2008;452:564–70.

[5] Khan J, Wei JS, Ringner M, Saal LH, Ladanyi M, Westermann F, et al. Classification and diagnostic prediction of cancers using gene expression profiling and artificial neural networks. Nat Med 2001;7:673–9.

[6] Schena M, Shalon D, Davis RW, Brown PO. Quantitative monitoring of gene expression patterns with a complementary DNA microarray. Science 1995;270:467–70.

[7] Macgregor PF. Gene expression in cancer: the application of microarrays. Exp Rev Mol Diagn 2003;3:185–200.

[8] Wang Z, Gerstein M, Snyder M. RNA-Seq: a revolutionary tool for transcriptomics. Nat Rev Genet 2009;10:57–63.

[9] Wu Y, Wang X, Wu F, Huang R, Xue F, Liang G, et al. Transcriptome profiling of the cancer, adjacent non-tumor and distant normal tissues from a colorectal cancer patient by deep sequencing. PloS One 2012;7:e41001.

[10] Kim SC, Jung Y, Park J, Cho S, Seo C, Kim J, et al. A high-dimensional, deep-sequencing study of lung adenocarcinoma in female never-smokers. PloS One 2013;8:e55596.

[11] Sinicropi D, Qu K, Collin F, Crager M, Liu ML, Pelham RJ, et al. Whole transcriptome RNA-Seq analysis of breast cancer recurrence risk using formalin-fixed paraffin-embedded tumor tissue. PloS One 2012;7:e40092.

[12] Yang L, Duff MO, Graveley BR, Carmichael GG, Chen LL. Genomewide characterization of non-polyadenylated RNAs. Genome Biol 2011;12:R16.

[13] Gutschner T, Diederichs S. The hallmarks of cancer: a long non-coding RNA point of view. RNA Biol 2012;9:703–19.

[14] Spizzo R, Almeida MI, Colombatti A, Calin GA. Long non-coding RNAs and cancer: a new frontier of translational research?. Oncogene 2012;31:4577–87

[15] Levin JZ, Yassour M, Adiconis X, Nusbaum C, Thompson DA, Friedman N, et al. Comprehensive comparative analysis of strand-specific RNA sequencing methods. Nat Methods 2010;7:709–15.

[16] Ozsolak F, Milos PM. RNA sequencing: advances, challenges and opportunities. Nat Rev Genet 2011;12:87–98.

[17] Liu L, Li Y, Li S, Hu N, He Y, Pong R, et al. Comparison of next-generation sequencing systems. J Biomed Biotechnol 2012;6:251364.

[18] Quail MA, Smith M, Coupland P, Otto TD, Harris SR, Connor TR, et al. A tale of three next generation sequencing platforms: comparison of Ion Torrent, Pacific Biosciences and Illumina MiSeq sequencers. BMC Genomics 2012;13:341.

[19] Ozsolak F, Platt AR, Jones DR, Reifenberger JG, Sass LE, McInerney P, et al. Direct RNA sequencing. Nature 2009;461:814–8.

[20] Schadt EE, Turner S, Kasarskis A. A window into third-generation sequencing. Hum Mol Genet 2010;19:R227–40.

[21] Wu TD, Watanabe CK. GMAP: a genomic mapping and alignment program for mRNA and EST sequences. Bioinformatics 2005;21:1859–75.

[22] Wu TD, Nacu S. Fast and SNP-tolerant detection of complex variants and splicing in short reads. Bioinformatics 2010;26:873–81.

[23] Trapnell C, Pachter L, Salzberg SL. TopHat: discovering splice junctions with RNA-Seq. Bioinformatics 2009;25:1105–11.

[24] Li H, Handsaker B, Wysoker A, Fennell T, Ruan J, Homer N, et al. The Sequence Alignment/Map format and SAMtools. Bioinformatics 2009;25:2078–9.

[25] HTSeq: Analysing high-throughput sequencing data with Python. Available from: <http://www-huber.embl.de/users/anders/HTSeq/doc/overview.html>.

[26] Trapnell C, Williams BA, Pertea G, Mortazavi A, Kwan G, van Baren MJ, et al. Transcript assembly and quantification by RNA-Seq reveals unannotated transcripts and isoform switching during cell differentiation. Nat Biotechnol 2010;28(5):511–5.

[27] Anders S, Huber W. Differential expression analysis for sequence count data. Genome Biol 2010;11:R106.

[28] Robinson MD, McCarthy DJ, Smyth GK. edgeR: a Bioconductor package for differential expression analysis of digital gene expression data. Bioinformatics 2010;26:139–40.

[29] Mortazavi A, Williams BA, McCue K, Schaeffer L, Wold B. Mapping and quantifying mammalian transcriptomes by RNA-Seq. Nat Methods 2008;5:621–8.

[30] Robinson MD, Oshlack A. A scaling normalization method for differential expression analysis of RNA-seq data. Genome Biol 2010;11:R25.

[31] Young MD, Wakefield MJ, Smyth GK, Oshlack A. Gene ontology analysis for RNA-seq: accounting for selection bias. Genome Biol 2010;11:R14.

[32] Gutschner T, Diederichs S. The hallmarks of cancer: A long non-coding RNA point of view. RNA Biol 2012;9:6.

[33] Spizzo R, Almeida MI, Colombatti A, Calin GA. Long non-coding RNAs and cancer: a new frontier of translational research? Oncogene 2012;31:4577–87.

[34] Prensner JR, Iyer MK, Balbin OA, Dhanasekaran SM, Cao Q, Brenner JC, et al. Transcriptome sequencing across a prostate cancer cohort identifies PCAT-1, an unannotated lincRNA implicated in disease progression. Nat Biotechnol 2011;29:742–9.

[35] Croce CM. Causes and consequences of microRNA dysregulation in cancer. Nat Rev Genet 2009;10:704–14.

[36] Calin GA, Croce CM. MicroRNA signatures in human cancers. Nat Rev Cancer 2006;6:857–66.

[37] Kozomara A, Griffiths-Jones S. miRBase: integrating microRNA annotation and deep-sequencing data. Nucleic Acids Res 2011;39: D152–7.

[38] Lee LW, Zhang S, Etheridge A, Ma L, Martin D, Galas D, et al. Complexity of the microRNA repertoire revealed by next-generation sequencing. RNA 2010;16:2170–80.

[39] Neilsen CT, Goodall GJ, Bracken CP. IsomiRs – the overlooked repertoire in the dynamic microRNAome. Trends Genet 2012;28:544–9.

[40] Cloonan N, Wani S, Xu Q, Gu J, Lea K, Heater S, et al. MicroRNAs and their isomiRs function cooperatively to target common biological pathways. Genome Biol 2011;12:R126.

[41] Nowell PC. The minute chromosome (Phl) in chronic granulocytic leukemia. Blut 1962;8:65–6.

[42] Douglass EC, Valentine M, Etcubanas E, Parham D, Webber BL, Houghton PJ, et al. A specific chromosomal abnormality in rhabdomyosarcoma. Cytogenet Cell Genet 1987;45:148–55.

[43] Zhao Q, Caballero OL, Levy S, Stevenson BJ, Iseli C, de Souza SJ, et al. Transcriptome-guided characterization of genomic rearrangements in a breast cancer cell line. Proc Natl Acad Sci USA 2009;106:1886–91.

[44] Maher CA, Kumar-Sinha C, Cao X, Kalyana-Sundaram S, Han B, Jing X, et al. Transcriptome sequencing to detect gene fusions in cancer. Nature 2009;458:97–101.

[45] Steidl C, Shah SP, Woolcock BW, Rui L, Kawahara M, Farinha P, et al. MHC class II transactivator CIITA is a recurrent gene fusion partner in lymphoid cancers. Nature 2011;471:377–81.

[46] Roberts KG, Morin RD, Zhang J, Hirst M, Zhao Y, Su X, et al. Genetic alterations activating kinase and cytokine receptor signaling in high-risk acute lymphoblastic leukemia. Cancer Cell 2012;22:153–66.

[47] Maher CA, Palanisamy N, Brenner JC, Cao X, Kalyana-Sundaram S, Luo S, et al. Chimeric transcript discovery by paired-end transcriptome sequencing. Proc Natl Acad Sci USA 2009;106:12353–8.

[48] Zhang Y, Gong M, Yuan H, Park HG, Frierson HF, Li H. Chimeric transcript generated by cis-splicing of adjacent genes regulates prostate cancer cell proliferation. Cancer Discov 2012;2:598–607.

[49] Rickman DS, Pflueger D, Moss B, VanDoren VE, Chen CX, de la Taille A, et al. SLC45A3-ELK4 is a novel and frequent erythroblast transformation-specific fusion transcript in prostate cancer. Cancer Res 2009;69:2734–8.
[50] Jiang Z, Jhunjhunwala S, Liu J, Haverty PM, Kennemer MI, Guan Y, et al. The effects of hepatitis B virus integration into the genomes of hepatocellular carcinoma patients. Genome Res 2012;22: 593–601.
[51] Sung WK, Zheng H, Li S, Chen R, Liu X, Li Y, et al. Genome-wide survey of recurrent HBV integration in hepatocellular carcinoma. Nat Genet 2012;44(7):765–9.
[52] Valle L, Serena-Acedo T, Liyanarachchi S, Hampel H, Comeras I, Li Z, et al. Germline allele-specific expression of TGFBR1 confers an increased risk of colorectal cancer. Science 2008;321:1361–5.
[53] Chen X, Weaver J, Bove BA, Vanderveer LA, Weil SC, Miron A, et al. Allelic imbalance in BRCA1 and BRCA2 gene expression is associated with an increased breast cancer risk. Hum Mol Genet 2008;17:1336–48.
[54] Tuch BB, Laborde RR, Xu X, Gu J, Chung CB, Monighetti CK, et al. Tumor transcriptome sequencing reveals allelic expression imbalances associated with copy number alterations. PloS One 2010;5:e9317.
[55] Ha G, Roth A, Lai D, Bashashati A, Ding J, Goya R, et al. Integrative analysis of genome-wide loss of heterozygosity and monoallelic expression at nucleotide resolution reveals disrupted pathways in triple-negative breast cancer. Genome Res 2012;22:1995–2007.
[56] Li G, Bahn JH, Lee JH, Peng G, Chen Z, Nelson SF, et al. Identification of allele-specific alternative mRNA processing via transcriptome sequencing. Nucleic Acids Res 2012; 40(13):e104.
[57] Poliseno L. Pseudogenes: newly discovered players in human cancer. Sci Signal 2012;5:re5.
[58] Salmena L, Poliseno L, Tay Y, Kats L, Pandolfi PP. A ceRNA hypothesis: the Rosetta Stone of a hidden RNA language? Cell 2011;146:353–8.
[59] Kalyana-Sundaram S, Kumar-Sinha C, Shankar S, Robinson DR, Wu YM, Cao X, et al. Expressed pseudogenes in the transcriptional landscape of human cancers. Cell 2012;149:1622–34.
[60] Pan Q, Shai O, Lee LJ, Frey BJ, Blencowe BJ. Deep surveying of alternative splicing complexity in the human transcriptome by high-throughput sequencing. Nat Genet 2008;40:1413–5.
[61] Venables JP, Klinck R, Koh C, Gervais-Bird J, Bramard A, Inkel L, et al. Cancer-associated regulation of alternative splicing. Nat Struct Mol Biol 2009;16:670–6.
[62] Langer W, Sohler F, Leder G, Beckmann G, Seidel H, Grone J, et al. Exon array analysis using re-defined probe sets results in reliable identification of alternatively spliced genes in non-small cell lung cancer. BMC Genomics 2010;11:676.
[63] Sveen A, Agesen TH, Nesbakken A, Rognum TO, Lothe RA, Skotheim RI. Transcriptome instability in colorectal cancer identified by exon microarray analyses: Associations with splicing factor expression levels and patient survival. Genome Med 2011;3:32.
[64] Cheung HC, Baggerly KA, Tsavachidis S, Bachinski LL, Neubauer VL, Nixon TJ, et al. Global analysis of aberrant pre-mRNA splicing in glioblastoma using exon expression arrays. BMC Genomics 2008;9:216.
[65] Shapiro IM, Cheng AW, Flytzanis NC, Balsamo M, Condeelis JS, Oktay MH, et al. An EMT-driven alternative splicing program occurs in human breast cancer and modulates cellular phenotype. PLoS Genet 2011;7:e1002218.
[66] Gardina PJ, Clark TA, Shimada B, Staples MK, Yang Q, Veitch J, et al. Alternative splicing and differential gene expression in colon cancer detected by a whole genome exon array. BMC Genomics 2006;7:325.
[67] Dutertre M, Lacroix-Triki M, Driouch K, de la Grange P, Gratadou L, Beck S, et al. Exon-based clustering of murine breast tumor transcriptomes reveals alternative exons whose expression is associated with metastasis. Cancer Res 2010;70: 896–905.
[68] Bemmo A, Dias C, Rose AA, Russo C, Siegel P, Majewski J. Exon-level transcriptome profiling in murine breast cancer reveals splicing changes specific to tumors with different metastatic abilities. PloS One 2010;5:e11981.
[69] Guo X, Chen QR, Song YK, Wei JS, Khan J. Exon array analysis reveals neuroblastoma tumors have distinct alternative splicing patterns according to stage and MYCN amplification status. BMC Med Genomics 2011;4:35.
[70] Misquitta-Ali CM, Cheng E, O'Hanlon D, Liu N, McGlade CJ, Tsao MS, et al. Global profiling and molecular characterization of alternative splicing events misregulated in lung cancer. Mol Cell Biol 2011;31:138–50.
[71] Derti A, Garrett-Engele P, Macisaac KD, Stevens RC, Sriram S, Chen R, et al. A quantitative atlas of polyadenylation in five mammals. Genome Res 2012;22:1173–83.
[72] Mayr C, Bartel DP. Widespread shortening of 3′UTRs by alternative cleavage and polyadenylation activates oncogenes in cancer cells. Cell 2009;138:673–84.
[73] Dominissini D, Moshitch-Moshkovitz S, Amariglio N, Rechavi G. Adenosine-to-inosine RNA editing meets cancer. Carcinogenesis 2011;32:1569–77.
[74] Galeano F, Tomaselli S, Locatelli F, Gallo A, A-to-I RNA. editing: the "ADAR" side of human cancer. Sem Cell Devel Biol 2012;23(3):244–50.
[75] Chen L, Li Y, Lin CH, Chan TH, Chow RK, Song Y, et al. Recoding RNA editing of AZIN1 predisposes to hepatocellular carcinoma. Nat Med 2013;19:209–16.
[76] Guttman M, Garber M, Levin JZ, Donaghey J, Robinson J, Adiconis X, et al. *Ab initio* reconstruction of cell type-specific transcriptomes in mouse reveals the conserved multi-exonic structure of lincRNAs. Nat Biotechnol 2010;28:503–10.
[77] Martin JA, Wang Z. Next-generation transcriptome assembly. Nat Rev Genet 2011;12:671–82.
[78] Shah SP, Kobel M, Senz J, Morin RD, Clarke BA, Wiegand KC, et al. Mutation of FOXL2 in granulosa-cell tumors of the ovary. N Engl J Med 2009;360:2719–29.
[79] Bass B, Hundley H, Li JB, Peng Z, Pickrell J, Xiao XG, et al. The difficult calls in RNA editing. Interviewed by H Craig Mak. Nat Biotechnol 2012;30:1207–9.
[80] Wei JS, Johansson P, Chen L, et al. Massively parallel sequencing reveals an accumulation of de novo mutations and an activating mutation of LPAR1 in a patient with metastatic neuroblastoma. PloS One (in press).

[81] Akavia UD, Litvin O, Kim J, Sanchez-Garcia F, Kotliar D, Causton HC, et al. An integrated approach to uncover drivers of cancer. Cell 2010;143:1005−17.

[82] Oren M, Rotter V. Mutant p53 gain-of-function in cancer. Cold Spring Harbor Perspect Biol 2010;2:a001107.

[83] Ng S, Collisson EA, Sokolov A, Goldstein T, Gonzalez-Perez A, Lopez-Bigas N, et al. PARADIGM-SHIFT predicts the function of mutations in multiple cancers using pathway impact analysis. Bioinformatics 2012;28:i640−6.

[84] Roychowdhury S, Iyer MK, Robinson DR, Lonigro RJ, Wu YM, Cao X, et al. Personalized oncology through integrative high-throughput sequencing: a pilot study. Sci Translat Med 2011;3(111):111ra121.

[85] Meyer UA, Zanger UM, Schwab M. Omics and drug response. Annu Rev Pharmacol Toxicol 2013;53:475−502.

[86] Manolio TA. Genomewide association studies and assessment of the risk of disease. N Engl J Med 2010;363:166−76.

[87] Barrie ES, Smith RM, Sanford JC, Sadee W. mRNA transcript diversity creates new opportunities for pharmacological intervention. Mol Pharmacol 2012;81:620−30.

[88] Consortium ICG. ICGC cancer genome projects. Available from: <http://www.icgc.org/>.

[89] National Cancer Institute. The Cancer Genome Atlas. Available from: <http://cancergenome.nih.gov/>.

[90] National Cancer Institute. Therapeutically Applicable Research to Generate Effective Treatments. Available from: <http://ocg.cancer.gov/>.

[91] Forbes SA, Bindal N, Bamford S, Cole C, Kok CY, Beare D, et al. COSMIC: mining complete cancer genomes in the Catalogue of Somatic Mutations in Cancer. Nucleic Acids Res 2011;39: D945−50.

[92] Cerami E, Gao J, Dogrusoz U, Gross BE, Sumer SO, Aksoy BA, et al. The cBio cancer genomics portal: an open platform for exploring multidimensional cancer genomics data. Cancer Discov 2012;2:401−4.

[93] Robinson DR, Wu YM, Kalyana-Sundaram S, Cao X, Lonigro RJ, Sung YS, et al. Identification of recurrent NAB2-STAT6 gene fusions in solitary fibrous tumor by integrative sequencing. Nat Genet 2013;45:180−5.

[94] McLeod HL. Cancer pharmacogenomics: early promise, but concerted effort needed. Science 2013;339:1563−6.

[95] Gerlinger M, Rowan AJ, Horswell S, Larkin J, Endesfelder D, Gronroos E, et al. Intratumor heterogeneity and branched evolution revealed by multiregion sequencing. N Engl J Med 2012;366:883−92.

[96] Nekrutenko A, Taylor J. Next-generation sequencing data interpretation: enhancing reproducibility and accessibility. Nat Rev Genet 2012;13:667−72.

Cancer Genomics

Tissue Microarrays in Studying Gynecological Cancers

Cécile Le Page[1]
Anne-Marie Mes-Masson[1,2]
Anthony M. Magliocco[3]

[1]Centre hospitalier de l'Université de Montréal, Institut du Cancer de Montréal, Montreal, QC, Canada
[2]Université de Montréal, Department of Medicine, Montreal, QC, Canada
[3]H. Lee Moffitt Cancer Center and Research Institute, Tampa, FL, USA

Contents

Cancer Genomics. DOI: http://dx.doi.org/10.1016/B978-0-12-396967-5.00005-0

Key Concepts

- TMAs are cost- and time-effective means of validating cancer biological markers identified by genomic approaches
- TMAs allow the analysis of hundreds of samples simultaneously and enable the conservation of priceless tissue
- TMAs allow the validation and analysis of several known and putative markers simultaneously and enable more reliable correlative analysis between markers
- The TMA has a large number of applications in cancer and pathology research providing a more comprehensible assessment of the tumor biology
- Good quality standards are required for the optimal utilization and analysis of TMAs

INTRODUCTION

Tumorigenesis is caused by the accumulation of genetic and epigenetic changes that give rise to different molecular abnormalities in numerous genes and proteins in cancer cells. Genetic changes can be single mutations, chromosome rearrangements or amplification resulting in aberrant protein expression. Some of these protein or gene modifications are silent in the sense that they have no impact on cell functions that contribute to tumorigenesis, while other so-called driver events may induce specific functional alterations that contribute to the pathogenesis of cancer cells. The identification of such molecular changes often occurs in cell lines or a small number of tissues and, subsequently, necessitates the analysis of a large number of patient samples to determine their relative abundance in a particular tumor type.

The investigation of pathogenesis has been revolutionized by the introduction of high-throughput technologies, such as DNA and tissue microarrays (TMAs). The TMA technology allows the transposition of punches of tissues from donor to a recipient block, resulting in a recipient block which can contain several hundreds of small tissue cores from many donor specimens. Hence, this method allows a rapid and simultaneous investigation of one or a few molecular markers in hundreds of tumor samples by means of *in situ* molecular assays such as immunohistochemistry (IHC) and fluorescence *in situ* hybridization (FISH). In addition, advanced computer image analysis methods, such as fractal dimension calculations, can also be applied [1]. As the recipient block can be sectioned multiple times, this allows a parallel molecular profiling of specimens for multiple *in situ* marker types (DNA, RNA or protein) on multiple sections of the same array. This technology therefore allows a more efficient and comprehensive assessment of tumor biology, and enables the identification of tumor makers more rapidly and at a lower cost compared to traditional "single tissue block" methods.

Another useful application of TMAs is for comparative studies to evaluate and qualify tissue specimens [2], to perform antibody screening or diminish the intra- and interlaboratory variability in staining methods and results [3]. Even if tissue immunostaining on whole tissue section is a common laboratory practice, it still entails high slide-to-slide and batch variability effects, which are mitigated in TMAs. TMAs can also pool samples from different sources to diminish variables caused by interlaboratory variables related to such variables as antigen retrieval, staining time, antibody source, antibody concentration and in the interpretation of staining results. Therefore, TMAs offer a useful method to control the quality and uniformity of results.

The technique was first described in 1986 by Battifora who reported a multitumor tissue block method in which 1-mm-thick cylindrical cores of different tissues were embedded in one paraffin block and from which numerous sections were cut and individually stained by immunohistochemistry [4]. Shortly after, the array format was described by Wan et al. in 1987 [5]. However, it took 10 years to develop a device that allows the technique to perform accurate arrays and be easily and reproducibly usable in research and pathology laboratories [6]. Since then a dramatic increase in the popularity and utility of the technique has been observed, which is reflected by the number of publications per year in which TMAs are used (Figure 5.1).

In its most common form, a tissue microarray is built from several cylindrical cores of tissues, withdrawn from either a formalin-fixed, paraffin-embedded (FFPE) block or from fixed or frozen tissues. Cores are vertically placed into an empty grid identified recipient paraffin block. The constructed recipient block can be horizontally sectioned in 3–5 μm sections allowing the production of numerous replicated slices. The thin layers of tissues are

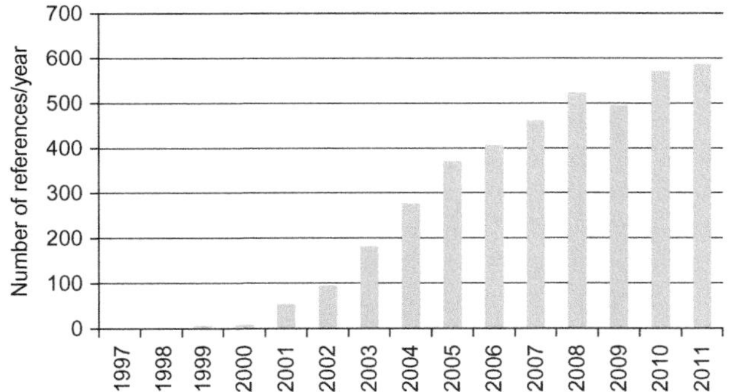

FIGURE 5.1 Frequency of tissue microarray studies reported in Pubmed*. The Pubmed database was used to search for the keywords "tissue" and "microarray" and "immunohistochemistry" or "fluorescence". References including gene microarray studies without tissue microarray were manually excluded. * http://www.ncbi.nlm.nih.gov/pubmed.

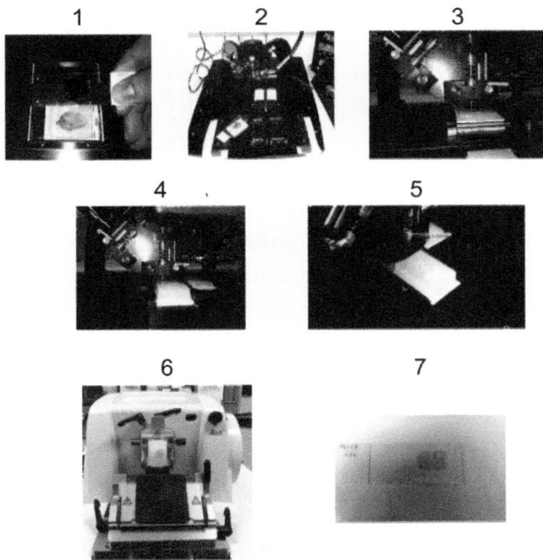

FIGURE 5.2 Construction of a tissue microarray. (1) The donor block is aligned with the corresponding H&E slide to select an area of interest. (2) An empty recipient paraffin block is positioned on the arrayer prior to receiving cores from donor blocks. (3) A cylindrical core is punched from the FFPE donor block. (4) The tumor core is vertically transferred to the new recipient block. (5) Image of a new TMA block after completion. (6) Horizontal sectioning of the new TMA with a microtome. (7) Image of a TMA slide after H&E staining.

then placed individually on a microscopic glass slide for subsequent analysis (Figure 5.2).

To introduce the readers to the essential basis and challenges of tissue microarray, this chapter provides a short overview of this increasingly popular technology for validating cancer biomarkers derived from genomics studies by focusing on several technical aspects of tissue microarray construction and analysis.

MATERIALS AND TISSUES

Instruments

It is important to know that the construction of tissue microarrays requires a substantial financial investment in equipment as well as in human resources, as experienced technicians or histotechnologists are required to prepare and section the TMAs properly.

Arrayers

The main device, the microarrayer, is a high-precision punching instrument (see Figure 5.2). Several suppliers now offer microarrayers with prices ranging from \$10 000 to more than \$100 000. The microarrayer can be manual or automated. One of the earliest instruments was supplied by Beecher Instruments Inc., although several different manufacturers now offer a variety of manual, semi-automated and automated versions (see for example http://www.pathologydevices.com/TMArrayer.htm). Differences between the manual and automated arrayers include features related to the positioning of the punches, software (none typically associated with the manual version), and speed and ease of replicate block production. However, an important advantage of the manual arrayer is the ability to control visually the depth of tissue punches, which is a significant advantage when working with tissues of different thicknesses.

Microscope

Another costly instrument required is a high-resolution microscope for selection of tissue areas, histological validation of the TMA and downstream analyses. Because the analysis of individual cores is a time-consuming step, several software packages have been developed which greatly facilitate high-throughput analyses. However, automated analysis involves additional cost and training, including the purchase of a slide scanner able to digitize slides at a very high resolution [7].

Microtome

Once a TMA is constructed, it is sectioned in thin paraffin layers (3– 6 μm), each of which contains the arrayed tissues cores, then layers are mounted on microscope glass slides. Cutting the TMA block is a delicate and important step. The number of sections cut from a TMA will impact the number of studies that can be done. While, ideally, one wants to maximize the number of slides produced from a given TMA, in order to maximize the use of tissue, caution is required as thinner tissue layers are more difficult to manipulate and thus easily wasted. Considerable skill is required for reliable sectioning of TMA blocks, also timing should be considered to remove the possibility of wasting material due to "block facing" and "antigen decay" on unstained slides. As TMA blocks become older there is "core fall out" as individual cores become exhausted. Poorly cut or incomplete TMA sections can still be used as controls or as experimental sections for working up novel assays.

Immunostaining Facilities or an Immunostainer

The quality of an assay is highly dependent on providing stringent quality control over all steps of the staining protocol. Automated staining robots have removed human variation from staining by automating reagent dispensing and, in some cases, the antigen retrieval process. However, these devices cannot overcome any deficiencies in specimen preparation prior to staining or problems with the primary reagents, such as antibody quality and activity. Automated immunohistochemical stainers are available from several suppliers (Leica Microsystem Covertiles, Ventana's Benchmark and Discovery, DAKO Autostainer Link, Biogenex Optimax, and others).

Image Analysis Devices and Software

With the increased usage of TMAs, numerous commercial sources of digital image acquisition and software have become available to enable computer-assisted analysis of target expression. These include Aperio, Caliper, HistoRx, Definiens, Ventana, Visiomorph and others.

Tissues

Even if tissue microarrays are most commonly constructed from FFPE blocks, TMAs can also be constructed from FFPE biopsies, FFPE cell pellets or from OCT frozen tissues [8–10]. The quality of proteins and nucleic acids in stored tissues is dependent on several factors that influence the quality and efficiency of subsequent experimentation. These factors can include time to tissue processing (warm and cold ischemia), fixation conditions, storage duration, and storage conditions. It has often been reported that tissue processing within 30 minutes is preferred for most applications to avoid loss of antigenicity and alteration of nucleic acids in the tissue [11]. Loss of antigenicity can also be observed in tissues stored for a long period, especially in non-optimal light and temperature conditions, and this can contribute to an important bias in the data obtained [12–15]. Within a given cohort, it is essential to verify through statistical analysis that the age of tissues and staining intensity are not correlated.

Tissue fixation is an important factor that can affect results generated from TMAs. The most common fixative for histologic tissues is an aqueous aldehyde-based solution, either fomaldehyde (4%) or glutaraldehyde (2–3%), which creates covalent bounds between macromolecules to maintain the tissue architecture and integrity of cellular components [16]. However, tissue fixation may also affect antigenicity and cause damage to nucleic acids. The chemical fixative can modify aldehyde groups of proteins but also surrounding soluble proteins in the tissue can be cross-linked to the cellular proteins and lead to an allosteric modification of epitopes. The advantage of formaldehyde over glutaraldehyde fixation is the reversibility of cross-linkages by heating. Another important advantage of the formaldehyde fixative is that membrane proteins are not destroyed during the process of fixation [17]. However, glutaraldehyde allows better morphological preservation. The process of fixation and antigen masking with aldehyde-based fixative is progressive. With increased fixation time there is a proportional decrease of immunoreactivity due to increased antigen masking. However, an overly short fixation time can lead to incomplete fixation particularly in the central portion of the tissue and thereby subsequently creates a heterogeneous immunoreactivity. Furthermore, exposure to alcohol during the processing steps leads to protein denaturation if formalin fixation is incomplete. Therefore, more moderate fixation conditions are recommended in order to maximize the use of the TMA. Clinically, it has been recommended to use a minimum of 6–8 hours of fixation to obtain a reliable estrogen receptor (ER) staining in breast cancer tissues and a maximum of 48–72 hours [18]. Several buffered formalin preparations are used in histology, but neutral buffered formalin (NBF) is the most common since it yields relatively good nucleic acids quality compared to other buffers. However, it has been reported that other fixatives may be more advantageous for the detection of certain specific antigens [19]. Perhaps disturbingly, the cellular localization of different stains has been shown to vary using different fixatives, implying that artifacts have been introduced during tissue processing [19]. Alternative fixatives are mostly alcohol or zinc based. More recently, attention has focused on a new generation of molecular fixatives that improve the quality of embedded nucleic acids such as RNAlater (however, protein antigenicity may be sacrificed). It is therefore recommended to consider carefully the appropriate fixative for all proteins of interest and/or test immunogenicity of the proteins in different fixatives before preparing the tissues to be used in TMAs. Pragmatically, however, many TMAs are constructed for pathology-based archival tissues where the best practice is at least to document the method of fixation used in the clinical setting. An issue sometimes arises in TMA construction due to variation in embedding paraffin hardness making construction and cutting of a TMA difficult. If this situation arises, we suggest re-embedding the donor blocks with fresh paraffin of the same formulation used for the host block. An additional problem arises in large studies that include specimens from multiple institutions spanning many years. We have observed significant differences in antigenicity associated with institution of specimen origin, and age of specimen. These changes are probably attributable to pre-analytical handling conditions (such as time to fixation and formulation of fixation) and changes in laboratory tissue handling procedures over time. Consequently, it is important to record these variables and possibly adjust for them during data analysis.

Ethics and Regulatory Issues

As many TMAs are produced from human tissues, it is essential prior to construction to have obtained the appropriate approval from the designated research ethics board. Several laws and guidelines need to be considered when dealing with human tissues, and international standards include such references as the Helsinki declaration (www.wma.net) and the UNESCO policy (www.unesco.org). In Canada, the Tri-Council Policy Statement: Ethical Conduct for Research Involving Humans (TCPS) is the joint research ethics policy statement of the federal research agencies that fund health research. This policy has recently been updated (www.pre.ethics.gc.ca/eng/policy-politique/initiatives/tcps2-eptc2/Default/). Whether participants providing tissues need to be prospectively consented, can be retrospectively consented, or whether anonymized tissues without consent can be used is determined largely by jurisdictional conventions. Consent should always be obtained using qualified personnel, and all laboratory members who use human biological specimens should be adequately trained, including around issues related to patient privacy and confidentiality. In the USA, the Belmont report led to the Federal Policy for the Protection of Human Subjects, informally known as the "Common Rule". The Office for Human Research Protections (OHRP) was also established within HHS.

Tissues removed for diagnostic purposes are an excellent source of material for research study and TMA construction. However, these materials are governed by regulations, laws and policies overseeing diagnostic clinical laboratories. In the USA, these include rules developed by CLIA (Clinical Laboratory Improvement Amendments) and certification bodies such as the College of American Pathologists. Some of these regulations restrict use of diagnostic tissue for research purposes up to 2 years and require clinical laboratories to maintain specimens for at least 10 years. Specimens may also be subject to local state and provincial laws. There is a lack of clarity regarding "ownership" of tissues but, in general, diagnostic tissues are assumed to be under the custodianship of the medical director of the diagnostic laboratory processing the specimen. Diagnostic material has special status as it may be needed for medico-legal reasons, or for additional testing (for example to assess a new integral marker necessary for enrollment into a clinical trial).

Design of Microarrays

The optimal design should take into consideration the study goal and the end user. If the array is visually scored, placing the cores in asymmetrical sub-arrays will considerably ease the interpretation of the staining analysis by orienting the user when viewing cores under the microscope (Figure 5.3). In addition, even if a large number of tissue cores may render the tissue array more efficient, too many cores can introduce staining variations, particularly on the edges of the array. For this reason, we do not generally recommend the construction of very large TMA with more than 500 cores but rather divide cases into sub-arrays in separate TMA blocks. If the cohort under study has to be divided over several TMAs, it is best to stain all slides at one time in order to

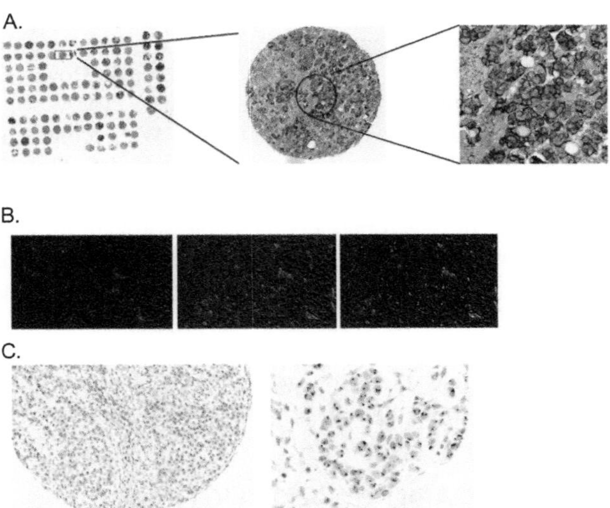

FIGURE 5.3 (A) Example of an ovarian cancer TMA stained with anti-E-cadherin antibody. Left image represents 1× magnification. Middle image represents a 20× magnification of cytoplasmic and membrane of E-cadherin. Right image represents a 200× magnification used for scoring analysis. (B) Representative images of immunofluorescence on ovarian cancer tissue. Blue DAPI signal in nuclei. Image 20×. Left image: Macrophage CD68+ staining (red). Middle image: CD206+ staining (green). Right image: Merge staining showing the co-localization of CD68 and CD206 markers (orange). Here CD206 expression is almost exclusively co-localized with CD68 staining. (C) Example of SISH staining. Low (left) and high (right) magnification.

minimize experimental variability. The introduction of positive and negative staining controls should be included whenever possible and may include different tissue types or cell lines with known profiles for the marker of interest. In addition, genetically modified cell lines that either under- or overexpress a marker of interest can be incorporated into the TMA. However, cell lines tend to stain differently from fixed tissue so, in the case of tumor cell lines, xenograft grown in immunocompromised animals can serve as excellent controls. Including the same controls on TMAs for a given cohort provides an additional means of ensuring that staining across different slides remains comparable. It is advantageous to include identical specimens across multiple TMAs to facilitate interpretation and enable normalization of data. It is also helpful to include additional controls of different organ tissue (such as normal liver) to help orient arrays. Orientation becomes progressively problematic as arrays are used and cores "drop out". Some investigators have noted that staining may have spatial variation across the array (stronger at top or sides) due to some manual methods of applying staining reagents. Consequently, balanced random construction designs are recommended to mitigate these effects.

A map should be designed to record the position of each tissue core within the TMA and scoring should be tied to this map. The scoring can then be linked to the clinical parameters of interest for subsequent analysis. Microsoft's Excel program is a simple way to store the map design, scoring results and clinical data. However, this platform may not be suitable for large TMAs with extended and complex associated data. We recommend cautious use of Excel as it is easy to corrupt data through accidental "sorting" of columns in data sheets. Some groups have designed publicly available software to facilitate the merger of TMA data and clinicopathological parameters [20,21] (http://genome-www.stanford.edu/TMA/combiner). Most recent automated scanners are supplied with software that automatically collects tissue images and links them to comprehensive database files that are easily linked to image analysis software.

METHODS AND CONSTRUCTION

Selection of Tissues and Cores

Most of the work involved in the construction of a TMA is spent during the process of tissue selection. Not only does it requires a large amount of time, depending on the size of the TMA, but also it is a crucial step involving technical and ethical issues. The data obtained from the analysis of a TMA are highly dependent on the quality of the TMA, including tissue quality, quality of core selection and design of the TMA. The selection of tissues is defined by the purpose of the study and should be able to answer the study question(s) with statistical robustness. For this purpose, the analytical variables should be carefully considered. Since human tissues are precious material, such selection will limit the number of cases in the TMA without decreasing the statistical robustness of the study. This includes parameters such as number of cases to be selected with specific clinical variables [22]. This is particularly important for outcome-based or progression-based disease TMA. For example, in the analysis of a potential prognosis marker, the TMA should include enough cases with good and worse prognosis, but each group of patients should have similar clinical characteristics, such as same range of grade, stages and age.

Before the physical construction of a TMA, a list of the selected tissues is generated. Then, all tissue blocks of the appropriate specimens are collected from the archive and reviewed by a pathologist in order to select the optimal donor blocks and appropriate punching areas. Depending on the endpoint of the study or the marker of interest, different tissue areas can be selected. When selecting the punching area, it would also be pertinent to record the tumor location of the punches in the tissue, i.e. center of the tumor, peripheral edge of the tumor, near a necrotic or inflammatory area, etc. This will help to determine if the staining of interest is dependent on the tumor heterogeneity or tumor location. It should also be considered that antigen preservation is often better near the edge of the specimen due to more rapid exposure to fixative. In general, a marker of true clinical value should display a uniform staining in the tumor and this could be assessed by the use of TMA with different tissue areas from several specimens.

The main critique of TMA in comparison of whole tissue slides is the small amount of tissue to be analyzed, which may not be representative of the whole tissue. Core size and number will depend on the tissue type and marker. Few studies have demonstrated a good concordance between results obtained with only one or two cores of 0.6 mm and whole tissue slides from ovarian or breast tissues [23–26]. However, for some markers, such as Ki-67, exhibiting strong tissue heterogeneity, additional cores should be included. Heterogeneous markers may be prone to false negative results on TMA analysis. In addition, a technical problem associated with TMA analysis is the tissue lost or folding during the sectioning or the staining. For ovarian or breast cancer TMA, we recommend that at least two cores per tissue should be considered, particularly if the cohort under study is small and each individual in the cohort is necessary to maintain a statistical robustness. Additional cores also enable an assessment of tumor heterogeneity and perhaps a more accurate estimate of antigen expression than single cores lending statistical approaches such as calculating mean expression. Core sizes usually range from 0.6 mm to 2 mm, with 0.6 mm being the "gold standard" for breast and ovarian cancer tissues. The main advantage of a smaller core is obviously a smaller depletion of the original tissue specimens and the possibility of building TMA slides with a greater number of cases per TMA slide. In addition, larger cores make the sections more difficult to manipulate and, especially in the case of breast tissue that may have a high fat content, cores are easily lost during processing when larger than 0.6 mm. Several websites describe the production of TMAs, and a visual construction of TMA with a manual arrayer can be viewed at: www.jove.com/video/3620/production-tissue-microarrays-immunohistochemistry-staining.

Cutting Slides

Once the TMA is constructed, the new paraffin block containing tissue cores can be cut into thin paraffin layers and transferred onto microscope glass slides. This step is technically delicate and, when poorly executed, can result in tissue loss, folding of thin sections and even shift cores from their initial location on the paraffin block. Sections are traditionally cut with a microtome and transferred to a water bath to collect TMA sections on the microscope slide. An alternative technique commonly used to decrease the frequency of tissue loss during sectioning is the tape transfer method which consists of using an adhesive tape placed over the face of the paraffin block prior to sectioning to stabilize the section during cutting and subsequent transfer to a microscope glass slide [27]. However increased non-specific staining of specific markers has been associated with this technique [28].

A single TMA block can potentially create 200 sections. However, since all donor tissue specimens are not of the same depth, the first and last sections are usually not appropriate for staining. In practice, 50 to 150 working sections are generated from one TMA block with 90% of cores in good condition.

An additional issue is antigen stability. It is recommended that immunohistochemical stains be performed on as freshly cut tissue sections as possible because antigens can decay due to oxidation. It is possible to reduce the extent of oxidation using techniques such as storing slides in nitrogen gas, dipping them in paraffin, or wrapping and freezing them. It is unfortunate that each time an array is sectioned there is a potential for tissue loss due to "trimming", consequently, it is advisable to batch stain a series of sections from a TMA to ensure maximum usage of tissue. We also recommend saving poorly cut or incomplete sections for use as pilot testing or controls.

Antigen Retrieval

Antigen masking that occurs during tissue fixation needs to be retrieved to recover the tissue antigenicity. Several methods have been implemented over the years and their performance is variable and dependent on the applications [29]. In particular, the antigen retrieval method is antigen/antibody dependent. The most common methods, which work for the majority of proteins, are 10 mM citrate buffer (pH 6) and EDTA or EGTA (pH 9) buffer with incubations between 80 and 100°C. Although several factors affect the efficiency of antigen retrieval, such as type of buffer, pH, time, the most important factor is heat, which induces hydrolytic breaks of intra- and intermolecular cross-links [30]. More recently, Namimatsu et al. have reported that a 0.05% citraconic anhydride solution at neutral pH may work as an efficient method for antigens difficult to stain such as CD4, cyclin D1 on breast cancer tissues [31,32]. Several suppliers offer pre-made retrieval solutions (Dako, Vector Labs, R&D Systems and Novocastra to mention only some). In addition, clinical manufacturers have developed robotic machinery to standardize the retrieval process (e.g. PT link from DAKO and Ventana Benchmark which has built-in retrieval protocols).

Staining

Several applications can be performed with a paraffin-embedded TMA. The most common application is immunohistochemistry (IHC) with one or two markers, immunofluorescence with multiple markers or fluorescence *in situ* hybridization (FISH). IHC involves a primary and at least one secondary antibody and a chromogenic substrate for detection. Several companies now offer kits to ensure reliable and reproducible staining. The most common kits are manufactured by DAKO Inc. While robust and reliable, this method cannot be used for the detection of more than two markers in the same tissue. Another major criticism of IHC is that analysis is generally semi-quantitative at best.

An alternative approach to protein labeling by IHC is immunofluorescence (IF) labeling with fluorophores. In recent years, there has not been an increasing number of reports describing the IF labeling of TMA sections compared to IHC because this approach is technically and financially challenging. A major technical challenge often observed, but poorly described, is the autofluorescence of cellular structures in FFPE tissues [33]. Several methodologies have been developed to deal with this problem [34]. Pretreatment of tissue slides with sodium borohydride and glycine, and inclusion of normal goat and normal donkey sera as a blocking step between antibodies, has also been proposed [35]. The reagent performance to mask autofluorescence depends on the fixative used and tissue type under study. Sudan Black B 0.1% in 70% ethanol has been successfully used on ovarian cancer TMAs, for example [36].

Another aspect to consider carefully is the choice of fluorophore combination when two or more are used. Several parameters should be considered and, ideally, the spectra of fluorophores should be well separated. As the level of protein expression affects the methodology, it is recommended that least expressed proteins be analyzed with the brightest fluorophores. Brightness index values are not absolute and can vary depending on the antibody, the antigen, the instrument, the staining protocol, the cell type, and other factors. BioLegend offers a brightness index guide to assist the choice (http://www.biolegend.com/multicolor_staining#data). The combination of primary antibody and fluorophores can also lead to different levels of background and even the sequence of hybridization involving different fluorophores should also be tested to reduce the autofluorescent background.

In addition to overcome technical challenges, quantitative IF labeling and analysis of FFPE material, particularly for TMA, needs an assisted fluorescence imaging system. Scanned images can be processed and quantified using sophisticated software, although defining parameters within these systems is not trivial. The major advantage of IF labeling is obviously the quantitative aspect of the technique and the possibility to perform multicolor labeling, analyzing several molecular candidates or identifying cellular subtypes on the same tissue specimen. Using such techniques Mote et al. were able to visualize the differential expression of progesterone receptors A and B isoforms on endometrial tissues during menstrual cycle [33]. Multiple antigen detection using IF approaches not only addresses aspects of material depletion but also allows concomitant and quantitative analysis of several candidates within the same cells. A fluorescent approach also provides the possibility of superior marker subcellular localization based on the co-staining of proteins of interest and a localization marker.

In Situ Hybridization and Fluorescence *In Situ* Hybridization (ISH and FISH)

Tissue sections also lend themselves to *in situ* hybridization approaches to visualize nucleic acid changes including chromosomal structural changes such as gene amplification, deletions, rearrangements and chromosome changes such as polysomy [37]. Additional methods can also visualize RNA. Further, pathogenic DNA or RNA can also be targeted to visualize infectious agents such as human papilloma virus (HPV) or Epstein-Barr virus (EBV) [38]. In general, labeled DNA probes are hybridized to tissue sections using appropriate temperatures to denature and re-anneal target DNA or RNA. After washing, the bound probes are visualized using antibodies linked to enzymes causing signal development. Signals may be fluorescent (FISH), chromogenic (CISH) or precipitated silver (SISH) [39]. Some of the automated staining devices (such as the Ventana Platforms, www.ventana.com) are capable of performing the entire deparaffinization, antigen retrieval and *in situ* hybridization automatically. Multiple probes can be used simultaneously with FISH and CISH enabling translocations and gene amplifications to be detected on a single slide. When SISH is used, generally multiple slides are used to calculate gene ratios to determine presence of amplification, deletion and polysomy. FISH has the advantage of multiplexing but the disadvantages of lack of signal stability and requirement of fluorescent microscopy. In addition, FISH is difficult to impossible to interpret manually on standard TMAs if an automated imaging acquisition device is not used. SISH and CISH have the advantage of signal stability and ease of visibility using standard bright-field microscopy, enabling manual observation and interpretation.

QUALITY CONTROL AND ANALYSIS

Slide Conservation and Storage

TMA blocks and sections are a non-sustainable resource and thereby a very precious material. To save material and avoid loss during the sectioning step, some laboratories may decide to cut a large number of sections at once. However, tissues deteriorate rapidly due to oxidation in contact with air, which can affect antigenicity of some proteins as reported with cytokeratins, PR and HER2 and ER on breast cancer tissues [13,40–42]. To prevent tissue oxidation, cut sections can be paraffin coated and stored in a dessicator at 4°C until further use but, to ensure the preservation of antigenicity [13], each antibody of interest should be cross-verified on fresh and stored tissues.

Staining and Antibody Quality

An important consideration in IHC and IF assays is the antibody quality and titer. The antibody staining should be specific and reproducible (Table 5.1). Specificity should be assessed, preferably by the same technique as the one being used for the study, i.e. IHC or IF, with the appropriate controls or blocking peptides. These can consist of formalin-fixed paraffin-embedded cell line pellets [9], cell line xenografts, or tissues with known expression patterns. However, validation in IHC can be challenging since appropriate negative and positive controls are not always available. Concomitant validation of the antibody

TABLE 5.1 Summary Table of Problems and Troubleshooting Encountered with TMA Staining and Scoring

Problems Encountered with TMA and IHC	Troubleshooting
Tumor heterogeneity	Increase the number of cores per tissue
Loss of tissue cores	Increase the number of cores per tissue
Artifactual staining localization	Fixation time
	Type of fixative
	Optimization of antibody concentration and incubation time
	Time of specimen processing
Antigen retrieval	Antigen retrieval buffer and pH
Non-specific staining	Specificity of antibodies, introduce positive and negative control tissues
	Optimization of antibody concentration and incubation time
Weak staining	Optimization of antibody concentration and incubation time
	Type of fixative
	Age of tissue specimen
	Time of specimen processing
Objective and tedious manual scoring	Virtual microscopy

specificity by western-blot analysis in well-characterized cell lines is also encouraged. Unfortunately, in the past, antibody validation has not consistently been considered as a standard for publication which, in part, explains the number of inconsistent reports in the literature related to ovarian cancer IHC analysis [43]. For more details about antibody validation methods see Bordeaux et al. [44].

Optimal results are obtained with an appropriate antibody concentration. Titration of the antibody determines the lowest dilution that gives the strongest signal with the lowest background. Although this is a general rule for secondary antibodies, it may not be directly applied to primary antibodies. The titer of antibody may affect the endpoint data, particularly when scoring is associated with clinical parameters or patient outcomes. Low concentration of antibody will highlight differences in the range of high expression while high antibody concentration will discriminate differences in the low range of expression. An interesting example was provided by the study of McCabe et al. [45] who showed that when high concentrations of HER2 and p53 antibodies were used, the lowest staining of these proteins was associated with poor survival. In contrast, when low antibody concentrations were used a stronger staining was associated with poor survival. This particular example is striking and has also to do with the complexity of the breast cancer disease which includes several subtypes with different clinical outcomes.

In IF, where labeling combinations are used, each combination of antibodies should be tested to ensure that there is no cross-reactivity. When using mouse secondary fluorescence labeled antibodies, the Mouse-on-Mouse blocking agent (Vector Labs) permits the consecutive use of several mouse secondary antibodies without cross-reactivity. Investigators should be aware that antibody performance may vary dramatically from batch to batch and decay over time. Consequently, it is advised that, for a TMA study, a single lot of antibody should be used and staining performed in a batch whenever possible.

Scoring and Analysis

The visual scoring of TMA can be very laborious. The traditional visualization with a microscope may take up to 2 hours of analysis per 100 cores, depending on the staining complexity and marker location. In this regard, digital imaging has tremendously facilitated this step.

There are two main methods for scoring – visual and automated. For visual scoring, the simplest is the semi-quantitative evaluation, usually by visual inspection of the staining intensity, giving each core a global value from 0 to 3. This type of scoring provides categorical data relatively easy to analyze and compare to clinical parameters. When staining is not homogeneous a percentage of stained tissue area can also be evaluated. A more complex approach takes into account both staining intensity and stained area, giving rise to a staining index called the "Quick score" that provides continuous values for statistical analyses [46,47]. The choice of scoring method will depend on the marker. For example, analyses of ovarian cancer tissues using antibodies directed against Ki67 or p53 are usually evaluated by the percentage of positive nuclei [48], while WT1 is scored by intensity and percentage of staining [49,50]. Because this type of analysis is subjective and arbitrary, at least two independent observers are necessary to reduce bias of personal interpretation and inter-observer variability should be assessed [51] by an intraclass correlation method. The Intraclass Correlation Coefficient (ICC), Cohen's kappa statistic, the Fleiss kappa and the concordance correlation coefficient have been proposed as suitable measures of agreement among non-exchangeable observers [52]. In addition, to reach a good scoring consensus between observers, classifier parameters should be well defined between observers to maintain consistency in the scoring.

When there is discrepancy between observers scoring should be reviewed. Another concern when evaluating scoring is the interlaboratory variability [53]. There are numerous studies with conflicting results as to the prognostic value of individual biomarkers in the same malignancy such as ovarian cancer [43]. To address this issue, several groups have proposed standardized scoring methods for commonly used specific markers such as p53, ER and PR in breast cancer tissues [54,55].

The automated method requires investment of capital into digital microscopy including a scanning device and image analysis software. There are now numerous commercial systems available that are designed for the quantification of IHC or IF staining [7,56,57]. This approach has the advantage of being more reproducible compared to the manual scoring once algorithms are defined in the software prior to analysis, more rapid and, importantly, give rise to continuous data instead of categorical data.

Before initiating the study analysis, there are important analytical parameters to verify. First, when several cores from the same patient are incorporated in the TMA, a core-to-core correlation test should be assessed to ensure that the number of cores is representative of a whole tissue section or that the tumor heterogeneity is represented. Second, since the TMA is constructed from archival tissue blocks, it is important to verify that the antigenicity of the epitope of interest has not been lost over time. A correlation test between age of tissue and scoring should be performed to ensure there is no statistical bias associated with archival tissues.

Advanced digital acquisition techniques may allow normalization of data prior to analysis. This could be accomplished by taking ratios of signal in cancer cells to internal controls such as stromal elements or adjacent benign tissues, or adjusting the signals for variables such as block age. It may also be advantageous to transform the data prior to data analysis using log transformation, or other methods that mitigate the effects of extreme data outliers on the analysis.

To analyze scoring data and correlation with clinical parameters, several statistical tests can be appropriate depending on a number of factors such as distribution of values, type of values (nominal, ordinal, categorical or continuous) [58]. A biostatistician can suggest the best test and should be consulted for deeper analysis. To visualize the survival distribution and the marker of interest, the most common tests are Kaplan–Meier plots, logistic regression or Cox regression analysis. In addition, it is also important that data are reported in a comprehensive manner to allow critical evaluation and comparison between studies. To this purpose, the NCI-European Organization for Research and Treatment of Cancer working group has developed guidelines, called REMARK, for the reporting of tumor marker studies [59]. There are also some powerful data analysis tools such as X-Tile available from the Rimm Laboratory that allow exploration of multiple cut points in continuous data to define optimal expression levels associated with key clinical endpoints. In general, when novel cut points are identified using discovery methods, it is incumbent on the investigator to validate the method on an alternate specimen collection TMA.

CONCLUSION

TMAs are a very effective approach for high-throughput molecular analysis of pathological tissues supporting the identification of new potential diagnostic and prognostic markers in human cancers. TMAs are now widely used for all types of tissue-based research, particularly in oncology. They offer a range of applications in research and quality assessment. TMAs have led to a significant acceleration of basic research findings towards clinical applications. Because TMAs have a limited representation of tumor heterogeneity, biomarkers that score significantly on this platform may represent the most robust of indicators. However, TMAs remain a research tool and should not be used as a diagnostic/prognostic tool at the individual patient level, for which analysis of a greater amount of tissue will ultimately be required. The migration of promising predictive marker assays from TMAs into routine clinical use will require careful evaluation of the staining characteristics of the selected marker in whole tissue sections with development of precise quantification or analytical measurement approaches.

GLOSSARY

Fluorescence *in situ* hybridization (FISH) is a cytogenetic technique that is used to detect and localize DNA sequences on chromosomes.

High-throughput screening Type of technique allowing a researcher to conduct quickly millions of chemical, genetic or pharmacological tests.

Immunohistochemistry (IHC) is a technique allowing the colorimetric detection of antigens in cells or tissue sections with antibodies.

Tissue microarray (TMA) is a technology allowing the transposition of punches of tissue cores from tissue donor blocks, arranged in an array fashion.

ABBREVIATIONS

FISH Fluorescence *in situ* hybridization
FFPE Formalin-fixed, paraffin-embedded tissue
IHC Immunohistochemistry
OCT Optimum cutting temperature
TMA Tissue microarray

REFERENCES

[1] Tambasco M, Eliasziw M, Magliocco AM. Morphologic complexity of epithelial architecture for predicting invasive breast cancer survival. J Transl Med 2010;8:140.

[2] Le Page C, Kobel M, de Ladurantaye M, Rahimi K, Madore J, Babinszky S, et al. Specimen quality evaluation from Canadian Biobanks participating in the COEUR repository. Biopreserv Biobank 2013;11:83–93.

[3] Hsu FD, Nielsen TO, Alkushi A, Dupuis B, Huntsman D, Liu CL, et al. Tissue microarrays are an effective quality assurance tool for diagnostic immunohistochemistry. Mod Pathol 2002;15:1374–80.

[4] Battifora H. The multitumor (sausage) tissue block: novel method for immunohistochemical antibody testing. Lab Invest 1986;55:244–8.

[5] Wan WH, Fortuna MB, Furmanski P. A rapid and efficient method for testing immunohistochemical reactivity of monoclonal antibodies against multiple tissue samples simultaneously. J Immunol Methods 1987;103:121–9.

[6] Kononen J, Bubendorf L, Kallioniemi A, Barlund M, Schraml P, Leighton S, et al. Tissue microarrays for high-throughput molecular profiling of tumor specimens. Nat Med 1998;4:844–7.

[7] Conway C, Dobson L, O'Grady A, Kay E, Costello S, O'Shea D. Virtual microscopy as an enabler of automated/quantitative assessment of protein expression in TMAs. Histochem Cell Biol 2008;130:447–63.

[8] Zhou L, Hodeib M, Abad JD, Mendoza L, Kore AR, Hu Z. New tissue microarray technology for analyses of gene expression in frozen pathological samples. BioTechniques 2007;43:101–5.

[9] Zietarska M, Madore J, Diallo JS, Diallo JS, Delvoye N, Saad F, et al. A novel method of cell embedding for tissue microarrays. Histopathology 2010;57:323–9.

[10] Schoenberg Fejzo M, Slamon DJ. Frozen tumor tissue microarray technology for analysis of tumor RNA, DNA, and proteins. Am J Pathol 2001;159:1645–50.

[11] Friede A, Grossman R, Hunt R, Li RM, Stren S, editors. National biospecimen network blueprint. Durham, NC: Constella Group, Inc; 2003.

[12] Xie R, Chung JY, Ylaya K, Williams RL, Guerrero N, Nakatsuka N, et al. Factors influencing the degradation of archival formalin-fixed paraffin-embedded tissue sections. J Histochem Cytochem 2011;59:356–65.

[13] DiVito KA, Charette LA, Rimm DL, Camp RL. Long-term preservation of antigenicity on tissue microarrays. Lab Invest 2004;84:1071–8.

[14] Barres V, Ouellet V, Lafontaine J, Tonin PN, Provencher DM, Mes-Masson AM. An essential role for Ran GTPase in epithelial ovarian cancer cell survival. Mol Cancer 2010;9:272.

[15] Karlsson C, Karlsson MG. Effects of long-term storage on the detection of proteins, DNA, and mRNA in tissue microarray slides. J Histochem Cytochem 2011;59:1113–21.

[16] Fox CH, Johnson FB, Whiting J, Roller PP. Formaldehyde fixation. J Histochem Cytochem 1985;33:845–53.

[17] Hayat MA. Chemical fixation. Principles and techniques of electron microscopy: biological applications. CRC Press; 1989. p. 1–74.

[18] Goldstein NS, Ferkowicz M, Odish E, Mani A, Hastah F. Minimum formalin fixation time for consistent estrogen receptor immunohistochemical staining of invasive breast carcinoma. Am J Clin Pathol 2003;120:86–92.

[19] Paavilainen L, Edvinsson A, Asplund A, Hober S, Kampf C, Ponten F, et al. The impact of tissue fixatives on morphology and antibody-based protein profiling in tissues and cells. J Histochem Cytochem 2010;58:237–46.

[20] Liu CL, Prapong W, Natkunam Y, Alizadeh A, Montgomery K, Gilks CB, et al. Software tools for high-throughput analysis and archiving of immunohistochemistry staining data obtained with tissue microarrays. Am J Pathol 2002;161:1557–65.

[21] Liu CL, Montgomery KD, Natkunam Y, West RB, Nielsen TO, Cheang MC, et al. TMA-Combiner, a simple software tool to permit analysis of replicate cores on tissue microarrays. Mod Pathol 2005;18:1641–8.

[22] Kajdacsy-Balla A, Geynisman JM, Macias V, Setty S, Nanaji NM, Berman JJ, et al. Practical aspects of planning, building, and interpreting tissue microarrays: the Cooperative Prostate Cancer Tissue Resource experience. J Mol Histol 2007;38:113–21.

[23] Camp RL, Charette LA, Rimm DL. Validation of tissue microarray technology in breast carcinoma. Lab Invest 2000;80:1943–9.

[24] Rosen DG, Huang X, Deavers MT, Malpica A, Silva EG, Liu J. Validation of tissue microarray technology in ovarian carcinoma. Mod Pathol 2004;17:790–7.

[25] Torhorst J, Bucher C, Kononen J, Haas P, Zuber M, Kochli OR, et al. Tissue microarrays for rapid linking of molecular changes to clinical endpoints. Am J Pathol 2001;159:2249–56.

[26] Hecht JL, Kotsopoulos J, Gates MA, Hankinson SE, Tworoger SS. Validation of tissue microarray technology in ovarian cancer: results from the Nurses' Health Study. Cancer Epidemiol Biomarkers Prev 2008;17:3043–50.

[27] Rimm DL, Camp RL, Charette LA, Olsen DA, Provost E. Amplification of tissue by construction of tissue microarrays. Exp Mol Pathol 2001;70:255–64.

[28] Catchpoole D, Mackie N, McIver S, Chetcuti A, Henwood A, Graf N, et al. Tape transfer sectioning of tissue microarrays introduces nonspecific immunohistochemical staining artifacts. Biotech Histochem 2011;86:421–8.

[29] D'Amico F, Skarmoutsou E, Stivala F. State of the art in antigen retrieval for immunohistochemistry. J Immunol Methods 2009;341:1–18.

[30] Shi SR, Key ME, Kalra KL. Antigen retrieval in formalin-fixed, paraffin-embedded tissues: an enhancement method for immunohistochemical staining based on microwave oven heating of tissue sections. J Histochem Cytochem 1991;39:741–8.

[31] Leong AS, Haffajee Z. Citraconic anhydride: a new antigen retrieval solution. Pathology 2010;42:77–81.

[32] Namimatsu S, Ghazizadeh M, Sugisaki Y. Reversing the effects of formalin fixation with citraconic anhydride and heat: a universal antigen retrieval method. J Histochem Cytochem 2005;53:3–11.

[33] Mote PA, Balleine RL, McGowan EM, Clarke CL. Colocalization of progesterone receptors A and B by dual immunofluorescent histochemistry in human endometrium during the menstrual cycle. J Clin Endocrinol Metab 1999;84:2963–71.

[34] Baschong W, Suetterlin R, Laeng RH. Control of autofluorescence of archival formaldehyde-fixed, paraffin-embedded tissue in confocal laser scanning microscopy (CLSM). J Histochem Cytochem 2001;49:1565–72.

[35] Szeszel MK, Crisman CL, Crow L, McMullen S, Major JM, Natarajan L, et al. Quantifying estrogen and progesterone receptor expression in breast cancer by digital imaging. J Histochem Cytochem 2005;53:753–62.
[36] Le Page C, Marineau A, Bonza PK, Rahimi K, Cyr L, Labouda I, et al. BTN3A2 expression in epithelial ovarian cancer is associated with higher tumor infiltrating T cells and a better prognosis. PLoS One 2012;7:e38541.
[37] Brown LA, Huntsman D. Fluorescent in situ hybridization on tissue microarrays: challenges and solutions. J Mol Histol 2007;38:151–7.
[38] Zeng WJ, Li Y, Fei HL, Cheng XD, Ye F, Lu WG, et al. The value of p16ink4a expression by fluorescence in situ hybridization in triage for high risk HPV positive in cervical cancer screening. Gynecol Oncol 2011;120:84–8.
[39] Dietel M, Ellis IO, Hofler H, Kreipe H, Moch H, Dankof A, et al. Comparison of automated silver enhanced in situ hybridisation (SISH) and fluorescence ISH (FISH) for the validation of HER2 gene status in breast carcinoma according to the guidelines of the American Society of Clinical Oncology and the College of American Pathologists. Virchows Arch 2007;451:19–25.
[40] Bertheau P, Cazals-Hatem D, Meignin V, de Roquancourt A, Verola O, Lesourd A, et al. Variability of immunohistochemical reactivity on stored paraffin slides. J Clin Pathol 1998;51:370–4.
[41] Fergenbaum JH, Garcia-Closas M, Hewitt SM, Lissowska J, Sakoda LC, Sherman ME. Loss of antigenicity in stored sections of breast cancer tissue microarrays. Cancer Epidemiol Biomarkers Prev 2004;13:667–72.
[42] Mirlacher M, Kasper M, Storz M, Knecht Y, Durmuller U, Simon R, et al. Influence of slide aging on results of translational research studies using immunohistochemistry. Mod Pathol 2004;17:1414–20.
[43] Le Page C, Huntsman D, Provencher D, Mes-Masson AM. Predictive and prognostic protein biomarkers in epithelial ovarian cancer: recommendation for future studies. Cancers 2010;2:913–54.
[44] Bordeaux J, Welsh A, Agarwal S, Killiam E, Baquero M, Hanna J, et al. Antibody validation. BioTechniques 2010;48:197–209.
[45] McCabe A, Dolled-Filhart M, Camp RL, Rimm DL. Automated quantitative analysis (AQUA) of in situ protein expression, antibody concentration, and prognosis. J Natl Cancer Inst 2005;97:1808–15.
[46] Allred DC, Clark GM, Elledge R, Fugua SA, Brown RW, Chamness GC, et al. Association of p53 protein expression with tumor cell proliferation rate and clinical outcome in node-negative breast cancer. J Natl Cancer Inst 1993;85:200–6.
[47] Detre S, Saclani Jotti G, Dowsett MA. A "quickscore" method for immunohistochemical semiquantitation: validation for oestrogen receptor in breast carcinomas. J Clin Pathol 1995;48:876–8.
[48] Lassus H, Leminen A, Lundin J, Lehtovirta P, Butzow R. Distinct subtypes of serous ovarian carcinoma identified by p53 determination. Gynecol Oncol 2003;91:504–12.
[49] Kobel M, Kalloger SE, Boyd N, McKinney S, Mehl E, Palmer C, et al. Ovarian carcinoma subtypes are different diseases: implications for biomarker studies. PLoS Med 2008;5:e232.
[50] Vermeij R, de Bock GH, Leffers N, Ten Hoor KA, Schulze U, Hollema H, et al. Tumor-infiltrating cytotoxic T lymphocytes as independent prognostic factor in epithelial ovarian cancer with Wilms tumor protein 1 overexpression. J Immunother 2011;34:516–23.
[51] Adams EJ, Green JA, Clark AH, Youngson JH. Comparison of different scoring systems for immunohistochemical staining. J Clin Pathol 1999;52:75–7.
[52] Nickerson CA. A Note on "A concordance correlation coefficient to evaluate reproducibility". Biometrics 1997;53:1503–7.
[53] Rhodes A, Jasani B, Barnes DM, Bobrow LG, Miller KD. Reliability of immunohistochemical demonstration of oestrogen receptors in routine practice: interlaboratory variance in the sensitivity of detection and evaluation of scoring systems. J Clin Pathol 2000;53:125–30.
[54] Charafe-Jauffret E, Tarpin C, Bardou VJ, Bertucci F, Ginestier C, Braud AC, et al. Immunophenotypic analysis of inflammatory breast cancers: identification of an 'inflammatory signature'. J Pathol 2004;202:265–73.
[55] Phillips T, Murray G, Wakamiya K, Askaa J, Huang D, Welcher R, et al. Development of standard estrogen and progesterone receptor immunohistochemical assays for selection of patients for antihormonal therapy. Appl Immunohistochem Mol Morphol 2007;15:325–31.
[56] Rojo MG, Garcia GB, Mateos CP, Garcia JG, Vicente MC. Critical comparison of 31 commercially available digital slide systems in pathology. Int J Surg Pathol 2006;14:285–305.
[57] Cregger M, Berger AJ, Rimm DL. Immunohistochemistry and quantitative analysis of protein expression. Arch Pathol Lab Med 2006;130:1026–30.
[58] Ryder EF, Robakiewicz P. Statistics for the molecular biologist: group comparisons. Curr Protoc Mol Biol 2001; Appendix 3: Appendix 3I.
[59] McShane LM, Altman DG, Sauerbrei W, Taube SE, Gion M, Clark GM. Reporting recommendations for tumor marker prognostic studies. J Clin Oncol 2005;23:9067–72.

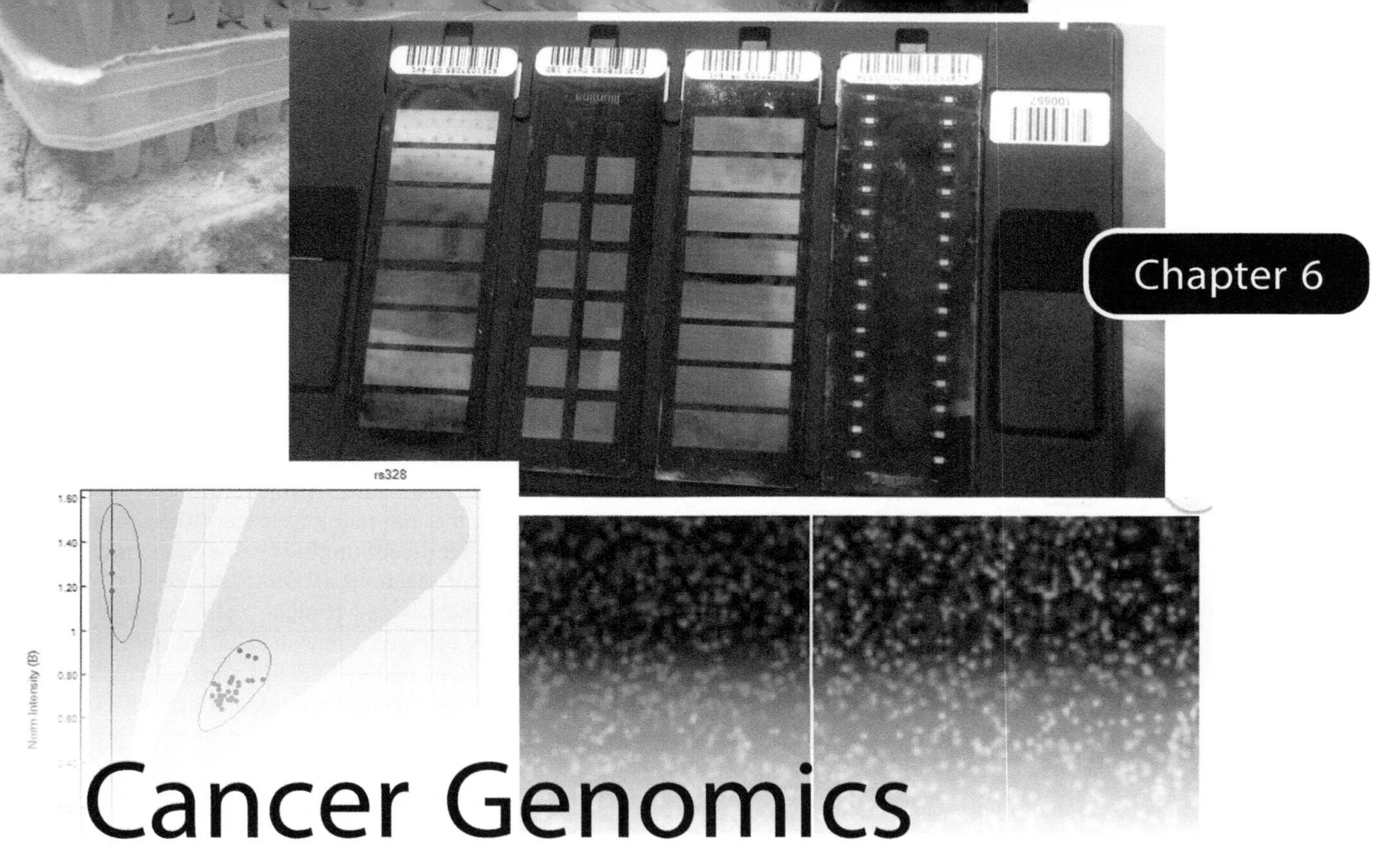

Cancer Genomics

Cancer Pharmacogenomics in Children

Shahrad Rod Rassekh[1,2]
Colin J.D. Ross[2,3,4]

[1]Department of Pediatrics, Division of Pediatric Hematology/Oncology/ BMT, University of British Columbia, Vancouver, BC, Canada
[2]Child and Family Research Institute, British Columbia's Children's Hospital, Vancouver, BC, Canada
[3]Department of Pediatrics, Division of Translational Therapeutics, University of British Columbia, Vancouver, BC, Canada
[4]Centre for Molecular Medicine and Therapeutics, University of British Columbia, Vancouver, BC, Canada

Contents

Cancer Genomics. DOI: http://dx.doi.org/10.1016/B978-0-12-396967-5.00006-2

Key Concepts

- Adverse drug reactions are common in pediatric oncology with approximately 40% having life-threatening or irreversible morbidity
- There is a large variation among children receiving the same chemotherapy at the same doses with respect to outcomes
- Pharmacogenetics is the study of host genetic factors that may explain why two children treated with the same medication may have significantly different outcomes
- Thiopurine methyltransferase (*TPMT*) has been shown to play a significant role in both toxicity to thiopurines, as well as to cisplatin-induced hearing loss
- Cisplatin-induced hearing loss is common and several genes have been identified that are associated with susceptibility to hearing loss
- Genes involved in the folate metabolism pathway appear to play an important role in methotrexate-induced toxicity
- Vincristine-induced neurotoxicity is common and variation in the CYP3A5 enzyme that metabolizes vinca alkaloids may explain the variation in individual toxicity
- Cardiac toxicity is the most serious complication of treatment of anthracycline and several genetic variants have been identified that may be associated with toxicity
- With new advances in technology and worldwide collaborations that are underway, pharmacogenomics has the potential to improve significantly the safety and effectiveness of pediatric cancer therapies in the future

INTRODUCTION

History of Adverse Drug Reactions in Pediatric Oncology

Hippocrates established the concept of personalized medicine over 2000 years ago, observing that physicians must "give different [drugs] to different patients" [1]. By understanding the specific determinants of disease and wellness in an individual, we can tailor the diagnosis, treatment and prevention of disease to avoid harm and maximize well-being. Pediatric cancer was once considered a universally fatal disease. Recently, there have been significant improvements in the treatment of pediatric cancer with the use of multimodal therapies including surgery, chemotherapy and radiation therapy. Pediatric cancer is now curable for a majority of children, with an overall 5-year survival rate of more than 82% [2]. Unfortunately the life-saving treatments that are required to cure the cancer have also resulted in significant morbidity or even mortality due to adverse drug reactions (ADRs). A paradox of modern drug development is that clinical trials investigate the safety and efficacy of medications at standard doses in large populations without taking into account individual patient differences. Consequently, individual variability in response to a medication can lead to severe or even life-threatening ADRs.

Adverse Drug Reactions

Despite the use of standard dosing, life-threatening or permanently disabling ADRs occur in children receiving therapy [3]. ADRs may be evident within minutes, days, months, or even years following exposure and may range from a minor skin rash to life-threatening heart failure or liver injury. ADRs can have serious consequences for patients, including death, permanent disability, treatment cessation, or reduced adherence to medications. Due to the historically poor outcomes of diseases such as cancer, many ADRs have been regarded as unavoidable complications of therapy. However, the risk factors for severe ADRs remain largely unclear, and the selection of specific drugs for an individual patient is often an empiric exercise [4]. This highlights the fact that one of the most difficult issues faced in clinical practice is deciding which drug and which dose is best for each individual.

Approximately 22% of all hospital admissions for pediatric cancer patients are caused by ADRs [5]. Nearly three-quarters of childhood cancer survivors suffer an ADR related to their therapy, and an alarming 40% of these ADRs can be classified as life threatening or permanently disabling [6]. Moreover, ADRs in pediatric oncology can result in significant treatment modifications and delays in therapy, which can significantly impact outcome. The significant individual variability in response to some medications suggests that individual differences must exist that are responsible for this variation. Non-genetic factors that can contribute to an individual's response to a medication include age, organ function, gut absorption, and drug interactions. However, genomic differences are thought to account for a significant proportion of the variability of drug response [7–11].

PHARMACOGENOMICS

There are many factors that influence the positive and negative effects that a medication will have on any given individual. Non-genetic factors that play an important role in drug toxicity include age, underlying organ function, and drug interactions. It is estimated that for many medications, genetic factors may account for more than half of the variability in drug response [7–11]. Recent technological advances, including next-generation DNA sequencing, have helped pharmacogenomics become a clinical reality. Genomic medicine has been described as "the use of information from genomes (from humans and other organisms) and their derivatives (RNA, proteins, and metabolites) to guide medical decision making" [12]. Sequencing of the human genome and the International HapMap Project have advanced this field forward at an unparalleled rate and have created the opportunity for the study of host variability that contributes to different responses to chemotherapy [13–16]. The overall goal of pharmacogenomics is to use genetic testing to identify patients at high risk of drug toxicity, or at risk of low efficacy, and use this information to modify treatment. Treatment modifications in pediatric cancer might include dose modifications, use of alternative chemotherapeutic drugs, or the concomitant administration of protective agents. To be clinically useful, pharmacogenomic tests must be accessible and affordable with a rapid turnaround time.

Pharmacogenomics of Thiopurine-Induced Toxicity in Children

The most well developed example of pharmacogenetic testing in childhood cancer is the study of the metabolism of thiopurines. The thiopurines, which include 6-mercaptopurine (6-MP), 6-thioguanine (6-TG), and azathioprine, are all metabolized by thiopurine-methyltransferase (*TPMT*) and this gene has been extensively investigated. Genetic polymorphisms in the *TPMT* locus have been identified with heterozygotes having intermediate activity of the enzyme and homozygotes having very low TPMT activity [17,18]. Moreover, the genotype–phenotype correlation is highly concordant and over 24 low-functional genetic variants have been identified. The two most common variants, *TPMT**2 and *3, account for more than 95% of defective TPMT activity in patients [19–22].

The treatment of acute lymphoblastic leukemia (ALL) relies heavily on the use of both 6-MP and 6-TG [23], both of which are inactive pro-drugs that require multistep activation into active thioguanine nucleotides (Figure 6.1). TPMT is involved in the inactivation of these active metabolites, normally functioning to reduce the activity of these drugs. It has been shown that approximately 10% of people have reduced TPMT activity levels, and 0.3% of the population has no detectable levels of TPMT activity [19,24–26]. Consequently, patients with low TPMT enzyme activity exhibit reduced drug inactivation and increased toxicity. The most common toxicity in patients with low TPMT activity is severe myelotoxicity, which can result in prolonged cytopenia, significant morbidity from infections and transfusional requirements, and even death [27–29]. It should be noted that low-functioning variants of *TPMT* do not explain all cases of myelotoxicity and that other genetic polymorphisms are likely to be involved in thiopurine-induced cytopenias [30].

TPMT variants may also play a very significant role in the efficacy of thiopurines. For example, it has been shown that patients with low TPMT activity exhibit better treatment response as measured by minimal residual disease testing than those with normal functioning variants of *TPMT* [31]. It is not known whether *TPMT* variants can play a role in the overall efficacy outcomes of overall survival and event-free survival.

There is also evidence that *TPMT* genetic variants are an important predictor of secondary malignancy. In a St Jude's ALL trial, children with low TPMT activity were found to be at increased risk of therapy-associated acute myeloid leukemia (tAML) [32]. This finding was replicated in a Nordic Society of Pediatric Hematology and Oncology trial, in which patients with low activity *TPMT* variants had an increased risk of tAML when compared to those with normal TPMT activity (1% ± 1%, vs 9% ± 6%, *p*-value 0.002) [33]. Low *TPMT* activity levels have also been reported in association with secondary brain tumors in the St Jude's Total Therapy XII trial, which utilized cranial radiation given concurrently with 6-MP. In this trial, the 8-year cumulative incidence of secondary brain tumors was 42.9% (± 20.6%) in patients with low *TPMT* activity variants versus only 4.7% for those with normal TPMT activity [34]. By contrast, German investigators in the Berlin–Frankfurt–Munster (BFM) ALL clinical trials group did not find an association between *TPMT* low activity genotype and second malignancy investigating 105 subjects who developed a second malignancy compared to 814 patients with ALL who did not develop a second malignancy. They found a similar distribution of *TPMT* variants in the two groups and concluded that *TPMT* does not play a major role in the etiology of second malignancy after treatment for childhood ALL using a BFM strategy [35]. It should be noted that differences in dosing of 6-MP (lower doses of 6-MP in maintenance on the BFM trials) and differences in timing of other concurrent therapies (such as concurrent radiation or methotrexate dosing) may explain the different outcomes when comparing the BFM trials with the Nordic and St Jude's cohorts. A recent paper investigating the

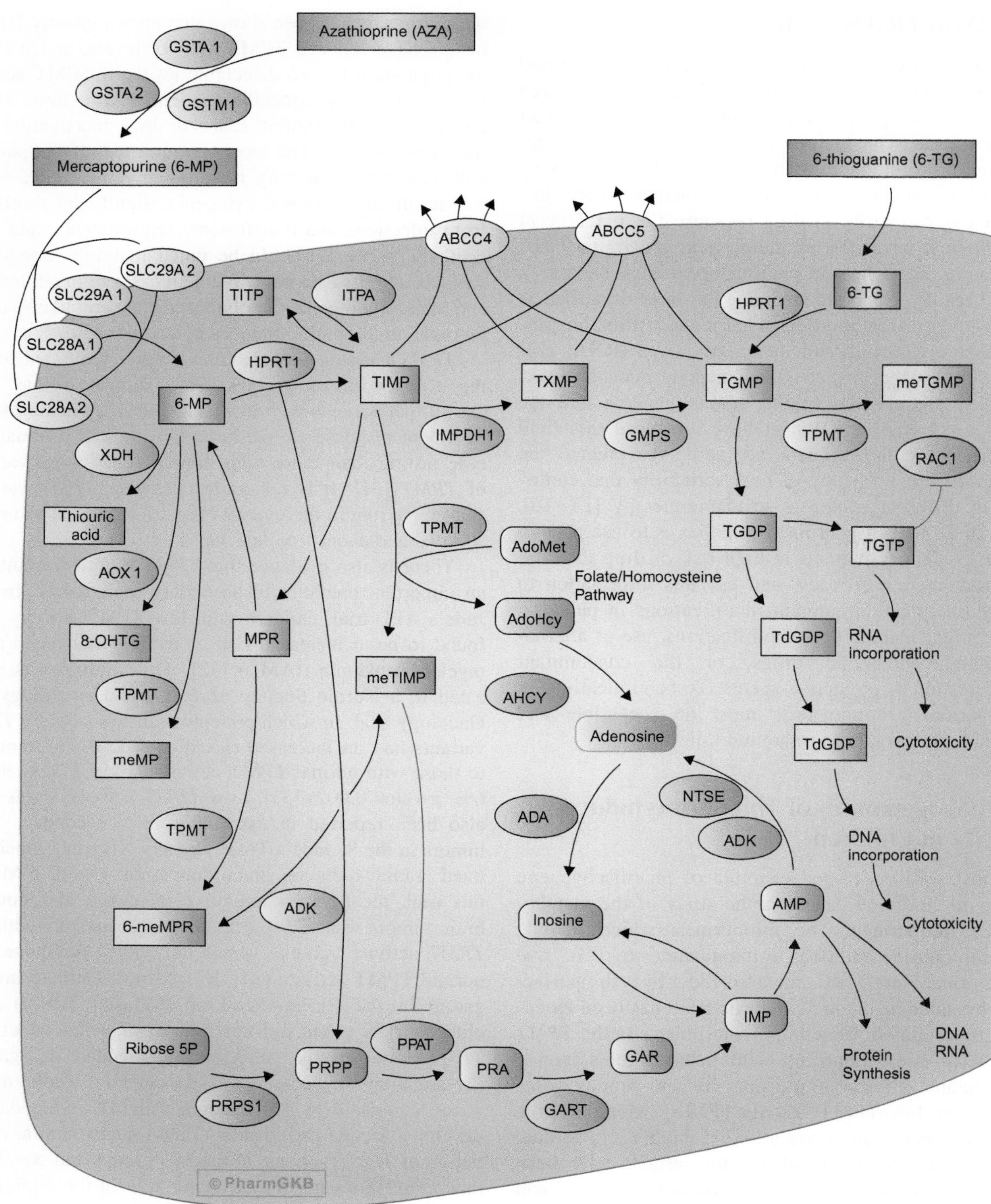

FIGURE 6.1 The thiopurine drugs are medications used in the treatment of acute lymphoblastic leukemia, 6-mercaptopurine (6-MP) is a pro-drug that requires activation into cytotoxic metabolites. However, synthesis of active metabolites is in competition with inactivation pathways catalyzed by thiopurine methyltransferase (*TPMT*). Zaza G, Cheok M, Krynetskaia N, Thorn C, Stocco G, Hebert JM, et al. Thiopurine pathway. *Pharmacogenet Genomics* 2010;20:573–4.

Nordic cohort further supported this hypothesis by showing that increased intensity and duration of 6-MP in combination with methotrexate was associated with an increased risk of second malignancy in those with low-functioning *TPMT* variants [36].

Despite all of this evidence for the important role of *TPMT* genotype in both toxicity and possibly efficacy, upfront genetic testing for *TPMT* variants is not yet being used routinely at most institutions. However, there are an increasing number of recommendations from organizations such as the Food and Drug Administraton (FDA), as well as international groups, suggesting that *TPMT* testing should be considered and that dose adjustments of thiopurines can be made using pharmacogenetic results [37–39]. The high sensitivity, specificity, positive and negative predictive values of *TPMT* genotyping suggest that this testing is worthy of adoption into routine clinical practice [37]. Studies have shown that prospective testing of *TPMT* genotype for dose adjustments are effective in reducing 6-MP toxicity without compromising toxicity [38]. Current Children's Oncology Group trials suggest that optional TPMT testing during induction therapy can be considered and, in the case of significant or unexpected prolonged cytopenias, that TPMT testing is recommended. However, the cost of routine pharmacogenetics testing is an important consideration and the current studies of cost-effectiveness of TPMT testing shows conflicting results and that, although testing often provides important clinical information, whether it is cost-effective is debatable at the present time [39,40].

Pharmacogenomics of Methotrexate-Induced Toxicity in Children

Methotrexate is an important medication that is used in the treatment of childhood leukemia, brain tumors, and osteosarcoma [40]. The anticancer effect of methotrexate works by depleting endogenous cellular folate through the inhibition of dihydrofolatereductase (*DHFR*) and blocking the metabolism of folic acid, leading to cell death in rapidly dividing cells. Methotrexate is a pro-drug, which requires intracellular polyglutamation for maximum cytotoxic effects (Figure 6.2) [41–44].

A variety of ADRs caused by methotrexate have been reported, including short-term problems such as mucositis, diarrhea, nausea and vomiting, hepatitis, seizures and skin reactions as well as long-term toxicities such as leucoencephalopathy and learning issues, such as attention difficulties [45–48]. The mechanism of how methotrexate causes mucositis is not yet fully known, and the potential role of genetic susceptibility factors also remains unknown. Several studies have shown that polymorphisms associated with increased toxicity appear to be associated with worse outcome as measured by event-free survival (EFS). One study identified that a non-synonymous variant in the *SLC19A1* (Solute carrier family 19 (folate transporter), member 1) gene (rs1051266) is associated with a worse prognosis in terms of event-free survival, despite higher plasma concentrations of methotrexate [49]. This same association of *SLC19A1* rs1051266 and methotrexate concentration was also observed in adult rheumatoid arthritis patients [50]; however, the association was not found in a separate pediatric study of 53 ALL patients [51].

One of the more concerning toxic effects of methotrexate is leucoencephalopathy, which can range from mild to severe and can result in significant long-term neurologic sequelae in anywhere from 0 to 69% of subjects [47,52–54]. The reason for this extreme variability in incidence of this toxicity in the literature is likely due to several factors, such as timing and dosing of methotrexate as well as due to different approaches to investigating and defining leucoencephalopathy. Methotrexate-induced leucoencephalopathy is considered worse in children treated under the age of 5 years and in those concomitantly or previously treated with radiation therapy [54]. Given the impact that leucoencephalopathy plays in long-term quality of life, there have been several groups of researchers interested in this important toxicity, resulting in numerous reports of pharmacogenetic associations with methotrexate-induced leucoencephalopathy. Genes involved in the metabolism of folate and methionine are thought to play an important role in methotrexate neurotoxicity and the methylene-tetrahydrofolate reductase (*MTHFR*) gene has been frequently investigated. In a study of 68 subjects with primary CNS lymphoma, a low activity variant of *MTHFR* (Ala222Val, rs1801133) was predictive of toxicity ($p = 0.013$) [55]. In a separate case report, a patient homozygous for this variant developed significant toxicity and delayed clearance of methotrexate [56]. Additional evidence comes from a study showing that low functioning *MTHFR* variants are associated with mucositis [57]. A second polymorphism within the *MTHFR* gene (A1298C) has been found to be associated with attention deficit hyperactivity disorder (ADHD). This variant is associated with the inattentive symptoms that are common in ALL survivors [58]. However, to complicate matters, several other studies have shown no correlation of *MTHFR* with methotrexate-induced toxicity [51,59–61].

A Canadian study, which was later replicated in a Children's Oncology Group (COG) cohort of children, found that *MTHFR* might play a significant role in methotrexate efficacy [59,62]. The initial single-center study found that in 201 children with ALL, those with either the *MTHFR* T677A or A1298C haplotypes had a lower probability of EFS in multivariate analysis (hazard

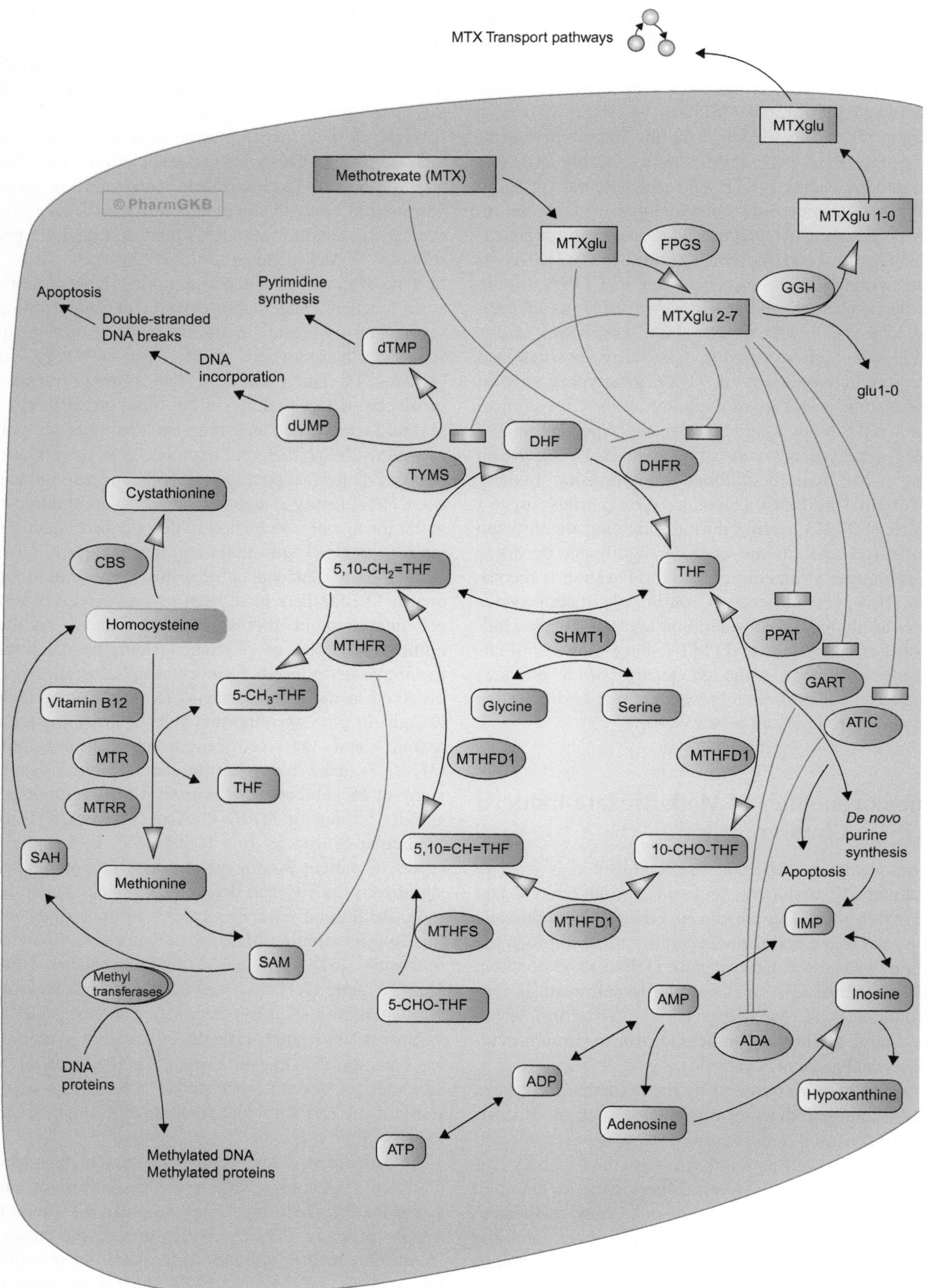

FIGURE 6.2 Methotrexate is medication used in the treatment of a variety of cancers as well as non-malignant diseases and is a folate analog that interacts with numerous enzymes in the folate pathway. Mikkelsen TS, Thorn CF, Yang JJ, Ulrich CM, French D, Zaza G, et al. PharmGKB summary: methotrexate pathway. *Pharmacogenet Genomics* 2011;21:679–86.

ratio = 2.2, 95% C.I. 0.9–5.6) [62]. In the subsequent COG ALL study, 520 subjects were genotyped for two common functional polymorphisms (C677T and A1298G). The C677T variant was significantly associated with relapse even when controlling for important covariates (hazard ratio = 1.82, $p = 0.008$). This finding was, in fact, more predictive of relapse than day 7 bone marrow response [59].

Other genes that may be critical in methotrexate toxicity include the ABC superfamily proteins *ABCB1*, *ABCC1–3*, and *ABCG2*. These genes are components of the methotrexate efflux pathway and have been specifically studied as genes of interest in methotrexate toxicity [63,64]. A study in children demonstrated that methotrexate encephalopathy was more frequent in patients homozygous for the *ABCB1* 3435 "TT" genotype versus "CC" or "CT" genotype patients [46]. A further study of 60 subjects with osteosarcoma identified an association of methotrexate toxicity with specific variants of *MTHFR* (A1958G, $p = 0.038$), ABCB1 (T3435C, $p = 0.027$), and *ABCC2* (T3563A, $p = 0.028$) [65]. However, a recent Spanish study of children with ALL did not replicate the associations with *MTHFR* or *ABCB1* [66]. In a large, well-designed, genome-wide study of 434 children with ALL, Trevino et al. found that toxicity was associated with methotrexate plasma concentrations and identified a strong genetic association with two highly linked variants of the solute carrier organic anion transporter family member 1B1 (*SLCO1B1*) (rs11045879 and rs4149081) with methotrexate clearance ($p < 2 \times 10^{-9}$) and replicated these in an independent cohort of 206 children ($p < 0.018$). In the replication cohort, these *SLC01B1* variants were highly associated with methotrexate-induced gastrointestinal toxicity (odds ratios 15.3–16.4; $p = 0.03–0.004$) [67]. This association with *SLCO1B1* was also replicated in the Spanish study of 115 children with ALL, mentioned above (rs11045879, $p = 0.03$) [66].

Pharmacogenetics appears to play a significant role in both the toxicity and efficacy of methotrexate. Further studies are needed in order to investigate the role of all the above genes as well as other undiscovered associations that could be used in a multigene model in order to delineate better who will develop toxicity. Once this is done, then subsequent studies will be needed to investigate whether dose modification can be used to maximize the therapeutic effect of methotrexate, while attempting to minimize harm.

Pharmacogenomics of Cisplatin-Induced Toxicity

Cisplatin is used in both children and adults for the treatment of solid tumors, but the use of this agent is limited by the high incidence of ADRs including cytopenias, irreversible hearing loss, peripheral neuropathy and nephrotoxicity [68–70]. Cisplatin causes a typical pattern of hearing loss that can be monitored with audiograms, with an initial decline in the high frequency ranges, which then progresses into the speech frequencies. The hearing loss continues to progress with further doses, and higher cumulative doses are associated with increased hearing loss. Ototoxicity is very common and it is estimated that 41–61% of children treated with cisplatin have hearing loss [71–76]. This toxicity is especially important in young children, as hearing loss considerably increases a child's risk of learning difficulties and social–emotional problems as well as generating significant costs to the healthcare system [74,77,78]. The large inter-individual variability in response to cisplatin is suggestive that genetic variation may be involved in cisplatin ADRs. Further evidence of the potential role of genetics are the different rates of toxicity noted in people of different ethnicities. Although there have been no studies investigating this with regards to ototoxicity, a higher rate of nephrotoxicity is observed in African–American patients compared to Caucasians (47.6% vs 8.3%, $p = 0.007$) [79].

The initial focus of genes implicated in cisplatin ototoxicity focused on genes involved in oxidative stress, as free radicals within the inner ear have been implicated in causing the toxicity due to cisplatin. There has been significant interest in the glutathione S-transferase (*GST*) gene family, given the important role it plays in protection against oxidative stress. Several adult studies have investigated variants in *GST* with varied results. A study of 39 adults found a significant association between cisplatin-induced hearing loss and expression of *GSTM3* ($p = 0.02$, uncorrected for multiple testing), but not with *GSTM1*, *GSTT1*, or *GSTP1*. Approximately 18% of patients with normal hearing after therapy carried the *GSTM3*B* allele versus 2.5% of those with hearing loss post-cisplatin [80]. In a second large series, the *GSTP1* variant was associated with protection against hearing loss due to cisplatin with an odds ratio of 4.21 (95% CI, 1.99–8.88) in 173 adult testicular cancer survivors. Overall cumulative cisplatin dose and certain *GST* genotypes (105-Val-GSTP1 homozygotes) were found to be statistically associated with hearing loss [81].

Studies investigating other forms of drug-induced hearing loss have been used to create potential candidate genes for the study of cisplatin-induced hearing loss. The common group of antibiotics from the aminoglycoside family have been extensively investigated for their association with hearing loss. The megalin gene has been shown to play a potential role in aminoglycoside-induced hearing loss, which led to investigation of its role in cisplatin ototoxicity. A small adult study identified one variant which was significantly associated with hearing loss with an odds ratio of 3.45 [82].

A study of adolescents with osteosarcoma investigated eight single nucleotide polymorphisms (SNPs) in six genes involved in platinum-DNA adduct repair (*ERCC2, XPC, XPA, ERCC1, ERCC4,* and *ERCC5*) and identified one variant in the *ERCC2* gene that was associated with a slightly improved survival outcome [83,84]. Unfortunately, this study did not have consistently reliable audiologic data and therefore only a weak association was found for the *XPC* gene (OR = 17.6 for homozygous risk allele, CI = 1.1–266.78, $p = 0.042$). Studies in adults with non-small cell lung tumors which are treated with cisplatin have identified two common SNPs in *ERCC1* that have been correlated with increased toxicity and survival [85,86].

The largest study of cisplatin pharmacogenomics in children was recently performed by the Canadian Pharmacogenomics Network for Drug Safety (CPNDS) and this study identified two genes that were associated with cisplatin-induced hearing loss [87,88]. This large candidate gene study of 220 drug metabolism genes revealed a strong association of cisplatin-associated hearing loss with loss-of-function genetic variants in thiopurine methyltransferase (*TPMT*) (odds ratio 10.9–18.0) and a moderate association with catechol-O-methyl transferase (*COMT*) (odds ratio 2.5–5.5). The findings were initially identified in a cohort of children with cancer from British Columbia and replicated in a second cohort of children from across Canada. The greater number of risk alleles in *TPMT* and *COMT* that were carried by patients, the more rapid the decline in hearing. Those who carried three or all four risk alleles often had clinically significant hearing loss with only a single dose of cisplatin. Potential mechanisms of toxicity associated with variants in these genes include increased efficiency of cisplatin cross-linking, as well as a possible role of the methionine pathway through a common substrate, S-adenosylmethionine [88].

Pharmacogenomics of Vincristine-Induced Toxicity

Vincristine is one of the most commonly used pediatric chemotherapeutic agents and is used to treat both hematologic malignancies and solid tumors. It is a natural alkaloid that interferes with the assembly of microtubule structures leading to cell apoptosis. A significant dose-limiting ADR associated with vincristine is peripheral neuropathy. Vincristine-induced peripheral neuropathy occurs in 4–98% of children [89,90]. This large variation is likely due to different definitions and reported severity of vincristine toxicity. This neuropathy can include sensory neuropathy (paresthesias, burning or pain), motor neuropathy, vocal cord paralysis, paralytic ileus, and autonomic neuropathies [89–93]. The mechanism of vincristine-induced neuropathy has not yet been fully elucidated and there is significant debate as to whether it is dose related. Studies have suggested that vincristine-induced neuropathy occurs in a dose-dependent fashion with peak vincristine plasma concentrations correlating with neuropathy [94–97]. In contrast, a pharmacokinetic study in children with leukemia found no association with toxicity and vincristine area-under-the-concentration–time curve [98].

Preliminary evidence that genetics may play a significant role in susceptibility to vincristine toxicity comes from the widely differing rates of toxicity between patients of different ancestries. Specifically, when compared to patients of African ancestry, patients of European ancestry had overall greater toxicity, more missed doses due to toxicity, and greater severity of toxicity with an average neurotoxicity grade (using the Common Terminology Criteria for Adverse Events version 3) of 2.72 versus 1, respectively [99]. These results suggest that genetic differences in the vincristine pathway may differ between patients of European and African ancestry, and these differences may be important in conferring susceptibility to vincristine-induced toxicity. *CYP3A5*, for example, has been shown to metabolize vinca alkyloids [100–103] and significant differences in the genetic variation in *CYP3A5* are found in patients of African and Caucasian ancestry [103]. Over 80% of African–Americans carry at least one functional copy of the *CYP3A5* allele, compared to only 10–20% of Caucasians [104]. In a recent study of vincristine-induced neuropathy and *CYP3A5* expression, a 100% rate of toxicity was observed in those patients who were *CYP3A5* non-expressors, compared to those who were *CYP3A5* expressors, who had an 88% rate of toxicity [91]. This suggests that this genetic variation in *CYP3A5* is in part linked with susceptibility to vincristine-induced neuropathy, but also that there are likely additional factors that are important, and that further research is required.

Pharmacogenomics of Anthracycline-Induced Cardiac Toxicity

Anthracyclines are a class of chemotherapeutic agents that are vital in the treatment of both solid and hematologic malignancies in childhood. Over 60% of all childhood cancer patients receive anthracyclines as part of their treatment, and use of anthracyclines is thought to have contributed greatly to improved cure rates for cancer in children [105]. The most significant toxicity of anthracyclines is cardiomyopathy, leading to heart failure in up to 16% of subjects, and asymptomatic heart dysfunction in up to 57% of subjects [106–108]. The frequent use of anthracyclines contributes significantly to the fact that survivors of childhood cancer are eight times more likely to die of heart disease, and 15 times more likely to experience heart failure than their contemporaries who did not have cancer [109,110]. The toxicity associated with anthracyclines can occur at the time of administration, as well as long after therapy has been completed, with heart failure being the second most common cause of death in survivors after secondary malignancy [111].

The two most consistent risk factors for anthracycline-induced cardiac toxicity are higher cumulative lifetime anthracycline dose and radiation therapy involving the heart. Additional risk factors include higher individual dose, shorter infusion time, younger age, longer time since treatment, and female gender [106,108,112,113]. There is a large inter-individual variability with some individuals having toxicity with minimal doses of anthracyclines, while others tolerate very high cumulative doses, suggesting a significant genetic component to the risk of cardiotoxicity [114].

Several studies have been conducted to identify genetic variants that may be predictive of anthracycline-induced cardiotoxicity. In a study of adults with non-Hodgkin lymphoma (NHL), genetic variants in 82 candidate genes were examined in 87 cases and 363 well matched control patients and identified five genes in three subunits of the NAD(P)H oxidase (*NCF4*, *RAC2*, *CYBA*) and two doxorubicin transporters (*MRP1/ABCC1* and *MRP2/ABCC2*) that were significantly associated with anthracycline-induced cardiotoxicity [115]. Variants in *ABCC1*, *ABCC2*, *RAC2*, and *CYBA* were associated with acute toxicity, while variants in *NCF4* were associated with chronic toxicity, which the authors defined as heart failure occurring at any point after the third cycle of therapy. This association with *NCF4* was replicated in a second study of 106 adults with NHL [116]. The first study of anthracycline-induced cardiac toxicity in children investigated 30 cases of heart failure with 115 controls, and used questionnaires and self-reports as part of a nested case-control study within the Childhood Cancer Survivor Study cohort [114]. This study looked specifically at two polymorphisms, *NQO1*2* (rs1800566) and *CBR3* V244M (rs1056892), and found no statistically significant results. However, there was a trend for association of the *CBR3* V244M polymorphism with anthracycline-induced cardiac toxicity with an odds ratio of 8.16 in multivariate analysis ($p = 0.056$ for the homozygous G/G compared to the A/A variant). A second pediatric study examined several genes in 76 children treated for ALL who were evaluated for late toxicity by echocardiograms and electrocardiograms. This study found an intronic variant (rs10836235) in the catalase gene (*CAT*) that was associated with toxicity, while a known functional variant in the promoter region was not associated [117]. A recent report from Hungary investigated nine polymorphisms in the *ABCC1* gene in 235 pediatric patients with ALL and identified a variant (rs374527) that was associated with lower left ventricular fractional shortening after chemotherapy [118]. The *ABCC1* gene was also identified previously in adult patients with non-Hodgkin lymphoma [115].

The largest study in children to date investigating anthracycline-induced cardiac toxicity was carried out across Canada through the CPNDS in an initial cohort of 156 children in British Columbia, with replication of the findings in a second Canadian cohort of 188 children. An analysis of 2977 genetic variants in 220 key drug biotransformation genes identified a highly significant association of anthracycline-induced cardiac toxicity with a synonymous coding variant within the solute carrier family 28 (*SLC28A3*) gene (rs7853758) with a protective odds ratio of 0.35 ($p = 1.8 \times 10^{-5}$). This newly discovered association was further replicated in a third cohort of 96 children from Amsterdam [119]. This study also identified several additional highly associated variants in genes, including drug transporters such as *SLC28A1*, *ABCB1*, *ABCB4*, *ABCC1* as well as an enzyme of the glucuronidation pathway, UDP-glucuronosyltransferase 1 (*UGT1A6*).

NEED FOR COLLABORATION, ACTIVE ADR SURVEILLANCE AND FUTURE DIRECTIONS

The safety of medications is an international concern that has attracted the attention of policy makers and clinicians worldwide. Removing medications from the market that have been associated with serious ADRs is not an optimal solution because this can leave seriously ill patients without therapeutic options. Rather, the solution lies in finding the mechanisms underlying specific ADRs so that these medications can continue to be used in patients for whom there is therapeutic benefit and enable better management or prevention of reaction among those at high risk of an ADR.

The absence of effective regulatory surveillance and the rarity of many drug-induced severe ADRs make it difficult for any one research group to accrue sufficient numbers of patients to determine potential susceptibility factors. Even in countries such as Canada, where ADR reporting is mandatory for pharmaceutical manufacturers, less than 5–20% of ADRs are reported [118,120,121]. Thus, national and international collaborations are necessary to identify and investigate sufficient numbers of patients with a specific ADR accurately to identify and validate ADR-susceptibility factors.

Many significant associations between genetic variations and differential drug responses have been found in children with cancer. With continued technological advancement and reduction in the costs of genome-wide analyses, there is a growing opportunity to identify the key factors that confer increased susceptibility or protection from specific severe ADRs. With the ability to do large genome-wide association and sequencing studies, there is a strong need for extremely well phenotyped cohorts of cases and controls for ADR analysis. The large variation in toxicity rates among many of the studies investigating similar medications highlights the importance of consistent clinical definitions and precise phenotyping. Much of the recent focus in pharmacogenomics research has been on improvements in genomic analysis technologies, yet an often overlooked but crucial component is needed improvements in accurate phenotyping.

Given the large number of patients required in genome-wide studies and the relatively small number of cases of pediatric cancer, large national and international collaborations are needed to ascertain enough subjects to achieve sufficient statistical power to distinguish real findings from stochastic noise. For example, the Canadian Pharmacogenomic Network for Drug Safety (CPNDS) and the C17 Research Network are focused on active surveillance of chemotherapeutic ADRs for enrollment into large Canada-wide patient cohorts from almost all the centers in Canada treating children with cancer [87,122,123]. Similar cohorts are being recruited in countries around the world. It is critical to collaborate in order to replicate findings and to ensure that research findings are effectively translated into clinical practice for the benefit of patients worldwide. Moreover, genetic diversity between different ancestries will be important to take into account when looking at the applicability of tests in different populations, which leads to an even greater need for international collaborations truly to understand how to deliver chemotherapy safely to all children of all different ethnicities. This can be made possible, for example, through the international collaborative groups such as the Children's Oncology Group (COG) and the Société Internationale d'Oncologie Pediatrique (SIOP). Both groups are now conducting pharmacogenomic research through currently open clinical trials that are collecting DNA and detailed outcome measures. Large cohort studies, such as the recently launched St Jude's Life Study, will prospectively test adults who were treated for cancer as children, and may provide valuable insight into the long-term outcomes and toxicity as survivors age. In the St Jude's Life Study, adults who were treated as children and who have survived over 10 years are being brought back for thorough testing and toxicity assessments and DNA is being collected for genomic studies [124].

A first step in pharmacogenetics is to identify the variants responsible for poor outcomes. These findings must be followed with individualized treatment plans that allow for alteration of therapy in order to change outcomes. For example, a dose of thiopurine chemotherapy could be reduced in patients with low TPMT activity if tested prior to initiation of therapy, leading to reduced toxicity but without a reduction in efficacy. Unfortunately, for many other agents, a reduction of dose may not be possible without reducing efficacy and, in these situations, other changes to overall treatment may be required. An alternative to reducing the chemotherapy dose is to use a protective agent. Several ongoing clinical trials are investigating the use of sodium thiosulfate to prevent cisplatin-induced hearing loss, or dexrazoxane to prevent anthracycline-induced cardiotoxicity. There are concerns regarding the safety of these protective agents, including concerns of tumor protection, as well as increased toxicities, such as second malignancies with dexrazoxane. In the future, the pharmacogenetic identification of patients at very high risk for specific ADRs may be an effective strategy to test the effectiveness of these protective agents. For example, predictive pharmacogenetic testing may provide significant benefit by identifying children at the highest risk of cardiac toxicity, enabling limited use of dexrazoxane exclusively within this subgroup of children.

As we study pharmacogenetics, it is becoming evident that, for many chemotherapeutic drugs, there will be multiple genes involved that are each critical in determining outcome. Therefore, current studies are investigating the role of multiple genes that each contribute to an individual's response to therapy, and novel methodologies for analysis of gene–gene interactions are required [125]. In addition, many clinical, non-genetic, factors may have important roles in influencing patient outcomes, even in those who are genetically at higher or lower susceptibility to drug-induced toxicity. A complex interplay likely exists between the host genome, the cancer genome, and the environment. The study of this interplay will require complex modeling of both genetic and non-genetic factors in order to create effective prediction models of an individual's response to chemotherapy. Notably, these studies may also provide insight into both drug toxicity and efficacy [16,59,62,65]. For example, a model that incorporates both clinical and genetic risk factors of anthracycline-induced cardiac toxicity has been developed, which places individuals in one of three risk categories: high risk, intermediate risk, or low risk of drug toxicity. As seen in Figure 6.3, those patients at high risk have very high rates of cardiac toxicity (over 80%) as compared to those patients predicted to be at low risk (under 5% risk of toxicity), which comprises almost half of the population of children receiving this drug [119].

A final area of future research that needs to be developed involves determining the interplay between tumor

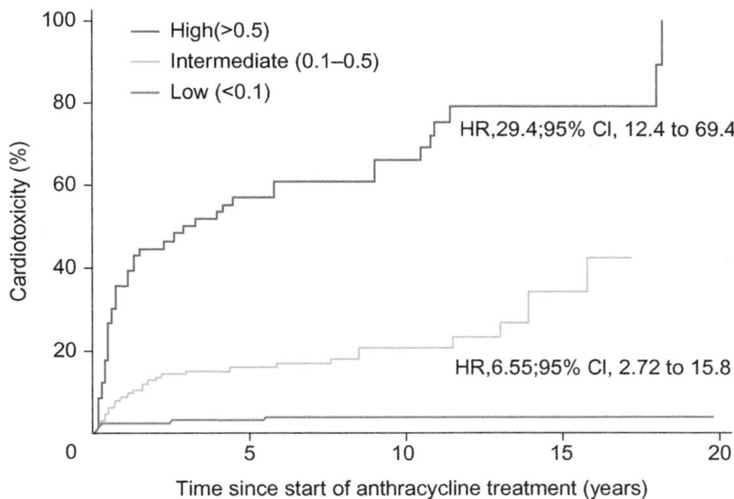

FIGURE 6.3 Model of risk of anthracycline cardiac toxicity, which utilizes both genetic information in combination with clinical risk factors in order to predict the risk of poor cardiac outcome post-therapy [119].

and host genomes. Much cancer research conducted to date has been restricted to either the cancer or host genome, but incorporating both genomes will provide the most predictive models of drug toxicity and efficacy for an individual patient.

CONCLUSION

There have been many exciting findings showing the importance of pharmacogenetics in childhood cancer and many important and clinically useful discoveries have been made. Some genes and their impact on toxicity have been extensively studied with decades of follow-up supporting research, such as *TPMT* in mercaptopurine toxicity and efficacy in childhood ALL, while the research into many other adverse reactions is still in its infancy. Ongoing research is needed better to predict and manage a patient's susceptibility to serious adverse reactions to cancer treatment. This work will lead to improvements not only to pediatric cancer cure rates, but also to better quality of life for childhood cancer survivors by reducing treatment-related toxicities.

ACKNOWLEDGMENTS

The authors have no conflicts of interest to declare. The authors would like to acknowledge all the investigators, trainees and staff of the Canadian Pharmacogenomic Network for Drug Safety (CPNDS) who help support our group's research into the pharmacogenomics of drug toxicity. The funding support of the C17 Research Network, Childhood Cancer Foundation – Candlelighter's Canada, Canadian Institute of Health Research, Canada Foundation for Innovation, and Genome British Columbia has been invaluable in our studies of pharmacogenomics in pediatric cancer.

GLOSSARY

Adverse drug event Harm associated with the use of a medication at any dose (normal dose, overdosing, or underdosing) or any harm associated with a drug such as events associated with discontinuation of the drug.

Adverse drug reaction Harm associated with the use of a medication at normal dosage during normal use. This is one form of an adverse drug event.

Ancestry Inherited properties associated with the descendants of one individual.

Anthracycline Class of drug used in cancer chemotherapy derived from *Streptomyces* bacterium.

Cardiomyopathy The measurable deterioration of the function of the myocardium usually leading to heart failure.

Genome-wide association Investigation of genetic variation throughout the entire genome to identify differences among individuals.

Homozygous Identical alleles of a gene are found on both homologous chromosomes.

Leucoencephalopathy Disease of the white matter of the brain, often due to localized areas of cell death.

Mucositis Inflammation and ulceration of the mucous membranes lining the digestive tract.

Negative predictive value The proportion of those truly without the condition that are correctly identified with a negative test result (ratio of true negatives to combined true and false negatives).

Neuropathy A functional deterioration or pathological change in one or more of the peripheral nerves.

Neurotoxicity Damage to the central and/or peripheral nervous system.

Pharmacogenetics Genetic differences that can influence an individual's response to a particular drug.

Pharmacogenomics The branch of pharmacology that deals with influence of genetic variation on drug response in individuals. It is sometimes referred to being the whole genome application of pharmacogenetics.

Polymorphism Natural variations in a gene, DNA sequence, or chromosome that have no adverse effects on the individual and occur with fairly high frequency in the general population.

Positive predictive value The proportion of true positives with the condition that are correctly identified with a positive test result (ratio of true positives to combined true and false positives).

Sensitivity The proportion of those with the condition who correctly test positive for it (ratio of true positives to true positives and false negatives).

Specificity The proportion of those without the condition who correctly test negative for it (ratio of true negatives to true negatives and false positives).

Synonymous coding variant A genetic variation in the open reading frame of a gene that does not alter the amino acid composition of the protein produced from the gene.

ABBREVIATIONS

6-MP 6-Mercaptopurine
ADHD Attention deficit hyperactivity disorder
ADR Adverse drug reaction
ALL Acute lymphoblastic leukemia
BFM Berlin–Frankfurt–Munich
COG Children's Oncology Group
CPNDS Canadian Pharmacogenomics Network for Drug Safety
EFS Event-free survival
DHFR Dihydrofolate reductase
GST Glutathione S-transferase
MTHFR Methylene-tetrahydrofolate reductase
SLC19A1 Solute carrier family 19
SIOP Société Internationale d'Oncologie Pediatrique
TAML Therapy-associated acute myeloid leukemia
TPMT Thiopurine-methyltransferase

REFERENCES

[1] Sykiotis GP, Kalliolias GD, Papavassiliou AG. Pharmacogenetic principles in the Hippocratic writings. J Clin Pharmacol 2005;45:1218–20.

[2] Ellison LF, De P, Mery LS, Grundy PE. Canadian Cancer Society Steering Committee C. Canadian cancer statistics at a glance: cancer in children. Can Med Assoc J 2009;180:422–4.

[3] WHO. International drug monitoring: the role of the hospital. Geneva, Switzerland: World Health Organization; 1966.

[4] Kalow W, Tang BK, Endrenyi L. Hypothesis: comparisons of inter- and intra-individual variations can substitute for twin studies in drug research. Pharmacogen 1998;8:283–9.

[5] Mitchell AA, Lacouture PG, Sheehan JE, Kauffman RE, Shapiro S. Adverse drug-reactions in children leading to hospital admission. Pediatrics 1988;82:24–9.

[6] Geenen MM, Cardous-Ubbink MC, Kremer LCM, van den Bos C, van der Pal HJ, Heinen RC, et al. Medical assessment of adverse health outcomes in long-term survivors of childhood cancer. J Am Med Assoc 2007;297:2705–15.

[7] Impicciatore M. Pharmacogenomic can give children safer medicines. Arch Dis Child 2003;88:366.

[8] Jaja C, Rothstein M. Pharmacogenomics. New York: John Wiley and Sons, Inc; 2003.

[9] Kling J. USFDA contemplates collection of pharmacogenomic data. Nat Biotechnol 2003;21:590.

[10] Classen DC, Pestotnik SL, Evans RS, Lloyd JF, Burke JP. Adverse drug events in hospitalized patients – Excess length of stay, extra costs, and attributable mortality. J Am Med Assoc 1997;277:301–6.

[11] Kalow W, Tang BK, Endrenyi L. Hypothesis: Comparisons of inter- and intra-individual variations can substitute for twin studies in drug research. Pharmacogenetics 1998;8:283–9.

[12] Ginsburg GS, Willard HF. Genomic and personalized medicine: foundations and applications. Translat Res 2009;154:277–87.

[13] Ansari M, Krajinovic M. Pharmacogenomics in cancer treatment defining genetic bases for inter-individual differences in responses to chemotherapy. Curr Opin Pediatr 2007;19:15–22.

[14] Venter JC, Adams MD, Myers EW, Li PW, Mural RJ, Sutton GG, et al. The sequence of the human genome. Science 2001;291:1304–51.

[15] Lander ES, Linton LM, Birren B, Nusbaum C, Zody MC, Baldwin J, et al. Initial sequencing and analysis of the human genome. Nature 2001;409:860–921.

[16] Paugh SW, Stocco G, Evans WE. Pharmacogenomics in pediatric leukemia. Curr Opin Pediatr 2010;22:703–10.

[17] Wang L, Weinshilboum R. Thiopurine S-methyltransferase pharmacogenetics: insights, challenges and future directions. Oncogene 2006;25:1629–38.

[18] Cheok MH, Evans WE. Acute lymphoblastic leukaemia: a model for the pharmacogenomics of cancer therapy. Nat Rev Cancer 2006;6:117–29.

[19] Schaeffeler E, Fischer C, Brockmeier D, Wernet D, Moerike K, Eichelbaum M, et al. Comprehensive analysis of thiopurine S-methyltransferase phenotype–genotype correlation in a large population of German Caucasians and identification of novel TPMT variants. Pharmacogenetics 2004;14:407–17.

[20] Schaeffeler E, Lang T, Zanger UM, Eichelbaum M, Schwab M. High-throughput genotyping of thiopurine S-methyltransferase by denaturing HPLC. Clin Chem 2001;47:548–55.

[21] Yates CR, Krynetski EY, Loennechen T, Fessing MY, Tai HL, Pui CH, et al. Molecular diagnosis of thiopurine S-methyltransferase deficiency: Genetic basis for azathioprine and mercaptopurine intolerance. Ann Internl Med 1997;126:608–14.

[22] Coulthard SA, Howell C, Robson J, Hall AG. The relationship between thiopurine methyltransferase activity and genotype in blasts from patients with acute leukemia. Blood 1998;92:2856–62.

[23] Pui CH, Evans WE. Drug therapy – Treatment of acute lymphoblastic leukemia. N Engl J Med 2006;354:166–78.

[24] Krynetski EY, Tai HL, Yates CR, Fessing MY, Loennechen T, Schuetz JD, et al. Genetic polymorphism of thiopurine S-methyltransferase: clinical importance and molecular mechanisms. Pharmacogenetics 1996;6:279–90.

[25] Krynetski EY, Evans WE. Pharmacogenetics as a molecular basis for individualized drug therapy: The thiopurine S-methyltransferase paradigm. Pharm Res 1999;16:342–9.

[26] McLeod HL, Lin JS, Scott EP, Pui CH, Evans WE. Thiopurine methyltransferase activity in American white subjects and black subjects. Clin Pharmacol Ther 1994;55:15–20.

[27] McLeod HL, Miller DR, Evans WE. Azathioprine-induced myelosuppression in thiopurine methyltransferase deficient heart-transplant recipient. Lancet 1993;341:1151.

[28] Evans WE, Horner M, Chu YQ, Kalwinsky D, Roberts WM. Altered mercaptopurine metabolism, toxic effects, and dosage requirement in a thiopurine methyltransferase-deficient child with acute lymphocytic-leukemia. J Pediatr 1991;119:985–9.

[29] Schutz E, Gummert J, Mohr F, Oellerich M. Azathioprine-induced myelosuppression in thiopurine methyltransferase deficient heart-transplant recipient. Lancet 1993;34:436.

[30] Stocco G, Cheok MH, Crews KR, Dervieux T, French D, Pei D, et al. Genetic polymorphism of inosine triphosphate pyrophosphatase is a determinant of mercaptopurine metabolism and toxicity during treatment for acute lymphoblastic leukemia. Clin Pharmacol Ther 2009;85:164–72.

[31] Stanulla M, Schaeffeler E, Flohr T, Cario G, Schrauder A, Zimmermann M, et al. Thiopurine methyltransferase (TPMT) genotype and early treatment response to mercaptopurine in childhood acute lymphoblastic leukemia. J Am Med Assoc 2005;293:1485–9.

[32] Relling MV, Yanishevski Y, Nemec J, Evans WE, Boyett JM, Behm FG, et al. Etoposide and antimetabolite pharmacology in patients who develop secondary acute myeloid leukemia. Leukemia 1998;12:346–52.

[33] Thomsen JB, Schroder H, Kristinsson J, Madsen B, Szumlanski C, Weinshilboum R, et al. Possible carcinogenic effect of 6-mercaptopurine on bone marrow stem cells – Relation to thiopurine metabolism. Cancer 1999;86:1080–6.

[34] Relling MV, Rubnitz JE, Rivera GK, Boyett JM, Hancock ML, Felix CA, et al. High incidence of secondary brain tumours after radiotherapy and antimetabolites. Lancet 1999;354:34–9.

[35] Stanulla M, Schaeffeler E, Moricke A, Coulthard SA, Cario G, Schrauder A, et al. Thiopurine methyltransferase genetics is not a major risk factor for secondary malignant neoplasms after treatment of childhood acute lymphoblastic leukemia on Berlin–Frankfurt–Munster protocols. Blood 2009;114:1314–8.

[36] Schmiegelow K, Al-Modhwahi I, Andersen MK, Behrendtz M, Forestier E, Hasle H, et al. Methotrexate/6-mercaptopurine maintenance therapy influences the risk of a second malignant neoplasm after childhood acute lymphoblastic leukemia: results from the NOPHO ALL-92 study. Blood 2009;113:6077–84.

[37] Haga SB, Thummel KE, Burke W. Adding pharmacogenetics information to drug labels: lessons learned. Pharmacogenet Genomics 2006;16:847–54.

[38] Relling MV, Gardner EE, Sandborn WJ, Schmiegelow K, Pui CH, Yee SW, et al. Clinical Pharmacogenetics Implementation Consortium Guidelines for thiopurine methyltransferase genotype and thiopurine dosing. Clin Pharmacol Ther 2011;89:387–91.

[39] Swen JJ, Nijenhuis M, de Boer A, Grandia L, Maitland-van der Zee AH, Mulder H, et al. Pharmacogenetics: from bench to byte – an update of guidelines. Clin Pharmacol Ther 2011;89:662–73.

[40] Cheng KKF. Association of plasma methotrexate, neutropenia, hepatic dysfunction, nausea/vomiting and oral mucositis in children with cancer. Eur J Cancer Care 2008;17:306–11.

[41] Jolivet J, Schilsky RL, Bailey BD, Drake JC, Chabner BA. Synthesis, retention, and biological activity of methotrexate polyglutamates in cultured human breast cancer cells. J Clin Invest 1982;70:351–60.

[42] Chabner BA, Allegra CJ, Curt GA, Clendeninn NJ, Baram J, Koizumi S, et al. Polyglutamation of methotrexate – is methotrexate a prodrug. J Clin Invest 1985;76:907–12.

[43] Zhao RB, Goldman ID. Resistance to antifolates. Oncogene 2003;22:7431–57.

[44] Fabre I, Fabre G, Goldman ID. Polyglutamylation, an important element in methotrexate cytotoxicity and selectivity in tumor versus murine granulocytic progenitor cells *in vitro*. Cancer Res 1984;44:3190–5.

[45] Shimasaki N, Mori T, Samejinia R, Sato R, Shimada H, Yahagi N, et al. Effects of methylenetetrahydrofolate reductase and reduced folate carrier 1 polymorphisms on high-dose methotrexate-induced toxicities in children with acute lymphoblastic leukemia or lymphoma. J Pediatr Hematol Oncol 2006;28:64–8.

[46] Robaey P, Krajinovic M, Marcoux S, Moghrabi A. Pharmacogenetics of the neurodevelopmental impact of anticancer chemotherapy. Develop Disabil Res Rev 2008;14:211–20.

[47] Jaksic W, Veljkovic D, Pozza C, Lewis I. Methotrexate-induced leukoencephalopathy reversed by aminophylline and high-dose folinic acid. Acta Haematol 2004;111:230–2.

[48] Elting LS, Cooksley C, Chambers M, Cantor SB. The burdens of cancer therapy – Clinical and economic outcomes of chemotherapy-induced mucositis. Cancer 2003;98:1531–9.

[49] Laverdiere C, Chiasson S, Costea I, Moghrabi A, Krajinovic M. Polymorphism G(80)A in the reduced folate carrier gene and its relationship to methotrexate plasma levels and outcome of childhood acute lymphoblastic leukemia. Blood 2002;100:3832–4.

[50] Dervieux T, Kremer J, Lein DO, Capps R, Barham R, Meyer G, et al. Contribution of common polymorphisms in reduced folate carrier and gamma-glutamylhydrolase to methotrexate polyglutamate levels in patients with rheumatoid arthritis. Pharmacogenetics 2004;14:733–9.

[51] Kishi S, Griener J, Cheng C, Das S, Cook EH, Pei D, et al. Homocysteine, pharmacogenetics, and neurotoxicity in children with leukemia. J Clin Oncol 2003;21:3084–91.

[52] Ziereisen F, Dan B, Azzi N, Ferster A, Damry N, Christophe C. Reversible acute methotrexate leukoencephalopathy: atypical brain MR imaging features. Pediatr Radiol 2006;36:205–12.

[53] Reddick WE, Glass JO, Helton KJ, Langston JW, Li CS, Pui CH. A quantitative MR imaging assessment of leukoencephalopathy in children treated for acute lymphoblastic leukemia without irradiation. Am J Neuroradiol 2005;26:2371–7.

[54] Reddick WE, Glass JO, Helton KJ, Langston JW, Xiong X, Wu S, et al. Prevalence of leukoencephalopathy in children treated for acute lymphoblastic leukemia with high-dose methotrexate. Am J Neuroradiol 2005;26:1263–9.

[55] Linnebank M, Moskau S, Jurgens A, Simon M, Semmler A, Orlopp K, et al. Association of genetic variants of methionine metabolism with methotrexate-induced CNS white matter changes in patients with primary CNS lymphoma. Neuro Oncol 2009;11:2–8.

[56] Muller J, Kralovanszky J, Adleff V, Pap E, Nemeth K, Komlosi V, et al. Toxic encephalopathy and delayed MTX clearance after high-dose methotrexate therapy in a child homozygous for the MTHFR C677T polymorphism. Anticancer Res 2008;28:3051–4.

[57] Ulrich CM, Yasui Y, Storb R, Schubert MM, Wagner JL, Bigler J, et al. Pharmacogenetics of methotrexate: toxicity among marrow transplantation patients varies with the methylenetetrahydrofolate reductase C677T polymorphism. Blood 2001;98:231–4.

[58] Krull KR, Brouwers P, Jain N, Zhang L, Bomgaars L, Dreyer Z, et al. Folate pathway genetic polymorphisms are related to

attention disorders in childhood leukemia survivors. J Pediatr 2008;152:101–5.

[59] Aplenc R, Thompson J, Han P, La M, Zhao H, Lange B, et al. Methylenetetrahydrofolate reductase polymorphisms and therapy response in pediatric acute lymphoblastic leukemia. Cancer Res 2005;65:2482–7.

[60] Costea I, Moghrabi A, Laverdiere C, Graziani A, Krajinovic M. Folate cycle gene variants and chemotherapy toxicity in pediatric patients with acute lymphoblastic leukemia. Haematologica 2006;91:1113–6.

[61] Ruiz-Argueles GJ, Coconi-Linares LN, Garces-Eisele J, Reyes-Nunez V. Methotrexate-induced mucositis in acute leukemia patients is not associated with the MTHFR 677T allele in Mexico. Hematology 2007;12:387–91.

[62] Krajinovic M, Lemieux-Blanchard E, Chiasson S, Primeau M, Costea I, Moghrabi A. Role of polymorphisms in MTHFR and MTHFD1 genes in the outcome of childhood acute lymphoblastic leukemia. Pharmacogenomics J 2004;4:66–72.

[63] Chen ZS, Lee K, Walther S, Raftogianis RB, Kuwano M, Zeng H, et al. Analysis of methotrexate and folate transport by multidrug resistance protein 4 (ABCC4): MRP4 is a component of the methotrexate efflux system. Cancer Res 2002;62:3144–50.

[64] Zeng H, Chen ZS, Belinsky MG, Rea PA, Kruh GD. Transport of methotrexate (MTX) and folates by multidrug resistance protein (MRP) 3 and MRP1: Effect of polyglutamylation on MTX transport. Cancer Res 2001;61:7225–32.

[65] Windsor RE, Strauss SJ, Kallis C, Wood NE, WJ. S. Germline genetic polymorphisms may influence chemotherapy response and disease outcome in osteosarcoma. Cancer 2012;118:1856–67.

[66] Lopez-Lopez E, Martin-Guerrero I, Ballesteros J, Pinan MA, Garcia-Miguel P, Navajas A, et al. Polymorphisms of the SLCO1B1 gene predict methotrexate-related toxicity in childhood acute lymphoblastic leukemia. Pediatr Blood Cancer 2011;57:612–9.

[67] Trevino LR, Shimasaki N, Yang WJ, Panetta JC, Cheng C, Pei D, et al. Germline genetic variation in an organic anion transporter polypeptide associated with methotrexate pharmacokinetics and clinical effects. J Clin Oncol 2009;27:5972–8.

[68] Brock P, Bellman S. Ototoxicity of cisplatinum. Br J Cancer 1991;63:159–60.

[69] McWhinney SR, Goldberg RM, McLeod HL. Platinum neurotoxicity pharmacogenetics. Molec Cancer Ther 2009;8:10–6.

[70] Yao X, Panichpisal K, Kurtzman N, Nugent K. Cisplatin nephrotoxicity: A review. Am J Med Sci 2007;334:115–24.

[71] Bokemeyer C, Berger CC, Hartmann JT, Kollmannsberger C, Schmoll HJ, Kuczyk MA, et al. Analysis of risk factors for cisplatin-induced ototoxicity in patients with testicular cancer. Br J Cancer 1998;77:1355–62.

[72] Brock PR, Yeomans EC, Bellman SC, Pritchard J. Cisplatin therapy in infants – short and long-term morbidity. Br J Cancer 1992;66:S36–40.

[73] Coradini PP, Cigana L, Selistre SGA, Rosito LS, Brunetto AL. Ototoxicity from cisplatin therapy in childhood cancer. J Pediatr Hematol Oncol 2007;29:355–60.

[74] Knight KRG, Kraemer DF, Neuwelt EA. Ototoxicity in children receiving platinum chemotherapy: Underestimating a commonly occurring toxicity that may influence academic and social development. J Clin Oncol 2005;23:8588–96.

[75] Li Y, Womer RB, Silber JH. Predicting cisplatin ototoxicity in children: the influence of age and the cumulative dose. Eur J Cancer 2004;40:2445–51.

[76] Kushner BH, Budnick A, Kramer K, Modak S, Cheung N-KV. Ototoxicity from high-dose use of platinum compounds in patients with neuroblastoma. Cancer 2006;107:417–22.

[77] Bess FH, Dodd-Murphy J, Parker RA. Children with minimal sensorineural hearing loss: Prevalence, educational performance, and functional status. Ear Hearing 1998;19:339–54.

[78] Dionne F, Mitton C, Rassekh R, Brooks B, Ross C, Hayden M, et al. Economic impact of a genetic test for cisplatin-induced ototoxicity. Pharmacogenomics J 2012;12:205–13.

[79] Shord SS, Thompson DM, Krempl GA, Hanigan MH. Effect of concurrent medications on cisplatin-induced nephrotoxicity in patients with head and neck cancer. Anticancer Drugs 2006; 17:207–15.

[80] Peters U, Preisler-Adams S, Hebeisen A, Hahn M, Seifert E, Lanvers C, et al. Glutathione S-transferase genetic polymorphisms and individual sensitivity to the ototoxic effect of cisplatin. Anticancer Drugs 2000;11:639–43.

[81] Oldenburg J, Kraggerud SM, Cvancarova M, Lothe RA, Fossa SD. Cisplatin-induced long-term hearing impairment is associated with specific glutathione S-transferase genotypes in testicular cancer survivors. J Clin Oncol 2007;25:708–14.

[82] Riedemann L, Lanvers C, Deuster D, Peters U, Boos J, Jurgens H, et al. Megalin genetic polymorphisms and individual sensitivity to the ototoxic effect of cisplatin. Pharmacogenomics J 2008; 8:23–8.

[83] Caronia D, Patino-Garcia A, Milne RL, Zalacain-Diez M, Pita G, Alonso MR, et al. Common variations in ERCC2 are associated with response to cisplatin chemotherapy and clinical outcome in osteosarcoma patients. Pharmacogenomics J 2009;9:347–53.

[84] Martin LP, Hamilton TC, Schilder RJ. Platinum resistance: The role of DNA repair pathways. Clin Cancer Res 2008;14:1291–5.

[85] Zhou W, Gurubhagavatula S, Liu G, Park S, Neuberg DS, Wain JC, et al. Excision repair cross-complementation group 1 polymorphism predicts overall survival in advanced non-small cell lung cancer patients treated with platinum-based chemotherapy. Clin Cancer Res 2004;10:4939–43.

[86] Suk R, Gurubhagavatula S, Park S, Zhou W, Su L, Lynch TJ, et al. Polymorphisms in ERCC1 and grade 3 or 4 toxicity in non-small cell lung cancer patients. Clin Cancer Res 2005; 11:1534–8.

[87] Carleton BC, Poole RL, Smith MA, Leeder J, Ghannadan R, Ross C, et al. Adverse drug reaction active surveillance: developing a national network in Canada's children's hospitals. Pharmacoepidemiol Drug Safety 2009;18:713–21.

[88] Ross CJ, Katzov-Eckert H, Dube M, Brooks B, Rassekh SR, Barhdadi A, et al. Genetic variants in TPMT and COMT are associated with hearing loss in children receiving cisplatin chemotherapy. Nat Genet 2009;41:1345–9.

[89] Gidding CEM, Meeuwsen-de Boer GJ, Koopmans P, Uges DRA, Kamps WA, de Graaf SSN. Vincristine pharmacokinetics after repetitive dosing in children. Cancer Chemother Pharmacol 1999;44:203–9.

[90] Tisdale JE, Miller DA. Drug-induced diseases. Prevention, detection, and management. Bethesda, MD: American Society of Health-System Pharmacists; 2005.

[91] Tarlaci S. Vincristine-induced fatal neuropathy in non-Hodgkin's lymphoma. Neurotoxicology 2008;29:748–9.

[92] Egbelakin A, Ferguson MJ, MacGill EA, Lehmann AS, Topletz AR, Quinney SK, et al. Increased risk of vincristine neurotoxicity associated with low CYP3A5 expression genotype in children with acute lymphoblastic leukemia. Pediatr Blood Cancer 2011;56:361–7.

[93] Kuruvilla G, Perry S, Wilson B, El-Hakim H. The natural history of vincristine-induced laryngeal paralysis in children. Arch Otolaryngol Head Neck Surg 2009;135:101–5.

[94] Callizot N, Andriambeloson E, Glass J, Revel M, Ferro P, Cirillo R, et al. Interleukin-6 protects against paclitaxel, cisplatin and vincristine-induced neuropathies without impairing chemotherapeutic activity. Cancer Chemother Pharmacol 2008;62:995–1007.

[95] Verstappen CCP, Koeppen S, Heimans JJ, Huijgens PC, Scheulen ME, Strumberg D, et al. Dose-related vincristine-induced peripheral neuropathy with unexpected off-therapy worsening. Neurology 2005;64:1076–7.

[96] Dougherty PM, Cata JP, Burton AW, Vu K, Weng HR. Dysfunction in multiple primary afferent fiber subtypes revealed by quantitative sensory testing in patients with chronic vincristine-induced pain. J Pain Symptom Manage 2007;33:166–79.

[97] Groninger E, Meeuwsen-de Boer T, Koopmans P, Uges D, Sluiter W, Veerman A, et al. Vincristine pharmacokinetics and response to vincristine monotherapy in an up-front window study of the Dutch Childhood Leukaemia Study Group (DCLSG). Eur J Cancer 2005;41:98–103.

[98] Crom WR, Degraaf SSN, Synold T, Uges DR, Bloemhof H, Rivera G, et al. Pharmacokinetics of vincristine in children and adolescents with acute lymphocytic-leukemia. J Pediatr 1994;125:642–9.

[99] Renbarger JL, McCammack KC, Rouse CE, Hall SD. Effect of race on vincristine-associated neurotoxicity in pediatric acute lymphoblastic leukemia patients. Pediatr Blood Cancer 2008;50:769–71.

[100] Dennison JB, Kulanthaivel P, Barbuch RJ, Renbarger JL, Ehlhardt WJ, Hall SD. Selective metabolism of vincristine in vitro by CYP3A5. Drug Metab Dispos 2006;34:1317–27.

[101] Dennison JB, Jones DR, Renbarger JL, Hall SD. Effect of CYP3A5 expression on vincristine metabolism with human liver microsomes. J Pharmacol Exp Ther 2007;321:553–63.

[102] Kajita J, Kuwabara T, Kobayashi H, Kobayashi S. CYP3A4 is mainly responsibile for the metabolism of a new vinca alkaloid, vinorelbine, in human liver microsomes. Drug Metab Dispos 2000;28:1121–7.

[103] Kuehl P, Zhang J, Lin Y, Lamba J, Assem M, Schuetz J, et al. Sequence diversity in CYP3A promoters and characterization of the genetic basis of polymorphic CYP3A5 expression. Nat Genet 2001;27:383–91.

[104] Xie HG, Wood AJJ, Kim RB, Stein CM, Wilkinson GR. Genetic variability in CYP3A5 and its possible consequences. Pharmacogenomics 2004;5:243–72.

[105] van Dalen EC, van der Pal HJH, Kok WEM, Caron HN, Kremer LCM. Clinical heart failure in a cohort of children treated with anthracyclines: A long-term follow-up study. Eur J Cancer 2006;42:3191–8.

[106] Kremer LCM, van Dalen EC, Offringa M, Voute PA. Frequency and risk factors of anthracycline-induced clinical heart failure in children: a systematic review. Ann Oncol 2002;13:503–12.

[107] Lipshultz SE, Colan SD, Gelber RD, Perezatayde AR, Sallan SE, Sanders SP. Late cardiac effects of doxorubicin therapy for acute lymphoblastic leukemia in childhood. N Engl J Med 1991;324:808–15.

[108] Von Hoff DD, Layard MW, Basa P, Davis HL, Von Hoff AL, Rozencweig M, et al. Risk-factors for doxorubicin-induced congestive heart-failure. Ann Intern Med 1979;91:710–807.

[109] Oeffinger KC, Mertens AC, Sklar CA, Kawashima T, Hudson MM, Meadows AT, et al. Chronic health conditions in adult survivors of childhood cancer. N Engl J Med 2006;355:1572–82.

[110] Mertens AC, Yasui Y, Neglia JP, Potter JD, Nesbit ME, Ruccione K, et al. Late mortality experience in five-year survivors of childhood and adolescent cancer: The Childhood Cancer Survivor Study. J Clin Oncol 2001;19:3163–72.

[111] Reulen RC, Winter DL, Frobisher C, Lancashire ER, Stiller CA, Jenney ME, et al. Long-term cause-specific mortality among survivors of childhood cancer. J Am Med Assoc 2010;304:172–9.

[112] Kremer LCM, van der Pal HJH, Offringa M, van Dalen EC, Voute PA. Frequency and risk factors of subclinical cardiotoxicity after anthracycline therapy in children: a systematic review. Ann Oncol 2002;13:819–29.

[113] Trachtenberg BH, Landy DC, Franco VI, Henkel JM, Pearson EJ, Miller TL, et al. Anthracycline-associated cardiotoxicity in survivors of childhood cancer. Pediatr Cardiol 2011;32:342–53.

[114] Wojnowski L, Kulle B, Schirmer M, Schluter G, Schmidt A, Rosenberger A, et al. NAD(P)H oxidase and multidrug resistance protein genetic polymorphisms are associated with doxorubicin-induced cardiotoxicity. Circulation 2005;112:3754–62.

[115] Rossi D, Rasi S, Franceschetti S, Capello D, Castelli A, De Paoli L, et al. Analysis of the host pharmacogenetic background for prediction of outcome and toxicity in diffuse large B-cell lymphoma treated with R-CHOP21. Leukemia 2009;23: 1118–26.

[116] Blanco JG, Leisenring WM, Gonzalez-Covarrubias VM, Kawashima TI, Davies SM, Relling MV, et al. Genetic polymorphisms in the carbonyl reductase 3 gene CBR3 and the NAD(P)H : Quinone oxidoreductase 1 gene NQO1 in patients who developed anthracycline-related congestive heart failure after childhood cancer. Cancer 2008;112:2789–95.

[117] Rajic V, Aplenc R, Debeljak M, Prestor VV, Karas-Kuzelicki N, Mlinaric-Rascan I, et al. Influence of the polymorphism in candidate genes on late cardiac damage in patients treated due to acute leukemia in childhood. Leukemia Lymphoma 2009;50:1693–8.

[118] Visscher H, Ross CJD, Rassekh SR, Barhdadi A, Dube MP, Al-Saloos H, et al. Pharmacogenomic prediction of anthracycline-induced cardiotoxicity in children. J Clin Oncol 2012;30:1422–8.

[119] Mittmann N, Knowles SR, Gomez M, Fish JS, Cartotto R, Shear NH. Evaluation of the extent of under-reporting of serious adverse drug reactions: the case of toxic epidermal necrolysis. Drug Safety 2004;27:477–87.

[120] Rzany B, Mockenhaupt M, Baur S, Schroder W, Stocker U, Mueller J, et al. Epidemiology of erythema exsudativum multiforme majus, Stevens-Johnson syndrome, and toxic epidermal necrolysis in Germany (1990–1992): structure and results of a population-based registry. J Clin Epidemiol 1996;49:769–73.

[121] Stephenson T. How children's responses to drugs differ from adults. Br J Clin Pharmacol 2005;59:670–3.

[122] Ross CJD, Visscher H, Sistonen J, Brunham LR, Pussegoda K, Loo TT, et al. The Canadian Pharmacogenomics Network for Drug Safety: A model for safety pharmacology. Thyroid 2010;20:681–7.

[123] Carleton B. Demonstrating utility of pharmacogenetics in pediatric populations: methodological considerations. Clin Pharmacol Ther 2010;88:757–9.

[124] Hudson MM, Ness KK, Nolan VG, Armstrong GT, Green DM, Morris EB, et al. Prospective medical assessment of adults surviving childhood cancer: study design, cohort characteristics, and feasibility of the St Jude Lifetime Cohort Study. Pediatr Blood Cancer 2011;56:825–36.

[125] Eichelbaum M, Ingelman-Sundberg M, Evans WE. Pharmacogenomics and individualized drug therapy. Ann Rev Med 2006;57:119–37.

Cancer Genomics

Biomarker Discovery and Development through Genomics

Janet E. Dancey

Ontario Institute for Cancer Research, Toronto, ON; NCIC Clinical Trials Group, Queen's University, Kingston, ON, Canada

Contents

Cancer Genomics. DOI: http://dx.doi.org/10.1016/B978-0-12-396967-5.00007-4

Key Concepts

- Molecular diagnostics are increasingly important in clinical research and management to stratify or identify molecularly profiled patient cohorts for targeted therapies, to modify the dose of a therapeutic agent, and to assess early response to therapy or monitor patients
- Key considerations for clinical biomarker studies include study design, pre-planned hypotheses, patient and specimen characteristics, assay methods, bioinformatics and statistical analysis methods
- To establish clinical validity and clinical utility a biomarker must be capable of being reproducibly quantified in independent laboratories by independent personnel so that there is confidence in the results that are to guide clinical decisions

USES OF BIOMARKERS IN CANCER RESEARCH AND CANCER CARE

Biomarkers are objectively measurable indicators of biological states [1]. In healthcare, biomarkers can provide information on the presence and classification of disease, or susceptibility to disease in an individual, or predict or monitor patient response to therapeutic interventions. The evaluation of new biomarkers in cancer medicine is directed toward improving diagnosis or treatment, and thus improving health outcomes and reducing the social and economic impact of the disease.

The evaluation of biomarkers has accelerated due to advances in genomics research and associated technologies. Such research has resulted in improvements in our understanding of the genetic basis of cancer, the identification of individual differences in outcomes from disease and response to treatments, and allowed for tailoring of diagnostic testing, treatment and monitoring of individual patients [2]. Although there are many possible reasons to evaluate biomarkers, including diagnosis and monitoring of disease, much of cancer biomarker research and development is focused on the identification of prognostic and predictive markers. Markers that are associated with survival or other clinical endpoints independently of any specific treatment are classified as prognostic markers [3]. Markers that are associated with clinical benefit but also correlate the effectiveness of a particular treatment are designated predictive markers (also known as effect modifier biomarkers) and thus can be used to select patients for therapy [3,4]. Certain biomarkers can be both prognostic and predictive. For example, the amplification of human epidermal growth factor receptor-2 (HER2) in breast cancer is associated with poor prognosis with conventional treatment but predicts for benefit from the HER2 monoclonal antibody trastuzumab [5,6]. Biomarkers may also predict risk of harm as pharmacogenomics markers are used to identify the risk of organ-based toxicities or altered metabolism and/or responses to therapies [3,4]. Biomarkers that correlate with patient outcomes and response to treatment are of particular interest to improve the management of cancer patients.

Despite thousands of biomarker publications, there are relatively few biomarkers that have been taken up by the clinical community to guide treatment decisions [3,7]. Potentially useful biomarkers address relevant clinical questions and there is confidence in the evidence that patients do better based on using the marker to inform management decisions. Such confidence arises when there is a strong association between marker, treatment and outcome. Such associations are demonstrated through research studies that are first and foremost reproducible, being based on appropriate clinical study designs [8]. However, most published biomarker studies share common deficits that erode reproducibility, including relatively small sample sizes; the assessment of relatively few markers; and may use samples, assays, reagents and analytical approaches that introduce sources of variability that result in failure of initial results to be confirmed in subsequent clinical studies [9].

An "omics"-based test uses multiple measurements of nucleic acids, proteins or metabolites and a computational model employed as the clinical test [10]. As multiple components can be measured simultaneously using these tests, many hope that the promise that "omics" science holds for medicine will be the development of better biomarkers for patient management. These multianalyte profiles may better correlate with individual patient's outcomes. However, the development of "omic" technologies magnifies the potential weaknesses of biomarker studies. Specifically, the complexity of these technologies and the resulting high-dimensional data require rigorous statistical, bioinformatics, laboratory, and clinical procedures to develop and validate these tests and evaluate their clinical usefulness.

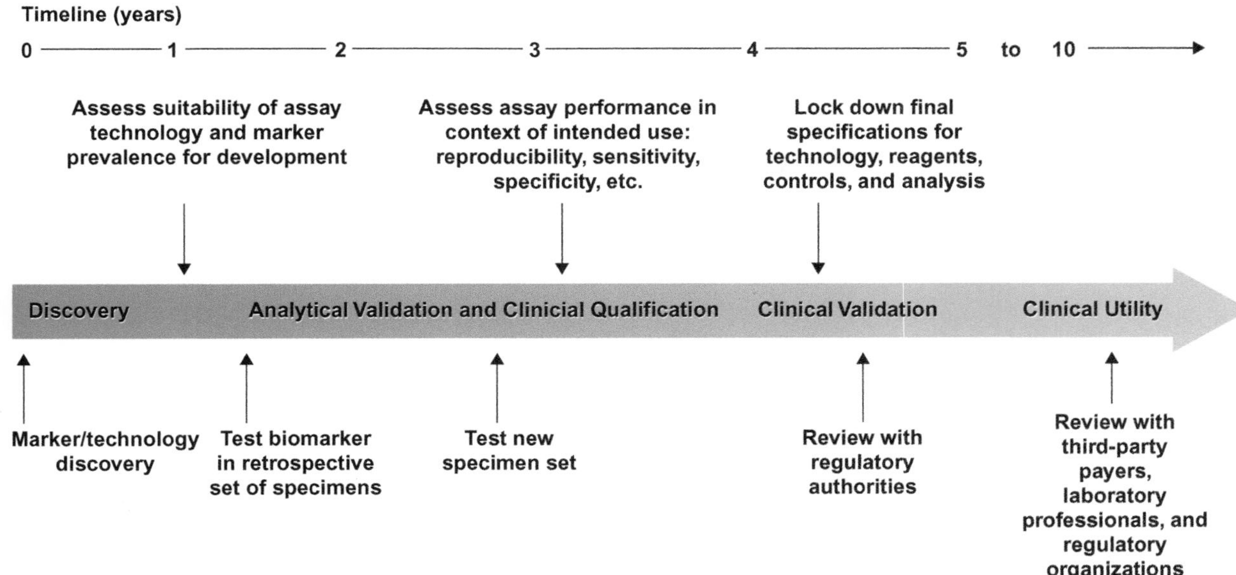

FIGURE 7.1 Overview of biomarker discovery and development.

The generation of robust and meaningful data is key to successful biomarker studies. The development of a prognostic or predictive assay that can guide treatment for cancer patients is usually an iterative process and is summarized in Figure 7.1. A process that begins with the discovery of a molecular characteristic or signature and a preliminary correlation with clinical outcome [11]. Subsequent steps involve optimization of the assay (including technical or analytical validation of the assay), clinical validation or evidence that the result of the analytically validated assay correlates with the clinical outcome of interest, and assessment of clinical utility for the intended use, i.e. the use of the assay results in a better outcome than standard methods or treatment. The further along the spectrum towards clinical uptake the biomarker is positioned, the greater the need to optimize and validate the biomarker assays [12]. Table 7.1 outlines key differences between the early and late phases of biomarker development.

In this chapter, we review the key considerations required to design and conduct biomarker studies that will yield results that can inform future clinical research and clinical practice with confidence. Key to the multistep process is the smooth migration from discovery platform, through verification studies and clinical assay development, to final clinical diagnostics platforms required for clinical trials and clinical application. Therefore, we will provide guidance in the necessary steps to navigate between these stages of biomarker discovery and implementation.

BIOMARKER DISCOVERY AND QUALIFICATION

Biomarker discovery is the phase of research during which the differential expression of markers between disease states is first identified. Discovery research often involves the use of advanced technologies to search for nucleic acid or protein changes that may be useful targets for therapy or correlate with therapeutic response and/or overall patient outcome. The product of the discovery phase is generally the list of markers that are found to be differentially expressed within patients with divergent clinical outcomes [13,14]. Often results derived from omic discovery efforts are referred to as "candidate biomarkers" because of the inherent false-discovery rates that arise from analyses of high dimensional data [13,14]. Candidate biomarkers identified from the discovery process can be added to candidates drawn from the literature, from other discovery approaches or sources. At the end of the discovery phase, there should be reasonable certainty that the correlation between the markers and outcomes is "true" and not due to bias or random error associated with the sample set, technology or analysis.

Most discovery efforts are conducted with biological materials selected to maximize the detection of meaningful correlations between biomarkers and clinical outcomes of interest while minimizing the sample number required for analysis. The samples that are employed in a study should reflect the patient population of interest, be of sufficient quality, quantity and number to be evaluated

TABLE 7.1 Differences between Biomarker Discovery and Clinical Development Phases

	Discovery	Clinical Development
Preanalytic		
Samples	Fresh-frozen	FFPE
	Samples of convenience	Samples of convenience
	(bias by site)	(bias by trial or locale)
Analysis		
Approach	Usually unbiased	Focused on analyte in tissue/matrix
Power	To detect variant	To reduce false negatives/false positives
Runs	Batched	Individual tests
Sensitivity/specificity	May have relatively high false discovery rate	Requires high sensitivity/specificity
Risk to patient	Minimal	High as treatment may be determined based on test result
Postanalytic		
	Not regulated	Health Authorities regulate laboratory and device

Adapted from [3]

successfully with the technology of interest. As there are numerous preanalytical factors that may confound analysis and interpretation [15,16], the reliability of the data derived from these various technology platforms should be assessed to determine the impact of sample quality and consistency of sample acquisition and preparation, as the majority of methodological errors arise due to these factors [16–19]. The highest quality samples can be found from biobanks that adhere to internationally recognized biobanking standards. Information on biobanks can be found from a number of resources, some of which are listed in Table 7.2.

Ideal samples are those that come from cases and controls that differ with respect to the condition of interest, have been collected and stored in a standardized fashion, are of sufficient quantity and quality to be evaluated using the proposed assay technology, and are associated with relevant biological and clinical data. In a discovery effort for a diagnostic marker for breast cancer, for example, the cases would be women with confirmed breast cancer, and the controls would be confirmed to be free of breast cancer, but would otherwise be identical to the cases (e.g. with respect to age, menopausal status, ethnicity, which are all related to the risk of development of breast cancer). To discover a prognostic marker to identify patients with early breast cancer destined to relapse, the cases are women initially diagnosed with early breast cancer with disease relapse and the controls are those without relapse. However, case and control samples should be matched for known prognostic factors such as stage, grade, and hormone receptor status and there should be complete clinical outcome data. Thus, the sample set should reflect the intended use of the biomarker and be associated with any additional and relevant data that might help determine whether candidate biomarkers demonstrate an improvement over current biomarkers.

The advantage of using omic discovery technologies is that they can concurrently assess multiple biomarkers of potential relevance. Because thousands of candidate biomarkers may be generated, and because of the extensive variability of host and disease factors, ideally, biomarkers would be assayed across thousands of samples, to minimize false positive and negative correlations that may occur through error, bias or overfitting. Efforts to develop novel biomarkers must balance the need for in-depth and comprehensive sample analysis and the need for large sample numbers to generate statistically and clinically meaningful correlations. In addition, to facilitate translation to clinical laboratories which do not currently use omic discovery platforms, biomarkers will eventually need to be measured using assays and samples commonly used in clinical laboratories. To resolve these conflicting needs, it is useful to use materials and technologies in the discovery phase that will identify candidate biomarkers, then, in subsequent phases, develop methods and samples that more closely adhere to a current clinical laboratory test [13,14].

As omic technology platforms generate large data sets, systems biology-based analyses are required to generate signatures that correlate with treatment outcome [14]. As mentioned previously, for the specific

TABLE 7.2 Useful Resources

Biobanking resources	Biobanking and Biomolecular Resources Research Infrastructure (BBMRI) provides a biobank registry for researchers. Registered biobanks are added to a web-catalog www.bbmri.eu Biorepositories and Biospecimens Research Branch National Cancer Institute has established Best Practices and a Biospecimen Science Initiative to determine preanalytical and analytical factors that impact assay performance and results http://biospecimens.cancer.gov/default.asp Biobank Resource Centre is an online resource for biobanks across Canada http://www.biobanking.ca/ UK Human Tissue Authority (HTA) provides training on the regulations and licensing process of research biobanks in the UK www.hta.gov.uk International Society for Biospecimen and Environmental Repositories publications on best practices for biobanks http://www.isber.org/wg/bs/
Regulatory guidance documents for test development	The FDA has a website that provides information on the process and regulations surrounding approval and marketing of devices http://www.fda.gov/downloads/Drugs/GuidanceComplianceRegulatoryInformation/Guidances/UCM332181.pdf http://www.fda.gov/downloads/drugs/scienceresearch/researchareas/pharmacogenetics/ucm116689.pdf http://www.fda.gov/medicaldevices/deviceregulationandguidance/guidancedocuments/ucm262292.htm National Institute of Standards and Technology (NIST) is the federal technology agency that works with industry to develop and apply technology, measurements, and standards. NIST has catalogs of reference standards and data http://srmcatalog.nist.gov Center for Disease Control (CDC) – Laboratory Practice Program – http://www.cdc.gov/mmwr/pdf/rr/rr5806.pdf CDC Evaluation of Genomic Applications in Practice and Prevention (EGAPP) Working Group was established to support the development of a systematic process for assessing the available evidence regarding the validity and utility of rapidly emerging genetic tests for clinical practice http://www.egappreviews.org/ The Cancer Diagnosis Program created templates to document an assay's analytical performance that are posted on the CDP website http://cdp.cancer.gov
Intellectual property and patent Issues	Gene Cards and GeneIP identify intellectual property related to specific genes that may be the basis for a molecular diagnostic. http://www.genecards.org http://geneip.xennexinc.com
Assessment of clinical utility	Agency for Healthcare Research and Quality (AHRQ) in collaboration with Evidence Practice Centers and the Journal of General Internal Medicine developed and published a Methods Guide for Medical Test Reviews http://effectivehealthcare.ahrq.gov/index.cfm/search-for-guides-reviews-andreports/?pageaction = displayproduct&productid = 1088 National Comprehensive Cancer Network (NCCN) have developed white papers for molecular testing and clinical utility of tumor markers in oncology Molecular Testing White Paper; NCCN Molecular Testing White Paper: Effectiveness, Efficiency, and Reimbursement [20] NCCN Task Force Report: Evaluating the Clinical Utility of Tumor Markers in Oncology http://www.nccn.org/JNCCN/supplements/PDF/TumorMarkers_Task_Force_Report.full.pdf
Reporting biomarker studies	REporting recommendations for tumor MARKer prognostic studies (REMARK) prognostic biomarker reporting guidance for publications [11] Standards for Reporting of Diagnostic Accuracy (STARD) initiative is to improve the accuracy and completeness of reporting of studies of diagnostic accuracy, to allow readers to assess the potential for bias in the study (internal validity) and to evaluate its generalizability (external validity) Biospecimen reporting for improved study quality (BRISQ) recommendations delineate the information on handling of biospecimens to be reported in a thorough, accurate, and standardized manner to improve the quality of research utilizing human tissues [21]

technology employed and biological material analyzed, it is important to know the strengths, weaknesses, and robustness of the readout of each platform and sample format. Specific strategies for data collection, annotation, storage, statistical analysis, and interpretation should be established first to guide such studies. For example, performance controls for both the instruments and the assay need to be performed to assure reproducibility of the data. Key parameters include the achievable depth and extent of genomic, transcriptome or proteome coverage. In addition, transformation of high-dimensional sequencing data into identifiable genes, transcripts and proteins requires considerable bioinformatics analyses. Currently, software packages automate analysis of nucleic acid or peptide sequences as well as the database searches required to yield protein, transcript or gene identifications. However, different software algorithms yield different results. Thus, to ensure accurate results, quality control measures are important at each step.

After generating a list of candidate biomarkers, it is important to evaluate the reliability of identified biomarkers to enhance confidence in marker–referred to as outcome correlations. Identification of important genes, transcripts or proteins of interest should be verified by using an alternative method. In many cases, omics technologies will provide a set of biomarker candidates that can be verified by a robust conventional analytical method [14]. For example, a mutation within a gene can be verified by Sanger sequencing, copy number variation by *in situ* hybridization, and protein expression by immunohistochemistry. Only a subset of candidate markers may be verified depending on the rigor of the discovery methods and the number of candidates identified. The subset of candidate biomarkers is usually based on existing biological knowledge of the function and relevance of the candidate biomarker in the disease of interest, technical performance of the assay and the availability of reagents for biomarker verification. In the end, only the protein or nucleic acid variants validated with high statistical confidence should be further evaluated.

Candidate biomarkers also need to be qualified. Qualification refers to the process of confirming the consistency of association between marker and disease or outcome of interest. The first step in the qualification process is to confirm that the gene candidate exhibiting differential expression is also identified using alternative, targeted assay methods. For example, if overexpression of a gene product is identified by gene expression array analysis to correlate with an outcome of interest, the subsequent demonstration that the correlation with outcome persists, when overexpression is demonstrated through a different assay method, increases the likelihood that the correlation is real and not limited to a specific assay system. In the second step of the qualification process, the differential expression of candidate biomarkers and outcomes is confirmed in a second sample set. The second sample set should have a larger number of cases and controls that reflect the variation in the population where the biomarker testing is intended to be used. The replication of results within the discovery samples and datasets enhances confidence in the reliability of the initial analytical methods and correlations. Biomarker–clinical outcome correlations replicated in data generated from independent samples support the robustness and potential generalizability of the observed correlations. This is a key step in moving candidate markers to the next phase of development (see Figure 7.1). These latter efforts are necessary not only to confirm the initial findings but also to establish the prevalence of the molecular marker in a sample set from the intended clinical-use population.

Important Disclosures during Biomarker Qualification

When candidate omics-based tests from the discovery phase are intended for further clinical development, several criteria should be disclosed to enable independent assessment and verification of the findings. The Institute of Medicine (IOM) 2012 report on omic technologies [10] details a series of recommendations:

a. Candidate omics-based tests should be confirmed using an independent set of samples, not used in generation of the computational model and, when feasible, blinded to any outcome or other phenotypic data until after the computational procedures have been locked down and the candidate omics-based test has been applied to the samples; b. Data and metadata used for development of the candidate omics-based test should be made available in an independently managed database (such as dbGaP) in standard format; c. Computer code and fully specified computational procedures used for development of the candidate omics-based test should be made sustainably available; and d. A candidate omics-based test should be defined precisely, including the molecular measurements, the computational procedures, and the intended clinical use of the test, in anticipation of the test validation phase.

Such disclosures allow for independent verification which supports the reliability of the results and the potential value of the marker test.

ASSAY VALIDATION

Successful clinical validation of a biomarker requires both a good assay and a robust validation strategy [4,22]. To provide meaningful and interpretable information, assays should: (1) measure what they claim to measure, (2) be reproducible, and (3) produce results that are statistically meaningful [7]. In addition, it is not sufficient to validate biomarkers on unique experimental platforms beyond the means or expertise of a routine clinical molecular diagnostics laboratory [23]. To be useful, a biomarker assay should be compatible with general clinical laboratory practices. In this phase of biomarker development, the research grade assay's analytical performance characteristics are evaluated and optimized so that they can be tested in samples that reflect the targeted population [3,14,24]. The end goal of this phase is to optimize the biomarker assay, streamline processing and enhance throughput to ensure a sound platform for subsequent clinical validation and eventual uptake of biomarker testing in clinical practice [25,26]. Therefore, fully evaluated and validated biomarkers and assays require the research assay is refined to meet the standards required for clinical tests. A biomarker development plan should consider all aspects of this process from discovery to intended clinical use and commercialization. A guidance document for the validation of bioanalytical assays is available through the United States Food and Drug Administration (FDA) website http://www.fda.gov/cder/guidance/.

Assay analytical validation involves a systematic evaluation of the processes required to demonstrate that a particular technique is reliable for its intended purpose [24]. Conceptually, analytical validation involves a series of assessments to determine possible sources of variability in results that may occur due to materials, methods, technologies, laboratory settings and other sources. The steps taken to optimize the assay should focus on reducing variability to enhance the ability to identify the true signal from noise [27]. Once the limitations of performance for the assay are determined, these can be evaluated for the suitability of the assay for clinical validation studies and as a clinical laboratory test. For established biomarkers, the performance characteristic limits are determined by several factors, including clinical validation studies, clinical laboratory requirements, published professional recommendations, proficiency testing findings or goals set by regulatory agencies [14]. For novel biomarkers such as those generated from omic discovery technologies, such performance specifications may not be available [25]. However, comparable analytical performance to an established similar assay system should be expected.

The Phases of Assay Analytical Validation

Five phases of assay analytical validation have been described [14,28,29]. The selection of the candidate assay occurs in the first stage. During stage 2, the goal is to assemble reagents and components and write the method validation plan. Stage 3 is the experimental phase of performance verification that will lead to a standard operating procedure. In stage 4, there is further assessment of the assay in the clinical context and identification of patient sampling issues, such as collection, storage and stability that may impact on results. Stage 5 is where the assay enters clinical laboratory use, and quality control (QC) monitoring, proficiency testing and batch-to-batch QC issues are fully explored [12,30].

Depending on the assay, it may not be possible to validate all possible analytical parameters, particularly for omic tests. However, different omic technologies have variable error rates [31], thus, a validation plan for evaluating and defining acceptable variability for the assays is required. The following parameters should be assessed and addressed with statistical significance indicated when feasible [28,29,32]:

1 Specificity and selectivity reflect the ability of the assay to differentiate and quantify the test article in the context of the bioassay components. Generation of control materials is a major challenge. For clinical development, control materials should be manufactured in sufficient quantities to provide a sustained source.
2 Accuracy reflects closeness of the test results to the true value. Often this is very difficult to ascertain for biological assays as it requires an independent true measure of this variable.
3 Imprecision (random error) for the analytical method is characterized by repeatedly measuring control materials, and these data are used to calculate reference ranges for control materials.
4 Precision (intra- and inter-assay) reflects how close values are upon replicate measurement, performed either within the same assay or in independent assays.
5 Calibration/standard curve (upper and lower limits of quantification) is the range of the standard curve that can be used to quantify test values.
6 Detection limit is the lowest value that can be detected above the established negative or background value.
7 Robustness is how well the assay transfers to another laboratory and/or another instrument within the same laboratory.
8 Preanalytical variation is the characterization and control of the preanalytical variation for the analyte to be measured correctly [14]. Standardization of sample collection procedures and appreciation of the limitations of the assay in various physiologic and pathologic conditions allow this variability to be minimized.

Presently, there are no regulatory or professional guidelines that stipulate the minimum requirements for analytical validation of assays or quality certification of laboratories that perform biomarker assays if there is no intent to use results to guide patient management or to support a regulatory approval of the laboratory test or diagnostic device. However, quality control is important for research discovery and is critical for moving biomarkers through a development pipeline aimed toward clinical application [28,29]. Good Laboratory Practice (GLP) guidelines for non-clinical laboratories include some general assay and laboratory quality control measures such as: (1) the development of standard operating procedures (SOPs) for all laboratory procedures and processes, including not only assay methodologies but also sample receipt, processing, and storage, personnel training, equipment maintenance/calibration, data management, and repository activities [33]; (2) the investment in the time and funds to develop qualified/validated assays; (3) the establishment of reference standards whenever possible and creating master lots and/or cell banks for all standards [34]; and (4) the establishment of statistically supported quality parameters for the reference cell lines, which can be as pass/fail criteria for the assay performance [28,29].

CLINICAL VALIDITY

Clinical validation refers to how well the assay result relates to the clinical outcome of interest [35]. To establish clinical validity, investigators should show that a positive test result is associated with a particular clinical endpoint or event, and a negative test result is not associated with that outcome in the population the test is intended to be used. For a predictive marker, the clinical endpoint may be related to either survival (i.e. overall survival or a progression-free interval) or response to therapy. To show clinical validity, the assay must characterize the biomarkers in specimens collected in routine clinical practice from patients within the clinical context and intended use of the biomarker. These results can then be translated into clinical sensitivity and specificity, and may be used to define statistical cutoffs by means of receiver operating characteristic (ROC) curves or other analyses [36].

To report outcomes of biomarker defined patient groups, clinical researchers generally calculate and report survival curves, hazard ratios with confidence intervals and *p*-values. Although these are important statistics for showing differences in outcomes within a study population, more relevant analyses for determining clinical validity of tests focus on diagnostic accuracy and diagnostic predictability. Diagnostic accuracy establishes how well the test discriminates between those patients with and without the parameter or outcome of interest. Diagnostic predictability establishes the ability of the test to predict the presence or absence of the parameter, or clinical outcome of interest for a given test result. A study may report impressive hazard ratios, *p*-values and narrow confidence intervals, but the biomarker may not be clinically useful for patient management decisions due to relatively poor diagnostic accuracy and/or predictability compared to available tests or management approaches [7].

Diagnostic accuracy is determined by calculating the test's sensitivity, specificity, likelihood ratio and ROC curve [14,36]. Sensitivity and specificity are the proportion of subjects with the clinical parameter that are correctly classified as such by the examined test. For example, immunohistochemistry (IHC) and fluorescence *in situ* hybridization (FISH) are established methods for detecting HER2 overexpression in formalin-fixed, paraffin wax-embedded tissue in diagnostic laboratories. There is a wide array of IHC antibodies with varying sensitivities available for HER2 testing, FDA-approved HercepTest IHC kit and PathVysion FISH kit are among the most widely used assay systems. A study comparing the sensitivity (and specificity) of the CBE356 antibody and HercepTest to predict gene amplification by PathVysion FISH and the prognostic value of each test in terms of overall survival reported sensitivity (and specificity) for CBE356 and HercepTest were 94% (89%) and 91% (66%), respectively. All three tests strongly correlated with prognosis in patients with breast cancer [37]. The likelihood ratio and ROC curve are derived from sensitivity and specificity values [38]. The likelihood ratio denotes the value of the test for increasing certainty about the clinical parameter of interest. For a predictive marker, it is the likelihood that a given test result would be expected in a patient that will benefit from a particular treatment compared to the likelihood that that same result would be expected in a patient that will not benefit from the treatment. The ROC curve enables the comparison of the diagnostic accuracy of tests and the determination of appropriate cut-points, depending on the clinical use of the test.

Diagnostic predictability is determined by calculating the positive and negative predictive values. The positive predictive value is the proportion of patients with positive test results that have the parameter or outcome of interest, and the negative predictive value defines the proportion of patients with a negative test that do not have the parameter or outcome. The predictive values of a test vary with the prevalence of the clinical parameter in the population examined.

Retrospective and Prospective Clinical Validation

Clinical validation studies may be conducted retrospectively or prospectively and on samples collected from clinical trials, tissue banks or other sources [7,39]. For early studies assessing for clinical validity, investigators may use a research grade assay and "samples of convenience"; for example, a collection of samples from the tumor of interest that may or may not have come from the specific patient population of interest, or been collected in a standardized fashion. Such samples may have biases that affect interpretation of the clinical value of the assay, including biases in specimen collection methods, differences in treatment, concomitant medications, co-morbidities, and means and timing of tumor assessments. In addition, the assays used may not have optimal performance characteristics. In contrast to these early studies, a formal clinical validation study requires a defined validation protocol with a specified analysis plan [28,29]. As part of that plan, the biomarker and assay should be specified for omic tests and the bioinformatics analysis validated and standardized [10]. Samples collected from a clinical trial are preferred as the patient population, treatment and sample collection, as well as the collection of the required clinical, sample and laboratory data, can be assured. It is possible to validate clinically an assay in a prospective–retrospective fashion if an adequate number of samples can be obtained from relevant clinical trials that have already been completed or other sources and for which the outcome data are available [7]. Importantly, the assay must be performed blinded with respect to patient outcome [7,10]. If a prospective clinical trial is required for clinical validation and the assay is considered investigational, an Investigational Device Exemption (IDE) should be filed with the FDA prior to trial commencement [40].

Intellectual property issues surrounding the use of a biological marker also need to be considered during the development of a biomarker test. For example, patents may exist that restrict the use of a biological marker for clinical management, and/or prevent the technology from being adapted to a clinical setting due to costs associated with licensing. The patent holders may require laboratories to purchase a sublicense prior to testing for the gene, transcript or protein in question, or can block laboratories from performing the assay, and mandate that all assays be performed in the company's clinical laboratories [41]. Therefore, it is imperative that investigators, and their institutions' technology-transfer offices, determine whether a potential patent holder for a molecular diagnostic has been granted licenses that might block the use of that diagnostic in clinical trials.

CLINICAL UTILITY

Clinical utility is defined as evidence of improved clinical outcomes, and a test's usefulness with regard to added value to patient management and decision making compared with the standard of care [42,43]. Clinical utility also refers to how useful the assay is for directing treatment. A test might have clinical validity by virtue of distinguishing the outcome of two subgroups within a population, perhaps with a high degree of statistical significance and reasonable diagnostic accuracy but, if this difference is not large and/or does not add to current decision making, the test will not be used. In this case, the test has clinical validity, but not clinical utility. Clinical utility can be determined with a prospective clinical trial, or with more than one prospective–retrospective trial or research study [39]. For the assessment of clinical utility, the assay should be both analytically and clinically validated. The demonstration of clinical utility for an omics-based test is fundamentally similar to that of biomarker tests, although one must pay special attention to the challenges specific to omics-based test development in the discovery and test validation phases, as noted in previous sections.

The process of gathering evidence to support clinical utility begins before a test is introduced into clinical practice, and continues following its clinical uptake. It is important to note that regulatory agencies with oversight for review and approval of devices such as the FDA, and certification of clinical laboratories such as Clinical Laboratory Improvements Amendments (CLIA), do not necessarily require evidence of clinical utility for their evaluations of clinical tests. FDA review of a biomarker test has focused principally on analytical and clinical validity, but not on demonstration of clinical utility [10]. CLIA laboratory certification ensures the quality of testing within the clinical laboratory. Therefore, FDA approval or clearance or CLIA certification does not necessarily imply that the use of the test results improves patient outcomes. It can take many years after a test is introduced into clinical practice to demonstrate clinical utility of the test. The level of evidence necessary, prior to clinical use, depends on disease setting, and varies among clinicians, patients, third party payers and policy makers [39,42,43].

Guidelines for Assessing Clinical Utility

Regulatory agencies do not necessarily approve tests based on clinical utility; however, guidelines committees and third party technical assessment panels perform opinion- or evidence-based analyses of available data to make recommendations regarding whether selected tumor biomarker tests should be used clinically, and whether they should be supported by reimbursement. Various methods have been proposed to establish the clinical utility of cancer biomarkers [38,44–47]. These have been used in technology assessment programs by organizations such as the National Institute for Health and Clinical Excellence (NICE) (http://www.nice.org.uk) in the UK, and the Center for Medical Technology Policy (www.cmtpnet.org), the Agency for Healthcare Research and Quality (www.ahrq.gov), the Centers for Disease

Control and Prevention's Evaluation of Genomic Applications in Practice and Prevention (eGAPP; www.egappreviews.org/), and the Blue Cross Blue Shield Technology Evaluation Center (BCBS TEC; http://www.bcbs.com/blueresources/tec/) in the USA. These organizations follow predefined criteria in an effort to assess the utility of a test along with its associated companion therapeutic(s) in an unbiased way. For example, the BCBS TEC assesses the utility of treatments and diagnostics on the following parameters: (1) the technology must be approved by the appropriate governmental regulatory bodies; (2) the scientific evidence must enable conclusions about the effect of the technology on health outcomes; (3) the technology must improve the net health outcome; (4) the technology must be as beneficial as established alternatives; and (5) the improvement must be expected to be attainable outside clinical trial research investigations. Reports provided by these evaluation committees are important as these documents delineate the concerns of the payers providing reimbursements for both the diagnostic test and the companion treatment.

INCORPORATION OF BIOMARKERS INTO CLINICAL TRIAL DESIGN

Much of clinical research generally involves clinical trials (phases 0–IV) of cancer treatments, which often incorporate biomarkers. Descriptions of the phases of cancer treatment trials and the types of biomarkers are summarized in Table 7.3. Over recent years, the role of biomarkers in clinical trials has expanded to include research tools to demonstrate mechanism of action to surrogate endpoints of clinical efficacy. Biomarkers may enable selection of patients most likely to respond to drugs, monitoring of treatment efficacy or identify early signs of toxicity [2,3,9].

Designing Clinical Trials to Determine Clinical Utility of Biomarkers

Although phases and corresponding designs of cancer treatment trials are well defined, clinical trial methods for assessing clinical utility of biomarkers are continually evolving [48]. There are three pathways to generate evidence of clinical utility: (1) prospective–retrospective studies using archived specimens from previously conducted clinical trials or research studies that address the intended use of the omics-based test; (2) prospective clinical trials that directly address the utility of the omics-based test, where either the test does not direct patient management; or (3) the test does direct patient management [7,10]. For assays intended to direct patient management in clinical trials, assays and methods must comply with regulations and professional standards covering clinical laboratory testing.

The National Cancer Institute has proposed a classification of integral, integrated and research biomarkers to reflect their role in clinical trials, the regulatory requirements and degree of analytical and clinical validation required to conduct the study [8]. Integral markers are used to determine patient eligibility; to assign patients to treatment; for risk stratification if such stratification leads

TABLE 7.3 Phases of Cancer Clinical Trials and Types of Biomarkers

Phase	Description	Examples of Biomarkers
0	Trials in which pharmacokinetic and pharmacodynamic effects of a drug are evaluated to enable comparisons with the same compound or agent in preclinical animal models. The dosing in phase 0 trials is low enough that the biological effect of an agent on its intended target can be detected without risk of serious adverse events	Modulation of protein target in surrogate tissues such as skin or blood
I	Trials are designed to determine the maximum tolerated dose of an agent with respect to its safety and toxicity profile without establishing efficacy in patients within a particular type of cancer	Demonstration of mechanism of drug action through assessment of proteins in tumor, functional imaging or toxicity
II	Trials in which an agent is dosed generally at or close to the maximum tolerated dose to ascertain the anticancer activity of the agent in different malignant diseases	Predictive markers of drug activity, assessment of tumor burden and response by measuring tumor size
III	Large, randomized trials that are intended to confirm the efficacy of an agent in a particular disease context compared with the standard-of-care therapy	Predictive markers that correlate with benefit or risk of toxicity
IV	Post-market trials that assess drug safety in large population sets and potential efficacy in new indications	Predictive and/or prognostic markers

to different treatments; for risk classification. Such trials are designed to show clinical utility and, if successful, support uptake into clinical practice. Because the assay is performed for medical decision making, a Clinical Laboratory Improvement Amendments-certified laboratory is generally required for assay and an IDE may be required by the FDA. Many countries have similar regulatory requirements. Integrated markers are performed on all or a statistical subset of patients, but are not used for medical decision making. These also require a high degree of analytical and clinical validation to support their incorporation into clinical trials. Research markers include all other assays; often referred to as correlative research.

The decision to move an integral prognostic or predictive biomarker forward into a clinical trial requires prior evidence of clinical validity. Even though clinical need should drive the decision to develop a biomarker assay into a test suitable for use in medical management, results from studies supporting the ability of the prognostic or predictive marker to distinguish subgroups of patients with different outcomes are still necessary [7]. The magnitude of the difference in treatment effect expected with use of the marker should be sufficient to support using different treatments for marker-positive and -negative patients. Analytical and clinical validation of the marker assay should be completed prior to clinical trial initiation to ensure that the assay is suitable for use throughout the trial. Assays should not change during the clinical study or clinical trial without a protocol amendment and discussion with the appropriate regulatory agencies. The clinical protocol for the prospective evaluation of the biomarker must include the important aspects related to use of the assay, including the impact on patient eligibility, the statistical design (including implications for power and control of type I error if subgroup analyses are planned) and monitoring for trial futility [7]. In addition, the protocol and/or accompanying laboratory manual should describe the logistical details such as sample preparation, timing, preparation, shipping and transmission of assay results. The magnitude of the benefit obtained from using the marker/assay should be stated prior to analysis, such that patients and clinicians would find the assay useful for deciding between treatment options. The magnitude of difference required to change medical management is also context dependent and will vary based on the clinical situation, the availability, effectiveness and toxicities of therapies and the resource and financial impact of using the biomarker [7,20,49].

Various trial designs have been proposed for assessing clinical utility [48]. These include marker-stratified designs and Bayesian adaptive designs [48,50–53]. In marker-stratified designs, the marker is measured in all of the patients. The patients are then stratified into marker-negative or -positive groups, and each group is randomized to receive either the experimental or the standard therapy. The strength of this design is the possibility for retrospective assessment of a different statistical endpoint or the use of different biomarkers, for example, enrichment designs that include only a defined marker group can be an efficient method for assessing a therapy in a biomarker-positive subgroup, such designs do not allow a complete assessment of the performance of the marker (i.e. the drug's effect on biomarker-negative patients). Bayesian designs start with an initial model of how the assay will perform relative to the primary outcome. As the trial progresses and data are obtained, the model is updated. Although such designs are being used in early drug development clinical trials, their efficiency relative to other designs is still under evaluation. Full descriptions of tumor biomarker trial designs have been published elsewhere, most of which are suitable for the evaluation of new omics-based tests [51,54,55].

Ultimately, the clinical use of the assay requires that it be performed in an appropriately accredited clinical laboratory. If the biomarker assay is investigational, application for its use in the clinical trial should be made to the regulatory agency. The IOM report on omic technologies [10] recommends the following steps for tests used in clinical trials conducted in the USA:

1 The candidate omics-based test and its intended use should be discussed with the FDA prior to initiation of validation studies.
2 Test validation should be performed in a CLIA-certified clinical laboratory, beginning with a defined candidate omics-based test from the discovery phase.
3 The CLIA-certified laboratory should design, optimize, validate, and implement the omics-based test under current clinical laboratory standards.
4 If the omics-based test will be performed in more than one CLIA-certified laboratory for a clinical trial, analytical validation and CLIA requirements for the same omics-based test should be met by each laboratory, working with the primary laboratory.

CONCLUSION

For biomarker studies to provide meaningful and interpretable information, it is critical that they be conducted in a manner that is as scientifically sound and methodologically rigorous as possible to minimize waste of resources and to generate creditable results. Regardless of methods and technologies used, key criteria for successful biomarker discovery efforts are: (1) a specific, well-defined, and relevant question that addresses a clinical need; (2) the assembly of high quality samples from clinically well-defined individuals; (3) a study sample size that provides sufficient statistical power to identify gene or protein variations that address the question under investigation; and (4) a defined method of data analysis conducted with adequate statistical or bioinformatics rigor. Researchers should also understand the potential constraints and limitations imposed on specimens, as well as the technology and the study design to generate datasets that are statistically meaningful. As the quality of clinical decisions can depend on the quality of biomarker data, appropriate analytical validation of biomarker assays, early in the development phase of a biomarker intended for clinical use, is essential. The extent of assay validation is often determined by the proposed purpose of the biomarker. Input of laboratory scientists, clinicians, pathologists and biostatisticians may be invaluable at all stages of the research; beginning with defining the question, sample requirements, assay development, and statistical analysis. A biomarker development plan aimed at developing a clinical test must consider trade-offs of cost, time, size of the study, quality of results, availability of technology, as well as consider the potential for return on the research and development investment. A constantly growing number of research groups entering the "omics" arena is a positive trend but without proper guidance in the discovery phase these efforts may result in clinically inadequate biological markers being advanced at great cost. With in-depth understanding of the issues associated with the biomarker discovery and development processes, including all those involved from the first step of sample selection to the final regulatory and clinical acceptance of the test, such research can result in high-quality data that will contribute to real progress in the adoption of omic technologies and analyses, which ultimately will benefit cancer patients by optimizing their treatment and care.

GLOSSARY

Analytical validation Traditionally, the assessment of a diagnostic test and measurement of its performance characteristics, for example, the range of conditions under which the assay will give reproducible and accurate data. With respect to omics, this term refers to the assessment of a test's ability to measure accurately and reliably the analyte(s) of interest in the clinical laboratory, as well as in specimens representative of the population of interest.

Accuracy The closeness of agreement between the value that is accepted either as a conventional true value or an accepted reference value and the value found experimentally.

Biomarker A biomarker is, according to the US National Institutes of Health, a characteristic that is objectively measured and evaluated as an indicator of normal biological processes, pathogenic processes, or pharmacologic responses to a therapeutic intervention.

Calibration curve A functional relationship between the analyte concentration in the standards (calibrators) and the measured response. The calibration curve is used to estimate the analyte concentration in test samples by dose interpolation.

Calibration standards Samples having a known concentration of analyte that are used in an assay to gauge the relationship between system responses (for example, absorbance units) and concentrations of an analyte.

Clinical laboratory improvements amendments (CLIA) United States federal regulatory standards that apply to all clinical laboratory testing.

Clinical utility Evidence of improved measurable clinical outcomes, usefulness and added value to patient management decision making compared with current management.

Clinical/biological validation The assessment of a test's ability to predict accurately and reliably the clinically defined pathological state or phenotype of interest.

Cross-validation A statistical method for preliminary confirmation of a computational model's performance using a single dataset, by dividing the data into multiple segments, and iteratively fitting the model to all but one segment and then evaluating its performance on the remaining segment.

Diagnostic biomarker A biomarker that identifies the presence or absence of pathological state or phenotype of interest.

Effect modifier A measure that identifies patients most likely to be sensitive or resistant to a specific treatment regimen or agent. An effect modifier is particularly useful when that measure can be used to identify the subgroup of patients for whom treatment will have a clinically meaningfully favorable benefit-to-risk profile.

Genomic biomarkers Biomarkers based on the analysis of DNA (deoxyribonucleic acid) profiles, especially the analysis of SNPs (single nucleotide polymorphisms).

Good clinical practice (GCP) An international quality standard that is provided by International Conference on Harmonization (ICH), an international body that defines standards, which governments can transpose into regulations for clinical trials involving human subjects.

Good laboratory practice (GLP) In the experimental (non-clinical) research arena, good laboratory practice specifically refers to a quality system of management controls for research laboratories and organizations to try to ensure the uniformity, consistency, reliability, reproducibility, quality, and integrity of chemical (including pharmaceuticals) non-clinical safety tests; from physio-chemical properties through acute to chronic toxicity tests.

High-dimensional data Large datasets characterized by the presence of many more predictor variables than observations, such as datasets that result from measurements of hundreds to thousands of molecules in a relatively small number of biological samples. The analysis of such datasets requires appropriate computing power and statistical methods.

Limit of detection The lowest concentration of analyte for which the response can be reliably distinguished from background noise.

Lower limit of quantitation The lowest concentration (amount) of analyte in a test sample that can be determined quantitatively with suitable accuracy (mean bias) and precision.

Metabolomic biomarkers A biomarker based on the analysis of metabolites (metabolites are the intermediates and products of metabolism).

Omics Scientific disciplines comprising study of related sets of biological molecules. Examples of omics disciplines include genomics, transcriptomics, proteomics, metabolomics, and epigenomics.

Omics-based test An assay composed of or derived from many molecular measurements and interpreted by a fully specified computational model to produce a clinically actionable result.

Overfitting Occurs when the model-fitting process unintentionally exploits characteristics of the data that are due to noise, experimental artifacts, or other chance effects that are not shared between datasets, rather than to the underlying biology that is shared between datasets. Overfitting leads to a statistical or computational model that exhibits very good performance on the particular dataset on which it is fit, but poor performance on other datasets. Although not unique to omics research, the chance of overfitting increases when the model has a large number of measurements relative to the number of samples.

Preanalytical variables Aspects of sample collection and handling that need to be standardized and documented prior to test development and use.

Predictive biomarker A biomarker that correlates with the effectiveness of treatment, which is sometimes also referred to as an "effect modifier" of treatment.

Precision The closeness of agreement (degree of scatter) between a series of measurements obtained from multiple sampling of the same homogeneous sample under the prescribed conditions. Precision may be considered at three levels: repeatability, intermediate precision and reproducibility.

Prognostic biomarker A biomarker correlated with a clinical outcome in the setting of natural history or a standard of care regimen; it is a variable used to estimate the risk of or time to clinical outcomes.

Proteomic biomarkers A biomarker based on the analysis of the protein profiles.

Quality control (QC) samples Pre-study validation and in-study samples having a known concentration (nominal) of analyte that are treated as unknowns in an assay. During pre-study validation, QC samples are used to generate information to demonstrate the method is suitable for its intended purpose. During in-study runs, QC values are used as the basis for accepting and rejecting bioanalytical method batches.

Reference material Defined by the National Institute of Standards and Technology as a material that is homogeneous and stable enough to be fit for use in a measurement process. The USP and the National Formulary list certified reference materials that meet the standards of the International Organization for Standardization (ISO) for reference materials in their ISO 17025 and ISO Guide 34 for reference material producers.

Repeatability The precision under the same operating conditions over a short interval of time. Repeatability is also termed as intra-batch or intra-run precision.

Reproducibility Precision of repeated measurements between laboratories and is termed inter-laboratory precision. Usually applies to collaborative studies that involve the standardization of a bioanalytical method across multiple laboratories.

Selectivity The extent to which a bioanalytical method can measure particular analyte(s) in a complex mixture without interference from other components of the mixture.

Sensitivity Analytical sensitivity is the ability of an analytical method to detect small quantities of the measured component. Diagnostic sensitivity is the probability that the individual will test positive when an individual has the disease.

Specificity Analytical specificity is the ability unequivocally to measure the analyte in the presence of other components that may be expected to be present in the biological specimen, including impurities, metabolites and endogenous matrix components. Diagnostic specificity is the probability that the individual will test negative when that individual does not have the disease.

Statistics and bioinformatics validation Verifying that the omics-based test can perform its intended task. Ideally, this involves assuring that the test can accurately predict the clinical outcome of interest in an independent set of samples that were not used in developing the test. Such validation is particularly important as omics tests typically involve computational models whose parameters can be overfit in any single dataset, leading to an overly optimistic sense of the test's accuracy.

Transcriptomics biomarkers A biomarker based on the analysis of RNA expression profiles.

ABBREVIATIONS

AHRQ Agency for Healthcare Research and Quality
BCBS TEC Blue Cross Blue Shield Technology Evaluation Center
CDC Centers for Disease Control and Prevention
EGAPP Evaluation of Genomic Applications in Practice and Prevention
CLIA Clinical Laboratory Improvements Amendment
CMPT Center for Medical Technology Policy
FDA Food and Drug Administration
NICE National Institute for Health and Clinical Excellence
IDE Investigational Device Exemption
GCP Good Clinical Practice
GLP Good Laboratory Practice
ICH International Conference on Harmonization
QC Quality control
ROC Receiver operating control
SOP Standard operating procedures

REFERENCES

[1] Biomarkers Definitions Working Group. Biomarkers and surrogate endpoints: preferred definitions and conceptual framework. Clin Pharmacol Ther 2001;69:89–95.

[2] Tan DS, Thomas GV, Garrett MD, Banerji U, de Bono JS, Kaye SB, et al. Biomarker-driven early clinical trials in oncology: a paradigm shift in drug development. Cancer J 2009;15:406–20.

[3] Williams PM, Lively TG, Jessup JM, Conley BA. Bridging the gap: moving predictive and prognostic assays from research to clinical use. Clin Cancer Res 2012;18:1531–9.

[4] Wagner JA, Williams SA, Webster CJ. Biomarkers and surrogate end points for fit-for-purpose development and regulatory evaluation of new drugs. Clin Pharmacol Ther 2007;81:104–7.

[5] Mass RD, Press MF, Anderson S, Cobleigh MA, Vogel CL, Dybdal N, et al. Evaluation of clinical outcomes according to HER2 detection by fluorescence in situ hybridization in women with metastatic breast cancer treated with trastuzumab. Clin Breast Cancer 2005;6:240–6.

[6] Jacobs TW, Gown AM, Yaziji H, Barnes MJ, Schnitt SJ. Specificity of HercepTest in determining HER-2/neu status of breast cancers using the United States Food and Drug Administration-approved scoring system. J Clin Oncol 1999;17:1983–7.

[7] McShane LM, Hayes DF. Publication of tumor marker research results: the necessity for complete and transparent reporting. J Clin Oncol 2012;30:4223–32.

[8] Schilsky RL, Doroshow JH, Leblanc M, Conley BA. Development and use of integral assays in clinical trials. Clin Cancer Res 2012;18:1540–6.

[9] McShane LM, Hunsberger S, Adjei AA. Effective incorporation of biomarkers into phase II trials. Clin Cancer Res 2009;15:1898–905.

[10] IOM. Evolution of translational omics: lessons learned and the path forward. Washington, DC: Institute of Medicine; 2012.

[11] McShane LM, Altman DG, Sauerbrei W, Taube SE, Gion M, Clark GM. REporting recommendations for tumor MARKer prognostic studies (REMARK). Breast Cancer Res Treat 2006;100:229–35.

[12] Lee JW, Figeys D, Vasilescu J. Biomarker assay translation from discovery to clinical studies in cancer drug development: quantification of emerging protein biomarkers. Adv Cancer Res 2007;96:269–98.

[13] Rifai N, Gerszten RE. Biomarker discovery and validation. Clin Chem 2006;52:1635–7.

[14] Rifai N, Gillette MA, Carr SA. Protein biomarker discovery and validation: the long and uncertain path to clinical utility. Nat Biotechnol 2006;24:971–83.

[15] Hewitt SM, Badve SS, True LD. Impact of preanalytic factors on the design and application of integral biomarkers for directing patient therapy. Clin Cancer Res 2012;18:1524–30.

[16] Tuefferd M, De Bondt A, Van Den Wyngaert I, Talloen W, Verbeke T, Carvalho B, et al. Genome-wide copy number alterations detection in fresh frozen and matched FFPE samples using SNP 6.0 arrays. Genes Chromosomes Cancer 2008;47:957–64.

[17] Silberring J, Ciborowski P. Biomarker discovery and clinical proteomics. Trends Anal Chem 2010;29:128.

[18] Pieragostino D, Petrucci F, Del Boccio P, Mantini D, Lugaresi A, Tiberio S, et al. Pre-analytical factors in clinical proteomics investigations: impact of ex vivo protein modifications for multiple sclerosis biomarker discovery. J Proteomics 2010;73:579–92.

[19] Christin C, Hoefsloot HC, Smilde AK, Hoekman B, Suits F, Bischoff R, et al. A critical assessment of feature selection methods for biomarker discovery in clinical proteomics. Mol Cell Proteomics 2013;12:263–76.

[20] Engstrom PF, Bloom MG, Demetri GD, Febbo PG, Goeckeler W, Ladanyi M, et al. NCCN molecular testing white paper: effectiveness, efficiency, and reimbursement. J Natl Compr Canc Netw 2011;9(Suppl. 6):S1–16.

[21] Moore HM, Kelly A, McShane LM, Vaught J. Biospecimen reporting for improved study quality (BRISQ). Clin Chim Acta 2012;413:1305.

[22] Pepe MS, Etzioni R, Feng Z, Potter JD, Thompson ML, Thornquist M, et al. Phases of biomarker development for early detection of cancer. J Natl Cancer Inst 2001;93:1054–61.

[23] Marchio C, Dowsett M, Reis-Filho JS. Revisiting the technical validation of tumour biomarker assays: how to open a Pandora's box. BMC Med 2011;9:41.

[24] Shah RB, Tawakkul MA, Khan MA. Process analytical technology: chemometric analysis of Raman and near infra-red spectroscopic data for predicting physical properties of extended release matrix tablets. J Pharm Sci 2007;96:1356–65.

[25] Lee JW, Weiner RS, Sailstad JM, Bowsher RR, Knuth DW, O'Brien PJ, et al. Method validation and measurement of biomarkers in nonclinical and clinical samples in drug development: a conference report. Pharm Res 2005;22:499–511.

[26] Wang J, Lee J, Burns D, Doherty D, Brunner L, Peterson M, et al. "Fit-for-purpose" method validation and application of a biomarker (C-terminal telopeptides of type 1 collagen) in denosumab clinical studies. Am Assoc Pharm Sci J 2009;11:385–94.

[27] Nowatzke W, Woolf E. Best practices during bioanalytical method validation for the characterization of assay reagents and the evaluation of analyte stability in assay standards, quality controls, and study samples. Am Assoc Pharm Sci J 2007;9: E117–22.

[28] Cummings J, Ward TH, Dive C. Fit-for-purpose biomarker method validation in anticancer drug development. Drug Discov Today 2010;15:816–25.

[29] Cummings J, Raynaud F, Jones L, Sugar R, Dive C. Fit-for-purpose biomarker method validation for application in clinical trials of anticancer drugs. Br J Cancer 2010;103:1313–7.

[30] Lee JW, Devanarayan V, Barrett YC, Weiner R, Allinson J, Fountain S, et al. Fit-for-purpose method development and validation for successful biomarker measurement. Pharm Res 2006;23: 312–28.

[31] Nothnagel M, Herrmann A, Wolf A, Schreiber S, Platzer M, Siebert R, et al. Technology-specific error signatures in the 1000 Genomes Project data. Hum Genet 2011;130:505–16.

[32] Smith WC, Sittampalam GS. Conceptual and statistical issues in the validation of analytic dilution assays for pharmaceutical applications. J Biopharm Stat 1998;8:509–32.

[33] Adamo JE, Bauer G, Berro M, Burnett BK, Hartman KA, Masiello LM, et al. A roadmap for academic health centers to establish good laboratory practice-compliant infrastructure. Acad Med 2012;87:279–84.

[34] Barker PE. Cancer biomarker validation: standards and process: roles for the National Institute of Standards and Technology (NIST). Ann NY Acad Sci 2003;983:142–50.

[35] Poste G, Carbone DP, Parkinson DR, Verweij J, Hewitt SM, Jessup JM. Leveling the playing field: bringing development of biomarkers and molecular diagnostics up to the standards for drug development. Clin Cancer Res 2012;18:1515–23.

[36] Steyerberg EW, Pencina MJ, Lingsma HF, Kattan MW, Vickers AJ, Van Calster B. Assessing the incremental value of diagnostic and prognostic markers: a review and illustration. Eur J Clin Invest 2012;42:216–28.

[37] Ainsworth R, Bartlett JM, Going JJ, Mallon EA, Forsyth A, Richmond J, et al. IHC for Her2 with CBE356 antibody is a more accurate predictor of Her2 gene amplification by FISH than HercepTest in breast carcinoma. J Clin Pathol 2005;58:1086–90.

[38] Deeks JJ, Altman DG. Diagnostic tests 4: likelihood ratios. Br Med J 2004;329:168–9.
[39] Simon RM, Paik S, Hayes DF. Use of archived specimens in evaluation of prognostic and predictive biomarkers. J Natl Cancer Inst 2009;101:1446–52.
[40] Meshinchi S, Hunger SP, Aplenc R, Adamson PC, Jessup JM. Lessons learned from the investigational device exemption review of Children's Oncology Group trial AAML1031. Clin Cancer Res 2012;18:1547–54.
[41] Jessup JM. Interlaboratory assay reproducibility study for loss of heterozygosity on chromosome 18 (18q LOH) in colon cancer. J Clin Oncol 2009;27: suppl: abstr 4052.
[42] Khoury MJ, McBride CM, Schully SD, Ioannidis JP, Feero WG, Janssens AC, et al. The Scientific Foundation for personal genomics: recommendations from a National Institutes of Health-Centers for Disease Control and Prevention multidisciplinary workshop. Genet Med 2009;11:559–67.
[43] Teutsch SM, Bradley LA, Palomaki GE, Haddow JE, Piper M, Calonge N, et al. The evaluation of genomic applications in practice and prevention (EGAPP) initiative: methods of the EGAPP Working Group. Genet Med 2009;11:3–14.
[44] Altman DG, Lyman GH. Methodological challenges in the evaluation of prognostic factors in breast cancer. Breast Cancer Res Treat 1998;52:289–303.
[45] Altman DG, Riley RD. Primer: an evidence-based approach to prognostic markers. Nat Clin Pract Oncol 2005;2:466–72.
[46] Hayes DF, Bast RC, Desch CE, Fritsche H, Kemeny NE, Jessup JM, et al. Tumor marker utility grading system: a framework to evaluate clinical utility of tumor markers. J Natl Cancer Inst 1996;88:1456–66.
[47] Simon R. Clinical trial designs for evaluating the medical utility of prognostic and predictive biomarkers in oncology. Per Med 2010;7:33–47.
[48] Gosho M, Nagashima K, Sato Y. Study designs and statistical analyses for biomarker research. Sensors (Basel) 2012;12:8966–86.
[49] Febbo PG, Ladanyi M, Aldape KD, De Marzo AM, Hammond ME, Hayes DF, et al. NCCN Task Force report: Evaluating the clinical utility of tumor markers in oncology. J Natl Compr Canc Netw 2011;9(Suppl. 5):S1–32; quiz S33.
[50] Freidlin B, Korn EL. Biomarker-adaptive clinical trial designs. Pharmacogenomics 2010;11:1679–82.
[51] Freidlin B, McShane LM, Korn EL. Randomized clinical trials with biomarkers: design issues. J Natl Cancer Inst 2010;102:152–60.
[52] Freidlin B, Jiang W, Simon R. The cross-validated adaptive signature design. Clin Cancer Res 2010;16:691–8.
[53] Zhou X, Liu S, Kim ES, Herbst RS, Lee JJ. Bayesian adaptive design for targeted therapy development in lung cancer – a step toward personalized medicine. Clin Trials 2008;5:181–93.
[54] Baker SG, Kramer BS, Sargent DJ, Bonetti M. Biomarkers, subgroup evaluation, and clinical trial design. Discov Med 2012;13:187–92.
[55] Mandrekar SJ, Sargent DJ. Design of clinical trials for biomarker research in oncology. Clin Invest (Lond) 2011;1:1629–36.

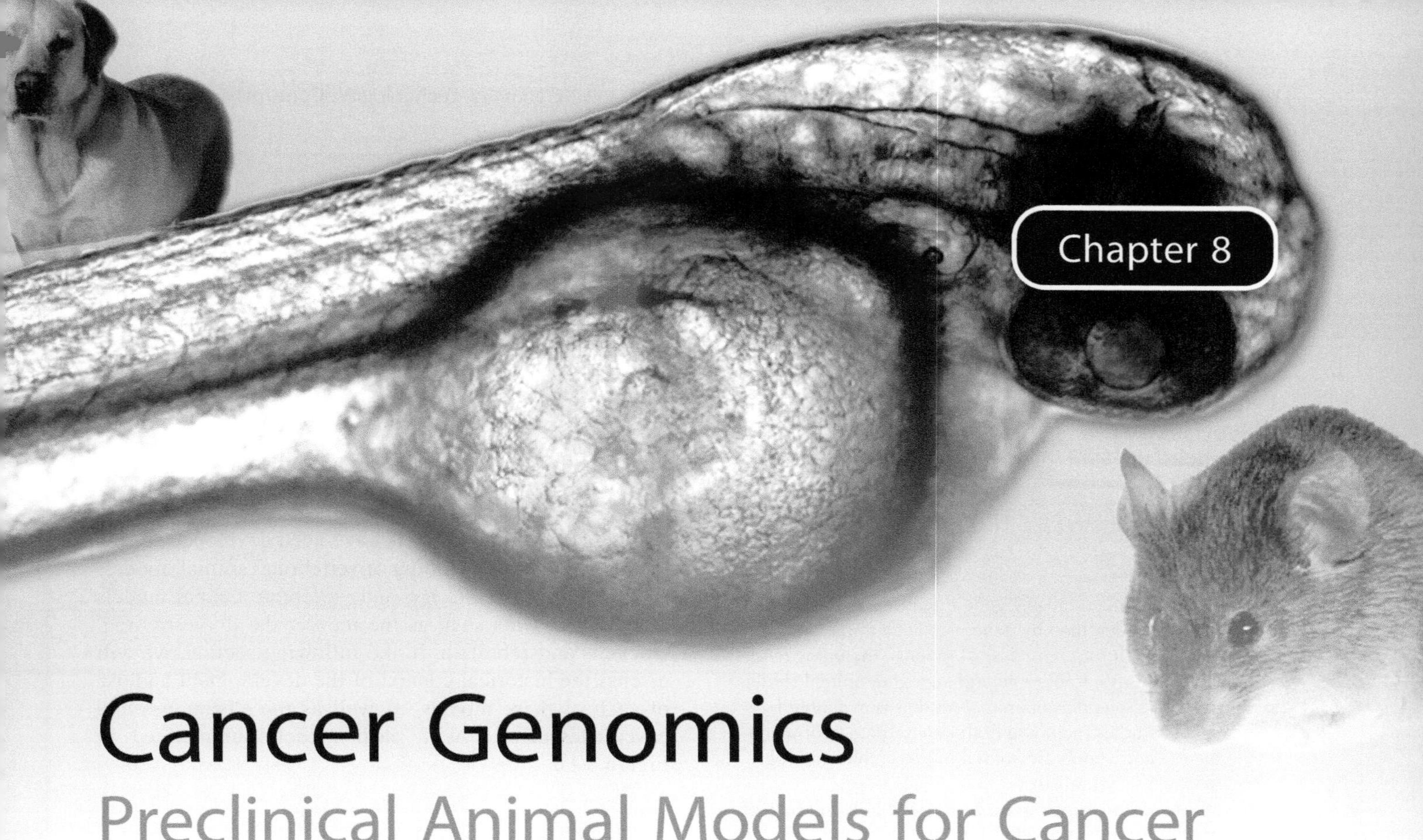

Cancer Genomics

Preclinical Animal Models for Cancer Genomics

Jason N. Berman[1]
Priscilla P.L. Chiu[2]
Graham Dellaire[3]

[1]*Departments of Pediatrics, Microbiology and Immunology, and Pathology and IWK Health Centre, Dalhousie University, Halifax, NS, Canada*
[2]*Department of Surgery, The Hospital for Sick Children, Toronto, ON, Canada*
[3]*Departments of Pathology and Biochemistry & Molecular Biology, Dalhousie University, Halifax, NS, Canada*

Contents

Cancer Genomics. DOI: http://dx.doi.org/10.1016/B978-0-12-396967-5.00008-6

Key Concepts

- Putative cancer-causing gene mutations and genetic pathways identified by genomics need to be validated using one or more preclinical models, including tissue culture assays, transgenic and xenograft animal studies
- A diverse selection of animal models is available for cancer studies including both invertebrate (worm and the fly) and vertebrate animal models (chicken, dog, mouse and zebrafish)
- Several criteria need to be considered in the choice of preclinical animal models for the validation of cancer genome studies, including conservation of genetic pathways, ease of use and genetic manipulation, adaptability to high-throughput screening, suitability for xenograft studies, and relative cost
- Genetically engineered mouse models (GEMMs) are the most widely used transgenic animal model for cancer research, and involve the introduction of genetic material into the germline of the mouse. Similarly, transgenic zebrafish have also grown in popularity as a cancer model due to their large numbers, high degree of genetic conservation with mice and humans, and their relatively low cost
- Non-germline genetically engineered mouse models (nGEMMs), are more rapid to produce and cheaper to maintain than traditional GEMMs. They are often generated as a byproduct of making a transgenic animal, and are a chimera of engineered and non-engineered cells. As such, tumors that form in nGEMMs arise in the context of normal tissues, better mimicking the natural state of cancer
- Human tumor xenograft animal models represent an essential experimental system for the study and validation of newly identified genetic mutations that play a role in cancer development, progression and treatment response. The mouse, chicken chorioallantoic membrane, and zebrafish xenograft models are the most common and are also in the development and validation of molecular targeted therapies for cancer
- When considering human xenograft animal models, there are a number of criteria that will determine the appropriate choice of model. These include whether to xenograft tumor cells in an orthotopic or heterotopic manner, and whether to use primary patient tumor material or cell lines. One must also consider tumor heterogeneity as well as whether an appropriate tumor niche exists to supply the stromal interactions between the tumor material and the xenograft host, which is often by necessity in an immunocompromised animal

HISTORY OF THE DEVELOPMENT OF ANIMAL MODELS FOR CANCER RESEARCH

A number of animal models have been developed for the study of cancer, including invertebrate animal models such as the worm and fly, and vertebrate animal models including rodents such as the mouse, the domestic dog, chickens and zebrafish. In the following section, we will discuss the historical context of the development and use of each of these models, as well as the advantages and caveats associated with each model (summarized in Figure 8.1).

Rodent Animal Models

For the past century, scientists have devised animal models to study human diseases to overcome the moral and ethical barriers involved with human experimentation. In the early part of the twentieth century, large animals were used to "graft" human cancers that exhibited "autonomous growth" due to the ease of implanting and retrieving the human cells. However, these experiments ultimately failed due to the lack of recognition at that time of the significant immune-based, xenogeneic barriers across different species. Later, the discovery that rat sarcoma cells survived in chick embryos revealed that specific conditions within the embryo supported the growth of cross-species tumor cells [1]. By the late 1930s, the identification of an "immune-privileged" site in the anterior chamber of the eye of guinea pigs and rabbits to support the growth of the human tumor cells [2] spawned interest in expanding the repertoire of human tumors to be grown and investigated using this "xenograft" model.

To overcome the immunological barriers associated with xenotransplantation in immune-competent hosts, scientists began to use new tools for xenografts after T lymphocytes were identified as the cellular effectors of immune-based xeno-rejection. Experimentally induced T lymphopenia by thymectomy (surgical removal of the thymus) was used to generate immunodeficient mice for human cancer xenografts with antilymphocyte globulin to deplete mature T lymphocytes [3]. In the 1960s, identification of spontaneous immunodeficient strains of mice became the natural successors as xenograft hosts.

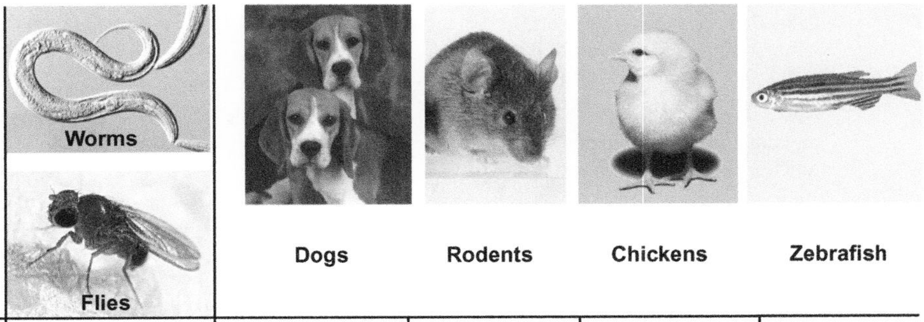

Animal Models	Invertebrates	Vertebrates			
	Worms / Flies	Dogs	Rodents	Chickens	Zebrafish
Criteria					
• Conservation of genetic/biochemical pathways vs human	+	+++	+++	++	++
• Suitable model for:					
Genetic screens	+++	NS*	++	NS	+++
GWAS studies	NS	+++	++	NS	++
Xenograft studies	NS	NS*	+++	++	+++
Chemical-genetic screens	+++	NS*	++	NS	+++
Chemically induced tumor studies	+++	NS*	++	+	+++
• Availability of viral & spontaneous tumor models	+	+++	+++	++	+++
• Availability of transgenic models	++	NS*	+++	NS	+++
• Ease of genetic manipulation	+++	+	+++	+	+++
• High-throughput screening capability	+++	NS	NS	+**	++
• Relative cost of each model	$	$$$$	$$$	$$	$$

NS = not suitable; * not suitable due to ethical issues; ** refers to the use of the chorioallantoic membrane assay only

FIGURE 8.1 Common animal models used in cancer research.

The T-lymphocyte-deficient *nude* mouse was initially identified in 1962 in the virology lab of Dr N R Grist [4] and the thymic defect further characterized by Pantelouris and Hair [5]. This athymic strain permitted for the first time successful, large-scale transplantation of primary human tumors [6] and cancer cell lines into a mouse host [7]. As more immune-deficient mouse strains were identified by spontaneous (e.g. *Prkdc*^{scid}, *beige*) and targeted (e.g. *rag-1*^{null}, *rag-2*^{null}) genetic mutations, these strains became the preferred hosts for human xenografts. Further to enhance human cell engraftment, combinations of immunodeficient strains (e.g. *nude/scid, NOD/scid, NOD/scid-il2r*γ^{null}) or antibody treatment of immunodeficient strains to deplete immune cells subsets (e.g. antimouse CD122 or IL-2 receptor β-chain to target NK cells) resulted in greater immune-depletion. Table 8.1 summarizes the immunological and cellular features of the most commonly used immunodeficient strains for the

TABLE 8.1 Comparisons of Cellular and Serological Immune Deficiencies of Immunodeficient Mouse Strains[1]

Cell	Nude	SCID	NS	NSG
B cells	Defective maturation	Pre-B and mature B cells absent[2]	Pre-B and mature B cells absent	Pre-B and mature B cells absent
T cells	Low or absent (mature and functional cells)	Pre-T and mature T cells absent[2]	Pre-T and mature T cells absent	Pre-T and mature T cells absent
NK cells	High	Normal	Decreased	Decreased
Macrophages	Normal	Normal	Normal number but NOD-derived sirpα haplotype enhances human cell engraftment[3]	Normal number but sirpα haplotype enhances human cell engraftment[3]
LAK cells	Normal	Normal	Normal	Decreased
Serum immunoglobulin	IgG low IgM normal IgA low	IgG low IgM low IgA low	IgG low IgM low IgA low	IgG low IgM low IgA low
Lymphoid organs	Athymic	Small thymus Small lymphoid organs	Small thymus (unless thymoma) Small lymphoid organs	Small thymus Small lymphoid organs
Radiation sensitivity	Normal radiosensitivity	More radiosensitive	More radiosensitive	More radiosensitive
Chemotherapy sensitivity	More hardy	More sensitive	More sensitive	More sensitive
Cost	$	$	$$	$$$

NS: NOD/scid; NSG: NOD/scid.*il2r*^{null}; SCID: CB17-*scid*
[1]*Modified from [8].*
[2]*Although SCID mice lack mature T cells and B cells, this deficiency is variable as aged SCID mice can have some T and B cells due to the "leakiness" of the genetic defect [9].*
[3]*[10].*

xenotransplantation of a variety of human hematological [11–15] and solid organ [16–20] cancers.

In parallel, the discovery of oncogenic viruses in mice that predisposed them to a high incidence of cancers was made in the mid-1950s, which illustrated the contribution of genes to oncogenesis and ushered in a new frontier in cancer research. With the discovery of oncogenes or tumor suppressor genes that promoted or suppressed carcinogenesis respectively, along with the development of recombination genetics as the tool to manipulate these genes within mouse germlines and not just in cell lines, powerful new murine cancer models were generated starting in the 1980s [21,22]. Cancer-prone mice generated through the transgenic overexpression of oncogenes or inactivation of tumor suppressor genes became widely used models to study the biology of cancers by dissecting the disease down to single gene contributions and allowing the progression of the disease to be observed over time. For therapeutic studies, the transgenic model allowed researchers to intervene in the progression of the cancer by targeting gene-specific pathways at specific stages of cancer development. Moreover, transgenic mice could be bred to other transgenic lines in order to generate more complex models of cancers. As a result, transgenic cancer models became a popular tool for researchers of cancer mechanisms as well as cancer therapies.

Other Animal Models

Although mice are arguably the "gold standard" and most commonly employed animal model for cancer studies, other vertebrates including the chicken, dog and, more recently, the zebrafish have grown in popularity. These vertebrate models have become more attractive in the last decade due to the sequencing of their genomes and well-defined genetics (e.g. dog pedigrees) [23], the growing repertoire of *in vitro* assays using these models (e.g. the chick chorioallantoic membrane assay [24]), as well as the availability of new tools for both reverse genetic approaches and the development of transgenic animals, especially in the case of the zebrafish [25]. In addition, invertebrates such as the fruit fly and the nematode *C. elegans* have also been employed as genetic models for the study of important cancer pathways [26,27]. In this section, we will review some of these alternative models for cancer biology and genomics.

Zebrafish

While recognized for a number of decades as a powerful vertebrate model for genetic studies of embryonic development and organogenesis, the zebrafish has more recently emerged as a highly efficient vertebrate tool for studying the pathogenesis of human diseases, including cancer [25,28]. The growth of the zebrafish as a cancer model has paralleled the recognition of cancer as a "genetic disease". The high fecundity and short generation times of the zebrafish embryo facilitate the generation of transgenic lines incorporating human oncogenes and tumor suppressors, which enable studies of tumor formation and maintenance in zebrafish [28–34]. The application of a number of carcinogenic agents has provided models of spontaneous tumor formation [35–37]. Large-scale mutagenesis screens in zebrafish that can identify novel genetic pathways in cancer progression have also been fruitful [32,37–41]. The first transgenic model of cancer in the zebrafish was a model of B-precursor acute lymphoblastic leukemia (ALL) published in 2003 and the first xenograft studies did not emerge until 2005 [42–44]. Transgenic strategies have since been well developed in the zebrafish. These strategies employ both tissue-specific and ubiquitous promoters, fluorescent reporters, such as green fluorescent protein (GFP), as well as the ability to induce gene expression on demand by employing approaches already pioneered in mice, including Cre-lox, tamoxifen-inducible and Tet-on/off systems [32,45–51]. More recently, the advent of zinc finger nuclease (ZFN) and transcription activator-like effector nuclease (TALEN) technologies has permitted much more facile genetic manipulation allowing the generation of permanent knockouts of human tumor suppressor genes [52–56]. Zebrafish transgenic lines share the advantage with rodent models of being vertebrates, but surpass rodents with respect to generation of large numbers of offspring, external fertilization, rapid embryonic development and optical clarity [25,46,57]. Xenograft approaches in the zebrafish are also gaining traction, as it is now possible to transplant human cancer cell lines and primary tumor samples into zebrafish embryos, which essentially behave as *in vivo* incubators in which to study the interaction of cancer cells with their microenvironment and/or their response to therapeutics [44].

Dog

The domestic dog, represented by over 400 breeds with specific physical characteristics and disease spectra, has become increasingly recognized as a unique and powerful animal model for the study of the genomic basis of disease including cancer [23]. Dogs develop cancer spontaneously, with a similar spectrum of cancers to those in humans, their tumors appear histologically similar to human tumors, and respond similarly to radiation and chemotherapy [58,59]. Cancers common in the dog in order of prevalence are: mammary carcinoma (52% of all tumors in female dogs); non-melanoma skin cancer; soft tissue sarcomas; testicular cancer; melanoma; non-Hodgkin lymphoma (NHL); head/neck cancer; and bone tumors [60]. In addition, specific cancers have been shown to be enriched in particular breeds as a byproduct of population bottle-necks and strong artificial selection during breed development. For example, T-cell lymphoma is more prevalent in Irish wolfhounds than in Cocker spaniels, but the opposite is true for B-cell lymphomas [58]. Dogs also share a similar environment and in many cases eat similar types of foods as their human companions, which may explain in part, the similar cancer spectrum and speak to the importance of environmental factors in cancer onset and progression. An additional advantage is that given the shorter lifespan of dogs and reduced average disease-free time of survival following cancer treatment (18 months versus 5–7 years in humans), canine clinical trials can be conducted over a shorter time frame than is possible in humans [59]. However, it is the unique genetic heritage of the dog that is the single most powerful aspect of this animal model for cancer genomics.

The artificial genetic bottle-necks that occurred during breed development, most of which were imposed within the last 200 years, create a unique situation where linkage disequilibrium (LD), the non-random association of two or more genetic loci, is a 100-fold more prevalent in dogs than in humans [13]. As a result, it has been possible to employ simple linkage mapping in the dog to determine mutations associated with several cancers found in specific breeds. For example, simple linkage mapping of microsatellite repeats was used to determine that mutations in the *folliculin* (*FLCN*) gene were responsible for a rare renal cancer in German shepherds [61,62]. The unique genetic heritage of the dog also leads to long LD haplotype clusters within breeds and, as a consequence, genome-wide association (GWA) studies can be accomplished with as few as 10 000 single nucleotide polymorphisms (SNPs) versus millions of SNPs in human studies [13]. However, since it has been at least 15 000 years since dogs were domesticated and most breeds are derived from what appear to be wolves, shorter LD clusters exist among dog breeds. With the sequencing of the dog genome in 2005 [23], GWA studies have been greatly advanced and coupled with the unique genetics of the domestic dog, it is now possible to efficiently identify cancer susceptibility genes using a two-stage mapping approach. For example, a GWA study using SNP-typing can be first employed to define an initial region of LD associated with a given cancer using a single breed where that cancer is highly prevalent (e.g. osteosarcoma in rotteweilers and greyhounds) [63], and then confirmation and narrowing of the region of LD can be done comparing SNP data across multiple breeds [23].

Despite the clear advantage of the dog as an animal model for the study of spontaneously derived cancers, the dog is less suited to reverse genetic approaches using mutagens to induce cancers, or to studies aimed to validate gene mutations that would induce cancers in these companion animals. These kinds of experiments are not only impractical, given relatively long breeding and maturation time of dogs, but the ethical and moral objections to these experiments as well as the cost to maintain these animals are considerable. As a result, the validation of genetic drivers of cancer identified through human genomic studies will continue to employ more standard animal models such as the mouse and zebrafish, which are the primary focus of this chapter.

Chicken

The chicken may seem an unlikely animal model for the study of cancer. However, this animal holds an important place in establishing cancer biology as a field of study over 100 years ago. Peyton Rous in 1911 published a paper describing a "filterable agent" isolated from a sarcoma of the hen that allowed the transmission of the cancer between animals, thus ushering in the field of tumor virology [61]. This virus, later coined the Rous sarcoma virus (RSV), and its study not only shed light on the etiology of avian sarcoma but led to a greater understanding of the role of retroviruses in cancer, in addition to setting the stage for the development of tissue culture and embryonic assays using the chicken as a cancer model [65]. Other notable achievements of the chicken as a cancer model include the development of the first assays for oncogenic transformation, the identification of the first viral oncogene, *v-src*, and its cellular origin, and aiding in the understanding of how retroviruses contribute to hematological malignancies [66]. In addition, the first successful anticancer vaccine was developed to immunize poultry against the Marek disease virus that causes T-cell lymphoma in chickens, decades before human vaccination against hepatitis B and human papilloma virus (HPV)-induced liver and cervical cancer, respectively, was achieved [67–69].

In the last decade, the chicken has been an important animal model for the study of molecular mechanisms of cancer, as well as solid tumor development and metastasis. For example, the chicken DT40 B lymphoblast cell line, which exhibits high rates of homologous recombination, is a favored model for the somatic knockout of genes for reverse genetic studies of gene function in DNA repair and cell cycle regulation [70]. More recently, the chicken has also proved to be an interesting model for epithelial ovarian cancer, which is relatively common in chickens, occurring in up to 7.4% of White Leghorn laying hens, and shares many traits with human ovarian cancer, including similar staging, histology and patterns of metastasis [71,72]. Finally, as we will discuss in depth later in this chapter, the chick chorioallantoic membrane assay, in which human tumors are engrafted in the developing chicken embryo, has proven to be a powerful model for the study of metastasis and in the screening of anti-angiogenic compounds [73].

With the sequencing of the chicken genome in 2004 [74] and the subsequent comparison of the genomes of different chicken breeds, it is likely that the chicken will continue to be a useful model for the determination of cancer susceptibility genes. Like the dog, the artificial selection and resulting variation between breeds of chickens can be exploited for gene mapping. As just one example, the variation between the Cornish and White Leghorn breeds was recently exploited to map SNPs in the gene encoding NK-lysin, an antibacterial cationic peptide with anticancer properties, resulting in new clues as to differences in innate immunity between the breeds [43].

Invertebrate Models

Despite the popularity of the mouse as an animal model for cancer studies, on the other end of the evolutionary spectrum, invertebrates including the fly, *Drosophila melanogaster*, and the nematode worm, *Caenorhabditis elegans*, have also been used to study oncogenic pathways. As metazoans, the fly and worm have a distinct advantage over unicellular eukaryotes such as yeast, as cancer is a pathological process of multicellular organisms. In addition, many key cellular processes such as programmed cell death (apoptosis), cell-cycle control, and cellular differentiation and cell migration, all of which are subverted in tumorigenesis, have been extensively studied and, in some cases, pioneered in these invertebrate models [26,75]. For example, *C. elegans* was instrumental in the identification of the genetic regulation of apoptosis through pioneering work in the 1980s by Robert Horovits, John Sulston and Sydney Brenner (reviewed in [26]), which consequently earned them the 2002 Nobel Prize for Physiology or Medicine. Several of the genes identified in these studies, such as

CED-3 and *CED-9*, the worm homologs of mammalian caspases and Bcl2-like genes, were subsequently found to play important roles in cancer development, as well as tumor response to chemotherapy and radiation [26,76]. The fly and the worm have also been instrumental in the discovery of RNA interference, a discovery which also garnered a Nobel Prize in 2006 [77]. By enabling the massively parallel interrogation of loss of gene function for 100s or 1000s of genes, this technique has revolutionized genetic screens for modifiers of cancer and other diseases [78]. The fly and worm are ideal animal models for chemical and genetic high-throughput screens due to their size, short life span, ability to generate large numbers of offspring, and their optical clarity which facilitates imaging-based screens.

Although the fruit fly does not seem like a natural model for cancer research, spontaneous tumors arising in flies were first described as early as 1918 [79], and a large number of tumor suppressor genes and oncogenes were identified in *Drosophila* using genetic screens for cellular overgrowth phenotypes in the late 1970s and in the 1980s (reviewed in [80]). One such phenotype was the overgrowth of neurons associated with mutations in the *Notch* gene, so called due to the notched wing phenotype in haploinsufficient individuals. In mammals, the Notch pathway regulates cell differentiation, proliferation and apoptosis and is implicated in both leukemia and in solid tumor development [80,81]. The molecular details of the *Ras* proto-oncogene pathway were also first elucidated in *Drosophila,* where this pathway regulates photoreceptor development in the eye, and in *C. elegans* where the *KRAS* gene homolog, *let-60,* plays a role in vulval development [80,82]. Another gene identified in genetic screens in *Drosophila* was *Hedgehog*, which plays a role in cellular differentiation, proliferation and apoptosis. Aberrant Hedgehog signaling promotes leukemia as well as solid cancers derived from many tissues, including skin (basal cell carcinoma), breast, lung and colon [80,83].

The fly and the worm have also been used to explore the molecular pathways controlling oncogenes (or tumor suppressors) first identified in human cancers, including adenomatous polyposis coli (*APC*), tuberous sclerosis 1 (*TSC1*), wingless type MMTV integration site family members (Wnts), and the *RET* proto-oncogene [80,84]. As discussed previously, given the adaptability of these invertebrate models to high-throughput screens, they have also consequently been used for chemical genetic screens to identify inhibitors of these conserved oncogene pathways [84]. For example, using an RNA interference (RNAi)-based chemical genetic screen in *Drosophila*, Gonsalves and colleagues identified novel inhibitors of the Wnt pathway, which proved to have activity against human colon cancer cells [85]. Despite these successes, it is much easier to identify specific small-molecule inhibitors of human oncogenes by employing human tumor and cancer cell lines, where the potential pitfall of poor conservation of drug–gene interactions between species is not a factor. Nonetheless, both the fly and the worm have and will continue to contribute to the elucidation of fundamental and highly conserved molecular genetic pathways involved in cancer development.

TRANSGENIC ANIMAL MODELS

Genetic manipulation of invertebrate genomes in *Drosophila* and *C. elegans* has been widely used to study oncogenic pathways which, at times, have been generated serendipitously as a byproduct of genetic screens. However, these models sometimes lack the evolutionary conservation of gene function, and/or complexity required to model a cancer in a way that is relevant for the development of human cancer therapies. As a result, genetic models for cancer in rodents and other vertebrates were developed for more comprehensive studies of cancer mechanisms and in therapeutic drug discovery. In the following section, we will discuss transgenic models in mice and zebrafish, which represent two of the most widely used vertebrate animal models.

Transgenic Mice – GEMMs and nGEMMs

Genetically engineered mouse models (GEMMs) are by far the most commonly used and abundant models for cancer research today. The widespread availability of the technology to manipulate the mouse germline has made GEMMs essential tools for cancer research, although the transgenic technology is patented by Harvard University and licensed by DuPont and requires contractual agreements for its use. The role of tumor suppressor genes in cancer is studied through inactivation of the gene in knockout mice, whereas transgenic mice overexpressing oncogenes yield the cancer phenotype. Transgenic expression of the human oncogene, *C-MYC*, under the control of the murine mammary tumor virus (MMTV) promoter was shown in the first transgenic model of breast cancer that cancer can be initiated solely by the aberrant activation of an oncogene [22]. Since then, the roles of numerous oncogenes and tumor suppressor genes involved in cancer development have been discovered. Simplistically defined, oncogenes encode proteins that regulate cell division by increasing the activity of transcription factors, signal transduction proteins or growth factor receptors. Tumor suppressor genes were generally found to be regulators of cell cycle arrest, cell death or DNA repair mechanisms. In the absence of these proteins, cells can proliferate unchecked, accumulating additional DNA mutations eventually to acquire a malignant phenotype. The most studied tumor suppressor gene is *TP53*, which encodes the p53 protein that is a critical regulator of the G_1/S cell cycle check point, and p53 knockout mice spontaneously develop tumors (mostly lymphomas) by 6 months of age [86].

The techniques for generating GEMMs can create potential challenges for investigators as the gene targeted for disruption in the knockout model may be one that is vital for embryogenesis, thus producing an embryonic lethal phenotype during development. Similarly, systemic and ubiquitous overexpression of the transgene can render the model useless (or of questionable fidelity) for cancer research if the mouse develops and succumbs from tumors elsewhere than the one of research interest. Therefore, the investigator needs to consider carefully the anticipated effects of gene function before deciding on the approach used for GEMM development.

Genetically altered mice can be derived either by injection of cloned DNA into the male pronucleus or fertilized egg or by the injection of genes into embryonic stem (ES) cell lines. The latter technique allows for cloned genes to integrate into specific regions in the genome by homologous recombination, as well as by "random" integration in the genome by non-homologous recombination via the DNA ends. Homologous recombination is employed to "knockout" genes but can also "knock-in" modified genes that affect gene function or protein structure. However, as cloned genes injected into the ES cell lines may also integrate randomly into the genome, genetic constructs that enable homologous recombinants to be distinguished from random integration are required. Typically, selectable markers inserted into random integration sites in ES cells are not expressed due to in-frame stop codons, which are removed during the process of homologous recombination at the true target site, thus allowing easy identification of site-specific gene integration. Importantly, to ensure that protein function is completely abrogated, the selectable marker gene should be integrated into the genetic position that encodes the protein's most critical domain, which will generate a loss-of-function allele. Following gene integration, ES cells are introduced into the blastocysts of a day 3.5 pregnant female prior to implantation of the chimeric blastocyst into a pseudo-pregnant female host. At birth, the resultant mice do not uniformly express the genetic background of the ES cells because the cells from the blastocyst do not carry the same genetic mutations (a feature that is harnessed for the generation of non-germline GEMMs as described later). Therefore, only those chimeric mice that harbor the desired genetic mutations in their germline are selected and bred in order to generate homozygous offspring that carry both copies of the manipulated gene. As this process requires the generation of multiple chimeras and genetic screening of the offspring, the time required to generate transgenic lines using this approach can take several years.

Alternatively, transgenic mice can be generated by the injection of plasmid DNA vectors (100–200 copies of the gene of interest) into the fertilized oocyte or the male pronucleus. The plasmid must contain all of the essential elements for the gene's expression including tissue-specific (and, in some cases, inducible) promoters/enhancers, open reading frames, RNA splice sites and polyadenylation sequences. For larger gene fragments (up to 200 kb), bacterial artificial chromosomes (BAC) have become the favored vector for the introduction of transgenes due to their consistency and stability, and the wide availability of mouse BAC libraries that are commercially available. The choice of promoter is important to ensure appropriate location for the expression of the gene of interest, especially for cancer research where therapy studies may be targeting a specific cancer cell population in the niche (see below). The injected embryos are implanted into the oviducts of a pseudo-pregnant female and all of the progeny are expected to contain copies of the gene. However, transgene expression and, therefore, cancer progression in these models can be affected by the number of copies of the gene integrated into the genome and the integration site(s) within the genome. Larger genes may only integrate once into the genome but smaller genes (<50 kb) may integrate into multiple sites or as a concatamer (up to 50 copies of the transgene arranged in a head-to-tail manner) in a single site. The progeny that most stably expresses the transgene are the "founders" and are then bred to maintain the transgenic line.

Injection of genetic material into the fertilized egg results in the integration into the genome at the one cell stage. This results in systemic gene expression in the "oncomouse" that may not be the desired effect. Therefore, tissue-specific promoters are required for organ-specific transgenic cancer models and many organ systems now have well-studied promoters for oncogene expression (Table 8.2). Such a focused approach allows for a relatively limited "field effect" of the transgene in order to study the cancer in a more representative and organ-specific site [94]. For an added level of complexity, the use of inducible promoters temporally to regulate transgene expression (expression of the transgene is "induced" by the promoter response to drugs, such as tetracycline promoter response to tetracycline added to the mouse diet) can be used to synchronize transgene transcription to detect the earliest events in tumor development, such as breast cancer progression from its initial ductal proliferation morphology [95]. While oncogene activation promotes cancer development in transgenic models, it is clear that the latency in cancer development observed in various transgenic lines is due to the

TABLE 8.2 Promoters Used to Express Transgenes in Specific Tissue Sites in Mice

Tissue	Promoter	References
Breast/mammary gland	MMTV-LTR	[87]
Immature thymocytes	LCK proximal promoter	[88]
Mature thymocytes/T cells	LCK distal promoter	[89]
Prostate	PB	[90,91]
Brain	FGF1B	[92]
Pancreas	RIP-Tag	[93]

FGF1B: fibroblast growth factor 1B; LCK: lymphocyte protein tyrosine kinase; MMTV-LTR: murine mammary tumor virus long terminal repeats; PB: probasin; RIP-Tag, rat insulin promoter large T antigen

additional, stochastic development of a mutation of a tumor suppressor gene. The converse also occurs in knockout mouse models that exhibit stochastic oncogene activation. These effects were demonstrated in double mutant GEMMs where the onset of cancer was accelerated compared to single mutants [96,97]. Therefore, GEMMs provide important insights into the genetic requirements for the clinical progression of cancer.

Controlling the timing and location of gene disruption in GEMMs may be vital if tumor suppressor gene function is essential for embryonic development. One strategy to avoid such embryonic lethal outcomes is to use conditional knockouts in which the timing and location of gene disruption is controlled using the Cre-loxP system [98], which is patented by the DuPont Company. The "floxed" transgenic construct involves flanking the essential components of the target gene (e.g. the gene promoter or critical exons for protein function) with LoxP elements. LoxP are bacterial sequences targeted by the bacterial Cre recombinase for recombination, thereby affecting the floxed gene's expression when the LoxP transgenic mice are infected with virus carrying the Cre gene [99] or crossed with mice carrying the bacterial Cre recombinase transgene [100]. Cre-mediated recombination of the floxed gene in the progeny modulates gene expression by splicing out the essential exon or bringing the promoter to the gene. While the Cre-loxP system can be used to activate genes, most commonly it is used for abrogating gene expression. Further regulation of Cre activity by placing it under the control of an inducible or tissue-specific promoter destines gene activation or inactivation to a specific time or tissue. Additionally, the yeast analogue to the bacterial Cre-lox Precombination system, Flp-Frt [101], can be added such that combined Cre–Flp transgenics can affect multiple, independent genetic recombination events. Such conditional GEMM lines are now available through academic and commercial suppliers. These GEMMs have generated more sophisticated models of cancer than traditional transgenic models, allowing the complexity of cancer development to be better captured experimentally. Conditional GEMMs have been used in the study of premalignant and malignant progression of various cancers including colorectal [69], breast [43], lung [102] and brain cancers [103].

While standard germline GEMMs are the predominant transgenic cancer models, concerns have been raised regarding their applicability to the study of human cancers since few human cancers arise from germline genetic mutations as compared to the more common sporadic mutations in somatic cells. Even with inducible promoters and conditional gene expression systems that provide temporal and spatial control to restrict transgene-mediated effects to a particular developmental stage or tissue (respectively), the homogeneity of the genetically engineered cells expressing the transgene may not always faithfully model cancer in a clinically relevant way. For example, genetic "field effects" can arise as a consequence of uniform gene expression in a field of cells, which may affect cell–stromal interactions that regulate tumor growth in a manner that does not occur in sporadic cancer [104]. More practically, the time and cost involved in generating germline GEMMs are potentially limiting especially for therapeutic screens.

To achieve "high-throughput" cancer models, *ex vivo* genetic manipulation of either adult stem/progenitor cells before reintroduction into an adult mouse (transplantation model) or ES cells prior to injection into mouse blastocysts (chimeric model) has garnered renewed interest for cancer researchers. Typically, the engineered cells are not uniformly distributed in the organism but rather sporadically and adjacent to normal, non-engineered cells. As such, this can make them difficult to identify before they exhibit an overt cancer phenotype. However, one can identify the developing cancer cells by using genetically encoded visible markers (such as GFP, which is visible under fluorescence imaging). In these so-called non-germline GEMMs (nGEMMs), tumors arise in the context of normal tissues, better mimicking the natural state of cancer development in humans than traditional germline GEMMs. Additionally, nGEMMs can also be generated quickly and less expensively because all of the genetic manipulation of the stem cells or ES cells is accomplished *ex vivo* and, as a result, one can avoid the cost of maintaining a large colony for animal husbandry, which is normally required to maintain germline GEMMs. This approach has already been used to generate lung [105], blood [106], breast [43] and liver [107] cancer models. Despite these advantages, limitations of transplantation and chimeric nGEMMs include the need for careful validation of multiallelic expression in each chimeric host, as well as extensive monitoring of ES cell pluripotency following genetic manipulation to ensure that genetically engineered ES cells can populate the organ or tissue being studied. As a result, investigators must functionally validate each genetic element following the generation of their models to ensure that their model is representative and remains clinically relevant.

Recent evidence suggests that cell-extrinsic processes are important contributors to cancer development [108,109]. Whereas traditional GEMMs model cell-intrinsic effects of oncogenes or tumor suppressor genes in tumorigenesis, a complementary use of GEMMs and nGEMMs in cancer research is in defining the niche or microenvironmental factors that promote or inhibit tumor growth or metastasis. Genetic targeting of the tumor environment to change local cytokine concentration [110], cell adhesion [111], stromal factors [112] and vascular network [113] has been used to determine the elements essential for cancer cell viability and circulation. Elucidation of the niche factors using these models will help to devise novel therapeutic strategies to limit cancer cell survival within the niche or restrict their ability for migration and metastases in human cancers. Resources for obtaining or generating transgenic and other cancer mouse models are listed in Table 8.3.

Transgenic Zebrafish

Transgenic approaches have been well developed in the zebrafish both for generating fluorescently labeled cells and tissues using cell-specific promoters and for overexpressing oncogenic driver mutations [25,45,114–116]. Cell-specific promoters enable exclusive expression in a given cell lineage or tissue, which may be beneficial for a particular oncogenic phenotype, as well as for restricting generalized toxicity and early mortality from more ubiquitous expression. The first transgenic cancer model in the zebrafish employed the *rag2* lymphoid-specific promoter to drive expression of the mouse *c-Myc* oncogene resulting in a clonal proliferation of immature T cells arising from the thymus, reminiscent of T-cell acute lymphoblastic leukemia [42]. Despite being cell-specific and resulting in the desired phenotype, the potency of the *C-myc* oncogene did not permit these fish to survive to sexual maturity to perpetuate the line, prompting the development of inducible gene expression systems for use in zebrafish. Subsequently, the Cre-lox system employed first in mice (described above), was adapted and used successfully to generate transgenic zebrafish models of ALL [47] and myeloid disease [48,117,118]. Typically, these systems involve the engineering of zebrafish embryos to encode an oncogene that may be inverted relative to its promoter or

TABLE 8.3 Online Resources for Cell and Animal Model Studies

Resources	Online Information
Cell lines	Non-profit
	American Tissue Culture Collection www.atcc.org European Collection of Cell Cultures www.hpacultures.org.uk/collections/ecacc.jsp Coriell Cell Repository(USA) ccr.coriell.org DSMZ-German Collection of Microorganisms and Cell Cultures http://www.dsmz.de/
	Commercial
	Invitrogen www.invitrogen.com Sigma Aldrich www.sigmaaldrich.com
Animal models	
Mice	
Transgenic mouse strains	Non-profit
	National Institutes of Health (USA) www.nih.gov/science/models/mouse/resources/hcc.html National Cancer Institute (USA) cancermodels.nci.nih.gov/ Medical Research Council Harwell (UK) www.har.mrc.ac.uk The Jackson Laboratory (USA) cre.jax.org/introduction.html
	Commercial
	Taconic www.taconic.com/KORmice Amplicon Express www.ampliconexpress.com Charles River www.criver.com
Immunodeficient and humanized mouse strains	Non-profit
	The Jackson Laboratory (USA) jaxservices.jax.org/invivo/pdx.html Case Western Reserve University (USA) cancer.case.edu/sharedresources/athymic/
	Commercial
	Harlan www.harlan.com
Zebrafish	
Zebrafish transgenic lines and resources	Non-profit

(Continued)

TABLE 8.3 (Continued)

Resources	Online Information
	ZFIN: Zebrafish Model Organism Database (USA) http://zfin.org/ Zebrafish International Resource Centre http://zebrafish.org
	Commercial
	Transposagen (provide TALENs for zebrafish model generation) http://transposagenbio.com

followed by a strong stop transcription signal, preventing expression, and flanked by Loxp sites. Expression of the oncogene is induced by injection of the embryo with an mRNA encoding the *Cre* recombinase that inverts or activates the gene via recombination between the Loxp sites, placing it under control of its promoter. Another strategy, which is particularly useful in juvenile and adult zebrafish and provides temporal control, is to cross the zebrafish line carrying the Loxp-flanked oncogene with another transgenic fish expressing *Cre* under an inducible promoter, such as the heat shock promoter. Similarly, tetracycline (Tet-on) and estrogen responsive (Tamoxifen-inducible) systems have also been employed successfully in the zebrafish [32,45,49–51,118]. However, it should be noted that inducible oncogene expression is not always desirable or capable of faithfully recapitulating a cancer phenotype. For example, despite the success of the *rag2* promoter in T-ALL generation in zebrafish, it was ineffective at generating B-precursor ALL in the case of the *TEL–AML1* fusion oncogene whereas, by contrast, a robust phenotype was observed using the ubiquitous *β-actin* promoter [119]. The authors of this study questioned whether these findings implied that the cell of origin in this type of leukemia antedates the B-precursor population expressing *rag2*. A number of zebrafish transgenic models have emerged using cell-specific or ubiquitous promoters, with or without inducible elements to model a host of human malignancies including leukemias [32,42,47,48,117,119], melanomas [120,122], sarcomas [31,121] and pancreatic tumors [123]. Generation of these models has been greatly facilitated in some cases by the use of efficient cloning systems that allow the insertion of transgenes into the genome at site-specific DNA breaks induced by the *I-Sce* meganuclease [124] or via the *Tol2* transposon system [125]. In particular, the Gateway cloning system (Invitrogen, Carlsbad, CA), which permits rapid recombination-based cloning of promoters, target genes, 3′ fluorescent molecular tags, and *Tol2* repeats in one single reaction, has greatly accelerated the generation of *Tol2* transposon-mediated transgenic zebrafish [126,127].

While effective, generation of germline transgenic zebrafish remains time-consuming. Each transgenic construct is injected into one-cell stage embryos and this F0 generation is reared to sexual maturity, which occurs approximately 3–4 months post-fertilization. Following the outcrossing of these animals to wild-type fish, the resulting F1 offspring can be screened for the presence of green or red fluorescence, if a gene encoding a fluorescent marker has been incorporated into the DNA construct. Confirmation by genomic polymerase chain reaction (PCR) on single embryos for the expression of the transgene identifies germline F1 embryos and their F0 founder parent. The F1 generation can then be intercrossed to generate homozygous F2 fish. To overcome the need for multiple generations to generate stable transgenic lines, the Langenau laboratory has recently shown the feasibility of using first generation mosaic zebrafish to model human cancers, analogous to the non-germline genetically engineered mouse models (nGEMMs). A number of constructs including different oncogenes and fluorescent markers can also be injected concurrently and will integrate by non-homologous recombination into the fish genome as concatamers [45]. This strategy has been particularly effective at evaluating collaborating mutations, such as *KRAS* and *TP53* in embryonal rhabdomyosarcoma [128] and *NOTCH*, *C-MYC* and GFP in T-ALL [129]. Moreover, as was described for nGEMMs, these mosaic fish may more closely replicate the somatic nature of most human cancers.

XENOGRAFT ANIMAL MODELS

Human tumor xenograft animal models represent an essential experimental system for the study and validation of newly identified genetic mutations that play a role in cancer development, progression and treatment response. In particular, xenografts are extensively used in both cancer drug discovery and in preclinical studies of tumor–gene interactions that are required to develop and validate targeted molecular therapies for cancer. This section will discuss the xenotransplantation of human tumor cells in the mouse, chicken and zebrafish, which are the most commonly used xenograft animal models (summarized in Figure 8.2).

Mouse Xenografts

Human cancer xenotransplantation into mice was made possible by the discovery of spontaneously immune-deficient mouse strains [130] allowing scientists to abandon time-consuming and laborious experimental models of immune-deficient mice. Over time, increasingly more immune-deficient strains would be discovered and developed to improve this model. While initial experiments largely focused on the technical aspects and feasibility of xenotransplantation for a variety of human cancers, efforts were eventually directed towards evaluating cancer biology, therapies and the mechanisms involved in mediating drug resistance. The functional heterogeneity of human cancer cell subsets in xenotransplantation formed the basis for a new model of cancer, now known as the cancer stem cell model, in which cancer cell functional properties recapitulates the functional hierarchy of normal biological systems. Additionally, humanized mice are now widely available to make the xenograft model as biologically close to the human disease as possible. Therefore, the murine xenograft model serves as a powerful and functionally validated tool to interrogate human cancers, revealing new insights that revolutionize the way we think about cancer itself.

Following the report and characterization of the T-lymphocyte deficient, athymic *nude* mouse in Scotland [4,5] researchers began to implant, inject and passage primary human cancers in *nude* mice – first in Europe [6,131,132] and then in the USA at the National Cancer Institute [7,133]. It was observed that only some of the cancers would engraft, as the *nude* defect encoded by the mutant *Foxn1* gene only affected the cellular immune response [129]. The *Scid* mouse, first reported in 1983 [9], was found to be even more immunodeficient than the *nude* mouse as the DNA repair defect encoded by the *Prkdc*^{scid} mutation [134] affected cellular as well as humoral immunity by impairing the antigen receptor gene recombination events required for B- and T-lymphocyte development. The C.B-17 *Scid*, the most immune-deficient and widely used *Scid* strain, generated blood [135], ovarian [136], lung [137], skin [138] and colorectal [139] cancer xenografts with substantial success. To achieve even greater immune-deficiency, *beige/nude/xid* (BNX) mice were bred to combine impaired NK cell activity (from the *beige* mouse background) [140] with the absence of T lymphocytes (from *nude*) and B lymphocytes (from *xid* or X-linked immunodeficiency [141]). Not as widely available, BNX mice were used for human hematopoietic [142] and skin cancer [143] xenografts. However, immune deficiency in *Scid* and BNX mice was not a permanent trait as evidence of cellular and humoral immunity was detected depending on the age and genetic background of the mice [144].

To abrogate further innate immune function, Leonard Shultz at the Jackson Laboratory developed the *NOD/scid* congenic strain to combine the lymphocyte deficiency of *Scid* with the non-obese diabetic (*NOD*) strain-derived defects in innate immunity, including NK cell dysfunction, complement deficiency and macrophage differentiation defects [145]. An example of a *NOD/scid* mouse xenografted with a human breast tumor is shown in Figure 8.2A. Treatment with anti-CD122 (IL-2 receptor β-chain) antibody to deplete NK cells further [146] rendered the *NOD/scid* (NS122) even better for cancer xenotransplantation [15,147]. While the strain was clearly more immune deficient than C.B-17 *Scid*, *NOD/scid* mice developed spontaneous endogenous thymic lymphomas starting at 6 months of age, limiting their use for long-term xenograft studies. Their susceptibility to lymphoma development was in part due to the lack of innate immune surveillance against malignant cells combined with the DNA repair defect. As a result, engineered genetic mutations of the recombination enzymes required for antigen receptor gene rearrangement, *rag-1* and *rag-2* [148,149], were the natural targets for lymphocyte depletion bred into the NOD background [150,151]. More recently, *NOD/scid* mice lacking the gene encoding the IL-2R γ-chain, also known as the cytokine receptor common γ-chain (*NOD/scid/il2r* γ^{null} or NSG), were developed by Shultz's group and the RIKEN laboratory in Japan [152,153] to improve engraftment without the high incidence of endogenous lymphoma seen in *NOD/scid* mice. With increased longevity, advanced disease xenograft models have also proved useful in studying therapies directed against late-stage cancers (reviewed in [154]). Finally, the introduction of human genes into *NOD/scid* congenics has further enhanced the ability to replicate the human tumor environment in murine xenografts [155].

With improved and increased xenografts, the spectrum of human cancer xenografts expanded to include blood, breast, brain, colorectal, soft tissue sarcoma, pancreatic, oropharyngeal, lung and gonadal cancers. It is through the modeling of human cancers in murine xenografts that the variable capacity of cell subsets to initiate cancer was functionally characterized and described. The first cancer to feature prominently this proposed functional model of cancer was acute myeloblastic leukemia (AML) [11], where the rare CD34 + CD38− subset was uniquely able to initiate leukemia in the *Scid* xenograft compared to other AML subsets. The CD34 + CD38− cells were labeled as leukemia stem cells as the leukemia xenograft was initiated and propagated uniquely by this stem cell-like CD34 + CD38− population. This model bore a strong resemblance to the process involved in normal hematopoiesis, as the cancer stem cells mirrored the hierarchy of hematopoiesis and the hematopoietic stem cells' unique capacity to regenerate the blood system [156]. Based on this model, "unique" cancer stem cells were functionally characterized in xenografts for a wide spectrum of human cancers, including breast [19], brain [16], colorectal [17], esophageal [127], lung [157] and liver [158] cancers. However, not all cancer stem cells were rare in frequency and the phenotypic markers used to

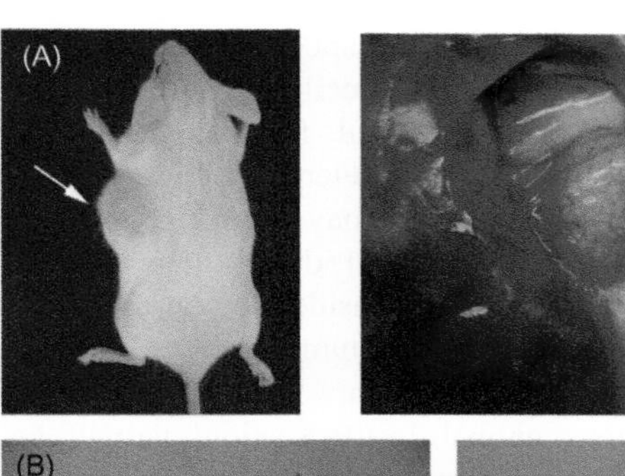

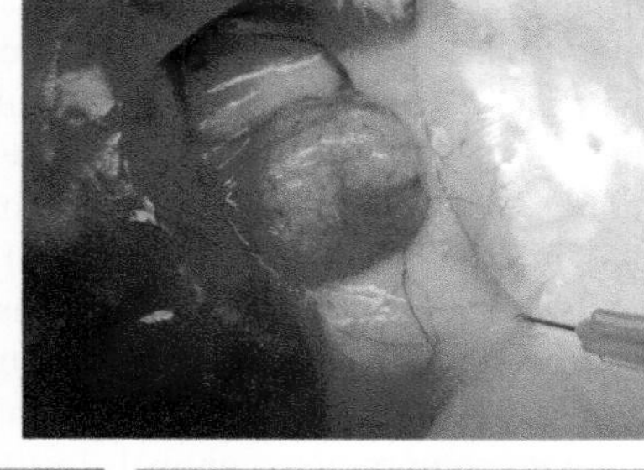
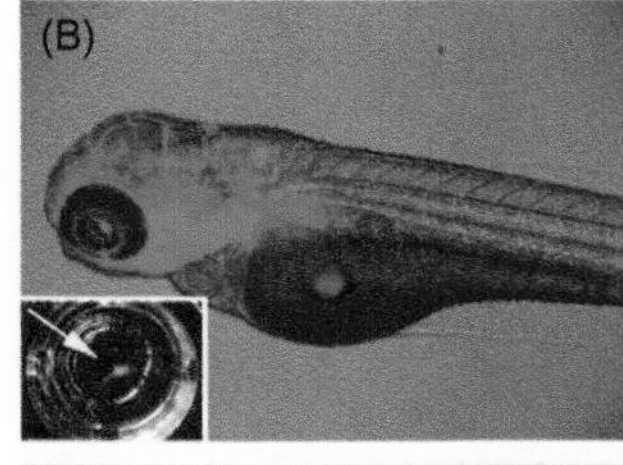

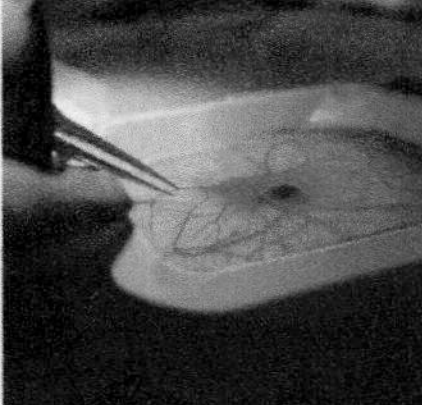
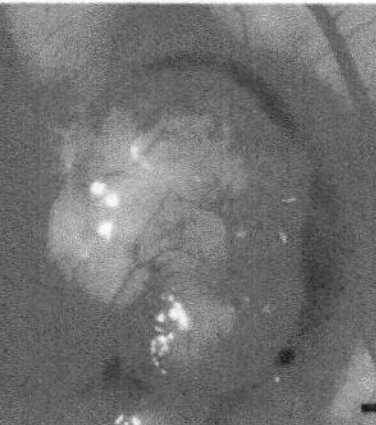

FIGURE 8.2 Human tumor xenograft animal models. (A) Mouse xenograft model. The left panel depicts a NOD/scid immunocompromised mouse carrying a breast tumor xenograft on its left flank (white arrow), which is shown after dissection of the animal in the right panel (images courtesy of Dr Paola Marcato, Dept Pathology, Dalhousie University). (B) Zebrafish xenograft model. Shown in the two panels are a 48-hour post-fertilization (hpf) zebrafish transgenic *fli-GFP* embryo (left) xenografted with Ewing's sarcoma cells (red), and the same embryo 72 hours later (right panel) (i.e. 96 hpf). The blood vessels of the fish appear green due to expression of a green fluorescent transgene (GFP) driven by a promoter derived from the endothelial cell-specific gene *fli1*. The inset depicts the same embryo within one well of a 96-well plate (white arrow). (C) Chicken chorioallantoic membrane xenograft model. In the panel at the left is a chicken egg with a window cut in it to allow access for the engraftment of human tumor cells. The middle panel shows the same chicken embryo *ex ovo*. The right panel depicts a close-up image of a xenografted human fibrosarcoma tumor growing on the chorioallantoic membrane of the chicken embryo. Scale bar = 1 mm. Images courtesy of Dr John Lewis, University of Alberta.

select for the cancer stem cell subset were not always the same across patient samples, although the functional identification of the cancer stem cell subset through xenografts correlated with a specific gene signature [159]. Hence, an emphasis was placed on functional identification of cancer stem cells rather than cell phenotype alone.

The clinical impact of the cancer stem cell model was also demonstrated using the xenografts, as it was proposed that cancer stem cells contributed to cancer relapse following chemotherapies that target rapidly proliferating "bulk" cancer cells while leaving slowly cycling cancer stem cells untouched [160]. Murine xenografts treated with chemotherapeutic agents further demonstrated the drug resistance of cancer stem cells [15], thereby providing a useful model to evaluate the molecular mechanisms mediating drug resistance as xenografts can be harvested and re-xenotransplanted. However, this model has generated controversy as some human cancers can be successfully xenotransplanted with as few as a single cell in the most immune deficient NSG strain [161], bringing into question the universality of the cancer stem cell model for human cancers and whether the hierarchy that it depicts truly exists in human disease or is an artifact of the model itself [162].

Chicken Chorioallantoic Membrane Model

The chicken embryo has also been used for human tumor xenografts. This model is called the chick chorioallantoic membrane assay (CAM) [24]. The CAM is a double-layered mesoderm that develops over 4–10 days post-fertilization (dpf) in the chick embryo and contains an extensive network of blood vessels that connects to the developing embryo's circulation [24]. The CAM assay was originally developed in 1911 by Rous and Murphy and involves removal of a small area of shell prior to the microinjection of tumor cells under the CAM of a 10 dpf chicken embryo [64,163]. The tumor cells then readily engraft forming a tumor mass, after which both angiogenesis and metastasis can be evaluated histologically or via light microscopy through the hole in the eggshell up to 17 dpf [24,163,164] (see Figure 8.2C). This model system is amenable to medium- to high-throughput drug screening for antitumor or anti-angiogenic compounds, and does not require immunosuppression as the early chick embryo (similarly to the zebrafish embryo, which we will discuss below) is naturally immunocompromised preventing xenograft rejection [24]. Nonetheless, care must be taken to avoid non-specific inflammatory reactions when engrafting later stage embryos [24]. The chief disadvantages of this model are the relative lack of chicken-specific antibodies needed to characterize changes in host tissues in response to metastasis, and the fact that the assay must be stopped before hatching of the chick at 17 dpf, thus preventing evaluation of macroscopic metastases in the organs in the chick [24].

Zebrafish Xenografts

Xenotransplantation studies in zebrafish were first conducted in 2005, when Lee et al. showed engraftment of human metastatic melanoma cells in zebrafish embryos at the blastula stage [165]. Subsequently, human tumor cell lines including breast cancer, leukemia, sarcoma have been grown *in vivo* within zebrafish embryos following xenotransplantation [44,166–169] (see

Figure 8.2B). In contrast to mouse xenografts, immunosuppression is not necessary early in development to prevent rejection as zebrafish embryos lack a fully functional immune system until 28 days post-fertilization [170–172]. An initial hurdle in the development of the zebrafish xenograft platform was the difference in temperature requirement for the maintenance of human cells compared with zebrafish embryos, since the latter are usually maintained at 28°C [173] while human cells are usually incubated at 37°C, reflecting human body temperature. Several groups have shown that human cells kept at temperatures as low as 31°C are still able to proliferate and form colonies in soft agar clonogenic assays [43,168,174]. Moreover, results from different groups, including our laboratory, suggest that zebrafish embryos kept at 35°C develop normally, while xenotransplanted human melanoma and leukemia cells still readily proliferate [166,168,171]. Thus, higher zebrafish maintenance temperatures of 34–35°C are most typically used when evaluating biological characteristics of xenotransplanted human cells. Developmental stage and location of injection of human cells have also been explored. While locations of injection range from the duct of Cuvier to the fourth ventricle of the central nervous system and time points from the early blastula to 30 days post-fertilization have been performed, the yolk sac and 24–48 hours post-fertilization (hpf) are generally accepted as the ideal anatomic location and developmental stage for injection, respectively [44,166,171,172,174,175]. Xenotransplantation of human cells into adult zebrafish has also been successfully undertaken, with the advantage of fully developed organ systems including the vasculature [176]. However, the adaptive immune system is also mature at this stage and so immunosuppressive measures such as gamma irradiation or steroid treatment are necessary to promote engraftment. An example of a zebrafish embryo following xenotransplantation with human tumor cells is shown in Figure 8.2B.

Important Considerations in Choosing a Xenograft Approach

As discussed above, human tumor xenografts represent an important step in the progression of genomic studies to preclinical studies of the role of newly identified genetic mutations in cancer, as well as in the definition of drug–tumor interactions that are the prelude to the development of targeted cancer therapies. However, beyond the genetic conservation of the molecular pathways to be studied using a particular xenograft animal model, there are a number of other criteria that must be considered when choosing the appropriate approach and model. These include whether to xenograft tumor cells in an orthotopic or heterotopic manner, and whether to use primary patient tumor material or cell lines. In addition, when using patient tumor material, for example from a biopsy, one must also consider heterogeneity within the tumor and sampling errors that may dramatically affect the interpretation and analysis of drug–tumor interactions. Finally, one must also consider tumor niche and stromal interactions between the tumor material and the xenograft host, which is almost always by necessity an immunocompromised animal. In this section, we will discuss in detail these considerations in the choice of xenograft model.

Orthotopic vs Heterotopic Xenografts

The site of human cell engraftment in the xenograft model is worthy of careful consideration for any scientific experiment. Orthotopic xenotransplantation means the human cells are implanted into the same physiological site in the animal (e.g. breast cancer cells implanted into the mouse breast tissue) whereas human cells implanted into other sites (e.g. intraperitoneal injection of breast cancer cells) is heterotopic xenotransplantation. This difference in location for cell implantation may not only affect the engraftment efficiency (i.e. the number of cells required for the cancer to develop) and viability of these cells (especially if the cells are dependent on the site's ability to support the growth of the cells and hence the term "niche" is used for studying orthotopic engraftment to determine site-specific features or requirements) but, in the case of therapeutic studies, the delivery and efficacy of the drug may also be affected by the xenotransplantation site. For those using xenograft models to assess metastatic disease, cancer cells implanted into heterotopic sites may fail to mimic the usual pattern of metastatic spread of the human disease, whereas orthotopic xenotransplants, although sometimes extremely difficult and technically challenging to accomplish (e.g. pancreatic cancer xenografts by ultrasound-guided injection), appear to reproduce the metastatic spread of the human cancers [136,177].

Nonetheless, the most widely used mouse xenograft model is heterotopic xenotransplantation of human cancer cells or cell lines, usually in the soft tissue of immunodeficient mice by subcutaneous injection. For cancer cell lines transfected with bioluminescent (e.g. beta-lactamase, *lacZ*) or fluorescent (green fluorescent protein, GFP) marker, subcutaneous or intramuscular transplantation sites allow investigators both to measure directly tumor growth (as the tumors are palpable and can be measured in all three dimensions) and facilitate detection and image-based quantification by luminescence or fluorescence. This model is particularly useful for therapy evaluation as the effects of drugs on tumor growth can be assessed easily and serially in a live model.

When niche-dependent factors or metastatic spread is required, orthotopic xenografts provide the most reliable model to assess tissue-specific effects on tumor cells as well as drug effects on cancer cells within the microenvironment. Such details are essential for drug discovery as factors within the microenvironment that support tumor growth or tumor spread can be harnessed as potential drug targets. An additional advantage of orthotopic xenografts is a slightly improved efficiency of engraftment compared to heterotopic models. Where input cell numbers play a critical and limiting factor, orthotopic xenotransplants may yield the highest efficiency.

The zebrafish provides a unique tool for visualization of tumor cell behavior and interaction with host cells. Injection into zebrafish embryos is generally heterotopic with injection into the yolk sac or duct of Cuvier/perivitelline space between the yolk and periderm near the heart and close to the developing subintestinal vessels [172] or into the common cardinal vein [169]. However, human glioblastoma multiforme cell lines can be injected and successfully engraft into the hindbrain ventricle of the central nervous system, representing an orthotopic xenograft model in the zebrafish embryo [174]. While embryonic xenografts at approximately 48 hours post-fertilization have been employed most frequently due to advantages of visualization in transparent embryos and no need for immunosuppression at this developmental stage, adult zebrafish xenografts have also been successful. Again, these have been heterotopic and predominantly by injection into the peritoneal cavity [176]. Xenotransplantation in these more mature hosts requires immunosuppression either by irradiation or dexamethasone treatment [122,176,178,179]. In both embryonic and adult zebrafish xenograft studies, significantly fewer cells are required compared with mice, on the order of 50–100 cells for embryos and up to 300 cells for adult fish. In this way, the zebrafish model may be more advantageous for studies using primary material available in limited quantities [180,181].

Xenotransplantation of Cell Lines vs Primary Cancers

Human cancer cell lines have been the workhorses for cancer studies. These are widely available commercially and from non-profit cell repositories (see Table 8.3), and are employed for cancer genomics, proteomic and high-throughput drug screening analyses. However, a persistent criticism of this approach in therapeutic studies is the lack of correlation between drug effects observed in cell lines compared to those from primary human cancers. Some of these discrepancies may be due to the fact that cell lines maintained under defined growth conditions for long periods of time select for cells that are genetically and epigenetically adapted for those specific culture conditions and are relatively dissimilar to the *in situ* microenvironment represented in the human disease. Such cell line "selection" may lack the metabolic, genetic and epigenetic features seen within the complexity of a solid tumor in the context of its niche [182]. Additionally, cancer cell lines may undergo genetic drift that is divergent from genetic evolution observed within *in situ* human cancers [148]. However, the abundance of cell lines remains a critical factor for their continued use in the study of human cancers as primary patient tumors are difficult to obtain and maintain, limited in repertoire and quantity and expensive to harvest and bank on a routine basis.

While most zebrafish xenograft studies have utilized human cancer cell lines, a small number of reports have successfully engrafted primary human cells. Human pancreas, colon and stomach carcinoma have been engrafted in the yolk of 48-hour zebrafish embryos, and tumor cells were observed to invade the embryo and form micrometastases as early as 24 hours post-injection (hpi) [175]. Importantly, this approach discriminated between infiltrating pancreatic adenocarcinoma and non-invasive chronic pancreatitis. Similarly, both human leukemia cell lines and samples derived from the bone marrow of patients with leukemia have been successfully xenografted into zebrafish embryos. Transplanted K562 erythroleukemia and NB4 acute promyelocytic leukemia cell lines proliferated and migrated throughout the fish and could be followed *in vivo* for several days [166,168] as did patient-derived T- and B-immunophenotype ALL [183] and multiple myeloma samples [184]. In AML, engraftment was demonstrated from CD34 + putative leukemia stem cells but not from CD34 − cells, suggesting that the zebrafish may reflect human disease biology to a similar extent that mouse models do, and thus like mice, represent a valuable tool for studying tumor stem cells [11,156,185].

Clonal Representation of Primary Human Cancers in Xenografts

Human cancers undergo genetic evolution over time and following chemotherapy treatments [147,186,187]. Therefore, an essential element for murine xenografts generated with primary patient cancer cells is faithfully to represent this clonal progression if genetic and/or therapeutic screening are the objectives of the xenotransplantation studies. While the genomics or genomic data are still being generated by deep sequencing analyses for large cohorts of cancer samples, recent reports with limited sample sizes suggest that human cancer cells passaged serially through xenografts were comparable to their primary human breast, colon and blood cancers in mutational profile [188,189]. This remains an important feature of the xenograft model as a tool for medical research as this allows the xenotransplantation model to serve as a validated model in preclinical drug screening and efficacy studies.

Niche/Stromal Interactions and the Humanization of Animal Models

Despite the severe immune deficiencies of the strains used for xenotransplantation, niche-specific barriers were still identified as significant obstacles for the successful and consistent engraftment of some human cells. More recently, "humanization" of these immunodeficient strains by transgenic expression of human leukocyte antigen (HLA) molecules, cytokines and growth factors, or co-injection of human cells with mesenchymal stem cells or stromal cells, has further tailored the murine niches to support specifically the engraftment of human cells (reviewed in [155]). Many of these strains and techniques have now become more widely available through specific investigators or institutions. For reference, Table 8.3 lists a number of online resources from which information can be obtained for currently available immunodeficient and humanized mouse strains.

Transparent zebrafish embryos provide the opportunity to visualize directly the interaction of transplanted human cancer cells with the host microenvironment. A number of transgenic lines are available and can facilitate these studies [125]. The *Tg(fli1-eGFP)* lines in which the *fli1* promoter is able to drive the expression of EGFP in all blood vessels throughout embryogenesis can be used for observing the migration of cancer cells through the vasculature as well as for analyzing tumor-induced angiogenesis (see Figure 8.2B) [190]. Similarly, *Tg(mpx:GFP)* line labels neutrophils [191,192] and the *Tg(mpeg1:eGFP)* labels macrophages [193]. *Tg(cd41:eGFP)* labels both platelets and hematopoietic stem cells [194,195]. These transgenic lines can be incorporated into xenograft studies to evaluate the interaction of these microenvironmental components on the xenografted tumor cells and the impact of the host's inflammatory response on tumor cell behavior. In adults, mutant transparent zebrafish lines have been engineered to permit tracking of transplanted cells down to the single cell level [196]. These fish provide a valuable tool for evaluation of cancer stem cell homing and migration of micrometastases.

CHOOSING AN ANIMAL MODEL

Advantages and Drawbacks of Various Animal Models

It is clear from the presented data that there is no ideal animal model for human cancers, and each model has its benefits and drawbacks (summarized in Figure 8.1). For example, the dog and the chicken are not easily manipulated genetically, yet spontaneous tumors in dogs resemble human malignancy to a greater degree than tumors in mice or zebrafish, and the genetic heritage of dog breeds make them ideal for GWA studies for cancer susceptibility alleles. Unquestionably, mice and zebrafish are two models that are very well-suited to transgenic and reverse genetic approaches, the chemical induction of cancers, as well as human tumor xenograft experiments. Yet even between these two models there are dramatic differences in cost, degree of gene conservation, and ease of use for high-throughput chemical or genetic screens, with the zebrafish being cheaper and a higher-throughput animal model at the cost of being a less genetically conserved model to humans than mice. For dogs, the moral and ethical issues surrounding chemical screens and xenografts make them wholly unsuited to these kinds of experiments. Invertebrate models, such as the fly and worm, do not share these concerns, benefit from well documented and powerful tools for genetic manipulation and, due to their size and short life cycle, are also excellent candidate organisms for high-throughput genetic and chemical screens. However, human tumor xenotransplantation in invertebrates is not feasible and their evolutionary distance from humans puts the conclusions drawn from invertebrate models under greater scrutiny, requiring extensive validation in mammalian preclinical models such as the mouse. Since the two most relevant animal models for the validation of gain-of-function mutations, cancer susceptibility alleles, and oncogenic translocation gene products identified by cancer genomics studies are the mouse and the zebrafish, in this chapter, we have primarily focused on these two animal models.

Xenografts vs Transgenic Models in Mice and Zebrafish

In regard to technologies, xenograft models are expensive, labor-intensive and occasionally impossible to generate due to intrinsic or extrinsic elements necessary for successful engraftment. However, the ability to assay a patient-specific drug response provides a powerful incentive for clinicians, as well as scientists, to develop xenograft models that can be used to predict clinical outcome, especially in cases of high-risk cancers such as lung cancer and pediatric T-cell acute lymphoblastic leukemia. GEMMs are initially expensive, slow to develop and potentially difficult to maintain (especially if lethal as homozygotes) but, once established, they represent a powerful tool to study single gene effects in cancer biology, for the validation of gain-of-function mutations identified by cancer genomics, and when transgenic or other genetically manipulated strains are combined, can closely resemble the complexity and natural progression of the human disease as additional mutations can spontaneously occur within the model. GEMMs are immune-competent and can be used for therapy studies where an intact immune system may be required to achieve treatment effects, whereas cancer xenografts in immunodeficient hosts cannot.

Despite the general conservation of oncogenic pathways between zebrafish and humans, some key genes have not been identified, such as the apparent absence of the *INK4a/ARF* tumor suppressor gene [197]. In addition, many of the tools and assays that have evolved for the interrogation of gene function, particularly at the protein level, remain in their developmental infancy for application to the zebrafish. Thus, the ability fully to recapitulate the genetic complexity of human tumors and provide the in-depth analysis currently available in the mouse model may not be possible using zebrafish at present and may remain limited in the future. These challenges have been overcome to some degree with the recent advent of zebrafish xenograft platforms.

CONCLUSIONS

Cancer genome studies continue to generate an ever-growing list of susceptibility alleles, and genetic mutations associated with cancers of various origins. As discussed in this chapter, animal models have and will continue to play a critical role in the study of how cancer-associated genetic changes contribute to cancer development, progression and treatment response. In particular, they are an essential means of translating genomic and *in vitro* tissue culture studies into preclinical studies of molecular targeted therapies for cancers. The choice of animal model is complex and multifactorial, as each animal model has its inherent advantages and disadvantages.

The mouse represents the "gold-standard" for the study of cancer, due to its close approximation to humans as a mammalian model that is easily manipulated genetically, and is amenable to both tumor xenografts and transgenic approaches. The chicken, historically a breakthrough animal model for the discovery of tumor viruses, is now being employed as an embryonic animal tumor model for drug screening and the study of angiogenesis and metastasis. Other vertebrate animal models include the domestic dog, a model highly amenable to GWA studies of cancer susceptibility given the unique genetics of these companion animals and the zebrafish, which is ideally suited to transgenic and xenograft approaches and is relatively inexpensive. Within these model systems, transgenic and xenograft strategies offer different opportunities. For example, transgenic mice require considerable resources to generate and maintain but once generated provide a permanent experimental system for studying cancer processes that preserves tumor–immune system interactions. By contrast, primary human tumor xenografts in mice, which are ideal for the study of tumor–drug interactions for personalized therapy, rely on finite tumor samples and employ immune-compromised animals, limiting the study of the interaction of immune cells with tumor cells in a context representative of what occurs in humans. Some of these limitations can now be overcome by the more time- and cost-efficient use of the zebrafish for transgenic and xenograft studies, but the true scope and utility of the fish model is still to be fully realized.

Finally, invertebrate animal models such as the fly and worm are the least expensive, and are highly suited to the discovery of modifiers of cancer-related pathways and, of all the models, are the most amenable to high-throughput chemical and genetic screens. However, invertebrate models are less ideal for the discovery and development of cancer therapeutics due to much lower gene conservation as compared to mammals or other vertebrates, which can reduce the ability of scientists to translate results in these models to patient treatment. In the end, no single animal model is suitable for studying every aspect of cancer. However, through the judicious choice of animal model and/or combination of these models, experiments can progress from gene discovery to preclinical development of molecular targeted agents, and the promise of cancer genomics to enable personalized cancer treatment may be realized.

GLOSSARY

Animal model An animal that is used to study disease processes, such as cancer.

Chimera A term used in genetics to describe an animal such as a mouse that is composed of a mixture of wild-type and genetically engineered cells that are derived from different animals.

Chorioallantoic membrane A double-layered mesoderm that develops over 4–10 days post-fertilization in the chick embryo and contains an extensive network of blood vessels that connects to the developing embryo's circulation.

Germline A term used in genetics to describe the cells within an animal that contribute to gamete production and therefore can pass on genetic information to the offspring.

Haploinsufficient A term in genetics used to describe a condition in diploid organisms where a single functional copy of a gene allele is incapable of producing enough gene product to allow the normal growth or development of the organism resulting often in a disease phenotype.

Heterotopic A term used to describe the engraftment or transplantation of cells to an abnormal or atypical anatomical location (e.g. lung cells transplanted under the epidermis).

Homozygous A term in genetics that refers to an individual or animal that has two identical copies of a particular gene allele.

Immunocompromised An animal or individual that is incapable of mounting an effective immune response.

Immune-privileged A term usually referring to an anatomical site in the body that is not subjected to immune surveillance and thus cells transplanted to that site or tissue are not rejected by the immune system.

Linkage disequilibrium A term in genetics that refers to the non-random association of two or more genetic loci within individuals in a population, which may or may not be on the same chromosome.

Mosaic A term used in genetics to refer to an animal composed of two or more cytogenetically distinct subpopulations of cells that were derived from the same zygote. This term is often used to describe the results of transgene experiments involving injection of genetic material into mouse and zebrafish single-cell embryos (i.e. the zygote).

Non-germline A term used in genetics to describe the cells within an animal that do not contribute to gamete production and therefore do not pass on genetic information to the offspring.

Orthotopic A term used to describe the engraftment or transplantation of cells to the same anatomical location from which they were derived (e.g. mammary cells transplanted to the mammary fat pad).

Preclinical A term that usually refers to experimentation or medical treatment in animals, with the aim of collecting data in support of the safety and efficacy of a new treatment prior to the initiation of clinical trials in humans.

Stroma A structural network of cells, blood vessels, and extracellular material that forms the connective tissue underlying tissue and organ formation.

Transgenic A term that is used to describe an animal that has been genetically engineered to express other genetic material, referred to as a transgene, that is not normally found in that animal or in that chromosomal position if the gene is derived from an endogenous sequence.

Tumor niche The specialized microenvironment in which a tumor grows, that includes stromal fibroblasts, blood vessels and immune cells.

Xenotransplantation Cross-species transplantation of cells from one organism to another (e.g. transplantation of human leukemia cells in the mouse).

ABBREVIATIONS

ALL Acute lymphoblastic leukemia
AML Acute myeloblastic leukemia
APC Adenomatous polyposis coli
BAC Bacterial artificial chromosome
BNX beige/nude/xid
CAM Chorioallantoic membrane
ES Embryonic stem
FLCN Folliculin
GEMM Genetically engineered mouse model
GFP Green fluorescent protein
GWA Genome-wide association
HLA Human leukocyte antigen
HPF Hours post-fertilization
HPV Human papilloma virus
LAK Lymphokine-activated killer
LD Linkage disequilibrium
MMTV Murine mammary tumor virus
nGEMM Non-germline genetically engineered mouse model
NK Natural killer
NOD Non-obese diabetic
NHL Non-Hodgkin lymphoma
PCR Polymerase chain reaction
RNAi RNA interference
SCID Severe combined immunodeficiency
SNP Small nucleotide polymorphism
TALEN Transcription activator-like effector nuclease
TSC1 Tuberous sclerosis 1
WNT Wingless type MMTV integration site
XID X-linked immunodeficiency
ZFN Zinc finger nucleases

REFERENCES

[1] Murphy JB. Transplantability of tissues to the embryo of foreign species. Its bearing on questions of tissue specificity and tumor immunity. J Exp Med 1913;17:482–93.

[2] Greene H. The heterologous transplantation of mouse tumors induced in vitro. Cancer Res 1946;6:396–402.

[3] Quayle J. Ability of CEA blood levels to reflect tumour burden: a study in a human xenograft model. Br J Cancer 1982;46:220–7.

[4] Flanagan S. 'Nude', a new hairless gene with pleiotropic effects in the mouse. Genet Res 1966;8:295–309.

[5] Pantelouris EM, Hair J. Thymus dysgenesis in nude (nu nu) mice. J Embryol Exp Morphol 1970;24:615–23.

[6] Povlsen C, Rygaard J. Heterotransplantation of human adenocarcinomas of the colon and rectum to the mouse mutant Nude. A study of nine consecutive transplantations. Acta Pathol Microbiol Scand A 1971;79:159–69.

[7] Giovanella B, Stehlin J, Williams LJ. Heterotransplantation of human malignant tumors in "nude" thymusless mice. II. Malignant tumors induced by injection of cell cultures derived from human solid tumors. J Natl Cancer Inst 1974;52:921–30.

[8] Clarke R, Dickson R. Animal models of tumor onset, growth, and metastasis. In: Bertino J, editor. Encyclopedia of cancer. New York: Academic Press; 1997.

[9] Bosma G, Custer R. A severe combined immunodeficiency mutation in the mouse. Nature 1983;301:527–30.

[10] Theocharides A, Jin L, Cheng P, Prasolava TK, Malko AV, Ho JM, et al. Disruption of SIRPα signaling in macrophages eliminates human acute myeloid leukemia stem cells in xenografts. J Exp Med 2012;209:1883–99.

[11] Lapidot T, Sirard C, Vormoor J, Murdoch B, Hoang T, Caceres-Cortes J, et al. A cell initiating human acute myeloid leukaemia after transplantation into SCID mice. Nature 1994;367:645–8.

[12] Cox C, Martin H, Kearns P, Virgo P, Evely R, Blair A. Characterization of a progenitor cell population in childhood T-cell acute lymphoblastic leukemia. Blood 2007;109:674–82.

[13] Thiollier C, Lopez C, Gerby B, Ignacimouttou C, Poglio S, Duffourd Y, et al. Characterization of novel genomic alterations and therapeutic approaches using acute megakaryoblastic leukemia xenograft models. J Exp Med 2012;209:2017–31.

[14] Lock RB, Liem N, Farnsworth ML, Milross CG, Xue C, Tajbakhsh M, et al. The nonobese diabetic/severe combined immunodeficient (NOD/SCID) mouse model of childhood acute lymphoblastic leukemia reveals intrinsic differences in biologic characteristics at diagnosis and relapse. Blood 2002;99:4100–8.

[15] Chiu P, Jiang H, Dick JE. Leukemia-initiating cells in human T-lymphoblastic leukemia exhibit glucocorticoid resistance. Blood 2010;116:5268–79.

[16] Singh S, Hawkins C, Clarke I, Squire JA, Bayani J, Hide T, et al. Identification of human brain tumour intiating cells. Nature 2004;432:396–401.

[17] O'Brien C, Pollett A, Gallinger S, Dick J. A human colon cancer cell capable of initiating tumour growth in immunodeficient mice. Nature 2007;445:106–10.

[18] Ricci-Vitiani L, Lombardi D, Pilozzi E, Biffoni M, Todaro M, Peschle C, et al. Identification and expansion of human colon-cancer-initiating cells. Nature 2007;445:111–5.

[19] Al-Hajj M, Wicha MS, Benito-Hernandez A, Morrison SJ, Clarke MF. Prospective identification of tumorigenic breast cancer cells. Proc Natl Acad Sci USA 2003;100:3983–8.

[20] Quintana E, Shackleton M, Sabel MS, Fullen DR, Johnson TM, Morrison SJ. Efficient tumour formation by single human melanoma cells. Nature 2008;456:593–9.

[21] Brinster RL, Chen HY, Messing A, van Dyke T, Levine AJ, Palmiter RD. Transgenic mice harboring SV40 T-antigen genes develop characteristic brain tumors. Cell 1984;37:367–79.

[22] Stewart TA, Pattengale PK, Leder P. Spontaneous mammary adenocarcinomas in transgenic mice that carry and express MTV/myc fusion genes. Cell 1984;38:627–37.

[23] Karlsson EK, Lindblad-Toh K. Leader of the pack: gene mapping in dogs and other model organisms. Nat Rev Genet 2008;9:713–25.

[24] Cimpean AM, Ribatti D, Raica M. The chick embryo chorioallantoic membrane as a model to study tumor metastasis. Angiogenesis 2008;11:311–9.

[25] Stoletov K, Klemke R. Catch of the day: zebrafish as a human cancer model. Oncogene 2008;27:4509–20.

[26] Potts MB, Cameron S. Cell lineage and cell death: *Caenorhabditis elegans* and cancer research. Nat Rev Cancer 2011;11:50–8.

[27] Rudrapatna VA, Cagan RL, Das TK. *Drosophila* cancer models. Dev Dyn 2012;241:107–18.

[28] Payne E, Look T. Zebrafish modelling of leukaemias. Br J Haematol 2009;146:247–56.

[29] Amatruda JF, Shepard JL, Stern HM, Zon LI. Zebrafish as a cancer model system. Cancer Cell 2002;1:229–31.

[30] Berghmans S, Jette C, Langenau D, Hsu K, Stewart R, Look T, et al. Making waves in cancer research: new models in the zebrafish. BioTechniques 2005;39:227–37.

[31] Langenau DM, Keefe MD, Storer NY, Guyon JR, Kutok JL, Le X, et al. Effects of RAS on the genesis of embryonal rhabdomyosarcoma. Genes Dev 2007;21:1382–95.

[32] Gutierrez A, Grebliunaite R, Feng H, Kozakewich E, Zhu S, Guo F, et al. Pten mediates Myc oncogene dependence in a conditional zebrafish model of T cell acute lymphoblastic leukemia. J Exp Med 2011;208:1595–603.

[33] Parng C, Seng WL, Semino C, McGrath P. Zebrafish: a preclinical model for drug screening. Assay Drug Dev Technol 2002;1:41–8.

[34] Amatruda JF, Patton EE. Genetic models of cancer in zebrafish. Int Rev Cell Mol Biol 2008;271:1–34.

[35] Spitsbergen JM, Tsai HW, Reddy A, Miller T, Arbogast D, Hendricks JD, et al. Neoplasia in zebrafish (Danio rerio) treated with 7,12-dimethylbenz[a]anthracene by two exposure routes at different developmental stages. Toxicol Pathol 2000;28:705–15.

[36] Spitsbergen JM, Tsai HW, Reddy A, Miller T, Arbogast D, Hendricks JD, et al. Neoplasia in zebrafish (Danio rerio) treated with N-methyl-N′-nitro-N-nitrosoguanidine by three exposure routes at different developmental stages. Toxicol Pathol 2000;28:716–25.

[37] Stern HM, Zon LI. Cancer genetics and drug discovery in the zebrafish. Nat Rev Cancer 2003;3:533–9.

[38] Mullins MC, Hammerschmidt M, Haffter P, Nusslein-Volhard C. Large-scale mutagenesis in the zebrafish: in search of genes controlling development in a vertebrate. Curr Biol 1994;4:189–202.

[39] Zon LI, Peterson RT. In vivo drug discovery in the zebrafish. Nat Rev Drug Discov 2005;4:35–44.

[40] Amsterdam A, Hopkins N. Mutagenesis strategies in zebrafish for identifying genes involved in development and disease. Trends Genet 2006;22:473–8.

[41] Soucy TA, Smith PG, Milhollen MA, Berger AJ, Gavin JM, Adhikari S, et al. An inhibitor of NEDD8-activating enzyme as a new approach to treat cancer. Nature 2009;458:732–6.

[42] Langenau DM, Traver D, Ferrando AA, Kutok JL, Aster JC, Kanki JP, et al. Myc-induced T cell leukemia in transgenic zebrafish. Science 2003;299:887–90.

[43] Wu M, Jung L, Cooper AB, Fleet C, Chen L, Breault L, et al. Dissecting genetic requirements of human breast tumorigenesis in a tissue transgenic model of human breast cancer in mice. Proc Natl Acad Sci USA 2009;106:7022–7.

[44] Konantz M, Balci TB, Hartwig UF, Dellaire G, Andre MC, Berman JN, et al. Zebrafish xenografts as a tool for *in vivo* studies on human cancer. Ann NY Acad Sci 2012;1266:124–37.

[45] Moore FE, Langenau DM. Through the looking glass: visualizing leukemia growth, migration, and engraftment using fluorescent transgenic zebrafish. Adv Hematol 2012;2012:478164.

[46] Berman J, Hsu K, Look AT. Zebrafish as a model organism for blood diseases. Br J Haematol 2003;123:568–76.

[47] Langenau DM, Feng H, Berghmans S, Kanki JP, Kutok JL, Look AT. Cre/lox-regulated transgenic zebrafish model with conditional myc-induced T cell acute lymphoblastic leukemia. Proc Natl Acad Sci USA 2005;102:6068–73.

[48] Forrester AM, Grabher C, McBride ER, Boyd ER, Vigerstad MH, Edgar A, et al. NUP98-HOXA9-transgenic zebrafish develop a myeloproliferative neoplasm and provide new insight into mechanisms of myeloid leukaemogenesis. Br J Haematol 2011;155:167–81.

[49] Campbell LJ, Willoughby JJ, Jensen AM. Two types of tet-on transgenic lines for doxycycline-inducible gene expression in zebrafish rod photoreceptors and a gateway-based tet-on toolkit. PLoS One 2012;7:e51270.

[50] Li Z, Huang X, Zhan H, Zeng Z, Li C, Spitsbergen JM, et al. Inducible and repressable oncogene-addicted hepatocellular carcinoma in Tet-on xmrk transgenic zebrafish. J Hepatol 2012;56:419–25.

[51] Li Z, Zheng W, Wang Z, Zeng Z, Zhan H, Li C, et al. A transgenic zebrafish liver tumor model with inducible Myc expression reveals conserved Myc signatures with mammalian liver tumors. Dis Model Mech 2012;6(2):414–23.

[52] Foley JE, Maeder ML, Pearlberg J, Joung JK, Peterson RT, Yeh JR. Targeted mutagenesis in zebrafish using customized zinc-finger nucleases. Nat Protoc 2009;4:1855–67.

[53] Foley JE, Yeh JR, Maeder ML, Reyon D, Sander JD, Peterson RT, et al. Rapid mutation of endogenous zebrafish genes using zinc finger nucleases made by Oligomerized Pool ENgineering (OPEN). PLoS One 2009;4:e4348.

[54] Cade L, Reyon D, Hwang WY, Tsai SQ, Patel S, Khayter C, et al. Highly efficient generation of heritable zebrafish gene mutations

using homo- and heterodimeric TALENs. Nucleic Acids Res 2012;40:8001–10.

[55] Moore FE, Reyon D, Sander JD, Martinez SA, Blackburn JS, Khayter C, et al. Improved somatic mutagenesis in zebrafish using transcription activator-like effector nucleases (TALENs). PLoS One 2012;7:e37877.

[56] Sander JD, Cade L, Khayter C, Reyon D, Peterson RT, Joung JK, et al. Targeted gene disruption in somatic zebrafish cells using engineered TALENs. Nat Biotechnol 2011;29:697–8.

[57] Detrich III HW, Westerfield M, Zon LI. Overview of the zebrafish system. Methods Cell Biol 1999;59:3–10.

[58] Rowell JL, McCarthy DO, Alvarez CE. Dog models of naturally occurring cancer. Trends Mol Med 2011;17:380–8.

[59] Paoloni M, Khanna C. Translation of new cancer treatments from pet dogs to humans. Nat Rev Cancer 2008;8:147–56.

[60] Vail DM, MacEwen EG. Spontaneously occurring tumors of companion animals as models for human cancer. Cancer Invest 2000;18:781–92.

[61] Jonasdottir TJ, Mellersh CS, Moe L, Heggebo R, Gamlem H, Ostrander EA, et al. Genetic mapping of a naturally occurring hereditary renal cancer syndrome in dogs. Proc Natl Acad Sci USA 2000;97:4132–7.

[62] Lingaas F, Comstock KE, Kirkness EF, Sorensen A, Aarskaug T, Hitte C, et al. A mutation in the canine BHD gene is associated with hereditary multifocal renal cystadenocarcinoma and nodular dermatofibrosis in the German Shepherd dog. Hum Mol Genet 2003;12:3043–53.

[63] Rosenberger JA, Pablo NV, Crawford PC. Prevalence of and intrinsic risk factors for appendicular osteosarcoma in dogs: 179 cases (1996–2005). J Am Vet Med Assoc 2007;231:1076–80.

[64] Rous P. A sarcoma of the fowl transmissible by an agent separable from the tumor cells. J Exp Med 1911;13:397–9.

[65] Rubin H. The early history of tumor virology: Rous, RIF, and RAV. Proc Natl Acad Sci USA 2011;108:14389–96.

[66] Plachy J, Hejnar J. Chicken cells – oncogene transformation, immortalization and more. Folia Biol (Praha) 2002;48:126–38.

[67] Purchase HG. Prevention of Marek's disease: a review. Cancer Res 1976;36:696–700.

[68] Ross NL. T-cell transformation by Marek's disease virus. Trends Microbiol 1999;7:22–9.

[69] Shibata H, Toyama K, Shioya H, Ito M, Hirota M, Hasegawa S, et al. Rapid colorectal adenoma formation initiated by conditional targeting of the Apc gene. Science 1997;278:120–3.

[70] Yamazoe M, Sonoda E, Hochegger H, Takeda S. Reverse genetic studies of the DNA damage response in the chicken B lymphocyte line DT40. DNA Repair (Amst) 2004;3:1175–85.

[71] Barua A, Bitterman P, Abramowicz JS, Dirks AL, Bahr JM, Hales DB, et al. Histopathology of ovarian tumors in laying hens: a preclinical model of human ovarian cancer. Int J Gynecol Cancer 2009;19:531–9.

[72] Lim W, Kim HS, Jeong W, Ahn SE, Kim J, Kim YB, et al. SERPINB3 in the chicken model of ovarian cancer: a prognostic factor for platinum resistance and survival in patients with epithelial ovarian cancer. PLoS One 2012;7:e49869.

[73] Tufan AC, Satiroglu-Tufan NL. The chick embryo chorioallantoic membrane as a model system for the study of tumor angiogenesis, invasion and development of anti-angiogenic agents. Curr Cancer Drug Targets 2005;5:249–66.

[74] International Chicken Genome Sequencing Consortium. Sequence and comparative analysis of the chicken genome provide unique perspectives on vertebrate evolution. Nature 2004;432:695-716.

[75] Polesello C, Roch F, Gobert V, Haenlin M, Waltzer L. Modeling cancers in *Drosophila*. Prog Mol Biol Transl Sci 2011;100:51–82.

[76] Zhivotovsky B, Orrenius S. Carcinogenesis and apoptosis: paradigms and paradoxes. Carcinogenesis 2006;27:1939–45.

[77] Zamore PD. RNA interference: big applause for silencing in Stockholm. Cell 2006;127:1083–6.

[78] Mohr SE, Perrimon N. RNAi screening: new approaches, understandings, and organisms. Wiley Interdiscip Rev RNA 2011;3:145–58.

[79] Stark MB. An hereditary tumor in *Drosophila*. J Cancer Res 1918;3:279–301.

[80] Potter CJ, Turenchalk GS, Xu T. *Drosophila* in cancer research. An expanding role. Trends Genet 2000;16:33–9.

[81] Ranganathan P, Weaver KL, Capobianco AJ. Notch signalling in solid tumours: a little bit of everything but not all the time. Nat Rev Cancer 2011;11:338–51.

[82] Sternberg PW, Han M. Genetics of RAS signaling in *C. elegans*. Trends Genet 1998;14:466–72.

[83] Chen K, Ahmed S, Adeyi O, Dick J, Ghanekar A. Human solid tumor xenografts in immunodeficient mice are vulnerable to lymphomagenesis associated with Epstein-Barr virus. PLoS One 2012;7:e39294.

[84] Gladstone M, Su TT. Chemical genetics and drug screening in *Drosophila* cancer models. J Genet Genomics 2011;38:497–504.

[85] Gonsalves FC, Klein K, Carson BB, Katz S, Ekas LA, Evans S, et al. An RNAi-based chemical genetic screen identifies three small-molecule inhibitors of the Wnt/wingless signaling pathway. Proc Natl Acad Sci USA 2011;108:5954–63.

[86] Donehower LA, Harvey M, Slagle BL, McArthur MJ, Montgomery CA, Butel JS, et al. Mice deficient for p53 are developmentally normal but susceptible to spontaneous tumours. Nature 1992;356:215–21.

[87] Stewart TA, Pattengale PK, Leder P. Spontaneous mammary adenocarcinomas in transgenic mice that carry and express MTV/myc fusion genes. Cell 1984;38:627–37.

[88] Allen JM, Forbush KA, Perlmutter RM. Functional dissection of the lack proximal promoter. Mol Cell Biol 1992;12:2758–68.

[89] Wildin RS, Wang HU, Forbrush KA, Perlmutter RM. Functional dissection of the murine lck distal promoter. J Immunol 1995;155:1286–95.

[90] Kasper S. Survey of genetically engineered mouse models for prostate cancer: analyzing the molecular basis of prostate cancer development, progression, and metastasis. J Cell Biochem 2005;94:279–97.

[91] Zhang J, Gao N, DeGraff DJ, Yu X, Sun Q, Case TC, et al. Characterization of cis elements of the probasin promoter necessary for prostate-specific gene expression. Prostate 2010;70:934–51.

[92] Chiu I, Touhalisky K, Liu Y, Yates A, Frostholm A. Tumorigenesis in transgenic mice in which the SV40 T antigen is driven by the brain-specific FGF1 promoter. Oncogene 2000;19:6229–39.

[93] Qian F, Hanahan D, Weissman IL. L-selectin can facilitate metastasis to lymph nodes in a transgenic mouse model of carcinogenesis. Proc Natl Acad Sci USA 2001;98:3976–81.

[94] Adams JM, Cory S. Transgenic models of tumor development. Science 1991;254:1161–7.

[95] Ewald D, Li M, Efrat S, Auer G, Wall RJ, Furth PA, et al. Time-sensitive reversal of hyperplasia in transgenic mice expressing SV40 T antigen. Science 1996;273:1384–6.
[96] Landesman-Bollag E, Channavajhala P, Cardiff R, Seldin D. p53 deficiency and misexpression of protein kinase CK2alpha collaborate in the development of thymic lymphomas in mice. Oncogene 1998;16:2965–74.
[97] McCormack S, Weaver Z, Deming S, Natarajan N, Torri J, Johnson MD, et al. Myc/p53 interactions in transgenic mouse mammary development, tumorigenesis and chromosomal instability. Oncogene 1998;16: 2755–66.
[98] Sauer B, Henderson N. Site-specific DNA recombination in mammalian cells by the Cre recombinase of bacteriophage P1. Proc Natl Acad Sci USA 1988;85:5166–70.
[99] Meuwissen R, Linn S, van der Valk M, Mooi W, Berns A. Mouse model for lung tumorigenesis through Cre/lox controlled sporadic activation of the K-Ras oncogene. Oncogene 2001;20:6551–8.
[100] Gu H, Marth J, Orban P, Mossmann H, Rajewsky K. Deletion of a DNA polymerase beta gene segment in T cells using cell type-specific gene targeting. Science 1994;265:103–6.
[101] McLeod M, Craft S, Broach JR. Identification of the crossover site during FLP-mediated recombination in the *Saccaromyces cerevisiae* plasmid 2μm circle. Mol Cell Biol 1986;6:3357–67.
[102] Tuveson D, Shaw A, Willis N, Silver DP, Jackson EL, Chang S, et al. Endogenous oncogenic K-ras(G12D) stimulates proliferation and widespread neoplastic and developmental defects. Cancer Cell 2004;5:375–87.
[103] Marino S, Vooijs M, van der Gulden H, Jonkers J, Berns A. Induction of medulloblastomas in p53-null mutant mice by somatic inactivation of Rb in the external granular layer cells of the cerebellum. Genes Dev 2000;14:994–1004.
[104] Trimboli A, Cantemir-Stone C, Li F, Wallace JA, Merchant A, Creasap N, et al. Pten in stromal fibroblasts suppresses mammary epithelial tumours. Nature 2009;461:1084–91.
[105] Zhou Y, Rideout WR, Zi T, Bressel A, Reddypalli S, Rancourt R, et al. Chimeric mouse tumor models reveal differences in pathway activation between ERBB family- and KRAS-dependent lung adenocarcinomas. Nat Biotechnol 2010;28:71–8.
[106] Pear WS, Miller JP, Xu L, Pui JC, Soffer B, Quackenbush RC, et al. Efficient and rapid induction of a chronic myelogenous leukemia-like myeloproliferative disease in mice receiving P210 bcr/abl-transduced bone marrow. Blood 1998;92:3780–92.
[107] Zender L, Spector M, Xue W, Flemming P, Cordon-Cardo C, Silke J, et al. Identification and validation of oncogenes in liver cancer using an integrative oncogenomic approach. Cell 2006;125:1253–67.
[108] Bhowmick NA, Neilson EG, Moses HL. Stromal fibroblasts in cancer initiation and progression. Nature 2004;432:332–7.
[109] Hanahan D, Weinberg RA. Hallmarks of cancer: the next generation. Cell 2011;144:646–74.
[110] Huber S, Gagliani N, Zenewicz LA, Huber FJ, Bosurgi L, Hu B, et al. IL-22BP is regulated by the inflammasome and modulates tumorigenesis in the intestine. Nature 2012;491:259–63.
[111] Qian F, Hanahan D, Weissman I. L-selectin can facilitate metastasis to lymph nodes in a transgenic mouse model of carcinogenesis. Proc Natl Acad Sci USA 2001;98:3976–81.
[112] Hill R, Song Y, Cardiff R, van Dyke T. Selective evolution of stromal mesenchyme with p53 loss in response to epithelial tumorigenesis. Cell 2005;123:1001–11.
[113] Nunez-Cruz S, Gimotty PA, Guerra MW, Connolly DC, Wu YQ, DeAngelis RA, et al. Genetic and pharmacologic inhibition of complement impairs endothelial cell function and ablates ovarian cancer neovascularization. Neoplasia 2012;14:994–1004.
[114] Amacher SL. Transcriptional regulation during zebrafish embryogenesis. Curr Opin Genet Dev 1999;9:548–52.
[115] Jessen JR, Willett CE, Lin S. Artificial chromosome transgenesis reveals long-distance negative regulation of rag1 in zebrafish. Nat Genet 1999;23:15–6.
[116] Carradice D, Lieschke GJ. Zebrafish in hematology: sushi or science? Blood 2008;111:3331–42.
[117] Yeh JR, Munson KM, Chao YL, Peterson QP, Macrae CA, Peterson RT. AML1-ETO reprograms hematopoietic cell fate by downregulating scl expression. Development 2008;135:401–10.
[118] Le X, Langenau DM, Keefe MD, Kutok JL, Neuberg DS, Zon LI. Heat shock-inducible Cre/Lox approaches to induce diverse types of tumors and hyperplasia in transgenic zebrafish. Proc Natl Acad Sci USA 2007;104:9410–5.
[119] Sabaawy HE, Azuma M, Embree LJ, Tsai HJ, Starost MF, Hickstein DD. TEL-AML1 transgenic zebrafish model of precursor B cell acute lymphoblastic leukemia. Proc Natl Acad Sci USA 2006;103:15166–71.
[120] Dovey M, White RM, Zon LI. Oncogenic NRAS cooperates with p53 loss to generate melanoma in zebrafish. Zebrafish 2009;6:397–404.
[121] Leacock SW, Basse AN, Chandler GL, Kirk AM, Rakheja D, Amatruda JF. A zebrafish transgenic model of Ewing's sarcoma reveals conserved mediators of EWS-FLI1 tumorigenesis. Dis Model Mech 2012;5:95–106.
[122] Patton EE, Widlund HR, Kutok JL, Kopani KR, Amatruda JF, Murphey RD, et al. BRAF mutations are sufficient to promote nevi formation and cooperate with p53 in the genesis of melanoma. Curr Biol 2005;15:249–54.
[123] Park SW, Davison JM, Rhee J, Hruban RH, Maitra A, Leach SD. Oncogenic KRAS induces progenitor cell expansion and malignant transformation in zebrafish exocrine pancreas. Gastroenterology 2008;134:2080–90.
[124] Thermes V, Grabher C, Ristoratore F, Bourrat F, Choulika A, Wittbrodt J, et al. I-SceI meganuclease mediates highly efficient transgenesis in fish. Mech Dev 2002;118:91–8.
[125] Kawakami K. Transposon tools and methods in zebrafish. Dev Dyn 2005;234:244–54.
[126] Villefranc JA, Amigo J, Lawson ND. Gateway compatible vectors for analysis of gene function in the zebrafish. Dev Dyn 2007;236:3077–87.
[127] Tang K, Dai Y, Tong M, Chan YP, Kwan PS, Fu L, et al. A $CD90^+$ tumor-initiating cell population with an aggressive signature and metastatic capacity in esophageal cancer. Cancer Res 2013;73(7):2322–32.
[128] Langenau DM, Keefe MD, Storer NY, Jette CA, Smith ACH, Ceol CJ, et al. Co-injection strategies to modify radiation sensitivity and tumor initiation in transgenic zebrafish. Oncogene 2008;27:4242–8.
[129] Blackburn C, Augustine C, Li R, Harvey RP, Malin MA, Boyd RL, et al. The nu gene acts cell-autonomously and is required for differentiation of thymic epithelial progenitors. Proc Natl Acad Sci USA 1996;93:5742–6.
[130] Povlsen C, Fialkow P, Klein E, Klein G, Rygaard J, Wiener F. Growth and antigenic properties of a biopsy-derived Burkitt's lymphoma in thymus-less (nude) mice. Int J Cancer 1973;11:30–9.

[131] Rygaard J, Povlsen C. Heterotransplantation of a human malignant tumour to "Nude" mice. Acta Pathol Microbiol Scand 1969;77:758–60.

[132] Povlsen C, Rygaard J. Heterotransplantation of human epidermoid carcinomas to the mouse mutant nude. Acta Pathol Microbiol Scand A 1972;80:713–7.

[133] Giovanella B, Yim S, Stehlin J, Williams LJ. Development of invasive tumors in the "nude" mouse after injection of cultured human melanoma cells. J Natl Cancer Inst 1972;48:1531–3.

[134] Fulop G, Phillips RA. The scid mutation in mice causes a general defect in DNA repair. Nature 1990;347:479–82.

[135] De Lord C, Clutterbuck R, Titley J, Ormerod M, Gordon-Smith T, Millar J, et al. Growth of primary human acute leukemia in severe combined immunodeficient mice. Exp Hematol 1991;19:991–3.

[136] Fu X, Hoffman R. Human ovarian carcinoma metastatic models constructed in nude mice by orthotopic transplantation of histologically-intact patient specimens. Anticancer Res 1993;13:283–6.

[137] Williams K, Ghosh R, Giridhar P, Gu G, Case T, Belcher SM, et al. Inhibition of stathmin1 accelerates the metastatic process. Cancer Res 2012;72:5407–17.

[138] Juhasz I, Albelda S, Elder D, Murphy GF, Adachi K, Herlyn D, et al. Growth and invasion of human melanomas in human skin grafted to immunodeficient mice. Am J Pathol 1993;143:528–37.

[139] Takahashi H, Nakada T, Puisieux I. Inhibition of human colon cancer growth by antibody-directed human LAK cells in SCID mice. Science 1993;259(5100):1460–3.

[140] Pflumio F, Fonteneau P, Gavériaux C, Cammisuli S, Loor F. The C57BL/6 nude, beige mouse: a model of combined T cell and NK effector cell immunodeficiency. Cell Immunol 1989;120:218–29.

[141] Amsbaugh DF, Hansen CT, Prescott B, Stashak PW, Barthold DR, Baker PJ. Genetic control of the antibody response to type III pneumococcal polysaccharide in mice: I. Evidence that an X-linked gene plays a decisive role in determining responsiveness. J Exp Med 1972;136:931–49.

[142] Kamel-Reid S, Dick J. Engraftment of immune-deficient mice with human hematopoietic stem cells. Science 1988;242:1706–9.

[143] Mulé J, Jicha D, Aebersold P, Travis W, Rosenberg S. Disseminated human malignant melanoma in congenitally immune-deficient (bg/nu/xid) mice. J Natl Cancer Inst 1991;83:350–5.

[144] Nonoyama S, Smith FO, Bernstein ID, Ochs HD. Strain-dependent leakiness of mice with severe combined immune deficiency. J Immunol 1993;150:3817–24.

[145] Shultz LD, Schweitzer PA, Christianson SW, Gott B, Schweitzer IB, Tennent B, et al. Multiple defects in innate and adaptive immunologic function in NOD/LtSz-scid mice. J Immunol 1995;154:180–91.

[146] Tanaka T, Kitamura F, Nagasaka Y, Kuida K, Suwa H, Miyasaka M. Selective long-term elimination of natural killer cells in vivo by an anti-interleukin 2 receptor β chain monoclonal antibody in mice. J Exp Med 1993;178:1103–7.

[147] Notta F, Mullighan CG, Wang JCY, Poeppl A, Doulatov S, Phillips LA, et al. Evolution of human BCR-ABL1 lymphoblastic leukaemia-initiating cells. Nature 2011;471:254.

[148] Masramon L, Vendrell E, Tarafa G, Capella G, Miro R, Ribas M, et al. Genetic instability and divergence of clonal populations in colon cancer cells in vitro. J Cell Sci 2006;119:1477–82.

[149] Oettinger M, Schatz DG, Gorka C, Baltimore D. RAG-1 and RAG-2, adjacent genes that synergistically activate V(D)J recombination. Science 1990;248:1517–23.

[150] Shultz L, Lang P, Christianson S, Gott B, Lyons B, Umeda S, et al. NOD/LtSz-Rag1null mice: an immunodeficient and radioresistant model for engraftment of human hematolymphoid cells, HIV infection, and adoptive transfer of NOD mouse diabetogenic T cells. J Immunol 2000;164:2496–507.

[151] Söderstrøm I, Bergman M, Colucci F, Lejon K, Bergqvist I, Holmberg D. Establishment and characterization of RAG-2 deficient non-obese diabetic mice. Scand J Immunol 1996;43:525–30.

[152] Ishikawa F, Yasukawa M, Lyons B, Yoshida S, Miyamoto T, Yoshimoto G, et al. Development of functional human blood and immune systems in NOD/SCID/IL2 receptor γ chain^{null} mice. Blood 2005;106:1565–73.

[153] Shultz LD, Lyons BL, Burzenski LM, Gott B, Chen X, Chaleff S, et al. Human lymphoid and myeloid cell development in NOD/LtSz-scid IL2Rγ^{null} mice engrafted with mobilized human hemopoietic stem cells. J Immunol 2005;174:6477–89.

[154] Francia G, Cruz-Munoz W, Man S, Xu P, Kerbel RS. Mouse models of advanced spontaneous metastasis for experimental therapeutics. Nat Rev Cancer 2011;11:135–44.

[155] Pearson T, Greiner D, Shultz L. Humanized SCID mouse models for biomedical research. Curr Top Microbiol Immunol 2008;324:25–51.

[156] Bonnet D, Dick J. Human acute myeloid leukemia is organized as a hierarchy that originates from a primitive hematopoietic cell. Nat Med 1997;3:730–7.

[157] Eramo A, Lotti F, Sette G, Pilozzi E, Biffoni M, Di Virgilio A, et al. Identification and expansion of the tumorigenic lung cancer stem cell population. Cell Death Differ 2008;15:504–14.

[158] Yang Z, Ho D, Ng M, Lau CK, Yu WC, Ngai P, et al. Significance of CD90^{+} cancer stem cells in human liver cancer. Cancer Cell 2008;13:153–66.

[159] Eppert K, Takenaka K, Lechman E, Waldron L, Nilsson B, van Galen P, et al. Stem cell gene expression programs influence clinical outcome in human leukemia. Nat Med 2011;17:1086–93.

[160] Wang J, Dick J. Cancer stem cells: lessons from leukemia. Trends Cell Biol 2005;15:494–501.

[161] Quintana E, Piskounova E, Shackleton M, Weinberg D, Eskiocak U, Fullen DR, et al. Human melanoma metastasis in NSG mice correlates with clinical outcome in patients. Science Translat Med 2012;4:149–59.

[162] Eaves CJ. Cancer stem cells: here, there, everywhere? Nature 2008;456:581–2.

[163] Ossowski L, Reich E. Experimental model for quantitative study of metastasis. Cancer Res 1980;40:2300–9.

[164] Scher C, Haudenschild C, Klagsbrun M. The chick chorioallantoic membrane as a model system for the study of tissue invasion by viral transformed cells. Cell 1976;8:373–82.

[165] Lee LM, Seftor EA, Bonde G, Cornell RA, Hendrix MJ. The fate of human malignant melanoma cells transplanted into zebrafish embryos: assessment of migration and cell division in the absence of tumor formation. Dev Dyn 2005;233:1560–70.

[166] Corkery DP, Dellaire G, Berman JN. Leukaemia xenotransplantation in zebrafish – chemotherapy response assay *in vivo*. Br J Haematol 2011;153:786–9.

[167] Veinotte CJ, Corkery D, Dellaire G, El_Naggar A, Sinclair K, Bernstein ML, et al. Using zebrafish xenotransplantation to study the role of Y-Box binding protein (YB-1) in the metastasis of Ewing family tumors. American Academy of Cancer Research Annual Meeting, Abstract 1398; 2012.

[168] Pruvot B, Jacquel A, Droin N, et al. 2011 Leukemia cell xenograft in zebrafish embryo for investigating drug efficacy. Haematologica 2011;96(4):6126.

[169] Stoletov K, Kato H, Zardouzian E, Kelber J, Yang J, Shattil S, et al. Visualizing extravasation dynamics of metastatic tumor cells. J Cell Sci 2010;123:2332–41.

[170] Lam SH, Chua HL, Gong Z, Lam TJ, Sin YM. Development and maturation of the immune system in zebrafish, *Danio rerio*: a gene expression profiling, in situ hybridization and immunological study. Dev Comp Immunol 2004;28:9–28.

[171] Haldi M, Ton C, Seng WL, McGrath P. Human melanoma cells transplanted into zebrafish proliferate, migrate, produce melanin, form masses and stimulate angiogenesis in zebrafish. Angiogenesis 2006;9:139–51.

[172] Nicoli S, Ribatti D, Cotelli F, Presta M. Mammalian tumor xenografts induce neovascularization in zebrafish embryos. Cancer Res 2007;67:2927–31.

[173] Detrich III HW, Westerfield M, Zon LI. The zebrafish: cellular and developmental biology, part A. Preface. Methods Cell Biol 2010;100: xiii

[174] Geiger GA, Fu W, Kao GD. Temozolomide-mediated radiosensitization of human glioma cells in a zebrafish embryonic system. Cancer Res 2008;68:3396–404.

[175] Marques IJ, Weiss FU, Vlecken DH, Nitsche C, Bakkers J, Lagendijk AK, et al. Metastatic behaviour of primary human tumours in a zebrafish xenotransplantation model. BMC Cancer 2009;9:128.

[176] Stoletov K, Montel V, Lester RD, Gonias SL, Klemke R. High-resolution imaging of the dynamic tumor cell vascular interface in transparent zebrafish. Proc Natl Acad Sci USA 2007;104:17406–11.

[177] Huynh AS, Abrahams DF, Torres MS, Baldwin MK, Gillies RJ, Morse DL. Development of an orthotopic human pancreatic cancer xenograft model using ultrasound guided injection of cells. PLoS One 2011;6:e20330.

[178] Traver D, Paw BH, Poss KD, Penberthy WT, Lin S, Zon LI. Transplantation and *in vivo* imaging of multilineage engraftment in zebrafish bloodless mutants. Nat Immunol 2003;4:1238–46.

[179] Berman J, Hsu K, Liu TX, et al. A transgenic zebrafish model of Hoxa9 and Meis1 mediated myeloid leukemias. 6th International Conference on Zebrafish Development and Genetics; 2004.

[180] Smith AC, Raimondi AR, Salthouse CD, Ignatius MS, Blackburn JS, Mizgirev IV, et al. High-throughput cell transplantation establishes that tumor-initiating cells are abundant in zebrafish T-cell acute lymphoblastic leukemia. Blood 2010;115:3296–303.

[181] Eguiara A, Holgado O, Beloqui I, Abalde L, Sanchez Y, Callol C, et al. Xenografts in zebrafish embryos as a rapid functional assay for breast cancer stem-like cell identification. Cell Cycle 2011;10:3751–7.

[182] Lichtenstein AV. Cancer: evolutionary, genetic and epigenetic aspects. Clin Epigenet 2010;1:85–100.

[183] Corkery D, Balci TB, Berman JN, Dellaire G. Evaluating novel targeted therapeutic agents in T-cell acute lymphoblastic leukemia using a zebrafish xenotransplantation model. American Academy of Cancer Research Annual Meeting, Abstract 4251; 2012.

[184] Ma D, Ramachandran A, Chang Y, Lin J, Zhang W, Tai YT, et al. Differences in the angiogenic response and subsequent growth of plasma cells from myeloma and MGUS patients xenografted into zebrafish embryos. Blood (ASH Annual Meeting Abstracts) 2012;120(21):2912.

[185] Wang JC, Lapidot T, Cashman JD, Doedens M, Addy L, Sutherland DR, et al. High level engraftment of NOD/SCID mice by primitive normal and leukemic hematopoietic cells from patients with chronic myeloid leukemia in chronic phase. Blood 1998;91:2406–14.

[186] Walter MJ, Shen D, Ding L, Shao J, Koboldt DC, Chen K, et al. Clonal architecture of secondary acute myeloid leukemia. N Engl J Med 2012;366:1090–8.

[187] Ding L, Ley TJ, Larson DE, Miller CA, Koboldt DC, Welch JS, et al. Clonal evolution in relapsed acute myeloid leukaemia revealed by whole-genome sequencing. Nature 2011;481:506–10.

[188] Schmitz M, Breithaupt P, Scheidegger N, Cario G, Bonapace L, Meissner B, et al. Xenografts of highly resistant leukemia recapitulate the clonal composition of the leukemogenic compartment. Blood 2011;118:1854–64.

[189] Harismendy O, Schwab R, Bao L, Olson J, Rozenzhak S, Kotsopoulos SK, et al. Detection of low prevalence somatic mutations in solid tumors with ultra-deep targeted sequencing. Genome Biol 2011;12:R124.

[190] Lawson ND, Weinstein BM. *In vivo* imaging of embryonic vascular development using transgenic zebrafish. Dev Biol 2002;248:307–18.

[191] Renshaw SA, Loynes CA, Trushell DM, Elworthy S, Ingham PW, Whyte MK. A transgenic zebrafish model of neutrophilic inflammation. Blood 2006;108:3976–8.

[192] He S, Lamers GE, Beenakker JW, Cui C, Ghotra VP, Danen EH, et al. Neutrophil-mediated experimental metastasis is enhanced by VEGFR inhibition in a zebrafish xenograft model. J Pathol 2012;227:431–45.

[193] Ellett F, Pase L, Hayman JW, Andrianopoulos A, Lieschke GJ. mpeg1 promoter transgenes direct macrophage-lineage expression in zebrafish. Blood 2011;117:e49–56.

[194] Lin HF, Traver D, Zhu H, Dooley K, Paw BH, Zon LI, et al. Analysis of thrombocyte development in CD41-GFP transgenic zebrafish. Blood 2005;106:3803–10.

[195] Ma D, Zhang J, Lin HF, Italiano J, Handin RI. The identification and characterization of zebrafish hematopoietic stem cells. Blood 2011;118:289–97.

[196] White RM, Sessa A, Burke C, Bowman T, LeBlanc J, Ceol C, et al. Transparent adult zebrafish as a tool for in vivo transplantation analysis. Cell Stem Cell 2008;2:183–9.

[197] Sharpless NE. INK4a/ARF: a multifunctional tumor suppressor locus. Mutat Res 2005;576:22–38.

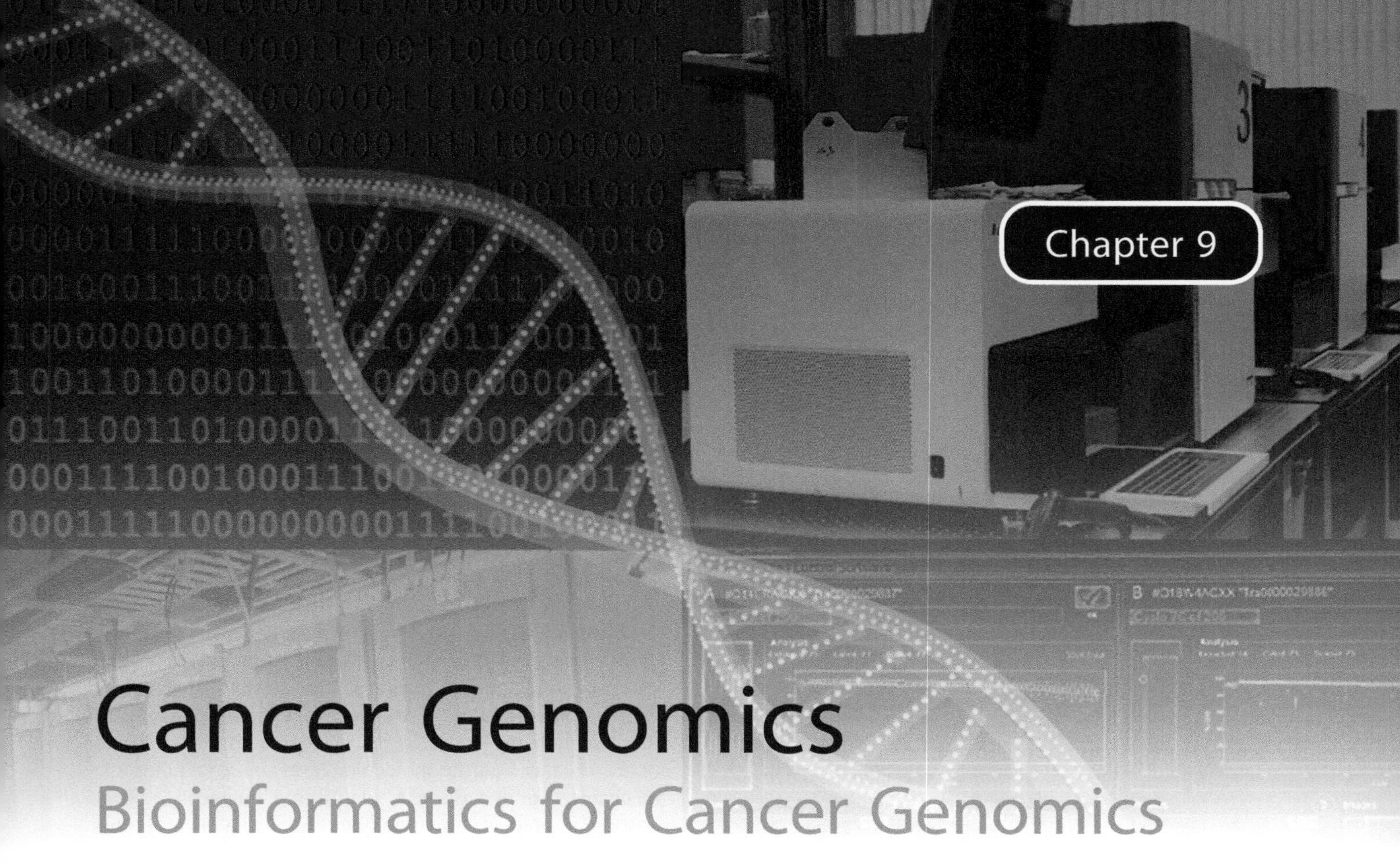

Cancer Genomics

Bioinformatics for Cancer Genomics

Katayoon Kasaian
Yvonne Y. Li
Steven J.M. Jones

Canada's Michael Smith Genome Sciences Centre, British Columbia Cancer Agency, Vancouver, BC, Canada

Contents

Cancer Genomics. DOI: http://dx.doi.org/10.1016/B978-0-12-396967-5.00009-8

Key Concepts

- Advances in DNA sequencing techniques have transformed the field of cancer biology. Using next-generation sequencing (NGS) technologies, we are now able to examine cancers on the molecular level and identify somatic mutations that have accrued during tumorigenesis. Analysis of these alterations can lead to the identification of molecular pathway(s) that are driving the disease
- Large volumes of data are generated as part of even the smallest cancer genomic experiments. Bioinformatic methods have provided solutions for easy data storage and access as well as sharing of data among different laboratories regardless of their geographical locations
- Different types of data can be generated as part of a cancer genomic experiment. Among these are whole genome, exome, transcriptome and epigenome sequencing datasets. Each type of experiment has its unique strengths and limitations; the most suitable technique for a particular experiment will depend on the scientific question in mind
- Somatic mutations in the tumor can be identified using the aligned or assembled sequence reads. These aberrations could involve only one base pair, i.e. single nucleotide variant (SNV), or could cover larger areas of the genome. Examples of such include indels, copy number variants (CNVs) and large structural variants (SVs)
- In the genomic analysis of a cancer sample, the challenge is to discriminate between driver and passenger mutations in a typically large pool of somatic variations. Driver mutations are those that give the tumor the capability to grow and divide without control while the passengers are merely the byproduct of the unstable cancer genome. Computational tools have been developed to aid this process; however, functional analysis of putative driver mutations is still a necessity
- The integration of different data types and mutation calls, with clinical information, as well as other "omics" datasets, will provide a comprehensive understanding of an individual's cancer and the disease pathway(s)

INTRODUCTION

Bioinformatics is an interdisciplinary field of study where computer science, statistics and information technology are applied to the field of biology. Advances in molecular biology tools and techniques have provided biologists with an unprecedented opportunity to gather large amounts of data. Bioinformatics offers not only a way to manage, store and easily access this information but also a way to visualize and analyze efficiently the data, enabling us to draw biologically significant conclusions.

Since the completion of the Human Genome Project (HGP), there has been a revolution in genomic technologies, particularly DNA sequencing techniques. Next-generation sequencing (NGS) technologies have transformed the field of cancer genomics; a complete human genome can now be sequenced at a high depth of coverage for a fraction of the cost and time it would take only a few years ago. Unravelling the genomic abnormalities that lead to cancer, potential therapeutic targets and the mechanisms behind tumor response or resistance to a particular treatment modality is integral to the advancement of cancer medicine. Therefore, the ultimate goal of the cancer genomics field is to explore fully the potential of these sequencing technologies in characterizing different types of cancer on the molecular level, understanding the mechanism of the disease, identifying diagnostic, prognostic and predictive markers and, finally, translating this knowledge into patient-based therapies. Computational biology and bioinformatic techniques provide solutions for examining complete genetic material of a cancer sample for every type of mutation including SNVs, insertions and deletions (indels), CNVs as well as SVs. The utility of the vast parallel sequencing machines is not, however, limited to analyzing the genome. The epigenome, the transcriptome, the proteome and the metabolome of a cell can all be investigated through these high-throughput technologies (Figure 9.1).

Bioinformatics helps to understand complex biological systems by systematically analyzing large biological datasets and by providing the necessary techniques for integrating different data types. This enables us to derive a global view of the healthy state of a cell and to identify how these are altered in the disease state. In this chapter, we focus on the bioinformatic algorithms and software applied in cancer genomics, particularly those used in the analysis of next-generation sequencing datasets. The algorithms and software discussed in this chapter are only illustrative examples of the more widely used tools and techniques in the field rather than a comprehensive survey.

DATA TYPES IN CANCER GENOMICS

Whole Genome and Exome Sequence Data

Cancers arise due to mutations that provide the cell with a growth advantage. In sporadic or non-familial cases of

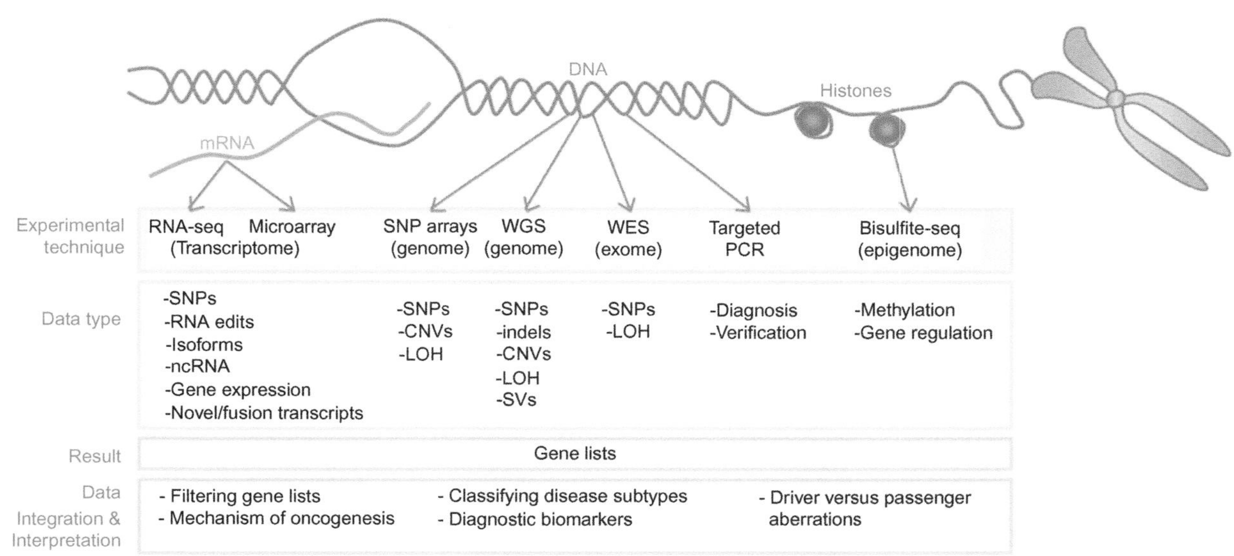

FIGURE 9.1 Through the application of high-throughput sequencing technologies, the genome, the epigenome and the transcriptome can be examined in great detail, providing a comprehensive picture of the state of health or any alterations leading to disease. Such experiments allow the identification of both small and large variations in individual samples.

cancer, these somatic events can be identified through the comparison of cancer and normal genomes of a patient. Whole genome shotgun sequencing provides the sequence of the complete DNA of a sample. These sequence data can be examined for the presence of various somatic alterations such as single nucleotide mutations, insertions and deletions, changes in copy number and large structural variations such as inversions, duplications, translocations and gene fusions. Although, whole genome sequencing provides the complete genetic landscape of a cell, most of the effort of the research community in cancer genomics has focused on identifying alterations in the coding regions of the genome. These are the mutations that change the sequence of the proteins and are thus more likely to alter protein structure and function and lead to cellular malfunctioning. This, along with the fact that whole genome sequencing is still not affordable enough to be carried out for individual patients in clinical settings or even in every research laboratory, has made whole exome sequencing an appealing alternative. Sequencing only the exons provides the information in the complete coding region of the genome at a high depth and for a lower cost than whole genome sequencing. Currently, the sequencing technologies provide such high sequence coverage that multiple exome libraries can be indexed, pooled and sequenced in a single experiment without losing any information while decreasing the cost even further. Whole exome sequence data can still unveil small mutations such as SNVs and indels. Recently, a few tools have been developed that promise the identification of regions of copy number loss or gain as well as coding structural variations from the exome capture data. However, most of the progress to date in finding somatic CNVs and SVs has been the result of whole genome sequencing experiments. This is due to change as more advanced algorithms and software become available.

Examining the cell's DNA provides a static view of the mutations that could potentially be disrupting protein functions. However, cells are dynamic entities, transcribing and translating the genetic information into protein products in accordance to their needs. Studying the dynamic profile of the cell through transcriptome sequencing or characterizing the protein collection of the cell can serve as a powerful tool for identifying disrupted pathways in a disease state.

Whole Transcriptome Sequence Data

It has long been known that there is a global change in the expression of genes in cancer cells compared with their normal counterparts. Some of these alterations, such as changes in the expression of oncogenes and tumor suppressors, will be drivers of the disease while others are the result of the malfunctioning cell and the fragile cancer genome. Using NGS technologies, the complete transcriptome of a cell can now be sequenced, providing a digital count of the expression of all genes. Through whole transcriptome sequencing, also referred to as RNA-seq, expressed mutations such as SNVs and indels are identified. *De novo* assembly of transcriptome data can also serve as a powerful tool for identifying events such as novel transcripts, skipped exons, retained introns or novel

TABLE 9.1 Advantages and Disadvantages of Different Data Types in Cancer Genomics

Data	Analysis Type	Advantages	Disadvantages
Whole genome shotgun sequencing	CNVs Indels SNVs SVs	Comprehensive interrogation of mutations	Costly No information on expression status
Whole exome capture sequencing	Indels SNVs	Cost efficient	Restricted to known annotations Detects only small coding mutations
Whole transcriptome shotgun sequencing	Expression Indels SNVs SVs	Cost efficient Digital gene expression Detects novel events	Detects only expressed alterations

splicing events. Differential expression analysis between malignant and adjacent normal tissues can shed light on the altered pathways in the disease and help in developing diagnostic and prognostic panels. However, such analysis in cancer genomics is hindered due to the typically limited access to neighboring matched normal tissue. Patient's blood usually serves as the normal sample and though it is a good reference for the tumor genome, the expression profile of the blood cells will be entirely different from that of a solid tumor, for example. Different data types in cancer genomics have varying strengths and limitations, generating as many different datasets as possible and their integration is the most promising solution in deciphering cancer signatures (Table 9.1).

Proteomic Data

Various high-throughput techniques such as protein microarrays and mass spectrometry have been developed for studying the complete collection of a cell's proteins, often referred to as the proteome. Proteomic analysis of a biological sample can unveil all the proteins present, their amount, specific post-translation modifications and all protein–protein interactions. Through such analyses of cancer and matched normal tissues or various cancer subtypes, one can identify diagnostic and prognostic biomarkers as well as novel drug targets. Our knowledge of the human proteome, however, has lagged behind the efforts such as the HGP which decoded the sequence of almost the entire genome. Understanding the structure and function of proteins is an important step in cancer genomics, leading to conclusions about the function of mutated proteins, whether they contribute to disease initiation and progression and how they can be targeted. The Human Proteome Project, launched in 2011, aims to identify the structure and function of at least one protein product of each protein-coding gene (http://www.hupo.org/research/hpp/) [1]. Such efforts combined with improvements in DNA sequencing technologies and integrative analyses tools will pave the way for delivering targeted therapies to patients.

Epigenomic Data

Next-generation sequencing technologies have also enabled the study of the epigenome, the transcriptional control of the cell. Mutations of several epigenetic enzymes are found in various cancers and thus there is increasing evidence that changes in the epigenome and the resultant alterations in the expressional profile of the cell could be the cause of many diseases including cancers. Examining the pattern of epigenetic marks associated with both the DNA and histone proteins throughout the whole genome of the cancer and matched normal tissue can provide a profound understanding of the changes leading to the disease state. Chromatin immunoprecipitation followed by sequencing (CHIP-seq) [2], with higher throughput and better sensitivity than CHIP-on-chip [3], provides a genome-wide view of specific DNA–protein interactions including histone modification marks. Profiling the methylation state of the genome is also now possible through techniques coupled with high-throughput sequencing [4]. These methods are divided into those which enrich for methyl-DNA [5–7], those which utilize methylation-dependent restriction enzymes [8,9] and the third category which is based on direct bisulfite conversion [10–16].

Data generation has arguably become the easiest and the most efficient step in studying a cancer genome. The challenge now is to analyze the sheer volume of data generated by the high-throughput technologies and to be able to integrate different types of mutational datasets such as SNVs, indels, CNVs, SVs, expression profiles and epigenetic alterations in order to draw a biologically correct and meaningful conclusion about the underlying cause of the disease and how best to treat it.

DATA MANAGEMENT

A vast amount of information is gathered as part of cancer genomic initiatives such as The Cancer Genome Atlas (TCGA), the Pediatric Cancer Genome Project (PCGP), International Cancer Genome Consortium (ICGC) [17], Cancer Genome Project (CGP), Cancer Cell Line Project (http://www.sanger.ac.uk/genetics/CGP/CellLines/), Cancer Genome Characterization Initiative (CGCI), Therapeutically Applicable Research to Generate Effective Treatments (TARGET) and the Cancer Cell Line Encyclopedia (CCLE) (http://www.broadinstitute.org/ccle/home) [18]. These large-scale initiatives are not restricted to studying diseases such as cancer but also include projects that examine the healthy population; examples include the HapMap [19], the 1000 Genomes Project [20], Personal Genome Project and the Human Genome Structural Variation Project [21]. In order to meet the collective goal of characterizing different cancers on the molecular level and identifying therapeutic targets, there is a need for efficient and rigorous management of the large and complex datasets generated mostly by the NGS technologies. Data collected as part of these studies include clinical data, raw genomic data as well as processed data such as short read alignment files and variant calls. An efficient data management strategy should allow easy data storage and access, as well as the ability to link different types of data. Cancer Genomic Hub (CGHub) (https://cghub.ucsc.edu/index.html), a national data center built by the University of California Santa Cruz (UCSC), is an example of such a centralized data storage system. It serves as the data repository center for TCGA and other related cancer genomic efforts; it is an automated resource that gives researchers access to sequence read alignment files as well as any available metadata.

In addition to raw and aligned data repositories, databases populated with derived information from large genomic studies are of considerable importance to the field. Great examples of these databases include dbSNP [22] where common variants in the general population identified through the HapMap [19], the 1000 Genomes Project [20] as well as other validated experiments are stored, and the COSMIC database [23] which stores curated somatic mutations identified in human cancers.

Tools such as NCBI's Entrez database browser [24] and BioMart Central Portal [25] provide a single point of access to several databases; this facilitates sharing as well as integration of data in the research community. Web services have also become indispensable tools in bioinformatics. Services provided by host institutes provide data and application access to scientists located anywhere in the world without having locally to install and maintain these resources.

The need for genomic data management is not restricted to large initiatives but also to smaller genome centers. Setting up automated pipelines for analyzing the data immediately after their production will ensure their efficient processing in a timely manner. Quality control and processing steps such as alignment of short reads to the reference genome, *de novo* assembly and variant calling would be part of such pipelines. These centers also benefit from efficient and robust data storage systems. Databases designed specifically for storage of cancer genomic variant calls will minimize redundancy, enable data integration with other databases and facilitate query of large data collections. As a result, such repositories can serve as powerful knowledge resources. They facilitate studies by which the profile of a specific point mutation, for instance, over thousands of sequenced cancer and normal libraries could provide the basis for a statistically based conclusion about its role in cancer and whether the affected gene could serve as a biomarker.

DATA ANALYSIS

Even more pronounced than the need for data storage in cancer genomics is the ever increasing need for powerful computational resources and efficient algorithms for data analysis and visualization.

Computational Resources

Currently, a single high-throughput DNA sequencing run using NGS technologies can produce 200 giga base pairs (Gbp) of sequence data. A tremendous amount of computational power is thus needed in every step of analysis, from the collection of raw data, to the alignment of reads to the 3 billion base pairs of the human reference genome, to variant calling and, finally, assembly. High-performance computing (HPC) facilities equipped with a cluster of high-speed computer nodes and multipetabyte storage systems can provide the hardware and the support needed for large-scale genomic projects. The HPC facility housed at the Michael Smith Genome Sciences Centre (http://www.bcgsc.ca/) with over 8000 cores/16 000 threads and 8.5 PB of storage system is an example of such an HPC system in Canada. Distributed and parallel computing, cloud computing and graphics processing unit (GPU) computing [26] are some of the examples of techniques and technologies that are moving to the spotlight as the genomics field is awaiting the arrival of single molecule sequencing technologies.

Analysis Algorithms and Tools

Sequence Alignment and Assembly

NGS technologies produce large numbers of short reads in a relatively short period of time. Application of these technologies in cancer genomics depends on the ability to reconstruct the complete genome from these reads with great accuracy in a time- and memory-efficient manner. Generally, two options exist: one is to align the reads to the reference genome and the other is to perform a *de novo* assembly.

Simply put, alignment refers to the task of finding the location in the complete genome where a sequence read was generated from. This is in essence a string-matching problem. There already exist various algorithms and software for solving this problem, all however facing a tradeoff between accuracy and speed. Although the standard Smith–Waterman algorithm [27], widely used for the alignment of longer reads, provides the most optimal solution, it becomes computationally intractable when working with a large number of short sequence reads. As a result, a growing number of algorithms for the alignment of NGS reads to the human reference genome have been implemented. The most widely used techniques fall into two broad categories, hash table- and suffix tree-based algorithms. The former constructs a hash data structure for indexing the sequence data and fast look up of strings. These hash tables can be constructed from either the reference genome or the set of sequence reads. The unindexed set is subsequently used to scan the hash table for matches and to identify the approximate location of each read. Finally, dynamic programming algorithms such as Smith–Waterman and its improved equivalents are used to find the exact placement of each read. Some of the more widely used aligners in this category include MAQ [28], SOAP [29] and SHRiMP [30]. The second category of aligners uses a different set of data structures for indexing the sequence data. These generally rely on indexing the reference genome using suffix trees, suffix arrays or the more widely used Ferragina Manzini (FM)-index, a compressed data structure based on the Burrows–Wheeler Transform (BWT). Suffix trees, data structures, where all the suffixes of a string are stored, and suffix arrays, arrays of integers corresponding to the starting position of all possible substrings, are not efficient for indexing large search spaces such as the human reference genome. Ferragina and Manzini demonstrated that suffix arrays built from the BWT transformed strings are more efficient than arrays built from the non-transformed strings [31]. As a consequence, the FM-index is the data structure of choice in the majority of current short-read aligners. They allow for rapid substring search and are generally more memory efficient than suffix trees and arrays. Examples of such aligners include Bowtie [32], BWA [33] and SOAP2 [34].

The often ignored limitation of the current aligners is their intended use for the alignment of normal sequence reads only. These tools are not yet optimized for the alignment of reads generated from cancer samples which could potentially contain large insertions and deletions, as well as evidence of other structural variations such as duplications, translocations and gene fusions. In addition, as the sequencing technologies improve and the reads become longer, the majority of the tools for alignment of short reads to the reference will not be applicable anymore. There will be a need for specialized software to map optimally longer reads, perhaps containing indels or other structural variations, to the reference genome.

An alternative option to the alignment process is *de novo* assembly of sequence reads. Such an approach allows for the identification of highly diverged DNA regions in the sequenced sample compared with the reference genome. The techniques used in assembling longer reads produced by the early sequencing technologies, such as Sanger, generally involved finding areas of overlap between reads and extending those into longer contigs. Shorter reads and higher coverage produced by the NGS technologies, however, make such algorithms computationally inefficient, if not unfeasible. Currently, the more widely used assemblers make use of the de Bruijn graph data structure where all possible substrings of size k are stored in the nodes of the graph and each edge indicates an overlap of size $k-1$ between the two connecting nodes [35–38]. Traversing such a graph built from raw sequence reads will yield a collection of contigs representing the sample's sequence. *De novo* assembly techniques are not yet as computationally efficient as alignment of reads directly to the reference genome and, hence, not yet as widely used. Currently, the genomic analysis of a cancer and its matched normal involves separate alignment of each sample; this is followed by variant calling and the identification of novel somatic mutations in the tumor tissue. With advances in assembly algorithms as well as increases in read length and insert sizes of paired-end libraries, it is conceivable that *de novo* assembly of tumor and normal genomes will eliminate the need for alignment to a reference. As a result, this approach can provide more comprehensive insights into each individual's unique genomic landscape and pave the way for more personalized diagnosis and treatment options.

Discovery of Point Mutations

The alignment and/or assembly results are subsequently explored for the presence of any type of somatic mutation including single nucleotide variants. The majority of the early SNV detection tools [39–41] rely on setting arbitrary thresholds for variables such as sequence coverage, read mapping quality, base quality and distance between mismatched bases in order to filter out technical noise and identify the positions that show true variability from the reference. These tools, however, are best suited for the analysis of normal samples and detection of germline variations where, for example, a heterozygote SNV would be expected to have variant allele frequency of 50% while, in a homozygote position, the variant base would be observed at 100% frequency. When analyzing tumor samples, contamination with adjacent normal tissue, the presence of multiple clonal populations within the tumor, as well as tumor aneuploidy can result in single nucleotide variants that are observed at any frequency. Probability model-based tools designed specifically for the detection of variants in cancer samples have been developed; these identify the most likely genotype at each position based on a probabilistic model for allelic distribution [42,43]. The dependence of all these tools on separate analysis of cancer and normal samples followed by their pair-wise subtraction has, however, deemed them as suboptimal in detecting somatic mutations. Recent developments in simultaneous analysis of matched sample pairs have resulted in more confident somatic mutation calls by calculating the likelihood of genotype differences between the two genomes, at all locations [39,44–46]. These algorithms allow for the detection of true somatic mutations which lack strong support in the tumor sequence data and distinguish them from false somatic events with weak support in the normal sequence data. The current state of cancer genomics requires the verification of computationally detected variant calls in their corresponding samples using orthogonal methods. In future, such verification may no longer be needed should advances in sequencing technologies and analysis tools lead to near optimal quality of reads and genotype calls.

Identification of Indels

Detecting small insertions and deletions (indels) from NGS short read products has proved more challenging than detecting single nucleotide variants. This is mainly attributed to the limitations of current aligners which, by default, allow a set number of small mismatches between a read and the reference, typically with no gaps, leading to misalignment or no alignment of reads spanning indels. Using split-read approaches, the Pindel software aims to detect large deletions and medium-size insertions from pair-end datasets [47]. Mapping short reads to repetitive regions and tandem repeats typically poses difficulties, leading to low sensitivity and specificity of the majority of indel detection tools. Gapped [33,48] and paired-end alignments [28,33,47] are methods that can improve detection sensitivity; parameters such as the number of reads supporting an indel, mapping and base qualities as well as presence or absence of homopolymer regions should be taken into account when estimating the true positive probabilities [39,40]. Dindel [49], the 1000 Genomes Project indel-caller [20], uses local realignment of reads to increase the accuracy of indel detection rate. Dindel accepts a list of potential indels and SNP calls as input, identifies all candidate haplotypes surrounding these sites and realigns reads to all the candidates in order to identify true events [49]. One limitation of Dindel, however, is its dependence on the sensitivity of the aligner that provides the initial list of potential insertion and deletions.

Indels, having the potential to alter or completely eliminate a protein's function, are the second most abundant type of variation in the human genome after SNVs [50]. As a result, there is a great need for development of robust probabilistic algorithms for detecting somatic indels from paired cancer and normal samples.

Structural Variation Detection

Structural alterations including large insertions and deletions, duplications, inversions, translocations and gene fusions have been associated with various cancer types [51]. Before the advent of NGS technologies, cytogenetics, karyotyping and fluorescent *in situ* hybridization, as well as array-bases techniques such as SNP arrays and array comparative genomic hybridization (CGH), were used in detecting large SVs. However, the emergence of next-generation sequencing technologies and the corresponding analysis tools has enabled the detection of various SVs including copy-neutral events and the corresponding break points at a much higher resolution and with greater accuracy.

Paired-end sequencing protocols, where the two ends of a single DNA molecule are read, allow the detection of SVs in the genomic data. Since the order and orientation of read pairs and the insert size distribution are known, any deviation from these expectations in the alignment might suggest a variation in the sample. Several tools have been developed which detect read pair anomalies and infer specific SVs [52–56]. However, we now know that the majority of structural variations are found in duplicated regions of the genome [21,57], regions that pose the most difficulty during the alignment process. As a result, alignment-based SV detection may result in many false positives while missing true events. An alternative to examining the alignment data for finding anomalies is to assemble the sequence reads *de novo* and compare the resultant contigs with the reference genome [58]. As the reads get longer, the assembly of individual genomes becomes more feasible and detection of SVs will have higher sensitivity and specificity.

Large deletions and amplifications, at times encompassing chromosome arms or whole chromosomes, lead to changes in number of gene copies and, in some cases, their expression levels. These structural variations are often collectively referred to as copy number variations. Variations in gene copy number can be detected using single-end as well as paired-end reads. Given the assumption that the whole genome is sampled uniformly and reads are generated with equal probability, depth of coverage can serve as a quantitative measure of copy number [59,60]. This assumption is not strictly correct, however. GC content, for instance, introduces bias during the sequencing experiment [61], while challenges such as alignment of short reads to repetitive regions of the reference genome leads to computational biases. Various techniques have been employed in identifying somatic CNVs by correcting these deviations from the expected distribution [62–64]. Several repositories containing somatic copy numbers from cancer datasets are publicly available [23,65–67]; these resources can facilitate the analysis of variation calls in one or multiple samples.

Expression Analysis

High-throughput sequencing of the complete transcriptome offers a few advantages over the more traditional means of expression analysis such as oligonucleotide microarray technologies. All expressed entities including novel transcripts, novel isoforms and non-human transcripts are sampled in these surveys of the whole transcriptome as opposed to microarray experiments which are restricted to known genes and annotations. Digital analysis of the transcriptome also increases both specificity and sensitivity; the high coverage that can be achieved through these experiments enables the identification of genes with even the lowest expression levels. Identifying differentially expressed genes or specific isoforms between malignant and normal states can reveal pathways which, when altered, might lead to tumorigenesis. Differential expression analysis can also identify subtypes of a disease and subsequently aid in finding diagnostic [68,69] and prognostic markers [70,71]. The discovery of novel subtypes of a disease can be accomplished through unsupervised clustering techniques such as hierarchical clustering of the expression data [72], use of self-organizing maps [73] or non-negative matrix factorization [74,75].

Expression analysis is not restricted to the cell's messenger RNA. Small non-coding transcripts such as miRNAs can also be subjected to high-throughput sequencing and analysis. Integration of protein-coding gene expression profiles with miRNA expression, promoter methylation and copy number variation data can provide clues to which genes are silenced and thus may function in tumor suppression, and which are overexpressed and might be acting as oncogenes.

Analytical Tools

In addition to computational infrastructure and software, programming languages and toolkits designed specifically for working with biological data are required for efficient analysis of genomic datasets. Analysis of variant calls, most likely stored in different file formats across a network, and their integration with externally available resources will require tools that allow easy data management and processing. Perl, Python, Java, C++ and R are some of the popular scripting and programming languages used in cancer genomics. Several toolkits designed specifically for biological data are developed for each of these languages; some of these include BioPerl (http://www.bioperl.org) [76], BioPython (http://biopython.org) [77], BioJava (http://biojava.org) [78], NCBI's C++ toolkit (http://ncbi.nih.gov/IEB/ToolBox/CPP_DOC/) and R BioConductor (http://bioconductor.org) [79] which provides powerful capabilities for statistical analysis of biological data. The majority of these resources are released under an Open Source license and thus can serve as building blocks for larger and more complex tools and software [80].

Visualizing the genomic data is an integral part of comprehensive investigation of cancer genomes. Tools such as genome browsers provide a graphical interface where the complete human genome as well as all available annotations of it can be explored. The widely used genome browsers in the field include the University of California Santa Cruz (http://genome.ucsc.edu) [81] and Ensembl browsers [82]. Various types of sequence data tracks can also be uploaded and visualized using tools such as Integrative Genomics Viewer (IGV) [83] and Circos [84]. The recently released cBio Cancer Genomics Portal [85] provides a platform for integrating and visualizing various types of cancer genomic data available through initiatives such as TCGA and ICGC; instances of the portal can also be installed locally (Table 9.2).

DATA INTERPRETATION

Over just a few years, the cancer genomics community has made great progress in developing algorithms and software for detecting various types of mutation from whole genome and transcriptome datasets. As improvements are made to sequencing technologies and detection tools, the most challenging task becomes mining the large and diverse mutational profiles that are generated in even a single patient experiment for mutations which contribute to disease initiation and progression.

TABLE 9.2 List of Bioinformatics Resources for Cancer Genomics

Description	Tools	URL
Programming languages & biology-specific modules	Perl, BioPerl	http://www.bioperl.org
	Python, BioPython	http://biopython.org
	Java, BioJava	http://biojava.org
	C++, NCBI's C++ toolkit	http://ncbi.nih.gov/IEB/ToolBox/CPP_DOC/
	R, BioConductor	http://bioconductor.org
Genome browsers & annotation resources	UCSC	http://genome.ucsc.edu/
	Ensembl	http://www.ensembl.org
	NCBI Map Viewer	http://www.ncbi.nlm.nih.gov/projects/mapview/
Sequence data analysis toolkits	SAMtools	http://samtools.sourceforge.net/
	GATK	http://www.broadinstitute.org/gsa/wiki/
Population polymorphism data	dbSNP	http://www.ncbi.nlm.nih.gov/projects/SNP/
	1000 Genomes Project	http://www.1000genomes.org/home
	Database of Genomic Variants	http://projects.tcag.ca/variation/
Repositories of cancer gene & mutations	COSMIC	http://www.sanger.ac.uk/genetics/CGP/cosmic/
	Cancer Gene Census	http://www.sanger.ac.uk/genetics/CGP/Census/
	OMIM	http://www.ncbi.nlm.nih.gov/omim/
	Mitelman Database	http://cgap.nci.nih.gov/Chromosomes/Mitelman
Data from cancer genomic initiatives	TCGA	http://cancergenome.nih.gov/
	ICGC	http://www.icgc.org/
	PCGP	http://www.pediatriccancergenomeproject.org/site/
	Cancer Cell Line Project	http://www.sanger.ac.uk/genetics/CGP/CellLines/
	CGCI	http://cgap.nci.nih.gov/cgci.html
	CCLE	http://www.broadinstitute.org/ccle/home
Gene expression resources	GEO	http://www.ncbi.nlm.nih.gov/geo/
	ArrayExpress	http://www.ebi.ac.uk/arrayexpress/
	Oncomine	https://www.oncomine.org/resource/login.html
Viewers	IGV	http://www.broadinstitute.org/igv/
	Circos	http://circos.ca/
	Genome Browsers, e.g. UCSC	http://genome.ucsc.edu/
Pathway analysis tools & databases	KEGG	http://www.genome.jp/kegg/
	DAVID	http://david.abcc.ncifcrf.gov/
	IPA	http://www.ingenuity.com/
	Pathway Interaction Database	http://pid.nci.nih.gov/
	Reactome	http://www.reactome.org
	STRING	http://string-db.org/
Integrative data analysis	PARADIGM	http://sbenz.github.com/Paradigm/
	cBio Cancer Genomics Portal	http://www.cbioportal.org
Resources for therapeutic intervention	DrugBank	http://www.drugbank.ca/
	TTD	http://bidd.nus.edu.sg/group/ttd/
	PharmGKB	http://www.pharmgkb.org/

The list of putative cancer-related somatic mutations can be refined using publicly available reference databases. These include repositories where variations in the healthy population are curated such as dbSNP [22], Database of Genomic Variants [86] and 1000 Genomes Project datasets [20], as well as databases where known cancer genes and their mutations are stored. Examples of such include: the COSMIC (Catalogue of Somatic Mutations in Cancer) [23], an open source database containing somatic mutations and copy number alterations associated with cancers; OMIM (Online Mendelian Inheritance in Man), which collates information on familial cancer genes and susceptibility loci; Cancer Gene Census [87], which catalogs all genes shown to be causally implicated in cancers; and the Mitelman database of chromosome aberrations and gene fusions in cancer [88]. The common assumption when using these databases is that variations that are commonly found in the general population are less likely to contribute to diseases such as cancer, while genes recurrently mutated in various cancer types are potential oncogenes and tumor suppressors.

When interpreting genomic data, it is imperative to be aware of potential confounding factors. The presence of circulating tumor cells in a normal blood sample, normal cell contamination in a tumor sample or a heterogeneous tumor sample with various different subclonal populations can lead to false positive and false negative mutation calls. Bioinformatic algorithms have been developed to estimate and correct for the amount of contamination and to determine more sensitively the copy number variation (CNANorm) [89], SNVs (MutationSeq) [46], or loss of heterozygosity (APOLLOH) [90]. In some cases, manual inspection of tumor and normal sequence files using tools such as IGV [83] can help in eliminating false positive somatic calls. The next step following the computational discovery of candidate mutations typically entails verification of those events in their corresponding sample(s). This step identifies the variant calls that were falsely identified as somatic due to sequencing or computational errors. Mutation verification usually involves the amplification of the potential variant site in both cancer and matched normal tissues using polymerase chain reaction (PCR) techniques followed by Sanger or next-generation sequencing (Figure 9.2).

The number and profile of somatic mutations demonstrate great variability in different cancers [91]. A high number of mutations are seen in cancers such as melanomas [92] whereas some pediatric cancers exhibit a very low number of alterations [93]. Regardless, all somatic mutations can be categorized into "drivers", changes that are responsible for disease pathogenesis and tumor evolution, or "passengers" which are simply the byproduct of the unstable cancer genome and provide no growth advantage to tumor cells [91]. Distinguishing these two types of entities is critical given that passenger mutations play no functional role in disease initiation, progression or maintenance and thus treatment(s) targeting them may prove ineffective. Although, gene(s) that are recurrently mutated in a large cohort of samples are easily distinguishable as drivers, the task becomes more difficult when examining a single or only a few samples. Computational techniques are developed that aim at distinguishing drivers and passengers *in silico* prior to the more labor-intensive and time-consuming procedures of functional validation in the wet lab. Since it is believed that mutations which result in changes in protein structure and function are more likely to act as drivers of cancer, the majority of the focus to date has been on mutations that affect protein-coding regions of the genome. A common computational strategy in examining the functional role of a somatic mutation is to determine its location with respect to functional domains and key amino acid residues in the protein product using resources such as UniProt. Software tools such as PolyPhen [94], MutationAssessor [95] and SIFT [96] use evolutionary conservation of gene sequences as well as homology to provide a likelihood score for the deleterious effect of a point mutation on protein structure and function. The general assumption that driver mutations providing growth advantage for the tumor must be under evolutionary positive selection while the passengers are less likely to be selected for has given rise to algorithms aiming to quantify the selection pressure on mutated genes in cancer samples. The algorithm developed by Greenman et al. [97] estimates the background rate of mutation in each gene using the set of silent mutations; any gene with more than the expected number of non-silent mutations might then be concluded to be under positive selection.

In addition to sequence-level analysis, genomic aberrations can be investigated on a structural level. This approach is limited to genes that have solved three-dimensional protein structures in the Protein Data Bank (PDB) [98] but can be extended to include genes with homology models, predicted protein structures of genes based on a solved structure of a similar-sequence gene. Repositories like SWISS-MODEL [99] have automatically built homology models for all known human genes where a solved structure of a similar gene exists. The protein and mutation can be visualized using molecular graphics software such as PyMOL (http://www.pymol.org/) or SwissPdb Viewer [100] and can provide a more realistic view of how the mutation might affect the catalytic site, nearby key residues, or protein–protein binding residues such as substrate/cofactor binding or dimerization sites. Conclusions based on structural analysis, however, are limited in that they do not take into consideration the

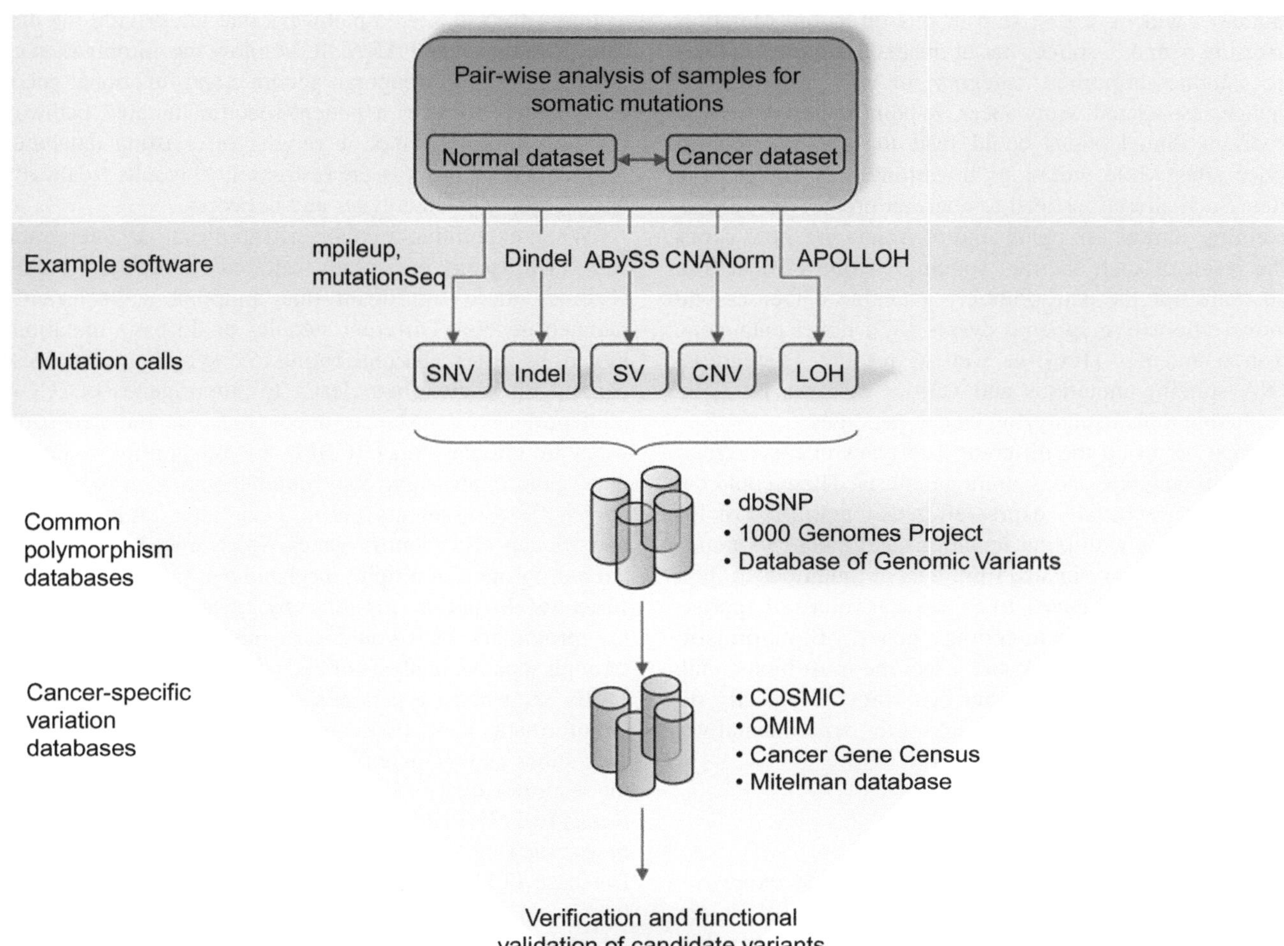

FIGURE 9.2 Filtering the identified somatic mutations in a cancer sample using publicly available databases of common genetic polymorphisms as well as known cancer-specific variants can narrow down the potentially long list of candidates to the most likely drivers.

actual protein conformation in the cell or residues that are key in retaining protein stability and/or play a role in protein–protein interactions.

Determining the deleterious effect(s) of mutations spanning more than one base pair is typically an easier task. Indels that occur in-frame add or remove at least one amino acid residue while out-of-frame indels can completely disrupt the sequence of the gene and hence the protein product. Inversions, duplications and translocations with break points in the middle of a gene can also alter the protein function. Changes in gene copy numbers and expression might be more difficult to interpret in whole genome and transcriptome studies. A single cancer sample could potentially have thousands of genes which are up- or downregulated compared with the normal tissue. Similar to SNPs, statistical approaches have been developed for identifying larger alterations such as CNVs that serve as disease drivers [101]. Integration of all mutational and expression data promises to provide a global view of the altered pathways and identify mutations that are indispensable to the cancer cell.

Although the majority of somatic mutations in any cancer sample fall outside the protein-coding regions, the scientific community for the most part has focused on examining the protein-coding changes. This is attributed not to a lack of interest but perhaps to a lack of comprehensive knowledge about the regulatory elements of the genome. Although it is more challenging to identify the role these non-coding mutations might be playing in cancer, integration of mutation data with known regulatory sites from resources such as the ENCODE data [102], TRANSFAC [103], JASPAR [104] and ORegAnno [105] databases will be a first step in this direction. Mutations affecting 5′ and 3′ splice sites at the exon–intron junctions are another important category of non-protein-coding changes associated with cancer. A point mutation in these invariant dinucleotides could lead to loss of canonical splice sites while mutations in introns may create novel sites. Such alterations lead to aberrant protein products by retaining introns, skipping and/or producing new exons. The result of such aberrant splicing will be evident from the transcriptome data. Publicly available databases with known alternative splicing events [82], novel pathogenic exon boundaries [106], as well as resources integrating RNA splicing mutations and various diseases [107] are invaluable tools in analyzing cancer genomes.

Analysis of all the different data types in cancer genomics will generate one or more gene lists. These could catalog the differentially expressed genes, amplified or lost genes or genes with point mutations and indels, to name a few. They can vary in size from tens of mutations to thousands of amplified genes, to expression values of approximately 20 000 protein-coding genes. Bioinformatics analysis allows us to rank and select the most biologically interesting set of genes, cluster them into biologically significant subgroups, or select a subset for experimental verification or functional studies.

DATA INTEGRATION

Integration of all somatic mutation calls and expression data from a cancer sample plays an important role in creating a molecular pathway hypothesis of aberrations driving the tumor. The biology of a cell is complex and involves processes and controls of those processes on the genomic, epigenomic, transcriptomic and proteomic levels (Figure 9.3). Molecules in the cell are not isolated, but are part of a collective "system" of interacting parts [108], and aberrations in one molecule can perturb the whole system. Network construction and modeling allows the simulation of complex biological pathways including their temporal aspects. For instance, a human colon cancer cell network was built using cell signaling pathways to model and predict the phenotypic fate of the cell, and could predict changes to cell proliferation rates by inhibiting certain genes or proteins [109].

Data integration can be performed by intersecting different gene lists or by visualizing them together on a genome plot like Circos (Figure 9.4). Interrogating the affected genes for enrichment of Gene Ontology (GO) terms [110] or particular gene sets [111] along with pathway analysis using repositories and tools such as KEGG (the Kyoto Encyclopedia of Genes and Genomes) [112], DAVID (the Database for Annotation, Visualization and Integrated Discovery) [113] and IPA (Ingenuity® Systems Pathway Analysis, www.ingenuity.com) can provide evidence for the altered pathways that are driving the disease. Tools such as PARADIGM allow the combination of all types of high-throughput genomic and functional genomic datasets to infer a patient-specific mutated pathway [114]. A limiting factor in relying on existing databases for pathway analysis is the restriction of results to already known and annotated genes and networks.

When examining a cohort of samples, data integration and identifying commonly altered pathway(s) often becomes more significant than pinpointing recurrently mutated gene(s). Different samples could have mutations in various genes, all contributing to one biological process which, when disturbed, leads to tumorigenesis. TCGA pilot project, for instance, uncovered core mutated pathways in glioblastomas (GBM) by integrating sequence data, gene expression, copy number variation as well as epigenetic assessments [115]. Data integration in these cohorts can also identify genes which may be frequently altered through multiple mechanisms (point mutations, structural disruption, loss of copy, or hypermethylation of the promoter) but would not otherwise be identified through separate analysis of each data type.

By sequencing a patient's tumor and normal genomes, bioinformatic methods enable not only the discovery of aberrations and the potential oncogenic mechanism but also the determination of actionable targets and potential therapies [116,117]. Integration of mutation calls with drug databases such as DrugBank [118] or Therapeutic Target Database (TTD) [119], which curate drug–target interactions from the literature, will identify those mutations that might be inhibited or activated by the already approved drugs. The drug response can also be predicted *in silico* using resources with known pharmacogenetic information such as the Pharmacogenomics Knowledgebase (PharmGKB) [120]. However, the effectiveness of drug prediction methods depends on the completeness and accuracy of the existing knowledge resources. Some databases are populated with semi-manually curated data while others contain fully automated literature-mined interactions. In the latter cases, data integrity must be carefully monitored as auto-curation can lead to erroneous entries. The context in which the data in the database were generated should also be considered during the analysis. DrugBank, for example, stores all proteins known to bind to a specific drug, and thus contains results from biochemical binding assays that may not persist in more complex environments such as the human body; as such, protein–drug interactions identified in DrugBank may not always represent viable drug candidates for the treatment of cancer patients. In contrast, TTD focuses on collating target information for drugs currently in clinical trials but only contains primary targets annotated for most drugs. In short, the available bioinformatic databases and tools contain a wealth of information that can provide the foundation for data interpretation. However, the

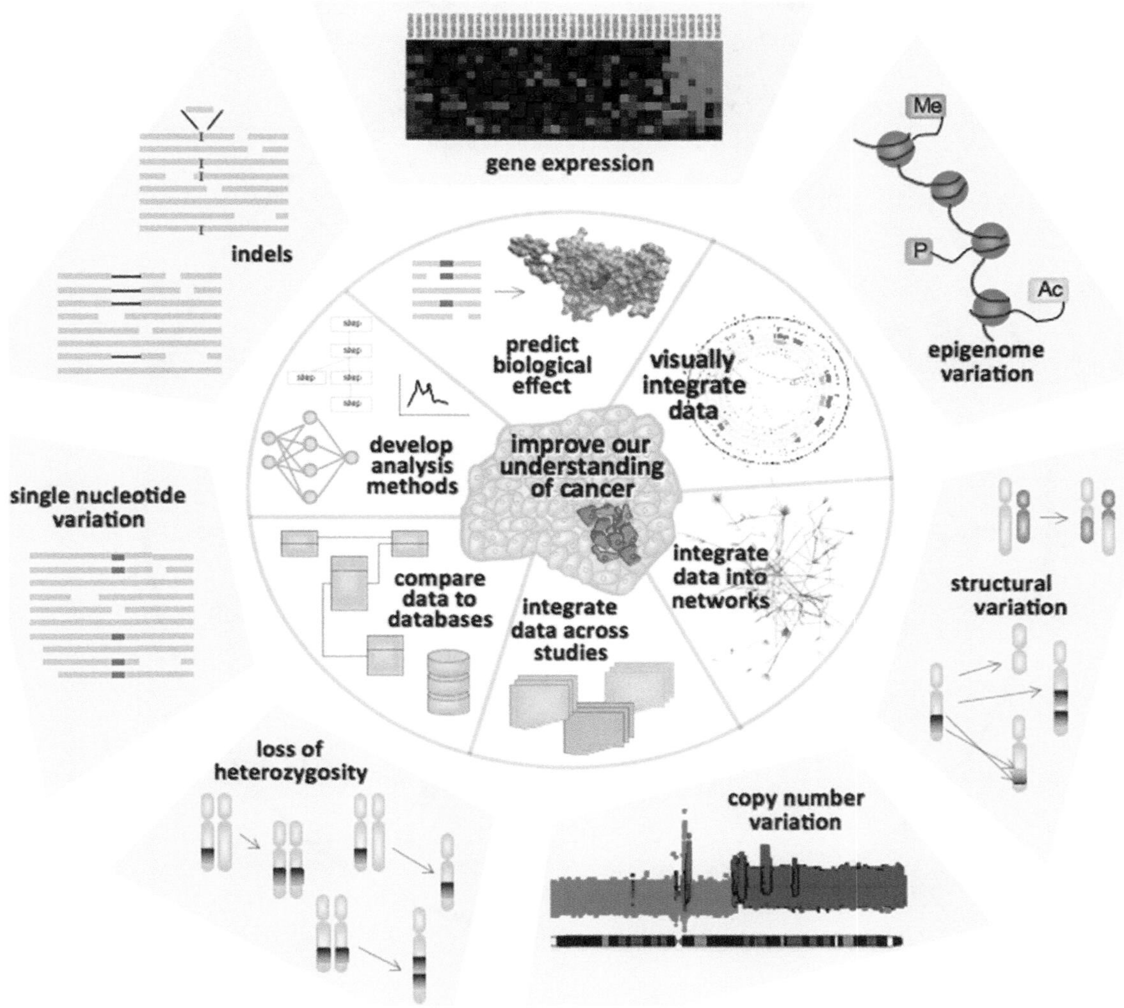

FIGURE 9.3 Identifying the perturbed pathways and networks driving the disease is integral to better understanding the drivers of cancer, exploring potential therapeutic targets and predicting drug response. Integration of data generated from both normal and cancer samples promises the unbiased examination of a cell's interconnected network of molecules.

accuracy and biases of these resources must be taken into consideration during analysis, especially when considering therapeutic intervention.

Data integration is not only applicable across all data types in one experiment but can also be applied to data across multiple platforms or analyses. Laboratories around the world are generating and analyzing data using various experimental designs, technologies and techniques. Bioinformatic methods have become key players for integration of these data. An example of such is the meta-analysis of microarray gene expression studies where the development of rigorous normalization methods can allow the vast amounts of gene expression data to be compared across platforms [121]. Data integration also requires the development of unified ontologies precisely to define and categorize biological concepts and data properties (http://www.geneontology.org/) [110].

Each data type independently can provide an overview of the alterations in a cancer genome. However, a more comprehensive view of the state of cell is obtained through their integration. Such a comprehensive analysis of diseased and normal cellular networks combined with computational modeling techniques would enable the identification of targets that can be inhibited in combination, or perhaps in a particular order, to maximize tumor cell death while sparing the healthy cells. For example, Iadevaia et al. modeled the IGF receptor signaling network and simulated the effects of inhibiting individual and combinations of targets. They predicted that optimal inhibition of MEK and mTOR would kill cancer cells but suboptimal inhibition would actually

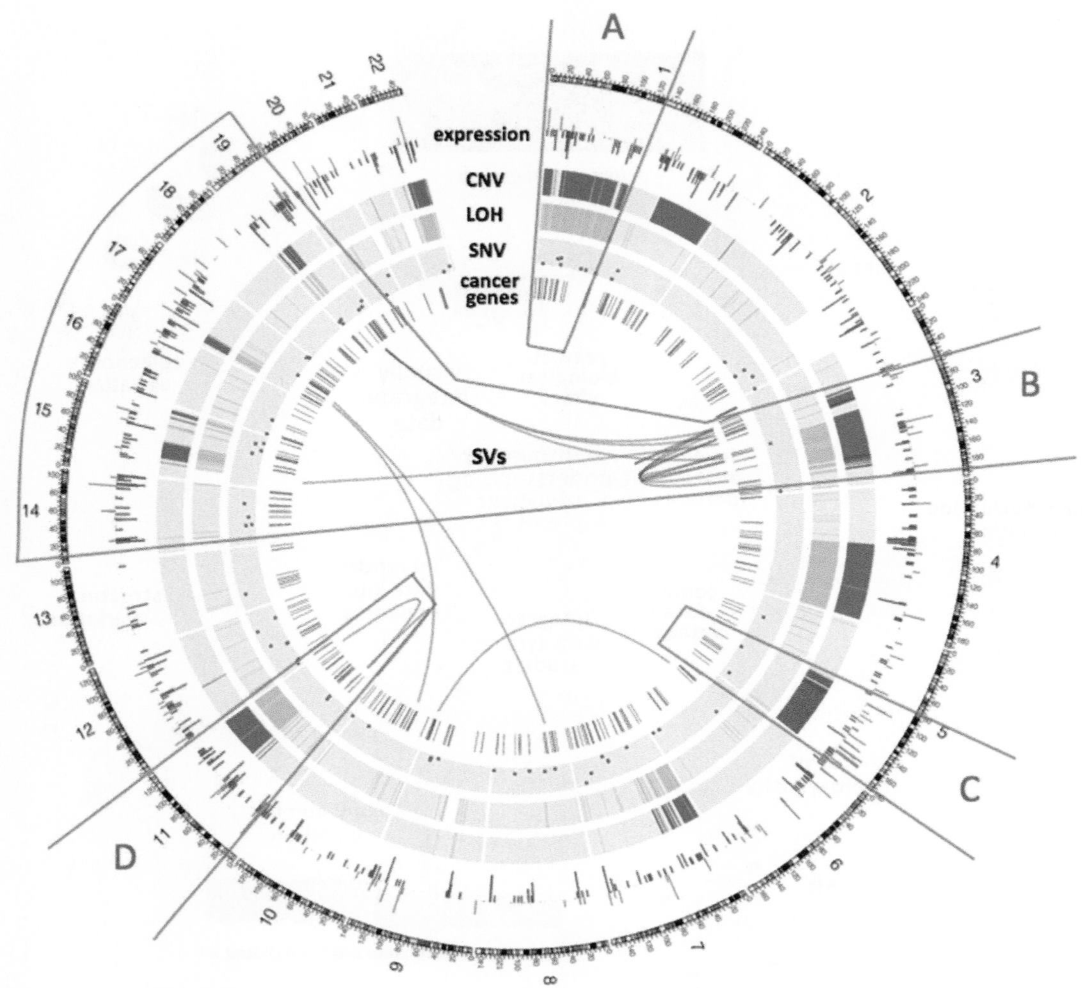

FIGURE 9.4 Data integration enables the interpretation of mutation calls and their biological significance. A hypothetical tumor-versus-normal dataset is plotted with Circos visualization software, showing examples of how integration can help interpret biological data. (A) An area with copy number loss (green CNV) and loss of heterozygosity (green LOH) agreeing with a one-copy deletion in the tumor genome. In this region, SNVs (red dots), which were heterozygous in the normal genome, may have become homozygous in the tumor. If only an inactivated allele of a tumor suppressor gene remains, for example, this SNV may contribute to oncogenesis. (B) A region showing varying copy number and a high density of structural variants may indicate extreme DNA rearrangement events such as chromothripsis. (C) Though copy number events do not necessarily correlate with expression results everywhere, regions where the two data types agree (red: CNV gain and high expression) may be of more importance and could provide a mechanistic basis for potential driver genes. (D) Integrating data can also help to verify if the data generated using different techniques agree with each other. In this region, for example, a deletion event (SV) correlates well with a loss of copy number (green CNV). In addition to the experimentally generated data, external datasets can also be included in such visualization plots. In this graph, one track for known cancer genes is added to the plot.

increase cancer cell viability. They subsequently validated this in cell proliferation assays [122].

CONCLUSION

Personalized cancer medicine, perhaps more appropriately termed "genomics medicine", aims to diagnose individuals based on their unique mutational profiles and provide therapies targeting those alterations while making predictions about the patient's response to treatment. Prior to the advent of NGS technologies, malignant tumors were diagnosed based on their anatomical site and histology characteristics; treatment options would then be based on patient's age, sex and/or family history. We now know that each cancer type is a collection of molecularly heterogeneous subtypes, and thus should not be treated identically. The power of NGS technologies in examining individual cancer genomes has given us the opportunity to identify different mutational profiles in patients with the same cancer type. For instance, *BRAF* activating mutations are found in 40–60% of melanoma patients while the rest carry wild-type copies of the gene [123]. Such knowledge allows for stratification of patients and targeted treatments; in this case, for instance, those with the *BRAF* mutation benefit from the drug vemurafenib while the rest do not. This ensures administration of the right treatment to the right individuals and prevents unnecessary negative side effects in patients not likely to benefit from the drug. Genomic studies have also provided the opportunity to study rare cancers for which no treatment options exist. Tumors which are rare with respect to the tissue type and anatomical site they occur in could be quite common on the molecular level, harboring mutations in known cancer genes and pathways [116]. Such analyses allow for the delivery of already approved drugs to patients for whom no therapy options exist. With further advancements and the decreasing cost, high-throughput and comprehensive studies are starting to become tractable for clinical use [124]. Cancer genomic studies, especially large-scale initiatives such as TCGA, will provide us with a comprehensive overview of all somatic mutations in different cancer types and subtypes. These efforts will help in identifying and cataloging the key players in the disease. Maintaining databases for storing and linking genomic aberrations, clinicopathological characteristics, treatments and outcome of the disease for each patient will become inevitable in the near future. These databases need to be easily accessible by the scientific community and clinicians in order to advance research even further, thus one day enabling early diagnosis and targeted therapies in the clinic. Recently developed tools such as GeneInsight [125] provide the necessary communication gateways between research centers, medical laboratories and practicing clinicians, enabling the delivery of personalized cancer medicine.

Although bioinformatic tools that can more readily automate the data analysis and integration processes are currently in development, it will take much work in the future automatically to predict or annotate the somatic aberrations at a clinically usable level. To date, most studies have involved expert opinion where a panel of experts would discuss and choose the most actionable target(s) [124]. Exploiting the information from mutational analysis of cancers in developing therapeutic strategies also involves functional validation studies. Gene knockout methods, small interfering RNA (siRNA) experiments and selective overexpression of genes in model organisms, established cell lines and cell cultures can provide the opportunity for elucidating the functional role of mutations. However, to keep in pace with computational cancer genomics, these functional screening techniques need to be efficient and high-throughput.

GLOSSARY

Actionable target A mutated protein which is shown to be causally involved in disease initiation, progression and/or maintenance and which can be directly targeted by drugs.

Alignment Alignment is the process of mapping short sequence reads to the reference genome.

Assembly Assembly in genomics refers to the merging of short sequence reads into longer pieces of DNA contigs.

Bioinformatics The application of computer technology and programming to the field of biology.

Cancer genomics The field of study which examines cancers on the molecular level in the hope of identifying cancer-causing mutations and pathways.

Contig A contiguous sequence of nucleotides that is the result of assembling and joining overlapping reads; contigs provide the consensus sequence of the source DNA/RNA.

Copy number variation (CNV) Two copies of every gene are present in human cells, one inherited from each parent. Gain or loss of large regions of the genome leads to the gain or loss of gene copies. These are referred to as copy number variations.

Coverage Sequence coverage is the number of reads that span a specific position in the genome. The higher the coverage, the more confident the genotype calls.

Data integration Different data types (and mutation calls) from whole genome, exome, transcriptome or epigenome experiments are generated in cancer genomic studies. Combining the analysis results from all experiments rather than separate analysis of each can yield conclusive results on the state of the sample under study.

DNA sequencing Tools and techniques used for finding the order of nucleotide base pairs in a DNA molecule.

Driver mutations These are the somatic mutations in cancers which are responsible for providing the cell with growth advantage. As a result, they contribute to disease initiation, progression and maintenance.

Exome sequencing Sequencing only the protein-coding regions of the genome, the set of complete exons.

Expression analysis Refers to the analysis of the expression level of all or a subset of genes. Expression level can be quantified using the number of transcribed copies of each gene.

Functional validation Refers to proving the functional role of a driver mutation at the bench.

Genome The complete set of genetic material in a cell.

Germline mutations Mutations which are present in the gametes and are passed from one generation the next.

High-throughput sequencing Technologies which parallelize the sequencing of whole genomes by reading the sequence of millions of short DNA segments simultaneously.

Indels Indels refer to small insertions and deletions in the genome.

Next-generation sequencing technologies These are the novel technologies which have enabled high-throughput sequencing of complete genomes. Refer to high-throughput sequencing.

Paired-end sequencing The sequencing technique where the two ends of a DNA molecule of a known size are sequenced.

Paired-sample analysis Somatic mutation detection techniques which consider the tumor and matched normal samples simultaneously when identifying the most probable somatic aberrations.

Passenger mutations Any cancer cell could have hundreds of mutated genes, not all are contributing to the disease however. The majority are the result of the unstable cancer genome and do not provide the cell with any growth advantage. These are referred to as passenger mutations.

Pathway analysis Refers to identifying a pathway, a common cellular process governed by a complex of proteins and other molecules, which is affected by the accumulated mutations in the cell.

RNA-seq Also referred to as whole transcriptome shotgun sequencing (WTSS). It is the technique for finding the sequence of the transcribed regions of the genome. cDNA is made using the cell's RNA as template and is then sequenced.

Single nucleotide polymorphism (SNP) These are the variations that are present in the general healthy population. Any two individuals can have thousands, if not more, single nucleotides that vary in their genomes.

Single nucleotide variation (SNV) These are single nucleotide differences that are not present at high frequency in the population and might play a causal role in various diseases.

Somatic mutation Mutations which could be present in any cell type of the body except the germ cells (sperms and eggs). These mutations are not passed from one generation to the next.

Structural variation These refer to variations that encompass large areas of the genome and could include large insertions and deletions, inversions, duplications, translocations and gene fusions.

Transcriptome Complete set of transcribed molecules in a cell.

Verification Mutations identified through computational approaches need to be confirmed in the corresponding samples using techniques such as targeted capture and re-sequencing. This process is referred to as verification.

ABBREVIATIONS

AML Acute myeloid leukemia
BWA Burrows–Wheeler aligner
CCLE Cancer Cell Line Encyclopedia
CGCI Cancer Genome Characterization Initiative
CGH Comparative genomic hybridization
CGHub Cancer Genomic Hub
CGP Cancer Genome Project
CNV Copy number variation
COSMIC Catalogue Of Somatic Mutations In Cancer
DbSNP Database of Single Nucleotide Polymorphisms
ENCODE Encyclopedia of DNA Elements
FDA Food and Drug Administration
Gbp Giga base pair
GBM Glioblastomas
GO Gene ontology
GPU Graphics processing unit
HGP Human Genome Project
HPC High performance computing
ICGC International Cancer Genome Consortium
IGF Insulin-like growth factor
IGV Integrative Genomics Viewer
IPA Ingenuity Pathway Analysis
KEGG Kyoto Encyclopedia of Genes and Genomes
MAQ Mapping and assembly with quality
miRNA MicroRNA
NGS Next-generation sequencing
OMIM Online Mendelian Inheritance in Man
OregAnno Open regulatory annotation
PCGP Pediatric Cancer Genome Project
PCR Polymerase chain reaction
PDB Protein Data Bank
PharmGKB Pharmacogenomics Knowledgebase
PolyPhen Polymorphism phenotyping
SHRiMP Short read mapping package
SIFT Sorting Intolerant From Tolerant
siRNA Small interfering RNA
SNP Single nucleotide polymorphism
SNV Single nucleotide variation
SOAP Short Oligonucleotide Alignment Program
SV Structural variation
TARGET Therapeutically Applicable Research to Generate Effective Treatments
TCGA The Cancer Genome Atlas
TRANSFAC Transcription Factor Database
TTD Therapeutic Target Database
UCSC University of California Santa Cruz
WES Whole exome sequencing
WGS Whole genome sequencing
WGSS Whole genome shotgun sequencing
WTSS Whole transcriptome shotgun sequencing

REFERENCES

[1] A gene-centric human proteome project: HUPO – the Human Proteome organization. Mol Cell Proteomics 2010;9:427–9.

[2] Robertson G, Hirst M, Bainbridge M, Bilenky M, Zhao Y, Zeng T, et al. Genome-wide profiles of STAT1 DNA association using chromatin immunoprecipitation and massively parallel sequencing. Nat Methods 2007;4:651–7.

[3] Lee TI, Johnstone SE, Young RA. Chromatin immunoprecipitation and microarray-based analysis of protein location. Nat Protoc 2006;1:729–48.

[4] Hirst M, Marra MA. Next generation sequencing based approaches to epigenomics. Brief Funct Genomics 2010;9:455–65.

[5] Wilson IM, Davies JJ, Weber M, Brown CJ, Alvarez CE, MacAulay C, et al. Epigenomics: mapping the methylome. Cell Cycle 2006;5:155–8.

[6] Jacinto FV, Ballestar E, Esteller M. Methyl-DNA immunoprecipitation (MeDIP): hunting down the DNA methylome. BioTechniques 2008;44:35–43.

[7] Serre D, Lee BH, Ting AH. MBD-isolated genome sequencing provides a high-throughput and comprehensive survey of DNA methylation in the human genome. Nucleic Acids Res 2010;38:391–9.

[8] Harris RA, Wang T, Coarfa C, Nagarajan RP, Hong C, Downey SL, et al. Comparison of sequencing-based methods to profile DNA methylation and identification of monoallelic epigenetic modifications. Nat Biotechnol 2010;28:1097–105.

[9] Maunakea AK, Nagarajan RP, Bilenky M, Ballinger TJ, D'Souza C, Fouse SD, et al. Conserved role of intragenic DNA methylation in regulating alternative promoters. Nature 2010;466 (7303):253–7.

[10] Cokus SJ, Feng S, Zhang X, Chen Z, Merriman Z, Haudenschild CD, et al. Shotgun bisulphite sequencing of the *Arabidopsis* genome reveals DNA methylation patterning. Nature 2008;452:215–9.

[11] Lister R, O'Malley RC, Tonti-Filippini J, Gregory BD, Berry CC, Millar AH, et al. Highly integrated single-base resolution maps of the epigenome in *Arabidopsis*. Cell 2008;133:523–36.

[12] Lister R, Pelizzola M, Dowen RH, Hawkins D, Hon G, Tonti-Filippini J, et al. Human DNA methylomes at base resolution show widespread epigenomic differences. Nature 2009;462:315–22.

[13] Meissner A, Gnirke A, Bell GW, Ramsahoye B, Lander ES, Jaenisch R. Reduced representation bisulfite sequencing for comparative high-resolution DNA methylation analysis. Nucleic Acids Res 2005;33:5868–77.

[14] Deng J, Shoemaker R, Xie B, Gore A, LeProust EM, Antosiewicz-Bourget J, et al. Targeted bisulfite sequencing reveals changes in DNA methylation associated with nuclear reprogramming. Nat Biotechnol 2009;27:353–60.

[15] Ball MP, Li JB, Gao Y, Lee JH, LeProust EM, Park IH, et al. Targeted and genome-scale strategies reveal gene-body methylation signatures in human cells. Nat Biotechnol 2009;27:361–8.

[16] Smith ZD, Gu H, Bock C, Gnirke A, Meissner A. High-throughput bisulfite sequencing in mammalian genomes. Methods 2009;48:226–32.

[17] International Cancer Genome Consortium. International network of cancer genome projects. Nature 2010;464:993–8.

[18] Barretina J, Caponigro G, Stransky N, Venkatesan K, Margolin AA, Kim S, et al. The Cancer Cell Line Encyclopedia enables predictive modelling of anticancer drug sensitivity. Nature 2012;483:603–7.

[19] International HapMap Consortium. The International HapMap Project. Nature 2003;426:789–96.

[20] 1000 Genomes Project Consortium. A map of human genome variation from population-scale sequencing. Nature 2010;467:1061–73.

[21] Human Genome Structural Variation Working Group, Eichler EE, Nickerson DA, Altshuler D, Bowcock AM, Brooks LD, et al. Completing the map of human genetic variation. Nature 2007;447: 161–5.

[22] Sherry ST, Ward MH, Kholodov M, Baker J, Phan L, Smigielski EM, et al. dbSNP: the NCBI database of genetic variation. Nucleic Acids Res 2001;29:308–11.

[23] Forbes SA, Bhamra G, Bamford S, Dawson E, Kok C, Clements J, et al. The Catalogue of Somatic Mutations in Cancer (COSMIC). Curr Protoc Hum Genet 2008 [Chapter 10:Unit 10.11].

[24] Weisemann JM, Boguski MS, Ouellette BF. Sequence databases: integrated information retrieval and data submission. Curr Protoc Hum Genet 2001 [Chapter 6:Unit 6.7].

[25] Guberman JM, Ai J, Arnaiz O, Baran J, Blake A, Baldock R, et al. BioMart Central Portal: an open database network for the biological community. Database (Oxford) 2011;2011:bar041.

[26] Schadt EE, Linderman MD, Sorenson J, Lee L, Nolan GP. Computational solutions to large-scale data management and analysis. Nat Rev Genet 2010;11:647–57.

[27] Smith TF, Waterman MS. Identification of common molecular subsequences. J Mol Biol 1981;147:195–7.

[28] Li H, Ruan J, Durbin R. Mapping short DNA sequencing reads and calling variants using mapping quality scores. Genome Res 2008;18:1851–8.

[29] Li R, Li Y, Kristiansen K, Wang J. SOAP: short oligonucleotide alignment program. Bioinformatics 2008;24:713–4.

[30] Rumble SM, Lacroute P, Dalca AV, Fiume M, Sidow A, Brudno M. SHRiMP: accurate mapping of short color-space reads. PLoS Comput Biol 2009;5:e1000386.

[31] Ferragina P, Manzini G. Opportunistic data structures with applications. Proceedings of the 41st Symposium on Foundation of Computer Science 2000;390–8.

[32] Langmead B, Trapnell C, Pop M, Salzberg SL. Ultrafast and memory-efficient alignment of short DNA sequences to the human genome. Genome Biol 2009;10:R25.

[33] Li H, Durbin R. Fast and accurate short read alignment with Burrows–Wheeler transform. Bioinformatics 2009;25:1754–60.

[34] Li R, Yu C, Li Y, Lam TW, Yiu SM, Kristiansen K, et al. SOAP2: an improved ultrafast tool for short read alignment. Bioinformatics 2009;25:1966–7.

[35] Zerbino DR, Birney E. Velvet: algorithms for de novo short read assembly using de Bruijn graphs. Genome Res 2008;18:821–9.

[36] Chaisson MJ, Pevzner PA. Short read fragment assembly of bacterial genomes. Genome Res 2008;18:324–30.

[37] Butler J, MacCallum I, Kleber M, Shlyakhter IA, Belmonte MK, Lander ES, et al. ALLPATHS: de novo assembly of whole-genome shotgun microreads. Genome Res 2008;18:810–20.

[38] Simpson JT, Wong K, Jackman SD, Schein JE, Jones SJ, Birol I. ABySS: a parallel assembler for short read sequence data. Genome Res 2009;19:1117–23.

[39] Koboldt DC, Chen K, Wylie T, Larson DE, McLellan MD, Mardis ER, et al. VarScan: variant detection in massively parallel sequencing of individual and pooled samples. Bioinformatics 2009;25: 2283–5.

[40] Li H, Handsaker B, Wysoker A, Fennell T, Ruan J, Homer N, et al. The Sequence Alignment/Map format and SAMtools. Bioinformatics 2009;25:2078–9.

[41] McKenna A, Hanna M, Banks E, Sivachenko A, Cibulskis K, Kernytsky A, et al. The Genome Analysis Toolkit: a MapReduce framework for analyzing next-generation DNA sequencing data. Genome Res 2010;20:1297–303.

[42] Goya R, Sun MG, Morin RD, Leung G, Ha G, Wiegand KC, et al. SNVMix: predicting single nucleotide variants from next-generation sequencing of tumors. Bioinformatics 2010;26: 730–6.

[43] Li R, Li Y, Fang X, Yang H, Wang J, Kristiansen K, et al. SNP detection for massively parallel whole-genome resequencing. Genome Res 2009;19:1124–32.

[44] Roth A, Ding J, Morin R, Crisan A, Ha G, Giuliany R, et al. JointSNVMix: a probabilistic model for accurate detection of somatic mutations in normal/tumour paired next-generation sequencing data. Bioinformatics 2012;28:907–13.

[45] Larson DE, Harris CC, Chen K, Koboldt DC, Abbott TE, Dooling DJ, et al. SomaticSniper: identification of somatic point mutations in whole genome sequencing data. Bioinformatics 2012;28:311–7.

[46] Ding J, Bashashati A, Roth A, Oloumi A, Tse K, Zeng T, et al. Feature-based classifiers for somatic mutation detection in tumour-normal paired sequencing data. Bioinformatics 2012;28:167–75.

[47] Ye K, Schulz MH, Long Q, Apweiler R, Ning Z. Pindel: a pattern growth approach to detect break points of large deletions and medium sized insertions from paired-end short reads. Bioinformatics 2009;25:2865–71.

[48] Homer N, Merriman B, Nelson SF. BFAST: an alignment tool for large scale genome resequencing. PLoS One 2009;4:e7767.

[49] Albers CA, Lunter G, MacArthur DG, McVean G, Ouwehand WH, Durbin R. Dindel: accurate indel calls from short-read data. Genome Res 2011;21:961–73.

[50] Mullaney JM, Mills RE, Pittard WS, Devine SE. Small insertions and deletions (INDELs) in human genomes. Hum Mol Genet 2010;19:R131–6.

[51] Futreal PA, Coin L, Marshall M, Down T, Hubbard T, Wooster R, et al. A census of human cancer genes. Nat Rev Cancer 2004;4: 177–83.

[52] Korbel JO, Urban AE, Affourtit JP, Godwin B, Grubert F, Simons JF, et al. Paired-end mapping reveals extensive structural variation in the human genome. Science 2007;318:420–6.

[53] Korbel JO, Abyzov A, Mu XJ, Carriero N, Cayting P, Zhang Z, et al. PEMer: a computational framework with simulation-based error models for inferring genomic structural variants from massive paired-end sequencing data. Genome Biol 2009;10:R23.

[54] Hormozdiari F, Alkan C, Eichler EE, Sahinalp SC. Combinatorial algorithms for structural variation detection in high-throughput sequenced genomes. Genome Res 2009;19:1270–8.

[55] Chen K, Wallis JW, McLellan MD, Larson DE, Kalicki JM, Pohl CS, et al. BreakDancer: an algorithm for high-resolution mapping of genomic structural variation. Nat Methods 2009;6:677–81.

[56] McKernan KJ, Peckham HE, Costa GL, McLaughlin SF, Fu Y, Tsung EF, et al. Sequence and structural variation in a human genome uncovered by short-read, massively parallel ligation sequencing using two-base encoding. Genome Res 2009;19:1527–41.

[57] Kidd JM, Cooper GM, Donahue WF, Hayden HS, Sampas N, Graves T, et al. Mapping and sequencing of structural variation from eight human genomes. Nature 2008;453:56–64.

[58] Robertson G, Schein J, Chiu R, Corbett R, Field M, Jackman SD, et al. *De novo* assembly and analysis of RNA-seq data. Nat Methods 2010;7:909–12.

[59] Xie C, Tammi MT. CNV-seq, a new method to detect copy number variation using high-throughput sequencing. BMC Bioinformatics 2009;10:80.

[60] Chiang DY, Getz G, Jaffe DB, O'Kelly MJT, Zhao X, Carter SL, et al. High-resolution mapping of copy-number alterations with massively parallel sequencing. Nat Methods 2009;6:99–103.

[61] Bentley DR, Balasubramanian S, Swerdlow HP, Smith GP, Milton J, Brown CG, et al. Accurate whole human genome sequencing using reversible terminator chemistry. Nature 2008;456:53–9.

[62] Yoon S, Xuan Z, Makarov V, Ye K, Sebat J. Sensitive and accurate detection of copy number variants using read depth of coverage. Genome Res 2009;19:1586–92.

[63] Campbell PJ, Stephens PJ, Pleasance ED, O'Meara S, Li H, Santarius T, et al. Identification of somatically acquired rearrangements in cancer using genome-wide massively parallel paired-end sequencing. Nat Genet 2008;40:722–9.

[64] Alkan C, Kidd JM, Marques-Bonet T, Aksay G, Antonacci F, Hormozdiari F, et al. Personalized copy number and segmental duplication maps using next-generation sequencing. Nat Genet 2009;41:1061–7.

[65] Edgar R, Domrachev M, Lash AE. Gene Expression Omnibus: NCBI gene expression and hybridization array data repository. Nucleic Acids Res 2002;30:207–10.

[66] Barrett T, Troup DB, Wilhite SE, Ledoux P, Rudnev D, Evangelista C, et al. NCBI GEO: archive for high-throughput functional genomic data. Nucleic Acids Res 2009;37:D885–90.

[67] Beroukhim R, Mermel CH, Porter D, Wei G, Raychaudhuri S, Donovan J, et al. The landscape of somatic copy-number alteration across human cancers. Nature 2010;463:899–905.

[68] Golub TR, Slonim DK, Tamayo P, Huard C, Gaasenbeek M, Mesirov JP, et al. Molecular classification of cancer: class discovery and class prediction by gene expression monitoring. Science 1999;286:531–7.

[69] Alizadeh AA, Eisen MB, Davis RE, Ma C, Lossos IS, Rosenwald A, et al. Distinct types of diffuse large B-cell lymphoma identified by gene expression profiling. Nature 2000;403:503–11.

[70] van de Vijver MJ, He YD, van't Veer LJ, et al. A gene-expression signature as a predictor of survival in breast cancer. N Engl J Med 2002;347:1999–2009.

[71] Paik S, Shak S, Tang G, Kim C, Baker J, Cronin M, Dai H, Hart AAM, Voskuil DW, et al. A multigene assay to predict recurrence of tamoxifen-treated, node-negative breast cancer. N Engl J Med 2004;351:2817–26.

[72] Eisen MB, Spellman PT, Brown PO, Botstein D. Cluster analysis and display of genome-wide expression patterns. Proc Natl Acad Sci USA 1998;95:14863–8.

[73] Tamayo P, Slonim D, Mesirov J, Zhu Q, Kitareewan S, Dmitrovsky E, et al. Interpreting patterns of gene expression with self-organizing maps: methods and application to hematopoietic differentiation. Proc Natl Acad Sci USA 1999;96:2907–12.

[74] Kim PM, Tidor B. Subsystem identification through dimensionality reduction of large-scale gene expression data. Genome Res 2003;13:1706–18.

[75] Brunet JP, Tamayo P, Golub TR, Mesirov JP. Metagenes and molecular pattern discovery using matrix factorization. Proc Natl Acad Sci USA 2004;101:4164–9.

[76] Stajich JE, Block D, Boulez K, Brenner SE, Chervitz SA, Dagdigian C, et al. The Bioperl toolkit: Perl modules for the life sciences. Genome Res 2002;12:1611–8.

[77] Cock PJ, Antao T, Chang JT, Chapman BA, Cox CJ, Dalke A, et al. Biopython: freely available Python tools for computational molecular biology and bioinformatics. Bioinformatics 2009;25:1422–3.

[78] Holland RC, Down TA, Pocock M, Prlic A, Huen D, James K, et al. BioJava: an open-source framework for bioinformatics. Bioinformatics 2008;24:2096–7.

[79] Gentleman RC, Carey VJ, Bates DM, Bolstad B, Dettling M, Dudoit S, et al. Bioconductor: open software development for computational biology and bioinformatics. Genome Biol 2004;5:R80.

[80] Stajich JE, Lapp H. Open source tools and toolkits for bioinformatics: significance, and where are we? Brief Bioinform 2006;7:287–96.

[81] Dreszer TR, Karolchik D, Zweig AS, Hinrichs AS, Raney BJ, Kuhn RM, et al. The UCSC Genome Browser database: extensions and updates 2011. Nucleic Acids Res 2012;40: D918–23.

[82] Flicek P, Amode MR, Barrell D, Beal K, Brent S, Carvalho-Silva D, et al. Ensembl 2012. Nucleic Acids Res 2012;40:D84–90.

[83] Thorvaldsdottir H, Robinson JT, Mesirov JP. Integrative Genomics Viewer (IGV): high-performance genomics data visualization and exploration. Brief Bioinform 2012.

[84] Krzywinski M, Schein J, Birol I, Connors J, Gascoyne R, Horsman D, et al. Circos: an information aesthetic for comparative genomics. Genome Res 2009;19:1639–45.

[85] Cerami E, Gao J, Dogrusoz U, Gross BE, Sumer SO, Aksoy BA, et al. The cBio cancer genomics portal: an open platform for exploring multidimensional cancer genomics data. Cancer Discov 2012;2:401–4.

[86] Iafrate AJ, Feuk L, Rivera MN, Listewnik ML, Donahoe PK, Qi Y, et al. Detection of large-scale variation in the human genome. Nat Genet 2004;36:949–51.

[87] Futreal PA, Coin L, Marshall M, Down T, Hubbard T, Wooster R, et al. A census of human cancer genes. Nat Rev Cancer 2004;4:177–83.

[88] Mitelman F, Johansson B, Mertens F. The impact of translocations and gene fusions on cancer causation. Nat Rev Cancer 2007;7:233–45.

[89] Gusnanto A, Wood HM, Pawitan Y, Rabbitts P, Berri S. Correcting for cancer genome size and tumour cell content enables better estimation of copy number alterations from next-generation sequence data. Bioinformatics 2012;28: 40–7.

[90] Ha G, Roth A, Lai D, Bashashati A, Ding J, Goya R, et al. Integrative analysis of genome-wide loss of heterozygosity and mono-allelic expression at nucleotide resolution reveals disrupted pathways in triple negative breast cancer. Genome Res 2012;22:1995–2007.

[91] Greenman C, Stephens P, Smith R, Dalgliesh GL, Hunter C, Bignell G, et al. Patterns of somatic mutation in human cancer genomes. Nature 2007;446:153–8.

[92] Nikolaev SI, Rimoldi D, Iseli C, Valsesia A, Robyr D, Gehrig C, et al. Exome sequencing identifies recurrent somatic MAP2K1 and MAP2K2 mutations in melanoma. Nat Genet 2011;44:133–9.

[93] Downing JR, Wilson RK, Zhang J, Mardis ER, Pui CH, Ding L, et al. The Pediatric Cancer Genome Project. Nat Genet 2012;44: 619–22.

[94] Adzhubei IA, Schmidt S, Peshkin L, Ramensky VE, Gerasimova A, Bork P, et al. A method and server for predicting damaging missense mutations. Nat Methods 2010;7:248–9.

[95] Reva B, Antipin Y, Sander C. Determinants of protein function revealed by combinatorial entropy optimization. Genome Biol 2007;8:R232.

[96] Ng PC, Henikoff S. Predicting deleterious amino acid substitutions. Genome Res 2001;11:863–74.

[97] Greenman C, Wooster R, Futreal PA, Stratton MR, Easton DF. Statistical analysis of pathogenicity of somatic mutations in cancer. Genetics 2006;173:2187–98.

[98] Berman HM, Westbrook J, Feng Z, Gilliland G, Bhat TN, Weissig H, et al. The Protein Data Bank. Nucleic Acids Res 2000;28:235–42.

[99] Kiefer F, Arnold K, Kunzli M, Bordoli L, Schwede T. The SWISS-MODEL Repository and associated resources. Nucleic Acids Res 2009;37:D387–92.

[100] Guex N, Peitsch MC. SWISS-MODEL and the Swiss-PdbViewer: an environment for comparative protein modeling. Electrophoresis 1997;18:2714–23.

[101] Beroukhim R, Getz G, Nghiemphu L, Barretina J, Hsueh T, Linhart D, et al. Assessing the significance of chromosomal aberrations in cancer: methodology and application to glioma. Proc Natl Acad Sci USA 2007;104:20007–12.

[102] ENCODE Project Consortium. The ENCODE (ENCyclopedia Of DNA Elements) Project. Science 2004;306:636–40.

[103] Wingender E. The TRANSFAC project as an example of framework technology that supports the analysis of genomic regulation. Brief Bioinform 2008;9:326–32.

[104] Portales-Casamar E, Thongjuea S, Kwon AT, Arenillas D, Zhao X, Valen E, et al. JASPAR 2010: the greatly expanded open-access database of transcription factor binding profiles. Nucleic Acids Res 2010;38:D105–10.

[105] Griffith OL, Montgomery SB, Bernier B, Chu B, Kasaian K, Aerts S, et al. ORegAnno: an open-access community-driven resource for regulatory annotation. Nucleic Acids Res 2008;36: D107–13.

[106] Buratti E, Chivers M, Hwang G, Vorechovsky I. DBASS3 and DBASS5: databases of aberrant 3′- and 5′-splice sites. Nucleic Acids Res 2011;39:D86–91.

[107] Wang J, Zhang J, Li K, Zhao W, Cui Q. SpliceDisease database: linking RNA splicing and disease. Nucleic Acids Res 2012;40: D1055–9.

[108] Wang E. A roadmap of cancer systems biology. In: Wang E, editor. Cancer systems biology. Boca Raton, FL: CRC Press; 2010.

[109] Christopher R, Dhiman A, Fox J, Gendelman R, Haberitcher T, Kagle D, et al. Data-driven computer simulation of human cancer cell. Ann NY Acad Sci 2004;1020:132–53.

[110] Ashburner M, Ball CA, Blake JA, Botstein D, Butler H, Cherry JM, et al. Gene ontology: tool for the unification of biology. The Gene Ontology Consortium. Nat Genet 2000;25:25–9.

[111] Subramanian A, Tamayo P, Mootha VK, Mukherjee S, Ebert BL, Gillette MA, et al. Gene set enrichment analysis: a knowledge-based approach for interpreting genome-wide expression profiles. Proc Natl Acad Sci USA 2005;102:15545–50.

[112] Kanehisa M, Goto S, Hattori M, Aoki-Kinoshita KF, Itoh M, Kawashima S, et al. From genomics to chemical genomics: new developments in KEGG. Nucleic Acids Res 2006;34:D354–7.

[113] Huang da W, Sherman BT, Lempicki RA. Systematic and integrative analysis of large gene lists using DAVID bioinformatics resources. Nat Protoc 2009;4:44–57.

[114] Vaske CJ, Benz SC, Sanborn JZ, Earl D, Szeto C, Zhu J, et al. Inference of patient-specific pathway activities from multi-dimensional cancer genomics data using PARADIGM. Bioinformatics 2010;26:i237–45.

[115] Cancer Genome Atlas Research Network. Comprehensive genomic characterization defines human glioblastoma genes and core pathways. Nature 2008;455:1061–8.

[116] Jones SJ, Laskin J, Li YY, Griffith OL, An J, Bilenky M, et al. Evolution of an adenocarcinoma in response to selection by targeted kinase inhibitors. Genome Biol 2010;11:R82.

[117] Wagle N, Berger MF, Davis MJ, Blumenstiel B, Defelice M, Pochanard P, et al. High-throughput detection of actionable genomic alterations in clinical tumor samples by targeted, massively parallel sequencing. Cancer Discov 2012;2:82–93.

[118] Wishart DS, Knox C, Guo AC, Shrivastava S, Hassanali M, Stothard P, et al. DrugBank: a comprehensive resource for *in silico* drug discovery and exploration. Nucleic Acids Res 2006;34:D668–72.

[119] Chen X, Ji ZL, Chen YZ. TTD: Therapeutic Target Database. Nucleic Acids Res 2002;30:412–5.

[120] Hewett M, Oliver DE, Rubin DL, Easton KL, Stuart JM, Altman RB, et al. PharmGKB: the Pharmacogenetics Knowledge Base. Nucleic Acids Res 2002;30:163–5.

[121] Ramasamy A, Mondry A, Holmes CC, Altman DG. Key issues in conducting a meta-analysis of gene expression microarray datasets. PLoS Med 2008;5:e184.

[122] Iadevaia S, Lu Y, Morales FC, Mills GB, Ram PT. Identification of optimal drug combinations targeting cellular networks: integrating phospho-proteomics and computational network analysis. Cancer Res 2010;70:6704–14.

[123] Flaherty KT, Puzanov I, Kim KB, Ribas A, McArthur GA, Sosman JA, et al. Inhibition of mutated, activated BRAF in metastatic melanoma. N Engl J Med 2010;363:809–19.

[124] Roychowdhury S, Iyer MK, Robinson DR, Lonigro RJ, Wu YM, Cao X, et al. Personalized oncology through integrative high-throughput sequencing: a pilot study. Sci Transl Med 2011;3:111–21.

[125] Aronson SJ, Clark EH, Babb LJ, Baxter S, Farwell LM, Funke BH, et al. The GeneInsight Suite: a platform to support laboratory and provider use of DNA-based genetic testing. Hum Mutat 2011;32:532–6.

Chapter 10

Cancer Genomics

Genomic Resource Projects

Matthew Parker
Erin Hedlund
Jinghui Zhang

Computational Biology Department, St Jude Children's Research Hospital Memphis, TN, USA

Contents

Cancer Genomics. DOI: http://dx.doi.org/10.1016/B978-0-12-396967-5.00010-4

Key Concepts

- Genomic data are widely available online for both bioinformaticians who want to analyze raw data and biologists/clinicians seeking biologically meaningful conclusions from the analyses
- Large consortia/projects have been launched with the goal of sequencing all types of cancer from pediatric to adult; these groups are producing data at an incredible rate
- Genomic data exist in four main categories: (1) raw, (2) processed/normalized, (3) interpreted, and (4) summarized within and across diseases
- Data can be downloaded from large data warehouses or from data portals with varying access restrictions; as such, these data portals provide a means to view interpreted data
- A number of on/off-line tools exist to view and analyze genomic data
- To determine the spectrum of disease-related mutations, which help inform clinical decisions, aggregation of a wide variety of genomic knowledge from as many sources as possible is necessary

INTRODUCTION: THE GENOMIC DATA GOLDMINE

Cancer is a genetic disease caused by alterations in DNA. Ever since the discovery of *HRAS* G12 mutations 30 years ago [1,2], identification of somatic mutations has provided important insight into the initiation, progression and prognosis of cancer. With the completion of the reference human genome sequence [3] and advances in sequencing and computing technologies, a systematic survey of the cancer genome landscape truly began in earnest in 2005. The first studies in the modern genomic era initially took the approach of targeted candidate gene re-sequencing using polymerase chain reaction (PCR) and Sanger sequencing. Notable examples include sequencing of the protein kinase gene family in multiple cancer types [4–6], sequencing of candidate genes chosen based on genetic lesions identified by copy number changes, gene expression profiling, and genes with a functional role in known disease pathways [7–10]. Some of the studies were ultimately expanded to analyze the entire spectrum of protein-coding genes in the human genome using these approaches [11–13].

With the advent of next-generation sequencing (NGS) technology, the comprehensive investigation of somatic alterations of the entire cancer genome at the base-pair resolution has become feasible. The drastic reduction of sequencing cost coupled with enhanced sequencing capacity of NGS has made it the technology of choice for ongoing genome resource projects like the Cancer Genome Project (CGP) and The Cancer Genome Atlas (TCGA). NGS also catalyzed the formation of new initiatives for genome-wide characterization of somatic lesions in cancer such as the International Cancer Genome Consortium (ICGC) and the St Jude Children's Research Hospital–Washington University Pediatric Cancer Genome Project (PCGP). Together, these cancer genome sequencing efforts are expected not only to yield an unparalleled view of the altered signaling pathways in adult and pediatric cancer, but also the identity of new gene targets against which novel therapeutics can be developed.

One common feature of the cancer genome sequencing projects is the huge volume of multidimensional data, including sequencing data, copy number variation (CNV), methylation, gene expression and microRNA expression. The raw data files required to carry out these analyses are accessible via public repository databases like the National Center for Biotechnology Information (NCBI) and the European Bioinformatics Institute (EBI). These raw data are of most interest to bioinformaticians who can subject them to reanalysis with new algorithms that may result in the identification of additional novel genetic lesions or may provide new biological insights from a different perspective than the initial scientific report of the data. In addition, the raw data can be used to test novel computational methods that provide further opportunities for reanalysis of individual datasets or meta-analyses across datasets. The overarching theme being that data generated to answer one specific question have the potential to be repurposed to answer another.

Although the final analytical results of the genetic lesions found by efforts of the genome resource projects are usually made available as large supplementary files appended to the primary publications, access to the processed data via a data portal is extremely important for both biomedical researchers and computational biologists. Different types of web-based data access portals have been developed to meet the diverse needs of the research community. Several cancer genome sequencing projects such as TCGA, ICGC and PCGP have developed their own data portals that allow researchers to access both the open and controlled datasets. The project data portals also provide user interfaces for gene-orientated queries as well as complex queries that integrate genomic, clinical and functional information. A third type of web portal collects data across multiple projects, performs reanalysis and quality control (QC) to ensure consistency and integrity across multiple projects, and provides query and graphical interfaces that support cross-project data interrogation. Through the use of these portals, biomedical researchers who are not specialized in computational analysis can now access processed data for cross-study comparison, for carrying out aggregated analysis across multiple disease cohorts, and for getting information on genes and pathways that may not have been the main focus of the primary publications.

The generation of genomic data is increasing at an exponential rate and these data archives and portals will preserve the data for many years to come. As our knowledge and experience in analyzing these data grows it will be possible to go back and reanalyze them with new algorithms that could potentially extract additional biologically meaningful information from the data.

Box 10.1 Types of Genomic Data

Next-Generation Sequencing (NGS)
Second-generation, or Next-Gen, sequencing that allows for the discovery of multiple types of genomic aberrations: single nucleotide variations, structural variation, small insertions/deletions, copy number variation, loss-of-heterozygosity and structural variations. NGS can be applied to either the whole genome, the exome or to the transcriptome.

Whole Genome Sequencing (WGS)
The most complete DNA sequencing, covering the majority of the genome: exons, introns, and intergenic regions.

Exome Capture Sequencing (ECS)
This method involves capture of just the coding exons of the human genome using specially designed probes which are complementary to these exons. The DNA of interest is hybridized to these probes and all other DNA washed away.

Targeted Sequencing
Like exon capture, targeted sequencing uses probes but, in this technique, they are designed to specific regions of the genome that are of interest, e.g. mutated genes. Usually targeted sequencing is used for the purpose of validation and high-throughput screening of recurring mutations.

RNA-Seq (Transcriptome Sequencing)
Next-generation sequencing of cDNA from all transcribed mRNAs or a subset using capture techniques described above. This allows for differential expression analysis like microarray-based methods (see below) but, in addition, facilitates the discovery of SNVs, SVs, and novel isoforms/exons.

miRNA-Seq
Like RNA sequencing small non-coding RNAs can be prepared from cells and sequenced. MiRNA-seq sequencing data provide the nucleotide sequence in addition to the expression levels.

ChIP-seq
This technique uses next-generation sequencing to discover the DNA bound by proteins of interest. This is an extension of chromatin immunoprecipitation (ChIP) where specific interactions between proteins and DNA are investigated. Using NGS enables a global view to be discerned.

Chromosome Conformation Capture (3C)
This technique allows researchers to examine the higher order structure of the genome (the chromatin) by cross-linking the physical interactions of chromatin with formaldehyde followed by enzymatic digestion and ligation with the frequency of ligation of two restriction fragments a measure of the frequency of interaction within the nucleus.

Chromosome Conformation Capture Carbon-Copy (5C)
5C is an extension of 3C using NGS technology to identify the ligation products, using ligation-mediated amplification to copy and amplify a subset of the 3C library followed by detection using NGS [14]. 5C is allowing three-dimensional interaction maps of the genome to be generated uncovering long-range interactions of promoters and distal elements that can potentially affect gene regulation [15].

Microarrays
Microarrays contain small sequences of thousands of genes or other genomic regions embedded on a solid surface such as a glass slide, which is subsequently hybridized with DNA or RNA. While less expensive and faster than NGS technologies, they offer only one or two data types per array type.

SNP Array
This is a DNA microarray (DNA probes immobilized on a glass slide/chip) used to determine single nucleotide changes as well as copy number variation in the genome. It can also be used to estimate copy number variations.

RNA Expression
RNA expression arrays ("microarrays") measure the differences in expression of genes from two different populations of samples. RNA "probes" are, like DNA microarrays, spotted onto glass slides or chips.

miRNA Expression
miRNA microarrays have been developed to measure the expression of these small non-coding RNAs.

RPPA
Antibody-based, reverse phase protein arrays for protein expression levels and phosphorylation state. Measures levels of phosphorylated isoforms.

KEY LARGE-SCALE CANCER GENOMICS PROJECTS

A number of large genomics projects have been launched with the aim of cataloging the somatic changes that lead to the development of cancer. These projects have a number of features in common:

- Sequencing multiple cancer types
- Sequencing cancer tissue and a matched non-cancer sample
- Sequencing to a high (bp) resolution with deep coverage
- Ensuring all samples have rigorous accompanying clinical information
- Data released to the public (with restrictions).

These projects differ in the types of sequencing being undertaken, with some projects choosing whole genome sequencing (WGS), and some exome. These data are often supplemented with further genome-wide analyses such as RNA-seq, gene expression, methylation, and genotyping.

The Cancer Genome Atlas (TCGA)

The Cancer Genome Atlas (TCGA) project was initiated jointly by the United States National Cancer Institute (NCI) and National Human Genome Research Institute (NHGRI), and aims to catalog and discover major cancer-causing somatic lesions in over 20 types of adult cancers. The overarching goal being to improve our ability to diagnose, treat and prevent cancer.

Nearly 20 research institutes have participated in TCGA through biospecimen collection, genome characterization, genome sequencing, genome data analysis, data coordination and, most recently, proteome characterization. To ensure scientific rigor and comprehensiveness of cancer genome profiling, DNA copy number alteration, messenger RNA expression, microRNA expression and CpG methylation were characterized using multiple complementary platforms and the resulting genomic data were analyzed by multiple computational algorithms. All analytical results for genome characterization and sequence mutations were deposited in standard common formats in the TCGA Data Coordination Center (DCC) at http://cancergenome.nih.gov/dataportal/.

TCGA took a phased-in strategy that began with a 3-year pilot project in 2006 targeting glioblastoma, lung cancer and ovarian cancer. The first published TCGA study focused on glioblastoma (World Health Organization grade IV), a deadly cancer with a median survival of approximately 1 year with generally poor responses to all therapeutic modalities. Recurrent somatic lesions were identified in three key signaling pathways, PI-3K/RAS, TP53 and RB. Furthermore, an integrative analysis of mutation, DNA methylation and clinical treatment data identified a link between MGMT promoter methylation and a hypermutator phenotype caused by mismatch repair deficiency in treated glioblastomas [8]. The next publication was a study concerning ovarian cancer, identifying subgroups based on transcription, microRNA and methylation profiling. *TP53* mutations were found in almost all tumors (96%) while mutations with low prevalence but high significance were found in *NF1*, *BRCA1*, *BRCA2*, *RB1* and *CDK12* along with 113 significant focal DNA copy number aberrations [16].

One important component of TCGA is the development of infrastructure that provides public access to genomic data through the Data Coordinating Center and the TCGA Data Portal, enabling researchers anywhere in the world to make and validate important discoveries. Over 100 peer-reviewed publications have been authored by investigators who are not part of TCGA Research Network but whose work is based on TCGA data.

In parallel with TCGA, NCI launched the Therapeutically Applicable Research to Generate Effective Treatments (TARGET) Initiative in 2006 which applied genomic technologies to identify new drug targets in high-risk childhood cancers. The initiative's goal is to characterize completely the genome, transcriptome, and epigenome of 100 to 200 cancer specimens from patients with five pediatric cancers: acute lymphoblastic leukemia (ALL), acute myeloid leukemia (AML), Wilms tumor, neuroblastoma and osteosarcoma. Recurrent oncogenic mutations and novel translocations resulting in activation of JAK/STAT signaling pathway were found in a subgroup of patients with a transcription profile similar to those that have BCR−ABL translocation. Genome characterization data for TARGET can be downloaded from the TARGET data portal (http://target.cancer.gov/dataportal/).

For both TCGA and TARGET, the sequence trace files generated through Sanger sequencing technologies are stored in the NCBI's Trace Archive; while the aligned reads (BAM files) from second- and third-generation sequencing technologies are contained in NCI's Cancer Genomics Hub (CGHub) for TCGA, and NCBI's Sequence Read Archives (SRA) for TARGET.

International Cancer Genome Consortium (ICGC)

Like TCGA, the ICGC is a collaboration between many research groups and aims to sequence 25 000 cancer genomes, supplementing these with epigenomic and transcriptomic studies for each case over a 10-year period [17]. As well as coordinating multiple smaller disparate sequencing projects around the globe, the ICGC also includes the data from the TCGA project described above. The consortium hopes to collect data from at least

50 cancer types having clinical and societal importance, and aims to create a resource of rapidly updated and freely available cancer genomic data. The ICGC provides a forum for collaboration and coordinates current large-scale projects for the elucidation of the genomic changes in cancer. All participants have agreed upon common standards for sample collection and storage that protects the identities of the donors.

This is the largest consortium of its type in the world and the number of samples that are to be sequenced is impressive, because it was calculated that 500 samples per tumor type would be required to have the statistical power to detect variations from the normal sequence occurring at a 5% or greater frequency in the tumor sample. However, this number may have to be reduced for rare tumors or increased for heterogeneous tumors that may contain different populations, "clones", of cancer cells as well as normal and immune cells.

Data are stored in local databases at each of the consortium members' institutions and a web portal, using BioMart, allows access to these datasets via a single interface (see below) alongside public databases such as Ensembl, Reactome, the Kyoto Encyclopedia of Genes and Genomes (KEGG), and the Catalogue of Somatic Mutations in Cancer (COSMIC). Currently, data for 6590 donors including samples from 31 different malignancies of adulthood and childhood are available. The ICGC Data Portal offers a gene search tool that directs users to the datasets available with aberrations in the particular gene of interest. It integrates information from COSMIC and other online sources and directs the user to other databases of interest. The main use of this portal is downloading interpreted data in text files for further offline analysis. It is also possible to perform rudimentary pathway and affected gene analyses and build more complex queries if desired.

Some of the notable findings of this project include *SF3B1* mutations in low-grade myelodysplasia [18] as well as frequent *NOTCH1*, *XPO1*, *MYD88* and *KLHL6* mutations in chronic lymphocytic leukemia (CLL) [19]. In particular, the CLL study integrated extensive clinical information and showed a significant reduction in survival for those individuals harboring *NOTCH1* mutations. These were found to be gain-of-function mutations in *NOTCH1* and *MYD88* and, as such, could be targeted by inhibitors. Thus, this work highlights how genomic studies can lead to potential new therapeutic approaches in treating this disease.

The St Jude Children's Research Hospital–Washington University Pediatric Cancer Genome Project (PCGP)

The St Jude Children's Research Hospital–Washington University Pediatric Cancer Genome Project (PCGP) is a privately funded initiative for identifying the somatic mutations that drive the initiation and biological and clinical behavior of pediatric cancer [20]. It was launched in 2010 aiming to obtain 30-fold haploid coverage of the whole genome of 600 pediatric tumors and matched non-tumor DNA samples (1200 total genomes). The large scope of this pediatric-focused cancer sequencing effort is necessary to explore fully the genetic basis of the unique cancers seen in children because the spectrum of cancers occurring in the pediatric population is markedly different from that seen in adults. For example, the major brain and peripheral solid tumors that arise in children, including medulloblastoma, neuroblastoma, rhabdomyosarcoma, Ewing's sarcoma, osteosarcoma, and Wilms tumor, are exceedingly rare in adults. Similarly, the specific genetic subtypes of ALL, the most common malignancy in children, differ markedly between children and adults. The PCGP was thus specifically designed to complement the larger government-funded genomic efforts focused on adult cancers such as the TCGA and ICGC projects. Because structural variations (SVs) such as inter- and intra-chromosomal rearrangements are a common mechanism of mutagenesis in pediatric leukemias and solid tumors, PCGP chose a WGS approach instead of exome or transcriptome sequencing in order to detect the full spectrum of somatic lesions in pediatric cancers. Given the relative rarity of pediatric cancers coupled with the heterogeneity of tumor subtypes, the analysis of a large number of pediatric cancers from a specific subtype by WGS was unfeasible in the short term. In the PCGP, the pediatric cancer subtypes for which outcome (cure) with current treatment is poor and/or where there is a conspicuous lack of knowledge regarding the genetic basis of the disease were prioritized for sequencing. All samples were analyzed by SNP arrays for quality control and a subset of samples was also analyzed by transcriptome sequencing. In addition to the discovery cohort analyzed by WGS for each subtype of cancer, a validation cohort was analyzed by either exome or transcriptome sequencing to define mutation frequency in a combined larger cohort. Mouse models harboring the most significant mutations were employed in combination with gene expression and epigenetic profiling to aid in the understanding of the functional impact of these gene mutations. Major findings published in 2012 included identification of key pathways mutated in early T-cell precursor acute lymphoblastic leukemia (T-ALL) patients [21], the discovery of *SYK* as a novel drug target in retinoblastoma [22], the identification of recurrent *K27M* mutations in histone H3.3 present in 70% of the pediatric GBM but absent in adult GBM [23], as well as high-frequency *ATRX* mutations in neuroblastoma [24] and subgroup-specific mutations in medulloblastoma [25].

In May 2012, PCGP uploaded 260 tumor and germline DNA sequence files (520 in all) from 15 pediatric cancers to the European Bioinformatics Institute data portal (The European Genome-Phenome Archive, EGA), providing researchers with immediate access to both published and unpublished data. At the time of data release, this resulted in a more than doubling of the high-coverage human WGS data available to the scientific community. The immediate release of PCGP data is expected to catalyze research in pediatric malignancies and lead to improvements in our ability to diagnose, monitor and treat patients with targeted therapies aimed at a subset of the identified alterations.

OTHER GENOMICS PROJECTS OF NOTE

Catalogue of Somatic Mutations in Cancer (COSMIC) and dbSNP

COSMIC differs significantly from other projects we have described in this chapter in that it mined literature to catalog somatic mutations, SVs, and CNVs in genes that are causally, but not necessarily experimentally, implicated in the development of cancer. Alongside the mutation information, metadata on the sample from the original publication were standardized and recorded. This effort has produced a rich database of aberrations that can be interrogated by gene, sample, tissue type or mutation description [26]. As more data are collected these resources will no doubt be expanded to cover the full spectrum of genetic changes in cancer.

The most common type of genetic variations between individuals are single nucleotide polymorphisms (SNPs), which occur at a surprisingly high frequency (every 500–1000 bases) and vary greatly from individual to individual. dbSNP [27] aims to catalog these genetic variations, with the current build (dbSNP 137) of the database containing 22 508 883 SNPs that fall within a gene. dbSNP can therefore be used to filter out those germline polymorphisms that have been erroneously classified as somatic mutations in COSMIC. For example, we have carried out filtering using dbSNP 135 which contains 46 160 022 SNPs. Of the 39 867 SNVs listed in COSMIC, 6659 were found to overlap. Integrating databases in this manner enables more intelligent filtering and makes them more useful to cancer researchers.

Exome Variant Server (EVS)

Initiated by the National Heart, Lung and Blood Institute (NHLBI), this project aims to combine large amounts of exome sequencing data of diverse germline samples to uncover coding mutations that could affect disorders of the heart, lung and blood. SNPs are available from 6503 thoroughly phenotyped samples [28] for bulk download and online exploration. A web interface allows searching for SNPs in a gene and its upstream or downstream sequence giving information on variants discovered and their ethnic distribution. The sheer volume of germline data allows this resource to serve both as a tool for finding new polymorphisms affecting disease, as well as a control population for cancer research.

1000 Genomes Project

The 1000 Genomes Project aims to describe and characterize the variations found in human genomes [29] and will hopefully then serve as a way of investigating the relationship between genetic polymorphisms and phenotypes by sequencing 2600 individuals from 26 populations from around the world. Whole genomes will be sequenced from blood samples at low coverage and combined with array-based genotyping (SNP array) and supplemented with deep coverage exome sequencing. With this design, the 1000 Genome Project hopes to pick up where the first generation studies, which may have missed rare variants, left off. The 1000 Genomes Project aims to identify 95% of variants that reside in currently accessible genomic regions and are present at greater than or equal to 1% (i.e. a common polymorphism) in each of five major population groups: African, Ad Mixed American, East Asian, European, and South Asian. To examine those rare coding variants, the frequency threshold is reduced from 1% to 0.1%. Like the NHLBI Project, the 1000 Genomes Project will serve as a useful large-scale control group for detailed cancer genomic analysis.

DATA SOURCES

Cancer genomic data can be broadly split into four categories [30], which often determine how it is protected, stored, and distributed:

1 **Raw:** Data that are produced by genome sequencing machines/array analysis hardware without modification. In the case of NGS data, these are raw FASTQ files. Distribution of these data is rare.
2 **Processed/normalized:** This category of data has been subject to a small amount of computational post-processing. In the case of NGS data, this processing usually involves alignment to a reference genome to produce an aligned BAM file. The owners of these data often tightly control their distribution, which is handled by public data archives. For example, to access Category 2 data from the PCGP or TCGA, qualified users must apply to these groups and provide research proposals as to how the data will be stored and analyzed and what research question is to be answered by the applicant. This application is scrutinized to ensure that it meets the requirements of the data provider and is often subject to review by the data requester's own institution to ensure the requirements can be met.

3 **Interpreted:** This category of data often represents the conclusions reached from analysis of the Category 2 data. This includes SNVs, indels, SVs and other biologically relevant findings. Most Category 3 data are freely available after publication either from the publisher's website in the form of supplemental tables or through data access portals developed by sequencing projects to disseminate interpreted results.
4 **Summarized:** Cross-sample analyses that result in the discovery of significant events in the cohort under scrutiny. These are the data often reported in detail in the published findings of a genomics study.

Archives

Large data warehouses include the European Genome Phenome Archive [31] (EGA, http://www.ebi.ac.uk/ega), NCBI's Sequence Read Archive (SRA, http://www.ncbi.nlm.nih.gov/sra) and NCI's Cancer Genomics Hub (CGHub, https://cghub.ucsc.edu/). Access to the raw data require authorization. For both SRA and CGHub, users must apply for authorized access through NCBI's Database of Genes and Phenotypes [32] (dbGaP, http://www.ncbi.nlm.nih.gov/gap). These organizations provide permanent data storage primarily to enable three activities:

1 The reproduction of published results as part of due diligence within the community.
2 The discovery of previously unknown aberrations.
3 The development of new methods.

Data archiving is often required by the research grant funding the generation of the data, and when publishing studies involving large data files like in NGS, the receiving journal more often than not requires deposition of the data in a public archive. These requirements ensure that the public archive of genomic data continues to expand at an exponential rate.

dbGaP stores data from any study that investigates genotype/phenotype relationships such as genome-wide association studies (GWAS), clinical sequencing, disease sequencing, etc. It stores these data in a two-tiered system with some of the data being "open access", while other more sensitive data are stored under "controlled access". Data must be submitted, with all relevant metadata, and are then subject to an intensive review process by dbGaP to ensure consent from subjects and approval from the submitter's organization has been obtained. Open-access data can be browsed without restriction on the website, which includes the studies, associated documents, phenotypes and the genotype–phenotype analyses. Any of these categories of data stored on dbGaP may be categorized as controlled access on a per-study basis. EGA operates under a similar umbrella and contains a separate NGS warehouse storing raw Category 1/2 data in its short read archive with other data being stored under relevant databases that are separated from the phenotypic information. Each of these archives provides project/study landing pages that list the datasets contained within each project, the data access policies and how to obtain access if required. For example, at the time of writing, the PCGP project page (https://www.ebi.ac.uk/ega/dacs/EGAC00001000044) contains links to data for 314 samples contained in nine datasets. Clicking on one of these studies provides further information on the data if they are freely available to download, or contact information if access is controlled.

Many of the available files are incredibly large, for example BAM files for a whole genome are on average ≈90 Gb, depending on the sequencing coverage (Figure 10.2). These large files can take days to download depending upon the speed of the network infrastructure of the archive and the speed of the receiving institution's Internet and network backbone. Reliability is also an issue with large data downloads, though many archives have the ability to restart failed downloads where the user left off. Once the large data files are downloaded, they must be stored, which can be costly (Figure 10.2). Subsequent processing and analysis of these data requires high-performance computing facilities and fast network resources. For example, the Washington University St Louis genomic data center is 16 000 square feet in size and contains 120 racks of data analysis and storage servers. These high-performance systems allow complex calculations to be carried out simultaneously and are supported by a fast network infrastructure that can transfer large sequencing files to computational servers in minutes. Desktop computers can run these kinds of analyses, but slowly, requiring weeks to months to complete an analysis.

Portals

Because of the specialized knowledge and resources required to reanalyze publicly available Category 1 and 2 data, many of the large sequencing projects have developed data access portals as a way of disseminating interpreted data (Categories 3 and 4) in an interactive forum for *de novo* research, offering the ability to analyze the large datasets produced without the requirement of programming knowledge and specialized infrastructure. Publications resulting from the large sequencing projects often discuss the major findings but there is also a wealth of SNP, indel, SV, CNV and other genomic metrics not featured in the primary publication. These data can often be found as supplementary tables so that researchers with particular interests can browse the data in spreadsheets. They are additionally deposited on data portals that present the data in a more integrated fashion with other datasets and unpublished results. Specific questions such

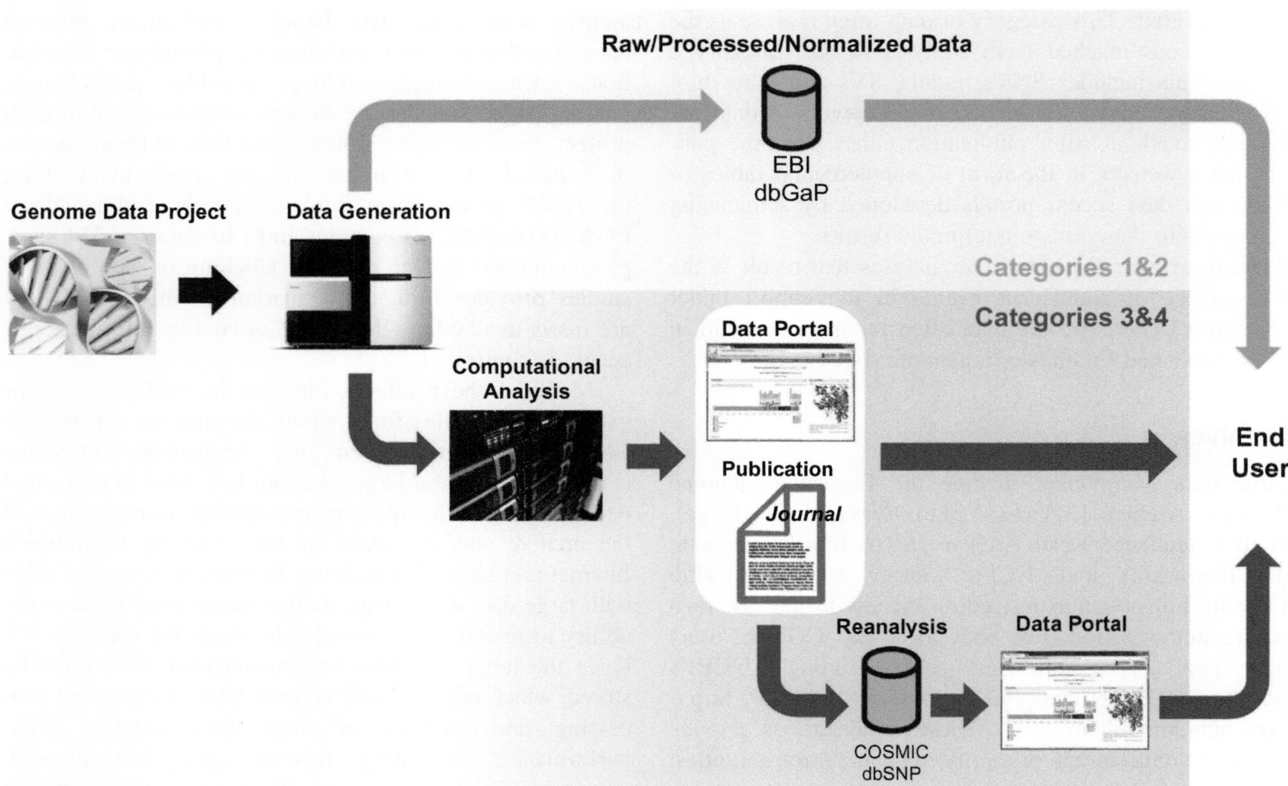

FIGURE 10.1 The flow of genomic data. Data produced by a genome sequencing project are minimally processed to create the data files that are either uploaded directly to data warehouses, such as dbGaP, for download by the end user, or are analyzed in-house to discover biologically meaningful aberrations. These data are summarized and published and/or uploaded to a genomic data portal for end users to access. Data can also be summarized by a third party to create additional databases and data portals.

as; "Is *RB1* mutated frequently in ovarian carcinoma?" are quickly answerable and often a one word query is all that is required to guide the user into more complex views where it is possible to explore networks, pathways, data summaries and additional data related to their original question. Offering these related data enables researchers to submerse themselves fully in the data, potentially leading to novel discoveries or stimulating questions the user would not have thought initially to ask.

All online portals attempt to summarize data in a visually pleasing manner, providing plots of complex information, and also offer downloads of raw data tables so that users may carry out their own analyses offline. A common set of features, implemented slightly differently in each case, can be found on most online data access portals:

- Gene search
- Disease search
- Display specific data type (i.e. SNV, CNV, SV)
- Display overlapping aberrations.

Data access portals differ from institution to institution with respect to the datasets and data types they contain. For example, the TCGA Data Portal features exclusively adult malignancies while the PCGP Explore portal provides data from childhood malignancies. It is common for the portals to provide a breakdown of what data are contained when the user first enters the site so it is clear what information can be queried. The types of data available vary from portal to portal; while most have sequence mutations, others supplement sequencing data with other analyses such as methylation, RNA expression, or protein levels assayed by high-throughput protein arrays.

Most of the portals allow bulk data downloads or programmatic access to the data through application programming interfaces (APIs). Bulk data download formats differ from project to project but contain similar information. For example, sequence mutations are often provided with additional information alongside the prerequisite chromosome, position and variant allele such as the amino acid change for a particular protein, the flanking sequence, the number of sequencing reads containing the variant, etc. It is of note that the level of validation, i.e. the confirmation of the aberration using an independent assay, can differ significantly from portal to portal and, unless explicitly stated, users should treat the data with caution to avoid overinterpretation of the results.

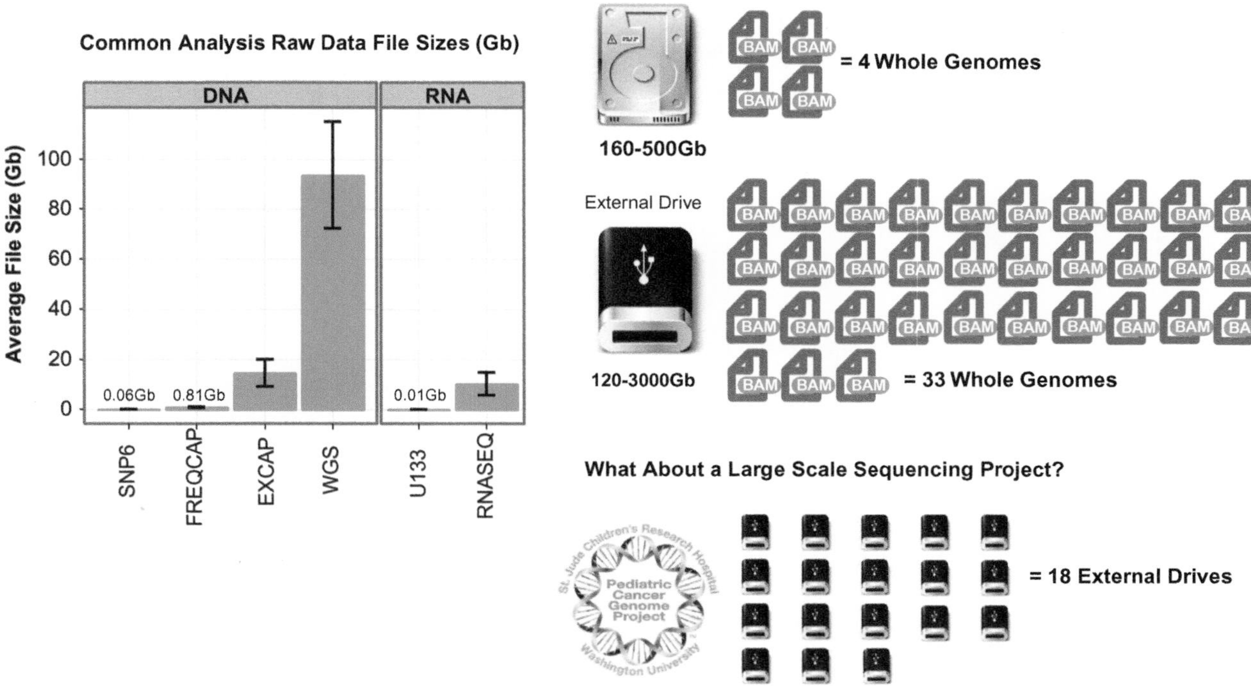

FIGURE 10.2 Data sizes. Genomic data can be very large depending on the type of analysis being performed. EXCAP = Exome capture, FREQCAP = Targeted capture, SNP6 = SNP 6.0 Array, WGS = Whole genome sequencing, RNASEQ = RNA (transcriptome) sequencing, U133 = Gene expression array. The average desktop computer drive is ≈ 500 Gb which means a user could store around four whole genomes, a large external drive could store 33 and a sequencing project such as the PCGP would require 18 external drives for storage of one copy of all the WGS.

The TCGA portal, for example, provides sequence mutation data as text files in "MAF" format, which includes the validation status and method along with other information on the mutation. We will discuss, in detail, specific examples of data analysis portals later in this chapter.

Processors

Genomic data can seem overwhelming and disparate in that sites from different institutions contain small pieces of the puzzle and trying to integrate all these data together is a major challenge for genomic research. Several groups have tried to address these concerns by processing sequencing data into unified resources; therefore, we are categorizing these organizations/services into "processors" of genomic data. There are two major types of genomic data processors, those that create repositories of disparate genomic data in one easily accessible location and those that provide tools to analyze disparate data *de novo*. Projects like COSMIC can be classified as serving the role of a data processor because they take published Category 3 data and create a powerful summary database of significant utility to the research community. Other organizations, such as BioMart, provide tools to access multiple data sources in one location, allowing integration of independent datasets. BioMart presents both a unified interface to databases around the world as well as tools to query and combine the data. Other data processing tools, like Galaxy, allow the computational analysis of public or private data in a web interface that gives biomedical researchers access to what is essentially a supercomputer in the cloud. Galaxy can be used to perform simple tasks, such as file formatting, to complex analyses like processing raw WGS data.

ONLINE/OFFLINE ANALYSIS TOOLS

A number of online data access portals and on/offline data analysis tools are available to integrate and analyze genomic data. In this section, we provide a summary of a large number of tools and the data that they contain (Table 10.1) as well as a more in-depth description of the best-in-class tools.

cBio Cancer Genomics Portal (MSKCC)

The cBio Cancer Genomics Portal currently contains published and preliminary data from TCGA and ICGC [33]

TABLE 10.1 Data Available via Online Data Access Portals. Summary of the Types of Data Contained within Each Data Portal

Data Portal	Institution	Data Available	SNV	CNV	SV	GV	mRNA Expression	miRNA Expression	Epigenetics	Proteomics	Metabolomics	Pathways	Raw data	Export	URL
PCGP "Explore"	St Jude Children's Research Hospital	53 cancer WGS & germline pairs	×	×	×	×	×					×	×	×	http://explore.pediatriccancergenomeproject.org/
Exome Variant Server	NHLBI	6503 germline exomes				×							×	×	http://evs.gs.washington.edu/EVS/
TCGA	NCI	6019 "downloadable" tumor samples	×	×			×	×	×			×	×	×	https://tcga-data.nci.nih.gov/tcga/
COSMIC	Sanger Center	4 991 081 experiments	×	×	×									×	http://www.sanger.ac.uk/genetics/CGP/cosmic/
cBio	MSKCC	5 published datasets & 15 provisional TCGA datasets	×	×			×	×	×	×		×		×	http://www.cbioportal.org/public-portal/
Oncomine	Compendia Bioscience	Subscription required for some features		×			×								https://www.oncomine.org/resource/login.html
G-DOC	Georgetown	5385 donors		×			×	×			×				https://gdoc.georgetown.edu/gdoc/
ICGC	Ontario Institute for Cancer Research	3561 donors	×	×	×		×	×	×			×	×	×	http://dcc.icgc.org/web/
Tumorscape	Broad	TCGA copy number data	×	×										×	http://www.broadinstitute.org/tumorscape/pages/portalHome.jsf
TARGET	NCI	No summary numbers provided	×	×			×	×	×				×		http://target.nci.nih.gov/dataMatrix/TARGET_DataMatrix.html

on which a user may carry out integrative genomic analysis of pre-computed results and *ad hoc* statistical testing. In addition to SNV, indel, CNV, and mRNA expression data displayed on most genomics portals, cBio also includes methylation and proteomics data such as protein and phosphoprotein levels. Another feature includes the ability to generate survival curves on the fly for a given query, quickly informing the user of any clinical significance. The authors of this portal have concentrated on a gene-centric view of the data to help consolidate the vast amount of information stored and allow complex analyses to be carried out with no computational knowledge. Once a gene or genes are selected for analysis, multiple tabs allow the user to browse all the data related to that gene or set of genes. Genes can be queried across cancers and mutual exclusivity or co-occurrence of aberrations can be examined in user-selected gene sets.

As well as simple one-gene searches, cBio also has the ability to build complex queries using a language dubbed "The Onco Query Language". For example, to display mutations and amplifications of *RB1* in all cancers, the user simply enters a single line "RB1: AMP MUT". To compare more than one gene, users can enter multiple genes and multiple restrictions to build a complex picture of the aberrations of interest. A summary of cases is displayed on the resulting query page and summaries for each individual cancer are shown below. When clicked, these links reveal a visualization termed "OncoPrint" that gives an immediate representation of the selected aberrations and their frequency. In addition, OncoPrints are editable in the browser and exportable in SVG format, allowing publication quality figures to be produced from the data.

Concentrating on a single disease shows a more in-depth analysis of the protein–protein interaction network, plots of expression along with CNV, survival analysis, a plot of the protein and its domains with mutations displayed, the protein array, an Integrative Genomics Viewer (IGV) viewer link (see below), downloads and the ability to bookmark your search for sharing or returning to it at a later date. Pathway networks are sourced from the pathway commons project [34] and are overlaid with genomic information allowing the user to determine quickly if other genes are disrupted in a related set of genes.

The source code is freely available and can be installed on a local machine or users can request an Amazon Virtual Machine (see below) that has been preloaded with the cBio portal to analyze their own data.

PCGP "Explore"

The PCGP (described above) is the largest pediatric cancer WGS project in the world. Its Category 2 data are available from EBI while the Category 3 and 4 data are available via a web portal called PCGP Explore (Figure 10.3).

PCGP Explore allows users to browse summary data such as CIRCOS plots, CNV heatmaps, gene expression heatmaps, summary data tables, and supplementary files for any or all PCGP published diseases. For all genes or a subset of the user's choosing, a matrix-like graphical summary of all aberrations can be customized to allow the selection of aberration types, disease types, patient groups, and pathway information. This matrix can be exported to Microsoft Excel for publication. Additional features include highly customizable gene expression boxplots, germline SNP searches, and R libraries that allow direct access to facilitate further statistical analysis of the data.

Explore provides users with the opportunity to search a gene of interest and view a downloadable and editable SVG image of all mutations found in that gene on a "Protein Paint", which also displays domains annotated by the conserved domain database [35]. In contrast to some of the other genome portals, a structural model of the protein is displayed, when available, allowing the user to highlight mutations of interest within the structure providing clues to functional impact. Beneath the images is a table of all aberrations found within that gene in all available PCGP samples, including SNVs, indels, SVs and CNVs.

Galaxy

Galaxy is an online library of executable tools that allows *de novo* manipulation and interactive analysis [36] of data that can be uploaded or chosen from a library of shared data available to Galaxy users. Historically, the user would have had to have knowledge of command line programming and data management skills to carry out the type of analyses Galaxy allows. Now, however, the biomedical scientist can carry out these analyses via a readily accessible web interface. The types of manipulation that can be carried out by Galaxy can be broadly split into three categories:

1 Queries
2 Sequence analysis tools
3 Output displays.

When a user submits an analysis for Galaxy to carry out, it is queued and run when computational resources become available (for small, simple analyses, this usually takes a minute or two). Multiple analyses, or jobs, are queued in the order that they were received. Like job submission systems used on high-performance computing clusters, Galaxy can also perform complex job management allowing the output of one analysis to be the input for another job in the queue. These jobs are stored in the

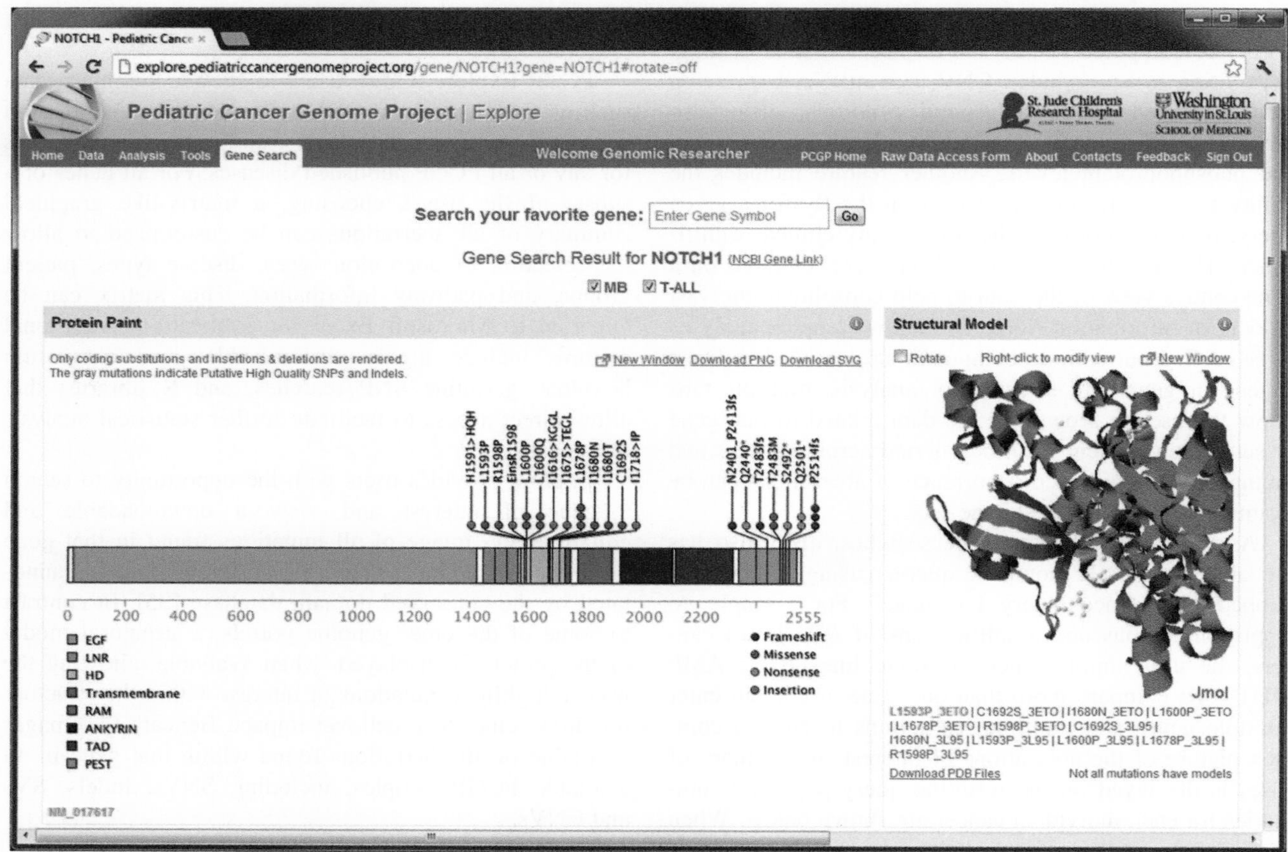

FIGURE 10.3 PCGP Explore. Explore displays data from the PCGP. In this example a "Gene Search" for NOTCH1 was performed. NOTCH1 was found to be the subject of aberrations in both T-cell precursor ALL and medulloblastoma. On the left, mutations are displayed on a schematic of the protein, "Protein Paint", which includes manually curated domains. On the right is a structural model of a region of NOTCH1 encompassing some of the residues that are mutated in these cancers. The mutations are highlighted in the protein structure allowing assessment of the mutation effect.

user's history and can be saved, shared and re-run. Additionally, these histories can be converted into "workflows" for analyses that are to be carried out multiple times on different sets of data.

Galaxy contains an enormous number of data formatting and analysis tools ranging from the simple, such as merging or adding columns to files, computing of an expression on every row in a file, converting formats, etc., to more complex operations such as converting coordinates from different versions of genomes (Figure 10.4), statistical analysis of data, and sequence alignments.

It is possible to perform complex command line operations in a graphical manner such as the annotations of SNVs and prediction of functional impact of missense mutations using SIFT [37]. One of the most powerful features of the site is the "NGS Toolbox" that allows users to run complex computational algorithms on raw NGS data. Analysis of BAM files is usually reserved for users of the Unix command line, but Galaxy provides a large selection of popular tools to perform analyses on these files, such as examining coverage, searching for and filtering genetic abnormalities such as SNVs and indels etc.

An extension of Galaxy's features has been undertaken by the biomedical developer community in an Apple App Store fashion, termed the "Galaxy Tool Shed", in which developers can create or convert tools that can be installed in a Galaxy instance on the web for use on their own datasets. At the time of writing, 2044 tools were available in the shed. The creation of the tool shed exponentially expands the scope of analyses that can be performed on Galaxy. It is also possible for peers to rate tools so that users can see what the best tools are for use with their analyses.

Galaxy has been described/utilized in more than 400 research articles including publications in top-tier journals where it has been utilized for analysis of raw ChIP-seq data [38] and RNA-seq mapping and analysis [39]. Data from NGS studies are also being deposited in Galaxy to facilitate third-party analysis [40]. In addition to being used on the web, Galaxy can also be installed locally to

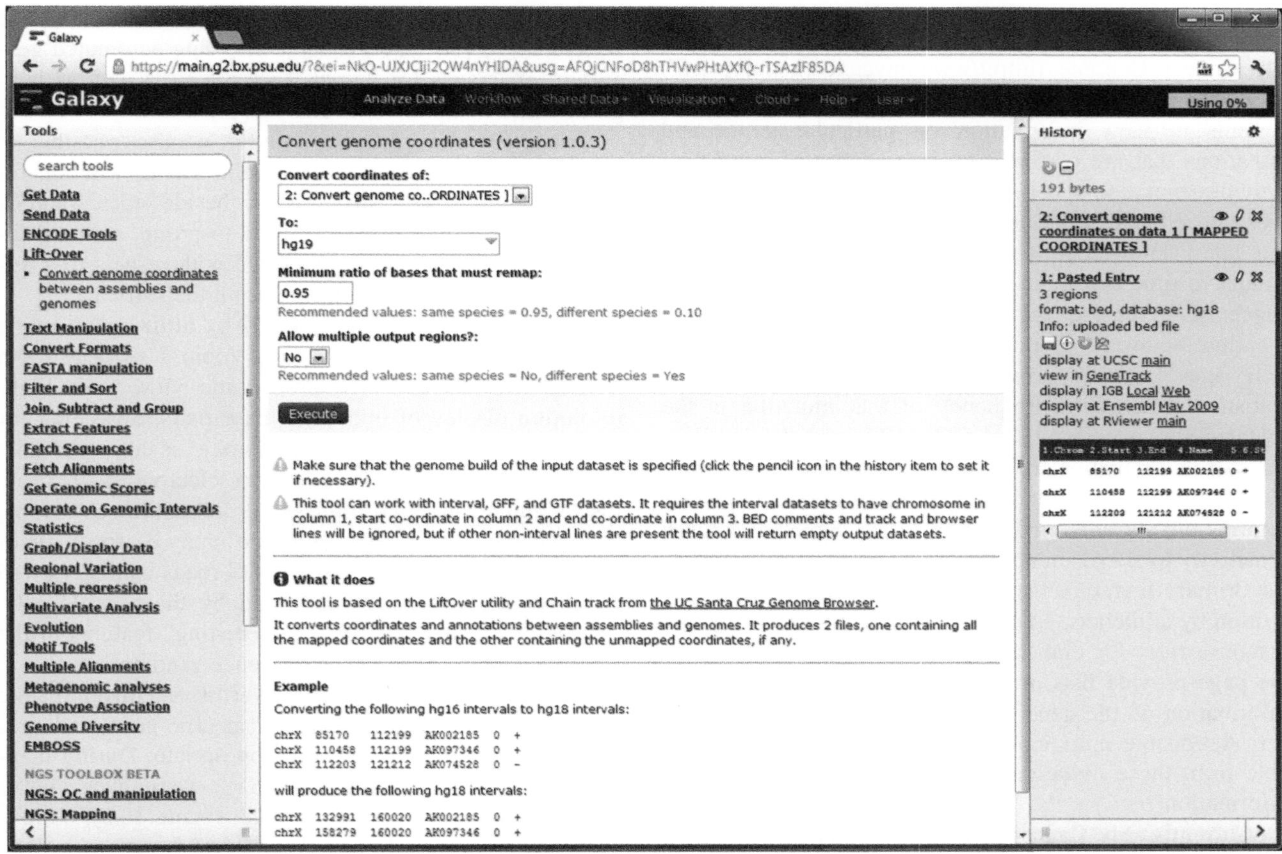

FIGURE 10.4 Galaxy. Galaxy is a web-based platform for the computational analysis of complex biological data. Galaxy is split into three sections: on the left the available "Tools" are listed while the right panel displays a "History" showing analyses that are either queued, running, or finished. The center panel displays both parameter options for the tool selected and analysis results. In this example, we have converted coordinates from an older to a newer genome build.

increase the speed of data analysis and provide more privacy for clinically sensitive data. Excellent walkthroughs of Galaxy's features are available to assist new users [41], and the site itself provides a variety of training videos.

BioMart

Biological databases can be disparate in their location and access style; often the user has to be familiar with a particular database/web-interface to get the most out of a resource, and gene identifiers frequently differ between databases making direct comparisons clumsy and unintuitive. To solve these issues, a database integration system has been created called BioMart [42]. BioMart provides a generic interface to enable queries of over 40 disparate biological databases around the world, providing users access to a variety of data without requiring them to learn the intricacies of each data center's interface. It can be used to get gene annotations or convert from one gene ID to another, problems that can be daunting and frustrating to many biomedical researchers.

BioMart enables querying across multiple data sources simultaneously – datasets that share common identifiers (e.g. Ensembl gene IDs) allow the linking of BioMarts with integrated queries on the same server and across servers around the globe.

Because of its power, BioMart has been integrated into a variety of large website frameworks like the ICGC Data Portal. Each of the centers in the consortium maintains their own BioMart server that can be seamlessly accessed from the ICGC portal alongside external databases like Ensembl and COSMIC. COSMIC has also populated their own instance of BioMart, termed COSMICMart, which has all of the somatic mutations contained within COSMIC as well as associated phenotypic data (http://www.sanger.ac.uk/genetics/CGP/cosmic/biomart/martview). This makes filtering COSMIC data extremely straightforward, allowing complex integrated queries to be constructed.

My Cancer Genome

My Cancer Genome (http://www.mycancergenome.org/) is a web resource that primarily enables clinicians and patients to view information on particular genes and mutations that are pertinent to their cancer of interest. It provides mutation-specific information on relevant therapies and clinical trials, thereby providing clinicians with the latest knowledge without them having to perform extensive time-consuming literature searches. Selecting a cancer gives the user the ability to select a gene and then a variant within it before proceeding to the gene information pages. Users are presented with a simple table listing statistics such as the frequency of that mutation in the selected disease (data from COSMIC) and the implications of that mutation for targeted therapeutics. For example, selecting Melanoma > BRAF > V600E quickly informs the physician that this mutation confers increased sensitivity to BRAF inhibitors and provides a reference to the primary literature. After a description of the mutation written by clinicians, a table lists the treatment agents and response rates for that treatment while tabs at the top of the page provide lists of clinical trials in the USA, more information on the gene, and its role in the selected cancer. Actionable mutations are, therefore, readily identifiable from those mutations for which there is no clinical information.

Currently, My Cancer Genome is a manually curated resource and has a limited number of mutations in its database. Experts in each disease create and edit sections of the site. Adding new data takes time, but it is an ever-expanding resource that will become increasingly important for personalized medicine as more mutations are investigated in clinical trials. It is databases like My Cancer Genome that will help to revolutionize cancer medicine and provide meaningful mutation information affecting patient care.

Standalone Genome Sequencing Viewers: IGV and More...

Although there is a wide range of NGS viewers available, one of the most powerful and comprehensive is the Integrative Genomics Viewer that allows cross-platform visualization of a broad range of sequencing and other genomic data such as gene expression [43].

IGV is an offline tool that can be downloaded from http://www.broadinstitute.org/igv and has been streamlined to run on almost all desktop computers independent of operating system (Windows, OS X and Unix). Although the tool runs locally, it is possible to import a wide range of data in a variety of formats available on local drives, websites or the cloud. Hundreds to thousands of samples can be viewed concurrently allowing very large-scale comparisons to be made. A command-line version of IGV is also available allowing automated generation of images for human assessment as well as integration into computational analysis pipelines.

Alongside the genomic data, it is also possible to import sample metadata such as clinical information for display in a color-coded matrix beside each sample. These annotations can also be used to group, sort and filter the data. The data are visualized with respect to a reference genome selectable from within the software.

Genome navigation is achieved by utilizing a Google Maps-like features such as pan and zoom. Users can zoom from a low-resolution whole-genome view to a high-resolution display of individual base pairs. Searching for genes or locations within the genome is also possible. Useful features include the ability to select views of "Gene Lists" which can be customized (or a "multilocus" view), allowing the user to view multiple analyses for a set of genes in an integrated manner. NGS reads can be viewed and SNV calls manually reviewed by the user. Finally, IGV automatically highlights "interesting" features of the reads which fail to match the reference genome.

The IGV viewer is a very powerful visualization suite but, for more detailed analyses of specific genetic aberrations, other tools may be more appropriate. During manual review of SNPs, indels and SVs, a sequence viewer like Bambino may be more suitable (Figure 10.5).

Bambino is a platform-independent viewer (as well as a variant calling suite) that reads BAM/SAM data and can be used to display the NGS reads of two samples (i.e. normal vs tumor) aligned to the reference genome [44]. It displays protein isoform information (RefSeq) allowing changes to the protein sequence to be displayed. One standout feature of this viewer is the display of read quality information where the background of each base pair is shaded dependent on quality score allowing for a more informed judgment on differences from the reference genome. Changes with respect to the reference are highlighted in red and indels can be readily distinguished. Other similar viewers include; Savant, Tablet, Artemis and EagleView. All viewers have their strengths and weaknesses and the decision on which viewer to utilize is dependent on the data type, platform, feature sets and personal preference.

Viewers such as IGV and Bambino allow for the human interpretation of genome-wide analyses as well as computational predictions to be examined and interrogated thoroughly.

CLOUD COMPUTING

Cloud computing has been gaining popularity lately with a number of companies offering storage of personal information on their servers via the Internet. Users can store

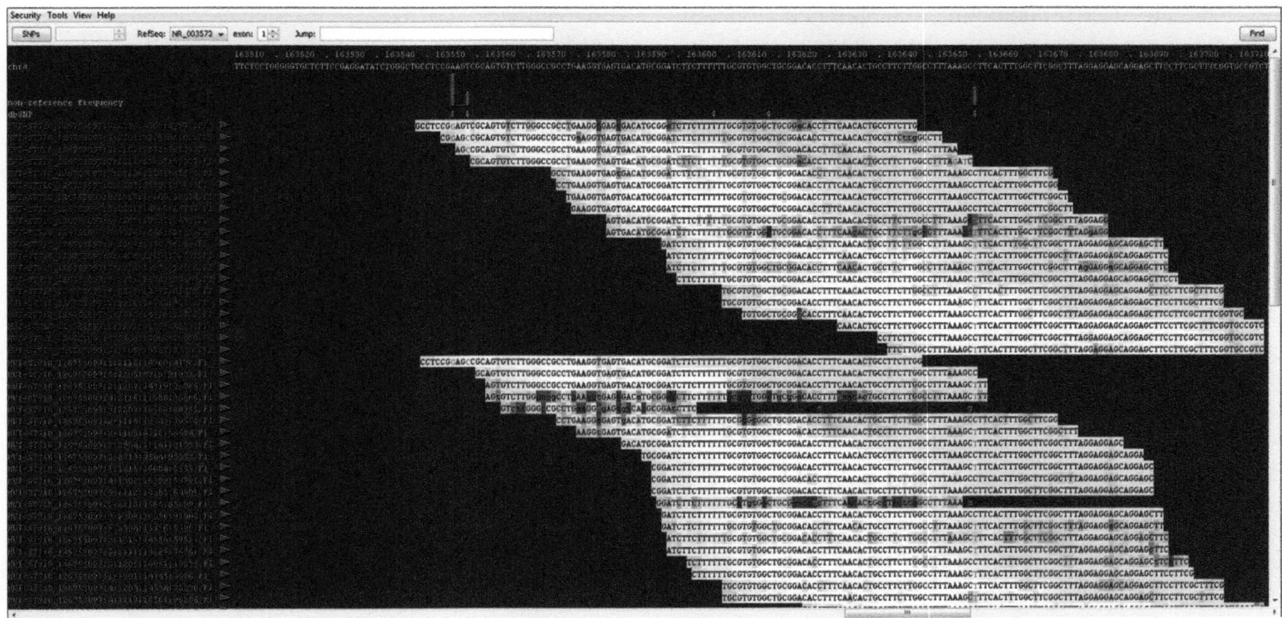

FIGURE 10.5 Bambino. Bambino is an NGS read viewer, suited for the manual review of computationally predicted genomic aberrations such as single nucleotide and structural variations.

Box 10.2 Common Genomic Formats

Here we will produce a box describing briefly the types of genomic formats common to portals and viewers (FASTQ, FASTA, SAM, BAM, MAF, VCF, BED, GFF, segmentation, Wiggle).

FASTA
A text file containing sequence information. Each sequence has an identifier line indicated by a greater than symbol (>) followed on the next line with the sequence data.

FASTQ
This text format is an extension of FASTA and facilitates the inclusion of sequencing quality alongside the sequence itself. FASTQ files usually contain 4 lines per sequence with the sequence ID denoted by an @ character, line 2 is the sequence, line 3 is denoted by a + and can optionally contain the sequence ID and line 4 encodes the quality values for each base in line 2.

SAM
Tab-delimited text file containing sequence alignment data.

BAM
A binary version of a SAM file.

BED
A tab-delimited text file that defines genomic features containing genomic position information (chromosome, start and end) and optional features of that region.

GFF
General feature format files are tab-delimited text files containing genomic regions and their features. There are several versions of this format.

MAF
The mutation alignment format is a tab-delimited text file containing lists of mutations. This format is required for TCGA mutation reports.

VCF
Variant call format files are used by the 1000 Genomes Project to record variants.

More information can be found at http://www.broadinstitute.org/software/igv/FileFormats

their photos, music, etc. for access on multiple devices, including smartphones, and often for free for a small amount of storage and pay-as-you-go for larger amounts. Computing in the cloud is quickly becoming a useful activity for biomedical researchers and it is now possible for individuals and research groups, without the infrastructure to store, manage, and analyze raw sequencing data. A number of services like Amazon Elastic Compute Cloud (EC2) or Windows Azure allow access to whole servers or "virtual machines" that can be customized based on the user's computing requirements. Like the cloud for storing personal data, these services are offered on a pay-as-you-go basis. Amazon EC2 allows users to pay only for the computation time they have used running their analyses. These services provide access to a shared pool of resources that can be utilized on an *ad hoc* basis dependent on the type of analyses being run. For example, analyses that could take 10 hours on a single machine could be split onto 10 servers with the results available in 1 hour. The highly scalable nature of cloud computing makes it an attractive service to researchers who do not have access to computing clusters.

One of the big advantages of cloud computing, besides access to computational resources, is reproducibility. A "snapshot" of a virtual machine can be taken, archived, and restored at any time with all data and software faithfully replicated allowing for reproducible computing. Some research groups are already providing their software in Amazon Virtual Machine snapshots so users can run the software with little configuration required. This ensures that the software will run smoothly with little user intervention, thus providing a more plug and play approach. The user simply requests the virtual machine and uploads data for analysis. cBio has, for example, implemented this with their cancer genomics portal allowing users to run their personal version of the portal for analysis and display of their own data.

The 1000 Genomes Project has also uploaded all current data (1700 genomes) to the Amazon cloud as a free public dataset. Users can analyze these data directly using Amazon's EC2 service without having to download the 200 TB of files stored there. Once a user has set up a computing server, the data are accessible and the user may run analyses freely on the data with the results being stored on the user's own server. Projects such as 1000 Genomes use technologies like the cloud to demonstrate their wide applicability and encourage development in this area.

BioLinux [45] (http://cloudbiolinux.org/) has been pre-configured to run on an Amazon EC2 cloud server or a desktop machine and includes popular NGS analysis software like BWA, Bowtie, bedtools, Picard, SAMtools, and Galaxy. This prevents the user from having to endure lengthy installation procedures that require Unix administration knowledge. Distributions like BioLinux provide a smoother entry into the bioinformatic analysis of sequencing data by providing detailed user guides and easily installable virtual machines that enable users to start to become familiar with running these tools from the command line.

Despite the increasing accessibility of the cloud for biomedical research, there are a number of roadblocks. It can be costly because raw NGS data files can be very large and disk space is expensive. Upload times for these files can also be very long (dependent on connection speeds); therefore, it may be more suitable for downstream analyses that often involve smaller files. However, the deposition of BAM files by large genomics projects could be a positive step towards easing these concerns. Restrictions are also placed on the number of transactions that can be made (i.e. the upload or download of a file). Patient privacy could also be an issue, especially if germline data are being analyzed. Clinical protocols that allow collection of patient material may not allow the analysis of the resulting data on what is, for all intents and purposes, a public server. Identification of a patient from NGS data, though, is extremely unlikely, especially if it has been diligently de-identified.

SYNOPSIS AND PROSPECTS (GENOMIC RESOURCES AND THE CLINIC)

Sequencing data generation is no longer a major hurdle to deciphering the cancer genome because of the release by large genome sequencing projects like TCGA, PCGP and ICGC of data from hundreds to thousands of cancer genomes and their matched normal (germline) samples. These data resources are growing exponentially as sequencing technology becomes more accessible. The high standards set by these projects ensure the data being released to the research community are of high quality and suitable for downstream analyses. Important discoveries have already been made using these data, and many more will appear in the coming decade. The cancer genome for the most part is still providing information that could lead to new treatment modalities and, above all, is changing the way we think about delivering cancer therapy. No longer will it be satisfactory to treat patients suffering from cancer of a particular tissue like the breast with one umbrella treatment regimen. Even now breast cancer patients are divided into large treatment subgroups based on hormone receptor status, and soon the standard of care may involve subdivision of patients into even smaller groups in light of recent genomic studies [46]. Clinical genomic sequencing will allow specific treatments to be tailored to the aberrations contained within the individual's genome. To realize this future of personalized medicine, genomic data resources first must be mined extensively to build databases that provide detailed information on specific aberrations and then their potential transformed into actionable treatments in the clinic. The numerous data sources and analysis tools we described in this chapter serve as the foundations for these databases, but we must first integrate this disparate stockpile of raw genomic information into something that is applicable in the clinic. Projects like My Cancer Genome have started the arduous task of manually populating such a database and, even though it is in its infancy, this resource sets an example for the integration of mutation data with actionable clinical information such as identifying drugs known to be effective against certain aberrations as well as providing details of relevant ongoing clinical trials. For NGS to become commonplace in clinical data analysis, pipelines must be standardized and the reproducibility of data must be held in the highest regard. With tools like Galaxy, whose history and workflow features allow the same analysis to be run on multiple datasets and pipelines that can be easily transferred between collaborators or other institutions, these goals seem attainable. As more and more biomedical researchers turn their hand to making sense of the wealth of cancer genomic data, the more understanding we will gain. New cloud computing initiatives such as the deposition of the 1000 Genomes data on the Amazon cloud service put high performance computing and complex data analysis algorithms increasingly at our fingertips.

Issues, however, still remain in the utilization of genome sequencing in the clinic, such as how much of the data should be reported to clinicians and ultimately the patients? Sequencing of the patient's genome may reveal aberrations not directly related to the current diagnosis. For instance, what should be done with information relating to a predisposition for heart disease or an elevated risk of mental illness? Often these data can be difficult to interpret and could lead to further expensive testing or unnecessary worry for the patient; however, not disclosing this information could deny the patient and clinician valuable information. These issues will need to be addressed alongside the need for better education for clinicians in the application of genomics to medical practice.

GLOSSARY

ChIP-seq This technique uses next-generation sequencing to discover the DNA bound by proteins of interest. This is an extension of chromatin immunoprecipitation (ChIP) where specific interactions between proteins and DNA are investigated.

Chromosome conformation capture (3C) This technique allows researchers to examine the higher order structure of the genome (the chromatin) by cross-linking the physical interactions of chromatin with formaldehyde followed by enzymatic digestion and ligation with the frequency of ligation of two restriction fragments a measure of the frequency of interaction within the nucleus.

Chromosome conformation capture carbon-copy (5C) 5C is an extension of 3C using NGS technology to identify the ligation products, using ligation-mediated amplification to copy and amplify a subset of the 3C library followed by detection using NGS14. 5C is allowing three-dimensional interaction maps of the genome to be generated uncovering long-range interactions of promoters and distal elements that can potentially affect gene regulation.

Cloud computing The use of computational resources that are delivered over a network, which is often the Internet. Third parties provide software or computing services that can be used on an "as needed" basis, providing the user with instant scalability if required. Examples include Amazon's EC2 cloud computing service and Microsoft Azure, both of which allow one to rent servers or clusters of servers to carry out complex calculations.

Data access control/policy Certain data/categories of data are often subject to controlled access. This is determined by the data producer and usually data access is granted after an application and verification process.

Data access portal A user-friendly web-based resource for accessing mainly category 3 and 4 data from large-scale genome sequencing projects.

Data integration Data integration takes many types of sequencing data, for example whole genome and transcriptome and attempts to combine them in a meaningful way. If a structural variation is found through whole genome sequencing of the DNA then integrating this with transcriptome sequencing could provide further evidence for the variation. Genomic data categories (the level of analysis usually dictates the category of data): (1) raw; (2) processed/normalized; (3) interpreted; and (4) summarized within and across diseases.

Data warehouse/archive A third-party provider of large-scale storage of category 1 and 2 data generated by genome sequencing projects. Data warehouses ensure universal accessibility to raw data under "open" or "controlled" access policies.

Exome capture sequencing (ECS) This method involves capture of just the coding exons of the human genome using specially designed probes which are complementary to these exons. The DNA of interest is hybridized to these probes and all other DNA washed away.

Microarrays Microarrays contain small sequences of thousands of genes or other genomic regions embedded on a solid surface such as a glass slide, which is subsequently hybridized with DNA or RNA. While less expensive and faster than NGS technologies, they offer only one or two data types per array type.

miRNA expression miRNA microarrays have been developed to measure the expression of these small non-coding RNAs.

miRNA-seq Like RNA sequencing small non-coding RNAs can be prepared from cells and sequenced. miRNA-seq sequencing data provide the nucleotide sequence in addition to the expression levels.

Next-generation sequencing (NGS) Modern high-throughput sequencing technologies that parallelize sequencing allowing thousands to millions of DNA molecules to be sequenced simultaneously, lowering the cost and making the sequencing of large genomes possible in weeks. Whole genome, exome, transcriptome and ChIP-seq all use next-generation sequencing technology.

Personalized medicine The adaption of therapeutic modalities to individual patients based on the use of their (mainly) genetic information. An example of this approach would be the use of mutation information to deliver a drug that is known to have activity against this mutation (e.g. patients with a *BRAF* V600E mutation respond more favorably to the BRAF inhibitor Vemurafenib).

Reverse phase protein array (RPPA) Antibody-based, reverse phase protein arrays for protein expression levels and phosphorylation state. Measures levels of phosphorylated isoforms.

RNA expression array RNA expression arrays ("microarrays") measure the differences in expression of genes from two different populations of samples. RNA "probes" are, like DNA microarrays, spotted onto glass slides or chips.

RNA-seq (transcriptome sequencing) Next-generation sequencing of cDNA from all transcribed mRNAs or a subset using capture techniques described above. This allows for differential expression analysis like microarray-based methods but, in addition, facilitates the discovery of SNVs, SVs, and novel isoforms/exons.

SNP array This is a DNA microarray (DNA probes immobilized on a glass slide/chip) used to determine single nucleotide changes as well as copy number variation in the genome. It can also be used to estimate copy number variations.

Targeted sequencing Like exon capture, targeted sequencing uses probes, but in this technique they are designed to specific regions of the genome that are of interest, e.g. mutated genes. Usually targeted sequencing is used for the purpose of validation and high-throughput screening of recurring mutations.

Whole genome sequencing (WGS) The most complete DNA sequencing, covering the majority of the genome: exons, introns, and intergenic regions.

ABBREVIATIONS

ALL Acute lymphoblastic leukemia
AML Acute myeloid leukemia
API Application programming interfaces
CGP Cancer Genome Project
CLL Chronic lymphocytic leukemia
CNV Copy number variation
COSMIC Catalogue of Somatic Mutations in Cancer
dbGaP Database of Genes and Phenotypes
DCC Data Coordination Center
EBI European Bioinformatics Institute
EC2 Elastic Compute Cloud
ECS Exome capture sequencing
EGA European Genome-Phenome Archive
EVS Exome variant server
GWAS Genome-wide association studies
ICGC International Cancer Genome Consortium
IGV Integrative Genomics Viewer
Indels Small insertions or deletions
KEGG Kyoto Encyclopedia of Genes and Genomes
LOH Loss-of-heterozygosity
MSKCC Memorial Sloan-Kettering Cancer Center
NCBI National Center for Biotechnology Information
NCI National Cancer Institute
NGS Next-generation sequencing
NHGRI National Human Genome Research Institute
NHLBI National Heart, Lung, and Blood Institute
PCGP Pediatric Cancer Genome Project
QC Quality control
RPPA Reverse phase protein array
SNP Single nucleotide polymorphism
SNV Single nucleotide variation
SRA Sequence read archive
T-ALL T-cell precursor ALL
TARGET Therapeutically applicable research to generate effective treatments
TCGA The Cancer Genome Atlas
WGS Whole genome sequencing

REFERENCES

[1] Reddy EP, Reynolds RK, Santos E, Barbacid M. A point mutation is responsible for the acquisition of transforming properties by the T24 human bladder carcinoma oncogene. Nature 1982;300:149–52.

[2] Tabin CJ, Bradley SM, Bargmann CI, Weinberg RA, Papageorge AG, Scolnick EM, et al. Mechanism of activation of a human oncogene. Nature 1982;300:143–9.

[3] International Human Genome Sequencing Consortium. Finishing the euchromatic sequence of the human genome. Nature 2004;431:931–45.

[4] Bardelli A, Parsons DW, Silliman N, Ptak J, Szabo S, Saha S, et al. Mutational analysis of the tyrosine kinome in colorectal cancers. Science 2003;300:949.

[5] Bignell G, Smith R, Hunter C, Stephens P, Davies H, Greenman C, et al. Sequence analysis of the protein kinase gene family in human testicular germ-cell tumors of adolescents and adults. Genes Chromosomes Cancer 2006;45:42–6.

[6] Davies H, Hunter C, Smith R, Stephens P, Greenman C, Bignell G, et al. Somatic mutations of the protein kinase gene family in human lung cancer. Cancer Res 2005;65:7591–5.

[7] Wood LD, Parsons DW, Jones S, Lin J, Sjoblom T, Leary RJ, et al. The genomic landscapes of human breast and colorectal cancers. Science 2007;318:1108–13.

[8] Cancer Genome Atlas Research Network. Comprehensive genomic characterization defines human glioblastoma genes and core pathways. Nature 2008;455:1061–8.

[9] Weir BA, Woo MS, Getz G, Perner S, Ding L, Beroukhim R, et al. Characterizing the cancer genome in lung adenocarcinoma. Nature 2007;450:893–8.

[10] Ding L, Getz G, Wheeler DA, Mardis ER, McLellan MD, Cibulskis K, et al. Somatic mutations affect key pathways in lung adenocarcinoma. Nature 2008;455:1069–75.

[11] Jones S, Zhang X, Parsons DW, Lin JC, Leary RJ, Angenendt P, et al. Core signaling pathways in human pancreatic cancers revealed by global genomic analyses. Science 2008;321:1801–6.

[12] Sjoblom T, Jones S, Wood LD, Parsons DW, Lin J, Barber TD, et al. The consensus coding sequences of human breast and colorectal cancers. Science 2006;314:268–74.

[13] Parsons DW, Li M, Zhang X, Jones S, Leary RJ, Lin JC, et al. The genetic landscape of the childhood cancer medulloblastoma. Science 2011;331:435–9.

[14] Dostie J, Richmond TA, Arnaout RA, Selzer RR, Lee WL, Honan TA, et al. Chromosome Conformation Capture Carbon Copy (5C): a massively parallel solution for mapping interactions between genomic elements. Genome Res 2006;16:1299–309.

[15] Sanyal A, Lajoie BR, Jain G, Dekker J. The long-range interaction landscape of gene promoters. Nature 2012;489:109–13.

[16] Cancer Genome Atlas Research Network. Integrated genomic analyses of ovarian carcinoma. Nature 2011;474:609–15.
[17] The International Cancer Genome Consortium. International network of cancer genome projects. Nature 2010;464:993–8.
[18] Papaemmanuil E, Cazzola M, Boultwood J, Malcovati L, Vyas P, Bowen D, et al. Somatic SF3B1 mutation in myelodysplasia with ring sideroblasts. N Engl J Med 2011;365:1384–95.
[19] Puente XS, Pinyol M, Quesada V, Conde L, Ordonez GR, Villamor N, et al. Whole-genome sequencing identifies recurrent mutations in chronic lymphocytic leukaemia. Nature 2011;475:101–5.
[20] Downing JR, Wilson RK, Zhang J, Mardis ER, Pui CH, Ding L, et al. The Pediatric Cancer Genome Project. Nat Genet 2012;44:619–22.
[21] Zhang J, Ding L, Holmfeldt L, Wu G, Heatley SL, Payne-Turner D, et al. The genetic basis of early T-cell precursor acute lymphoblastic leukaemia. Nature 2012;481:157–63.
[22] Zhang J, Benavente CA, McEvoy J, Flores-Otero J, Ding L, Chen X, et al. A novel retinoblastoma therapy from genomic and epigenetic analyses. Nature 2012;481:329–34.
[23] Wu G, Broniscer A, McEachron TA, Lu C, Paugh BS, Becksfort J, et al. Somatic histone H3 alterations in pediatric diffuse intrinsic pontine gliomas and non-brainstem glioblastomas. Nat Genet 2012;44:251–3.
[24] Cheung NK, Zhang J, Lu C, Parker M, Bahrami A, Tickoo SK, et al. Association of age at diagnosis and genetic mutations in patients with neuroblastoma. J Am Med Assoc 2012;307:1062–71.
[25] Robinson G, Parker M, Kranenburg TA, Lu C, Chen L, Ding L, et al. Novel mutations target distinct subgroups of medulloblastoma. Nature 2012;488:43–8.
[26] Reva B, Antipin Y, Sander C. Predicting the functional impact of protein mutations: application to cancer genomics. Nucleic Acids Res 2011;39:e118.
[27] Sherry ST, Ward MH, Kholodov M, Baker J, Phan L, Smigielski EM, et al. dbSNP: the NCBI database of genetic variation. Nucleic Acids Res 2001;29:308–11.
[28] NHLBI. Exome Variant Server. Vol. 2012 NHLBI Exome Sequencing Project 2012.
[29] The 1000 Genomes Project Consortium. A map of human genome variation from population-scale sequencing. Nature 2010;467:1061–73.
[30] Chin L, Hahn WC, Getz G, Meyerson M. Making sense of cancer genomic data. Genes Dev 2011;25:534–55.
[31] Leinonen R, Akhtar R, Birney E, Bower L, Cerdeno-Tarraga A, Cheng Y, et al. The European Nucleotide Archive. Nucleic Acids Res 2011;39:D28–31.
[32] Mailman MD, Feolo M, Jin Y, Kimura M, Tryka K, Bagoutdinov R, et al. The NCBI dbGaP database of genotypes and phenotypes. Nat Genet 2007;39:1181–6.
[33] Cerami E, Gao J, Dogrusoz U, Gross BE, Sumer SO, Aksoy BA, et al. The cBio cancer genomics portal: an open platform for exploring multidimensional cancer genomics data. Cancer Discov 2012;2:401–4.
[34] Cerami EG, Bader GD, Gross BE, Sander C. cPath: open source software for collecting, storing, and querying biological pathways. BMC Bioinformatics 2006;7:497.
[35] Marchler-Bauer A, Lu S, Anderson JB, Chitsaz F, Derbyshire MK, DeWeese-Scott C, et al. CDD: a Conserved Domain Database for the functional annotation of proteins. Nucleic Acids Res 2011;39: D225–9.
[36] Goecks J, Nekrutenko A, Taylor J. The Galaxy Team. Galaxy: a comprehensive approach for supporting accessible, reproducible, and transparent computational research in the life sciences. Genome Biol 2010;11:R86.
[37] Kumar P, Henikoff S, Ng PC. Predicting the effects of coding non-synonymous variants on protein function using the SIFT algorithm. Nat Protoc 2009;4:1073–81.
[38] Kadauke S, Udugama MI, Pawlicki JM, Achtman JC, Jain DP, Cheng Y, et al. Tissue-specific mitotic bookmarking by hematopoietic transcription factor GATA1. Cell 2012;150:725–37.
[39] Bhatt DM, Pandya-Jones A, Tong AJ, Barozzi I, Lissner MM, Natoli G, et al. Transcript dynamics of proinflammatory genes revealed by sequence analysis of subcellular RNA fractions. Cell 2012;150:279–90.
[40] Schuster SC, Miller W, Ratan A, Ratan LP, Giardine B, Kasson LR, et al. Complete Khoisan and Bantu genomes from southern Africa. Nature 2010;463:943–7.
[41] Blankenberg D, Von Kuster G, Coraor N, Ananda G, Lazarus R, Mangan M, et al. Galaxy: a web-based genome analysis tool for experimentalists. Curr Protoc Mol Biol 2010;10:1–21 Chapter 19, Unit 19.
[42] Smedley D, Haider S, Ballester B, Holland R, London D, Thorisson G, et al. BioMart – biological queries made easy. BMC Genomics 2009;10:22.
[43] Thorvaldsdottir H, Robinson JT, Mesirov JP. Integrative Genomics Viewer (IGV): high-performance genomics data visualization and exploration. Brief Bioinform 2012.
[44] Edmonson MN, Zhang J, Yan C, Finney RP, Meerzaman DM, Buetow K. Bambino: a variant detector and alignment viewer for next-generation sequencing data in the SAM/BAM format. Bioinformatics 2011;27:865–6.
[45] Krampis K, Booth T, Chapman B, Tiwari B, Bicak M, Field D, et al. Cloud BioLinux: pre-configured and on-demand bioinformatics computing for the genomics community. BMC Bioinformatics 2012;13:42.
[46] Shah SP, Roth A, Goya R, Oloumi A, Ha G, Zhao Y, et al. The clonal and mutational evolution spectrum of primary triple-negative breast cancers. Nature 2012;486:395–9.

Part 3

Hereditary Cancer Syndromes

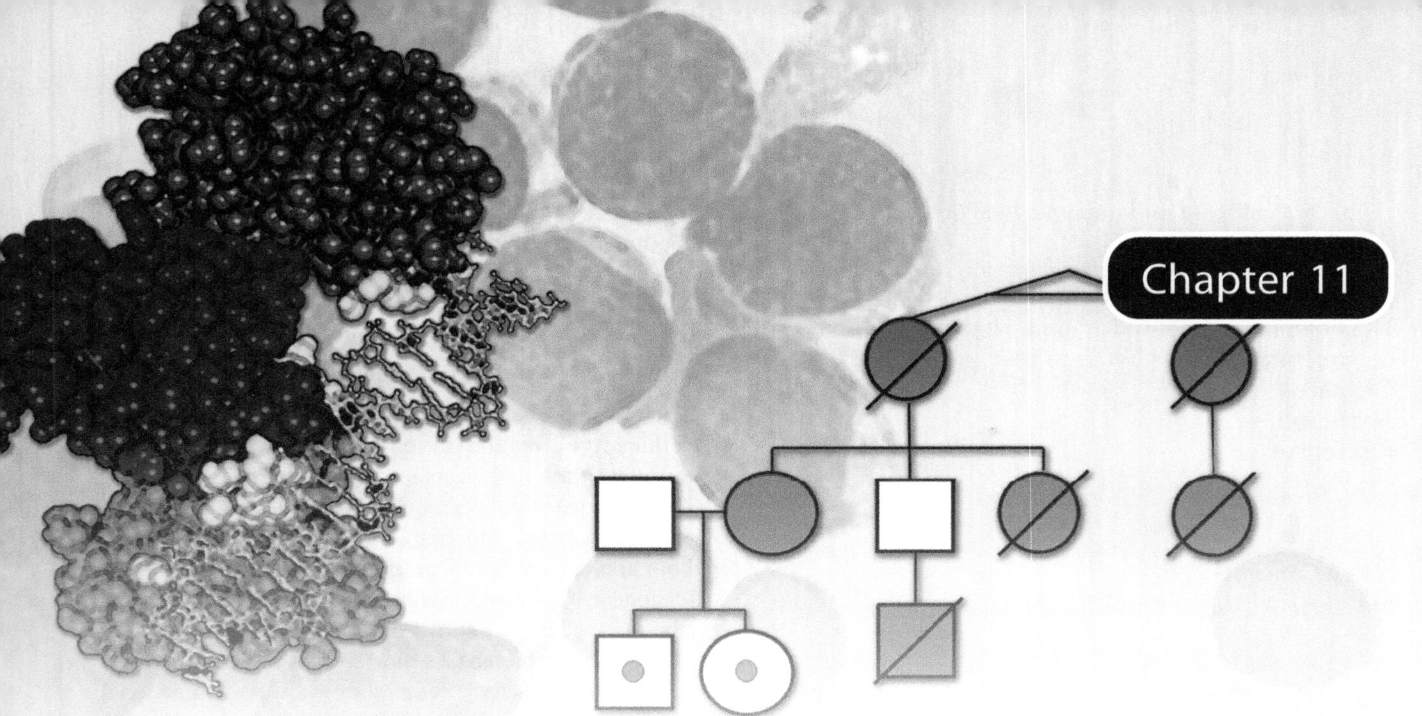

Cancer Genomics

Genetic Basis of Hereditary Cancer Syndromes

David Malkin

Division of Hematology/Oncology, The Hospital for Sick Children and Department of Pediatrics, University of Toronto, Toronto, ON, Canada

Contents

Cancer Genomics. DOI: http://dx.doi.org/10.1016/B978-0-12-396967-5.00011-6

Key Concepts

- At least 25% of childhood cancers are hereditary or familial
- Germline mutations of both tumor suppressor genes and oncogenes can be associated with cancer susceptibility
- Knudson's "two hit" hypothesis explains the genetic basis of tumor suppressor gene-associated cancer susceptibility
- Germline alterations of cancer susceptibility genes are necessary but not sufficient for the development of cancer
- Cancer may be the only recognizable phenotype in some cancer predisposition syndromes
- Patients with syndromes presenting with a spectrum of congenital anomalies commonly develop cancers as a result of an underlying gene mutation that is associated with both aberrant normal development and carcinogenesis
- Genetic testing for cancer susceptibility genes is feasible in both adults and children
- Clinical surveillance protocols can be implemented in genetically at-risk individuals for early cancer detection

INTRODUCTION

Cancer is the most common cause of disease-related death in children beyond the newborn period. Although childhood cancers account for only a small fraction of all human cancer [1], and while most childhood cancers occur sporadically and with a poorly defined etiology, hereditary or familial factors are evident in 25–40% of cases [2]. Genes found to be mutated and inherited in cancer predisposition syndromes are involved not only in the pathogenesis of inherited cancers but also in their sporadic counterparts. Furthermore, these genes almost universally play critical roles in human carcinogenesis even outside the context of childhood cancers. Cancer predisposition syndromes manifesting in childhood fall into two phenotypic categories: those with no non-malignant manifestations and those with coincident congenital anomalies. The former include hereditary retinoblastoma (RB), Li–Fraumeni syndrome and familial polyposis, while the latter category includes von Hippel Lindau disease, Beckwith–Wiedemann syndrome and Gorlin syndrome. Cancer predisposition is caused by constitutional alterations in one of three groups of genes:

1. Tumor suppressor genes encode proteins that control cellular proliferation by either inhibiting progression through the cell cycle or promoting apoptosis. Usually, one functional copy of the gene is sufficient to exert normal function. Functional inactivation of both alleles permits uncontrolled proliferation.
2. Oncogenes activate cellular proliferation and a mutation on one allele is sufficient for oncogenic activation.
3. DNA stability or repair genes are not directly involved in regulation of cell proliferation; however, functional disruption of these genes leads to a higher mutation rate across the genome, again leading to tumor formation.

Introduction of routine predictive genetic testing together with development and implementation of clinical surveillance guidelines has led to early tumor detection and improved survival for both children and adults with hereditary forms of cancer. This chapter addresses the diversity of molecular mechanisms in several prototypical childhood cancer predisposition syndromes. A comprehensive list of hereditary cancer syndromes, their associated phenotypes and genotypes is found in Table 11.1.

RETINOBLASTOMA

Retinoblastoma (RB) is a rare childhood tumor thought to arise in the embryonic retinal epithelium. RB occurs in approximately 1 in 15 000 births and accounts for 2.5–4.0% of all childhood cancers with 90% arising before 5 years of age [3]. RB is the prototype cancer caused by mutations of a tumor suppressor gene, *RB1*. RB occurs in both a heritable and a non-heritable form. Patients with non-heritable RB usually present with unilateral tumors and are older (median age 22 months) at diagnosis, whereas patients with heritable RB often develop bilateral or multifocal tumors and are younger (median age 11 months) at presentation. In approximately 40% of RB cases, the disease is inherited as an autosomal dominant trait, with a penetrance approaching 100% [4]. The remaining 60% of cases are sporadic (non-heritable). Fifteen percent of unilateral RB is heritable but by chance develops in only one eye. On the basis of this inheritance pattern, Knudson proposed the "two hit" hypothesis, which formed the basis for understanding (and discovery) of tumor suppressor genes [4]. The Knudson hypothesis proposes that two mutational events are necessary for tumor development. In heritable tumors, one mutation is

TABLE 11.1 Hereditary Syndromes Associated with Childhood Neoplasms

Syndrome	OMIM Entry	Major Tumor Types	Mode of Inheritance	Genes
Hereditary gastrointestinal malignancies				
Adenomatous polyposis of the colon	175100	Colon, thyroid, stomach, intestine, hepatoblastoma	Dominant	*APC*
Juvenile polyposis	174900	Gastrointestinal	Dominant	*SMAD4/ DPC4*
Peutz–Jeghers syndrome	175200	Intestinal, ovarian, pancreatic	Dominant	*STK11*
Genodermatoses with cancer predisposition				
Nevoid basal cell carcinoma syndrome	109400	Skin, medulloblastoma	Dominant	*PTCH*
Neurofibromatosis type 1	162200	Neurofibroma, optic pathway glioma, peripheral nerve sheath tumor	Dominant	*NF1*
Neurofibromatosis type 2	101000	Vestibular schwannoma	Dominant	*NF2*
Tuberous sclerosis	191100	Hamartoma, renal angiomyolipoma, renal cell carcinoma	Dominant	*TSC1/TSC2*
Xeroderma pigmentosum	278730, 278700, 278720, 278760, 278740, 278780, 278750, 133510	Skin, melanoma, leukemia	Recessive	*XPA, B, C, D, E, F G, POLH*
Rothmund–Thomson syndrome	268400	Skin, bone	Recessive	*RECQL4*
Leukemia/lymphoma predisposition syndromes				
Bloom syndrome	210900	Leukemia, lymphoma, skin	Recessive	*BLM*
Fanconi anemia	227650	Leukemia, squamous cell carcinoma, gynecological system	Recessive	*FANCA.B.C. D_2,E.F.G*
Schwachman Diamond syndrome	260400	Leukemia/myelodysplasia	Recessive	*SBDS*
Nijemegen breakage syndrome	251260	Lymphoma, medulloblastoma, glioma	Recessive	*NBS_1*
Ataxia teleangiectasia	208900	Leukemia, lymphoma	Recessive	*ATM*
Genitourinary cancer predisposition syndromes				
Simpson–Golabi–Behmel syndrome	312870	Embryonal tumors, Wilms tumor	X-linked	*GPC3*
von Hippel–Lindau syndrome	193300	Retinal and central nervous hemangioblastoma, pheochromocytoma, renal cell carcinoma	Dominant	*VHL*
Beckwith–Wiedemann syndrome	130650	Wilms tumor, hepatoblastoma, adrenal carcinoma, rhabdomyosarcoma	Dominant	*CDKN1C/ NSD1*
Wilms tumor syndrome	194070	Wilms tumor	Dominant	*WT1*
WAGR syndrome	194072	Wilms tumor, gonadoblastoma	Dominant	*WT1*
Costello syndrome	218040	Neuroblastoma, rhabdomyosarcoma, bladder carcinoma	Dominant	*H-Ras*
Central nervous system predisposition syndromes				
Retinoblastoma	180200	Retinoblastoma, osteosarcoma	Dominant	*RB1*

(*Continued*)

TABLE 11.1 (Continued)

Syndrome	OMIM Entry	Major Tumor Types	Mode of Inheritance	Genes
Rhabdoid predisposition syndrome	601607	Rhabdoid tumor, medulloblastoma, choroids plexus tumor		*SNF5/INI1*
Medulloblastoma predisposition	607035	Medulloblastoma	Dominant	*SUFU*
Sarcoma/bone cancer predisposition syndromes				
Li–Fraumeni syndrome	151623	Soft tissue sarcoma, osteosarcoma, breast, adrenocortical carcinoma, leukemia, brain tumor	Dominant	*TP53*
Multiple exostosis	133700,133701	Chondrosarcoma	Dominant	*EXT1/EXT2*
Werner syndrome	277700	Osteosarcoma, menigioma	Recessive	*WRN*
Endocrine cancer predisposition syndromes				
MEN1	131000	Pancreatic islet cell tumor, pituitary adenoma, parathyroid adenoma	Dominant	*MEN1*
MEN2	171400	Medullary thyroid carcinoma, pheochromocytoma, parathyroid hyperplasia	Dominant	*RET*

OMIM, Online Mendelian Inheritance in Man.

inherited through the germline and the second occurs in somatic cells, leading to their multifocal, early age of onset. In non-heritable tumors, both mutations occur in somatic cells, leading to their unifocal, later onset. Survivors of heritable retinoblastoma have a 100-fold increased risk of developing mesenchymal tumors such as osteogenic sarcoma, fibrosarcoma and melanoma later in life.

The *RB* gene maps to chromosome 13q14 [5]. *RB* consists of 27 exons and encodes pRB, a 105-kD nuclear phosphoprotein that plays a central role in the control of cell cycle regulation, particularly in determining transition from G_1 through S (DNA synthesis) phase in virtually all cell types [6].

In the developing retina, inactivation of the *RB* gene is necessary and sufficient for tumor formation. It is now clear, however, that these tumors develop as a result of a more complex interplay of aberrant expression of other cell cycle control genes. In particular, a tumor surveillance pathway mediated by Arf, MDM2, MDMX and p53 is activated after loss of pRB during development of the retina. In a small fraction of RB tumors, no *RB1* mutations are detected; in the majority of these, high-level amplification of the *MYCN* oncogene is observed suggesting a novel mechanism of tumorigenesis in the presence of non-mutated *RB1* genes [7]. Not only do these observations provide a provocative biologic mechanism for tumor formation in retinoblastoma, but they also point to potential molecular targets for developing novel therapeutic approaches to this tumor. For example, the MDM2/MDMX antagonist, Nutlin-3a, efficiently targets the p53 pathway and is effective as an ocular formulation in treating RB in orthotopic xenografts [8]. Whether these concepts might also lead to the development of chemopreventive approaches to RB in mutation carriers is an avenue open to further study.

The critical importance of early diagnosis and existence of effective surveillance guidelines highlight the role of genetic testing and counseling in patients and families with RB. Family history may not always be informative because of a high *de novo* mutation rate or incomplete penetrance of inherited mutations. Routine clinical testing is available for mutation detection. A multistep approach reported by Gallie and coworkers detected 89% (199/224) of mutations in bilaterally affected probands and both mutant alleles in 84% (112/134) of tumors from unilaterally affected probands [9]. It is recommended that patients at risk undergo ophthalmologic examination under anesthesia starting at birth until age 4 years. This surveillance is performed in conjunction with genetic counseling.

WILMS TUMOR

Wilms tumor (WT), or nephroblastoma, is an embryonal malignancy that arises from remnants of immature kidney

[10]. It affects approximately 1 in 7000 children, usually before the age of 6 years (median age at diagnosis, 3.5 years). Approximately 5–10% of children present with synchronous or metachronous bilateral tumors. WT typically presents as an asymptomatic abdominal mass, although a small fraction of children have symptoms such as hematuria or hypertension. Approximately 20% of children present with metastatic disease.

Unlike RB, patients with WT not infrequently present with non-neoplastic congenital anomalies, suggesting a close relationship between WT and aberrations of normal development. A peculiar feature of WT is its association with nephrogenic rests, foci of primitive but non-malignant cells whose persistence suggests a defect in kidney development. These precursor lesions are found within the normal kidney tissue of more than one third of children with WT. Nephrogenic rests may persist, regress spontaneously, or grow into large masses that simulate true WT and present a difficult diagnostic challenge [10]. Another intriguing feature of WT is its association with specific congenital abnormalities, including genitourinary anomalies, sporadic aniridia, mental retardation and hemihypertrophy. A genetic predisposition to WT is observed in two distinct disease syndromes with urogenital system malformations – the WAGR (Wilms tumor, aniridia, genitourinary abnormalities, mental retardation) syndrome [11] and the Denys–Drash syndrome (DDS) [12] – and in Beckwith–Wiedemann syndrome (BWS) [13].

The WAGR syndrome is associated with constitutional deletions of chromosome 11q13 [11]. The WAGR deletion encompasses a number of contiguous genes, including the aniridia gene *Pax6*. The cytogenetic observation in patients with WAGR was also important in cloning of the *WT1* gene at chromosome 11p13. *WT1* spans approximately 50 kb of DNA, containing 10 exons that encode the WT1 protein transcription factor. DDS, the second syndrome closely associated with this locus, is a rare association of WT, intersex disorders and progressive renal failure [12]. Virtually all patients with DDS carry germline *WT1* point mutations.

WT1 is altered in only 10% of Wilms tumors. This observation implies the existence of alternative loci in the etiology of this childhood renal malignancy. One such locus also resides on the short arm of chromosome 11, telomeric of *WT1*, at 11p15. This gene, designated *WT2*, is associated with BWS. Patients with BWS are at increased risk of developing Wilms tumor as well as other embryonic malignancies, including rhabdomyosarcoma (RMS), neuroblastoma and hepatoblastoma [13]. The putative *BWS* gene maps to chromosome 11p15 [14] and its complex structure is discussed later in this chapter. Using long-oligonucleotide array comparative genomic hybridization (array CGH), a novel gene termed *WTX* was identified on chromosome Xq11.1. *WTX* is inactivated in one third of WTs and tumors with *WTX* mutations lack *WT1* mutations [15]. A role of *WTX* in hereditary forms of WT has not been defined. Bilateral WT or a family history of WT occurs in 1–5% of patients. Although linkage studies have indicated that the gene for familial WT must be distinct from *WT1* and *WT2* and from the gene that predisposes to BWS, this gene has been neither cytogenetically localized nor isolated.

NON-SYNDROMIC TUMORS AND SUSCEPTIBILITY LOCI

Retinoblastoma and Wilms tumor represent two classic tumors whose inheritance patterns have been well established and for which the genetic basis of tumor susceptibility is generally understood. In both situations, as outlined above, they most commonly present in the absence of a family history of other tumors – that is, they are “tumor specific” within families. Further along in this chapter we will explore the genetic basis of a spectrum of syndromes that are recognizable by the diverse cancer phenotypes across generations and family members. However, it is also worth noting several common pediatric malignancies that, for the most part, do not appear to harbor striking germline alterations in susceptibility loci. These tumors represent an important subset of childhood cancers for which the advent of large-scale genome-wide sequencing may provide insight into novel mechanisms of susceptibility, distinct from the generally single-gene disorders of commonly cited cancer predisposition syndromes.

Neuroblastoma

Neuroblastoma (NB) is the most common soft tissue tumor of children. The embryonic neural crest gives rise to the peripheral nervous system including cranial and spinal sensory ganglia, the autonomic ganglia, the adrenal medulla and other paraendocrine cells distributed throughout the body – accounting for the widespread distribution of presenting and metastatic sites of NB.

A small subset of neuroblastomas (probably <10) is inherited in an autosomal dominant fashion. Until recently, the only gene definitively associated with neuroblastoma risk was *PHOX2*B, also linked to central apnea, or Ondine’s curse [16]. Utilizing high-resolution microarray and next-generation sequencing approaches, *de novo* and inherited missense mutations in the tyrosine kinase domain of the ALK (anaplastic lymphoma kinase) gene on chromosome 2p23 have been observed in many hereditary neuroblastoma families, as well as in sporadic cases [17]. Remarkably, different stages of disease present in different family members harboring the same constitutional ALK mutation – strongly suggesting the requirement of subsequent genetic events in somatic cells that determine biological aggressiveness. Neuroblastoma has occasionally been reported in the context of cancer syndromes such as Li–Fraumeni and Beckwith–Wiedemann syndromes for which specific susceptibility genes have been associated. However, in the absence of these familial cancer phenotypes, whole-exome, genome and transcriptome sequencing of neuroblastoma has identified only a small handful of putative pathogenic germline variants (*ALK*, *CHEK2*, *PINK1* and *BARD1*) [18].

Sarcomas of Childhood

While sarcomas account for approximately 20% of all childhood malignant tumors, outside of the context of a few susceptibility syndromes, constitutional genetic susceptibility does not appear to play a significant role in their etiology. Rhabdomyosarcoma is observed in hereditary cancer syndromes including Li–Fraumeni syndrome (see following section) in which carriers harbor constitutional mutations of the *TP53* tumor suppressor gene. The possible importance of the patched gene, *PTCH*, in the development of RMS is suggested by the finding that mice lacking this gene develop RMS. *PTCH* is mutated in the germline of patients with Gorlin syndrome, a disorder that includes predisposition to tumor (medulloblastoma) development. Strikingly, *PTCH* is shown to regulate another gene, *GLI*, which is found to be amplified in RMS and Gorlin syndrome-associated tumors. RMS itself has been rarely reported in Gorlin syndrome. Activation of the *HRAS* oncogene by heterozygous germline mutations predisposes to RMS in Costello syndrome [19]. Mutations in the CDKN1C gene complex on chromosome 11p are also associated with RMS in Beckwith–Wiedemann syndrome (see below), further highlighting the multiple molecular pathways associated with rhabdomyosarcomagenesis. Osteosarcoma is seen in the context of LFS, retinoblastoma and syndromes associated with mutations in the helicase gene complex. Despite the remarkable aneuploidy characteristic of somatic karyotypes (as measured by conventional chromosome spreads or whole-genome sequencing), no other osteosarcoma-specific susceptibility loci have been identified to date. Similarly, while Ewing sarcoma is occasionally reported to occur in multiple first- or second-degree relatives, it is not observed in the context of known familial cancer syndromes. Other much less common sarcomas, such as infantile fibrosarcoma, although characterized by a somatic t(12;15) translocation, are not known to be familial. Current large-scale microarray and genome-wide studies are likely to shed some light on the role of constitutional genetic alterations in "sporadic" or "familial" sarcomas. To date, however, no definitive clues as to their heritable etiology have been observed.

CANCER PREDISPOSITION SYNDROMES

Several hereditary cancer syndromes are associated with the occurrence of childhood as well as adult-onset neoplasms. Although it is beyond the scope of this chapter to describe them all, it is worthwhile to discuss a few to highlight the important molecular basis on which these disorders develop (Table 11.2). Several syndromes are described in each of two categories: (1) those not associated with other congenital anomalies; and (2) those for whom patients are typically identified by the presence of a spectrum of anomalies at or shortly after birth, their cancers only manifesting later in infancy or childhood.

Cancer Predisposition Syndromes with Malignant-Only Phenotype

Li–Fraumeni Syndrome

Li–Fraumeni familial cancer syndrome (LFS) is the prototypical familial cancer predisposition syndrome. The definition of classical LFS requires a proband with a sarcoma diagnosed before 45 years of age, a first-degree relative diagnosed as having any cancer when younger than 45 years and a first- or second-degree relative with a diagnosis of cancer when younger than 45 years or a sarcoma at any age [20]. The classic spectrum of tumors that includes soft-tissue sarcomas, osteosarcomas, breast cancer, brain tumors, leukemia and adrenocortical carcinoma (ACC) has been overwhelmingly substantiated by numerous subsequent studies, although other cancers, usually of particularly early age of onset, are also observed (Figure 11.1). Similar patterns of cancer that do not meet the classic definition have been coined Li–Fraumeni-like syndrome (LFS-L). The Chompret critieria for genetic testing for LFS were initially described in 2001 and updated in 2009. The sensitivity and specificity of the Chompret criteria are 82% and 58%, respectively, making it perhaps the most rigorous and relevant definition to justify *TP53* mutation analysis [21].

Germline alterations of the *TP53* tumor suppressor gene are associated with LFS [22,23]. These are primarily missense mutations that encode a stabilized mutant protein. The spectrum of germline *TP53* mutations is similar to that of somatic mutations found in a wide variety of tumors. Carriers are heterozygous for the mutation and, in tumors derived from these individuals, the second (wild-type) allele is frequently deleted or mutated, leading to functional inactivation. Several comprehensive databases document all reported germline (and somatic) *TP53* mutations and are of particular value in evaluating novel

TABLE 11.2 Clinical Criteria for Classic Li–Fraumeni Syndrome, LFS-Like Criteria and Chompret Criteria

Classic LFS criteria
Proband diagnosed with sarcoma before age 45
A first-degree relative with cancer diagnosed before age 45
A first- or second-degree relative on the same parental lineage with cancer diagnosed before age 45 or a sarcoma at any age
LFS-like critera (Birch)
Proband with any childhood cancer or sarcoma, brain tumor or adrenocortical cancer diagnosed before age 45
First- or second-degree relative with typical LFS cancer (sarcoma, breast cancer, brain tumor, leukemia or adrenocortical cancer) diagnosed at any age, AND
A first- or second-degree relative on the same side of the family with any cancer diagnosed under age 60
LFS-like criteria (Eeles)
Two first- or second-degree relatives with LFS-related malignancies at any age
Chompret criteria for LFS
Proband diagnosed with a narrow-spectrum cancer (sarcoma, brain tumor, breast cancer or adrenocortical carcinoma) before age 46 and at least one first- or second-degree relative with any cancer, except breast cancer if the proband has breast cancer
Proband with multiple primary tumors, two of which belong to the narrow spectrum and the first of which occurred before age 46, regardless of family history
Proband with adrenocortical cancer or choroid plexus carcinoma, regardless of age at diagnosis or family history

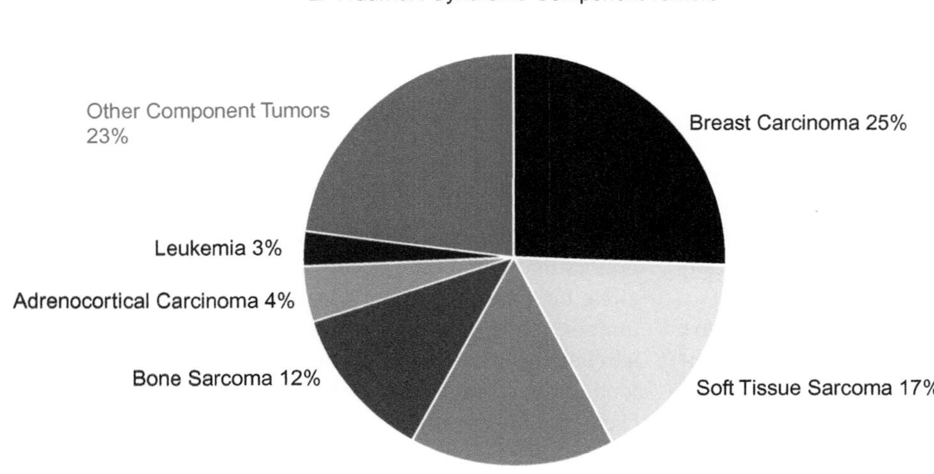

FIGURE 11.1 Pie graph demonstrating the frequency of different tumor types in Li–Fraumeni syndrome. Data from Nichols KE, Malkin D, Garber JE, Fraumeni JF Jr, Li FP. Germ-line p53 mutations predispose to a wide spectrum of early-onset cancers. Cancer Epidemiol Biomarkers Prev 2001;10:83–7.

mutations as well as phenotype–genotype correlations. Studies in the USA and in Europe have suggested that germline *p53* mutations occur at the rate of about 1:5000 individuals. In Brazil, a specific germline mutation at codon R337H (c.1010 G > A, genomic nucleotide number 17 588) in exon 10, within the oligomerization domain of p53, was first identified in children with adrenocortical carcinoma (ADC) in families with no reported history of cancer [24,25]. The allele frequency of R337H in the population of southeast and southern Brazil is about 15 times higher than any other single *p53* mutation associated with LFS [26]. In fact, the carrier rate (1:300) for this mutation is higher than any other known cancer susceptibility mutation of any gene worldwide. At pH in the low to normal physiological range (up to 7.5), the mutant protein forms normal oligomers and retains its suppressor function. However, at high physiologic pH, the histidine replacing arginine at codon 337 becomes deprotonated and is unable to donate a hydrogen bond critical for protein dimerization. This prevents p53 from assembling into a functional transcription factor. This unique biochemical feature might contribute to the particular features of R337H families, which often show incomplete penetrance and heterogeneous tumor patterns [27].

Analysis of tumor patterns in R337H carriers and their families reveals all the common features of LFS/LFL, clearly establishing that this mutant predisposes to a wide spectrum of multiple cancers. In R337H carriers, the penetrance at age 30 is less than 20% (compared to about 50% in "classical" LFS). However, the penetrance over lifetime is about 90%, similar to "classical" LFS.

The high prevalence of a rare mutation raised the question of a possible founder effect among Brazilian families' carriers of the same alteration. Using a dense panel of single nucleotide polymorphisms (SNPs) encompassing the whole *p53* gene revealed the presence of a rare haplotype, with a probability that the mutation arose independently on this haplotype of less than 10^{-8} [28], establishing the existence of a founder effect – the only known one in any gene predisposing to childhood cancers.

Approximately 75% of families with classic LFS have detectable *TP53* alterations. It is not clear whether the remainder are associated with the presence of modifier genes, promoter defects yielding abnormalities of p53 expression, or simply the result of weak phenotype–genotype correlations (i.e. the broad clinical definition encompasses families that are not actual members of LFS). The variability in type of cancer and age of onset within and between LFS families suggests that expression of modifier genes might influence the underlying mutant *TP53* genotype. Several of these have been described including those that accelerate age of tumor onset in *TP53* mutation carriers, such as an SNP in the promoter of the *MDM2* gene (SNP 309) that is involved in p53 degradation pathway [29], accelerated telomere attritition and increased constitutional copy number variation (CNV) [30], whereas others, such as a 16-bp duplication in *TP53* intron 3 (PIN3), delay tumor onset by up to 19 years [31]. Whole-genome sequencing (WGS) of children with sporadic medulloblastoma revealed a subset demonstrating characteristic evidence of chromothripsis, defined as a high frequency of intrachromosomal rearrangements localized to two to three chromosomes in each tumor. Remarkably, all patients whose tumors demonstrated chromothripsis were observed to have a germline *TP53* mutation; in fact, all were of the SHH medulloblastoma subgroup. This finding suggests an important role for early *TP53* mutations in chromothripsis and provides a possible genetic mechanism for pathogenesis of SHH medulloblastoma associated with LFS [32]. WGS has also uncovered another otherwise occult relationship with germline *TP53* mutations when it was reported in a WGS analysis of childhood acute lymphoblastic leukemia (ALL) that patients with hypodiploid ALL exhibited a high frequency of germline *TP53* mutations. The unique phenotype also suggests an important role of early TP53 mutation in leukemogenesis of this particular rare and aggressive subtype [33].

Until recently, options for intervention in LFS were thought to be limited, but two studies have clearly demonstrated the value of total body imaging using rapid sequence whole body MRI (with or without biochemical marker studies) in *TP53* mutation carriers (Table 11.3) [34,35]. Evaluation of these have proven to be effective in early tumor detection which leads to improved survival – offering hope for these patients that the combination of molecular testing with early clinical surveillance can interfere with the natural history of the disease.

Hereditary Paraganglioma Syndromes

Paragangliomas are benign non-catecholamine secreting tumors that commonly occur in the head and neck region, along the parasympathetic chain. Catecholamine-secreting tumors can develop along the sympathetic chain, in the adrenal medulla (pheochromocytoma) alongside the aortopulmonary vasculature, the organ of Zuckercandl, or even the bladder and vas deferens. Paragangliomas have an estimated population incidence of 1 in 30 000. However, in the presence of an underlying germline mutation in the succinyl dehydrogenase (*SDHx*) complex, the tumor rate is extraordinarily high with disease penetrance approaching 80% [36,37]. Nearly 30% of non-metastatic paragangliomas and pheochromocytomas are due to germline *SDHx* mutations and 44% of adults and 81% of pediatric patients with metastatic disease carry germline *SDHx* mutations [38]. Other tumors, including renal cell carcinoma, oncocytoma, papillary thyroid

TABLE 11.3 Clinical Surveillance Strategy for *TP53* Mutation Carriers

Tumor Type	Surveillance Strategy
Children	
Adrenocortical carcinoma	Ultrasound of abdomen and pelvis every 3–4 months
	Complete urinalysis every 3–4 months
	Bloodwork every 4 months: ESR, LDH, β-HCG, α-fetoprotein, 17-OH-progesterone, testosterone, DHEAS, androstenedione
Brain tumor	Annual MRI of the brain
Soft tissue and bone sarcoma	Annual total body MRI
Leukemia/lymphoma	Bloodwork every 4 months: CBC profile
Regular evaluation with family physician with close attention to any medical concerns or complaints	
Adults	
Breast cancer	Monthly breast self-examination starting at age 18 years
	Semiannual clinical breast exam starting at age 20–25 years or 5–10 years before the earliest known breast cancer in the family
	Annual mammogram and breast MRI screening starting at age 20–25 years, or individualized based on earliest age of onset in family
	Consider risk-reducing bilateral mastectomy
Brain tumor	Annual MRI of the brain
Soft tissue and bone sarcoma	Total body MRI to be used as a baseline
Colon cancer	Biennial colonoscopies beginning at age 40 years, or 10 years before the earliest known colon cancer in the family
Melanoma	Annual dermatology examination
Leukemia/lymphoma	CBC profile every 6 months for indications of leukemia/lymphoma
	Annual abdominal ultrasound
Regular evaluation with family physician with close attention to any medical concerns or complaints	

cancer, pituitary tumors, gastrointestinal stroma tumors (GIST) and, rarely, neuroblastoma are observed in *SDHx* mutation carriers.

Succinate dehydrogenase (SDH) is a component of respiratory Complex II in the mitochondria. This enzyme complex is responsible for converting succinate to fumarate as part of the Krebs cycle. SDH is composed of four distinct proteins called SDHA, SDHB, SDHC and SDHD [39]. A fifth gene called *SDHAF2*, or SDH assembly factor 2, is responsible for assembling all of the individual SDH proteins into a fully functioning enzyme complex [40]. Germline mutations in each of these *SDHx* genes may lead to development of paragangliomas or pheochromocytomas. Lack of a functioning SDH complex leads to increased succinate, with subsequent increases in hypoxia inducible factor (HIF) signaling and possible histone deregulation. Germline mutations in other genes such as *NF1*, *VHL*, *RET*, *TMEM127* and *MAX* have also been associated with the development of paragangliomas and pheochromocytomas (Figure 11.2). Based on gene expression and pathway analysis, these tumors can be divided into two different clusters which correspond to their underlying gene mutations: Cluster 1 (Cluster 1A: *SDHx*, Cluster 1B: *VHL*) associated with pseudohypoxia and aberrant VEGF signaling and Cluster 2 *(RET/NF1/TMEM127/MAX)* associated with aberrant kinase signaling pathways.

Alterations in each *SDHx* gene lead to different disease phenotypes and clinical presentations, as outlined in Table 11.4 [39]. To facilitate genetic diagnosis, risk assessment and treatment options, it is now possible to test for all the SDHx genes simultaneously.

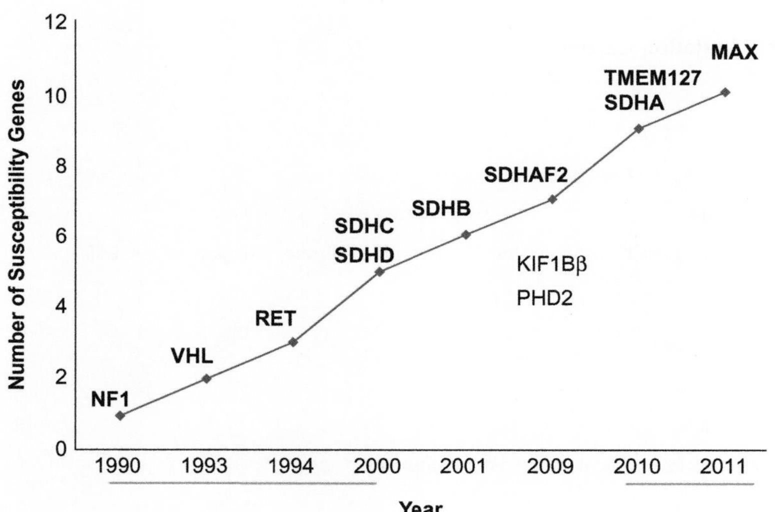

FIGURE 11.2 Accelerated discovery of genes associated with predisposition to hereditary pheochromocytoma/paraganglioma syndromes.

TABLE 11.4 SDHx Genotype:Phenotype Correlations

	PGL1	PGL2	PGL3	PGL4	PGLX
SDH Gene	*SDHD*	*SDH5 (SDHAF2)*	*SDHC*	*SDHB*	*SDHA*
Chromosomal location	11q23	11q11.3	1q21	1p35-36.1	5p15
Most common mutation	frameshift	point	nonsense	missense	missense
Head and neck PGLs	++	++	++	+	+
PCC (any abdominal)	+/−	−	+/−	++	++
Catecholamine secreting	+/−	−	+/−	++	?
Malignant	−	−	unknown	++	+
Associated with GIST	+	−	+	+	−
Associated with thyroid cancer	+	−	−	+	−
Associated with renal tumors (renal cell carcinoma and oncocytoma)	−	−	−	+	−
Associated with neuroblastoma	−	−	−	+	−

Regular surveillance can detect early tumors in patients with underlying germline *SDHx* mutations. As with surveillance in *TP53* mutation carriers, this is important so that smaller, asymptomatic SDH-deficient tumors can be removed before they transform to malignant and metastatic disease. Although no formalized screening guidelines exist, many clinicians will perform annual physical examinations, blood pressure checks (for hypertension due to increased catecholamines) and blood work for serum metabolites. Previously, urine catecholamines were examined from 24-hour urine specimens, but many have eliminated urine screening in favor of serum testing. Fractionated plasma metanephrines are the most sensitive and specific serum test for detecting secreting paragangliomas and pheochromocytomas [41]. Increased methoxytyramine, a metabolite of dopamine, seems to be helpful for predicting the likelihood of metastatic disease and for distinguishing SDH-related tumors from VHL-related tumors. However, testing of methoxytyramine remains difficult to obtain on a clinical basis.

Regular imaging has been demonstrated by several groups to be very effective at identifying SDH-related tumors, especially in the setting of negative biochemical results [42]. Screening approaches using rapid sequence whole body MRI in conjunction with urinary and/or fractionated plasma metanephrine levels are being used widely with abnormal MRI results (or biochemical results) being followed with positron emission tomography (PET) imaging for refining the anatomical location of the tumor.

DICER1 *Syndrome*

In addition to the many cancer predisposition syndromes previously described in the literature, as powerful genomic/genetic tools and improved recognition of cancer associations are used, "new" syndromes continue to be defined. One of these, *DICER1* syndrome, is a very recently characterized phenotypic association of distinctive dysontogenic hyperplastic or overtly malignant tumors. The most frequent of these is the rare childhood lung malignancy pleuropulmonary blastoma (PPB). A wide spectrum of other, primarily endocrine manifestations are evident: ovarian Sertoli–Leydig cell tumors (SLCT), nodular thyroid hyperplasia, pituitary blastoma, papillary and follicular thyroid carcinoma, cervical rhabdomyosarcoma, cystic nephroma and possibly Wilms tumor [43]. Germline mutations in *DICER1* have been identified in children and young adults affected with one or several of these tumors and somatic *DICER1* mutations have been variously identified in sporadic component tumors of the disorder. DICER1 is an endoribonuclease that processes hairpin precursor microRNAs (miRNAs) into short, functional miRNAS. Mature 5′ miRNAs as well as other components of the RNA-induced silencing complex (RISC) downregulate three targeted mRNAs [44].

Unlike the classical "two hit" mechanism associated with inactivation of tumor suppressor genes, the effect of *DICER1* loss of function appears to result primarily from an initial inactivating mutation that reduces by half the amount of wild-type DICER1 protein, while the second hit specifically knocks out production of 5′ mature miRNAs. Furthermore, penetrance of the mutations is highly variable and the explanation for the wide spectrum of both hyperplastic and malignant neoplasms is not clear. While many of the lesions in *DICER1* mutation carriers are relatively indolent or benign, the risk in childhood of some potentially lethal tumors such as PPB and pineoblastoma indicates a need for clinical surveillance particularly targeting the lungs, abdomen and brain. Such surveillance protocols have recently been developed but they, unlike those for other more thoroughly defined syndromes, are still in evolution.

Multiple Endocrine Neoplasia

The multiple endocrine neoplasia (MEN) disorders comprise at least three different diseases, MEN type 1, MEN type 2A, and MEN type 2B, which are all cancer predisposition syndromes that affect different endocrine organs. The most common features of MEN type 1 are parathyroid adenomas (about 90% of cases), pancreatic islet cell tumors (50–75% of cases) and pituitary adenomas (25–65% of cases) [45]. Twenty-eight percent of MEN1 mutation carriers demonstrate clinical or biochemical evidence of disease by age 15 years. MEN2A is associated with medullary thyroid carcinoma (MTC), parathyroid adenoma and pheochromocytoma. The risk of developing MTC in mutation carriers with MEN2A is 100%; prophylactic thyroidectomy is recommended before age 5 years in all children who are confirmed carriers. MEN2B is a related disorder, but with onset of the tumors in early infancy, ganglioneurinoma of the gastrointestinal tract and skeletal abnormalities.

MEN1 is caused by mutation in the tumor suppressor gene, *MEN1*; MEN2A and 2B are caused by mutations in the proto-oncogene *RET*. While 100% of *RET* mutation carriers will develop MTC, approximately 1–7% of patients with sporadic MTC harbor germline *RET* mutations. The pattern of mutations seen in MEN2 families does not follow the "two hit" hypothesis for tumor suppressor genes: the RET proto-oncogene is not inactivated and there is no loss of the second allele in the tumors. Thus, the predisposition to cancer in families with MEN2 is based on the inheritance of an activating mutation in the *RET* proto-oncogene. This unusual pattern of inheritance is almost unique in the field of hereditary cancer predisposition syndromes. Genetic testing is possible by direct mutation analysis of the 10 exons of the gene. Furthermore, well-established clinical surveillance tools exist for early detection of pheochromocytoma, papillary and medullary thyroid cancer and pancreatic and pituitary tumors associated with the MEN disorders.

Cancer Predisposition Syndromes with Coincident Congenital Anomalies

Beckwith–Wiedemann Syndrome

Beckwith–Wiedemann syndrome (BWS) occurs with a frequency of 1 in 13 700 births. BWS is associated with a wide spectrum of phenotypic stigmata including hemihyperplasia, exomphalos, macroglossia, gigantism and ear pits (posterior aspect of the pinna). Laboratory findings may include profound neonatal hypoglycemia,

TABLE 11.5 Beckwith–Wiedemann Syndrome Genetic and Epigenetic Subgroups

	DNA	RNA	Karyotype	Frequency (%)	Inheritance
A. Regional	Paternal 11p15 UPD		Normal lipis	10–20	Sporadic
	Disruption of *KCNQ10T1*		Duplication 11p15	1	Sporadic
			Transl/Inver	1	Sporadic
B. Domain 1	*H19* hypermethylation	*IGF2* LOI	Normal	2	Sporadic
	Normal *H19* methylation	*IGF2* LOI	Normal	25–50	Sporadic
C. Domain 2	*CDKN1C* mutation	KNQ1OT1 LOI	Normal	5–10	Sporadic
	CDKN1C mutation		Normal	25	AD
	KvDMR1 LOM		Normal	50	Sporadic
D. Other	Unknown	Unknown	Normal	5	AD
	Unknown		Normal	10–20	Sporadic

AD: autosomal dominant; LOI: loss of imprinting; LOM: loss of methylation; UPD: uniparental disomy.

polycythemia, hypocalcemia, hypertriglyceridemia, hypercholesterolemia and high serum α-fetoprotein (AFP) level. With increasing age, phenotypic and laboratory features of BWS become less pronounced. Although neurocognitive defects are not universal in BWS, early diagnosis of the condition is crucial to avoid deleterious neurologic effects of neonatal hypoglycemia and to initiate an appropriate screening protocol for tumor development. The increased risk for tumor formation in BWS patients is estimated at 7.5% and is further increased to 10% if hemihyperplasia is present. Tumors occurring with the highest frequency include Wilms tumor, hepatoblastoma, neuroblastoma, rhabdomyosarcoma and adrenocortical carcinoma.

The genetic basis of BWS is complex and it does not appear that a single gene is responsible for the phenotype. Various 11p15 chromosomal or molecular alterations have been associated with the BWS phenotype (Table 11.5) and its tumors [46]. Abnormalities in this region impact an imprinted domain, indicating that it is more likely that normal gene regulation in this part of chromosome 11p15 occurs in a regional manner and may depend on various interdependent factors or genes. These include the paternally expressed genes *IGF2* and *KCNQ1OT1* and the maternally expressed genes *H19*, *CDKN1C* and *KCNQ1*. Paternal uniparental disomy, in which two alleles are inherited from one parent (the father), has been reported in approximately 15% of sporadic BWS patients [47]. The insulin/IGF2 region is always represented in the uniparental disomy, although the extent of chromosomal involvement is highly variable. Alterations in allele-specific DNA methylation of *IGF2* and *H19* reflect this paternal imprinting phenomenon [47]. As with other cancer susceptibility syndromes, effective clinical surveillance protocols regularly identify tumors with demonstrable improved outcomes. Regular (every 3 month) AFP levels and abdominal/pelvic ultrasound are recommended until the affected child is approximately 9 years old and generally beyond the risk age for the associated tumors.

Gorlin Syndrome

Nevoid basal cell carcinoma syndrome, or Gorlin syndrome, is a rare autosomal dominant disorder characterized by multiple basal cell carcinomas, developmental defects including bifid ribs and other spine and rib abnormalities, palmar and plantar pits, odontogenic keratocysts and generalized overgrowth [48]. The sonic hedgehog (Shh) signaling pathway is a key regulator of development. Gorlin syndrome appears to be caused by germline mutations of the tumor suppressor gene *PTCH*, a receptor for Shh. Medulloblastoma develops in approximately 5% of patients with Gorlin syndrome, virtually always before 5 years of age. Furthermore, approximately 10% of patients diagnosed with medulloblastoma by the age of 2 years are found to have other phenotypic features consistent with Gorlin syndrome and harbor germline *PTCH* mutations [49]. Although Gorlin syndrome develops in individuals with germline mutations of *PTCH*, a subset of children with medulloblastoma harbor germline mutations of another gene, *SUFU*, in the Shh pathway, with accompanying loss of heterozygosity in the tumors.

Molecular and Clinical Surveillance for Cancer Predisposition in Children

As in tumors of adults, continued investigation of the molecular alterations that underlie tumor pathogenesis in children can be expected to provide insights that will lead to highly specific molecular therapeutic targets and that, in turn, should lead to more specific, more efficacious and less toxic therapies. Because tumors associated with highly penetrant genetic predispositions often occur early in life, such insights into pathogenesis also provide novel opportunities for the diagnosis of cancer in children. The evidence to justify the use and continued refinement of molecular analysis for tumor diagnosis, prognosis and development of novel therapeutic avenues for children with cancer is overwhelming. The use of molecular screening as a tool for identification of children at risk for the purpose of developing rational clinical surveillance guidelines is more controversial. Several issues are worth noting. Based on recommendations of the American Society of Clinical Oncology [50], genetic testing should only be undertaken with fully informed consent, including elements of risk assessment, psychological implications of test results (both benefits and risks), risks of employer or insurance discrimination, confidentiality issues and options and limitations of medical surveillance or prevention strategies. When children are not competent to give informed consent, the main consideration should be for the welfare of the child. Although screening for some mutations, such as *TP53*, *RET* or *RB*, is associated with clearly defined beneficial medical management decisions that lead to improved outcomes, it has been argued that presymptomatic identification of other gene mutations, such as in *DICER1* syndrome, are of less obvious clinical benefit. Regardless of the scenario, the complexity of the issues underlies the importance that referral of these patients and families is made to an experienced multidisciplinary team including oncologists, geneticists, psychologists and genetic counselors to facilitate the most appropriate management.

THE POTENTIAL IMPACT OF GENOME STUDIES ON HEREDITARY CANCER SYNDROMES

As is noted throughout this chapter, the vast majority of genes associated with hereditary cancer predisposition syndromes were identified using a variety of classical genetic techniques, including gene cloning (*RB1*), candidate gene approach (*TP53*-Li–Fraumeni syndrome) and reverse genetic approaches from functional characterization in syndrome-specific tumors (*DICER1* syndrome). To date, at least within the field of hereditary cancer syndromes with a strong pediatric foundation, whole-genome surveys have been uninformative with the notable exception to this being the discovery of germline *ALK* mutations in some familial neuroblastomas. However, at the time of writing of this chapter, several large-scale efforts are underway to determine the frequency and characteristics of novel germline mutations in genes that may be associated with susceptibility to many different childhood (and adult) cancer types. Within the adult cancer world, these efforts have generally revealed only evident risk alleles or allele clusters (such as those on chromosome 8q24 associated with increased risk to prostate cancer); whether higher resolution deep-sequencing efforts will uncover more clearly defined genotype–phenotype correlations remains to be determined. A major consideration in these efforts arises from the discovery of "incidental" constitutional genetic alterations. This is particularly relevant in the predisposition gene discovery in the context of childhood cancer. In this setting, the use of whole-genome/whole-exome platforms will necessarily identify genetic variants or deleterious mutations in genes entirely unrelated to the cancer susceptibility gene discovery exercise. Furthermore, for the most part, the child will not have been the one who provides consent either for research or clinical genetic testing. Thus, the potential impact of revealing information generated from the next-generation sequencing data on the child, as well as unaffected and untested, family members, is not insignificant. Virtually nothing is published yet exploring the psychological effect of predictive genetic testing in children; though evidence from studies of parents of children at risk generally support prudent use of genetic testing in the context of intensive pre- and post-testing genetic counseling.

GLOSSARY

Cancer predisposition syndrome A constellation of cancers in a family that is associated with hereditable alterations in cancer susceptibility genes.

Copy number variation A DNA sequence of greater than 1 kilobase in length that is duplicated or deleted in the human genome.

Knudson's two hit hypothesis A theory established by Dr Alfred Knudson to explain the pattern of tumor development in hereditary and sporadic retinoblastoma, predicting the existence of tumor suppressor genes.

Li–Fraumeni syndrome A cancer predisposition syndrome, originally described in 1969, caused by germline mutations in the *TP53* tumor suppressor gene and associated with a wide spectrum of early onset child and adult-onset cancers.

Succinyl dehydrogenase Enzyme complex responsible for converting succinate to fumarate as part of the Krebs cycle.

TP53 The most frequently altered gene in human cancer.

Tumor suppressor gene Gene that encodes a protein whose loss of function leads to aberrant cell cycle control.

ABBREVIATIONS

ACC Adrenocortical carcinoma
BWS Beckwith–Wiedemann syndrome
CGH Comparative genomic hybridization
CNV Copy number variation
CPS Cancer predisposition syndrome
DDS Denys–Drash syndrome

LFS Li–Fraumeni syndrome
MEN Multiple endocrine neoplasia
MTC Medullarythyroid carcinoma
NB Neuroblastoma
PPB Pleuropulmonary blastoma
RB Retinoblastoma
RMS Rhabdomyosarcoma
SDH Succinyldehydrogenase
Shh Sonic hedgehog
VHL von Hippel–Lindau disease
WT Wilms tumor

REFERENCES

[1] Landis SH, Murray T, Bolden S, Wingo PA. Cancer statistics. CA Cancer J Clin 1998;48:6–29.

[2] Knapke S, Nagarajan R, Correll J, Kent D, Burns K. Hereditary cancer risk assessment in a pediatric cancer follow up clinic. Pediatr Blood Cancer 2012;58:85–9.

[3] Devesa SS. The incidence of retinoblastoma. Am J Ophthalmol 1975;80:263–5.

[4] Knudson Jr AG, Hethcote HW, Brown BW. Mutation and childhood cancer: a probabilistic model for the incidence of retinoblastoma. Proc Natl Acad Sci USA 1975;72:5116–20.

[5] Squire J, Dryja TP, Dunn J, Goddard A, Hofmann T, Musarella M, et al. Cloning of the esterase D gene: a polymorphic gene probe closely linked to the rentinoblastoma locus on chromosome 13. Proc Natl Acad Sci USA 1986;83:6573–7.

[6] Friend SH, Bernards R, Rogelj S, Weinberg RA, Rapaport JM, Albert DM, et al. A human DNA segment with properties of the gene that predisposes to retinoblastoma and osteosarcoma. Nature 1986;323:643–6.

[7] Rushlow DE, Mol BM, Kennett JY, Yee S, Pajovic S, Thériault BL, et al. Characterisation of retinoblastoma without RB1 mutations: genomic, gene expression, and clinical studies. Lancet Oncol 2013;14:327–34.

[8] Brennan RC, Federico S, Bradley C, Zhang J, Flores-Otero J, Wilson M, et al. Targeting the p53 pathway in retinoblastoma with subconjunctival Nutlin-3a. Cancer Res 2011;71:4205–13.

[9] Richter S, Vandezande K, Chen N, Zhang K, Sutherland J, Anderson J, et al. Sensitive and efficient detection of RB1 gene mutations enhances care for families with retinoblastoma. Am J Hum Genet 2003;72:253–69.

[10] Beckwith JB, Kiviat NB, Bonadio JF. Nephrogenic rests, nephroblastomatosis, and the pathogenesis of Wilms' tumor. Pediatr Pathol 1990;10:1–36.

[11] Riccardi VM, Sujansky E, Smith AC, Francke U. Chromosomal imbalance in the Aniridia-Wilms' tumor association: 11p interstitial deletion. Pediatrics 1978;61:604–10.

[12] Mueller RF. The Denys–Drash syndrome. J Med Genet 1994;31:471–7.

[13] Sotelo-Avila C, Gonzalez-Crussi F, Fowler JW. Complete and incomplete forms of Beckwith–Wiedemann syndrome: their oncogenic potential. J Pediatr 1980;96:47–50.

[14] Ping AJ, Reeve AE, Law DJ, Young MR, Boehnke M, Feinberg AP. Genetic linkage of Beckwith–Wiedemann syndrome to 11p15. Am J Hum Genet 1989;44:720–3.

[15] Rivera MN, Kim WJ, Wells J, Driscoll DR, Brannigan BW, Han M, et al. An X chromosome gene, WTX, is commonly mutated in Wilms tumor. Science 2007;315:642–5.

[16] Trochet D, Bourdeaut F, Janoueix-Lerosey I, Deville A, de Pontual L, Schleiermacher G, et al. Germline mutations of the paired-like homeobox 2B (PHOX2B) gene in neuroblastoma. Am J Hum Genet 2004;74:761–4.

[17] Mossé YP, Laudenslager M, Longo L, Cole KA, Wood A, Attiyeh EF, et al. Identification of ALK as a major neuroblastoma susceptibility gene. Nature 2008;455:930–5.

[18] Pugh TJ, Morozova O, Attiyeh EF, Asgharzadeh S, Wei JS, Auclair D, et al. The genetic landscape of high-risk neuroblastoma. Nat Genet 2013;45:279–84.

[19] Gripp KW, Lin AE. Costello syndrome: a Ras/mitogen activated protein kinase pathway syndrome (rasopathy) resulting from HRAS germline mutations. Genet Med 2012;14:285–92.

[20] Li FP, Fraumeni Jr JF. Rhabdomyosarcoma in children: epidemiologic study and identification of a familial cancer syndrome. J Natl Cancer Inst 1969;43:1365–73.

[21] Tinat J, Bougeard G, Baert-Desurmont S, Vasseur S, Martin C, Bouvignies E, et al. 2009 version of the Chompret criteria for Li–Fraumeni syndrome. J Clin Oncol 2009;27:e108–9.

[22] Malkin D, Li FP, Strong LC, Fraumeni JF, Nelson CE, Kim DH, et al. Germ line p53 mutations in a familial syndrome of breast cancer, sarcomas, and other neoplasms. Science 1990;250: 1233–8.

[23] Srivastava S, Zou ZQ, Pirollo K, Blattner W, Chang EH. Germline transmission of a mutated p53 gene in a cancer-prone family with Li–Fraumeni syndrome. Nature 1990;348:747.

[24] Figueiredo BC, Sandrini R, Zambetti GP, Pereira RM, Cheng C, Liu W, et al. Penetrance of adrenocortical tumours associated with the germline TP53 R337H mutation. J Med Genet 2006;43:91–6.

[25] Ribeiro RC, Sandrini F, Figueiredo B, Zambetti GP, Michalkiewicz E, Lafferty AR, et al. An inherited p53 mutation that contributes in a tissue-specific manner to pediatric adrenal cortical carcinoma. Proc Natl Acad Sci USA 2001;98:9330–5.

[26] Achatz MI, Olivier M, Le Calvez F, Martel-Planche G, Lopes A, Rossi BM, et al. The TP53 mutation, R337H, is associated with Li-Fraumeni and Li-Fraumeni-like syndromes in Brazilian families. Cancer Lett 2007;245:96–102.

[27] DiGiammarino EL, Lee AS, Cadwell C, Zhang W, Bothner B, Ribeiro RC, et al. A novel mechanism of tumorigenesis involving pH-dependent destabilization of a mutant p53 tetramer. Nat Struct Biol 2002;9:12–6.

[28] Pinto EM, Billerbeck AE, Villares MC, Domenice S, Mendonça BB, Latronico AC. Founder effect for the highly prevalent R337H mutation of tumor suppressor p53 in Brazilian patients with adrenocortical tumors. Arq Bras Endocrinol Metabol 2004;48:647–50.

[29] Bougeard G, Baert-Desermont S, Tournier I, Vasseur S, Martin C, Brugieres L, et al. Impact of the MDM2 SNP 309 and p53 Arg72Pro polymorphism on age of onset in Li–Fraumeni syndrome. J Med Genet 2006;43:531–3.

[30] Shlien A, Marshall CR, Pienkowska M, Pienkowska M, Feuk L, Novokmet A, et al. Excessive genomic DNA copy number variation in the Li–Fraumeni cancer predisposition syndrome. Proc Natl Acad USA 2008;32:11264–9.

[31] Marcel V, Palmero EI, Falagan-Lotsch P, Martel-Planche G, Ashton-Prolla P, Olivier M, et al. TP53 PIN3 and SNP 309 polymorphisms as genetic modifiers in the Li–Fraumeni syndrome: impact on age at first diagnosis. J Med Genet 2009;46:766–72.

[32] Rausch T, Jones DT, Zapatka M, Stütz AM, Zichner T, Weischenfeldt J, et al. Genome sequencing of pediatric medulloblastoma links catastrophic DNA rearrangements with TP53 mutations. Cell 2012;148:59–71.

[33] Holmfeldt L, Wei L, Diaz-Flores E, Walsh M, Zhang J, Ding L, et al. The genomic landscape of hypodiploid acute lymphoblastic leukemia. Nat Genet 2013;45:242–52.

[34] Masciari S, Van den Abbeele AD, Diller LR, Rastarhuyeva I, Yap J, Schneider K, et al. F18-fluorodeoxyglucose-positron emission tomography/computed tomography screening in Li–Fraumeni syndrome. J Am Med Assoc 2008;299:1315–9.

[35] Villani A, Tabori U, Schiffman J, Shlien A, Beyene J, Druker H, et al. Biochemical and imaging surveillance in TP53 mutation carriers with Li–Fraumeni syndrome: a prospective observational study. Lancet Oncol 2011;12:559–67.

[36] Giminez-Roqueplo AP, Dahia PL, Robledo M. An update on the genetics of paraganglioma, pheochromocytoma and associated hereditary syndromes. Hormon Metab Res 2012;44:328–33.

[37] Benn DE, Giminez-Roqueplo AP, Reilly JR, Bertherat J, Burgess J, Byth K, et al. Clinical presentation and penetrance of pheochromocytoma/paraganglioma syndromes. J Clin Endocrinol Metab 2006;91:827–36.

[38] Gimenez-Roqueplo AP, Tischler AS. Pheochromocytoma and paraganglioma: progress on all fronts. Endocr Pathol 2012;23:1–3.

[39] Pasini B, Stratakis CA. SDH mutations in tumorigenesis and inherited endocrine tumors: lesson from the pheochromocytoma-paraganglioma syndromes. J Intern Med 2009;266:19–42.

[40] Hao HX, Khalimonchuk O, Shraders M, Dephoure N, Bayley JP, Kunst H, et al. SDH5, a gene required for flavination of succinate dehydrogenase is mutated in paraganglioma. Science 2009;325: 1139–42.

[41] Eisenhofer G, Lenders JW, Timmers H, Mannelli M, Grebe SK, Hofbauer LC, et al. Measurements of plasma methoxytyramine, normetanephrine, and metanephrine as discriminators of different hereditary forms of pheochromocytoma. Clin Chem 2011;57:411–20.

[42] Gimenez-Roqueplo AP, Caumont-Prim A, Houzard C, Hignette C, Hernigou A, Halimi P, et al. Imaging work-up for screening of paraganglioma and pheochromocytoma in SDHx mutation carriers: a multicenter prospective study from the PGL.EVA Investigators. J Clin Endocrinol Metab 2013;98:E162–73.

[43] Choong SC, Priest JR, Foulkes WD. Exploring the endocrine manifestations of DICER1 mutations. Trends Mol Med 2012;18:503–5.

[44] Bahubeshi A, Tischkowitz M, Foulkes WD. miRNA processing and human cancer: DICER1 cuts the mustard. Sci Transl Med 2011;3:111.

[45] Chandrasekharappa SC, Guru SC, Manickam P, Olufemi SE, Collins FS, Emmert-Buck MR, et al. Positional cloning of the gene for multiple endocrine neoplasia-type 1. Science 1997;276:404–7.

[46] Mannens M, Hoovers JM, Redeker E, Verjaal M, Feinberg AP, Little P, et al. Parental imprinting of human chromosome region 11p15.3-pter involved in the Beckwith–Wiedemann syndrome and various human neoplasia. Eur J Hum Genet 1994;2:3–23.

[47] Reik W, Brown KW, Schneid Y, Le Bouc Y, Bickmore W, Maher ER. Imprinting mutations in the Beckwith–Wiedemann syndrome suggested by altered imprinting pattern in the IGF2-H19 domain. Hum Mol Genet 1995;4:2379–85.

[48] Gorlin RJ. Nevoid basal-cell carcinoma syndrome. Medicine (Baltimore) 1987;66:98–113.

[49] Cowan R, Hoban P, Kelsey A, Birch JM, Gattamaneni R, Evans DG. The gene for the naevoid basal cell carcinoma syndrome acts as a tumour-suppressor gene in medulloblastoma. Br J Cancer 1997;76:141–5.

[50] Statement of the American Society of Clinical Oncology: genetic testing for cancer susceptibility. Adopted on February 20. J Clin Oncol 1996; 14:1730.

Part 4

Genomics of Adult and Pediatric Cancers

4.1 Adult Solid Tumors

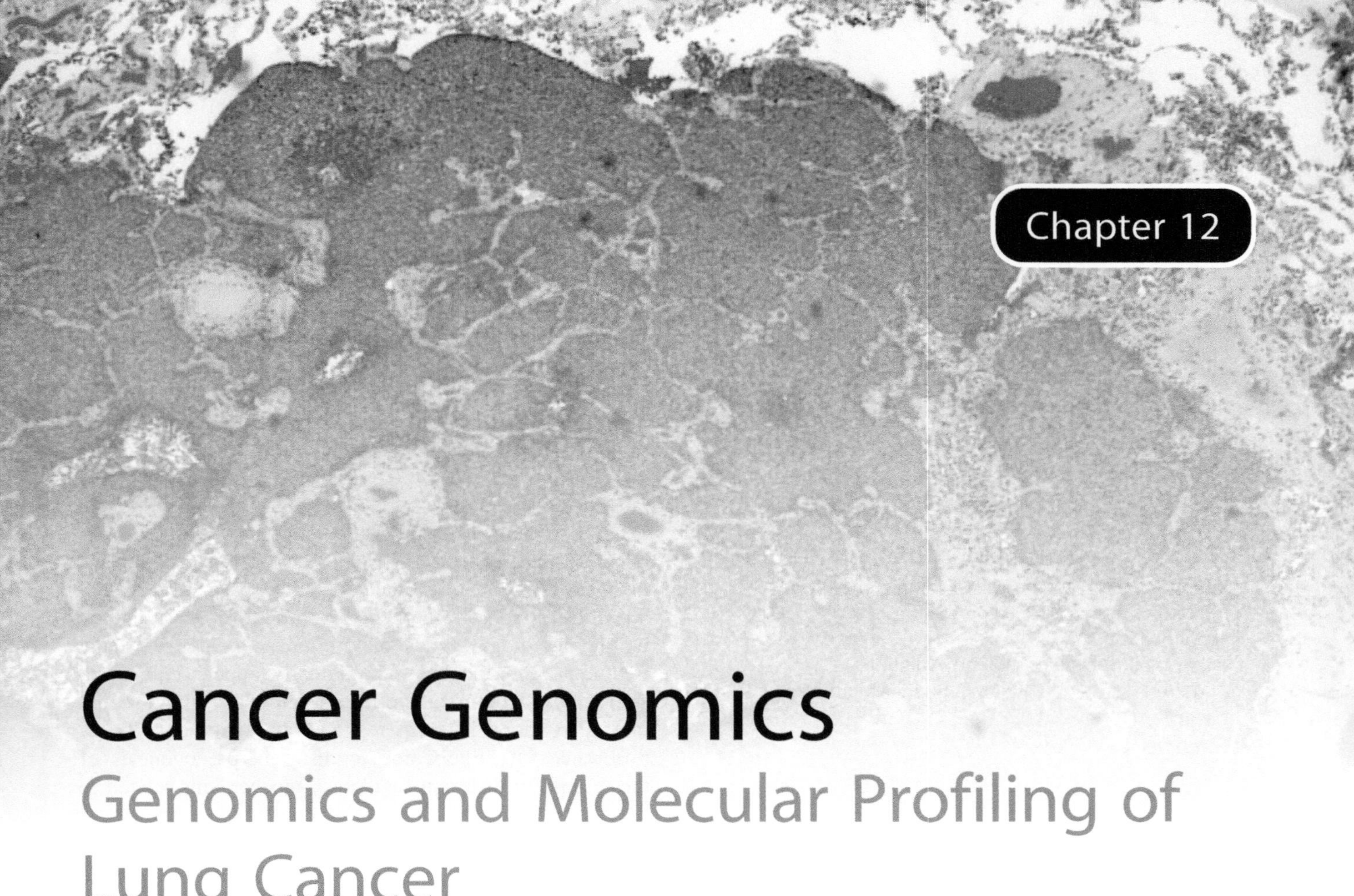

Cancer Genomics

Genomics and Molecular Profiling of Lung Cancer

Zhaolin Xu

Department of Pathology, Dalhousie University and Queen Elizabeth II Health Sciences Centre, Halifax NS, Canada

Contents

Cancer Genomics. DOI: http://dx.doi.org/10.1016/B978-0-12-396967-5.00012-8

Key Concepts

- Lung cancer is the leading cause of cancer-related death in both males and females worldwide and prognosis is poor. Overall, the 5-year survival rate is only about 16% which has not changed significantly in the past several decades
- Most lung cancer patients are in a locally advanced or late stage of the disease at the time of diagnosis and the prognosis by conventional treatment is often disappointing
- Approximately half of lung adenocarcinomas have identifiable gene mutations such as *KRAS, EFGR, ALK, BRAF, PI3KCA, Her-2, MET* etc. Most of these genes are considered as driving genes and mutually exclusive
- Targeting these mutant genes or gene products using specific agents has demonstrated effective responses, bringing a new direction of cancer treatment, i.e. personalized targeted therapy
- It is essential to identify the gene mutations in lung cancer tissue in order to achieve an efficient response from personalized targeted therapy

INTRODUCTION

Lung cancer is the most prevalent life-threatening cancer and remains the leading cause of cancer-related death worldwide [1,2]. The majority (about 85%) of lung cancers are non-small cell lung cancer (NSCLC) and the remaining are small cell carcinoma [3,4]. NSCLC is a group of lung cancers other than small cell carcinoma which includes adenocarcinoma, squamous cell carcinoma, large cell carcinoma and several others. Surgery is considered as one of the most effective treatments for NSCLC. However, most patients with NSCLC (about 70–80%) present with locally advanced or distant metastasis at the time of diagnosis and are usually not candidates for surgical treatment. For these patients, palliative chemotherapy and/or radiotherapy becomes the main treatment of choice. Although chemotherapy/radiotherapy may improve cancer-related symptoms, the prognosis is still very poor. An overall 5-year survival rate for all stages of lung cancer, regardless of subtypes, is about 16% (5-year survival for small cell carcinoma is only 6% and for NSCLC 17%) [2,5]. The survival rates have not significantly changed in the past several decades. Although refinements of therapy based upon unique histologic types may improve response to the treatment, tumors that histologically appear equivalent sometimes demonstrate different clinical behavior and may also show different response to treatments. These differences are likely due to different pathogenesis of the disease or different regulatory mechanisms related to the tumor growth. The pathogenesis of lung cancer involves alteration and accumulation of molecular abnormalities over time. The abnormalities can occur at different levels including the DNA sequence, gene transcripts, protein expression and cell signaling involved in cell proliferation, differentiation, apoptosis, and angiogenesis. These changes are related to alterations in normal genes due to mutations, deletions, amplifications or silencing. Recent major developments in the understanding of the molecular pathogenesis of lung cancer have led to new strategies for early detection, molecular profiling, and personalized targeted therapy. Detailed analysis of lung cancer at the molecular level provides the basis for a totally new direction of treatment which has resulted in much higher response rates, longer progression-free survival (PFS) and improved quality of life. Gene mutations/alterations and protein expressions identified by various methods such as DNA sequencing, reverse transcriptase polymerase chain reaction (RT-PCR), fluoresence *in situ* hybridization (FISH) and immunohistochemistry (IHC) have been shown to be useful in classifying tumors and formulating prognostic or predictive factors in various types of lung cancer. Using high-throughput analysis in gene and protein expression levels has made it possible to identify biomarkers as potential targets for personalized targeted therapy. Therefore, the treatment options in modern oncology have included tumor marker-based targeted therapies in addition to conventional methods. It is also crucial to keep seeking new tumor markers in order to improve further the clinical outcome of lung cancer through personalized targeted therapy.

EPIDEMIOLOGY

Lung cancer is the most common cancer in males worldwide (excluding skin cancer) followed by prostate cancer (more common in developed countries) and stomach cancer (particularly in developing countries), and it has been the most common cancer diagnosed each year since 1985 [2]. Globally, it is the fourth most common cancer in females behind breast cancer, cervical cancer, and colorectal cancer. About 1.35 million people were diagnosed with lung cancer worldwide in 2002 (12% of all invasive cancers) and, among those, 71% were males and 29% females [2]. The incidence rates are higher in more developed countries than in developing countries (61/100 000 vs 29/100 000 in males and 19/100 000 vs 10/100 000 in females) [5]. In North America, many European countries, Australia and New Zealand, the incidence rates of lung cancer in males reached a peak in the mid-1980s and have slightly declined since then [6–8]. However, among males in many Eastern European countries, and countries in southeastern Asia including Japan and China, and among females in most developed countries, the incidence rates of lung cancer either continue to increase or have recently reached a plateau [7–9]. Because of a long latency period, the patterns of the incidence closely reflect smoking prevalence from 20 to 30 years earlier [10–13]. It is estimated that about 7% of men and women, i.e. 1 in 14, will develop lung cancer during their lifetime.

Lung cancer is the leading cause of cancer-related deaths in males, and is ranked second among females behind breast cancer worldwide. In North America, some European countries and a number of other countries, lung cancer has overtaken breast cancer as the leading cause of cancer death among females [2]. The mortality of lung cancer accounted for about 18% of total cancer-related deaths, and 2% of all causes of deaths worldwide [14]. Lung cancer causes more deaths than the next three most common cancers combined (colon, breast and prostate cancers). Approximately 70% of lung cancer deaths occurred in males. The average age-standardized mortality rate of lung cancer, similar to the incidence, was about twice as high in more developed countries than in less developed countries (53/100 000 vs 26/100 000 among males and 15/100 000 vs 9/100 000 among females in 2002) [5]. It was estimated that more than 70% of lung cancer deaths worldwide were due to cigarette smoking [15]. The figure is even higher in more developed countries, reaching about 90% among males and 70% in females. In developing countries, as a contrast, about two thirds of lung cancer deaths in males and one fourth in females were due to cigarette smoking. Mortality rates in males have been decreasing over the last two to three decades worldwide. However, mortality rates in females are still increasing in most countries.

STANDARD TREATMENT OF LUNG CANCER AND PROGNOSIS

Following is a list of the treatments of choice for lung cancer in current clinical practice. However, the availability of the choices varies in each cancer center.

Surgery

Surgical removal of tumor tissue is an effective treatment for lung cancer. However, only a small percentage of patients have surgical indications because most patients are at an advanced stage of the disease at the time of diagnosis [16]. There are four common types of surgery used to treat lung cancer:

- Lobectomy: This is a standard surgical treatment for lung cancer to excise a whole lobe of the lung where the tumor is located
- Pneumonectomy: This is a surgical treatment to excise one lung for selected patients whose lung cancer cannot be resected adequately or safely by lobectomy
- Wedge resection: A type of surgery to remove a tumor and surrounding normal lung tissue for selected patients who cannot tolerate lobectomy or larger surgeries for a variety of reasons. Resection of a segment of a lobe is called a segmental resection
- Sleeve resection: This is a type of surgery to remove a lung tumor including a portion of proximal bronchus in an adjacent lobe to give a safe bronchial resection margin.

Radiation Therapy

High-energy X-rays or other types of radiation are used to kill cancer cells or keep them from growing. Two types of radiation therapy are commonly used in clinical practice: (1) external beam radiation therapy – which uses a machine external to the body to send radiation toward the cancer; (2) internal radiation therapy – whereby a radioactive substance is placed directly into or near the cancer. The former method is more commonly used than the latter for lung cancer patients.

Certain types of lung cancer are sensitive to radiation therapy such as small cell carcinoma. Radiation therapy also can be given as palliative treatment.

Chemotherapy

Chemotherapy is a type of treatment that uses drugs to stop or slow down the growth of cancer cells, either by killing the cells or by stopping them from dividing. Certain types of lung cancer are sensitive to chemotherapy such as small cell carcinoma. Chemotherapy can also be used as palliative, neoadjuvant (chemotherapy before surgery) or adjuvant (chemotherapy after surgery) therapy. Chemoprevention is a type of treatment using drugs to reduce the risk of developing cancer or to prevent cancer from recurrence.

Personalized Targeted Therapy

This is a type of treatment using drugs or other substances that recognize cancer cells with specific markers and attack these cancer cells without harming normal cells. Monoclonal antibodies, tyrosine kinase inhibitors and other molecules are used as targeted therapy for treatment of NSCLC. Monoclonal antibodies attach and kill the cancer cells, block their growth, or prevent them from spreading. For example, cetuximab binds to the epidermal growth factor receptor (EGFR) and prevents cancer cells from growing and dividing. Bevacizumab binds to the vascular endothelial growth factor (VEGF) and blocks the growth of new blood vessels within the tumor inhibiting tumor growth. Tyrosine kinase inhibitors block intercellular signals needed for tumors to grow. Several new drugs are available such as erlotinib and gefitinib that target *EGFR* and crizotinib, which targets the *EML4-ALK* fusion gene found in some cases of NSCLC.

Laser Therapy

This is a type of treatment which uses a laser beam to kill cancer cells.

Photodynamic Therapy

This therapy employs both non-toxic light-sensitive compounds (photosensitizers) and a certain type of laser light to kill cancer cells. The photosensitizer accumulates in cancer cells and becomes active when it is exposed to laser light. Fiberoptic tubes are then used to carry the laser light to the cancer cells, where the drug becomes active and kills the cells.

Cryotherapy

This type of treatment freezes and destroys neoplastic tissue, such as carcinoma *in situ*.

Electrocautery

This approach employs a probe or needle heated by an electric current to destroy abnormal tissue.

The prognosis of lung cancer in general is poor (Table 12.1) and is related to the clinical stage of the disease which is determined by the status of the primary tumor, lymph node involvement and distant metastasis (Table 12.2).

TABLE 12.1 5-Year Survival Rates for Lung Cancer in the USA from 2002 to 2008 [17]

Stage at Diagnosis	Stage Distribution (%)	5-Year Relative Survival (%)
Localized (cancer confined to primary site)	15	52.2
Regional (spread to regional lymph nodes)	22	25.1
Distant metastasis	56	3.7
Unknown (unstaged)	6	7.9

SOMATIC GENE MUTATIONS IN LUNG CANCER

Pathogenesis of Lung Cancer

The pathogenesis of lung cancer is a complicated process and is multifactorial. Among many influential factors, gene mutations are considered as driving forces responsible for the development of lung cancer. Changes in certain genes such as proinflammatory interleukin-8 and some DNA-repair genes observed in non-neoplastic lung tissue of smokers and also in lung cancer patients imply that the alterations in these genes probably precede epithelial clonal evolution and eventually lead to lung cancer development. Loss of heterozygosity of tumor suppressor genes at chromosomal region 3p21.3 (a member of the Ras association domain family), 3p14.2 (a fragile histidine triad gene), 9p21 (p16), and 17p13 (p53) are related to early events in the development of NSCLC [19]. While *KRAS* mutations are believed to be related to the development of smoking-related adenocarcinoma, mutations in the *EGFR* kinase domain occur early in the development of adenocarcinoma and are usually unrelated to smoking [20,21]. Many gene mutations or alterations such as *p53* mutations, *EGFR* mutations, aberrant methylation of promoter regions of genes, loss of heterozygosity, microsatellite instability and c-*Myc* amplification may also appear in normal lung tissue near lung cancers that likely increase the risks of recurrence and second primary tumors [19,21,22]. Methylation of *p16* is related to the development of squamous cell lung cancer and occurs early in the disease process such as in premalignant squamous cell lung lesions (e.g. dysplasia). In contrast, *p16* methylation occurs very rarely in adenocarcinoma

TABLE 12.2 TNM Staging System for Lung Cancer (7th Edition) [18]

Primary tumor			
T1	Tumor ≤3 cm diameter, surrounded by lung or visceral pleura, without invasion more proximal than lobar bronchus		
T1a	Tumor ≤2 cm in diameter		
T1b	Tumor >2 cm but ≤3 cm in diameter		
T2	Tumor >3 cm but ≤7 cm, or tumor with any of the following features: Involves main bronchus, ≥2 cm distal to carina Invades visceral pleura Associated with atelectasis or obstructive pneumonitis that extends to the hilar region but does not involve the entire lung		
T2a	Tumor >3 cm but ≤5 cm		
T2b	Tumor >5 cm but ≤7 cm		
T3	Tumor >7 cm or any of the following: Directly invades any of the following: chest wall, diaphragm, phrenic nerve, mediastinal pleura, parietal pericardium, main bronchus <2 cm from carina (without involvement of carina) Atelectasis or obstructive pneumonitis of the entire lung Separate tumor nodules in the same lobe		
T4	Tumor of any size that invades the mediastinum, heart, great vessels, trachea, recurrent laryngeal nerve, esophagus, vertebral body, carina, or with separate tumor nodules in a different ipsilateral lobe		
Regional lymph nodes (N)			
N0	No regional lymph node metastases		
N1	Metastasis in ipsilateral peribronchial and/or ipsilateral hilar lymph nodes and intrapulmonary nodes, including involvement by direct extension		
N2	Metastasis in ipsilateral mediastinal and/or subcarinal lymph node(s)		
N3	Metastasis in contralateral mediastinal, contralateral hilar, ipsilateral or contralateral scalene, or supraclavicular lymph node(s)		
Distant metastasis (M)			
M0	No distant metastasis		
M1	Distant metastasis		
M1a	Separate tumor nodule(s) in a contralateral lobe; tumor with pleural nodules or malignant pleural or pericardial effusion		
M1b	Distant metastasis (in extrathoracic organs)		
Stage groupings			
Stage IA	T1a–T1b	N0	M0
Stage IB	T2a	N0	M0
Stage IIA	T1a, T1b, T2a	N1	M0
	T2b	N0	M0
Stage IIB	T2b	N1	M0
	T3	N0	M0
Stage IIIA	T1a, T1b, T2a, T2b	N2	M0
	T3	N1, N2	M0
	T4	N0, N1	M0
Stage IIIB	T4	N2	M0
	Any T	N3	M0
Stage IV	Any T	Any N	M1a or M1b

[23,24]. Methylated *ASC-TMS1* is associated with the recurrence of lung cancer [25,26]. Some data have indicated that promoter methylation of various genes in stage I NSCLC was associated with recurrence after resection [27,28]. The *KRAS*, *PTEN*, phosphoinositide 3-kinase (PI3K), and cyclin-dependent kinase pathways have been implicated in the proliferation of bronchoalveolar stem cells and the latter could serve as the precursors of lung cancer cells [29,30]. However, the potential role of the stem cells in the development of lung cancer needs further clarification.

An association between single nucleotide polymorphism (SNP) variation at 15q24–15q25 and susceptibility to lung cancer has been identified recently [31]. The region of the SNP variation includes two genes encoding subunits of the nicotinic acetylcholine receptor alpha regulated by nicotine exposure [32,33]. Germline alterations in nucleotide excision repair genes, such as *ERCC1*, also increase lung cancer susceptibility, especially when accompanied by smoking [34,35]. On the other hand, increased expression of DNA synthesis and repair genes, including *RRM1* (the regulatory subunit of ribonucleotide reductase) and *ERCC1*, correlate with better prognosis in NSCLC, but no benefit from platinum-based chemotherapy [36,37].

Molecular Profiling in Lung Cancer

Lung cancer patients with a similar clinical stage and tumor histology sometimes have quite different responses to treatment and prognosis, which is likely due to molecular heterogeneity and the presence of various gene mutations. Lung cancer heterogeneity can be measured at a global level by DNA microarray techniques that profile the expressions of tens of thousands of genes simultaneously from tumor tissue. The technology was initially developed in 1995 for the analysis of cDNAs and then applied to the analysis of oligonucleotides in 1997 [38,39]. Microarray technology has been used for the global analysis of a large number of histologically diverse lung cancer specimens since 2001 [40]. Gene profiling can provide valuable information to guide the direction of the treatment, to predict tumor response to therapy and to improve the clinical outcome through personalized targeted therapy. In addition to gene profiling, microRNAs have been identified as important regulators of gene expression. MicroRNA expression is commonly deregulated in lung cancer and microRNA signatures associated with clinical outcome have been identified. Some microRNA profiling studies have indicated that there was a greater than 80% reduction in let-7 expression levels in 44–60% of NSCLC cells compared to normal lung epithelial cells and reduced expression of let-7 was associated with poor prognosis [41,42]. Loss of microRNA-128b, located on chromosome 3p, has been shown to correlate with the response to *EGFR* tyrosine kinase inhibitions in lung cancer patients [43]. Gene mutations, tumor-specific DNA alterations (such as loss of heterozygosity) and circulating tumor cells can also be identified in the peripheral blood of lung cancer patients. Relevant circulating DNA from lung cancer cells can be detected in the plasma or serum of patients, and the levels of DNA are associated with clinical outcomes. Some studies indicated that a decline in the number of circulating tumor cells was associated with response to treatment. Therefore, blood molecular profiling is also able to provide useful information with clinical implications.

Most targeted therapeutic agents are designed to target the activity of specific proteins such as tyrosine kinases. However, genomic profiling and mRNA level may not always correlate with protein expression and may not provide information on protein–protein interactions or post-translational modifications such as phosphorylation. Therefore, understanding the complexity of the protein signaling network and protein-based profiling is critical in developing effective targeted therapeutic agents. While IHC analysis remains the most widely used method for assessing individual proteins, high-throughput proteomic techniques, such as mass spectrometry and protein microarrays, make it possible to view overall signal transduction networks [44,45]. Serum mass spectrometry profiles may be able to distinguish lung cancer patients from normal controls and could predict survival differences among patients with adenocarcinoma [46,47]. Multiplexed analyses of dozens of cytokines and angiogenic factors can be achieved with a small amount of serum or plasma. One of the advantages of blood molecular profiling is that tumor response to the therapy could be evaluated by monitoring the protein levels repeatedly during treatment without the need of repeated tumor tissue biopsies.

COMMON GENE MUTATIONS AND CLINICAL SIGNIFICANCE

Lung cancer can be divided into a variety of subtypes based on their histologic appearance (Table 12.3) and a revised classification of lung adenocarcinoma has been proposed recently (Table 12.4). It is believed that all cancers are caused by or related to so-called driving gene(s) regardless of the histologic types. About one half of lung adenocarcinomas are found to have gene mutations which are often mutually exclusive. Following are some major gene mutations especially identified in lung adenocarcinoma.

KRAS

RAS genes were initially identified in the rat genome and their DNA sequences were also found in rat sarcoma-inducing retroviruses [50,51]. The *RAS* genes encode the

TABLE 12.3 2004 WHO Classification of Invasive Malignant Epithelial Lung Tumors [48]

Squamous cell carcinoma
Papillary
Clear cell
Small cell
Basaloid
Small cell carcinoma
Variant: combined small cell carcinoma
Adenocarcinoma
Adenocarcinoma, mixed subtype
Acinar adenocarcinoma
Papillary adenocarcinoma
Bronchioloalveolar carcinoma
Variants: non-mucinous mucinous, mixed non-mucinous and mucinous or indeterminate
Solid adenocarcinoma with mucin production
Variants:
Fetal adenocarcinoma
Mucinous ("colloid") carcinoma
Mucinous cystadenocarcinoma
Signet ring adenocarcinoma
Clear cell adenocarcinoma
Large cell carcinoma
Variants:
Large cell neuroendocrine carcinoma
Combined large cell neuroendocrine carcinoma
Basaloid carcinoma
Lymphoepithelioma-like carcinoma
Clear cell carcinoma
Large cell carcinoma with rhaboid phenotype
Adenosquamous carcinoma
Sarcomatoid carcinoma
Variants:
Pleomorphic carcinoma
Spindle cell carcinoma
Giant cell carcinoma
Carcinosarcoma
Pulmonary blastoma

(Continued)

TABLE 12.3 (Continued)

Carcinoid tumor
Typical carcinoid
Atypical carcinoid
Salivary gland tumors
Mucoepidermoid carcinoma
Adenoid cystic carcinoma
Epithelial-myoepithelial carcinoma

TABLE 12.4 International Association for the Study of Lung Cancer (IASLC), American Thoracic Society (ATS), European Respiratory Society (ERS) Classification of Lung Adenocarcinoma (2011) [49]

Preinvasive lesions
Atypical adenomatous hyperplasia
Adenocarcinoma in situ (≤3 cm, formerly BAC*)
Non-mucinous
Mucinous
Mixed mucinous/non-mucinous
Minimally invasive adenocarcinoma (≤3 cm lepidic predominant tumor with ≤5 mm invasion)
Non-mucinous
Mucinous
Mixed mucinous/non-mucinous
Invasive adenocarcinoma
Lepidic predominant (formerly non-mucinous BAC pattern, with >5 mm invasion)
Acinar predominant
Papillary predominant
Micropapillary predominant
Solid predominant with mucin production
Variants:
Invasive mucinous adenocarcinoma (formerly mucinous BAC)
Colloid
Fetal (low and high grade)
Enteric

**BAC: bronchioloalveolar carcinoma*

membrane-bound 21-kDa guanosine triphosphate (GTP)-binding proteins which are involved in several signaling pathways related to cell growth, differentiation, and apoptosis such as *MAPK* (mitogen-activated protein kinase), *PI3K* (phosphoinositide 3-kinase) and *STAT* (signal transducer and activator of transcription) signaling cascades. The human homologs of rat genes (Harvey ras or *HRAS*, Kirsten ras or *KRAS*, and neuroblastoma ras or *NRAS*) were also identified in the early 1980s [52,53]. About one third of all human cancers were found to have *RAS* mutations with *KRAS* mutation accounting for the majority and approximately 97% of *KRAS* mutations in lung cancer involve a point mutation in codons 12 or 13 [54,55]. The mutation results in replacement of an amino acid at position 12, 13, or 61 that causes impaired GTPase activity and constitutive activation of *RAS* signaling [54,55]. Interestingly, the vast majority of *RAS* mutations in lung adenocarcinomas are *KRAS* mutations which are uncommon in squamous cell carcinomas. The presence of *KRAS* mutations in lung cancer is a negative prognostic factor with a hazard ratio (HR) for death of 1.40 (95% confidence interval, 1.18−1.65) and an association with an overall poor clinical outcome [56]. In a randomized prospective clinical trial, patients with stage II−IIIA NSCLC were randomized to receive postoperative radiation therapy or chemoradiation therapy and *KRAS* mutations were evaluated as part of the study. In a total of 197 cases available for *KRAS* mutational analysis, the mutations were identified in 24% of the cases. A median survival in patients with wild-type *KRAS* was 42 months and in patients with *KRAS* mutations 25 months, with a risk ratio of 0.59 (wild type:mutant *KRAS*) [57]. However, univariate or multivariate analyses indicate that *KRAS* mutations apparently were not a significant prognostic factor for survival [58,59]. Tyrosine kinase inhibitors such as erlotinib or gefitinib are ineffective against *KRAS* mutant lung cancer. The response rate was lower in NSCLC patients treated with erlotinib plus chemotherapy than those treated with chemotherapy alone. No direct and effective tyrosine kinase inhibitors or other types of inhibitors are available at present time to target *KRAS* mutations.

EGFR

EGFR is a receptor tyrosine kinase and belongs to the HER/erbB family which includes *HER1* (EGFR/erbB1), *HER2* (neu, erbB2), *HER3* (erbB3), and *HER4* (erbB4). The *EGFR* gene is located on the short arm of chromosome 7 (7p11.2) and encodes a 170-kDa type I transmembrane growth factor receptor with tyrosine kinase activity. *EGFR* has an extracellular, cysteine-rich ligand-binding domain, a single α-helix transmembrane domain, a cytoplasmic tyrosine kinase domain and a carboxy-terminal signaling domain [60,61]. Once bound to ligand the tyrosine kinase becomes activated and results in autophosphorylation of the cytoplasmic domain of the receptor which, in turn, activates downstream signaling pathways such as the RAS-RAF-MEK-MAPK pathway, PI3K-AKT pathway, and signal transducer and activator of transcription (STAT) pathway. *EGFR* downstream signaling leads to cell proliferation, differentiation, decreased apoptosis, and angiogenesis [62,63]. Improper activation of the *EGFR* tyrosine kinase inhibits tumor cell apoptosis resulting in tumor progression and activated matrix metalloproteinases, which stimulate cell motility and invasion and promote metastasis [64,65]. Somatic mutations in the tyrosine kinase domain of *EGFR* can be classified into three types based on their nucleotide changes: deletions in exon 19 that result in the loss of four to six amino acids (E746 to S752), single-nucleotide point mutations occurring in exons 18, 20, 21, and insertions/duplications mostly in exon 20 [66]. About 85−90% of all *EGFR* mutations occur in exon 19 and exon 21 with a slightly higher frequency in the former (45%) than the latter (40−45%) [67].

EGFR mutations can be detected through direct DNA sequencing by polymerase chain reaction (PCR) to amplify genomic DNA that corresponds to exons 18 to 21 of the *EGFR* genes [68]. Direct sequencing may miss up to 30% of the mutations due to a variety of reasons, such as imbalanced PCR amplification, copy number of mutant alleles, and contamination of wild-type alleles of non-tumor cells [69]. Because of the relatively low sensitivity by direct sequencing, many other methods have been developed such as high-resolution melting analysis, peptide nucleic acid-locked clamping method, single-stranded conformational polymorphism and denaturing high-performance liquid chromatography analysis of PCR amplicons following digestion by CEL1 endonuclease, and co-amplification at lower denaturation temperature coupled with PCR [70,71].

The incidence of *EGFR* mutations varies geographically in patients with lung cancer. Some studies screened relatively large tumor samples and showed that the incidence of *EGFR* mutations is higher in the Asian population than non-Asian patients (32% vs 7%); higher in females than males (38% vs 10%); higher in never-smokers than smokers (47% vs 7%); and higher in adenocarcinoma than non-adenocarcinoma (30% vs 2%) [72].

ALK

Anaplastic lymphoma kinase (ALK) is a member of the insulin receptor family of receptor tyrosine kinases. Although *ALK* is normally expressed in the central nervous system, small intestine, and testis, translocations or rearrangement of the gene have been identified in several cancers including anaplastic large cell lymphoma, lung cancer, neuroblastoma, and inflammatory myofibroblastic tumor [73,74]. ALK gene translocation was initially identified in a subset of lymphoma in which there is chromosomal translocation of the ALK gene to fuse with nucleophosmin resulting in the expression of an oncogene that drives tumor growth [75]. A similar type of gene alteration was also found in NSCLC, especially lung adenocarcinoma with rearrangement of the gene on the short arm of chromosome 2 leading to the expression of a fusion protein between the N-terminal portion of echinoderm microtubule-associated protein-like 4 (EML4) and the C-terminal kinase domain of *ALK* [76]. This gene rearrangement results in aberrant activation of downstream oncogenic signaling pathways including MAP kinase, PI3 kinase, and STAT3-dependent pathways causing cell proliferation, stromal invasion, and apoptotic inhibition [77,78]. The ALK gene rearrangement in lung cancer was first detected by transfection of a cDNA expression library from NSCLC into 3T3 fibroblasts, resulting in selective transformation [77]. The EML4–ALK fusion gene is considered as a driver gene of tumorigenesis. The incidence rates vary from 1.6 to 11.6% with an average about 4–5% and its clinical and pathological characteristics are similar to those with EGFR mutations, i.e. more commonly seen in female patients, Asian ethnicity, light or never smoking history, and adenocarcinoma histology [79–81]. In a phase II study, 82 ALK-positive patients with 96% adenocarcinoma were found from approximately 1500 patients screened for ALK rearrangements [82]. Among the ALK-positive cases, 76% were non-smokers. The presence of the EML4–ALK gene rearrangement is mutually exclusive of other known driver gene mutations, including EGFR, KRAS, and BRAF. Some studies indicated that the *ALK* mutated tumors were more likely associated with "signet ring cell" histology, more prevalent in younger lung cancer patients, and less strongly associated with female gender than that in *EGFR* mutations [83]. There are several methods to detect *ALK* gene rearrangement including FISH using "breakapart" probes, which is considered the standard method; multiplexed RT-PCR to detect the fusion transcript; or IHC to detect aberrant *ALK* protein expression. RT-PCR and IHC seem to have less sensitivity than the FISH method. One study showed that only 20 of 29 patients who were *ALK* positive by FISH had the *EML4–ALK* fusion breakpoint confirmed by RT-PCR and 19 of 25 patients who were *ALK* positive by FISH showed *ALK* protein expression with positive IHC [82].

BRAF

BRAF is one of three members of the *RAF* kinase family: A-RAF, B-RAF, and RAF-1 (also known as c-RAF), which is a serine-threonine kinase that links *RAS* GTPases to downstream proteins of the *MAPK* family to control cell proliferation [84,85]. Somatic *BRAF* mutations were originally identified in melanomas, but are also present in other cancer types such as lung cancer, thyroid cancer and colon cancer. Some studies revealed that *BRAF* mutations were clustered within the P-loop (exon 11) and activation segment (exon 15) of the kinase domain. Among the various types of mutations, a single substitution of glutamic acid for valine at residue 600 (V600E, initially designated as V599E), which lies within the activation segment of the kinase domain, was observed in approximately 90% of cases with *BRAF* mutations [85,86]. *BRAF* mutations are found in 1–3% of NSCLC, most of which are adenocarcinomas [85,87]. Biologically, *BRAF* mutations are associated with increased kinase activity and lead to constitutive activation of *MAPK2* and *MAPK3*. The activating *BRAF* mutations can induce cell transformation and promote cell viability, cell proliferation, and tumorigenesis [85,87]. *BRAF* mutations are mutually exclusive to *KRAS, EGFR* and *ALK* mutations. One study showed that among 581 patients with NSCLC, five tumors (0.8%) had mutations in exon 15 and no mutation was discovered in exon 11 of the *BRAF* gene [88]. Two tumors had a point mutation of GTG to GAG at codon 600, which substituted glutamic acid for valine (V600E); two tumors had a point mutation of GAT to GGT at codon 594, which substituted glycine for aspartic acid (D594G); and one tumor had a point mutation without any amino acid substitution.

PIK3CA

PIK3CA encodes phosphatidylinositol 3-kinases (the *PI3K* protein family) which are lipid kinases that regenerate phosphatidylinositol-3-phosphate [89]. The latter is a key mediator between growth factor receptors and intracellular downstream signaling pathways. The main catalytic subunit of *PI3K* proteins is the p110α isoform. Mutations in this gene have been identified in many tumor types including 30% of glioblastomas and gastric cancers, and approximately 2% of NSCLCs [90]. Mutations of *PIK3CA* in NSCLC most frequently affect residues Glu542 and Glu545 in exon 9, which encode the catalytic domain [87,90]. *PIK3CA* mutations can occur in both squamous cell carcinoma and adenocarcinoma in a similar frequency [91]. In contrast to many other gene mutations which are mutually exclusive, *PIK3CA* mutations can be concurrent with other gene mutations including *EGFR*-mutated tumors [91]. *PIK3CA* mutations lead to activation of the AKT (protein kinase B) signaling pathway resulting in oncogenic cellular transformation and cell proliferation. *PIK3CA* amplification is more likely seen in male patients, in smokers, and in squamous cell carcinoma [92,93]. However, the oncogenic role of *PIK3CA* amplification in lung cancer is not quite clear.

HER2

HER2 protein (also known as ERBB-2) is a member of the *HER* family of receptor tyrosine kinases which forms homodimers or heterodimers with other members of the *HER* family [94]. *HER2* gene mutations occur in approximately 2% of NSCLC but *HER2* expression occurs in about 20% of the cases [95,96]. The mutations are mainly related to inframe insertions in exon 20, mostly involving the amino acid sequence Tyr-Val-Met-Ala at codon 776 which leads to constitutive activation of the receptor [96]. The clinical and pathologic characters of *HER2* mutations are similar to *EGFR* and other gene mutations, namely they are more commonly seen in people who have never smoked, female, and Asian patients, and are more often present in adenocarcinoma than other types of lung cancer [96–98]. *HER2* mutations are mutually exclusive with other gene mutations such as *EGFR*, *KRAS* and *ALK* in lung cancer.

MET

The *MET* gene is located on chromosome 7q21-q31 and encodes hepatocyte growth factor receptor (HGFR) which is a receptor tyrosine kinase [99]. Amplification of this gene is noted to be associated with secondary resistance to *EGFR* tyrosine kinase inhibitors through a kinase switch independent from *KRAS* and *EGFR* mutations. *MET* gene amplification is seen in 1.4–21.0% of patients with NSCLC not previously treated with *EGFR*-specific tyrosine kinase inhibitors and approximately 20% of tumors from patients with acquired resistance [100–103]. The differences in the percentage of patients with *MET* gene amplification might be related to differences in methods and cut-off values used to assess and define the amplification. *MET* gene amplification occurs in both squamous cell carcinoma and adenocarcinoma [102]. Some studies showed that *MET* amplification was associated with increased concentrations of phosphorylated *HGFR* that initiated tumorigenesis and led to cell proliferation, survival, motility and invasion, epithelial-mesenchymal transition and branched morphogenesis [100]. Small-hairpin RNA-mediated knock down of *HGFR* expression induced apoptosis and growth inhibition in cells with *MET* amplification [104]. These data suggest that *MET* amplification is necessary and sufficient to drive the proliferation of cancer cells. Mutations in the kinase domain of *MET* are uncommon in lung cancer compared to renal and gastric carcinomas. In a cohort of 188 lung adenocarcinomas, only three somatic *MET* mutations were identified, two in exon 13 encoding the juxtamembrane domain (Arg988del and Tyr1021Asn) and one in exon 18 encoding the kinase domain (Gly1260Cys) [104]. In addition, an intronic splice variant with exon 14 deletions has been reported in 2–3% of NSCLC tumors in Japanese cohorts [101,105]. However, tumor microarray expression analysis of *MET/HGF* in human cancers demonstrated that both *MET* and *HGF* are commonly expressed in human solid cancers, including lung cancer. One study showed that 72% (29 of 40) of lung cancers were found to express *MET*, and 40% (16 of 40) had *MET* receptor overexpression. Moreover, phospho-*MET* expression is found to be the highest in lung cancer, followed by ovarian, breast, renal, and colon cancers [106]. Although the biological importance of these mutations is uncertain, some studies have indicated that mutations of the *MET* gene resulted in delayed receptor downregulation and increased ligand-mediated proliferation [105], while amplification of *HGFR* correlated with a poor prognosis in surgically resected lung cancer [103].

PERSONALIZED TARGETED THERAPY IN LUNG CANCER

Selection of Targets

The first breakthrough towards personalized treatment happened in 2004 when the activating mutations in the tyrosine kinase domain of *EGFR* were identified in patients who had a dramatic response to the *EGFR* tyrosine kinase inhibitor, gefitinib [107,108]. Since then, personalized targeted therapy for lung cancer has become a focus of interest. Personalized targeted therapy is based on the identification of molecular abnormalities, especially gene mutations, in the cancer cells that ideally are the drivers in the development of disease. The targets for personalized targeted therapy should be selected for those mutated genes, with a goal that inhibition of mutated gene expression would halt cancer cell growth and progression. The key is to identify the specific driver genes for targeting as a number of genomic abnormalities may occur in a cancer cell but very few of these are appropriate for personalized targeted therapy [87]. This helps to explain why the number of successful clinical trials of molecular-targeted therapies in lung cancer is far fewer than the number of unsuccessful trials. The best way to evaluate a successful personalized targeted therapy is through a clinical trial to demonstrate a higher tumor response rate, prolonged PFS, or a better clinical outcome with improved overall survival in lung cancer patients whose tumors harbor the targeted mutation. By contrast, similar results would not be expected in patients without the mutation.

EGFR Tyrosine Kinase Inhibitors

Activating mutations in *EGFR* are the best-known target in NSCLC. There are two major classes of agents to inhibit *EGFR* activity, monoclonal antibodies and tyrosine kinase inhibitors. Cetuximab is an example of humanized mouse monoclonal antibody developed against the ligand-binding domain of *EGFR* as a competitive antagonist. Gefitinib (Iressa) and erlotinib (Tarceva) are tyrosine kinase inhibitors approved for the treatment of advanced or metastatic NSCLC. These agents have higher binding affinity for mutant *EGFR* than for wild type. They reversibly and specifically inhibit *EGFR* downstream signaling by binding to the ATP site on the kinase domain of the receptor. Although in an initial study using a tyrosine kinase inhibitor, gefitinib, to target *EGFR*, the tumor response rates in unselected patients were below 20%, which was inferior to standard first-line chemotherapy [109], several subsequent single-arm clinical studies have demonstrated that the tumor response rates to *EGFR* tyrosine kinase inhibitors in patients with *EGFR*-mutant tumors ranged from 55 to 78% [110–114].

The Iressa™ Pan-Asia Study (IPASS) was the first comparative study to confirm that gefitinib is superior to standard chemotherapy [114]. In this study, patients with lung adenocarcinoma who were of Asian ancestry and light or never-smokers were randomized to receive either gefitinib (250 mg daily) alone or paclitaxel (200 mg/m^2) plus carboplatin (5 times AUC). Among 437 tumor samples tested for *EGFR* mutations, 261 (59.7%) had positive results. Response rates in patients with *EGFR* mutant tumors were 71.2% in the gefitinib arm and 47.3% in the chemotherapy arm. Gefitinib significantly improved progression-free survival in patients with *EGFR*-mutant tumors (hazard ratio (HR) = 0.48, $p < 0.0001$) [114]. The importance of biomarker selection and personalized targeted therapy is further supported by other randomized studies in lung cancer patients with *EGFR* mutations to receive either *EGFR* tyrosine kinase inhibitors or chemotherapy. Higher response rates and longer PFS have again been demonstrated in the patients treated with *EGFR* tyrosine kinase inhibitors in these studies to confirm *EGFR* tyrosine kinase inhibitors to be a superior treatment of choice for advanced lung cancer patients with activating mutations in *EGFR*. The ISEL35 and BR-21 studies compared gefitinib and erlotinib, respectively, in lung cancer patients who had failed to respond to one or more chemotherapy regimens [115,116]. Both studies used *EGFR* protein expression and *EGFR* gene-copy number as biomarkers but, interestingly, had different results on the predictive value of the biomarkers. In the ISEL study, both biomarkers seemed to be marginally predictive (interaction test $p = 0.049$ and 0.045, respectively) but the biomarkers were not predictive in the BR-21 study (interaction test $p = 0.33$ and 0.12, respectively). Since the number of patients who harbored an *EGFR* mutation was small (26 in ISEL and 15 in BR-21), the results may not be meaningful. The INTEREST study compared gefitinib with docetaxel as second-line and third-line therapy, and none of the three biomarkers (*EGFR* protein expression, *EGFR* gene-copy number, *EGFR* mutations) were predictive of survival [117], which is probably also due to the small number of patients with *EGFR* mutations (44 patients in total). Conversely, in the IPASS study, *EGFR* mutations were an obvious predictive biomarker for both objective tumor response to gefinitib treatment and progression-free survival ($p < 0.001$), but not for overall survival ($p = 0.99$) [118]. There was also a significant difference between gefitinib and chemotherapy in patients with high *EGFR* gene-copy number which seemed to be predictive of PFS (HR = 0.66; 95% CI 0.50–0.88; $p = 0.005$). The predictive power of high *EGFR* gene-copy number could be explained by the concurrent presence of *EGFR* mutations which accounted for 78% of patients with high *EGFR* gene-copy number. By contrast, in patients with low *EGFR* gene-copy number, only 33% had *EGFR* mutations. Patients with both high gene-copy number and mutations in *EGFR* treated with gefitinib had a longer PFS (HR = 0.48; 95% CI 0.34–0.67), while chemotherapy was superior to gefitinib in patients with high gene-copy number without *EGFR* mutations (HR = 3.85; 95% CI 2.09–7.09) [118]. *EGFR* mutations should be the predictive biomarker for first-line *EGFR* tyrosine kinase inhibitors in patients with advanced stage of NSCLC. On the other hand, the prognostic or predictive value of *EGFR* mutations in early-stage NSCLC is less well defined. One study indicated that patients with resectable lung cancer who harbor *EGFR* mutations might have a better survival outcome than patients without *EGFR* mutations [119]. However, multivariate analysis only confirmed smoking status – and not mutations in *EGFR* – to be an independent prognostic factor in this group of patients. The prognostic or predictive value of *EGFR* mutations in early-stage NSCLC is also not consistent in some studies. In a randomized phase III study (BR-19), adjuvant gefitinib treatment was compared with placebo in an unselected patient population with resectable NSCLC. Although overall survival was similar between the two groups (HR = 1.23; $p = 0.136$), patients with *EGFR* mutations ($n = 76$) seemed to have a worse survival outcome with adjuvant gefitinib than with placebo (HR = 1.58; $p = 0.18$) [120]. One of the explanations is that the tumor biology of *EGFR* mutations in early-stage NSCLC might be different from advanced-stage disease.

The overall response rates of NSCLC with *EGFR* mutations treated with tyrosine kinase inhibitors are 50 to 100%, while response rates among patients with wild-type *EGFR* are 0 to 30% [109]. Among all *EGFR* mutations, some that respond particularly well to tyrosine kinase inhibitors include point mutations in exons 18 (G719A/C) and 21 (L858R and L861Q), and in-frame deletions in exon 19.

Inhibitor against EML4–ALK Fusion Gene

Crizotinib, initially launched as an inhibitor of the *MET* tyrosine kinase, is an effective drug to target NSCLC harboring an *ALK* gene rearrangement. Crizotinib inhibits phosphorylation of activated *ALK* and reduces cell proliferation. In a cohort of 82 NSCLC patients with a mean treatment duration of 6.4 months, the overall response rate to crizotinib was 57% (46 confirmed partial responses and one confirmed complete response) and 33% (27 patients) had a stable disease, and PFS was approximately 9.2 months. In this study, 77% of patients continued to receive crizotinib at the time of data cutoff, and the estimated probability of 6-month PFS was 72% [82]. In contrast, a recent meta-analysis showed that 6-month PFS in NSCLC patients following conventional second-line chemotherapy was only 27.2% [121]. Not all lung cancer patients with *EML4–ALK* gene rearrangements had dramatic benefit from crizotinib treatment. Among 82 lung cancer patients, six patients (7%) had disease progression at the time of their first staging scans; three patients (4%) had stable disease but concurrently developed new tumors consistent with progressive disease; two patients (3%) had an increase in the tumor size of more than 20% after two cycles of therapy; and one patient (1%) had tumor regression of 25% during cycle 1 of the treatment but subsequently had an increase in tumor size of more than 20% [82]. The mechanisms for resistance in this subset of patients are not fully understood. To evaluate further the effectiveness of crizotinib in lung cancer with *ALK* gene rearrangements, a randomized phase III clinical trial was initiated [122]. This study compared crizotinib treatment and standard of care in 318 advanced lung cancer patients with *ALK* gene rearrangements who had failed to respond to one prior treatment with platinum-based chemotherapy. The primary endpoint of this trial was to evaluate PFS and the results are also very encouraging. Despite the effectiveness of crizotinib in treating lung cancer patients with *ALK* gene rearrangements, development of acquired resistance caused by secondary mutations in the kinase domain of the *EML4-ALK* fusion transgene is a problem for achieving a better clinical outcome.

Inhibitors against MET Gene Mutation/Amplification

Tivantinib is a novel tyrosine kinase inhibitor targeted against *MET*. A randomized phase II study in patients with non-squamous cell histology comparing a combination of tivantinib and erlotinib with erlotinib alone showed an improvement in PFS (HR $= 0.61$, $p < 0.05$) and overall survival (HR $= 0.58$, $p < 0.05$) favoring the combination arm [123]. Small molecule inhibitors of *HGFR* kinase activity in mutant *MET* cells are being developed and early preclinical *MET* inhibitors that have been studied include geldanamycin, K252a, SU11274, and PHA-665752 [124–127]. These small molecule *MET* inhibitors have demonstrated an ability to inhibit cell growth and survival and to induce apoptosis by inhibiting *MET/HGF* signaling in various lung cancer cell lines. Different phases of clinical trials are underway.

Other Inhibitors against Gene Mutations

Various specific inhibitors of *BRAF* mutations, such as sorafenib, PLX4720, and AZ628, have recently been developed as cancer treatments [128]. Sorafenib was originally developed as a *BRAF* inhibitor, but is in fact a multikinase inhibitor of *RAF-1*, *BRAF*, *VEGFR 1*, *2*, and *3*, platelet-derived growth factor receptor, and CD117 (also known as c-Kit). *PIK3CA* inhibitors are still at a relatively early stage in development in clinical settings. The small molecule BEZ235 (Novartis, Basel, Switzerland) developed to target PI3K and mTOR proteins has demonstrated some antitumor activity in mice [129] and several PI3K inhibitors are in early clinical development; however, the response rate to single agents is low. Tumor cells harboring *HER2* mutations are sensitive to treatment with small molecule tyrosine kinase inhibitors such as lapatinib (HKI-272, Wyeth, Madison, NJ, USA) that target the kinase activity of *EGFR* and *HER2* [130], but seem insensitive to those that target *EGFR* alone. Some encouraging results have been seen in patients with lung adenocarcinoma with *HER2* mutations treated with tyrosine kinase inhibitors [131].

Acquired Resistance and Secondary Mutations

Despite effective molecular-targeted therapies, almost all treated tumors will eventually become resistant to treatment. The development of secondary mutations is one of the reasons patients fail to achieve complete cure. The most common secondary mutation is the T790M in *EGFR*, which is a substitution of methionine for threonine at codon 790 in exon 20 occurring in the original mutation, or alternatively an exon 19 deletion or an exon 21 L858R point mutation [132,133]. A single amino acid change of T790M in *EGFR* alters the conformation of the protein, resulting in interfering with the binding of *EGFR* tyrosine kinase inhibitors and thus restoring ATP binding to the levels of wild-type *EGFR* [132]. By increasing the affinity of the receptor for ATP, the T790M mutation reduces the effectiveness of tyrosine kinase inhibitors. The T790M mutation accounts for about 50–60% of acquired resistance to *EGFR* tyrosine kinase inhibitors. However, whether the T790M mutation pre-exists in a small percentage of cells or is acquired from exposure to *EGFR* tyrosine kinase inhibitors remains controversial. One study showed the presence of the T790M mutation in 28% of tumor samples with no previous exposure to tyrosine kinase inhibitors [79]. The prognostic value of the T790M mutation is also debatable. Patients with this mutation had a worse progression-free survival when treated with erlotinib than patients without the mutation [134]. However, lung cancer patients who were found to have the T790M mutation at the time of progression seemed to have slower progression of the disease. In a study of 93 patients with a tumor re-biopsy taken at the time of resistance to *EGFR* tyrosine kinase inhibitors, 62% of the cases were found to have the T790M mutation. Median post-progression survival for patients with the T790M mutation was 19 months, which is longer than the median overall survival of 12 months in patients without the T790M mutation ($p = 0.036$) [135]. In addition to the T790M mutation, there are several less common secondary resistance mutations including D761Y and L747S in exon 19, and T854A in exon 21, which arise in the background of sensitizing L858R mutations [136,137]. *MET* gene amplification accounts for about 20% of tyrosine kinase inhibitor resistance [100]. *ALK* inhibitor, crizotinib, related secondary mutations have also been found and the incidence of these resistant mutations is unclear.

Development of tyrosine kinase inhibitors specifically targeting these secondary mutations will improve the clinical outcome for those with acquired resistance. There are several other tyrosine kinase inhibitors such as the irreversible inhibitors CL-387, CL-785, HKI-272, and BIBW 2992 that are still in clinical development. These irreversible inhibitors may be effective against those lung cancers that have developed secondary resistance mutations [138].

Rapid development of personalized targeted therapy in the past several years has provided hopes for subsets of lung cancer patients and has also pointed to a new direction of treatment for non-small cell lung cancer especially for patients with lung adenocarcinoma. About 10–20% of all lung cancer patients harbor mutations in *EGFR* and 4–5% of the patients harbor mutations in *EML4–ALK*. These patients will benefit from *EGFR* tyrosine kinase inhibitors and *ALK* inhibitors, respectively. Testing for mutations for selected lung cancer patients for first-line *EGFR* tyrosine kinase inhibitor therapy is standard practice and selection of patients for personalized targeted therapy will also become a standard of routine clinical practice in the near future. Eventually all lung cancer patients should have their tumor molecular profiling done before starting treatment for the purpose of selecting the optimal treatment of choice and achieving the best clinical outcome.

GLOSSARY

Allele An allele is one of a number of alternative forms of the same gene or same genetic locus (generally a group of genes).

Amplicon A piece of DNA or RNA that is the source and/or product of natural or artificial amplification or replication events.

Angiogenesis The formation and differentiation of blood vessels.

Antagonist A chemical that acts within the body to reduce the physiological activity of another chemical substance.

Apoptosis A genetically determined process of cell self-destruction (programmed cell death) that is activated either by the presence of a stimulus or by the removal of a stimulus or suppressing agent and, when halted (as by genetic mutation), may result in uncontrolled cell growth and tumor formation.

Chemotherapy A general term for the use of a chemical agent to stop cancer cells from growing.

Chromosome A thread-like structure of nucleic acids and protein found in the nucleus of most living cells, carrying genetic information in the form of genes.

Codon A specific sequence of three consecutive nucleotides that is part of the genetic code and that specifies a particular amino acid in a protein or starts or stops protein synthesis – also called triplet.

Cohort A group of subjects who have shared a particular event together during a particular time span.

Differentiation The sum of the developmental processes whereby apparently unspecialized cells, tissues, and structures attain their adult form and function.

DNA sequence A succession of letters that indicate the order of nucleotides within a DNA (using GACT) molecule.

Domain Any of the three-dimensional subunits of a protein that together make up its tertiary structure, that are formed by folding its linear peptide chain.

Driving gene A gene as the driver of tumorigenesis.

EGFR Epidermal growth factor receptor is the cell-surface receptor for members of the epidermal growth factor family (EGF-family) of extracellular protein ligands.

Encode To specify the genetic code for.

Exon A nucleotide sequence in DNA that carries the code for the final mRNA molecule and thus defines a protein's amino acid sequence.

FISH Fluoresence *in situ* hybridization is a cytogenetic technique that is used to detect and localize the presence or absence of specific DNA sequences on chromosomes.

Gene mutation A permanent change in the DNA sequence that makes up a gene. Mutations range in size from a single DNA building block (DNA base) to a large segment of a chromosome.

Hazard ratio An expression of the hazard or chance of events occurring in the treatment arm as a ratio of the hazard of the events occurring in the control arm.

Heterodimer A macromolecular complex formed by two different macromolecules.

Heterozygosity Having different alleles at one or more corresponding chromosomal loci.

Homodimers A macromolecular complex formed by two identical molecules.

Homologs Also known as homologues indicating chromosome pairs of approximately the same length, centromere position, and staining pattern, with genes for the same characteristics at corresponding loci.

Immunohistochemistry The application of histochemical and immunologic methods to chemical analysis of living cells and tissues.

Isoform Any of two or more functionally similar proteins that have a similar but not identical amino acid sequence and are either encoded by different genes or by RNA transcripts from the same gene which have had different exons removed.

KRAS A protein encoded by the KRAS gene which performs an essential function in normal tissue signaling, and the mutation of a KRAS gene is an essential step in the development of many cancers.

Ligand A group, ion, or molecule coordinated to a central atom or molecule in a complex.

Lobectomy Excision of a whole lobe of the lung.

Lung cancer Malignancy tumors originally arising from lung tissue and the vast majority are bronchogenic in origin.

MAPK (mitogen-activated protein kinase) A serine/threonine-specific protein kinase involved in directing cellular responses to a diverse array of stimuli, such as mitogens, osmotic stress, heat shock and proinflammatory cytokines. They regulate cell proliferation, gene expression, differentiation, mitosis, cell survival, and apoptosis.

Mass spectrometry An instrumental method for identifying the chemical constitution of a substance by means of the separation of gaseous ions according to their differing mass and charge.

Metastasis The spread of a disease-producing agency (as cancer cells or bacteria) from the initial or primary site of disease to another part of the body.

Methylation The addition of a methyl group to a substrate or the substitution of an atom or group by a methyl group. It can be involved in modification of heavy metals, regulation of gene expression, regulation of protein function, and RNA metabolism.

Microarray A supporting material (such as a glass or plastic slide) onto which numerous molecules or molecular fragments usually of DNA or protein are attached in a regular pattern for use in biochemical or genetic analysis.

MicroRNA A class of post-transcriptional regulators. They are short about 22 nucleotide RNA sequences that bind to complementary sequences in the 3′ UTR of multiple target mRNAs, usually resulting in their silencing.

Molecular profiling A set of data exhibiting the significant features of something at a molecular level such as gene mutations.

Mortality The ratio of deaths in an area to the population of that area.

Multivariate analyses Based on the statistical principle of multivariate statistics, which involves observation and analysis of more than one statistical outcome variable at a time.

Neoplastic Relating to, or constituting a tumor or neoplasia.

Non-small cell lung cancer A group of subtypes of lung cancer which excludes small cell lung cancer.

Oligonucleotides A relatively short single-stranded nucleic-acid chain (as an oligodeoxynucleotide or oligoribonucleotide) usually consisting of up to approximately 20 nucleotides.

Oncogene A gene having the potential to cause a normal cell to become cancerous.

Palliative chemotherapy To use chemical agents to extend life and alleviate symptoms but not used as an attempt to cure cancer completely.

Personalized targeted therapy A type of medication that blocks the growth of cancer cells by interfering with specific targeted molecules needed for carcinogenesis and tumor growth that occurs in individual patients.

Pathogenesis The mechanism that causes the disease and can also describe the origin and development of the disease.

Phosphorylation The addition of phosphate to an organic compound through the action of a phosphorylase or kinase.

PI3K Phosphoinositide 3-kinase is an enzyme involved in cellular functions such as cell growth, proliferation, differentiation, motility, survival and intracellular trafficking.

Pneumonectomy Surgical treatment to excise one side of the lung.

Point mutation A gene mutation involving the substitution, addition, or deletion of a single nucleotide base.

Polymorphism Existence of a gene in several allelic forms.

Progression-free survival The length of time during and after treatment during which the disease being treated does not get worse.

Protein expressions A subcomponent of gene expression consisting of the stages after DNA has been transcribed to messenger RNA (mRNA), then translated into polypeptide chains, which are ultimately folded into proteins.

PTEN Phosphatase and tensin homolog, a tumor suppressor gene through the action of its phosphatase protein product to dephosphorylate phosphatidylinositol (3,4,16)-trisphosphate.

Radiation therapy The medical use of ionizing radiation to control or kill malignant cells.

Retrovirus A single-stranded RNA virus that produces reverse transcriptase by means of which DNA is produced using its RNA as a template and replicates in a host cell that is often tumorigenic.

Signal transduction It occurs when an extracellular signaling molecule activates a cell surface receptor. In turn, this receptor alters intracellular molecules creating a response.

SNP Single nucleotide polymorphism is a DNA sequence variation occurring when a single nucleotide (A, T, C or G) in the genome differs between paired chromosomes in a human.

Somatic mutation A change in the genetic structure that is neither inherited nor passed to offspring.

STAT Signal transducer and activator of transcription which regulates many aspects of cell growth, survival and differentiation.

Stem cells The biological cells able to divide (through mitosis) and differentiate into diverse specialized cell types and can self-renew to produce more stem cells.

Suppressor gene A gene that protects a cell from one step on the path to cancer. When this gene is mutated to cause a loss or reduction in its function, the cell can progress to cancer, usually in combination with other genetic changes.

Tumorigenesis The process involved in the production of a new tumor or tumors.

Tyrosine kinase An enzyme that can transfer a phosphate group from ATP to a protein in a cell.

Univariate analysis The simplest form of quantitative (statistical) analysis carried out with the description of a single variable and its attributes of the applicable unit of analysis.

Wedge resection A type of surgery to remove a tumor and surrounding normal lung tissue.

Wild type A phenotype, genotype, or gene that predominates in a natural population of organisms in contrast to that of mutant forms.

REFERENCES

[1] Boyle P, Levin B. World Cancer Report. Lyon: International Agency for Research on Cancer; 2008.

[2] Jemal A, Bray F, Center M, Ferlay J, Ward E, Forman D. Global cancer statistics. CA Cancer J Clin 2011;61:69–90.

[3] Jemal A, Siegel R, Ward E, Hao Y, Xu J, Murray T, et al. Cancer statistics, 2008. CA Cancer J Clin 2008;58:71–96.

[4] Herbst RS, Heymach JV, Lippman SM. Lung cancer. N Engl J Med 2008;359:1367–80.

[5] GLOBOCAN 2002: Cancer incidence, mortality and prevalence worldwide. IARC CancerBase No. 5; Version 2.0. IARC Press; 2004. Available at: <http://www-dep.iarc.fr/>.

[6] Devesa SS, Bray F, Vizcaino AP, Parkin DM. International lung cancer trends by histologic type: male:female differences diminishing and adenocarcinoma rates rising. Int J Cancer 2005;117: 294–9.

[7] Kamangar F, Dores GM, Anderson WF. Patterns of cancer incidence, mortality and prevalence across five continents: defining priorities to reduce cancer disparities in different geographic regions of the world. J Clin Oncol 2006;24:2137–50.

[8] Janssen-Heijnen ML, Coebergh JW. Trends in incidence and prognosis of the histological subtypes of lung cancer in North America, Australia, New Zealand and Europe. Lung Cancer 2001;31:123–37.

[9] Chen K, Wang PP, Sun B, Li Q, Perruccio A, Power D, et al. Twenty-year secular changes in sex specific lung cancer incidence rates in an urban Chinese population. Lung Cancer 2006;51: 13–9.

[10] Birring SS, Peake MD. Symptoms and the early diagnosis of lung cancer. Thorax 2005;60:268–9.

[11] Alberg AJ, Brock MV, Samet JM. Epidemiology of lung cancer: looking to the future. J Clin Oncol 2005;23:3175–85.

[12] Peace LR. A time correlation between cigarette smoking and lung cancer. Statistician 1985;34:371–81.

[13] Weiss W. Cigarette smoking and lung cancer trends. A light at the end of the tunnel? Chest 1997;111:1414–6.

[14] Mathers CD, Loncar D. Projections of global mortality and burden of disease from 2002 to 2030. PLoS Med 2006;3:e442.

[15] Ezzati M, Lopez AD. Estimates of global mortality attributable to smoking in 2000. Lancet 2003;362:847–52.

[16] Howlader N, Noone AM, Krapcho M, Neyman N, Aminou R, Waldron W, et al., editors. SEER Cancer Statistics Review, 1975–2008. Bethesda, MD: National Cancer Institute. Available at: <http://seer.cancer.gov/csr/1975_2008/>, based on November 2010 SEER data submission, posted to the SEER website; 2011.

[17] Howlader N, Noone AM, Krapcho M, Neyman N, Aminou R, Waldron W, et al., editors. SEER Cancer Statistics Review, 1975–2009 (Vintage 2009 Populations). Bethesda, MD: National Cancer Institute. Available at: <http://seer.cancer.gov/csr/1975_2009_pops09/>, based on November 2011 SEER data submission, posted to the SEER website; 2012.

[18] Edge SB, Byrd DR, Compton CC, Fritz AG, Greene FL, Trotti A. AJCC cancer staging manual. 7th ed. New York: Springer; 2010.

[19] Wistuba II, Mao L, Gazdar AF. Smoking molecular damage in bronchial epithelium. Oncogene 2002;21:7298–306.

[20] Westra WH. Early glandular neoplasia of the lung. Respir Res 2000;1:163–9.

[21] Tang X, Shigematsu H, Bekele BN, Roth JA, Minna JD, Hong WK, et al. EGFR tyrosine kinase domain mutations are detected in histologically normal respiratory epithelium in lung cancer patients. Cancer Res 2005;65:7568–72.

[22] Sato M, Shames DS, Gazdar AF, Minna JD. A translational view of the molecular pathogenesis of lung cancer. J Thorac Oncol 2007;2:327–43.

[23] Bhutani M, Pathak AK, Fan YH, Liu DD, Lee JJ, Tang H, et al. Oral epithelium as a surrogate tissue for assessing smoking-induced molecular alterations in the lungs. Cancer Prev Res 2008;1:39–44.

[24] Licchesi JD, Westra WH, Hooker CM, Herman JG. Promoter hypermethylation of hallmark cancer genes in atypical adenomatous hyperplasia of the lung. Clin Cancer Res 2008;14:2570–8.

[25] Belinsky SA, Liechty KC, Gentry FD, Wolf HJ, Rogers J, Vu K, et al. Promoter hypermethylation of multiple genes in sputum precedes lung cancer incidence in a high-risk cohort. Cancer Res 2006;66:3338–44.

[26] Machida EO, Brock MV, Hooker CM, Nakayama J, Ishida A, Amano J, et al. Hypermethylation of ASC/TMS1 is a sputum marker for late-stage lung cancer. Cancer Res 2006;66:6210–8.

[27] Brock MV, Hooker CM, Ota-Machida E, Han Y, Guo M, Ames S, et al. DNA methylation markers and early recurrence in stage I lung cancer. N Engl J Med 2008;358:1118–28.

[28] Kim JS, Kim JW, Han J, Shim YM, Park J, Kim DH. Cohypermethylation of p16 and FHIT promoters as a prognostic factor of recurrence in surgically resected stage I non-small cell lung cancer. Cancer Res 2006;66:4049–54.

[29] Yang Y, Iwanaga K, Raso MG, Wislez M, Hanna AE, Wieder ED, et al. Phosphatidylinositol 3-kinase mediates bronchioalveolar stem cell expansion in mouse models of oncogenic K-ras-induced lung cancer. PLoS One 2008;3(5):e2220.

[30] Yanagi S, Kishimoto H, Kawahara K, Sasaki T, Sasaki M, Nishio M, et al. Pten controls lung morphogenesis, bronchioalveolar stem cells, and onset of lung adenocarcinomas in mice. J Clin Invest 2007;117:2929–40.

[31] Amos CI, Wu X, Broderick P, Gorlov IP, Gu J, Eisen T, et al. Genome-wide association scan of tag SNPs identifies a susceptibility locus for lung cancer at 15q25.1. Nat Genet 2008;40:616–22.

[32] Hung RJ, McKay JD, Gaborieau V, Boffetta P, Hashibe M, Zaridze D, et al. A susceptibility locus for lung cancer maps to nicotinic acetylcholine receptor subunit genes on 15q25. Nature 2008;452:633–7.

[33] Thorgeirsson TE, Geller F, Sulem P, Rafnar T, Wiste A, Magnusson KP, et al. A variant associated with nicotine dependence, lung cancer and peripheral arterial disease. Nature 2008;452:638–42.

[34] Spitz MR, Wei Q, Dong Q, Amos CI, Wu X. Genetic susceptibility to lung cancer: the role of DNA damage and repair. Cancer Epidemiol Biomarkers Prev 2003;12:689–98.

[35] Yu D, Zhang X, Liu J, Yuan P, Tan W, Guo Y, et al. Characterization of functional excision repair cross-complementation

group 1 variants and their association with lung cancer risk and prognosis. Clin Cancer Res 2008;14:2878–86.

[36] Olaussen KA, Dunant A, Fouret P, Brambilla E, Andre F, Haddad V, et al. DNA repair by ERCC1 in non-small-cell lung cancer and cisplatin-based adjuvant chemotherapy. N Engl J Med 2006;355:983–91.

[37] Zheng Z, Chen T, Li X, Haura E, Sharma A, Bepler G. DNA synthesis and repair genes *RRM1* and *ERCC1* in lung cancer. N Engl J Med 2007;356:800–8.

[38] Schena M, Shalon D, Davis RW, Brown PO. Quantitative monitoring of gene expression patterns with a complementary DNA microarray. Science 1995;270:467–70.

[39] Wodicka L, Dong H, Mittmann M, Ho MH, Lockhart DJ. Genome-wide expression monitoring in *Saccharomyces cerevisiae*. Nat Biotechnol 1997;15:1359–67.

[40] Bhattacharjee A, Richards WG, Staunton J, Li C, Monti S, Vasa P, et al. Classification of human lung carcinomas by mRNA expression profiling reveals distinct adenocarcinoma subclasses. Proc Natl Acad Sci USA 2001;98:13790–5.

[41] Takamizawa J, Konishi H, Yanagisawa K, Tomida S, Osada H, Endoh H, et al. Reduced expression of the let-7 microRNAs in human lung cancers in association with shortened postoperative survival. Cancer Res 2004;64:3753–6.

[42] Calin GA, Croce CM. MicroRNA signatures in human cancers. Nat Rev Cancer 2006;6:857–66.

[43] Weiss GJ, Bemis LT, Nakajima E, Sugita M, Birks DK, Robinson WA, et al. EGFR regulation by microRNA in lung cancer: correlation with clinical response and survival to gefitinib and EGFR expression in cell lines. Ann Oncol 2008;19 (6):1053–9.

[44] Fenn JB, Mann M, Meng CK, Wong SF, Whitehouse CM. Electrospray ionization for mass spectrometry of large biomolecules. Science 1989;246:64–71.

[45] Patterson SD, Aebersold RH. Proteomics: the first decade and beyond. Nat Genet 2003;33:311–23.

[46] Chen G, Gharib TG, Wang H, Huang CC, Kuick R, Thomas DG, et al. Protein profiles associated with survival in lung adenocarcinoma. Proc Natl Acad Sci USA 2003;100:13537–42.

[47] Yanagisawa K, Shyr Y, Xu BJ, Massion PP, Larsen PH, White BC, et al. Proteomic patterns of tumour subsets in non-small-cell lung cancer. Lancet 2003;362:433–9.

[48] Travis WD, Brambilla E, Müller-Hermelink HK, Harris CC, editors. WHO classification of tumours: pathology and genetics of the lung, pleura, thymus and heart. Lyon: IARC Press; 2004.

[49] Travis WD, Brambilla E, Noguchi M, Nicholson AG, Geisinger KR, Yatabe Y, et al. International Association for the Study of Lung Cancer/American Thoracic Society/European Respiratory Society International Multidisciplinary Classification of Lung Adenocarcinoma. J Thorac Oncol 2011;6(2):244–85.

[50] Harvey JJ. An unidentified virus which causes the rapid production of tumours in mice. Nature 1964;204:1104–5.

[51] Ellis RW, Defeo D, Shih TY, Gonda MA, Young HA, Tsuchida N, et al. The p21 src genes of Harvey and Kirsten sarcoma viruses originate from divergent members of a family of normal vertebrate genes. Nature 1981;292:506–11.

[52] Shih C, Padhy LC, Murray M, Weinberg RA. Transforming genes of carcinomas and neuroblastomas introduced into mouse fibroblasts. Nature 1981;290:261–4.

[53] Shimizu K, Goldfarb M, Suard Y, Perucho M, Li Y, Kamata T, et al. Three human transforming genes are related to the viral RAS oncogenes. Proc Natl Acad Sci USA 1983;80:2112–6.

[54] Bos JL. Ras oncogenes in human cancer: a review. Cancer Res 1989;49:4682–9.

[55] Forbes S, Clements J, Dawson E, Bamford S, Webb T, Dogan A, et al. COSMIC 2005. Br J Cancer 2006;94:318–22.

[56] Mascaux C, Iannino N, Martin B, Paesmans M, Berghmans T, Dusart M, et al. The role of RAS oncogene in survival of patients with lung cancer: a systematic review of the literature with meta-analysis. Br J Cancer 2005;92:131–9.

[57] Schiller JH, Adak S, Feins RH, Keller SM, Fry WA, Livingston RB, et al. Lack of prognostic significance of p53 and K-ras mutations in primary resected non-small-cell lung cancer on E4592: a laboratory ancillary study on an eastern cooperative oncology group prospective randomized trial of postoperative adjuvant therapy. J Clin Oncol 2001;19:448–57.

[58] Tsao MS, Aviel-Ronen S, Ding K, Lau D, Liu N, Sakurada A, et al. Prognostic and predictive importance of p53 and ras for adjuvant chemotherapy in non small-cell lung cancer. J Clin Oncol 2007;25:5240–7.

[59] Winton T, Livingston R, Johnson D, Rigas J, Johnston M, Butts C, et al. Vinorelbine plus cisplatin vs. observation in resected non-small-cell lung cancer. N Engl J Med 2005;352:2589–97.

[60] Yarden Y, Sliwkowski MX. Untangling the ErbB signaling network. Nat Rev Mol Cell Biol 2001;2:127–37.

[61] Wells A. EGF receptor. Int J Biochem Cell Biol 1999;31:637–43.

[62] Jimeno A, Hidalgo M. Pharmacogenomics of epidermal growth factor receptor (EGFR) tyrosine kinase inhibitors. Biochim Biophys Acta 2006;1766:217–29.

[63] Mosesson Y, Yarden Y. Oncogenic growth factor receptors: implications for signal transduction therapy. Semin Cancer Biol 2004;14:262–70.

[64] Woodburn JR. The epidermal growth factor receptor and its inhibition in cancer therapy. Pharmacol Ther 1999;82:241–50.

[65] Ellerbroek SM, Halbleib JM, Benavidez M, Warmka JK, Wattenberg EV, Stack MS, et al. Phosphatidylinositol 3-kinase activity in epidermal growth factor-stimulated matrix metalloproteinase-9 production and cell surface association. Cancer Res 2001;61:1855–61.

[66] Shigematsu H, Gazdar AF. Somatic mutations of epidermal growth factor receptor signaling pathway in lung cancers. Int J Cancer 2006;118:257–62.

[67] Sakurada A, Shepherd FA, Tsao MS. Epidermal growth factor receptor tyrosine kinase inhibitors in lung cancer: impact of primary or secondary mutations. Clin Lung Cancer 2006;7(Suppl. 4): S138–44.

[68] Eberhard DA, Giaccone G, Johnson BE. Biomarkers of response to epidermal growth factor receptor inhibitors in non-small-cell lung cancer working group: standardization for use in the clinical trial setting. J Clin Oncol 2008;26:983–94.

[69] Zhu CQ, da Cunha Santos G, Ding K, Sakurada A, Cutz JC, Liu N, et al. Role of KRAS and EGFR as biomarkers of response to erlotinib in National Cancer Institute of Canada Clinical Trials Group Study BR.21. J Clin Oncol 2008;26:4268–75.

[70] Pao W, Ladanyi M. Epidermal growth factor receptor mutation testing in lung cancer: searching for the ideal method. Clin Cancer Res 2007;13:4954–5.

[71] Willmore-Payne C, Holden JA, Layfield LJ. Detection of epidermal growth factor receptor and human epidermal growth factor receptor 2 activating mutations in lung adenocarcinoma by high-resolution melting amplicon analysis: correlation with gene copy number, protein expression, and hormone receptor expression. Hum Pathol 2006;37:755–63.

[72] Yatabe Y, Mitsudomi T. Mutations of the epidermal growth factor receptor gene and related genes as determinants of epidermal growth factor receptor tyrosine kinase inhibitors sensitivity in lung cancer. Cancer Sci 2007;98:1817–24.

[73] Pulford K, Lamant L, Espinos E, Jiang Q, Xue L, Turturro F, et al. The emerging normal and disease-related roles of anaplastic lymphoma kinase. Cell Mol Life Sci 2004;61:2939–53.

[74] Kutok JL, Aster JC. Molecular biology of anaplastic lymphoma kinase-positive anaplastic large-cell lymphoma. J Clin Oncol 2002;20:3691–702.

[75] Shiota M, Mori S. The clinicopathological features of anaplastic large cell lymphomas expressing p80NPM/ALK. Leuk Lymphoma 1996;23:25–32.

[76] Perner S, Wagner PL, Demichelis F, Mehra R, Lafargue CJ, Moss BJ, et al. EML4-ALK fusion lung cancer: a rare acquired event. Neoplasia 2008;10:298–302.

[77] Soda M, Choi YL, Enomoto M, Takada S, Yamashita Y, Ishikawa S, et al. Identification of the transforming EML4-ALK fusion gene in non-small-cell lung cancer. Nature 2007;448:561–6.

[78] Amin HM, Lai R. Pathobiology of ALK^+ anaplastic large cell lymphoma. Blood 2007;110:2259–67.

[79] Maheswaran S, Sequist LV, Nagrath S, Ulkus L, Brannigan BA, Collura CV, et al. Detection of mutations in EGFR in circulating lung-cancer cells. N Engl J Med 2008;359:366–77.

[80] Wong DW, Leung EL, So KK, Tam IY, Sihoe AD, Cheng LC, et al. The EML4-ALK fusion gene is involved in various histologic types of lung cancers from nonsmokers with wild-type EGFR and KRAS. Cancer 2009;15:1723–33.

[81] Shaw AT, Yeap BY, Mino-Kenudson M, Digumarthy SR, Costa DB, Heist RS, et al. Clinical features and outcome of patients with non-small cell lung cancer who harbor EML4-ALK. J Clin Oncol 2009;27:4247–53.

[82] Kwak EL, Bang YJ, Camidge DR, Shaw AT, Solomon B, Maki RG, et al. Anaplastic lymphoma kinase inhibition in non-small cell lung cancer. N Engl J Med 2010;363:1693–703.

[83] Rodig SJ, Mino-Kenudson M, Dacic S, Yeap BY, Shaw A, Barletta JA, et al. Unique clinicopathologic features characterize ALK-rearranged lung adenocarcinoma in the Western populations. Clin Cancer Res 2009;15:5216–23.

[84] Leicht DT, Balan V, Kaplun A, Singh-Gupta V, Kaplun L, Dobson M, et al. Raf kinases: function, regulation and role in human cancer. Biochim Biophys Acta 2007;1773:1196–212.

[85] Davies H, Bignell GR, Cox C, Stephens P, Edkins S, Clegg S, et al. Mutations of the *BRAF* gene in human cancer. Nature 2002;417:949–54.

[86] Brose MS, Volpe P, Feldman M, Kumar M, Rishi I, Gerrero R, et al. BRAF RAS mutations in human lung cancer and melanoma. Cancer Res 2002;62(23):6997–7000.

[87] Ding L, Getz G, Wheeler DA, Mardis ER, McLellan MD, Cibulskis K, et al. Somatic mutations affect key pathways in lung adenocarcinoma. Nature 2008;455:1069–75.

[88] Kobayashi M, Sonobe M, Takahashi T, Yoshizawa A, Ishikawa M, Kikuchi R, et al. Clinical significance of BRAF gene mutations in patients with non-small cell lung cancer. Anticancer Res 2011;31:4619–24.

[89] Jimenez C, Jones DR, Rodríguez-Viciana P, Gonzales-Garcia A, Leonardo E, Wennstrom S, et al. Identification and characterization of a new oncogene derived from the regulatory subunit of phosphoinositide 3-kinase. EMBO J 1998;17:743–53.

[90] Samuels Y, Wang Z, Bardelli A, Silliman N, Ptak J, Szabo S, et al. High frequency of mutations of the PIK3CA gene in human cancers. Science 2004;304:554.

[91] Kawano O, Sasaki H, Endo K, Suzuki E, Haneda H, Yukiue H, et al. PIK3CA mutation status in Japanese lung cancer patients. Lung Cancer 2006;54:209–15.

[92] Okudela K, Suzuki M, Kageyama S, Bunai T, Nagura K, Igarashi H, et al. PIK3CA mutation and amplification in human lung cancer. Pathol Int 2007;57:664–71.

[93] Angulo B, Suarez-Gauthier A, Lopez-Rios F, Medina PP, Conde E, Tang M, et al. Expression signatures in lung cancer reveal a profile for EGFR-mutant tumours and identify selective PIK3CA overexpression by gene amplification. J Pathol 2008;214:347–56.

[94] Coussens L, Yang-Feng TL, Liao YC, Chen E, Gray A, McGrath J, et al. Tyrosine kinase receptor with extensive homology to EGF receptor shares chromosomal location with neu oncogene. Science 1985;230:1132–9.

[95] Heinmöller P, Gross C, Beyser K, Schmidtgen C, Maass G, Pedrocchi M, et al. HER2 status in non-small cell lung cancer: results from patient screening for enrollment to a phase II study of herceptin. Clin Cancer Res 2003;9:5238–43.

[96] Stephens P, Hunter C, Bignell G, Edkins S, Davies H, Teague J, et al. Lung cancer: intragenic ERBB2 kinase mutations in tumours. Nature 2004;431:525–6.

[97] Shigematsu H, Takahashi T, Nomura M, Majmudar K, Suzuki M, Lee H, et al. Somatic mutations of the HER2 kinase domain in lung adenocarcinomas. Cancer Res 2005;65:1642–6.

[98] Buttitta F, Barassi F, Fresu G, Felicioni L, Chella A, Paolizzi D, et al. Mutational analysis of the HER2 gene in lung tumors from Caucasian patients: mutations are mainly present in adenocarcinomas with bronchioloalveolar features. Int J Cancer 2006;119:2586–91.

[99] Bean J, Brennan C, Shih JY, Riely G, Viale A, Wang L, et al. MET amplification occurs with or without T790M mutations in EGFR mutant lung tumors with acquired resistance to gefitinib or erlotinib. Proc Natl Acad Sci USA 2007;104:20932–7.

[100] Engelman JA, Zejnullahu K, Mitsudomi T, Song Y, Hyland C, Park JO, et al. MET amplification leads to gefitinib resistance in lung cancer by activating ERBB3 signaling. Science 2007;316:1039–43.

[101] Onozato R, Kosaka T, Kuwano H, Sekido Y, Yatabe Y, Mitsudomi T. Activation of MET by gene amplification or by splice mutations deleting the juxtamembrane domain in primary resected lung cancers. J Thorac Oncol 2009;4:5–11.

[102] Beau-Faller M, Ruppert AM, Voegeli AC, Neuville A, Meyer N, et al. MET gene copy number in non-small cell lung cancer: molecular analysis in a targeted tyrosine kinase inhibitor naive cohort. J Thorac Oncol 2008;3:331–9.

[103] Cappuzzo F, Marchetti A, Skokan M, Rossi E, Gajapathy S, Felicioni L, et al. Increased MET gene copy number negatively

affects survival of surgically resected non-small-cell lung cancer patients. J Clin Oncol 2009;27:1667–74.

[104] Lutterbach B, Zeng Q, Davis LJ, Hatch H, Hang G, Kohl NE, et al. Lung cancer cell lines harboring *MET* gene amplification are dependent on Met for growth and survival. Cancer Res 2007;67:2081–8.

[105] Kong-Beltran M, Seshagiri S, Zha J, Zhu W, Bhawe K, Mendoza N, et al. Somatic mutations lead to an oncogenic deletion of met in lung cancer. Cancer Res 2006;66:283–9.

[106] Ma PC, Tretiakova MS, MacKinnon AC, Ramnath N, Johnson C, Dietrich S, et al. Expression and mutational analysis of MET in human solid cancers. Genes Chromosomes Cancer 2008;47:1025–37.

[107] Lynch TJ, Bell DW, Sordella R, Gurubhagavatula S, Okimoto RA, Brannigan BW, et al. Activating mutations in the epidermal growth factor receptor underlying responsiveness of non-small-cell lung cancer to gefitinib. N Engl J Med 2004;350:2129–39.

[108] Paez JG, Jänne PA, Lee JC, Tracy S, Greulich H, Gabriel S, et al. EGFR mutations in lung cancer: correlation with clinical response to gefitinib therapy. Science 2004;304:1497–500.

[109] Sequist LV, Bell DW, Lynch TJ, Haber DA. Molecular predictors of response to epidermal growth factor receptor antagonists in non-small-cell lung cancer. J Clin Oncol 2007;25:587–95.

[110] Inoue A, Kobayashi K, Usui K, Maemondo M, Okinaga S, Mikami I, et al. First-line gefitinib for patients with advanced non-small-cell lung cancer harboring epidermal growth factor receptor mutations without indication for chemotherapy. J Clin Oncol 2009;27:1394–400.

[111] Rosell R, Moran T, Queralt C, Porta R, Cardenal F, Camps C, et al. Screening for epidermal growth factor receptor mutations in lung cancer. N Engl J Med 2009;361:958–67.

[112] Tamura K, Okamoto I, Kashii T, Negoro S, Hirashima T, Kudoh S, et al. Multicentre prospective phase II trial of gefitinib for advanced non-small cell lung cancer with epidermal growth factor receptor mutations: results of the West Japan Thoracic Oncology Group trial (WJTOG0403). Br J Cancer 2008;98:907–14.

[113] Sutani A, Nagai Y, Udagawa K, Uchida Y, Koyama N, Murayama Y, et al. Gefitinib for non-small-cell lung cancer patients with epidermal growth factor receptor gene mutations screened by peptide nucleic acid-locked nucleic acid PCR clamp. Br J Cancer 2006;95:1483–9.

[114] Mok TS, Wu YL, Thongprasert S, Yang CH, Chu DT, Saijo N, et al. Gefitinib or carboplatin-paclitaxel in pulmonary adenocarcinoma. N Engl J Med 2009;361:947–57.

[115] Thatcher N, Chang A, Parikh P, Rodrigues Pereira J, Ciuleanu T, von Pawel J, et al. Gefitinib plus best supportive care in previously treated patients with refractory advanced non-small cell lung cancer: Results from a randomized, placebo-controlled, multicentre study (Iressa Survival Evaluation in Lung Cancer). Lancet 2005;366:1527–37.

[116] Shepherd FA, Rodrigues Pereira J, Tan EH, Hirsh V, Thongprasert S, et al. Erlotinib in previously treated non-small cell lung cancer. N Engl J Med 2005;353:123–32.

[117] Kim ES, Hirsh V, Mok T, Socinski MA, Gervais R, Wu YL, et al. Gefitinib versus docetaxel in previously treated non-small-cell lung cancer: a randomised phase III trial (INTEREST). Lancet 2008;372:1809–18.

[118] Fukuoka M, Wu YL, Thongprasert S, Sunpaweravong P, Leong T, Tsuboi M, et al. Biomarker analyses and final overall survival results from a phase III, randomized, open-label, first-line study of gefitinib versus carboplatin/paclitaxel in clinically selected patients with advanced non-small-cell lung cancer in Asia (IPASS). J Clin Oncol 2011;29(21):2866–74.

[119] Kosaka T, Yatabe Y, Onozato R, Kuwano H, Mitsudomi T. Prognostic implication of EGFR, KRAS, and TP53 gene mutations in a large cohort of Japanese patients with surgically treated lung adenocarcinoma. J Thorac Oncol 2009;4:22–9.

[120] Goss GD, Lorimer I, Tsao MS, O'Callaghan CJ, Ding K, Masters GA, et al. A phase III randomized, double-blind, placebo-controlled trial of the epidermal growth factor receptor inhibitor gefitinb in completely resected stage IB-IIIA non-small cell lung cancer (NSCLC): NCIC CTG BR.19 [abstract]. J Clin Oncol 2010;28(Suppl. 18): LBA7005.

[121] Di Maio M, Chiodini P, Georgoulias V, Hatzidaki D, Takeda K, Wachters FM, et al. Meta-analysis of single agent chemotherapy compared with combination chemotherapy as second line treatment for advanced non-small cell lung cancer. J Clin Oncol 2009;27:1836–43.

[122] Christensen JG, Zou HY, Arango ME, Li Q, Lee JH, McDonnell SR, et al. Cytoreductive antitumor activity of PF-2341066, a novel inhibitor of anaplastic lymphoma kinase and c-Met, in experimental models of anaplastic large cell lymphoma. Mol Cancer Ther 2007;6:3314–22.

[123] Schiller J, Akerley W, Brugger W, Ferrari D, Garmey EG, Gerber DE, et al. Results from ARQ 197-209: a global randomized placebo-controlled phase II clinical trial of erlotinib plus ARQ 197 versus erlotinib plus placebo in previously treated EGFR inhibitor-naive patients with locally advanced or metastatic non-small cell lung cancer (NSCLC) [abstract]. J Clin Oncol 2010;28(Suppl. 18):LBA7502.

[124] Ma PC, Jagadeeswaran R, Jagadeesh S, Tretia.kova MS, Nallasura V, Fox EA, et al. Functional expression and mutations of c-Met and its therapeutic inhibition with SU11274 and small interfering RNA in non-small cell lung cancer. Cancer Res 2005;65:1479–88.

[125] Sattler M, Pride YB, Ma P, Gramlich JL, Chu SC, Quinnan LA, et al. A novel small molecule met inhibitor induces apoptosis in cells transformed by the oncogenic TPR-MET tyrosine kinase. Cancer Res 2003;63:5462–9.

[126] Christensen JG, Schreck R, Burrows J, Kuruganti P, Chan E, Le P, et al. A selective small molecule inhibitor of c-Met kinase inhibits c-Met-dependent phenotypes in vitro and exhibits cytoreductive antitumor activity in vivo. Cancer Res 2003;63:7345–55.

[127] Ma PC, Schaefer E, Christensen JG, Salgia R. A selective small molecule c-MET Inhibitor, PHA665752, cooperates with rapamycin. Clin Cancer Res 2005;11:2312–9.

[128] Halilovic E, Solit DB. Therapeutic strategies for inhibiting oncogenic BRAF signaling. Curr Opin Pharmacol 2008;8(4):419–26.

[129] Engelman JA, Chen L, Tan X, Crosby K, Guimaraes AR, Upadhyay R, et al. Effective use of PI3K and MEK inhibitors to treat mutant Kras G12D and PIK3CA H1047R murine lung cancers. Nat Med 2008;14:1351–6.

[130] Wang SE, Narasanna A, Perez-Torres M, Xiang B, Wu FY, Yang S, et al. HER2 kinase domain mutation results in constitutive phosphorylation and activation of HER2 and EGFR and

resistance to EGFR tyrosine kinase inhibitors. Cancer Cell 2006;10:25–38.

[131] De Grève J, Teugels E, Geers C, Decoster L, Galdermans D, De Mey J, et al. Clinical activity of afatinib (BIBW 2992) in patients with lung adenocarcinoma with mutations in the kinase domain of HER2/neu. Lung Cancer 2012;76(1):123–7.

[132] Kobayashi S, Boggon TJ, Dayaram T, Janne PA, Kocher O, Meyerson M, et al. EGFR mutation and resistance of non-small-cell lung cancer to gefitinib. N Engl J Med 2005;352:786–92.

[133] Kosaka T, Yatabe Y, Endoh H, Yoshida K, Hida T, Tsuboi M, et al. Analysis of epidermal growth factor receptor gene mutation in patients with non-small-cell lung cancer and acquired resistance to gefitinib. Clin Cancer Res 2006;12:5764–9.

[134] Rosell R, Molina MA, Costa C, Simonetti S, Gimenez-Capitan J, Bertran-Alamillo J, et al. Pretreatment EGFR T790M mutation and BRCA1 mRNA expression in erlotinib-treated advanced non-small-cell lung cancer patients with EGFR mutation. Clin Cancer Res 2011;17:1160–8.

[135] Oxnard GR, Arcila ME, Sima CS, Riely GJ, Chmielecki J, Kris MG, et al. Acquired resistance to EGFR tyrosine kinase inhibitors in EGFR mutant lung cancer: distinct natural history of patients with tumor harboring the T790M mutation. Clin Cancer Res 2011;17:1616–22.

[136] Balak MN, Gong Y, Riely GJ, Somwar R, Li AR, Zakowski MF, et al. Novel D761Y and common secondary T790M mutations in epidermal growth factor receptor–mutant lung adenocarcinomas with acquired resistance to kinase inhibitors. Clin Cancer Res 2006;12:6494–501.

[137] Bean J, Riely GJ, Balak M, Marks JL, Ladanyi M, Miller VA, et al. Acquired resistance to epidermal growth factor receptor kinase inhibitors associated with a novel T854A mutation in a patient with EGFR-mutant lung adenocarcinoma. Clin Cancer Res 2008;14:7519–25.

[138] Sharma SV, Bell DW, Settleman J, Haber DA. Epidermal growth factor receptor mutations in lung cancer. Nat Rev Cancer 2007;7:169–81.

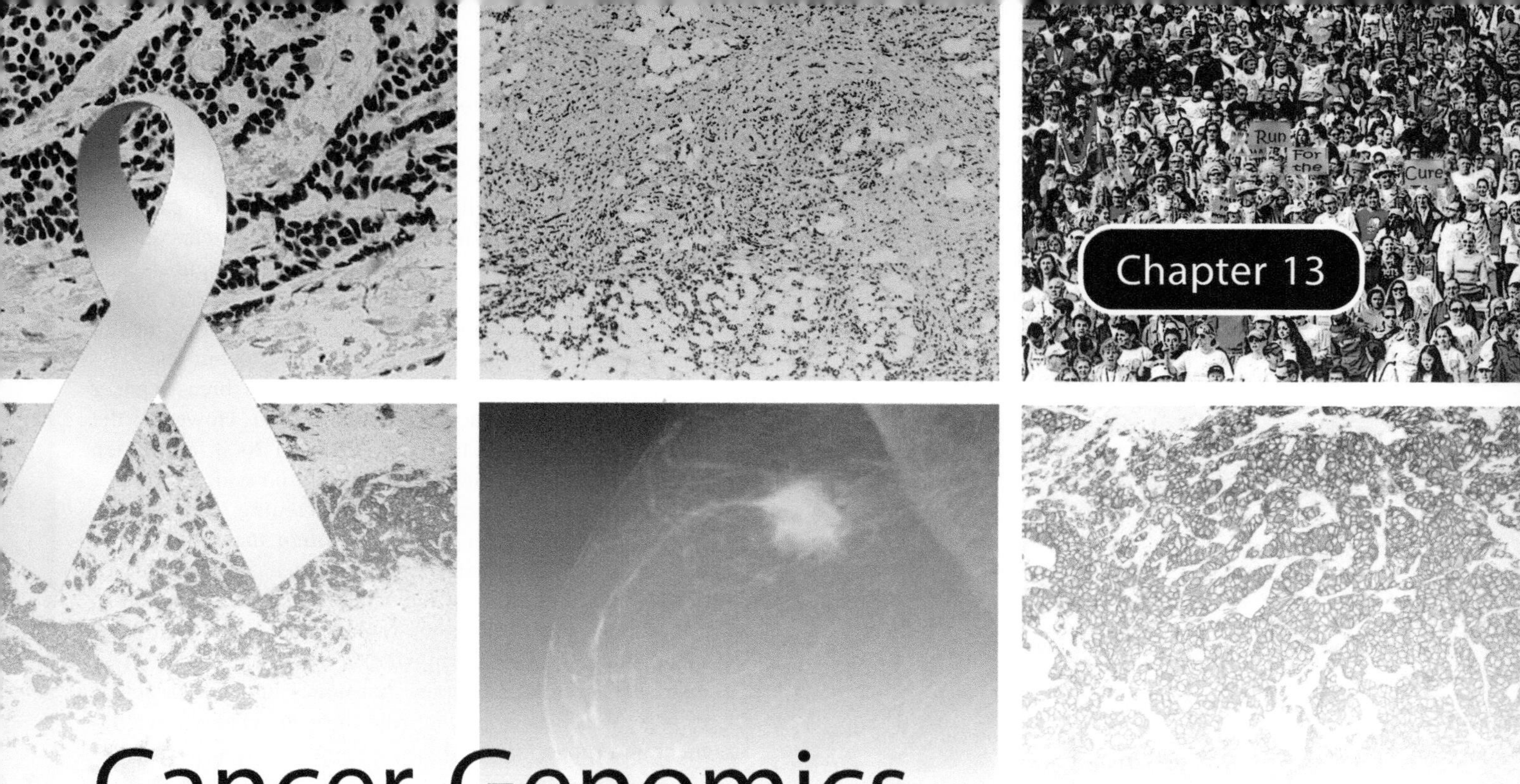

Cancer Genomics

Breast Cancer Genomics

Moamen Bydoun[1]
Paola Marcato[1]
Graham Dellaire[1,2]

[1]*Department of Pathology, Dalhousie University, Halifax, NS, Canada*
[2]*Department of Biochemistry and Molecular Biology, Dalhousie University, Halifax, NS, Canada*

Contents

Cancer Genomics. DOI: http://dx.doi.org/10.1016/B978-0-12-396967-5.00013-X

Key Concepts

- Breast cancer is the most common cancer affecting women worldwide and a major cause of cancer-related mortality
- The prediction of patient outcome as well as the development of new treatments lie at the heart of rapidly evolving breast cancer research
- Breast cancer is clinically assessed based on tumor size and spread, and hormone receptor status. New approaches stratify tumors based on the gene expression and genomic profiles of individual patients
- Gene expression profiling of individual tumors permitted subtyping of breast cancer and defined high- and low-risk groups
- Current genome-wide analysis offers new insights into the genetic basis of breast cancer which differentiate individual tumors based on single-nucleotide and/or gene copy variations while taking into account intra-tumoral variability. This appears to be the most robust prediction of prognosis and treatment response to date
- Ideally, combining patient data from genome-wide studies and gene expression analysis can be used to de-subtype breast cancer into discrete low-, intermediate- and high-risk groups
- Both gene expression and genome-wide analysis are the most patient-oriented approaches as they pave the way towards tailored treatment

INTRODUCTION

Breast Cancer Epidemiology

Breast cancer is the most common cancer affecting women worldwide, with approximately 1.5 million new cases every year, and it accounts for 16% of cancer deaths [1,2]. On average, 1 in 8 women will develop breast cancer during their lifetime [3]. Even with early detection, it remains the second most common cause of cancer-related death in females after lung cancer [4]. Despite the profound advances in cancer diagnosis and treatment over the last two decades, breast cancer is still associated with high mortality and remains a global health concern in both developed and developing countries.

Prior to the year 2000, the incidence of breast cancer had been gradually increasing since 1980. However, the rate reached a plateau between 2000 and 2003 for women of all ages, a time that coincides with the widespread use of mammograms as a standard screening method [5]. Notably, breast cancer incidence within the same period has dropped dramatically among women 50–69 years of age. Hence, this drop in incidence has been attributed to the reduced use of hormone replacement therapy in post-menopausal women during the early 2000s. As we will discuss later in the chapter, hormones and, in particular estrogen, play an important role in breast cancer development and progression [5].

Breast Anatomy and Development

The human breast is part of the integument and is located between the second and sixth ribs just anterior to the rib cage and intercostal muscles (Figure 13.1). On average, the breast contains about 20 lobes, each of which is divided into smaller lobules (known as terminal ductal-lobular units; TDLUs). The lobules are lined by glandular epithelium which are responsible for milk production. Breast ducts transport milk to the nipples [6].

The initial development of the primary human breast tissue takes place during the sixth week of embryogenesis. Until puberty, the human breast is still relatively immature and closely resembles fetal and infant breast tissue in terms of existing cell lineages and structural complexity. It includes branching ducts which terminate in terminal buds and few TDLUs. Complete tissue development occurs after puberty eventually to form a complex endocrinologic system. Post puberty, a new epithelial breast cell lineage appears and contributes to expanding the complexity of the breast. The further build-up of such complexity relies on factors such as hormone levels, pregnancy, diet, age, and family history [6]. When the breast is fully developed at the age of 18, it contains highly branching segmental and sub-segmental ducts and well-developed TDLUs [7].

Hormonal Regulation and Predisposition to Transformation

Although the steroid hormone estrogen is key for breast tissue development during puberty and pregnancy, it may also predispose to breast cancer. This relates to the role that estrogen plays in breast cell division as well as the

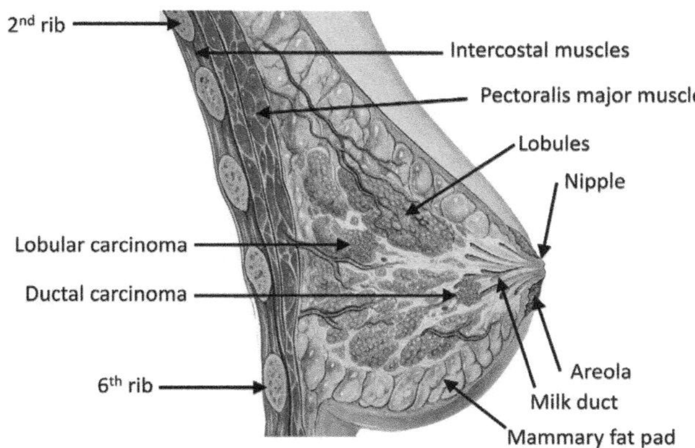

FIGURE 13.1 Basic anatomy of the human breast. The human breast is located between the second and sixth ribs anterior to the rib cage and intercostal muscles. Gland lobules (also known as terminal ductal-lobular units; TDLUs) are the structural units of the lobes of the breast. The lobules are lined by epithelial cells and are responsible for milk production. Milk is transported into the nipples through ducts. The majority of tumors arise from the epithelial lining of the ducts (ductal carcinoma); however, they can arise from the lobular region (lobular carcinoma). Adapted from: Breast anatomy normal scheme.png by Patrick J. Lynch, medical illustrator; C. Carl Jaffe, MD, cardiologist (Yale University Center for Advanced Instructional Media). Reproduced under the Creative Commons Attribution 3.0.

regulation of other hormones that contribute to the proliferation of mammary tissue. In addition, estrogen can support the growth of breast cancers that express receptors for estrogen. Not surprisingly, there have been extensive studies on the epidemiological and experimental relevance of estrogen in the etiology of breast cancer [8]. For instance, higher estrogen levels in serum correlate with increased breast cancer incidence in North American females compared to Asian females, which also suggests that genetic differences may affect hormone synthesis and predisposition to cancer [9,10]. On average, postmenopausal women who develop breast cancer have a 15% higher serum estrogen concentration than healthy women [11]. The complex development and plasticity of the breast, combined with responsiveness to estrogen, render it susceptible to disease occurrence. Since estrogen stimulates the development of cells within lobules and ducts, the latter become susceptible to cellular transformation and accumulation of mutations. If transformation favors cell growth, it may give rise to a rapidly proliferating tumor mass.

The majority of breast cancers (80–85%) arise from the epithelial cell layer of milk ducts, which is rich in estrogen receptors (ERs); these tumors are known as ductal carcinomas (see Figure 13.1). In contrast, lobular carcinomas, which are less prevalent (10–15%), arise from the lobular region of the breast. Other breast cancers (e.g. inflammatory breast cancer) are far less common and comprise the remaining 1–5%. Since the breast is a highly vascularized tissue, it forms a welcoming environment for tumor formation and progression. In addition, breast tissue contains mammary stem cells (located at the branching area of the duct) that replenish the cells of the lobules and ducts during a woman's lifetime. Although this is an important aspect of normal breast development and physiology, the same niche within mammary tissue that supports normal stem cells can also support tumor-initiating breast cancer stem cells.

BREAST CANCER DEVELOPMENT AND PROGRESSION: A STEP-WISE PHENOMENON

The development of an invasive carcinoma requires a series of genetic aberrations which initially give rise to a hyperplasia (increase in cell number) followed by an atypical hyperplasia (appearance of abnormal breast cells) and the formation of a carcinoma *in situ* (CIS) with a non-invasive phenotype. This non-invasive tumor mass remains confined to the lobules or milk ducts and does not invade surrounding tissues. As such, these tumors are referred to as lobular carcinoma *in situ* (LCIS) or ductal carcinoma *in situ* (DCIS), some of which are benign and never progress to the invasive stage even without treatment. However, in many cases, tumors can acquire additional genetic aberrations which result in progression into an invasive carcinoma (IC). The latter will ultimately lead to tumor cell dissemination via lymphatics or blood circulation leading to distant-site metastatic disease. Breast cancers tend to metastasize to organs such as the lymph nodes, lungs, bone, brain and skin, and these metastases are responsible for the majority of mortality and morbidity. The inability to determine which DCIS and LCIS will proceed to IC and which will remain *in situ*, necessitates that all patients receive treatment. Consequently, some patients may be over-treated.

For women with carcinoma *in situ*, surgical resection may be curative. It is generally accepted that patients with well-differentiated DCIS (retaining much of the ordered structure of normal tissue) have the best prognosis, while a poorly differentiated DCIS can progress to a grade III IDC if left untreated [12]. In addition to histological differences, it should be noted that genomic characteristics, which define a high grade IDC, are different from those of a well-differentiated DCIS and these genetic differences can be used to predict patient outcome and guide selection of treatment regimen [13,14].

Germline and Somatic Aberrations

Many factors have been shown to predispose women to the development of breast cancer, such as age (over 50), early menarche age, late pregnancy, late menopause, obesity and dietary factors, high-dose radiation early in life and nulliparity [15]. Nonetheless, family history and genetic background remain the top predisposing factors for developing breast cancer.

Women who have relatives affected by breast cancer have a higher risk of developing the disease compared to the general population. As a result, scientists studying breast cancer initially treated it as a solely hereditary disease, pursuing genetic variations in genes that might explain the susceptibility to breast cancer within families. To their surprise, genetic predisposition was later found to be linked to less than 10% of breast cancer cases [16]. Still, many susceptibility events were identified which included mutations in key genes such as *breast cancer 1* and *2* (*BRCA1* and *BRCA2*) and *tumor protein 53* (*TP53*) (the gene encoding the tumor suppressor p53) [17]. Mutations in these genes can lead to the loss of cell cycle control and the accumulation of DNA damage, both of which are prominent hallmarks of cancer as described by Hanahan and Weinberg [18]. Frame shift mutations in *BRCA1* and *BRCA2* result in expression of a truncated protein that is non-functional and capable of promoting genomic instability and chromosomal aberrations [19].

The absence of *BRCA1* or *BRCA2* mutations (termed BRCAx) in families with breast cancer history, predicts a lower grade disease with well-differentiated histology compared to *BRCA1/BRCA2* mutant families [20]. Patients with a germline mutation in *TP53* or *check point kinase 2* (*CHEK2*) develop Li–Fraumeni syndrome and are susceptible to many cancers. These individuals have a high risk of developing bilateral breast cancer at an early age [19,21]. These germline mutations can be classified into three subgroups based on their frequency and penetrance (magnitude of impact on patient): *high penetrance with low frequency* (*BRCA1*, *BRCA2*, *PTEN*, *TP53*, *STK11*), *low penetrance with low frequency* (*CHEK2*, *ATM*, *PALB2*, *BRIP1*) or *low penetrance with high frequency* (*FGFR2*, *LSP1*, *MAP3K1*, *TGFB1*, *TOX3*, single nucleotide variants (SNVs) at 2q35,16q12,8q) [22].

Although many germline and somatic mutations have been characterized in breast cancer, no single mutation has been described as "the" initiating event. For instance, pancreatic and lung cancers are often initiated by a mutation in *K-RAS*, while some hematological malignancies are driven by single gene fusion events (e.g. *PML–RARα* in promyelocytic leukemia and *BCR–ABL* in chronic myelogenous leukemia) [23]. Somatic aberrations identified in breast cancer include both copy number alterations (CNAs) and single nucleotide variants (SNVs). Examples of CNAs include amplifications in human epidermal growth factor receptor 2 (*HER2/ERBB2*), *CCND1*, *WIP1*, *GASC1* genes [22]. In the case of CNAs, which encompass amplifications or deletions of multiple and/or large chromosomal regions, such aberrations can be a reflection of a high genomic instability and, as such, are predictive of a poor prognosis in cancer patients. In contrast, patients with no or low genomic instability tend to have a better prognosis [24].

Similarly, common gene deletions in tumor suppressor genes such as *PTEN* and *TP53* have also been reported in breast cancers. It is important to realize that a mutation in one tumor suppressor allele is often insufficient for cancer initiation but it predisposes someone to cancer as mutation in the remaining wild-type allele can lead to loss of heterozygosity (LOH) in a population of cells driving tumor progression. Certain mutations correlate with commonly deregulated signaling pathways and transcriptomic profiles (as defined by breast cancer subtypes; discussed later) and can also predict patient outcome. For instance, a mutation in the tumor suppressor *PTEN* indicates a hyperactive PI3K/mTOR pathway and yields tumors with a high mitotic index [22]. In addition, high frequency of *TP53* mutations correlates with the most aggressive tumors and predicts poor outcome (Table 13.1) [21].

Although the contribution of the aforementioned genetic aberrations to breast cancer progression has been reasonably understood, it is not always as clear how other CNAs and genetic mutations (SNVs) contribute to patient outcome. As a result, much effort has been focused on correlating mutational profiles in breast tumors with treatment response and outcome, as we will discuss later. Finally, it should be noted that breast tumors are not homogeneous with respect to genetic mutations, and the underlying genomic aberrations within regions of the tumor can give rise to distinct tumor biology and behavior.

Epigenetic Deregulation

Although germline and somatic mutations underline the molecular basis of breast cancer development, accumulating evidence delineates the importance of epigenetic changes in breast cancer carcinogenesis. Epigenetic alterations are fundamentally different from mutational events in that they are more frequent, are reversible, and tend to be non-random (i.e. gene-specific). Cell transformation is associated with epigenetic events that predominantly involve DNA methylation. In general, cancer cells display global hypomethylation, while hypermethylation occurs at gene-specific sites [25]. Hypomethylation of CpG islands may lead to aberrant expression of genes that are normally silenced by methylation in a particular tissue or developmental stage. This can contribute to overexpression of oncogenes leading to neoplastic transformation and cancer development. In addition, genome-wide hypomethylation

TABLE 13.1 Overview of Common Genetic Mutations in Breast Cancer Susceptibility Genes

Germline		Somatic		
Single nucleotide variants	Copy number variations	Single nucleotide variants	Copy number aberrations	
			Gene amplifications	Gene deletions
BRCA1	MYCN	TP53	ERRB2	PTEN
BRCA2	MYB	PTEN	CCND1	PPP2R2A
TP53		RB1	WIP1	MTAP
PTEN		USH2A	GASC1	MAP2K4
STK11		MYO3A	PIK3CA	
CHEK2		RAFB1	AKT1/2	
ATM		NRAS	STAT3	
MLH1		EGFR	ZNF703	
MSH2/6		ITGA2	MYC	
PMS2		LAMA2/4	MDM2	
BLM		MYH2/3/4/11/14	CCNE1	
WRN		DNAH2/3/5/9/11	EMSY (c11or f30)	
RB1			PAK1	
CDH1			RSF1	
PALB2			PALB2	
BRIP1			BRIP1	
NBS1			FGFR2	
RAD50			LSP1	
			MAP3K1	
			TGFB1	
			TOX3	

promotes genomic instability via destabilizing chromosomal peri-centromere regions [26].

On the other hand, gene-specific hypermethylation can promote tumor development through the unwanted silencing of tumor suppressor genes, a phenomenon that is considered a hallmark of cancer [18]. Changes in DNA methylation can affect gene expression either directly through activation (or silencing) of a gene via hypomethylation (or hypermethylation) of its promoter, or indirectly by activation or silencing of other genes that in turn regulate the expression of a gene of interest. In breast cancer, genes directly affected by hypermethylation include tumor suppressor genes (e.g. *BRCA1*), metastasis-inhibitory genes (e.g. *CDH1*, *TIMP-3*), hormone receptor genes (e.g. *ER-α*, *PR*) and cell cycle control genes (e.g. p16INK4a (*CDKN2A*)) [27]. For example, *CDKN2A* is found silenced by methylation in 30% of breast tumors [28] resulting in hyperproliferative tumor proliferation. In addition, the loss of expression of the *estrogen receptor alpha* (*ER-α*) gene observed in the aggressive triple-negative breast cancers (TNBC) is due to promoter hypermethylation and not a mutational or gene deletion event [29]. The lack of ER indicates insensitivity to hormone therapy and an aggressive tumor phenotype. In addition hypermethylation-mediated loss of *CDH1* (encodes E-cadherin) and *TIMP-3* gene expression is associated with a metastatic tumor phenotype and poor patient survival [30,31]. Similarly, the *BRCA1* gene has been also reported to be subject to silencing by DNA hypermethylation [32], an event that can promote genomic instability and subsequent accumulation of DNA damage [33].

BREAST CANCER DIAGNOSIS AND PROGNOSIS: A HISTORICAL OVERVIEW

For the last three decades, new approaches for breast cancer treatment have been extensively proposed in the hope of improving disease outcome and prolonging patient survival. Nonetheless, traditional surgical tumor resection or mastectomy with/without radiation therapy remains the best treatment for breast cancer. Unfortunately, in patients with advanced metastatic disease, resection is not helpful or recommended especially since post-surgery trauma can promote aggressiveness of the residual tumor cells. As a result, chemotherapy is now the gold standard for patients with metastasizing breast cancer but is also an option for those with localized disease.

In principle, chemotherapy involves the administration of chemical agents that generally target proliferating cancer cells. Chemotherapeutic agents can be given in an adjuvant form post-primary treatment (surgery, anti-hormone therapy, radiation) to reduce the risk of recurrence or to offer palliative care in patients with advanced metastatic disease. In contrast, neoadjuvant chemotherapy is given before primary treatment to shrink the primary tumor and reduce the need for aggressive surgical resection (i.e. radical mastectomy). Being able to predict whether patients will benefit from or require adjuvant therapy or neoadjuvant regimen is a major leap forward towards personalized medicine.

Clinical Diagnosis and Prognosis

Histopathological Stage and Tumor Grade

Standard assessment of breast cancer patients relies on certain histological criteria as well as cytological characteristics of tumors [34]. Tumor grade assesses the histopathology of breast tumors by characterizing the differentiation status, nuclear pleomorphism and proliferative capacity (e.g. by determining the mitotic index and/or proliferation using S-phase markers such as Ki-67). The differentiation status evaluates how closely the tumor tissue resembles the normal tissue it arose from. For instance, a well-differentiated tumor has similar histological features to normal tissues and hence has a better prognosis compared to poorly differentiated or anaplastic tumors. Another method of breast cancer assessment is tumor stage which defines the behavior of a tumor based on tumor size and location, lymph node involvement, and presence/absence of metastasis (Figure 13.2). A stage 0 breast cancer represents a carcinoma *in situ* which is confined within the duct. Stage I tumors spread beyond the confines of the duct or lobule, and may reach the lymph nodes. Lymph node dissemination defines a stage II disease while further invasion of other tissues (skin, muscle) is stage III. Notably, there are some exceptions in this staging system especially in large-size tumors. For example, lymph node-negative tumors which are larger than 5 cm still classify as stage II. Tumors are upgraded to Stage IV only when distant metastasis is detected [35].

Receptor Statuses (ER, PR and HER2)

As part of the standard assessment of breast tumors, immunohistochemical analyses are used to characterize the expression of three receptors, namely the estrogen receptor (ER), progesterone receptor (PR), and the human epidermal growth factor receptor 2 (HER2) (see Figure 13.2). Up to 70–80% of breast cancer cases are positive for hormone receptors which allows the use of standardized hormone therapy (e.g. tamoxifen, which targets ER+ tumors). HER2 is found overexpressed in 10–15% of all cases where the *ERBB2* gene, which encodes HER2, is overexpressed or amplified [36]. HER2+ patient tumors are treated with anti-HER2 therapies, predominantly the humanized monoclonal anti-HER2 antibody (Herceptin; trastuzumab), which prevents HER2 dimerization and intracellular signaling. The remaining 15–25% of breast cancer cases lack expression of all three receptors and are referred to as triple-negative breast cancers. The latter do not respond to anti-hormone therapy or anti-HER2 drugs and, as a result, have a poor prognosis [37]. Current clinical practice often combines hormone receptor status with histological grade and tumor stage to predict treatment response and prognosis.

Although receptor-based classification offers an insight into treatment choice and patient outcome, it remains a crude assessment as patients may have different clinical outcomes and respond differently to therapy even though they have the same receptor status. In particular, receptor classification fails to take into account the underlying molecular and genomic basis of breast cancer. As a result, much effort has been focused on the characterization of gene expression changes in breast cancer that can be used to predict patient outcome and treatment response, which we will discuss in detail in the next section.

Gene Expression Signatures

When predicting the outcome of breast cancer, many factors are taken into account. Age, menopause, hormone receptor status, tumor grade and stage are all used to predict clinical outcome in breast cancer patients. Although these clinically based predictions have been useful in prediction of prognosis, choice of treatment, and risk of recurrence, they emerge with a limited predictive capacity when differentiating patients on an individual basis. In fact, two patients with similar clinico-pathological predictions may have different outcomes and respond

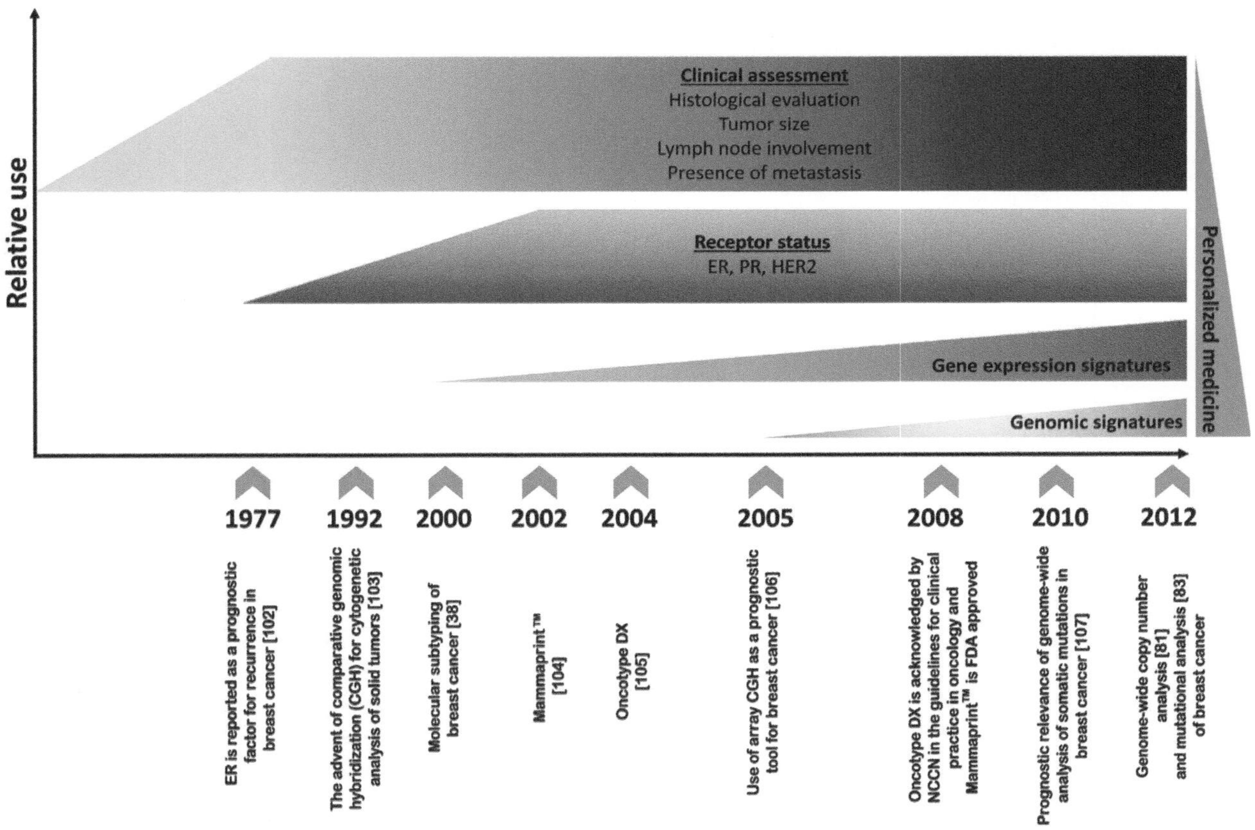

FIGURE 13.2 Timeline and selected key events in the evolution of breast cancer prognosis and predictive signatures. The x-axis represents key events and is not drawn to scale. The y-axis represents the overall use of a certain method clinically or preclinically.

differently to treatment. This, in part, relates to the heterogeneity of breast tumors among patients. As a result, recent advances in breast cancer research have utilized high-throughput microarray-based technologies to address the heterogeneity of patient tumors based on gene expression profiles (Figure 13.3). Gene expression profiling not only allows patients with tumors of similar grade or hormone status to be differentially stratified but can also be used to produce predictive and prognostic gene signatures based on the expression of a set of key regulatory genes. To date, over 50 breast cancer gene signatures have been proposed to predict patient survival, risk of recurrence, treatment response, and the chance of metastasis. The majority of these signatures also aim to define high- and low-risk groups of patients who may differentially benefit from adjuvant chemotherapy or radiation treatment.

Breast Cancer Intrinsic Subtypes and PAM 50

The analysis of gene expression using microarray-based techniques has allowed breast cancer researchers to undertake a more comprehensive approach towards profiling of breast cancers. In 2000, Perou et al. proposed a comprehensive molecular profiling of breast cancer that classified breast cancers into five distinct subtypes; *luminal A*, *luminal B*, *basal-like*, *HER2-enriched* and *claudin-low* (see Figure 13.3) [38]. These expression profiles give rise to differential tumor biology and distinct clinical outcomes [39,40]. These subtypes can be partially distinguished based on receptor status which may not be exclusive to a single subtype. In other words, although ER + tumors constitute the majority of the luminal A subtype, they overlap with other subtypes (e.g. luminal B). The same applies to TNBC which can be of basal-like or claudin-low subtypes, both of which have a poor prognosis. Hence, the terminology describing the breast cancer intrinsic subtypes and/or hormone receptor status has not always been clearly defined and care must be taken when navigating the literature describing the various breast cancer gene signatures.

The intrinsic classification was later refined by Parker et al. into a list of 50 genes (PAM50; prediction analysis of microarray), which was sufficient to distinguish the five subtypes [41]. More recently, a new intrinsic subtype has

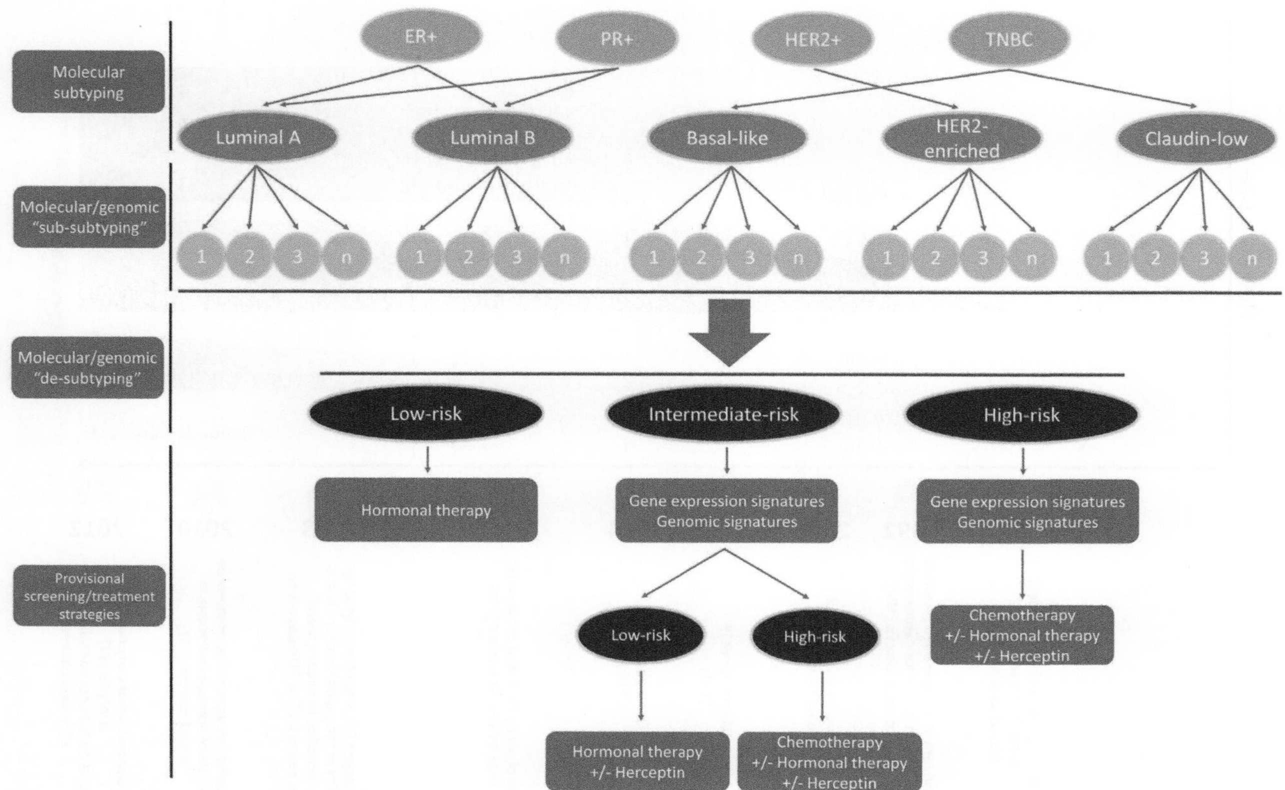

FIGURE 13.3 Current and prospective subtyping of breast cancer patients. The figure signifies the necessity of "de-subtyping" breast cancers into a three-level risk grouping (high, intermediate, and low) to combine all the current/upcoming subtypes.

been identified as *claudin-low* which predominantly consists of triple-negative tumors that express low levels of E-cadherin and display an intermediate prognosis between basal-like and luminal subtypes [42]. The claudin-low subtype is also known as the *mesenchymal subtype*, which reflects that fact that these tumors are enriched for a mesenchymal and tumor-initiating cell population [43,44].

Luminal A cancers tend to have a good prognosis predominantly because they are ER + and respond to hormone therapy. In contrast, basal-like and HER2-enriched patients tend to have a worse prognosis, yet they are predicted to respond well to adjuvant chemotherapy [45]. In a recent randomized trial, HER2-enriched tumors were found to be sensitive to anthracyclines over other chemotherapies [46] indicating that anthracyclines may be a better treatment option for this subtype. Later studies have further subdivided the five intrinsic subtypes (see Figure 13.3). For instance, basal-like and HER2-enriched tumors that are ER-negative can be further subdivided into four groups of different clinical outcome, based on differences in immune cell infiltration of the tumor and the expression of immune response pathway genes (*STAT1*, *SP110*, *NFKBI*, *IFI44*, *IFNGR1*, *ISGF3G*, and *IRF7*) [47].

ER-Positive Gene Signatures

The primary method of treatment of ER + patients is hormonal therapy through the administration of the estrogen receptor antagonist tamoxifen (Nolvadex). However, further adjuvant chemotherapy (e.g. anthracyclines, taxanes) is often recommended in higher risk ER + individuals who meet certain criteria, which include older age, tumor size (>1 cm), lymph node (LN) involvement, and high tumor grade (poorly differentiated). If patients have all risk factors, it is almost necessary for them to receive adjuvant therapy to reduce recurrence and improve event-free survival. However, in most cases, patients present with only one of these risk factors after which the decision to use adjuvant chemotherapy becomes less compelling or obvious. Although the cautious use of adjuvant therapy in intermediate-risk patients has dramatically increased their survival [48], the fear of over-treatment is becoming a concern, especially in terms of the cost of chemotherapy treatment and increased morbidity associated with treating patients who may not benefit from adjuvant therapy. The issue of over-treatment of breast cancer patients has been addressed by Gnant and Steger [48]. Fortunately, online tools such as Adjuvant! have been developed to aid physicians in deciding whether to proceed with adjuvant chemotherapy in post-surgery early ER + patients [49]. Although Adjuvant! and similar tools are useful in addressing over-treatment, their accuracy can still be improved.

Unlike tumor grade and receptor status, gene signatures may provide clinicians with a more reliable approach to predict prognosis and possible benefit of using adjuvant therapy. For example, 21-gene (Oncotype DX) and 70-gene (Mammaprint) signatures [50,51] are currently being used successfully in North America and parts of Europe to divide early stage ER + patients into those who are predicted to benefit from adjuvant chemotherapy and those who do not require further treatment [52,53]. In 2008, the National Cancer Comprehensive Network (NCCN) announced new updates for the use of Oncotype DX as part of the NCCN Clinical Practice Guidelines in Oncology. These guidelines consider the use of Oncotype DX in ER + /PR + , HER2-negative patients with poor to moderately differentiated tumors and a size between 0.6 and 1.0 cm [54]. Both signatures have predictive and prognostic potential especially in high-risk patients and have helped to increase disease-free survival of those who received anthracycline-based adjuvant chemotherapy [55,56]. However, the predictive value of these signatures can be extended to other chemotherapies including microtubule-targeting drugs such as taxanes, a class of drug typically used to treat HER2 + tumors [57].

More specific gene signatures are under development to predict prognosis in specific stages of ER + breast cancer but are more limited in their application to other breast tumors than the 21-gene and 70-gene signatures [58]. For instance, a microarray-based 97-gene genomic grade index (MapQuant DX) is capable of differentiating between grade II tumors which will behave as grade III tumors (high-risk group) or those which will behave as stage I tumors (low-risk group) [59,60]. The low-risk group can be treated without the need for chemotherapy. Some quantitative RT-PCR-based signatures have also been developed to examine small sets of genes or to verify the predictive value of other signatures. Using this approach Sánchez-Navarro et al. developed an 8-gene recurrence score for the prediction of 5-year metastasis-free survival which was derived from previous gene signatures including the 70-gene signature [61]. Similarly, Jerevall et al. combined the two-gene ratio (*HOXB13: IL17BR*) and a five-gene (*BUB1B*, *CENPA*, *NEK2*, *RACGAP1*, *RRM2*) molecular grade index (MGI) to produce the Breast Cancer Index, which was capable of identifying patients with low-risk ER + breast cancer who are treated with tamoxifen but do not require adjuvant chemotherapy [62,63] (see Table 13.1).

A major drawback of gene expression signatures is the turnover time of testing since patient samples can only be processed at a reference center and cannot be done in local hospitals. In addition, it is also challenging to obtain a biopsy from small tumors, especially in early stage ER + patients who eventually may not require adjuvant therapy. Although Oncotype DX is highly predictive of high-risk and low-risk patients and the potential benefit of chemotherapy, it remains elusive in some intermediate-risk cases where the benefit of chemotherapy is unclear. In 2006, the National Cancer Institute (NCI) performed the phase III clinical trial of TAILORx to assess the risk of recurrence and the need for chemotherapy in early stage LN-negative intermediate-risk ER + patients. This trial was set to determine whether hormone therapy alone or combined with adjuvant chemotherapy is more beneficial to patients with an intermediate Oncotype DX score between 11 and 25 [64]. A similar phase III clinical trial (MINDACT) assessed the applicability of the Mammaprint signature on ER + LN-positive patients to determine the benefit of adjuvant chemotherapy [65].

Interestingly, the aforementioned gene signatures have been primarily developed for patients with ER + ductal carcinomas. It has been assumed that these signatures could also be applied to the less common lobular carcinomas. However, recent studies suggest that lobular carcinomas have a different genomic and molecular profile from ductal carcinomas [66,67]. Hence, existing gene signatures must be verified and new signatures identified for lobular carcinoma patients.

HER2-Positive Gene Signatures

HER2 + tumors are a clinically challenging subgroup and can respond differently to therapy, reflecting the complex nature of these tumors. Since these tumors are often associated with a poor prognosis, almost all HER2 + patients receive chemotherapy along with Herceptin (trastuzumab) as a neoadjuvant or an adjuvant. In high-risk patients, Herceptin is combined with either Tykerb (lapatinib) or Perjeta (pertuzumab) [58]. HER2 + patients who are node positive can benefit from paclitaxel after receiving adjuvant doxorubicin plus cyclophosphamide therapy [68]. Notably, ER + patients who also overexpress HER2 can be treated with anti-HER2 drugs. Although HER2-targeted therapies can be effective, several lines of evidence show that a subset of patients develop resistance by acquiring specific genomic alterations [69]. As a result, gene signatures have been developed to predict resistance to anti-HER2 drugs [70,71]. Interestingly, patients who received Herceptin yet later developed recurrence had a different gene expression profile than those who did not recur [70]. These differentially expressed genes must be further studied to determine the mechanism of Herceptin resistance.

The 70-gene signature can be an independent prognostic indicator of outcome in HER2 + patients and allows the identification of a low-risk early stage group with a favorable long-term metastasis-free survival [72]. Moreover, a 158-gene signature identified three subtypes of HER2 + tumors with mixed stage, different histologic grade, and ER status of which one subtype predicted poor outcome and reduced long-term survival. The differentially expressed genes that gave rise to these predictions were mostly associated with processes of tumor invasion, metastasis and antitumor immunity (e.g. *CXCR4*, *PLAU*, *CX3CR1*, *TGFBR3*, and *STAT5*) [73]. Clearly, more effort is required to identify further the molecular and genomic basis of HER2 + tumors and their response to therapy, especially in patients who received Herceptin and failed to respond. Further validation of gene expression profiles in these patients may offer insights into understanding how Herceptin resistance will develop, and could be used to determine when adjuvant therapy is needed.

Triple-Negative Breast Cancer Gene Signatures

TNBCs do not express ER, PR or HER2 and hence fail to respond to anti-hormone therapy (e.g. tamoxifen) or Herceptin. Consequently, cytotoxic chemotherapy remains the main treatment modality for TNBC patients. Generally, TNBCs are poorly differentiated tumors with a poor prognosis, with the exception of medullary carcinoma and adenoid cystic carcinoma, which tend to have a favorable outcome [74,75]. Discrete gene signatures of TNBC have not been fully identified, predominantly because no low-risk group of patients is predicted to benefit from adjuvant chemotherapy. In addition, applying gene signatures of ER + tumors to TNBC showed limited success in predicting survival and treatment options [76].

Unlike ER + signatures, which often reflect the proliferative potential of a tumor, existing TNBC signatures are often related to the susceptibility of these tumors to antitumor immunity [76]. High expression of inflammatory and angiogenic genes is associated with poor prognosis. In contrast, high B-cell count and low interleukin 8 (IL-8) expression correlates with better prognosis [77]. A recent study identified a novel set of 14 genes which predict metastatic outcome and may help spare early stage TNBC patients from unneeded adjuvant therapy. This gene set is also linked to inflammatory cytokine regulation (e.g. *CXCL13*, *RGS4*) [78]. More recently, a microarray-based gene expression analysis of TNBC identified six subtypes: two *basal-like* tumors (BL1 and BL2); *immunomodulatory* (IM) tumors; *mesenchymal* (M) *tumors*; *mesenchymal stem-like* (MSL) tumors; and *luminal androgen receptor* (LAR) tumors. The gene signature of each subtype revolved around an activated "driver" growth factor/hormone pathway (e.g. PI3K/mTOR) and correlated with poor relapse-free survival. By cross-matching these signatures to established TNBC cell lines, the authors were able to predict the sensitivity of these subtypes to currently approved drugs. For instance, cell line models for M and MSL subtypes responded well to BEZ-235 (PI3K/mTOR inhibitor), while BL1 and BL2 subtype models were sensitive to cisplatin [79]. Turner et al. have utilized high-resolution microarray-based comparative genomic hybridization (aCGH) to examine gene expression and its relation to copy number variation. *The fibroblast growth factor receptor 2* (*FGFR2*) gene was found amplified in a small subgroup (4%) of TNBC, and *FGFR2*-overexpressing TNBC cell lines were sensitive to the FGFR inhibitor (PD173074) [80].

Collectively, these studies overwhelmingly demonstrate that TNBC is a heterogeneous disease and finding a robust gene signature that accurately stratifies patient risk and treatment response remains a challenging obstacle. Any gene signatures developed for TNBC must also undergo similar testing and validation to that of Oncotype DX and Mammaprint signatures used to stratify patients with ER + breast cancers (Table 13.2) [58]. Nonetheless, these studies demonstrate the power of gene expression analysis to uncover important molecular and cellular pathways for which new targeted therapies can be developed in the future.

Genomic Signatures

In the last five years, the ability of genome-wide analysis to generate information has been increasing at an ever-growing pace. Genome-wide analysis techniques based on next-generation sequencing have become faster, more reliable, and more easily applicable to clinical practice. These technologies are transforming how we view cancer development and progression by providing a means of documenting the underlying genetic basis of human cancers at single nucleotide resolution. As a result, multiple studies have been published revealing the diversity of the breast cancer genomic landscape which, in turn, has provided the opportunity for new approaches for therapeutic targeting.

Defining the breast cancer landscape using genome-wide analysis allows us to examine: single nucleotide variations (SNVs); single nucleotide polymorphisms (SNPs); chromosomal aberrations or translocations; somatic copy-number aberrations (CNAs); transcriptome profiles; and epigenetic changes. This is achieved by high-throughput high-resolution microarray techniques and parallel DNA or RNA sequencing, yielding valuable information which can be used to select cancer-causing alterations to guide treatment choice and/or predict treatment efficacy. Importantly, these technologies have revealed that the

TABLE 13.2 Gene Expression Signatures Used for Breast Cancer Diagnosis and Prognostic Testing

Signature	Test Type	Target Patient Cohort	Predictive Value	Gene List	Reference
Oncotype DX	qRT-PCR	Stage I or II ER + LN-negative	Recurrence risk Chance of metastasis	Ki67, STK15, Survivin, CCNB1, MYBL2, GSTM1, BAG1, CD68, ER, PR, BCL2, SCUBE2, MMP11, CTSL2, ERBB2, GRB7, ACTB, GAPDH, RPLPO, GUS, TFRC	51
Mammaprint	Microarray	Stage I or II ER + LN-negative and LN-positive (size <5 cm) some HER2 +	10-year metastasis-free survival	BBC3, EGLN1, ESM1, IGFBP5, FGF18, SCUBE2, TGFB3, WISP1, FLT1, STK32B, RASSF7, DCK, MELK, EXT1, GNAZ, EBF4, MTDH, PITRM1, QSCN6L1, CCNE2, ECT2, CENPA, LIN9, KNTC2, MCM6, NUSAP1, ORC6L, TSPYL5, RUNDC1, PRC1, RFC4, RECQL5, CDCA7, DTL, COL4A2, GPR180, GPR126, RTN4RL1, DIAPH3, CDC42BPA, PALM2, ALDH4A1, AYTL2, OXCT1, PECI, GMPS, GSTM3, SLC2A3, FLT, FGF18, COL4A2, GPR180, EGLN1, MMP9, LOC100288906, C9orf30, ZNF533, C16orf61, SERF1A, C20orf46, LOC730018, LOC100131053, AA555029_RC, LGP2, NMU, UCHL5, JHDM1D, AP2B1, MS4 A7, RAB6B	50
97-gene genomic grade index (MapQuant DX)	Microarray	Grade I, II and III ER +	Histologic behavior Relapse-free survival Risk of recurrence	LAMB2, FLJ23554, SIRT3, SESN1, FLJ20477, CDCA3, TIMELESS, TPX2, OIP5, CENPA, BBS1, H2AFZ, KPNA2, KIF11, HMGB3, CCNB2, CCNE2, KIF20A, DDX39, EXO1, KNTC2, RNASEH2A, SPAG5, POLQ, GTSE1, ONSON, FLJ20641, CDCA8, RACGAP1, ORMDL2, UBE2S, ESPL1, KIF4A, ASPM, DKFZp762E1312, NUSAP1, MCM10, C20orf24, FLJ10156, DC13, KNSL7, HSMPP8, RRM2, BUB1B, ZWINT, TTC10, CYBRD1, PRC1, C10orf3, CCT5, UBE2N, 13CDNA73, STARD13, FLJ21062, FLJ21827, CX3CR1, CDC25A, BLM, TTK, MYBL2, MELK, CDK2, FOXM1, CCNB1, STK6, BM039, BRRN1, PLK1, CDC2, MARS, FEN1, TRIP13, HMMR, AURKB, MCM4, BUB1, MCM2, MLF1IP, CENPF, SLC7A5, BIRC5, TROAP, CDC20, MAD2L1, NUDT1, GMPS, KIF2C, SHMT2, CENPE,DLG7, MKI67, CDKN3, CCNA2, LMNB1	59
8-gene recurrence score	qRT-PCR	Stage I or II ER + /PR +	Metastasis-free survival Risk of recurrence	DTL, ECT2, MTDH, PRC1, RFC4, SCUBE2, STK32B, ZNF533	61
14-gene distant metastasis signature	Microarray-verified (qRT-PCR-based)	TNBC	Chance of relapse Metastasis-free survival	RGS4, CXCL13, HAPLN1, HRBL, MATN1, PRTN3, SSX3, PRS28, EXOC7, ABO, CLIC5, RFXDC2, ZNF3, PRRG3	78
158-gene signature	Microarray	HER2 +, ER +/−, any stage or grade	Disease-free survival Benefit from anti-HER2 therapies	LTBP1, SPTBN2, GPR172A, ZNF609, LOC116143, ELMO3, ZNF281, PXDN, CASP6, PGM2L1, INHBA, IRF6, C12ORF29, GRHL1, KPNA3, LPGAT1, MGC40579, SH3YL1, COL8A1, S100A11, PEX11B, MED19, ZNF690, CD55, SENP5, COMP, SLK, NUDT9, KIAA1549, C7ORF25, DCBLD1,	73

(*Continued*)

TABLE 13.2 (Continued)

Signature	Test Type	Target Patient Cohort	Predictive Value	Gene List	Reference
				PLAU, GLRX2, AFTIPHILIN, HSPBP1, GRB7, DDEF2, LAD1, PKP3, CXCR4, PERLD1, TSPAN7, S100B, ANPEP, ARIH2, GPC3, SOX10, EDNRB, SH3KBP1, STON2, RBP4, C6ORF188, PDE2A, STAT5A, RNF166, MFAP4, STXBP1, SOCS1, ATP8B2, TIGAL, LONRD1, TGFBR3, MAP7D3, CYORF15B, GPX3, ALAS2, PLAC8, CLSTN3, HEMGN, NLGN2, HMHA1, CRMP1, FYN, ZAP70, FLNC, AXIN2, FGF2, LAT, NDN, RBM5, ARHGAP19, RASIP1, TF, ATOH8, MAP4, C6ORF190, SH3BP5, SNCA, RAC2, SLC38A5, PIM1, KIRREL, TMEFF1, CX3XR1, SDPR, PPP5C, CLCN7, MLXIPL, C17ORF68, SFRS5, MTHFR, AFP, DKFZP686E10196, DRCTNNB1A, IL17RD, PPP1R14A, EPB41, SNF1LK2, ITGA4, 8MAR, PRKCA, KIAA1754L, OSBPL3, PLXNA4B, SLC25A42, BIN3, EDG5, STARD4, HHIP, CD3D, LC16A7, ETS1, LTBP4, C14ORF49, CD69, PELI2, CDON, FAIM3, C11ORF57, RFT1, ABCD4, SEMA6A, CFL2, ADD3, FAM46A, DMN, WDHD1, ZNF276, FAM89A, PTPRCAP, LOC133308, UBE2G1, BMP4, TM2A, ST6GALNAC6, KIAA1826, PRSS12, KLF9, SEMA7A, TLE4, GAS7, GYPC, GLTSCR1, LOC90624, TCF7, DPYSL2, ST3GAL2, COLEC12	
PAM50	qRT-PCR and microarray	All breast cancer patients	Intrinsic subtyping Benefit of neoadjuvant therapy Overall survival	FOXC1, MIA, KNTC2, CEP55, ANLN, MELK, GPR160, TMEM45B, ESR1, FOXA1, ERBB2, GRB7, GFR4, BLVRA, BAG1, CDC20, CCNE1, ACTR3B, MYC, SFRP1, KRT14, KRT17, KRT5, MLPH, CCNB1, CDC6, TYMS, UBE2T, RRM2, MMP11, CXXC5, ORC6L, MDM2, KIF2C, PGR, MKI67, BCL2, EGFR, PHGDH, CDH3, NAT1, SLC39A6, MAPT, ,UBE2C, PTTG1, EXO1, CENPF, CDCA1, MYBL2, BIRC5	41

Oncotype DX and Mammaprint are the only signatures which have been approved for clinical use. Other signatures are still in preclinical stages and require further validation and feasibility analyses.

extent of divergence of breast tumors among individuals, as well as that during different stages of tumor development, is poorly characterized by traditional prognostic methods.

Copy Number Aberrations and Integrative Clustering

Since HER2 overexpression in breast cancer can result from gene amplification, recent genome-wide analysis focuses on examining how gene copy number alterations contribute to breast carcinogenesis and whether copy number-based classifications could be developed for breast tumors. In a large scale study, Curtis et al. examined CNAs in 2000 breast tumors of all subtypes using MLPA (multiplex ligation probe analysis). This technique detects genomic deletions and amplifications which can be later assigned as cancer-causing events. The study identified 45 regions with recurrent deletions or amplifications that largely affected gene expression and hence were considered as possible "driver" events. The most common events were deletions of *PTEN*, *PPP2R2A*, *MTAP*, *MAP2K4* and amplifications in *ERBB2*, *MYC*, *MDM2*, *CCND1*, *CCNE1* and *ZNF703*. Deletion in *PPP2R2A* correlated with highly mitotic ER + tumors within the luminal B subtype [81]. By combining copy number changes with gene expression patterns, Curtis et al. were able to derive 10 distinct groups, dubbed as integrative clusters, which have differential CNAs and corresponding gene expression signatures. These clusters had 10 distinct copy number aberrations and gene expression maps which translated into differential survival rates. Interestingly, cluster IV has the best prognosis as it is devoid of any CNAs and has strong antitumor immune signatures. In contrast, clusters V and X have the poorest prognosis as they correlate with aggressive HER2-enriched subtypes and high genomic instability [81]. Although, patient clustering based on copy number alterations and gene expression profiles can help explain differences in prognosis and response to treatment, these associations still remain elusive and, as a result, are inapplicable in a clinical setting today. It is crucial, however, to examine what role the identified CNAs play in cancer development and whether they can be independent predictors of prognosis or treatment response.

In a study which examined somatic CNAs and exon mutations of 100 breast tumors, Stephens et al. identified significant differences in the number of somatic mutations among breast cancer patients. Mutations and indels (insertions and deletions) were identified in many "suspect" genes (*AKT1*, *BRCA1*, *CDH1*, *GATA3*, *PIK3CA*, *PTEN*, *RB1* and *TP53*), but also in new potentially cancer-causing genes such as *AKT2*, *ARID1B*, *CASP8*, *CDKN1B*, *MAP3K1*, *MAP3K13*, *NCOR1*, *SMARCD1* and *TBX3*. These genes were considered driver genes since they were mutated more frequently than what is predicted by pure chance and have also been previously associated with cancer. Interestingly, these genes are part of the JUN kinase signaling pathway which might indicate the importance of the pathway as a therapy target used in combination with adjuvant therapies. One of the identified mutational signatures strongly correlated with a cytosine mutation in CpG dinucleotide regions. The degree of these substitutions strongly correlated with age in ER-negative patients but not in ER+ patients, delineating differential dynamics of mutation occurrence in breast cancer subtypes [30].

It is generally accepted that ER+ patients who do not respond to aromatase inhibitors tend to have Ki-67-high tumors. Using parallel genome sequencing of ER+ tumors before treatment, Ellis et al. showed that aromatase inhibitor resistance is strongly associated with a higher frequency of somatic mutations (*RUNX1*, *CBFB*, *MYH9*, *MLL3*, *TP53*, *SF3B1*, *MAP3K1* and *MAP2K4*) and genomic structure aberrations when compared to Ki67-low tumors [34]. Banerji et al. calculated the rate of mutation acquisition around one mutation per one million base pairs. Whole-exome sequencing of 103 breast cancer patients in Mexico and Vietnam confirmed many recurring mutations (*PIK3CA*, *AKT1*, *GATA1*, *MAP3K1* and *TP53*) but also identified new driver events such as a mutation in the transcription factor *CBFB*, *RUNX1* deletion, and a *MAGI3–AKT3* fusion event in TNBC patients. This fusion event promotes a constitutively active AKT/TOR pathway but is sensitive to AKT inhibitors [82].

The advent of genome-wide analysis of breast tumors has opened a new era of patient stratification and options for treatment. Combining copy number aberrations with expression signatures appears to be the most robust approach in breast cancer subtyping and the latest addendum towards personalized medicine. Nonetheless, these approaches must be verified and validated through randomized clinical trials to determine their robustness and applicability in standard clinical practice.

Intra-Tumoral Heterogeneity and Mutational Signatures

Intra-tumoral heterogeneity represents the pathological, molecular and genomic variability that exists among cells within the same tumor. To date, it remains a major obstacle against complete remission post-treatment. This heterogeneity further contributes to the necessity of introducing personalized care approaches to our healthcare systems. Using TNBC tumors, Shah et al. revealed that the frequency of a mutational event is variable within and among tumors. A mutation with a high clonal frequency means that it is present in the majority of cells within a tumor and is also indicative of being an early driver event. In contrast, passenger events tend to have lower clonal frequency as they accompany driver events but have no clear implication for tumor development. Not surprisingly, *TP53*, *PTEN* and *PIK3CA*, which are commonly mutated in breast cancers, had the highest clonal frequency delineating their important role in breast cancer development and progression [83]. Interestingly, some mutations in genes regulating the integrin signaling pathway (e.g. *MYO3A*) were observed. Although aberrations to these pathways have been implicated in cancer, no mutations have been associated with these genes [83]. Hence, further validation of these mutations is necessary to understand their contribution to breast cancer, especially TNBC, allowing us to develop target-specific inhibitors which can be given in combination with current chemotherapies. Lastly, through the sequencing of 21 breast cancer genomes, a particularly unique form of somatic mutations was uncovered, termed "kataegis", that describes a shower of clustered mutations in a single chromosomal locus that are predominately C to T transitions in the context of TpC dinucleotide sequences [84]. It was later discovered that enzymes of the AID/APOBEC family of cytosine deaminases were responsible for kataegis [85,86]. However, the prognostic value and the consequences of this kind of mutation spectrum in breast cancer remain to be fully characterized.

FUTURE OF BREAST CANCER PROGNOSIS AND TREATMENT

Feasibility of Molecular/Genomic Signatures in Clinical Practice

Apart from OncotypeDX and Mammaprint, gene expression and genome-wide analyses have been limited to research use only and have not been incorporated into clinical practice. Although these technologies are constantly being optimized for higher efficiency and cost-effectiveness, they are still expensive for population based-screening and, as a result, cannot be incorporated into our healthcare systems as standard procedures. Reduction in costs, decreased turnover times and multiplexing are all factors that can promote the introduction of genome-wide analysis into clinical practice, especially with regard to whole-genome sequencing. On average, it requires around 7 days to sequence, process and analyze one human genome. However, newer robust sequencing techniques are now able to perform full multiplex genome sequencing in less than 48 hours. A faster turnover time reduces testing and technician costs to the healthcare system and may aid in the introduction of high-throughput sequencing as a standard screening method for breast cancer patients in the future.

Another approach involves targeting high-risk patients only for genomic and molecular screens for the prediction of therapy choice and response. Such an approach requires "de-subtyping" of all breast cancer subtypes back into high-, intermediate- and low-risk groups (see Figure 13.3). As a result, high-risk patients can have more frequent follow-ups for early detection of relapse post-treatment. In contrast, genomic/gene expression signatures may not be necessary for low-risk patients who generally do not need additional chemotherapy. Importantly, this approach may prevent over-treatment, over-diagnosis, and ultimately reduce costs.

Non-Invasive Monitoring of Residual and Occult Disease

In the last three decades, there have been attempts to improve strategies for detection of occult breast cancer at earlier stages in the hope of improving diagnosis. These strategies involve standard mammography, diagnostic imaging (MRI; magnetic resonance imaging), as well as follow-up imaging for detection of recurrent and metastatic disease. Although the increased and standardized use of mammogram screening correlated with increased diagnosis of breast cancer, it has led to earlier detection of occult disease and reduced breast cancer mortality in women [87,88]. Interestingly, the contribution of mammography on patient mortality is not as significant when taking into account other factors such as increased awareness and improved treatments [89]. In fact, on average, more than 700 women must be screened for 10 consecutive years in order to prevent one breast cancer death [88]. As a result, the benefits of mammography are questionable. A possible explanation is that early tumors detected by mammography tend to be less aggressive in nature as evident by their histopathology [90] and gene expression profiles [91]. On the other hand, MRI is commonly used for assessment of breast cancers which were previously confirmed by biopsies. MRI allows preoperative detection of the regional spread of a lesion and can guide marginal excision or mastectomy [92]. Notably, multiple reports have revealed that MRI can produce false positives primarily due to inter-observer variability. In fact, MRI false positives give rise to 1–2% of unnecessary mastectomies [92,93].

Serum markers are also being used for detection of breast cancer in susceptible individuals and in post-treatment patient monitoring. The most common serum markers for breast cancer are the carcinoembryonic antigen (CEA) and the cancer antigen 27.29 (CA 27.29; MUC1 glycoprotein) [94]. In 2004, Cristofanilli et al. suggested that circulating tumor cells (CTCs) can be used as a marker for breast cancer diagnosis and prognosis, as well as for treatment monitoring post-chemotherapy or radiation [95]. Later studies confirmed the role of CTCs as prognostic indicators in patients with metastatic breast cancer [96–98] and they are superior to traditional imaging methods [99]. Further studies are required to establish how the CTC status can be utilized feasibly in a clinical setting. Circulating tumor DNA (ctDNA) has been also proven to be a highly sensitive prognostic marker of metastatic breast cancer and provides the earliest answer for treatment response when compared to traditional serum markers (i.e. CA 15–3) [100]. Cell-free DNA allows detection of cancer-causing genetic aberrations at a very high sensitivity hence determining the presence of occult tumor cells in primary sites or micro-metastasis containing dormant cells. Combination of data analyses from CTCs and ctDNA may provide additional insights to complement traditional imaging methods.

Need for Standardization and Personalized Medicine

Although high-throughput genome-wide studies provide us with the basis underlying the breast cancer landscape, more efforts are needed to filter down the data and decide what could be used routinely in clinical practice. It is obvious that some genomic events occur frequently and consistently in breast cancer. However, little is known about the thousands of rare and unique events that distinguish individual tumors and most likely give rise to differences in prognosis and treatment response. More studies are required to understand the functions of these genes and how they contribute to carcinogenesis, the outcome of which may be the development of targeted treatments and combination therapies tailored to individual tumors to achieve more desirable treatment responses.

The aforementioned genomic approach to personalized medicine will require genome-wide analysis of individual tumors followed by target validation. Although this is clinically unfeasible at this point, it offers hope that genome-wide precision medicine is an option to pursue in identification of "druggable" targets and more precise genome-based subtyping of breast cancer patients. It is important to realize that traditional pathological subtyping (receptor status, stage etc.) may not be replaced by genomic approaches. Instead, the latter may act as complementary prognostic tools for more accurate predictions and assignment of treatments.

ACKNOWLEDGMENTS

We would like to thank Dr Penny Barnes (Dept Pathology, Dalhousie University) for providing the histological images in the masthead of this chapter. This work was funded by operating grants to PM from the Canadian Institutes of Health Research (CIHR), and to GD from the Canadian Breast Cancer Foundation (CBCF) – Atlantic. Both PM and GD are Senior Scientists of the Beatrice Hunter Cancer Research Institute (BHCRI) and MB is supported by a trainee award from the BHCRI with funds provided by Cancer Care Nova Scotia as part of The Terry Fox Strategic Health Research Training Program in Cancer Research at CIHR.

GLOSSARY

Adjuvant chemotherapy Chemotherapy administered post-primary treatment (surgery, anti-hormone therapy, radiation) to reduce the risk of recurrence or to offer palliative care in patients with advanced metastatic disease.

Anthracyclines DNA intercalating class of antibiotics which are derived from the bacterium *Streptomyces* and are used as potent anticancer agents. The most common anthracycline is doxorubicin.

Carcinoma *in situ* Carcinoma which remains confined within the lobule or duct from which it arises.

Chemotherapy Administration of chemical agents that generally target proliferating cancer cells.

Circulating tumor cells (CTCs) Tumors cells that disseminate away from the primary tumor into the circulation and are often indicative of distant-site metastasis.

Circulating tumor DNA (ctDNA) Tumor DNA which sheds from the primary tumor upon cell lysis and can be detected by highly sensitive molecular techniques.

Clonal frequency of mutation The percentage of cells within a tumor which have a specific mutation relative to the total number of cells.

Copy number aberrations (CNAs) Gain or loss of certain regions of the genome which may cause gain or loss in gene copy number.

DNA methylation Addition of a methyl group to the cytosine base of DNA. The methylation of a regulatory DNA sequence can silence the expression of a given gene.

DNA microarray DNA chip which allows measurement of the expression levels of thousands of genes simultaneously. Microarrays form the basis of the gene expression profiling of breast cancer.

Driver event An event that is responsible for cancer initiation and/or progression.

Ductal carcinoma Cancer that arises from the cells which line the milk ducts.

Epidermal growth factor receptor (HER2) A cell surface receptor that binds to external growth stimuli (e.g. EGF; epidermal growth factors) and signals cell proliferation and inhibits cell death.

Epigenetic modification Non-genetic changes that affect gene expression and may be inherited.

Estrogen The primary steroid female hormone that is important for development and function of several organs primarily the human breast.

Estrogen receptor An intracellular receptor that activates cell growth and proliferation when it is bound by its ligand estrogen.

Event-free survival The proportion of people who are expected to remain free of a disease (the event) after treatment.

Gene expression profile Measurement of the expression of a well-defined (or genome-wide) set of genes that differentially classifies individual tumors.

Genome The entire genic and intragenic sequences of DNA in an organism.

Germline mutations Mutations that are inherited from parents and passed along to the offspring.

Hormone replacement therapy Supplement of hormones due to the lack or insufficiency of endogenous hormones as part of a medical treatment.

Hyperplasia An abnormal increase in cell number within an organ causing it to become enlarged.

Indels Insertions or deletions of a single nucleotide or whole sequences.

Integument General term describing the outer covering of an organism, e.g. human skin.

Intra-tumoral heterogeneity Represents the pathological, molecular and genomic variability that exists among cells within the same tumor.

Kataegis A pattern of clustered hypermutation found in some cancers including breast. The term kataegis is derived from the Greek word for "thunder". The mutations are predominately cytosine to thymine in the context of a TpC dinucleotide and the AID/APOBEC family of cytosine deaminases is responsible for the process of kataegis.

Ki-67 A nuclear antigen used commonly to determine proliferating cells within a tumor biopsy; it is a nucleolar protein that is highly expressed in S-phase of the cell cycle.

Lobe Gross anatomical extension of the breast which consists of lobule units.

Lobular carcinoma Cancer that arises from the milk-producing glandular cells in the lobules.

Lobule Glandular part of the lobe and is responsible for milk production.

Loss of heterozygosity (LOH) The loss of the wild-type allele when the other allele is mutated or lost (e.g. p53 gene in cancer).

Lymphatic system Part of the circulatory system consisting of a network of connecting vessels that transfer a clear fluid called lymph throughout the body. The lymphatic system drains lymph from tissues back into the blood.

Mammography Imaging technique that delivers low-dose X-ray radiation for the early detection and diagnosis of breast cancer.

Menarche age Age at which the first menstrual cycle (or first menstrual bleeding) occurs in females.

Menopause Age (40–50) at which the monthly menstrual cycles cease and signifies the end of the fertile phase of a woman's life.

Mitotic index Measure of the proliferative capacity of cells and is defined as the ratio of the number of cells undergoing mitosis over the total number of cells.

Morbidity Physical state of being ill, diseased or unhealthy.

Mortality The state of being susceptible to death; mortality rate = the rate of death.

Neoadjuvant chemotherapy Chemotherapy that is given before primary treatment to shrink the primary tumor and reduce the need for aggressive surgical resection.

Nulliparity The state of not having offspring; i.e. childless.

Oncogenes Genes which contribute to cell proliferation and survival once their expression is amplified or their function is altered (activating mutation, gene amplification, epigenetic modification, post-translational stability).

Overall survival The proportion of people who are expected to live after a certain amount of time (e.g. 5 or 10 years). Overall survival takes into account death from any cause which may be unrelated to the disease.

Passenger mutation A mutation that results from increased genomic instability and does not contribute to the progression of cancer.

Penetrance of a mutation Magnitude of impact on patient outcome.

Personalized medicine An approach by which medical treatments are tailored to individual patients for optimal treatment responses.

Predictive value Overall capacity of a method to predict an outcome (e.g. survival, recurrence risk, treatment response etc.).

Progesterone The primary steroid female hormone required for the female menstrual cycle as well as different stages of embryogenesis.

Progesterone receptor (PR) A nuclear receptor that can act as a transcription factor when it is activated by its ligand progesterone.

Risk of recurrence Chance of a cancer patient to present with recurring disease post-treatment.

Sample turnover time Time required for processing of patient sample from time of receiving to the availability of the results.

Single nucleotide polymorphism (SNP) A single base pair substitution which occurs at a high frequency in the human population, often contributes to inter-individual DNA sequence variability and does not contribute to disease.

Single nucleotide variation (SNV) Mutation that affects a single base pair in the genome and may play a role in disease.

Somatic mutations DNA mutations that are acquired during a person's life time which can be due to endogenous metabolism and/or exogenous exposure to carcinogens or radiation.

Taxanes Microtubule-inhibiting class of drugs which are widely used as anticancer agents. They are derived from yew plants (genus *Taxus*). The most common taxanes are docetaxel and paclitaxel.

Terminal ductal-lobular units (TDLUs) The basic functional units of the breast which are composed of extralobular and intralobular ducts. The TDLUs provide the conduit from which the milk drains from the lobules to allow secretion from the nipple.

Triple-negative breast cancer A subtype of breast cancer that lacks the expression of all three receptors (ER, PR, and HER2) and predicts a poor prognosis as it is not responsive to anti-hormone therapies.

Tumor grade Histological assessment of the differentiation state of a tumor tissue, i.e. how much it resembles the normal tissue it arose from. Well-differentiated tumor: tumor closely resembles normal tissue, poorly-differentiated: tumor does not represent the original normal tissue.

Tumor stage A numerical assessment that reflects tumor size and degree of spread (i.e. local invasion and metastasis). It is considered as the best predictor of prognosis.

Tumor suppressor genes Genes whose loss (or inactivation) can contribute to cell proliferation and survival and can predispose a cell to cancerous transformation.

Whole exome sequencing Selective sequencing of the gene coding regions of the genome which offers a cheaper and faster alternative to whole-genome sequencing. This technique sequences cDNA that is derived from total RNA and hence lack intragenic sequences.

Whole genome sequencing Sequencing of all the DNA present in an organism's genome.

ABBREVIATIONS

ACGH Array-based comparative genomic hybridization
CA 27.29 Cancer antigen 27.29
CEA Carcinoembryonic antigen
CIS Carcinoma *in situ*
CNA Copy number aberration
CTC Circulating tumor cells
CtDNA Circulating tumor DNA
DCIS Ductal carcinoma *in situ*
DNA Deoxyribonucleic acid
ER Estrogen receptor
HER2 Epidermal growth factor receptor 2
IC Invasive carcinoma
IDC Invasive ductal carcinoma
IM Immunomodulatory
LAR Luminal androgen receptor
LCIS Lobular carcinoma *in situ*
LOH Loss of heterozygosity
M Mesenchymal
MGI Molecular grade index
MLPA Multiplex ligation probe analysis
MRI Magnetic resonance imaging
MSL Mesenchymal stem-like
mTOR Murine target of rapamycin
NCCN National Cancer Comprehensive Network
PAM50 Prediction analysis of microarrays 50
PI3K Phosphoinositide 3-kinase
PR Progesterone receptor

qRT-PCR Reverse transcriptase quantitative polymerase chain reaction
RNA Ribonucleic acid
SNP Single nucleotide polymorphism
SNV Single nucleotide variation
TAILORx Trial assigning individualized options for treatment
TDLU Terminal ductal-lobular units
TNBC Triple-negative breast cancer

REFERENCES

[1] GLOBOCAN. The GLOBOCAN project: fact sheets. <http://globocan.iarc.fr/factsheets/populations/factsheet.asp?uno = 900#WOMEN>; 2008

[2] Key TJ, Verkasalo PK, Banks E. Epidemiology of breast cancer. Lancet Oncol 2001;2:133–40.

[3] Howlader N, Noone AM, Krapcho M, Neyman N, Aminou R, Waldron W, editors. SEER Cancer Statistics Review, 1975–2010. Bethesda, MD: National Cancer Institute. Available at: <http://seer.cancer.gov/csr/1975_2010/>, based on November 2012 SEER data submission, posted to the SEER web site, April 2013.

[4] Smith RA, Brooks D, Cokkinides V, Saslow D, Brawley OW. Cancer screening in the United States, 2013: a review of current American Cancer Society guidelines, current issues in cancer screening, and new guidance on cervical cancer screening and lung cancer screening. CA Cancer J Clin 2013;63:88–105.

[5] Jemal A, Siegel R, Ward E, Murray T, Xu J, Thun MJ. Cancer statistics. CA Cancer J Clin 2007;57:43–66.

[6] Russo J, Russo IH. Development of the human breast. Maturitas 2004;49:2–15.

[7] Petersen OW, Polyak K. Stem cells in the human breast. Cold Spring Harb Perpect Biol 2010;2:a003160.

[8] Henderson BE, Ross RK, Bernstein L. Estrogens as a cause of human cancer: The Richard and Hinda Rosenthal Foundation Award Lecture. Cancer Res 1988;48:246–53.

[9] Shimizu H, Ross RK, Bernstein L, Pike MC, Henderson BE. Serum oestrogen levels in postmenopausal women: comparison of American whites and Japanese in Japan. Br J Cancer 1990;62:451–3.

[10] Bernstein L, Yuan JM, Ross RK, Pike MC, Hanisch R, Lobo R, et al. Serum hormone levels in pre-menopausal Chinese women in Shanghai and white women in Los Angeles: results from two breast cancer case-control studies. Cancer Causes Control 1990;1:51–8.

[11] Thomas HV, Reeves GK, Key TJ. Endogenous estrogen and post-menopausal breast cancer: a quantitative review. Cancer Causes Control 1997;8:922–8.

[12] Simpson PT, Reis-Filho JS, Gale T, Lakhani SR. Molecular evolution of breast cancer. J Pathol 2005;205:248–54.

[13] Knudsen ES, Pajak TF, Qeenan M, McClendon AK, Armon BD, Schwartz GF, et al. Retinoblastoma and phosphate and tensin homolog tumor suppressors: impact on ductal carcinoma *in situ* progression. J Natl Cancer Inst 2012;104:1825–36.

[14] Rakovitch E, Nofech-Mozes S, Hanna W, Narod S, Thiruchelvam D, Saskin R, et al. HER2/neu and Ki-67 expression predict non-invasive recurrence following breast-conserving therapy for ductal carcinoma *in situ*. Br J Cancer 2012;106:1160–5.

[15] Lipworth L. Epidemiology of breast cancer. Eur J Cancer Prev 1995;4:7–30.

[16] Sutcliffe S, Pharoah PD, Easton DF, Ponder BA. Ovarian and breast cancer risks to women in families with two or more cases of ovarian cancer. Int J Cancer 2000;87:110–7.

[17] Gayther SA, Pharoah PD, Ponder BA. The genetics of inherited breast cancer. J Mammary Gland Biol 1998;3:365–76.

[18] Hanahan D, Weinberg RA. Hallmarks of cancer: the next generation. Cell 2011;144:646–74.

[19] Charpentier A, Aldaz CM. In: Coleman WB, Tsongalis GJ, editors. The molecular basis of breast carcinogenesis in the molecular basis of human cancer. Totowa: Humana Press; 2002. p. 347–63.

[20] Lakhani SR, Gusterson BA, Jacquemier J, Sloane JP, Anderson TJ, van de Vijver MJ, et al. The pathology of familial breast cancer: Histological features of cancers in families not attributable to mutations in BRCA1 and BRCA2. Clin Cancer Res 2000;6:782–9.

[21] Melhem-Bertrandt A, Bojadzieva J, Ready KJ, Obeid E, Liu DD, Gutierrez-Barrera AM, et al. Early onset HER2-positive breast cancer is associated with germline TP53 mutations. Cancer 2012;118:908–13.

[22] Hirshfield KM, Rebbeck TR, Levine AJ. Germline mutations and polymorphisms in the origins of cancers in women. J Oncol 2010;297671.

[23] Weinberg RA. The biology of cancer. New York: Garland Publishing Group; 2007.

[24] Habermann JK, Doering J, Hautaniemi S, Roblick UJ, Bundgen NK, Nicorici D, et al. The gene expression signature of genomic instability in breast cancer is an independent predictor of clinical outcome. Int J Cancer 2009;124:1552–64.

[25] Herman JG, Baylin SB. Gene silencing in cancer in association with promoter hypermethylation. N Engl J Med 2003;349:2042–54.

[26] Eden A, Gaudet F, Waghmare A, Jaenisch R. Chromosomal instability and tumors promoted by DNA hypomethylation. Science 2003;300:455.

[27] Yan PS, Shi H, Rahmatpanah F, Hsiau TH, Hsiau AH, Leu YW, et al. Differential distribution of DNA methylation within the RASSF1A CpG island in breast cancer. Cancer Res 2003;63:6178–86.

[28] Herman JG, Merlo A, Mao L, Lapidus RG, Issa JP, Davidson NE, et al. Inactivation of the CDKN2/p16/MTS1 gene is frequently associated with aberrant DNA methylation in all common human cancers. Cancer Res 1995;55:4525–30.

[29] Lapidus RG, Ferguson AT, Ottaviano YL, Parl FF, Smith Hs, Weitzman SA, et al. Methylation of estrogen and progesterone receptor gene 5′ CpG islands correlates with lack of estrogen and progesterone receptor gene expression in breast tumors. Clin Cancer Res 1996;2:805–10.

[30] Singhai R, Patil VW, Jaiswal SR, Patil SD, Tayade MB, Patil AV. E-Cadherin as a diagnostic biomarker in breast cancer. N Am J Med Sci 2011;3:227–33.

[31] Mylona E, Magkou C, Giannopoulou I, Agrogiannis G, Markaki S, Keramopoulos A, et al. Expression of tissue inhibitor of matrix metalloproteinases (TIMP)-3 protein in invasive breast carcinoma: relation to tumor phenotype and clinical outcome. Breast Cancer Res 2006;8:R57.

[32] Rice JC, Massey-Brown KS, Futscher BW. Aberrant methylation of the BRCA1 CpG island promoter is associated with decreased BRCA1 mRNA in sporadic breast cancer cells. Oncogene 1998;17:1807–12.
[33] Esteller M. Relevance of DNA methylation in the management of cancer. Lancet Oncol 2003;4:351–8.
[34] Ellis MJ, Ding L, Shen D, Luo J, Suman VJ, Wallis JW, et al. Whole-genome analysis informs breast cancer response to aromatase inhibition. Nature 2012;486:353–60.
[35] Elston CW, Ellis IO, Pinder SE. Pathological prognostic factors in breast cancer. Crit Rev Oncol Hematol 1999;31:209–23.
[36] Konecny G, Pauletti G, Pegram M, Untch M, Dandekar S, Aguilar Z, et al. Quantitative association between HER-2/neu and steroid hormone receptors in hormone receptor-positive primary breast cancer. J Natl Cancer Inst 2003;95:142–53.
[37] Dawson SJ, Provenzano E, Caldas C. Triple negative breast cancers: clinical and prognostic implications. Eur J Cancer 2009;45 (Suppl 1):27–40.
[38] Perou CM, Sorlie T, Eisen MB, van de Rijn M, Jeffrey SS, Rees CA, et al. Molecular portraits of human breast tumours. Nature 2000;406:747–52.
[39] Sorlie T, Perou CM, Tibshirani R, Aas T, Geisler S, Johnsen H, et al. Gene expression patterns of breast carcinomas distinguish tumor subclasses with clinical implications. Proc Natl Acad Sci USA 2001;98:10869–74.
[40] Sorlie T, Tibshirani R, Parker J, Hastie T, Marron JS, Nobel A, et al. Repeated observation of breast tumor subtypes in independent gene expression data sets. Proc Natl Acad Sci USA 2003;100:8418–23.
[41] Parker JS, Mullins M, Cheang MC, Leung S, Voduc D, Vickery T, et al. Supervised risk predictor of breast cancer based on intrinsic subtypes. J Clin Oncol 2009;27:1160–7.
[42] Prat A, Parker JS, Karginova O, Fan C, Livasy C, Herschkowitz JI, et al. Phenotypic and molecular characterization of the claudin-low intrinsic subtype of breast cancer. Breast Cancer Res 2010;12:R68.
[43] Hennessy BT, Gonzalez-Angulo AM, Stemke-Hale K, Gilcrease MZ, Krishnamurthy S, Lee JS, et al. Characterization of a naturally occurring breast cancer subset enriched in epithelial-to-mesenchymal transition and stem cell characteristics. Cancer Res 2009;69:4116–24.
[44] Bruna A, Greenwood W, Le Quesne J, Teschendorff A, Miranda-Saavedra D, Rueda OM, et al. TGFβ induces the formation of tumour initiating cells in claudin low breast cancer. Nat Commun 2012;3:1055.
[45] Prat A, Perou CM. Deconstructing the molecular portraits of breast cancer. Mol Oncol 2011;5:5–23.
[46] Cheang MC, Voduc KD, Tu D, Jiang S, Leung S, Chia SK, et al. Responsiveness of intrinsic subtypes to adjuvant anthracycline substitution in the NCIC.CTG MA.5 randomized trial. Clin Cancer Res 2012;18:2402–12.
[47] Teschendorff AE, Miremadi A, Pinder SE, Ellis IO, Caldas C. An immune response gene expression module identifes a good prognosis subtype in estrogen receptor negative breast cancer. Genome Biol 2007;8:R157.
[48] Gnant M, Steger GG. Fighting overtreatment in adjuvant breast cancer therapy. Lancet 2009;374:2029–30.
[49] Olivotto IA, Bajdik CD, Ravdin PM, Speers CH, Coldman AJ, Norris BD, et al. Population-based validation of the prognostic model ADJUVANT! for early breast cancer. J Clin Oncol 2005;23:2716–25.
[50] van de Vijver MJ, He YD, van't Veer LJ, Dai H, Hart AAM, Voskuil DW, et al. A gene expression signature as a predictor of survival in breast cancer. N Engl J Med 2002;347:1999–2009.
[51] Paik S, Shak S, Tang G, Kim C, Baker J, Cronin M, et al. A multigene assay to predict recurrence of tamoxifen-treated, node-negative breast cancer. N Engl J Med 2004;351:2817–26.
[52] Albain KS, Barlow WE, Shak S, Hortobagyi GN, Livingston RB, Yeh IT, et al. Prognostic and predictive value of the 21-gene recurrence score assay in postmenopausal women with node-positive, oestrogen-receptor-positive breast cancer on chemotherapy: a retrospective analysis of a randomised trial. Lancet Oncol 2010;11:55–65.
[53] Mook S, Schmidt MK, Viale G, Pruneri G, Eekhout I, Floore A, et al. The 70-gene prognosis signature predicts disease outcome in breast cancer patients with 1–3 positive lymph nodes in an independent validation study. Breast Cancer Res Treat 2009;116: 295–302.
[54] NCCN Updates Breast Cancer Guidelines; 2008. Available at: <http://www.nccn.org>. Updated April 2013.
[55] Knauer M, Mook S, Rutgers EJ, Bender RA, Hauptmann M, van de Vijver MJ, et al. The predictive value of the 70-gene signature for adjuvant chemotherapy in early breast cancer. Breast Cancer Res Treat 2010;120:655–61.
[56] Paik S, Tang G, Shak S, Kim C, Baker J, Kim W, et al. Gene expression and benefit of chemotherapy in women with node-negative, estrogen receptor-positive breast cancer. J Clin Oncol 2006;24:3726–34.
[57] Baselga J, Seidman AD, Rosen PP, Norton L. HER2 overexpression and paclitaxel sensitivity in breast cancer: therapeutic implications. Oncology 1997;11(3 Suppl 2):43–8.
[58] Arranz EE, Vara JÁ, Gámez-Pozo A, Zamora P. Gene signatures in breast cancer: current and future uses. Transl Oncol 2012;5:398–403.
[59] Sotiriou C, Wirapati P, Loi S, Harris A, Fox S, Smeds J, et al. Gene expression profiling in breast cancer: understanding the molecular basis of histologic grade to improve prognosis. J Natl Cancer Inst 2006;98:262–72.
[60] Loi S, Haibe-Kains B, Desmedt C, Lallemand F, Tutt AM, Gillet C, et al. Definition of clinically distinct molecular subtypes in estrogen receptor-positive breast carcinomas through genomic grade. J Clin Oncol 2007;25:1239–46.
[61] Sánchez-Navarro I, Gámez-Pozo A, Pinto A, Hardisson D, Madero R, Lopez R, et al. An 8-gene qRT-PCR-based gene expression score that has prognostic value in early breast cancer. BMC Cancer 2010;10:336.
[62] Jerevall PL, Brommesson S, Strand C, Gruvberger-Saal S, Malmstrom P, Nordenskjold B, et al. Exploring the two-gene ratio in breast cancer – independent roles for HOXB13 and IL17BR in prediction of clinical outcome. Breast Cancer Res Treat 2008;107:225–34.
[63] Jerevall PL, Ma XJ, Li H, Salunga R, Kesty NC, Erlander MG, et al. Prognostic utility of HOXB13:IL17BR and molecular grade index in early-stage breast cancer patients from the Stockholm trial. Br J Cancer 2011;104:1762–9.
[64] National Institutes of Health 2006–2013. Hormone therapy with or without combination chemotherapy in treating women who have

undergone surgery for node-negative breast cancer (The TAILORx Trial). NCT00310180. Available at: <http://clinicaltrials.gov/ct2/show/NCT00310180?term = NCT00310180&rank = 1>.
[65] National Institutes of Health 2006–2013. Genetic testing or clinical assessment in determining the need for chemotherapy in women with breast cancer that involves no more than 3 lymph nodes (The MINDACT Trial). NCT00433589. Available at: <http://clinicaltrials.gov/ct2/show/NCT00433589?term=nct00433589&rank=1>.
[66] Gruel N, Lucchesi C, Raynal V, Rodrigues MJ, Pierron G, Goudefroye R, et al. Lobular invasive carcinoma of the breast is a molecular entity distinct from luminal invasive ductal carcinoma. Eur J Cancer 2010;46:2399–407.
[67] Bertucci F, Orsetti B, Nègre V, Finetti P, Rouge C, Ahomadegbe JC, et al. Lobular and ductal carcinomas of the breast have distinct genomic and expression profiles. Oncogene 2008;27:5359–72.
[68] Hayes DF, Thor AD, Dressler LG, Weaver D, Edgerton S, Cowan D, et al. HER2 and response to paclitaxel in node-positive breast cancer. N Engl J Med 2007;357:1496–506.
[69] Minuti G, Cappuzzo F, Duchnowska R, Jassem J, Fabi A, O'Brien T, et al. Increased MET and HGF gene copy numbers are associated with trastuzumab failure in HER2-positive metastatic breast cancer. Br J Cancer 2012;107:793–9.
[70] Khoury T, Kanehira K, Wang D, Ademuyiwa F, Mojica W, Cheney R, et al. Breast carcinoma with amplified HER2: a gene expression signature specific for trastuzumab resistance and poor prognosis. Mod Pathol 2010;23:1364–78.
[71] Vegran F, Boidot R, Coudert B, Fumoleau P, Arnould L, Garnier J, et al. Gene expression profile and response to trastuzumab-docetaxel-based treatment in breast carcinoma. Br J Cancer 2009;101:1357–64.
[72] Knauer M, Cardoso F, Wesseling J, Bedard PL, Linn SC, Rutgers EJT, et al. Identification of a low-risk subgroup of HER-2-positive breast cancer by the 70-gene prognosis signature. Br J Cancer 2010;103:1788–93.
[73] Staaf J, Ringner M, Vallon-Christersson J, Jonsson G, Bendahl PO, Holm K, et al. Identification of subtypes in human epidermal growth factor receptor 2-positive breast cancer reveals a gene signature prognostic of outcome. J Clin Oncol 2010;28:1813–20.
[74] Martinez SR, Beal SH, Canter RJ, Chen SL, Khatri VP, Bold RJ. Medullary carcinoma of the breast: a population-based perspective. Med Oncol 2011;28:738–44.
[75] Boujelbene N, Khabir A, Jeanneret Sozzi W, Mirimanoff RO, Khanfir K. Clinical review – breast adenoid cystic carcinoma. Breast 2012;21:124–7.
[76] Desmedt C, Haibe-Kains B, Wirapati P, Buyse M, Larsimont D, Bontempi G, et al. Biological processes associated with breast cancer clinical outcome depend on the molecular subtypes. Clin Cancer Res 2008;14:5158–65.
[77] Rody A, Karn T, Liedtke C, Pusztai L, Ruckhaeberle E, Hanker L, et al. A clinically relevant gene signature in triple negative and basal-like breast cancer. Breast Cancer Res 2011;13:R97.
[78] Yau C, Esserman L, Moore DH, Waldman F, Sninsky J, Benz CC. A multigene predictor of metastatic outcome in early stage hormone receptornegative and triple-negative breast cancer. Breast Cancer Res 2010;12:R85.
[79] Lehmann BD, Bauer JA, Chen X, Sanders ME, Chakravarthy AB, Shyr Y, et al. Identification of human triple-negative breast cancer subtypes and preclinical models for selection of targeted therapies. J Clin Invest 2011;121:2750–67.
[80] Turner N, Lambros MB, Horlings HM, Pearson A, Sharpe R, Natrajan R, et al. Integrative molecular profiling of triple negative breast cancers identifies amplicon drivers and potential therapeutic targets. Oncogene 2010;29:2013–23.
[81] Curtis C, Shah SP, Chin SF, Turashvili G, Rueda OM, Dunning MJ, et al. The genomic and transcriptomic architecture of 2,000 breast tumours reveals novel subgroups. Nature 2012;486:346–52.
[82] Banerji S, Cibulskis K, Rangel-Escareno C, Brown KK, Carter SL, Frederick AM, et al. Sequence analysis of mutations and translocations across breast cancer subtypes. Nature 2012;486:405–9.
[83] Shah SP, Roth A, Goya R, Oloumi A, Ha G, Zhao Y, et al. The clonal and mutational evolution spectrum of primary triple-negative breast cancers. Nature 2012;486:395–9.
[84] Nik-Zainal S, Alexandrov LB, Wedge DC, Van Loo P, Greenman CD, Raine K, et al. Mutational processes molding the genomes of 21 breast cancers. Cell 2012;149:979–93.
[85] Lada AG, Dhar A, Boissy RJ, Hirano M, Rubel AA, Rogozin IB, et al. AID/APOBEC cytosine deaminase induces genome-wide kataegis. Biol Direct 2012;7:47.
[86] Taylor BJ, Nik-Zainal S, Wu YL, Stebbings LA, Raine K, Campbell PJ, et al. DNA deaminases induce break-associated mutation showers with implication of APOBEC3B and 3A in breast cancer kataegis. eLife 2013;2:e00534.
[87] Nelson HD, Tyne K, Naik A, Bougatsos C, Chan BK, Humphrey L. Screening for breast cancer: an update for the U.S. Preventive Services Task Force. Ann Intern Med 2009;151(10):727–37.
[88] The Canadian Task Force on Preventive Health Care. Recommendations on screening for breast cancer in average-risk women aged 40–74 years. C Med Assoc. J 2011;183:1991–2001.
[89] Autier P, Boniol M, Gavin A, Vatten LJ. Breast cancer mortality in neighbouring European countries with different levels of screening but similar access to treatment: trend analysis of WHO mortality database. Br Med J 2011;343:d4411.
[90] Groenendijk RP, Bult P, Tewarie L, Peer PG, van der Sluis RF, Ruers TJ, et al. Screen-detected breast cancers have a lower mitotic activity index. Br J Cancer 2000;82:381–4.
[91] Houssami N, Hayes DF. Review of preoperative magnetic resonance imaging (MRI) in breast cancer: should MRI be performed on all women with newly diagnosed, early stage breast cancer? CA Cancer J Clin 2009;59:290–302.
[92] Crispo A, Barba M, D'Aiuto G, De Laurentiis M, Grimaldi M, Rinaldo M, et al. Molecular profiles of screen detected vs. symptomatic breast cancer and their impact on survival: results from a clinical series. BMC Cancer 2013;13:15.
[93] Miller BT, Abbott AM, Tuttle TM. The influence of preoperative MRI on breast cancer treatment. Ann Surg Oncol 2012;19:536–40.
[94] Bast Jr RC, Ravdin P, Hayes DF, Bates S, Fritsche H, Jessup JM, et al. 2000 update of recommendations for the use of tumor markers in breast and colorectal cancer: clinical practice guidelines of the American Society of Clinical Oncology. J Clin Oncol 2001;19:1865–78.
[95] Cristofanilli M, Budd GT, Ellis MJ, Stopeck A, Matera J, Miller MC, et al. Circulating tumor cells, disease progression, and survival in metastatic breast cancer. N Engl J Med 2004;351:781–91.

[96] Consoli F, Grisanti S, Amoroso V, Almici C, Verardi R, Marini M, et al. Circulating tumor cells as predictors of prognosis in metastatic breast cancer: clinical application outside a clinical trial. Tumori 2011;97:737–42.

[97] Wallwiener M, Hartkopf AD, Baccelli I, Riethdorf S, Schott S, Pantel K, et al. The prognostic impact of circulating tumor cells in subtypes of metastatic breast cancer. Breast Cancer Res Treat 2013;137:503–10.

[98] Zhang L, Riethdorf S, Wu G, Wang T, Yang K, Peng G, et al. Meta-analysis of the prognostic value of circulating tumor cells in breast cancer. Clin Cancer Res 2012;18:5701–10.

[99] Budd GT, Cristofanilli M, Ellis MJ, Stopeck A, Borden E, Miller MC, et al. Circulating tumor cells versus imaging – predicting overall survival in metastatic breast cancer. Clin Cancer Res 2006;12:6403–9.

[100] Dawson SJ, Tsui DW, Murtaza M, Biggs H, Rueda OM, Chin SF, et al. Analysis of circulating tumor DNA to monitor metastatic breast cancer. N Engl J Med 2013;368:1199–209.

[101] Stephens PJ, Tarpey PS, Davies H, Van Loo P, Greenman C, Wedge DC, et al. The landscape of cancer genes and mutational processes in breast cancer. Nature 2012;486:400–4.

[102] Knight WA, Livingston RB, Gregory EJ, McGuire WL. Estrogen receptor as an independent prognostic factor for early recurrence in breast cancer. Cancer Res 1977;37(12):4669–71.

[103] Kallioniemi A, Kallioniemi OP, Sudar D, Rutovitz D, Gray JW, Waldman F, et al. Comparative genomic hybridization for molecular cytogenetic analysis of solid tumors. Science 1992;258(5083):818–21.

[104] van't Veer LJ, Dai H, van de Vijver MJ, He YD, Hart AA, Mao M, et al. Gene expression profiling predicts clinical outcome of breast cancer. Nature 2002;415(6871):530–6.

[105] Paik S, Shak S, Tang G, Kim C, Baker J, Cronin M, et al. A multigene assay to predict recurrence of tamoxifen-treated, node-negative breast cancer. N Engl J Med 2004;351 (27):2817–26.

[106] Callagy G, Pharoah P, Chin SF, Sangan T, Daigo Y, Jackson L, et al. Identification and validation of prognostic markers in breast cancer with the complementary use of array-CGH and tissue microarrays. J Pathol 2005;205(3):388–96.

[107] Kan Z, Jaiswal BS, Stinson J, Janakiraman V, Bhatt D, Stern HM, et al. Diverse somatic mutation patterns and pathway alterations in human cancers. Nature 2010;466(7308):869–73.

Cancer Genomics

Prostate Cancer Genomics as a Driver of Personalized Medicine

Michael Fraser[1]
Alejandro Berlin[1,2]
Veronique Ouellet[3]
Fred Saad[3,4]
Robert G. Bristow[1,2]

[1]*Ontario Cancer Institute and Princess Margaret Cancer Center (University Health Network), Toronto, ON, Canada*

[2]*Departments of Radiation Oncology and Medical Biophysics, University of Toronto, Toronto, ON, Canada*

[3]*Centre de Recherche du Centre hospitalier de l'Université de Montréal, Montreal, QC, Canada*

[4]*Department of Surgery, Centre hospitalier de l'Université de Montréal, Montreal, QC, Canada*

Contents

Cancer Genomics. DOI: http://dx.doi.org/10.1016/B978-0-12-396967-5.00014-1

Key Concepts

- Current prognostic factors based on clinical T-category, PSA, and Gleason score can be used to stratify cohorts of patients with localized prostate cancer into low-, intermediate-, and high-risk of recurrence and the development of metastatic disease. However, these factors cannot determine the risk for an individual man
- Additional prognostic biomarkers, based on tumor genetics, are urgently required for personalized prostate cancer medicine
- Genome-wide association studies have identified several prostate cancer susceptibility loci, which may be useful in population-wide screening programs
- Recent efforts to characterize the genomic landscape of prostate cancer have found a relatively low number of overall and recurrent point mutations. Several recurrent copy number alterations (e.g. Nkx3.1, c-Myc, and others) have been identified, and some of these may be useful prognostic biomarkers for failure of local therapy
- Recurrent gene fusions, such as *TMPRSS2–ERG*, are very common in prostate cancer, and fusion-positive cancers may possess unique biology relative to fusion-negative cancers. In particular, early-onset cancers (<55 years) appear to possess a unique set of androgen-dependent gene fusions and other structural rearrangements, relative to late-onset cancers. However, the prognostic significance of particular fusions, including *TMPRSS2–ERG*, is still unclear
- Prostate cancers harboring *BRCA1/2* mutations are unique in that they are at substantially higher risk of possessing occult metastases and progressing to high-risk disease (particularly those with *BRCA2* mutations). However, *BRCA1/2*-mutated tumors may also be amenable to personalized medicine using PARP inhibitors, which selectively kill *BRCA1/2*-mutant cells that are deficient in homologous recombination-mediated DNA damage repair
- Prostate cancer is a uniquely multifocal and potentially multiclonal disease, and intra-focal heterogeneity may complicate the search for effective biomarkers of treatment response and/or disease aggressiveness. A robust assessment of the genomics of multiclonal disease, and of patient-matched primary and metastatic disease, is required to characterize more completely the contribution of differential tumor foci to disease progression and prognosis
- Significant efforts are currently underway to sequence the genomes of prostate cancers within several unique patient cohorts, including efforts by The Cancer Genome Atlas (TCGA) and the International Cancer Genome Consortium (ICGC), as well as other groups

THE NEED FOR GENOMICS IN PROSTATE CANCER PROGNOSIS

Prostate cancer (CaP) is the most commonly diagnosed malignancy in men in the Western world, with 1 in 6 men diagnosed with CaP and 1 in 34 dying of metastatic disease. In North America and Europe, over 500 000 cases are diagnosed annually [1,2].

Treatment options for CaP depend on the TNM-staging of the disease. Using the prognostic variables of T-category, serum prostate specific antigen (PSA), and pathologic Gleason score (GS), men with localized CaP are placed in low-, intermediate- and high-risk groupings. These risk groups predict for biochemical relapse (also referred to as biochemical relapse-free rate (bRFR) and based on post-treatment increases in PSA) and prostate cancer-specific survival (PCSS) [3].

In low- or intermediate-risk CaP which is localized to the prostate (Figure 14.1), treatments such as active surveillance, radical prostatectomy, and either external beam radiotherapy (RT) or brachytherapy are used. The choice of treatment will depend on patient preference and other considerations (e.g. operative risk and co-morbidities). In other patients, radical prostatectomy with positive surgical margins, and/or a rising PSA with initial curative intent, is followed by adjuvant or salvage RT to the prostatic fossa. Patients with high-risk disease are managed with surgery alone, or with surgery (or RT) in combination with androgen-deprivation therapy due to the increased risk of subclinical metastases. Finally, for men who develop metastatic or recurrent disease, palliative non-curative treatment has led to improved progression-free survival and includes continuous androgen deprivation, chemotherapy using docetaxel and prednisone, secondary hormonal manipulation using enzalutamide or abiraterone, systemic radionuclides (Ra223) and immunotherapy (Sipulcel-T). Further improvement in pain symptomology can be achieved with targeted, palliative RT (8 Gy single dose or 20–30 Gy in daily fractions) [1,4]. At present, there are few, if any, biomarkers to differentiate the use of one systemic agent versus another, either as a single or combined treatment [5].

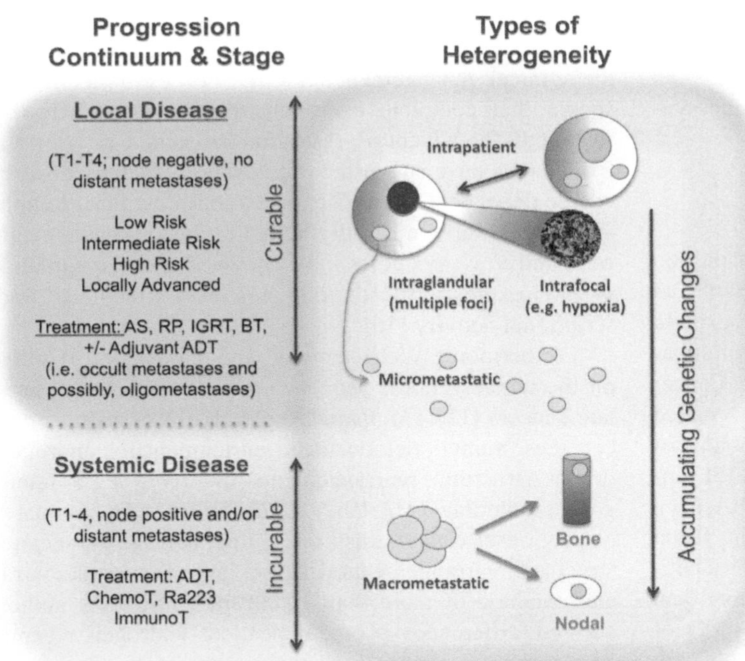

FIGURE 14.1 A continuum of genomic aberrations in prostate cancer. Prostate cancer is driven by a set of genetic and microenvironmental events that act sequentially to drive progression from localized, potentially curable disease towards distantly metastatic, androgen-resistant, and ultimately lethal cancer. Recurrent early events that precede therapeutic interventions include amplification and/or overexpression of c-Myc, deletion of Nkx3.1, androgen-dependent ETS gene fusions such as *TMPRSS2–ERG*, SPINK1 overexpression (in ETS fusion-negative disease), ERK/MAP kinase activation, and others. While recurrent point mutations are rare in prostate cancer, recurrent mutations of SPOP and TP53 have been observed in several studies. Additional aberrations may arise during the course of treatment, particularly in cancers treated with hormone therapy, where selection for androgen receptor mutations and amplification may hasten the progression to androgen-insensitive disease.

However, the currently used clinical prognostic factors of T-category, PSA and Gleason score explain only a moderate proportion of the observed variation in clinical outcome [2]. The use of PSA alone to determine the clinical course in otherwise "clinically silent" disease needs to be buttressed with biomarkers based on tumor biology [1]. For example, biochemical relapse can range from 20 to 60% in intermediate- and high-risk patients treated with precision radiotherapy or surgery alone [6,7]. There are no tests currently used in the clinic that can differentiate patients who will be cured by local therapy alone versus patients who need combined modality treatments as a means of intensification therapy due to predicted local and/or systemic resistance. Further complicating the issue for personalized medicine is the fact that many low-risk CaPs are indolent and their over-treatment can result in significant morbidity. For example, up to two thirds of indolent CaPs can be followed without treatment when reliably triaged to active surveillance alone, thereby preventing gastrointestinal and genitourinary side effects of RT or surgery. However, one third of these low-risk patients are being inaccurately classified as low risk and will transit to intermediate-risk disease within 7–10 years [8]. Here again, on an individualized basis, there are no tests that predict *a priori* which patients with indolent CaP are at additional risk. Therefore, better predictors of treatment outcome and patient prognosis are required to individualize CaP treatment and to provide the optimal therapy with minimal side effects.

The combination of technological advances and large directed efforts has recently led to several comprehensive studies of cancer genomes. These have greatly accelerated the search for genetic mutations implicated in both etiology and progression toward aggressive and metastatic disease. In the 1990s, DNA sequencing using Sanger chemistry evolved from vertical gel-electrophoresis (with a capacity of $\approx$1000 nucleotide bases per day) to capillary instruments yielding over 1 million bases per day. This capacity has kept improving at a breathtaking pace, with current instruments generating in excess of 10 billion bases per day. These advances have been accompanied by drastic decreases in cost. Estimated reagent costs for a whole genome sequence and associated bioinformatics are now approximately $10–20 000 (US): a 10 000-fold decrease since the first genome was sequenced. We now discuss the potential benefit of these technologies towards the personalized, clinical management of prostate cancers.

GENETIC ARCHITECTURE OF PROSTATE CANCER AND CLINICAL OUTCOME

As detailed above, the current clinical management for CaP reflects a limited implementation of genomics in characterizing tumor biology. Key areas of research that will be reviewed include genomic studies of: (1) genome-wide association studies for prostate cancer risk;

(2) indolent versus non-indolent localized cancers, including castrate-resistant metastatic cancers; and (3) the special case of *BRCA2*-mutated tumors, which are an exemplar of personalized medicine whereby genetic synthetic lethality can be used for targeted therapy.

Genetics of Prostate Cancer Risk

Genome-wide association studies (GWAS) have identified multiple common genetic variants associated with an increased risk of prostate cancer (CaP), but these explain less than one third of the heritability. Recently, the multinational PRACTICAL consortium (Prostate Cancer Association Group to Investigate Cancer Associated Alterations in the Genome) has summarized common prostate cancer susceptibility alleles by genotyping 211 155 single nucleotide polymorphisms (SNPs) in blood DNA from 25 074 prostate cancer cases and 24 272 controls. From this, 23 new prostate cancer susceptibility loci were identified at genome-wide significance ($p < 5 \times 10^{-8}$). More than 70 prostate cancer susceptibility loci, explaining ~30% of prostate cancer risk, have now been identified. Patients with these SNPs in their germline have a 4.7-fold higher risk than the average of the population for developing prostate cancer [9]. In a related study, a meta-analysis of four large GWAS confirmed the association of a CaP susceptibility locus (rs11672691 on chromosome 19), which was not only associated with risk but also an aggressive CaP subtype with poor prognosis [10]. Such information may be used by physicians to intensify current screening protocols using PSA and MRI or ultrasound-based lesion identification (and MRI-guided biopsies) within these higher-risk populations.

Genetics of Localized Disease through to Metastases

Genetic mutations and rearrangements in CaP lead to the inactivation of tumor suppressor genes and activation of oncogenes. These play a central role in the initiation and progression of CaP. The most common genetic changes include deletion of the *PTEN* (phosphatase and tensin homolog deleted on chromosome 10) tumor suppressor gene and translocations involving the ETS family of transcription factors and the *TMPRSS2* locus, leading to *TMPRSS2−ERG* fusions and increased transcription [11−13]. Individually, each of these changes has been found in upwards of 50−75% of CaP specimens [12]. Other common genetic changes include: deletions of *ATBF1*; amplification of *c-MYC*; androgen receptor gene amplification; *RAF*, *KLF5* and *KL6* deletion. Deletions have also been reported for the *RB1*, *NKX3-1*, *E-cadherin*, *p16INK4A*, *p27KIP*, and *SMAD4* genes [14]. Other microdeletions exist in more than 20% of CaP patients at 1q, 5q, 20q, and 22q [15].

There are good precedents that justify the expectation that a better understanding of CaP genomics will improve the clinical management of CaP. For example, having shown that complete androgen ablation is an effective treatment for advanced castration-resistant CaP, correlative studies have suggested that hormone therapy efficacy can be dependent on *ERG* expression status. For example, AR expression is markedly higher in fusion-positive cancers, and *in vitro* studies have shown that downregulation of *ERG* expression leads to decreased AR-mediated transcriptional activity [16].

Furthermore, Weischenfeldt and colleagues reported on the whole-genome sequencing of 11 early-onset prostate cancers [17]. Compared to elderly-onset prostate cancer, these tumors demonstrated enrichment for androgen-driven structural rearrangements involving ETS family genes, including *TMPRSS2−ERG*. This was in contrast to elderly-onset cancers displaying non-androgen-associated structural rearrangements. Using a large tissue microarray and database of more than 10 000 patients, these authors showed differences in age-dependent androgen receptor levels and a prevalence of rearrangements in early-stage prostate cancers, which starts to substantiate a potential novel pathway of progression in early-stage versus elderly-associated prostate cancers [17]. However, it should be noted that inactivation of genes associated with aggressive prostate cancers does not have to be linked to ETS fusions. For example, biallelic *PTEN* inactivation, by either homozygous deletion or deletion of one allele and mutation of the other, occurs in most *PTEN*-defective cancers and characterizes a particularly aggressive subset of metastatic and hormone-refractory prostate cancers; this is independent of fusion status [18].

These examples prove that new treatments can be guided by fundamental molecular observations. Additionally, reports already suggest that alterations in *NKX3-1*, *TMPRSS2−ERG*, *PTEN* and *c-Myc* gene status (alone or in combination) can be associated with clinical outcome [19−21]. The number and complexity of genomic alterations increase between primary prostate cancers and lethal castrate-resistant disease. In a sequencing study of 45 patients, over 40% of castrate-resistant prostate cancers (CRPCs) had genomic alterations involving the androgen receptor gene (*AR*), including *AR* copy number gain (24% of CRPCs) or *AR* point mutation (20% of CRPCs). Other recurrent mutations included transmembrane protease, serine 2 gene (*TMPRSS2*)−v-ets erythroblastosis virus E26 oncogene homolog (avian) gene (*ERG*) fusion (44%); phosphatase and tensin homolog gene (*PTEN*) loss (44%); tumor protein p53 gene (*TP53*) mutation (40%); retinoblastoma gene (*RB*) loss (28%); v-myc myelocytomatosis viral oncogene homolog (avian) gene (*MYC*) gain (12%); and phosphatidylinositol-4,5-bisphosphate 3-kinase, catalytic subunit α gene (*PIK3CA*) mutation (4%) [22]. There was a high incidence of genomic alterations involving key genes important for DNA repair, including breast cancer 2, early onset gene (*BRCA2*) loss (12%) and ataxia telangiectasia mutated gene (*ATM*) mutations (8%). The implication of having functional defects in DNA repair genes is discussed below.

DNA epigenetic modification also plays a role in CaP (e.g. methlylation of *glutathione S-transferase pi* (*GSTpi*)) and may be associated with disease progression [23,24]. In a recent high-throughput sequencing study investigating genome-wide DNA methylation patterns in 51 tumor and 53 benign prostate samples, more than 147 000 cancer-associated epigenetic alterations were observed [25]. Global methylation patterns in the genome showed significant differences based on the presence or absence of a *TMPRSS2-ERG* rearrangement. Using functional studies, the authors proposed that hypermethylation of microRNA (miR-26a) represents an alternative pathway of *ERG* rearrangement-independent EZH2 activation. This salient observation of increased differential methylation events in fusion-negative tumors now helps to explain the pathogenesis of prostate cancer in up to 50% of patients who do not harbor fusions [25].

The Special Case of BRCA1/2-Mutated Tumors

Men who harbor mutations in genes involved in DNA repair and genome surveillance (e.g. *BRCA2*, *NBS1*) have an increased risk for prostate cancer and poor prognosis [26,27]. Of interest, these patients do not seem to have increased toxicity when treated with DNA-damaging radiotherapy or chemotherapy, thus there are no reasons to use surgery over RT in this population [28]. Instead, these patients appear to be at risk for increased occult metastases and a rapid progression to castrate-resistant metastatic CaP. In one of the largest studies to date, eight BRCA1 carriers, 61 BRCA2 carriers, and 1940 non-carriers were evaluated for associations between BRCA1/2 status and other CaP prognostic factors. Patients with germline *BRCA2* mutations were more frequently associated with Gleason ≥ 8, T3/T4 stage, nodal involvement, and metastases at diagnosis than CaP in non-carriers. Prostate cancer cause-specific survival was significantly longer in non-carriers than in carriers (15.7 vs 8.6 years), and therefore BRCA2 status should be considered when tailoring clinical management for these patients [29].

However, BRCA2-related CaP may be one of the best examples of personalized medicine, as the DNA repair defects in these tumor cells provide the opportunity for treatment with inhibitors of the PARP protein, which has been reported selectively to kill *BRCA1/2*-defective ovarian and breast cancers that are HR defective. The PARP1 and PARP2 proteins are required for repair of DNA single-strand breaks (SSB) and, in cells treated with small molecule inhibitors of PARP (PARPi), unrepaired SSBs at replication forks are converted into DSBs which require HR-mediated repair to offset cell lethality. Tumor cells lacking HR function (e.g. deficient in BRCA1 or BRCA2 expression) are exquisitely sensitive to PARPi due to their inability to repair replication-associated DSBs; this form of targeted therapy is referred to as "synthetic cell lethality". Early results in *BRCA*-mutated CaP with PARP inhibitors are promising [30]. There is also interest in developing biomarkers that predict functional losses in DNA repair activity, in addition to mutations in DNA repair genes, better to predict clinical PARPi efficacy in sporadic CaP that may exhibit DNA repair defects of unknown origin [15,22,31].

DEFINING HETEROGENEITY IN PROSTATE CANCER: MULTIFOCALITY AND MULTICLONALITY

CaP is unique in that it is a multifocal cancer with clonal subpopulations and varied molecular abnormalities that can determine whether cancers are relatively indolent or aggressively metastatic [32]. Heterogeneity exists both within and between patients (see Figure 14.1). It is therefore critical to define genetic heterogeneity that exists within a given prostate gland and, indeed within a given focus of cancer, as pathological staging criteria do not adequately account for heterogeneity of this type. Indeed, the clinical success of detailed disease typing in breast cancer (e.g. ER/PR/HER2 status) strongly suggests that a more complete understanding of disease heterogeneity may be of great utility in stratifying patients and ultimately in the individualization of therapy.

"Aggressive" prostate cancer can be defined as those cancers that harbor biology associated with local resistance to RT or surgery and/or a capacity for distant metastases that increase patient lethality and decrease prostate cancer-specific survival. However, it is not clear whether there exists one "lethal" clone within a given prostate gland that ultimately seeds distant metastases, or whether a "field defect" within the prostate as a whole gives rise to multiple lethal clones. Recent studies have identified specific lesions found in primary vs metastatic disease [33] and, although the number of patient-matched primary/metastatic tumors analyzed to date is small, these studies have the potential to shed light on the origins of metastasis in CaP. Similarly, it remains unclear whether differential metastasis to bone vs soft tissue can be predicted *a priori* based on the genetic signature of a tumor.

Intra-prostatic heterogeneity may also affect the clinical course for a given patient. While current prostate biopsy protocols allow the identification of unique focal regions of disease within a prostate with differing Gleason scores, these foci may be further subcategorized by genomics with differential *TMPRSS2–ERG* fusion, *PTEN*, *c-MYC* and *NKX3.1* status [20,23]. Marked intrafocal heterogeneity challenges the concept that *TMPRSS2–ERG* fusion events are always an early step in prostate cancer development. The marked heterogeneity of CaP also compromises the concept of analyzing ERG status in diagnostic needle core biopsies as a basis for making treatment decisions [34]. However, unbiased genome-wide approaches (e.g. whole-genome sequencing) to characterize the individual foci within a single patient's tumor are currently lacking. Indeed, recent data using a combination of gene copy number and targeted sequencing analyses show that tumors with identical Gleason scores may exhibit profound genetic heterogeneity [33].

Taken together, the above data begin to help explain the relatively large disparity in clinical outcomes for patients with identically staged disease. This has recently been underscored in other cancer types (e.g. renal cancer), in which single needle biopsies did not adequately account for the uneven distribution of alterations throughout the genetically heterogeneous tumor [35]. Likewise, the contribution of heterogeneity and genomic instability within adjacent stroma and normal prostate epithelium [36] to prostate tumor biology is also unclear. Intra-ductal carcinoma of the prostate (IDC-P), which occurs in approximately 20–30% of CaP foci, is strongly prognostic for early biochemical relapse, disease-free survival, and metastatic failure rate, independent of Gleason score in patients undergoing radiotherapy for CaP [37], although the genetic basis for this aggressive pathologic subtype is unclear.

CaP also displays large heterogeneity between patients, which may reflect the currently limited understanding of the genetic subtypes of the disease. For example, the incidence rates of the *TMPRSS2–ERG* fusion are highly variable between patient cohorts, with Caucasian cohorts showing a far higher incidence of the fusion than Asian cohorts [38–40]. On the contrary, castrate-resistant CaP (CRPC) displays far less variability in *TMPRSS2–ERG* fusion and *PTEN* status, perhaps suggesting a lower degree of heterogeneity in CRPC vs primary CaP. Indeed, the observations that *TMPRSS2–ERG* fusions, alterations in androgen receptor signaling and the PI3K/PTEN/AKT axis are nearly universal in metastatic CaP [33] support this hypothesis.

While pathologic Gleason grade is a valuable staging tool for prostate cancer, it is unclear whether Gleason grades are a surrogate for cancer genotype; that is, is there a common genetic pattern that underlies Gleason grade 3 vs 4 vs 5? Furthermore, if such a putative genotype exists, it is not clear what genetic heterogeneity would exist in the context of the overall Gleason score. Penney and colleagues recently identified an mRNA signature based on comparing tumors with Gleason score ≤ 6 to those with Gleason score ≥ 8, and applied that signature to a Gleason score $= 7$ test cohort to discriminate lethal cases [41]. This gene expression signature significantly improved the ability to predict lethal prostate cancer, compared with discrimination of the cases as either Gleason $3 + 4$ or $4 + 3$. This suggests that it may be possible to delineate genomic signatures that synergize with pathologic Gleason score for improved patient stratification. Very recent data have shown that the number of copy number alterations (CNAs) seen in predominantly Gleason 4 tumors is significantly increased relative to predominantly Gleason 3 tumors [42]. Despite this, it is clear from other large-scale genomics efforts [33] that there is considerable genetic variation within tumors with equivalent Gleason scores, implying that additional tools will be required to stratify patients more effectively.

GENOME-WIDE SEQUENCING AND RNA EXPRESSION STUDIES: DEVELOPING SIGNATURES OF CLINICAL UTILITY

In more unbiased discovery approaches, a set of DNA-based biomarkers called GEMCap (for Genomic Evaluators of Metastatic Prostate Cancer) was used with the Kattan nomogram to improve the prediction of recurrence following surgery [43]. Finally, another recent unbiased approach that used a combination of array comparative genomic hybridization (aCGH), focused exon re-sequencing, as well as RNA and miRNA expression profiles, enabled the observation of novel CNAs and gene signatures that were prognostic following radical prostatectomy [33].

Efforts to sequence the entire prostate cancer genome have, to date, been limited (Table 14.1). Berger and colleagues performed whole-genome sequencing of seven prostate cancers from patients with intermediate- or high-risk disease [46]. They identified a relatively low incidence of non-synonymous mutations (median of 20 per patient), and a highly variable incidence of rearrangements and translocations. Furthermore, they showed that rearrangement breakpoints could be stratified based on *TMPRSS2–ERG* fusion status, suggesting a distinct underlying biology of ERG fusion-positive vs -negative cancers. The same group subsequently sequenced 57 prostate cancers (55 primary cancers and two neuroendocrine metastases), and identified a phenomenon termed "chromoplexy", in which multiple interdependent DNA translocations and deletions occur over relatively few generations, in a kind of "punctuated evolution" of

TABLE 14.1 Summary of Previous Prostate Cancer Sequencing Reports

Reference	Sequencing Methodologies	Sample Number and Source	Summary of Major Findings
[33]	Targeted exon re-sequencing	218 primary and metastatic prostate cancers, prostate cancer cell lines and xenografts	Identification of *NCOA2* as putative oncogene (11% of cases) Robust classification of low- and high-risk disease based on DNA copy number Identification of TMPRSS2–ERG-associated genetic aberrations
[44]	Targeted exon re-sequencing	8 metastatic prostate cancers and matched normal tissue from 6 patients	Identification of somatic alterations that may contribute to lethal disease (e.g. *MTOR, BRCA2, CHD5*, and others)
[45]	Whole-exome sequencing	23 tumors from 16 lethal, metastatic prostate cancers and 3 high-grade primary cancers	Relatively high numbers of somatic mutations in advanced prostate cancers, but very few recurrently mutated genes Identified a subset of cancers with "hypermutated" genomes, which may be linked to therapeutic resistance
[46]	Whole-genome sequencing (30 ×)	7 radical prostatectomies; Gleason score ≥7	Identified novel CaP-associated rearrangements (e.g. *MAGI2, CADM2*) and point mutations (e.g. *HSP-1*) Stratification of differential rearrangement breakpoints based on TMPRSS2–ERG status
[47]	Whole-exome sequencing	50 lethal, metastatic castrate-resistant prostate cancers; 11 high-grade localized prostate cancers	Demonstrated a low incidence of point mutations even in heavily treated, advanced disease Confirmed a monoclonal origin of lethal cancers Identified several mechanisms of disruption of androgen receptor signaling in castrate-resistant disease Defined novel TMPRSS2–ERG-positive disease subtype based on CHD1 deletion.
[48]	Genome-wide methylated DNA immunoprecipitation sequencing (MeDIP)	51 prostate tumors, 53 benign prostate	Global methylation patterns differ based on TMPRSS2–ERG fusion status Identified miR-26a as a putative regulator of EZH2 which has been linked to invasiveness and metastasis in fusion-negative tumors
[49]	Whole-exome sequencing; whole-transcriptome sequencing	112 tumor/normal tissue pairs	Identified novel recurrent somatic mutations in *MED12, FOXA1*, and *SPOP* *SPOP*-mutated tumors define ETS fusion-negative cancers
[17]	Whole-genome sequencing (50 ×)	11 early-onset (<50 years) prostate cancers	Characterization of androgen-driven structural rearrangements in early-onset disease, suggesting a differential mechanism of tumorigenesis vs late-onset
[42]	Whole-genome sequencing (61 ×)	55 primary cancers, 2 neuroendocrine metastases	Developed concept of "chromoplexy"; interdependent DNA translocations and deletions occurring over relatively few generations (i.e. *punctuated evolution)* Characterization of the oncogenic evolution of prostate cancer

prostate cancer [42]. This is distinct from "chromothripsis", in which thousands of rearrangements occur during a single dramatic event during carcinogenesis (Figure 14.2) [50]. However, whether chromoplexy generally characterizes prostate cancer remains unclear.

Using whole-transcriptome sequencing, Beltran and colleagues profiled 37 prostate tumors, including seven neuroendocrine prostate cancers (NEPC) [51]. NEPC cases showed significant gene amplification and overexpression of *AURKA* (Aurora Kinase A), relative to

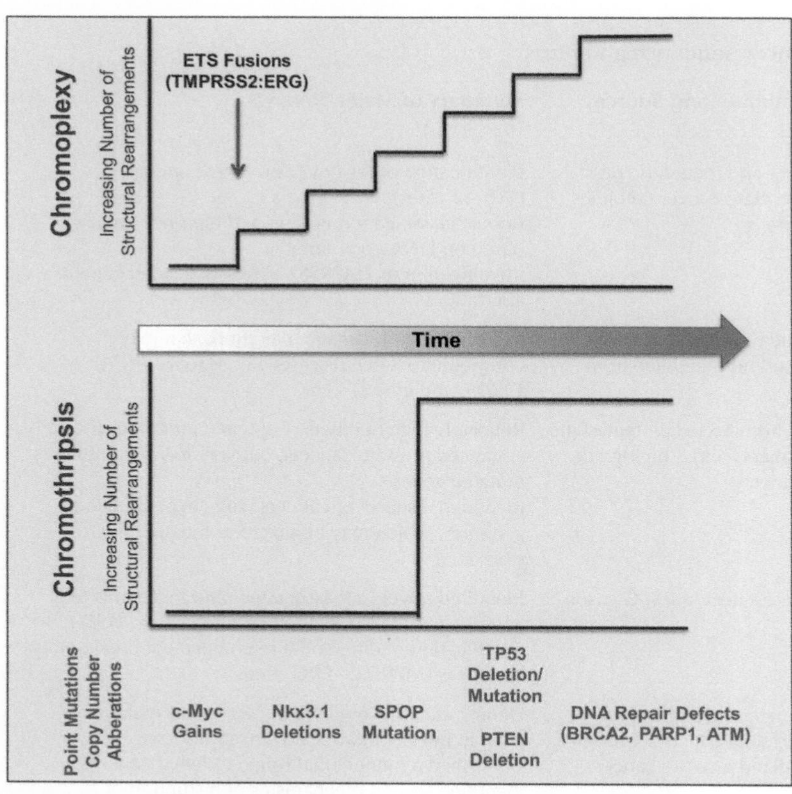

FIGURE 14.2 Chromoplexy, chromothripsis, and prostate cancer progression. Garraway and colleagues recently described a phenomenon termed "chromoplexy", whereby multiple interdependent genetic rearrangements occur in a punctuated manner, ultimately giving rise to a clinically apparent tumor [42]. This is in contrast to the phenomenon of "chromothripsis", recently described by Campbell and colleagues in chronic lymphocytic leukemia, as a single massive focal rearrangement of one or a few chromosomes [50]. While evidence of chromothripsis has been found in several cancer types, to date there is no evidence of its widespread occurrence in prostate cancer. Mutations and amplification/deletion of key cancer "driver" genes occur over the background of this overall genomic rearrangement, producing the genomic landscape that ultimately characterizes the tumor.

prostate adenocarcinoma. In a broader cohort, the same group demonstrated overexpression of *AURKA* in 76% of NEPC, 12% of adenocarcinomas, and in 0% of benign prostate tissue. This confirms the utility of using transcriptome analyses to identify potentially targetable lesions, such as *AURKA* overexpression.

Similarly, using matched whole-genome and whole-transcriptome sequencing, as well as aCGH, Wu and colleagues recently profiled a rare hybrid luminal-neuroendocrine form of prostate cancer, and confirmed that its genomic profile contained signatures of both cell types [52]. Thus, while disease subtypes such as NEPC are relatively rare, the potential for identification of novel mutations and rearrangements and for the individualization of therapy via next-generation sequencing-based assays is an intriguing corollary of these studies.

A number of genome-wide, large-scale studies of cancer genomes have also revealed tremendous heterogeneity within and among cancers of different histological origin, and imply the existence of a large repertoire of previously unknown cancer susceptibility genes. For example, a mutational screen of 518 protein kinases in 210 human cancers of various origins identified many previously unidentified cancer mutations [53]. Using a less directed approach, another study used extensive transcriptome, genome and epigenome characterization of 209 glioblastomas, as well as sequencing of 601 candidate cancer genes in a subset of these tumors, and revealed three signaling pathways important in this disease: p53, Rb, and the phosphatidylinositol 3-kinase (PI3K) pathway, which are pathways that are commonly disrupted in a variety of cancers [54]. Similarly, a survey of 623 genes in 188 human lung adenocarcinomas identified over 1000 somatic mutations linked to tumor development [55]. In yet another study, the analysis of 20 857 transcripts (representing 18 191 genes) in 11 breast and 11 colorectal tumors found a handful of commonly mutated gene "mountains" and a much larger number of gene "hills" mutated at low frequency; which again identified the PI3K pathway as a major contributor to carcinogenesis [56]. Parallel results were obtained in pancreatic cancer and glioblastoma [57,58]. Furthermore, with advances in massively parallel sequencing technology, whole cancer genome sequencing has revealed a strikingly high abundance of somatic mutations in each cancer genome, with over 10 000 acquired mutations in some single tumors [55,59–61]. Taken together, these studies illustrate the profound genetic heterogeneity found not just within prostate tumors but in all forms of cancer, which clearly represents one of the major challenges to personalized therapy.

Despite the seemingly insurmountable challenge of tumor heterogeneity, the myriad of reported genetic alterations in cancers not only can confer a biological advantage to the tumor but may actually constitute genetic changes that represent actionable therapeutic targets (e.g. *PIK3CA*, *BRAF*, *NF1*, *KDR*, and *PIK3R1*). Additional insight derived from various cancer genome studies includes correlations between cancer mutations and prognosis, such as *IDH1* and *IDH2* mutations in several types of gliomas [58,62] and in a small subset of prostate cancers [63]. In a lung cancer genome study [55], clinical correlations existed between the number of mutations and clinical grade, mutations in specific genes and risk factors (i.e. *EGFR* mutations and smoking), and pathological subtypes (i.e. negative correlation for *LRPB1*, *TP53*, and *INHBA* with acinar, papillary and BAC subtypes). These preliminary studies have both identified specific challenges and new opportunities that will translate to clinical benefits.

As outlined above, a commonly mutated pathway that represents a valid therapeutic target in many cancers is the PI3K pathway. With respect to CaP, Gerald and colleagues demonstrated that frequent alterations occur in the PI3K/AKT/PTEN axis; with alterations occurring in 42% of primary prostate tumors, and in 100% of metastases [33]. In addition to providing valuable insights into the biological processes that underlie the development and differentiation of prostate tumors, data such as these provide strong support for the clinical validation of PI3K pathway inhibitors in treating CaP. Future studies using more detailed approaches are likely to refine further our understanding of which pathways may be most effectively targeted for the treatment of both primary and metastatic prostate cancer. Moreover, recent studies have validated the feasibility of obtaining genome-wide next-generation sequencing data for a given patient within 3–4 weeks [64], including DNA and RNA isolation, sequencing, and clinical review of key findings. Approaches such as this would allow for individualized, biomarker-driven therapy, based on next-generation sequencing, within a clinically relevant timeframe.

CONCLUSIONS: TRANSFERRING GENOMIC INFORMATION INTO CLINICAL TESTS

Through deep sequencing and multimodal genomic studies (i.e. DNA, RNA, non-coding RNA), it has become feasible to define the genomic landscape of prostate cancer with incredible detail. When combined with clinical data, the value of these studies is multiplied by allowing the collective data to be mined for novel biological markers of patient outcome and treatment response. To that end, the Canadian Prostate Cancer Genome Network (CPC-GENE) is sequencing 500 prostate cancer genomes from 300 patients with intermediate-risk disease, each of whom has at least 5 years of clinical follow-up. This will permit a robust assessment of genetic changes associated with treatment failure in potentially curable, organ-confined disease. Furthermore, a focus on multiple regions of disease from the same prostate will allow us to investigate the potential implications of intra-prostatic heterogeneity on treatment outcome. Additional groups within TCGA and the ICGC are currently sequencing other prostate cancer cohorts, and thus the next few years should bring an enormous increase in our understanding of the fundamental genomic alterations that characterize the disease (Table 14.2).

As discussed above, Garraway and colleagues sequenced the genomes of 57 prostate tumors and matched normal tissues to characterize interdependent gene rearrangements and deletions during oncogenesis and progression. They termed this interdependence "chromoplexy", in which multiple cancer genes are coordinately disrupted over a few events. If this relates to outcome, the concept of chromoplexy may allow for better triaging of patients into risk groups that have the least or greatest chance of metastasis during progression and drive individualized treatment based on their "chromoplexy" [42]. These changes may be amenable to less complex testing and thus more widely applied using standard clinical platforms such as immunohistochemistry (IHC) or fluorescence *in situ* hybridization (FISH), which can be used on primary patient biopsies to inform treatment management. Indeed, a Canadian transnational project is underway (The Terry Fox Research Institute–Canadian Prostate Cancer Biomarker Network) which will create tissue microarrays on more than 2000 localized prostate cancer patients who underwent radical prostatectomy, precision radiotherapy or active surveillance. Most importantly, this unique tissue resource is accompanied by detailed clinico-pathological data including patient outcome.

If the CaP susceptibility genes and prognostic gene signatures derived from various genome-wide studies can be distilled down to information that can be tested on such arrays by FISH and IHC, it is highly likely that large-scale genomic studies will inform clinical practice. These assays could then ensure that important genomic changes within a patient's cancer can be transferred to hospitals worldwide and drive new standards of care in the treatment for prostate cancer.

ACKNOWLEDGMENTS

This work was supported by a grant from the Prostate Cancer Canada and Movember Foundation to the CPC-GENE project, the

TABLE 14.2 International Prostate ICGC Programs

Host Country	Scientific Questions	Sample Types	Platforms	Unique Features
Canada	Identify genetic factors that predict outcome in patients with intermediate risk CaP who underwent radical prostatectomy or image-guided radiotherapy Identify genetic factors that predict outcome in patients with castrate-resistant prostate cancer Evaluate intra- and inter-prostate genetic heterogeneity	Fresh frozen research biopsy Fresh frozen and FFPE radical prostatectomy	Illumina HiSeq 2000	All patients have minimum of 5y clinical follow-up for outcome studies Prospective re-consenting allows for collection of matched normal tissue Whole-genome sequencing of material taken from a unique bank of fresh-frozen pre-treatment biopsies
France	Characterization of genomic factors of prostate cancer in French Caucasian and African-Caribbean men			Capitalizes on existing clinical and epidemiological databases of French Caucasian (PROGENE) and French African-Caribbean (KARU) men.
Germany	Identify novel genetic alterations in early-onset prostate cancers (< 50 y.a.)	Fresh frozen radical prostatectomy	Illumina HiSeq 2000	Early onset prostate cancer may have a distinct genomic profile; may be a strong familial component with higher proportion of hereditary cancers
United Kingdom	Comparison of prostate cancer genomes from men in different ethnic groups (Caucasian, Chinese, Malaysian, Indian) to gain insights into disease etiology Compare genomes of castrate-resistant disease, cancers that ultimately went on to become castrate-resistant, and cancers from patients with low vs. intermediate vs. high risk disease Assessment of prostate cancer intra-prostate and intra-cancer heterogeneity	Fresh frozen and FFPE radical prostatectomy TURPs-extracted prostate tissue	Illumina HiSeq 2000	Fresh frozen prostate cancer metastases available from warm autopsy program Tissue available from several ethnic groups
United States (The Cancer Genome Atlas)	Analyze correlation between Gleason score and other tumor characteristics to better define aggressiveness Examine samples from majority populations and compare to underrepresented populations, such as African Americans who have the highest incidence rate of prostate cancer Integrate genomic information with the proteomic data	Fresh frozen radical prostatectomy	Multiple platforms	Multimodal Data: CNV, SNV, Methylome, RNA expression

Ontario Institute for Cancer Research, Terry Fox Research Institute in partnership with the Canadian Partnership Against Cancer to the Canadian Prostate Cancer Biomarker Network and the Princess Margaret Cancer Center Foundation. This research was also funded in part by the Ontario Ministry of Health and Long Term Care. The views expressed do not necessarily reflect those of the Ontario Ministry of Health and Long Term Care. RGB is a Canadian Cancer Society Research Scientist.

GLOSSARY

Adenocarcinoma Epithelial cell cancer, occurring in a gland.

Androgen deprivation A therapy for prostate cancer involving the use of drugs that block either the synthesis or function of androgen hormones, which are often required for prostate cancer cell survival.

Brachytherapy A form of radiotherapy in which a radioactive source is placed adjacent to or within the area requiring treatment.

Castrate-resistant prostate cancer (CRPC) An advanced, incurable form of prostate cancer which has ceased to respond to androgen-deprivation therapy.

Chromoplexy The occurrence of multiple, interdependent DNA translocations and deletions occurring over relatively few generations.

Chromothripsis The occurrence of thousands of DNA rearrangements during a single, massive event during the development of cancer.

Copy number alteration (CNA) A class of genomic structural variation characterized by gain or loss of one or more copies of a section of DNA, distinguished from normal copy number variation (CNV) by its association with a pathological state (e.g. cancer).

DNA methylation A chemical modification of DNA bases (usually cytosine), which can influence the rate of transcription of specific genes.

External beam radiotherapy Delivery of ionizing radiation to the prostate gland using an external radiation source.

Genome-wide association study (GWAS) A protocol in which single nucleotide variants (SNVs) are identified in the genome of individuals that either do or do not have a particular cancer (or other disease). SNVs that occur more frequently in those that have the disease than those that do not may be important for predicting disease onset

Gleason score A description of the morphological differentiation of prostate cancer, as determined by a pathologist under the microscope. The Gleason score represents the sum of two Gleason grades, ranging from 1 to 5 (where 1 is well-differentiated and 5 is very poorly differentiated), representing the most common and the second most common grade within an area of tumor.

Intra-ductal carcinoma of the prostate (IDC-P) A phenomenon observed in approximately 25% of prostate cancers in which cancer cells invade the prostatic ducts. IDC-P is strongly associated with high-grade cancers, and may be an independent prognostic factor for relapse and metastasis.

Nomogram A graphical calculator representing the relationship between three or more variables. In the context of prostate cancer, nomograms such as the Kattan nomogram are prognostic tools that can be used to decide which treatment approaches will result in the greatest benefit for men at various stages of prostate cancer.

Prostate specific antigen (PSA) A protein secreted by the prostate gland in response to androgen stimulation. The presence of PSA in the blood can be an indicator of the presence of prostate cancer.

Radical prostatectomy Surgical removal of the entire prostate gland.

Sanger sequencing A protocol involving the use of dideoxynucleotide-based chain termination for sequencing short strands of DNA. Sanger sequencing is serial, in that only a single DNA sequence can be determined in any given experimental reaction.

Synthetic lethality A phenomenon whereby simultaneous disruption of two proteins results in cell death, whereas disruption of either gene individually does not.

TMPRSS2–ERG fusion A fusion of two genes on chromosome 21 (*TMPRSS2* and *ERG*), which is observed in approximately 40–50% of prostate cancers.

TNM staging A system that describes the extent of a cancer based on the size of the tumor (T), the presence of tumor cells in the regional lymph nodes (N), and the presence or absence of distant metastases (M).

Whole-genome sequencing A protocol involving sequencing of short fragments of the entire genome. Modern whole-genome sequencing involves massively parallel sequencing of millions of DNA fragments in a single reaction.

Whole-transcriptome sequencing Massively parallel sequencing of cDNA (reverse transcribed mRNA) to gain information about the RNA content of a tumor.

ABBREVIATIONS

ACGH Array comparative genomic hybridization
AR Androgen receptor
BRCA1/2 Breast cancer type 1/2 susceptibility protein
BRFR Biochemical relapse-free rate
CaP Prostate cancer
CPC-GENE Canadian Prostate Cancer Genome Network
ERG ETS-related gene
ETS Erythroblast transformation-specific
FISH Fluorescence *in situ* hybridization
GEMCap Genomic evaluators of metastatic prostate cancer
GWAS Genome-wide association study
Gy Gray
ICGC International Cancer Genome Consortium
IDC-P Intra-ductal carcinoma of the prostate
miRNA MicroRNA
MRI Magnetic resonance imaging
NBS1 Nijmegen breakage syndrome 1
NEPC Neuroendocrine prostate cancer
PARP Poly(ADP)-ribose polymerase
PSA Prostate-specific antigen
PTEN Phosphatase and tensin homolog
RB1 Retinoblastoma 1
RT Radiotherapy
SPOP Speckle-type POZ protein
TCGA The Cancer Genome Atlas
TMPRSS2 Transmembrane protease, serine 2
TP53 Tumor protein 53

REFERENCES

[1] Nichol AM, Warde P, Bristow RG. Optimal treatment of intermediate-risk prostate carcinoma with radiotherapy: clinical and translational issues. Cancer 2005;104:891–905.

[2] D'Amico AV, Cote K, Loffredo M, Renshaw AA, Chen MH. Pretreatment predictors of time to cancer specific death after prostate specific antigen failure. J Urol 2003;169:1320–4.

[3] D'Amico AV, Moul J, Carroll PR, Sun L, Lubeck D, Chen MH. Cancer-specific mortality after surgery or radiation for patients with clinically localized prostate cancer managed during the prostate-specific antigen era. J Clin Oncol 2003;21:2163–72.

[4] Mukherji D, Eichholz A, De Bono JS. Management of metastatic castration-resistant prostate cancer: recent advances. Drugs 2012;72:1011–28.

[5] Yap TA, Swanton C, de Bono JS. Personalization of prostate cancer prevention and therapy: are clinically qualified biomarkers in the horizon? EPMA J 2012;3:3.

[6] Grimm P, Billiet I, Bostwick D, Dicker AP, Frank S, Immerzeel J, et al. Comparative analysis of prostate-specific antigen free survival outcomes for patients with low, intermediate and high risk prostate cancer treatment by radical therapy. Results from the

Prostate Cancer Results Study Group. Br J Urol Internatl 2012;109 (Suppl. 1):22–9.

[7] Kollmeier MA, Zelefsky MJ. How to select the optimal therapy for early-stage prostate cancer. Crit Rev Oncol Hematology 2012;83:225–34.

[8] Klotz L. Active surveillance for prostate cancer: overview and update. Curr Treat Options Oncol 2013;14:97–108.

[9] Eeles RA, Olama AA, Benlloch S, Saunders EJ, Leongamornlert DA, Tymrakiewicz M, et al. Identification of 23 new prostate cancer susceptibility loci using the iCOGS custom genotyping array. Nat Genet 2013;45:385–91.

[10] Amin Al Olama A, Kote-Jarai Z, Schumacher FR, Wiklund F, Berndt SI, Benlloch S, et al. meta-analysis of genome-wide association studies to identify prostate cancer susceptibility loci associated with aggressive and non-aggressive disease. Hum Mol Genet 2013;22:408–15.

[11] Bismar TA, Yoshimoto M, Vollmer RT, Duan Q, Firszt M, Corcos J, et al. PTEN genomic deletion is an early event associated with ERG gene rearrangements in prostate cancer. Br J Urol Internatl 2011;107:477–85.

[12] Tomlins SA, Rhodes DR, Perner S, Dhanasekaran SM, Mehra R, Sun XW, et al. Recurrent fusion of TMPRSS2 and ETS transcription factor genes in prostate cancer. Science 2005;310:644–8.

[13] Squire JA. TMPRSS2-ERG and PTEN loss in prostate cancer. Nat Genet 2009;41:509–10.

[14] Ishkanian AS, Zafarana G, Thoms J, Bristow RG. Array CGH as a potential predictor of radiocurability in intermediate risk prostate cancer. Acta Oncol 2010;49:888–94.

[15] Ishkanian AS, Mallof CA, Ho J, Meng A, Albert M, Syed A, et al. High-resolution array CGH identifies novel regions of genomic alteration in intermediate-risk prostate cancer. Prostate 2009;69:1091–100.

[16] Minner S, Enodien M, Sirma H, Luebke AM, Krohn A, Mayer PS, et al. ERG status is unrelated to PSA recurrence in radically operated prostate cancer in the absence of antihormonal therapy. Clin Cancer Res 2011;17:5878–88.

[17] Weischenfeldt J, Simon R, Feuerbach L, Schlangen K, Weichenhan D, Minner S, et al. Integrative genomic analyses reveal an androgen-driven somatic alteration landscape in early-onset prostate cancer. Cancer Cell 2013;23:159–70.

[18] Krohn A, Diedler T, Burkhardt L, Mayer PS, De Silva C, Meyer-Kornblum M, et al. Genomic deletion of PTEN is associated with tumor progression and early PSA recurrence in ERG fusion-positive and fusion-negative prostate cancer. Am J Pathol 2012;181:401–12.

[19] Liu W, Laitinen S, Khan S, Vihinen M, Kowalski J, Yu G, et al. Copy number analysis indicates monoclonal origin of lethal metastatic prostate cancer. Nat Med 2009;15:559–65.

[20] Locke JA, Zafarana G, Ishkanian AS, Milosevic M, Thoms J, Have CL, et al. NKX3.1 haploinsufficiency is prognostic for prostate cancer relapse following surgery or image-guided radiotherapy. Clin Cancer Res 2012;18:308–16.

[21] Zafarana G, Ishkanian AS, Malloff CA, Locke JA, Sykes J, Thoms J, et al. Copy number alterations of c-MYC and PTEN are prognostic factors for relapse after prostate cancer radiotherapy. Cancer 2012;118:4053–62.

[22] Beltran H, Yelensky R, Frampton GM, Park K, Downing SR, MacDonald TY, et al. Targeted next-generation sequencing of advanced prostate cancer identifies potential therapeutic targets and disease heterogeneity. Eur Urol 2013;63:920–6.

[23] Cooper CS, Foster CS. Concepts of epigenetics in prostate cancer development. Br J Cancer 2009;100:240–5.

[24] Kron K, Pethe V, Briollais L, Sadikovic B, Ozcelik H, Sunderji A, et al. Discovery of novel hypermethylated genes in prostate cancer using genomic CpG island microarrays. PloS One 2009;4: e4830.

[25] Borno ST, Fischer A, Kerick M, Falth M, Laible M, Brase JC, et al. Genome-wide DNA methylation events in TMPRSS2-ERG fusion-negative prostate cancers implicate an EZH2-dependent mechanism with miR-26a hypermethylation. Cancer Discov 2012;2:1024–35.

[26] Cybulski C, Wokolorczyk D, Kluzniak W, Jakubowska A, Gorski B, Gronwald J, et al. An inherited NBN mutation is associated with poor prognosis prostate cancer. Br J Cancer 2013;108:461–8.

[27] Narod SA, Neuhausen S, Vichodez G, Armel S, Lynch HT, Ghadirian P, et al. Rapid progression of prostate cancer in men with a BRCA2 mutation. Br J Cancer 2008;99:371–4.

[28] Vesprini D, Narod SA, Trachtenberg J, Crook J, Jalali F, Preiner J, et al. The therapeutic ratio is preserved for radiotherapy or cisplatin treatment in BRCA2-mutated prostate cancers. Can Urol Assoc J 2011;5:E31–5.

[29] Castro E, Goh C, Olmos D, Saunders E, Leongamornlert D, Tymrakiewicz M, et al. Germline BRCA mutations are associated with higher risk of nodal involvement, distant metastasis, and poor survival outcomes in prostate cancer. J Clin Oncol 2013;31:1748–57.

[30] Sandhu SK, Omlin A, Hylands L, Miranda S, Barber LJ, Riisnaes R, et al. Poly (ADP-ribose) polymerase (PARP) inhibitors for the treatment of advanced germline BRCA2 mutant prostate cancer. Ann Oncol 2013;24:1416–8.

[31] Fraser M, Zhao H, Luoto KR, Lundin C, Coackley C, Chan N, et al. PTEN deletion in prostate cancer cells does not associate with loss of RAD51 function: implications for radiotherapy and chemotherapy. Clin Cancer Res 2012;18:1015–27.

[32] Van der Kwast TH, Evans A, Lockwood G, Tkachuk D, Bostwick DG, Epstein JL, et al. Variability in diagnostic opinion among pathologists for single small atypical foci in prostate biopsies. Am J Surg Pathol 2010;34:169–77.

[33] Taylor BS, Schultz N, Hieronymus H, Gopalan A, Xiao Y, Carver BS, et al. Integrative genomic profiling of human prostate cancer. Cancer Cell 2010;18:11–22.

[34] Minner S, Gartner M, Freudenthaler F, Bauer M, Kluth M, Salomon G, et al. Marked heterogeneity of ERG expression in large primary prostate cancers. Modern Pathol 2013;26:106–16.

[35] Gerlinger M, Rowan AJ, Horswell S, Larkin J, Endesfelder D, Gronroos E, et al. Intratumor heterogeneity and branched evolution revealed by multiregion sequencing. N Engl J Med 2012;366:883–92.

[36] Joshua AM, Vukovic B, Braude I, Hussein S, Zielenska M, Srigley J, et al. Telomere attrition in isolated high-grade prostatic intraepithelial neoplasia and surrounding stroma is predictive of prostate cancer. Neoplasia 2007;9:81–9.

[37] Van der Kwast T, Al Daoud N, Collette L, Sykes J, Thoms J, Milosevic M, et al. Biopsy diagnosis of intraductal carcinoma is prognostic in intermediate and high risk prostate cancer patients treated by radiotherapy. Eur J Cancer 2012;48:1318–25.

[38] Lee SE, Kim DS, Lee WK, Park HZ, Lee CJ, Doo SH, et al. Application of the Epstein criteria for prediction of clinically insignificant prostate cancer in Korean men. Br J Urol Internatl 2010;105:1526–30.

[39] Miyagi Y, Sasaki T, Fujinami K, Sano J, Senga Y, Miura T, et al. ETS family-associated gene fusions in Japanese prostate cancer: analysis of 194 radical prostatectomy samples. Modern Pathol 2010;23:1492–8.

[40] Mao X, Yu Y, Boyd LK, Ren G, Lin D, Chaplin T, et al. Distinct genomic alterations in prostate cancers in Chinese and Western populations suggest alternative pathways of prostate carcinogenesis. Cancer Res 2010;70:5207–12.

[41] Penney KL, Sinnott JA, Fall K, Pawitan Y, Hoshida Y, Kraft P, et al. mRNA expression signature of Gleason grade predicts lethal prostate cancer. J Clin Oncol 2011;29:2391–6.

[42] Baca SC, Prandi D, Lawrence MS, Mosquera JM, Romanel A, Drier Y, et al. Punctuated evolution of prostate cancer genomes. Cell 2013;153:666–77.

[43] Paris PL, Weinberg V, Albo G, Roy R, Burke C, Simko J, et al. A group of genome-based biomarkers that add to a Kattan nomogram for predicting progression in men with high-risk prostate cancer. Clin Cancer Res 2010;16:195–202.

[44] Robbins CM, Tembe WA, Baker A, Sinari S, Moses TY, Beckstrom-Sternberg S, et al. Copy number and targeted mutational analysis reveals novel somatic events in metastatic prostate tumors. Genome Res 2011;21:47–55.

[45] Kumar A, White TA, MacKenzie AP, Clegg N, Lee C, Dumpit RF, et al. Exome sequencing identifies a spectrum of mutation frequencies in advanced and lethal prostate cancers. Proc Natl Acad Sci 2011;41:17087–92.

[46] Berger MF, Lawrence MS, Demichelis F, Drier Y, Cibulskis K, Sivachenko AY, et al. The genomic complexity of primary human prostate cancer. Nature 2011;470:214–20.

[47] Grasso CS, Wu Y-M, Robinson DR, Cao X, Dhanasekaran SM, Khan AP, et al. The mutational landscape of lethal castration-resistant prostate cancer. Nature 2012;487:239–43.

[48] Börno ST, Fischer A, Kerick M, Fälth M, Laible M, Brase JC, et al. Genome-wide DNA methylation events in *TMPRSS2–ERG* fusion-negative prostate cancers implicate an EZH2-dependent mechanism with *miR-26a* hypermethylation. Cancer Discov 2012;2:1024–35.

[49] Barbieri CE, Baca SC, Lawrence MS, Demichelis F, Blattner M, Theurillat J-P, et al. Exome sequencing identifies recurrent *SPOP*, *FOXA1* and *MED12* mutations in prostate cancer. Nat Genet 2012;44:685–9.

[50] Stephens PJ, Greenman CD, Fu B, Yang F, Bignell GR, Mudie LJ, et al. Massive genomic rearrangement acquired in a single catastrophic event during cancer development. Cell 2011;144: 27–40.

[51] Beltran H, Rickman DS, Park K, Chae SS, Sboner A, MacDonald TY, et al. Molecular characterization of neuroendocrine prostate cancer and identification of new drug targets. Cancer Discov 2011;1:487–95.

[52] Wu C, Wyatt AW, Lapuk AV, McPherson A, McConeghy BJ, Bell RH, et al. Integrated genome and transcriptome sequencing identifies a novel form of hybrid and aggressive prostate cancer. J Pathol 2012;227:53–61.

[53] Greenman C, Stephens P, Smith R, Dalgliesh GL, Hunter C, Bignell G, et al. Patterns of somatic mutation in human cancer genomes. Nature 2007;446:153–8.

[54] The Cancer Genome Atlas Research Network. Comprehensive genomic characterization defines human glioblastoma genes and core pathways. Nature 2008;455:1061–8.

[55] Ding L, Getz G, Wheeler DA, Mardis ER, McLellan MD, Cibulskis K, et al. Somatic mutations affect key pathways in lung adenocarcinoma. Nature 2008;455:1069–75.

[56] Wood LD, Parsons DW, Lin J, Sjoblom T, Leary RJ, et al. The genomic landscapes of human breast and colorectal cancers. Science 2007;318:1108–13.

[57] Jones S, Zhang X, Parsons DW, Lin JC, Leary RJ, Angenendt P, et al. Core signaling pathways in human pancreatic cancers revealed by global genomic analyses. Science 2008;321:1801–6.

[58] Parsons DW, Jones S, Zhang X, Lin JC, Leary RJ, Angenendt P, et al. An integrated genomic analysis of human glioblastoma multiforme. Science 2008;321:1807–12.

[59] Ley TJ, Mardis ER, Ding L, Fulton B, McLellan MD, Chen K, et al. DNA sequencing of a cytogenetically normal acute myeloid leukaemia genome. Nature 2008;456:66–72.

[60] Mardis ER, Ding L, Dooling DJ, Larson DE, McLellan MD, Chen K, et al. Recurring mutations found by sequencing an acute myeloid leukemia genome. N Engl J Med 2009;361:1058–66.

[61] Shah SP, Morin RD, Khattra J, Prentice L, Pugh T, Burleigh A, et al. Mutational evolution in a lobular breast tumour profiled at single nucleotide resolution. Nature 2009;461:809–13.

[62] Yan H, Parsons DW, Jin G, McLendon R, Rasheed BA, Yuan W, et al. IDH1 and IDH2 mutations in gliomas. N Engl J Med 2009;360:765–73.

[63] Ghiam AF, Cairns RA, Thoms J, Dal Pra A, Ahmed O, Meng A, et al. IDH mutation status in prostate cancer. Oncogene 2012;31:3826.

[64] Roychowdhury S, Iyer MK, Robinson DR, Lonigro RJ, Wu YM, Cao X, et al. Personalized oncology through integrative high-throughput sequencing: a pilot study. Sci Translat Med 2011;3:111–21.

Chapter 15

Cancer Genomics

Genomics of Colorectal Cancer

Daniel Gaston[1]
Carman Giacomantonio[2]

[1]Department of Pathology, Dalhousie University, Halifax, NS, Canada
[2]Departments of Pathology and Surgery, Dalhousie University, Halifax, NS, Canada

Contents

Cancer Genomics. DOI: http://dx.doi.org/10.1016/B978-0-12-396967-5.00015-3

Key Concepts

- Loss of function of mismatch repair genes (*MLH1, MSH2, MSH6, PMS1, PMS2*) leads to microsatellite instability and colorectal cancer (CRC). Germline mutations in these genes is a cause of Lynch syndrome, also known as hereditary non-polyposis colon cancer (HNPCC)
- *APC*, through the Wnt pathway, regulates the levels of β-catenin. Loss of *APC* function occurs in both sporadic CRC and familial adenomatous polyposis (FAP) colon cancer, and results in an abnormal increase in β-catenin levels and the inappropriate activation and overexpression of a number of genes that promote tumor growth and proliferation
- Colon tumors can be classified as one of three main subtypes based on epigenetic status and genomic instability: (1) tumors with microsatellite instability (MSI), which have the best prognosis; (2) tumors without microsatellite instability (non-MSI) that tend to be more aggressive and feature chromosomal instability (CIN); and (3) tumors that exhibit the CpG island methylator phenotype (CIMP) which, in the absence of MSI or CIN, has the poorest prognosis. However, these subtypes are not necessarily mutually exclusive and tumors that exhibit both the CIMP and CIN phenotypes can occur
- Activating mutations in *KRAS* lead to activation of the EGFR pathway. Tumors that are positive for *KRAS* mutations do not respond to treatment with EGFR-inhibiting therapies
- Frequently mutated genes in colorectal cancer tend to cluster in several common pathways such as DNA repair, Wnt signaling, TGF-β, MAPK, or Akt/mTOR. The genes and pathways are commonly involved in cell growth, proliferation, and resistance to apoptosis
- High-throughput sequencing technologies and genomics are providing an increasingly detailed and more complex picture of the colon cancer genome. These studies have uncovered new information about methylation patterns and transcriptional profiles of genes in colorectal cancer, and have been useful in defining new phenotypes (e.g. the CpG island methylator subtype of CRC). However, the translation of these findings into clinical care remains a challenge
- While still in early stages, non-coding RNA and transcriptional profiling (RNA-seq) hold promise for generating new colorectal cancer diagnostics signatures

INTRODUCTION

Colorectal cancer (CRC) is the third most common cause of cancer mortality in women and fourth in men [1]. It is fitting therefore that The Cancer Genome Atlas (TCGA) published a comprehensive characterization of the genetics of CRC as their third publication [2]. Because CRC is often diagnosed at a late stage, and the early detection of cancer dramatically increases survival, the identification of genetic risk factors in CRC is of the utmost importance.

The most widely recognized pathway describing CRC progression is the adenoma–carcinoma sequence [3] (Figure 15.1): beginning as benign polyps/dysplastic lesions before progressing to advanced adenoma, and finally to invasive carcinoma. A variety of molecular pathways and genes are involved in the progression, which typically occurs over years or decades [4].

Carcinomas that remain confined to the colon wall are curable with surgery, while most (73%) of those that progress to stage III tumors (metastasize to regional lymph nodes) are treatable with a combination of surgery, chemotherapy, and radiotherapy [5,6].

Progression of this type falls into two main categories defined by particular molecular features: tumors with microsatellite instability (MSI), where strand slippage coupled with defects in mismatch repair cause growth or retraction in the number of repeats within microsatellite regions, and those without microsatellite instability (non-MSI). Non-MSI, or microsatellite stable (MSS), tumors generally have a poor prognosis and tend to be more aggressive, often featuring chromosomal instability (CIN). CIN is characterized by translocations, inversions, and copy number alterations, whereas MSI tumors, which account for 10–15% of cases, are less aggressive and have both better clinical outcomes and prognosis.

In addition to the adenoma–carcinoma sequence, the "serrated pathway" [7–9] has come to be recognized more recently. Serrated adenomas occur as a subset of serrated polyps, which also include benign hyperplastic polyps. Serrated adenomas are distinguished based on their architecture, which is described in more detail elsewhere [10] but, in brief, the colon polyp features a serrated gland architecture along with changes characteristic of adenoma. Small serrated adenomas may be indistinguishable from hyperplastic polyps under the microscope.

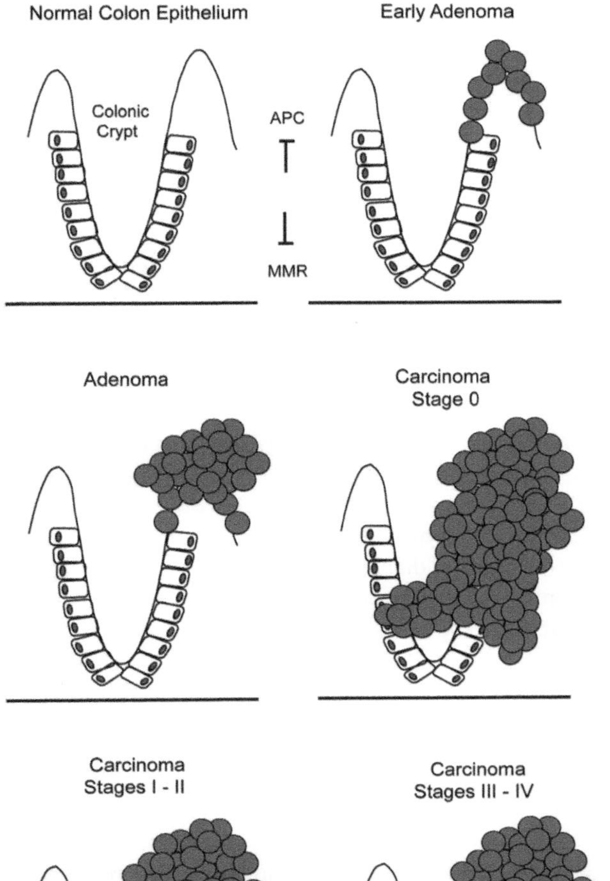

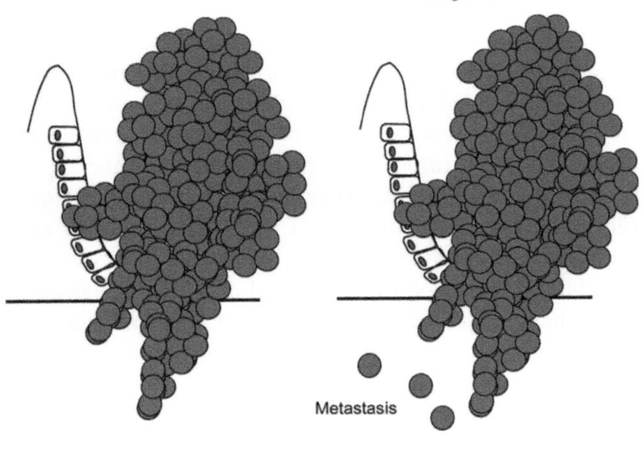

FIGURE 15.1 The adenoma–carcinoma sequence of colorectal cancer. Loss of function of either *APC* or mismatch repair genes (*MLH1*, *MSH2*, etc.) leads to the initial hyperproliferation of epithelial cells of the colonic lining and an early adenomatous polyp or lesion. Subsequent somatic mutations in other genes such as *KRAS* act as drivers in the progression to larger adenomas and eventually carcinoma. Carcinoma may then progress through various stages leading to eventual metastasis to other organs and lymph nodes.

CIN is found in 85% of sporadic CRC cases and MSI in the remaining 15% [9]. In these sporadic MSI cases, the instability itself is due to silencing of genes in the mismatch repair (MMR) pathway (see below). Silencing occurs through methylation of CpG islands in the promoter region of these genes, a phenomenon typically overlapping with the CpG island methylator phenotype (CIMP) [2,7,9,11–13], described in more detail later.

As with other cancers, CRC is categorized by a series of stages (0–IV), known as staging. These stages reflect the degree of invasiveness of the tumor locally, lymph node involvement, and whether the tumor has metastasized elsewhere in the body. CRC is staged through a mixture of preoperative diagnostic methods (biopsy, imaging) and what is found during the surgery itself (pathologic staging). The most widely used staging method measures three criteria: tumor size, lymph node, and metastasis involvement (referred to as the TNM staging system) to determine the overall stage, which is summarized below:

- **Stage 0:** Tumors have not yet penetrated the bowel wall
- **Stage I:** Tumors have already invaded several layers into the muscular wall of the colon or rectum
- **Stage II:** Tumors have penetrated the colon or rectal wall completely and may have begun to invade nearby tissue as well
- **Stage III:** Tumor has spread to regional lymph nodes
- **Stage IV:** Metastasis to nearby organs has occurred, most often the liver or lungs.

A great deal of research and clinical trials has focused on defining prognostic indicators for treatment in stage III tumors as many CRC cases are not diagnosed until this late stage. Five-year survival rates for CRC are approximately 64%, with treatment consisting of surgery followed by chemotherapy and possibly radiation depending on the extent of tumor growth [1,14].

Sporadic cases of CRC are the most common, accounting for 70–80% of cases, with the remaining 20–30% having some familial component [15–18], but only a small proportion belonging to a defined heritable syndrome. Familial cases are defined as two or more CRC cases between first-degree relatives. Risk factors for sporadic CRC include age, gender, environmental factors (such as diet and exercise), and inflammatory bowel disease. While sporadic CRC comprises the majority of all CRC cases, the lifetime average risk within the general population is only 5%, with much higher rates in Western and developed nations compared to other regions [1].

Familial colorectal cancer and CRC syndromes make up only 5–7% of all CRC cases; however, the identification of the genetics underlying them led to the early characterization of the most common genetic mechanisms in CRC. Due to the importance in determining the major genetic factors of CRC, we will first focus on these hereditary syndromes.

COLORECTAL CANCER SYNDROMES

While upwards of 30% of CRC cases have some evidence for an underlying familial/hereditary component, only

5–7% belong to a defined hereditary syndrome. The CRC syndromes are characterized by the inheritance of strongly penetrant, Mendelian variants that lead to a high CRC risk in families, often affecting individuals at much younger ages than seen in the general population [16,17,19,20]. The study of families affected by CRC syndromes led to the first major discoveries of the genetics underlying colorectal cancer initiation and progression.

Hereditary CRC syndromes have been traditionally divided between the polyposis and non-polyposis syndromes. The most common non-polyposis syndrome, and most common CRC syndrome in general, is Lynch syndrome, while the most common of the polyposis syndromes is familial adenomatous polyposis (FAP) [17].

Families that show clear genetic inheritance patterns of a CRC syndrome have been amenable to genetic linkage studies. Genetic linkage studies are able to use well-resolved family trees, accurate phenotypic data for family members (affected/non-affected status), a model of the mode of genetic inheritance (recessive, dominant, complete versus incomplete penetrance), and genetic markers (restriction fragment length polymorphisms (RFLPs) or single nucleotide polymorphisms (SNPs)), to identify the region of the genome linked to the phenotype under study. In the case of recessively inherited disorders, genetic linkage identifies a region of autozygosity, where a region on both chromosomes has been inherited from a common ancestor. Depending on the size of the family and the number/density of genetic markers, the region identified can contain a few dozen to hundreds or thousands of genes. Despite these limitations, and prior to more recent "omics" advances, linkage analysis has led to the discovery of countless disease-causing genes that are inherited in a Mendelian fashion. With respect to the study of hereditary CRC syndromes, genetic linkage studies have identified many genes, generally involved in DNA repair or signal transduction, which can lead to CRC when one or both copies of the gene do not function properly. Consequently, many of these same genes have also been implicated in familial or sporadic CRC cases (see later).

Lynch Syndrome and the Mismatch Repair System

Lynch syndrome, also called hereditary non-polyposis colorectal cancer (HNPCC), is an autosomal dominant form of hereditary CRC and the most common of the CRC syndromes [21]. Although the specific diagnostic guidelines used vary by location (reviewed in [17]), Lynch syndrome is characterized by early average age of onset (43 years) compared to the general population (65 years), shorter timespan for development of carcinomas from small adenomas (2–3 years versus 10 or more), higher risk of additional CRC after surgical resection, increased risk of additional non-colonic cancers, poorly differentiated pathology, increased survival when stage-matched, a tendency for microsatellite instability (MSI), and a tendency for tumors to be localized in the proximal colon.

At the genetic level, Lynch syndrome is caused by a germline loss-of-function mutation in one allele of one of the mismatch repair (MMR) genes: *MLH1* [22], *MSH2* [23], *MSH6*, *PMS1*, or *PMS2* [22,24]. Somatic loss of heterozygosity (LOH), where the remaining functional allele is lost, initiates tumorigenesis. Mutations in *MSH2* and *MLH1* account for 60 and 30% of Lynch syndrome cases, respectively, while mutations in *MSH6* account for most of the remaining 10% of cases. Both screening for MSI and immunohistochemistry for MMR genes can be used as molecular diagnostics of Lynch syndrome in conjunction with other phenotypic factors, with immunohistochemistry being the simplest and most widely available molecular pathology test [9,25].

During DNA replication of microsatellites, DNA slippage can result in incorrect base pairing, a DNA lesion normally repaired by MMR proteins. However, when MMR is defective, this slippage can lead to nucleotide gains or losses within the repeats. While most microsatellites are located outside of genes, some genes do contain repeats within functional coding regions. In this context, MSI can result in frameshifts and loss of function, which can affect tumor suppressor genes such as the transforming growth factor beta receptor type 2 (*TGFBR2*) and BCL2-associated X protein (*BAX*) [26–28]. This occurs in both Lynch syndrome and sporadic CRC cases with MSI.

The identification of *MLH1* and *MSH2* as genes causing Lynch syndrome was done through a combination of linkage mapping in affected families and evolutionary reasoning based on conservation of gene function between species. Previous studies had shown that a significant number of Lynch syndrome patients' tumors had microsatellite instability and mismatch repair defects [29,30]. Genes responsible for mismatch repair in *E. coli* had already been discovered and two of them, *MutS* and *MutL*, had been shown to cause MSI when non-functional in both *E. coli* and yeast models [31,32].

Based on these data, the human homologs of these genes were identified. *MSH2*, the human homolog of *MutS*, was located within one of the candidate regions identified in linkage studies [23,33] and the homolog of *MutL*, *MLH1* within another [22,34]. *PMS1* and *PMS2* were also identified as homologs of *MutL* and mutations in both genes were shown to cause Lynch syndrome [22,24]. These important gene discoveries, which have affected the health of countless CRC patients worldwide, are a perfect illustration of the importance of basic science research to medical science. Without both the basic discoveries in yeast and the application of evolutionary biology to target and guide the research, these genes would have taken much longer to identify and at a significantly higher cost.

In addition to straightforward LOF mutations in these genes, inherited hypermethylation of *MLH1* and *MSH2* has been observed, also resulting in gene silencing equivalent to a heterozygous LOF mutation [35,36]. Deletions of the 3′ region of *EPCAM/TACSD1*, the gene immediately upstream of *MSH2*, results in transcriptional read through and inactivation of *MSH2* [17,37,38]. Methylation of the *MLH1* promoter, and its resulting inactivation, is also observed in sporadic CRC cases and occurs in approximately 15% of all CRCs [39,40].

Familial Adenomatous Polyposis

Familial adenomatous polyposis (FAP) is also an autosomal dominant condition with complete (100%) penetrance. Individuals with FAP develop hundreds to thousands of adenomatous polyps, which occur primarily in the lower colon in a process that begins in young adulthood. The probability that one or more of these polyps will become malignant is nearly 100%, with extra-colonic polyps and tumors being extremely common [17]. Attenuated familial adenomatous polyposis (AFAP) is similar to FAP, although affected individuals generally develop fewer than 100 colonic polyps. These polyps, known as adenomas, typically begin to appear in the patient's teens or early 20s, and progress to carcinomas by their 40s or 50s.

Both FAP and AFAP are caused by germline truncating mutations in one allele of the *APC* gene (adenomatous polyposis coli). As in Lynch syndrome, somatic LOH of the remaining functional allele leads to tumorigenesis. Restriction fragment length polymorphism-based mapping studies of affected families had indicated the genetic mutation causing FAP was located on the q arm of chromosome 5 [41–43]. Further positional cloning experiments eventually identified truncating mutations in *APC* [44–47]. Unlike Lynch syndrome, FAP and AFAP do not exhibit MSI, instead they exhibit chromosomal instability and recurrent somatic loss of specific chromosomal regions (described in more detail later).

APC is part of the protein complex that regulates the degradation of β-catenin, the terminal signaling molecule of the Wnt pathway. Mutations in *APC* that prevent it from properly joining this complex, or result in the gene being lost in its entirety, lead to an increase in β-catenin levels and the inappropriate activation and overexpression of a number of genes that promote growth and proliferation [48–50]. It is in this role that *APC* acts as a tumor suppressor gene, and *APC* mutations are also commonly observed in sporadic cases of CRC.

Other Hereditary CRC Syndromes

MUTYH-associated polyposis (MAP) is an autosomal recessive polyposis syndrome [51] caused by mutations in the *MUTYH* gene (also known as *MYH*). Unlike the autosomal dominant syndromes (Lynch, FAP, AFAP), MAP is caused by the inheritance of two loss-of-function variants, one affecting each allele, in *MYH*. Human *MYH* is the homolog of another *E. coli* DNA repair enzyme, *MutY*. *MutY* performs base-excision repair following oxidative damage of guanine-containing DNA nucleotides [52]. While its incidence rate is unknown, it is estimated to account for up to 1% of CRC cases [53]. MAP may present similarly to AFAP and displays a broad phenotypic range [54,55].

Recessive loss-of-function (LOF) mutations in *MYH* result in G:C to T:A transversions that occur frequently in the genes *APC*, *KRAS* (Kirsten rat sarcoma viral oncogene homolog), and *BRCA1/2* [54,56], genes already known to play a role both in hereditary and non-hereditary CRCs as well as other cancers.

Other polyposis syndromes that carry significant lifetime risk of CRC are juvenile polyposis syndrome (JPS), which carries a lifetime risk of between 10 and 38% [17], and Peutz–Jeghers syndrome (PJS), carrying a 39% lifetime risk [57]. Between 40 and 50% of JPS cases carry germline mutations in the genes *SMAD4*, *BMPR1A*, or *ENG* [58], all three of which are part of the TGF-β/SMAD pathway. Between 50 and 60% of PJS patients carry germline mutations in *LKB1* (also known as *STK11*) [59].

GENOMIC FEATURES OF COLORECTAL CANCER AND GENOME-WIDE ANALYSES

Non-hereditary CRC, usually described as sporadic, may have a familial component (higher than normal frequency of first-degree relatives with CRC or other cancers), but is separate from the more traditional syndromes described previously. Although familial or hereditary CRC account for the minority of total CRC cases compared to sporadic cancers, their power to identify causative germline mutations has also been informative for understanding sporadic CRC. The identification of *APC* and the MMR genes that cause FAP, AFAP, and Lynch syndrome has led to the design of custom microarray screens that can be used to look for mutations in known CRC genes. This has led to the characterization of a broader spectrum of possible mutations in these genes, which predispose individuals to CRC, such as the common D132H mutation of *MLH1* [60].

Because of the weak familial component observed in some sporadic CRC cases, the identification of potentially heritable risk loci is crucial. The identification of risk loci for sporadic CRC has taken two primary approaches: genome-wide association studies (GWAS) or, more recently, large-scale genomics studies using targeted and whole-genome screens of patients. These approaches find genomic markers strongly associated with the disease state (e.g. often the case in GWAS) or specific genes that are mutated more often than expected in CRC cases than in controls.

Sporadic CRC most often proceeds by the traditional adenoma–carcinoma sequence and, much like in FAP, the majority exhibit chromosomal rather than microsatellite instability. Mutation or loss of *APC* is followed by mutation or loss of either *KRAS* or *TP63* (tumor protein p63), followed by increased chromosomal instability and, finally, progression to carcinoma [49,61,62]. A less common tumor progression for sporadic CRC is the more recently described serrated pathway (described above) [7–9]. This pathway features an early mutation in the *BRAF* (B-Raf) gene, resulting in the CpG island methylator phenotype (CIMP), where subsequent hypermethylation of oncogenes drives tumorigenesis [63,64]. *BRAF* mutations are seen in 10–15% of sporadic CRC cases, the majority of which exhibit this phenotype [65]. *MLH1* is silenced in roughly half of these *BRAF*-mutation positive tumors, which produces microsatellite instability as described previously [11,12,66,67]. *BRAF* wild-type (wt) MSI tumors have a good prognosis, both *BRAF*-mutated MSI and *BRAF*-wt microsatellite stable (MSS) tumors have an intermediate prognosis, and *BRAF*-mutated MSS tumors have the most unfavorable prognosis [20,68].

Outside of serrated pathway CRC cases, many sporadic cases of CRC feature a change in the methylation status of *MLH1*. Epigenetic silencing of *MLH1* results in a typical MMR defect as described previously in Lynch syndrome [47,66]. Some sporadic CRC cases have shown hereditary *MLH1* variants, which may indicate a susceptibility to epigenetic silencing [69].

Tumors tend to acquire mutations at an elevated rate compared to non-cancerous tissues. These somatic mutations have been divided into two classes: driver mutations and passenger mutations. Driver mutations tend to occur in different individuals more frequently than expected by chance, occur in "cancer genes", often confer some selective advantage to the tumor, and lineages with these mutations outcompete other cancer cells in the tumor. Passenger mutations, on the other hand, offer no selective advantage and are selectively neutral, being carried along by driver mutations. Driver mutations provide key insights into the molecular mechanisms of cancer progression, can serve as biological markers for prognosis and chemotherapy resistance, and may represent possible therapeutic targets. While passenger mutations are typically selectively neutral, it is possible that, under future conditions, they could be exposed to selective forces and become driver mutations. Passenger mutations can also act as targets of treatment or become deleterious [70]. Further functional studies can also reassess passenger mutations, as this pool of mutations may also harbor previously unrecognized candidate drivers of tumor development [71].

This complex landscape of driver/passenger mutations and frequent or recurrent genomic events (e.g. structural variation and epigenetic changes) is not unique to sporadic CRC. However, given the much greater number of sporadic cases, they have proven the most useful for high-throughput genomic studies by allowing easy accrual of large sample sizes. Nonetheless, many of the genomic features described below for sporadic CRC, can also apply to somatic changes that occur in tumors in patients with hereditary CRC syndromes.

As discussed briefly above, methylation and gene silencing of *MLH1* is also a frequent occurrence in sporadic CRC, and hypermethylation of *MSH2* has also been described. Indeed hypermethylation of CpG islands constitutes a significant phenotype or subtype of CRC (i.e. CIMP as discussed above). These epigenetic changes in the genome are major driving events in cancer development and progression and, in the next section, we will describe the epigenomics of CRC.

Epigenomics

In up to 15% of sporadic CRC cases, somatic silencing of both alleles of the *MLH1* gene occurs due to methylation of the promoter region, leading to inactivation of the mismatch repair machinery similarly to that of Lynch syndrome [47,66,72]. Heritable hypermethylation of *MLH1* and *MSH2* have also been reported [35,36,73,74] in both sporadic and Lynch syndrome cases. Deletions of the 3′ exon of *EPCAM* have also been shown to produce this hypermethylated phenotype (CIMP) [37,38].

With the advent of genome level techniques for interrogating methylation status, as opposed to individual CpG islands, there has been greater interest in identifying epigenetic signatures that can further differentiate CRC subtypes on a molecular basis and provide both predictive and prognostic markers. This hypermethylation leads to transcriptional silencing of genes and, combined with other genetic changes, contributes to the overall dysregulation of multiple pathways crucial to tumorigenesis.

A recent study [13] examined genome-wide methylation patterns in 125 tumor samples and 29 adjacent tissues. Statistical clustering analysis produced four distinct groups based on presence or absence of the CIMP methylator phenotype. The CIMP-High and CIMP-Low categories are consistent with previous data [7]; however, they split the non-CIMP category into two subcategories, one predominantly made up of tumors in the distal colon with a high *TP53* mutation rate, while the other category was predominantly made up of rectal tumors. As expected, CIMP-High tumors were significantly associated with *MLH1* hypermethylation (and hypermethylation of other cancer-specific genes) as well as the *BRAF* V600E mutation. The CIMP-Low category was enriched for *KRAS* mutations and hypermethylation of a subset of loci seen in the CIMP-High cases. With the exception of *BRAF* mutations, which are unique to the CIMP-High phenotype in these studies, loci are enriched in particular CIMP subtypes (as identified in these studies) but not unique [7,13]. Nonetheless, combinations of genes, forming methylation profiles, were able to differentiate CIMP tumors from the non-CIMP tumors, and CIMP-High from CIMP-Low [13]. It is important to note that CIMP categories are not exclusive with other phenotypic categories and, indeed, overlap significantly with (and may be caused by) other phenotypes such as MSI and *BRAF* mutation status for example. These overlapping phenotypes are also relevant in relation to CRC patient outcome, as it has been noted that there are differences in prognosis between patients with tumors that are CIMP-only (which have the poorest prognosis), CIMP + CIN (which have an intermediate prognosis), or exhibit MSI (which have the best prognosis) [75].

Genome-Wide Association Studies of Colorectal Cancer Susceptibility

Genome-wide association studies (GWAS) have been leveraged to study a range of complex phenotypes and human diseases. For non-familial studies, they are able to identify the strongest statistically significant markers (typically microsatellite repeats or SNPs) associated with the phenotype under study (i.e. cancer). Genes that are in linkage with these markers are all candidates for linkage to the phenotype. GWAS can be highly informative, but require large sample sizes with minimal familial relationships between samples (e.g. unidentified consanguineous relationships could confound analysis) and should be restricted to a specific population, as different populations are expected to have both shared and unique risk factors.

Many loci have been identified that confer some risk for the development of CRC, and some specific genes have been clearly linked. It is easiest to identify candidate genes when they are already known to function in relevant CRC pathways of tumorigenesis or to interact with and/or modulate the action of known CRC genes. A large GWAS meta-analysis, combining the data from multiple GWAS together, identified significant loci associated with the gene coding for nucleic acid binding protein 1 (*NABP1*), laminin gamma 1 (*LAMC1*), cyclin D2 (*CCND2*), and T-Box 3 (*TBX3*) [76]. *TBX3* is a target of Wnt signaling and *NADP1* is involved in DNA repair, consistent with the pathology of other CRC genes. Other potential risk loci for sporadic CRC that have been identified from individual GWAS include *CDKN1A*, *POLD3*, *SHROOM2* [77], *SLC5A9*, and *DUSP4* [78]. As with most GWAS, the associated risk factors for individual loci are often quite small and population-specific, and it is currently unclear how many of the recently identified loci may play a role in cancer initiation or progression.

Exome/Genome Sequencing and the Identification of Common Somatic Mutations

In recent years, whole-genome sequencing has dropped in cost dramatically. In addition, many sequencing providers have also developed targeted sequencing protocols to sequence selectively the exons of known protein-coding genes, as well as a growing selection of microRNAs (miRNA), long non-coding RNAs (lncRNA), and other functional RNA species. This targeted sequencing (exome-seq) examines a fraction of the whole genome (1–2%) and focuses on the discovery of single nucleotide variants (SNVs) and small insertion/deletion (indel) events in the known coding regions of the genome at a fraction of the cost. These advances have led to many large-scale sequencing efforts and the development of international consortia doing targeted large-scale sequencing of a variety of cancer types, such as The Cancer Genome Atlas (TCGA), which aims to provide high resolution, genomic profiles of 20 different cancer types using a mixture of high-throughput "omics" techniques including transcriptomics and whole-genome/exome sequencing. These large-scale projects aim to provide comprehensive profiles of commonly recurring somatic mutations. Genes that acquire mutations more commonly than expected by chance are presumed to be potential driver mutations [79].

Recently, TCGA, building on previous high-throughput sequencing projects in CRC [80–82], published their comprehensive findings in colon cancer [2]. Their study included results from exome-seq in 224 tumor-normal pairs, somatic copy number alterations from SNP microarrays, low-coverage whole-genome sequencing, and transcriptome profiling.

These genomic profiling studies have, as expected, identified recurrent driver mutations in many of the same genes that cause hereditary CRC syndromes and genes already identified as recurrent somatic variants. In addition to cataloging commonly recurring somatic mutations, The Cancer Genome Atlas project was also able to define two classes of CRC based on their mutational signature, one of which they labeled as hypermutated (16% of cases, >12 non-silent mutations per 10^6 base pairs, median number of total mutations = 728). Most (77%) of the hypermutated tumors also featured high MSI, of these most also showed *MLH1* methylation and/or one or more mutations in mismatch repair genes (*MLH1*, *MLH3*, *MSH2*, *MSH3*, *MSH6*, *PMS2*) or DNA polymerase epsilon (*POLE*). This set of features was not generally observed in the non-hypermutated cases. For the non-hypermutated tumors, which comprised the bulk of cases, the most commonly mutated genes were *APC*, *TP53*, *KRAS*, *PIK3CA*, *FBXW7*, *SMAD4*, *TCF7L2*, *NRAS*, *CTNNB1*, *SMAD2*, *FAM123B (WTX)*, *ATM*, *ARID1A*, and *SOX9*. The hypermutated cases contained mutations predominantly in *ACVR2A*, *APC*, *TGFBR2*, *MSH3*, *MSH6*, *SLC9A9*, *TCF7L2*, and *BRAF*. While *APC* was one of the most frequently mutated genes in both hypermutated and non-hypermutated tumors it, along with *TP53*, were considerably underrepresented in the hypermutated cases compared to the non-hypermutated. Although distinguishing between driver and passenger mutations in large-scale exome or genome sequencing projects is difficult, those that occur more frequently than expected by chance in large sample sizes, as is the case here, are considered strong candidates for being drivers.

Other exome-sequencing projects have identified other candidate driver mutations, in addition to those discussed above, in sporadic CRC cases including *FANCM*, *LAMB4*, *PTCHD3*, *LAMC3*, *TREX2*, and *NOTCH3*, all of which were found as germline mutations in patients with corresponding LOH in tumor cells. Taken in total these recurrently mutated genes tend to share several features (Table 15.1), notably key pathways that are dysregulated in many cancers including the Wnt signaling pathway (*SMAD2*, *SMAD4*, *APC*, *CTNNB1*, *TCF7L2*, *TP53*), mismatch repair (as summarized above), the adherens junction pathway (*SMAD2*, *SMAD4*, *CTNNB1*, *TCF7L2*, *TGFBR2*), TGF-β signaling (*SMAD2*, *SMAD4*, *ACVR2A*, *TGFBR2*), MAPK signaling (*NRAS*, *TGFBR2*, *TP53*, *KRAS*, *BRAF*), and others. The impact of some of these genes and pathways is discussed in more detail later.

The similarity between the recurrent somatic mutations [83] in these experiments, along with which phenotype they cause (CIMP, MSI, non-MSI/MSS), combined with the general proportion of each phenotypic subcategory relative to all characterized tumors, closely parallels that observed in hereditary and familial CRCs. Furthermore, the chromosomal regions affected by copy number alterations (CNAs) (i.e. gains and losses) or translocations in CRC also tended to disrupt or alter the expression of similar tumor suppressor and oncogenes found mutated in hereditary CRC and other cancers [2].

Structural Variation

Structural variation encompasses a broad group of large genomic abnormalities including copy number alterations involving gene gains or losses, as well as chromosomal translocations, inversions, and gene fusions. These types of abnormalities are common in cancer cells including CRC. Indeed *APC*-induced CRCs commonly feature CIN as described previously. Many of these variants have been mapped to locations with frequent copy number variations (CNVs) in the human genome that are presumed to be functionally neutral [84]. These structural variations are identified by both FISH (fluorescence *in situ* hybridization) and array-based methods, although paired-end (whole genome) sequencing based methods have become increasingly popular. These whole genome-based methods provide increased resolution and the ability to directly identify gene breakpoints.

The commonly occurring (80%) CIN phenotype in sporadic CRC is associated with poor prognosis [3,9,85,86]. Several genomic “hotspots” where recurrent instability (gain/loss, rearrangement) has been observed have been identified, including human chromosomes 5q, 8p, 17p, and 18q [87–90]. CIN contributes to tumorigenesis by promoting aneuploidy as well as LOH, and is itself driven by underlying genetic events. A number of genes have been identified that can cause the CIN phenotype in human tumors including *PLK1*, *APC*, *TERC*, *TP53*, and *BRCA1/2* [86]. These genes fall into three major categories concerning their molecular role: genes involved in: (1) chromosomal segregation (i.e. *PLK1*, *APC*); (2) telomere regulation (*TERC*); and (3) DNA damage response (*BRCA1/2*, *TP53*).

While the CIN phenotype and the gene pathways that regulate chromosomal stability were once thought to be mutually exclusive with MSI and the CpG island methylator phenotype (CIMP), there is now evidence of overlap between the three categories, even if only in a small percentage of cases and at low levels (as discussed above) [90]. These findings highlight how genomics methods are defining new, previously unknown molecular subtypes of cancers, potentially opening the doors to new therapies and treatment guidelines.

Transcriptomics

Transcriptomics covers several different technologies but, in all cases, the focus is on actively transcribed genes, both protein-coding and non-coding RNA (ncRNA). RNA sequencing (RNA-seq), which is high-throughput sequencing of isolated RNA molecules, provides a wealth

TABLE 15.1 Genes Commonly Mutated in Colorectal Cancer*

Gene	Cellular Pathway(s)	Associated Hereditary CRC Syndrome
ACVR2A	TGF-β signaling	
APC	Wnt signaling, actin cytoskeleton regulation	Familial adenomatous polyposis (FAP) and attentuated FAP (AFAP)
ATM	DNA repair, cell cycle	
BMPR1A	TGF-β signaling, Wnt signaling	Juvenile polyposis syndrome (JPS)
BRAF	mTOR signaling, FGFR signaling, MAPK signaling, ERK signaling, actin cytoskeleton regulation, NGF signaling	
BRCA1	Cell cycle, DNA repair	
BRCA2	Cell cycle, DNA repair	
CCND2	Cell cycle, Wnt signaling, p53 signaling	
CDKN1A	Cell cycle, DNA repair, p53 signaling	
CTNNB1	Wnt signaling, adherens junction	
DUSP4	MAPK signaling	
ENG	TGF-β signaling	Juvenile polyposis syndrome (JPS)
FAM123B	Wnt signaling	
FANCM	DNA repair	
KRAS	VEGF signaling, MAPK signaling, actin cytoskeleton regulation	
LAMB4	ECM-receptor interaction, focal adhesion	
LAMC1	ECM-receptor interaction, focal adhesion	
LAMC3	ECM-receptor interaction, focal adhesion	
MLH1	Mismatch repair	Lynch syndrome (HNPCC)
MLH3	Mismatch repair	Lynch syndrome (HNPCC)
MSH2	Mismatch repair	Lynch syndrome (HNPCC)
MSH3	Mismatch repair	Lynch syndrome (HNPCC)
MSH6	Mismatch repair	Lynch syndrome (HNPCC)
MYH	DNA repair	Lynch syndrome (HNPCC)
NOTCH3	NOTCH signaling	
NRAS	VEGF signaling, MAPK signaling, actin cytoskeleton regulation	
PIK3CA	mTOR signaling, VEGF signaling, EGFR signaling, ERBB2 signaling	
PLK1	Cell cycle	
PMS1	Mismatch repair	
PMS2	Mismatch repair	
POLD3	Cell cycle, DNA repair	
POLE	Cell cycle, DNA repair	
SLC5A9	Transmembrane transport	
SLC9A9	Transmembrane transport	
SMAD2	SMAD signaling, Wnt signaling, TGF-β signaling, adherens junction	
SMAD4	SMAD signaling, Wnt signaling, TGF-β signaling, adherens junction	Juvenile polyposis syndrome (JPS)

(*Continued*)

TABLE 15.1 (Continued)

Gene	Cellular Pathway(s)	Associated Hereditary CRC Syndrome
STK11 (LKB1)	AMPK signaling, mTOR signaling, PI3 kinase signaling	Peutz–Jeghers syndrome (PJS)
TCF7L2	Adherens junction, Wnt signaling	
TGFBR2	TGF-β signaling, SMAD signaling, MAPK signaling, adherens junction	
TP53	MAPK signaling, Wnt signaling, NOTCH signaling, DNA repair, apoptosis and cell cycle control	
TP63	NOTCH signaling, DNA repair, apoptosis and cell cycle control	
TREX2	DNA repair	
WTX	Wnt signaling	

**As with many other cancer types, there is a preponderance of genes involved in signaling cascades such as Wnt, MAPK, and TGF-β, as well as DNA repair and cell cycle associated genes*

of information including comparative gene expression levels, allele-specific expression, direct sequence data (identification of genetic variants), changes in copy number, mRNA isoform abundance, and gene fusions (i.e. by identification of fusion transcripts). As with exome-seq and whole-genome sequencing, the advent of RNA-seq is providing much broader and comprehensive views of cancer genomics, although more focused and targeted approaches are also useful (e.g. microarrays or targeted RNA-seq) with many providers now producing curated "panels" for cancer research.

In the last decade, the identification of new classes of ncRNA and expansion of the number of known ncRNA genes has been driven largely by RNA-seq. Both lncRNA and miRNA are being increasingly investigated for their role in cancer progression, including in colorectal cancer. Transcriptional profiling provides a dynamic view of the cancer genome and the effect of mutations and chromosomal instabilities. These profiles, as well as providing active information about the molecular landscape of the cell, may also be useful as prognostic markers and for stratification and subtyping of cancers.

Non-Coding RNA and Colorectal Cancer

First identified in the early 1990s [91], miRNAs are small RNA molecules (~20 bases) that possess bases complementary with one or more mRNA molecules to which they can bind, acting as transcriptional and post-transcriptional modifiers of gene expression [92]. As sequencing of the total complement of RNA in a cell or tissue has become more affordable, attention has turned to deciphering the role (if any) of ncRNAs in normal biology and disease. In addition to miRNA, lncRNA and long non-coding intergenic RNA (lincRNA) have also gained wider attention. Thought to also be involved in the regulation of gene expression, knowledge of all three RNA species remains relatively limited when compared to protein-coding genes. Nevertheless, large-scale projects continue to identify new ncRNA genes and potential targets. In particular, much of the focus for research in ncRNA (and transcriptomics in general) has been on identifying molecular profiles of CRC subtypes or as prognostic indicators.

MicroRNA profiling has been demonstrated to differentiate between normal tissue, polyps, and colorectal tumors [93] but, because they can target multiple genes, their effect can be widespread throughout the genome. They may act in analogous fashions to known CRC oncogenes, with some evidence of miRNAs influencing MSI status [94], or in completely novel fashions. While a number of miRNAs have been proposed as specific markers of early-stage CRC or as prognostic markers, the exact biological significance and role of most ncRNAs remains unclear [93,95,96]. While these studies have yet to produce potentially clinically actionable results, this burgeoning field of research is growing dramatically.

IMPACT OF COMMON SOMATIC MUTATIONS ON MOLECULAR PATHWAYS IN COLORECTAL CANCER

Whether due to a hereditary CRC syndrome, familial susceptibility, or a sporadic cause, once the cancer is initiated, tumors acquire somatic mutations over time. Within the tumor mass, competition results in both proliferation and natural selection of cancer cells with specific gene mutations that allow them to adapt to changing growth conditions, and a hallmark of cancer development and progression that drives the acquisition of new mutations is genomic instability. For these reasons, there has been a great deal of interest in surveying the somatic mutational landscape of various cancer types in order to define subcategories, enhance prognosis, and to guide treatment. High-throughput genomic studies like the exome/whole-genome studies discussed previously have greatly increased our understanding of the mutational landscape of CRC.

Due to their common recurrence, there are several genes mutated and molecular pathways subverted in CRC (see Table 15.1) that warrant further discussion of their biological context. Perhaps the most important of these genes is *KRAS*, mutations in which occur in approximately 40% of all CRC cases (both sporadic and inherited). *KRAS*, a GTPase, is involved in a wide variety of signaling pathways, including the epithelial growth factor (EGFR) and the mitogen-activated kinase (MAPK) pathways, which influence cell proliferation and are actionable target pathways for chemotherapy. In colorectal cancer, *KRAS* mutations that occur on their own in the absence of *APC* mutations generally do not result in cancer formation [97]. More importantly, activating mutations of *KRAS* represent an important biomarker for whether *EGFR* inhibitor therapy will be effective (see below). Other members of the ras-gene family (*HRAS* and *NRAS*), which also function as downstream members of various growth factor signaling pathways, have also been identified as recurrent somatic mutations in CRC, although at lower frequencies (5% for *NRAS*) [39].

BRAF, another gene activated by ras-family proteins, and a regulator of the MAPK pathway, is also recurrently mutated and has been identified as a key molecular player in the serrated pathway and CIMP subtype (described above) [64]. The *BRAF* V600E mutation in particular lies within the active domain of *BRAF*, and results in enhanced *BRAF* activity that promotes increased cell division and growth factor secretion.

Mutations in *PIK3CA* appear in up to 25% of CRC cases [39], a gene whose protein product, when active, increases production of the second messenger phosphatidylinositol-3,4,5-triphosphate (PIP3). PIP3 mediates cell proliferation, growth and survival, which is a recurring theme associated with oncogene activation. *PTEN* (phosphatase and tensin homolog), a known tumor suppressor in many cancers, also acts on PIP3, with inactivating mutations or loss of gene expression in 10–20% of CRC cases [39]. Both genes act as regulators of the *Akt/PKB* (protein kinase B) signaling pathway. *Akt* ultimately (but indirectly) activates *mTOR* (mammalian target of rapamycin), a serine/threonine protein kinase that is an important mediator of cell survival, proliferation, and growth. As with many other cancers, gene inactivation or LOH of *TP53* is common (up to 70%). Similarly LOH at 18q (70%) affecting *SMAD2*, *SMAD3*, *SMAD4*, and/or *TGF-β* type II receptor (*TGFβIIR*) is also a common genetic event in CRC, all of which are genes in the TGF-β signaling pathway (discussed below) [39].

In CRC with MSI, many genes containing microsatellite repeats in their coding region acquire frameshift mutations leading to loss of function. The most important of these is likely somatic mutations in *TGFβIIR*. The TGF-β signaling pathway is an important mediator of cell proliferation, survival, growth, and apoptosis. In the context of CRC, the TGF-β pathway inhibits proliferation within the colon epithelium [98,99], and thus disruption of normal TGF-β signaling (e.g. by loss of *TGFβIIR*) is an important part of the carcinogenetic pathway in MSI tumors that promotes tumor progression. Other genes affected by mutations in repetitive DNA as a result of MSI include DNA repair (*MLH3*, *MSH6*, *MSH3*, *CHK1*), cell cycle and apoptosis genes (*BAX*, *PTEN*, *FAS*), and genetic regulators of proliferation (*GRB1*, *TCF4*, *WISP3*) [28].

MOLECULAR DIAGNOSTICS AND TREATMENT GUIDANCE

Currently, the most widely used screen tool for colorectal cancer is the fecal occult blood (FOB) test, which has low sensitivity and low specificity for CRC and requires follow-up tests including colonoscopy. As a result, there is considerable effort in working towards translating many of the genomics findings discussed in this chapter into clinically actionable tests with higher specificity and sensitivity for CRC, with some early success. Translating genetic/genomic findings from research studies into clinically relevant diagnostics and treatment guidelines is a difficult task, particularly as new technologies, such as high-throughput sequencing, have become more heavily used and produce large volumes of potentially clinically actionable data, although sequence-based tests can also provide high false-positive signals if not controlled carefully. These studies have led to the identification of biological markers of particular CRC stages and subtypes, the discovery of a number of prognostic markers that indicate CRC tumor progression and patient outcome, and have helped identify predictive (or therapeutic) markers that can be used to guide treatment based on their ability to predict tumor response to a given therapy.

The identification of hereditary genetic mutations causing CRC syndromes has provided some of the largest and most immediate benefits in affected families. Instead of routine surveillance in all family members, genetic tests can identify only those who carry the relevant genetic variation. Carriers of identified risk loci have various levels of surveillance and screening by colonoscopy, upper endoscopy, and small bowel imaging as required based on risk. These forms of surveillance, often begin as a child, teenager, or young adult, and their frequency can vary from 1 to 3 years based on the initial age of surveillance and depending on the particular CRC syndrome in question [17].

Mismatch repair gene mutational status in sporadic CRC cases is an important indicator of prognosis, with tumors driven by defects in MMR having lower recurrence rates, longer time to recurrence, and improved survival rates [100]. In particular, MSI in CRC has been associated with improved survival in stage II cases after complete resection [101].

With respect to signaling pathways mediating growth, the EGFR pathway is a common treatment target in cancer, and anti-EGFR antibody therapies such as cetuximab and panitumumab, or small molecule inhibitors of EGFR such as gefitinib or erlotinib, are used in CRC treatment. However, activating mutations in *KRAS*, found in approximately 40% of colorectal cancers, are indicative of treatment failure in patients treated with EGFR-targeting therapies [102–104]. *KRAS* mutations alone do not fully explain EGFR-targeted therapy response in CRC patients, as only 20–30% of CRC cases respond to EGFR-targeted therapy, including wild-type *KRAS* cases. One explanation may be complex feedback regulation of EGFR signaling via other common mutations in CRC. For example, CRCs carrying the activating V600E mutation in *BRAF* do not respond to *BRAF* inhibitors such as vemurafenib, which has otherwise been successful in the treatment of melanomas harboring the *BRAF* (V600E) mutation [105]. The reason that vemurafenib treatment failed to inhibit the growth of *BRAF* (V600E) carrying CRCs was that *BRAF* inhibition induced the activation of EGFR via a feedback mechanism that promoted cell proliferation. However, the combined treatment of these CRCs with inhibitors of both *EGFR* and *BRAF* was found to be successful [105].

Even when treatments have been successful, mutation and gene expression profiling may also improve patient outcomes. For example, 5-FU (fluorouracil), a common adjuvant chemotherapy used for the treatment of stage III CRC, has been investigated for benefit in treatment of stage II cases after surgical resection, although the results have proven controversial. Since clinicians are concerned with over-treating patients with chemotherapy such as 5-FU when surgical resection may be curative, several diagnostic assay panels have been developed based on quantitative gene expression profiling data to predict when adjuvant therapy is warranted [106]. Through such studies, new indicators of response to 5-FU have been identified, including defects in mismatch repair, the high-CIMP methylator phenotype, and thymidylate synthase expression, all of which are associated benefits from 5-FU adjuvant therapy in stage II CRC [107].

Despite the potential of genomic studies, the majority of proposed biomarkers derived from genome-scale projects have yet to be validated in multiple studies or in large clinical trials. Work will continue to try to validate many of these, including CD133+ status, a proposed marker of colorectal cancer stem cells [108–110], which has been linked with poor clinical outcomes, but whose status remains controversial [111,112].

CONCLUSIONS AND FUTURE DIRECTIONS

Despite being a relatively small proportion of the total number of cases, the study of familial colorectal cancer has provided early insights into the genetic mechanisms of this cancer. However, the bulk of CRC cases are sporadic and their etiology is multifactorial based on a complex interaction of dietary, environmental and subtle genetic factors that remain poorly defined. With advances in technology and decreasing costs in the genomics space, large-scale studies of sporadic cancers have provided further insights into the genetic basis of this complex disease. As a result, more than ever before, we have an increasingly detailed and complicated picture of the genetic landscape of CRC, which has resulted in the ability to subtype this disease based on molecular features instead of purely gross morphological changes.

As with many other cancers, mutations affecting the regulation of key pathways for cell proliferation, division, and survival are the most frequently observed, and most important in terms of treatment and prognosis. Many of the genes that drive CRC are involved in β-catenin regulation and the Wnt pathway; for instance *APC* is a key regulator of β-catenin and its normal function prevents its translocation to the nucleus by activating its degradation. Other genes, like *KRAS* and *BRAF*, acquire activating mutations that trigger the oncogenic and constitutive activation of EGFR and the MAP kinases. When mutated, these genes are also key prognostic indicators for treatment outcome for both chemotherapeutic agents and EGFR-targeting therapies.

Genomic technologies and their clinical application in defining the mutational landscape of CRCs have been useful in characterizing and refining the molecular subtypes of this disease. Although the more traditional, standard subtyping based on MSI status and mutations in a few key genes (the MMR genes, *BRAF*, *APC*, *KRAS*) is still the predominate means of stratifying CRC patients, more complex subtyping will eventually enter into clinical practice. This is a trend not unique to CRC and is extending to a variety of other cancers. Using genomic analyses, we have already extended the subtypes of CRC beyond MSI and CIN to include the CpG island methylator phenotype, which itself can be further subdivided into CIMP-High and CIMP-Low categories to help guide therapy. Future advances in genomics and, in particular, advances in the area of modified genome sequencing approaches such as methylation-specific sequencing, will enable cancer genomes to be interrogated to an ever increasing degree, allowing even further stratification. Indeed, the relatively recent explosion in high-throughput, genome scale genetic and epigenetic studies in large numbers of tumors and matched tissues is driving a revolution in how we study and classify colorectal cancer.

The translation of genomic research findings to clinical care, with a particular emphasis on prognostic indicators, is and will remain an area of intense focus in the coming decade. MicroRNA expression profiles hold a lot of promise in the construction of screening panels and as prognostic signatures, providing a broad view of changes in molecular signaling pathways and of malignant dysregulation. Similarly, thanks to work on the detection of circulating fetal DNA, advances in the ability to detect and sequence circulating tumor DNA promises a rapid and non-invasive means routinely to follow patient therapeutic response, as well as providing early detection of cancer relapse. In addition, circulating tumor DNA can be sequenced to identify shifts in the population structure of tumors and the emergence of resistant populations. Surveys of this type during treatment could be used to switch therapies when necessary to prevent such emergence. Finally, as advances in high-throughput sequencing methods drive down costs, and our experience with appropriate workflows (including bioinformatics) and possible false-positive rates grows, these technologies will begin to enter wider clinical practice. And with the clinical adoption of genomic medicine, it is hoped that cancer-related mortality will be greatly reduced for all cancers including CRC.

GLOSSARY

Adenoma A benign tumor of glandular origin. Adenomas are a possible precursor of malignancy. In individuals with heritable colorectal cancer syndromes, monitoring for adenoma formation is an important strategy for early diagnosis.

Adjuvant A pharmaceutical delivered in conjunction with another drug whose action it modifies. 5-Fluorouracil is a common adjuvant administered with other chemotherapy agents during treatment of some colorectal cancers

Allele One of (usually two) versions of a gene located at the same locus (position) on a chromosome. Alleles may be identical (homozygous) or variant (heterozygous).

Allele frequency The proportion of an allele observed within a population. The allele frequency is simply the ratio of all copies of a variant allele observed within the population divided by the chromosomal population size.

APC The adenomatous polyposis coli gene. A tumor suppressor that acts as a regulator of β-catenin. Truncating germline mutations in APC cause both familial adenomatous polyposis and attenuated familial adenomatous polyposis.

Autosome Any human chromosome other than the X and Y sex chromosomes.

Autozygosity A genomic region of homozygosity where both strands of DNA were inherited from a common ancestor.

Biomarker A portmanteau of the words biological and marker. Biomarkers are a measurable characteristic used to indicate the status of some process, generally normal versus a pathological state.

Carcinoma A cancerous tumor derived from epithelial cells.

Chromosomal instability One of many genomic instabilities observed in cancer. It is characterized by large-scale rearrangements of regions of a chromosome. Includes duplications and deletions as well as translocations.

Copy number variation The gain or loss of copy numbers of a gene from the normal two alleles. Copy number variation can occur through the loss of one or both alleles (resulting in hemizygosity) or gene duplications. Copy number variation can be neutral or pathogenic. Neutral variation adds a new level of polymorphism to the human genome.

DNA sequencing The process of determining the exact order of nucleotides in a fragment (or molecules) of DNA.

Driver mutations Driver mutations, in contrast to passenger mutations, are acquired somatic mutations in tumor cells that typically offer a degree of selective advantage. Driver mutations play critical roles in tumor progression and proliferation or offer resistance to chemotherapeutic agents.

Dysplasia An abnormality of tissue development characterized by an overabundance of immature cells versus mature and properly differentiated cells.

Epigenomics The genome-wide analysis of epigenetic features. Most often involves the description of modified DNA or DNA associated proteins occupying promoter regions which affect gene expression, turning them off or on. Epigenetic alterations do not result in changes to a DNA sequence but are chemical modifications that affect the interaction of DNA with chromatin, histones, and transcription factors.

Exome The portion of the human genome that is made up of the exons of protein-coding genes. Includes untranslated regions but not introns (although partial sequencing of intronic regions occurs due to the nature of exon capture). Is estimated to only make up approximately 1−2% of the human genome.

Familial adenomatous polyposis A type of inherited colorectal cancer syndrome caused by defects in the *APC* gene. Features chromosomal instability and the development of polyps within the colon beginning at a young age. An attenuated form (AFAP) is also caused by defects in *APC* but features fewer polyps.

Fecal occult blood Blood in the feces that is not readily apparent. FOB tests are aimed at detecting globin, DNA, or other factors to indicate the presence of blood.

Genome The totality of genetic information of an organism. In humans, includes both the nuclear and mitochondrial genomes.

Genomics The study of genomes. Includes genome-wide analysis of structure, sequence, function, and evolution.

Genome-wide association studies A method for using genome-wide polymorphism data (usually single nucleotide polymorphisms)

and large groups of patients and controls to identify loci linked to a disease.

High-throughout sequencing DNA sequencing technologies characterized by their ability to sequence thousands to millions of DNA fragments in parallel.

Hyperplasia An increase in the proliferation (and sometimes size) of cells in a tissue.

Immunohistochemistry The use of antibodies to stain and visualize the presence of proteins in tissue sections. Antibodies can be conjugated to enzymes (e.g. horseradish peroxidase) or fluorescently tagged.

Indel A portmanteau of the words insertion and deletion. Used to describe small insertion or deletion events that occur within a gene. May include complex cases where multiple events have occurred at the same position.

Linkage Genetic markers that are physically close to one another on a chromosome tend not to be separated from one another, or broken up, during recombination. These markers (often SNPs) are said to be in linkage. Linkage analysis can be performed in suitable pedigrees to identify regions of a chromosome with a greater degree of linkage than expected and that is preferentially shared among disease carriers than unaffected members of the family. It is a powerful tool for reducing the search space from a whole genome to part of a chromosome in disease genetics research.

Locus A chromosomal location.

Loss of heterozygosity Loss of heterozygosity results in the loss of a functional tumor suppressor in a two hit manner. The first deleterious mutation in a single allele results in a heterozygous state. In individuals with a hereditary syndrome, the first hit is a germline mutation. Most often the second hit is a deletion of the remaining functional allele although often LOH is used as a more general term to also describe the case where the second hit is a second deleterious mutation. The end result of both cases is the lack of a functional gene product.

Lynch syndrome A hereditary colorectal syndrome caused by defects in mismatch repair enzymes, resulting in microsatellite instability.

Meta-analysis An experimental design where the results or data of multiple experiments are combined together and analyzed.

Microsatellite instability Found in a subset of colorectal cancer, MSI is caused by defects in the mismatch repair pathway. Strand slippage in microsatellite repeats during DNA replication causes expansion/contraction of microsatellite repeat regions which can cause frameshifts and loss of function in genes with such repeats in their coding regions.

Mismatch repair The molecular means of recognizing and repairing the inappropriate insertion, deletion, or misincorporation of incorrect base pairs during DNA replication. Mismatch repair is also used for repairing certain kinds of DNA damage. Defects in certain mismatch repair enzymes are a major cause of certain types of colorectal cancer, in particular Lynch syndrome.

mRNA Messenger RNA. RNA that is transcribed from a gene and which will be transcribed into protein.

Metastasis When a tumor spreads from its original location to elsewhere in the body.

Mutation A change in a gene, chromosome, or region of a chromosome.

ncRNA Non-coding RNA. Non-coding RNA refers to RNA genes that do not encode for protein. A broad class of genes with several subsets including the more common tRNA. New classes with presumed regulatory roles include microRNA (small, approximately 20 bp) and long non-coding RNA.

Passenger mutations The subset of acquired somatic mutations in tumor cells that are presumed to be selectively neutral. These mutations "hitch hike" along with driver mutations during selective sweeps and clonal expansion of tumor cells.

Polymorphism When more than one allele exists in a population for a particular locus. Strictly speaking, a polymorphism should be a neutral variation and does not include disease-causing mutations. Includes variation at a single nucleotide, indels, or larger chromosomal variation.

Polyp An abnormal growth of tissue that protrudes from a mucous membrane.

Resection The partial surgical removal of an organ.

RNA-seq A term used to describe a range of technologies used to sequence RNA molecules, either directly or, more commonly, RNA that has been first converted in to double-stranded DNA by reverse transcription. RNA-seq is crucial to the field of transcriptomics.

Serrated A pathological term that, in this case, is used to describe the appearance of certain types of polyps. Serrated applies to the sawtooth-like appearance of this type of polyp.

Single nucleotide polymorphism A type of polymorphism that involves only a single nucleotide.

Somatic mutation A mutation occurring in a non-germline cell. Somatic cells are all cells in the body that are not part of the germline.

Staging The process of assigning a tumor to its correct pathological stage using various tests and diagnostic indicators. More advanced stages of cancer have worse prognoses and may have already involved lymph nodes or metastasis to other organs.

Transcriptional profile A higher-order view of gene expression in a disease or tissues. A profile may be generated from any of a number of technologies (microarray, RNA-seq) and describes the active transcriptional state of a cell, tissue, or tumor.

Transcriptomics The sequencing and analysis of gene expression. While this can apply to the use of microarray technology, most often it is used when referring to the use of sequencing-based methods. Used to study messenger RNA as well as non-coding RNA.

Tumorigenesis The process by which a tumor forms.

ABBREVIATIONS

AFAP Attenuated familial adenomatous polyposis
CIMP CpG island methylator phenotype
CIN Chromosomal instability
CRC Colorectal cancer
FAP Familial adenomatous polyposis
FOB Fecal occult blood
GWAS Genome-wide association studies
lncRNA Long non-coding RNA
lincRNA Long non-coding intergenic RNA
LOH Loss of heterozygosity
LOF Loss of function

miRNA MicroRNA
MMR Mismatch repair
MSI Microsatellite instability
MSS Microsatellite stable
RFLP Restriction fragment length polymorphism
SNP Single nucleotide polymorphism
SNV Single nucleotide variant
TCGA The Cancer Genome Atlas
TNM Tumor size, lymph node, and metastasis

REFERENCES

[1] Jemal A, Bray F, Center M, Ferlay J, Ward E, Forman D. Global cancer statistics. CA Cancer J Clin 2011;61:69–90.

[2] The Cancer Genome Atlas Network. Comprehensive molecular characterization of human colon and rectal cancer. Nature 2012; 487:330–7.

[3] Fearon ER, Vogelstein B. A genetic model for colorectal tumorigenesis. Cell 1990;61:759–67.

[4] Cunningham D, Atkin W, Lenz HJ, Lynch HT, Minsky B, Nordinger B, et al. Colorectal cancer. Lancet 2010;375:1030–47.

[5] André T, Boni C, Mounedji-Boudiaf L, Navarro M, Tabernero J, Hickish T, et al. Oxaliplatin, fluorouracil, and leucovorin as adjuvant treatment for colon cancer. N Engl J Med 2004;350: 2343–51.

[6] Libutti S, Salz L, Willett C. Cancer of the colon. In: DeVita VT, Hellman S, Rosenberg SA, editors. Cancer: principles and practice of oncology. 9th ed. Philadelphia, PA: Lippincott Williams & Wilkins; 2011. p. 1084–126.

[7] Shen L, Toyota M, Kondo Y, Lin E, Zhang L, Guo Y, et al. Integrated genetic and epigenetic analysis identifies three different subclasses of colon cancer. Proc Natl Acad Sci USA 2007;104:18654–9.

[8] Young J, Jenkins M, Parry S, Young B, Nancarrow D, English D, et al. Serrated pathway colorectal cancer in the population: genetic consideration. Gut 2007;56:1453–9.

[9] Cunningham D, Atkin W, Lenz HJ, Lynch HT, Minsky B, Nordinger B, et al. Colorectal cancer. Lancet 2010;375:1030–47.

[10] Noffsinger A. Serrated polyps and colorectal cancer: new pathway to malignancy. Annu Rev Pathol 2009;4:343–64.

[11] Toyota M, Ahuja N, Ohe-Toyota M, Herman J, Baylin S, Issa J. CpG island methylator phenotype in colorectal cancer. Proc Natl Acad Sci USA 1999;96:8681–6.

[12] Toyota M, Ho C, Ahuja N, Jair K, Li Q, Ohe-Toyota M, et al. Identification of differentially methylated sequences in colorectal cancer by methylated CpG island amplification. Cancer Res 1999;59(10):2307–12.

[13] Hinoue T, Weisenberger D, Lange C, Shen H, Byun HM, Van Den Berg D, et al. Genome-scale analysis of aberrant DNA methylation in colorectal cancer. Genome Res 2012;22:271–82.

[14] American Cancer Society. Cancer Facts & Figures 2013. Atlanta: American Cancer Society; 2013.

[15] Kastrinos F, Syngal S. Recently identified colon cancer predispositions: MYH and MSH6 mutations. Semin Oncol 2007;34:418–24.

[16] Patel S, Ahnen D. Familial colon cancer syndromes: an update of a rapidly evolving field. Curr Gastroenterol Rep 2012;14:428–38.

[17] Gala M, Chung D. Hereditary colon cancer syndromes. Semin Oncol 2011;38:490–9.

[18] Power D, Gloglowski E, Lipkin S. Clinical genetics of hereditary colorectal cancer. Hematol Oncol Clin N Am 2010;24:837–59.

[19] Johns L, Houlston R. A systematic review and meta-analysis of familial colorectal cancer risk. Am J Gastroenterol 2001;96: 2992–3003.

[20] Samowitz W, Sweeney C, Herrick J, Albertsen H, Levin TR, Murtaugh MA, et al. Poor survival associated with the BRAF V600E mutation in microsatellite-stable colon cancers. Cancer Res 2005;65:6063–9.

[21] Lynch H, de la Chapelle A. Hereditary colorectal cancer. N Engl J Med 2003;348:919–32.

[22] Papadopoulos N, Nicolaides NC, Wei YF, Ruben SM, Carter KC, Rosen CA, et al. Mutation of a mutL homolog in hereditary colon cancer. Science 1994;263:1625–9.

[23] Leach FS, Nicolaides NC, Papadopoulos N, Liu B, Jen J, Parsons R, et al. Mutations of a mutS homolog in hereditary nonpolyposis colorectal cancer. Cell 1993;75:1215–25.

[24] Nicolaides NC, Papadopoulos N, Liu B, Wei YF, Carter KC, Ruben SM, et al. Mutations of two PMS homologues in hereditary nonpolyposis colon cancer. Nature 1994;371:75–80.

[25] Umar A, Boland CR, Terdiman JP, Syngal S, de la Chapelle A, Ruschoff J, et al. Revised Bethesda guidelines for hereditary nonpolyposis colorectal cancer (Lynch syndrome) and microsatellite instability. J Natl Cancer Inst 2004;96:261–8.

[26] Perucho M. Cancer of the microsatellite mutator phenotype. Biol Chem 1996;377:675–84.

[27] Peltomäki P. Deficient DNA mismatch repair: a common etiologic factor for colon cancer. Hum Mol Genet 2001;10:735–40.

[28] Boland C, Goel A. Microsatellite instability in colorectal cancer. Gastroenterology 2010;138:2073–2087000.

[29] Aaltonen LA, Peltomaki P, Leach FS, Sistonen P, Pylkkanen L, Mecklin JP, et al. Clues to the pathogenesis of familial colorectal cancer. Science 1993;260:812–6.

[30] Ionov Y, Peinado MA, Malkhosyan S, Shibata D, Perucho M. Ubiquitous somatic mutations in simple repeated sequences reveal a new mechanism for colonic carcinogenesis. Nature 1993;363:558–61.

[31] Levinson G, Gutman GA. Slipped-strand mispairing: a major mechanism for DNA sequence evolution. Mol Biol Evol 1987;4:203–21.

[32] Strand M, Prolla TA, Liskay RM, Petes TD. Destabilization of tracts of simple repetitive DNA in yeast by mutations affecting DNA mismatch repair. Nature 1993;365:274–6.

[33] Fishel R, Lescoe M, Rao M, Copeland NG, Jenkins NA, Garber J, et al. The human mutator gene homolog MSH2 and its association with hereditary nonpolyposis colon cancer. Cell 1993;75: 1027–38.

[34] Lindblom A, Tannergard P, Werelius B, Nordenskjold M. Genetic mapping of a second locus predisposing to hereditary nonpolyposis colon cancer. Nat Genet 1993;5:279–82.

[35] Gazzoli I, Loda M, Garber J, Syngal S, Kolodner R. A hereditary nonpolyposis colorectal carcinoma case associated with hypermethylation of the MLH1 gene in normal tissue and loss of heterozygosity of the unmethylated allele in the resulting microsatellite instability-high tumor. Cancer Res 2002;62:3925–8.

[36] Miyakura Y, Sugano K, Akasu T, Yoshida T, Maekawa M, Saitoh S, et al. Extensive but hemiallelic methylation of the hMLH1 promoter region in early-onset sporadic colon cancers with microsatellite instability. Clin Gastroenterol Hepatol 2004;2:147–56.

[37] Ligtenberg M, Kuiper R, Chan T, Goossens M, Hebeda KM, Voorendt M, et al. Heritable somatic methylation and inactivation of MSH2 in families with Lynch syndrome due to deletion of the 3′ exons of TACSTD1. Nat Genet 2009;41:112–7.

[38] Niessen R, Hofstra R, Westers H, Ligtenberg MJ, Kooi K, Jager PO, et al. Germline hypermethylation of MLH1 and EPCAM deletions are a frequent cause of Lynch syndrome. Genes Chromosomes Cancer 2009;48:737–44.

[39] Fearon E. Molecular genetics of colorectal cancer. Annu Rev Pathol 2011;6:479–507.

[40] Vilar E, Gruber SB. Microsatellite instability in colorectal cancer – the stable evidence. Nat Rev Clin Oncol 2010;7:153–62.

[41] Herrera L, Kakati S, Gibas L, Pietrzak E, Sandberg AA. Gardner syndrome in a man with an interstitial deletion of 5q. Am J Med Genet 1986;25:473–6.

[42] Nakamura Y, Lathrop M, Leppert M, Dobbs M, Wasmuth J, Wolff E, et al. Localization of the genetic defect in familial adenomatous polyposis within a small region of chromosome 5. Am J Hum Genet 1988;43:638–44.

[43] Bodmer WF, Bailey CJ, Bodmer J, Bussey HJ, Ellis A, Gorman P, et al. Localization of the gene for familial adenomatous polyposis on chromosome 5. Nature 1987;328:614–6.

[44] Groden J, Thliveris A, Samowitz W, Carlson M, Gelbert L, Albertson H, et al. Identification and characterization of the familial adenomatous polyposis coli gene. Cell 1991;66:589–600.

[45] Joslyn G, Carlson M, Thliveris A, Albertsen H, Gelbert L, Samowitz W, et al. Identification of deletion mutations and three new genes at the familial polyposis locus. Cell 1991;66:601–13.

[46] Kinzler KW, Nilbert MC, Su LK, Vogelstein B, Bryan TM, Levy DB, et al. Identification of FAP locus genes from chromosome 5q21. Science 1991;253:661–5.

[47] Herman J, Umar A, Polyak K, Graff JR, Ahuja N, Issa JP, et al. Incidence and functional consequences of hMLH1 promoter hypermethylation in colorectal carcinoma. Proc Natl Acad Sci USA 1998;95:6870–5.

[48] Nathke IS. The adenomatous polyposis coli protein. Mol Pathol 1999;52:169–73.

[49] Morin PJ, Sparks AB, Korinek V, Barker N, Clevers H, Vogelstein B, et al. Activation of β-catenin-Tcf signaling in colon cancer by mutations in β-catenin or APC. Science 1997;275:1787–90.

[50] Chung D. The genetic basis of colorectal cancer: insights into critical pathways of tumorigenesis. Gastroenterology 2000;119: 854–65.

[51] Al-Tassan N, Chmiel N, Maynard J, Fleming N, Livingston AL, Williams GT, et al. Inherited variants of MYH associated with somatic G:C→T:A mutations in colorectal tumors. Nat Genet 2002;30:227–32.

[52] Slupska M, Baikalov C, Luther W, Chiang J, Wei Y, Miller J. Cloning and sequencing a human homolog (hMYH) of the *Escherichia coli* mutY gene whose function is required for the repair of oxidative DNA damage. J Bacteriol 1996;178:3885–92.

[53] Croitoru M, Cleary S, Di Nicola N, Manno M, Selander T, Aronson M, et al. Association between biallelic and monoallelic germline MYH gene mutations and colorectal cancer risk. J Natl Cancer Inst 2004;96:1631–4.

[54] Sieber OM, Lipton L, Crabtree M, Heinimann K, Fidalgo P, Phillips RK, et al. Multiple colorectal adenomas, classic adenomatous polyposis, and germ-line mutations in MYH. N Engl J Med 2003;348:791–9.

[55] Sampson J, Dolwani S, Jones S, Eccles D, Ellis A, Evans DG, et al. Autosomal recessive colorectal adenomatous polyposis due to inherited mutations of MYH. Lancet 2003;362:39–41.

[56] Nghiem Y, Cabrera M, Cupples C, Miller J. The mutY gene: a mutator locus in *Escherichia coli* that generates G.C–T.A transversions. Proc Natl Acad Sci USA 1988;85:2709–13.

[57] Giardiello FM, Brensinger JD, Tersmette AC, Goodman SN, Petersen GM, Booker SV, et al. Very high risk of cancer in familial Peutz-Jeghers syndrome. Gastroenterology 2000;119:1447–53.

[58] Sweet K, Willis J, Zhou X-P, Gallione C, Sawada T, Alhopuro P, et al. Molecular classification of patients with unexplained hamartomatous and hyperplastic polyposis. J Am Med Assoc 2005;294:2465–73.

[59] Zbuk K, Eng C. Hamartomatous polyposis syndromes. Nat Clin Practice Gastroenterol Hepatol 2007;4:492–502.

[60] Lipkin S, Rozek L, Rennert G, Yang W, Chen PC, Hacia J, et al. The MLH1 D132H variant is associated with susceptibility to sporadic colorectal cancer. Nat Genet 2004;36:694–9.

[61] Miyoshi Y, Nagase H, Ando H, Horii A, Ichii S, Nakatsuru S, et al. Somatic mutations of the APC gene in colorectal tumors: mutation cluster region in the APC gene. Hum Mol Genet 1992;1:229–33.

[62] Powell SM, Zilz N, Beazer-Barclay Y, Bryan TM, Hamilton SR, Thibodeau SN, et al. APC mutations occur early during colorectal tumorigenesis. Nature 1992;359:235–7.

[63] Bond C, Umapathy A, Buttenshaw R, Wockner L, Leggett B, Whitehall V. Chromosomal instability in BRAF mutant, microsatellite stable colorectal cancers. PloS One 2012;7:10.

[64] Ogino S, Nosho K, Kirkner G, Kawasaki T, Meyerhardt JA, Loda M, et al. CpG island methylator phenotype, microsatellite instability, BRAF mutation and clinical outcome in colon cancer. Gut 2009;58:90–6.

[65] Rajagopalan H, Bardelli A, Lengauer C, Kinzler K, Vogelstein B, Velculescu V. Tumorigenesis: RAF/RAS oncogenes and mismatch-repair status. Nature 2002;418:934.

[66] Kane M, Loda M, Gaida G, Lipman J, Mishra R, Goldman H, et al. Methylation of the hMLH1 promoter correlates with lack of expression of hMLH1 in sporadic colon tumors and mismatch repair-defective human tumor cell lines. Cancer Res 1997;57(5):808–11.

[67] Koinuma K, Shitoh K, Miyakura Y, Furukawa T, Yamashita Y, Ota J, et al. Mutations of BRAF are associated with extensive hMLH1 promoter methylation in sporadic colorectal carcinomas. Interml J Cancer 2004;108:237–42.

[68] Ogino S, Shima K, Meyerhardt J, McCleary NJ, Ng K, Hollis D, et al. Predictive and prognostic roles of BRAF mutation in stage III colon cancer: results from intergroup trial CALGB 89803. Clin Cancer Res 2012;18:890–900.

[69] Chen H, Taylor N, Sotamaa K, Mutch DG, Powell MA, Schmidt AP, et al. Evidence for heritable predisposition to epigenetic silencing of MLH1. Internatl J Cancer 2007;120: 1684–8.

[70] McFarland CD, Korolev KS, Kryukov GV, Sunyaev SR, Mirny LA. Impact of deleterious passenger mutations on cancer progression. Proc Natl Acad Sci USA 2013;110:2910–5.

[71] Zhang L, Komurov K, Wright WE, Shay JW. Identification of novel driver tumor suppressors through functional interrogation of putative passenger mutations in colorectal cancer. Int J Cancer 2013;132:732–7.

[72] Veigl M, Kasturi L, Olechnowicz J, Ma AH, Lutterbaugh JD, Periyasamy S, et al. Biallelic inactivation of hMLH1 by epigenetic gene silencing, a novel mechanism causing human MSI cancers. Proc Natl Acad Sci USA 1998;95:8698–702.

[73] Hitchins M, Williams R, Cheong K, Halani N, Lin VA, Packham D, et al. MLH1 germline epimutations as a factor in hereditary nonpolyposis colorectal cancer. Gastroenterology 2005;129:1392–9.

[74] Suter CM, Martin DI, Ward RL. Germline epimutation of MLH1 in individuals with multiple cancers. Nat Genet 2004;36:497–501.

[75] Simons CC, Hughes LA, Smits KM, Khalid-de Bakker CA, de Bruïne AP, Carvalho B, et al. A novel classification of colorectal tumors based on microsatellite instability, the CpG island methylator phenotype and chromosomal instability: implications for prognosis. Ann Oncol 2013;8:2048–56.

[76] Peters U, Jiao S, Schumacher F, Hutter CM, Aragaki AK, Baron JA, et al. Identification of genetic susceptibility loci for colorectal tumors in a genome-wide meta-analysis. Gastroenterology 2013;144:799–807.

[77] Dunlop M, Dobbins S, Farrington S, Jones AM, Palles C, Whiffin N, et al. Common variation near CDKN1A, POLD3 and SHROOM2 influences colorectal cancer risk. Nat Genet 2012; 44:770–6.

[78] Fernandez-Rozadilla C, Cazier J-B, Tomlinson IP, Carvajal-Carmona LG, Palles C, Lamas MJ, et al. A colorectal cancer genome-wide association study in a Spanish cohort identifies two variants associated with colorectal cancer risk at 1p33 and 8p12. BMC Genomics 2013;14:55.

[79] Stratton M, Campbell P, Futreal P. The cancer genome. Nature 2009;458:719–24.

[80] Wood L, Parsons D, Jones S, Lin J, Sjöblom T, Leary RJ, et al. The genomic landscapes of human breast and colorectal cancers. Science 2007;318: 1108–13.

[81] Sjöblom T, Jones S, Wood L, Parsons DW, Lin J, Barber TD, et al. The consensus coding sequences of human breast and colorectal cancers. Science 2006;314:268–74.

[82] Bass A, Lawrence M, Brace L, Ramos AH, Drier Y, Cibulskis K, et al. Genomic sequencing of colorectal adenocarcinomas identifies a recurrent VTI1A-TCF7L2 fusion. Nat Genet 2011;43: 964–8.

[83] Smith CG, Naven M, Harris R, Colley J, West H, Li N, et al. Exome re-sequencing identifies potential tumor suppressor genes that predispose to colorectal cancer. Hum Mutat 2013;34(7): 1026–34.

[84] Camps J, Grade M, Nguyen Q, Hormann P, Becker S, Hummon AB, et al. Chromosomal breakpoints in primary colon cancer cluster at sites of structural variants in the genome. Cancer Res 2008;68:1284–95.

[85] Lengauer C, Kinzler K, Vogelstein B. Genetic instability in colorectal cancers. Nature 1997;386:623–7.

[86] Pino M, Chung D. The chromosomal instability pathway in colon cancer. Gastroenterology 2010;138:2059–72.

[87] Vogelstein B, Fearon E, Hamilton S, Kern SE, Preisinger AC, Leppert M, et al. Genetic alterations during colorectal-tumor development. N Engl J Med 1988;319:525–32.

[88] Lengauer C, Kinzler K, Vogelstein B. Genetic instabilities in human cancers. Nature 1998;396:643–9.

[89] Lothe R, Peltomäki P, Meling G, Meling GI, Aaltonen LA, Nyström-Lahiti M, et al. Genomic instability in colorectal cancer: relationship to clinicopathological variables and family history. Cancer Res 1993;53:5849–52.

[90] Trautmann K, Terdiman J, French A, Roydasgupta R, Sein N, Kakar S, et al. Chromosomal instability in microsatellite-unstable and stable colon cancer. Clin Cancer 2006;12:6379–85.

[91] Lee R, Feinbaum R, Ambros V. The C. elegans heterochronic gene *lin-4* encodes small RNAs with antisense complementarity to *lin-14*. Cell 1993;75(5):843–54.

[92] Chen K, Rajewsky N. The evolution of gene regulation by transcription factors and microRNAs. Nat Rev Genet 2007;8: 93–103.

[93] Oberg A, French A, Sarver A, Subramanian S, Morlan BW, Riska SM, et al. miRNA expression in colon polyps provides evidence for a multihit model of colon cancer. PloS One 2011;6:6.

[94] Earle J, Luthra R, Romans A, Abraham R, Ensor J, Yao H, et al. Association of microRNA expression with microsatellite instability status in colorectal adenocarcinoma. J Mol Diagn 2010;12: 433–40.

[95] Cekaite L, Rantala J, Bruun J, Guriby M, Agesen TH, Danielsen SA, et al. MiR-9, -31, and -182 deregulation promote proliferation and tumor cell survival in colon cancer. Neoplasia (NY) 2012;14:868–79.

[96] Smits K, Paranjape T, Nallur S, Wouters KA, Weijenberg MP, Schouten LJ, et al. A let-7 microRNA SNP in the KRAS 3′UTR is prognostic in early-stage colorectal cancer. Clin Cancer Res 2011;17:7723–31.

[97] Vogelstein B, Kinzler K. Cancer genes and the pathways they control. Nat Med 2004;10:789–99.

[98] Markowitz S, Bertagnolli M. Molecular origins of cancer: Molecular basis of colorectal cancer. N Engl J Med 2009; 361:2449–60.

[99] Parsons R, Myeroff LL, Liu B, Wilson JK, Markowitz SD, Kinzler KW, et al. Microsatellite instability and mutations of the transforming growth factor beta type II receptor gene in colorectal cancer. Cancer Res 1995;55:5548–50.

[100] Sinicrope F, Foster N, Thibodeau S, Marsoni S, Monges G, Labianca R, et al. DNA mismatch repair status and colon cancer recurrence and survival in clinical trials of 5-fluorouracil-based adjuvant therapy. J Natl Cancer Inst 2011;103:863–75.

[101] Merok MA, Ahlquist T, Royrvik EC, Tufteland KF, Hekktoen M, Sjo OH, et al. Microsatellite instability has a positive prognostic impact on stage II colorectal cancer after complete resection: results from a large, consecutive Norwegian series. Ann Oncol 2012;24:1274–82.

[102] Dijkstra J, Heideman D, Meijer G, Boers JE, 't Hart NA, Diebold J, et al. KRAS mutation analysis on low percentage of colon cancer cells: the importance of quality assurance. Virchows Arch 2013;462:39–46.

[103] Lièvre A, Bachet J-B, Le Corre D, Boige V, Landi B, Emile JF, et al. KRAS mutation status is predictive of response to cetuximab therapy in colorectal cancer. Cancer Res 2006;66:3992–5.

[104] Misale S, Yaeger R, Hobor S, Scala E, Janakiraman M, Liska D, et al. Emergence of KRAS mutations and acquired resistance to anti-EGFR therapy in colorectal cancer. Nature 2012;486: 532–6.

[105] Prahallad A, Sun C, Huang S, Di Nicolantonio F, Salazar R, Zecchin D, et al. Unresponsiveness of colon cancer to BRAF (V600E) inhibition through feedback activation of EGFR. Nature 2012;483:100–3.

[106] Kelley R, Venook A. Prognostic and predictive markers in stage II colon cancer: is there a role for gene expression profiling? Clin Colorectal Cancer 2011;10:73–80.

[107] Donada M, Bonin S, Barbazza R, Pettirosso D, Stanta G. Management of stage II colon cancer – the use of molecular biomarkers for adjuvant therapy decision. BMC Gastroenterol 2013;13:36.

[108] Chen S, Song X, Chen Z, Li X, Li M, Liu H, et al. CD133 expression and the prognosis of colorectal cancer: a systematic review and meta-analysis. PloS One 2013;8:2.

[109] O'Brien C, Pollett A, Gallinger S, Dick J. A human colon cancer cell capable of initiating tumour growth in immunodeficient mice. Nature 2007;445:106–10.

[110] Ricci-Vitiani L, Lombardi D, Pilozzi E, Biffoni M, Todaro M, Peschle C, et al. Identification and expansion of human colon-cancer-initiating cells. Nature 2007;445:111–5.

[111] Horst D, Kriegl L, Engel J, Kirchner T, Jung A. CD133 expression is an independent prognostic marker for low survival in colorectal cancer. Br J Cancer 2008;99:1285–9.

[112] Lugli A, Iezzi G, Hostettler I, Muraro MG, Mele V, Tornillo L, et al. Prognostic impact of the expression of putative cancer stem cell markers CD133, CD166, CD44s, EpCAM, and ALDH1 in colorectal cancer. Br J Cancer 2010;103: 382–90.

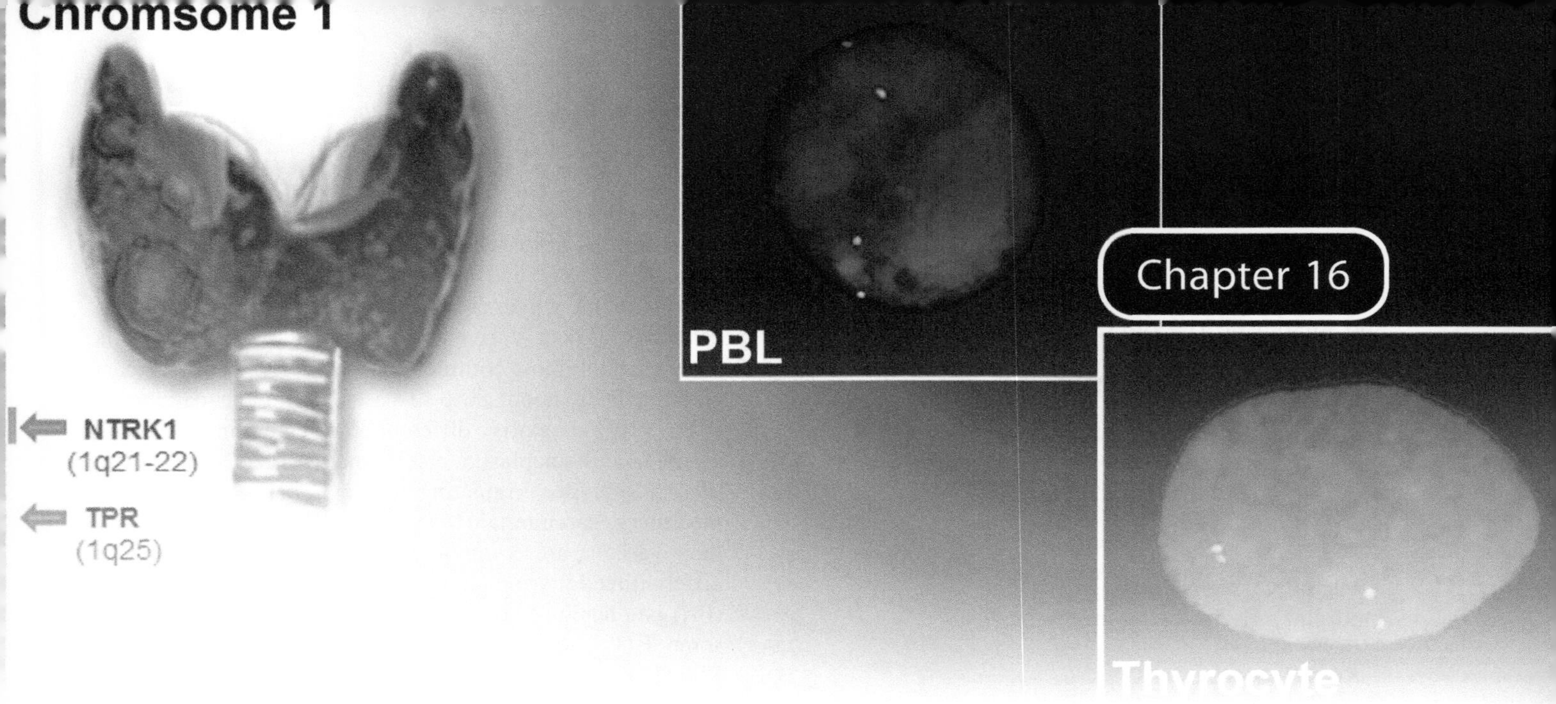

Chapter 16

Cancer Genomics

Thyroid Cancer

Angela Greco[1]
Claudia Miranda[1]
Maria Grazia Borrello[1]
Marco A. Pierotti[2]

[1]*Molecular Mechanisms Unit, Department of Experimental Oncology and Molecular Medicine, Fondazione IRCCS Istituto Nazionale dei Tumori, Milan, Italy*

[2]*Scientific Directorate, Fondazione IRCCS Istituto Nazionale dei Tumori, Milan, Italy*

Contents

Cancer Genomics. DOI: http://dx.doi.org/10.1016/B978-0-12-396967-5.00016-5

Key Concepts

- Different tumor histotypes originate from thyroid
- High proneness of thyroid epithelium to chromosomal rearrangements
- Structure and mechanisms of action of the different thyroid tumor-associated oncogenes
- PTC-associated oncogenes are produced by chromosome rearrangements involving tyrosine kinase receptor gene
- High frequency of *BRAFV600E* mutation in PTC
- Frequent deregulation of MAPK pathway in PTC
- FTC feature *RAS* gene mutations and *PAX8/PPPRγ* rearrangement
- The most aggressive thyroid tumors, PDTC and ATC, feature some genetic alterations typical of PTC and FTC, as well as frequent deregulation of the PTEN/PI3K/AKT pathway
- *RET* mutations are the most frequent genetic alteration in MTC
- Novel players in thyroid carcinogenesis are being unveiled by gene and miRNA expression studies

INTRODUCTION

Thyroid cancer (TC) is a relatively rare disease, accounting for approximately 1% of all new cases of cancer each year. Nevertheless, it represents the most common endocrine malignancy and its incidence has been increasing over the last few decades [1–3]. The reasons for this increase are unclear, but they are in part related to the improvement of diagnostic tools that allow early diagnosis.

According to the cell of origin, TCs are divided into those derived from follicular epithelial cells, that produce thyroid hormone, and those originating from parafollicular, or C cells, which are responsible for the production of calcitonin. Those derived from follicular epithelial cells are by far the most common, and comprise papillary carcinoma (PTC, about 80% of cases), follicular carcinoma (FTC, 10%), poorly differentiated carcinoma (PDTC, 4–6%), and anaplastic carcinoma (ATC, 2–5%). The second subtype, originating in the C cells, is known as medullary carcinoma (MTC, 5–10%). PTC, FTC and their variants are commonly grouped under the name of differentiated thyroid cancers (DTCs). Follicular adenoma (FA) is a benign tumor that may represent a precursor for some FTC. Less differentiated thyroid cancers, namely PDTC and ATC, could arise either *de novo* from thyrocytes or through de-differentiation of the DTCs [4,5].

The cornerstone of TC management remains surgery, followed by thyroid stimulating hormone (TSH) suppression and ablative doses of radioiodine as adjuvant treatment of DTC, treatments which do not play a curative role in MTC. In ATC, surgery is often not technically feasible.

DTCs have usually a very good long-term prognosis; nevertheless about 20% of patients will fail therapy due to local or distant tumor recurrence, failure to absorb radioiodine or, in rare cases, due to progression of well-differentiated tumors to poorly differentiated or undifferentiated carcinomas.

The last two decades have been marked by significant insights that have revolutionized our understanding of TC at the molecular level. The genetic lesions associated with this tumor type have been in part elucidated. More recently, global gene and miRNA expression analyses have provided new insight into global pathophysiologic changes in TC. On the whole, the dissection of the genetic events governing thyroid carcinogenesis has improved our understanding of the molecular basis of TC and provided opportunities for developing novel therapeutic approaches. In this chapter, we review the genetic lesions associated with the different thyroid tumor histotypes (Table 16.1). The functional consequences and the intracellular pathways affected by the different genetic alterations are illustrated in Figure 16.1. A schematic representation of different thyroid tumor types originating from a follicular cell or C cell is depicted in Figure 16.2.

PAPILLARY THYROID CARCINOMA (PTC)

PTC is the most frequent thyroid tumor type, accounting for approximately 80% of all thyroid malignancies [4]. It is associated with exposure to external radiation, as documented by the increased incidence of TC observed among

TABLE 16.1 Genetic Lesions Associated with Different Thyroid Tumor Histotypes

Tumor Histotype	Genetic Lesion
Papillary thyroid carcinoma	*RET* rearrangements
	NTRK1 rearrangements
	ALK rearrangements
	AKAP9/BRAF rearrangements
	BRAF mutations
	RAS mutations
Follicular thyroid carcinoma	*RAS* mutations
	PAX8/PPARγ rearrangements
	mt DNA mutations
	GRIM-19 mutations
	PIK3CA mutations
	PIK3CA copy gain
	PTEN mutations
	AKT mutation
Poorly differentiated thyroid carcinoma	*RAS* mutations
	BRAF mutations
	RET rearrangements
	PIK3CA mutations
	AKT mutation
	TP53 mutations
	CTNNB1 mutations
Anaplastic thyroid carcinoma	*RAS* mutations
	BRAF mutations
	PIK3CA mutations
	PIK3CA copy gain
	PTEN mutations
	AKT mutation
	TP53 mutations
	ALK mutations
	CTNNB1 mutations
Hereditary medullary thyroid carcinoma	*RET* germline mutations
(MEN2A, MEN2B, FMTC)	
Sporadic medullary thyroid carcinoma	*RET* mutations
	RAS mutations

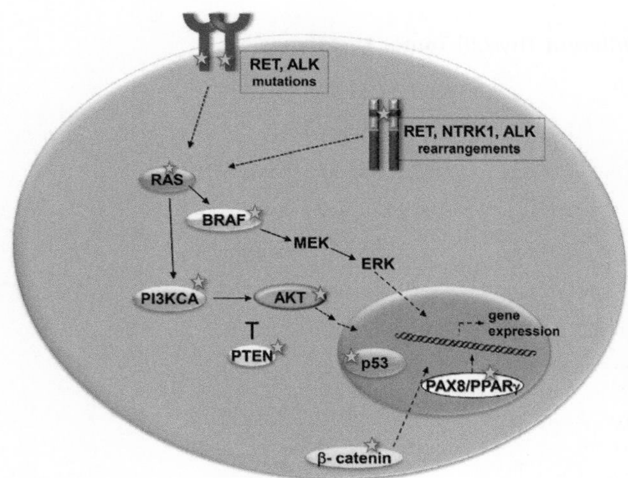

FIGURE 16.1 Molecular alterations (stars) and deregulated intracellular pathways in thyroid cancer.

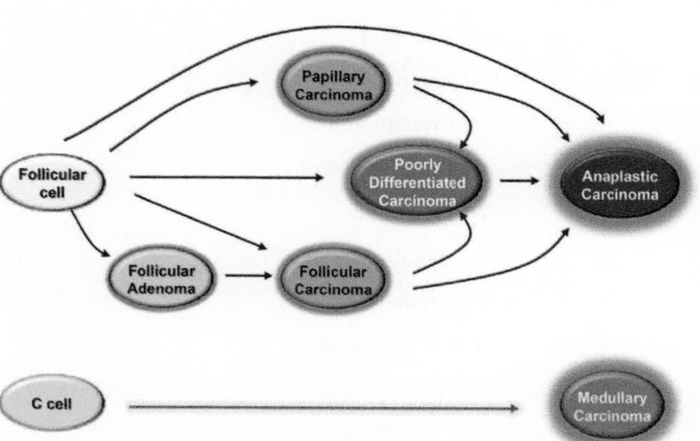

FIGURE 16.2 Schematic representation of different thyroid tumor types originating from follicular cell or C cell.

the atomic bomb survivors of Hiroshima and Nagasaki, and after the nuclear accident at Chernobyl [6,7]. PTCs are generally slow-growing tumors but they often spread to regional lymph nodes, frequently presenting as metastases at the time of diagnosis.

Classic PTC is microscopically characterized by a papillary architecture and the presence of follicular cells with "ground glass" nuclei that exhibit irregularities in shape. Another typical PTC feature is the presence of psammoma bodies, found in 40–50% of cases. Although the prognosis of classic PTC is excellent, patient outcome is influenced by patient age, tumor size and presence of distant metastases.

In addition to classic PTC, several histological variants, differing in morphological pattern as well as in prognosis have been described. The most common are: papillary thyroid microcarcinoma (PTMC), follicular, oncocytic, clear cells, diffuse sclerosing, tall cell, columnar cell, solid and cribriform. PTMC is defined as a tumor measuring <1 cm in diameter, often found incidentally, and is considered the early stage of PTC. It has an indolent course and very favorable prognosis despite the presence of multifocality within the thyroid and synchronous nodal local-regional spread [8]. For most variants, the prognosis is similar to that of classical PTC. Other rare variants such as tall cell, columnar cell and solid variant, are characterized by an aggressive behavior and worse outcome, with the primary tumor often presenting with extrathyroid extension and vascular invasion.

The last two decades have been marked by significant expansion in the understanding of the molecular basis of PTC. It is now apparent that more than 70% of this tumor type is characterized by genetic lesions leading to the activation of the MAPK signaling pathway [9] (see Figure 16.1). These alterations include chromosomal rearrangements and point mutations. In addition to rearrangements involving rearranged during transfection *(RET)* and neurotrophic tyrosine kinase receptor type 1 *(NTRK1)* genes, identified more than 20 years ago [9], rearrangements of the *BRAF* and the *ALK* genes have also been recently reported [10,11]. The most common point mutations detected in PTC involve *BRAF* as well as *RAS* genes. PTC-associated genetic lesions are generally mutually exclusive. However, simultaneous presence of *RET/PTC* rearrangements and *BRAF* mutations, although rare, has been reported in patients with recurrent disease [12].

High Frequency of Oncogenic Chromosomal Rearrangements in PTC

Chromosomal rearrangements producing fusion oncogenes have been considered for a long time a hallmark of human leukemia and sarcoma. Among epithelial tumors, neoplasias arising from thyrocytes, in particular PTC, represented an exception insofar as they are associated with frequent oncogenic chromosomal rearrangements. In the last few years, oncogenic chromosomal rearrangements have also been detected in other epithelial tumors such as prostate, lung, and breast cancer [13]. In the following sections, we will describe the common oncogenic rearrangements occurring in PTC. In addition to *RET/PTC* and *TRK* oncogenes, discovered more than 20 years ago [9], chromosomal rearrangements involving *BRAF* and *ALK* genes have been more recently reported [10,11].

RET/PTC, *TRK* and *EML4/ALK* oncogenes share a common mechanism of action. In each case, the chromosomal rearrangement results in the fusion of sequences encoding a receptor tyrosine kinase (TK) domain fused to "activating" sequences from different donor genes.

The donor genes contain dimerization domains, which confer on the oncoproteins constitutive dimerization and tyrosine kinase activity. In the case of *AKAP9/BRAF* oncogene, the lack of the *BRAF* autoinhibitory domain confers constitutive, elevated, kinase activity.

Despite the high frequency of chromosomal rearrangements in PTC and, to a lesser extent, in FTC, the molecular basis underlying the predisposition of thyrocytes to undergo chromosome rearrangements is not completely understood.

The high proneness of thyroid epithelium to chromosomal rearrangements might reflect a peculiar sensitivity of thyrocytes to ionizing radiation, and/or the intrinsic capacity of a cell to repair radiation-induced DNA double-strand breaks [14]. In this respect, it has been proposed that, following DNA damage, thyrocytes would be more prone to aberrant DNA repair than to apoptosis [15], and this would increase the likelihood of gene rearrangements.

Several reports have proposed that the spatial proximity of translocation-prone gene loci may favor gene rearrangements. In fact, proximity between rearranging gene pairs such as *RET* and *H4*, and *NTRK1* and *TPR*, has been reported in interphase thyroid nuclei [16,17], possibly as a result of the three-dimensional organization of the chromosomal domains within these nuclei [18]. In this model, the spatial organization of chromosomal domains in a given tissue may favor the formation of specific oncogenes through chromosomal translocations between neighboring regions following radiation, and/or by facilitating non-homologous end-joining of close free DNA ends within or between adjacent chromosomes [14].

As will be detailed later, the majority of rearrangements found in PTC are intrachromosomal inversions. Gandhi et al. [19] also suggest that this predominance of intrachromosomal inversions can be explained by nuclear architecture, and demonstrated that loci involved in intrachromosomal rearrangements are located more centrally within their chromosome territory.

RET/PTC Rearrangements

RET/PTC oncogenes were the first genetic lesion identified in PTC [20]. They are produced by rearrangements of the *RET* proto-oncogene, on chromosome 10q11.2, encoding a receptor type tyrosine kinase later on identified as the signaling component of a multimolecular complex that includes ligands of the glial cell line-derived neurotrophic factor (GDNF) family and their co-receptors (GFRα1–4) [21]. Ontogenically, the *RET* gene is essential for development of the sympathetic, parasympathetic and enteric neurons, kidney and male germ cells [22]. Germline loss-of-function mutations in the *RET* gene lead to impaired formation of the enteric nervous system and congenital aganglionosis of the colon (Hirschsprung's disease), while germline activating point mutations are causally related to the hereditary forms of MTC, such as multiple endocrine neoplasia (MEN)2A and 2B, and familial MTC (FMTC) [22].

A significant fraction (20–30%) of PTC chromosomal rearrangements juxtapose the RET intracellular domains (including the TK domain and the C-terminal tail) to the 5′ end of several heterologous genes (donor genes) producing chimeric *RET/PTC* oncogenes with constitutive TK activity (reviewed in [9] and listed in Table 16.2). *RET/PTC1* and *RET/PTC3* oncogenes are the most frequent, accounting together for more than 90% of all the rearrangements. Both *RET/PTC1* and *RET/PTC3* oncogenes are produced by a paracentric inversion, as the relative activating genes, *CCDC6* and *NCOA4*, as well as *RET*, reside on chromosome 10q. The other, less frequent *RET/PTC* oncogenes are produced by interchromosomal rearrangements (Table 16.2). RET/PTC oncoproteins display constitutive dimerization and tyrosine kinase activity, thus triggering neoplastic transformation. A further consequence of *RET/PTC* rearrangement is the structural/functional alteration of one copy of the involved donor gene, possibly contributing to neoplastic transformation. For example, a tumor suppressor function has been suggested for some fusion partners of *RET* such as *H4/CCDC6* and *PRKARIA*.

RET/PTC oncogenes are considered an early event playing a causative role in the pathogenesis of PTC. This hypothesis is based on several lines of evidence: (1) *RET/PTC* expression is present in microscopic and occult PTCs [23]; (2) *RET/PTC1* expression is sufficient to cause diagnostic PTC-associated changes in the nuclear envelope and chromatin structure of normal human primary thyrocytes [24]; and (3) *RET/PTCs* induce morphological transformation of PC-CL3 rat thyroid epithelial cells [25].

PTCs harboring *RET/PTC* rearrangements display predominantly the classical papillary variant. Patients have a high rate of lymph node metastases and possibly a more favorable prognosis, especially when *RET/PTC1* oncogene is present [26].

RET/PTC incidence is higher in tumors from pediatric patients and from patients with a history of radiation exposure. For example, cases of PTC post-Chernobyl and from atomic bomb survivors show an increased frequency of chromosomal rearrangements (mainly *RET/PTC*), and a decrease in point mutations (mainly in *BRAF*) [7,27].

The mechanism by which *RET/PTC* oncogenes trigger transformation has been elucidated (reviewed in [22,28,29]). Constitutive *RET/PTC* dimerization, mediated by the activating portion, results in constitutive phosphorylation at crucial tyrosine residues such as Y900 and Y905 in the kinase A loop, and Y1062, the major RET docking site, and Y1015 in the intracellular domain. Y900 and Y905 are docking sites binding Grb7 and SH2Bβ1. Y1015, a docking site for PLCγ, is essential for kidney

TABLE 16.2 PTC-Associated Oncogenes Derived from Chromosomal Rearrangements

Oncogene	Rearranging Genes and Chromosome Location				Rearrangement Type
*RET/PTC1**	*RET*	10q11.2	*CCD6*	10q21	Inversion
RET/PTC2			*PRKARIA*	17q23	Translocation
*RET/PTC3**			*NCOA4*	10q11.2	Inversion
RET/PTC4			*NCOA4*	10q11.2	Inversion
RET/PTC5			*GOLGAS*	14q	Translocation
RET/PTC6			*TRIM24*	7q32–34	Translocation
RET/PTC7			*TRIM33*	1p13	Translocation
RET/PTC8			*KTN1*	14q22.1	Translocation
RET/PTC9			*RFG9*	18q21–22	Translocation
ELKS-RET			*ELKS*	6p21	Translocation
PCM1-RET			*PCM1*	8p21-22	Translocation
RFP-RET			*RFP*	6p21	Translocation
HOOK3-RET			*HOOK3*	8p11.21	Translocation
TRK	*NTRK1*	1q21–22	*TPM3*	1q22–23	Inversion
TRK-T1			*TPR*	1q25	Inversion
TRK-T2			*TPR*	1q25	Inversion
TRK-T3			*TFG*	3q11–12	Translocation
AKAP9-BRAF	*BRAF*	7q34	*AKAP9*	7q21–q22	Inversion
EML4-ALK	*ALK*	2p23	*EML4*	2p21	Inversion

* Given that PTC1 and PTC3 account for 90% of all *RET* rearrangements, chromosomal inversions are the most frequent chromosomal rearrangements in PTC.

development and for full *RET/PTC* transforming activity. Y1062 binds multiple proteins including SHCA, SHCC, IRS1/2, FRS2, DOK1/4/5 and Enigma and is essential for *RET* transforming activity and also for development.

Downstream to the cited adaptor proteins, *RET* activates RAS/ERK, PI3K/AKT, and NFκB pathways [22,28,29]. Moreover, a crucial role for β-catenin in *RET* neoplastic signaling in PTC has been recently suggested [30,31]. Besides directly altering cell signaling by relying on already available signaling molecules, *RET* activates a transcriptional program, including the transcription of inflammatory molecules and other TK receptors, such as MET/HGFR [32]. It was recently shown that RET/PTC1 signaling is involved in cross-talk with Met at both the transcriptional and signaling levels to drive thyrocyte neoplastic transformation [30].

TRK Rearrangements

TRK oncogenes arise from rearrangements of the *NTRK1* gene (also known as TRKA) on chromosome 1. *NTRK1* was originally isolated from a human colon carcinoma as a transforming oncogene activated by a somatic rearrangement that fused *TPM3* (non-muscle tropomyosin 3) gene to the kinase domain of a novel tyrosine kinase receptor. Cloning and characterization of the full length receptor gene led eventually to the identification of *NTRK1* as the high affinity receptor for nerve growth factor (reviewed in [33]). *NTRK1* exerts a critical role in the development and maturation of the central and peripheral nervous system; in addition, it stimulates proliferation of a number of cell types such as lymphocytes, keratinocytes and prostate cells. Somatic rearrangements of the *NTRK1* gene in PTC are less common than those involving the *RET* gene; their frequency does not exceed 12%. The different *TRK* oncogenes so far identified are listed in Table 16.2 and reviewed in [34]. *TRK* oncogene, which is identical to that first isolated from colon carcinoma, and containing sequences from the *TPM3* gene on chromosome 1q22-23, is the most frequent one. *TRK-T1* and *TRK-T2* derive from rearrangement between *NTRK1* and different portions of the *TPR* gene on chromosome 1.

TRK-T3 is activated by fusion of *NTRK1* with a novel gene on chromosome 3q11-12, first identified through the characterization of this translocation, and designated as TRK fused gene (*TFG*). The majority of *TRK* oncogenes are produced by inversion on chromosome 1 (see Table 16.2). TRK oncoprotein dimerization mediated by the activating sequences promotes the constitutive phosphorylation of five tyrosine residues crucial for *NTRK1* activity (Y490, Y670, Y674, Y675, and Y785) which, in turn, interact with *PLC-γ*, *SHC*, *FRS2*, *FRS3*, *IRS1* and *IRS2* and activate intracellular signaling pathways.

The association of *NTRK1* rearrangements with radiation is not clearly defined. Unlike *RET*, for which an increased frequency of gene rearrangements was observed in radiation-associated PTCs, the frequency of *NTRK1* rearrangements in these tumors was similar to that of sporadic tumors [35,36].

Since the number of PTCs observed to carry *TRK* oncogenes is limited, no unequivocal data concerning the correlation of *NTRK1* rearrangements with clinicopathological features are available. However, this may be in part due to the fact that *TRK* rearrangement is not included in the genotyping of PTCs, and often PTCs carrying *RET* and *NTRK1* rearrangement are considered as a unique group.

Transgenic mice carrying *TRK-T1* oncogene under the control of thyroglobulin promoter (Tg-TRK-T1 mice) develop thyroid hyperplasia and papillary carcinoma [37], thus suggesting that *TRK* oncogenes represent an early event in the process of thyroid carcinogenesis. Crossing of Tg-TRK-T1 mice with $p27^{kip1}$-deficient mice increased the penetrance of thyroid cancer and shortens the latency period of tumor incidence, indicating that *TRK-T1* needs cooperation with tumor suppressor genes to transform thyroid epithelium [38].

ALK Rearrangements

The anaplastic lymphoma kinase (ALK) is a receptor tyrosine kinase first identified as the product of a gene rearrangement in anaplastic large cell lymphoma and subsequently found to be rearranged, mutated, or amplified in a further series of tumors (reviewed in [39]). The most common *ALK* genetic alterations are chromosomal translocations or inversions involving the chromosome 2p23, where the *ALK* gene is located, resulting in chimeric oncoproteins with constitutive ALK tyrosine kinase activity. In addition, *ALK* is activated by point mutations in neuroblastoma. Recently, *ALK* rearrangements have been detected in 10 out of 19 PTCs from atomic bomb survivors characterized by the absence of known genetic alterations [11]. In about half of the cases, the fusion partner was *EML4*, also located on chromosome 2 and the same gene involved in oncogenic rearrangements in non-small cell lung carcinoma.

AKAP9/BRAF Rearrangements

Mutations that activate BRAF kinase are the most frequently oncogenic mutation in PTC, as described in the next section. In 11% of PTCs with recent history of radiation exposure, *BRAF* has been found activated by a chromosomal rearrangement, namely a paracentric inversion of chromosome 7q, that generates a chimeric oncoprotein AKAP9 and the BRAF kinase domain lacking its autoinhibitory domain. The AKAP9/BRAF oncoprotein displays elevated kinase activity and transforming activity comparable to BRAFV600E, which is the common mutation found in melanoma [10].

BRAF Mutations

BRAF (v-Raf murine sarcoma viral oncogene homolog B1) is one of the three mammalian isoforms of a serine–threonine kinase that, upon activation by RAS binding and translocation to the cell membrane, activate MEK and thus the MAPK cascade.

BRAF is one of the most common human oncogenes, mutated in 7–9% of all malignant solid tumors [40], with the highest prevalence (60%) in melanoma. Although more than 40 mutations have been identified, the most frequent mutation, accounting for over 90% of all *BRAF* mutations, is a thymidine to adenine transversion at nucleotide 1799 (T1799A) resulting in a valine-to-glutamate substitution at residue 600 (V600E). Such mutation destabilizes the inactive conformation of BRAF protein leading to constitutive activation and constitutive phosphorylation of downstream effectors of the MAPK pathway [41]. After melanoma, PTC is the second most common human malignancy where *BRAF* mutations are frequent, being found in 29–87% of cases (reviewed in [42]). Although the *BRAFV600E* mutation is the most prevalent in PTC, other rare alterations contribute to *BRAF* oncogenic activation in this malignancy. These include the lysine-to-glutamine K601Qpoint mutation, small inframe insertions or deletions surrounding codon 600, and the *AKAP9/BRAF* chromosome rearrangement described in the previous section.

BRAFV600E mutation is highly prevalent in PTC with classical histology and in tall cell variant tumors. Numerous studies investigating the clinical significance of *BRAF* mutation in PTC have reported association with aggressive tumor features, advanced tumor stage at presentation, tumor recurrence and metastasis. *BRAFV600E* has been proposed as an independent predictor of tumor recurrence, and PTCs carrying *BRAFV600E* mutation exhibit decreased expression of genes involved in thyroid hormone biosynthesis, including sodium iodide symporter, decreased ability to trap radioiodine and consequent treatment failure [42].

Transgenic mice with thyroid-specific expression of *BRAFV600E* developed PTC closely recapitulating those seen in human tumors [43], supporting a role of *BRAF* oncogene in tumor initiation and differentiation.

RAS Mutations

Human *HRAS*, *KRAS* and *NRAS* genes encode highly related G-proteins that reside in the inner surface of the cell membrane and transmit signals arising from cell-membrane receptor tyrosine kinases and G-protein-coupled receptors along the MAPK, PI3K/AKT and other signaling pathways. Activating point mutations leading to constitutive activation typically affect codons 12, 13 and 61 of the *RAS* genes [44]. In thyroid cancer, *RAS* mutations are most prevalent in FTC as will be discussed in the corresponding section. However, they are also present in 10–20% of PTC, particularly in the follicular variant [45].

FOLLICULAR THYROID CARCINOMA (FTC)

FTC is characterized by evidence of follicular cell differentiation but lacking the diagnostic nuclear features of PTC [4]. It is usually encapsulated and solitary and it may develop from follicular adenoma. FTC is more common in geographic regions of iodine insufficiency and metastasizes almost exclusively via the blood stream. Compared to PTC, cervical lymph node involvement is less common at diagnosis while distant metastases, mainly involving bone and lung, can be found in 20% of cases. Invasiveness of the capsule and vascular invasion are the microscopic characteristics of FTC. FTC includes conventional and oncocytic/Hurthle cell carcinoma (HCC) variants, differing on biological and clinical grounds, leading some groups to classify HCC as a distinct pathologic entity. HCC is defined as a tumor composed of 75% or greater Hurthle cells (large polygonal cells characterized by eosinophilic granular cytoplasm, rich in mitochondria) with complete capsular or vascular invasion [46]. The tumor cells typically present an abundant granular, eosinophilic cytoplasm derived from the presence of a large number of mitochondria. Moreover, 30% of HCC cases present with cervical lymph node involvement and/or distant metastases.

FTC displays a more aggressive course and a poorer prognosis when compared with PTC; 50% of patients with a widely invasive FTC die, whereas patients with a minimally invasive FTC have an excellent prognosis [47].

RAS Oncogenes

Oncogenic mutations of the three members of the *RAS* gene family are among the first genetic lesions originally identified in tumors of the thyroid follicular epithelium. However, some discrepancies remain with regard to the overall frequency of *RAS* mutations and their prevalence in specific thyroid tumors. *RAS* mutations have been detected in up to 50% of FTCs, 40% of FAs, and 25% of HCCs with the following relative frequency: *NRAS* > *HRAS* > *KRAS*. They are also present in 10–20% of PTCs belonging to the follicular variant at a high frequency in PDTC and ATC (reviewed in [48]). The presence of *RAS* mutations in adenomas suggests that *RAS* activation may represent an early step in thyroid carcinogenesis. This hypothesis is supported by the finding that transgenic mice with targeted overexpression of the *Tg-NRAS-Q61K* oncogene developed thyroid follicular neoplasms (11% FAs, about 40% invasive FTCs, in cases with a mixed papillary/follicular morphology). Such studies also suggested that *RAS* oncogenes may predispose tumor to de-differentiation and metastatic behavior, as about 25% of the Tg-NRAS-Q61K carcinomas displayed poorly differentiated areas with evidence of vascular invasion and distant metastases [49]. Accordingly, a significant correlation between *RAS* mutations, metastases and poor prognosis has been found in patients [50,51]. *RAS* activation may be insufficient to induce malignant growth but may predispose tumors to the acquisition of further genetic or epigenetic alterations that lead to a fully transformed phenotype. This concept is in keeping with the observation that *RAS* mutations may affect chromosomal stability *in vitro* [52].

PAX8/PPARγ Rearrangements

The *PAX8/PPARγ* fusion oncogene results from the t(2;3)(q13;p25) translocation that fuses PAX8, a transcription factor essential for thyroid differentiation, to most of the peroxisome proliferator-activated receptor γ (PPARγ) [53]. The *PAX8/PPARγ* oncogene is found in 30–35% of FTCs, and also in a subset (2–13%) of FA [54].

With respect to the mechanisms activated by *PAX8/PPARγ*, both gain- and loss-of-function activities have been suggested. Deregulation of the normal functions of *PAX8* and *PPARγ* as well as the unique transcriptional activities of the fusion oncoprotein have been reported, and it has been suggested that *PAX8/PPARγ*-stimulated growth could depend at least in part on loss of *PPARγ* functions caused by the inhibition of wild-type protein [48]. The concept that *PPARγ* downregulation or inhibition may be a key event in thyroid carcinogenesis is corroborated by the evidence of frequent *PPARγ* downregulation in translocation-negative papillary or follicular thyroid tumors [55].

The importance of *PPARγ* deregulation in FTC is further underscored by the detection of a novel *CREB3L2/PPARγ* fusion gene generated by t(3;7)(q34;p25) rearrangement in <3% of FTC [56].

The *PAX8/PPARγ* rearrangement and *RAS* point mutations rarely overlap in the same tumor suggesting that they represent distinct pathogenetic pathways in the development of FTC. In keeping with this, gene expression studies have shown that *PAX8/PPARγ*-positive FTCs display a distinct transcriptional signature [57].

FTC also features deregulation of the PTEN/PI3K/AKT pathway, as will be detailed in the PDTC and ATC paragraph.

Mitochondrial Defects

Defects in mitochondrial proteins encoded by both nuclear and mitochondrial genes have been detected in HCCs. Sequencing of mtDNA identified potentially pathogenic mutations in subunits of complex I (NADH-ubiquinone oxidoreductase) have been found in 53% of HCCs. Immunohistochemical analyses have recently confirmed a specific lack of complex I, which was expressed at <5% of the level determined in surrounding non-cancerous tissue [58,59]. Moreover, mutations in the nuclear gene *GRIM-19*, coding for a dual function protein involved in complex I mitochondrial function and cell death, have been found in 15% of HCCs [60]. Overall, these findings support the concept of the existence of defects in the energy production machinery in oncocytic thyroid tumor cells and suggest that their increased mitochondrial content might be a compensatory mechanism. Interestingly, lack of complex I has been proposed to prevent tumor cells from undergoing apoptosis [59].

POORLY DIFFERENTIATED THYROID CARCINOMA (PDTC) AND ANAPLASTIC THYROID CARCINOMA (ATC)

PDTC and ATC represent the most aggressive thyroid tumors. They can represent the terminal stage of the de-differentiation process of PTC or FTC or can arise *de novo* [4].

PDTC is a rare tumor with higher incidence in areas having iodine deficiency. It behaves intermediately between DTC and ATC. PDTC can appear as a unique rapidly growing thyroid with or without cervical node involvement; distant metastases can be also present. Three different histological patterns have been described: insular, trabecular and solid. The mean 5-year survival of PDTC patients is less than 50%.

ATC is one of the most malignant tumors in humans, with a median survival of 4–12 months from the time of diagnosis. At the clinical level, it is characterized by a large thyroid mass, rapidly growing and infiltrating surrounding tissues and muscles. The larynx, esophagus and trachea may also be involved in up to 70% of patients. Distant metastases at lung, bone and brain can also occur at presentation or during the course of the illness. The clinical course of the disease can be so rapid that surgical procedures performed are often useless. The molecular alterations of PDTC and ATC have been recently reviewed [61] and are summarized below.

Genetic Alterations Shared with DTC

According to the notion that they can arise from well-differentiated precursor lesions both PDTC and ATC display some of the alterations associated with DTC. PDTC carry mutations of *RAS* (46–55%), of *BRAF* (12–17%), as well as *RET* rearrangements (13%). In ATC, *RAS* mutations (6–52%) as well as *RAF* mutations (25–29%) are present [62]. The rare presence of *RET/PTC*, *TRK* and *PAX8/PPPRγ* oncogenes in PDTC and ATC suggests that tumors with these alterations do not usually progress to more aggressive forms.

Alterations of the PTEN/PI3K/AKT Pathway

The PTEN/PI3K/AKT pathway plays a central role in many cellular processes, including growth, motility and apoptosis, and is frequently upregulated in human cancer. Enhanced signaling by this pathway in cancer occurs through a variety of mechanisms, including genetic mutations, epigenetic mutations altering gene and protein expression, and post-translational modifications. The importance of this pathway in thyroid tumorigenesis has recently been recognized. Overall, alterations of this pathway are believed to promote thyroid cell transformation to FTC and to be involved in the progression to PDTC and ATC [63,64]. Mutations of the *PIK3CA* gene, encoding the catalytic subunit of PI3K, have been found in ATC, PDTC and FTC, and in related recurrent or metastatic lesions, and are associated with tumor aggressiveness. Genetic copy gain of the *PIK3CA* gene was found in approximately 30 and 50% of FTC and ATC, respectively. Whether mutations and amplification of the *PIK3CA* gene are sufficient to promote thyroid cancer *in vivo* has not been investigated yet.

PTEN is a negative regulator of the PI3K pathway by preventing AKT activation. The first evidence of *PTEN* involvement in thyroid tumors was provided by the high frequency of follicular thyroid neoplasia in patients affected by Cowden syndrome, caused by germline loss-of-function mutations of *PTEN*. Loss of *PTEN* activity is frequent also in sporadic thyroid tumors. Point mutations are rare, and occur prevalently in FTC and ATC. Loss of heterozygosity of *PTEN* gene is common in thyroid tumors. *PTEN* expression is silenced by promoter methylation in FTC and ATC, with the highest methylation in the latter. Overall, loss of *PTEN* activity is proposed to cooperate with other genetic lesions in promoting thyroid tumorigenesis [63].

The serine-threonine kinase AKT1 is a key effector of PI3K and is frequently mutated in cancer. The most prevalent mutation E17K, proven to cause constitutive AKT1 activation, has been recently found in FTC, ATC, PDTC and in metastatic lesions from tumors of different histotype, also in association with other genetic lesions, thus suggesting a role in tumor progression [65].

TP53 Mutations

The *TP53* gene encodes a tumor suppressor protein involved in cell cycle regulation, senescence, apoptosis and it is frequently inactivated in human cancer [66]. In thyroid carcinoma, *TP53* mutations represent a late event and are associated with tumor progression and aggressiveness. *TP53* mutations are rare or absent in DTC. On the contrary, they have been found in 26% of PDTC and in a very high proportion of ATC, although with variable frequency (60–83%) in different studies [61].

ALK Mutations

Mutations of the *ALK* gene, leading to oncogenic activation, are commonly associated with neuroblastoma [39] and, recently, two novel mutations of the ALK receptor tyrosine kinase (TK) have been detected in 11% of ATCs [67]. These mutations were not found in adjacent matching normal tissue, led to an increase in ALK kinase activity, activation of the PI3K/AKT and MAP kinase pathways, and exhibited potent *in vitro* transforming activity. Of note, in the same study, no *ALK* mutations were detected in FTC and PTC samples [67]. The presence of *ALK* mutations in ATC and *ALK* rearrangements in radiation-associated PTC underlines the occurrence of two different mechanisms of *ALK* oncogenic activation in two different follicular thyroid cell malignancies.

β-Catenin Mutations

The β-catenin protein is encoded by the *CTNNB1* gene and is involved in cell adhesion and transcription, depending on its cellular localization [68].The cell adhesion function involves interaction with cell membrane cadherins at adherens junctions. Cellular abundance of β-catenin is constitutively regulated by proteosomal degradation. Inhibition of such degradation by the Wnt pathway stabilizes β-catenin by allowing its nuclear translocation and transcription function. In up to 25% of PDTC and 65% of ATC, mutations in β-catenin lead to its nuclear accumulation by preventing its ubiquitination and degradation, which is consistent with Wnt pathway activation, thus enhancing its tumor-promoting activity [61]. Alternatively, the activity of β-catenin can be deregulated by the loss of E-cadherin expression, an event frequently observed in PDTC [69].

EXPRESSION PROFILE STUDIES IN EPITHELIAL THYROID TUMORS

Gene Expression

High-throughput gene expression studies have provided useful tools for diagnosis and molecular classification, as well as for identification of prognostic markers, thus offering the possibility of tailoring better treatment regimens for cancer. In the last few years, several groups have determined gene expression profiles of thyroid tumors, and identified signatures associated with the different histotypes. A multigene expression profile was found that could discriminate between benign and malignant lesions and, among the latter, was specifically associated with the follicular or papillary histotype of TC. Overall, a general prevalence of upregulated genes was reported in PTC compared to FTC, whereas a larger number of genes are underexpressed in FTC compared to PTC, consistent with loss-of-heterozygosity rates in these different forms of TC [70].

The set of genes capable of discriminating PTC from normal thyroid includes genes underexpressed in PTC that are related to specialized thyroid functions, such as thyroid peroxidase and sodium iodide symporter, the latter associated with the presence of *BRAF* mutations [71]. This is consistent with the fact that most malignant thyroid tumors are hypofunctioning in trapping iodine and producing thyroid hormone. Among the upregulated genes, adhesion genes have been found as the most frequent gene ontology class; genes involved in signal transduction (*MET*, *CITED1* or *S100A6*), cell cycle regulators (*SFN* and *CCND1*) were also found in different studies [72,73]. Although PTCs display common gene expression features, induced by alternative mutations leading to MAPK pathway activation, specific gene signatures for the different genetic lesions have also been identified. In particular, differences have been identified between *BRAF*-positive and *BRAF*-negative PTCs, indicating profound changes in biology related to this oncogene [71].

Different groups have investigated signatures capable of distinguishing radio-induced from sporadic PTCs. In the first study, no gene signature separating sporadic from post-Chernobyl PTCs was identified; however, further analyses showed differences in the expression of genes involved in response to H_2O_2 or γ-radiation and in homologous recombination [73].

In contrast to PTC, much less is known regarding the gene expression profile of FTC. The majority of the existing gene expression analyses have considered the diagnostic aspect and defined gene signatures differentiating FTC from FA, and have identified a combination of discriminating genes, including *CCND2*, *PCSK2*, and *PLAB*. Changes in gene expression have also been observed in *PAX8/PPAR*γ-bearing FTCs, whereas *RAS*-related TC gene signatures have yet to be specified [71,74]. A combination of five genes has been proposed to discriminate between PTC and FTC expression profiles [70].

PDTCs are characterized by upregulation of genes related to cell cycle progression and involved in MAPK and TGFβ pathways [71].

The gene signature of ATC is characterized by upregulation of genes involved in cell cycle progression and chromosome segregation. ATC also displays downregulation of genes involved in cell adhesion, which may lead to tumor invasion and metastasis [71].

MicroRNA Expression

MicroRNAs (miRNAs) are phylogenetically conserved single-stranded RNA molecules of 19–25 nt involved in modulation of gene expression via translational blockade or transcript degradation. miRNA involvement in human cancer has been extensively characterized; they represent a new layer of gene regulation, and their expression profiles may represent a potential tool for the diagnosis of malignancies as well as determining prognosis and therapeutic response [75].

The main miRNA alteration in cancer cells is abnormal expression. Accordingly, several studies have shown deregulation of miRNA expression in TC [76]. Furthermore, functional studies support a crucial role for miRNAs in TC development [62]. Aberrant miRNA expression profiles that distinguish among PTC, FTC, ATC, and normal thyroid tissue have been already suggested [76]. However, while the aberrant expression of specific miRNAs is established, the involvement of other miRNAs in TC histotypes is still under study. Generally, several onco-miRs, defined as miRNAs upregulated in cancer and targeting genes antagonizing tumor progression (e.g. miR-221 -222, -21, -155), and several tumor-suppressive miRNAs, defined as downregulated in tumor progression and that target genes encoding oncogenic factors (e.g. miR200, let-7, miR-30) are both involved in thyroid carcinogenesis. However, changes in the expression of tumor-suppressive miRNAs are more associated with ATC.

Moreover, the existence of circulating miRNAs in the blood of cancer patients has recently raised the possibility that the profile of miRNAs isolated from peripheral blood may provide a novel source of tumor markers [77]. The first of these studies has been undertaken in PTC patients, and several diagnostic miRNA biomarkers have been suggested [78].

MEDULLARY THYROID CARCINOMA (MTC)

MTC is a neuroendocrine tumor which is sporadic in about 75% of cases. In the other 25%, it occurs as a component of the autosomal dominant familial MEN 2 syndromes. MEN 2, first described by Sipple [79], includes three disorders: MEN 2A, MEN 2B, and familial MTC (FMTC) [80]. MTC is typically located in the middle third of the lobe (where C cells are most represented) and can be unilateral in cases of sporadic tumors or multiple and bilateral in hereditary MTC, and commonly not capsulated. Up to 50% of patients present with cervical node involvement, and up to 15% with distant metastases. MTC is characterized by nests or trabeculae of polygonal, round or spindle cells, separated by fibrovascular stroma. Tumor cells express calcitonin carcinoembryonic antigen, other neuroendocrine markers such as chromogranin A and synaptophysin, and, in some cases, TTF-1 and low molecular weight keratins. Calcitonin (CT) levels have a diagnostic and prognostic value. Occult distant metastases after thyroidectomy and neck dissection are responsible for the persistence of high levels of CT. Five and ten year survival rates are 83 and 74%, respectively, and related to age, male gender, tumor extension, and presence of distant metastases.

RET Mutations

The *RET* gene, already described in detail above, is crucial in MTC being mutated in the germline in virtually all hereditary MTC cases, and somatically in 30–50% of sporadic MTCs. A spectrum of *RET* mutations has been identified in MEN2 families. The MEN2 RET database (http://www.arup.utah.edu/database/MEN2/MEN2_welcome.php) is a repository for MEN2-associated RET sequence variation and a reference for RET genotype/MEN2 phenotype correlations [81]. In 98% of MEN 2A, the mutations affect one of the five cysteines in the extracellular cysteine-rich domain of RET: codons 609, 611, 618, 620, and 634 [82]. These "gain-of-function" mutations result in ligand-independent receptor dimerization and constitutive activation. Codon 634 is the most commonly affected, being mutated in 80% of MEN2A cases. In FMTC, mutations affect either the already described extracellular cysteines or the intracellular region of *RET* (codons 768, 790, 791, 804, 806, and 891 are mainly involved) [22]. Most MEN 2B patients (95% of cases) carry the M918T-mutation in the RET kinase domain, causing receptor autophosphorylation and activation, resulting in a shift in substrate specificity; the remaining fraction harbors the A883F substitution or other rare mutations. Altered RET signaling triggered by these *RET* mutants is responsible for more aggressive MTC and for the other pathologies associated with MEN 2B [83].

The presence of a known MEN2-associated *RET* mutation in the germline of an MTC patient identifies hereditary MTC disease, thus allowing preclinical identification of family members at risk for MTC, as well as providing information about the risk for the proband to develop other tumors associated with MEN2 syndromes. In fact, the *RET* mutation type, assessed by *RET* genetic testing, guides clinical decisions, as different *RET* mutants have been associated with different risk profiles. Consequently, prophylactic thyroidectomy for asymptomatic mutation carriers is recommended at different ages according to the *RET* mutation-associated risk level [84].

RET oncogenes are able to drive MTC formation, as demonstrated by transgenic mice models. Mice expressing *RET-C634R* (MEN2A mutant) or *RET-M918T* (MEN2B mutant) under the control of CT gene promoter, developed MTC [85]. However, the existence of additional genetic alterations occurring at somatic level and cooperating with *RET* mutants has been suggested to be required for full development of MTC. This is supported by the fact that MTC in FMTC patients displaying germline *RET* mutations may be monoclonal [86] and also by mouse model experiments, as knock-in of the *M918T* mutation into mouse endogenous *RET* gene caused C-cell hyperplasia but not MTC [87]. A secondary genetic event may target the *RET* gene itself by duplication of the mutant allele or deletion of the wild-type (wt) allele [88], or involve alterations at different chromosomes [89].

In summary, hereditary MTC is a model of genetically determined cancer in which both diagnostic and therapeutic strategies rely on the identification of specific mutations.

Nonetheless, a variable phenotypic expression within and between families harboring the same *RET* mutation suggests the existence of modifier genes. This is also suggested by transgenic mice studies showing that genetic background may strongly affect the MTC phenotype with tumor penetrance varying from 0 to 98% in different mouse strains [90].

The genetic lesions associated with sporadic MTC are less defined. Also, in this case, *RET* gene activation is involved, as oncogenic *RET* mutations have been detected in up to 50% of cases. *RET-M918T* somatic mutation, the same associated with MEN2B, is the most frequent *RET* mutation in sporadic MTC, and is associated with a more aggressive disease [91].

RAS Mutations

The involvement of *RAS* mutations, limited to the sporadic form of MTC, has now been recognized. Specifically *HRAS* and *KRAS* mutations at codons 61 and 12 occur in sporadic MTC. No *NRAS* or *BRAF* mutations have been detected in the same large patient series. The fraction of *RAS* mutation-positive MTC varies from 17 to 68% of the *RET*-negative MTC cases. However, all the analyses suggest that the activation of *RAS* and *RET* represents alternative genetic events in sporadic MTC tumorigenesis [92,93].

CONCLUSIONS

In the last few decades, the identification of the genetic lesions associated with the different types of TC has partially unveiled the molecular mechanisms governing thyroid carcinogenesis. The accumulation of knowledge has impacted the clinical management of patients, through the generation of novel diagnostic/prognostic molecular tools, and the design of novel therapeutic approaches targeting the pathways deregulated in TC. Nevertheless, the key events of thyroid carcinogenesis remain to be fully elucidated. We believe that, in the next few years, the integration of studies employing different high-throughput techniques will aid in the full definition of the genomic landscape of TC.

GLOSSARY

AKT A protein-serine-threonine kinase that is activated by phosphorylation in response to growth factors or insulin, playing a major role in cell metabolism, growth, and survival.

Anaplastic thyroid carcinoma (ATC) A highly malignant thyroid tumor composed in part or wholly of undifferentiated cells.

B-RAF Cytoplasmic protein with serine/threonine activity, involved in the regulation of the MAP kinase/ERKs signaling pathway, which affects cell division, differentiation, and secretion. The *BRAF* gene was shown to be mutated in human cancers.

C cell (parafollicular cell) A cell type of the thyroid gland, situated between or within the walls of follicles, producing and secreting calcitonin.

Chromosomal rearrangements Structural rearrangements caused by chromosome breakage followed by reconstitution in an abnormal combination, resulting in inversion, translocation or deletion.

Follicular thyroid carcinoma (FTC) Thyroid neoplasm, not belonging to papillary thyroid carcinoma, with evidence of capsular and/or vascular invasion.

Medullary thyroid carcinoma (MTC) Thyroid cancer of aggressive nature originating from C cells and in close association with multiple endocrine neoplasia syndromes.

MicroRNA Small double-stranded, non-protein coding RNAs, 21–25 nucleotides in length involved in repressing translation of target RNA by binding to homologous 3′ UTR region as an imperfect match.

Multiple endocrine neoplasia type 2 (MEN2) Group of medical disorders associated with tumors of the endocrine system and transmitted in an autosomal dominant pattern, in which gain-of-function mutations of the *RET* gene frequently play a causative role.

***NTRK1*(TRKA)** Proto-oncogene encoding a tyrosine protein kinase that serves as a receptor for nerve growth factor (NGF), which primarily regulates growth, differentiation, and programmed cell death of neurons in both the peripheral and central nervous system.

Oncogene Gene, usually playing a role in cell growth and proliferation, whose gain-of-function alteration leads to neoplastic cell transformation.

Point mutation Alteration in DNA sequence caused by a single nucleotide base change, insertion, or deletion resulting in silent, missense or nonsense mutations.

Papillary thyroid carcinoma (PTC) Thyroid neoplasm deriving from the thyroid follicular epithelial cells, displaying papillary architecture and characteristic nuclear features.

RAS Family of guanine nucleotide-binding proteins involved in cell signaling pathways, cell growth and apoptosis that have been frequently found mutated in human cancers.

Receptor tyrosine kinases Class of transmembrane proteins that have intrinsic protein–tyrosine kinase activity and act as receptors for growth factors, neurotrophic factors, and other extracellular signaling molecules.

RET Proto-oncogene encoding a tyrosine kinase receptor for members of the glial cell line-derived neurotrophic factor (GDNF) family.

RET/PTC Activated version of the *RET* proto-oncogene, found in papillary thyroid carcinoma.

Thyrocytes Epithelial cells (also called follicular cells) of the thyroid gland that are responsible for the production and secretion of thyroid hormones, that is, thyroxine (T4) and triiodothyronine (T3).

ABBREVIATIONS

ATC Anaplastic thyroid carcinoma
DTC Differentiated thyroid cancer
FA Follicular adenoma
FMTC Familial medullary thyroid carcinoma
FTC Follicular thyroid carcinoma
HCC Hurthle cell carcinoma
MEN Multiple endocrine neoplasia
MTC Medullary thyroid carcinoma
PDTC Poorly differentiated thyroid carcinoma
PTC Papillary thyroid carcinoma
TC Thyroid cancer
TK Tyrosine kinase

Due to the large number of cited papers, we were not able to cite all individual references, and we apologize to the authors whose references are not cited. In many cases we refer the readers to reviews and references therein.

REFERENCES

[1] Davies L, Weleh HG. Increasing incidence of thyroid cancer in the United States, 1973–2002. J Am Med Assoc 2006;295: 2164–7.

[2] Burgess JR, Tucker P. Incidence trends for papillary thyroid carcinoma and their correlation with thyroid surgery and thyroid fine-needle aspirate cytology. Thyroid 2006;16:47–53.

[3] Colonna M, Guizard AV, Schvartz C, Velten M, Raverdy N, Molinie F, et al. A time trend analysis of papillary and follicular cancers as a function of tumour size: a study of data from six cancer registries in France (1983–2000). Eur J Cancer 2007;43: 891–900.

[4] Delellis RA, Lloyd RV, Heitz PU, Eng C. Pathology and genetics of tumors of endocrine organs. In: IARC Press, editor. World Health Organization classification of tumours. Lyon: IARC Press; 2004.

[5] Nikiforov YE, Nikiforova MN. Molecular genetics and diagnosis of thyroid cancer. Nat Rev Endocrinol 2011;7:569–80.

[6] Nakachi K, Hayashi T, Hamatani K, Eguchi H, Kusunoki Y. Sixty years of follow-up of Hiroshima and Nagasaki survivors: current progress in molecular epidemiology studies. Mutat Res 2008;659: 109–17.

[7] Nikiforov YE. Radiation-induced thyroid cancer: what we have learned from Chernobyl. Endocr Pathol 2006;17:307–17.

[8] Sakorafas GH, Giotakis J, Stafyla V. Papillary thyroid microcarcinoma: a surgical perspective. Cancer Treat Rev 2005;31:423–38.

[9] Greco A, Borrello MG, Miranda C, Degl'Innocenti D, Pierotti MA. Molecular pathology of differentiated thyroid cancer. Q J Nucl Med Mol Imaging 2009;53:440–53.

[10] Ciampi R, Knauf JA, Kerler R, Gandhi M, Zhu Z, Nikiforova MN, et al. Oncogenic AKAP9-BRAF fusion is a novel mechanism of MAPK pathway activation in thyroid cancer. J Clin Invest 2005;115:94–101.

[11] Hamatani K, Mukai M, Takahashi K, Hayashi Y, Nakachi K, Kusunoki Y. Rearranged anaplastic lymphoma kinase ((ALK)) gene in adult-onset papillary thyroid cancer among atomic-bomb survivors. Thyroid 2012;11:1153–9.

[12] Henderson YC, Shellenberger TD, Williams MD, El-Naggar AK, Frederick MJ, Cieply KM, et al. High rate of BRAF and RET/PTC dual mutations associated with recurrent papillary thyroid carcinoma. Clin Cancer Res 2009;15:485–91.

[13] Zhang H, Oliveira AM. Fusion genes in epithelial neoplasia. J Clin Pathol 2010;63:4–11.

[14] Pierotti MA, Greco A. Oncogenic rearrangements of the NTRK1/NGF receptor. Cancer Lett 2006;232:90–8.

[15] Yang TT, Namba H, Hara T, Takmura N, Nagayama Y, Fukata S, et al. p53 induced by ionizing radiation mediates DNA end-jointing activity, but not apoptosis of thyroid cells. Oncogene 1997;14:1511–9.

[16] Nikiforova MN, Stringer JR, Blough R, Medvedovic M, Fagin JA, Nikiforov YE. Proximity of chromosomal loci that participate in radiation-induced rearrangements in human cells. Science 2000;290:138–41.

[17] Roccato E, Bressan P, Sabatella G, Rumio C, Vizzotto L, Pierotti MA, et al. Proximity of TPR and NTRK1 rearranging loci in human thyrocytes. Cancer Res 2005;65:2572–6.

[18] Gandhi M, Medvedovic M, Stringer JR, Nikiforov YE. Interphase chromosome folding determines spatial proximity of genes

participating in carcinogenic RET/PTC rearrangements. Oncogene 2006;25:2360–6.

[19] Gandhi MS, Stringer JR, Nikiforova MN, Medvedovic M, Nikiforov YE. Gene position within chromosome territories correlates with their involvement in distinct rearrangement types in thyroid cancer cells. Genes Chromosomes Cancer 2009;48: 222–8.

[20] Fusco A, Grieco M, Santoro M, Berlingieri MT, Pilotti S, Pierotti MA, et al. A new oncogene in human papillary thyroid carcinomas and their lymph-nodal metastases. Nature 1987;328:170–2.

[21] Airaksinen MS, Saarma M. The GDNF family: signalling, biological functions and therapeutic value. Nat Rev Neurosci 2002;3:383–94.

[22] Arighi E, Borrello MG, Sariola H. RET tyrosine kinase signaling in development and cancer. Cytokine Growth Factor Rev 2005;16:441–67.

[23] Viglietto G, Chiappetta G, Martinez-Tello FJ, Fukunaga FH, Tallini G, Rigopoulou D, et al. RET/PTC oncogene activation is an early event in thyroid carcinogenesis. Oncogene 1995;11:1207–10.

[24] Fischer AH, Bond JA, Taysavang P, Battles OE, Wynford-Thomas D. Papillary thyroid carcinoma oncogene (RET/PTC) alters the nuclear envelope and chromatin structure. Am J Pathol 1998;153:1443–50.

[25] Jhiang SM, Sagartz JE, Tong Q, Parker-Thornburg J, Capen CC, Cho JY, et al. Targeted expression of the ret/PTC1 oncogene induces papillary thyroid carcinomas. Endocrinology 1996;137:375–8.

[26] Adeniran AJ, Zhu Z, Gandhi M, Steward DL, Fidler JP, Giordano TJ, et al. Correlation between genetic alterations and microscopic features, clinical manifestations, and prognostic characteristics of thyroid papillary carcinomas. Am J Surg Pathol 2006;30:216–22.

[27] Hamatani K, Eguchi H, Ito R, Mukai M, Takahashi K, Taga M, et al. RET/PTC rearrangements preferentially occurred in papillary thyroid cancer among atomic bomb survivors exposed to high radiation dose. Cancer Res 2008;68:7176–82.

[28] Asai N, Jijiwa M, Enomoto A, Kawai K, Maeda K, Ichiahara M, et al. Ret receptor signaling: dysfunction in thyroid cancer and Hirschsprung's disease. Pathol Int 2006;56:164–72.

[29] Castellone MD, Santoro M. Dysregulated RET signaling in thyroid cancer. Endocrinol Metab Clin N Am 2008;37:363–74.

[30] Cassinelli G, Favini E, Degl'Innocenti D, Salvi A, De Petro G, Pierotti MA, et al. RET/PTC1-driven neoplastic transformation and proinvasive phenotype of human thyrocytes involve Met induction and beta-catenin nuclear translocation. Neoplasia 2009;11:10–21.

[31] Castellone MD, De Falco V, Rao DM, Bellelli R, Muthu M, Basolo F, et al. The β-catenin axis integrates multiple signals downstream from RET/papillary thyroid carcinoma leading to cell proliferation. Cancer Res 2009;69:1867–76.

[32] Borrello MG, Alberti L, Fischer A, Degl'Innocenti D, Ferrario C, Gariboldi M, et al. Induction of a proinflammatory programme in normal human thyrocytes by the RET/PTC1 oncogene. Proc Natl Acad Sci USA 2005;102:14825–30.

[33] Kaplan DR, Miller FD. Neurotrophin signal transduction in the nervous system. Curr Opin Neurobiol 2000;10:381–91.

[34] Greco A, Miranda C, Pierotti MA. Rearrangements of NTRK1 gene in papillary thyroid carcinoma. Mol Cell Endocrinol 2009;321:44–9.

[35] Rabes HM, Demidchik EP, Sidorow JD, Lengfelder E, Beimfohr C, Hoelzel D, et al. Pattern of radiation-induced RET and NTRK1 rearrangements in 191 post-Chernobyl papillary thyroid carcinomas: biological, phenotypic, and clinical implications. Clin Cancer Res 2000;6:1093–103.

[36] Bounacer A, Schlumberger M, Wicker R, Du-Villard JA, Caillou B, Sarasin A, et al. Search for NTRK1 proto-oncogene rearrangements in human thyroid tumours originated after therapeutic radiation. Br J Cancer 2000;82:308–14.

[37] Russell JP, Powell DJ, Cunnane M, Greco A, Portella G, Santoro M, et al. The *TRK-T1* fusion protein induces neoplastic transformation of thyroid epithelium. Oncogene 2000;19:5729–35.

[38] Fedele M, Palmieri D, Chiappetta G, Pasquinelli R, De Martino I, Arra C, et al. Impairment of the p27kip1 function enhances thyroid carcinogenesis in TRK-T1 transgenic mice. Endocr Relat Cancer 2009;16:483–90.

[39] Barreca A, Lasorsa E, Riera L, Machiorlatti R, Piva R, Ponzoni M, et al. Anaplastic lymphoma kinase in human cancer. J Mol Endocrinol 2011;47:R11–23.

[40] Davies H, Bignell GR, Cox C, Stephens P, Edkins S, Clegg S, et al. Mutations of the BRAF gene in human cancer. Nature 2002;417:949–54.

[41] Wan PT, Garnett MJ, Roe SM, Lee S, Niculescu-Duvaz D, Good VM, et al. Mechanism of activation of the RAF-ERK signaling pathway by oncogenic mutations of B-RAF. Cell 2004;116:855–67.

[42] Xing M. BRAF mutation in papillary thyroid cancer: pathogenic role, molecular bases, and clinical implications. Endocr Rev 2007;28:742–62.

[43] Knauf JA, Ma X, Smith EP, Zhang L, Mitsutake N, Liao XH, et al. Targeted expression of BRAFV600E in thyroid cells of transgenic mice results in papillary thyroid cancers that undergo dedifferentiation. Cancer Res 2005;65:4238–45.

[44] Malumbres M, Barbacid M. RAS oncogenes: the first 30 years. Nat Rev Cancer 2003;3:459–65.

[45] Zhu Z, Gandhi M, Nikiforova MN, Fischer AH, Nikiforov YE. Molecular profile and clinical-pathologic features of the follicular variant of papillary thyroid carcinoma. An unusually high prevalence of ras mutations. Am J Clin Pathol 2003;120:71–7.

[46] Rosai J, Carcangiu ML, Delellis RA. Tumors of the thyroid gland. Atlas of tumor pathology. Washington, DC: Armed Forces Institute of Pathology; 1992. p. 161–82

[47] Sobrinho-Simoes M, Eloy C, Magalhaes J, Lobo C, Amaro T. Follicular thyroid carcinoma. Mod Pathol 2011;24(Suppl. 2): S10–8.

[48] Nikiforova MN, Nikiforov YE. Molecular genetics of thyroid cancer: implications for diagnosis, treatment and prognosis. Expert Rev Mol Diagn 2008;8:83–95.

[49] Vitagliano D, Portella G, Troncone G, Francione A, Rossi C, Bruno A, et al. Thyroid targeting of the N-ras(Gln61Lys) oncogene in transgenic mice results in follicular tumors that progress to poorly differentiated carcinomas. Oncogene 2006;25:5467–74.

[50] Manenti G, Pilotti S, Re FC, Della Porta G, Pierotti MA. Selective activation of *ras* oncogenes in follicular and undifferentiated thyroid carcinomas. Eur J Cancer 1994;30A:987–93.

[51] Garcia-Rostan G, Zhao H, Camp RL, Pollan M, Herrero A, Pardo J, et al. Ras mutations are associated with aggressive

tumor phenotypes and poor prognosis in thyroid cancer. J Clin Oncol 2003;21:3226–35.
[52] Saavedra HI, Knauf JA, Shirokawa JM, Wang J, Ouyang B, Elisei R, et al. The RAS oncogene induces genomic instability in thyroid PCCL3 cells via the MAPK pathway. Oncogene 2000;19:3948–54.
[53] Kroll TG, Sarraf P, Chen CJ, Mueller E, Spiegelman BM, Fletcher JA. PAX8-PPARγ1 fusion oncogene in human thyroid carcinoma. Science 2000;289:1357–60.
[54] Nikiforova MN, Lynch RA, Biddinger PW, Alexander EK, Dorn GW, Tallini G, et al. RAS point mutations and PAX8-PPAR gamma rearrangement in thyroid tumors: evidence for distinct molecular pathways in thyroid follicular carcinoma. J Clin Endocrinol Metab 2003;88:2318–26.
[55] Placzkowski KA, Reddi HV, Grebe SK, Eberhardt NL, McIver B. The role of the PAX8/PPARγ fusion oncogene in thyroid cancer. PPAR Res 2008;2008:672829.
[56] Lui WO, Zeng L, Rehrmann V, Deshpande S, Tretiakova M, Kaplan EL, et al. CREB3L2-PPARγ fusion mutation identifies a thyroid signaling pathway regulated by intramembrane proteolysis. Cancer Res 2008;68:7156–64.
[57] Lui WO, Foukakis T, Liden J, Thoppe SR, Dwight T, Höög A, et al. Expression profiling reveals a distinct transcription signature in follicular thyroid carcinomas with a PAX8-PPAR(γ) fusion oncogene. Oncogene 2005;24:1467–76.
[58] Gasparre G, Porcelli AM, Bonora E, Pennisi LF, Toller M, Iommarini L, et al. Disruptive mitochondrial DNA mutations in complex I subunits are markers of oncocytic phenotype in thyroid tumors. Proc Natl Acad Sci USA 2007;104:9001–6.
[59] Zimmermann FA, Mayr JA, Neureiter D, Feichtinger R, Alinger B, Jones ND, et al. Lack of complex I is associated with oncocytic thyroid tumours. Br J Cancer 2009;100:1434–7.
[60] Maximo V, Botelho T, Capela J, Soares P, Lima J, Taveira A, et al. Somatic and germline mutation in GRIM-19, a dual function gene involved in mitochondrial metabolism and cell death, is linked to mitochondrion-rich (Hurthle cell) tumours of the thyroid. Br J Cancer 2005;92:1892–8.
[61] Soares P, Lima J, Preto A, Castro P, Vinagre J, Celestino R, et al. Genetic alterations in poorly differentiated and undifferentiated thyroid carcinomas. Curr Genomics 2011;12:609–17.
[62] Pallante P, Visone R, Croce CM, Fusco A. Deregulation of microRNA expression in follicular cell-derived human thyroid carcinomas. Endocr Relat Cancer 2010;17:F91–104.
[63] Xing M. Genetic alterations in the phosphatidylinositol-3 kinase/Akt pathway in thyroid cancer. Thyroid 2010;20:697–706.
[64] Saji M, Ringel MD. The PI3K-Akt-mTOR pathway in initiation and progression of thyroid tumors. Mol Cell Endocrinol 2010;321:20–8.
[65] Ricarte-Filho JC, Ryder M, Chitale DA, Rivera M, Heguy A, Ladanyi M, et al. Mutational profile of advanced primary and metastatic radioactive iodine-refractory thyroid cancers reveals distinct pathogenetic roles for BRAF, PIK3CA, and AKT1. Cancer Res 2009;69:4885–93.
[66] Levine AJ, Oren M. The first 30 years of p53: growing ever more complex. Nat Rev Cancer 2009;9:749–58.
[67] Murugan AK, Xing M. Anaplastic thyroid cancers harbor novel oncogenic mutations of the ALK gene. Cancer Res 2011;71: 4403–11.
[68] Gavert N, Ben-Ze'ev A. β-Catenin signaling in biological control and cancer. J Cell Biochem 2007;102:820–8.
[69] Rocha AS, Soares P, Fonseca E, Cameselle-Teijeiro J, Oliveira MC, Sobrinho-Simoes M. E-cadherin loss rather than β-catenin alterations is a common feature of poorly differentiated thyroid carcinomas. Histopathology 2003;42:580–7.
[70] Aldred MA, Huang Y, Liyanarachchi S, Pellegata NS, Gimm O, Jhiang S, et al. Papillary and follicular thyroid carcinomas show distinctly different microarray expression profiles and can be distinguished by a minimum of five genes. J Clin Oncol 2004;22:3531–9.
[71] Rusinek D, Szpak-Ulczok S, Jarzab B. Gene expression profile of human thyroid cancer in relation to its mutational status. J Mol Endocrinol 2011;47:R91–103.
[72] Jarzab B, Wiench M, Fujarewicz K, Simek K, Jarzab M, Oczko-Wojciechowska M, et al. Gene expression profile of papillary thyroid cancer: sources of variability and diagnostic implications. Cancer Res 2005;65:1587–97.
[73] Maenhaut C, Detours V, Dom G, Handkiewicz-Junak D, Oczko-Wojciechowska M, Jarzab B. Gene expression profiles for radiation-induced thyroid cancer. Clin Oncol (R Coll Radiol) 2011;23:282–8.
[74] Freitas BC, Cerutti JM. Genetic markers differentiating follicular thyroid carcinoma from benign lesions. Mol Cell Endocrinol 2010;321:77–85.
[75] Di LG, Croce CM. Roles of small RNAs in tumor formation. Trends Mol Med 2010;16:257–67.
[76] Braun J, Huttelmaier S. Pathogenic mechanisms of deregulated microRNA expression in thyroid carcinomas of follicular origin. Thyroid Res 2011;4(Suppl. 1):S1.
[77] Kosaka N, Iguchi H, Ochiya T. Circulating microRNA in body fluid: a new potential biomarker for cancer diagnosis and prognosis. Cancer Sci 2010;101:2087–92.
[78] Yu S, Liu Y, Wang J, Guo Z, Zhang Q, Yu F, et al. Circulating microRNA profiles as potential biomarkers for diagnosis of papillary thyroid carcinoma. J Clin Endocrinol Metab 2012;97:2084–92.
[79] Sipple JH. The association of pheochromocytoma with carcinoma of the thyroid gland. Am J Med 1961;31:163–6.
[80] Marx SJ. Molecular genetics of multiple endocrine neoplasia types 1 and 2. Nat Rev Cancer 2005;5:367–75.
[81] Margraf RL, Crockett DK, Krautscheid PM, Seamons R, Calderon FR, Wittwer CT, et al. Multiple endocrine neoplasia type 2 RET protooncogene database: repository of MEN2-associated RET sequence variation and reference for genotype/phenotype correlations. Hum Mutat 2009;30(4):548–56.
[82] Eng C, Clayton D, Schuffenecker I, Lenoir G, Cote G, Gagel RF, et al. The relationship between specific RET proto-oncogene mutations and disease phenotype in multiple endocrine neoplasia type 2. International RET mutation consortium analysis. J Am Med Assoc 1996;276:1575–9.
[83] Santoro M, Carlomagno F, Romano A, Bottaro DP, Dathan NA, Grieco M, et al. Activation of *RET* as a dominant transforming gene by germline mutations of MEN2A and MEN2B. Science 1995;267:381–3.
[84] Frank-Raue K, Rondot S, Raue F. Molecular genetics and phenomics of RET mutations: impact on prognosis of MTC. Mol Cell Endocrinol 2010;322:2–7.

[85] Acton DS, Velthuyzen D, Lips CJ, Hoppener JW. Multiple endocrine neoplasia type 2B mutation in human RET oncogene induces medullary thyroid carcinoma in transgenic mice. Oncogene 2000;19:3121–5.

[86] Gagel RF, Marx SJ. Multiple endocrine neoplasia. In: Kronenberg HM, Melamed J, Polonsky K, Larsen P, editors. Williams Textbook of Endocrinology. 11th ed. Philadelphia: Saunders, Elsevier Science; 2008. p. 1705–46.

[87] Smith-Hicks CL, Sizer KC, Powers JF, Tischler AS, Costantini F. C-cell hyperplasia, pheochromocytoma and sympathoadrenal malformation in a mouse model of multiple endocrine neoplasia type 2B. EMBO J 2000;19:612–22.

[88] Huang SC, Torres-Cruz J, Pack SD, Kock CA, Vortmeyer AO, Mannan P, et al. Amplification and overexpression of mutant RET in multiple endocrine neoplasia type 2-associated medullary thyroid carcinoma. J Clin Endocrinol Metab 2003;88:459–63.

[89] Ye L, Santarpia L, Cote GJ, El-Naggar AK, Gagel RF. High resolution array-comparative genomic hybridization profiling reveals deoxyribonucleic acid copy number alterations associated with medullary thyroid carcinoma. J Clin Endocrinol Metab 2008;93:4367–72.

[90] Cranston AN, Ponder BA. Modulation of medullary thyroid carcinoma penetrance suggests the presence of modifier genes in a RET transgenic mouse model. Cancer Res 2003;63:4777–80.

[91] Elisei R, Cosci B, Romei C, Bottici V, Renzini G, Molinaro E, et al. Prognostic significance of somatic RET oncogene mutations in sporadic medullary thyroid cancer: a 10-year follow-up study. J Clin Endocrinol Metab 2008;93:682–7.

[92] Moura MM, Cavaco BM, Pinto AE, Leite V. High prevalence of RAS mutations in RET-negative sporadic medullary thyroid carcinomas. J Clin Endocrinol Metab 2011;96:E863–8.

[93] Ciampi R, Mian C, Fugazzola L, Cosci B, Romei C, Barollo S, et al. Evidence of low prevalence of RAS mutations in a large medullary thyroid cancer series. Thyroid 2013;23(1):50–7.

Part 4.2

Age Non-Specific Tumors and Hematopoietic Malignancies

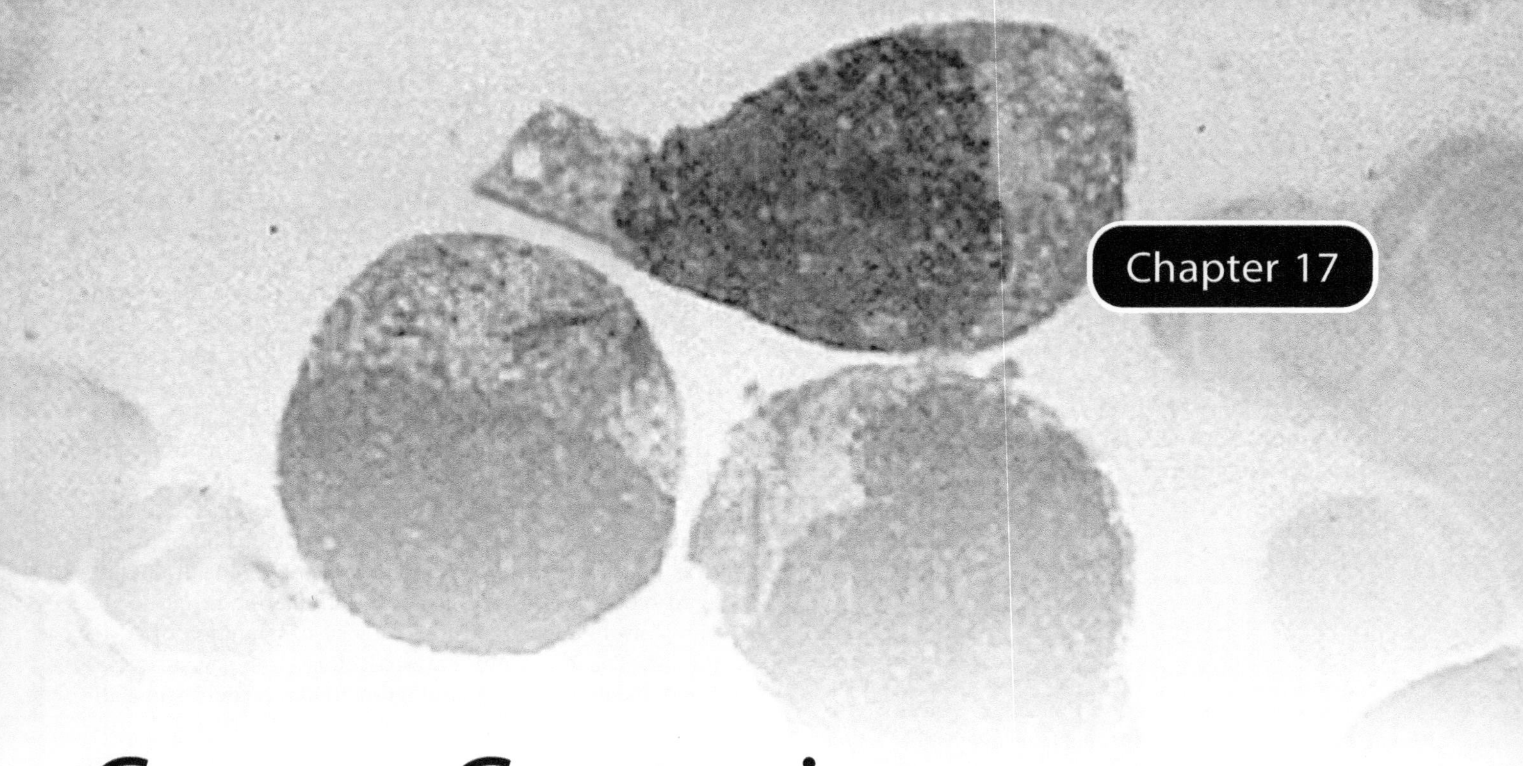

Cancer Genomics
Acute Myeloid Leukemia

Robert J. Arceci[1]
Jason N. Berman[2]
Soheil Meshinchi[3]

[1]*Children's Center for Cancer and Blood Disorders, Hematology/Oncology, The Ron Matricaria Institute of Molecular Medicine at Phoenix Children's Hospital, Department of Child Health, University of Arizona, College of Medicine, Phoenix AZ, USA*

[2]*Departments of Pediatrics, Microbiology and Immunology, and Pathology and IWK Health Centre, Dalhousie University, Halifax, NS, Canada*

[3]*Clinical Research Division, Fred Hutchinson Cancer Research Center, Department of Pediatrics, University of Washington School of Medicine, Seattle, WA, USA*

Contents

Cancer Genomics. DOI: http://dx.doi.org/10.1016/B978-0-12-396967-5.00017-7

Key Concepts

- AML is characterized by a large number of heterogeneous chromosomal alterations and selective gene mutations
- The frequency and genes mutated in AML in children differ in many instances with those identified in adults
- Common pathways that are affected by chromosomal alterations and gene mutations involve those regulating cell cycle, differentiation, proliferation, DNA repair and chromatin structure
- Whole genome and RNA sequencing have led to a more comprehensive characterization of AML resulting in the recognition of convergent pathways that are more therapeutically targetable

INTRODUCTION

Acute myeloid leukemia (AML) is a heterogeneous and complex disease that represents the culmination of interaction between genetic and epigenetic alterations in the hematopoietic progenitors, leading to dysregulation of multiple critical signal transduction pathways. The result is the increased proliferation, survival and accumulation of leukemic cells at the expense of normal hematopoiesis leading to anemia, thrombocytopenia and neutropenia.

A single step process does not appear to be responsible for AML pathogenesis. For example, *in vivo* studies have demonstrated that common cytogenetic alterations, such as t(8;21) or inv(16), that are highly associated with AML, are not sufficient for the disease pathogenesis. These findings led to the hypothesis that evolution of AML may require multiple, genetic changes or "hits" and that the disease may require cooperation between two or more alterations [1].

One such model suggests that AML is the result of cooperation between two classes of genetic alterations where the initial event leads to maturation arrest [1]. Initial transforming events include genomic alterations such as core binding factor (CBF) or mixed-lineage leukemia (MLL) translocations that are referred to as *Type II mutations* that are thought to arise in the hematopoietic stem/progenitor cells. Hematopoietic precursors that develop such genetic alterations would have self-renewal properties, but lack proliferative potential and represent a minor clone in the hematopoietic progenitor population. However, they would potentially persist as a minor clone for extended periods of time.

For example, multiple studies have demonstrated that, in teenage patients with t(8;21) AML, the fusion transcript was able to be detected in the patients' neonatal Guthrie card blood spots, demonstrating intrauterine origin of the transforming event and long latency of evolution of AML [2–4]. The maturation-arrested clone would persist as a subclinical population until a time that it acquired an additional genetic "hit" that would provide a proliferative advantage. The proliferative change, referred to as *Type I mutation*, includes activating mutations such as those affecting FLT3/ITD and c-KIT receptor tyrosine kinases. Hematopoietic cells with concomitant Type I and Type II mutations would have both self-renewal capacity as well as a significant proliferative advantage, thus having the capacity to outcompete normal hematopoietic cells, resulting in normal hematopoietic insufficiency. *In vivo* studies have provided support for this cooperation model; for instance, transduction of cDNAs encoding individual mutations (i.e. genes representing Type I or Type II mutations) did not result in an AML phenotype. However, simultaneous introduction of genes encoding Type I and Type II mutations led to rapid evolution of leukemia [5].

Recent technologic advances have allowed more complete interrogation of the AML transcriptome, genome and epigenome, leading to the identification of an increasing number of alterations associated with AML and suggesting that the process that leads to AML pathogenesis may prove to be even more complex. Current data suggest that additional genomic and epigenomic events in leukemic cells may lead to further alteration of the leukemic clone and introduce genomic heterogeneity and clonal evolution within the leukemic cell population.

EPIDEMIOLOGY

With the exception of an increased incidence of AML in infants, the incidence of AML is relatively constant until early adolescence when it continues to rise slowly through young adulthood. After 50 years of age, the incidence rises dramatically, mostly due to secondary MDS/AML until after about 80 years of age, when the incidence declines [6].

The incidence of AML also varies according to race and ethnicity. For example, in the USA, the annual incidence of AML in Hispanic children and young adults is nine per million, which is mainly secondary to the increase of acute promyelocytic leukemia (APL) [7,8]. The United States Surveillance, Epidemiology and End Results (SEER) database demonstrates similar incidence rates between Caucasian and African-American youth [9]. The rates of AML cases from Japan, Australia and Zimbabwe are 8, 8 and 11 per million annually [10].

ETIOLOGY

Mutagenic Environmental Exposures

Exposure to ionizing radiation after the atomic bomb detonations in Hiroshima and Nagasaki during World War II resulted in a 20-fold increase in the incidence of AML that peaked 6–8 years after exposure [11–14]. A variety of chemical exposures have been linked to the development of AML in adults, most notably petroleum products, benzene, herbicides and pesticides [15–19]. Secondary AML as a result of exposure to genotoxic chemotherapeutic agents has also been extensively documented [20–24]. Of note, there has been no definitive evidence that exposure to ultrasound or electromagnetic fields, such as observed from electrical power lines, increases the incidence of AML [25,26]. Similarly, there have been no compelling data reported to support viral causes for AML.

Inherited Predisposition

Down syndrome (DS) represents the most common inherited condition with the highest incidence of developing leukemia with a 10- to 20-fold increased risk of leukemia. Acute megakaryoblastic leukemia (AMKL) during the first 4 years of life represents the most common subtype [27]. AMKL is classified as M7 subtype of AML by FAB morphologic classification. Of note, approximately 10% of neonates with DS have a "transient myeloproliferative disorder" (TMD) that mimics AMKL but usually resolves spontaneously in 4–6 weeks. However, approximately 20–30% of these patients subsequently develop true AMKL. While it is unclear how the presence of trisomy 21 increases the risk of leukemia in these patients, both TMD and AMKL share somatic mutations of the GATA1 transcription factor in nearly all cases [28].

Several bone marrow failure syndromes (BMFS) also have an increased risk of developing AML. These include Fanconi anemia (defects in DNA repair), Kostmann syndrome (defects in neutrophil elastase *ELA2*), Shwachman–Diamond Syndrome (ribosomal processing *SBDS* gene), Diamond–Blackfan anemia (defects in ribosomal small and large subunit proteins) and dyskeratosis congenita (X-lined mutations in the dyskerin or autosomal recessive forms with mutations in genes regulating telomere maintenance and RNA processing). Other inherited disorders with a predisposition to develop AML include congenital amegakaryocytic thrombocytopenia (defects in the *CFFA2* gene and thrombopoietin receptor gene, *c-mpl*), autosomal dominant macrothrombocytopenia (Fechtner syndrome, *MYH9* gene), and familial platelet disorder with propensity to myeloid malignancy (FPD/AML, germline mutations in the *RUNX1/CEBPA* gene) [29–36].

In addition, mutations in the neurofibromin gene, encoding an RAS-inactivating GTPase, are responsible for neurofibromatosis type I, which is associated with an increased incidence of juvenile myelomonocytic leukemia (JMML) and AML [37,38]. In Noonan syndrome, mutations in the *PTPN11* gene, which encodes a SHP-2 tyrosine phosphatase, leads to increased incidence of JMML [39,40]. Mutations in *CBL*, encoding an E3 ubiquitin ligase, have been reported to be associated with a dominant inheritance of a Noonan syndrome-like disorder and predisposition to JMML [41–43]. Patients with Li–Fraumeni syndrome and Bloom syndrome have inherited defects in DNA repair genes *p53* and the *BLM* helicase gene, respectively, and are associated with a propensity to develop leukemia, including AML [29,44–47].

The increased incidence of AML can vary from nearly 100% concordance in monozygotic or identical twins to approximately 20% in fraternal twins up until about 6 years of age, after which the incidence appears to decrease to that of the general population [48–50]. Non-twin, familial cases of AML are rare and often associated with constitutional translocations, such as t(7;20), t(3;6) or monosomy 7 [51–53].

GENETIC AND EPIGENETIC ALTERATIONS

Disease-associated alterations in AML can be grouped into several broad categories as outlined below. Large structural alterations of the chromosomes identified in leukemic cells, including duplications, deletions and translocations, are usually identified by standard karyotype analysis in patients with AML. Specific karyotypic abnormalities remain key characteristics for diagnosis as well as for defining prognosis in AML.

In addition, disease-associated mutations of specific genes that mediate hematopoietic cell differentiation, survival and proliferation have been identified. These mutated genes commonly modify key signal transduction pathways that, in turn, can contribute to AML pathogenesis. Such mutations have come to define AML biology as they are prevalent, provide prognostic information and act as potential targets for directed therapy. Additional areas of investigation are also emerging as major contributors in the development of AML, including alterations of non-coding RNA (miRNA) expression and function as well as epigenetic alterations due to aberrant DNA methylation and/or altered histone modifications.

Chromosomal Alterations

Karyotypic Alterations

Cytogenetic alterations have been the cornerstone of diagnosis in AML for several decades and have provided valuable tools for appropriate diagnosis of AML and provide clinically meaningful prognostic tools for risk-based therapy allocation. Although a large number of chromosomal alterations are identified in AML, the majority of pediatric AML cases fall into six specific cytogenetic categories, where 25% have CBF AML (t(8;21) or inv(16)), 12% have t(15;17), 20% have rearrangements involving the *MLL* gene and 20% are without karyotypic abnormalities (normal karyotype, NK). In addition to karyotypic alterations, disease-associated mutations are identified in AML, with the highest prevalence in those with normal karyotype; >90% pediatric AML cases appear to have at least one genomic alteration.

Although non-random numerical and structural chromosome aberrations have been identified in approximately 80% of pediatric AML, there is significant age-related variability in the prevalence and the type of cytogenetic alterations in childhood AML (Figure 17.1). AML in infants (<1 year of age) is most commonly defined by the presence of MLL translocations. In addition, there is notable paucity of CBF AML, and rarely no karyotypic alterations (normal karyotype, NK). The prevalence of MLL decreases over the first decade of life to that seen in adults. Similarly, the prevalence of CBF AML and NK AML increases from virtually undetectable in the first year of life to that seen in young adults by age 15. Given the high association of specific karyotypic alterations in AML, and its association with outcome, the most recent WHO AML classification is primarily based on recurrent cytogenetic alterations where, for example, detection of t(8;21) or inv(16) is sufficient for confirmed diagnosis of AML regardless of marrow blast percentage [54].

Chimeric Transcription Factors

Chimeric transcription factors are generally produced as a result of chromosomal translocations that fuse a DNA-binding domain of a transcriptional activator to a transcriptional repressor. The transcriptionally active domain is thereby redirected to target genes of the transcriptional activator, many of which function in myeloid differentiation.

The t(8;21) translocation involves the DNA-binding domain of the transcriptional activator AML1 (RUNX1, CBFA2) on chromosome 21q22 that is fused to the transcriptional repressor ETO (CBF2T1) on chromosome 8q22. The t(8;21) translocation is associated with characteristic clinical, morphologic and immunophenotypic characteristics, such as chloromas that can occur in about 20% of cases [55,56]. Over 80% of cases with t(8;21) are classified as FAB (French–American–British) type M2. However, of cases classified as FAB M2, only approximately 40% will be found to harbor t(8;21) [56]. The immunophenotype of cases with t(8;21) is also commonly characterized by the expression of the B-cell antigen, CD19, and the natural killer (NK) cell antigen, CD56 [57]. In both adults and children, t(8;21) is associated with a favorable prognosis. Improved outcomes for adult patients with AML and t(8;21) have been shown to be associated with dose intensity of cytarabine treatment [58].

Chromosomal abnormalities involving chromosome 16, such as inv(16) or t(16;16), disrupt the *CBFB* gene at chromosome band 16q22 and often fuse it to the *MYH11* gene at chromosome band 16p13. *CBFB* encodes the beta subunit of the core binding factor (CBF) transcriptional complex, which normally heterodimerizes with *AML1* (or *CBFA2*, which encodes the alpha subunit of CBF). The fusion of *CBFB* with the myosin heavy chain gene, *MYH11*, results in decreased CBF transcriptional activity. The inv(16) and t(16;16) abnormalities are closely linked to the FAB M4Eo subtype, and both confer a favorable prognosis for both adults and children with AML [59,60].

AML with t(15;17) is associated with FAB M3 subtype or APL. The t(15;17) translocation results in a fusion protein that combines the retinoid acid receptor alpha and PML; this fusion protein confers an increased sensitivity to the differentiating effects of pharmacologic doses of all-*trans* retinoic acid (ATRA). Since the introduction of ATRA as part of chemotherapy-based treatment, outcomes for patients with APL have improved significantly, with an approximately 75% overall survival [61]. Much less common translocations, involving RAR alpha, can also result in APL (e.g. t(11;17) that involves the *PLZF* gene) [62]. Identification of cases with t(11;17) is important because of their decreased sensitivity to ATRA.

MLL Gene Rearrangements

MLL-rearranged AML represents a heterogeneous group of leukemias with variable outcomes depending on the type of rearrangement [63]. Translocations of chromosomal band 11q23 involving the *MLL* gene, including most AML secondary to epipodophyllotoxin exposure (e.g. etoposide), are frequently associated with monocytic differentiation (FAB M4 and M5) and often have an unfavorable prognosis [63]. In an international study examining outcomes for a large number of young patients, overall survival was 100%, 63%, 27% and 22% for patients with AML characterized by t(1;11), t(9;11), t(4;11), and t(6;11), respectively [64]. AML with t(10;11)

(A)

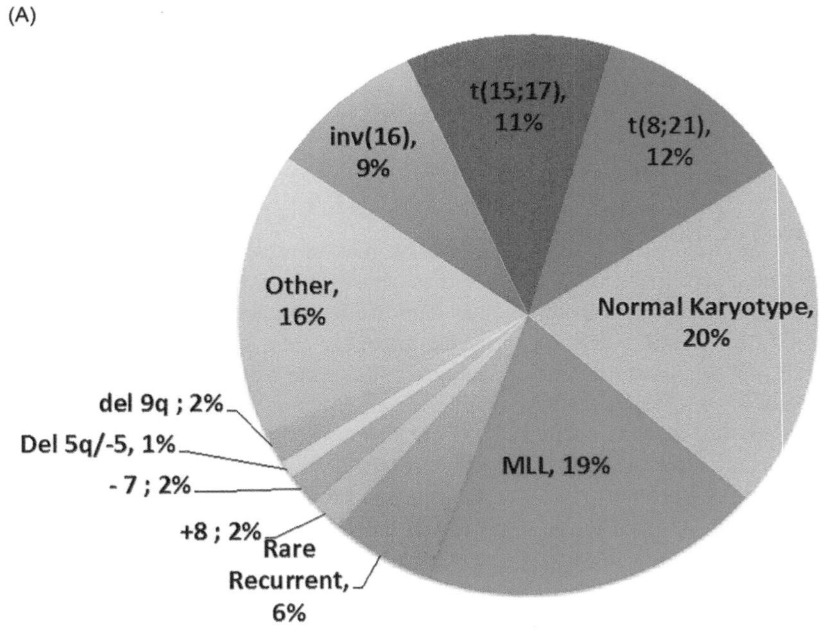

(B)

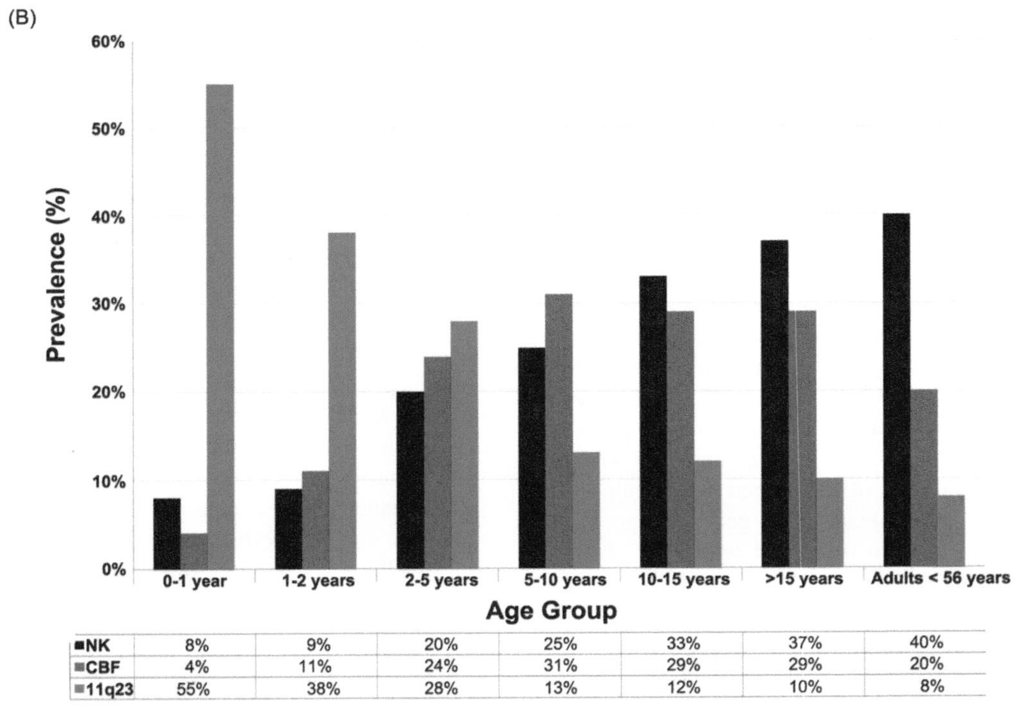

	0-1 year	1-2 years	2-5 years	5-10 years	10-15 years	>15 years	Adults < 56 years
NK	8%	9%	20%	25%	33%	37%	40%
CBF	4%	11%	24%	31%	29%	29%	20%
11q23	55%	38%	28%	13%	12%	10%	8%

FIGURE 17.1 (A) Percentages of chromosomal alterations in pediatric AML and (B) frequency of some common subtypes according to age.

is characterized by a fusion of the *CALM* gene on chromosome 11 with the *AF10* gene [65]. Prognosis appears poor for cases with t(10;11) [66]. The variability in outcomes and lack of definitive data that demonstrate the superiority of hematopoietic stem cell transplant (HSCT) for patients with *MLL*-rearranged AML have led most cooperative trials not to recommend allogeneic HSCT in first remission [67,68].

The t(1;22)(p13;q13) translocation is essentially restricted to AMKL and occurs in as many as one third of

AMKL cases in children. In leukemias with t(1;22), the *OTT (RBM15)* gene on chromosome 1 is fused to the *MAL (MKL1)* gene on chromosome 22 [69,70]. Once associated with a very poor prognosis [71], AMKL with the t(1;22) now has an improved outcome with contemporary treatment; thus, HSCT may not be indicated in these patients [72,73].

Miscellaneous Chromosomal Abnormalities

Chromosomal abnormalities associated with poor prognosis in patients with AML include those involving chromosome 7 (monosomy 7, but not necessarily del(7q)), chromosome 5 (monosomy 5 and del(5q)) and the long arm of chromosome 3 (inv(3)(q21;q26) or t(3;3)(q21;q26)) [74,75]. These cytogenetic subgroups are also associated with poor prognosis in both children and adults with AML, although abnormalities of the long arm of chromosome 3q and 5q are extremely rare in children with AML [76].

Sub-Karyotypic Chromosomal Abnormalities

Small structural alterations, including deletions or duplications, can escape detection by conventional cytogenetics but be detected using specialized techniques such as fluorescence *in situ* hybridization (FISH). Loss of heterozygosity (LOH) that is generally associated with deletions can lead to haplo-insufficiency that in turn may contribute to the development of AML. Single nucleotide polymorphism (SNP) genotyping and copy number evaluation have the capacity to detect genomic alterations of even smaller magnitude. For example, these methodologies have identified regions of copy-neutral LOH (CN-LOH), also referred to as acquired uniparental disomy (aUPD) [77–79]. aUPD arises from a homologous recombination-mediated process by which a homozygous state is achieved after an initial acquisition of a heterozygous mutation. For example, during the DNA repair process, a mutant allele is used as a template, and the wild-type allele is converted to that of the mutant one, resulting in a homozygous state and altering the gene dosage of that mutation. aUPD has been described with *FLT3/ITD* and *CEBPA* mutations [80,81].

SOMATIC MUTATIONS AND THEIR CONTRIBUTION TO SURVIVAL IN CHILDHOOD AML

Somatic mutations in genes known to contribute to hematopoiesis have been identified in a significant proportion of AML, and the presence of these mutations has been shown to be associated with clinical outcome. New discovery initiatives including the NCI-sponsored Therapeutically Applicable Research to Generate Effective Treatments (TARGET) AML initiative have broadened our knowledge of the mutations involved in malignant myeloid hematopoiesis, although true prevalence and clinical significance of these novel findings are yet to be defined.

DISEASE-ASSOCIATED MUTATIONS IN SIGNAL TRANSDUCTION PATHWAYS

Chromosomal translocations described above are considered to cause differentiation arrest. A broad body of data has demonstrated that such alterations are not sufficient for frank leukemic transformation, and additional events are required for the evolution of leukemic phenotype. Activating mutations of the kinase-mediated RAS/MAP signal transduction pathways have been demonstrated to provide the proliferative signal for the arrested clone and lead to leukemic progression. Regulation of cellular processes is initiated by extracellular signals (cytokines) whose actions are mediated through various signal transduction pathways, resulting in cellular differentiation, proliferation and survival. Stimulation of specific membrane bound receptors by their native ligand leads to the activation of their intracellular kinase moieties resulting in dysregulation of the signal transduction pathways. Currently, six common mutations are implicated in AML pathogenesis, including mutations in *FLT3*, *NPM1*, *CEBPA*, *RAS*, *c-KIT* and *WT1* genes.

FLT3 *Mutations*

The most commonly mutated gene in childhood AML is FMS-like tyrosine kinase 3 (*FLT3*). FLT3, a receptor tyrosine kinase (RTK), is a membrane-bound receptor with an intrinsic tyrosine kinase activity that is highly expressed in hematopoietic stem/progenitor cells and regulates hematopoietic differentiation and proliferation. *FLT3* is expressed at variable levels in all myeloid malignancies. Mutations in the *FLT3* gene lead to constitutive activation of the receptor kinase activity and can be due to the internal tandem duplication (*FLT3/ITD*) of the juxtamembrane domain coding sequence, or a missense mutation in the activation loop domain (*FLT3/ALM*) [82,83]. Internal tandem duplication (*FLT3/ITD*) mutations are detected in 12% of all children with AML. There is a significant age-associated variation that increases from nearly none in the first year of life to nearly 20% by age 10 years, with a similar prevalence reported in adult AML (Figure 17.2).

FLT3 mutations lead to constitutive activation of the FLT3 kinase moiety, although by different mechanisms that in turn lead to different functional and clinical endpoints [83,84]. For instance, there are significant biological differences between the two types of *FLT3* mutations [85–88] as *FLT3/ITD*s and *FLT3/ALM*s appear to promote activation of different downstream effectors and

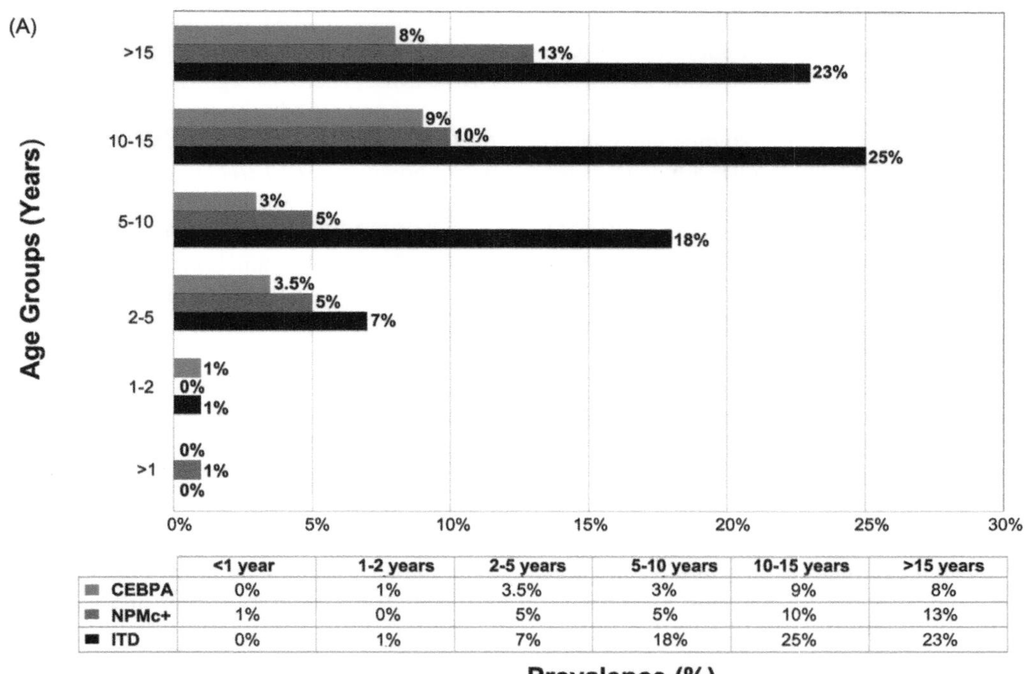

	<1 year	1-2 years	2-5 years	5-10 years	10-15 years	>15 years
CEBPA	0%	1%	3.5%	3%	9%	8%
NPMc+	1%	0%	5%	5%	10%	13%
ITD	0%	1%	7%	18%	25%	23%

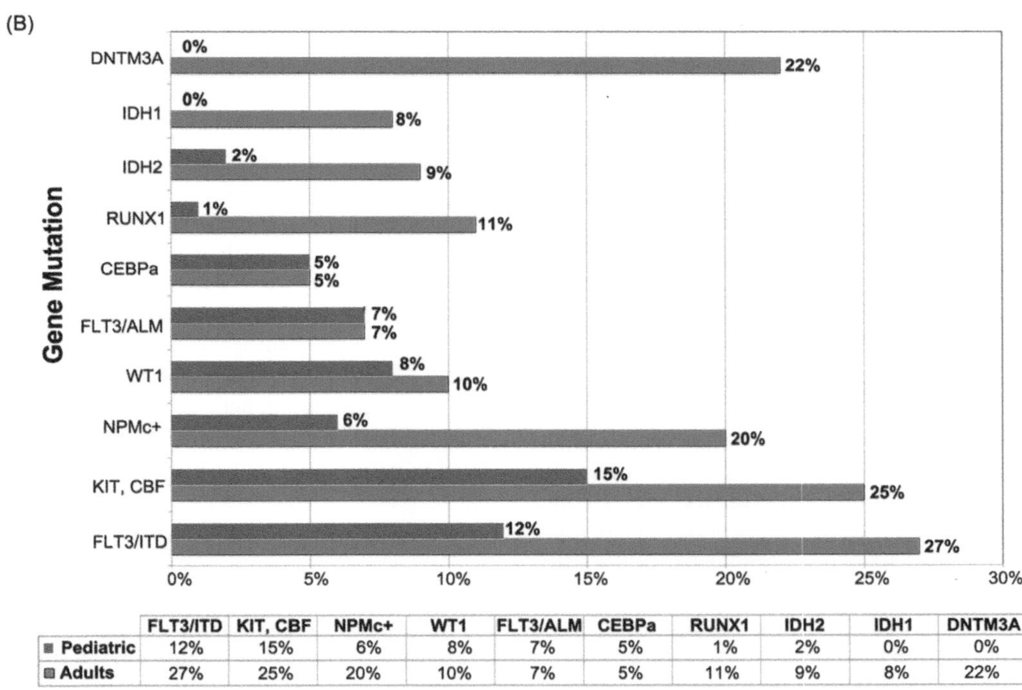

	FLT3/ITD	KIT, CBF	NPMc+	WT1	FLT3/ALM	CEBPa	RUNX1	IDH2	IDH1	DNTM3A
Pediatric	12%	15%	6%	8%	7%	5%	1%	2%	0%	0%
Adults	27%	25%	20%	10%	7%	5%	11%	9%	8%	22%

FIGURE 17.2 (A) Frequency of specific gene mutations by age and (B) in adults vs children.

biological responses [85]. Animal studies have demonstrated that mice harboring *FLT3/ITDs* primarily develop an oligoclonal myeloproliferative disorder, while mice harboring *FLT3/ALM* are more likely to develop oligoclonal lymphoid disorders [86]. The biologic differences between *FLT3/ITD* and *FLT3/ALM* are further highlighted in the clinical setting, where the presence of *FLT3/ITD* has been associated with significant leukocytosis, increased induction failure, high rate of post-remission relapse and overall poor survival with conventional chemotherapy, whereas those with *FLT3/ALM* have similar characteristics and outcomes as those without *FLT3* mutations [89,90]. Several studies have shown that patients with *FLT3/ITD* who receive allogeneic stem cell transplantation in CR have an improved outcome, thus providing a target-based therapy allocation for this high-risk cohort of patients [89,91,92]. Further, novel kinase inhibitors have shown efficacy in inducing a high rate of remission in patients with *FLT3/ITD*, although long-term benefits of such interventions are yet to be determined [93]. Again, in contrast to *FLT3/ITD*, *FLT3/ALM* does not seem to be similarly amenable to inhibition by kinase inhibitors [94–96].

c-KIT *Mutations*

c-KIT is a proto-oncogene located on chromosome band 4q11-12 which, similar to *FLT3*, encodes a transmembrane protein that is a member of the type III receptor tyrosine kinase family [97]. Mutations in the *c-KIT* gene lead to cytokine independent growth and increased drug resistance [98,99]. Mutations of *c-KIT* are frequent in CBF AML (25%), tend to cluster within the activation loop (exon 17) and a region of the extracellular domain integral to receptor dimerization (exon 8) [97], and were reported to be associated with worse clinical outcome in adults [100–104]. Although adult data suggested that the presence of *c-KIT* mutations are associated with worse clinical outcome, a large-scale study conducted in a pediatric cooperative trial demonstrated a prevalence of *c-KIT* mutations of nearly 20% but lack of clinical significance in pediatric CBF AML [105].

*Nucleophosmin (*NPM1*) Mutations*

NPM1 encodes a ubiquitously expressed and highly conserved phosphoprotein that is primarily localized to the nucleolus, but shuttles rapidly between the nucleus and cytoplasm in its role as a molecular chaperone. Abnormalities in *NPM1* have been implicated in the pathogenesis of various hematopoietic malignancies, including anaplastic large cell lymphoma with t(2;5) translocation, resulting in expression of the NPM1–ALK fusion protein. Mutations in *NPM1* are common in AML, with a prevalence of 30% in adult and 8–10% in pediatric AML [106–109]. *NPM1* mutations appear to be more prevalent in AML with normal karyotype, with a prevalence of nearly 40–50% in adults and 20% in pediatric AML. Disease-associated mutations, characterized by four base insertions in exon 12 of the *NPM* gene, lead to impaired nuclear localization of the nucleophosmin protein. Presence of *NPM* mutations portends favorable outcome with reduced relapse risk and improved survival; those with *NPM* mutations have a similar outcome as those with CBF AML. There is significant overlap between *NPM* mutations and *FLT3/ITD*, as nearly one third of those with *NPM* mutations also have *FLT3/ITD*. In adult AML, it has been shown that *FLT3/ITD* retains its high-risk status in the presence of *NPM* mutations. In pediatric AML, the presence of *NPM1* mutations in the presence of *FLT3/ITD* has been reported to ameliorate partially the poor prognosis conferred by *FLT3/ITD* alone [108].

CEBPA *Mutations*

CCAAT/enhancer binding protein-alpha (C/EBPA) is encoded by the *CEBPA* gene and functions as a transcription factor that regulates proliferation and terminal granulocytic differentiation. In the hematopoietic system, C/EBPA is expressed in cells of myelomonocytic origin and regulates the expression of genes that regulate granulocyte differentiation. Mutations in the *CEBPA* gene occur in 4% of childhood AML cases and the overwhelming majority of cases with these mutations occur in a bi-allelic manner, in which two distinct mutations occur, one in the N-terminal domain (NTD) and the second in the opposite allele affecting the bZip domain [110]. *NTD* mutation results in an early termination codon, which truncates the p42 protein, allowing intact translation of only the p30 protein. In a bi-allelic model, the bZip mutation in the opposite allele disrupts the bZip domains of its protein products, resulting in loss of function. Such a bi-allelic mutation would result in the expression of p30 as the only functional protein, which was shown to be sufficient for development of AML [111]. This model would argue that the two mutations are required to evolve in a bi-allelic manner for AML pathogenesis. COG studies have demonstrated that *CEBPA* mutations tend to occur in patients with normal cytogenetics and are associated with decreased relapse risk as well as improved survival compared to patients without a mutation [110].

RAS *Mutations*

Mutations in the *N-ras* gene occur in approximately 10–30% of adult AML cases [112,113] and 10% of pediatric AML [113]. Point mutations frequently occur in codons 12, 13, and 61 [114], and prevent the hydrolysis of *RAS–GTP*, keeping *RAS* in the "on" or activated state. The coincident occurrence of both an *FLT3* and *ras* mutation in the same case is very rare, as might be expected, since there may be no *a priori* selective advantage to two mutations affecting the same pathway. However, unlike *FLT3/ITD* mutations, *RAS* mutations are not prognostically important [113]. With both *FLT3* and *ras* mutations, there may be discordance between diagnostic and relapse specimens. Thus, a patient with either a *ras* or *FLT3* mutation at diagnosis may relapse without the mutation (or vice versa). This suggests that these mutations are not the primary oncogenic events in all cases.

*Wilms Tumor-1 (*WT1*) Gene Expression and Mutation*

The *WT1* gene, located on chromosome 11p13, encodes a zinc-finger protein that exists in multiple isoforms and functions as a transcription factor. *WT1* is expressed primarily in tissues of the developing genitourinary and hematopoietic systems, and mutations in *WT1* occur in both syndrome-associated and sporadic cases of nephroblastoma (Wilms tumor), the most common childhood renal malignancy. *WT1* is also expressed in CD34 + hematopoietic progenitors, and is overexpressed in a subset of acute leukemias, including AML, ALL, and CML in blast crisis [115]. Somatic mutations of the *WT1* gene have been reported to have variable prognostic significance [116]. Initial studies in adults suggested that the presence of *WT1* mutations was an independent prognostic factor for poor outcome. However, recent studies, including a comprehensive evaluation of pediatric AML, have identified *WT* mutations in 8–10% of patients; further, it was demonstrated that a strong association between *WT1* mutations and *FLT3/ITD* existed. In that study, although the presence of *WT1* mutation was associated with worse outcome, it was shown that the prognostic significance of *WT1* mutation was due to *FLT3/ITD;* however, in patients without *FLT3/ITD*, the presence of a *WT1* mutation did not portend poor outcome [105]. In contrast, Hollink et al. have reported that *WT1* mutations represent an independent poor risk factor [64]. Furthermore, although expression of *WT1* has been used as a marker of disease response and emergence of MRD [117,118], no significant association of *WT1* expression was observed in a pediatric trial study [119]. Comprehensive evaluation of *WT1* expression within specific risk groups has not been performed in childhood AML and currently is not utilized in risk group stratification.

GATA-1 *Mutations*

GATA-1 mutations are present in nearly all children with Down syndrome and either transient myeloproliferative disease (TMD) or AMKL, but not in patients without Down syndrome and AMKL [120–123]. GATA-1 is a transcription factor that is required for normal development of erythroid cells, megakaryocytes, eosinophils, and mast cells. *GATA-1* mutations appear to confer increased sensitivity to cytarabine, possibly by decreasing cytidine deaminase expression and thus providing a possible explanation for the superior outcomes for children with Down syndrome and M7 AML [124].

Mutations of Genes Affecting Chromatin Structure and Miscellaneous Signaling Pathways

Genomic alterations in the *TET2* gene (4q24) have been demonstrated in myeloid malignancies in adults. The prevalence of these mutations appears to be approximately 20% in MDS, 10–20% in myeloproliferative disease and 12% in adult AML. Data from COG studies show that these mutations are uncommon in children with a prevalence of about 3%. In adults, *TET2* mutations have been associated with decreased overall survival compared to patients with AML characterized by WT *TET2* [125]; however, the prognostic impact is unclear in children. Similarly, mutations of genes affecting other cellular pathways in AML, such as *SHIP*, *PTPN11*, *RUNX1* and *IDH1* and *IDH2*, are relatively rare in young patients with AML. Mutations in *IDH1* and *IDH2* in adults with AML are particularly intriguing in that they confer a new enzymatic activity on the enzyme product that results in altered cellular metabolism. These mutations have been associated with a neutral or somewhat poorer prognosis [126–128].

The presence of the Philadelphia chromosome and/or *BCR–ABL* fusion in children and young adults with AML most likely represents chronic myelogenous leukemia (CML) that has transformed to AML rather than *de novo* AML.

Recent advances in sequencing have enabled interrogation of the entire leukemic genome for identification of novel disease-associated mutations. Studies by Ley et al. initially provided the feasibility of WGS in AML [129] and went on to identify novel somatic mutations that were prevalent and were associated with clinical outcome in adult disease [130,131]. Despite the high prevalence of these novel mutations in adult AML, Ho et al. demonstrated that these mutations are extremely rare in childhood AML [132], providing the rationale for similar whole genome interrogation of AML genome in childhood AML. Recent efforts by the TARGET AML initiative have provided some insight into the mutations in childhood AML with identification of potential mutations in childhood AML [133].

Insights from Genome-Wide Analysis

Whole-genome and exon sequencing studies have provided important information on both the identification of novel gene mutations as well as providing key information on the number and types of gene mutations occurring in individual cases of AML. The latter observations have also allowed for the assessment of how various genes could contribute to alterations of specific cell survival, proliferation and drug resistance or sensitivity pathways. The importance of such approaches is further demonstrated by the fact that 20–50% (children compared to adults) of AML cases have normal karyotypes and many have no evidence for any of the known mutated genes thought to be important for AML [134–136].

The Cancer Genome Atlas (TCGA) report provides several important insights [134]. Using high-resolution SNP arrays, few copy number changes were observed except in samples of high-risk AML. These data suggest that, unlike some solid tumors, extensive disruption of chromosomes (e.g. chromothripsis) does not occur as part of AML [137]. While only 23 genes were observed to be mutated relatively frequently, most were only found mutated in just a few cases. However, mutations appeared to cluster into common categories, including transcription factor fusions, myeloid-specific transcription factor genes, NPM1, tumor suppressor genes, DNA methylation-related genes, signaling genes, chromatin-modifying genes, cohesion-complex genes and splicesosome-complex related genes. Of note, mutations of miRNAs were observed in 4% of the samples. Epigenetic profiling revealed distinctive patterns of CpG methylation that clustered with cytogenetic subtypes, but also increased methylation levels in samples with mutated *IDH1* and *IDH2*. In contrast, loss of CpG methylation was observed in samples with mutations of *NPM1*, *DNMT3A* and *FLT3*. The majority of methylation changes occurred in gene-coding regions. The highest concordance of an inverse relationship of CpG methylation and gene expressions was observed in the APL subtype. Thus, while the total number of observed genetic, transcription, and epigenetic changes may be in the hundreds or low thousands in terms of events, most of the changes are associated with a small number of key proliferative, differentiation and survival pathways.

Epigenetic Changes

An increasing body of evidence has demonstrated that epigenetic modifications may contribute to the pathogenesis of leukemia. The term epigenetics refers to heritable changes of DNA and chromatin that modify gene expression and cellular physiology other than alterations in the DNA sequence. Epigenetic changes that underlie the development of leukemia can be in one of two major categories: changes in the DNA methylation state and alterations in histone or other chromatin-modifying proteins. These two processes collaborate with one another and can lead to activation or repression of gene expression. Emerging evidence shows that altered methylation as well as changes in histone modification regulates both normal and malignant hematopoiesis [138–142].

Regulators of epigenetic patterning have been observed more commonly in adult compared to pediatric AML. For instance, while mutations of *TET2* on chromosome 4q24 have been reported in about 12% of adult AML (about 23% in normal cytogenetic AML) and portend a poor prognosis [126,143,144], these mutations occur in only about 6% of pediatric AML and may also be similarly associated with poor prognosis [132,145,146]. Mutations of the *IDH1* (isocitrate dehydrogenase) gene interestingly produce a gain-of-function activity that in turn results in increased levels of 2-hydroxyglutarate and aberrantly hypermethylated promoter CpG sites [140,147,148]. While *IDH1* mutations have been reported in about 8% of young adults with AML, and are often associated with the presence of *NPM1* mutations, they appear to occur in up to 10% of pediatric AML [149–151].

Poor outcome has been associated with *IDH1* mutations in adults, but this has not been demonstrated in multivariate analyses from pediatric trials [149–151]. A minor SNP at rs11554137 associated with *IDH1*, found in about 10% of pediatric patients with AML, could also not be correlated with outcome [152]. While mutations of the DNA methyltransferase encoding gene, *DNMT3A*, may occur in up to 20% of adult AML [130,153–155], such mutations are exceedingly rare in pediatric AML [132,156].

Evidence linking alterations in DNA methylation in malignancies has prompted comprehensive evaluations of DNA methylation patterns in AML in an attempt to identify specific methylation profiles that can be used for risk identification as well as identification of potential targets for epigenetically directed therapies. One study in adult AML used DNA methylation patterns to group patients into 16 different cohorts based on their methylation pattern [140]. The methylation pattern segregated with specific mutation and cytogenetic groups. This study also identified a 15-gene methylation classifier that appeared to predict overall survival. The contribution of alterations in DNA methylation patterning and their clinical significance are not yet well defined in childhood AML, but such evaluations are currently underway.

Histone lysine acetylation is catalyzed by histone acetyltransferases (HATs) and associated with transcriptionally active chromatin. Several transcriptional co-activators, including CREB binding protein (CBP), have been shown to possess HAT activity. Transcriptional co-repressor complexes, such as nuclear co-repressor 1 (NCOR1), exist in protein complexes containing histone deacetylase (HDAC) activity and can modify their activity. As acetylation and methylation status of the histone lysine residues helps to determine whether chromatin is transcriptionally active or repressed, altered histone modification patterns have been shown to underlie in part the development of AML by altering expression of oncogenes. As these histone alterations in methylation and acetylation have been found to make a significant contribution to myeloid transformation and proliferation, they also represent potential targets for epigenetic therapy [157].

Transcriptional Alterations

The consequences of chromosome rearrangements, gene mutations and epigenetic changes can result in distinct RNA expression patterns that have been, in some instances, associated with specific AML subtypes and outcomes.

Multidrug resistance has been associated with rapid drug efflux mediated by a family of drug transporters, including the multidrug resistance gene 1 (*MDR1*) that encodes P-glycoprotein (PgP). Other proteins of the same family of multidrug resistance such as MRP1 (multidrug resistance-associated protein 1) and LRP (lung resistance protein) also contribute to drug efflux and resistance to chemotherapy. Unlike the evidence in adults with AML, *MDR* gene expression in pediatric patients has not been definitively associated with poor clinical outcomes. Sievers et al. demonstrated a prevalence of 13% for the expression of PgP in a group of 130 pediatric AML patients treated on the Children's Cancer Group study CCG-2891, but the expression of this transporter was not significantly correlated with outcome [158]. Another pediatric study demonstrated that *MDR1* expression was not higher in patients with relapsed AML [159]. The Pediatric Oncology Group (POG) 9421 trial randomized the use of cyclosporine A as an inhibitor of *MDR1* but did not demonstrate any advantage in the group receiving MDR inhibition [160].

The expression level of *WT1* has been implicated in pathogenesis and prognosis in AML. Although *WT1* expression level at the time of diagnosis and during remission has been correlated with clinical outcome, such findings have not been uniformly observed [117,119]. Telomerase activity has been implicated in leukemogenesis; however, there are data to suggest that telomerase activity has prognostic significance, although it could potentially represent a therapeutic target [161]. *BAALC* (brain and acute leukemia, cytoplasmic) is a gene whose elevated expression has also been implicated in the adverse outcome of adults with AML [162]. High levels of expression of *AF1q*, an *MLL* fusion partner, have been reported to be associated with poor outcome [163]. At this time, these and most single gene associations have not been definitively validated and, thus, are not used in current cooperative group clinical trials for stratifying patients to particular treatments.

Similarly, although total cellular RNA, including microRNA expression, signatures can accurately discriminate cases with specific chromosomal translocations and identify novel subsets of AML [164–168], independent studies have not always been able to identify an expression pattern with consistent prognostic significance [169,170]. In part, the discrepancies identified in different results may arise from different methodologies and analytical methods, as well as the leukemia content of the samples being tested.

Some of the differences observed in gene expression studies are also beginning to be understood in terms of epigenetic regulation of chromatin function. The finding of a similar expression profile between an AML subset with the methylation and silencing of the *CEBPA* promoter and another characterized by *CEBPA* mutations demonstrates the critical relationship between epigenetic changes and gene expression in AML [171,172]. As an example, the distinctive myeloid/T-lymphoid characteristics of this subtype have been linked to the finding that decreased *CEBPA* expression leads to increased expression of T-lineage genes in hematopoietic precursors [142,172]. Larger, prospective studies using gene expression profiling along with independent cohort confirmation are likely to be required to establish a useful role of genomic profiling in risk identification in AML.

INTEGRATION OF MOLECULAR DATASETS AND PATHWAY DIRECTED TREATMENT

An important goal of the integration of various molecular data, such as genome-wide sequencing, epigenetic changes and altered transcription, is that a more complete description of the biology of a patient's leukemia will be able to be leveraged to determine which therapy would result in an optimal clinical response and outcome.

Using a more restricted set of only 18 genes for mutational analysis in 398 patients under the age of 60 years, Patel et al. identified at least one somatic alteration in 97% of patients [135]. Of note, patients had been randomized to receive either high-dose or standard-dose daunorubicin during induction treatment following their diagnosis. When clinical outcomes were examined, a reduced overall survival was associated with the presence of *FLT3/ITD* (37%), partial duplication of *MLL* (*MLL–PTD*, 5%) and mutations of *ASXL1* (additional sex combs-like 1) and *PHF6* (plant homeodomain finger 6). *DNMT3A* mutations were frequently observed concomitantly with mutations of *NPM1* and *FLT3*, and *FLT3* mutations were also more frequently observed in combination with *NPM1* mutations.

Better overall survival was associated with the presence of *CEBPA* (3%) and *IDH2* mutations (8%); of note, both of these genes encode chromatin-modulating proteins. Surprisingly, the usual favorable prognostic impact of *NPM1* mutation was found to be restricted to those patients whose AML also had *IDH1* or *IDH2* mutations. Of further interest, a positive effect on survival of receiving high-dose daunorubicin was observed in patients with *DNMT3A* or *NPM1* mutations or translocations of *MLL*, while such an effect was not observed for other patients.

While such analyses are important, they are limited by the focus on only a small subset of the key genes that could have either leukemia driver or modulator functions or potentially no direct impact on the leukemia physiology. This study also did not include an analysis of epigenetic or transcriptional patterns. Nevertheless, the association of specific gene combinations with response to different therapies is certainly a beginning in the generation of predictive signatures that can be linked to selective treatments.

CONCLUSION

The future of integrative genomics biologically to guide therapy of individual patients within a clinically relevant time frame will require optimization of tumor sampling, sequencing methodologies, integrative analyses that have a high probability of predicting responses and, of course, clinically testing new therapeutic agents. In addition, the ability to assess rare clones that would predict specific resistance at the time of diagnosis should provide the means to introduce targeted therapies early in treatment to avoid the emergence of resistant disease. The leukemias, in part because of their relatively easy accessibility, may be able to lead the way.

GLOSSARY

Dominant oncogenes Genes that when mutated result in an activation of a key protein or pathway, usually directly, such as a mutation in a tyrosine kinase receptor that results in the constitutive activation of that receptor.

Epigenetics Modifications to DNA, such as cytosine methylation, or to chromatin, such as histone methylation or phosphorylation.

Leukemia stem cell Also referred to as a leukemia initiating cell, refers to a cell that is self-renewing and can give rise to leukemia when transplanted into an immunodeficient mouse as a xenograft.

Tumor suppressor genes Genes that when inactivated contribute to malignant transformation.

Type I mutation Genes mutated in AML that involve proliferation of leukemia cells.

Type II mutation Genes mutated in AML that restrict the differentiation potential of leukemia cells.

ABBREVIATIONS

AF1q Gene fusion partner with MLL also called MLLT11 for myeloid/lymphoid or mixed-lineage leukemia (trithorax homolog, Drosophila); translocated to 11

AMKL Acute megakaryoblastic leukemia

ASXL1 Additional sex combs-like 1

BAALC Brain and acute leukemia, cytoplasmic

BMFS Bone marrow failure syndrome

CBF Core binding factor

CBL Proto-oncogene derived from Cas-Br-M (murine) ecotropic retroviral transforming sequence

CEBPA CCAAT/enhancer binding protein (C/EBP), alpha

CFFA2 Gene mutated in congenital amegakaryocytic thrombocytopenia

DNMT DNA methyl transferase

ELA2 Elastase gene

FLT FMS-like tyrosine kinase

FMS Feline sarcoma virus

IDH Isocitrate dehydrogenase

JMML Juvenile myelomonocytic leukemia

KIT From the v-kit Hardy-Zuckerman 4 feline sarcoma viral oncogene

MDR Multidrug resistance

MLL Mixed-lineage leukemia

MPL Myeloproliferative leukemia virus oncogene

MYH9 Myosin heavy chain gene involved in autosomal dominant macrothrombocytopenia

NPM Nucleophosmin

PgP P-glycoprotein

PHF6 Plant homeodomain finger 6

PTPN11 Protein tyrosine phosphatase, non-receptor type 11 often mutated in JMML; also called SHP2

RUNX1 Runt-related transcription factor 1 gene also called aml1 for acute myeloid leukemia 1 gene

SBDS Shwachman–Bodian–Diamond syndrome

SNP Single nucleotide polymorphism

TET Gene name derived from ten-eleven translocation-1

TMD Transient myeloproliferative syndrome seen in infants with Down syndrome

UPD Uniparental disomy

WT Wilms tumor

REFERENCES

[1] Gilliland DG, Jordan CT, Felix CA. The molecular basis of leukemia. Hematology (Am Soc Hematol Educ Program) 2004:80–97.

[2] Wiemels JL, Xiao Z, Buffler PA, Maia AT, Ma X, Dicks BM, et al. In utero origin of t(8;21) AML1-ETO translocations in childhood acute myeloid leukemia. Blood 2002;99:3801–5.

[3] Mahmoud HH, Ridge SA, Behm FG, Pui CH, Ford AM, Raimondi SC, et al. Intrauterine monoclonal origin of neonatal concordant acute lymphoblastic leukemia in monozygotic twins. Med Pediatr Oncol 1995;24:77–81.

[4] Greaves MF. Stem cell origins of leukaemia and curability. Br J Cancer 1993;67:413–23.

[5] Schessl C, Rawat VP, Cusan M, Deshpande A, Kohl TM, Rosten PM, et al. The AML1-ETO fusion gene and the FLT3 length mutation collaborate in inducing acute leukemia in mice. J Clin Invest 2005;115:2159–68.

[6] Juliusson G, Antunovic P, Derolf A, Lehmann S, Mollgard L, Stockelberg D, et al. Age and acute myeloid leukemia: real world data on decision to treat and outcomes from the Swedish Acute Leukemia Registry. Blood 2009;113:4179–87.

[7] Douer D, Preston-Martin S, Chang E, Nichols PW, Watkins KJ, Levine AM. High frequency of acute promyelocytic leukemia among Latinos with acute myeloid leukemia. Blood 1996;87:308–13.

[8] Glazer ER, Perkins CI, Young Jr. JL, Schlag RD, Campleman SL, Wright WE. Cancer among Hispanic children in California,

1988–1994: comparison with non-Hispanic white children. Cancer 1999;86:1070–9.

[9] Oblimersen: Augmerosen, BCL-2 antisense oligonucleotide – Genta, G 3139, GC 3139, oblimersen sodium. Drugs R D 2007;8:321–34.

[10] Stiller CA. Epidemiology and genetics of childhood cancer. Oncogene 2004;23:6429–44.

[11] Ichimaru M, Ishimaru T, Belsky JL. Incidence of leukemia in atomic bomb survivors belonging to a fixed cohort in Hiroshima and Nagasaki. 1950–1971: radiation dose, years after exposure, age at exposure, and type of leukemia. J Radiat Res 1978;19:262–82.

[12] Kato H, Schull WJ. Studies of the mortality of A-bomb survivors: mortality, 1950–1978: Part I. Cancer mortality. Radiat Res 1982;90:395–432.

[13] Kato I, Tajima K, Hirose K, Nakagawa N, Kuroishi T, Tominaga S. A descriptive epidemiological study of hematopoietic neoplasms in Japan. Jpn J Clin Oncol 1985;15:347–64.

[14] Shimizu Y, Schull WI, Kato H. Cancer risk among atomic bomb survivors: the RERF Life Span Study. J Am Med Assoc 1990;264:601–4.

[15] Korte JE, Hertz-Picciotto I, Schulz MR, Ball LM, Duell EJ. The contribution of benzene to smoking-induced leukemia. Environ Health Perspect 2000;108:333–9.

[16] McBride ML. Childhood cancer and environmental contaminants. Can J Public Health 1998;89(Suppl. 1):S53–62, S58–68.

[17] van Waalwijk van Doorn-Khosrovani SB, Janssen J, Maas LM, Godschalk RW, Nijhuis JG, van Schooten FJ. Dietary flavonoids induce MLL translocations in primary human $CD34^{+}$cells. Carcinogenesis 2007;28:1703–9.

[18] Yin SN, Hayes RB, Linet MS, Li GL, Dosemeci M, Travis LB, et al. A cohort study of cancer among benzene-exposed workers in China: overall results. Am J Ind Med 1996;29:227–35.

[19] Mills PK, Zahm SH. Organophosphate pesticide residues in urine of farmworkers and their children in Fresno County, California. Am J Ind Med 2001;40:571–7.

[20] Barnard DR, Alonzo TA, Gerbing RB, Lange B, Woods WG. Comparison of childhood myelodysplastic syndrome, AML FAB M6 or M7, CCG 2891: report from the children's oncology group. Pediatr Blood Cancer 2007;49:17–22.

[21] Barnard DR, Woods WG. Treatment-related myelodysplastic syndrome/acute myeloid leukemia in survivors of childhood cancer – an update. Leuk Lymphoma 2005;46:651–63.

[22] Millot F, Esperou H, Bordigoni P, Dalle JH, Michallet M, Michel G, et al. Allogeneic bone marrow transplantation for chronic myeloid leukemia in childhood: a report from the Societe Francaise de Greffe de Moelle et de Therapie Cellulaire (SFGM-TC). Bone Marrow Transplant 2003;32:993–9.

[23] Stine KC, Saylors RL, Sawyer JR, Becton DL. Secondary acute myelogenous leukemia following safe exposure to etoposide. J Clin Oncol 1997;15:1583–6.

[24] Duffner PK, Krischer JP, Horowitz ME, Cohen ME, Burger PC, Friedman HS, et al. Second malignancies in young children with primary brain tumors following treatment with prolonged postoperative chemotherapy and delayed irradiation: a Pediatric Oncology Group study. Ann Neurol 1998;44:313–6.

[25] Linet MS, Hatch EE, Kleinerman RA, Robison LL, Kaune WT, Friedman DR, et al. Residential exposure to magnetic fields and acute lymphoblastic leukemia in children. N Engl J Med 1997;337:1–7.

[26] Armstrong B, Theriault G, Guenel P, Deadman J, Goldberg M, Heroux P. Association between exposure to pulsed electromagnetic fields and cancer in electric utility workers in Quebec, Canada, and France. Am J Epidemiol 1994;140:805–20.

[27] Ross JA, Spector LG, Robison LL, Olshan AF. Epidemiology of leukemia in children with Down syndrome. Pediatr Blood Cancer 2005;44:8–12.

[28] Wechsler J, Greene M, McDevitt MA, Anastasi J, Karp JE, Le Beau MM, et al. Acquired mutations in GATA1 in the megakaryoblastic leukemia of Down syndrome. Nat Genet 2002;32:148–52.

[29] Alter BP, Giri N, Savage SA, Peters JA, Loud JT, Leathwood L, et al. Malignancies and survival patterns in the National Cancer Institute inherited bone marrow failure syndromes cohort study. Br J Haematol 2010;150:179–88.

[30] Tamary H, Nishri D, Yacobovich J, Zilber R, Dgany O, Krasnov T, et al. Frequency and natural history of inherited bone marrow failure syndromes: the Israeli Inherited Bone Marrow Failure Registry. Haematologica 2010;95:1300–7.

[31] Geddis AE. Congenital amegakaryocytic thrombocytopenia and thrombocytopenia with absent radii. Hematol Oncol Clin North Am 2009;23:321–31.

[32] Owen C, Barnett M, Fitzgibbon J. Familial myelodysplasia and acute myeloid leukaemia – a review. Br J Haematol 2008;140:123–32.

[33] Owen CJ, Toze CL, Koochin A, Forrest DL, Smith CA, Stevens JM, et al. Five new pedigrees with inherited RUNX1 mutations causing familial platelet disorder with propensity to myeloid malignancy. Blood 2008;112:4639–45.

[34] Matheny CJ, Speck ME, Cushing PR, Zhou Y, Corpora T, Regan M, et al. Disease mutations in RUNX1 and RUNX2 create nonfunctional, dominant-negative, or hypomorphic alleles. EMBO J 2007;26:1163–75.

[35] Rheingold SR. Acute myeloid leukemia in a child with hereditary thrombocytopenia. Pediatr Blood Cancer 2007;48:105–7.

[36] Walker LC, Stevens J, Campbell H, Corbett R, Spearing R, Heaton D, et al. A novel inherited mutation of the transcription factor RUNX1 causes thrombocytopenia and may predispose to acute myeloid leukaemia. Br J Haematol 2002;117:878–81.

[37] Lauchle JO, Braun BS, Loh ML, Shannon K. Inherited predispositions and hyperactive Ras in myeloid leukemogenesis. Pediatr Blood Cancer 2006;46:579–85.

[38] de Vries AC, Zwaan CM, van den Heuvel-Eibrink MM. Molecular basis of juvenile myelomonocytic leukemia. Haematologica 2010;95:179–82.

[39] Tartaglia M, Niemeyer CM, Fragale A, Song X, Buechner J, Jung A, et al. Somatic mutations in PTPN11 in juvenile myelomonocytic leukemia, myelodysplastic syndromes and acute myeloid leukemia. Nat Genet 2003;34:148–50.

[40] Loh ML, Vattikuti S, Schubbert S, Reynolds MG, Carlson E, Lieuw KH, et al. Mutations in PTPN11 implicate the SHP-2 phosphatase in leukemogenesis. Blood 2004;103:2325–31.

[41] Shiba N, Kato M, Park MJ, Sanada M, Ito E, Fukushima K, et al. CBL mutations in juvenile myelomonocytic leukemia and pediatric myelodysplastic syndrome. Leukemia 2010;24:1090–2.

[42] Muramatsu H, Makishima H, Jankowska AM, Cazzolli H, O'Keefe C, Yoshida N, et al. Mutations of an E3 ubiquitin ligase c-Cbl but not TET2 mutations are pathogenic in juvenile myelomonocytic leukemia. Blood 2010;115:1969–75.

[43] Niemeyer CM, Kang MW, Shin DH, Furlan I, Erlacher M, Bunin NJ, et al. Germline CBL mutations cause developmental abnormalities and predispose to juvenile myelomonocytic leukemia. Nat Genet 2010;42:794–800.

[44] Horwitz M. The genetics of familial leukemia. Leukemia 1997;11:1347–59.

[45] Poppe B, Van Limbergen H, Van Roy N, Vandecruys E, De Paepe A, Benoit Y, et al. Chromosomal aberrations in Bloom syndrome patients with myeloid malignancies. Cancer Genet Cytogenet 2001;128:39–42.

[46] Lynch HT, Weisenburger DD, Quinn-Laquer B, Snyder CL, Lynch JF, Lipkin SM, et al. Family with acute myelocytic leukemia, breast, ovarian, and gastrointestinal cancer. Cancer Genet Cytogenet 2002;137:8–14.

[47] Alter BP. Bone marrow failure syndromes in children. Pediatr Clin North Am 2002;49:973–88.

[48] Greaves MF, Maia AT, Wiemels JL, Ford AM. Leukemia in twins: lessons in natural history. Blood 2003;102: 2321–33.

[49] Greaves M. Pre-natal origins of childhood leukemia. Rev Clin Exp Hematol 2003;7:233–45.

[50] Wiemels J. Chromosomal translocations in childhood leukemia: natural history, mechanisms, and epidemiology. J Natl Cancer Inst Monogr 2008;2008(39):87–90.

[51] Markkanen A, Ruutu T, Rasi V, Franssila K, Knuutila S, de la Chapelle A. Constitutional translocation t(3;6)(p14;p11) in a family with hematologic malignancies. Cancer Genet Cytogenet 1987;25:87–95.

[52] Minelli A, Maserati E, Giudici G, Tosi S, Olivieri C, Bonvini L, et al. Familial partial monosomy 7 and myelodysplasia: different parental origin of the monosomy 7 suggests action of a mutator gene. Cancer Genet Cytogenet 2001;124:147–51.

[53] Riccardi VM, Humbert JR, Peakman D. Acute leukemia associated with trisomy 8 mosaicism and a familial translocation 46, XY,t(7;20)(p13;p12). Am J Med Genet 1978;2:15–21.

[54] Vardiman JW, Harris NL, Brunning RD. The World Health Organization (WHO) classification of the myeloid neoplasms. Blood 2002;100:2292–302.

[55] Brown LM, Daeschner CD, Timms J, Crow W. Granulocytic sarcoma in childhood acute myelogenous leukemia. Pediatr Neurol 1989;5:173–8.

[56] Rubnitz JE, Raimondi SC, Halbert AR, Tong X, Srivastava DK, Razzouk BI, et al. Characteristics and outcome of t(8;21)-positive childhood acute myeloid leukemia: a single institution's experience. Leukemia 2002;16:2072–7.

[57] Hurwitz CA, Raimondi SC, Head D, Krance R, Mirro J, Kalwinsky DK, et al. Distinctive immunophenotypic features of t (8;21)(q22;q22) acute myeloblastic leukemia in children. Blood 1992;80:3182–8.

[58] Byrd JC, Dodge RK, Carroll A, Baer MR, Edwards C, Stamberg J, et al. Patients with t(8;21)(q22;q22) and acute myeloid leukemia have superior failure-free and overall survival when repetitive cycles of high-dose cytarabine are administered. J Clin Oncol 1999;17:3767–75.

[59] Lie SO, Abrahamsson J, Clausen N, Forestier E, Hasle H, Hovi L, et al. Long-term results in children with AML: NOPHO-AML study group – report of three consecutive trials. Leukemia 2005;19:2090–100.

[60] Lange BJ, Smith FO, Feusner J, Barnard DR, Dinndorf P, Feig S, et al. Outcomes in CCG-2961, a children's oncology group phase 3 trial for untreated pediatric acute myeloid leukemia: a report from the children's oncology group. Blood 2008;111:1044–53.

[61] Bally C, Fadlallah J, Leverger G, Bertrand Y, Robert A, Baruchel A, et al. Outcome of acute promyelocytic leukemia (APL) in children and adolescents: an analysis in two consecutive trials of the European APL group. J Clin Oncol 2012;30:1641–6.

[62] Licht JD, Chomienne C, Goy A, Chen A, Scott AA, Head DR, et al. Clinical and molecular characterization of a rare syndrome of acute promyelocytic leukemia associated with translocation (11;17). Blood 1995;85:1083–94.

[63] Balgobind BV, van den Heuvel-Eibrink MM, Menezes RX, Reinhardt D, Hollink IH, Arentsen-Peters ST, et al. Evaluation of gene expression signatures predictive for cytogenetic and molecular subtypes of pediatric acute myeloid leukemia. Haematologica 2011;96(2):221–30.

[64] Hollink IH, van den Heuvel-Eibrink MM, Zimmermann M, Balgobind BV, Arentsen-Peters ST, Alders M, et al. Clinical relevance of Wilms tumor 1 gene mutations in childhood acute myeloid leukemia. Blood 2009;113:5951–60.

[65] Carlson KM, Vignon C, Bohlander S, Martinez-Climent JA, Le Beau MM, Rowley JD. Identification and molecular characterization of CALM/AF10fusion products in T cell acute lymphoblastic leukemia and acute myeloid leukemia. Leukemia 2000; 14:100–4.

[66] Dreyling MH, Schrader K, Fonatsch C, Schlegelberger B, Haase D, Schoch C, et al. MLL and CALM are fused to AF10 in morphologically distinct subsets of acute leukemia with translocation t(10;11): both rearrangements are associated with a poor prognosis. Blood 1998;91:4662–7.

[67] Oliansky DM, Rizzo JD, Aplan PD, Arceci RJ, Leone L, Ravindranath Y, et al. The role of cytotoxic therapy with hematopoietic stem cell transplantation in the therapy of acute myeloid leukemia in children: an evidence-based review. Biol Blood Marrow Transplant 2007;13:1–25.

[68] Appelbaum FR, Rosenblum D, Arceci RJ, Carroll WL, Breitfeld PP, Forman SJ, et al. End points to establish the efficacy of new agents in the treatment of acute leukemia. Blood 2007;109:1810–6.

[69] Ma Z, Morris SW, Valentine V, Li M, Herbrick JA, Cui X, et al. Fusion of two novel genes, RBM15 and MKL1, in the t(1;22) (p13;q13) of acute megakaryoblastic leukemia. Nat Genet 2001;28:220–1.

[70] Mercher T, Coniat MB, Monni R, Mauchauffe M, Nguyen Khac F, Gressin L, et al. Involvement of a human gene related to the Drosophila spen gene in the recurrent t(1;22) translocation of acute megakaryocytic leukemia. Proc Natl Acad Sci USA 2001;98:5776–9.

[71] Bernstein J, Dastugue N, Haas OA, Harbott J, Heerema NA, Huret JL, et al. Nineteen cases of the t(1;22)(p13;q13) acute megakaryblastic leukaemia of infants/children and a review of 39 cases: report from a t(1;22) study group. Leukemia 2000;14:216–8.

[72] Duchayne E, Fenneteau O, Pages MP, Sainty D, Arnoule C, Dastuque N, et al. Acute megakaryoblastic leukaemia: a national clinical and biological study of 53 adult and childhood cases by the Groupe Francais d'Hematologie Cellulaire (GFHC). Leuk Lymphoma 2003;44:49–58.

[73] Inaba H, Londero M, Maurer SH, Onciu M, Ge Y, Taub JW, et al. Acute megakaryoblastic leukemia without GATA1 mutation after transient myeloproliferative disorder in an infant without Down syndrome. J Clin Oncol 2011;29:e230–3.

[74] Grimwade D, Walker H, Oliver F, Wheatley K, Harrison C, Harrison G, et al. The importance of diagnostic cytogenetics on outcome in AML: analysis of 1,612 patients entered into the MRC AML 10 trial. The medical research council adult and children's Leukaemia working parties. Blood 1998;92:2322–33.

[75] Lugthart S, Groschel S, Beverloo HB, Kayser S, Valk PJ, Zelderen-Bhola SL, et al. Clinical, molecular, and prognostic significance of WHO type inv(3)(q21q26.2)/t(3;3)(q21;q26.2) and various other 3q abnormalities in acute myeloid leukemia. J Clin Oncol 2010;28:3890–8.

[76] Stevens RF, Hann IM, Wheatley K, Gray RG. Marked improvements in outcome with chemotherapy alone in paediatric acute myeloid leukemia: results of the United Kingdom Medical Research Council's 10th AML trial. MRC Childhood Leukaemia working party. Br J Haematol 1998;101:130–40.

[77] Raghavan M, Lillington DM, Skoulakis S, Debernardi S, Chaplin T, Foot NJ, et al. Genome-wide single nucleotide polymorphism analysis reveals frequent partial uniparental disomy due to somatic recombination in acute myeloid leukemias. Cancer Res 2005;65:375–8.

[78] Raghavan M, Smith LL, Lillington DM, Chaplin T, Kakkas I, Molloy G, et al. Segmental uniparental disomy is a commonly acquired genetic event in relapsed acute myeloid leukemia. Blood 2008;112:814–21.

[79] Gupta M, Raghavan M, Gale RE, Chelala C, Allen C, Molloy G, et al. Novel regions of acquired uniparental disomy discovered in acute myeloid leukemia. Genes Chromosomes Cancer 2008;47:729–39.

[80] Walter MJ, Payton JE, Ries RE, Shannon WD, Deshmukh H, Zhao Y, et al. Acquired copy number alterations in adult acute myeloid leukemia genomes. Proc Natl `Acad Sci USA 2009;106:12950–5.

[81] Radtke I, Mullighan CG, Ishii M, Su X, Cheng J, Ma J, et al. Genomic analysis reveals few genetic alterations in pediatric acute myeloid leukemia. Proc Natl Acad Sci USA 2009;106:12944–9.

[82] Nakao M, Yokota S, Iwai T, Kaneko H, Horiike S, Kashima K, et al. Internal tandem duplication of the flt3 gene found in acute myeloid leukemia. Leukemia 1996;10:1911–8.

[83] Yamamoto Y, Kiyoi H, Nakano Y, Suzuki R, Kodera Y, Miyawaki S, et al. Activating mutation of D835 within the activation loop of FLT3 in human hematologic malignancies. Blood 2001;97:2434–9.

[84] Thiede C, Steudel C, Mohr B, Schaich M, Schakel U, Platzbecker U, et al. Analysis of FLT3-activating mutations in 979 patients with acute myelogenous leukemia: association with FAB subtypes and identification of subgroups with poor prognosis. Blood 2002;99:4326–35.

[85] Choudhary C, Schwable J, Brandts C, Tickenbrock L, Sargin B, Kindler T, et al. AML-associated Flt3 kinase domain mutations show signal transduction differences compared with Flt3 ITD mutations. Blood 2005;106:265–73.

[86] Grundler R, Thiede C, Miething C, Steudel C, Peschel C, Duyster J. Sensitivity toward tyrosine kinase inhibitors varies between different activating mutations of the FLT3 receptor. Blood 2003;102:646–51.

[87] Vempati S, Reindl C, Wolf U, Kern R, Petropoulos K, Naidu VM, et al. Transformation by oncogenic mutants and ligand-dependent activation of FLT3 wild-type requires the tyrosine residues 589 and 591. Clin Cancer Res 2008;14:4437–45.

[88] Rocnik JL, Okabe R, Yu JC, Lee BH, Giese N, Schenkein DP, et al. Roles of tyrosine 589 and 591 in STAT5 activation and transformation mediated by FLT3-ITD. Blood 2006;108:1339–45.

[89] Meshinchi S, Alonzo TA, Stirewalt DL, Zwaan M, Zimmerman M, Reinhardt D, et al. Clinical implications of FLT3 mutations in pediatric AML. Blood 2006;108:3654–61.

[90] Gamis AS, Alonzo TA, Gerbing RB, Aplenc R, Sung L, Meshinchi S, et al. Remission rates in childhood acute myeloid leukemia (AML) utilizing a dose-intensive induction regimen with or without gemtuzumab ozogamicin (GO): initial results from the children's oncology group Phase III Trial, AAML0531. Blood (ASH Annual Meeting Abstracts) 2010;116(21):182.

[91] Meshinchi S, Arceci RJ, Sanders JE, et al. Role of allogeneic stem cell transplantation in FLT3/ITD-positive AML. Blood 2006;108:400–1. Author reply 401.

[92] Bornhauser M, Illmer T, Schaich M, Soucek S, Ehninger G, Thiede C. Improved outcome after stem-cell transplantation in FLT3/ITD-positive AML. Blood 2007;109:2264–5. Author reply 2265.

[93] Ravandi F, Cortes JE, Jones D, Faderl S, Garcia-Manero G, Konopleva MY, et al. Phase I/II study of combination therapy with sorafenib, idarubicin, and cytarabine in younger patients with acute myeloid leukemia. J Clin Oncol 2010;28:1856–62.

[94] Kelly LM, Yu JC, Boulton CL, Apatira M, Li J, Sullivan CM, et al. CT53518, a novel selective FLT3 antagonist for the treatment of acute myelogenous leukemia (AML). Cancer Cell 2002;1:421–32.

[95] Weisberg E, Boulton C, Kelly LM, Manley P, Fabbro D, Meyer T, et al. Inhibition of mutant FLT3 receptors in leukemia cells by the small molecule tyrosine kinase inhibitor PKC412. Cancer Cell 2002;1:433–43.

[96] Spiekermann K, Dirschinger RJ, Schwab R, Bagrintseva K, Faber F, Buske C, et al. The protein tyrosine kinase inhibitor SU5614 inhibits FLT3 and induces growth arrest and apoptosis in AML-derived cell lines expressing a constitutively activated FLT3. Blood 2003;101:1494–504.

[97] Renneville A, Roumier C, Biggio V, Nibourel O, Boissel N, Fenaux P, et al. Cooperating gene mutations in acute myeloid leukemia: a review of the literature. Leukemia 2008;22:915–31.

[98] Ning ZQ, Li J, McGuinness M, Arceci RJ. STAT3 activation is required for Asp(816) mutant c-Kit induced tumorigenicity. Oncogene 2001;20:4528–36.

[99] Ning ZQ, Li J, Arceci RJ. Signal transducer and activator of transcription 3 activation is required for Asp(816) mutant c-Kit-mediated cytokine-independent survival and proliferation in human leukemia cells. Blood 2001;97:3559–67.

[100] Care RS, Valk PJ, Goodeve AC, Abu-Duhier FM, Geertsma-Kleinekoort WM, Wilson GA, et al. Incidence and prognosis of c-KIT and FLT3 mutations in core binding factor (CBF) acute myeloid leukaemias. Br J Haematol 2003;121:775–7.

[101] Nanri T, Matsuno N, Kawakita T, Suzushima H, Kawano F, Mitsuya H, et al. Mutations in the receptor tyrosine kinase

pathway are associated with clinical outcome in patients with acute myeloblastic leukemia harboring t(8;21)(q22;q22). Leukemia 2005;19:1361–6.

[102] Schnittger S, Kohl TM, Haferlach T, Kern W, Hiddemann W, Spiekermann K, et al. KIT-D816 mutations in AML1-ETO-positive AML are associated with impaired event-free and overall survival. Blood 2006;107:1791–9.

[103] Cairoli R, Beghini A, Grillo G, Nadali G, Elice F, Ripamonti CB, et al. Prognostic impact of c-KIT mutations in core binding factor leukemias: an Italian retrospective study. Blood 2006;107:3463–8.

[104] Paschka P, Marcucci G, Ruppert AS, Mrozek K, Chen H, Kittles RA, et al. Adverse prognostic significance of KIT mutations in adult acute myeloid leukemia with inv(16) and t(8;21): a Cancer and Leukemia Group B Study. J Clin Oncol 2006; 24:3904–11.

[105] Pollard JA, Alonzo TA, Gerbing RB, Ho PA, Zeng R, Ravindranath Y, et al. Prevalence and prognostic significance of KIT mutations in pediatric patients with core binding factor AML enrolled on serial pediatric cooperative trials for de novo AML. Blood 2010;115:2372–9.

[106] Dohner K, Schlenk RF, Habdank M, Scholl C, Rucker FG, Corbacioglu A, et al. Mutant nucleophosmin (NPM1) predicts favorable prognosis in younger adults with acute myeloid leukemia and normal cytogenetics: interaction with other gene mutations. Blood 2005;106:3740–6.

[107] Gale RE, Hills R, Pizzey AR, Kottaridis PD, Swirsky D, Gilkes AF, et al. Relationship between FLT3 mutation status, biologic characteristics, and response to targeted therapy in acute promyelocytic leukemia. Blood 2005;106:3768–76.

[108] Hollink IH, Zwaan CM, Zimmermann M, Arentsen-Peters TC, Pieters R, Cloos J, et al. Favorable prognostic impact of NPM1 gene mutations in childhood acute myeloid leukemia, with emphasis on cytogenetically normal AML. Leukemia 2009;23:262–70.

[109] Brown P, McIntyre E, Rau R, Meshinchi S, Lacayo N, Dahl G, et al. The incidence and clinical significance of nucleophosmin mutations in childhood AML. Blood 2007;110:979–85.

[110] Ho PA, Alonzo TA, Gerbing RB, Pollard J, Stirewalt DL, Hurwitz C, et al. Prevalence and prognostic implications of CEBPA mutations in pediatric acute myeloid leukemia (AML): a report from the Children's Oncology Group. Blood 2009;113:6558–66.

[111] Kirstetter P, Schuster MB, Bereshchenko O, Moore S, Dvinge H, Kurz E, et al. Modeling of C/EBPalpha mutant acute myeloid leukemia reveals a common expression signature of committed myeloid leukemia-initiating cells. Cancer Cell 2008;13:299–310.

[112] Radich JP, Kopecky KJ, Willman CL, Weick J, Head D, Appelbaum F, et al. N-ras mutations in adult de novo acute myelogenous leukemia: prevalence and clinical significance. Blood 1990;76:801–7.

[113] Berman JN, Gerbing RB, Alonzo TA, Ho PA, Miller K, Hurwitz C, et al. Prevalence and clinical implications of NRAS mutations in childhood AML: a report from the Children's Oncology Group. Leukemia 2011;25:1039–42.

[114] Byrne JL, Marshall CJ. The molecular pathophysiology of myeloid leukaemias: ras revisited. Br J Haematol 1998;100:256–64.

[115] Bergmann L, Maurer U, Weidmann E. Wilms tumor gene expression in acute myeloid leukemias. Leuk Lymphoma 1997;25:435–43.

[116] Mrozek K, Marcucci G, Paschka P, Whitman SP, Bloomfield CD. Clinical relevance of mutations and gene-expression changes in adult acute myeloid leukemia with normal cytogenetics: are we ready for a prognostically prioritized molecular classification?. Blood 2007;109:431–48.

[117] Lapillonne H, Renneville A, Auvrignon A, Flamant C, Blaise A, Perot C, et al. High WT1 expression after induction therapy predicts high risk of relapse and death in pediatric acute myeloid leukemia. J Clin Oncol 2006;24:1507–15.

[118] Bergmann L, Miething C, Maurer U, Brieger J, Karakas T, Weidmann E, et al. High levels of Wilms' tumor gene (wt1) mRNA in acute myeloid leukemias are associated with a worse long-term outcome. Blood 1997;90:1217–25.

[119] Noronha SA, Farrar JE, Alonzo TA, Gerbing RB, Lacayo NJ, Dahl GV, et al. WT1 expression at diagnosis does not predict survival in pediatric AML: a report from the Children's Oncology Group. Pediatr Blood Cancer 2009;53:1136–9.

[120] Groet J, McElwaine S, Spinelli M, Rinaldi A, Burtscher I, Mulligan C, et al. Acquired mutations in GATA1 in neonates with down's syndrome with transient myeloid disorder. Lancet 2003;361:1617–20.

[121] Hitzler J, Zipursky A. GATA 1 mutations as clonal markers of minimal residual disease in acute megakaryoblastic leukemia of Down syndrome – a new tool with significant potential applications. Leuk Res 2005;29:1239–40.

[122] Hitzler JK, Cheung J, Li Y, Scherer SW, Zipursky A. GATA1 mutations in transient leukemia and acute megakaryoblastic leukemia of Down syndrome. Blood 2003;101:4301–4.

[123] Rainis L, Bercovich D, Strehl S, Teigler-Schlegel A, Stark B, Trka J, et al. Mutations in exon 2 of GATA1 are early events in megakaryocytic malignancies associated with trisomy 21. Blood 2003;102:981–6.

[124] Ge Y, Jensen TL, Stout ML, Flatley RM, Grohar PJ, Ravindranath Y, et al. The role of cytidine deaminase and GATA1 mutations in the increased cytosine arabinoside sensitivity of Down syndrome myeloblasts and leukemia cell lines. Cancer Res 2004;64:728–35.

[125] Shih AH, Abdel-Wahab O, Patel JP, Levine RL. The role of mutations in epigenetic regulators in myeloid malignancies. Nat Rev Cancer 2012;12:599–612.

[126] Abbas S, Lugthart S, Kavelaars FG, Schelen A, Koenders JE, Zeilemaker A, et al. Acquired mutations in the genes encoding IDH1 and IDH2 both are recurrent aberrations in acute myeloid leukemia: prevalence and prognostic value. Blood 2010;116:2122–6.

[127] Marcucci G, Maharry K, Wu YZ, Radmacher MD, Mrozek K, Margeson D, et al. IDH1 and IDH2 gene mutations identify novel molecular subsets within de novo cytogenetically normal acute myeloid leukemia: a Cancer and Leukemia Group B study. J Clin Oncol 2010;28:2348–55.

[128] Paschka P, Schlenk RF, Gaidzik VI, Habdank M, Kronke J, Bullinger L, et al. IDH1 and IDH2 mutations are frequent genetic alterations in acute myeloid leukemia and confer adverse prognosis in cytogenetically normal acute myeloid leukemia with NPM1 mutation without FLT3 internal tandem duplication. J Clin Oncol 2010;28:3636–43.

[129] Ley TJ, Mardis ER, Ding L, Fulton B, McLellan MD, Chen K, et al. DNA sequencing of a cytogenetically normal acute myeloid leukaemia genome. Nature 2008;456:66–72.

[130] Ley TJ, Ding L, Walter MJ, McLellan MD, Lamprecht T, Larson DE, et al. DNMT3A mutations in acute myeloid leukemia. N Engl J Med 2010;363(25):2424–33.

[131] Mardis ER, Ding L, Dooling DJ, Larson DE, McLellan MD, Chen K, et al. Recurring mutations found by sequencing an acute myeloid leukemia genome. N Engl J Med 2009;361:1058–66.

[132] Ho PA, Kutny MA, Alonzo TA, Gerbing RB, Joaquin J, Raimondi SC, et al. Leukemic mutations in the methylation-associated genes DNMT3A and IDH2 are rare events in pediatric AML: a report from the Children's Oncology Group. Pediatr Blood Cancer 2011;57:204–9.

[133] Meshinchi S, Hunger SP, Aplenc R, Adamson PC, Jessup JM. Lessons learned from the investigational device exemption review of Children's Oncology Group trial AAML1031. Clin Cancer Res 2012;18:1547–54.

[134] The Cancer Genome Atlas Research Network. Genomic and epigenomic landscapes of adult de novo acute myeloid leukemia. N Engl J Med 2013;368(22):2059–74.

[135] Patel JP, Gonen M, Figueroa ME, Fernandez H, Sun Z, Racevskis J, et al. Prognostic relevance of integrated genetic profiling in acute myeloid leukemia. N Engl J Med 2012;366:1079–89.

[136] Pui CH, Carroll WL, Meshinchi S, Arceci RJ. Biology, risk stratification, and therapy of pediatric acute leukemias: an update. J Clin Oncol 2011;29(5):551–65.

[137] Stephens PJ, Greenman CD, Fu B, Yang F, Bignell GR, Mudie LJ, et al. Massive genomic rearrangement acquired in a single catastrophic event during cancer development. Cell 2011;144:27–40.

[138] Ji H, Ehrlich LI, Seita J, Murakami P, Doi A, Lindau P, et al. Comprehensive methylome map of lineage commitment from haematopoietic progenitors. Nature 2010;467(7313):338–42.

[139] Rice KL, Hormaeche I, Licht JD. Epigenetic regulation of normal and malignant hematopoiesis. Oncogene 2007;26:6697–714.

[140] Figueroa ME, Lugthart S, Li Y, Erpelinck-Verschueren C, Deng X, Christos PJ, et al. DNA methylation signatures identify biologically distinct subtypes in acute myeloid leukemia. Cancer Cell 2010;17:13–27.

[141] Figueroa ME, Reimers M, Thompson RF, Ye K, Li Y, Selzer RR, et al. An integrative genomic and epigenomic approach for the study of transcriptional regulation. PLoS One 2008;3:e1882.

[142] Figueroa ME, Wouters BJ, Skrabanek L, Glass J, Li Y, Erpelinck-Verschueren CA, et al. Genome-wide epigenetic analysis delineates a biologically distinct immature acute leukemia with myeloid/T-lymphoid features. Blood 2009;113:2795–804.

[143] Armand P, Kim HT, Zhang MJ, Perez WS, Dal Cin PS, Klumpp TR, et al. Classifying cytogenetics in patients with acute myelogenous leukemia in complete remission undergoing allogeneic transplantation: a Center for International Blood and Marrow Transplant Research study. Biol Blood Marrow Transplant 2012;18:280–8.

[144] Metzeler KH, Maharry K, Radmacher MD, Mrozek K, Margeson D, Becker H, et al. TET2 mutations improve the new European LeukemiaNet risk classification of acute myeloid leukemia: a Cancer and Leukemia Group B study. J Clin Oncol 2011;29:1373–81.

[145] Kuhn MW, Radtke I, Bullinger L, Goorha S, Cheng J, Edelmann J, et al. High-resolution genomic profiling of adult and pediatric core-binding factor acute myeloid leukemia reveals new recurrent genomic alterations. Blood 2012;119:e67–75.

[146] Langemeijer SM, Jansen JH, Hooijer J, van Hoogen P, Stevens-Linders E, Massop M, et al. TET2 mutations in childhood leukemia. Leukemia 2011;25:189–92.

[147] Gross S, Cairns RA, Minden MD, Driggers EM, Bittinger MA, Jang HG, et al. Cancer-associated metabolite 2-hydroxyglutarate accumulates in acute myelogenous leukemia with isocitrate dehydrogenase 1 and 2 mutations. J Exp Med 2010;207:339–44.

[148] Lu C, Ward PS, Kapoor GS, Rohle D, Turcan S, Abdel-Wahab O, et al. IDH mutation impairs histone demethylation and results in a block to cell differentiation. Nature 2012;483:474–8.

[149] Green CL, Evans CM, Hills RK, Burnett AK, Linch DC, Gale RE. The prognostic significance of IDH1 mutations in younger adult patients with acute myeloid leukemia is dependent on FLT3/ITD status. Blood 2010;116:2779–82.

[150] Damm F, Heuser M, Morgan M, Wagner K, Gorlich K, Grosshennig A, et al. Integrative prognostic risk score in acute myeloid leukemia with normal karyotype. Blood 2011;117:4561–8.

[151] Ho PA, Alonzo TA, Kopecky KJ, Miller KL, Kuhn J, Zeng R, et al. Molecular alterations of the IDH1 gene in AML: a Children's Oncology Group and Southwest Oncology Group study. Leukemia 2010;24:909–13.

[152] Ho PA, Kopecky KJ, Alonzo TA, Gerbing RB, Miller KL, Kuhn J, et al. Prognostic implications of the IDH1 synonymous SNP rs11554137 in pediatric and adult AML: a report from the Children's Oncology Group and SWOG. Blood 2011;118:4561–6.

[153] Marcucci G, Metzeler KH, Schwind S, Becker H, Maharry K, Mrozek K, et al. Age-related prognostic impact of different types of DNMT3A mutations in adults with primary cytogenetically normal acute myeloid leukemia. J Clin Oncol 2012;30:742–50.

[154] Renneville A, Boissel N, Nibourel O, Berthon C, Helevaut N, Gardin C, et al. Prognostic significance of DNA methyltransferase 3A mutations in cytogenetically normal acute myeloid leukemia: a study by the Acute Leukemia French Association. Leukemia 2012;26:1247–54.

[155] Shah MY, Licht JD. DNMT3A mutations in acute myeloid leukemia. Nat Genet 2011;43:289–90.

[156] Hollink IH, Feng Q, Danen-van Oorschot AA, Arentsen-Peters ST, Verboon LJ, Zhang P, et al. Low frequency of DNMT3A mutations in pediatric AML, and the identification of the OCI-AML3 cell line as an in vitro model. Leukemia 2012;26:371–3.

[157] Ernst T, Chase AJ, Score J, Hildago-Curtis CE, Bryant C, Jones AV, et al. Inactivating mutations of the histone methyltransferase gene EZH2 in myeloid disorders. Nat Genet 2010;42:722–6.

[158] Sievers EL, Loken MR. Detection of minimal residual disease in acute myelogenous leukemia. J Pediatr Hematol Oncol 1995;17:123–33.

[159] den Boer ML, Pieters R, Kazemier KM, Rottier MM, Zwaan CM, Kaspers GJ, et al. Relationship between major vault protein/lung resistance protein, multidrug resistance-associated protein, P-glycoprotein expression, and drug resistance in childhood leukemia. Blood 1998;91:2092–8.

[160] Becton D, Dahl GV, Ravindranath Y, Chang MN, Behm FG, Raimondi SC, et al. Randomized use of cyclosporin A (CsA) to modulate P-glycoprotein in children with AML in remission: Pediatric Oncology Group Study 9421. Blood 2006;107:1315–24.

[161] Verstovsek S, Manshouri T, Smith FO, Giles FJ, Cortes J, Estey E, et al. Telomerase activity is prognostic in pediatric patients with acute myeloid leukemia: comparison with adult acute myeloid leukemia. Cancer 2003;97:2212–7.

[162] Baldus CD, Thiede C, Soucek S, Bloomfield CD, Thiel E, Ehninger G. BAALC expression and FLT3 internal tandem duplication mutations in acute myeloid leukemia patients with normal cytogenetics: prognostic implications. J Clin Oncol 2006;24:790–7.

[163] Tse W, Meshinchi S, Alonzo TA, Stirewalt DL, Gerbing RB, Woods WG, et al. Elevated expression of the AF1q gene, an MLL fusion partner, is an independent adverse prognostic factor in pediatric acute myeloid leukemia. Blood 2004;104:3058–63.

[164] Lacayo NJ, Meshinchi S, Kinnunen P, Yu R, Wang Y, Stuber CM, et al. Gene expression profiles at diagnosis in de novo childhood AML patients identify FLT3 mutations with good clinical outcomes. Blood 2004;104:2646–54.

[165] Wouters BJ, Lowenberg B, Delwel R. A decade of genome-wide gene expression profiling in acute myeloid leukemia: flashback and prospects. Blood 2009;113:291–8.

[166] Wilson CS, Davidson GS, Martin SB, Andries E, Potter J, Harvey R, et al. Gene expression profiling of adult acute myeloid leukemia identifies novel biologic clusters for risk classification and outcome prediction. Blood 2006;108:685–96.

[167] Wang Y, Li Z, He C, Wang D, Yuan X, Chen J, et al. MicroRNAs expression signatures are associated with lineage and survival in acute leukemias. Blood Cells Mol Dis 2010;44:191–7.

[168] Balgobind BV, Raimondi SC, Harbott J, Zimmermann M, Alonzo TA, Auvrignon A, et al. Novel prognostic subgroups in childhood 11q23/MLL-rearranged acute myeloid leukemia: results of an international retrospective study. Blood 2009;114:2489–96.

[169] Valk PJ, Verhaak RG, Beijen MA, Erpelinck CA, Barjesteh van Waalwijk van Doorn-Khosrovani S, Boer JM, et al. Prognostically useful gene-expression profiles in acute myeloid leukemia. N Engl J Med 2004;350:1617–28.

[170] Yagi T, Morimoto A, Eguchi M, Hibi S, Sako M, Ishii E, et al. Identification of a gene expression signature associated with pediatric AML prognosis. Blood 2003;102:1849–56.

[171] Wouters BJ, Jorda MA, Keeshan K, Louwers I, Erpelinck-Verschueren CA, Tielemans D, et al. Distinct gene expression profiles of acute myeloid/T-lymphoid leukemia with silenced CEBPA and mutations in NOTCH1. Blood 2007;110:3706–14.

[172] Wouters BJ, Koss C, Delwel R. Gene expression profiling for improved dissection of acute leukemia: a recently identified immature myeloid/T-lymphoid subgroup as an example. Blood Cells Mol Dis 2008;40:395–400.

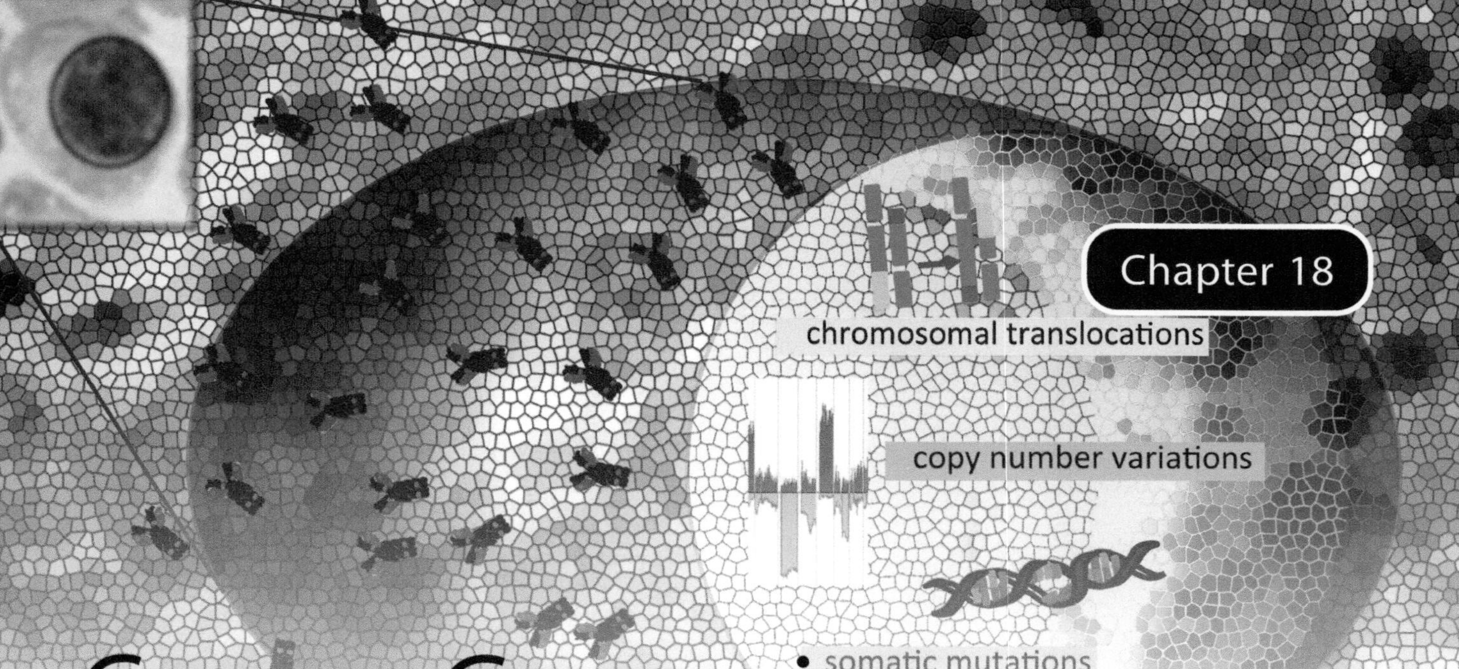

Cancer Genomics

Genomics in Multiple Myeloma: From Conventional Cytogenetics to Novel Whole-Genome Sequencing Approaches

Francesca Cottini[1]
Kenneth C. Anderson[1]
Giovanni Tonon[2]

[1]*Department of Medical Oncology, Jerome Lipper Multiple Myeloma Center, Dana-Farber Cancer Institute and Department of Medicine, Harvard Medical School, Boston, MA, USA*

[2]*Division of Molecular Oncology, Functional Genomics of Cancer Unit, San Raffaele Scientific Institute, Milan, Italy*

Contents

Cancer Genomics. DOI: http://dx.doi.org/10.1016/B978-0-12-396967-5.00018-9

Key Concepts

- Multiple myeloma (MM) is a B-cell disorder characterized by the slow proliferation of clonal aberrant plasma cells in the bone marrow. It is clinically associated with the presence of a monoclonal protein in the blood and in the urine, together with signs of end-organ damage such as bone osteolytic lesions, renal failure, and bone marrow infiltration (leukopenia, anemia and thrombocytopenia)
- Multiple myeloma derives from a pre-malignant lesion called monoclonal gammopathy of undetermined significance (MGUS), which is present in 1% of adults and progresses to malignant multiple myeloma at a rate of 1% per year
- Multiple myeloma karyotypes are complex with several numerical and structural aberrations, making MM genomes more similar to those found in epithelial tumors and the blast phase of chronic myelogenous leukemia than to those in other hematopoietic malignancies
- Based on the karyotype, MM patients can be divided into two subgroups: hyperdiploid cases with more than 46 but less than 76 chromosomes (34–50% of MM), and non-hyperdiploid cases which include patients with a hypodiploid (up to 44–45 chromosomes), pseudodiploid (44/45 or 46/47 chromosomes with gains or losses), or near-tetraploid karyotype. The hyperdiploid karyotype is associated with a better prognosis since patients belonging to this group have a lower incidence of primary immunoglobulin heavy chain (IgH) translocations and chromosome 13 deletions (see below)
- Primary immunoglobulin heavy chain translocations occur in up to 50% of cases of MGUS and MM, juxtaposing known oncogenes such as *CCND1* or *CCND3* (t(11;14)), fibroblast growth factor receptor 3 (*FGFR3*) combined with the nuclear protein *MMSET* (t(4;14)), and *MAF* (t(14;16) or t(14;20)) to the immunoglobulin transcriptional regulatory regions
- Secondary translocations happen without involvement of an immunoglobulin locus during late stage disease, and contribute to progression by dysregulating other oncogenes, such as *MYC*
- MM genomes are characterized by numerous and recurrent copy number variations (CNVs), resulting in genomic gains or losses. The most frequent numerical abnormalities are deletion of chromosome 13, deletion of chromosome 17p13, and amplification of chromosome 1q21, all associated with a worse patient outcome
- Common reported mutations in MM include genes involved in protein translation and RNA processing, histone methylation, NF-κB signaling, and MAP kinase pathway. Activating mutations in *KRAS* have been proposed as molecular markers of the transition from MGUS to myeloma. In contrast, *TP53* mutations are rare at diagnosis, while associated with disease progression and evolution to extramedullary myeloma or secondary plasma cell leukemia (sPCL)
- Pre-malignant MGUS cells share several genomic alterations with active MM including primary translocations and karyotype abnormalities. However, MM cells have a higher incidence of monosomy 13, show activating mutations of *KRAS,* and also have greater myc mRNA levels to promote proliferation
- Plasma cell leukemia (PCL) occurs when malignant plasma cells become independent from bone marrow stromal cell support and move to the bloodstream compartment. PCL can arise *de novo* or evolve from MM. *TP53* inactivation, p14ARF promoter methylation and *PTEN* loss are crucial for extramedullary tumor expansion and hence have a putative role in the development of PCL from MM

INTRODUCTION

Multiple myeloma (MM) is characterized by the clonal proliferation and accumulation of malignant plasma cells in the bone marrow (BM), capable of producing immunoglobulins or immunoglobulin light chains. Myeloma has an incidence of 20 000 cases per year in the USA, which increases with age. The median age at diagnosis is 70 years and males are more commonly affected than females. It is often preceded by a condition called monoclonal gammopathy of undetermined significance (MGUS), vastly more common than myeloma, occurring in 1% of the population older than 50 years and in up to 10% of individuals older than 75 years. MM can progress to an aggressive stage defined as plasma cell leukemia (PCL), with circulating peripheral blood malignant plasma cells which are less dependent on the bone marrow milieu [1,2]. The BM microenvironment is indeed essential for the survival of intramedullary MM cells. In particular, direct binding to bone marrow stromal cells (BMSCs) and extracellular matrix (ECM) via cell-surface adhesion molecules allows MM cells to survive, migrate, and develop resistance to treatment. Moreover, BMSCs release various cytokines such as interleukin 6 (IL-6), insulin-like growth factor 1 (IGF-1), and vascular endothelial growth factor (VEGF), which further enhance MM growth [1]. Clinically, MM is defined by the presence of more than 10% clonal plasma cells in the BM, together with a serum or urine monoclonal protein (M protein) and evidence of end-organ damage (bone lesions, hypercalcemia, renal failure and anemia), due to the underlying plasma cell disorder. Bone pain is the most commonly reported symptom, often associated with a concomitant pathological fracture or lytic lesion due to an imbalance between the enhanced activity of the osteoclasts and the reduction in number and function of the osteoblasts. Hypercalcemia can result from mobilization of calcium from bones, causing lethargy, weakness, depression, and confusion. Renal failure occurs in nearly 25% of myeloma patients. In particular, the excess of light chains overloads tubular cells, leading to proteinuria (usually just light chains – Bence Jones proteins), glycosuria, aminoaciduria and isosthenuria (type 2 proximal renal tubular acidosis). MM patients can also present anemia, thrombocytopenia and leukopenia as a result of bone marrow failure due to the replacement of normal marrow by the expanding tumor cells, which also release factors capable of inhibiting hematopoiesis. MM patients also have a higher susceptibility to bacterial infections. Indeed, despite the presence of an M spike, hypogammaglobulinemia occurs due to a decreased production and an increased destruction of normal antibodies. Hyperviscosity symptoms such as headache, fatigue, visual disturbances, and retinopathy can develop, depending on the physical properties of the M component (most commonly with IgM, IgG3, and IgA paraproteins). Moreover, sensory neuropathy can ensue as a result of the hyperproduction of specific IgM subtypes or via infiltration of peripheral nerves by amyloid, an extracellular accumulation of light chains in beta-sheets conformation. The most important differential diagnosis in patients with myeloma involves their distinction from individuals with MGUS, since MGUS is more common than myeloma and is a benign condition. Non-IgG subtype, abnormal kappa/lambda free light chain ratio, and serum M protein >1.5 g/dL are associated with a higher incidence of progression of MGUS to myeloma. The clinical evaluation of patients with myeloma includes a careful physical examination, a complete blood count to assess the presence of anemia, as well as serum calcium, urea nitrogen, creatinine, and uric acid levels to detect signs of renal failure. Chest and bone radiographs may show lytic lesions or diffuse osteopenia, while an increased serum alkaline phosphatase is associated with extensive bone formation. Protein electrophoresis and immunoelectrophoresis are useful for detecting and characterizing M spikes. Usually, the serum M component is IgG in more than half of the cases; in the other cases it is IgA (25% of patients), IgD (1%) or simply composed of just light chains. Patient stratification is based on the International Staging System (ISS), measuring serum levels of β_2-microglobulin and albumin. Stage I is considered when β_2-microglobulin <3.5 mg/L and albumin >3.5 g/dL, while stage III is diagnosed when β_2-microglobulin >5.5 mg/L. The treatment of MM patients is based on the usage of a systemic therapy to control disease progression and symptomatic supportive care to prevent or treat disease complications such as anemia, hypercalcemia, osteopenia, or spinal compression. The initial standard treatment for newly diagnosed myeloma is dependent on whether or not the patient is a candidate for high-dose chemotherapy with autologous stem cell transplant. In newly diagnosed patients who are transplant candidates, newer agents (bortezomib, a proteasome inhibitor, and lenalidomide, an immunomodulatory derivative of thalidomide) combined with pulsed glucocorticoids have now become standard of care as induction therapy, achieving >90% response rate while allowing for successful harvesting of peripheral blood stem cells for transplantation. In patients who are not transplant candidates, besides the options available for transplant candidates, therapy consists of intermittent pulses of an alkylating agent, melphalan, with prednisone. Relapsed myeloma can be treated with novel agents including lenalidomide and bortezomib, in combination with dexamethasone and liposomal doxorubicin, with 60% of partial responses and 10–15% of complete responses. The median overall survival of patients with myeloma is nowadays 7–8 years, with subsets of younger patients surviving more than 10 years [2].

Different clinical manifestations and outcomes of MM can be linked to its wide spectrum of genomic abnormalities, gene expression profiles, and single gene mutations, suggesting that MM is not a single linear disease, but rather can originate from diverse separate abnormalities. Therefore, delineating the molecular pathogenesis of MM is essential in order to develop targeted therapies, already widely used in other hematological (chronic myeloid leukemia, CML) and solid cancers, to allow personalized therapies and improve outcome. In this chapter, we will briefly describe MM-associated chromosomal imbalances (genomic gains and losses), balanced chromosomal rearrangements, and somatic mutations.

COMPLEX KARYOTYPES IN MULTIPLE MYELOMA

MM karyotypes are usually complex with several numerical and structural aberrations, showing more similarities to solid cancers than other hematological disorders. Karyotypic analysis performed by conventional cytogenetics underestimates the number of patients with chromosomal abnormalities in MM, given the low proliferation rate of malignant plasma cells. Nevertheless, cytogenetics has been pivotal in the identification of two MM patient subgroups: hyperdiploid cases with more than 46 but less than 76 chromosomes (34–50% of MM according to different statistics); and non-hyperdiploid cases which include patients with hypodiploid (up to 44–45 chromosomes), pseudodiploid (44/45 or 46/47 chromosomes with gains or losses), and near-tetraploid karyotype [3,4]. Patients with hyperdiploidy are more frequently older, male, of IgG kappa type, and have lytic bone disease. Hyperdiploid karyotype is associated with a better prognosis than the hypodiploid variant [4]. The hyperdiploid variant is typically characterized by gain of the odd-numbered chromosomes. In particular, based on several previous reports and on the recent compendium of MM-associated chromosomal abnormalities by Morgan and coworkers, various combinations of trisomies, affecting chromosomes 15, 9, 5, 19, 3, 11, 7 and 21 (ranked by decreasing order) are present in patients with hyperdiploid karyotype, leading to the deregulation of multiple genes [5,6]. However, hyperdiploid MM is also a heterogeneous group: individuals with additional chr11 or chr1q gains or chr13 loss, for example, have a poor prognosis despite this favorable karyotype [5].

BALANCED CHROMOSOMAL REARRANGEMENTS IN MULTIPLE MYELOMA

Traditionally, the hypodiploid group is associated with poorer overall survival, while the hyperdiploid group presents a better outcome [4]. In fact, non-hyperdiploid patients tend to have a higher incidence of primary immunoglobulin heavy chain (IgH) translocations (85% versus <30%) and chromosome 13 deletions (>70% vs 40%) [7]. Patients who harbor chromosome translocations show overexpression of genes (including *CCND1*, *CCND3*, *MAF*, *MAFB*, *WHSC1* – also called *MMSET* – and *FGFR3*) via their juxtaposition to the immunoglobulin heavy chain locus. IgH translocations in MM are considered primary and early events. Indeed, they are thought to originate in terminally differentiated B cells from errors during physiologic DNA recombination, mostly at the time of isotype class switching, the process necessary to change the isotype specific class of immunoglobulins, in order to create IgM, IgD, IgG, IgA or IgE or, rarely, during somatic hypermutation, the mechanism aimed at generating the variable regions of immunoglobulins [8].

Translocation t(11;14)

The most frequent translocation in MM is t(11;14)(q13;q32), which is present in about 20% of MM and also present in MGUS. MM cases with t(11;14) are CD20 positive, show lymphoplasmacytic morphology, and are hyposecretory [9]. Patients with t(11;14) also commonly bear a near diploid DNA content [8]. Moreover, the majority of IgM MM cases have t(11;14) translocation. Translocation t(11;14) is also a hallmark of mantle cell lymphoma (MCL). In MCL, breakpoints are at VDJ recombination sites originating during somatic hypermutation, while translocations in MM are due to errors during isotype class switching recombination in 60% of cases, with breakpoints at the IgH switch region or during somatic hypermutation in the remaining 40% of MM [9].

The t(11;14) translocation leads to upregulation of cyclin D1 (*CCDN1*). In normal cells, cyclin D1 accumulates and activates its kinase partners, the cyclin-dependent kinases (CDK)-4 or -6, leading to phosphorylation and inactivation of pRb tumor suppressor protein, thereby allowing G1-S transition. Cyclin D1 is also downstream of the mammalian target of rapamycin, the threonine kinase mTOR. Upregulation of CCDN1 is associated with a favorable outcome in most series and low rate of progression to plasma cell leukemia [8,10,11]. These observations are correlated with a lower proliferative index of cyclin D1-expressing MM cells, despite the role of CCDN1 in cell cycle progression [8]. Apart from cyclin D1, other cyclin D proteins are also commonly deregulated in MM, often in the absence of chromosomal translocations, by mechanisms currently unknown. There is an increased expression of one of the three CCND proteins in virtually all MGUS and MM cases, even though these proteins are not expressed in normal B lymphocytes or plasma cells. Indeed, in addition to *CCND1* rearrangements, chromosomal translocations have also been identified for the other cyclins, specifically in 2% of MM (t(6;14) for *CCND3*), and in less than 1% of MM (t(12;14) for *CCND2*). Moreover, as suggested by Kuehl and Bergsagel, specific patient subgroups, in the absence of evident genetic lesions affecting cyclins, nevertheless present a peculiar and specific dysregulation of CCND1 (31%), CCND2 (19%), CCND1 + CCND2 (19%) or CCND3 (2%) [12,13]. Therefore, disruption of G1-S checkpoint control apparently exerts a fundamental role in promoting tumor onset and growth, often in association with *MYC* activation, amplification or translocations, and with *RAS* mutations.

Translocation t(4;14)

The second most common translocation pairs *MMSET/WHSC1*, a nuclear SET DOMAIN protein encoded on 4p16, with *FGFR3* (fibroblast growth factor receptor), an oncogenic tyrosine kinase receptor. The t(4;14) translocation is often present alongside chromosome 13 monosomies or deletions and is associated with poor clinical outcome [11,14,15]. The t(4;14)(p16;q32) translocation has also been reported in MGUS, but is present more frequently in patients with smoldering or active MM [16]. This translocation occurs in about 15% of MM, and all the breakpoints are located in the IgH switch region. MMSET is always expressed in t(4;14) MM, while 23–30% of t(4:14) cases do not demonstrate FGFR3 overexpression, suggesting that MMSET is the main target of this translocation.

MMSET is a lysine histone methyltransferase affecting H3K4, H3K27, H3K36 and H4K20. Lysine methylation is fundamental to regulate chromatin conformation and accessibility to elicit gene transcription [17]. Overexpression of MMSET in MM cells causes global changes in chromatin status, inducing an increase in H3K36 dimethylation and trimethylation along with a reduction of H3K27 methylation, a phenotype usually associated with gene repression [18]. Overexpression of MMSET has a role in cellular adhesion, clonogenic growth, and tumorigenicity in MM, as demonstrated by Lauring et al. [19]. Several proteins are in turn modulated by MMSET. Among them, pro-apoptotic proteins (BAX, BCL2 and caspase 6), cell cycle proteins (cyclin E2, E2F2, TP53INP1, and CDC25A), DNA repair proteins (ATM, E2F2, and GADD45A), and integrin-mediated signaling proteins (CDC42 and integrin alpha-L) are regulated by MMSET. Therefore, MMSET upregulation has a widespread role in modulating the expression of crucial transcripts in MM pathogenesis [18].

FGFR3 is the other partner of t(4;14) translocation. FGFR3 is a tyrosine kinase receptor which undergoes dimerization and tyrosine autophosphorylation when activated, resulting in cell proliferation or differentiation. It has an oncogenic function in MM by signaling through RAS-MAPK and STAT pathways and thereby promoting IL-6 independent survival [20,21]. *FGFR3* activating mutations have also been reported, specifically in patients with the t(4;14) translocation, that are mutually exclusive with the *RAS* mutation. Indeed, both alterations lead to constitutive activation of the mitogen-activated protein kinase (MAPK) pathway.

MAF Translocations

The *MAF* translocation group includes t(14;16) with overexpression of *c-MAF* in 5% of patients, t(14;20) which deregulates *MAFB* (2% of patients), and t(8;16) involving *MAFA* (<1%). The incidence of t(14;16) is higher in MM cell lines (25%) [22]. These translocations are relatively rare in MM and are associated with poor prognosis [15]. In contrast to MM, t(14;20) in MGUS/SMM does not appear to impact prognosis. c-Maf is the cellular homolog of v-maf, the transforming gene of the avian retrovirus AS42 first isolated from a spontaneous fibrosarcoma in chicken. c-MAF, MAFA and MAFB are members of the large family of basic leucine zipper transcription factors that includes also JUN, FOS and AP1. These proteins can form heterodimers and are involved in proliferation, differentiation, and responsiveness to IL-6. c-Maf elicits MM growth by transactivating cyclin D2, a promoter of cell cycle progression, and also upregulates integrin beta7 and C-C chemokine receptor-1 (CCR1) [23,24]. Increased integrin-beta7 expression on plasma cells correlates with poor survival in newly diagnosed MM treated with different regimens. Along with other integrins, it mediates cell adhesion-mediated drug resistance (CAM-DR) [25]. Moreover, integrin beta7/E-cadherin interactions enhance production of VEGF and mediate adhesion to bone marrow stromal cells as well [23].

CHROMOSOMAL IMBALANCES IN MULTIPLE MYELOMA

MM genomes are characterized by numerous and recurrent copy number variations (CNVs) that can involve focal genomic loci or wider regions, resulting in genomic gains or losses. Array comparative genomic hybridization (aCGH) is a molecular cytogenetic technique which is more sensitive than conventional cytogenetics and fluorescence *in situ* hybridization (FISH) analysis to detect these focal events, scanning the entire genome for copy number aberrations in a single step [26,27].

Copy Number Variations with Prognostic Significance

The most frequent recurrent numerical abnormalities present in both hyperdiploid and hypodiploid groups with prognostic significance include deletion of chromosome 13, deletion of chromosome 17p13, and amplification of chromosome 1q21 (Figure 18.1) [28,29].

Deletion of Chromosome 13

Deletion of chromosome 13 is present in 50–60% of newly diagnosed MM, with complete monosomy in 85% of cases and small deletions in the remaining patients [30]. It is more frequent in the non-hyperdiploid group (>70%) in comparison to the hyperdiploid group (35%). The incidence of deletion of chr13 ranges from 20% of MGUS cases (usually in only a subset of tumor cells) to nearly 70% of PCL or MM cell lines [30]. When present in MGUS and MM cases, deletion of chromosome 13 is often associated with balanced translocations such as t(4;14) or t(14;16), suggesting its cooperative role in MM progression [31].

The true target of the deletion is still a mystery, even though this same region is also deleted in chronic lymphocytic leukemia. The minimum deleted region (MDR) is at 13q14, where the *RB1* (retinoblastoma) genomic

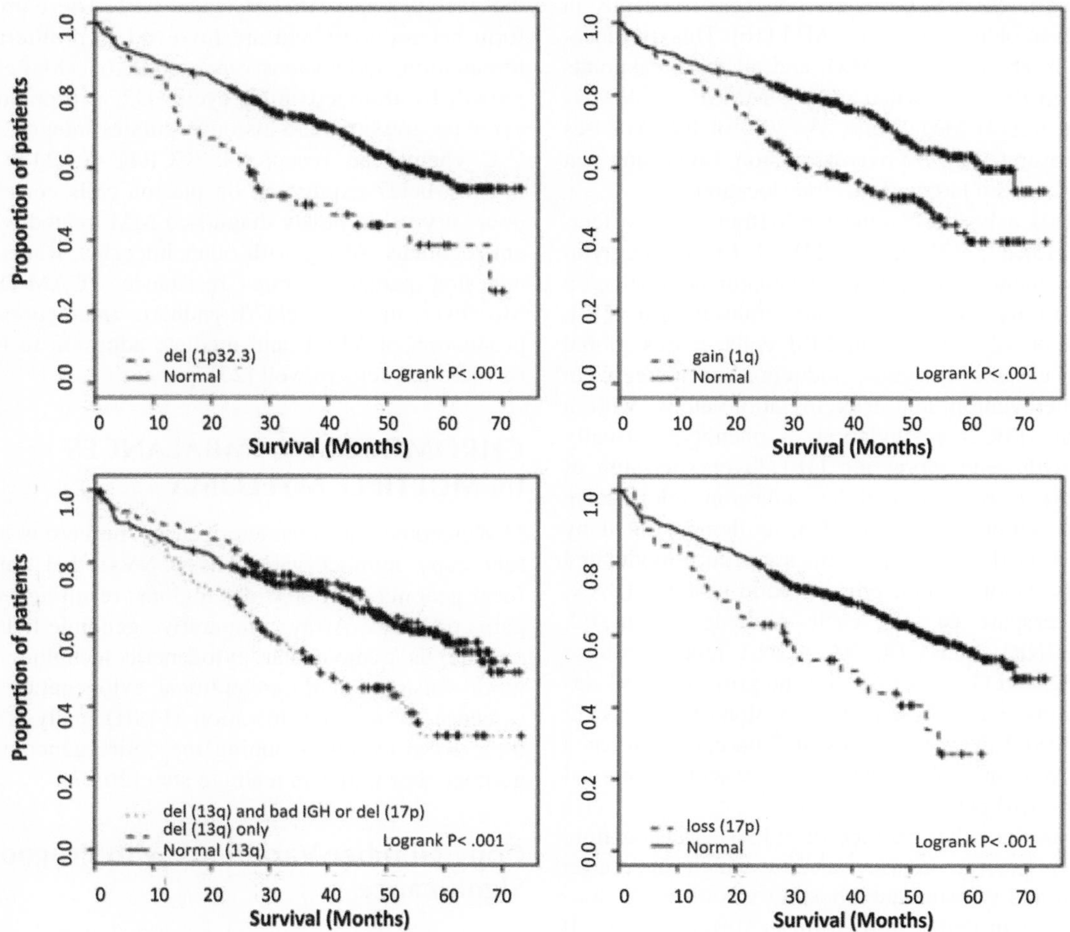

FIGURE 18.1 Clinical outcomes associated with chromosomal abnormalities. Survival varies with the chromosomal imbalances including deletion of short arm of chromosome 1, gain of long arm of chromosome 1, deletion of long arm of chromosome 13, and loss of 17p region (modified from [6]).

locus is located. However, since bi-allelic deletions as well as *RB1* mutations in MM are rare at diagnosis and are acquired during disease progression, *RB1* is probably not an isolated lesion or the main target of this genomic loss [32]. Indeed, located in the same deleted locus is the miR-15a/16-1 cluster, which has a fundamental role in the development of other lymphoid malignancies, such as chronic lymphocytic leukemia and mantle cell lymphoma as well as solid tumors (breast, prostate, and non-small cell lung carcinoma). These miRNAs control expression of proteins including cyclin D1, cyclin D2 and BCL2. Their role in MM pathogenesis has been studied by Roccaro et al. who showed *in vitro* and in the context of the bone marrow milieu that miRNA 15a/16-1 regulates proliferation and growth of MM cells by inhibiting AKT, MAP-kinases, and NF-κB pathways [33]. Other genes located in the commonly deleted area include *EBPL*, *RNASEH2B* and *RCBTB2*, but their function is unknown.

Deletion of chromosome 13 has traditionally been associated with poor prognosis, evidenced by reduced event-free survival (EFS) and overall survival (OS) [34–36]; however, recent studies showed that bortezomib-based regimens can overcome the adverse impact conferred by this abnormality [37–39]. One possible confounder is also its close association with the t(4;14)(p16;q32) and hypodiploid karyotype [34]; therefore, the prognostic value of deletion of chromosome 13 could also be indirect.

Deletion of Chromosome 17p

Another important chromosomal abnormality is the deletion of chromosome 17p involving the *TP53* locus, which is absent in MGUS, rare in newly diagnosed MM (5–10%), and more common in relapsed and refractory MM (20–40%). It is often associated with *TP53* inactivating mutations [40,41]. This abnormality has a strong negative impact on prognosis. Indeed, it is particularly common in early relapsed patients after both conventional chemotherapy and high-dose therapy with autologous stem cell transplant [29]. Several therapeutic regimens are under investigation to overcome this adverse prognostic factor, although their effectiveness is as yet unclear. A recent study by the NCRI Haematological Oncology study group showed that induction therapy with thalidomide is associated with improved response rates (RR), but not OS, in patients with 17p-deleted MM [42]. Evaluation of RR and clinical outcomes in the HOVON-65/GMMG-HD4 trial demonstrated that patients with del(17p) MM greatly benefited from regimens containing versus lacking bortezomib. Indeed, the median progression-free survival (PFS) extended from 12.0 to 26.2 months ($p = 0.024$), and the 3-year OS increased from 17 to 69% ($p = 0.028$) in non-bortezomib treated versus bortezomib-treated patients [43], suggesting that bortezomib induces apoptosis, at least in part, via a p53-independent mechanism. However, patients with del(17p) MM still have a worse prognosis compared with those patients without this cytogenetic aberration. This observation likely implies that long-term administration of bortezomib can improve, but not fully overcome, the adverse outcome associated with del(17p). Novel drugs to treat patients with del(17p) MM are hence urgently needed.

Chromosome 1 Abnormalities

Chromosome 1 abnormalities are also very common in MM, including deletions of chromosome 1p or gains of chromosome 1q. Deletions of 1p have been identified in approximately 7–40% of MM cases using cytogenetics, FISH, and comparative genomic hybridization; when present, this abnormality is associated with poor prognosis, with shorter PFS and OS [44,45]. A rare reciprocal translocation involving *MYC*, t(1;8)(p12;q24), has also been reported in a few MM cases. There are different minimally deleted regions (MDRs) in the p arm that include several potential candidate tumor suppressor genes such as: *TP73*, a p53-family protein (1p35-p36); LAPTM5 (lysosome-associated protein, transmembrane-5), a hematopoietic-specific transcript induced by retinoic acid on 1p34; *CDKN2C* (1p32), *MTF2* and *TMED5* (located at 1p22.1-21.2, the most common MDR in this area, deleted in 22.5% of cases); and *FAM46C* (1p12), that has also been found mutated in 13% of cases (see below) [6]. Deletions of the tumor suppressor *CDKN2C* (p18) have been identified in up to 40% of MM cell lines, 15% MM, 10.3% SMM, and 4.5% MGUS. Deletions affecting this locus are strongly associated with short overall survival (OS) in patients treated with ASCT, as reported by Boyd et al. [45,46]. CDKN2C belongs to the family of cyclin-dependent kinase inhibitors and interacts with CDK4 or CDK6 to regulate G1/S phase, in the context of wild-type RB1 function. This study of Boyd pointed out that aberrancies of *FAM46C* locus or *FAM46C* mutations are also associated with shortened survival (median OS 25.7 months vs 51.3 months, $p = 0.004$) [46]. The biologic function of FAM46C remains unknown. Finally, other proteins deleted might also modulate MM growth and progression including LAPTM5, which is involved in myeloid and B-cell differentiation, and MTF2, which is part of the polycomb repressive complex two (PRC2) and implicated in embryonic stem (ES) cell pluripotency and programming.

The amplification of the 1q21 region is also frequent, detected by FISH in about 40% of newly diagnosed MM and in 70% of relapsed MM. Gain of 1q has an adverse effect on OS, independent of other chromosomal aberrations [47,48]. In particular, +1q21 (3 copies) and +1q21 (>3 copies) are independent predictors for adverse PFS and OS, and the outcome of patients is strictly linked to the copy number variation of +1q21 found in MM cells. Indeed, patients presenting with more than 3 copies of +1q21 have a shorter median PFS and OS, whereas 3 copies or fewer of +1q21 confer a less severe outcome [43]. Additionally, the evolution toward active MM is more likely in patients with smoldering MM and 1q21, where smoldering MM is a condition characterized by more than 10% of plasma cells in the bone without evidence of end-organ damage [49]. Several genes are located on the 1q arm, specifically in the 1q21.1-q23.3 region. Among them, one of the candidate genes is *CKS1B*, a member of the Cks/Suc1 family of small proteins (9–18 kDa) that binds cyclin-dependent protein kinases thereby regulating their function [47]. *PSMD4*, one of the proteasome genes associated with reduced bortezomib response, is located in the same hotspot. Both high *PSMD4* expression levels and high 1q21 copy number adversely affect clinical outcome. Patients whose MM has high proteasome mRNA activity, evidenced by high *PSMD4* expression, have a worse prognosis [50]. *MUC1*, *MCL1*, *PDZK1*, *IL6R* and *BCL9* genes are also located in the 1q region. Soluble MUC1 correlates with tumor burden and contributes to cellular growth via beta-catenin and NF-κB pathway activation [51,52]. Myeloid cell factor-1 (MCL1), an anti-apoptotic member of the Bcl-2 family, is required for MM survival; conversely, its cleavage by bortezomib mediates apoptotic death [53]. *PDZK1* amplification has been associated with drug resistance to melphalan [54]. IL6R is crucial for MM signaling pathways mediated by interleukin-6 and BLC9 enhances beta-catenin-mediated transcriptional activity and thereby promotes MM proliferation, migration and invasion [55].

Other Common Regions of Copy Number Variations

Chromosome 8 Abnormalities

Chromosome 8 abnormalities represent an important hotspot in MM as well. Indeed, *MYC* oncogene maps to chromosome 8p24. *MYC* rearrangements are rare in MGUS and smoldering MM, but are present in about 15% of active MM, in 45% of relapsed and refractory MM and in practically all MM cell lines. *MYC* rearrangements are complex and involve unbalanced translocations and insertions (often affecting three chromosomes) or small duplications, amplifications, and inversions [56]. Moreover, they are often initially present in a subset of tumor cells, which tend to gain survival advantages to become the predominant tumor clone. Whether *MYC* rearrangements are an early event in MGUS to MM progression or not remains debatable. Even though *MYC* rearrangements are acquired late during disease progression and secondary to non-Ig locus translocation, *MYC* activation signature is present in two thirds of newly diagnosed MM [57]. Moreover, *MYC* increased expression from normal PCs to MGUS to MM suggests that *MYC* activity can trigger malignant transformation even in the absence of translocation, which may therefore represent a later event [57]. Importantly, the role of *MYC* in MM has been studied *in vivo* using a transgenic mouse model. *MYC* activation in the germinal center B cells of the so-called Vk*MYC mice causes an indolent form of MM, recapitulating the biological and clinical features of the human disease [58]. Finally, given the widespread activation of *MYC* in MM patients, *MYC* modulation by drugs may represent a novel, potentially very effective therapeutic approach in MM. In particular, one strategy involves the use of BRD4 inhibitors. Indeed, BRD4 directly regulates *MYC* gene expression, and the release of BRD4 protein from the *MYC* locus causes *MYC* transcriptional silencing. BRD4 inhibitors, tested in *in vitro* and *in vivo* models of MM, heavily change the MYC transcriptome and block MM growth, thus representing a good option for reducing *MYC* activation in MM [59].

Chromosome 11 Abnormalities

Up to 40% of MM present with gains of chromosome 11 or have increased copy numbers at the *CCND1* locus with or without t(11:14), the most common balanced translocation in MM. In particular, patients can be divided into three groups based on cyclin D1 expression: high expression with t(11;14)(q13;q32); intermediate expression with trisomy 11; and very weak expression with no abnormality [8,34]. In addition, the 11q22 locus could also be the site of homozygous deletion, leading to the loss of *BIRC2* and *BIRC3* genes, alongside a few others in the same locus [60,61]. Deletion of this particular area is associated with activation of the NF-κB pathway, increased levels of NIK protein (involved in both canonical and non-canonical NF-κB activation and that can also be site of translocation and amplification, see below), and constitutive NFKB2 processing.

Chromosome 4, 14, 16 and X Abnormalities

Copy number variations of chromosomes 4, 14, and 16 are often related to translocations (see above).

Chromosome 4 frequently demonstrates deletion of the telomere of the short arm as a consequence of the unbalanced t(4;14) translocation that abrogates FGFR3 expression (see above). Losses affecting chromosome 14 are found in 12% of MM patients, involving the whole chromosome, or smaller regions. Disruption of the *WWOX* locus, which is located at the fragile site FRA16D of chromosome 16, is commonly associated with t(14:16) [62]. WWOX is an oxidoreductase involved in protein degradation, transcription, and RNA splicing. It is also pro-apoptotic, both through the tumor necrosis factor (TNF) apoptotic pathway and via direct physical interaction with p53 and p73 [63]. Moreover, homozygous deletion of a nearby small region including *CYLD* has been reported [5]. *CYLD*, which is also mutated in 21% of patients with loss of heterozygosity, is a putative tumor suppressor gene and negative regulator of the non-canonical NF-κB pathway [62]. Therefore, chromosome 16 aberrancies can lead in some cases to the combined loss of two tumor suppressor genes (*WWOX* and *CYLD*). The presence of del(16q) is associated with shorter overall survival, with a significantly worse outcome in combination with either t(4;14) or del(17p) [62].

Chromosome 6 can be deleted or amplified. Deletions on chromosome 6 are detected in up to 35% of MM (6q is commonly involved). The most commonly deleted region is centered at 6q25.3. *TIAM2*, a guanine nucleotide exchange factor that can promote proliferation and invasion, and *PARK2*, a component of a multiprotein E3 ubiquitin ligase complex that mediates the targeting of substrate proteins for proteasomal degradation, are among the genes located in this area. *PARK2* is also mutated in familial cases of Parkinson disease and shares functional similarities to *LRRK2*, which is mutated in MM as well. Gain of chromosome 6p is also frequent, and the MDR from 6p22.3 to p21.31 includes *JARID2*, which interacts with polycomb repressive complex 2; and *SCUBE3*, which is involved in TGF-beta signaling [5,6].

Chromosome X: homozygous deletions of Xp11.2 are present in up to 30% of MM patients and involve *UTX*, which encodes a histone H3 lysine 27 (H3K27) demethylase. *UTX* is also mutated in 10% of MM and in other

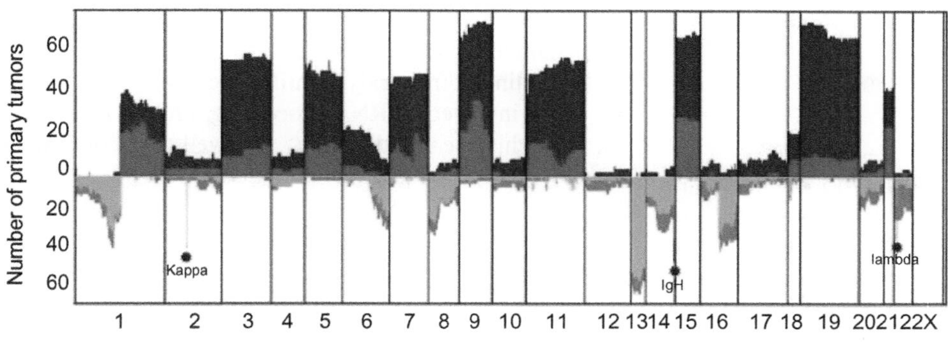

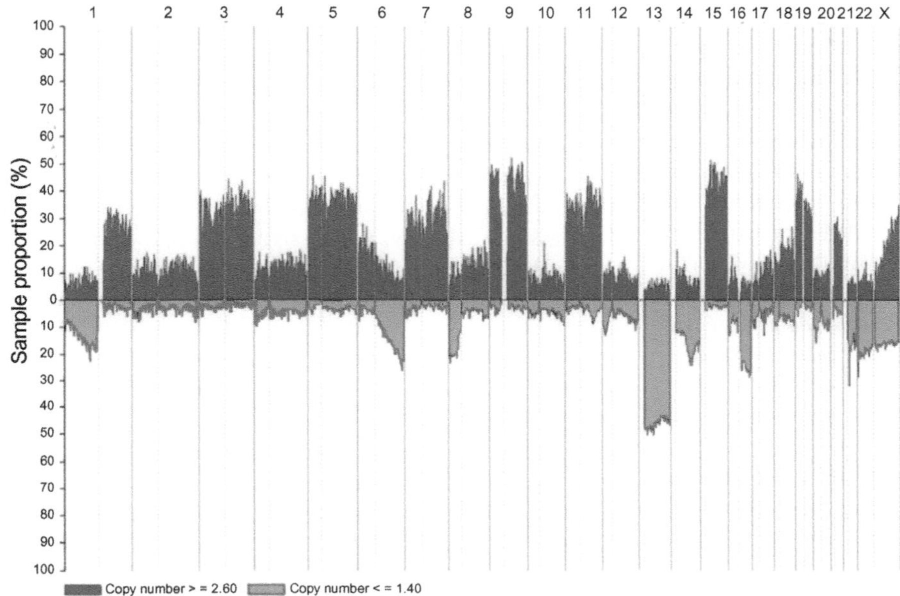

FIGURE 18.2 Summary plots of copy number variations in MM patient samples. The figure shows two summaries of genomic profiles of MM modified from Carrasco et al. [5] and Walker et al. [6]. Genomic gains are in red and genomic losses are in green. All the chromosomes are represented along the x-axis, while the frequency of abnormalities is along the y-axis. Both tumor collections present overlapping genomic copy number changes.

cancers albeit with a lower percentage [64,65]. *UTX* is one of a limited number of genes on the X chromosome that escapes X inactivation in females and has a male counterpart, *UTY*. Therefore, bi-allelic mutations may be necessary for oncogenesis [64]. Of note, Vk*-Myc mice may also have recurrent (~50%) homozygous deletions of *Kdm6a/Utx* locus.

Rare Copy Number Variations

Chromosomes 12, 20 and 22 also show abnormalities. In particular, chromosome 12 frequently demonstrates deletions at the 12p13.1-p13.2 locus, including *CDKN1B*, a negative regulator of cell cycle and at the 12q23.2 locus, where *IGF1* (insulin-like growth factor 1), *PAH* and *ASCL1* are located, while chromosome 20 and 22 can present both deletions and interstitial gains. The biological and clinical significance of these alterations is currently unknown.

Figure 18.2 presents a summary of the most common copy number variations in multiple myeloma [5,6].

SOMATIC MUTATIONS

Cancer development is due to the acquisition of different classes of DNA sequence changes. In particular, during cancer progression there is an accumulation of somatic, heritable variations that are transmitted division after division, and are selected by the neoplastic clones to promote cancer progression and adaptation. Somatic mutations can be considered "drivers" when they confer growth advantage or other cancer-related phenotypes, or "passengers" when they appear in the tumor as a consequence of errors arising during DNA replication but are functionally inconsequential. Given the staggering amount of data emerging from whole-genome sequencing studies, it is thus fundamental to separate the driver from the passenger mutations. Several reports have started to identify genes mutated in MM. Commonly mutated genes are *RAS*, *FGFR3* and *TP53*.

RAS and FGFR3 Mutations

RAS mutations at the "classical" codons 12, 13 or 61 are present in about 30–50% of newly diagnosed MM but are rarely seen in MGUS (less than 5%), suggesting that they are a possible transforming event. The prevalence of *RAS* mutations tends to increase with disease progression and development of PCL. Either *NRAS* or *KRAS* mutations can be present, though they are mutually exclusive. The preponderance of *NRAS* or *KRAS* mutations varies according to different studies [66,67]. Some emerging evidence suggests that *KRAS* and *NRAS* mutations may emerge at different times during disease progression and it seems that only *KRAS* and not *NRAS* mutations are significantly associated with shorter overall survival and progression-free survival [68,69]. *RAS* mutations are more common in t(11;14) MM or in MM that expresses cyclin D1. In contrast, they are absent in t(4;14) MM that constitutively activates the MAPK pathway via FGFR3 upregulation. In addition, *RAS* mutations determine the intra- or extramedullary fate of cancer cells: MM cells without *RAS* mutations accumulate in the bone marrow, while clones endowed with *RAS* mutations are more often extramedullary [70].

FGFR3-activating mutations occur in a fraction of MM cases with t(4;14) [71] and they are mutually exclusive with *RAS* and *BRAF* mutations.

TP53 Mutations

TP53 mutations are not common in intramedullary MM, but are more frequent in advanced phases of disease. *TP53* mutations happen in one third of patients with hemizygous deletion of chromosome 17 but are absent in all the other groups of patients. The majority of *TP53* mutations are located at the DNA-binding domain, in residues important for the direct contact with DNA or for the folding and stabilization of the DNA-binding domain itself. In contrast, in the vast majority of del(17p) hemizygous patients, normal p53 protein is present and functional [41].

Novel Mutations

Recently, the technology of next-generation sequencing has been applied to MM. The first whole-genome sequencing of MM was performed on 38 patients [65]. This study confirmed the presence of somatic mutations in *KRAS* and *NRAS* (50% patients), as well as *TP53* in 8% of cases. Moreover, two point mutations were identified in *CCND1*, commonly translocated in MM but never mutated in other cancers.

Interestingly, this study identified new mutations affecting genes involved in RNA processing, protein homeostasis, and histone methylation, as well as confirming mutations in NF-κB pathway proteins.

NF-κB Mutations

Mutations and copy number changes in genes belonging to the NF-κB pathway have been identified by two different groups [60,61] and confirmed by the recent whole-genome sequencing project [65]. The NF-κB family of transcription factors is composed of NFkB1 (p50 and its precursor p105), NFkB2 (p52 and its precursor p100), RelA (p65), RelB and c-Rel, that can activate two different pathways, the classical (or canonical) and the alternative (or non-canonical) cascade. In general, NF-κB mutations in B-lymphoid malignancies involve genes in the classical pathway, which signal through B-cell receptor [72]. However, in MM, point mutations implicate genes important in both canonical and non-canonical NF-κB pathways (*BTRC, CARD11, CYLD, IKBIP, IKBKB, MAP3K1, MAP3K14, RIPK4, TLR4, TNFRSF1A, TRAF3, NFKB2, CD40, TACI*). These mutations are present in 17% of MM cases and in up to 42% of MM cell lines. *NIK* upregulation in particular is common in patients with an active NF-κB signature. Indeed, more than 50% of NF-κB-mutated MM and one third of MM cell lines have mutations in *TRAF3*, a protein responsible for NIK protein stability, along with TRAF2 and BIRC2/BIRC3. *BIRC2/BIRC3* deletion or *TRAF2* aberrations also increase NIK protein stability.

In addition, *NIK* overexpression can also be due to amplification or chromosomal translocation involving *NIK* itself, leading to activation of both non-canonical and canonical pathways. Finally, gain-of-function mutations or overexpression of *NFkB2, NFkB1, CD40, LTBR*, and *TACI* also activate the NF-κB pathway, resulting in pathway addiction. *CARD11* mutations, discovered also in diffuse large B-cell lymphoma, can also alter the complex formation required for NF-κB signaling downstream of the antigen receptors [65].

Mutations in RNA Processing Genes

RNA processing is a brand new field in cancer biology, starting with the identification of aberrancies in microRNA production and, more recently, highlighted by the discovery of mutations involving multiple components of the RNA splicing machinery such as *U2AF35, ZRSR2, SRSF2* and *SF3B1* in myelodysplastic syndromes and other cancers [73].

DIS3 is mutated in 10–18% of MM [65,74]. It encodes a highly conserved RNA exonuclease, which serves as the catalytic component of the exosome complex involved in regulation of processing and abundance of different RNA species. The four observed mutations occur at highly conserved regions and cluster within the RNA domain facing the enzyme's catalytic pocket. Thus, *DIS3* mutations might deregulate protein translation and mRNA processing. Even though the precise role of DIS3 in MM pathogenesis is still unknown, patients whose MM bears *DIS3* mutations often exhibit deletion of the remaining *DIS3* allele, suggesting that these mutations can complete the loss of function of the protein. However, other evidence suggests that one of the mutations can represent a gain of function. Another interesting finding is that *DIS3* mutations are associated with deletions of the *RB1* region, as reported recently [74]. Whether these events are collaborative and may elicit a synergic effect in MM growth and progression is still unknown. In addition, mutation of *DIS3* may be enriched in MM with either a t(4;14) or t(11;14) [74].

FAM46C gene is mutated in 13% of patients, and its genomic locus is deleted in 10–15% of cases. *FAM46C* deletion is associated with a worse prognosis [46]. It belongs to the Ntase fold protein superfamily, which transfers NMP from NTP to an acceptor hydroxyl group of protein or nucleic acid. FAM46C function is still unknown, but GSEA algorithms define a concomitant expression of FMA46C with a set of ribosomal proteins known to be tightly co-regulated and involved in initiation and elongation of protein translation. Similarly, FAM46C can also behave as an mRNA stability factor [46,65].

Mutations Involving Genes Important for Protein Homeostasis

Protein homeostasis is a crucial process for normal and malignant plasma cells because the enormous rate of production of immunoglobulins can lead to ER stress and apoptosis if not properly controlled. In addition, proteasome inhibitors are very effective MM therapies and act, at least in part, by interfering with unfolded protein response (UPR) and protein degradation [75–77]. Therefore, it is not surprising that mutations in key proteins of this process have been discovered in MM. According to a recent sequencing paper, two patients have mutations in the unfolded protein response gene *XBP1* [65]. XBP1 is a transcription factor that regulates genes responsible for protein folding and degradation in the endoplasmic reticulum, thereby controlling ER stress [78]. This protein is also essential for plasma cell differentiation and survival, since it maintains ER quality control. Overexpression of the spliced form of XBP1 has been shown to cause an MM-like syndrome in mice; conversely, Xbp1 −/− mice are devoid of plasma cells [79,80]. XBP1 is highly expressed in MM, and patients with a low XBP1 spliced/unspliced ratio have significantly better overall survival [81]. The XBP1 pathway and, in particular, the XBP1 spliced form, are fundamental for the growth of MM cells, since knockdown of XBP1 or inhibition of its splicing directly results in MM apoptosis. Moreover, blocking the splicing of XBP1 enhances the sensitivity of MM cells to other ER stress-inducing drugs, such as bortezomib and HSP90 inhibitors, further confirming the important role of UPR in MM [82]. Therefore, activating mutations in *XBP1* might constitutively activate XBP1 signaling and splicing to favor MM growth.

In addition to *XBP1*, *IRF4* and *PRDM1* have been found mutated in MM. These two proteins play critical roles in plasma cell differentiation and survival. IRF4 is a transcriptor factor abundant in terminally differentiated plasma cells and essential during B-cell differentiation [83]. MM cells are dependent on IRF4 by a mechanism described as "non-oncogene addiction". Indeed, IRF4, which can also be induced by MYC or rarely translocated (t(6;14)(p25,q32)), is hence able to activate the expression of genes normally silenced in plasma cells, leading to enhanced cellular growth [84,85]. The mutations discovered are located in the DNA-binding domain, possibly leading to modification in its transcriptional activity. PRDM1 or Blimp-1 promotes plasmacytic differentiation by repressing PAX5, thereby allowing expression of XBP1 [86]. When constitutively active, it can maintain XBP1 expression.

Another mutated gene involved in protein homeostasis in the context of neuronal development is *LRRK2*. This protein is a serine–threonine kinase that phosphorylates translation initiation factor 4E-binding protein (4EBP1) and MPKK. It is mutated in 8% of MM patients, and is associated with a predisposition to Parkinson disease. Nothing is known about the role of LRRK2 in MM; however, it stands to reason to hypothesize a function in the modulation of the endocytic vesicular trafficking and autophagy pathways, as described in other cellular systems [87].

Mutations in BRAF

BRAF is a serine–threonine protein kinase in the MAP kinase/ERK signaling pathway. Four percent of MM patients (1/38 from the initial study and 7/241 and 2/22 from additional screenings [65,74]) have *BRAF* mutations, and this gene is also frequently mutated in melanoma, thyroid, and colorectal carcinomas [88]. K601N and V600E substitutions, reported in 50% of melanoma patients, were described in MM patients, as well as a new mutation (G469A). *BRAF* mutations, in particular the V600E, induce constitutive signaling through the mitogen-activated protein kinase (MAPK) pathway, thereby stimulating cancer cell proliferation and survival. Importantly, BRAF inhibitors that block the active conformation of the BRAF kinase induce objective responses and improvement of OS in melanoma [89]. However, they are associated with the concomitant onset of well-differentiated cutaneous squamous-cell carcinomas and keratoacanthomas, due to hyperactivation of BRAF in wild-type cells that harbor upstream pathway activation including oncogenic RAS or upregulated receptor tyrosine kinases. This paradoxical MAPK-pathway activation must be considered when introducing BRAF inhibitor therapeutic regimens in MM [90,91].

Mutations in Histone-Modifying Genes

Another gene family frequently mutated in MM involves histone-modifying enzymes. Chapman et al. identified mutations in histone methyltransferases including *MLL, MLL2, MLL3, WHSC1/MMSET* and *WHSC1L1* as well as in the histone demethylases *UTX* [65]. In addition, the study from Walker et al. discovered other mutated histone-modifying genes such as *ACTL7B, ASXL1, CHAF1B, ESCO1, KDM2A, KDM4B, MLL5, MYST1, PCGF5, SETD1B, SETD6* and *TOX* [74]. These abnormalities play important roles in MM biology. ESCO1 and MYST1 are histone acetyltransferases. KDM2A, KDM4B, and KDM6A are all demethylases that act at CpG islands and have a role in epigenetic silencing, while MLL5, SETD1B, and SETD6 are histone methyltransferases at different histone sites. ASXL1 is a member of the Polycomb group of proteins, which are necessary for the maintenance of stable repression of genomic loci. Chromatin assembly factor IB (CHAFIB) is required for the assembly of histone octamers onto newly replicated DNA. TOX protein contains an HMG box DNA-binding domain and is involved in chromatin assembly, transcription, and replication. Since *WHSC1/MMSET* is frequently translocated in MM and *UTX* is deleted in up to 30% of cases, epigenetic control regulating histone modification can repress or elicit translation of keynote proteins in MM [6,64]. However, a detailed knowledge of the function of the single proteins in MM is still missing.

Somatic Mutations Stratified by Chromosomal Translocations

Despite a similar clinical phenotype, MM demonstrates a variable pattern of chromosomal imbalances and rearrangements. To obtain a precise understanding of MM pathogenesis, MM subsets sharing similar chromosomal abnormalities should be compared both for their biology and clinical phenotypes. For example, a recent study has compared somatic mutations in t(4;14) and t(11;14) MM [74]. At diagnosis patients with t(4;14) MM have an increased mutational load compared to t(11;14) MM. Interestingly, only 3% of mutations were seen in both groups. Genes mutated with comparable frequencies in both translocation groups included *NRAS, KRAS, BRAF* and *DIS3*, as well as *DNAH5*, a member of the axonemal dynein family mutated in the majority of cases of primary ciliary dyskinesia. Furthermore, two genes were recurrently mutated in the t(4;14) group (*ABCA13* and *PRKD2*), while eight different genes were reported in the t(11;14) subset (*ACTG1, EFNB2, F8, LRR1Q1, PCLO, PLD1, SEC31A* and *SSPO*). Given the exclusivity of these mutations, the authors searched for distinct deregulated pathways between the two groups, using Gene Ontology analysis. This analysis showed an enrichment for genes involved in chromatin organization (*KDM2A, KDM4B, KDM6A, MLL, MLL2, MLL5, MYST1, PCGF5, SETD1B, SETD6*), cytoskeleton remodeling, and microtubule and actin filament-based processes in the t(4;14) group; while genes involved in protein phosphorylation, phosphate metabolism, and Ras signaling (*NRAS, KRAS* and *BRAF*) were enriched in the t(11;14) group. As mentioned previously, RAS/MAPK pathway mutations are mutually exclusive in MM, since this pathway can be deregulated in presence of alteration of many of these abnormalities. Other mutated genes present at lower frequency in both groups include genes involved in the regulation of plasma cell biology and the NF-κB pathway (*PRDM1, PAX2, PAX7, IL6R, IRF4, IL10RA, BCL6, MAX, MYC, CYLD, MALT1, BMP2K, CARD6, TNIP1, IRF5, MKI67* and *LTB*), as well as cell cycle related genes (*ATM, ATR, CCAR1, CDC14B, PTPLAD1, PTPN21, PTPRU, TP53, TACC2, FAT1* and *FAT4*). In addition to DNAH5, other dynein-associated proteins are also mutated (DNAH2, DNAH3, DNAH10, DNAH11, DYNC2H1, TXNDC2, CCDC154, WWC1, TRPS1, RPRG and ODF1), but their role in MM is at present unclear.

This same study also demonstrated that specific mutations are sometimes present in minor subclones, which can later become dominant by gaining other mutations that confer growth and survival advantages [92]. For example, subclonal populations sharing a single common driver mutation (*RAS* or *ATM*) can acquire additional mutations that increase the fitness of the clone itself. This subclonal structure is also present in other hematological malignancies (AML, acute lymphoblastic leukemia) and in solid tumors (renal, breast and prostate carcinomas) [93–95].

GENOMIC ALTERATIONS IN MGUS AND PCL

Plasma cell dyscrasia spans the spectrum from MGUS to smoldering MM to active MM and finally to PCL. Karyotypic analyses are rarely informative in MGUS due to the low fraction of proliferating cells and the low number of monoclonal plasma cells. However, by FISH the prevalence of hyperdiploidy is similar to MM (42% in MGUS and 50% in MM). Also the prevalence of the three primary IgH rearrangements [t(6;14), t(11;14), and t(14;16)] is similar. Most primary IgH rearrangements result from an aberrant IgH switch recombination event in germinal center B cells, suggesting that primary IgH translocations probably represent an early genetic event in myelomagenesis [96]. Chromosome 13 deletion is rare in MGUS, but its presence in t(11;14) and t(6;14) MGUS may indicate transition to MM [97]. Another possible genetic marker indicative for progression of MGUS to MM is the presence of *RAS* mutation, identified in only 5% of MGUS but more frequent in MM [67,70]. *MYC* translocation, considered to be a secondary translocation because it occurs only in MM and in general is not Ig locus-associated, can also be a transforming event, initially resulting in deregulated expression, and later on in the development of the disease after genomic rearrangements [98,99].

In the future, whole-genome sequencing approaches will better define the MGUS–MM transition, since somatic mutations acquired over time can be tracked to delineate their role during development of active MM from MGUS, as well as acquisition of drug resistance and relapse of disease.

In a recent study, the genome of a t(4:14) MM-bearing patient was explored using whole-genome sequencing at diagnosis, first relapse, second relapse and progression to secondary plasma cell leukemia (sPCL) [100]. Alterations that persist at all the stages of disease can be considered fundamental for MM survival. Among them, *AFF1*, *ATXN1*, *CNGA3*, *COL2A1*, *CSMD3*, *KRT9*, *LRRC4C*, *MAGI1*, *MYPN*, and *RNF145* mutations were identified. The role of *AFF1* in cancer persistence was suggested: AFF1 is involved in RNA polymerase II-mediated transcription and co-localizes with MLL at the *HOXA9* promoter, thereby leading to *HOXA9A* upregulation [65]. In contrast, mutations present only at relapse or after leukemic transformation can be implicated in neoplastic progression to PCL. Two genes of particular interest are *RB1* and *ZKSCAN3*. Indeed, the patient at diagnosis had a heterozygous deletion of *RB1* locus followed by a truncating mutation in the other copy of the gene, leading to complete loss of Rb1 protein at the PCL stage. ZKSCAN3 is also involved in G1/S checkpoint, since it modulates the levels of CCND2 and other cyclin D proteins. Finally, following disease stages in a single patient also allowed for identification of mutations present only in particular phases of the disease, suggesting that all clonal progenitors are present at diagnosis and that different subclones can become dominant and subsequently disappear under certain conditions due to treatment selection or clonal evolution. Another possibility is the presence of multiple subclones that evolve independently into discrete foci in the bone marrow.

Apart from the mutations identified above, distinct genomic abnormalities are reported in PCL. PCL is the final leukemic progression of MM, when plasma cells become independent from bone marrow stromal cell support and can survive in the bloodstream. It can be the first presentation of disease or represent the evolution of MM, when it is called secondary PCL (sPCL). Primary PCL (pPCL) has a better prognosis than sPCL. Although pPCL and sPCL have overlapping genetic characteristics, they are not identical. pPCL is often non-hyperdiploid, while sPCL tumors, like MM, are sometimes hyperdiploid.

14q32 (IgH) translocations are ubiquitous in PCL; in primary PCL they are exclusively targeted at 11q13 (CCND1), while in secondary PCL, the 14q32 rearrangements observed reflect the spectrum of translocations observed in MM. *TP53* deletions or mutations are more common in PCL than in MM. Impairment of the TP53 tumor suppressor pathway via *TP53* mutations and deletions or p14ARF promoter methylation appears to be crucial for extramedullary tumor expansion. Another alteration with a putative role in transition from MM to sPCL is *PTEN* loss. PTEN is tyrosine phosphatase, involved in several pathways and in cell cycle regulation. Indeed, *PTEN* deletions are reported in 20–33% of sPCL cases, a significantly higher incidence than in MM (6%) and pPCL (8%) [96].

Figure 18.3 and Table 18.1 summarize the types of genomic abnormalities and their frequencies at the different stages of disease progression from MGUS through MM to PCL.

Finally, sequential genomic analyses can also define mechanisms of drug resistance. For example, studies to delineate the molecular mechanism of bortezomib resistance demonstrate acquisition *in vitro* of an Ala49Thr mutation residing in a highly conserved bortezomib-binding pocket in the proteasome β5-subunit (PSMB5) protein, occurring in association with a dramatic overexpression (up to 60-fold) of PSMB5 but not of the other proteasome subunits, like PSMB6, PSMB7 and PSMA7 [101]. This mutation has not been reported in patients so far.

To conclude, such longitudinal sequencing studies, when paired with clinical data, will yield further insights into the development and progression of MM and also provide new opportunities for molecular diagnosis and therapeutics.

GLOSSARY

Aneuploidy Aneuploidy is a condition where the number of chromosomes in the cells is not the typical one for the species. An increase in the normal number of chromosomes is called hyperdiplody, while a reduction is named hypodiplody.

Array comparative genomic hybridization (aCGH) A technique based on the hybridization of differentially labeled DNA

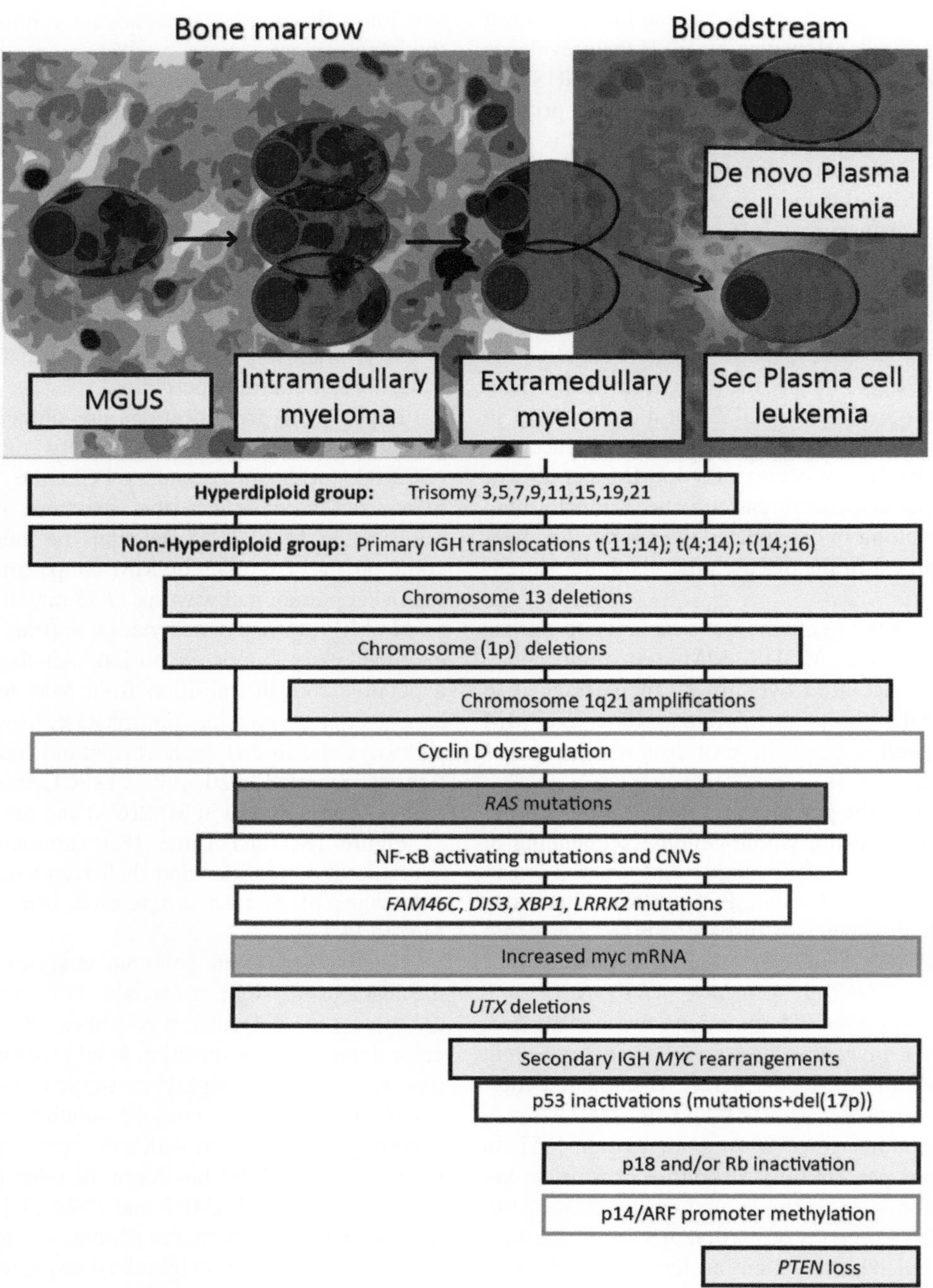

FIGURE 18.3 Types of genomic abnormalities and their frequencies at the different stages of disease progression from MGUS through MM to PCL.

(usually from normal and pathological sample DNAs) aimed to detect DNA gains and losses based on changes in signal ratios.

Autologous stem cell transplant (ASCT) A medical procedure in which hematopoietic stem cells from an individual are isolated, stored, and infused in the same individual after high-dose chemotherapy or radiation therapy to speed up and improve immune system recovering.

Autophagy Autophagy is a catabolic process involving the degradation of intracellular components in the lysosomal compartment, in order to reallocate nutrients from unnecessary processes to more essential ones. It maintains the balance between the synthesis, degradation, and subsequent recycling of cellular products and is exploited by MM cells to survive in critical starving conditions, although when maximally activated, autophagy could lead to apoptosis.

Bone marrow Bone marrow is a soft, gelatinous tissue which fills the cavities of the bones and is the site where blood cell biogenesis takes place.

TABLE 18.1 Types of Genomic Abnormalities and their Frequencies at the Different Stages of Disease Progression from MGUS through MM to PCL

Genetic Lesion	MGUS (%)	MM (%)	PCL (%)
Hyperdiploidy	50	50	20
t(11;14)	5–10	20	25–60
t(4;14)	2–3	15	15–25
MAF translocations		5	15–35
Del(13q)	20	50–60	60–80
Del(1p)	4	7–40	
Chr 1q21 amplification		40–70	
Cyclin D dysregulation	60	80	
RAS mutations	<5	30–50	30
FAM46C, DIS3, XPB1, LRRK2 mutations		8–30	
NF-κB activating mutations and CNVs		15–20	
Increased myc mRNA	0–5	60	
IgH *MYC* rearrangements	1–2	15	30–50
UTX deletions		30	
p53 inactivations (mutations + del(17p))	5	10–20	20–80
p18 and/or Rb inactivation		<5	25–30
p14 promoter methylation		<5	25–30
PTEN loss	0	<2	8–33

Bortezomib A drug used in multiple myeloma treatment regimens, which blocks proteasome activity leading to accumulation of misfolded proteins and thereby multiple myeloma growth inhibition and death.

Chromosomal rearrangement Chromosomal rearrangements are changes in the structure of a normal chromosome. They are mediated by a breakage in the DNA at two different locations, followed by an erroneous rejoining, which creates a different chromosomal configuration.

The different types of rearrangements include:

- Deletion: removal of a DNA sequence
- Insertion: addition of a DNA sequence
- Balanced translocation: a reciprocal fusion of a chromosome to another, which generates two derivative chromosomes with no apparent loss (or gain) of chromosomal sequences
- Unbalanced translocation: a translocation that generates a derivative chromosome(s) with missing chromosomal sequences
- Inversion: intra-chromosomal breaking at two different sites, followed by flipping of the segment and repair of the ends.

Copy number variations (CNVs) CNVs are small DNA structural changes, which include insertions, deletions, duplications and complex multisite variants, ranging from a few bases to megabases in size. CNVs can influence gene expression and gene dosage, being related to cancer susceptibility or progression.

Cytogenetics Branch of genetics aimed at studying number and structure abnormalities of chromosomes.

Event-free survival Time interval after treatment, during which there are no objective signs of disease relapse.

Fluorescence *in situ* hybridization (FISH) Cytogenetic technique used to identify and localize specific DNA sequences on metaphase and interphase spreads, through the use of fluorescently labeled probes.

Gene set enrichment analysis (GSEA) Method of analysis of gene expression profiling data which compares two sample groups to identify predetermined gene subsets concomitantly dysregulated in one group versus the other.

Histone demethylases Enzymes which remove methyl groups from histones and other proteins.

Immunoglobulin A protein produced by plasma cells and B lymphocytes, which plays a role in the immune response to identify and neutralize pathogens. Antibodies are composed of two large heavy chains and two small light chains and present a constant region and a variable region, which recognizes pathogen antigens.

Karyotype Karyotype is the number and appearance of chromosomes in the nucleus of cells. A normal human karyotype includes 22 pairs of autosomal chromosomes and one pair of sex chromosomes (46,XX in females or 46,XY in males). Abnormal karyotypes can be due to abnormalities in the number of chromosomes (aneuploidy) or in the structure of chromosomes.

Monoclonal gammopathy of undetermined significance (MGUS) A condition characterized by the presence of a monoclonal paraprotein in the blood, less than 10% of plasma cells in the bone marrow and no clinical symptoms. It can evolve to active multiple myeloma at a rate of 1% per year.

miRNA A small non-coding RNA molecule of 22 nucleotides average length, which regulates transcriptional and post-transcriptional gene expression and protein translation.

Minimal deleted region A region of the genome that is concomitantly lost in several samples.

Multiple myeloma Multiple myeloma is a clonal proliferation of neoplastic plasma cells in the bone marrow, which synthesize abnormal amounts of immunoglobulin or immunoglobulin fragments.

Overall survival (OS) The percentage of patients enrolled in a particular study who are still alive after they were diagnosed with or treated for a disease, at a specific time point.

Plasma cell A plasma cell is a fully differentiated, mature lymphocyte of the B-cell lineage, which produces and secretes antibodies.

Plasma cell leukemia Presence of more than 20% of malignant plasma cells in the bloodstream.

Progression-free survival (PFS) The percentage of patients enrolled in a particular study who are free of disease after a particular treatment, at a specific time point.

Proteasome Protein complexes, which degrade unneeded or misfolded proteins by proteolysis.

Response rate (RR) Percentage of patients who respond to a specific treatment or fulfill the specific criteria of response established for a particular clinical trial.

RNA processing A biological process aimed at generating mature mRNA, or a functional tRNA or rRNA from the primary transcript. Processing of pre-mRNA involves the steps of capping the 5′ end, polyadenylation of the 3′ end, splicing and sometimes RNA editing.

Smoldering myeloma A condition characterized by more than 10% of plasma cells in the bone without evidence of end-organ damage

Somatic point mutations A change of the nucleotide sequence of the genome of a cell.

There are different types of point mutations:

- Silent mutations, which code for the same (or a sufficiently similar) amino acid
- Missense mutations, which code for a different amino acid
- Nonsense mutations, which code for a stop codon and truncate the protein.

Subclonal heterogeneity Presence of different clonal populations (with variable percentage) with diverse genetic background within the malignant cells of the same individual.

Unfolded protein response (UPR) A cellular conserved stress response which is activated by the accumulation of unfolded or misfolded proteins in the lumen of the endoplasmic reticulum. UPR initially restores normal cell function by halting protein translation and activating signaling pathways which lead to the synthesis of chaperones to help protein folding. However, if the cellular unbalance is too prolonged, the UPR induces apoptosis.

Whole-exome sequencing (WES) A sequencing method by which the exons from an entire genome are sequenced.

Whole-genome sequencing (WGS) A sequencing method by which an entire genome is sequenced.

ABBREVIATIONS

aCGH Array comparative genomic hybridization
ASCT Autologous stem cell transplant
BM Bone marrow
CNV Copy number variation
EFS Event-free survival
FISH Fluorescence *in situ* hybridization
GSEA Gene set enrichment analysis
IgH Immunoglobulin heavy chain
MDR Minimum deleted region
MGUS Monoclonal gammopathy of undetermined significance
MM Multiple myeloma
OS Overall survival
PFS Progression-free survival
PCs Plasma cells
PCL Plasma cell leukemia
RR Response rate
UPR Unfolded protein response

REFERENCES

[1] Hideshima T, Mitsiades C, Tonon G, Richardson PG, Anderson KC. Understanding multiple myeloma pathogenesis in the bone marrow to identify new therapeutic targets. Nat Rev Cancer 2007;7:585–98.

[2] Palumbo A, Anderson K. Multiple myeloma. N Engl J Med 2011;364:1046–60.

[3] Gould J, Alexanian R, Goodacre A, Pathak S, Hecht B, Barlogie B, et al. Plasma cell karyotype in multiple myeloma. Blood 1988;453–6.

[4] Smadja NV, Bastard C, Brigaudeau C, Leroux D, Fruchart C, Groupe Français de Cytogénétique Hématologique. Hypodiploidy is a major prognostic factor in multiple myeloma. Blood 2001;98:2229–38.

[5] Carrasco DR, Tonon G, Huang Y, Zhang Y, Sinha R, Feng B, et al. High-resolution genomic profiles define distinct clinico-pathogenetic subgroups of multiple myeloma patients. Cancer Cell 2006;
9:313–25.

[6] Walker B, Leone PE, Chiecchio L, Dickens NJ, Jenner MW, Boyd KD, et al. A compendium of myeloma-associated chromosomal copy number abnormalities and their prognostic value. Blood 2010;116:e56–65.

[7] Fonseca R, Debes-Marun CS, Picken EB, Dewald GW, Bryant SC, Winkler JM, et al. The recurrent IgH translocations are highly associated with nonhyperdiploid variant multiple myeloma. Blood 2003;102:2562–7.

[8] Fonseca R, Blood EA, Oken MM, Kyle RA, Dewald GW, Bailey RJ, et al. Myeloma and the t(11;14)(q13;q32); evidence for a biologically defined unique subset of patients. Blood 2002;99: 3735–41.

[9] Hoyer JD, Hanson CA, Fonseca R, Greipp PR, Dewald GW, Kurtin PJ, et al. The (11;14)(q13;q32) translocation in multiple myeloma. A morphologic and immunohistochemical study. Am J Clin Pathol 2000;113:831–7.

[10] Soverini S, Cavo M, Cellini C, Terragna C, Zamagni E, Ruggeri D, et al. Cyclin D1 overexpression is a favorable prognostic variable

for newly diagnosed multiple myeloma patients treated with high-dose chemotherapy and single or double autologous transplantation. Blood 2003;102:1588–94.

[11] Gertz M, Lacy MQ, Dispenzieri A, Greipp PR, Litzow MR, Henderson KJ, et al. Clinical implications of t(11;14)(q13;q32), t(4;14)(p16.3;q32), and -17p13 in myeloma patients treated with high-dose therapy. Blood 2005;106:2837–40.

[12] Shaughnessy J, Gabrea A, Qi Y, Brents L, Zhan F, Tian E, et al. Cyclin D3 at 6p21 is dysregulated by recurrent chromosomal translocations to immunoglobulin loci in multiple myeloma. Blood 2001;98:217–23.

[13] Bergsagel PL, Kuehl WM, Zhan F, Sawyer J, Barlogie B, Shaughnessy J, et al. Cyclin D dysregulation: an early and unifying pathogenic event in multiple myeloma. Blood 2005;106:296–303.

[14] Keats JJ, Reiman T, Maxwell CA, Taylor BJ, Larratt LM, Mant MJ, et al. In multiple myeloma, t(4;14)(p16;q32) is an adverse prognostic factor irrespective of FGFR3 expression. Blood 2003;101:1520–9.

[15] Fonseca R, Blood E, Rue M, Harrington D, Oken MM, Kyle RA, et al. Clinical and biologic implications of recurrent genomic aberrations in myeloma. Blood 2003;101:4569–75.

[16] Avet-Loiseau H, Facon T, Daviet A, Godon C, Rapp MJ, Harousseau JL, et al. 14q32 translocations and monosomy 13 observed in monoclonal gammopathy of undetermined significance delineate a multistep process for the oncogenesis of multiple myeloma. Cancer Res 1999;59:4546–50.

[17] Marango J, Shimoyama M, Nishio H, Meyer JA, Min DJ, Sirulnik A, et al. The MMSET protein is a histone methyltransferase with characteristics of a transcriptional corepressor. Blood 2008;111:3145–54.

[18] Martinez-Garcia E, Popovic R, Min DJ, Sweet SM, Thomas PM, Zamdborg L, et al. The MMSET histone methyl transferase switches global histone methylation and alters gene expression in t(4;14) multiple myeloma cells. Blood 2011;117:211–20.

[19] Lauring J, Abukhdeir AM, Konishi H, Garay JP, Gustin JP, Wang Q, et al. The multiple myeloma associated MMSET gene contributes to cellular adhesion, clonogenic growth, and tumorigenicity. Blood 2008;111:856–64.

[20] Chesi M, Brents LA, Ely SA, Bais C, Robbiani DF, Mesri EA, et al. Activated fibroblast growth factor receptor 3 is an oncogene that contributes to tumor progression in multiple myeloma. Blood 2001;97:729–36.

[21] Masih-Khan E, Trudel S, Heise C, Li Z, Paterson J, Nadeem V, et al. MIP-α (CCL3) is a downstream target of FGFR3 and RAS-MAPK signaling in multiple myeloma. Blood 2006;108:3465–71.

[22] Chesi M, Bergsagel PL, Shonukan OO, Martelli ML, Brents LA, Chen T, et al. Frequent dysregulation of the c-maf proto-oncogene at 16q23 by translocation to an Ig locus in multiple myeloma. Blood 1998;91:4457–63.

[23] Hurt EM, Wiestner A, Rosenwald A, Shaffer AL, Campo E, Grogan T, et al. Overexpression of c-maf is a frequent oncogenic event in multiple myeloma that promotes proliferation and pathological interactions with bone marrow stroma. Cancer Cell 2004;5:191–9.

[24] Chang H, Qi Q, Xu W, Patterson B. c-Maf nuclear oncoprotein is frequently expressed in multiple myeloma. Leukemia 2007;21:1572–4.

[25] Neri P, Ren L, Azab AK, Brentnall M, Gratton K, Klimowicz AC, et al. Integrin β7-mediated regulation of multiple myeloma cell adhesion, migration, and invasion. Blood 2011;117:6202–13.

[26] Davies JJ, Wilson IM, Lam WL. Array CGH technologies and their applications to cancer genomes. Chromosome Res 2005;13:237–48.

[27] Inazawa J, Inoue J, Imoto I. Comparative genomic hybridization (CGH)-arrays pave the way for identification of novel cancer-related genes. Cancer Sci 2004;95:559–63.

[28] Avet-Loiseau H, Li C, Magrangeas F, Gouraud W, Charbonnel C, Harousseau JL, et al. Prognostic significance of copy-number alterations in multiple myeloma. J Clin Oncol 2009;27:4585–90.

[29] Sawyer JR. The prognostic significance of cytogenetics and molecular profiling in multiple myeloma. Cancer Genet 2011;204:3–12.

[30] Avet-Loiseau H, Daviet A, Sauner S, Bataille R. Chromosome 13 abnormalities in multiple myeloma are mostly monosomy 13. Br J Haematol 2000;111:1116–7.

[31] Avet-Loiseau H, Li J, Morineau N. Monosomy 13 is associated with the transition of monoclonal gammopathy of undetermined significance to multiple myeloma. Blood 1999;94:2583–9.

[32] Juge-Morineau N, Harousseau JL, Amiot M, Bataille R. The retinoblastoma susceptibility gene RB-1 in multiple myeloma. Leukemia Lymphoma 1997;24:229–37.

[33] Roccaro AM, Sacco A, Thompson B, Leleu X, Azab AK, Azab F, et al. MicroRNAs 15a and 16 regulate tumor proliferation in multiple myeloma. Blood 2009;113:6669–80.

[34] Tricot G, Barlogie B, Jagannath S, Bracy D, Mattox S, Vesole DH, et al. Poor prognosis in multiple myeloma is associated only with partial or complete deletions of chromosome 13 or abnormalities involving 11q and not with other karyotype abnormalities. Blood 1995;86:4250–6.

[35] Gutiérrez NC, Castellanos MV, Martín ML, Mateos MV, Hernández JM, Fernández M, et al. Prognostic and biological implications of genetic abnormalities in multiple myeloma undergoing autologous stem cell transplantation: t(4;14) is the most relevant adverse prognostic factor, whereas RB deletion as a unique abnormality is not associated with adverse prognosis. Leukemia 2007;21:143–50.

[36] Chng WJ, Santana-Dávila R, Van Wier SA, Ahmann GJ, Jalal SM, Bergsagel PL, et al. Prognostic factors for hyperdiploid-myeloma: effects of chromosome 13 deletions and IgH translocations. Leukemia 2006;20:807–13.

[37] Mateos M-V, Hernandez JM, Hernandez MT, Gutierrez N-C, Palomera L, Fuertes M, et al. Bortezomib plus melphalan and prednisone in elderly untreated patients with multiple myeloma: results of a multicenter phase 1/2 study. Blood 2006;108:2165–72.

[38] Chang H, Trieu Y, Qi X, Xu W, Stewart KA, Reece D. Bortezomib therapy response is independent of cytogenetic abnormalities in relapsed/refractory multiple myeloma. Leukemia Res 2007;31:779–82.

[39] Jagannath S, Richardson PG, Sonneveld P, Schuster MW, Irwin D, Stadtmauer EA, et al. Bortezomib appears to overcome the poor prognosis conferred by chromosome 13 deletion in phase 2 and 3 trials. Leukemia 2007;21:151–7.

[40] Neri A, Baldini L, Trecca D, Cro L, Polli E, Maiolo AT. P53 gene mutations in multiple myeloma are associated with advanced forms of malignancy. Blood 1993;81:128–35.

[41] Lodé L, Eveillard M, Trichet V, Soussi T, Wuillème S, Richebourg S, et al. Mutations in TP53 are exclusively associated with del(17p) in multiple myeloma. Haematologica 2010;95:1973–6.

[42] Boyd K, Ross F, Tapper W, Chiecchio L, Dagrada G, Konn ZJ, et al. The clinical impact and molecular biology of del(17p) in multiple myeloma treated with conventional or thalidomide based therapy. Genes Chromosomes Cancer 2011;774:765–74.

[43] Neben K, Lokhorst HM, Jauch A, Bertsch U, Hielscher T, van der Holt B, et al. Administration of bortezomib before and after autologous stem cell transplantation improves outcome in multiple myeloma patients with deletion 17p. Blood 2012;119:940–8.

[44] Qazilbash MH, Saliba RM, Ahmed B, Parikh G, Mendoza F, Ashraf N, et al. Deletion of the short arm of chromosome 1 (del 1p) is a strong predictor of poor outcome in myeloma patients undergoing an autotransplant. Biol Blood Marrow Transplant 2007;13:1066–72.

[45] Leone PE, Walker BA, Jenner MW, Chiecchio L, Dagrada G, Protheroe RK, et al. Deletions of CDKN2C in multiple myeloma: biological and clinical implications. Clin Cancer Res 2008;14: 6033–41.

[46] Boyd KD, Ross FM, Walker BA, Wardell CP, Tapper WJ, Chiecchio L, et al. Mapping of chromosome 1p deletions in myeloma identifies FAM46C at 1p12 and CDKN2C at 1p32.3 as being genes in regions associated with adverse survival. Clin Cancer Res 2011;17:7776–84.

[47] Shaughnessy J. Amplification and overexpression of CKS1B at chromosome band 1q21 is associated with reduced levels of p27Kip1 and an aggressive clinical course in multiple myeloma. Hematology 2005;10(Suppl. 1):117–26.

[48] Wu KL, Beverloo B, Lokhorst HM, Segeren CM, van der Holt B, Steijaert MM, et al. Abnormalities of chromosome 1p/q are highly associated with chromosome 13/13q deletions and are an adverse prognostic factor for the outcome of high-dose chemotherapy in patients with multiple myeloma. Br J Haematol 2007;136:615–23.

[49] Hanamura I, Stewart JP, Huang Y, Zhan F, Santra M, Sawyer JR, et al. Frequent gain of chromosome band 1q21 in plasma-cell dyscrasias detected by fluorescence in situ hybridization: incidence increases from MGUS to relapsed myeloma and is related to prognosis and disease progression following tandem stem-cell transplantation. Blood 2006;108:1724–32.

[50] Shaughnessy JD, Qu P, Usmani S, Heuck CJ, Zhang Q, Zhou Y, et al. Pharmacogenomics of bortezomib test-dosing identifies hyperexpression of proteasome genes, especially PSMD4, as novel high-risk feature in myeloma treated with Total Therapy 3. Blood 2011;118:3512–24.

[51] Treon SP, Maimonis P, Bua D, Young G, Raje N, Mollick J, et al. Elevated soluble MUC1 levels and decreased anti-MUC1 antibody levels in patients with multiple myeloma. Blood 2000;96:3147–53.

[52] Kawano T, Ahmad R, Nogi H, Agata N, Anderson K, Kufe D. MUC1 oncoprotein promotes growth and survival of human multiple myeloma cells. Int J Oncol 2008;33:153–9.

[53] Podar K, Gouill SL, Zhang J, Opferman JT, Zorn E, Tai YT, et al. A pivotal role for Mcl-1 in Bortezomib-induced apoptosis. Oncogene 2008;27:721–31.

[54] Inoue J, Otsuki T, Hirasawa A, Imoto I, Matsuo Y, Shimizu S, et al. Overexpression of PDZK1 within the 1q12-q22 amplicon is likely to be associated with drug-resistance phenotype in multiple myeloma. Am J Pathol 2004;165:71–81.

[55] Mani M, Carrasco DE, Zhang Y, Takada K, Gatt ME, Dutta-Simmons J, et al. BCL9 promotes tumor progression by conferring enhanced proliferative, metastatic, and angiogenic properties to cancer cells. Cancer Res 2009;69:7577–86.

[56] Dib A, Gabrea A, Glebov OK, Bergsagel PL, Kuehl WM. Characterization of MYC translocations in multiple myeloma cell lines. J Natl Cancer Inst Monogr 2008;2008(39):25–31.

[57] Chng WJ, Huang GF, Chung TH, Ng SB, Gonzalez-Paz N, Troska-Price T, et al. Clinical and biological implications of MYC activation: a common difference between MGUS and newly diagnosed multiple myeloma. Leukemia 2011;25:1026–35.

[58] Chesi M, Robbiani DF, Sebag M, Chng WJ, Affer M, Tiedemann R, et al. AID-dependent activation of a MYC transgene induces multiple myeloma in a conditional mouse model of post-germinal center malignancies. Cancer Cell 2008;13:167–80.

[59] Delmore JE, Issa GC, Lemieux ME, Rahl PB, Shi J, Jacobs HM, et al. BET bromodomain inhibition as a therapeutic strategy to target c-Myc. Cell 2011;146:904–17.

[60] Annunziata CM, Davis RE, Demchenko Y, Bellamy W, Gabrea A, Zhan F, et al. Frequent engagement of the classical and alternative NF-κB pathways by diverse genetic abnormalities in multiple myeloma. Cancer Cell 2007;12:115–30.

[61] Keats JJ, Fonseca R, Chesi M, Schop R, Baker A, Chng WJ, et al. Promiscuous mutations activate the noncanonical NF-κB pathway in multiple myeloma. Cancer Cell 2007;12:131–44.

[62] Jenner MW, Leone PE, Walker BA, Ross FM, Johnson DC, Gonzalez D, et al. Gene mapping and expression analysis of 16q loss of heterozygosity identifies WWOX and CYLD as being important in determining clinical outcome in multiple myeloma. Blood 2007;110:3291–300.

[63] Aqeilan RI, Pekarsky Y, Herrero JJ, Palamarchuk A, Letofsky J, Druck T, et al. Functional association between Wwox tumor suppressor protein and p73, a p53 homolog. Proc Natl Acad Sci USA 2004;101:4401–6.

[64] van Haaften G, Dalgliesh GL, Davies H, Chen L, Bignell G, Greenman C, et al. Somatic mutations of the histone H3K27 demethylase gene UTX in human cancer. Nat Genet 2009;41:521–3.

[65] Chapman M, Lawrence MS, Keats JJ, Cibulskis K, Sougnez C, Schinzel AC, et al. Initial genome sequencing and analysis of multiple myeloma. Nature 2011;471:467–72.

[66] Liu P, Leong T, Quam L, Billadeau D, Kay NE, Greipp P, et al. Activating mutations of N- and K-ras in multiple myeloma show different clinical associations: analysis of the Eastern Cooperative Oncology Group Phase III Trial. Blood 1996;88:2699–706.

[67] Bezieau S, Devilder MC, Avet-Loiseau H, Mellerin MP, Puthier D, Pennarun E, et al. High incidence of N and K-Ras activating mutations in multiple myeloma and primary plasma cell leukemia at diagnosis. Hum Mutat 2001;18:212–24.

[68] Chng WJ, Gonzalez-Paz N, Price-Troska T, Jacobus S, Rajkumar SV, Oken MM, et al. Clinical and biological significance of RAS mutations in multiple myeloma. Leukemia 2008;22:2280–4.

[69] Steinbrunn T, Stühmer T, Gattenlöhner S, Rosenwald A, Mottok A, Unzicker C, et al. Mutated RAS and constitutively activated Akt delineate distinct oncogenic pathways, which independently contribute to multiple myeloma cell survival. Blood 2011; 117:1998–2004.

[70] Rasmussen T, Kuehl M, Lodahl M, Johnsen HE, Dahl IM. Possible roles for activating RAS mutations in the MGUS to MM transition and in the intramedullary to extramedullary transition in some plasma cell tumors. Blood 2005;105: 317–23.

[71] Intini D, Baldini L, Fabris S, Lombardi L, Ciceri G, Maiolo AT, et al. Analysis of FGFR3 gene mutations in multiple myeloma patients with t(4;14). Br J Haematol 2001;114:362–4.
[72] Demchenko M, Kuehl WM. A critical role for the NFkB pathway in multiple myeloma. Oncotarget 2010;1:59–68.
[73] Papaemmanuil E, Cazzola M, Boultwood J, Malcovati L, Vyas P, Bowen D, et al. Somatic SF3B1 mutation in myelodysplasia with ring sideroblasts. N Engl J Med 2011;365:1384–95.
[74] Walker B, Wardell CP, Melchor L, Hulkki S, Potter NE, Johnson DC, et al. Intraclonal heterogeneity and distinct molecular mechanisms characterize the development of t(4;14) and t(11;14) myeloma. Blood 2012;120:1077–86.
[75] Adams J. The proteasome: a suitable antineoplastic target. Nat Rev Cancer 2004;4:349–60.
[76] Hideshima T, Bradner JE, Chauhan D, Anderson KC. Intracellular protein degradation and its therapeutic implications. Clin Cancer Res 2005;11:8530–3.
[77] Hideshima T, Richardson P, Chauhan D, Palombella VJ, Elliott PJ, Adams J, et al. The proteasome inhibitor PS-341 inhibits growth, induces apoptosis, and overcomes drug resistance in human multiple myeloma cells. Cancer Res 2001;61:3071–76.
[78] Lee A-H, Iwakoshi NN, Anderson KC, Glimcher LH. Proteasome inhibitors disrupt the unfolded protein response in myeloma cells. Proc Natl Acad Sci USA 2003;100:9946–51.
[79] Carrasco DR, Sukhdeo K, Protopopova M, Sinha R, Enos M, Carrasco DE, et al. The differentiation and stress response factor XBP-1 drives multiple myeloma pathogenesis. Cancer Cell 2007;11:349–60.
[80] Reimold M, Iwakoshi NN, Manis J, Vallabhajosyula P, Szomolanyi-Tsuda E, Gravallese EM, et al. Plasma cell differentiation requires the transcription factor XBP-1. Nature 2001;412: 300–7.
[81] Bagratuni T, Wu P, Gonzalez de Castro D, Davenport EL, Dickens NJ, Walker BA, et al. XBP1s levels are implicated in the biology and outcome of myeloma mediating different clinical outcomes to thalidomide-based treatments. Blood 2010;116: 250–3.
[82] Mimura N, Fulciniti M, Gorgun G, Tai YT, Cirstea D, Santo L, et al. Blockade of XBP1 splicing by inhibition of IRE1α is a promising therapeutic option in multiple myeloma. Blood 2012;119:5772–81.
[83] Mittrücker HW, Matsuyama T, Grossman A, Kündig TM, Potter J, Shahinian A, et al. Requirement for the transcription factor LSIRF/IRF4 for mature B and T lymphocyte function. Science 1997;275:540–3.
[84] Iida S, Rao PH, Butler M, Corradini P, Boccadoro M, Klein B, et al. Deregulation of MUM1/IRF4 by chromosomal translocation in multiple myeloma. Nat Genet 1999;17:226–30.
[85] Shaffer AL, Emre NC, Lamy L, Ngo VN, Wright G, Xiao W, et al. IRF4 addiction in multiple myeloma. Nature 2008;454:226–31.
[86] Shapiro-Shelef M, Lin KI, McHeyzer-Williams LJ, Liao J, McHeyzer-Williams MG, Calame K. Blimp-1 is required for the formation of immunoglobulin secreting plasma cells and pre-plasma memory B cells. Immunity 2003;19:607–20.
[87] Alegre-Abarrategui J, Christian H, Lufino MM, Mutihac R, Venda LL, Ansorge O, et al. LRRK2 regulates autophagic activity and localizes to specific membrane microdomains in a novel human genomic reporter cellular model. Hum Molec Genet 2009;18:4022–34.
[88] Davies H, Bignell GR, Cox C, Stephens P, Edkins S, Clegg S, et al. Mutations of the BRAF gene in human cancer. Nature 2002;417:949–54.
[89] Sosman J, Kim KB, Schuchter L, Gonzalez R, Pavlick AC, Weber JS, et al. Survival in BRAF V600-mutant advanced melanoma treated with vemurafenib. N Engl J Med 2012;366:707–14.
[90] Oberholzer P, Kee D, Dziunycz P, Sucker A, Kamsukom N, Jones R, et al. RAS mutations are associated with the development of cutaneous squamous cell tumors in patients treated with RAF inhibitors. J Clin Oncol 2012;30:316–21.
[91] Su F, Viros A, Milagre C, Trunzer K, Bollag G, Spleiss O, et al. RAS mutations in cutaneous squamous-cell carcinomas in patients treated with BRAF inhibitors. N Engl J Med 2012;207–15.
[92] Keats JJ, Chesi M, Egan JB, Garbitt VM, Palmer SE, Braggio E, et al. Clonal competition with alternating dominance in multiple myeloma. Blood 2012;120:1067–76.
[93] Gerstung M, Beisel C, Rechsteiner M, Wild P, Schraml P, Moch H, et al. Reliable detection of subclonal single-nucleotide variants in tumour cell populations. Nat Commun 2012;3:811.
[94] Shah SP, Morin RD, Khattra J, Prentice L, Pugh T, Burleigh A, et al. Mutational evolution in a lobular breast tumour profiled at single nucleotide resolution. Nature 2009;461:809–13.
[95] Campbell PJ, Yachida S, Mudie LJ, Stephens PJ, Pleasance ED, Stebbings LA, et al. The patterns and dynamics of genomic instability in metastatic pancreatic cancer. Nature 2010;467: 1109–13.
[96] Tiedemann RE, Gonzalez-Paz N, Kyle RA, Santana-Davila R, Price-Troska T, Van Wier SA, et al. Genetic aberrations and survival in plasma cell leukemia. Leukemia 2008;22:1044–52.
[97] Chiecchio L, Dagrada GP, Ibrahim AH, Dachs Cabanas E, Protheroe RK, Stockley DM, et al. Timing of acquisition of deletion 13 in plasma cell dyscrasias is dependent on genetic context. Haematologica 2009;94:1708–13.
[98] Chiecchio L, Dagrada GP, White HE, Towsend MR, Protheroe RKM, Cheung KL, et al. Frequent upregulation of MYC in plasma cell leukemia. Genes Chromosom Cancer 2009;48: 624–636.
[99] Shou Y, Martelli ML, Gabrea A, Qi Y, Brents LA, Roschke A, et al. Diverse karyotypic abnormalities of the c-myc locus associated with c-myc dysregulation and tumor progression in multiple myeloma. Proc Natl Acad Sci USA 2000;97:228–33.
[100] Egan JB, Shi CX, Tembe W, Christoforides A, Kurdoglu A, Sinari S, et al. Whole genome sequencing of multiple myeloma from diagnosis to plasma cell leukemia reveals genomic initiating events, evolution and clonal tides. Blood 2012;120:1060–6.
[101] Oerlemans R, Franke NE, Assaraf YG, Cloos J, van Zantwijk I, Berkers CR, et al. Molecular basis of bortezomib resistance: proteasome subunit beta5 (PSMB5) gene mutation and overexpression of PSMB5 protein. Blood 2008;112:2489–99.

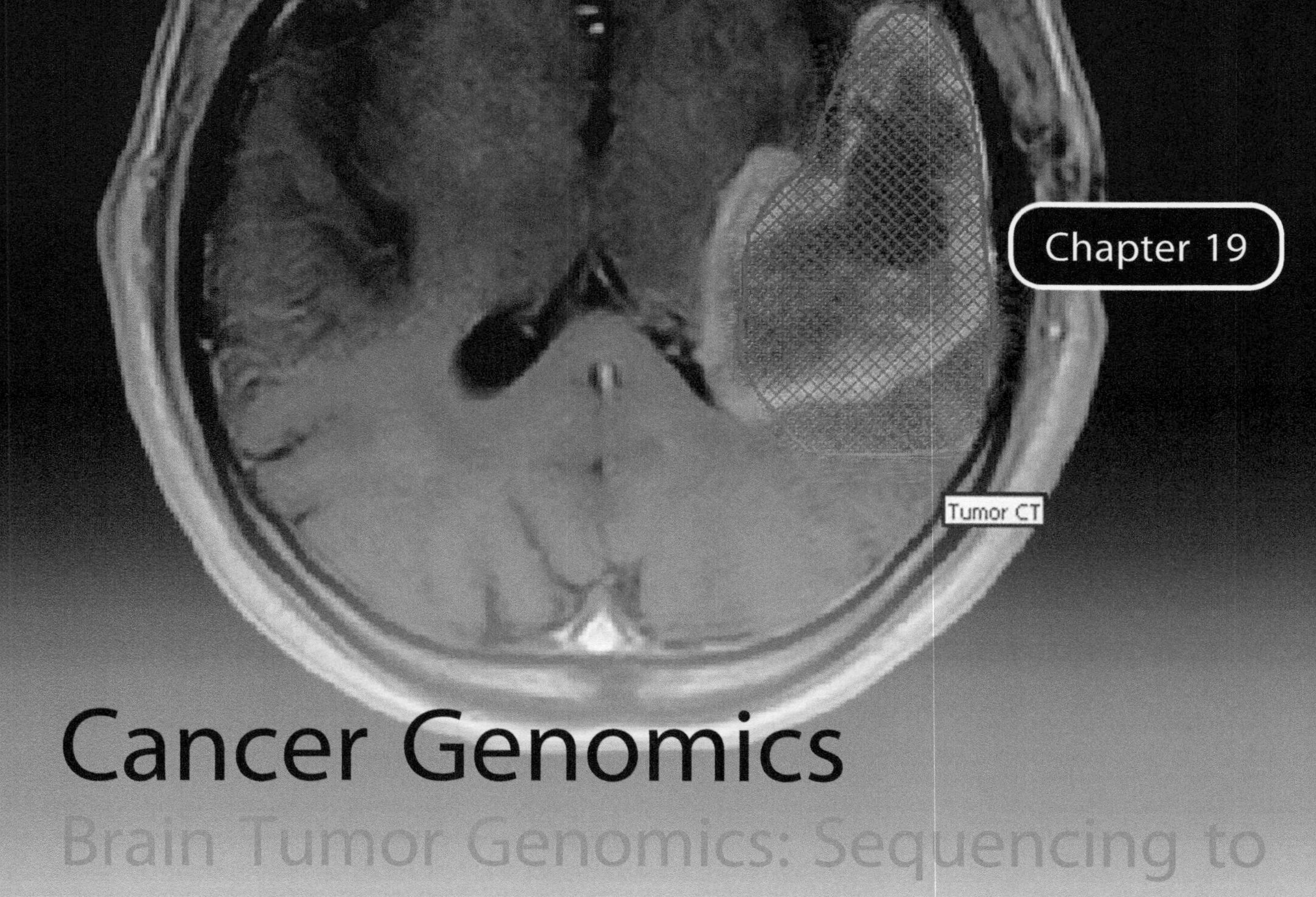

Chapter 19

Cancer Genomics

Brain Tumor Genomics: Sequencing to Clinical Utility

Branavan Manoranjan[1,2,3]
Parvez Vora[1]
Chitra Venugopal[1]
Sheila K. Singh[1,2,3,4]

[1]*McMaster Stem Cell and Cancer Research Institute, McMaster University, Hamilton, ON, Canada*
[2]*Michael G. DeGroote School of Medicine, McMaster University, Hamilton, ON, Canada*
[3]*Department of Biochemistry and Biomedical Sciences, McMaster University, Hamilton, ON, Canada*
[4]*Surgery, Faculty of Health Sciences, McMaster University, Hamilton, ON, Canada*

Contents

Cancer Genomics. DOI: http://dx.doi.org/10.1016/B978-0-12-396967-5.00019-0

Key Concepts

Gliomas

- Originate from glial cells of the central nervous system
- Histopathological subtypes include: astrocytoma, oligodendroglioma, oligoastrocytoma
- WHO grades include benign grades I and II to aggressive, poor outcome grades III and IV
- Glioblastoma is an astrocytic grade IV glioma that presents with histological features of necrosis or microvascular proliferation
- Molecular classification of glioblastoma consists of the following subgroups: proneural (focal amplifications in *PDGFRA* and point mutations in *IDH1*), neural (overexpression of neuronal markers), classical (overexpression of *EGFR*), and mesenchymal (mutations in *NF1*)
- Current treatment is guided by methylation status of *MGMT*, as those patients who have a hypermethylated *MGMT* promoter will respond to temozolomide treatment, extending median survival by 3 months

Medulloblastoma

- Represent the most frequent malignant pediatric brain tumor
- WHO classification of grade IV tumor that includes the following histopathological subtypes: classical, desmoplastic/nodular, medulloblastoma with extensive nodularity, anaplastic, and large cell
- Molecular classification consists of four subgroups: Group 1 (Wnt, best prognosis), Group 2 (Shh, intermediate prognosis), Group 3 (worst clinical outcome associated with *C-MYC* amplification, metastasis and recurrence), and Group 4 (intermediate–poor outcome)

Ependymoma

- Third most common pediatric brain tumor
- Found within the supratentorium, posterior fossa, and spinal cord but more common in posterior fossa in pediatric population, while supratentorial and spinal tumors are present more often in adults
- WHO classification includes grades I–III
- Molecular classification of posterior fossa pediatric ependymomas include two subtypes: Group A (younger patients, balanced genome, upregulation of *LAMA2*, and more likely to exhibit recurrence, metastasis, and death) and Group B (older children, improved clinical outcome, upregulation of *NELL2*)

Pediatric glioblastoma

- Histologically identical to adult glioblastoma but differs in several genomic alterations including minimal *TP53*, *IDH1*, and *IDH2* mutations in pediatric glioblastoma
- Recently associated with recurring gain-of-function heterozygous mutations in *H3F3A*

Diffuse intrinsic pontine glioma

- Not surgically removed due to its location in brainstem
- Recently associated with mutations in *H3F3A* and *HIST1H3B* suggesting possible epigenetic therapeutic targets

INTRODUCTION TO BRAIN TUMORS

Brain tumors are one of the deadliest forms of cancer, responsible for an estimated 13 700 deaths for the year 2012 in the USA alone [1]. These tumors are also the leading cause of childhood cancer mortality [2]. Brain tumors comprise a diverse group of neoplasms, which include primary tumors that arise from different kind of cells in the central nervous system (CNS) and secondary tumors that metastasize to the brain from a distant primary cancer. Brain tumors may present with generalized symptoms including headache, nausea, visual disturbances and vomiting which may vary in duration and frequency depending on the type of tumor. They may also present with focal symptoms and signs which reflect the intracranial location of the tumor, including seizures, hemiparesis and aphasia [3]. As these tumors perturb or invade the central nervous system, they can generate debilitating deficits and progressive symptoms that can incur a decreased quality of life for brain tumor patients. Here, we provide an overview of major primary brain tumors, discuss their general classification and review the latest insights into the genomic alterations that underlie their development.

GLIOMAS

Clinical Presentation and Epidemiology

Gliomas are histologically heterogeneous and invasive brain tumors that originate from glial cells of the central nervous system and differ in location, morphological features, and tendency for progression and therapy response. According to the latest statistical report by the Central Brain Tumor registry of the USA (CBTRUS), gliomas account for 30% of all primary brain and CNS tumors and 80% of malignant tumors [1]. Although pediatric gliomas resemble their adult counterparts in histology, they have distinct clinical and biological differences, which have emerged from recent publications of genome-wide data from pediatric gliomas [4–7]. Thus, they will be discussed separately below. Nonetheless, clinical presentation in children generally resembles that of adults, with insidious progression from low-grade gliomas to high-grade (malignant) gliomas. The most common presenting symptoms and signs are progressive headaches, seizures and focal neurological deficits.

Glioblastomas (GBMs), which account for the majority of gliomas (54%), are one of the most aggressive human cancers, feared for its near uniformly fatal prognosis [1,8,9]. GBM usually occurs in the cerebral hemispheres of adults, but can also occur in young children and infants. Its growth is rapid and infiltrative, and diagnostic pathological features include nuclear pleomorphism, microvascular proliferation and necrosis [10–13]. For adults with newly diagnosed GBM, median survival rarely extends beyond 12 months [8], with inexorable recurrence and generally poor responses to all therapeutic modalities. Even recent breakthroughs with adjuvant chemotherapeutic agents such as temozolomide will only garner an additional 3 months of survival [14]. For those facing the diagnosis, there are few biomarkers of favorable prognosis and, accordingly, few therapies strongly influencing disease outcome [9].

Classifying Gliomas

Historical Perspective

American neurosurgeons Harvey Cushing and Percival Bailey were among the first to establish a classification system for brain tumors. Published in 1926, their book *A Classification of the Tumors of the Glioma Group on a Histogenetic Basis with a Correlated Study of Prognosis* formed the basis of modern day neuro-oncology [15]. In later years, various other groups proposed different classification systems to define human brain tumors more clearly [16–19], but failed to define a universally accepted method.

In the early 1950s, the World Health Organization (WHO) started an initiative to establish a classification and grading system of human tumors based on histopathological and clinical diagnostic criteria. Based on the consensus of an international working group, the WHO aimed to publish classification criteria for nervous system tumors that would be accepted and used worldwide. The first edition, published in 1979, was based primarily on the microscopic characteristics of the tumor, including morphologically identifying cell types by light microscopy and analyzing histological patterns with conventional staining techniques [20]. In the second edition, published in 1993, the WHO working group made several major alterations and also added several new tumor entities based on detailed immunohistochemical and ultrastructural techniques [21]. With increasing knowledge of the genetic basis of human tumors, the third edition, published in 2000, included genomic alterations and gene expression patterns in addition to histological criteria. It also included sections on epidemiology, clinical symptoms, prognosis and predictive factors [22].

The latest (fourth) edition, published in 2007 [23], widely benefited from the focus of research on cytogenetics (e.g. chromosomal aberrations) and advanced molecular genetics (e.g. genome sequencing). Such an approach not only classifies brain tumors based on histopathology, gene expression patterns and molecular signature, but also subclassifies specific tumors based on their intertumoral heterogeneity.

Current Histopathological Classification and Grading System of Gliomas

The current universally accepted WHO system classifies malignant gliomas mainly by histopathology, using their morphological resemblance to different glial cell types: astrocytomas (related to astrocytes), oligodendrogliomas (related to oligodendrocytes) and oligoastrocytomas (mixture of both types). The tumors are further divided by tumor grade (I to IV) based on additional histological features present in the tumor. Grade I gliomas, often considered benign, are curable with surgery and generally do not progress to higher grade tumors. Grade II and grade III gliomas are more invasive, evolve into higher-grade tumors and have a poor outcome. The GBMs define grade IV tumors, with the most invasive pathology and dismal prognosis [23].

GBM, an astrocytic grade IV glioma, is diagnosed based on morphological resemblance to the astrocytic cell lineage and presence of extensive necrosis or microvascular proliferation [24]. Apart from the WHO classification system, GBM can also be categorized into two groups: primary GBMs as neoplasms in patients with no history of diagnosis of lower grade gliomas, and secondary GBMs as cancers that have clinical evidence of malignant progression from a pre-existing low-grade astrocytic tumor [25]. However, based on histopathological findings alone, primary and secondary GBMs are indistinguishable [26]. Moreover, the WHO classification of GBM lacked prognostic direction or prediction of response to therapy. This warranted a more systematic examination of the GBM genome that may define further molecular subclasses and, consequently, allow stratification of treatment. In order better to understand the GBM genome, various groups did genome-wide profiling studies to identify gene signatures that drive tumorigenesis [27–31].

The most systematic gene expression and copy number profiling of GBM was done by The Cancer Genome Atlas (TCGA) Research Network which led to gene expression-based molecular classification of GBM into proneural, neural, classical and mesenchymal subtypes [9,32].

Systematic Sequencing of Glioblastoma Genome

As with most cancerous cells, a GBM tumor cell arises from a lineage of mitotic cell divisions of a single fertilized egg. However, the cancer cell acquires a set of differences in its genome during these mitotic cell divisions, leading to genetic changes like chromosomal aberrations, nucleotide substitutions and epigenetic modifications.

With the aim to catalog genetic abnormalities in human cancer cells, The Cancer Genome Atlas (TCGA) pilot project started a systematic and comprehensive screen of genomic changes in a large GBM cohort of 206 patient samples. TCGA used an integrated large-scale multidimensional analysis to evaluate data generated primarily from gene expression profiling, single-nucleotide polymorphism genotyping and copy number variation in GBM tissue samples. Using targeted Sanger sequencing, 601 genes were analyzed in 91 samples and the most significantly mutated genes found were *TP53*, *PTEN*, *PDGFRA*, *IDH1*, *NF1*, *EGFR* and *RB1*. Furthermore, a group within the TCGA network subclassified GBM molecularly into proneural, neural, classical and mesenchymal subtypes based on the expression of signature genes [9].

Verhaak et al. identified clinically relevant subtypes of GBM characterized by abnormalities in *PDGFRA*, *IDH1*, *EGFR*, and *NF1*, based on TCGA's research data [9]. Validation of these subtypes was done using gene expression profiles of 260 GBM samples compiled from previously published papers [33–36].

Proneural

Focal amplifications in *PDGFRA* and point mutations in *IDH1* defined the proneural subtype of GBM. Many of these tumors had *TP53* mutations (54%), along with high levels of oligodendrocytic lineage genes, such as *PDGFRA*, *NKX2.2* and *OLIG2*.

Neural

The neural subtype, as the name suggests, is marked by overexpression of neuronal markers including *NEFL*, *GABRA1*, *SYT1*, and *SLC12A5*. Mutations in *TP53*, *EGFR* and *PTEN* were also seen but none of them had significantly higher or lower rates to define the subtype.

Classical

The main features of this subtype include high *EGFR* expression along with frequent chromosome 7 amplification (93%) and focal 9p21.3 homozygous deletion spanning *Ink4a/ARF* locus. Absence of *TP53*, *NF1*, *PDGFRA*, or *IDH1* mutations/amplifications also mark the classical subtype.

Mesenchymal

The mesenchymal subtype is characterized by mutation in the tumor suppressor gene *NF1* (37%) as well as high expression of *CHI3L1* and *MET* as previously shown [33].

The following section details the functional significance and the role of key defining oncogenes. Ultimately, this molecular subclassification may lead future scientists and clinicians to engineer more individualized drug treatment for GBM patients.

Common Mutations in GBM

Decades of DNA sequencing efforts have identified hundreds of germline and somatically mutated genes in the human GBM genome. The most frequently observed mutated genes include *TP53*, *IDH1*, *PDGFRA*, *PTEN*, *NF1*, *EGFR* and *RB1* (Table 19.1).

TP53: The *p53* tumor suppressor gene (*TP53*) is the most frequently mutated gene in human cancers. Loss of *p53* either through mutation or deletion appears to be a relatively early event during the progression of low-grade astrocytoma to a high-grade GBM [26,37]. The TCGA study found 42% of their GBM samples had *p53* mutations [32]. *p53* is known to regulate apoptosis and cell cycle arrest [38,39]. In addition, *p53* responds to stress signals by regulating mitochondrial respiration [40], inducing autophagy [41] and suppressing the self-renewal of adult neural stem cells [42].

PTEN: Phosphatase and tensin homolog (*PTEN*), present on chromosome 10q region, is a tumor suppressor gene mutated or deleted in nearly 40–50% of gliomas [32,37]. Loss or mutation of *PTEN* is characteristic of an aggressive subclass of GBM and primarily leads to deregulation of the phosphatidylinositol 3-kinase (PI3K) signaling pathway. The amino terminal domain of *PTEN* is important in regulating cell migration and invasion by directly dephosphorylating focal adhesion kinases [43].

EGFR: Approximately 20% of GBM patients have *EGFR* amplifications [32]. *EGFR* mutation is more commonly seen in primary ($>60\%$) than in secondary glioblastomas ($<10\%$) [44,45]. The *EGFRvIII* is the most common variant mutated with deletion of exons 2–7. It leads to constitutive activation of the EGF receptor and enhanced cell proliferation by downregulation of p27

TABLE 19.1 Distribution of Frequently Mutated Genes across GBM Subtypes.

Gene	Proneural (n = 37)	Neural (n = 19)	Classical (n = 22)	Mesenchymal (n = 38)	Total No. of Mutations
TP53	20 (54%)	4 (21%)	0 (0%)	12 (32%)	36
PTEN	6 (16%)	4 (21%)	5 (23%)	12 (32%)	27
NF1	2 (5%)	3 (16%)	1 (5%)	14 (37%)	20
EGFR	6 (16%)	5 (26%)	7 (32%)	2 (5%)	20
IDH1	11 (30%)	1 (5%)	0 (0%)	0 (0%)	12
PIK3R1	7 (19%)	2 (11%)	1 (5%)	0 (0%)	10
RB1	1 (3%)	1 (5%)	0 (0%)	5 (13%)	7
ERBB2	2 (5%)	3 (16%)	1 (5%)	1 (3%)	7
EGFRvIII	1 (3%)	0 (0%)	5 (23%)	1 (3%)	7
PIK3CA	3 (8%)	1 (5%)	1 (5%)	1 (3%)	6
PDGFRA	4 (11%)	0 (0%)	0 (0%)	0 (0%)	4

From [9]

through activation of the phosphatidylinositol 3-kinase (PI3K)/AKT pathway [46,47].

***NF1*:** Neurofibromin 1 (NF1), a potent tumor suppressor, is inactivated in 21% of GBM patients [32]. Excessive proteasomal degradation and genetic loss may contribute to the inactivation of *NF1* [48,49] which, in turn, leads to deregulation of the Ras signaling pathway. NF1-deficient mouse astrocytes are shown to have increased cell proliferation and migration mediated through the Ras signaling pathway [50].

***IDH1*:** Recurrent mutation in the gene encoding isocitrate dehydrogenase 1 (*IDH1*), a cytosolic protein, has been recently found to be present in approximately 12% of glioblastomas [48]. Interestingly, *IDH1* mutations were found to be present in 50–80% of secondary glioblastomas (those arising from low-grade gliomas), suggesting that these may drive the progression of a lower grade astrocytoma to a glioblastoma [51]. *IDH1* is known to catalyze the oxidative decarboxylation of isocitrate to α-ketoglutarate, leading to production of NADPH in the tricarboxylic acid (TCA) cycle [52].

***PDGFRα*:** Nearly 13% of all glioblastomas show amplification of platelet-derived growth factor receptor alpha (*PDGFRα*) [9,33]. Overexpression and gene amplification of *PDGFRα* occur mostly in low-grade gliomas and secondary GBMs [37,53,54]. Not much is known, in general, about the *in vivo* molecular mechanisms by which PDGFRα promotes GBM growth and invasion. However, PDGF, the PDGFRα ligand, is a well-known secreted mitogen, and is associated with various processes including proliferation, cellular migration, development, and angiogenesis. Hence, it is likely that PDGF not only has an autocrine effect on producing cells but also potential paracrine effects on tumor cells.

***RB1*:** Retinoblastoma 1 (*RB1*), located on the long arm of chromosome 13, is commonly deregulated in many tumors, and is found to be deleted in 11–14% of GBM patients [32,55]. RB1 protein controls the cell cycle regulation complex that governs the G1–S transition of cells, thus allowing for mitosis and cell division.

Epigenetic Alterations in Human Gliomas

Decades of genomic research have revealed a list of varied structural changes in genes, like gene deletions, insertions, amplifications, rearrangements and point mutations that lead to gliomagenesis. However, recent advances in the emerging field of epigenetics have proposed further genomic abnormalities such as aberrant DNA methylation and post-translational histone modification as molecular mechanisms that lead to tumor formation.

DNA Methylation

Global DNA hypomethylation, accompanied by the hypermethylation of promoter CpG islands associated with tumor suppressor genes, is widely implicated in the development and progression of gliomas [56]. Genomic hypomethylation is common in GBMs and can affect up to an estimated 10 million CpG dinucleotides per haploid tumor genome [57]. It may result in genomic instability by activating oncogenes and promoting tumorigenesis. On the other hand, hypermethylation is shown to occur frequently in genes associated with tumor suppression [58],

DNA repair [59], cell invasion [60] and migration [61]. In particular, hypermethylation in the promoter of O6-methylguanine-DNA methyltransferase (MGMT), which results in epigenetic silencing of *MGMT*, is present in 35–45% of glioma patients [62]. Interestingly, silencing of *MGMT* was shown to be associated with longer survival of GBM patients treated with radiation and undergoing chemotherapy with alkylating agents like temozolomide [63]. As an emerging prognostic and predictive value of methylation status of MGMT promoter was discovered, it is now used as an important clinical marker in brain tumors.

Furthermore, the TCGA working group recently identified a glioma CpG island methylator phenotype (G-CIMP) based on characterization of cancer-specific CpG island hypermethylation of genome across a discovery set of 272 TCGA GBM samples [64]. G-CIMP was found to be closely correlated with *IDH1* mutations, an association that may be explained by the fact that IDH proteins regulate levels of α-ketoglutarate (α-KG) and 2-hydroxyglutarate (2-HG), important for regulation of DNA and histone methylation [65].

Histone Modification

A number of histone regulatory genes, including histone deacetylases (*HDAC2* and *HDAC9*), histone demethylases (*JMJD1A* and *JMKD1B*) and histone methyltransferases (*SET7*, *SETD7*, *MLL*, *MLL3* and *MLL4*), have been found to be mutated in GBM patients [48]. In addition, immunohistochemical analysis of glioma samples shows strong nuclear expression of *HDAC1* and *HDAC2* and their expression levels are shown to increase during tumor progression and recurrence [66]. Alterations of histone modification patterns can affect the structure and integrity of the genome, which may promote tumorigenesis [67]. Unfortunately, the current knowledge on the functional significance of histone modifications in gliomas is very limited.

Heterogeneity

GBMs have marked intertumoral and intratumoral heterogeneity on cellular as well as genomic levels. Recognition of intertumoral heterogeneity has largely contributed to the molecular classification of GBM as discussed earlier. However, the high degree of intratumoral heterogeneity present in GBM is poorly understood. At cellular level, tumor cells have been proposed to fall into at least two classes: brain tumor-initiating cells (BTICs) and their more differentiated progeny. Based on the "cancer stem cell" model, it has been hypothesized that GBM arises from a subpopulation of BTICs (also called GBM stem cells) that initiate and maintain the tumor bulk, and are exclusively capable of driving brain tumor formation in an animal model [68,69]. BTICs are able to self-renew and give rise to phenotypically diverse non-BTIC cells with limited division properties that compose the tumor bulk. Adding to the complexity of heterogeneity, BTICs exist as a lineage of self-renewing cell types that may convert into each other within one GBM [70]. Identification of the complete BTIC-driven cellular hierarchy is an unresolved issue, as markers that exclusively define each lineage within the tumor cell hierarchy are still lacking.

At the genetic level, one of the most important markers expressed at different proportions in a GBM is the mutated receptor *EGFRvIII* [46]. This mutation was recently shown to be responsible for maintaining intratumoral heterogeneity in GBM with the help of interleukin 6 (IL-6) and leukemia inhibitory factor (LIF) cytokines [71]. Furthermore, the power of next-generation ultra-deep sequencing technology now makes it possible to characterize mutational heterogeneity within a GBM. Researchers are now trying to identify subpopulations of cells within the tumor which exclusively mark the presence of crucial mutations driving the tumorigenesis. Recent studies have identified variable frequencies of specific *EGFR*, *PIK3CA*, *PTEN* and *TP53* base mutations between two foci within the same primary GBM tumor as well as two foci within the second recurrence [72]. However, further investigation of intratumoral heterogeneity is warranted in order to understand the driving factors of GBM development, and ultimately to provide new avenues for personalized therapeutic interventions.

Current and Future Therapeutic Approaches

Despite advances in understanding the genetics behind glioma and more specifically GBM, prognosis still remains dismal. The current gold-standard treatment consists of surgery followed by fractionated radiotherapy with concomitant temozolomide (TMZ), then adjuvant TMZ. This has helped to postpone progression and extend overall survival from 12.1 months to 14.6 months [73]. However, biological predictors of chemosensitivity were identified in a small subset of GBM patients surviving beyond two years, likely due to a methylated MGMT promoter allowing for enhanced TMZ efficacy [63].

Different therapeutic strategies explored to improve survival include drug delivery enhancement, use of multi-targeted inhibitors and discovery of new targets. One idea is to investigate a cancer gene therapy approach by manipulating microRNA expression using small molecule compounds, which may serve as a novel targeted therapeutic strategy for GBM [74,75]. Intratumoral heterogeneity, at the pathological as well as the genetic level, represents a major obstacle to the design of efficacious therapies [76,77]. An alternative approach to overcome this problem is to use multiple inhibitors targeting several signaling pathways that may otherwise induce GBM resistance [78]. Inactivating the oncogene along with its downstream signaling molecules may have a greater impact on the progression of tumors. Also, identification of predictive biomarkers of resistance/response rates of individual treatment regimens can be helpful to selected patient groups with favorable molecular features, such as the aforementioned patients with MGMT promoter hypermethylation who are undergoing TMZ treatment [63]. In addition, combination of anti-angiogenic drugs with current cytotoxic therapies has shown modestly encouraging results, although further optimization of these drug regimens is still needed.

Regrettably, with all the rapid development in the field of neuro-oncology in the last decade, we still cannot offer a cure to GBM patients. However, patients now have an extended progression-free survival and relatively improved quality of life. Further incremental research that aims to uncover the molecular and genetic drivers of tumor heterogeneity will allow scientists to develop targeted therapies that disarm the invasive spread of GBM.

MEDULLOBLASTOMA

Classifying Medulloblastoma

Historical Perspective

Medulloblastoma (MB) represents the most frequent malignant pediatric brain tumor. James Homer Wright identified MB as a distinct central nervous system (CNS) tumor in 1910 [79]. At the time, Wright believed MB to arise from restricted neuronal precursors termed "neuroblasts". Shortly thereafter, in 1925, Percival Bailey and Harvey Cushing described the presence of glial and neuronal cells in MB suggesting the cell of origin to be of a more primitive embryonic neuroepithelial cell, which they termed "medulloblast", resulting in the conceptual link between neural stem cells (NSCs) and MB ontogeny [80]. More recent molecular classifications of MB have re-conceptualized the heterogeneity that exists within pathological subtypes by giving context to the role of key developmental signaling pathways in MB pathogenesis [81–86]. This work has lent greater support to the presence of a primitive cerebellar stem or precursor cell of origin.

Clinical Presentation and Epidemiology

Medulloblastoma accounts for 18% of pediatric intracranial tumors, with 350 new diagnoses each year in the USA. The incidence peaks at two timepoints during childhood: 3–4 years and 8–9 years of age [1]. Clinically, patients with medulloblastoma commonly present with symptoms associated with increased intracranial pressure, including headaches (especially upon walking), vomiting, irritability, lethargy, and ataxia [87]. The most common physical examination findings are ataxia, hyperreflexia, cranial nerve VI deficits, and papilledema. Computed tomography usually shows a hyperdense mass arising from the vermis (80%), eventually calcified (<20%), obstructing the fourth ventricle and frequently leading to hydrocephalus [87] On T1, tumors might be moderately hypointense, whereas T2-weighted images show mild hyperintensity. Furthermore, leptomeningeal metastases may occur in cranial or spinal subarachnoid spaces and must therefore be carefully assessed [87].

Current Histopathological Classification and Grading System

The 2007 WHO classification of CNS tumors [23] separated MB (grade IV tumor) into several histopathological subtypes: classical, desmoplastic/nodular (D/N), MB with extensive nodularity (MBEN), anaplastic MB, and large cell MB. Although these subtypes speak to the cellular and morphological heterogeneity of the bulk tumor, alluding to the presence of various cellular origins or a common primitive cellular source, the clinical and prognostic utility of these histopathological subtypes remains limited.

Systematic Sequencing of Medulloblastoma Genome

The more recent classification of MB based on multiple genomic platforms has re-conceptualized the heterogeneity that exists within pathological subgroups while also giving context to the role of key developmental signaling pathways in regulating MB tumorigenesis [81–86]. The recent consensus for this new molecular classification system consists of four subgroups, each distinct in terms of prognosis and predicted therapeutic response [86]. Groups 1 and 2 are characterized by upregulation of genes in the Wnt (7–8% of patients) or sonic hedgehog (Shh) (28–32% of patients) pathways, respectively (Figure 19.1).These two subgroups are separated from each other and other subgroups on principal components analysis and both are associated with improved clinical outcomes, when compared to Groups 3 (26–27% of patients) and 4 (34–38% of patients), which are characterized by a greater propensity for metastatic disease and poor clinical outcomes [83–86,88–91]. Although recent transgenic murine models have identified amplification of *C-MYC* signaling to characterize Group 3 MBs, [92,93] both Group 3 and 4 MBs are collectively known as

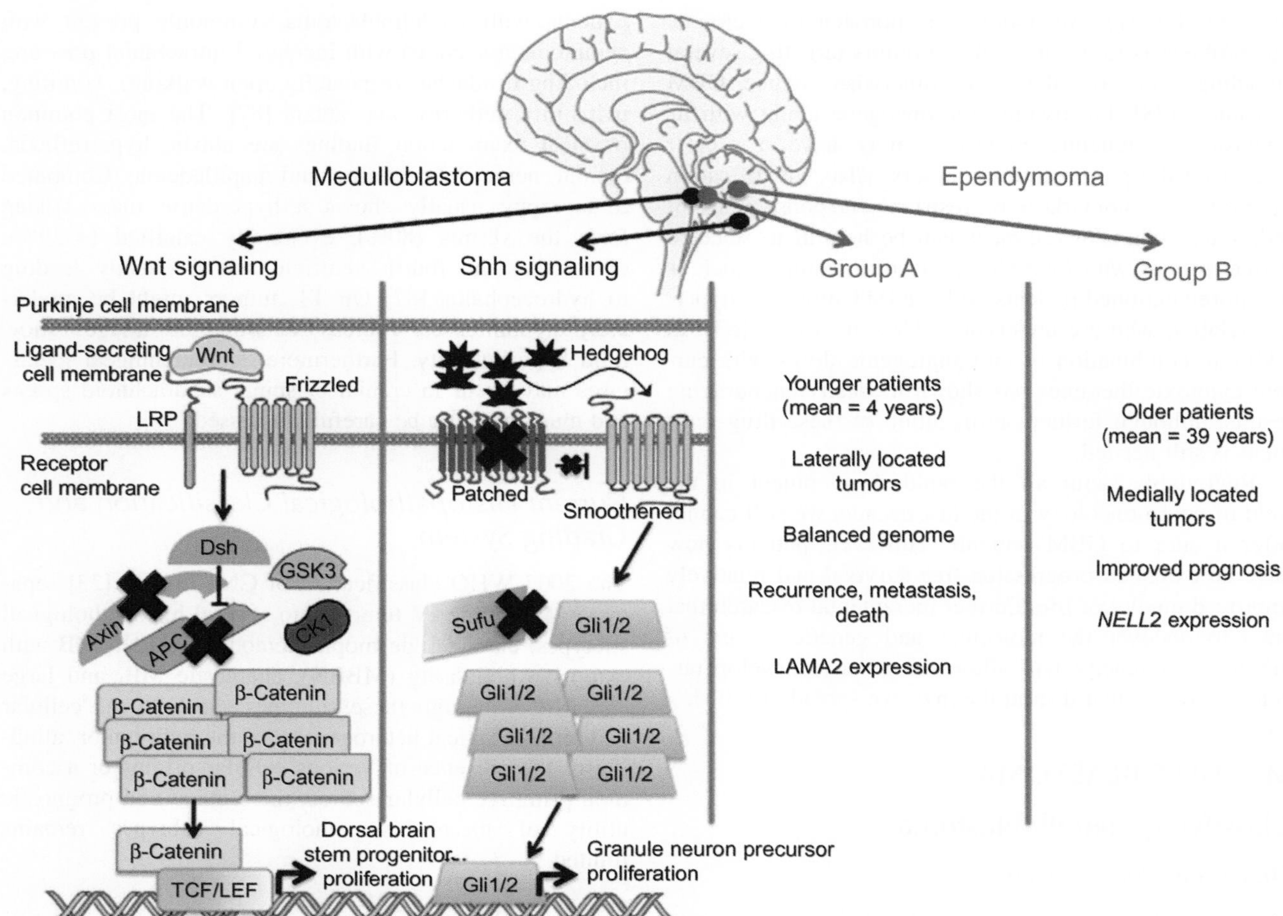

FIGURE 19.1 Posterior fossa ependymoma subgroups and known developmental signaling pathways during medulloblastoma tumorigenesis. Posterior fossa pediatric ependymomas are categorized into two subgroups: (1) Group A poor outcome patients associated with laterally located tumors and the expression of *LAMA2* and (2) Group B improved outcome patients associated with medial tumors and the expression of *NELL2*. Two well-characterized signaling pathways associated with medulloblastoma subgroups A and B are Wnt and Shh, respectively. During normal cerebellar development, activated canonical Wnt signaling regulates the proliferation of cerebellar stem cells. However, mutations in *APC* or *CTNNB1*, which code for β-catenin, result in the accumulation of cytosolic β-catenin and activation of target genes, promoting the proliferation of dorsal brain stem progenitors and the formation of MB. Shh signaling regulates the migration of granule neuron precursors in a controlled fashion during cerebellar development. However, mutations in Shh signaling pathway receptors (*PTCH*) or downstream inhibitors (*SUFU*) may lead to hyperstimulation of the pathway, resulting in the uncontrolled proliferation of granule neuron precursors and MB tumorigenesis.

"non-Shh/Wnt" subgroups, as they remain aggressive and refractory to current treatment modalities [94].

Common Mutations in Medulloblastoma

Wnt subgroup

Germline mutations in the Wnt pathway inhibitor *APC* serve as a predisposing factor to Turcot syndrome, which includes a proclivity to MB. With regards to sporadic cases of MB, somatic mutations of *CTNNB1* encoding β-catenin have served as a key marker. The germline and somatic data are in keeping with an etiological role for canonical Wnt signaling in the pathogenesis of this subgroup, leading to the nomenclature of "Wnt subgroup MB". Aside from a "Wnt pathway gene signature," these tumors are also distinguished by cytogenetic anomalies associated with monosomy 6 [85,86]. Recent whole-exome and deep sequencing of the MB genome has added additional genetic anomalies to the Wnt subgroup involving novel recurrent mutations in an RNA helicase gene, *DDX3X*, and the nuclear co-repressor complex genes *GPS2* and *BCOR* [95,96].

***CTNNB1*:** The encoded protein β-catenin is a member of adherens junction proteins, required for the creation and maintenance of epithelial cell layers. Specifically, this protein anchors the actin cytoskeleton and is responsible for transmitting the contact inhibition signal resulting in the cessation of cell division once the epithelial sheet is confluent. Moreover, the encoded protein is also the main effector protein of the Wnt signaling pathway, a developmentally conserved pathway required for the regulation of stem cell populations.

***APC*:** *APC* (adenomatous polyposis coli) functions as a tumor suppressor and inhibitor of the Wnt signaling pathway. At the cellular level, *APC* is involved in cell migration, adhesion, transcriptional activation, and apoptosis. Pathologic mutations are often clustered in a short region designated the mutation cluster region, which results in a truncated protein product.

***DDX3X*:** DEAD box proteins are putative RNA helicases, implicated in several cellular processes involving changes to the RNA secondary structure such as translation initiation, nuclear and mitochondrial splicing, and ribosome and spliceosome assembly.

***GPS2*:** The encoded protein is involved in G protein-mitogen-activated protein kinase (MAPK) signaling and, when overexpressed in mammalian cells, this gene could potently suppress RAS- and MAPK-mediated signals, suggesting an inhibitory function for this gene.

***BCOR*:** *BCOR* was initially shown to interact with co-repressor of BCL6 as a mechanism for promoting apoptosis. Additional interactions involve class I and II HDACs, suggesting a novel link between the two HDAC classes.

Shh subgroup

Germline mutations in the Shh receptor, *PTCH*, are associated with Gorlin syndrome, which includes a predisposition to MB. Similarly, individuals with germline mutations of the Shh inhibitor, *SUFU*, are also predisposed to MB. Sporadic cases of MB within the Shh subgroup are mostly associated with somatic mutations of *PTCH*, *SMO*, and *SUFU* along with amplifications of *GLI1* and *GLI2* [85,86]. Cytogenetic anomalies unique to the Shh subgroup include deletion of chromosome 9q, as the *PTCH* gene is interestingly located at chromosome 9q22 [85,86].

***PTCH*:** The *PTCH* gene is the receptor for the secreted hedgehog molecules. Functionally, the gene is a tumor suppressor and is frequently mutated in a number of malignancies.

***SUFU*:** Suppressor of fused (*SUFU*) is a negative downstream regulator of the hedgehog signaling pathway.

***SMO*:** SMO is a G protein-coupled receptor that interacts with the PTCH protein. Upon activation by the hedgehog/PTCH complex, SMO transduces intracellular signals leading to transcription of hedgehog target genes.

***GLI1*, *GLI2*:** The encoded transcription factors are activated by the sonic hedgehog signaling pathway and bind DNA through zinc finger motifs. The activity and nuclear localization of *Gli1* is negatively regulated by *p53*. Both, *Gli1* and *Gli2* function to regulate stem cell proliferation in normal and malignant populations.

Group 3

Group 3 tumors are primarily characterized by *c-MYC* amplification but not *MYCN* amplification [83,85]. Secondary genes of interest include *OTX2*, which is amplified and overexpressed in Group 3 and 4 tumors [85]. Moreover, a number of genes implicated in retinal development are also overexpressed in Group 3 tumors; however, their functional significance remains to be established [83–85]. RNA-seq of Group 3 tumors has recently identified recurrent translocations of *PVT1* including *PVT1-MYC* and *PVT1-NDRG1*, which arise through chromothripsis [97]. Cytogenetic anomalies include more common gains of chromosome 1q, and/or loss of 5q and 10q when compared to Group 4 tumors [85,86]. Interestingly, the strongest evidence for a clear subset within a subgroup in MB is found in Group 3, in which Group 3α tumors contain *MYC* amplification and are associated with high risk of recurrence and death, whereas Group 3β tumors do not harbor *MYC* amplifications and are associated with a clinical outcome similar to Group 4 patients, albeit relatively poor [84].

***MYC*:** The protein encoded by this gene is a transcription factor involved in cell cycle progression, apoptosis, and cellular transformation. Mutations, overexpression, rearrangement, and translocation of *MYC* have been associated with several malignancies.

***OTX2*:** The encoded protein acts as a transcription factor essential for brain, craniofacial, and sensory organ development. Pseudogenes are known to exist on chromosomes 2 and 9.

***PVT1*:** This is a long non-coding RNA gene involved in several translocations in various malignancies. Transcription of *PVT1* is regulated by MYC.

***NDRG1*:** This gene is a member of the N-myc downregulated gene family that belongs to the alpha/beta hydroxylase superfamily. Functionally, the protein is involved in stress/hormone responses, cell growth, and differentiation. *NDRG1* is also critical for p53-mediated caspase activation and apoptosis.

Group 4

Group 4 MBs remain as the least characterized subgroup. Although overexpression of genes involved in neuronal differentiation and development has been identified, the genetic and clinical relevance of these genes remains speculative [81–86]. The most significant genetic finding associated with this subgroup has come from an investigation of 1087 unique MBs and their somatic copy number aberrations, in which a tandem duplication of *SNCAIP*, a gene associated with Parkinson disease, was found to be restricted to Group 4 tumors [97]. Cytogenetic anomalies include isochromosome 17q, which is more common in Group 4 (66%) than Group 3 tumors (26%), and loss of the X chromosome, which is seen in 80% of females with Group 4 MBs [81–86].

***SNCAIP*:** The encoded gene product contains several protein–protein interaction domains (ankyrin-like repeats, ATP/GTP-binding motif).

Epigenetic Alterations in Human Medulloblastoma

The advent of high-resolution genomic arrays has enabled the profiling of large cohorts of MB cases to identify novel recurrent amplifications and homozygous deletions targeting genes involved in chromatin modification, particularly histone 3 lysine 9 (H3K9). Deregulation of the histone code has been implicated in MB tumorigenesis through recurrent homozygous deletions of the H3K9 methyltransferase *EHMT1*, high-level amplifications of the H3K9 demethylase *JMJD2C*, and global reduction in H3K9 methylation levels [98]. Recent whole-genome sequencing of DNA from MB patients has also uncovered several distinct components of the epigenetic machinery within distinct MB molecular subgroups. Novel regulators of H3K27 and H3K4 trimethylation in Groups 3 and 4 include *KDM6A* and *ZMYM3*, whereas *SMARCA4* and *CREBBP* function as *CTNNB1*-associated chromatin remodelers in the Wnt subgroup [95,96]. The functional and potential therapeutic significance of these epigenetic regulators remains to be investigated.

***EHMT1*:** The encoded protein functions as a histone methyltransferase that is part of the E2F6 complex, which represses transcription. Through H3K9 methylation *EHMT1* may silence *MYC*- and *E2F*-target genes, suggesting a role in G0/G1 cell cycle transition.

***JMJD2C*:** This protein is a member of the Jumonji domain 2 (JMJD2) family of nuclear proteins and functions as a trimethylation specific demethylase.

***KDM6A*:** This gene is located on the X chromosome and the encoded protein functions as a histone H3 tri/dimethylation demethylase.

***ZMYM3*:** This highly conserved gene is located on the X chromosome, subject to X inactivation, and most abundantly expressed in the brain. The encoded protein functions as a component of histone deacetylase-containing multiprotein complex.

***SMARCA4*:** The protein product is a member of the SWI/SNF family, and therefore has helicase and ATPase activities required for epigenetic regulation of gene expression.

***CREBBP*:** This gene is ubiquitously expressed and functions as a transcriptional co-activator for many transcription factors. Although the protein was initially shown to bind cAMP-response element binding protein (CREB), it is now recognized as a regulator of embryonic development, cell growth, and homeostasis. More importantly, since *CREBBP* has intrinsic histone acetyltransferase activity and behaves as a scaffold for stabilizing additional protein interactions with the transcription complex, it is able to couple chromatin remodeling to transcription factor recognition.

Treatment

Current treatment for MB consists of surgical removal of the tumor, followed by adjuvant radio- and chemotherapy. However, the drive for novel therapeutics based on the characterization of aberrant genomic and molecular pathway phenotypes remains best illustrated in MB. Countless agents have shown promising results in preclinical studies. For example, HDAC inhibitors from several classes including short-chain fatty acids (i.e. valproic acid), natural cyclic peptides (i.e. romidepsin), and aminobenzamides (i.e. MS-275) have been shown to inhibit MB cell proliferation at various efficacies and are being pursued in phase I studies [99]. Shh inhibitors represent the most significant class of agents in the pipeline for treating MB patients, specifically for those with Shh subgroup tumors. Given the limited clinical utility of cyclopamine due to its selective upstream inhibition for those patients with downstream pathway activation (mutations in *Gli* or *SUFU*), several small molecule inhibitors of Gli1-mediated transcription have been identified (GANT58, GANT61) [100]. These compounds have demonstrated strong preclinical efficacy and are currently being studied through early-phase clinical trials. Another aspect of therapeutic targeting involves the tumor vasculature by inhibiting or destructing abnormal blood vessels through anti-angiogenic or vascular disrupting agents. The principal angiogenic factors produced by MB are VEGF (vascular endothelial growth factor) and bFGF (basic fibroblast growth factor), which are thought to promote angiogenesis through stimulating the activity of integrins, $\alpha_v\beta_3$ and $\alpha_v\beta_5$ on endothelial cells [101,102]. Integrin activation results in the propagation of intracellular signals, which maintain endothelial cell survival and enhance proliferation and motility resulting in increased tumor vasculature [102]. Consequently, specific VEGF inhibitors such as bevacizumab and sunitinib are currently in early-phase clinical trials for MB patients. However, since VEGF and other targets for anti-angiogenic therapy are also critical for both physiologic and pathologic angiogenesis, several clinical side effects of anti-VEGF therapy are now beginning to emerge [101,103]. One area of particular concern relates to the creation of a hypoxic environment following anti-angiogenic therapy that may then serve as a viable niche for medulloblastoma stem cells [104]. Consequently, it is imperative that novel subgroup-specific molecular agents consider the complexity of the tumor niche when targeting the vasculature, signaling pathway phenotype, or distinct genomic alterations.

Heterogeneity

Since rare MB stem cell populations typically comprise a minority of cells within a heterogeneous tumor and these cells may be underrepresented on bulk tumor analyses, it is possible that very low transcript levels identify stem cell regulatory genes when profiling bulk MBs. Consequently, current molecular profiling techniques may not truly account for those genes preferentially expressed within the MB stem cell. Moreover, a recent report by Gerlinger et al. has shed light on intratumor heterogeneity, which may lead to an underestimation of the tumor genomics landscape portrayed from single tumor biopsy or resection samples, which is the current practice for MB molecular profiling [105]. These findings have major implications for the study of MB as the field is currently being driven by various gene expression platforms that do not account for intratumor heterogeneity. Consequently, a tumor that may appear to be a Group 2 Shh-driven MB in one region based on transcriptome analysis may actually be a Group 3 or Group 4 MB within other regions of the tumor. Moreover, with the recent report of intratumor heterogeneity, the validity and true clinical utility of the current MB molecular subtypes may be brought into question. Future research efforts and model systems when studying MB must account for its heterogeneous nature. One such model is the BTIC.

The BTIC model system when applied to MB pathogenesis provides a framework to study the heterogeneity of MB. The interplay between MB stem cells and their tumor niche offers researchers with multiple perspectives in regards to tumor biology and the differential regulation of specific subsets of tumor cells by developmental signaling pathways. Such heterogeneous regulation of distinct cell populations by signaling pathways has been well established by Wang et al. [106] who proposed a cell–cell interaction model based on CD133 marker expression and Shh receptor phenotype in MB stem cells. CD133+ MB stem cells displayed a gene expression pattern similar to cerebellar granule neuron precursors in that they contained the Shh signaling machinery and Shh receptor gene expression, whereas CD133− non-MB stem cells resembled gene expression similar to Purkinje cells of normal development by serving as Shh-secreting cells. Consequently, it is imperative also to investigate the differential gene expression pattern of distinct cell populations within the bulk tumor for the development of selective and targeted therapies.

OTHER BRAIN TUMORS

Ependymoma

Ependymoma is the third most common pediatric brain tumor, representing approximately 9% of primary brain tumors in children, with an estimated incidence of 200 per year. The median age of diagnosis of pediatric patients is 5 years [1]. Anatomically, ependymomas occur throughout the CNS, including the supratentorium, posterior fossa, and spinal cord. In the pediatric population, the posterior fossa is the most frequent site of tumorigenesis with 70% of cases occurring in the fourth ventricle, whereas supratentorial and spinal tumors present more often in adults [107,108]. According to the WHO, ependymoma may be classified into grades I–III. Although there are distinct anatomical locations for the pathogenesis of ependymoma, histologically these tumors remain indistinguishable. In keeping with poor histological parameters and a high mortality rate of 45%, novel genomic markers have been investigated further to delineate genomic subtypes and therapeutic targets [107,108].

The most frequent chromosomal abnormalities in ependymoma include losses on chromosome 6q, 17p, and 22q and gains on 1q and 9q [108]. Given the rarity of somatic mutations involving *PTEN* (10q), *TP53* (17p), and *NF2* (22q) in pediatric intracranial ependymoma, driver genes have yet to be identified within regions of genetic alteration using mutational analyses [108]. However, additional anomalies that have been discovered include focal amplifications of *EGFR*, *MYCN*, *NOTCH1*, and *NOTCH4* along with deletions of *CDKN2A* and *SULT4A1* [109–111]. Interestingly, a recurrent observation in several tumor cohorts is that up to 50% of posterior fossa cases have a balanced genomic profile [112,113]. Consequently, the genetic landscape of ependymoma may appear to be heterogeneous with some tumors exhibiting frequent gross numerical chromosomal alterations, while others consist of only single focal aberrations, or even a balanced genome.

In an attempt to merge genomics with clinical utility, a recent transcriptional profiling of two large independent cohorts of ependymomas has revealed the presence of two demographically, transcriptionaly, genetically, and clinically distinct groups of pediatric posterior fossa ependymomas [113] (see Figure 19.1). Group A patients are younger, associated with laterally located tumors with a balanced genome, and are much more likely to exhibit recurrence, metastasis, and death when compared to Group B patients. Consequently, Group B patients may be treated less aggressively than Group A patients, while novel adjuvant therapies are still required for Group A patients. The most striking candidate marker genes for distinguishing the two molecular subgroups are upregulation of *LAMA2* in Group A and *NELL2* in Group B. Given the poor outcome in Group A patients, additional markers include previously reported biomarkers of dismal clinical outcome: *CHI3L1, TNC, VEGF, EGFR, ERRB4*, and *BIRC5*. It is evident with these data that transcriptional profiling may provide a tool for prospective prognostication and stratification of pediatric posterior fossa ependymoma patients.

The treatment of ependymoma is moving towards molecularly guided therapies as several multikinase inhibitors targeting *PDGFR* and *VEGFR* are in various stages of development. *ERBB2* remains one of the more promising and yet controversial targets, since the functional significance of *ERBB2* overexpression or amplification in predicting tyrosine kinase inhibitor sensitivity remains to be explored. Nevertheless, investigations by the Collaborative Ependymoma Research Network (CERN) are currently underway for phase II studies treating recurrent ependymoma pediatric patients with the *ERBB1/2* inhibitor lapatanib and *VEGFR* inhibitor bevacizumab [114]. The role of HDAC inhibitors is also an area of great interest for establishing novel therapeutics. *In vitro* studies of the HDAC inhibitor, trichostatin A, in ependymoma cell lines has demonstrated a reduction in cell proliferation associated with G2/M arrest, DNA damage, and induction of apoptosis [115]. Continued study of the complex mechanisms responsible for the initiation and maintenance of ependymoma is required for translating genomic discoveries into therapies with high clinical utility.

***MYCN*:** The protein product of this gene contains a basic helix-loop-helix (bHLH) domain that enables MYCN to bind DNA through dimerizing with another bHLH protein.

***NOTCH1, NOTCH4*:** The Notch family of proteins is part of the type 1 transmembrane protein family. These proteins play a critical role in many developmental processes by controlling cell fate decisions by functioning through the evolutionarily conserved Notch signaling pathway.

***CDKN2A*:** This is an important tumor suppressor gene. The gene has many transcript variants which differ in the first exon. Two isoforms function as inhibitors of CDK4 kinase. Another transcript contains an alternate open reading frame (ARF) that specifies a protein structurally unrelated to the products of the other variants. This protein product functions as a stabilizer of p53 as it may interact with and sequester the E3 ubiquitin ligase, MDM2, which is responsible for the degradation of p53.

***SULT4A1*:** The encoded protein is a brain-specific sulfotransferase involved in the metabolism of neurotransmitters.

***LAMA2*:** Laminin is an extracellular protein that serves as a major component of the basement membrane. The protein functions to mediate the attachment, migration, and cytoarchitecture of normal and malignant cellular processes. *LAMA2* encodes for the alpha2 chain, which constitutes one of the subunits of laminin2 and laminin4.

***NELL2*:** The encoded protein is a glycoprotein that acts as a homotrimer in the cytoplasm. Functional studies suggest a role in neural cell growth and differentiation.

***CHI3L1*:** This gene encodes a glycoprotein member of the glycosyl hydrolase 18 family, which contains a family of chitinases that function to catalyze the hydrolysis of chitin. This particular gene encodes for a protein that lacks chitinase activity and is secreted by activated macrophages implicating inflammation and tissue remodeling.

***TNC*:** The functional protein is an extracellular matrix protein that is implicated in guidance of migrating neurons along with axons during development, synaptic plasticity, and neuronal regeneration.

***BIRC5*:** This gene is a member of the inhibitor of apoptosis (IAP) gene family, which encodes negative regulatory proteins that prevent apoptotic cell death. Although most IAP family members contain multiple baculovirus IAP repeats (BIR) domains, *BIRC5* encodes proteins with only a single BIR domain.

Pediatric Glioblastoma

Pediatric glioblastoma (GBM) accounts for approximately 3% of all childhood brain tumors [1]. Although pediatric GBM is histologically identical to adult GBM, several genomic alterations distinguish the two tumors [5]. For example, *IDH1* or *IDH2* mutations commonly found in adult GBM are rare in pre-adolescent children [5]. Similarly, *TP53* mutations, a hallmark feature of adult gliomas, are a less common event in early childhood GBM cases [116]. However, with the advent of whole-exome sequencing, unique genomic anomalies specific to pediatric GBM have been identified for the first time [7,117]. Recurring gain-of-function heterozygous mutations in the *H3F3A* gene, which encodes histone H3.3, have been shown to regulate telomere maintenance and/or heterochromatin stability (Figure 19.2). Although these mutations suggest epigenetic regulators play a critical role in the pathogenesis of pediatric GBMs, the clinical utility of these mutations remains to be investigated. As novel genomic regulators of malignant transformation continue to be discovered, so do new therapeutic modalities in treating pediatric GBM. The integrin inhibitors represent a promising area of research as integrins, a family of cell adhesion receptors, have been shown to maintain invasive properties of aggressive malignancies [118]. Gamma

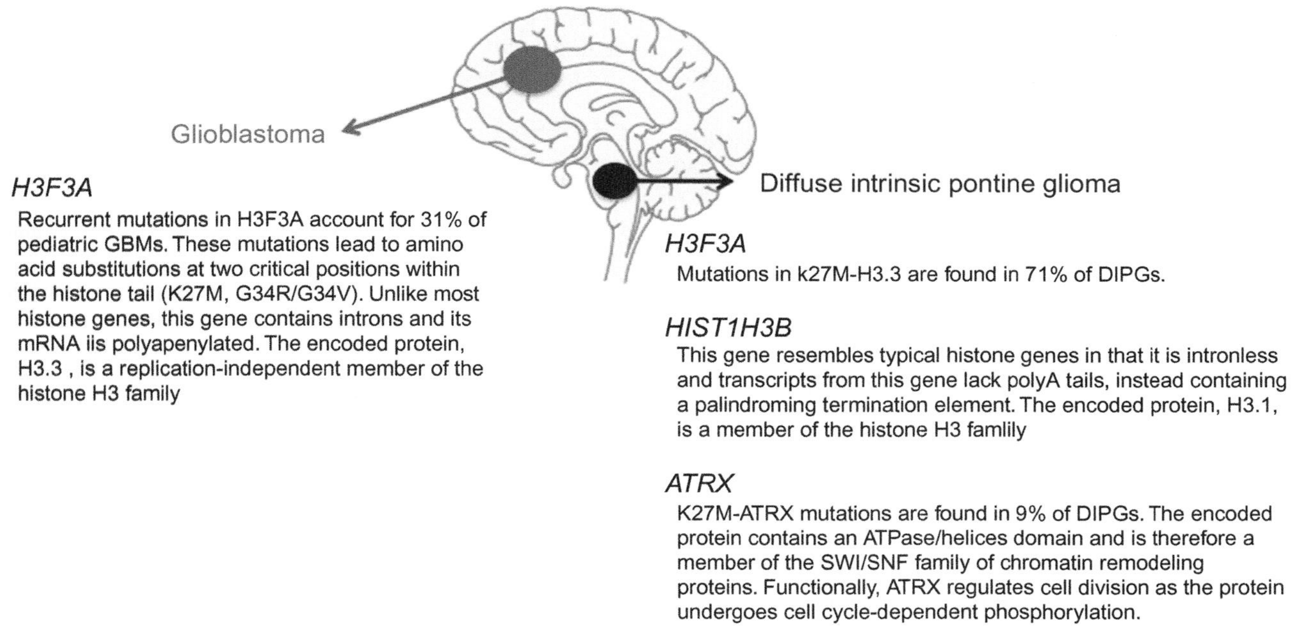

FIGURE 19.2 Localization and novel hallmark genetic mutations associated with high-grade pediatric gliomas. Pediatric glioblastoma (GBM) occur more frequently in the cerebral cortex whereas diffuse intrinsic pontine gliomas (DIPGs) are localized to the brainstem. Although mutations characteristic of adult gliomas are not present in pediatric high-grade gliomas, recent whole-exome sequencing has identified mutations in the following histone-modifying genes: *H3F3A* in pediatric GBM and DIPG along with *HIST1H3B* in DIPG.

secretase inhibitors are also currently being developed to block Notch signaling, which has been described as an essential promoter of GBM cell survival. Lastly, similar to adult GBM, temozolomide treatment may be a novel mechanism for targeted therapy as *MGMT* promoter methylation in pediatric GBM has been shown to correlate with improved progression-free survival [119].

Diffuse Intrinsic Pontine Glioma

Diffuse intrinsic pontine glioma (DIPG) is an anatomical variant of high-grade pediatric glioma, which has remained a surgical challenge for several decades due to its location in the neurologically delicate brainstem. Brainstem gliomas account for approximately 20% of all CNS tumors among children younger than 15 years of age [120]. The median age at presentation for all brainstem gliomas in children is 6–7 years, with males and females equally affected [120]. Clinically, the most common presentation is that of a mass arising in the pons and causing diffuse enlargement. Patients typically report a short duration of symptoms, with a median of one month and have a triad of signs including cranial neuropathy, long tract, and cerebellar deficits. Hydrocephalus and metastatic disease from diffuse tumors of the pons are rare [120].

Similar to pediatric GBM, whole-genome sequencing has only recently identified novel mutations in the *H3F3A* and *HIST1H3B* genes, which encode histone H3.3 and H3.1, respectively (see Figure 19.2) [117,121]. Histones are basic nuclear proteins responsible for the nucleosome structure of chromosomes. As the nucleosome is formed from DNA being wrapped in repeating units around an octamer consisting of two molecules of each core histone (H2A, H2B, H3, and H4), mutations in histone-coding genes provide an epigenetic mechanism of tumor formation. Additional mutations in *TP53* and *ATRX* have also been identified, in keeping with the findings of pediatric GBM cases [117,121]. The functional significance of these mutations with respect to clinical outcome, histological differences, along with intra- and intertumoral heterogeneity remains to be established. Current treatment for DIPG maintains no established role for chemotherapy with radiotherapy serving as the standard treatment, albeit palliative in nature as only 10% of children survive beyond 2 years [122]. Nevertheless, novel agents as part of a sequenced radio-chemotherapy regimen are continuously being investigated with the *EGFR* inhibitor cetuximab or HDAC inhibitor vorinostat providing two such examples [123].

***H3F3A*:** Unlike most histone genes, this gene contains introns and its mRNA is polyadenylated. The encoded protein, H3.3, is a replication-independent member of the histone H3 family.

***HIST1H3B*:** This gene is found in the large histone gene cluster on chromosome 6p22-p21.3. The gene resembles typical histone genes in that it is intronless and transcripts from this gene lack poly-A tails, instead containing a palindroming termination element. The encoded protein, H3.1, is a member of the histone H3 family.

***ATRX*:** The encoded protein contains an ATPase/helices domain and is therefore a member of the SWI/SNF family of chromatin remodeling proteins. Mutations in these genes also provide the possibility for epigenetic regulation of the tumor's genetic landscape as diverse changes in DNA methylation patterns have been reported with *ATRX* mutations, suggesting a link between chromatin remodeling, DNA methylation, and gene expression. Moreover, ATRX may also serve as a regulator of cell division as the protein undergoes cell cycle-dependent phosphorylation.

CONCLUSION

The field of brain tumor genomics has drastically evolved over the past 100 years, with several key discoveries having been made in only the past decade. With the advent of molecular subtyping and deep genome sequencing of malignant tissue, the identification of additional classification systems rooted in clinical outcome and risk stratification is just around the corner. However, the heterogeneous nature of brain tumors remains a burden to be reckoned with as recent reports have shed light on the extent of intratumoral heterogeneity within solid cancers. Given the urgent desire for targeted therapies and the observation of a heterogeneous genomic landscape, other frameworks and model systems should be investigated for exploring the dynamic nature of brain tumors. One such model system is that of the cancer stem cell (CSC) or brain tumor-initiating cell (BTIC). Since rare stem cell populations typically comprise a minority of cells within a heterogeneous tumor and these cells may be underrepresented on bulk tumor analyses, it is possible that very low transcript levels identify critical CSC regulatory genes when profiling bulk tumors. Consequently, current molecular profiling techniques may not truly account for those genes preferentially expressed within the CSC population. Moreover, the CSC model when applied to tumorigenesis provides a framework to study the interplay between CSCs and their tumor niche, offering researchers multiple perspectives with regard to tumor biology and differential gene expression patterns of specific subsets of tumor cells. Although primary human CSC cultures are technically challenging and provide a limited number of cells for data acquisition, future transcriptional analyses of stem cell genes may benefit from tissue banks consisting of prospectively sorted CSC RNA. Consequently, continued study of larger numbers of human CSCs and specifically BTIC specimens could eventually elucidate key stem cell regulatory genes, their pathways, and overall molecular mechanisms that could provide novel therapeutic targets for brain tumors.

GLOSSARY

Chromothripsis A phenomenon of complex inter- and intra-chromosomal rearrangements seen in cancer cells.

Hemiparesis A condition of muscular weakness causing inability to move one side of the body.

Supratentorium A region of the brain located above tentorium cerebelli.

ABBREVIATIONS

2-HG 2-Hydroxyglutarate
α-KG α-Ketoglutarate
BTIC Brain tumor-initiating cell
CSC Cancer stem cell
CNS Central nervous system
DIPG Diffuse intrinsic pontine glioma
EGFR Epidermal growth factor receptor
GBM Glioblastoma
G-CIMP CpG island methylator phenotype
HDAC Histone deacetylase
IDH1 Isocitrate dehydrogenase 1
LIF Leukemia inhibitory factor
MB Medulloblastoma
MGMT O6-methylguanine-DNA methyltransferase
NF1 Neurofibromin 1
PDGFRα Platelet-derived growth factor receptor alpha
PTEN Phosphatase and tensin homolog
RB1 Retinoblastoma 1
TCGA The Cancer Genome Atlas
TCA Tricarboxylic acid
TP53 p53 tumor suppressor gene
WHO World Health Organization

REFERENCES

[1] CBTRUS. Primary brain and central nervous system tumors diagnosed in the United States in 2004–2008. http://www.cbtrus.org/2012-NPCR-SEER/CBTRUS_Report_2004-2008_3-23-2012pdf; 2012.

[2] Pollack IF. Brain tumors in children. N Engl J Med 1994;331:1500–7.

[3] DeAngelis LM. Brain tumors. N Engl J Med 2001;344:114–23.

[4] Bax DA, Mackay A, Little SE, Carvalho D, Viana-Pereira M, Tamber N, et al. A distinct spectrum of copy number aberrations in pediatric high-grade gliomas. Clin Cancer Res 2010;16:3368–77.

[5] Paugh BS, Qu C, Jones C, Liu Z, Adamowicz-Brice M, Zhang J, et al. Integrated molecular genetic profiling of pediatric high-grade gliomas reveals key differences with the adult disease. J Clin Oncol 2010;28:3061–8.

[6] Qu HQ, Jacob K, Fatet S, Ge B, Barnett D, Delattre O, et al. Genome-wide profiling using single-nucleotide polymorphism arrays identifies novel chromosomal imbalances in pediatric glioblastomas. Neuro-oncology 2010;12:153–63.

[7] Schwartzentruber J, Korshunov A, Liu XY, Jones DT, Pfaff E, Jacob K, et al. Driver mutations in histone H3.3 and chromatin remodelling genes in paediatric glioblastoma. Nature 2012;482:226–31.

[8] Ohgaki H, Kleihues P. Epidemiology and etiology of gliomas. Acta Neuropathol 2005;109:93–108.

[9] Verhaak RG, Hoadley KA, Purdom E, Wang V, Qi Y, Wilkerson MD, et al. Integrated genomic analysis identifies clinically relevant subtypes of glioblastoma characterized by abnormalities in PDGFRA, IDH1, EGFR, and NF1. Cancer Cell 2010;17:98–110.

[10] Maher EA, Furnari FB, Bachoo RM, Rowitch DH, Louis DN, Cavenee WK, et al. Malignant glioma: genetics and biology of a grave matter. Genes Devel 2001;15:1311–33.

[11] Wechsler-Reya R, Scott MP. The developmental biology of brain tumors. Annu Rev Neurosci 2001;24:385–428.

[12] Zhu Y, Parada LF. The molecular and genetic basis of neurological tumours. Nat Rev Cancer 2002;2:616–26.

[13] Huse JT, Holland EC. Targeting brain cancer: advances in the molecular pathology of malignant glioma and medulloblastoma. Nat Rev Cancer 2010;10:319–31.

[14] Stupp R, Mason WP, van den Bent MJ, Weller M, Fisher B, Taphoorn MJ, et al. Radiotherapy plus concomitant and adjuvant temozolomide for glioblastoma. N Engl J Med 2005;352:987–96.

[15] Bailey P, Cushing H. A classification of the tumors of the glioma group on a histogenetic basis with a correlated study of prognosis. Philadelphia, PA: Lippincott; 1926.

[16] Roussy PG, Oberling C. La metaplasie epitheliale et ses rapports avec les tumeurs. Ann Surg 1931;93:90–8.

[17] Kernohan JW, Mabon RF, et al. A simplified classification of the gliomas. Proc Staff Meet Mayo Clin 1949;24:71–5.

[18] Ringertz N. Grading of gliomas. Acta Pathol Microbiol Scand 1950;27:51–64.

[19] Daumas-Duport C, Scheithauer B, O'Fallon J, Kelly P. Grading of astrocytomas. A simple and reproducible method. Cancer 1988;62:2152–65.

[20] Zülch KJ, World Health Organization. Histological typing of tumours of the central nervous system. Geneva: World Health Organization; 1979.

[21] Kleihues P, Burger PC, Scheithauer BW, Zülch KJ. Histological typing of tumours of the central nervous system. Berlin, New York: Springer-Verlag; 1993.

[22] Kleihues P, Cavenee WK, International Agency for Research on Cancer. Pathology and genetics of tumours of the nervous system. Lyon: IARC Press; 2000.

[23] Louis DN, International Agency for Research on Cancer, World Health Organization. WHO classification of tumours of the central nervous system. Lyon: International Agency for Research on Cancer; 2007.

[24] Miller CR, Perry A. Glioblastoma. Arch Pathol Lab Med 2007;131:397–406.

[25] Kleihues P, Ohgaki H. Primary and secondary glioblastomas: from concept to clinical diagnosis. Neuro-oncology 1999;1:44–51.

[26] Ohgaki H, Kleihues P. Genetic pathways to primary and secondary glioblastoma. Am J Pathol 2007;170:1445–53.

[27] Shai R, Shi T, Kremen TJ, Horvath S, Liau M, Cloughesy TF, et al. Gene expression profiling identifies molecular subtypes of gliomas. Oncogene 2003;22:4918–23.

[28] Freije WA, Castro-Vargas FE, Fang Z, Horvath S, Cloughesy T, Liau LM, et al. Gene expression profiling of gliomas strongly predicts survival. Cancer Res 2004;64:6503–10.

[29] Liang Y, Diehn M, Watson N, Bollen AW, Aldape KD, Nicholas MK, et al. Gene expression profiling reveals molecularly and clinically distinct subtypes of glioblastoma multiforme. Proc Natl Acad Sci USA 2005;102:5814–9.

[30] Maher EA, Brennan C, Wen PY, Durso L, Ligon KL, Richardson A, et al. Marked genomic differences characterize primary and secondary glioblastoma subtypes and identify two distinct molecular and clinical secondary glioblastoma entities. Cancer Res 2006;66:11502–13.

[31] Ruano Y, Mollejo M, Ribalta T, Fiaño C, Camacho FI, Gómez E, et al. Identification of novel candidate target genes in amplicons of glioblastoma multiforme tumors detected by expression and CGH microarray profiling. Mol Cancer 2006;5:39.

[32] TCGA CGARN. Comprehensive genomic characterization defines human glioblastoma genes and core pathways. Nature 2008; 455:1061–8.

[33] Phillips HS, Kharbanda S, Chen R, Forrest WF, Soriano RH, Wu TD, et al. Molecular subclasses of high-grade glioma predict prognosis, delineate a pattern of disease progression, and resemble stages in neurogenesis. Cancer Cell 2006;9:157–73.

[34] Sun L, Hui AM, Su Q, Vortmeyer A, Kotliarov Y, Pastorino S, et al. Neuronal and glioma-derived stem cell factor induces angiogenesis within the brain. Cancer Cell 2006;9:287–300.

[35] Beroukhim R, Getz G, Nghiemphu L, Barretina J, Hsueh T, Linhart D, et al. Assessing the significance of chromosomal aberrations in cancer: methodology and application to glioma. Proc Natl Acad Sci USA 2007;104:20007–12.

[36] Murat A, Migliavacca E, Gorlia T, Lambiv WL, Shay T, Hamou MF, et al. Stem cell-related "self-renewal" signature and high epidermal growth factor receptor expression associated with resistance to concomitant chemoradiotherapy in glioblastoma. J Clin Oncol 2008;26:3015–24.

[37] Furnari FB, Fenton T, Bachoo RM, Mukasa A, Stommel JM, Stegh A, et al. Malignant astrocytic glioma: genetics, biology, and paths to treatment. Genes Dev 2007;21:2683–710.

[38] Kastan MB, Zhan Q, el-Deiry WS, Carrier F, Jacks T, Walsh WV, et al. A mammalian cell cycle checkpoint pathway utilizing p53 and GADD45 is defective in ataxia-telangiectasia. Cell 1992;71:587–97.

[39] Clarke AR, Purdie CA, Harrison DJ, Morris RG, Bird CC, Hooper ML, et al. Thymocyte apoptosis induced by p53-dependent and independent pathways. Nature 1993;362:849–52.

[40] Matoba S, Kang JG, Patino WD, Wragg A, Boehm M, Gavrilova O, et al. p53 regulates mitochondrial respiration. Science 2006;312:1650–3.

[41] Crighton D, Wilkinson S, O'Prey J, Syed N, Smith P, Harrison PR, et al. DRAM, a p53-induced modulator of autophagy, is critical for apoptosis. Cell 2006;126:121–34.

[42] Meletis K, Wirta V, Hede SM, Nister M, Lundeberg J, Frisen J. p53 suppresses the self-renewal of adult neural stem cells. Development 2006;133:363–9.

[43] Tamura M, Gu J, Matsumoto K, Aota S, Parsons R, Yamada KM. Inhibition of cell migration, spreading, and focal adhesions by tumor suppressor PTEN. Science 1998;280:1614–7.

[44] Ekstrand AJ, Sugawa N, James CD, Collins VP. Amplified and rearranged epidermal growth factor receptor genes in human glioblastomas reveal deletions of sequences encoding portions of the N- and/or C-terminal tails. Proc Natl Acad Sci USA 1992; 89:4309–13.

[45] Watanabe K, Tachibana O, Sata K, Yonekawa Y, Kleihues P, Ohgaki H. Overexpression of the EGF receptor and p53 mutations are mutually exclusive in the evolution of primary and secondary glioblastomas. Brain Pathol 1996;6:217–23 [discussion 223–214]

[46] Huang HS, Nagane M, Klingbeil CK, Lin H, Nishikawa R, Ji XD, et al. The enhanced tumorigenic activity of a mutant epidermal growth factor receptor common in human cancers is mediated by threshold levels of constitutive tyrosine phosphorylation and unattenuated signaling. J Biol Chem 1997;272:2927–35.

[47] Narita Y, Nagane M, Mishima K, Huang HJ, Furnari FB, Cavenee WK. Mutant epidermal growth factor receptor signaling down-regulates p27 through activation of the phosphatidylinositol 3-kinase/Akt pathway in glioblastomas. Cancer Res 2002;62:6764–9.

[48] Parsons DW, Jones S, Zhang X, Lin JC, Leary RJ, Angenendt P, et al. An integrated genomic analysis of human glioblastoma multiforme. Science 2008;321:1807–12.

[49] McGillicuddy LT, Fromm JA, Hollstein PE, Kubek S, Beroukhim R, De Raedt T, et al. Proteasomal and genetic inactivation of the NF1 tumor suppressor in gliomagenesis. Cancer Cell 2009;16:44–54.

[50] Banerjee S, Crouse NR, Emnett RJ, Gianino SM, Gutmann DH. Neurofibromatosis-1 regulates mTOR-mediated astrocyte growth and glioma formation in a TSC/Rheb-independent manner. Proc Natl Acad Sci USA 2011;108:15996–6001.

[51] Ichimura K, Pearson DM, Kocialkowski S, Bäcklund LM, Chan R, Jones DT, et al. IDH1 mutations are present in the majority of common adult gliomas but rare in primary glioblastomas. Neuro-oncology 2009;11:341–7.

[52] Raimundo N, Baysal BE, Shadel GS. Revisiting the TCA cycle: signaling to tumor formation. Trends Mol Med 2011;17:641–9.

[53] Smith JS, Wang XY, Qian J, Hosek SM, Scheithauer BW, Jenkins RB, et al. Amplification of the platelet-derived growth factor receptor-A (PDGFRA) gene occurs in oligodendrogliomas with grade IV anaplastic features. J Neuropathol Exp Neurol 2000;59:495–503.

[54] Martinho O, Longatto-Filho A, Lambros MB, Martins A, Pinheiro C, Silva A, et al. Expression, mutation and copy number analysis of platelet-derived growth factor receptor A (PDGFRA) and its ligand PDGFA in gliomas. Br J Cancer 2009;101:973–82.

[55] Ichimura K, Schmidt EE, Goike HM, Collins VP. Human glioblastomas with no alterations of the CDKN2A (p16INK4A, MTS1) and CDK4 genes have frequent mutations of the retinoblastoma gene. Oncogene 1996;13:1065–72.

[56] Kim TY, Zhong S, Fields CR, Kim JH, Robertson KD. Epigenomic profiling reveals novel and frequent targets of aberrant DNA methylation-mediated silencing in malignant glioma. Cancer Res 2006;66:7490–501.

[57] Cadieux B, Ching TT, VandenBerg SR, Costello JF. Genome-wide hypomethylation in human glioblastomas associated with specific copy number alteration, methylenetetrahydrofolate reductase allele status, and increased proliferation. Cancer Res 2006;66:8469–76.

[58] Agnihotri S, Wolf A, Munoz DM, Smith CJ, Gajadhar A, Restrepo A, et al. A GATA4-regulated tumor suppressor network represses formation of malignant human astrocytomas. J Exp Med 2011;208:689–702.

[59] Nakamura M, Watanabe T, Yonekawa Y, Kleihues P, Ohgaki H. Promoter methylation of the DNA repair gene MGMT in astrocytomas is frequently associated with G:C → A:T mutations of the TP53 tumor suppressor gene. Carcinogenesis 2001;22:1715–9.

[60] Lindemann C, Hackmann O, Delic S, Schmidt N, Reifenberger G, Riemenschneider MJ. SOCS3 promoter methylation is mutually exclusive to EGFR amplification in gliomas and promotes glioma cell invasion through STAT3 and FAK activation. Acta Neuropathol 2011;122:241–51.

[61] Alonso MM, Diez-Valle R, Manterola L, Rubio A, Liu D, Cortes-Santiago N, et al. Genetic and epigenetic modifications of Sox2 contribute to the invasive phenotype of malignant gliomas. PLoS One 2011;6:e26740.

[62] Esteller M, Hamilton SR, Burger PC, Baylin SB, Herman JG. Inactivation of the DNA repair gene O6-methylguanine-DNA methyltransferase by promoter hypermethylation is a common event in primary human neoplasia. Cancer Res 1999;59:793–7.

[63] Hegi ME, Diserens AC, Gorlia T, Hamou MF, de Tribolet N, Weller M, et al. MGMT gene silencing and benefit from temozolomide in glioblastoma. N Engl J Med 2005;352:997–1003.

[64] Noushmehr H, Weisenberger DJ, Diefes K, Phillips HS, Pujara K, Berman BP, et al. Identification of a CpG island methylator phenotype that defines a distinct subgroup of glioma. Cancer Cell 2010;17:510–22.

[65] Lu C, Ward PS, Kapoor GS, Rohle D, Turcan S, Abdel-Wahab O, et al. IDH mutation impairs histone demethylation and results in a block to cell differentiation. Nature 2012;483:474–8.

[66] Campos B, Bermejo JL, Han L, Felsberg J, Ahmadi R, Grabe N, et al. Expression of nuclear receptor corepressors and class I histone deacetylases in astrocytic gliomas. Cancer Sci 2011;102:387–92.

[67] Esteller M. Cancer epigenomics: DNA methylomes and histone-modification maps. Nat Rev Genet 2007;8:286–98.

[68] Singh SK, Clarke ID, Terasaki M, Bonn VE, Hawkins C, Squire J, et al. Identification of a cancer stem cell in human brain tumors. Cancer Res 2003;63:5821–8.

[69] Singh SK, Hawkins C, Clarke ID, Squire JA, Bayani J, Hide T, et al. Identification of human brain tumour initiating cells. Nature 2004;432:396–401.

[70] Chen R, Nishimura MC, Bumbaca SM, Kharbanda S, Forrest WF, Kasman IM, et al. A hierarchy of self-renewing tumor-initiating cell types in glioblastoma. Cancer Cell 2010;17:362–75.

[71] Inda MM, Bonavia R, Mukasa A, Narita Y, Sah DW, Vandenberg S, et al. Tumor heterogeneity is an active process maintained by a mutant EGFR-induced cytokine circuit in glioblastoma. Genes Dev 2010;24:1731–45.

[72] Nickel GC, Barnholtz-Sloan J, Gould MP, McMahon S, Cohen A, Adams MD, et al. Characterizing mutational heterogeneity in a glioblastoma patient with double recurrence. PLoS One 2012;7: e35262.

[73] Stupp R, Hegi ME, Mason WP, van den Bent MJ, Taphoorn MJ, Janzer RC, et al. Effects of radiotherapy with concomitant and adjuvant temozolomide versus radiotherapy alone on survival in glioblastoma in a randomised phase III study: 5-year analysis of the EORTC-NCIC trial. Lancet Oncol 2009;10:459–66.

[74] Lu Z, Liu M, Stribinskis V, Klinge CM, Ramos KS, Colburn NH, et al. MicroRNA-21 promotes cell transformation by targeting the programmed cell death 4 gene. Oncogene 2008;27:4373–9.

[75] Silber J, Lim DA, Petritsch C, Persson AI, Maunakea AK, Yu M, et al. miR-124 and miR-137 inhibit proliferation of glioblastoma multiforme cells and induce differentiation of brain tumor stem cells. BMC Med 2008;6:14.

[76] Burger PC, Vogel FS, Green SB, Strike TA. Glioblastoma multiforme and anaplastic astrocytoma. Pathologic criteria and prognostic implications. Cancer 1985;56:1106–11.

[77] Ohgaki H, Kleihues P. Population-based studies on incidence, survival rates, and genetic alterations in astrocytic and oligodendroglial gliomas. J Neuropathol Exp Neurol 2005;64:479–89.

[78] Wen PY, Kesari S, Drappatz J. Malignant gliomas: strategies to increase the effectiveness of targeted molecular treatment. Expert Rev Anticancer Ther 2006;6:733–54.

[79] Wright JH. Neurocytoma or neuroblastoma, a kind of tumor not generally recognized. J Exp Med 1910;12:556–61.

[80] Bailey P, Cushing H. Medulloblastoma cerebelli: a common type of midcerebellar glioma of childhood. Arch Neurol Psychiatr 1925;14:192.

[81] Pomeroy SL, Tamayo P, Gaasenbeek M, Sturla LM, Angelo M, McLaughlin ME, et al. Prediction of central nervous system embryonal tumour outcome based on gene expression. Nature 2002;415:436–42.

[82] Thompson MC, Fuller C, Hogg TL, Dalton J, Finkelstein D, Lau CC, et al. Genomics identifies medulloblastoma subgroups that are enriched for specific genetic alterations. J Clin Oncol 2006;24:1924–31.

[83] Kool M, Koster J, Bunt J, Hasselt N, Lakeman A, van Sluis P, et al. Integrated genomics identifies five medulloblastoma subtypes with distinct genetic profiles, pathway signatures and clinicopathological features. PLoS One 2008;3:e3088.

[84] Cho YJ, Tsherniak A, Tamayo P, Santagata S, Ligon A, Greulich H, et al. Itegrative genomic analysis of medulloblastoma identifies a molecular subgroup that drives poor clinical outcome. J Clin Oncol 2011;29:1424–30.

[85] Northcott PA, Korshunov A, Witt H, Hielscher T, Eberhart CG, Mack S, et al. Medulloblastoma comprises four distinct molecular variants. J Clin Oncol 2011;29:1408–14.

[86] Taylor MD, Northcott PA, Korshunov A, Remke M, Cho YJ, Clifford SC, et al. Molecular subgroups of medulloblastoma: the current consensus. Acta Neuropathol 2012;123:465–72.

[87] Bourdeaut F, Miquel C, Alapetite C, Roujeau T, Doz F. Medulloblastomas: update on a heterogeneous disease. Curr Opin Oncol 2011;23:630–7.

[88] Ellison DW, Onilude OE, Lindsey JC, Lusher ME, Weston CL, Taylor RE, et al. β-Catenin status predicts a favorable outcome in childhood medulloblastoma: the United Kingdom Children's Cancer Study Group Brain Tumour Committee. J Clin Oncol 2005;23:7951–7.

[89] Clifford SC, Lusher ME, Lindsey JC, Langdon JA, Gilbertson RJ, Straughton D, et al. Wnt/Wingless pathway activation and chromosome 6 loss characterize a distinct molecular sub-group of medulloblastomas associated with a favorable prognosis. Cell Cycle 2006;5:2666–70.

[90] Fattet S, Haberler C, Legoix P, Varlet P, Lellouch-Tubiana A, Lair S, et al. Beta-catenin status in paediatric medulloblastomas: correlation of immunohistochemical expression with mutational status, genetic profiles, and clinical characteristics. J Pathol 2009;218:86–94.

[91] Kool M, Korshunov A, Remke M, Jones DT, Schlanstein M, Northcott PA, et al. Molecular subgroups of medulloblastoma: an international meta-analysis of transcriptome, genetic aberrations, and clinical data of WNT, SHH, Group 3, and Group 4 medulloblastomas. Acta Neuropathol 2012;123:473–84.

[92] Kawauchi D, Robinson G, Uziel T, Gibson P, Rehg J, Gao C, et al. A mouse model of the most aggressive subgroup of human medulloblastoma. Cancer Cell 2012;21:168–80.

[93] Pei Y, Moore CE, Wang J, Tewari AK, Eroshkin A, Cho YJ, et al. An animal model of MYC-driven medulloblastoma. Cancer Cell 2012;21:155–67.

[94] Ellison DW. Childhood medulloblastoma: novel approaches to the classification of a heterogeneous disease. Acta Neuropathologica 2010;120:305–16.

[95] Pugh TJ, Weeraratne SD, Archer TC, Pomeranz Krummel DA, Auclair D, Bochicchio J, et al. Medulloblastoma exome sequencing uncovers subtype-specific somatic mutations. Nature 2012;488:106–10.

[96] Robinson G, Parker M, Kranenburg TA, Lu C, Chen X, Ding L, et al. Novel mutations target distinct subgroups of medulloblastoma. Nature 2012;488:43–8.

[97] Northcott PA, Shih DJ, Peacock J, Garzia L, Morrissy AS, Zichner T, et al. Subgroup-specific structural variation across 1,000 medulloblastoma genomes. Nature 2012;488:49–56.

[98] Northcott PA, Nakahara Y, Wu X, Feuk L, Ellison DW, Croul S, et al. Multiple recurrent genetic events converge on control of histone lysine methylation in medulloblastoma. Nat Genet 2009;41:465–72.

[99] Furchert SE, Lanvers-Kaminsky C, Juurgens H, Jung M, Loidl A, Fruhwald MC. Inhibitors of histone deacetylases as potential therapeutic tools for high-risk embryonal tumors of the nervous system of childhood. Int J Cancer 2007;120:1787–94.

[100] Lauth M, Bergstrom A, Shimokawa T, Toftgard R. Inhibition of GLI-mediated transcription and tumor cell growth by small-molecule antagonists. Proc Natl Acad Sci USA 2007;104:8455–60.

[101] Huber H, Eggert A, Janss AJ, Wiewrodt R, Zhao H, Sutton LN, et al. Angiogenic profile of childhood primitive neuroectodermal brain tumours/medulloblastomas. Eur J Cancer 2001;37:2064–72.

[102] MacDonald TJ, Taga T, Shimada H, Tabrizi P, Zlokovic BV, Cheresh DA, et al. Preferential susceptibility of brain tumors to the antiangiogenic effects of an alpha(v) integrin antagonist. Neurosurgery 2001;48:151–7.

[103] Seaman S, Stevens J, Yang MY, Logsdon D, Graff-Cherry St C, Croix B. Genes that distinguish physiological and pathological angiogenesis. Cancer Cell 2007;11:539–54.

[104] Eyler CE, Rich JN. Survival of the fittest: cancer stem cells in therapeutic resistance and angiogenesis. J Clin Oncol 2008; 26:2839–45.

[105] Gerlinger M, Rowan AJ, Horswell S, Larkin J, Endesfelder D, Gronroos E, et al. Intratumor heterogeneity and branched evolution revealed by multiregion sequencing. N Engl J Med 2012;366:883–92.

[106] Wang X, Venugopal C, Manoranjan B, McFarlane N, O'Farrell E, Nolte S, et al. Sonic hedgehog regulates Bmi1 in human medulloblastoma brain tumor-initiating cells. Oncogene 2012;31:187–99.

[107] Kilday JP, Rahman R, Dyer S, Ridley L, Lowe J, Coyle B, et al. Pediatric ependymoma: biological perspectives. Mol Cancer Res 2009;7:765–86.

[108] Mack SC, Taylor MD. The genetic and epigenetic basis of ependymoma. Childs Nerv Syst 2009;25:1195–201.

[109] Mendrzyk F, Korshunov A, Benner A, Toedt G, Pfister S, Radlwimmer B, et al. Identification of gains on 1q and epidermal growth factor receptor overexpression as independent prognostic markers in intracranial ependymoma. Clin Cancer Res 2006;12:2070–9.

[110] Modena P, Lualdi E, Facchinetti F, Veltman J, Reid JF, Minardi S, et al. Identification of tumor-specific molecular signatures in intracranial ependymoma and association with clinical characteristics. J Clin Oncol 2006;24:5223–33.

[111] Puget S, Grill J, Valent A, Bieche I, Dantas-Barbosa C, Kauffmann A, et al. Candidate genes on chromosome 9q33-34 involved in the progression of childhood ependymomas. J Clin Oncol 2009;27:1884–92.

[112] Korshunov A, Witt H, Hielscher T, Benner A, Remke M, Ryzhova M, et al. Molecular staging of intracranial ependymoma in children and adults. J Clin Oncol 2010;28:3182–90.

[113] Witt H, Mack SC, Ryzhova M, Bender S, Sill M, Isserlin R, et al. Delineation of two clinically and molecularly distinct subgroups of posterior fossa ependymoma. Cancer Cell 2011;20:143–57.

[114] Shonka NA. Targets for therapy in ependymoma. Target Oncol 2011;6:163–9.

[115] Rahman R, Osteso-Ibanez T, Hirst RA, Levesley J, Kilday JP, Quinn S, et al. Histone deacetylase inhibition attenuates cell growth with associated telomerase inhibition in high-grade childhood brain tumor cells. Mol Cancer Ther 2010;9:2568–81.

[116] Pollack IF, Finkelstein SD, Burnham J, Holmes EJ, Hamilton RL, Yates AJ, et al. Age and TP53 mutation frequency in childhood malignant gliomas: results in a multi-institutional cohort. Cancer Res 2001;61:7404–7.

[117] Wu G, Broniscer A, McEachron TA, Lu C, Paugh BS, Becksfort J, et al. Somatic histone H3 alterations in pediatric diffuse intrinsic pontine gliomas and non-brainstem glioblastomas. Nat Genet 2012;44:251–3.

[118] Desgrosellier JS, Cheresh DA. Integrins in cancer: biological implications and therapeutic opportunities. Nat Rev Cancer 2010;10:9–22.

[119] Donson AM, Addo-Yobo SO, Handler MH, Gore L, Foreman NK. MGMT promoter methylation correlates with survival benefit and sensitivity to temozolomide in pediatric glioblastoma. Pediatr Blood Cancer 2007;48:403–7.

[120] Donaldson SS, Laningham F, Fisher PG. Advances toward an understanding of brainstem gliomas. J Clin Oncol 2006;24:1266–72.

[121] Khuong-Quang DA, Buczkowicz P, Rakopoulos P, Liu XY, Fontebasso AM, Bouffet E, et al. K27M mutation in histone H3.3 defines clinically and biologically distinct subgroups of pediatric diffuse intrinsic pontine gliomas. Acta Neuropathol 2012;124:439–47.

[122] Hargrave D, Bartels U, Bouffet E. Diffuse brainstem glioma in children: critical review of clinical trials. Lancet Oncol 2006;7:241–8.

[123] MacDonald TJ, Aguilera D, Kramm CM. Treatment of high-grade glioma in children and adolescents. Neuro-oncology 2011;13:1049–58.

Part 4.3

Pediatric Cancers

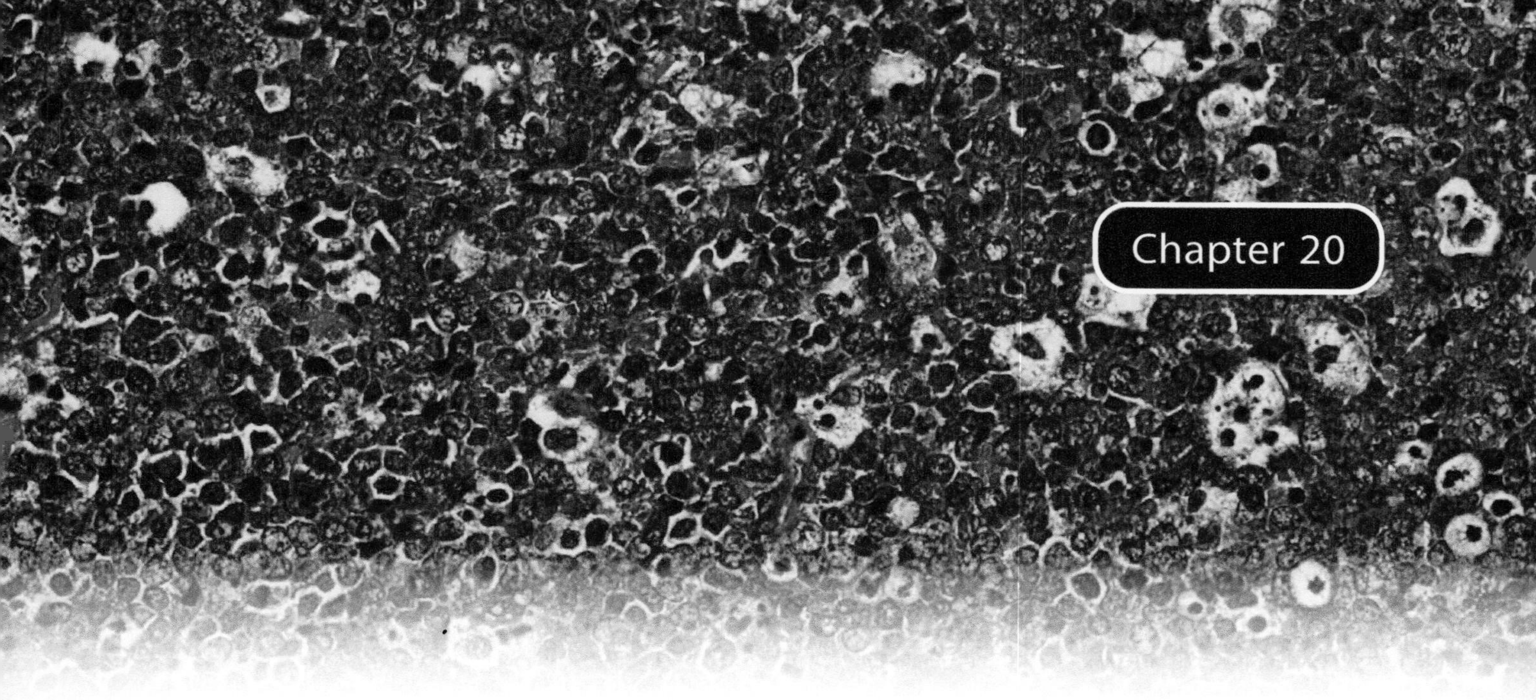

Cancer Genomics

Genomic Basis of Pediatric Lymphomas

Nathanael G. Bailey
Thomas Gross
Megan S. Lim

Department of Pathology, University of Michigan, Ann Arbor, MI, USA
Department of Pediatrics, The Ohio State University, Nationwide Children's Hospital Columbus, OH, USA

Contents

Cancer Genomics. DOI: http://dx.doi.org/10.1016/B978-0-12-396967-5.00020-7

Key Concepts

- Lymphomas are the third-most common group of tumors in children
- Lymphomas may be classified into a variety of subtypes with different clinical behavior
- Genomic profiling studies have augmented our understanding of lymphoma and have demonstrated different groups with similar morphology
- Although pediatric and adult lymphomas exhibit some similarity at the genomic level, some genetic aberrations are more common in the pediatric population
- The identification of genes and pathways implicated in lymphomagenesis may allow targeted therapies to be employed
- Further studies are needed to increase our understanding of the genetic basis of lymphoma, particularly in the pediatric population

INTRODUCTION

Mature lymphomas are a varied group of malignancies that account for approximately 11% of childhood cancer in the USA and are the third most common group of tumors in the pediatric population [1]. Mature lymphomas may be further subclassified clinically as Hodgkin and non-Hodgkin lymphomas; the non-Hodgkin lymphoma (NHL) category is itself heterogeneous and comprises neoplasms of both B-cell and T-cell origin. Lymphomas are uncommon in infants, and the incidence increases with age. The age-adjusted annual incidence of lymphoma in children under the age of 14 is approximately 17 per 1 million, nearly evenly divided between Hodgkin and non-Hodgkin categories. The incidence of lymphoma increases substantially in the adolescent population (patients 15–19 years of age), when Hodgkin lymphoma becomes predominant [1].

The spectrum of NHL that occurs in children differs markedly from that of adults. Generally, childhood lymphomas consist of aggressive subtypes (for example, Burkitt lymphoma, diffuse large B-cell lymphoma, anaplastic large cell lymphoma, lymphoblastic lymphoma), while the indolent lymphomas frequently identified in adults (for example, chronic lymphocytic leukemia/small lymphocytic lymphoma, low-grade follicular lymphoma) are very uncommon [2]. In this chapter, we provide an updated summary of the large-scale genomic studies of common B-cell and T-cell lymphomas that occur in children and young adolescents and highlight differences with those of adult lymphomas and stress the importance of continued investigations of lymphomas of the pediatric population.

OVERVIEW OF B-CELL DEVELOPMENT

B cells are part of the adaptive immune system, and their major function is the production of antibodies that specifically recognize foreign antigens in order to facilitate their elimination from the body. B-cell development begins in the bone marrow where initial V(D)J rearrangements of the immunoglobulin genes occur through double-stranded DNA breaks mediated by RAG1 and RAG2. The combinatorial diversity generated by V(D)J rearrangement, in addition to non-templated addition of random nucleotides by the action of TdT, generates a broad repertoire of immunoglobulins with sufficient variation to recognize many foreign antigens. The membrane-bound immunoglobulin, with CD79A/B, forms the B-cell receptor (BCR). Mature, antigen inexperienced, naïve B cells then leave the bone marrow and enter the circulation. BCL2 pathways inhibiting apoptosis are upregulated in this B-cell subset. Upon exposure to antigen, these naïve, IgM- and IgD-expressing B cells take part in the T-cell dependent germinal center reaction. The germinal center is a specialized structure within the lymph node where the antigenic affinity of the immunoglobulin expressed by the B cells is increased through a process known as affinity maturation. Germinal center B cells upregulate BCL6 expression, downregulate BCL2, proliferate rapidly, and undergo somatic hypermutation, a process mediated by activation-induced deaminase (AID). Through somatic hypermutation of the previously rearranged immunoglobulin genes, greater diversity is generated. Many of these newly hypermutated immunoglobulin genes are non-functional or less specific for the antigen than the unmutated forms; progeny expressing these immunoglobulins undergo apoptosis, which is facilitated by the downregulation of BCL2. The progeny B cells expressing immunoglobulin with increased antigenic affinity are selected for survival. Many of these cells undergo AID-mediated class switching of the heavy chain, leading to IgG, IgA, or IgE expression. The antigen-selected cells upregulate IRF4, which leads to downregulation of BCL6, and the cells proceed to differentiate into memory B cells or plasma cells [3]. Plasma cell differentiation requires downregulation of PAX5 and expression of BLIMP1. Understanding normal B-cell development is relevant to the pathogenesis of B-cell neoplasia, as B-cell lymphomas exhibit phenotypes comparable to normal B cells at various stages of development, and many genetic aberrancies present in

B-cell lymphomas may be due to errors that are secondary to V(D)J rearrangement, somatic hypermutation, and class switch recombination (e.g. *IG* rearrangements, hypermutation of *BCL6*, etc.) [4,5].

LYMPHOBLASTIC LYMPHOMA

Immature neoplasms of T-cell and B-cell precursors are called either acute lymphoblastic leukemias (ALL) or lymphoblastic lymphomas (LBL), depending upon whether they primarily involve extramedullary sites (LBL) or have significant bone marrow involvement (ALL). An arbitrary cut-off of 25% bone marrow blasts has been used to distinguish between the two diagnoses [6]. Precursor lymphoid neoplasms are categorized based upon their B-cell or T-cell lineage commitment. Within a lineage, ALL and LBL are considered to be biologically equivalent in the WHO framework [6].

B-Lymphoblastic Lymphoma

B-ALL/LBL typically has a leukemic clinical presentation and constitutes a minority of cases of LBL [7]. In contrast with T-LBL, B-LBL rarely has mediastinal involvement and frequently exhibits an extranodal distribution [8]. B-ALL/LBL is subcategorized by cytogenetic abnormality which provides for risk stratification. B-ALL/LBLs with high hyperdiploidy or ets variant 6/runt-related transcription factor 1 fusion *(ETV6-RUNX1)* fusion are considered "good risk" [9]. Those cases with hypodiploidy, myeloid/lymphoid or mixed-lineage leukemia (trithorax homolog, Drosophila) *(MLL)* rearrangements, and breakpoint cluster region/c-abl oncogene 1, non-receptor tyrosine kinase fusion *(BCR–ABL1)* fusion tend to have poor outcomes [9]. Two other cytogenetic subgroups are defined by the WHO: those with interleukin 3 (colony-stimulating factor, multiple)/immunoglobulin heavy chain locus fusion *(IL3–IGH@)*, and those with transcription factor 3 (E2A immunoglobulin enhancer binding factors E12/E47)/pre-B-cell leukemia homeobox 1 fusion *(TCF3–PBX1)*[6], both of which appear to have intermediate risk with modern treatment approaches [9].

Recent genome-wide studies have further expanded the spectrum of mutations associated with B-ALL/LBL. Paired box 5 *(PAX5)*, a transcription factor necessary for B-lineage commitment, is deleted or inactivated through mutation in $\approx$30% of B-ALL cases [10]. In an additional 10% of cases, deletions of other genes necessary for B-cell development are present (early B-cell factor 1 *(EBF1)*, IKAROS family zinc finger 1 *(IKZF1)*, IKAROS family zinc finger 3 *(IKZF3)*, B-cell linker *(BLNK)*, lymphoid enhancer-binding factor 1 *(LEF1)*, and *(TCF3)*) [10]. *IKZF1* deletion is present in the majority of *BCR-ABL1* fusion-positive B-ALL/LBL [11] and its deletion is associated with a poor prognosis in *BCR–ABL1* fusion-negative cases [12]. Many *IKZF1*-deleted B-ALLs without *BCR–ABL1* fusion exhibit gene expression very similar to B-ALLs with *BCR–ABL1* fusion [12]. Many of these B-ALLs with "*BCR–ABL1*-like" gene expression and *IKZF1* alterations have translocations leading to upregulation of cytokine receptor-like factor 2 *(CRLF2)* [13–15] and mutations in Janus kinases 1, 2, or 3 *(JAK1, JAK2,* or *JAK3)* [13,16].

Patients with B-ALL/LBL with *BCR–ABL1* fusion or "*BCR–ABL1*-like" gene expression have high-risk disease. Tyrosine kinase inhibitors (TKI) that are effective in chronic myelogenous leukemia (imatinib, dasatinib, nilotinib, sorafenib, ponatinib) are being extensively evaluated in *BCR–ABL1*-positive B-ALL/LBL. As "*BCR–ABL1*-like" cases frequently have *JAK* mutations, *JAK* inhibitors such as lestaurtinib and ruxolitinib are beginning to be explored in trials.

T-Lymphoblastic Lymphoma

T-ALL/LBL only accounts for around 15% of childhood ALL; however, it constitutes the majority (85–90%) of LBL, commonly manifesting as a mediastinal mass with bulky disease [6]. T-ALL/LBL frequently has rearrangements involving the T-cell receptor (TCR) loci [17]. Many of these rearrangements juxtapose transcription factors and the enhancer/promoter region of the TCR locus, leading to inappropriate overexpression of the transcription factor. Examples include rearrangements involving T-cell leukemia homeobox 1 *(TLX1 (HOX11))*, T-cell leukemia homeobox 3 *(TLX3 (HOX11L2))*, T-cell acute lymphocytic leukemia 1 *(TAL1 (SCL))*, LIM domain only 2 (rhombotin-like 1) *(LMO2)*, and lymphoblastic leukemia derived sequence 1 *(LYL1)* [18–22]. Gene expression profiling has identified subgroups of T-ALL/LBL distinguished by transcription factor overexpression. T-ALL/LBL with overexpression of *LYL1* exhibit a pro-T, CD4/CD8 double-negative phenotype, *TLX1* overexpressing cases have an early cortical CD4/CD8 double-positive phenotype, and *TAL1* overexpressing cases have a late cortical double-positive phenotype with expression of surface CD3 [23]. *TLX1* overexpressing cases appear to have a favorable prognosis [23–25], while expression of *TAL1* or *LYL1* was associated with poorer outcomes [23]. Deregulation of these transcription factors frequently occurs in the absence of cytogenetically recognizable translocations [23], and abnormalities identified by classical cytogenetics are not especially predictive of outcome [26], in contrast with B-ALL/LBL.

Additional oncogenic gene fusions seen in T-ALL/LBL include activation of tyrosine kinases through TCR beta locus and lymphocyte-specific protein tyrosine kinase fusions *(TRB@–LCK)* [27] and chimeric fusions of *ABL1* and nucleoporin 214 kDa *(NUP214)* [28]. *NUP214–ABL1* fusions are associated with overexpression of either *TLX1* or *TLX3* and deletions of cyclin-dependent kinase inhibitor 2A/B *(CDKN2A/B (p16INK4a/p15INK4b))* [28]; deletions of *CDKN2A/B* are generally very common in T-ALL/LBL, occurring in approximately 70% of cases [10].

Mutations of notch 1 *(NOTCH1)* are frequent in T-ALL/LBL, occurring in greater than 50% of cases and in multiple gene expression subgroups [29]. The mutations typically involve the heterodimerization domain or the C-terminal PEST domain of the protein, and lead to increased NOTCH signaling. Mutations of F-box and WD repeat domain containing 7, E3 ubiquitin protein ligase *(FBXW7)* are seen in approximately 30% of T-ALL/LBL and stabilize the intracellular domain of NOTCH1, which is crucial to its signaling [30,31]. The presence of *NOTCH1/FBXW7* mutations appears to be associated with improved outcomes [32,33].

Inactivating mutations of PHD finger protein 6 *(PHF6)*, located at Xq26, have been identified in approximately 16% of pediatric and 38% of adult T-ALL/LBL. *PHF6* mutations are associated with abnormal expression of *TLX1* and *TLX3* and occur nearly exclusively in males. These inactivating mutations may partly explain the male predominance seen in T-ALL/LGL [34].

A limited number of studies have compared T-ALL and T-LBL at the genetic level, generally supporting that the diseases are largely genetically similar, although some differences have been suggested [35–37]. *MLL* may be overexpressed preferentially in LBL cases [36], and losses of the neurofibromin 1 *(NF1)* locus may be more frequent in LBL than ALL [35]. A zebrafish model has led to the recognition that human T-LBLs exhibit increased B-cell CLL/lymphoma 2 *(BCL2)* expression relative to T-ALLs, with impaired vascular invasion [38]. In this model, AKT activation facilitated dissemination of T-LBL, suggesting a role for anti-BCL2 and AKT inhibitors in T-LBL [38].

The elucidation of the genetic pathways that are dysregulated in T-LBL holds promise in contributing to treatment. The implication of NOTCH signaling pathways in T-ALL/LBL pathogenesis suggests that drugs such as gamma-secretase inhibitors (GSI) may be effective. However, *FBXW7* mutations appear to impart resistance to GSI therapy, potentially limiting their utility [31]. Investigational compounds targeting Notch1 or gamma secretase include BMS-906024, MK0752, and PF-03084014. These compounds, along with AKT inhibitors (afuresertib, MK2206) have begun to be assessed in clinical trials.

The majority of the genomic work on lymphoblastic neoplasia has been performed on leukemic samples. While many of the findings are likely largely translatable to LBL, further studies of LBL specimens are necessary to determine if therapeutically relevant differences exist between LBL and ALL.

MATURE B-CELL LYMPHOMAS

In the pediatric population, the most common mature B-cell lymphomas are Burkitt lymphoma (BL) and diffuse large B-cell lymphoma (DLBCL). These are both aggressive neoplasms which are considered to be distinct pathologic entities by the World Health Organization classification [6]. DLBCL is the most common lymphoma of adults [39] and is a morphologic diagnosis of exclusion: a diffuse proliferation of large- to medium-sized B cells that does not meet diagnostic criteria for another lymphoma subtype is DLBCL, not otherwise specified [6]. As might be expected, this diagnostic category does not represent a discrete biologic entity, and outcomes are heterogeneous. In contrast with the marked heterogeneity present within DLBCL, BL has long been recognized as a discrete pathologic entity [40], albeit one with clinical and epidemiologic variants [41]. Prototypical BL has a characteristic childhood onset, typical morphologic features, and v-myc myelocytomatosis viral oncogene homolog (avian) *(MYC)* rearrangements as a characteristic genetic lesion [6]. BL is highly aggressive, but is frequently cured with intensive chemotherapy.

Diffuse Large B-Cell Lymphoma

A seminal study by Alizadeh et al. examined the heterogeneity of DLBCL through gene expression analysis using cDNA microarrays [42]. Two major subgroups of DLBCL could be identified based upon gene expression patterns: a germinal center B-cell-like group (GCB) and an activated B-cell-like group (ABC) (Figure 20.1). In the adult population, approximately 45–50% of DLBCL have the GCB signature, with ABC cases accounting for approximately 30–40% [43,44]. Not only is DLBCL genetically heterogeneous, the differences are clinically relevant: the ABC subgroup has a significantly decreased overall survival than does GCB DLBCL in patients given standard anthracycline-based chemotherapy (CHOP-like regimens) [42]. This prognostic impact was confirmed in a subsequent study [44] and was shown to retain its importance in patients additionally treated with rituximab [43]. Immunohistochemical classification of pediatric DLBCL suggested that they are predominantly of the GCB subtype [45,46], which has been confirmed in subsequent gene expression profiling studies [47,48].

The predictive power of the GCB/ABC classifier may be approximated by examining the expression of as few as six genes *(LMO2*, B-cell CLL/lymphoma 6 *(BCL6)*, *BCL2*, cyclin D2 *(CCND2)*, chemokine (C-C motif) ligand 3 *(CCL3)* fibronectin 1 *(FN1))* [49]. Expression of

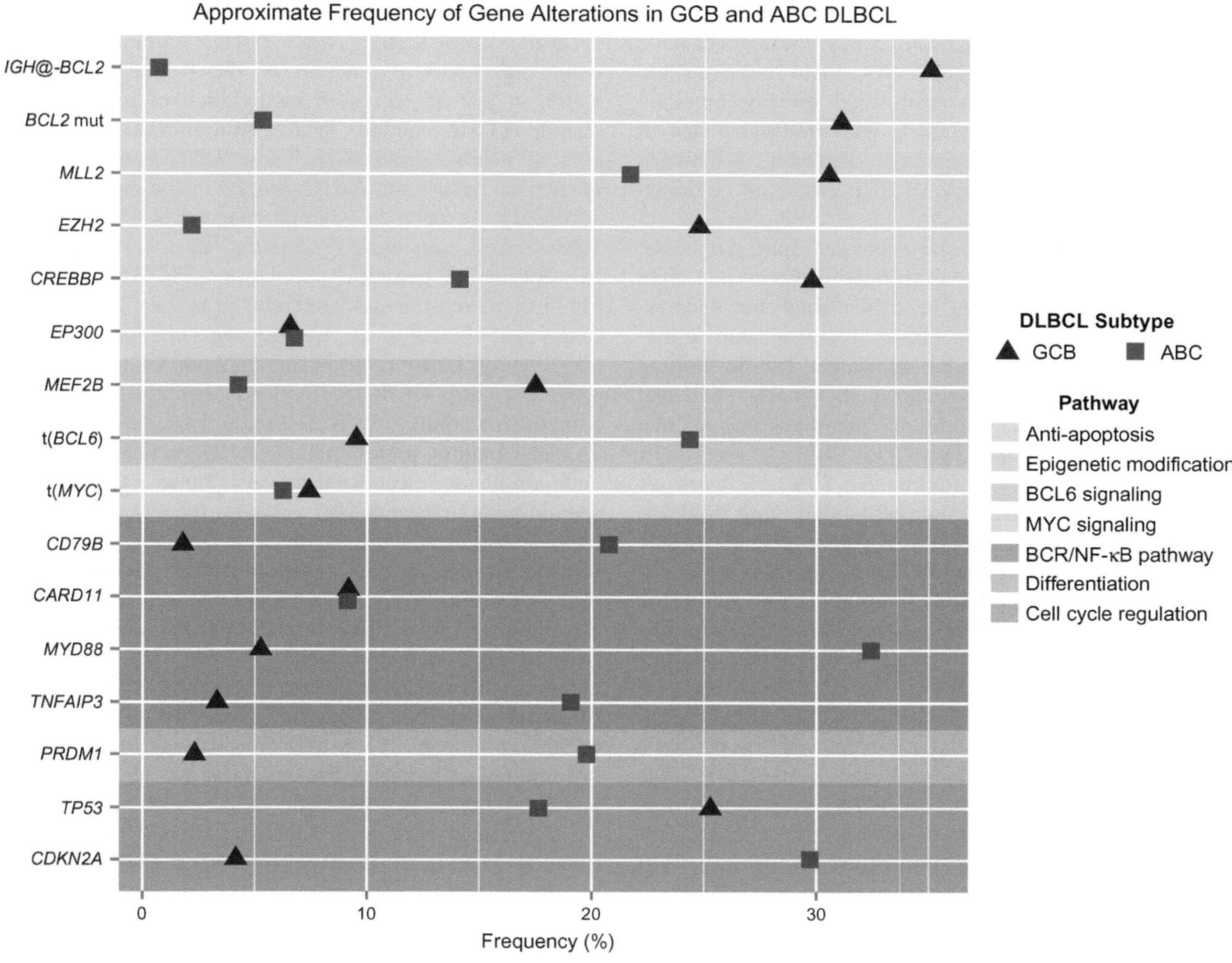

FIGURE 20.1 Approximate frequency of genetic alterations in germinal center B-cell (GCB) and activated B-cell (ABC) diffuse large B-cell lymphoma. Mut: mutation; t: translocation. Data for the figure was compiled from multiple sources [44,51,52,54,57,65–67,70,71,75,115].

LMO2 and *BCL6* is associated with the GCB gene signature and a favorable prognosis, whereas *BCL2, CCL3,* and *CCND2* are overexpressed in the ABC subtype and associated with an unfavorable prognosis. *FN1* overexpression was favorable and was among genes originally associated with a "lymph node" gene signature [42]. *FN1* is likely expressed by non-lymphoid background stromal cells [49]. Indeed, gene expression profiles of background nonmalignant cells have been shown to be prognostic in DLBCL, and *FN1* expression is correlated with the favorable "stromal-1" gene signature [43]. Notably, the role of the stromal microenvironment in the prognosis of pediatric DLBCL has not been investigated.

Array comparative genomic hybridization (aCGH) studies have demonstrated various copy number abnormalities (CNAs) in DLBCL. GCB and ABC DLBCL tend to harbor different CNAs: GCB DLBCL is associated with gains of amplification of 2p and gains of 12q12, and ABC DLBCL frequently has trisomy 3, gains of 3q and 18q21-q22, losses of 6q [50,51]. These CNAs are frequently associated with over- or underexpression of genes that may be involved in the pathogenesis of these neoplasms, including v-relreticuloendotheliosis viral oncogene homolog (avian) *(REL)* at 2p, Mdm2, p53 E3 ubiquitin protein ligase homolog (mouse) *(MDM2)* at 12q, forkhead box P1 *(FOXP1)* at 3q, *BCL2* at 18q, and tumor necrosis factor, alpha-induced protein 3 *(TNFAIP3 (A20))* at 6q [51,52]. aCGH has identified deletions of 4p14 and 19q13.32 in pediatric lymphomas that have not been identified in adult DLBCL [47]. These deletions may downregulate ras homolog family member H *(RHOH)* and/or amyloid beta (A4) precursor protein-binding, family B, member 2 *(APBB2)* at 4p14 and BCL2-associated X protein *(BAX)* and/or BCL2 binding component 3 *(BBC3)* at 19q [47]. However, further study is needed to determine what role, if any, these genes have in pediatric lymphomagenesis. As might be expected, aCGH demonstrates an overall trend toward increased genomic complexity in DLBCL with increasing age at diagnosis; however, no clear age demarcation between pediatric-type and adult-type DLBCL cases can be drawn on this basis [48].

Although DLBCL can be separated into prognostic subgroups based upon gene expression and recurrent copy number abnormalities have been identified, the underlying genetic changes responsible for malignant transformation are only beginning to be elucidated. The developing understanding of the molecular pathogenesis of B-cell lymphomas has begun to identify pathways that contain promising targets for therapy.

A large fraction (approximately 30%) of adult *de novo* GCB-like DLBCL are associated with t(14;18) [53,54]. This rearrangement, also common in low-grade adult follicular lymphomas, leads to *IGH@−BCL2* fusion and subsequent BCL2 overexpression, inhibiting apoptosis. *IGH@−BCL2* is uncommon in the ABC-like category [53]; however, overexpression of *BCL2* is frequent in ABC DLBCL, due to *BCL2* amplification or other mechanisms [55]. In contrast with adult GCB DLBCL, t(14;18) is virtually never seen in pediatric DLBCL cases [46].

Somatic hypermutation of *BCL6*, a transcriptional repressor whose expression is required for germinal center formation, is an especially common event in GCB DLBCL [57]. These mutations can involve *BCL6* autoregulatory sites and lead to deregulated overexpression of BCL6 [58]. *BCL6* rearrangements, on the other hand, are more frequently associated with ABC DLBCL; these rearrangements also appear to upregulate BCL6 expression [57].

Translocations juxtaposing immunoglobulin loci with interferon regulatory factor 4 *(IRF4)* have been recently described in a subset (~15%) of pediatric GCB DLBCL [60]. This translocation causes upregulation of *IRF4*. The exact mechanism of this translocation in lymphomagenesis is unclear. *IRF4* normally suppresses *BCL6* and activates *BLIMP1*; however, in lymphomas with *IG−IRF4* fusion, *BCL6* expression is retained, and *BLIMP1* is not expressed. *BCL6* mutations are also common in lymphomas with this fusion, and the mutated *BCL6* gene in conjunction with the *IG−IRF4* fusion likely deregulates normal IRF4 function [60]. In contrast with pediatric DLBCL, *IG−IRF4* fusion appears to be uncommon in adults [48,60].

Recently, mutations in genes involved in epigenetic regulation have been discovered to be common events in DLBCL. Recurrent heterozygous somatic mutations of enhancer of zeste homolog 2 (Drosophila) *(EZH2)* atY641 have been identified in approximately 25% of GCB-like DLBCL [61]. *EZH2* Y641 mutations appear, in conjunction with the action of the residual wild-type allele, to lead to increased methylation of lysine 27 on histone H3, subsequently leading to transcriptional repression of nearby genes [62]. These *EZH2* mutations frequently co-occur with *IGH@−BCL2* fusion and may be a secondary event in lymphomagenesis [63]. Like *IGH@−BCL2* fusion, *EZH2* mutations are rare to nonexistent in the pediatric population [48]. Myeloid/lymphoid or mixed-lineage leukemia 2 *(MLL2)*, another histone methyltransferase, is mutated in approximately 20−30% of DLBCL, occurring in both GCB and ABC types [64−66]. DLBCLs frequently contain deletions or somatic mutations of CREB binding protein *(CREBBP)* and E1A binding protein p300 *(EP300)*, which are histone and non-histone acetyltransferases. These mutations are seen in both GCB and ABC subtypes [66,67]. The heterozygous loss of function of these genes may have broad implications for gene transcription. Early data suggest that BCL6 may be upregulated and p53 downregulated in the presence of CREBBP and EP300 haploinsufficiency [67]. Myocyte enhancer factor 2B *(MEF2B)*, a calcium-dependent transcription factor involved with recruitment of histone modifying enzymes such as CREBBP and EP300, is mutated in approximately 10−15% of DLBCL, presumably with consequences similar to that of *CREBBP* and *EP300* mutations [64−66].

A common feature of ABC-like DLBCL is constitutive activation of the nuclear factor-kappaB (NF-κB) signaling pathway [68]. In a subset of ABC DLBCL (~10%), this pathway may be driven by activating mutations in caspase recruitment domain family, member 11 *(CARD11)*, a scaffolding protein necessary for NF-κB pathway activation by B-cell receptor (BCR) signaling [69]. Other ABC DLBCLs have wild-type *CARD11*, but exhibit "chronic active" BCR signaling, which may be due in some cases to mutant CD79B and CD79A, which are BCR subunits [70]. Activating mutations of myeloid differentiation primary response gene [88] *(MYD88)*, a toll/interleukin-1 receptor mediator, occur in approximately 30% of ABC DLBCLs and provide another pathway for NF-κB activation [71]. NF-κB upregulation in ABC DLBCL may also be due to frequent (~20%) inactivating mutations of *TNFAIP3* (*A20*), a negative regulator of the pathway [52,72]. In addition to NF-κB activation, many ABC DLBCLs harbor inactivating mutations or deletions of PR domain containing 1, with ZNF domain *(PRDM1 (BLIMP1))*, which blocks terminal B-cell differentiation into plasma cells [73−75] and deletions of the tumor suppressor *CDKN2A (p16INK4a)* [51].

The discovery of the genetic events underlying DLBCL pathogenesis has led to investigation of drugs targeting the altered pathways and genes. Anti-BCL2 molecules such as oblimersen (G3139) have been developed and are being evaluated in clinical trials [56]. BCL6 is inactivated by acetylation, suggesting that histone deacetylation inhibitor (HDACI) therapy may have efficacy in tumors with *BCL6* mutation or translocation [59]. In addition, lymphomas with *MEF2B*, *CREBBP*, and *EP300* alterations are promising candidates for HDACI therapy. Numerous HDACI compounds are being evaluated in clinical trials for lymphoma (e.g. romidepsin, panobinostat, entinostat, abexinostat, vorinostat, etc.). Inhibition of methyltransferase activity is attractive in lymphomas with *EZH2* or *MLL2* mutations, and specific inhibition of *EZH2* with GSK126 has shown promising results in mouse models of *EZH2*-mutated DLBCL [76]. NF-κB signaling may be decreased by blocking IκBα degradation through inhibition of either neddylation or the protesome [77], possibly using drugs such as MLN4924, bortezomib, carfilzomib, orixazomib. The BCR signaling pathway includes several tyrosine kinases (BTK, SYK, PRKCB) that may be targetable by drugs such as ibrutinib, AVL-292, fostamatinib, and enzastaurin. Compounds that target the PI3K pathway, also activated by BCR signaling [77], are being evaluated in clinical trials (pictrelisib, CAL 101, BKM 120, SAR245408).

Burkitt Lymphoma

As previously mentioned, classical BL has characteristic morphology, immunophenotype, and *MYC* translocations; however, examples of high-grade B-cell lymphomas with some, but not all, features of classical Burkitt lymphoma exist. In the past, these cases have been classified as Burkitt-like [78] or atypical Burkitt lymphoma [79]. Some of these tumors are associated with *MYC* translocations, which are frequently present in DLBCL as well as BL (Figure 20.2) [80]. Two large gene expression studies published in 2006 contributed to the understanding of BL and its relationship with DLBCL [81,82]. BL was found to be genetically simple, with *MYC* rearrangements to *IG* loci. On the other hand, *MYC* rearrangements in DLBCL generally involve non-immunoglobulin gene partners, and DLBCL exhibits more chromosomal abnormalities [82]. Some lymphomas were identified that could not be easily classified as either BL or DLBCL on a gene expression basis, with some features of both entities [81,82]. It is now recognized that in a subset of high-grade B-cell lymphomas, specific designation as either BL or DLBCL is impossible – these are considered to be unclassifiable high-grade lymphomas with features intermediate between BL and DLBCL [6]. Some of these tumors (referred to as "double-hit" lymphomas) contain, in addition to rearrangement of the *MYC* locus, *IGH@–BCL2* or *BCL6* rearrangements. These lymphomas have a particularly poor prognosis; however, "double-hit" lymphomas in the pediatric population are rare to non-existent [83,84]. Although MYC dysregulation is thought to be the primary oncogenic driver in BL, other genes are frequently mutated in BL [85–87]. These include *TCF3* and inhibitor of DNA binding 3 *(ID3)*; mutations in either of these genes lead to increased TCF3 activity that appears to increase PI3 kinase (PI3K) signaling and survival [85–87]. Additionally, cyclin D3 *(CCND3)* mutations are relatively common in BL, which lead to progression through the cell cycle [87]. Of note, many genes commonly mutated in ABC DLBCL *(BCL2, MYD88, CD79B, EZH2)* do not appear to be mutational targets in BL [87], and *ID3* mutations are rare to absent in DLBCL [85–87]. These results suggest that inhibition of MYC and PI3K signaling pathways may be of benefit to patients with BL.

Given the increased incidence of lymphoma with aging, most genomic studies of lymphoma have focused on the adult population, although pediatric BL cases were included in the studies by Hummel et al. [82] and Dave et al. [81]. Distinction between BL and DLBCL may be difficult in the pediatric population, and these neoplasms may be more closely related in children than in adults. Pediatric DLBCL has increased proliferation rates, increased MYC protein expression and incidence of *MYC* rearrangements, and decreased BCL2 expression when compared with adult DLBCL, features that are all similar to BL [45,47,88]. The findings of two gene expression studies also support this relationship. Klapper et al. analyzed 65 cases of aggressive mature non-Hodgkin lymphoma in children <18 years of age [89]. They found that 31% of cases morphologically defined as DLBCL were reclassified on a molecular basis as BL, a rate much higher than that seen in adults. On the other hand, the vast majority of morphologic BL cases had a BL gene signature [89]. In contrast, Deffenbacher et al. studied 70 cases of pediatric BL and DLBCL and found that approximately 20% of both morphologic BL and DLBCL cases were reclassified into the other group by gene expression profiling [47]. The reason for this difference is unclear but may be due to geographic differences in morphologic classification of BL and DLBCL in the USA and Germany. Taken together, the results suggest that morphologic and genomic classification of BL and DLBCL are more likely to be discrepant in pediatric lymphomas than in adults, and classification by conventional criteria may be challenging. Both studies support that pediatric BL does not differ substantially from classical BL in adults, re-emphasizing the relatively homogeneous nature of this disease [47,89].

The accumulated evidence suggests that aggressive mature B-cell lymphomas in children are a more uniform, more genetically simple group of tumors than are those in adults. Morphologically and genetically, distinction between DLBCL and BL may be difficult in children.

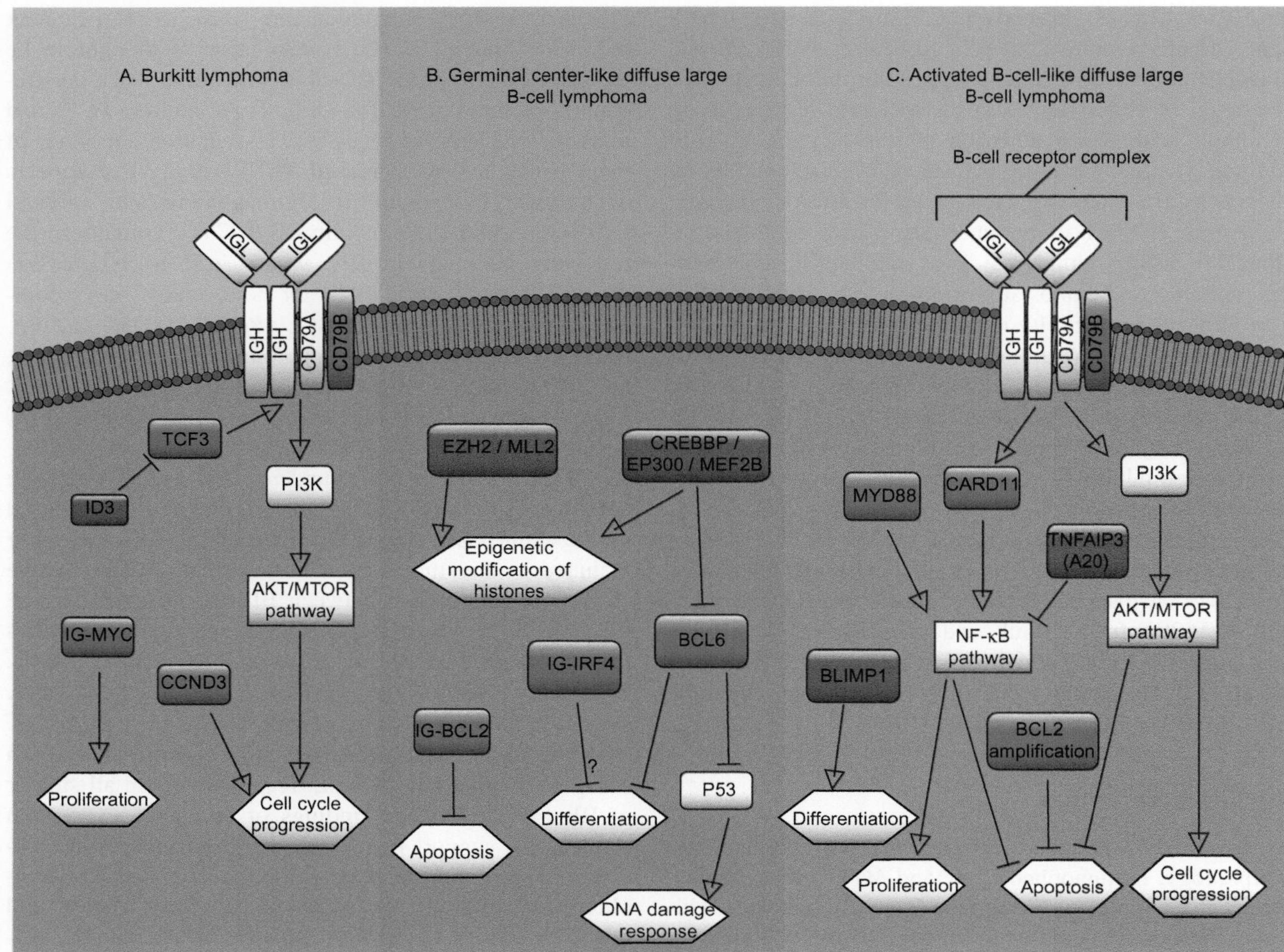

FIGURE 20.2 Pathways commonly implicated in aggressive B-cell lymphomagenesis. Activating mutations are designated by a green symbol, and inactivating mutations are designated by a red symbol. (A) Burkitt lymphoma is characterized by activation of *MYC*, through upregulation due to *IG-MYC* fusion. Mutations of *ID3* and *TCF3* lead to increased B-cell receptor (BCR) signaling, and activation of the PI3K/AKT/MTOR pathway. Activating mutations of *CCND3* lead to increased cell cycle progression. (B) Germinal center-like diffuse large B-cell lymphomas (GCB DLBCL) have a variety of recurrent genetic aberrations. *IG–BCL2* fusion generates upregulation of BCL2, an antiapoptotic molecule. *BCL6* is a master regulator of the germinal center reaction, and activating mutations have a variety of effects, including decreased response to DNA damage and decreased differentiation. *BCL6* may also be disregulated by mutations involving *CREBBP*, *EP300*, and *MEF2B*, through decreased acetylation. Epigenetic dysregulation of histones appears important in GCB DLBCL, as the methyltransferases *EZH2* and *MLL2* are frequently mutated, along with *CREBBP*, *EP300*, and *MEF2B*, which are also involved in histone acetylation. (C) Activated B-cell-like diffuse large B-cell lymphoma (ABC DLBCL) commonly has increased NF-κB signaling due to mutations in *CARD11*, *MYD88*, *TNFAIP3* or "chronic active" B-cell receptor signaling, which may be secondary to *CD79B* mutation. *BCL2* amplifications are common in ABC DLBCL, as are inactivating mutations in *PRDM1 (BLIMP1)*, leading to decreased apoptosis and differentiation, respectively. Many of these pathways are potential therapeutic targets.

Discrimination between these entities may currently be of somewhat lower importance in children than in adults. Therapy for BL and DLBCL is essentially the same in the pediatric population, whereas in adult patients therapies differ dramatically. In adults, DLBCL is generally treated with R-CHOP (rituximab, cyclophosphamide, doxorubicin, vincristine, and prednisone), while a variety of more intensive regimens are used for BL (e.g. CODOX-M/IVAC (cyclophosphamide, vincristine, doxorubicin, high dose methotrexate/ifosfamide, etoposide, high dose cytarabine), EPOCH-R (etoposide, prednisone, vincristine, cyclophosphamide, doxorubicin, rituximab), R-HyperCVAD (rituximab, hyperfractionated cyclophosphamide, vincristine, doxorubicin, dexamethasone, methotrexate, cytarabine), etc.). BL does not adequately respond to CHOP-based regimens [90]. In addition, children with aggressive mature B-cell lymphomas generally have much better survival than is seen in adults [91–93]. However, therapies targeting specific pathways hold the promise of reduced toxicity compared with conventional chemotherapy, and a subset of children with B-cell lymphomas still have suboptimal clinical outcomes. Therefore, new therapeutic agents are still needed in this population. The genetic basis of pediatric lymphoma is less well characterized currently than that of adults; further large-scale genomic studies are necessary to identify novel therapeutic targets in these aggressive neoplasms.

ANAPLASTIC LARGE CELL LYMPHOMA

T-cell lymphomas arise from cells of the innate and adaptive immune system. Of the 20 distinct peripheral T-cell lymphoma entities currently recognized by the World Health Organization [6], anaplastic large cell lymphoma (ALCL), anaplastic lymphoma kinase (ALK) positive represents the most common subtype in children and young adolescents [6]. ALCL is a distinct subtype of aggressive CD30 positive peripheral T-cell NHL that makes up about 10–15% of pediatric NHL [94,95]. In the pediatric population, primary cutaneous and ALK-negative ALCL are very rare [96]. The majority of pediatric ALCL (90–95%) expresses ALK due to specific genetic events that lead to overexpression and constitutive activation of ALK [97]. In about 80% of tumors, ALK overexpression arises due to a t(2;5)(p23;q35) translocation that juxtaposes the *ALK* gene on chromosome 2p23 to the nucleophosmin *(NPM)* gene on 5q35. This fusion gene encodes a chimeric, constitutively activated tyrosine kinase, NPM–ALK, consisting of the N-terminal portion of the *NPM* fused to the catalytic domain of *ALK*. The fusion with NPM results in activation of the ALK kinase domain and its expression in a deregulated and ectopic manner [94,98]. Over 19 alternative fusion partner genes for ALK have been identified and involve clathrin heavy polypeptide-like gene *(CLTC)*, non-muscle tropomyosin 3 gene *(TPM3)*, non-muscle tropomyosin 4 *(TPM4)*, moesin (MSN), Trk fusion gene *(TFG)*, cysteinyl-tRNAsynthetase *(CARS)*, non-muscle myosin heavy chain *(MYH9)*, RAN binding protein *(RANBP2)*, and 5-aminoimidazole-4-carboxamide ribonucleotideformyl-transferase/IMPcyclohydrolase gene *(ATIC)*. Most of the partner proteins contain an oligomerization domain that confers the fusion protein with constitutive activation of the ALK kinase [99].

Gene expression studies of ALCLs are relatively sparse due to the rarity of the disease. The largest study performed on 32 patients with systemic ALCL (25 ALK-positive and 7 ALK-negative) and five cell lines demonstrated the existence of two groups corresponding to morphologic subgroups (common vs small cell and mixed variants) [100]. Supervised hierarchical clustering analysis of ALK-positive and ALK-negative ALCLs demonstrated subsets of genes including *BCL6*, protein tyrosine phosphatase, non-receptor type 12 *(PTPN12)*, CCAAT/enhancer binding protein (C/EBP), beta-*(CEBPB)*, and serpin peptidase inhibitor, clade A (alpha-1 antiproteinase, antitrypsin), member 1 *(SERPINA1)* to be overexpressed in ALK-positive ALCL. Other functional groups of genes that were overrepresented in gene ontology (GO) terms in ALK-positive ALCL included those related to immunologic functions (immune response, response to pathogen) and the NF-κB/I-kB cascade, leukocyte transendothelial migration (integrin, alpha M *(ITGAM)*, paxillin *(PXN)*, junctional adhesion molecule 3 *(JAM3)*), focal adhesion and adherens junction (catenin (cadherin-associated protein), beta 1 *(CTNNB1)*, ras-related C3 botulinum toxin substrate 1 *(RAC1)*). Interestingly, many of these are recapitulated in the proteomic signature of ALCL [101,102]. The molecular signature of ALK-negative ALCLs did not provide any clues to the molecular mechanisms of the tumor subtype suggesting that it was composed of heterogeneous molecular entities. Gene expression profiling (GEP) studies of ALK-positive ALCL-derived cell lines combined with functional RNAi screening revealed the transcription factors *CEBPB* and BCL2-related protein A1 *(BCL2A1)* as critical genes required for cell transformation or maintenance of survival [103,104]. The downstream signaling pathways induced by the variant partner gene products are not known, although GEP studies comparing TPM3-ALK-positive and NPM-ALK-positive ALCL indicate that there may be pathways that are shared [105].

Genomic profiling of 74 ALCLs including 43 ALK-positive and 31 ALK-negative cases by comparative genomic hybridization (CGH) revealed genomic imbalances in 25 (58%) of ALK-positive ALCL and 20 (65%) of ALK-negative ALCLs [106]. Interestingly, ALK-positive cases with *NPM-ALK* or variant *ALK* translocations showed a similar profile of secondary genetic alterations, including gains of 17p and 17q24-qter, and losses of 4q13-q21 and 11q14. ALK-negative cases demonstrated more frequent gains of 1q and 6p21. Some abnormalities such as gains of chromosome 7 and 6q and loss of 13q were observed in both ALK-positive and ALK-negative cases. It is unknown whether there were specific alterations seen in pediatric ALCLs vs adult cases.

MicroRNAs are non-coding RNAs that regulate the expression of target proteins by binding at the 3′-UTR regions and inhibiting translation [107]. The role of microRNAs in regulation of key cellular processes such as differentiation and proliferation has been highlighted by numerous profiling studies which indicate that they serve as tumor-specific biomarkers and show the presence of unique miRNA signatures associated with tumor subtypes. Recent miRNA profiling studies of ALCLs indeed demonstrated distinct profiles of "oncomirs" that were able to distinguish ALK-positive ALCLs from ALK-negative subtypes [108]. Importantly, five members of the

TABLE 20.1 Possible Therapeutic Targets in Lymphoma

Lymphoma	Age Group	Mutation/Pathway	Possible Targeted Therapies
BL	Adult & pediatric	*IG-MYC*	Antisense molecules
	Adult & pediatric	*TCF3/ID3*	PI3K inhibition
	Adult & pediatric	*CCND3*	CDK4/6 inhibition
GCB DLBCL	Adult > pediatric	*BCL6* mutation/rearrangement	Histone deacetylase inhibitor
	Adult	*IG-BCL2*	BCL2 inhibitor
	Adult	*EZH2/MLL2* mutation	Demethylating agents
	?	*CREBBP/EP300/MEF2B* mutation	Histone deacetylase inhibitor
	Pediatric > adult	*IG-MYC*	Antisense molecules
	Pediatric	*IG-IRF4*	?
ABC DLBCL	Adult > pediatric	NF-κB activation due to mutations in *CARD11, MYD88, CD79B, TNFAIP3.*	Protesome inhibition
			Rituximab
			Neddylation inhibition
	?	*CREBBP/EP300/MEF2B* mutation	Histone deacetylase inhibitor
ALCL	Adult/pediatric	*ALK* rearrangement	ALK inhibition

BL: Burkitt lymphoma; GCB DLBCL: germinal center-like diffuse large B-cell lymphoma; ABC DLBCL: activated B-cell-like diffuse large B-cell lymphoma; ALCL: Anaplastic large cell lymphoma

MYC-regulated miR-17-92 cluster located on chromosome 13q31.3 (miR-17-1, miR-18a, miR-20a), or its paralogue on chromosome Xq26 (miR-363, miR-106a, and miR-20b) were highly expressed in ALK-positive ALCLs. Similarly, miRNAs that were downregulated in ALK-positive ALCL are negative regulators of proteins previously implicated with ALK-driven oncogenesis. These include Bcl2-like protein Mcl1 (miR-29b and miR-101) and mTOR (miR-101) [108,109].

Somatic mutations are rare in anaplastic large cell lymphoma; however, most of the studies reporting single nucleotide variants have been performed using directed Sanger sequencing of candidate genes. Monoallelic and biallelic mutations of the perforin *(PRF1)* gene have been reported in some cases of childhood lymphomas. Direct sequencing identified six different mutations in 12 patients (27.3%) of 44 patients with t(2;5)-positive ALCL. Two of the mutations were novel mutations: 529C > T and I471 > G, while four had been previously described. The incidence of *PRF1* mutations was found to be significantly higher in patients with ALCL compared with 400 control subjects, among whom only heterozygous A91V was observed in 41 subjects (10.2%) (chi-square test, 10.9; $p < 0.01$) [110]. *PRF1* mutations have been described in other NHL and are thought to result in defective perforin-mediated cytotoxicity due to abnormal conformational changes induced by the A91V mutation [111]. Amplification of sonic hedgehog (SHH) gene in a subset of ALK positive ALCL [112] has been shown to lead to deregulation of the sonic hedgehog signaling pathway.

Given the successes of TKI therapy in chronic myelogenous leukemia [113] and ALK-driven lung cancer [114], inhibition of ALK signaling is a promising therapeutic approach in ALK-positive ALCL. TKIs with anti-ALK activity are currently being assessed in clinical trials (crizotinib, AP26113). With the advent of more targeted therapy, it may be possible effectively to treat the mutational events underlying ALCL pathogenesis while reducing side effects and adverse long-term consequences of therapy (Table 20.1).

CONCLUSIONS AND FUTURE STUDIES

Pediatric lymphomas are biologically and clinically distinct from those of the adult population. Due to their rarity, a limited number of large-scale genomic studies utilizing CGH, gene expression profiling, miRNA profiling and next-generation sequencing strategies have been performed. However, the preliminary data do support the notion that despite some overlap in the genetic landscape between the pediatric and adult population, there are distinct genetic abnormalities in lymphomas of the pediatric population. Important questions pertaining to the genetic mechanisms involved in the genesis of pediatric follicular lymphomas and diffuse large B-cell lymphomas will be important to pursue. Whether tumors such as ALK-positive ALCLs in which there is a clear driving oncogene harbor "passenger, modifier" genetic alterations also remain to be determined. Next-generation sequencing studies of pediatric lymphomas will undoubtedly reveal novel genetic alterations that will impact our understanding of their pathogenesis and aid in development of targeted therapies.

GLOSSARY

Acetyltransferase An enzyme that moves an acetyl group from an acetyl donor to an acceptor. Acetylation is important in regulating the function of proteins and can affect gene expression through histone modification.

Acute lymphoblastic leukemia A neoplasm of immature B or T cells that exhibits extensive blood/bone marrow involvement.

Anaplastic large cell lymphoma T-cell lymphoma typically characterized by large cell size, CD30 expression, and *ALK* rearrangement.

Burkitt lymphoma Highly proliferative mature B-cell lymphoma, typically characterized by *MYC* translocations.

Diffuse large B-cell lymphoma Genetically heterogeneous group of mature B-cell lymphomas with large cell size and aggressive clinical behavior.

Epigenetics Regulation of gene expression or phenotype through modifications that do not alter the sequence of DNA.

Lymphoblastic lymphoma A neoplasm of immature B or T cells that exhibits extramedullary involvement (does not extensively involve the bone marrow or blood).

Methyltransferase An enzyme that transfers a methyl group from a methyl donor to an acceptor molecule. Methylation of DNA and/or histones is important in epigenetic regulation of gene expression.

Proteosome An intracellular complex that degrades damaged or unnecessary proteins that have been targeted for elimination.

Transcription factor A protein that binds to a DNA sequence to regulate its transcription to RNA.

Tyrosine kinase An enzyme that transfers a phosphate group from ATP to tyrosine residues, typically contributing to activation of a signal transduction pathway.

ABBREVIATIONS

ABC Activated B-cell
ALCL Anaplastic large cell lymphoma
ALL Acute lymphoblastic leukemia
BCR B-cell receptor
BL Burkitt lymphoma
DLBCL Diffuse large B-cell lymphoma
GCB Germinal center B-cell
LBL Lymphoblastic lymphoma
HDACI Histone deacetylase inhibitor
NHL Non-Hodgkin lymphoma

REFERENCES

[1] SEER Cancer Statistics Review, 1975–2009 (Vintage 2009 Populations) [database on the Internet]. Bethesda, MD: National Cancer Institute; 2012. Available from: <http://seer.cancer.gov/csr/1975_2009_pops09/>.

[2] Morton LM, Wang SS, Devesa SS, Hartge P, Weisenburger DD, Linet MS. Lymphoma incidence patterns by WHO subtype in the United States, 1992–2001. Blood 2006;107:265–76.

[3] Victora GD, Nussenzweig MC. Germinal centers. Annu Rev Immunol 2012;30:429–57.

[4] Nussenzweig A, Nussenzweig MC. Origin of chromosomal translocations in lymphoid cancer. Cell 2010;141:27–38.

[5] Shaffer III AL, Young RM, Staudt LM. Pathogenesis of human B cell lymphomas. Annu Rev Immunol 2012;30:565–610.

[6] Swerdlow SH, Campo E, Harris NL, Jaffe ES, Pileri SA, Stein H, et al. WHO classification of tumours of haematopoietic and lymphoid tissues. 4th ed. Lyon: IARC; 2008.

[7] Cossman J, Chused TM, Fisher RI, Magrath I, Bollum F, Jaffe ES. Diversity of immunological phenotypes of lymphoblastic lymphoma. Cancer Res 1983;43:4486–90.

[8] Lin P, Jones D, Dorfman DM, Medeiros LJ. Precursor B-cell lymphoblastic lymphoma: a predominantly extranodal tumor with low propensity for leukemic involvement. Am J Surg Pathol 2000;24:1480–90.

[9] Moorman AV, Ensor HM, Richards SM, Chilton L, Schwab C, Kinsey SE, et al. Prognostic effect of chromosomal abnormalities in childhood B-cell precursor acute lymphoblastic leukaemia: results from the UK Medical Research Council ALL97/99 randomised trial. Lancet Oncol 2010;11:429–38.

[10] Mullighan CG, Goorha S, Radtke I, Miller CB, Coustan-Smith E, Dalton JD, et al. Genome-wide analysis of genetic alterations in acute lymphoblastic leukaemia. Nature 2007;446:758–64.

[11] Mullighan CG, Miller CB, Radtke I, Phillips LA, Dalton J, Ma J, et al. BCR-ABL1 lymphoblastic leukaemia is characterized by the deletion of Ikaros. Nature 2008;453:110–4.

[12] Mullighan CG, Su X, Zhang J, Radtke I, Phillips LA, Miller CB, et al. Deletion of IKZF1 and prognosis in acute lymphoblastic leukemia. N Engl J Med 2009;360:470–80.

[13] Harvey RC, Mullighan CG, Chen IM, Wharton W, Mikhail FM, Carroll AJ, et al. Rearrangement of CRLF2 is associated with mutation of JAK kinases, alteration of IKZF1, Hispanic/Latino ethnicity, and a poor outcome in pediatric B-progenitor acute lymphoblastic leukemia. Blood 2010;115:5312–21.

[14] Mullighan CG, Collins-Underwood JR, Phillips LA, Loudin MG, Liu W, Zhang J, et al. Rearrangement of CRLF2 in B-progenitor- and Down syndrome-associated acute lymphoblastic leukemia. Nat Genet 2009;41:1243–6.

[15] Russell LJ, Capasso M, Vater I, Akasaka T, Bernard OA, Calasanz MJ, et al. Deregulated expression of cytokine receptor gene, CRLF2, is involved in lymphoid transformation in B-cell precursor acute lymphoblastic leukemia. Blood 2009;114: 2688–98.

[16] Mullighan CG, Zhang J, Harvey RC, Collins-Underwood JR, Schulman BA, Phillips LA, et al. JAK mutations in high-risk childhood acute lymphoblastic leukemia. Proc Natl Acad Sci USA 2009;106:9414–8.

[17] Graux C, Cools J, Michaux L, Vandenberghe P, Hagemeijer A. Cytogenetics and molecular genetics of T-cell acute lymphoblastic leukemia: from thymocyte to lymphoblast. Leukemia 2006;20: 1496–510.

[18] Bernard OA, Busson-LeConiat M, Ballerini P, Mauchauffé M, Della Valle V, Monni R, et al. A new recurrent and specific cryptic translocation, t(5;14)(q35;q32), is associated with expression of the Hox11L2 gene in T acute lymphoblastic leukemia. Leukemia 2001;15:1495–504.

[19] Finger LR, Kagan J, Christopher G, Kurtzberg J, Hershfield MS, Nowell PC, et al. Involvement of the TCL5 gene on human chromosome 1 in T-cell leukemia and melanoma. Proc Natl Acad Sci USA 1989;86:5039–43.

[20] Hatano M, Roberts CW, Minden M, Crist WM, Korsmeyer SJ. Deregulation of a homeobox gene, HOX11, by the t(10;14) in T cell leukemia. Science 1991;253:79–82.

[21] Mellentin JD, Smith SD, Cleary ML. lyl-1, a novel gene altered by chromosomal translocation in T cell leukemia, codes for a protein with a helix-loop-helix DNA binding motif. Cell 1989;58:77–83.

[22] Royer-Pokora B, Loos U, Ludwig WD. TTG-2, a new gene encoding a cysteine-rich protein with the LIM motif, is overexpressed in acute T-cell leukaemia with the t(11;14)(p13;q11). Oncogene 1991;6:1887–93.

[23] Ferrando AA, Neuberg DS, Staunton J, Loh ML, Huard C, Raimondi SC, et al. Gene expression signatures define novel oncogenic pathways in T cell acute lymphoblastic leukemia. Cancer Cell 2002;1:75–87.

[24] Bergeron J, Clappier E, Radford I, Buzyn A, Millien C, Soler G, et al. Prognostic and oncogenic relevance of TLX1/HOX11 expression level in T-ALLs. Blood 2007;110:2324–30.

[25] Ferrando AA, Neuberg DS, Dodge RK, Paietta E, Larson RA, Wiernik PH, et al. Prognostic importance of TLX1 (HOX11) oncogene expression in adults with T-cell acute lymphoblastic leukaemia. Lancet 2004;363:535–6.

[26] Karrman K, Forestier E, Heyman M, Andersen MK, Autio K, Blennow E, et al. Clinical and cytogenetic features of a population-based consecutive series of 285 pediatric T-cell acute lymphoblastic leukemias: rare T-cell receptor gene rearrangements are associated with poor outcome. Genes Chromosomes Cancer 2009;48:795–805.

[27] Tycko B, Smith SD, Sklar J. Chromosomal translocations joining LCK and TCRB loci in human T cell leukemia. J Exp Med 1991;174:867–73.

[28] Graux C, Cools J, Melotte C, Quentmeier H, Ferrando A, Levine R, et al. Fusion of NUP214 to ABL1 on amplified episomes in T-cell acute lymphoblastic leukemia. Nat Genet 2004;36:1084–9.

[29] Weng AP, Ferrando AA, Lee W, Morris JP, Silverman LB, Sanchez-Irizarry C, et al. Activating mutations of NOTCH1 in human T cell acute lymphoblastic leukemia. Science 2004;306: 269–71.

[30] Malyukova A, Dohda T, von der Lehr N, Akhoondi S, Corcoran M, Heyman M, et al. The tumor suppressor gene hCDC4 is frequently mutated in human T-cell acute lymphoblastic leukemia with functional consequences for Notch signaling. Cancer Res 2007;67:5611–6.

[31] O'Neil J, Grim J, Strack P, Rao S, Tibbitts D, Winter C, et al. FBW7 mutations in leukemic cells mediate NOTCH pathway activation and resistance to gamma-secretase inhibitors. J Exp Med 2007;204:1813–24.

[32] Callens C, Baleydier F, Lengline E, Ben Abdelali R, Petit A, Villarese P, et al. Clinical impact of NOTCH1 and/or FBXW7 mutations, FLASH deletion, and TCR status in pediatric T-cell lymphoblastic lymphoma. J Clin Oncol 2012;30:1966–73.

[33] Jenkinson S, Koo K, Mansour MR, Goulden N, Vora A, Mitchell C, et al. Impact of NOTCH1/FBXW7 mutations on outcome in pediatric T-cell acute lymphoblastic leukemia patients treated on the MRC UKALL 2003 trial. Leukemia 2013;27:41–7.

[34] Van Vlierberghe P, Palomero T, Khiabanian H, Van der Meulen J, Castillo M, Van Roy N, et al. PHF6 mutations in T-cell acute lymphoblastic leukemia. Nat Genet 2010;42:338–42.

[35] Basso K, Mussolin L, Lettieri A, Brahmachary M, Lim WK, Califano A, et al. T-cell lymphoblastic lymphoma shows differences and similarities with T-cell acute lymphoblastic leukemia by genomic and gene expression analyses. Genes Chromosomes Cancer 2011;50:1063–75.

[36] Raetz EA, Perkins SL, Bhojwani D, Smock K, Philip M, Carroll WL, et al. Gene expression profiling reveals intrinsic differences between T-cell acute lymphoblastic leukemia and T-cell lymphoblastic lymphoma. Pediatr Blood Cancer 2006;47:130–40.

[37] Uyttebroeck A, Vanhentenrijk V, Hagemeijer A, Boeckx N, Renard M, Wlodarska I, et al. Is there a difference in childhood T-cell acute lymphoblastic leukaemia and T-cell lymphoblastic lymphoma? Leuk Lymphoma 2007;48:1745–54.

[38] Feng H, Stachura DL, White RM, Gutierrez A, Zhang L, Sanda T, et al. T-lymphoblastic lymphoma cells express high levels of BCL2, S1P1, and ICAM1, leading to a blockade of tumor cell intravasation. Cancer Cell 2010;18:353–66.

[39] Armitage JO, Weisenburger DD. New approach to classifying non-Hodgkin's lymphomas: clinical features of the major histologic subtypes. Non-Hodgkin's lymphoma classification project. J Clin Oncol 1998;16:2780–95.

[40] Berard C, O'Conor GT, Thomas LB, Torloni H. Histopathological definition of Burkitt's tumour. Bull World Health Organ 1969;40:601–7.

[41] Wright DH. What is Burkitt's lymphoma and when is it endemic? Blood 1999;93:758.

[42] Alizadeh AA, Eisen MB, Davis RE, Ma C, Lossos IS, Rosenwald A, et al. Distinct types of diffuse large B-cell lymphoma identified by gene expression profiling. Nature 2000;403:503–11.

[43] Lenz G, Wright G, Dave SS, Xiao W, Powell J, Zhao H, et al. Stromal gene signatures in large-B-cell lymphomas. N Engl J Med 2008;359:2313–23.

[44] Rosenwald A, Wright G, Chan WC, Connors JM, Campo E, Fisher RI, et al. The use of molecular profiling to predict survival after chemotherapy for diffuse large-B-cell lymphoma. N Engl J Med 2002;346:1937–47.

[45] Miles RR, Raphael M, McCarthy K, Wotherspoon A, Lones MA, Terrier-Lacombe MJ, et al. Pediatric diffuse large B-cell lymphoma demonstrates a high proliferation index, frequent c-Myc protein expression, and a high incidence of germinal center subtype: Report of the French-American-British (FAB) international study group. Pediatr Blood Cancer 2008;51:369–74.

[46] Oschlies I, Klapper W, Zimmermann M, Krams M, Wacker HH, Burkhardt B, et al. Diffuse large B-cell lymphoma in pediatric patients belongs predominantly to the germinal-center type B-cell lymphomas: a clinicopathologic analysis of cases included in the German BFM (Berlin-Frankfurt-Munster) Multicenter Trial. Blood 2006;107:4047–52.

[47] Deffenbacher KE, Iqbal J, Sanger W, Shen Y, Lachel C, Liu Z, et al. Molecular distinctions between pediatric and adult mature B-cell non-Hodgkin lymphomas identified through genomic profiling. Blood 2012;119:3757–66.

[48] Klapper W, Kreuz M, Kohler CW, Burkhardt B, Szczepanowski M, Salaverria I, et al. Patient age at diagnosis is associated with the molecular characteristics of diffuse large B-cell lymphoma. Blood 2012;119:1882–7.

[49] Lossos IS, Czerwinski DK, Alizadeh AA, Wechser MA, Tibshirani R, Botstein D, et al. Prediction of survival in diffuse large-B-cell lymphoma based on the expression of six genes. N Engl J Med 2004;350:1828–37.

[50] Bea S, Zettl A, Wright G, Salaverria I, Jehn P, Moreno V, et al. Diffuse large B-cell lymphoma subgroups have distinct genetic profiles that influence tumor biology and improve gene-expression-based survival prediction. Blood 2005;106:3183–90.

[51] Lenz G, Wright GW, Emre NC, Kohlhammer H, Dave SS, Davis RE, et al. Molecular subtypes of diffuse large B-cell lymphoma arise by distinct genetic pathways. Proc Natl Acad Sci USA 2008;105:13520–5.

[52] Compagno M, Lim WK, Grunn A, Nandula SV, Brahmachary M, Shen Q, et al. Mutations of multiple genes cause deregulation of NF-κB in diffuse large B-cell lymphoma. Nature 2009;459:717–21.

[53] Huang JZ, Sanger WG, Greiner TC, Staudt LM, Weisenburger DD, Pickering DL, et al. The t(14;18) defines a unique subset of diffuse large B-cell lymphoma with a germinal center B-cell gene expression profile. Blood 2002;99:2285–90.

[54] Iqbal J, Sanger WG, Horsman DE, Rosenwald A, Pickering DL, Dave B, et al. BCL2 translocation defines a unique tumor subset within the germinal center B-cell-like diffuse large B-cell lymphoma. Am J Pathol 2004;165:159–66.

[55] Iqbal J, Neppalli VT, Wright G, Dave BJ, Horsman DE, Rosenwald A, et al. BCL2 expression is a prognostic marker for the activated B-cell-like type of diffuse large B-cell lymphoma. J Clin Oncol 2006;24:961–8.

[56] Kang MH, Reynolds CP. Bcl-2 inhibitors: targeting mitochondrial apoptotic pathways in cancer therapy. Clin Cancer Res 2009;15:1126–32.

[57] Iqbal J, Greiner TC, Patel K, Dave BJ, Smith L, Ji J, et al. Distinctive patterns of BCL6 molecular alterations and their functional consequences in different subgroups of diffuse large B-cell lymphoma. Leukemia 2007;21:2332–43.

[58] Pasqualucci L, Migliazza A, Basso K, Houldsworth J, Chaganti RS, Dalla-Favera R. Mutations of the BCL6 proto-oncogene disrupt its negative autoregulation in diffuse large B-cell lymphoma. Blood 2003;101:2914–23.

[59] Bereshchenko OR, Gu W, Dalla-Favera R. Acetylation inactivates the transcriptional repressor BCL6. Nat Genet 2002;32:606–13.

[60] Salaverria I, Philipp C, Oschlies I, Kohler CW, Kreuz M, Szczepanowski M, et al. Translocations activating IRF4 identify a subtype of germinal center-derived B-cell lymphoma affecting predominantly children and young adults. Blood 2011;118:139–47.

[61] Morin RD, Johnson NA, Severson TM, Mungall AJ, An J, Goya R, et al. Somatic mutations altering EZH2 (Tyr641) in follicular and diffuse large B-cell lymphomas of germinal-center origin. Nat Genet 2010;42:181–5.

[62] Sneeringer CJ, Scott MP, Kuntz KW, Knutson SK, Pollock RM, Richon VM, et al. Coordinated activities of wild-type plus mutant EZH2 drive tumor-associated hypertrimethylation of lysine 27 on histone H3 (H3K27) in human B-cell lymphomas. Proc Natl Acad Sci USA 2010;107:20980–5.

[63] Ryan RJ, Nitta M, Borger D, Zukerberg LR, Ferry JA, Harris NL, et al. EZH2 codon 641 mutations are common in BCL2-rearranged germinal center B cell lymphomas. PLoS One 2011;6:e28585.

[64] Lohr JG, Stojanov P, Lawrence MS, Auclair D, Chapuy B, Sougnez C, et al. Discovery and prioritization of somatic mutations in diffuse large B-cell lymphoma (DLBCL) by whole-exome sequencing. Proc Natl Acad Sci USA 2012;109:3879–84.

[65] Morin RD, Mendez-Lago M, Mungall AJ, Goya R, Mungall KL, Corbett RD, et al. Frequent mutation of histone-modifying genes in non-Hodgkin lymphoma. Nature 2011;476:298–303.

[66] Pasqualucci L, Trifonov V, Fabbri G, Ma J, Rossi D, Chiarenza A, et al. Analysis of the coding genome of diffuse large B-cell lymphoma. Nat Genet 2011;43:830–7.

[67] Pasqualucci L, Dominguez-Sola D, Chiarenza A, Fabbri G, Grunn A, Trifonov V, et al. Inactivating mutations of acetyltransferase genes in B-cell lymphoma. Nature 2011;471:189–95.

[68] Davis RE, Brown KD, Siebenlist U, Staudt LM. Constitutive nuclear factor κB activity is required for survival of activated B cell-like diffuse large B cell lymphoma cells. J Exp Med 2001;194:1861–74.

[69] Lenz G, Davis RE, Ngo VN, Lam L, George TC, Wright GW, et al. Oncogenic CARD11 mutations in human diffuse large B cell lymphoma. Science 2008;319:1676–9.

[70] Davis RE, Ngo VN, Lenz G, Tolar P, Young RM, Romesser PB, et al. Chronic active B-cell-receptor signalling in diffuse large B-cell lymphoma. Nature 2010;463:88–92.

[71] Ngo VN, Young RM, Schmitz R, Jhavar S, Xiao W, Lim KH, et al. Oncogenically active MYD88 mutations in human lymphoma. Nature 2011;470:115–9.

[72] Kato M, Sanada M, Kato I, Sato Y, Takita J, Takeuchi K, et al. Frequent inactivation of A20 in B-cell lymphomas. Nature 2009;459:712–6.

[73] Calado DP, Zhang B, Srinivasan L, Sasaki Y, Seagal J, Unitt C, et al. Constitutive canonical NF-κB activation cooperates with disruption of BLIMP1 in the pathogenesis of activated B cell-like diffuse large cell lymphoma. Cancer Cell 2010;18:580–9.

[74] Mandelbaum J, Bhagat G, Tang H, Mo T, Brahmachary M, Shen Q, et al. BLIMP1 is a tumor suppressor gene frequently disrupted in activated B cell-like diffuse large B cell lymphoma. Cancer Cell 2010;18:568–79.

[75] Pasqualucci L, Compagno M, Houldsworth J, Monti S, Grunn A, Nandula SV, et al. Inactivation of the PRDM1/BLIMP1 gene in diffuse large B cell lymphoma. J Exp Med 2006;203:311–7.

[76] McCabe MT, Ott HM, Ganji G, Korenchuk S, Thompson C, Van Aller GS, et al. EZH2 inhibition as a therapeutic strategy for lymphoma with EZH2-activating mutations. Nature 2012;492:108–12.
[77] Lenz G, Staudt LM. Aggressive lymphomas. N Engl J Med 2010;362:1417–29.
[78] Harris NL, Jaffe ES, Stein H, Banks PM, Chan JK, Cleary ML, et al. A revised European-American classification of lymphoid neoplasms: a proposal from the International Lymphoma Study Group. Blood 1994;84:1361–92.
[79] Jaffe ES. World Health Organization. Pathology and genetics of tumours of haematopoietic and lymphoid tissues. Lyon: IARC Press; 2001.
[80] Kramer MH, Hermans J, Wijburg E, Philippo K, Geelen E, van Krieken JH, et al. Clinical relevance of BCL2, BCL6, and MYC rearrangements in diffuse large B-cell lymphoma. Blood 1998;92:3152–62.
[81] Dave SS, Fu K, Wright GW, Lam LT, Kluin P, Boerma EJ, et al. Molecular diagnosis of Burkitt's lymphoma. N Engl J Med 2006;354:2431–42.
[82] Hummel M, Bentink S, Berger H, Klapper W, Wessendorf S, Barth TF, et al. A biologic definition of Burkitt's lymphoma from transcriptional and genomic profiling. N Engl J Med 2006;354:2419–30.
[83] Aukema SM, Siebert R, Schuuring E, van Imhoff GW, Kluin-Nelemans HC, Boerma EJ, et al. Double-hit B-cell lymphomas. Blood 2011;117:2319–31.
[84] Snuderl M, Kolman OK, Chen YB, Hsu JJ, Ackerman AM, Dal Cin P, et al. B-cell lymphomas with concurrent IGH-BCL2 and MYC rearrangements are aggressive neoplasms with clinical and pathologic features distinct from Burkitt lymphoma and diffuse large B-cell lymphoma. Am J Surg Pathol 2010;34:327–40.
[85] Love C, Sun Z, Jima D, Li G, Zhang J, Miles R, et al. The genetic landscape of mutations in Burkitt lymphoma. Nat Genet 2012;44:1321–5.
[86] Richter J, Schlesner M, Hoffmann S, Kreuz M, Leich E, Burkhardt B, et al. Recurrent mutation of the ID3 gene in Burkitt lymphoma identified by integrated genome, exome and transcriptome sequencing. Nat Genet 2012;44:1316–20.
[87] Schmitz R, Young RM, Ceribelli M, Jhavar S, Xiao W, Zhang M, et al. Burkitt lymphoma pathogenesis and therapeutic targets from structural and functional genomics. Nature 2012;490:116–20.
[88] Poirel HA, Cairo MS, Heerema NA, Swansbury J, Aupérin A, Launay E, et al. Specific cytogenetic abnormalities are associated with a significantly inferior outcome in children and adolescents with mature B-cell non-Hodgkin's lymphoma: results of the FAB/LMB 96 international study. Leukemia 2009;23:323–31.
[89] Klapper W, Szczepanowski M, Burkhardt B, Berger H, Rosolowski M, Bentink S, et al. Molecular profiling of pediatric mature B-cell lymphoma treated in population-based prospective clinical trials. Blood 2008;112:1374–81.
[90] Smeland S, Blystad AK, Kvaloy SO, Ikonomou IM, Delabie J, Kvalheim G, et al. Treatment of Burkitt's/Burkitt-like lymphoma in adolescents and adults: a 20-year experience from the Norwegian Radium Hospital with the use of three successive regimens. Ann Oncol 2004;15:1072–8.
[91] Cairo MS, Gerrard M, Sposto R, Auperin A, Pinkerton CR, Michon J, et al. Results of a randomized international study of high-risk central nervous system B non-Hodgkin lymphoma and B acute lymphoblastic leukemia in children and adolescents. Blood 2007;109:2736–43.
[92] Gerrard M, Cairo MS, Weston C, Auperin A, Pinkerton R, Lambilliote A, et al. Excellent survival following two courses of COPAD chemotherapy in children and adolescents with resected localized B-cell non-Hodgkin's lymphoma: results of the FAB/LMB 96 international study. Br J Haematol 2008;141:840–7.
[93] Patte C, Auperin A, Gerrard M, Michon J, Pinkerton R, Sposto R, et al. Results of the randomized international FAB/LMB96 trial for intermediate risk B-cell non-Hodgkin lymphoma in children and adolescents: it is possible to reduce treatment for the early responding patients. Blood 2007;109:2773–80.
[94] Drexler HG, Gignac SM, von Wasielewski R, Werner M, Dirks WG. Pathobiology of NPM-ALK and variant fusion genes in anaplastic large cell lymphoma and other lymphomas. Leukemia 2000;14:1533–59.
[95] Perkins SL. Work-up and diagnosis of pediatric non-Hodgkin's lymphomas. Pediatr Dev Pathol 2000;3:374–90.
[96] Tomaszewski MM, Moad JC, Lupton GP. Primary cutaneous Ki-1(CD30) positive anaplastic large cell lymphoma in childhood. J Am Acad Dermatol 1999;40:857–61.
[97] Kutok JL, Aster JC. Molecular biology of anaplastic lymphoma kinase-positive anaplastic large-cell lymphoma. J Clin Oncol 2002;20:3691–702.
[98] Pulford K, Morris SW, Turturro F. Anaplastic lymphoma kinase proteins in growth control and cancer. J Cell Physiol 2004;199:330–58.
[99] Amin HM, Lai R. Pathobiology of ALK^+ anaplastic large-cell lymphoma. Blood 2007;110:2259–67.
[100] Lamant L, de Reynies A, Duplantier MM, Rickman DS, Sabourdy F, Giuriato S, et al. Gene-expression profiling of systemic anaplastic large-cell lymphoma reveals differences based on ALK status and two distinct morphologic ALK^+ subtypes. Blood 2007;109:2156–64.
[101] Fanale M, Bartlett NL, Forero-Torres A, Rosenblatt J, Horning SJ, Franklin AR, et al. The antibody-drug conjugate brentuximab vedotin (SGN-35) induced multiple objective responses in patients with relapsed or refractory CD30-positive lymphomas in a phase 1 weekly dosing study. Blood (ASH Annual Meeting Abstracts) 2009;114:2731.
[102] Lim MS, Carlson ML, Crockett DK, Fillmore GC, Abbott DR, Elenitoba-Johnson OF, et al. The proteomic signature of NPM/ALK reveals deregulation of multiple cellular pathways. Blood 2009;114:1585–95.
[103] Piva R, Pellegrino E, Mattioli M, Agnelli L, Lombardi L, Boccalatte F, et al. Functional validation of the anaplastic lymphoma kinase signature identifies CEBPB and BCL2A1 as critical target genes. J Clin Invest 2006;116:3171–82.
[104] Chiarle R, Voena C, Ambrogio C, Piva R, Inghirami G. The anaplastic lymphoma kinase in the pathogenesis of cancer. Nat Rev Cancer 2008;8:11–23.
[105] Bohling SD, Jenson SD, Crockett DK, Schumacher JA, Elenitoba-Johnson KS, Lim MS. Analysis of gene expression profile of TPM3-ALK positive anaplastic large cell lymphoma reveals overlapping and unique patterns with that of NPM-ALK

positive anaplastic large cell lymphoma. Leuk Res 2008;32: 383–93.

[106] Salaverria I, Bea S, Lopez-Guillermo A, Lespinet V, Pinyol M, Burkhardt B, et al. Genomic profiling reveals different genetic aberrations in systemic ALK-positive and ALK-negative anaplastic large cell lymphomas. Br J Haematol 2008;140:516–26.

[107] Bartel DP. MicroRNAs: genomics, biogenesis, mechanism, and function. Cell 2004;116:281–97.

[108] Merkel O, Hamacher F, Laimer D, Sifft E, Trajanoski Z, Scheideler M, et al. Identification of differential and functionally active miRNAs in both anaplastic lymphoma kinase $(ALK)^+$ and ALK^- anaplastic large-cell lymphoma. Proc Natl Acad Sci USA 2010;107:16228–33.

[109] Desjobert C, Renalier MH, Bergalet J, Dejean E, Joseph N, Kruczynski A, et al. MiR-29a down-regulation in ALK-positive anaplastic large cell lymphomas contributes to apoptosis blockade through MCL-1 overexpression. Blood 2011;117:6627–37.

[110] Cannella S, Santoro A, Bruno G, Pillon M, Mussolin L, Mangili G, et al. Germline mutations of the perforin gene are a frequent occurrence in childhood anaplastic large cell lymphoma. Cancer 2007;109:2566–71.

[111] Clementi R, Locatelli F, Dupre L, Garaventa A, Emmi L, Bregni M, et al. A proportion of patients with lymphoma may harbor mutations of the perforin gene. Blood 2005;105:4424–8.

[112] Singh RR, Cho-Vega JH, Davuluri Y, Ma S, Kasbidi F, Milito C, et al. Sonic hedgehog signaling pathway is activated in ALK-positive anaplastic large cell lymphoma. Cancer Res 2009;69:2550–8.

[113] Druker BJ, Talpaz M, Resta DJ, Peng B, Buchdunger E, Ford JM, et al. Efficacy and safety of a specific inhibitor of the BCR-ABL tyrosine kinase in chronic myeloid leukemia. N Engl J Med 2001;344:1031–7.

[114] Kwak EL, Bang YJ, Camidge DR, Shaw AT, Solomon B, Maki RG, et al. Anaplastic lymphoma kinase inhibition in non-small-cell lung cancer. N Engl J Med 2010;363:1693–703.

[115] Young KH, Leroy K, Moller MB, Colleoni GW, Sánchez-Beato M, Kerbauy FR, et al. Structural profiles of TP53 gene mutations predict clinical outcome in diffuse large B-cell lymphoma: an international collaborative study. Blood 2008;112:3088–98.

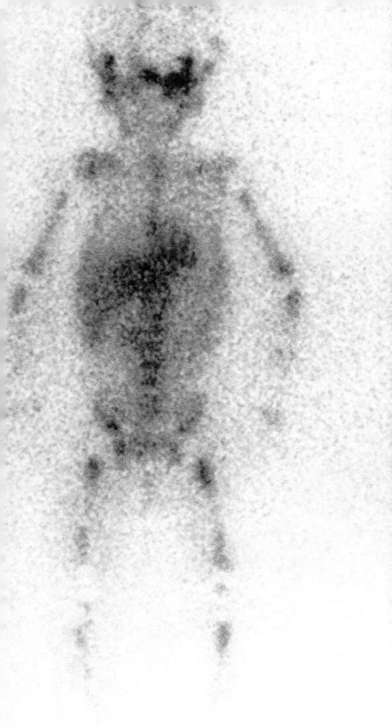

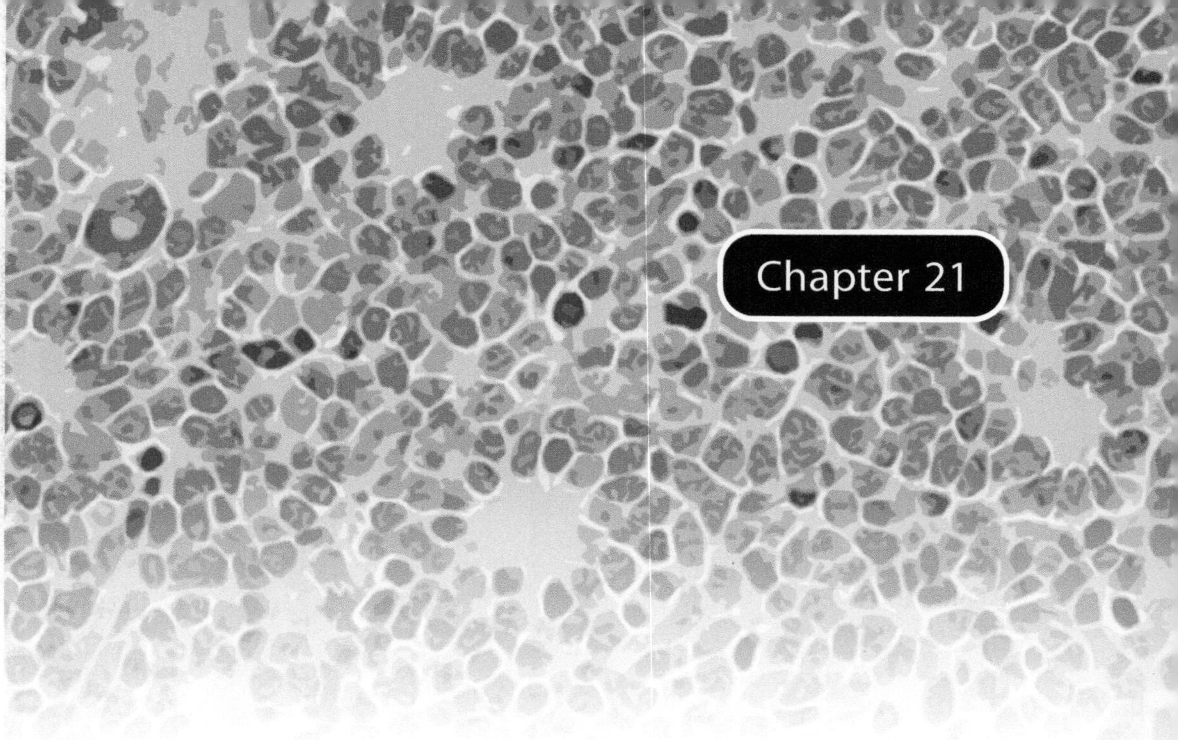

Chapter 21

Cancer Genomics

Neuroblastoma

Daniel A. Morgenstern[1,2]
Meredith S. Irwin[2]

[1]Paediatric Oncology, Great Ormond Street Hospital, London, UK;
[2]Department of Paediatrics, Hospital for Sick Children and University of Toronto, Toronto, ON, Canada

Contents

Cancer Genomics. DOI: http://dx.doi.org/10.1016/B978-0-12-396967-5.00021-9

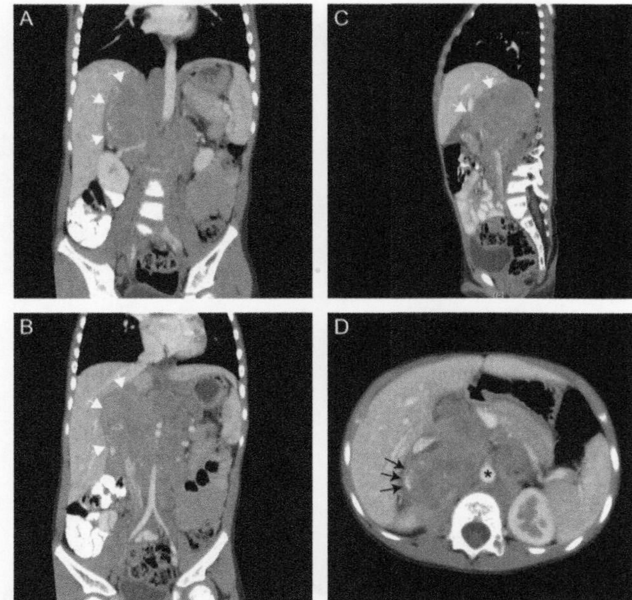

FIGURE 21.1 CT images of a typical neuroblastoma arising from the right adrenal (A and B, coronal views; C, sagittal; white arrowheads indicate tumor margin). In the axial section (D), calcifications within the tumor can be seen (black arrows) and the tumor shows typical encasement of major blood vessels including the aorta (asterisk).

Key Concepts

- Risk stratification involving patient, biological and, increasingly, genetic factors is crucial to the management of neuroblastoma, a tumor with a hugely variable natural history
- Amplification of *MYCN* was one of the earliest identified genetic markers in oncology and remains an important predictor of poor outcome
- Somatic mutations in neuroblastoma are rare, with mutations in *ALK* (the most commonly implicated gene discovered to date) found in only 8% of tumors
- Chromosomal abnormalities (particularly segmental, rather than numerical) are associated with poor outcome and are of particular relevance in predicting prognosis and determining treatment decisions in patients with localized disease
- In the future, it is hoped that the integration of multiple sources of genetic information (including whole-genome sequencing, mRNA and miRNA microarrays and epigenetic analysis) will lead to a better understanding of neuroblastoma biology, more precise improved risk stratification and potentially lead to the identification of novel therapeutic targets

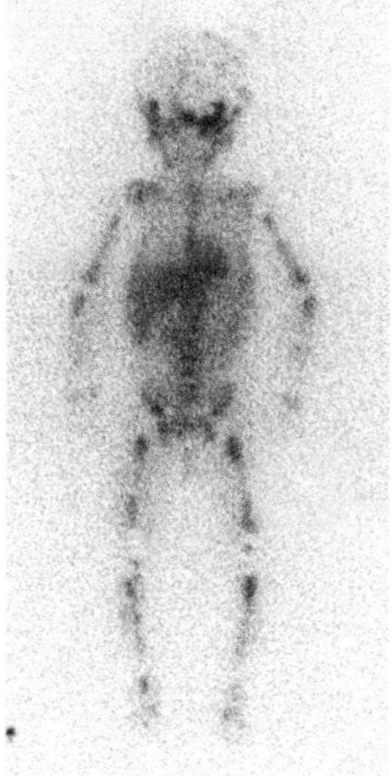

FIGURE 21.2 MIBG (*meta*-iodobenzylguanadine) nuclear medicine image of a patient with metastatic neuroblastoma showing widespread bony metastases throughout the skeleton. Uptake in liver, heart, bladder and salivary glands is physiological.

INTRODUCTION

Neuroblastoma is the most frequent extra-cranial solid tumor of childhood and most common malignancy in infants [1]. Average annual age-adjusted incidence is nine per million children, equating to approximately 650 new cases each year in the USA [1]. Reflecting its developmental origins, 40% of cases are diagnosed during the first 3 months of life and the median age of diagnosis is 17 months. Neuroblastoma is a neural crest-derived tumor that typically arises within tissues of the sympathetic nervous system, most commonly the adrenal glands or paraspinal sympathetic chain (Figures 21.1 and 21.3). Disseminated disease is detected in approximately one half of cases and clinical presentations vary depending on the pattern of disease (Figure 21.2). Metastatic spread to bone, bone marrow and liver results in symptoms including pancytopenia, irritability, weight loss and

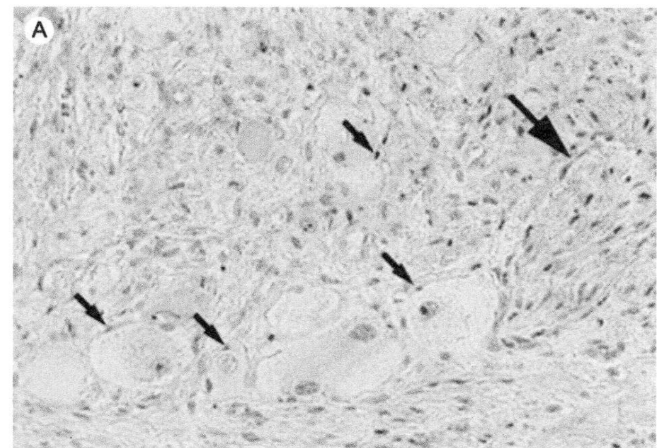

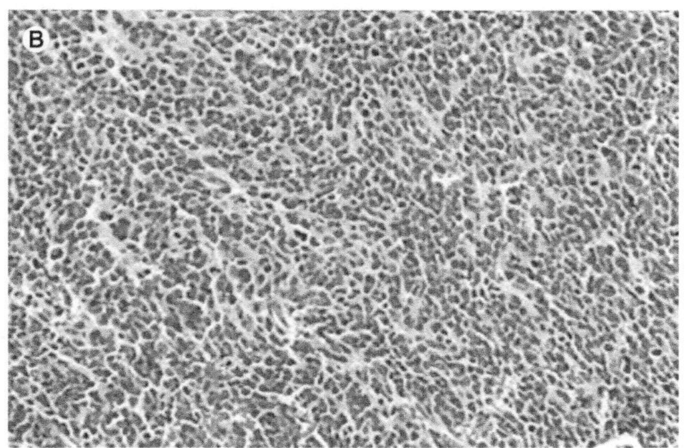

FIGURE 21.3 Representative histological appearances of neuroblastoma showing Schwann cells and ganglion cells (arrows) in stroma-rich neuroblastoma (A) and sheets of densely packed small round blue cells in stroma-poor neuroblastoma (B). Reproduced with permission from Maris JM, Hogarty MD, Bagatell R, Cohn SL. Neuroblastoma. Lancet 2007;369:2106–20.

bone pain [2]. Neuroblastoma has a fascinating and varied natural history. Disease in infants often regresses spontaneously even when initially widespread (stage 4S /4M disease). However, the outlook for older children with metastatic disease is poor. Despite intensive multimodality treatment including cytotoxic chemotherapy, surgery, myeloablative chemotherapy and stem cell rescue, radiotherapy, differentiation therapy and, introduced most recently, immunotherapy, 5-year overall survival rates remain at less than 50% [3]. Given the diverse clinical behavior of neuroblastoma, risk stratification plays a crucial role in determining management and has evolved with the identification of clinical and biological prognostic factors. In the mid-1980s, neuroblastoma became one of the first cancers in which a defined molecular marker (amplification of the *MYCN* oncogene) was incorporated into routine clinical practice [4–6].

The largest cooperative pediatric oncology groups from North America, Europe, and Japan formed the International Neuroblastoma Risk Group (INRG) task force to identify clinical and biological prognostics that determine risk of relapse and guide therapy. The most recent INRG stratification includes clinical stage, patient age, tumor histology, *MYCN* amplification status and markers of other chromosomal abnormalities (ploidy and 11q loss of heterozygosity (LOH)) [7,8]. Based on these factors neuroblastoma patients can be subclassified into low-, intermediate- and high-risk groups (Table 21.1). Overall survival (OS) for low-, intermediate- and high-risk groups is 95, 90, and 40–50%, respectively (Figure 21.4). Efforts are ongoing to develop novel molecular and genetic markers to allow more accurate prediction of disease response through, for example, defining an "ultra-high-risk" group of patients whose disease is predicted to be refractory to standard high-risk induction chemotherapy. In addition, a thorough understanding of the genetics of neuroblastoma is critical to the identification of novel therapeutic targets. This chapter examines the current state of knowledge of neuroblastoma genetics, from the genes associated with rare cases of familial neuroblastoma to the genetic determinants of sporadic disease. Major advances in high-density DNA- and RNA-based microarrays, whole-genome sequencing (WGS), and non-coding or microRNA and epigenetic profiling have led to significant insights into the molecular pathogenesis of neuroblastoma, host predisposition, potential therapeutic targets, and more precise risk prediction to guide treatments.

GERMLINE PREDISPOSITION

Familial Neuroblastoma

Although familial neuroblastoma comprises only 1–2% of all cases, investigation of the underlying genetic abnormalities in these families has yielded valuable insights into the pathogenesis of neuroblastoma and led to the characterization of novel molecular targets that are also relevant in sporadic disease. The first neuroblastoma predisposition gene identified was the paired-like homeobox 2B (*PHOX2B*), which is located at 4p12 and encodes a transcription factor involved in the regulation of neural development [9]. Loss-of-function mutations including missense, frameshift and non-polyalanine expansion mutations have been detected in 6–10% of familial neuroblastoma cases [10–12]. *PHOX2B* is also commonly mutated in cases of congenital central hypoventilation syndrome (CCHS, previously known as Ondine's curse) and such patients are at increased risk of neuroblastoma. In addition to CCHS, patients with *PHOX2B* mutations are at risk for other neurocristopathies including Hirschsprung's disease. In *Phox2b−/−* mice, and those engineered to overexpress one of the CCHS non-polyalanine expansion mutations, the autonomic ganglia fail to form properly [13,14]; however,

TABLE 21.1 International Neuroblastoma Risk Group (INRG) Risk Stratification for Neuroblastoma

<table>
<tr><th>INRG Stage</th><th>Age (Months)</th><th>Histologic Category</th><th>Grade</th><th>MYCN</th><th>11q Aberration</th><th>Ploidy</th><th>Risk Group</th></tr>
<tr><td>L1/L2</td><td></td><td>GN maturing; GNB intermixed</td><td></td><td></td><td></td><td></td><td>A Very low</td></tr>
<tr><td rowspan="2">L1</td><td rowspan="2"></td><td rowspan="2">Any, except GN maturing or GNB intermixed</td><td rowspan="2"></td><td>NA</td><td></td><td></td><td>B Very low</td></tr>
<tr><td>Amp</td><td></td><td></td><td>K High</td></tr>
<tr><td rowspan="6">L2</td><td rowspan="2"><18</td><td rowspan="2">Any, except GN maturing or GNB intermixed</td><td rowspan="2"></td><td rowspan="2">NA</td><td>No</td><td></td><td>D Low</td></tr>
<tr><td>Yes</td><td></td><td>G Intermediate</td></tr>
<tr><td rowspan="4">≥18</td><td rowspan="4">GNB nodular
Neuroblastoma</td><td rowspan="2">Differentiating</td><td rowspan="2">NA</td><td>No</td><td></td><td>E Low</td></tr>
<tr><td>Yes</td><td></td><td rowspan="2">H Intermediate</td></tr>
<tr><td>Poorly differentiated or undifferentiated</td><td>NA</td><td></td><td></td></tr>
<tr><td></td><td>Amp</td><td></td><td></td><td>N High</td></tr>
<tr><td rowspan="5">M</td><td><18</td><td></td><td></td><td>NA</td><td></td><td>Hyperdiploid</td><td>F Low</td></tr>
<tr><td><12</td><td></td><td></td><td>NA</td><td></td><td>Diploid</td><td>I Intermediate</td></tr>
<tr><td>12 to <18</td><td></td><td></td><td>NA</td><td></td><td>Diploid</td><td>J Intermediate</td></tr>
<tr><td><18</td><td></td><td></td><td>Amp</td><td></td><td></td><td>O High</td></tr>
<tr><td>≥18</td><td></td><td></td><td></td><td></td><td></td><td>P High</td></tr>
<tr><td rowspan="3">MS</td><td rowspan="3"><18</td><td rowspan="3"></td><td rowspan="3"></td><td rowspan="2">NA</td><td>No</td><td></td><td>C Very low</td></tr>
<tr><td>Yes</td><td></td><td>Q High</td></tr>
<tr><td>Amp</td><td></td><td></td><td>R High</td></tr>
</table>

INRG stage is preoperative stage; L1/L2: localized or locoregional tumor; M: distant metastatic disease; MS: metastatic disease in infants with metastases confined to skin, liver and/or bone marrow (previously stage 4S); GN: ganglioneuroma; GNB: ganglioneuroblastoma; NA: non-amplified; Amp: *MYCN* amplified. Adapted from [7].

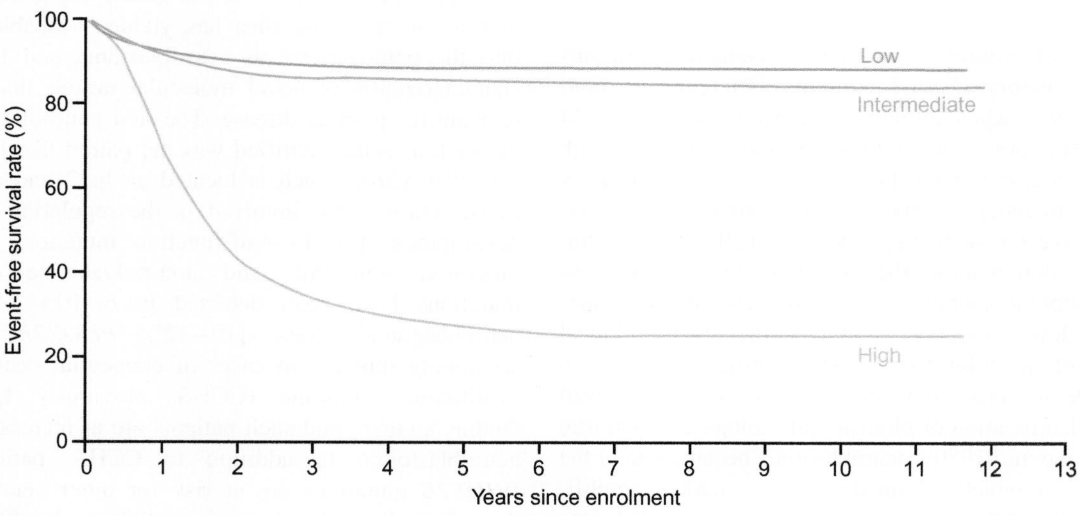

FIGURE 21.4 Kaplan–Meier curve showing event-free survival for neuroblastoma patients treated between 1986 and 2001 in Children's Cancer Group, Pediatric Oncology Group or Children's Oncology Group studies. Patients are classified at diagnosis as low-, intermediate- or high-risk based on clinical and biological features, including patient age, tumor stage, histology and *MYCN* amplification status. EFS for low- and intermediate-risk patients is approximately 85–90% at 10 years, but only 30% for those with high-risk disease. Reproduced with permission from Maris JM, Hogarty MD, Bagatell R, Cohn SL. Neuroblastoma. Lancet 2007;369:2106–20.

neural crest tumors have not been identified. *PHOX2B* mutations are not common in sporadic tumors.

Genome-wide single nucleotide polymorphism (SNP) linkage studies have subsequently been employed to explore hereditary neuroblastoma pedigrees and, in 2008, this led to the successful identification of *ALK* as a major familial neuroblastoma predisposition gene [15]. Initial linkage signal was located at chromosome 2p23-24, which is very close to the *MYCN* locus and, subsequently, three germline mutations in *ALK* were identified in eight neuroblastoma families [15]. *ALK* is a receptor tyrosine kinase and is implicated as an oncogene in a number of predominantly adult cancers in which chromosomal translocations lead to the constitutive activation of the ALK kinase domain. These include the nucleophosmin-ALK (*NPM–ALK*) fusion common in ALK-positive anaplastic large-cell lymphoma, echinoderm microtubule associated protein like 4-ALK (*EML4–ALK*) fusion in a subset of non-small cell lung cancers, and other fusions in squamous cell carcinomas and inflammatory myofibroblastic tumors [16]. Within these neuroblastoma families, the germline mutations in *ALK* (G1128A, R1192P and R1275Q) are all within the kinase domain and predicted to lead to kinase activation [15]. Although *ALK* mutations appear to account for the majority of familial neuroblastoma cases, it is clear that penetrance is incomplete and studies are ongoing to identify additional genetic variants that may affect the probability of an individual with a germline *ALK* mutation for developing neuroblastoma. An initial linkage analysis identified chromosome 2p23-p24, including the *ALK* locus itself, as most strongly associated, suggesting that penetrance may be determined by variations in the non-mutated *trans ALK* allele [17]. Although familial neuroblastoma is very rare, *ALK* mutations and amplifications have also been identified in more than 10% of sporadic NB tumors and the importance of *ALK* as an oncogenic driver and therapeutic target is discussed later in this chapter in the section on somatic mutations.

ALK and *PHOX2B* mutations account for the majority of familial neuroblastoma cases. However, neuroblastoma is found in association with other germline mutations in cancer predisposition syndromes including Li–Fraumeni (p53), Noonan (*PTPN11*), neurofibromatosis (*NF1*, *RAS*) and Beckwith–Wiedemann syndrome. It is likely that WGS efforts will lead to the identification of additional germline mutations in patients with a family history of neuroblastoma.

Predisposition to Sporadic Neuroblastoma

The underlying causes of the majority of neuroblastoma cases, which are sporadic rather than hereditary, remain to be elucidated. Epidemiological studies have failed to demonstrate any environmental causes and, given the developmental origins of the disease, it is likely that underlying genetic factors are, at least, partly responsible. Consequently, considerable research effort has been directed to identifying constitutional genomic variations that may contribute to increased neuroblastoma susceptibility. Unlike the rare mutations in *ALK* and *PHOX2B* described above, these variations are considerably more frequent but, individually, have a less dramatic impact on neuroblastoma risk (Figure 21.5). It is likely that the interplay between multiple genetic factors and potentially,

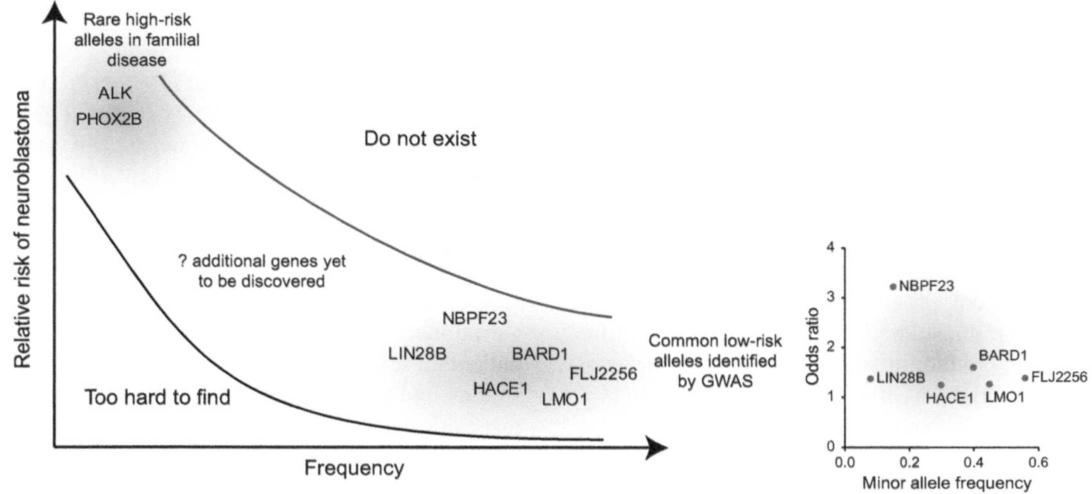

FIGURE 21.5 Currently identified neuroblastoma predisposition genes fall into two groups: the rare, high-risk mutations in *ALK* and *PHOX2B* associated with familial neuroblastoma, and the common, low-risk polymorphisms discovered through genome-wide association studies. At present it is unclear whether additional genes, which are less common but confer a higher risk of neuroblastoma, exist. Figure adapted from original concept relating to breast cancer susceptibility in [39].

as yet unidentified, environmental exposures may influence the initiation and progression of neuroblastoma.

Genome-wide association studies (GWAS) of constitutional DNA samples have identified numerous SNPs associated with different clinical features including specific risk groups. In a study of 1752 patients, Maris and colleagues identified an association of neuroblastoma with three SNPs at 6p22 [18]. For patients with neuroblastoma, these risk alleles were associated with a more aggressive disease phenotype (metastatic disease, *MYCN* amplification and increased relapse risk). The 6p22 locus contains two genes (*FLJ22536* and *FLJ44180*), although at present little is known about their function or potential role in neuroblastoma pathogenesis. A subsequent GWAS analysis of 2251 patients identified four SNPs within the LIM-domain-only 1 (*LMO1*) gene at 11p15.4 [19]. *LMO1* is one of a family of four transcriptional regulators required for development and strongly implicated in human cancer [20]. *LMO3* has been shown to have oncogenic properties in neuroblastoma as a result of interaction with the transcription factor HEN2 [21], while *LMO2* has been directly implicated in T-cell acute lymphoblastic leukemia in patients with X-linked severe combined immunodeficiency (SCID) treated with retroviral-mediated gene therapy, in which the gene being replaced inserts within the *LMO2* locus [22]. Examination of neuroblastoma tumor DNA revealed increased *LMO1* copy number gains as a result of duplications and this gain is associated with an aggressive disease phenotype (i.e. metastatic disease and older patient age) and worse patient outcome [19]. Interestingly *LMO1* aberrations were rarely detected in *MYCN*-amplified tumors. *LMO1* copy number alterations and SNPs are associated with increased *LMO1* expression in neuroblastoma tumors and cell lines, and depletion of *LMO1* by short-interfering RNA results in increased neuroblastoma cell proliferation. Taken together these findings suggest an oncogenic role for *LMO1* in neuroblastoma.

A subsequent study has reported additional associations of neuroblastoma with SNPs at 6q16 within the *HACE1* and *LIN28B* genes [23]. *HACE1* encodes an E3 ubiquitin protein ligase implicated as a tumor suppressor gene in Wilms tumors and a variety of other cancers [24]; in neuroblastoma, low expression is associated with high-risk disease and poor prognosis [23]. Specific substrates of *HACE1* in neuroblastoma have not yet been identified; however, *HACE1* ubiquitinates and targets the small GTPase Rac1, which is involved in cell motility, and *HACE1* knockdown enhances migration [25]. Since mutations in genes implicated in the Rho/Rac pathway have been identified in neuroblastoma [26], it is possible that *HACE1* ubiquitination of Rac1 may regulate migration and/or neuronal differentiation in neuroblastoma. *LIN28B* encodes one of two Lin28 proteins that act as repressors of the *let*-7 family of microRNAs and play important roles in stem cell biology and cancer development [27]. High levels of *LIN28B* expression are associated with metastatic disease and poor outcome in neuroblastoma and it has recently been demonstrated that LIN28B-mediated repression of *let*-7 miRNAs leads to increased MYCN protein levels and enhanced cellular proliferation [28]. In addition to the increased neuroblastoma risk associated with constitutional polymorphisms within the *LIN28B* locus, changes in *LIN28B* expression also appear to play an important role within neuroblastoma cells, with neuroblastoma tumors showing amplification and overexpression of *LIN28B* [28]. Knockdown of *LIN28B* leads to growth inhibition of neuroblastoma *in vitro*. Furthermore, murine transgenic models in which *Lin28B* is overexpressed in developing neural crest spontaneously develop neuroblastoma confirming a role for *LIN28B* as a neuroblastoma oncogene [28].

Most of the original GWAS examined DNA from neuroblastoma patients from all risk groups. A GWAS focusing specifically on high-risk neuroblastoma patients confirmed the finding of an association with SNPs at 6p22 and identified an additional region of interest with six SNPs at 2q35 associated with neuroblastoma risk [29]. The region of interest is located within introns of the *BARD1* (BRCA-1-associated RING domain-1) gene that encodes for a protein involved in stabilizing the tumor suppressor *BRCA1* [30]. Disease-causing mutations identified in *BRCA-1* in breast and ovarian cancers affect the ability of BRCA-1 to bind BARD-1. The functional significance of *BARD1* in neuroblastoma has since been demonstrated. Disease-associated SNPs result in the expression of a particular isoform, *BARD1β*, that has oncogenic properties and can inhibit apoptosis and stabilize Aurora kinase B [31]. Further susceptibility variants have been identified through GWAS approaches focusing instead on low-risk neuroblastoma cases. SNPs within dual specificity phosphatase 12 (*DUSP12)* at 1q23.3, dead box polypeptide 4 isoform (*DDX4)* and IL-31 receptor A precursor (*IL31RA)* at 5q11.2 and hydroxysteroid 17-beta dehydrogenase (*HSD17B12)* at 11p11.2 were all determined to be predisposition alleles associated with low-risk neuroblastoma [32].

In addition to identification of SNP risk alleles, GWAS analyses have been used to identify copy number variations (CNVs) in germline DNA of neuroblastoma patients. Previously, constitutional copy number alteration studies had identified specific CNVs associated with predisposition to non-malignant conditions including autoimmune diseases, autism and schizophrenia, and a general increase in CNVs was detected in patients with Li–Fraumeni syndrome [33]. In a study of CNVs, as detected by copy number analyses at 500 000 SNPs in approximately 1500 patients and more than 5000 controls, an association was detected between neuroblastoma and hemizygous deletion at 1q21.1, with deletion present in 15.6% of cases, but only 9.1% of controls, odds ratio of 2.5 [34]. Furthermore, the 1q21.1 deletion was detected in the germline of parents of affected patients. Interestingly, this region has sequence homology to several previously identified neuroblastoma breakpoint family (NBPF) genes, leading to the identification of a new NBPF family member, *NBPF23*. The precise role of these genes is not yet clear; however, *NBPF23* does show high levels of expression in fetal brain and sympathetic ganglia consistent with a potential role in sympathetic nervous system development. The founding member of the family, *NBPF1*, was discovered to be involved in a chromosomal translocation in a patient with neuroblastoma [35] and is located on chromosome 1p36 – a region that shows frequent loss of heterozygosity (LOH) in neuroblastoma (see below). It has recently been suggested that NBPF1 may form a trimolecular complex that includes the neuroblastoma tumor suppressor clusterin [36], although the functional significance in neuroblastoma pathogenesis remains unclear [37].

Overall, therefore, these GWAS approaches have identified a number of genes (including *LMO1*, *BARD1*, *LIN28B*, *NBPF23*) with potential roles in the pathogenesis of neuroblastoma. Many of the constitutional DNA polymorphisms identified are relatively common within both the neuroblastoma and control population, and the effect size or relative risk for any single genetic variation on neuroblastoma susceptibility is subtle. For most sporadic neuroblastoma cases, it is likely that a combination of underlying constitutional genomic aberrations interact with additional tumor-specific abnormalities to lead to the development of neuroblastoma and, specifically, the subgroups of neuroblastoma with divergent clinical presentations and outcomes. Future studies that integrate analyses of constitutional copy number aberrations or specific SNPs with tumor-specific mutations, mRNA or miRNA expression analyses will undoubtedly lead to models that may be more predictive of prognosis and ultimately response to therapies. Finally, recent studies also suggest that certain SNPs (e.g. *SPAG16*) may contribute to ethnic disparity in the outcome of patients with neuroblastoma, adding further complexity to the interpretation of whole-population GWAS studies that include heterogeneous racial backgrounds [38].

SOMATIC CHANGES IN NEUROBLASTOMA

Considerable effort has been expended in the hope of identifying recurrent genetic abnormalities in neuroblastoma that may provide insight into tractable therapeutic targets. However, in contrast to many adult carcinomas in which oncogenic mutations (e.g. *EGFR* in lung cancer, *KRAS* in colon cancer, *BRAF* in melanoma) are relatively common [39], frequent recurrent mutations have not been identified in neuroblastoma. Next-generation sequencing techniques have led to several recent publications that together have reported structural and sequence aberrations in more than 500 neuroblastoma tumors. Versteeg and colleagues performed whole-genome sequencing (WGS) on DNA from 87 neuroblastoma tumors and their matched germline (lymphocyte) DNA included samples from all patients of stages and risk groups. They identified structural variants consistent with chromotripsis or shredding of chromosomes in almost 20% of high stage tumors, which was associated with poor prognosis [26]. Structural variants were also detected in *ODZ3, PTPRD* and *CSMD1*, all of which are involved in neuritogenesis. Mutations were also identified in *ALK* (6%), *ATRX*, and *TIAM1*, a regulator of the cytoskeleton and neuritogenesis and rare mutations were detected in other neuritogenesis genes. Maris and colleagues focused their sequencing efforts on 240 high-risk neuroblastoma samples as part of the TARGET (Therapeutically Applicable Research to Generate Effective Treatments) initiative. Using a combination of whole-exome sequencing (WES), WGS and transcriptome sequencing they demonstrated a low exon mutation frequency of only 0.6 per Mb, markedly less than that of most adult cancers, and very few recurrently mutated genes [40]. The most commonly mutated genes included *ALK* (9.2% of cases), *PTPN11* (2.9%), *ATRX* (2.5%), *MYCN* (1.7%) and *NRAS* (0.83%), highlighting the rarity of recurrent mutations in neuroblastoma. Considerably more common are chromosomal aberrations, including numerical and segmental chromosomal abnormalities, variations in ploidy and amplifications, particularly of the *MYCN* gene locus. Additional mutations and deletions in other genes, including *ARID1A* and *ARID1B* have been detected using WGS [41]. Some of the mutated genes identified in these sequencing studies are discussed in more detail below.

ABNORMALITIES OF SPECIFIC GENES IN NEUROBLASTOMA

ALK

In addition to a role in rare familial neuroblastoma, abnormalities in *ALK* (anaplastic lymphoma kinase) have also been identified in sporadic neuroblastoma and represent the most common single gene mutation discovered to date [40]. Activating somatic mutations in *ALK* have been detected in up to 10% of primary tumors, with a further 4% having high-level focal amplification [15,42]. The most common *ALK* mutations in sporadic disease (R1275Q, F1174L and F1245V) are all located within the tyrosine kinase domain and act as oncogenic drivers both *in vitro* and *in vivo*[43,44]. In a murine model, targeted expression of ALKF1174L in developing neural crest induces the formation of tumors resembling human neuroblastoma [45,46]. In zebrafish, transgenic expression of *MYCN* together with the activated mutant *ALKF1174L* also results in neural crest tumors that resemble neuroblastoma [47]. In this and other models, tumors respond to pharmacological ALK inhibition and, consequently, there is considerable interest in ALK inhibitors for neuroblastoma therapy [48]. The small molecule dual MET and ALK antagonist crizotinib has recently been shown to be effective in the treatment of patients with ALK translocation-positive anaplastic large cell lymphomas and a subset of neuroblastoma tumors in a phase I pediatric clinical trial [49]. Based on preclinical data, it is likely that the response to crizotinib is mutation specific, and efforts are underway to identify novel strategies to inhibit ALK mutant proteins that are relatively resistant to crizotinib. Interestingly, while initial studies demonstrated no significant survival differences based on tumor *ALK* mutation status, survival of patients with the F1174L mutation is significantly worse than for those with R1275Q, perhaps relating to a correlation between F1174L mutation and *MYCN* amplification [50]. In addition to mutations, copy number abnormalities encompassing the *ALK* gene locus are present in up to 20% of neuroblastoma cases mainly as a result of gain of 2p or whole of chromosome 2. *ALK* gene copy number correlates with mRNA expression and, in turn, this is associated with poor survival, even in the presence of the wild-type gene [50]. A large study of more than 1500 tumors from the Children's Oncology Group is in progress to determine more definitively whether specific *ALK* mutations are associated with outcome or other clinical and biological risk factors.

ATRX

ATRX (alpha thalassemia/mental retardation syndrome X-linked) is a member of a family of chromatin remodeling proteins and plays a role in nucleosome assembly and maintenance of telomeres. WGS efforts identified mutations in 2.5–6% of sporadic neuroblastoma tumors [26,40] although, intriguingly, *ATRX* mutations are considerably more common in adolescent and young adults. In a cohort of 104 patients with advanced stage neuroblastoma, *ATRX* mutations were detected in 44% of tumors from patients aged ≥12 years, but only 9% for those aged 18 months to 12 years, and in 0% for infants aged <18 months [51]. Larger studies are underway to confirm this association of *ATRX* mutations with older patients, including adolescents and young adults with neuroblastoma – a subset of patients who rarely have *MYCN*-amplified tumors and often present with indolent and chronic disease but, ultimately, have a poor prognosis. *ATRX* mutations were associated with loss of nuclear ATRX protein, long telomeres and evidence of ALT (alternative lengthening of telomeres) pathway activation. ATRX-related changes in telomeres may therefore play a role in the more chronic clinical course seen in older neuroblastoma patients, and potentially suggests therapeutic approaches to target these aberrant telomere-related pathways in *ATRX* mutant tumors.

ARID1A/1B

Additional chromatin remodeling genes with a potential role in neuroblastoma have recently been identified. In a study in which whole-exome sequencing (WES) as well as WGS (subsets of cases analyzed by low and high coverage approaches) was performed, point mutations or intragenic deletions of *ARID1B* were identified in 7% of neuroblastoma tumors, with point mutations in *ARID1A* in a further 6% [41]. *ARID1A* and *ARID1B* aberrations were associated with early treatment failure and poor survival. Although functional studies in neuroblastoma have not yet been completed, mutations in *ARID1B* have been reported as oncogenic drivers in a number of adult cancers [52]. Aberrant expression of other proteins involved in chromatin remodeling and epigenetic regulation of gene expression in neuroblastoma are discussed in the section on epigenetics later in this chapter, and include *CASZ1*, *EZH2*, and *CHD5*. Interestingly, somatic mutations in the chromatin remodeling complex genes *H3.3*, *ATRX* and *DAXX* have recently been identified in pediatric glioblastoma [53].

PTPN11

The *PTPN11* gene encodes SHP-2, a protein tyrosine phosphatase (PTP) with a regulatory role in multiple intracellular signaling pathways. Germline mutations in *PTPN11* are present in approximately one half of patients with Noonan's syndrome and these individuals have an increased risk of a number of cancers, including neuroblastoma [54]. *PTPN11* mutations have also been detected in a small percentage of sporadic neuroblastoma [40,55], although the functional significance and any impact on prognosis have yet to be explored. The mechanisms by which SHP-2 phosphatase is involved in neuroblastoma pathogenesis have not been elucidated. However, clues to its role include the importance of SHP-2 signaling upstream of RAS, which is also mutated in a small subset of neuroblastoma tumors.

MYCN Amplification

MYCN was first identified in neuroblastoma cell lines by homology to viral *myc* and is a member of a highly conserved family of oncogenes that also includes *MYC*

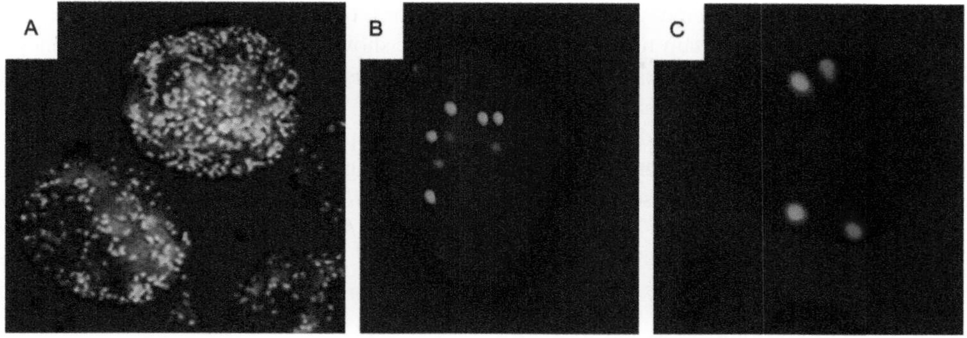

FIGURE 21.6 Fluorescence *in situ* hybridization (FISH) analysis of neuroblastoma tumor cells showing *MYCN* amplification (A), gain (B) or normal copy number (C). *MYCN* probe is green, *CEP2* probe (red) acts a control. From Wang M, Zhou C, Cai R, Li Y, Gong L. Copy number gain of *MYCN* gene is a recurrent genetic aberration and favorable prognostic factor in Chinese pediatric neuroblastoma patients. Diagn Pathol 2013;8:5. Reproduced under terms of the Creative Commons Attribution License © Wang et al.

(*c-myc*) and *MYCL* [4]. The MYCN protein is a helix-loop-helix/leucine zipper transcription factor that, through the formation of heterodimers with Max (Myc-associated factor X), binds to E-box sequences (CACGTG) and regulates a large number of downstream targets. The MYC proteins control multiple cellular processes including cell growth, proliferation and apoptosis, protein synthesis, differentiation and metabolism [56,57] and have been implicated as an oncogene in many pediatric and adult malignancies. *MYCN* amplification (MNA) is present in 20−25% of neuroblastoma tumors, and close to 40% of high-risk tumors (Figure 21.6) [5,6]. In children with metastatic disease over the age of 18 months, the presence of MNA is not an independent predictor for event-free survival (EFS). In contrast, MNA has particular prognostic importance in infants with metastatic disease − patients with non-MNA tumors have a 3-year EFS of 93% compared to only 10% for those with MNA tumors [58] − and is a crucial consideration for risk stratification and treatment allocation in current neuroblastoma protocols. For tumors without MNA, there has been controversy over the importance of the level of expression of MYCN mRNA or protein. Recent data using a 157-gene signature comprising MYCN-regulated targets demonstrated that this signature identifies poor prognosis patients with both MNA and non-MNA tumors [59]. These non-MNA tumors with poor outcome had low levels of *MYCN* mRNA, but high levels of MYCN protein (presumably related to protein stabilization), thus suggesting an important oncogenic role for *MYCN* even in neuroblastoma tumors without amplification.

The important role of *MYCN* in driving tumorigenesis in neuroblastoma has been demonstrated in murine models in which MYCN transgene is overexpressed in developing neural crest through a construct using the tyrosine hydroxylase (TH) promoter. These *TH−MYCN* mice spontaneously develop neuroblastoma tumors and serve as a useful model for drug development [60]. *TH−MYCN* mice have also been crossed with other genetically engineered mice (including those expressing a mutant *ALK* transgene) to develop additional neuroblastoma models. For example, compound *TH−MYCN* caspase 8−/− mice develop enhanced bone marrow metastases [61]. At present, despite considerable effort, it has proved impossible directly to target *MYCN* with small molecule drugs, although inhibitors against aurora kinase and PI3 kinase [62,63] that destabilize the MYCN protein and BET bromodomain inhibitors that interfere with MYCN transcriptional activity [64] represent potential alternative therapeutic strategies.

CHROMOSOMAL ABNORMALITIES

Given the heterogeneous clinical behavior of neuroblastoma, it is not surprising that the tumor can harbor a wide variety of structural genetic abnormalities. Three genomic subtypes are recognized [65]. Type 1 tumors typically occur in infants, have a favorable outcome, a hyperdiploid or near triploid DNA content and do not show focal chromosomal abnormalities or *MYCN* amplification (MNA). Type 2A are intermediate-risk tumors with near diploidy or tetraploidy and no MNA. These tumors typically have gain of chromosome 17q or deletion of 11q. Type 2B comprises high-risk tumors with *MYCN* amplification and frequently associated 1p loss or 17q gain. Principal component analysis of gene expression profiles from primary tumors has identified a six-gene profile that discriminates four neuroblastoma clusters, three of which correlate with the genomic subtypes outlined above [66]. The fourth cluster was similar to other type 2A tumors but with higher expression of various nervous system developmental genes (such as Erbb receptors and SoxE family genes). Thus, type 2A tumors may include two underlying genetic subgroups. Initial studies focused on studies of regions of interest

(ROIs) with gain and loss of chromosome segments with the hypothesis that these loci might contain tumor suppressor genes or oncogenes. More recent studies suggest that the overall genomic pattern of chromosomal aberrations may in itself be predictive. Specific loci and data to support the prognostic significance of segmental chromosome aberrations (SCA) are discussed in the following sections.

17q Gain

Gain of the long arm of chromosome 17 is the most frequent cytogenetic abnormality in neuroblastoma [67] and is present in approximately 50% of tumors, including up to 90% of high-risk tumors [68]. 17q gain is associated with multiple adverse prognostic risk factors including stage 4 disease, age greater than 1 year, diploidy, *MYCN* amplification and 1p deletion, and consequently correlates with worse clinical outcomes [69]. Its presence is particularly discriminatory for outcome in low-stage tumors without MNA, for which reported 5-year OS is 100% in the absence of 17q gain, but only 52.5% if 17q gain is present [70]. As with other chromosomal aberrations, considerable effort has been directed to determining the potential active genes in this region (17q23.1-qter) and several have been proposed, including *PPM1D* located at 17q23.2 [71]. *PPM1D* encodes for a phosphatase (also known as WIP1) that is induced by p53, but negatively regulates p53 and consequently acts as an oncogene [72]. Amplification or mutation of *PPM1D* is implicated in breast and ovarian cancer. In neuroblastoma cells, knockdown using antisense oligonucleotides leads to decreased cell proliferation consistent with *PPM1D* having an oncogenic role [71]. Other genes of interest also located on 17q include *BIRC5* (survivin) [73–75] and the neighboring genes *NME1* and *NME2* [76–78]. The survivin inhibitor YM155 has shown promising preclinical efficacy in neuroblastoma cells [79].

11q Loss

11q LOH is present in approximately one third of primary neuroblastoma cases [80–82]. Like 17q gain, it is more common in stage 4 disease and predicts poor prognosis; however, 11q LOH is rarely associated with MNA. Similar to many SCA, 11q LOH may be particularly relevant in identifying low- or intermediate-risk patients who are at highest risk for relapse [7]. Integrated genomic profiling of tumors with 11q LOH (without 1p loss or MNA) indicates that there are two biological subgroups with different outcomes and gene expression profiles. Tumors with a favorable gene expression signature [83] have a similar overall gene expression profile regardless of 11q status, suggesting that 11q loss is compensated for, whereas tumors with an unfavorable signature and 11q LOH showed downregulation of genes located on 11q [81]. Thus, it is unclear if the poor prognosis associated with 11q LOH reflects simply the loss of a particular tumor suppressor gene or is also determined by other underlying genetic events. One candidate tumor suppressor is *TSLC1/IGSF4/CADM1* (tumor suppressor in lung cancer 1/immunoglobulin superfamily 4/cell adhesion molecule 1) located at 11q23 and previously implicated in non-small cell lung cancer. Reduced expression of *CADM1* is found in unfavorable neuroblastoma and correlates with adverse outcome, while overexpression in neuroblastoma cells *in vitro* leads to reduced proliferation [84–86]. Other potential TSG located within 11q include *IGSF4*, *SDHD*, *MLL* and *PP2R1B* [87].

1p Loss

1p deletion in neuroblastoma was first reported in 1977 [88] and like other SCA is a powerful predictor of poor outcome, especially in infants. For example, in a series of 89 cases, 3-year EFS was 100% for stage 1, 2 or 4S tumors without 1p loss, but only 34% for those with 1p deletion, with equivalent EFS of 53 and 0%, respectively, for stage 3 or 4 disease [89]. A larger analysis of more than 900 patients confirmed deletion of 1p in 23% of cases and this correlated with MNA and other high risk features such as metastatic disease, age over 1 year and unfavorable histology [80]. Within high-risk tumors, 1p loss is consequently a frequent finding. Interestingly, a recent analysis of 22 patients aged 10–24 years showed no 1p deletion, suggesting that (like *MYCN* amplification), 1p deletion may be infrequent in older children and adolescents [90]. One potentially relevant tumor suppressor gene located within the 1p locus is *CHD5*, a member of the chromodomain superfamily that affects proliferation and apoptosis via the p19Arf/p53 pathway [91]. Recent examination of neuroblastoma has demonstrated that while no somatic mutations or deletions of *CHD5* were found, *CHD5* expression may be regulated by epigenetic modifications. Specifically, methylation of the *CHD5* promoter was frequently detected in high-risk tumors with both 1p deletion and *MYCN* amplification [92]. Thus, following loss of one allele through 1p deletion, expression of the remaining *CHD5* allele may be further reduced by promoter methylation. However, as with other neuroblastoma SCA, multiple genes within the deleted region are likely to play a role in driving malignancy. Gene expression analysis of tumors with deletion of 1p have shown reduced mRNA levels of many genes located to this region, including *GNB1*, *CLSTN1*, *CDC42*, *MFN2*, *SLC35E2*, *CAPZB*, *TCEB3*, *VAMP3*, *RBAF600*, *STMN1*, *UBE4B*, *CORT*, *CASP9* and *CAMTA1* [87,93,94].

Segmental Chromosomal Abnormalities

Historically, chromosomal abnormalities have been identified using polymerase chain reaction (PCR)-based assays or fluorescence *in situ* hybridization (FISH) to probe individual chromosomes. The recent development of array CGH (comparative genomic hybridization) and, more recently, MLPA (multiplex ligation probe amplification) and SNP arrays, has permitted the whole genome to be characterized and has led to an improved understanding of how chromosomal abnormalities in neuroblastoma may affect prognosis. Array CGH analyses of primary neuroblastoma tumors demonstrates multiple profiles with variable outcomes [95]. Tumors with numerical chromosomal abnormalities (NCA) characterized by whole chromosome gains and losses, but no segmental chromosomal abnormalities, were associated with an excellent outcome even in patients aged over 18 months. Patients whose tumors had SCA alone, or combined SCA and NCA, had poorer outcomes and, in a multivariable analysis, the presence of SCA and *MYCN* amplification status were the strongest predictors of relapse. SCA have subsequently been confirmed as a strong predictor of poor outcome in infants with locally unresectable or metastatic *MYCN*non-amplified disease [96]. An analysis of the INRG dataset in which SCA were defined only in terms of 1p or 11q deletion or 17q gain showed a similar result in patients with *MYCN*non-amplified disease [97]. Interestingly, examination of paired tumor samples at diagnosis and relapse indicates that tumors that initially have only NCA may subsequently develop SCA at relapse [98], suggesting increasing chromosomal instability with tumor progression. The number of SCA also correlates with increasing age at diagnosis in diploid/tetraploid, *MYCN*non-amplified (i.e. biologically lower risk) tumors and this may contribute to the poorer outcomes of older patients [99]. In the future, cooperative group studies will likely prospectively study whether the presence of SCA can be used to distinguish clinically less favorable subsets of patients who would otherwise be classified as low or intermediate risk in order to assign appropriate risk-adapted therapies.

mRNA EXPRESSION AND SIGNATURES

Given that recurrent mutations are infrequent in neuroblastoma, the discovery of altered expression of genes and their protein products may lend insight into the understanding of neuroblastoma biology, as well as provide prognostic information and potential biomarkers to predict effective therapies in a more personalized or precision approach. In addition, minimal residual disease (MRD) detection of neuroblastoma-specific transcripts in the bone marrow or peripheral blood during and following completion of therapy has been used to predict prognosis.

Trk Genes

The neurotrophin receptors are a family of transmembrane receptor tyrosine kinases with important roles in neurodevelopment and differential expression in neuroblastoma [100]. The three main Trk receptors and their ligands are TrkA, which binds nerve growth factor (NGF); TrkB, and its ligand, brain-derived neurotrophic factor (BDNF); and TrkC, or neutrophin-3 (NT3). Of these, *TrkA* and *TrkC* are expressed during development of sympathetic adrenal neurons. Studies of neuroblastoma have demonstrated high levels of expression of both *TrkA* and *TrkC* in tumors with favorable biological characteristics – younger patients with localized disease and without *MYCN* amplification – and *TrkA/TrkC* expression correlates with favorable patient outcome [101–104]. By contrast, high levels of *TrkB* (and *BDNF*) are seen in biologically unfavorable tumors, particularly those with *MYCN* amplification, and *Trk B* expression is associated with drug resistance and angiogenesis [105–107]. To date, activating translocations or mutations involving any of the *Trk* genes have not been identified [100] and it is unclear exactly the roles differential *Trk* expression may play in the initiation and progression of neuroblastoma. Nevertheless, small molecule inhibitors of *Trk*, such as lestaurtininb (CEP-701) are being explored as potential therapeutic agents for neuroblastoma [108,109].

mRNA Expression Signatures

Although the management of neuroblastoma already relies on risk stratification using clinico-pathological criteria, it is clear that for some patients this is inadequate. Most patients with low/intermediate-risk disease have an excellent prognosis. However, there remain individual patients who have poor outcome and for whom this reduced therapy is inadequate. Similarly, within the large group of high-risk patients, while some respond well to induction therapy, many tumors are refractory and these individuals (often referred to as "ultra-high risk") would likely benefit from alternative therapies upfront. Numerous attempts have been made to use mRNA gene expression profiling to define more precisely risk groups that may be more predictive than current classifications. In one of the largest studies, using reverse transcription polymerase chain reaction (RT–PCR) to examine expression of 59 genes in neuroblastoma tumors from SIOPEN/GPOH studies (and subsequently validated using tumors from patients treated within COG), the multigene expression profile was shown to be an independent predictor of outcome even after controlling for currently used risk factors, with an odds ratio of 19.3 for OS and 3.96 for progression-free survival (PFS) [110]. In a separate study, using published microarray datasets, a 42-gene classifier was identified incorporating genes detected in at least two of the four datasets [111]. This multigene classifier includes many genes implicated in neuroblastoma biology (including *MYCN, NTRK1, NME1, CADM1, FYN, ODC1* and *WSB1*) and was again an independent predictor of poor outcome. A smaller study looking only at *MYCN*non-amplified metastatic tumors, used Affymetrix microarrays to determine gene expression profiles and reported a 55-gene signature predictive of poor outcome [112]. For example, among patients aged >12 months (all *MYCN* non-amplified, metastatic), PFS was 16% for those with "molecular high risk" disease versus 79% for those classified as low risk. The difficulties of incorporating expression signatures involving so many genes into routine clinical practice has also led to the development of more simplified models. It has recently been reported that the expression pattern of just three genes (*CHD5*, chromodomain helicase DNA-binding protein 5; *PAFAH1B1*, platelet-activating factor acetylhydrolase isoform 1B; and *NME1*, non-metastatic cells 1, protein expressed in) can be used to predict outcome [78], although the model awaits validation before it can be incorporated into clinical practice. Finally, gene expression profiling of tumors has also been used to explore differences in infiltrating host immune cells, particularly tumor-associated macrophages (TAMs). Expression of TAM-related genes (*CD33, CD16, IL6R, IL10* and *FCGR3*) forms part of a 14-gene classification score that correlates with outcome of patients with *MYCN*non-amplified metastatic disease [113]. COG is currently conducting a pilot study for high-risk patients that includes an aim to examine the feasibility of incorporating molecular profiling of *MYCN*non-amplified tumors using this 14-gene signature to identify patients with ultra-high-risk neuroblastoma.

miRNAs

Beyond control of gene transcription, it is becoming increasingly apparent that there is also a complex network of regulation at the level of mRNA and protein translation. Introns and other non-coding regions of chromosomes encode multiple microRNA genes which produce small 22–25 nucleotide RNAs that directly inhibit protein translation or target mRNAs for degradation [114]. Among those miRNAs implicated in neuroblastoma, the best characterized are *miR34a* and the *miR-17-92* cluster. The *miR34a* gene is a direct target of p53, and appears to act as a tumor suppressor gene, with ectopic expression of *miR34a* leading to cell cycle arrest and apoptosis [115]. Downstream targets of *miR34a* include cyclins and cyclin-dependent inhibitors (CDK4, CDK6, cyclin E2), the E2F transcription factors that induce cell cycle progression, and the anti-apoptotic protein BCL2. In neuroblastoma tumors with 1p loss, the more aggressive tumor behavior may be partly explained by reduced expression of *miR34a* located at 1p36.23 [116]. In contrast, the *miR-17-92* cluster (and paralogues miR-106a and miR-106b) has oncogenic properties. Transcription of these miRNAs is directly activated by *MYC* and they show increased levels of expression in *MYCN*-amplified neuroblastoma [117,118]. Downstream targets inhibited by *miR-17-92* include p21 (a cyclin-dependent kinase that negatively regulates cell cycle progression) and the pro-apoptotic BIM (Bcl-2 interacting mediator of cell death). *In vitro* and *in vivo* treatment with an antagomir, which blocks the function of *miR-17-92*, leads to upregulation of p21 and BIM with consequent apoptosis and inhibition of tumor growth [118]. Interestingly, the *miR-17-92* cluster also regulates multiple components of the TGF-β signaling pathway, suggesting additional roles in proliferation and cell adhesion [119]. Many other miRNAs are differentially expressed in *MYCN*-amplified and non-amplified neuroblastoma and may also play important functional roles in tumor behavior [120]. Among these is *let-7* that, as previously discussed, is downregulated by *LIN28B* leading to increased levels of MYCN protein [28]. Additional microRNAs have been identified and implicated in neuroblastoma cell proliferation, apoptosis, and differentiation. Furthermore, similar to mRNA, miRNA signatures have now been shown to have prognostic significance, and recent data demonstrated that a 25-miRNA signature could predict progression-free and overall survival for a cohort of 500 patients [121]. In addition, analyses were feasible both with frozen tissue and archived formalin fixed paraffin-embedded samples.

EPIGENETICS

In addition to genomic factors, gene expression is also regulated by epigenetic modifications, in particular DNA methylation (Figure 21.7). Both candidate gene and, more recently, whole genome-wide screening approaches have been used to identify and characterize epigenetic alterations in neuroblastoma. In total, approximately 75 genes have been reported to be epigenetically modified in neuroblastoma cell lines and/or primary samples [122]. An important initial finding was the frequent inactivation of the gene encoding the apoptotic protease caspase 8 in *MYCN*-amplified neuroblastomas. *CASP8*, located at 2q33, was completely methylated in 63% of *MYCN*-amplified tumors, but only one of 26 (4%) non-amplified [123]. Functional studies confirmed an association of loss of caspase 8 expression with resistance of neuroblastoma

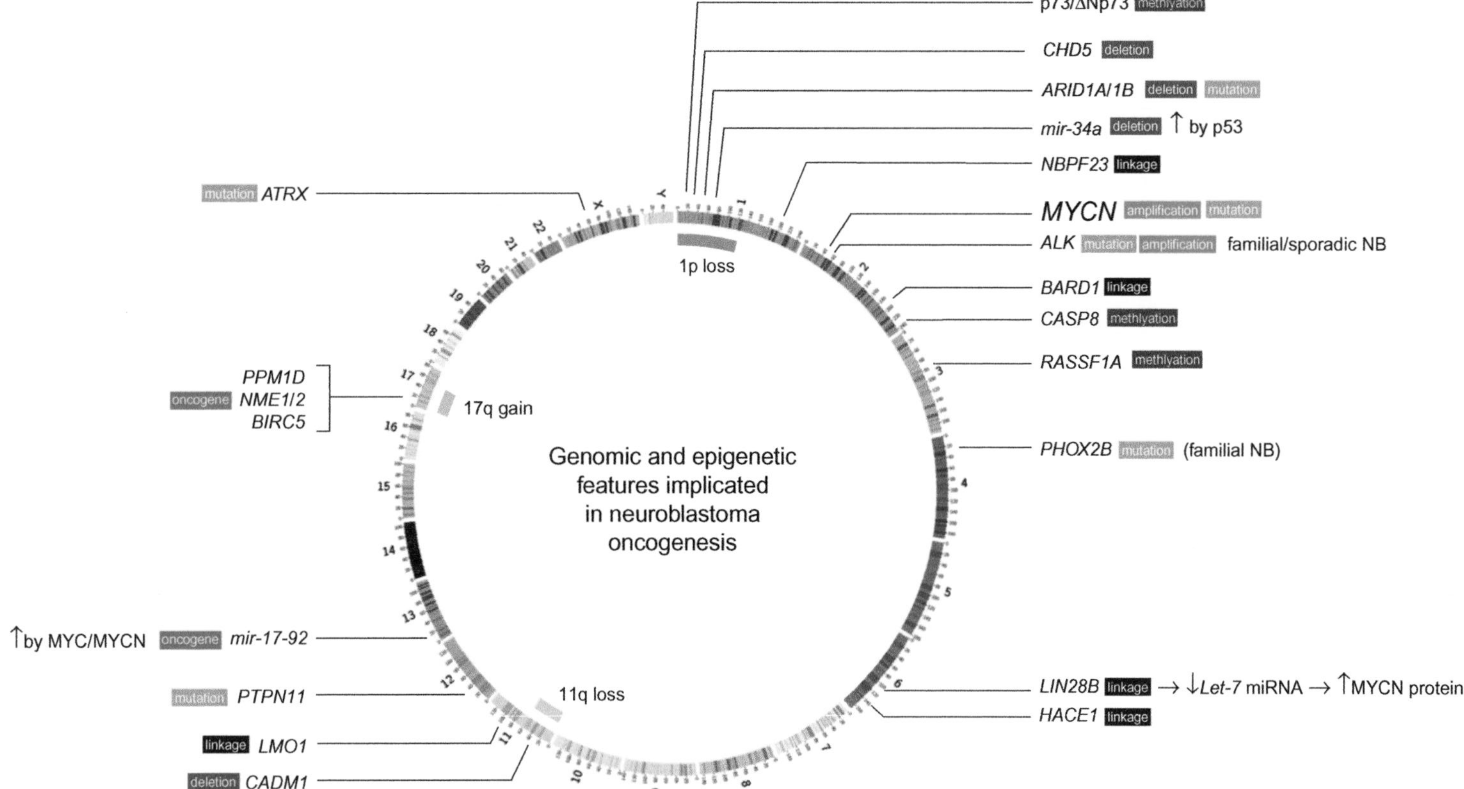

FIGURE 21.7 Summary of the major genomic and epigenetic features implicated in neuroblastoma oncogenesis identified to date. These include gene amplifications (pale green), mutations (orange), deletions (red), oncogenes (dark green) and epigenetic modifications such as methylation (grey). Gene loci identified through linkage analyses for neuroblastoma susceptibility genes are highlighted in blue.

cells to chemotherapy or death-receptor induced apoptosis. A subsequent study demonstrated that neuroblastoma tumors also frequently have epigenetic silencing of other apoptosis-related genes, particularly *RASSF1A*[124]. *RASSF1A* (RAS-association domain family 1 isoform A) interacts with multiple components of the RAS signaling pathway, but expression is silenced in 70–95% of tumors [122]. Numerous other genes that undergo DNA methylation in neuroblastoma have since been identified, with many such epigenetic changes correlating with patient outcome or risk factors such as patient age, tumor stage or *MYCN* amplification status [122]. These genes include the transcription factor *HOXA9*, cell cycle regulator *14.3.3σ* and apoptosis pathway receptor *DcR2*[125,126]. By contrast, the oncogenic truncated ΔNp73 isoform encoded by the *p73* gene is overexpressed in a subset of neuroblastoma tumors as a result of DNA demethylation and this correlates with adverse outcome [127,128].

In order to identify DNA methylation biomarkers that can be used for risk stratification and outcome prediction, it is likely that unbiased genome-wide screening techniques, including expression analyses following treatment with demethylating agents, promoter assays using affinity based capture, and methylation microarrays or sequencing following bisulfate treatment, will be required to identify additional genes (and signatures) regulated by methylation in neuroblastoma. To date, the majority of unbiased large-scale studies in neuroblastoma have relied upon treatment of neuroblastoma cells with demethylating agents, such as the histone deacetylase (HDAC) inhibitors trichostatin A and vorinostat. It is not yet clear whether epigenomic approaches will result in more robust risk stratification either independently or by integrating results with other high-throughput genomic studies such as mRNA or miRNA expression profiles or proteomic data. Currently, several early phase trials are exploring HDAC inhibitors alone or in combination with other therapies for the treatment of neuroblastoma.

HOST GENOMICS

In addition to the identification of germline predisposition alleles, it is likely that other host factors influence neuroblastoma development, response to therapy and susceptibility to treatment-related toxicities. With the incorporation of immunotherapy into the standard management of high-risk neuroblastoma, the role of the host immune system in determining response has become more apparent. Patients with high-risk disease benefit from an immunotherapy regimen that includes the Ch14.18 chimeric monoclonal antibody directed against a disialoganglioside (GD2) expressed on the surface of neuroblastoma cells [129]. However, response depends at least in part on the host immunogenotype, specifically, polymorphisms in the Fcγ receptor and natural killer (NK) cell KIR/KIR-ligand genotypes. Among patients treated with the murine anti-GD2 antibody (3F8), together with GM–CSF, improved progression-free survival significantly correlated with *FCGR2A*-R/R genotype, compared to *FCGR2A*-H/R or H/H, likely due to higher binding affinity of FcγR2A-R131 to the murine IgG3 3F8 antibody [130]. By contrast, for patients treated with a humanized antibody-cytokine construct (hu14.18-IL2), the opposite trend was apparent; patients with *FCGR2A*-H/H genotype responding more frequently than those with H/R or R/R [131]. These results likely reflect the different antibody constructs used and the differences in the cell types involved in antibody-dependent cell cytotoxicity (ADCC). This latter study also demonstrated the importance of NK cell genotypes, with all responding patients being mismatched for KIR/KIR-ligand, with no responses among KIR-matched patients [131]. As the immune response to neuroblastoma and the role of immunotherapy become better characterized, it is likely that additional features of the host immune genome will be identified as determinants of outcome and, specifically, response to immunotherapy. These predictors of response may then be incorporated into regimens aiming to personalize immunotherapy treatments.

Similarly, although pharmacogenomics has not been incorporated into routine management of neuroblastoma, it is likely to play an increasingly important role in identifying patients who may benefit from dose adjustments for particular chemotherapy agents or who are at risk of specific side effects. For example, patients with a particular polymorphism of the *UGT1A1* gene, which encodes for a glucoronyltransferase involved in drug metabolism, experience increased toxicity when treated with irinotecan [132] – a chemotherapy drug currently used to treat relapsed neuroblastoma. Genomic approaches are also showing promise in predicting ototoxicity induced by platinum-containing chemotherapy agents, which play an important role in initial management of neuroblastoma. Specific glutathione S-transferase (GST) genotypes are associated with increased risk of platinum-induced hearing loss [133,134]. GWAS that analyzed SNPs in 220 candidate drug metabolism genes identified ototoxicity risk alleles in catechol O-methyltransferase (*COMT*), and thiopurine S-methyltransferase (*TPMT*) [135]. Unbiased genome-wide approaches are currently underway in the hope of discovering additional determinants of susceptibility to ototoxicity, renal toxicity, anthracycline-induced cardiotoxicity, and second malignancies. However, the challenge will be how to incorporate knowledge about potential increased risk of specific toxicities against the need to ensure adequate antitumor therapy and further prospective validation studies will be required before personalizing therapy based on risk alleles for short- or long-term toxicities.

CONCLUSION

Current risk stratification involves the use of clinical and biological factors, including age, stage, histology, *MYCN* copy number, 11q LOH and ploidy that subclassify patients into high-, intermediate- and low-risk neuroblastoma. Biomarkers to identify patients with the poorest prognosis ("ultra-high risk") who will fail upfront therapy would facilitate a more personalized approach to therapy in which these patients could be offered novel therapies, perhaps related to specific genomic characteristics, such as crizotinib for patients whose tumors express mutant *ALK*. In addition, identifying the patients at lowest risk for relapse might allow for a reduction of chemotherapy that would potentially result in fewer short- and long-term side effects. The rapidly increasing availability and decreasing costs of high-density array-based technologies, next-generation sequencing, and mRNA, miRNA and epigenomic expression profiles have allowed the identification of tumor-specific genetic predictors of clinical phenotype (e.g. metastasis, clinical risk group), response to treatments and, in some cases, have identified biomarkers to support the use of targeted therapies (e.g. *ALK* status). GWAS have also identified risk alleles for neuroblastoma development and specific SNPs associated with different neuroblastoma clinical phenotypes and risk groups, as well as pharmacogenomic predictors of specific toxicities to chemotherapies. Given the lack of recurrently mutated driver mutations in neuroblastoma, mRNA- and miRNA-based signatures may provide more insight into pathways that could be targeted by kinase inhibitors and other agents. Furthermore, the emerging field of phosphoproteomics may also contribute to a better understanding of neuroblastoma intracellular signaling and identify novel therapeutic targets. Modeling genetic aberrations in mice and zebrafish, in addition to the use of well-characterized patient-specific cells and xenografts, will also be invaluable to study the roles of candidate genes and to perform preclinical studies of drugs that are predicted to target these pathways. Finally, the majority of current genomic studies are performed using primary tumor samples obtained at diagnosis. Recent studies in breast cancer, melanoma, and medulloblastoma demonstrate significant differences in the profiles of metastases in comparison to primary tumors [136,137]. Furthermore, in many cancers, additional mutations are acquired between diagnosis and relapse. Thus, DNA and RNA profiles of neuroblastoma tissue from different metastastic sites and relapsed samples should be evaluated. Integrated computational analyses that combine germline and tumor-specific genomic data will undoubtedly be invaluable, in combination with classical clinical and biological prognostic factors, to develop more precise risk stratification models. A better understanding of the genetic determinants or biomarkers of response to specific therapies will ultimately facilitate a more personalized approach to treatment of patients with newly diagnosed and relapsed neuroblastoma.

GLOSSARY

Chromotripsis Process of genomic rearrangement by which a cell may acquire multiple genetic mutations in a single step in which one or more chromosomes is shattered into multiple fragments and then reconstructed by the cell.

Copy number variations (CNV) Structural DNA abnormality in which the genome contains an abnormal number of copies of a particular section of DNA (including deletions or duplications).

Neuritogenesis The process of neural development in which neurons sprout and extend neurites to form axons and dendrites and create a functional neuronal network.

Neurocristopathy Developmental abnormality involving neural crest-derived tissues.

Segmental chromosomal abnormality (SCA) Genomic abnormality involving loss or gain of one or more individual parts of one or multiple chromosomes; in contrast to numerical chromosomal abnormality that involves loss or gain of whole chromosomes.

Single nucleotide polymorphism (SNP) A DNA sequence variation involving a single nucleotide difference between paired chromosomes.

ABBREVIATIONS

ALK Anaplastic lymphoma kinase
ARID1A/1B AT-rich interactive domain-containing protein 1A/1B
ATRX Alpha thalassemia/mental retardation syndrome X-linked
BARD1 BRCA-1 associated RING domain-1
CCHS Congenital central hypoventilation syndrome
CGH Comparative genomic hybridization
CNV Copy number variations
COG Children's Oncology Group
EFS Event-free survival
FISH Fluorescence *in situ* hybridization
GWAS Genome-wide association study
HACE1 HECT domain and ankyrin repeat containing E3 ubiquitin protein ligase 1
HDAC Histone deactylase
INRG International Neuroblastoma Risk Group
KIR Killer cell immunoglobulin-like receptor
LMO1 LIM-domain-only 1
LOH Loss of heterozygosity
MIBG *meta*-Iodobenzylguanadine
miR MicroRNA
MLPA Multiplex ligation probe amplification
MNA MYCN amplification
MYCN V-myc myelocytomatosis viral-related oncogene, neuroblastoma derived
NBPF23 Neuroblastoma breakpoint family 23
NCA Numerical chromosomal abnormality
NK cell Natural killer cell
OR Odds ratio
OS Overall survival
PFS Progression-free survival

PHOX2B Paired-like homeobox 2b
PI3K Phosphoinositide 3-kinase
PTPN11 Protein tyrosine phosphatase non-receptor type 11
ROI Region of interest
RT–PCR Reverse transcription polymerase chain reaction
SCA Segmental chromosomal abnormality
SCID Severe combined immunodeficiency
SHP2 Src homology phosphatase 2
SIOPEN International Society of Paediatric Oncology Europe Neuroblastoma
SNP Single nucleotide polymorphism
TAM Tumor-associated macrophage
TH Tyrosine hydroxylase
Trk Tyrosine receptor kinsase
WES Whole exome sequencing
WGS Whole genome sequencing

REFERENCES

[1] National Cancer Institute. Surveillance, Epidemiology and End Results Database. Available from: <http://seer.cancer.gov> [cited February 19, 2013].

[2] Maris JM. Recent advances in neuroblastoma. N Engl J Med 2010;362:2202–11.

[3] Matthay KK, Reynolds CP, Seeger RC, Shimada H, Adkins ES, Haas-Kogan D, et al. Long-term results for children with high-risk neuroblastoma treated on a randomized trial of myeloablative therapy followed by 13-*cis*-retinoic acid: a children's oncology group study. J Clin Oncol 2009;27:1007–13.

[4] Schwab M, Alitalo K, Klempnauer KH, Varmus HE, Bishop JM, Gilbert F, et al. Amplified DNA with limited homology to myc cellular oncogene is shared by human neuroblastoma cell lines and a neuroblastoma tumour. Nature 1983;305:245–8.

[5] Brodeur GM, Seeger RC, Schwab M, Varmus HE, Bishop JM. Amplification of N-myc in untreated human neuroblastomas correlates with advanced disease stage. Science 1984;224:1121–4.

[6] Seeger RC, Brodeur GM, Sather H, Dalton A, Siegel SE, Wong KY, et al. Association of multiple copies of the N-myc oncogene with rapid progression of neuroblastomas. N Engl J Med 1985;313:1111–6.

[7] Cohn SL, Pearson ADJ, London WB, Monclair T, Ambros PF, Brodeur GM, et al. The International Neuroblastoma Risk Group (INRG) classification system: an INRG Task Force report. J Clin Oncol 2009;27:289–97.

[8] Ambros PF, Ambros IM, Brodeur GM, Haber M, Khan J, Nakagawara A, et al. International consensus for neuroblastoma molecular diagnostics: report from the International Neuroblastoma Risk Group (INRG) Biology Committee. Br J Cancer 2009;100:1471–82.

[9] Brunet J-F, Pattyn A. Phox2 genes – from patterning to connectivity. Curr Opin Genet Devel 2002;12:435–40.

[10] Mossé YP, Laudenslager M, Khazi D, Carlisle AJ, Winter CL, Rappaport E, et al. Germline PHOX2B mutation in hereditary neuroblastoma. Am J Hum Genet 2004;75:727–30.

[11] Trochet D, Bourdeaut F, Janoueix-Lerosey I, Deville A, de Pontual L, Schleiermacher G, et al. Germline mutations of the paired-like homeobox 2B (PHOX2B) gene in neuroblastoma. Am J Hum Genet 2004;74:761–4.

[12] Raabe EH, Laudenslager M, Winter C, Wasserman N, Cole K, LaQuaglia M, et al. Prevalence and functional consequence of PHOX2B mutations in neuroblastoma. Oncogene 2008;27:469–76.

[13] Nagashimada M, Ohta H, Li C, Uesaka T, Brunet JF, Amiel J, et al. Autonomic neurocristopathy-associated mutations in PHOX2B dysregulate Sox10 expression. J Clin Invest 2012;122:3145–58.

[14] Pattyn A, Morin X, Cremer H, Goridis C, Brunet JF. The homeobox gene Phox2b is essential for the development of autonomic neural crest derivatives. Nature 1999;399:366–70.

[15] Mossé YP, Laudenslager M, Longo L, Cole KA, Wood A, Attiyeh EF, et al. Identification of ALK as a major familial neuroblastoma predisposition gene. Nature 2008;455:930–5.

[16] Mossé YP, Wood A, Maris JM. Inhibition of ALK signaling for cancer therapy. Clin Cancer Res 2009;15:5609–14.

[17] Devoto M, Specchia C, Laudenslager M, Longo L, Hakonarson H, Maris J, et al. Genome-wide linkage analysis to identify genetic modifiers of ALK mutation penetrance in familial neuroblastoma. Hum Hered 2011;71:135–9.

[18] Maris JM, Mossé YP, Bradfield JP, Hou C, Monni S, Scott RH, et al. Chromosome 6p22 locus associated with clinically aggressive neuroblastoma. N Engl J Med 2008;358:2585–93.

[19] Wang K, Diskin SJ, Zhang H, Attiyeh EF, Winter C, Hou C, et al. Integrative genomics identifies LMO1 as a neuroblastoma oncogene. Nature 2011;469:216–20.

[20] Matthews JM, Lester K, Joseph S, Curtis DJ. LIM-domain-only proteins in cancer. Nat Rev Cancer 2013;13:111–22.

[21] Aoyama M, Ozaki T, Inuzuka H, Tomotsune D, Hirato J, Okamoto Y, et al. LMO3 interacts with neuronal transcription factor, HEN2, and acts as an oncogene in neuroblastoma. Cancer Res 2005;65:4587–97.

[22] Hacein-Bey-Abina S, Kalle Von C, Schmidt M, McCormack MP, Wulffraat N, Leboulch P, et al. LMO2-associated clonal T cell proliferation in two patients after gene therapy for SCID-X1. Science 2003;302:415–9.

[23] Diskin SJ, Capasso M, Schnepp RW, Cole KA, Attiyeh EF, Hou C, et al. Common variation at 6q16 within HACE1 and LIN28B influences susceptibility to neuroblastoma. Nat Gen 2012;44:1126–30.

[24] Zhang L, Anglesio MS, O'Sullivan M, Zhang F, Yang G, Sarao R, et al. The E3 ligase HACE1 is a critical chromosome 6q21 tumor suppressor involved in multiple cancers. Nat Med 2007;13:1060–9.

[25] Castillo-Lluva S, Tan C-T, Daugaard M, Sorensen PHB, Malliri A. The tumour suppressor HACE1 controls cell migration by regulating Rac1 degradation. Oncogene 2013;32:1735–42.

[26] Molenaar JJ, Koster J, Zwijnenburg DA, van Sluis P, Valentijn LJ, van der Ploeg I, et al. Sequencing of neuroblastoma identifies chromothripsis and defects in neuritogenesis genes. Nature 2012;483:589–93.

[27] Piskounova E, Polytarchou C, Thornton JE, LaPierre RJ, Pothoulakis C, Hagan JP, et al. Lin28A and Lin28B inhibit let-7 microRNA biogenesis by distinct mechanisms. Cell 2011;147:1066–79.

[28] Molenaar JJ, Domingo-Fernández R, Ebus ME, Lindner S, Koster J, Drabek K, et al. LIN28B induces neuroblastoma and enhances MYCN levels via let-7 suppression. Nat Genet 2012;44:1199–206.

[29] Capasso M, Devoto M, Hou C, Asgharzadeh S, Glessner JT, Attiyeh EF, et al. Common variations in BARD1 influence susceptibility to high-risk neuroblastoma. Nat Gen 2009;41:718–23.

[30] Wu LC, Wang ZW, Tsan JT, Spillman MA, Phung A, Xu XL, et al. Identification of a RING protein that can interact in vivo with the BRCA1 gene product. Nat Genet 1996;14:430–40.

[31] Bosse KR, Diskin SJ, Cole KA, Wood AC, Schnepp RW, Norris G, et al. Common variation at BARD1 results in the expression of an oncogenic isoform that influences neuroblastoma susceptibility and oncogenicity. Cancer Res 2012;72:2068–78.

[32] Nguyen LB, Diskin SJ, Capasso M, Wang K, Diamond MA, Glessner J, et al. Phenotype restricted genome-wide association study using a gene-centric approach identifies three low-risk neuroblastoma susceptibility Loci. PLoS Genet 2011;7:e1002026.

[33] Shlien A, Tabori U, Marshall CR, Pienkowska M, Feuk L, Novokmet A, et al. Excessive genomic DNA copy number variation in the Li-Fraumeni cancer predisposition syndrome. Proc Natl Acad Sci USA 2008;105:11264–9.

[34] Diskin SJ, Hou C, Glessner JT, Attiyeh EF, Laudenslager M, Bosse K, et al. Copy number variation at 1q21.1 associated with neuroblastoma. Nature 2009;459:987–91.

[35] Vandepoele K, Andries V, Van Roy N, Staes K, Vandesompele J, Laureys G, et al. A constitutional translocation t(1;17)(p36.2;q11.2) in a neuroblastoma patient disrupts the human NBPF1 and ACCN1 genes. PLoS One 2008;3:e2207.

[36] Chayka O, Corvetta D, Dews M, Caccamo AE, Piotrowska I, Santilli G, et al. Clusterin, a haploinsufficient tumor suppressor gene in neuroblastomas. J Natl Cancer Inst 2009;101:663–77.

[37] Vandepoele K, Staes K, Andries V, van Roy F. Chibby interacts with NBPF1 and clusterin, two candidate tumor suppressors linked to neuroblastoma. Exp Cell Res 2010;316:1225–33.

[38] Gamazon ER, Pinto N, Konkashbaev A, Im HK, Diskin SJ, London WB, et al. Trans-population analysis of genetic mechanisms of ethnic disparities in neuroblastoma survival. J Natl Cancer Inst 2013;105:302–9.

[39] Harris TJR, McCormick F. The molecular pathology of cancer. Nat Rev Clin Oncol 2010;7:251–65.

[40] Pugh TJ, Morozova O, Attiyeh EF, Asgharzadeh S, Wei JS, Auclair D, et al. The genetic landscape of high-risk neuroblastoma. Nat Genet 2013;45:279–84.

[41] Sausen M, Leary RJ, Jones S, Wu J, Reynolds CP, Liu X, et al. Integrated genomic analyses identify ARID1A and ARID1B alterations in the childhood cancer neuroblastoma. Nat Genet 2013;45:12–7.

[42] Chen Y, Takita J, Choi YL, Kato M, Ohira M, Sanada M, et al. Oncogenic mutations of ALK kinase in neuroblastoma. Nature 2008;455:971–4.

[43] George RE, Sanda T, Hanna M, Fröhling S, Luther W, Zhang J, et al. Activating mutations in ALK provide a therapeutic target in neuroblastoma. Nature 2008;455:975–8.

[44] Chen QR, Song YK, Wei JS, Bilke S, Asgharzadeh S, Seeger RC, et al. An integrated cross-platform prognosis study on neuroblastoma patients. Genomics 2008;92:195–203.

[45] Heukamp LC, Thor T, Schramm A, De Preter K, Kumps C, De Wilde B, et al. Targeted expression of mutated ALK induces neuroblastoma in transgenic mice. Sci Transl Med 2012;4:141–91.

[46] Berry T, Luther W, Bhatnagar N, Jamin Y, Poon E, Sanda T, et al. The ALK(F1174L) mutation potentiates the oncogenic activity of MYCN in neuroblastoma. Cancer Cell 2012;22:117–30.

[47] Zhu S, Lee J-S, Guo F, Shin J, Perez-Atayde AR, Kutok JL, et al. Activated ALK collaborates with MYCN in neuroblastoma pathogenesis. Cancer Cell 2012;21:362–73.

[48] Carpenter EL, Mossé YP, Targeting ALK. in neuroblastoma – preclinical and clinical advancements. Nat Rev Clin Oncol 2012;9:391–9.

[49] Mossé YP, Lim MS, Voss SD, Wilner K, Ruffner K, Laliberte J, et al. Safety and activity of crizotinib for paediatric patients with refractory solid tumours or anaplastic large-cell lymphoma: a Children's Oncology Group phase 1 consortium study. Lancet Oncol 2013;14:472–80.

[50] De Brouwer S, Preter KD, Kumps C, Zabrocki P, Porcu M, Westerhout EM, et al. Meta-analysis of neuroblastomas reveals a skewed ALK mutation spectrum in tumors with MYCN amplification. Clin Cancer Res 2010;16:4353–62.

[51] Cheung NKV, Zhang J, Lu C, Parker M, Bahrami A, Tickoo SK, et al. Association of age at diagnosis and genetic mutations in patients with neuroblastoma. J Am Med Assoc 2012;307:1062–71.

[52] Shain AH, Pollack JR. The spectrum of SWI/SNF mutations, ubiquitous in human cancers. PLoS One 2013;8:e55119.

[53] Schwartzentruber J, Korshunov A, Liu X-Y, Jones DT, Pfaff E, Jacob K, et al. Driver mutations in histone H3.3 and chromatin remodelling genes in paediatric glioblastoma. Nature 2012;82:226–31.

[54] Mutesa L, Pierquin G, Janin N, Segers K, Thomée C, Provenzi M, et al. Germline PTPN11 missense mutation in a case of Noonan syndrome associated with mediastinal and retroperitoneal neuroblastic tumors. Cancer Genet Cytogenet 2008;182:40–2.

[55] Bentires-Alj M, Paez JG, David FS, Keilhack H, Halmos B, Naoki K, et al. Activating mutations of the Noonan syndrome-associated SHP2/PTPN11 gene in human solid tumors and adult acute myelogenous leukemia. Cancer Res 2004;64:8816–20.

[56] Eilers M, Eisenman RN. Myc's broad reach. Genes Devel 2008;22:2755–66.

[57] Westermark UK, Wilhelm M, Frenzel A, Henriksson MA. The MYCN oncogene and differentiation in neuroblastoma. Semin Cancer Biol 2011;21:256–66.

[58] Schmidt ML, Lukens JN, Seeger RC, Brodeur GM, Shimada H, Gerbing RB, et al. Biologic factors determine prognosis in infants with stage IV neuroblastoma: A prospective Children's Cancer Group study. J Clin Oncol 2000;18:1260–8.

[59] Valentijn LJ, Koster J, Haneveld F, Aissa RA, van Sluis P, Broekmans ME, et al. Functional MYCN signature predicts outcome of neuroblastoma irrespective of MYCN amplification. Proc Natl Acad Sci USA 2012;109:19190–5.

[60] Chesler L, Weiss WA. Genetically engineered murine models – contribution to our understanding of the genetics, molecular pathology and therapeutic targeting of neuroblastoma. Semin Cancer Biol 2011;21:245–55.

[61] Teitz T, Inoue M, Valentine MB, Zhu K, Rehg JE, Zhao W, et al. Th-MYCN mice with caspase-8 deficiency develop advanced neuroblastoma with bone marrow metastasis. Cancer Res 2013;73:4086–97.

[62] Otto T, Horn S, Brockmann M, Eilers U, Schüttrumpf L, Popov N, et al. Stabilization of N-Myc is a critical function of Aurora A in human neuroblastoma. Cancer Cell 2009;15:67–78.

[63] Chesler L, Schlieve C, Goldenberg DD, Kenney A, Kim G, McMillan A, et al. Inhibition of phosphatidylinositol 3-kinase destabilizes Mycn protein and blocks malignant progression in neuroblastoma. Cancer Res 2006;66:8139–46.

[64] Puissant A, Frumm SM, Alexe G, Bassil CF, Qi J, Chanthery YH, et al. Targeting MYCN in neuroblastoma by BET bromodomain inhibition. Cancer Discov 2013;3:308–23.

[65] Brodeur GM. Neuroblastoma: biological insights into a clinical enigma. Nat Rev Cancer 2003;3:203–16.

[66] Abel F, Dalevi D, Nethander M, Jörnsten R, De Preter K, Vermeulen J, et al. A 6-gene signature identifies four molecular subgroups of neuroblastoma. Cancer Cell Int 2011;11:9.

[67] Bown N, Cotterill S, Lastowska M, O'Neill S, Pearson AD, Plantaz D, et al. Gain of chromosome arm 17q and adverse outcome in patients with neuroblastoma. N Engl J Med 1999;340:1954–61.

[68] Meddeb M, Danglot G, Chudoba I, Vénuat AM, Bénard J, Avet-Loiseau H, et al. Additional copies of a 25 Mb chromosomal region originating from 17q23.1-17qter are present in 90% of high-grade neuroblastomas. Genes Chromosom Cancer 1996;17:156–65.

[69] Lastowska M, Cotterill S, Pearson AD, Roberts P, McGuckin A, Lewis I, et al. Gain of chromosome arm 17q predicts unfavourable outcome in neuroblastoma patients. UK Children's Cancer Study Group and the UK Cancer Cytogenetics Group. Eur J Cancer 1997;33:1627–33.

[70] Abel F, Ejeskär K, Kogner P, Martinsson T. Gain of chromosome arm 17q is associated with unfavourable prognosis in neuroblastoma, but does not involve mutations in the somatostatin receptor 2(SSTR2) gene at 17q24. Br J Cancer 1999;81:1402–9.

[71] Saito-Ohara F, Imoto I, Inoue J, Hosoi H, Nakagawara A, Sugimoto T, et al. PPM1D is a potential target for 17q gain in neuroblastoma. Cancer Res 2003;63:1876–83.

[72] Le Guezennec X, Bulavin DV. WIP1 phosphatase at the crossroads of cancer and aging. Trends Biochem Sci 2010;35:109–14.

[73] Lamers F, van der Ploeg I, Schild L, Ebus ME, Koster J, Hansen BR, et al. Knockdown of survivin (BIRC5) causes apoptosis in neuroblastoma via mitotic catastrophe. Endocr Relat Cancer 2011;18:657–68.

[74] Islam A, Kageyama H, Takada N, Kawamoto T, Takayasu H, Isogai E, et al. High expression of survivin, mapped to 17q25, is significantly associated with poor prognostic factors and promotes cell survival in human neuroblastoma. Oncogene 2000;19:617–23.

[75] Adida C, Berrebi D, Peuchmaur M, Reyes-Mugica M, Altieri DC. Anti-apoptosis gene, survivin, and prognosis of neuroblastoma. Lancet 1998;351:882–3.

[76] Godfried MB, Veenstra M, Sluis von P, Boon K, v Asperen R, Hermus MC, et al. The N-myc and c-myc downstream pathways include the chromosome 17q genes nm23-H1 and nm23-H2. Oncogene 2002;21:2097–101.

[77] Chandrasekharappa SC, Gross LA, King SE, Collins FS. The human NME2 gene lies within 18kb of NME1 in chromosome 17. Genes Chromosom. Cancer 1993;6:245–8.

[78] Garcia I, Mayol G, Ríos J, Domenech G, Cheung NK, Oberthuer A, et al. A three-gene expression signature model for risk stratification of patients with neuroblastoma. Clin Cancer Res 2012;18:2012–23.

[79] Lamers F, Schild L, Koster J, Versteeg R, Caron HN, Molenaar JJ. Targeted BIRC5 silencing using YM155 causes cell death in neuroblastoma cells with low ABCB1 expression. Eur J Cancer 2012;48:763–71.

[80] Attiyeh EF, London WB, Mossé YP, Wang Q, Winter C, Khazi D, et al. Chromosome 1p and 11q deletions and outcome in neuroblastoma. N Engl J Med 2005;353:2243–53.

[81] Fischer M, Bauer T, Oberthür A, Hero B, Theissen J, Ehrich M, et al. Integrated genomic profiling identifies two distinct molecular subtypes with divergent outcome in neuroblastoma with loss of chromosome 11q. Oncogene 2010;29:865–75.

[82] Guo C, White PS, Weiss MJ, Hogarty MD, Thompson PM, Stram DO, et al. Allelic deletion at 11q23 is common in MYCN single copy neuroblastomas. Oncogene 1999;18:4948–57.

[83] Oberthuer A, Berthold F, Warnat P, Hero B, Kahlert Y, Spitz R, et al. Customized oligonucleotide microarray gene expression-based classification of neuroblastoma patients outperforms current clinical risk stratification. J Clin Oncol 2006;24:5070–8.

[84] Ando K, Ohira M, Ozaki T, Nakagawa A, Akazawa K, Suenaga Y, et al. Expression of TSLC1, a candidate tumor suppressor gene mapped to chromosome 11q23, is downregulated in unfavorable neuroblastoma without promoter hypermethylation. Int J Cancer 2008;123:2087–94.

[85] Nowacki S, Skowron M, Oberthuer A, Fagin A, Voth H, Brors B, et al. Expression of the tumour suppressor gene CADM1 is associated with favourable outcome and inhibits cell survival in neuroblastoma. Oncogene 2008;27:3329–38.

[86] Michels E, Hoebeeck J, De Preter K, Schramm A, Brichard B, De Paepe A, et al. CADM1 is a strong neuroblastoma candidate gene that maps within a 3.72 Mb critical region of loss on 11q23. BMC Cancer 2008;8:173.

[87] Wang Q, Diskin S, Rappaport E, Attiyeh E, Mosse Y, Shue D, et al. Integrative genomics identifies distinct molecular classes of neuroblastoma and shows that multiple genes are targeted by regional alterations in DNA copy number. Cancer Res 2006;66:6050–62.

[88] Brodeur GM, Sekhon G, Goldstein MN. Chromosomal aberrations in human neuroblastomas. Cancer 1977;40:2256–63.

[89] Caron H, van Sluis P, de Kraker J, Bökkerink J, Egeler M, Laureys G, et al. Allelic loss of chromosome 1p as a predictor of unfavorable outcome in patients with neuroblastoma. N Engl J Med 1996;334:225–30.

[90] Castel V, Villamón E, Cañete A, Navarro S, Ruiz A, Melero C, et al. Neuroblastoma in adolescents: genetic and clinical characterisation. Clin Transl Oncol 2010;12:49–54.

[91] Bagchi A, Papazoglu C, Wu Y, Capurso D, Brodt M, Francis D, et al. CHD5 is a tumor suppressor at human 1p36. Cell 2007;128:459–75.

[92] Koyama H, Zhuang T, Light JE, Kolla V, Higashi M, McGrady PW, et al. Mechanisms of CHD5 inactivation in neuroblastomas. Clin Cancer Res 2012;18:1588–97.

[93] Janoueix-Lerosey I, Novikov E, Monteiro M, Gruel N, Schleiermacher G, Loriod B, et al. Gene expression profiling of 1p35-36 genes in neuroblastoma. Oncogene 2004;23:5912–22.

[94] Carén H, Ejeskär K, Fransson S, Hesson L, Latif F, Sjöberg RM, et al. A cluster of genes located in 1p36 are down-regulated in neuroblastomas with poor prognosis, but not due to CpG island methylation. Mol Cancer 2005;4:10.

[95] Janoueix-Lerosey I, Schleiermacher G, Michels E, Mosseri V, Ribeiro A, Lequin D, et al. Overall genomic pattern is a predictor of outcome in neuroblastoma. J Clin Oncol 2009;27:1026–33.

[96] Schleiermacher G, Michon J, Ribeiro A, Pierron G, Mosseri V, Rubie H, et al. Segmental chromosomal alterations lead to a higher risk of relapse in infants with MYCN-non-amplified localised unresectable/disseminated neuroblastoma (a SIOPEN collaborative study). Br J Cancer 2011;105:1940–8.

[97] Schleiermacher G, Mosseri V, London WB, Maris JM, Brodeur GM, Attiyeh E, et al. Segmental chromosomal alterations have prognostic impact in neuroblastoma: a report from the INRG project. Br J Cancer 2012;107:1418–22.

[98] Schleiermacher G, Janoueix-Lerosey I, Ribeiro A, Klijanienko J, Couturier J, Pierron G, et al. Accumulation of segmental alterations determines progression in neuroblastoma. J Clin Oncol 2010;28:3122–30.

[99] Souzaki R, Tajiri T, Teshiba R, Kinoshita Y, Yosue R, Kohashi K, et al. Correlation between the number of segmental chromosome aberrations and the age at diagnosis of diploid neuroblastomas without MYCN amplification. J Pediatr Surg 2011;46:2228–32.

[100] Brodeur GM, Minturn JE, Ho R, Simpson AM, Iyer R, Varela CR, et al. Trk receptor expression and inhibition in neuroblastomas. Clin Cancer Res 2009;15:3244–50.

[101] Nakagawara A, Arima-Nakagawara M, Scavarda NJ, Azar CG, Cantor AB, Brodeur GM. Association between high levels of expression of the TRK gene and favorable outcome in human neuroblastoma. N Engl J Med 1993;328:847–54.

[102] Nakagawara A, Arima M, Azar CG, Scavarda NJ, Brodeur GM. Inverse relationship between trk expression and N-myc amplification in human neuroblastomas. Cancer Res 1992;52:1364–8.

[103] Yamashiro DJ, Nakagawara A, Ikegaki N, Liu XG, Brodeur GM. Expression of TrkC in favorable human neuroblastomas. Oncogene 1996;12:37–41.

[104] Rydén M, Sehgal R, Dominici C, Schilling FH, Ibáñez CF, Kogner P. Expression of mRNA for the neurotrophin receptor trkC in neuroblastomas with favourable tumour stage and good prognosis. Br J Cancer 1996;74:773–9.

[105] Eggert A, Grotzer MA, Ikegaki N, Liu X-G, Evans AE, Brodeur GM. Expression of the neurotrophin receptor TrkA down-regulates expression and function of angiogenic stimulators in SH-SY5Y neuroblastoma cells. Cancer Res 2002;62:1802–8.

[106] Ho R, Eggert A, Hishiki T, Minturn JE, Ikegaki N, Foster P, et al. Resistance to chemotherapy mediated by TrkB in neuroblastomas. Cancer Res 2002;62:6462–6.

[107] Nakagawara A, Azar CG, Scavarda NJ, Brodeur GM. Expression and function of TRK-B and BDNF in human neuroblastomas. Mol Cell Biol 1994;14:759–67.

[108] Minturn JE, Evans AE, Villablanca JG, Yanik GA, Park JR, Shusterman S, et al. Phase I trial of lestaurtinib for children with refractory neuroblastoma: a new approaches to neuroblastoma therapy consortium study. Cancer Chemother Pharmacol 2011;68:1057–65.

[109] Norris RE, Minturn JE, Brodeur GM, Maris JM, Adamson PC. Preclinical evaluation of lestaurtinib (CEP-701) in combination with retinoids for neuroblastoma. Cancer Chemother Pharmacol 2011;68:1469–75.

[110] Vermeulen J, De Preter K, Naranjo A, Vercruysse L, Van Roy N, Hellemans J, et al. Predicting outcomes for children with neuroblastoma using a multigene-expression signature: a retrospective SIOPEN/COG/GPOH study. Lancet Oncol 2009;10:663–71.

[111] De Preter K, Vermeulen J, Brors B, Delattre O, Eggert A, Fischer M, et al. Accurate outcome prediction in neuroblastoma across independent data sets using a multigene signature. Clin Cancer Res 2010;16:1532–41.

[112] Asgharzadeh S, Pique-Regi R, Sposto R, Wang H, Yang Y, Shimada H, et al. Prognostic significance of gene expression profiles of metastatic neuroblastomas lacking MYCN gene amplification. J Natl Cancer Inst 2006;98:1193–203.

[113] Asgharzadeh S, Salo JA, Ji L, Oberthuer A, Fischer M, Berthold F, et al. Clinical significance of tumor-associated inflammatory cells in metastatic neuroblastoma. J Clin Oncol 2012;30:3525–32.

[114] Schulte JH, Horn S, Schlierf S, Schramm A, Heukamp LC, Christiansen H, et al. MicroRNAs in the pathogenesis of neuroblastoma. Cancer Lett 2009;274:10–5.

[115] He L, He X, Lim LP, de Stanchina E, Xuan Z, Liang Y, et al. A microRNA component of the p53 tumour suppressor network. Nature 2007;447:1130–4.

[116] Welch C, Chen Y, Stallings RL. MicroRNA-34a functions as a potential tumor suppressor by inducing apoptosis in neuroblastoma cells. Oncogene 2007;26:5017–22.

[117] Schulte JH, Horn S, Otto T, Samans B, Heukamp LC, Eilers UC, et al. MYCN regulates oncogenic MicroRNAs in neuroblastoma. Int J Cancer 2008;122:699–704.

[118] Fontana L, Fiori ME, Albini S, Cifaldi L, Giovinazzi S, Forloni M, et al. Antagomir-17-5p abolishes the growth of therapy-resistant neuroblastoma through p21 and BIM. PLoS One 2008;3:e2236.

[119] Mestdagh P, Boström A-K, Impens F, Fredlund E, Van Peer G, De Antonellis P, et al. The miR-17-92 microRNA cluster regulates multiple components of the TGF-β pathway in neuroblastoma. Mol Cell 2010;40:762–73.

[120] Chen Y, Stallings RL. Differential patterns of microRNA expression in neuroblastoma are correlated with prognosis, differentiation, and apoptosis. Cancer Res 2007;67:976–83.

[121] De Preter K, Mestdagh P, Vermeulen J, Zeka F, Naranjo A, Bray I, et al. miRNA expression profiling enables risk stratification in archived and fresh neuroblastoma tumor samples. Clin Cancer Res 2011;17:7684–92.

[122] Decock A, Ongenaert M, Vandesompele J, Speleman F. Neuroblastoma epigenetics: from candidate gene approaches to genome-wide screenings. Epigenetics 2011;6:962–70.

[123] Teitz T, Wei T, Valentine MB, Vanin EF, Grenet J, Valentine VA, et al. Caspase 8 is deleted or silenced preferentially in childhood neuroblastomas with amplification of MYCN. Nat Med 2000;6:529–35.

[124] Astuti D, Agathanggelou A, Honorio S, Dallol A, Martinsson T, Kogner P, et al. RASSF1A promoter region CpG island hypermethylation in phaeochromocytomas and neuroblastoma tumours. Oncogene 2001;20:7573–7.

[125] Banelli B, Gelvi I, Di Vinci A, Scaruffi P, Casciano I, Allemanni G, et al. Distinct CpG methylation profiles characterize different clinical groups of neuroblastic tumors. Oncogene 2005;24:5619–28.

[126] Alaminos M, Davalos V, Cheung NKV, Gerald WL, Esteller M. Clustering of gene hypermethylation associated with clinical risk groups in neuroblastoma. J Natl Cancer Inst 2004;96:1208–19.

[127] Casciano I, Mazzocco K, Boni L, Pagnan G, Banelli B, Allemanni G, et al. Expression of ΔNp73 is a molecular marker for adverse outcome in neuroblastoma patients. Cell Death Differ 2002;9:246–51.

[128] Casciano I, Banelli B, Croce M, Allemanni G, Ferrini S, Tonini GP, et al. Role of methylation in the control of ΔNp73 expression in neuroblastoma. Cell Death Differ 2002;9:343–5.

[129] Yu AL, Gilman AL, Ozkaynak MF, London WB, Kreissman SG, Chen HX, et al. Anti-GD2 antibody with GM-CSF, interleukin-2, and isotretinoin for neuroblastoma. N Engl J Med 2010;363:1324–34.

[130] Cheung NKV. FCGR2A polymorphism is correlated with clinical outcome after immunotherapy of neuroblastoma with anti-GD2 antibody and granulocyte macrophage colony-stimulating factor. J Clin Oncol 2006;24:2885–90.
[131] Delgado DC, Hank JA, Kolesar J, Lorentzen D, Gan J, Seo S, et al. Genotypes of NK cell KIR receptors, their ligands, and Fc γ receptors in the response of neuroblastoma patients to Hu14.18-IL2 immunotherapy. Cancer Res 2010;70:9554–61.
[132] Innocenti F, Undevia SD, Iyer L, Chen PX, Das S, Kocherginsky M, et al. Genetic variants in the UDP-glucuronosyltransferase 1A1 gene predict the risk of severe neutropenia of irinotecan. J Clin Oncol 2004;22:1382–8.
[133] Peters U, Preisler-Adams S, Hebeisen A, Hahn M, Seifert E, Lanvers C, et al. Glutathione S-transferase genetic polymorphisms and individual sensitivity to the ototoxic effect of cisplatin. Anticancer Drugs 2000;11:639–43.
[134] Oldenburg J, Kraggerud SM, Cvancarova M, Lothe RA, Fossa SD. Cisplatin-induced long-term hearing impairment is associated with specific glutathione S-transferase genotypes in testicular cancer survivors. J Clin Oncol 2007;25:708–14.
[135] Ross CJD, Katzov-Eckert H, Dubé MP, Brooks B, Rassekh SR, Barhdadi A, et al. Genetic variants in TPMT and COMT are associated with hearing loss in children receiving cisplatin chemotherapy. Nat Genet 2009;41:1345–9.
[136] Wu X, Northcott PA, Dubuc A, Dupuy AJ, Shih DJ, Witt H, et al. Clonal selection drives genetic divergence of metastatic medulloblastoma. Nature 2012;482:529–33.
[137] Nguyen DX, Massagué J. Genetic determinants of cancer metastasis. Nat Rev Genet 2007;8:341–52.

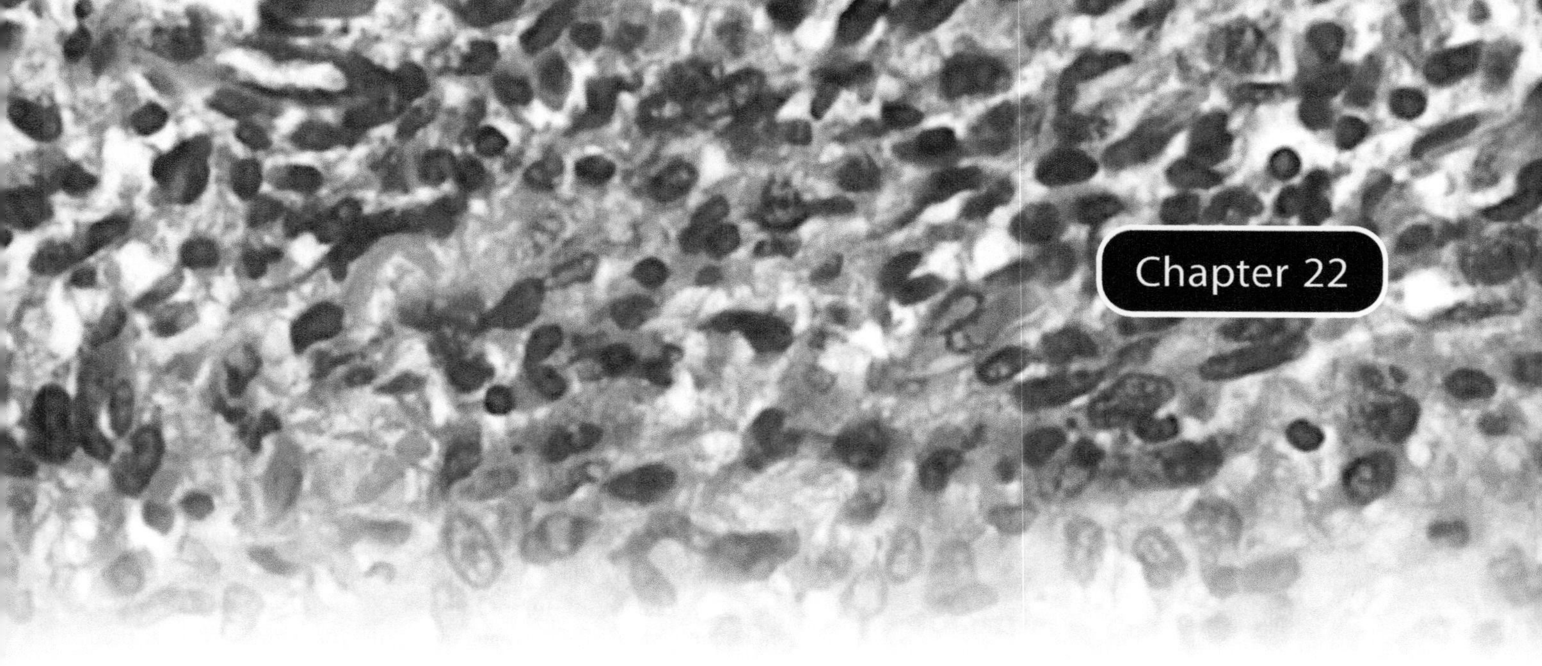

Cancer Genomics
Soft Tissue Sarcomas

Amal M. EL-Naggar
Gabriel Leprivier
Poul H. Sorensen

Department of Molecular Oncology, BC Cancer Research Center, Vancouver, BC; Department of Pathology and Laboratory Medicine, University of British Columbia, Vancouver, BC, Canada

Contents

Cancer Genomics. DOI: http://dx.doi.org/10.1016/B978-0-12-396967-5.00022-0

Key Concepts

- STSs are a rare heterogeneous group of cancers of mesenchymal origin
- STSs are more common in pediatric age groups
- Histological subtypes of STSs share common features
- Many STSs are highly aggressive tumors with a strong propensity for local recurrence and metastasis
- Metastatic spread represents the single-most powerful predictor of poor outcome in high-risk STSs
- Genetic and non-genetic factors play a role in sarcomagenesis
- RMS, SS, GIST, CFS, and EES are famous examples of STSs with simple karyotypic defects, while UPS, LMS, LPS, and extraosseous osteosarcomas are common examples of STSs with complex karyotypic defects

INTRODUCTION

Soft tissue sarcomas (STSs) are a relatively rare, heterogeneous group of non-epithelial aggressive cancers arising in mesenchymal tissues anywhere in the body, including extraskeletal tissues such as muscles, fat, blood vessels, nerves, and synovial tissues. The most frequent locations for STSs are the extremities (59%), the trunk (19%), the retroperitoneum (15%), and the head and neck (9%). STSs account for less than 1% of adult malignancies [1,2], but are relatively more common in children where they represent up to 15% of pediatric malignancies [2,3].

The majority of sarcomas are sporadic with unknown etiology. However, certain hereditary syndromes such as Li–Fraumeni syndrome, familial retinoblastoma, as well as environmental factors, such as ionizing radiation play a role in the etiology of specific types of sarcomas. Most cases associated with hereditary syndromes harbor mutations affecting tumor suppressor genes such as the *TP53* tumor suppressor gene, *RB1*, or *CHK2* [4–6], to support sarcomagenesis. Exposure to chemical carcinogens, such as phenoxyacetic acid in herbicides [7], and radiation [8] increases the risk of developing STSs. Of note, radiation-induced STSs are rare, difficult to treat, and associated with poor prognosis [9,10]. Radiation-induced DNA damage [11] or chromosomal instability as a result of radiation-induced alterations in telomere functions [12] may, at least, partially explain the pathophysiology of radiation-induced STSs. Because of rapidly evolving genomic and proteomic technologies, our understanding of the oncogenic mechanisms underlying sarcomagenesis is improving. STSs often arise *de novo*; however, in some instances, they develop from a pre-existing lesion, such as leiomyosarcoma arising in the presence of uterine fibroids (Figure 22.1), as previously reported [13].

STSs vary in their malignant potential. Some are highly malignant with high frequencies of pulmonary metastasis, such as pleomorphic STS with myogenic differentiation [14] as well as synovial sarcoma (SS), liposarcoma (LPS), leiomyosarcoma (LMS), and alveolar soft part sarcoma (ASPS) [15,16], while others have a more indolent course, such as congenital fibrosarcoma (CFS) [17] and giant cell tumor of tendon sheath (Figure 22.2A) [18,19]. Some benign soft tissue tumors may resemble malignancies, making them particularly challenging to diagnose (Figure 22.2B). For treatment purposes, clinicians require more detailed information regarding the biologic behavior of a tumor, indicated by the tumor grade; i.e. the degree of differentiation of tumor cells, ranging from well-differentiated (low grade), to moderately differentiated (intermediate grade) and, finally, to poorly differentiated (high grade) as the cells progressively lose the features seen in normal cells with the poorly differentiated cells having a worse prognosis. Tumors can be stratified from low to high grade. Stratification is based on various histopathologic features, such as the degree of cytologic atypia, the mitotic rate, and the presence or absence of necrosis [20]. Staging of STSs to detect precisely local tumor extent and distant spread is critical for determining the most effective treatment strategy. Indeed, two systems are available [21]: the Musculoskeletal Tumour Society (MSTS) staging system [22] and the GTNM (grading, tumor, nodes, metastases) staging system [23].

The management strategy for STSs includes chemotherapy followed by surgery and/or radiation [4]. Modern multiagent chemotherapy regimens have tremendously improved outcomes for patients with localized high-risk diseases such as rhabdomyosarcoma (RMS). However, the prognosis for patients with metastatic disease remains dismal [24–27]. Currently, the presence of metastatic

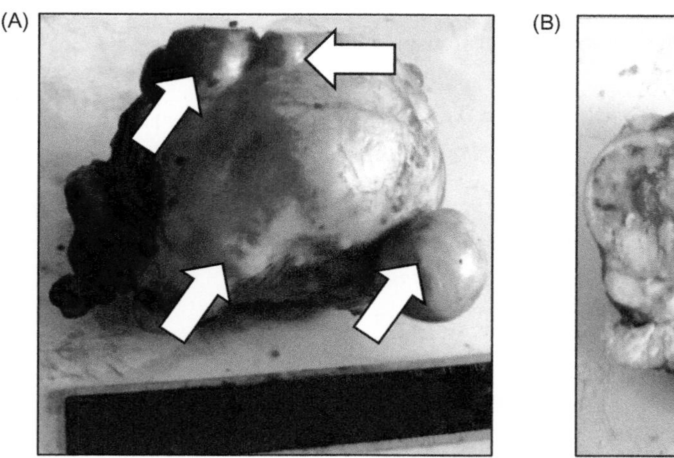
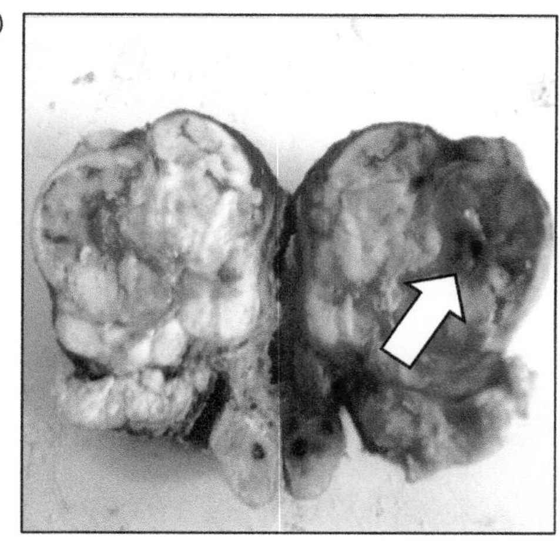

FIGURE 22.1 (A) Gross picture of the uterus with multiple leiomyomata (arrows). (B) A sagittal section of the uterus in (A) showing a leiomyosarcoma developed from a pre-existing uterine leiomyoma. Arrow points to the area with malignant transformation.

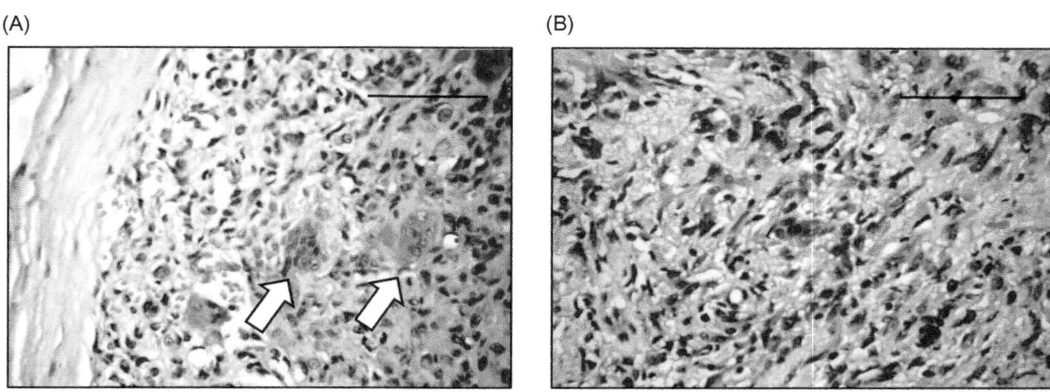

FIGURE 22.2 (A) Giant cell tumor of the tendon sheath. Photomicrograph of H&E-stained tissue section showing a variable collection of giant cells (arrows) admixed with mononuclear cells. (B) Spindle cell neoplasm of undetermined malignancy. Photomicrograph of H&E-stained tissue section showing spindle, bizarre shaped cells arranged in short fascicles without evidence of mitosis or necrosis. Scale bars, 50 μm.

disease is the most powerful predictor of poor outcome in sarcomas, and is responsible for >90% of cancer deaths overall [28]. A better understanding of the signaling pathways that impact metastatic spread has tremendous potential to reduce disease burden in childhood sarcoma.

GENETICS

Genetic alterations contribute significantly to STS pathogenesis. STSs can be categorized into two distinct genetic groups: (1) those with simple karyotypic defects, and (2) those with complex karyotypic defects [4,29,30].

STSs with Simple Karyotypic Defects

Simple karyotypic defects include recurrent chromosomal translocations, chromosomal amplifications, and specific oncogenic mutations. These genetic alterations may result in production of novel chimeric oncoproteins, altered gene expression, or dysregulation of signaling pathways via functional changes in specific signaling proteins [31]. The most common STSs associated with chromosomal translocations are listed in Table 22.1.

Recurrent disease-specific chromosomal translocations, often occurring in an otherwise chromosomally quiet background, are prototypical of STSs with simple karyotypes. Such translocations fuse the coding regions of two different partner genes to produce a so-called fusion transcript that encodes either an aberrant chimeric transcription factor or a constitutively active tyrosine kinase. Chimeric transcription factors typically harbor both a powerful transactivating domain and a DNA-binding domain, which can drive cellular transformation by altering the gene expression profile of transformed cells. For example, in extraosseous Ewing sarcoma (EES), the transcription regulatory domain of the *EWSR1* gene is fused

TABLE 22.1 The Most Common STS-Associated Chromosomal Translocations

Tumor	Cytogenetic Event	Molecular Event	Frequency	Diagnostic Utility?
Alveolar soft part sarcoma	t(X;17)(p11;q21)	*ASPL–TFE3* fusion	>90%	Yes
Angiomatoid fibrous histiocytoma	t(12;16)(q13;p11)	*FUS–ATF* fusion	?	Yes
Clear cell sarcoma	t(12;22)(q13;q12)	*EWS–ATF1* fusion	>75%	Yes
Congenital fibrosarcoma	t(12;15)(p13;q25)	*ETV6–NTRK3* fusion	>90%	Yes
Congenital mesoblastic nephroma	t(12;15)(p13;q25)	*ETV6–NTRK3* fusion	>90%	Yes
Dermatofibrosarcoma protuberans	t(17;22)(q22;q13)	*COL1A1–PDGFB* fusion	?	Yes
Desmoplastic small round cell tumor	t(11;22)(p13;q12)	*EWS–WT1* fusion	95%	Yes
Endometrial stromal tumor	t(7;17)(p15;q21)	*JAZF1–JJAZ1* fusion	30%	Yes
Extraosseous Ewing sarcoma	t(11;22)(q24;q12)	*EWS–FLI1* fusion	> 85%	Yes
	t(21;22)(q12;q12)	*EWS–ERG* fusion	10-15%	Yes
	t(2;22)(q33;q12)	*EWS–FEV* fusion	< 1%	Yes
	t(7;22)(p22;q12)	*EWS–ETV1* fusion	< 1%	Yes
	t(17;22)(q12;q12)	*EWS–E1AF* fusion	< 1%	Yes
	t(16;21)(p11;q22)	*FUS–ERG* fusion	?	?
Gastrointestinal stromal tumor		*KIT* or *PDGFRA* mutations	>85%	Yes
Inflammatory myofibroblastic tumor	t(2;var)(p23;var)	*ALK* fusion genes	>50%	Yes
Myxoid liposarcoma	t(12;16)(q13;p11)	*TLS–CHOP* fusion	95%	Yes
	t(12;22)(q13;q12)	EWS–CHOP fusion	5%	Yes
Rhabdoid tumor	Deletion of 22q	*INI1* inactivation	>90%	Yes
Alveolar rhabdomyosarcoma	t(2;13)(q35;q14)	*PAX3–FKHR* fusion	~55%	Yes
	t(1;13)(q36;q14)	*PAX7–FKHR* fusion	~20%	Yes
	t(X;3)(q13.1;q35)	*PAX3–AFX* fusion	?	?
	t(2;2)(q35;p23)	*PAX3–NCOA1* fusion	?	?
Synovial sarcoma	t(X;18)(p11;q11)	*SYT–SSX1 or SYT–SSX2* fusion	>90%	Yes

to the DNA-binding domain of members of the ETS family of transcription factors, including *FLI1* [32,33] or *ERG* [34], to generate a powerful transcriptional regulator. An example of a chimeric tyrosine kinase is the *ETV6–NTRK3* oncoprotein of CFS [35,36] and congenital mesoblastic nephroma [37,38].

Diverse experimental approaches have revealed functional roles for oncoproteins in cellular transformation and *trans*-differentiation [4,39–42]. The presence of recurrent genetic events such as chromosomal translocations are also a powerful diagnostic aid, as these alterations are unique to a particular subtype of sarcoma, such as the *ETV6–NTRK3* gene fusion in CFS [36,43] and congenital mesoblastic nephroma [37,38]. Some of the chromosomal translocations may also have prognostic value, such as the *PAX7–FOXO1A* fusion, reported to be associated with improved prognosis, compared with *PAX3–FOXO1A* fusions, in alveolar rhabdomyosarcoma (ARMS) [44–46].

Although less frequent, fusion oncoproteins may also result in growth factor overexpression, such as PDGFβ which is overexpressed in dermatofibrosarcoma protuberans (DFSP) due to the fusion of the *COL1A1* promoter to the *PDGFβ* coding sequence [47,48]. Chromosomal translocations may also affect signaling pathways, as occurs in inflammatory myofibroblastic tumors, which display constitutive activation of proliferation and survival pathways as a result of a chimeric fusion encoding a functional ALK receptor tyrosine kinase [49–51].

Specific oncogenic mutations have been reported in certain sarcoma subtypes, such as in gastrointestinal stromal tumor (GIST) with activating mutations in the c-KIT growth factor receptor, or in PDGFRA [52]. Chromosomal amplifications have also been reported in some sarcomas, such as well-differentiated liposarcoma [53,54]. In such cases, there is amplification of q14-15 of chromosome 12 resulting in overexpression of certain genes, including *MDM2* and *CDK4* [53]. Further, some STS subtypes such as rhabdoid tumors, a highly aggressive tumor subtype with poor outcome, malignant peripheral nerve sheath tumors and epithelioid sarcomas, show intergenic deletions with partial or complete loss of the *hSNF5/INI1* gene (chromosome 22q11.2), a central component of the SWI/SNF chromatin remodeling complex, resulting in cell cycle progression [55].

STSs with Complex Karyotypic Defects

While some STSs feature simple karyotypic defects, other STSs are characterized by high genomic instability, featuring multiple chromosomal duplications, deletions, and complex rearrangements. A complex karyotype is defined as the presence of five or more independent chromosome abnormalities at diagnosis [56,57]. *TP53* mutations are frequently observed, as well as loss of telomeres, thought to be responsible for repeated chromosome breakage and fusion during mitosis [58]. Additional genetic events have been identified, including mutations in oncogenes such as *RAS*, or tumor suppressor genes including *RB1*; however, their diagnostic utility remains questionable [59]. STS subtypes that are prototypes of this group include leiomyosarcoma [60] and undifferentiated pleomorphic sarcoma (UPS) [61].

HISTOLOGIC SUBTYPES OF STSs

Sarcomas include over 100 histologic subtypes that vary widely in their genetic abnormalities, pathogenesis, and clinical outcomes, of which some are common in adults, such as liposarcoma, while others are predominant in childhood such as rhabdomyosarcoma [62]. Examples of the most common STSs are listed in Table 22.2. We will discuss several common subtypes of STSs, including ARMS, SS, GIST, CFS, and EES as examples of STSs with simple karyotypic defects, and undifferentiated pleomorphic sarcoma (UPS), LMS, LPS, and extraosseous osteosarcomas as examples of STSs with complex karyotypic defects.

Rhabdomyosarcoma (RMS)

RMS is the most common STS in childhood populations, arising in limbs, the central axis, or head and neck areas, and comprising >50% of STSs in children. Two thirds of cases occur before the age of 6 years, with a second peak occurring during mid-adolescence. RMS has different histologic variants; embryonal and alveolar subtypes are the most common while botryoid and pleomorphic subtypes are less frequent. The embryonal RMS (ERMS) subtype (~70% of RMS cases) tends to affect infants and toddlers, whereas ARMS (~20% of RMS cases) affects all age groups. The botryoid and pleomorphic subtypes (~10% of RMS cases) commonly affect adults [62,63]. ARMS and ERMS most commonly occur as a mass, and the presenting signs and symptoms are related to the anatomic

TABLE 22.2 The Major Subtypes of STSs

STS Subtype	Tissue of Origin	Route of Spread	Age Predeliction
UPS	Pluripotential mesenchymal cells	Hematogenous	Older adults
LPS	Fatty tissue	Hematogenous	Adults
RMS	Striated muscles	Lymphatic	Children
LMS	Smooth muscles	Hematogenous	Adults/children
Hemangiosarcoma	Blood vessels	Hematogenous/lymphatic	Adults
SS	Synovial tissue	Lymphatic	Adolescents and young adults
Neurofibrosarcoma	Peripheral nerves	Hematogenous (rare)	Young and middle-aged adults
Chondrosarcoma	Mesenchymal precursors with chondrogenic potential	Hematogenous	Adults
EES	Proposed mesenchymal stem cells	Hematogenous	Children
Osteosarcoma	Proposed mesenchymal stem cells	Hematogenous	Children
GIST	Digestive tract (interstitial cells of Cajal)	Mesenteric spread/portal vein	Older adults

site of the primary tumor. The most common locations of primary disease are the head and neck region, genitourinary tract, and extremities [64,65], and survival is poor for patients with distant tumor spread [66,67].

The genetic mechanisms underlying the pathogenesis of the alveolar subtype are the characteristic chromosomal translocations t(2;13)(q35-37;q14) and t(1;13)(p36;q14), generating *PAX3−FOXO1A* and *PAX7−FOXO1A* gene fusions, respectively [68−74]. In addition, other amplification events have been described in ARMS which include the 2p24, 12q13-q14, and 13q31 chromosomal regions, of which the target genes include *MYCN*, *CDK4*, and *MIR17HG*, respectively [75−77]. While consistent chromosomal translocations of ARMS have been described, only random cytogenetic abnormality is reported in ERMS [78]. Indeed, it has been shown that ERMS frequently exhibits gains of chromosomes or chromosomal regions such as 2, 7, 8, 1, 12, 13q21, and 20, and losses of 1p35-36.3, 6, 9q22, 14q21-32, and 17 [79−81]. Of note, losses of 9q22 and 6, involving 6q26-2, affects putative tumor suppressor genes *PTCH* and *IGF-2*, respectively [82]. Further, elevated levels of wild-type *PAX3* or *PAX7* expression [83,84], and trisomy 2 have also been described in ERMS [85].

RMS cases are categorized into different risk groups; low, intermediate, and high risk, based on both pretreatment clinical and radiographic data, as well as surgical and pathological findings. Generally, the management strategy for RMS involves local control measures including surgical resection, if feasible, and radiotherapy in conjunction with systemic chemotherapy because of high metastatic potential [86]. The management strategies for adults with RMS are similar to those for children. Of interest, the treatment strategies designed based on prognostic stratification of patients show remarkable difference in their overall philosophy. For instance, the International Society of Pediatric Oncology Malignant Mesenchymal Tumor (MMT) Group prefers the use of chemotherapy as a front-line approach and its trials aim at avoiding, if possible, major surgical procedures and long-term effects of radiotherapy. By contrast, the goal of the Soft Tissue Sarcoma (STS) Committee of the Children's Oncology Group (COG) (COG-STS) is to apply local control measures for non-metastatic cases soon after the initial operation or biopsy [87,88].

Currently, the emergence of targeted therapy has profoundly changed the overall outcome for patients and this is true for many diseases. In the context of rhabdomyosarcoma, small molecule inhibitors include targeting tyrosine kinases such as IGF-1 receptor inhibitor [89,90], anti-angiogenic drugs such as bevacizumab (Avastin) and sorafenib (Nexavar) [91], and mTOR inhibitors such as temsirolimus (Torisel) and everolimus (Afinitor) are approved for the treatment of advanced renal cell carcinoma [92]. Many of these agents are in clinical trials and showed potential benefits when used alone or in combination with other anticancer regimens, providing encouraging approaches for RMS treatment. Interestingly, an additional therapeutic focus of research is immune-mediated destruction of the *PAX−FOXO1* fusion oncoprotein by vaccination [66] or by kinase inhibitors [93]. For more information, readers are referred to more comprehensive reviews [94,95].

Synovial Sarcoma (SS)

Synovial sarcoma (SS) is an aggressive STS subtype, commonly arising from synovial tissue lining joint cavities of the extremities, which often affects adolescents and young adults rather than older individuals [96−98]. It accounts for approximately 8% of all soft tissue sarcomas [99].

Microscopically, SS features two morphologically different cell types, including those with epithelial-like morphology arranged in gland-like structures, and sheets of spindle-shaped mesenchymal cells. Three microscopic patterns of SS have been characterized which include biphasic (with both epithelial and spindle-shaped regions) and tumors with a monophasic pattern (predominantly showing spindle-shaped morphology) [100−102]. SS is characterized by a specific t(X;18)(q11.2;p11) chromosomal translocation involving the *SSX1* or *SSX2* gene on chromosome X and the *SYT* gene on chromosome 18, with the resulting chimeric transcripts expressing SYT−SSX1 or SYT−SSX2 fusion proteins in which the C-terminal amino acids of SYT are replaced by amino acids from the C-terminus of the SSX proteins [103−105].

Recently, it has been shown that *SS18−SSX* fusions are involved in a complex that represses *ATF2* target gene expression by recruiting polycomb repressor proteins and histone deacetylase 1 (HDAC1) to the DNA. Disruption of this complex by siRNA knockdown of components of the complex or treatment with HDAC inhibitors rescues target gene expression, restoring growth suppression and apoptosis [106], a crucial finding that may have a potential therapeutic impact.

Further, a new study has shown that the altered BAF complex, harboring the SS18−SSX fusion protein in place of the wild-type SS18, is devoid of the tumor suppressor BAF47 (hSNF5). Indeed, the altered complex binds to the *Sox2* locus and reverses polycomb-mediated repression resulting in *Sox2* activation which is crucial for tumor proliferation. Of interest, restoration of the wild-type complexes by increasing the concentration of wild-type SS18 led to growth arrest [107], a finding that may be exploited for therapeutic purposes.

The prognosis of SS is generally poor with 50–70% of cases developing metastases. The management strategy involves wide resection followed by polychemotherapy with or without irradiation [108].

Gastrointestinal Stromal Tumor (GIST)

Gastrointestinal stromal tumor (GIST) is the most common mesenchymal-derived tumor arising from the digestive tract, and accounts for 1–3% of all gastrointestinal malignancies. It affects adults between 40 and 70 years old, and lacks gender predilection [109]. The predominant mechanism of pathogenesis is a gain-of-function mutation in the tyrosine kinase growth factor receptor, c-KIT (or CD117), first described by Hirota and his group [110], detectable by immunohistochemistery (IHC) for overexpression or by sequencing. The gain-of-function mutations of the c-KIT pathway are indispensible for GIST pathogenesis as they confer a growth advantage by permitting unchecked growth of tumor cells and evasion of apoptosis [110,111]. The most frequent KIT mutations, found in almost 75% of cases, are those affecting the intracellular and extracellular juxtamembrane domains (exon 8, 9, 11) and the activation loop of the kinase domain (exon 17) N822 residue, resulting in disruption of the autoinhibitory mechanism of KIT without altering the nucleotide binding domain [112,113]. For more information, readers are referred to review articles [114,115]. Of note, it has been shown that GIST cases lacking KIT mutations have intragenic activation mutations in the related receptor tyrosine kinase, platelet-derived growth factor receptor alpha (PDGFRA) [116,117].

Prognosis for localized GIST is favorable, as tumors can usually be managed with complete surgical excision [118]. However, GIST has a high recurrence rate and up to 40–90% of patients show hepatic or mesenteric implants (Figure 22.3), possibly due to tumor rupture [32]. The tyrosine kinase inhibitors (TKIs), including imatinib mesylate, have been successful for the treatment of GIST. These agents bind to the adenosine triphosphate (ATP)-binding pocket crucial for phosphorylation/activation of the growth factor receptor, thereby inhibiting tumor growth. TKIs are used preoperatively, and are particularly important for patients with unresectable or borderline resectable tumors, or patients at high risk of recurrence after surgical removal of the primary GIST tumor [119].

FIGURE 22.3 A recurrent gastrointestinal stromal tumor (GIST). The image shows the gross picture of a mesenteric nodule of a patient previously diagnosed with gastrointestinal stromal tumor (GIST) in the stomach. The cut section of the mesentric nodule displayed cystic spaces, and necrosis.

Congenital Fibrosarcoma (CFS)

Congenital (or infantile) fibrosarcoma (CFS) is a malignant neoplasm of fibroblasts that occurs in patients aged 2 years or younger. Contrary to the poor outcome associated with most STS subtypes, CFS has an excellent prognosis and a very low metastatic rate. Histologically, CFS is comprised of sheets of malignant spindle cells forming interlacing cords with focal collections of inflammatory cells. Areas of hemorrhage, necrosis, and calcifications are frequently seen [120]. Mitoses can often be readily identified.

CFS is characterized by a recurrent t(12;15)(p13;q25) rearrangement which fuses the *ETV6* (or *TEL*) gene from chromosome 12p13 with the chromosome 15q25 *NTRK3* neurotrophin-3 receptor gene (also known as *TRKC*), resulting in expression of *ETV6–NTRK3* chimeric transcripts which provide an important diagnostic marker of CFS [35,36]. *ETV6–NTRK3* fusion transcripts encode the helix-loop-helix (HLH) protein dimerization domain of ETV6 fused to the protein tyrosine kinase (PTK) domain of NTRK3. Indeed, the expressed chimeric PTK likely contributes to oncogenesis by dysregulation of NTRK3 signal transduction pathways, namely RAS-ERK and PI3K-AKT [35,121,122]. Of note, the ETV6–NTRK3 chimeric oncoprotein has potent transforming activity in fibroblasts and is also expressed in human secretory breast carcinoma [122].

Extraosseous Ewing Sarcoma (EES)

Extraosseous Ewing sarcoma (EES) is a rare, small, round, blue cell tumor of mesenchymal origin with the same histologic appearance and pathogenesis as ES of bone. It is usually seen in the second or third decade of life with a 1:2 female:male ratio. The most commonly affected sites include the lower extremity, head and neck, paravertebral region, and the pelvis [123,124]. Like ES of bone, EES is characterized by etiologic gene fusions of EWS to different members of the ETS transcription factor family. The expression of these chimeric EWS–ETS fusion proteins is pathognomonic of the disease [32,34,125,126]. Moreover, numerous studies have dissected how EWS–ETS fusion proteins impart malignant behavior to transformed cells [127–129].

Although these findings have revealed important insights into disease pathogenesis and have led to new molecular tools for diagnosis, they have not impacted treatment and outcome. EWS–ETS chimeric transcription factors are challenging to target, although several groups have used multiple approaches to reveal potentially promising downstream pathways [130–132]. Secondly, translocations occur in both localized and metastatic disease, and thus cannot, on their own, account for metastatic behavior [84,133]. The prognostic factors of EES are similar to primary osseous Ewing sarcomas and treatment strategies are similar in both, including adequate surgical resection, aggressive chemotherapy, and radiotherapy [134].

Undifferentiated Pleomorphic Sarcoma (UPS)

Undifferentiated pleomorphic sarcoma (UPS), including tumors previously classified as malignant fibrous histiocytoma (MFH), constitutes 20–70% of STS [14], and this entity was considered the most common STS in late adulthood. Basically, UPS commonly affects adults aged 50–70 with a higher propensity to affect men (1:2 female:male ratio) [135].

Intriguingly, re-evaluation of retrospective series of cases primarily diagnosed as MFH revealed that only a small minority of cases were truly eligible to be diagnosed as MFH while the majority of cases were discovered to be specific sarcoma subtypes other than MFH. Indeed, it is believed that UPS may originate either from a primitive pluripotential mesenchymal cell that exhibits a range of differentiation or from high-grade neoplasms of poor differentiation [14,136]. Of note, the diagnostic frequency of MFH is markedly diminished as a result of advanced molecular pathology assays and diagnostic techniques which provide powerful tools allowing reclassification of MFH cases.

UPS can arise anywhere in the body. However, the most frequent locations are in the deep soft tissue of the extremities and the retroperitoneum. The clinical presentation depends on the tumor location [20,137–139].

The specific genetic alterations associated with UPS are largely unknown. However, *C-MYC* amplifications, gains involving regions such as 8q21.3-qter and 9q32-qter, and losses involving multiple regions such as 13q21-q22, and 18q12-q22, have all been reported [140], reflecting the highly aggressive nature of this malignancy.

Leiomyosarcoma (LMS)

Leiomyosarcoma (LMS) is a highly malignant tumor arising from smooth muscles everywhere in the body, frequently affecting retroperitoneum, uterus, skin, superficial soft tissues, and the deep compartments of the extremities. These tumors account for 10–15% of all soft tissue sarcomas [141,142]. LMS is categorized into three major groups for prognostic and treatment purposes: somatic soft tissue LMS, cutaneous LMS, and vascular LMS; however, the three groups are histologically identical [143]. LMS are refractory tumors showing treatment resistance. Therefore, the prognosis is poor with low survival rates compared to other STSs [144]. Histologically, LMS exhibit areas of high cellularity, commonly arranged in fascicles. The malignant cells are characterized by abundant pink to deep red cytoplasm on hematoxylin and eosin (H&E) staining, with cigar-shaped, centrally located nuclei. However, these distinguishing features are lost in de-differentiated tumors [143].

The cytogenetic changes of LMS are highly complex with many areas of deletions and amplifications of which deletion of 13q14, the region harboring the retinoblastoma (*RB1*) tumor suppressor gene, is a common event. Further, the aggressive high-grade tumors often exhibit segmental deletions of 10q, 2p, and 12p, as well as gains in 1q and 17p [145–147]. In addition, deletion at the 11q22-q24 locus harboring the ataxia telangiectasia mutated gene (*ATM*) has been described in LMS [148]. Amplified regions have been described in LMS, such as the amplification of 12q12-q21, frequently observed in sarcomas, which harbors *CDK4*, *SAS* and oncogenes *GLI* and *MDM2* [149,150].

Liposarcoma (LPS)

In adults, LPS is one of the most frequent STS subtypes, arising from primitive mesenchymal cells and accounting for 14–18% of all soft-tissue sarcomas [151,152]. Microscopically, these tumors exhibit morphological diversity which is further correlated with their biologic behavior. LPS is classified into four main groups including: well-differentiated liposarcoma/de-differentiated liposarcoma (WD/DDLPS), myxoid/round cell liposarcoma (MLPS) and pleomorphic liposarcoma (PLPS) [153,154]. MLPS, commonly arising in the deep soft tissues such as the thigh or the retroperitoneum, represents the most common variant accounting for ~10% of all adult soft tissue sarcomas [135,155].

The prognosis of LPS depends on several factors including histologic features, site, and size of the LPS.

WDLPS is considered a malignancy of low-grade potential while DDLPS and PLPS are high-grade tumors with strong propensity for local recurrence and metastasis [156,157].

The cytogenetic changes of LPS vary based on the histologic subtype. For the WD/DDLPS subtype, the presence of supernumerary ring and/or giant rod chromosomes with amplified segments from the 12q13-15 region is characteristic, harboring several oncogenes including *HMGA2, MDM2, CDK4, HMGA2, TSPAN31, OS1, OS9, CHOP* and *GLI1* [154,158]. Further, amplifications involving *c-Jun*, and apoptosis signaling kinase 1 (*ASK1*), located on 1p32 and 6q23, respectively, have been described in DDLPS [159]. Similarly, PLPS exhibit complex karyotypes including gains: 1p, 1q21-q32, 2q, 3p, 3q, 5p12-p15, 5q, 6p21, 7p, 7q22, and losses involving 1q, 2q, 3p, 4q, 10q, 11q, 12p13, 13q14, 13q21-qter, 13q23-24, reviewed in [160–162]. Contrary to the complex karyotype of WD/DDLPS and PLPS, the defining cytogenetic event in MLPS is a balanced translocation, most commonly t(12;16)(q13;p11), fusing *FUS* (transcriptional regulatory domains interacting with the RNA polymerase II complex) with *DDIT3* (a DNA-binding leucine-zipper transcription factor that plays a role in cell cycle control and adipocytic differentiation) [163]. It was shown that the *FUS–DDIT3* fusion oncoprotein can induce a sarcoma phenotype in cell model systems [164].

Extraosseous Osteosarcoma

Extraosseous osteosarcomas are high-grade tumors, comprising nearly 1–1.2% of all osteosarcomas. The most commonly affected sites are the extremities, thorax, and the abdomen. Extraosseous and skeletal osteosarcomas have the same histologic variants: osteoblastic, chondroblastic, and fibroblastic. These tumors are chemoresistant, and the best treatment strategy is surgical resection. The prognosis of extraosseous osteosarcomas is poor, with 5-year survival ranging from 12 to 25% [165–168].

Extraosseous osteosarcomas are histologically pleiotropic, containing small and round, clear, multinucleated, spindled, epithelioid, plasmacytoid, and/or fusiform cells. Histopathologic diagnosis of osteosarcoma relies on the presence of osteoid – a homogeneous, pink, structureless extracellular material produced by the tumor cells. Both osteosarcoma of bone and extraosseous osteosarcomas lack consistent genetic abnormalities [169]. Several subtypes of osteosarcomas have been described which have in common the production of osteoid by the malignant cells and the potential for systemic metastasis [25,170].

The etiologies of skeletal and extraosseous osteosarcoma are poorly understood, and most cases are sporadic [171], though certain environmental and genetic factors have been associated with osteosarcoma [172,173]. *RB1* and *TP53* are two well-characterized genes implicated in osteosarcoma. *TP53* is commonly altered in osteosarcoma either by allelic loss, point mutations or gene rearrangements [174–176]. Up to 60% of high-grade osteosarcomas show *TP53* mutations, compared with 1% of low-grade osteosarcomas [177,178]. *RB1*, located at chromosome 13q14 encodes a 110-kDa protein that negatively regulates progression of the cell cycle from G0/G1 into S phase [179,180].

Treatment strategies for extraosseous osteosarcomas include surgery combined with pre- and postoperative chemotherapy. With this multidisciplinary approach, long-term survival has increased to 70%. Patients who have recurrent disease or metastatic lesions (typically in lungs) at diagnosis have a lower survival rate of 20% [181–183].

MOLECULAR DIAGNOSIS OF STSs

There is an expanding array of molecular assays with high diagnostic and prognostic predictive value for sarcomas, that can easily be conducted on fresh or formalin fixed, paraffin-embedded tissue biopsies. Traditional tests including conventional cytogenetics and karyotypic analysis, the previously standard method for diagnosing chromosomal translocations [184], have largely been replaced with the advent of fluorescence *in situ* hybridization (FISH) and reverse transcription polymerase chain reaction (RT-PCR)-based methods to detect these alterations on formalin fixed, paraffin-embedded sections. In fact, with new and emerging molecular diagnostic tools it has become increasingly easier to identify specific events such as chromosomal translocations (as in ES and ARMS), breakpoint regions, epigenetic changes, gene amplifications (as in WD/DDLPS), and deletions (as in rhabdoid tumors or well-differentiated spindle cell liposarcoma) [82,185,186].

Other molecular diagnostic tools have also emerged, including Sanger sequencing, pyrosequencing and high-resolution melting, providing information about specific molecular aberrations on gene level [187]. High-throughput next-generation sequencing (NGS) technology has become an indispensible approach to DNA sequencing, providing revolutionary methods for rapid DNA analysis at the single nucleotide level, and NGS is currently being adopted by many laboratories for routine diagnostic use. NGS represents a very reliable platform with high sensitivity and speed, permitting simultaneous scanning of complete sets of genes from individual patient samples, although affordability remains an issue [188,189]. Recent low- to medium-throughput NGS platforms may be more amenable to diagnostic laboratories, including the Ion Torrent PGM [190], Pacific Biosciences RS [191] and the Illumina MiSeq [192]. The adaptation of these platforms to the molecular diagnostics of STSs is currently under investigation. The use of NGS may enable detection in massively parallel sequencing of individual and pooled samples of genome-wide structural alterations such as copy number changes, chromosomal rearrangements, gene mutations, alternative splice forms, and DNA methylation status, thereby expanding our knowledge of the mutations, translocations, epigenetic alterations, and aberrant signaling pathways associated with specific sarcoma types. Of interest, NGS of the protein-coding regions of the human genome can provide precise transcriptome profiles (by RNA-seq), while other seq-based methods (ChIP-seq, methyl-seq, and DNase-seq) can generate genome-wide profiling of epigenetic marks and chromatin structure [155].

COMMONLY DEREGULATED SIGNALING PATHWAYS IN STSs

Identification of crucial signaling pathways altered in STSs have been central in understanding STS biology, although to date has provided only a few promising actionable therapeutic targets. Herein, we will highlight some of the commonly altered signaling pathways in STSs based on our current knowledge.

Insulin-Like Growth Factor 1 Receptor (IGF1R) Pathway

The IGF1R pathway has been identified as a major signaling pathway implicated in tumorigenesis of several sarcomas. Indeed, extensive molecular alterations of its components have been reported [193,194]. So far, three IGF1R ligands have been identified, namely, IGF1, IGF2, and insulin [195]. Upon binding to the IGF1R extracellular domain, these ligands induce the autophosphorylation of the receptor intracellular domain which, in turn, activates intracellular signaling cascades such as the phosphoinositide-3-kinase (PI3K)-AKT and RAS-mitogen activated protein kinase (MAPK) pathways [55]. The activation of such signaling pathways promotes cell proliferation and survival. Of note, IGF1R has been explored as a therapeutic target in ES and RMS and IGF1R inhibitors are currently undergoing evaluation in clinical trials [196].

More specifically, two clinical trials have included IGF1R as a therapeutic target in RMS, of which: (1) a phase I clinical trial, including two metastatic ARMS patients, using IGF1R-targeted antibody figitumumab, showed disease progression within 6 weeks of starting treatment [197]; and (2) a phase II clinical trial using IGF1R-targeted antibody R1507 showed radiological responses [198]. In terms of ES, a phase I clinical trial using IGF1R-targeted antibodies showed a promising therapeutic response [197,199,200]. However, more recent phase II clinical trials demonstrated that the therapeutic benefits of IGF1R antibodies persist for only a limited time period [98,201]. These studies are reviewed in [202].

The PI3K/AKT/mammalian Target of Rapamycin (mTOR) Pathway

The PI3K/AKT/mTOR pathway is involved in a broad range of fundamental cellular processes, including metabolism regulation, proliferation, survival, autophagy and control of protein synthesis [203,204]. Hyperactivation of the PI3K/AKT/mTOR pathway is a well-known deregulated mechanism central to oncogenic signaling in a large number of cancers [205]. Extensive studies and compelling evidence have reported a deregulation of the PI3K pathway in different kinds of sarcomas such as GIST [206,207]. Currently, phase 1 and 2 clinical trials of several PI3K inhibitors for use in STS treatment are ongoing [207].

For example, a phase II clinical trial conducted on 82 patients with advanced sarcoma using mTOR inhibitor, AP23573, showed partial response (PR) or stable disease (SD) in 39% of patients with bone sarcomas while SD was observed in 62, 25, and 22% of patients with leiomyosarcoma, liposarcoma, and other soft tissue sarcomas, respectively [208]. Another trial including 212 patients with advanced bone and soft tissue sarcomas treated with ridaforolimus showed enhanced median overall survival in a subset of patients [209]. For more information, readers are referred to a more comprehensive review [210].

mTOR is a serine/threonine kinase lying downstream of the PI3K/AKT pathway. In keeping with this, deregulation of this pathway, such as through inactivation of the AKT inhibitor phosphatase and tensin homolog (PTEN), has been reported as a major mechanism implicated in intrinsic activation of the mTOR pathway in sarcomas [211]. Other potential mechanisms contributing to mTOR overactivation have been reported such as loss of the tuberous sclerosis complex (TSC) tumor suppressor, a negative regulator of mTOR. Moreover, loss of the tumor suppressor liver kinase B1 (LKB1) protein, another negative regulator of mTOR through its control of adenine monophosphate-activated protein kinase (AMPK), also leads to hyperactivation of mTOR signaling [10,212]. Therefore, mTOR is a potentially promising therapeutic target in STSs. Of note, mTOR plays critical roles in regulating signaling pathways deregulated in a wide variety of STSs such as the signaling pathway for IGF-1R, which is hyperactivated in rhabdomyosarcoma, leiomyosarcoma, and synovial sarcoma [212,213]. Therefore, mTOR represents an attractive therapeutic target in these tumors.

Indeed, rapamycin, a well-known mTOR inhibitor, has been shown to be an effective anticancer agent in a wide range of preclinical models. Clinical trials with this agent alone and in combination with other anticancer agents have demonstrated potential clinical benefits in several types of sarcoma [92]. For example, in a metastatic myxoid chondrosarcoma case study, the use of the mTOR inhibitor sirolimus achieved a partial response [214]. Further, a phase II clinical trial conducted on 61 patients with recurrent or refractory sarcomas, involving STSs and bone sarcomas using everolimus showed a promising outcome response after 4 months with complete response (CR), PR, or SD in 13% of patients with STSs or bone sarcoma and 27% of patients with GIST [215]. Moreover, another phase II trial conducted using temsirolimus on 40 patients with advanced sarcomas demonstrated PR only in patients with undifferentiated fibrosarcoma and uterine leiomyosarcoma (5% of cases) [216]. In another cohort involving 212 patients with advanced STSs or bone sarcomas, a phase II clinical trial using ridaforolimus achieved a median overall survival of 40 weeks [217]. These studies are reviewed in [218].

Notch, Wnt, and Hedgehog (hh) Pathways

Aberrant activation of Notch, Wnt, and hh pathways have been unambiguously implicated in the development of a wide variety of malignant neoplasms [219–221]. The role of these pathways has been characterized in some STSs, notably RMS [222–227]. Of note, Notch has been found to promote invasion and metastasis of RMS potentially through induction of N-cadherin and α9-integrin [228], pointing to the Notch pathway as a promising target for the treatment of RMS.

While Notch signaling has oncogenic promoting effects in RMS as it promotes invasion and metastasis, the Wnt/β-catenin pathway has been shown to exert anti-oncogenic effects [229,230]. The Wnt/β-catenin pathway gets activated upon binding of the Wnt ligand to the Frizzled (Fz) receptor, resulting in the recruitment of the axin complex to the receptor. This event prevents axin-mediated degradation of β-catenin, therefore promoting accumulation of β-catenin which can then translocate to the nucleus to activate transcription of its target genes. Constitutive activation of the Wnt/β-catenin signaling pathway as a result of genetic alterations is commonly encountered in malignant tumors [231]. However, activation of the Wnt/β-catenin signaling pathway in rhabdomyosarcoma has been shown to promote myogenic differentiation through induction of MyoD transcription, a finding supporting the anti-oncogenic role of the Wnt/β-catenin signaling pathway in this disease [230].

One of the fundamental pathways implicated in modulation of Wnt activity and cancer is the hedgehog pathway. It has been shown that inhibition of sonic hedgehog (Shh) signaling led to Gli3-dependent reduction in Wnt-mediated transcriptional activation via antagonizing active forms of the Wnt transcriptional effector, β-catenin [232]. Furthermore, the inverse correlation between Wnt and hh activation in human gastric cancer has been reported, in which overexpression of Gli1 suppressed Wnt transcriptional activity and nuclear β-catenin accumulation [233].

The hedgehog pathway is critical for normal development. In addition, extensive studies and compelling evidence point to its potential contribution to tumorigenesis and metastasis. Classic activation of the hh pathway is initiated by the binding of the mature hh proteins to a transmembrane protein receptor called Patched-1 (PTCH1) receptor. Upon binding of the ligand, the inhibitory effect of PTCH1 on an Fz family-related transmembrane receptor called Smoothened (SMO) is released, thereby preventing proteasomal degradation of downstream Gli (glioma-associated oncogene homolog) transcription factors (Gli1, Gli2, and Gli3). Glis are subsequently translocated to the nucleus where they control transcription of hh target genes [234]. The hh signaling pathway has been implicated in the development of RMS. It has been shown that Ptch1-mutant mice, which exhibit overactive hh signaling, developed RMS with the tumor cells being resistant to apoptosis. Further reports demonstrated a consistent activation of the hh pathway in ERMS and translocation-negative ARMS. Of interest, hedgehog pathway inhibition with a smoothened inhibitor, IPI-926 [235], may represent a potentially new therapeutic strategy for hh-driven STSs.

CONCLUSIONS

STSs vary widely in their histologic appearance and behavior. Many etiological factors and genetic aberrations are likely involved in STS development. To date, and despite extensive studies, the etiology and biology of many STSs have yet to be elucidated. Diagnosis of STSs remains challenging owing to their undifferentiated appearances, or their resemblance to benign lesions, particularly those STSs with low-grade malignant potential. As a result of rapidly evolving genomic and proteomic technologies, our understanding of the oncogenic mechanisms underlying sarcomagenesis is markedly improved. STS research has extensive avenues and possibilities for investigation. Although insights into STS pathogenesis and new molecular diagnostic tools have emerged, these findings have failed to impact the prognosis for sarcoma patients with metastatic disease. Identifying factors which contribute to STS metastasis, and that can be targeted therapeutically, has tremendous potential to impact survival outcomes for patients with this group of diseases.

GLOSSARY

AKT A serine/threonine kinase serving as proto-oncogene.

Chromosomal translocation A chromosome abnormality caused by rearrangement of parts between non-homologous chromosomes. A gene fusion may be created when the translocation joins two otherwise separated genes and it is frequent in cancers.

C-kit A tyrosine kinase growth factor receptor, also known as CD117. Gain-of-function mutation is a major contributor to GIST pathogenesis.

IGF1R pathway A crucial cellular signaling pathway involving the IGF1 receptor (IGF1R) and IGF1 and 2 ligands that provides a potent survival and proliferative signals to stimulate growth in many different cell types.

mTOR A serine/threonine kinase downstream of PI3K.

MYCN A member of the MYC family of transcription factors which encodes a protein with a basic helix-loop-helix (bHLH) domain.

Oncogene A gene that has the potential to cause cancer, commonly playing a role in cell growth and proliferation. Its overexpression often leads to neoplastic cell transformation.

PDGFRA A gene that encodes a cell surface tyrosine kinase receptor, related to c-kit, for members of the platelet-derived growth factor family.

PI3K/AKT/mTOR pathway An intracellular signaling pathway involved in cellular proliferation, growth, survival and mobility and often exhibits alterations of its components in a wide variety of tumors, therefore representing an attractive target for anticancer therapy.

Soft tissue sarcomas (STSs) A heterogeneous group of malignancies of mesenchymal origin arising from anywhere in the body, usually aggressive.

Tyrosine kinase inhibitors (TKIs) These agents bind to the adenosine triphosphate (ATP)-binding pocket, crucial for phosphorylation and activation of the growth factor receptor, inhibiting tumor growth.

ABBREVIATIONS

ALK Anaplastic lymphoma kinase
AMPK Adenine monophosphate-activated protein kinase
ARMS Alveolar rhabdomyosarcoma
ASK1 Apoptosis signaling kinase 1
ASPS Alveolar soft part sarcoma
ATF2 Activating transcription factor 2
ATM Ataxia telangiectasia mutated
ATP Adenosine triphosphate
CD11 Cluster of differentiation
CDK Cyclin-dependent kinase
CFS Congenital fibrosarcoma
CHK2 Checkpoint kinase 2
COG-STS Children's Oncology Group–Soft Tissue Sarcoma (STS) Committee
COL1A1 Collagen, type I, alpha 1
CR Complete response
DDLPS De-differentiated liposarcoma
DFSP Dermatofibrosarcoma protuberans
Dvl Dishevelled
EES Extraosseous Ewing sarcoma
ES Ewing sarcoma
ETV6 ETS variant gene 6
EWSR1 Ewing sarcoma breakpoint region 1
FISH Fluorescence *in situ* hybridization
FKHR Forkhead homolog 1 rhabdomyosarcoma
FLI1 Friend leukemia virus integration 1
FOXO1 Forkhead box protein O1
FUS Fused in sarcoma
GIST Gastrointestinal stromal tumor
Gli Glioma-associated oncogene homolog
GSK-3β Glycogen synthase kinase 3
HDAC1 Histone deacetylase 1
hh Hedgehog
LKB1 Liver kinase B1
LS Liposarcoma
MAPK Mitogen activated protein kinase
MDM2 Murine double minute 2
MFH Malignant fibrous histiocytoma
MLPS Myxoid liposarcoma
MMT Malignant mesenchymal tumor
mTOR Mammalian target of rapamycin
NTRK3 Neutrophilic tyrosine kinase receptor, type3
PAX3 Paired box 3
PAX7 Paired box 7
PCR Polymerase chain reaction
PDGFRA Platelet-derived growth factor receptor alpha
PDGFβ Platelet-derived growth factor beta
PI3K Phosphatidylinositol 3-kinase
PLPS Pleomorphic liposarcoma
PR Partial response
PTEN Phosphatase and tensin homolog
RAS Rat sarcoma
RB1 Retinoblastoma 1
RMS Rhabdomyosarcoma
SD Stable disease
STSs Soft tissue sarcomas
TKIs Tyrosine kinase inhibitors
TLS Translocated in liposarcoma
TP53 Tumor protein 53
UPS Undifferentiated pleomorphic sarcoma
WDLPS Well-differentiated liposarcoma

REFERENCES

[1] Gutierrez JC, Perez EA, Moffat FL, Livingstone AS, Franceschi D, Koniaris LG. Should soft tissue sarcomas be treated at high-volume centers? An analysis of 4205 patients. Ann Surg 2007;245:952–8.

[2] Sinha S, Peach AH. Diagnosis and management of soft tissue sarcoma. Br Med J 2010;341:c7170.

[3] Loeb DM, Thornton K, Shokek O, Souid AK, Meyers PA, Gebhardt M. Pediatric soft tissue sarcomas. Surg Clin N Am 2008;88:615–27.

[4] Helman LJ, Meltzer P. Mechanisms of sarcoma development. Nat Rev Cancer 2003;3:685–94.

[5] Cohen RJ, Curtis RE, Inskip PD, Fraumeni Jr. JF. The risk of developing second cancers among survivors of childhood soft tissue sarcoma. Cancer 2005;103:2391–6.

[6] Lindor NM, Greene MH. The concise handbook of family cancer syndromes. Mayo Familial Cancer Program. J Natl Cancer Inst 1998;90:1039–71.

[7] Kogevinas M, Becher H, Benn T, Bertazzi PA, Boffetta P, Bueno-de-Mesquita HB, et al. Cancer mortality in workers exposed to phenoxy herbicides, chlorophenols, and dioxins. An expanded and updated international cohort study. Am J Epidemiol 1997;145:1061–75.

[8] Rubino C, Shamsaldin A, Le MG, Labbé M, Guinebretière JM, Chavaudra J, et al. Radiation dose and risk of soft tissue and bone sarcoma after breast cancer treatment. Breast Cancer Res Treat 2005;89:277–88.

[9] Riad S, Biau D, Holt GE, Werier J, Turcotte RE, Ferguson PC, et al. The clinical and functional outcome for patients with radiation-induced soft tissue sarcoma. Cancer 2012;118:2682–92.

[10] Zhang L, Anglesio MS, O'Sullivan M, Zhang F, Yang G, Sarao R, et al. The E3 ligase HACE1 is a critical chromosome 6q21 tumor suppressor involved in multiple cancers. Nat Med 2007;13:1060–9.

[11] Sterpone S, Cozzi R. Influence of XRCC1 Genetic polymorphisms on ionizing radiation-induced DNA damage and repair. J Nucleic Acids 2010;2010:780369

[12] Berardinelli F, Antoccia A, Buonsante R, Gerardi S, Cherubini R, De Nadal V, et al. The role of telomere length modulation in delayed chromosome instability induced by ionizing radiation in human primary fibroblasts. Environ Molec Mutagen 2013;54: 172–9.

[13] Walker CL, Stewart EA. Uterine fibroids: the elephant in the room. Science 2005;308:1589–92.

[14] Fletcher CD, Gustafson P, Rydholm A, Willen H, Akerman M. Clinicopathologic re-evaluation of 100 malignant fibrous histiocytomas: prognostic relevance of subclassification. J Clin Oncol 2001;19:3045–50.

[15] Billingsley KG, Burt ME, Jara E, Ginsberg RJ, Woodruff JM, Leung DH, et al. Pulmonary metastases from soft tissue sarcoma: analysis of patterns of diseases and postmetastasis survival. Ann Surg 1999;229:602–10, discussion 610–2.

[16] Park JH, Kang CH, Kim CH, Chae IJ, Park JH. Highly malignant soft tissue sarcoma of the extremity with a delayed diagnosis. World J Surg Oncol 2010;8:84.

[17] Sheng WQ, Hisaoka M, Okamoto S, Tanaka A, Meis-Kindblom JM, Kindblom LG, et al. Congenital-infantile fibrosarcoma. A clinicopathologic study of 10 cases and molecular detection of the ETV6-NTRK3 fusion transcripts using paraffin-embedded tissues. Am J Clin Pathol 2001;115:348–55.

[18] Kauffman SL, Stout AP. Histiocytic tumors (fibrous xanthoma and histiocytoma) in children. Cancer 1961;14:469–82.

[19] Walling AK, Gasser SI. Soft-tissue and bone tumors about the foot and ankle. Clin Sports Med 1994;13:909–38.

[20] Grimer R, Judson I, Peake D, Seddon B. Guidelines for the management of soft tissue sarcomas. Sarcoma 2010;2010:506182.

[21] Mankin HJ, Hornicek FJ. Diagnosis, classification, and management of soft tissue sarcomas. Cancer Control 2005;12:5–21.

[22] Peabody TD, Gibbs Jr CP, Simon MA. Evaluation and staging of musculoskeletal neoplasms. J Bone Jt Surg A 1998;80:1204–18.

[23] Pang LM, Roebuck DJ, Griffith JF, Kumta SM, Metreweli C. Alveolar soft-part sarcoma: a rare soft-tissue malignancy with distinctive clinical and radiological features. Pediatr Radiol 2001;31: 196–9.

[24] Bernstein ML, Devidas M, Lafreniere D, Souid AK, Meyers PA, Gebhardt M, et al. Intensive therapy with growth factor support for patients with Ewing tumor metastatic at diagnosis: Pediatric Oncology Group/Children's Cancer Group Phase II Study 9457--a report from the Children's Oncology Group. J Clin Oncol 2006;24:152–9.

[25] Granowetter L, Womer R, Devidas M, Krailo M, Wang C, Bernstein M, et al. Dose-intensified compared with standard chemotherapy for nonmetastatic Ewing sarcoma family of tumors: a Children's Oncology Group Study. J Clin Oncol 2009;27:2536–41.

[26] Walterhouse D, Watson A. Optimal management strategies for rhabdomyosarcoma in children. Paediatr Drugs 2007;9:391–400.

[27] Hegyi M, Felne Semsei A, Jakab Z, Antal I, Kiss J, et al. Results of the treatment of pediatric osteosarcoma in the Hungarian population. Magyar Onkol 2012;56:30–7 [In Hungarian].

[28] Gupta GP, Massague J. Cancer metastasis: building a framework. Cell 2006;127:679–95.

[29] US Cancer Statistics Working Group. United States Cancer Statistics: 1999 Incidence. Atlanta, GA: Department of Health and Human Services, Centers for Disease Control and Prevention and National Cancer Institute; 2002.

[30] Mitelman F, Johansson B, Mertens F. The impact of translocations and gene fusions on cancer causation. Nat Rev Cancer 2007;7:233–45.

[31] Mitelman F. Recurrent chromosome aberrations in cancer. Mutat Res 2000;462:247–53.

[32] Delattre O, Zucman J, Plougastel B, Desmaze C, Melot T, Peter M, et al. Gene fusion with an ETS DNA-binding domain caused by chromosome translocation in human tumours. Nature 1992;359: 162–5.

[33] May WA, Lessnick SL, Braun BS, Klemsz M, Lewis BC, Lunsford LB, et al. The Ewing's sarcoma EWS/FLI-1 fusion gene encodes a more potent transcriptional activator and is a more powerful transforming gene than FLI-1. Molec Cell Biol 1993;13:7393–8.

[34] Sorensen PH, Lessnick SL, Lopez-Terrada D, Liu XF, Triche TJ, Denny CT. A second Ewing's sarcoma translocation, t(21;22), fuses the EWS gene to another ETS-family transcription factor, ERG. Nat Genet 1994;6:146–51.

[35] Knezevich SR, McFadden DE, Tao W, Lim JF, Sorensen PH. A novel ETV6-NTRK3 gene fusion in congenital fibrosarcoma. Nat Genet 1998;18:184–7.

[36] Bourgeois JM, .Knezevich SR, Mathers JA, Sorensen PH. Molecular detection of the ETV6-NTRK3 gene fusion differentiates congenital fibrosarcoma from other childhood spindle cell tumors. Am J Surg Pathol 2000;24:937–46.

[37] Knezevich SR, Garnett MJ, Pysher TJ, Beckwith JB, Grundy PE, Sorensen PH. ETV6-NTRK3 gene fusions and trisomy 11 establish a histogenetic link between mesoblastic nephroma and congenital fibrosarcoma. Cancer Res 1998;58:5046–8.

[38] Rubin BP, Chen CJ, Morgan TW, Xiao S, Grier HE, Kozakewich HP, et al. Congenital mesoblastic nephroma t(12;15) is associated

with ETV6-NTRK3 gene fusion: cytogenetic and molecular relationship to congenital (infantile) fibrosarcoma. Am J Pathol 1998;153:1451–8.
[39] Riggi N, Cironi L, Suva ML, Stamenkovic I. Sarcomas: genetics, signaling, and cellular origins. Part 1: The fellowship of TET. J Pathol 2007;213:4–20.
[40] Liao W, Ho CY, Yan YL, Postlethwait J, Stainier DY. Hhex and scl function in parallel to regulate early endothelial and blood differentiation in zebrafish. Development 2000;127:4303–13.
[41] Tognon C, Garnett M, Kenward E, Kay R, Morrison K, Sorensen PH. The chimeric protein tyrosine kinase ETV6-NTRK3 requires both Ras-Erk1/2 and PI3-kinase-Akt signaling for fibroblast transformation. Cancer Res 2001;61:8909–16.
[42] Schwab CJ, Jones LR, Morrison H, Ryan SL, Yigittop H, Schouten JP, et al. Evaluation of multiplex ligation-dependent probe amplification as a method for the detection of copy number abnormalities in B-cell precursor acute lymphoblastic leukemia. Genes Chromosomes Cancer 2010;49:1104–13.
[43] Lannon CL, Sorensen PH. ETV6-NTRK3: a chimeric protein tyrosine kinase with transformation activity in multiple cell lineages. Semin Cancer Biol 2005;15:215–23.
[44] Kazanowska B, Reich A, Stegmaier S, Békássy AN, Leuschner I, Chybicka A, et al. Pax3-fkhr and pax7-fkhr fusion genes impact outcome of alveolar rhabdomyosarcoma in children. Fetal Pediatr Pathol 2007;26:17–31.
[45] Sorensen PH, Lynch JC, Qualman SJ, Tirabosco R, Lim JF, Maurer HM, et al. PAX3-FKHR and PAX7-FKHR gene fusions are prognostic indicators in alveolar rhabdomyosarcoma: a report from the children's oncology group. J Clin Oncol 2002;20: 2672–9.
[46] Kelly KM, Womer RB, Sorensen PH, Xiong QB, Barr FG. Common and variant gene fusions predict distinct clinical phenotypes in rhabdomyosarcoma. J Clin Oncol 1997;15:1831–6.
[47] Simon MP, Pedeutour F, Sirvent N, Grosgeorge J, Minoletti F, Coindre JM, et al. Deregulation of the platelet-derived growth factor B-chain gene via fusion with collagen gene COL1A1 in dermatofibrosarcoma protuberans and giant-cell fibroblastoma. Nat Genet 1997;15:95–8.
[48] O'Brien KP, Seroussi E, Dal Cin P, Sciot R, Mandahl N, Fletcher JA, et al. Various regions within the alpha-helical domain of the COL1A1 gene are fused to the second exon of the PDGFB gene in dermatofibrosarcomas and giant-cell fibroblastomas. Genes Chromosomes Cancer 1998;23:187–93.
[49] Butrynski JE, D'Adamo DR, Hornick JL, Dal Cin P, Antonescu CR, Jhanwar SC, et al. Crizotinib in ALK-rearranged inflammatory myofibroblastic tumor. N Engl J Med 2010;363:1727–33.
[50] Lawrence B, Perez-Atayde A, Hibbard MK, Rubin BP, Dal Cin P, Pinkus JL, et al. TPM3-ALK and TPM4-ALK oncogenes in inflammatory myofibroblastic tumors. Am J Pathol 2000;157:377–84.
[51] Griffin CA, Hawkins AL, Dvorak C, Henkle C, Ellingham T, Perlman EJ. Recurrent involvement of 2p23 in inflammatory myofibroblastic tumors. Cancer Res 1999;59:2776–80.
[52] Rubin BP, Blanke CD, Demetri GD, Dematteo RP, Fletcher CD, Goldblum JR, et al. Protocol for the examination of specimens from patients with gastrointestinal stromal tumor. Arch Pathol Lab Med 2010;134:165–70.
[53] Pedeutour F, Suijkerbuijk RF, Van Gaal J, Van de Klundert W, Coindre JM, Van Haelst A, et al. Chromosome 12 origin in rings and giant markers in well-differentiated liposarcoma. Cancer Genet Cytogenet 1993;66:133–4.
[54] Bovee JV, Hogendoorn PC. Molecular pathology of sarcomas: concepts and clinical implications. Virchows Arch 2010;456: 193–9.
[55] Quesada J, Amato R. The molecular biology of soft-tissue sarcomas and current trends in therapy. Sarcoma 2012;2012:849456.
[56] Guillou L, Aurias A. Soft tissue sarcomas with complex genomic profiles. Virchows Arch 2010;456:201–17.
[57] Reversade B, Escande-Beillard N, Dimopoulou A, Fischer B, Chng SC, Li Y, et al. Mutations in PYCR1 cause cutis laxa with progeroid features. Nat Genet 2009;41:1016–21.
[58] Murnane JP. Telomere loss as a mechanism for chromosome instability in human cancer. Cancer Res 2010;70:4255–9.
[59] Swords RT, Kelly KR, Smith PG, Garnsey JJ, Mahalingam D, Medina E, et al. Inhibition of NEDD8-activating enzyme: a novel approach for the treatment of acute myeloid leukemia. Blood 2010;115:3796–800.
[60] Albertson DG, Collins C, McCormick F, Gray JW. Chromosome aberrations in solid tumors. Nat Genet 2003;34:369–76.
[61] Lahat G, Tuvin D, Wei C, Wang WL, Pollock RE, Anaya DA, et al. Molecular prognosticators of complex karyotype soft tissue sarcoma outcome: a tissue microarray-based study. Ann Oncol 2010;21:1112–20.
[62] Parham DM, Maki RG, Albritton KH. Pediatric and adult sarcomas: pathologic and management issues. ASCO Annual Meeting, 2011 Educational Book, e15-e23. Category: Sarcoma.
[63] Parham DM, Ellison DA. Rhabdomyosarcomas in adults and children: an update. Arch Pathol Lab Med 2006;130:1454–65.
[64] Hazarika M, Jiang X, Liu Q, Lee SL, Ramchandani R, Garnett C, et al. Tasigna for chronic and accelerated phase Philadelphia chromosome-positive chronic myelogenous leukemia resistant to or intolerant of imatinib. Clin Cancer Res 2008;14:5325–31.
[65] Raney RB, Anderson JR, Barr FG, Donaldson SS, Pappo AS, Qualman SJ, et al. Rhabdomyosarcoma and undifferentiated sarcoma in the first two decades of life: a selective review of intergroup rhabdomyosarcoma study group experience and rationale for Intergroup Rhabdomyosarcoma Study V. J Pediatr Hematol Oncol 2001;23:215–20.
[66] Rodeberg D, Paidas C. Childhood rhabdomyosarcoma. Semin Pediatr Surg 2006;15:57–62.
[67] Gonsalves FC, Klein K, Carson BB, Katz S, Ekas LA, Evans S, et al. An RNAi-based chemical genetic screen identifies three small-molecule inhibitors of the Wnt/wingless signaling pathway. Proc Natl Acad Sci USA 2011;108:5954–63.
[68] Turc-Carel C, Lizard-Nacol S, Justrabo E, Favrot M, Philip T, Tabone E. Consistent chromosomal translocation in alveolar rhabdomyosarcoma. Cancer Genet Cytogenet 1986;19:361–2.
[69] Douglass EC, Valentine M, Etcubanas E, Parham D, Webber BL, Houghton PJ, et al. A specific chromosomal abnormality in rhabdomyosarcoma. Cytogenet Cell Genet 1987;45:148–55.
[70] Whang-Peng J, Knutsen T, Theil K, Horowitz ME, Triche T. Cytogenetic studies in subgroups of rhabdomyosarcoma. Genes Chromosomes Cancer 1992;5:299–310.

[71] Du S, Lawrence EJ, Strzelecki D, Rajput P, Xia SJ, Gottesman DM, et al. Co-expression of alternatively spliced forms of PAX3, PAX7, PAX3-FKHR and PAX7-FKHR with distinct DNA binding and transactivation properties in rhabdomyosarcoma. Internatl J Cancer 2005;115:85–92.

[72] Xia SJ, Barr FG. Chromosome translocations in sarcomas and the emergence of oncogenic transcription factors. Eur J Cancer 2005;41:2513–27.

[73] Lae M, Ahn EH, Mercado GE, Chuai S, Edgar M, Pawel BR, et al. Global gene expression profiling of PAX-FKHR fusion-positive alveolar and PAX-FKHR fusion-negative embryonal rhabdomyosarcomas. J Pathol 2007;212:143–51.

[74] Barr FG. Gene fusions involving PAX and FOX family members in alveolar rhabdomyosarcoma. Oncogene 2001;20:5736–46.

[75] Barr FG, Duan F, Smith LM, Gustafson D, Pitts M, Hammond S, et al. Genomic and clinical analyses of 2p24 and 12q13-q14 amplification in alveolar rhabdomyosarcoma: a report from the Children's Oncology Group. Genes Chromosomes Cancer 2009;48:661–72.

[76] Reichek JL, Duan F, Smith LM, Gustafson DM, O'Connor RS, Zhang C, et al. Genomic and clinical analysis of amplification of the 13q31 chromosomal region in alveolar rhabdomyosarcoma: a report from the Children's Oncology Group. Clin Cancer Res 2011;17:1463–73.

[77] Mercado GE, Xia SJ, Zhang C, Ahn EH, Gustafson DM, Laé M, et al. Identification of PAX3-FKHR-regulated genes differentially expressed between alveolar and embryonal rhabdomyosarcoma: focus on MYCN as a biologically relevant target. Genes Chromosomes Cancer 2008;47:510–20.

[78] Weber-Hall S, Anderson J, McManus A, Abe S, Nojima T, Pinkerton R, et al. Gains, losses, and amplification of genomic material in rhabdomyosarcoma analyzed by comparative genomic hybridization. Cancer Res 1996;56:3220–4.

[79] Bridge JA, Liu J, Weibolt V, Baker KS, Perry D, Kruger R, et al. Novel genomic imbalances in embryonal rhabdomyosarcoma revealed by comparative genomic hybridization and fluorescence in situ hybridization: an intergroup rhabdomyosarcoma study. Genes Chromosomes Cancer 2000;27:337–44.

[80] Ohyashiki K, Ohyashiki JH, Sandberg AA. Cytogenetic characterization of putative human myeloblastic leukemia cell lines (ML-1, -2, and -3): origin of the cells. Cancer Res 1986;46:3642–7.

[81] Polito P, Dal Cin P, Sciot R, Brock P, Van Eyken P, Van den Berghe H. Embryonal rhabdomyosarcoma with only numerical chromosome changes. Case report and review of the literature. Cancer Genet Cytogenet 1999;109:161–5.

[82] Bridge JA, Cushman-Vokoun AM. Molecular diagnostics of soft tissue tumors. Arch Pathol Lab Med 2011;135:588–601.

[83] Tiffin N, Williams RD, Shipley J, Pritchard-Jones K. PAX7 expression in embryonal rhabdomyosarcoma suggests an origin in muscle satellite cells. Br J Cancer 2003;89:327–32.

[84] Ginsberg JP, de Alava E, Ladanyi M, Wexler LH, Kovar H, Paulussen M, et al. EWS-FLI1 and EWS-ERG gene fusions are associated with similar clinical phenotypes in Ewing's sarcoma. J Clin Oncol 1999;17:1809–14.

[85] Wang-Wuu S, Soukup S, Ballard E, Gotwals B, Lampkin B. Chromosomal analysis of sixteen human rhabdomyosarcomas. Cancer Res 1988;48:983–7.

[86] Esiashvili N, Prabhu R, Kahn S, Paulino A. Current strategies and challenges in treatment of childhood rhabdomyosarcoma. J Radiat Oncol 2013;2:159–68.

[87] Gibson BE, Wheatley K, Hann IM, Stevens RF, Webb D, Hills RK, et al. Treatment strategy and long-term results in paediatric patients treated in consecutive UK AML trials. Leukemia 2005;19:2130–8.

[88] Taube C, Wei X, Swasey CH, Joetham A, Zarini S, Lively T, et al. Mast cells, Fc epsilon RI, and IL-13 are required for development of airway hyperresponsiveness after aerosolized allergen exposure in the absence of adjuvant. J Immunol 2004;172:6398–406.

[89] Maloney EK, McLaughlin JL, Dagdigian NE, Garrett LM, Connors KM, Zhou XM, et al. An anti-insulin-like growth factor I receptor antibody that is a potent inhibitor of cancer cell proliferation. Cancer Res 2003;63:5073–83.

[90] King ER, Wong KK. Insulin-like growth factor: current concepts and new developments in cancer therapy. Recent Pat Anticancer Drug Discov 2012;7:14–30.

[91] Wu H, Huang C, Chang D. Anti-angiogenic therapeutic drugs for treatment of human cancer. J Cancer Mol 2008;4:37–45.

[92] Agulnik M. New developments in mammalian target of rapamycin inhibitors for the treatment of sarcoma. Cancer 2012;118:1486–97.

[93] Zeng FY, Dong H, Cui J, Liu L, Chen T. Glycogen synthase kinase 3 regulates PAX3-FKHR-mediated cell proliferation in human alveolar rhabdomyosarcoma cells. Biochem Biophys Res Commun 2010;391:1049–55.

[94] Wachtel M, Schafer BW. Targets for cancer therapy in childhood sarcomas. Cancer Treat Rev 2010;36:318–27.

[95] Wang C. Childhood rhabdomyosarcoma: recent advances and prospective views. J Dent Res 2012;91:341–50.

[96] Brodsky JT, Burt ME, Hajdu SI, Casper ES, Brennan MF. Tendosynovial sarcoma. Clinicopathologic features, treatment, and prognosis. Cancer 1992;70:484–9.

[97] Ladenstein R, Treuner J, Koscielniak E, d'Oleire F, Keim M, Gadner H, et al. Synovial sarcoma of childhood and adolescence. Report of the German CWS-81 study. Cancer 1993;71:3647–55.

[98] Lager JJ, Lyden ER, Anderson JR, Pappo AS, Meyer WH, Breitfeld PP. Pooled analysis of phase II window studies in children with contemporary high-risk metastatic rhabdomyosarcoma: a report from the Soft Tissue Sarcoma Committee of the Children's Oncology Group. J Clin Oncol 2006;24:3415–22.

[99] Wang Y, Krivtsov AV, Sinha AU, North TE, Goessling W, Feng Z, et al. The Wnt/β-catenin pathway is required for the development of leukemia stem cells in AML. Science 2009;327:1650–3.

[100] Birdsall S, Osin P, Lu YJ, Fisher C, Shipley J. Synovial sarcoma specific translocation associated with both epithelial and spindle cell components. Internatl J Cancer 1999;82:605–8.

[101] Miettinen M, Lehto VP, Virtanen I. Monophasic synovial sarcoma of spindle-cell type. Epithelial differentiation as revealed by ultrastructural features, content of prekeratin and binding of peanut agglutinin. Virchows Archiv B 1983;44:187–99.

[102] Allander SV, Illei PB, Chen Y, Antonescu CR, Bittner M, Ladanyi M, et al. Expression profiling of synovial sarcoma by cDNA microarrays: association of ERBB2, IGFBP2, and ELF3 with epithelial differentiation. Am J Pathol 2002;161:1587–95.

[103] Shipley JM, Clark J, Crew AJ, Birdsall S, Rocques PJ, Gill S, et al. The t(X;18)(p11.2;q11.2) translocation found in human

synovial sarcomas involves two distinct loci on the X chromosome. Oncogene 1994;9:1447–53.

[104] Fisher C. Synovial sarcoma. Ann Diagnost Pathol 1998;2:401–21.

[105] Brett D, Whitehouse S, Antonson P, Shipley J, Cooper C, Goodwin G. The SYT protein involved in the t(X;18) synovial sarcoma translocation is a transcriptional activator localised in nuclear bodies. Hum Molec Genet 1997;6:1559–64.

[106] Su L, Sampaio AV, Jones KB, Pacheco M, Goytain A, Lin S, et al. Deconstruction of the SS18-SSX fusion oncoprotein complex: insights into disease etiology and therapeutics. Cancer Cell 2012;21:333–47.

[107] Kadoch C, Crabtree GR. Reversible disruption of mSWI/SNF (BAF) complexes by the SS18-SSX oncogenic fusion in synovial sarcoma. Cell 2013;153:71–85.

[108] Gerbaulet A, Wickenhauser C, Scholten J, Peschke K, Drube S, Horny HP, et al. Mast cell hyperplasia, B-cell malignancy, and intestinal inflammation in mice with conditional expression of a constitutively active kit. Blood 2011;117:2012–21.

[109] Sandrasegaran K, Rajesh A, Rydberg J, Rushing DA, Akisik FM, Henley JD. Gastrointestinal stromal tumors: clinical, radiologic, and pathologic features. Am J Roentgenol 2005;184:803–11.

[110] Hirota S, Isozaki K, Moriyama Y, Hashimoto K, Nishida T, Ishiguro S, et al. Gain-of-function mutations of c-kit in human gastrointestinal stromal tumors. Science 1998;279:577–80.

[111] Siehl J, Thiel E. C-kit, GIST, and imatinib. Recent Results Cancer Res 2007;176:145–51.

[112] Yoo SK, Starnes TW, Deng Q, Huttenlocher A. Lyn is a redox sensor that mediates leukocyte wound attraction *in vivo*. Nature 2011;480:109–12.

[113] Heinrich MC, Corless CL, Blanke CD, Demetri GD, Joensuu H, Roberts PJ, et al. Molecular correlates of imatinib resistance in gastrointestinal stromal tumors. J Clin Oncol 2006;24:4764–74.

[114] Fletcher JA, Rubin BP. KIT mutations in GIST. Curr Opin Genet Devel 2007;17:3–7.

[115] Feuring-Buske M, Gerhard B, Cashman J, Humphries RK, Eaves CJ, Hogge DE. Improved engraftment of human acute myeloid leukemia progenitor cells in beta 2-microglobulin-deficient NOD/SCID mice and in NOD/SCID mice transgenic for human growth factors. Leukemia 2003;17:760–3.

[116] Heinrich MC, Corless CL, Duensing A, McGreevey L, Chen CJ, Joseph N, et al. PDGFRA activating mutations in gastrointestinal stromal tumors. Science 2003;299:708–10.

[117] Burger H, den Bakker MA, Kros JM, van Tol H, de Bruin AM, Oosterhuis W, et al. Activating mutations in c-KIT and PDGFRalpha are exclusively found in gastrointestinal stromal tumors and not in other tumors overexpressing these imatinib mesylate target genes. Cancer Biol Ther 2005;4:1270–4.

[118] DeMatteo RP, Lewis JJ, Leung D, Mudan SS, Woodruff JM, Brennan MF. Two hundred gastrointestinal stromal tumors: recurrence patterns and prognostic factors for survival. Ann Surg 2000;231:51–8.

[119] Eisenberg BL, Trent JC. Adjuvant and neoadjuvant imatinib therapy: current role in the management of gastrointestinal stromal tumors. Internatl J Cancer 2011;129:2533–42.

[120] Corsi A, Boldrini R, Bosman C. Congenital-infantile fibrosarcoma: study of two cases and review of the literature. Tumori 1994;80:392–400.

[121] Punnett HH, Tomczak EZ, Pawel BR, de Chadarevian JP, Sorensen PH. ETV6-NTRK3 gene fusion in metastasizing congenital fibrosarcoma. Med Pediatr Oncol 2000;35:137–9.

[122] Tognon C, Knezevich SR, Huntsman D, Roskelley CD, Melnyk N, Mathers JA, et al. Expression of the ETV6-NTRK3 gene fusion as a primary event in human secretory breast carcinoma. Cancer Cell 2002;2:367–76.

[123] Narula MK, Gupta N, Anand R, Kapoor S. Extraosseous Ewing's sarcoma/primitive neuroectodermal tumor of the sacral nerve plexus. Ind J Radiol Imag 2009;19:151–4.

[124] Rud NP, Reiman HM, Pritchard DJ, Frassica FJ, Smithson WA. Extraosseous Ewing's sarcoma. A study of 42 cases. Cancer 1989;64:1548–53.

[125] Ordonez JL, Osuna D, Herrero D, de Alava E, Madoz-Gurpide J. Advances in Ewing's sarcoma research: where are we now and what lies ahead? Cancer Res 2009;69:7140–50.

[126] Arvand A, Denny CT. Biology of EWS/ETS fusions in Ewing's family tumors. Oncogene 2001;20:5747–54.

[127] Erkizan HV, Uversky VN, Toretsky JA. Oncogenic partnerships: EWS-FLI1 protein interactions initiate key pathways of Ewing's sarcoma. Clin Cancer Res 2010;16:4077–83.

[128] Ouchida M, Ohno T, Fujimura Y, Rao VN, Reddy ES. Loss of tumorigenicity of Ewing's sarcoma cells expressing antisense RNA to EWS-fusion transcripts. Oncogene 1995;11:1049–54.

[129] Tanaka K, Iwakuma T, Harimaya K, Sato H, Iwamoto Y. EWS-Fli1 antisense oligodeoxynucleotide inhibits proliferation of human Ewing's sarcoma and primitive neuroectodermal tumor cells. J Clin Invest 1997;99:239–47.

[130] Braunreiter CL, Hancock JD, Coffin CM, Boucher KM, Lessnick SL. Expression of EWS-ETS fusions in NIH3T3 cells reveals significant differences to Ewing's sarcoma. Cell Cycle 2006;5:2753–9.

[131] Prieur A, Tirode F, Cohen P, Delattre O. EWS/FLI-1 silencing and gene profiling of Ewing cells reveal downstream oncogenic pathways and a crucial role for repression of insulin-like growth factor binding protein 3. Mol Cell Biol 2004;24:7275–83.

[132] Smith R, Owen LA, Trem DJ, Wong JS, Whangbo JS, Golub TR, et al. Expression profiling of EWS/FLI identifies NKX2.2 as a critical target gene in Ewing's sarcoma. Cancer Cell 2006;9:405–16.

[133] Dubois S, Grier HE, Lessnick SL. Ewing's sarcoma. In: Orkin SH, Fisher DE, Look AT, Lux SE, Ginsburg D, Nathan DG, editors. Oncology of infancy and childhood. Philadelphia: Saunders Elsevier; 2009. p. 829–69.

[134] Tural D, Molinas Mandel N, Dervisoglu S, Dervisoglu S, Oner Dincbas F, Koca S, et al. Extraskeletal Ewing's sarcoma family of tumors in adults: prognostic factors and clinical outcome. Jap J Clin Oncol 2012;42:420–6.

[135] Fletcher CKU, Krishnan K, Mertens F. Pathology and genetics of tumours of soft tissue and bone. Lyon, France: IARC Press; 2002.

[136] Fletcher CD. Pleomorphic malignant fibrous histiocytoma: fact or fiction? A critical reappraisal based on 159 tumors diagnosed as pleomorphic sarcoma. Am J Surg Pathol 1992;16:213–28.

[137] Matushansky I, Charytonowicz E, Mills J, Siddiqi S, Hricik T, Cordon-Cardo C. MFH classification: differentiating undifferentiated pleomorphic sarcoma in the 21st Century. Expert Rev Anticancer Ther 2009;9:1135–44.

[138] Scott EW, Simon MC, Anastasi J, Singh H. Requirement of transcription factor PU.1 in the development of multiple hematopoietic lineages. Science 1994;265:1573–7.

[139] Kearney MM, Soule EH, Ivins JC. Malifnant fibrous histiocytoma: a retrospective study of 167 cases. Cancer 1980;45:167–78.

[140] Tarkkanen M, Larramendy ML, Bohling T, Serra M, Hattinger CM, Kivioja A, et al. Malignant fibrous histiocytoma of bone: analysis of genomic imbalances by comparative genomic hybridisation and C-MYC expression by immunohistochemistry. Eur J Cancer 2006;42:1172–80.

[141] Ravindranath Y, Abella E, Krischer JP, Wiley J, Inoue S, Harris M, et al. Acute myeloid leukemia (AML) in Down's syndrome is highly responsive to chemotherapy: experience on Pediatric Oncology Group AML Study 8498. Blood 1992;80:2210–4.

[142] Gustafson P, Willen H, Baldetorp B, Ferno M, Akerman M, Rydholm A. Soft tissue leiomyosarcoma. A population-based epidemiologic and prognostic study of 48 patients, including cellular DNA content. Cancer 1992;70:114–9.

[143] Weiss S, Goldblum J. Enzinger and Weiss's soft tissue tumors. 4th ed. Philadelphia: Mosby-Harcort; 2001.

[144] Mankin HJ, Casas-Ganem J, Kim JI, Gebhardt MC, Hornicek FJ, Zeegen EN. Leiomyosarcoma of somatic soft tissues. Clin Orthopaed Relat Res 2004;421:225–31.

[145] El-Rifai W, Sarlomo-Rikala M, Knuutila S, Miettinen M. DNA copy number changes in development and progression in leiomyosarcomas of soft tissues. Am J Pathol 1998;153:985–90.

[146] Hu J, Khanna V, Jones M, Surti U. Genomic alterations in uterine leiomyosarcomas: potential markers for clinical diagnosis and prognosis. Genes Chromosomes Cancer 2001;31:117–24.

[147] Meza-Zepeda LA, Kresse SH, Barragan-Polania AH, Bjerkehagen B, Ohnstad HO, Namløs HM, et al. Array comparative genomic hybridization reveals distinct DNA copy number differences between gastrointestinal stromal tumors and leiomyosarcomas. Cancer Res 2006;66:8984–93.

[148] Ul-Hassan A, Sisley K, Hughes D, Hammond DW, Robinson MH, Reed MW. Common genetic changes in leiomyosarcoma and gastrointestinal stromal tumour: implication for ataxia telangiectasia mutated involvement. Internatl J Exp Pathol 2009;90:549–57.

[149] Berner JM, Forus A, Elkahloun A, Meltzer PS, Fodstad O, Myklebost O. Separate amplified regions encompassing CDK4 and MDM2 in human sarcomas. Genes Chromosomes Cancer 1996;17:254–9.

[150] Ragazzini P, Gamberi G, Pazzaglia L, Serra M, Magagnoli G, Ponticelli F, et al. Amplification of CDK4, MDM2, SAS and GLI genes in leiomyosarcoma, alveolar and embryonal rhabdomyosarcoma. Histol Histopathol 2004;19:401–11.

[151] Kransdorf MJ. Malignant soft-tissue tumors in a large referral population: distribution of diagnoses by age, sex, and location. Am J Roentgenol 1995;164:129–34.

[152] Lee MO, Kim EH, Jang HJ, Park MN, Woo HJ, Han JY, et al. Effects of a single nucleotide polymorphism in the chicken NK-lysin gene on antimicrobial activity and cytotoxicity of cancer cells. Proc Natl Acad Sci USA 2012;109:12087–92.

[153] Kindblom LG. Lipomatous tumors – how we have reached our present views, what controversies remain and why we still face diagnostic problems: a tribute to Dr Franz Enzinger. Adv Anat Pathol 2006;13:279–85.

[154] Ley TJ, Ding L, Walter MJ, McLellan MD, Lamprecht T, Larson DE, et al. DNMT3A mutations in acute myeloid leukemia. N Engl J Med 2010;363:2424–33.

[155] Singer S, Nielsen T, Antonescu C. Molecular biology of soft tissue sarcoma. In: DeVita V, Lawrence T, Rosenberg S, editors. Cancer: principles & practice of oncology. 9th ed. Philadelphia: Lippincott Williams & Wilkins; 2011.

[156] Enzinger F, Weiss S. Soft tissue tumors. 3rd ed. St Louis: Mosby-Year Book; 1995. p. 431–66.

[157] Jelinek JS, Kransdorf MJ, Shmookler BM, Aboulafia AJ, Malawer MM. Liposarcoma of the extremities: MR and CT findings in the histologic subtypes. Radiology 1993;186: 455–9.

[158] Rieker RJ, Weitz J, Lehner B, Egerer G, Mueller A, Kasper B, et al. Genomic profiling reveals subsets of dedifferentiated liposarcoma to follow separate molecular pathways. Virchows Arch 2010;456:277–85.

[159] Chibon F, Mariani O, Derre J, Malinge S, Coindre JM, Guillou L, et al. A subgroup of malignant fibrous histiocytomas is associated with genetic changes similar to those of well-differentiated liposarcomas. Cancer Genet Cytogenet 2002;139:24–9.

[160] Taylor BS, Barretina J, Socci ND, DeCarolis P, Ladanyi M, Meyerson M, et al. Functional copy-number alterations in cancer. PloS One 2008;3:e3179.

[161] Idbaih A, Coindre JM, Derre J, Mariani O, Terrier P, Ranchère D, et al. Myxoid malignant fibrous histiocytoma and pleomorphic liposarcoma share very similar genomic imbalances. Lab Invest 2005;85:176–81.

[162] Schmidt H, Bartel F, Kappler M, Würl P, Lange H, Bache M, et al. Gains of 13q are correlated with a poor prognosis in liposarcoma. Mod Pathol 2005;18:638–44.

[163] Sandberg AA. Updates on the cytogenetics and molecular genetics of bone and soft tissue tumors: lipoma. Cancer Genet Cytogenet 2004;150:93–115.

[164] Schwarzbach MH, Koesters R, Germann A, Mechtersheimer G, Geisbill J, Winkler S, et al. Comparable transforming capacities and differential gene expression patterns of variant FUS/CHOP fusion transcripts derived from soft tissue liposarcomas. Oncogene 2004;23:6798–805.

[165] Ratnagiri R, Garg V, Chaturvedi R. Extra osseous osteosarcoma of the retroperitoneum: an unusual entity. J Cancer Res Ther 2012;8:424–6.

[166] Lee JS, Fetsch JF, Wasdhal DA, Lee BP, Pritchard DJ, Nascimento AG. A review of 40 patients with extraskeletal osteosarcoma. Cancer 1995;76:2253–9.

[167] Wurlitzer F, Ayala L, Romsdahl M. Extraosseous osteogenic sarcoma. Arch Surg 1972;105:691–5.

[168] Patel SR, Benjamin RS. Primary extraskeletal osteosarcoma – experience with chemotherapy. J Natl Cancer Inst 1995;87: 1331–3.

[169] Wu JM, Montgomery E. Classification and pathology of childhood sarcomas. Surg Clin N Am 2008;88:483–520.

[170] Gorlick R, Anderson P, Andrulis I, Arndt C, Beardsley GP, Bernstein M, et al. Biology of childhood osteogenic sarcoma and potential targets for therapeutic development: meeting summary. Clin Cancer Res 2003;9:5442–53.

[171] Fuchs B, Pritchard DJ. Etiology of osteosarcoma. Clin Orthopaed Relat Res 2002;397:40–52.

[172] Nellissery MJ, Padalecki SS, Brkanac Z, Singer FR, Roodman GD, Unni KK, et al. Evidence for a novel osteosarcoma tumor-suppressor gene in the chromosome 18 region genetically linked with Paget disease of bone. Am J Hum Genet 1998;63:817–24.

[173] Wick MR, Siegal GP, Unni KK, McLeod RA, Greditzer III HG. Sarcomas of bone complicating osteitis deformans (Paget's disease): fifty years' experience. Am J Surg Pathol 1981;5: 47–59.

[174] Siddiqui R, Onel K, Facio F, Nafa K, Diaz LR, Kauff N, et al. The TP53 mutational spectrum and frequency of CHEK2*1100delC in Li-Fraumeni-like kindreds. Familial Cancer 2005;4:177–81.

[175] Overholtzer M, Rao PH, Favis R, Lu XY, Elowitz MB, Barany F, et al. The presence of p53 mutations in human osteosarcomas correlates with high levels of genomic instability. Proc Natl Acad Sci USA 2003;100:11547–52.

[176] Castresana JS, Rubio MP, Gomez L, Kreicbergs A, Zetterberg A, Barrios C. Detection of TP53 gene mutations in human sarcomas. Eur J Cancer 1995;31A:735–8.

[177] Gokgoz N, Wunder JS, Mousses S, Eskandarian S, Bell RS, Andrulis IL. Comparison of p53 mutations in patients with localized osteosarcoma and metastatic osteosarcoma. Cancer 2001;92 (8):2181–9.

[178] Tsuchiya T, Sekine K, Hinohara S, Namiki T, Nobori T, Kaneko Y. Analysis of the p16INK4, p14ARF, p15, TP53, and MDM2 genes and their prognostic implications in osteosarcoma and Ewing sarcoma. Cancer Genet Cytogenet 2000;120:91–8.

[179] Grana X, Garriga J, Mayol X. Role of the retinoblastoma protein family, pRB, p107 and p130 in the negative control of cell growth. Oncogene 1998;17:3365–83.

[180] Weinberg RA. The retinoblastoma protein and cell cycle control. Cell 1995;81:323–30.

[181] Yonemoto T, Tatezaki S, Ishii T, Satoh T, Kimura H, Iwai N. Prognosis of osteosarcoma with pulmonary metastases at initial presentation is not dismal. Clin Orthopaed Relat Res 1998;349: 194–9.

[182] Bramer JA, van Linge JH, Grimer RJ, Scholten RJ. Prognostic factors in localized extremity osteosarcoma: a systematic review. Eur J Surg Oncol 2009;35:1030–6.

[183] Hegyi M, Semsei AF, Jakab Z, Antal I, Kiss J, Szendroi M, et al. Good prognosis of localized osteosarcoma in young patients treated with limb-salvage surgery and chemotherapy. Pediatr Blood Cancer 2011;57:415–22.

[184] Bickmore W. Karyotype analysis and chromosome banding. Encyclopedia of life sciences, Vol. 10. Chichester: John Wiley & Sons; 2005. p. 555–61.

[185] Kaltenboeck B, Wang C. Advances in real-time PCR: application to clinical laboratory diagnostics. Adv Clin Chem 2005;40:219–59.

[186] Leslie JD, Ariza-McNaughton L, Bermange AL, McAdow R, Johnson SL, Lewis J. Endothelial signaling by the Notch ligand Delta-like 4 restricts angiogenesis. Development 2007;134:839–44.

[187] Merkelbach-Bruse S, Wardelmann E, Kunstlinger H, Buttner R, Schildhaus HU. Molecular methods in the diagnosis of sarcoma. Der Pathol 2011;32:24–31 [In German].

[188] Meldrum C, Doyle MA, Tothill RW. Next-generation sequencing for cancer diagnostics: a practical perspective. Clin Biochem Rev 2011;32:177–95.

[189] Quail MA, Smith M, Coupland P, Otto TD, Harris SR, Connor TR, et al. A tale of three next generation sequencing platforms: comparison of Ion Torrent, Pacific Biosciences and Illumina MiSeq sequencers. BMC Genomics 2012;13:341.

[190] Rothberg JM, Hinz W, Rearick TM, Schultz J, Mileski W, Davey M, et al. An integrated semiconductor device enabling non-optical genome sequencing. Nature 2011;475:348–52.

[191] Marques IJ, Weiss FU, Vlecken DH, Nitsche C, Bakkers J, Lagendijk AK, et al. Metastatic behaviour of primary human tumours in a zebrafish xenotransplantation model. BMC Cancer 2009;9:128.

[192] Bentley VL, Corkery D, Veinotte CJ, Berman JN, Dellaire G. Zebrafish XT: A pre-clinical animal tumour model for the discovery of novel therapies for T-cell acute lymphoblastic leukemia. Terry Fox Research Institute, 4th Annual Research Meeting, Ottawa, Ontario; 2013.

[193] Rikhof B, de Jong S, Suurmeijer AJ, Meijer C, van der Graaf WT. The insulin-like growth factor system and sarcomas. J Pathol 2009;217:469–82.

[194] Zha J, Lackner MR. Targeting the insulin-like growth factor receptor-1R pathway for cancer therapy. Clin Cancer Res 2010;16:2512–7.

[195] Trajkovic-Arsic M, Kalideris E, Siveke JT. The role of insulin and IGF system in pancreatic cancer. J Molec Endocrinol 2013;50:R67–74.

[196] Olmos D, Martins AS, Jones RL, Alam S, Scurr M, Judson IR. Targeting the insulin-like growth factor 1 receptor in Ewing's sarcoma: reality and expectations. Sarcoma 2011;2011:402508.

[197] Olmos D, Postel-Vinay S, Molife LR, Okuno SH, Schuetze SM, Paccagnella ML, et al. Safety, pharmacokinetics, and preliminary activity of the anti-IGF-1R antibody figitumumab (CP-751,871) in patients with sarcoma and Ewing's sarcoma: a phase 1 expansion cohort study. Lancet Oncol 2010;11:129–35.

[198] Patel S, Pappo A, Crowley J, Reinke D, Eid S, Ritland S, et al. A SARC global collaborative phase II trial of R1507, a recombinant human monoclonal antibody to the insulin-like growth factor-1 receptor (IGF1R) in patients with recurrent or refractory sarcomas. J Clin Oncol 2009;27:15s, (suppl; abstr 10503).

[199] Tolcher AW, Sarantopoulos J, Patnaik A, Papadopoulos K, Lin CC, Rodon J, et al. Phase I, pharmacokinetic, and pharmacodynamic study of AMG 479, a fully human monoclonal antibody to insulin-like growth factor receptor 1. J Clin Oncol 2009;27:5800–7.

[200] Kurzrock R, Patnaik A, Aisner J, Warren T, Leong S, Benjamin R, et al. A phase I study of weekly R1507, a human monoclonal antibody insulin-like growth factor-I receptor antagonist, in patients with advanced solid tumors. Clin Cancer Res 2010;16:2458–65.

[201] Strausberg RL, Feingold EA, Grouse LH, Derge JG, Klausner RD, Collins FS, et al. Generation and initial analysis of more than 15,000 full-length human and mouse cDNA sequences. Proc Natl Acad Sci USA 2002;99:16899–903.

[202] Martins AS, Olmos D, Missiaglia E, Shipley J. Targeting the insulin-like growth factor pathway in rhabdomyosarcomas: rationale and future perspectives. Sarcoma 2011;2011:209736.

[203] Shanware NP, Bray K, Abraham RT. The PI3K, metabolic, and autophagy networks: interactive partners in cellular health and disease. Ann Rev Pharmacol Toxicol 2013;53:89–106.

[204] Steelman LS, Chappell WH, Abrams SL, Kempf RC, Long J, Laidler P, et al. Roles of the Raf/MEK/ERK and PI3K/PTEN/Akt/mTOR pathways in controlling growth and sensitivity to therapy-implications for cancer and aging. Aging 2011;3:192−222.

[205] Janku F, Wheler JJ, Naing A, Falchook GS, Hong DS, Stepanek VM, et al. PIK3CA mutation H1047R is associated with response to PI3K/AKT/mTOR signaling pathway inhibitors in early-phase clinical trials. Cancer Res 2013;73:276−84.

[206] Valkov A, Kilvaer TK, Sorbye SW, Donnem T, Smeland E, Bremnes RM, et al. The prognostic impact of Akt isoforms, PI3K and PTEN related to female steroid hormone receptors in soft tissue sarcomas. J Translat Med 2011;9:200.

[207] Patel S. Exploring novel therapeutic targets in GIST: Focus on the PI3K/Akt/mTOR pathway. Curr Oncol Rep 2013.

[208] Chawla S, Sankhala K, Chua V, et al. A phase 1 study of AP23573 (an mTOR inhibitor) in patients (pts) with advanced sarcomas. Presented at ASCO 2005 Abstr 9068.

[209] Chawla SP, Tolcher AW, Staddon AP, Schuetze S, D'Amato GZ, Blay JY, et al. Survival results with AP23573, a novel mTOR inhibitor, in patients (pts) with advanced soft tissue or bone sarcomas: Update of phase II trial. J Clin Oncol (Meeting Abstracts) 2007;25(Suppl):10076.

[210] Wu X, Zou Y, He X, Yuan R, Chen Y, Lan N, et al. Tumor-infiltrating mast cells in colorectal cancer as a poor prognostic factor. Int J Surg Pathol 2013;21:111−20.

[211] Vemulapalli S, Mita A, Alvarado Y, Sankhala K, Mita M. The emerging role of mammalian target of rapamycin inhibitors in the treatment of sarcomas. Target Oncol 2011;6:29−39.

[212] Wan X, Helman LJ. The biology behind mTOR inhibition in sarcoma. Oncologist 2007;12:1007−18.

[213] Xie Y, Skytting B, Nilsson G, Brodin B, Larsson O. Expression of insulin-like growth factor-1 receptor in synovial sarcoma: association with an aggressive phenotype. Cancer Res 1999;59: 3588−91.

[214] Merimsky O, Bernstein-Molho R, Sagi-Eisenberg R. Targeting the mammalian target of rapamycin in myxoid chondrosarcoma. Anticancer Drugs 2008;19:1019−21.

[215] Richter S, Pink D, Hohenberger P, Schuette H, Casali PG, Pustowka A, et al. Multicenter, triple-arm, single-stage, phase II trial to determine the efficacy and safety of everolimus (RAD001) in patients with refractory bone or soft tissue sarcomas including GIST. J Clin Oncol 2010;28(Suppl. 15): Abstract 10038.

[216] Okuno S, Bailey H, Mahoney MR, Adkins D, Maples W, Fitch T, et al. A phase 2 study of temsirolimus (CCI-779) in patients with soft tissue sarcomas: a study of the Mayo phase 2 consortium (P2C). Cancer 2011;117:3468−75.

[217] Chawla SP, Staddon AP, Baker LH, Schuetze SM, Tolcher AW, D'Amato GZ, et al. Phase II study of the mammalian target of rapamycin inhibitor ridaforolimus in patients with advanced bone and soft tissue sarcomas. J Clin Oncol 2012;30:78−84.

[218] Riedel RF. Systemic therapy for advanced soft tissue sarcomas: highlighting novel therapies and treatment approaches. Cancer 2012;118:1474−85.

[219] Gupta S, Takebe N, Lorusso P. Targeting the hedgehog pathway in cancer. Ther Adv Med Oncol 2010;2:237−50.

[220] Anastas JN, Moon RT. WNT signaling pathways as therapeutic targets in cancer. Nat Rev Cancer 2013;13:11−26.

[221] Hu YY, Zheng MH, Zhang R, Liang YM, Han H. Notch signaling pathway and cancer metastasis. Adv Exp Med Biol 2012;727:186−98.

[222] Ecke I, Petry F, Rosenberger A, Tauber S, Mönkemeyer S, Hess I, et al. Antitumor effects of a combined 5-aza-2′deoxycytidine and valproic acid treatment on rhabdomyosarcoma and medulloblastoma in Ptch mutant mice. Cancer Res 2009;69:887−95.

[223] Roma J, Almazan-Moga A, Sanchez de Toledo J, Gallego S. Notch, wnt, and hedgehog pathways in rhabdomyosarcoma: from single pathways to an integrated network. Sarcoma 2012;2012: 695603.

[224] Zibat A, Missiaglia E, Rosenberger A, Pritchard-Jones K, Shipley J, Hahn H, et al. Activation of the hedgehog pathway confers a poor prognosis in embryonal and fusion gene-negative alveolar rhabdomyosarcoma. Oncogene 2010;29:6323−30.

[225] Rota R, Ciarapica R, Miele L, Locatelli F. Notch signaling in pediatric soft tissue sarcomas. BMC Med 2012;10:141.

[226] Yamanaka H, Oue T, Uehara S, Fukuzawa M. Hedgehog signal inhibitor forskolin suppresses cell proliferation and tumor growth of human rhabdomyosarcoma xenograft. J Pediatr Surg 2011;46: 320−5.

[227] Pressey JG, Anderson JR, Crossman DK, Lynch JC, Barr FG. Hedgehog pathway activity in pediatric embryonal rhabdomyosarcoma and undifferentiated sarcoma: a report from the Children's Oncology Group. Pediatr Blood Cancer 2011;57: 930−8.

[228] Masia A, Almazan-Moga A, Velasco P, Reventós J, Torán N, Sánchez de Toledo J, et al. Notch-mediated induction of N-cadherin and α9-integrin confers higher invasive phenotype on rhabdomyosarcoma cells. Br J Cancer 2012;107:1374−83.

[229] Bouron-Dal Soglio D, Rougemont AL, Absi R, Giroux LM, Sanchez R, Barrette S, et al. Beta-catenin mutation does not seem to have an effect on the tumorigenesis of pediatric rhabdomyosarcomas. Pediatr Devel Pathol 2009;12:371−3.

[230] Singh S, Vinson C, Gurley CM, Nolen GT, Beggs ML, Nagarajan R, et al. Impaired Wnt signaling in embryonal rhabdomyosarcoma cells from p53/c-fos double mutant mice. Am J Pathol 2010;177:2055−66.

[231] Clevers H, Nusse R. Wnt/β-catenin signaling and disease. Cell 2012;149:1192−205.

[232] Ulloa F, Itasaki N, Briscoe J. Inhibitory Gli3 activity negatively regulates Wnt/β-catenin signaling. Curr Biol 2007;17:545−50.

[233] Matsuda K, Shimada A, Yoshida N, Ogawa A, Watanabe A, Yajima S, et al. Spontaneous improvement of hematologic abnormalities in patients having juvenile myelomonocytic leukemia with specific RAS mutations. Blood 2007;109: 5477−80.

[234] Liang DC, Liu HC, Yang CP, Jaing TH, Hung IJ, Yeh TC, et al. Cooperating gene mutations in childhood acute myeloid leukemia with special reference on mutations of ASXL1, TET2, IDH1, IDH2, and DNMT3A. Blood 2013;121:2988−95.

[235] Lau L, Supko JG, Blaney S, Hershon L, Seibel N, Krailo M, et al. A phase I and pharmacokinetic study of ecteinascidin-743 (Yondelis) in children with refractory solid tumors. A Children's Oncology Group study. Clin Cancer Res 2005;11:672−7.

[236] Jain S, Xu R, Prieto VG, Lee P. Molecular classification of soft tissue sarcomas and its clinical applications. Internatl J Clin Exp Pathol 2010;3:416−28.

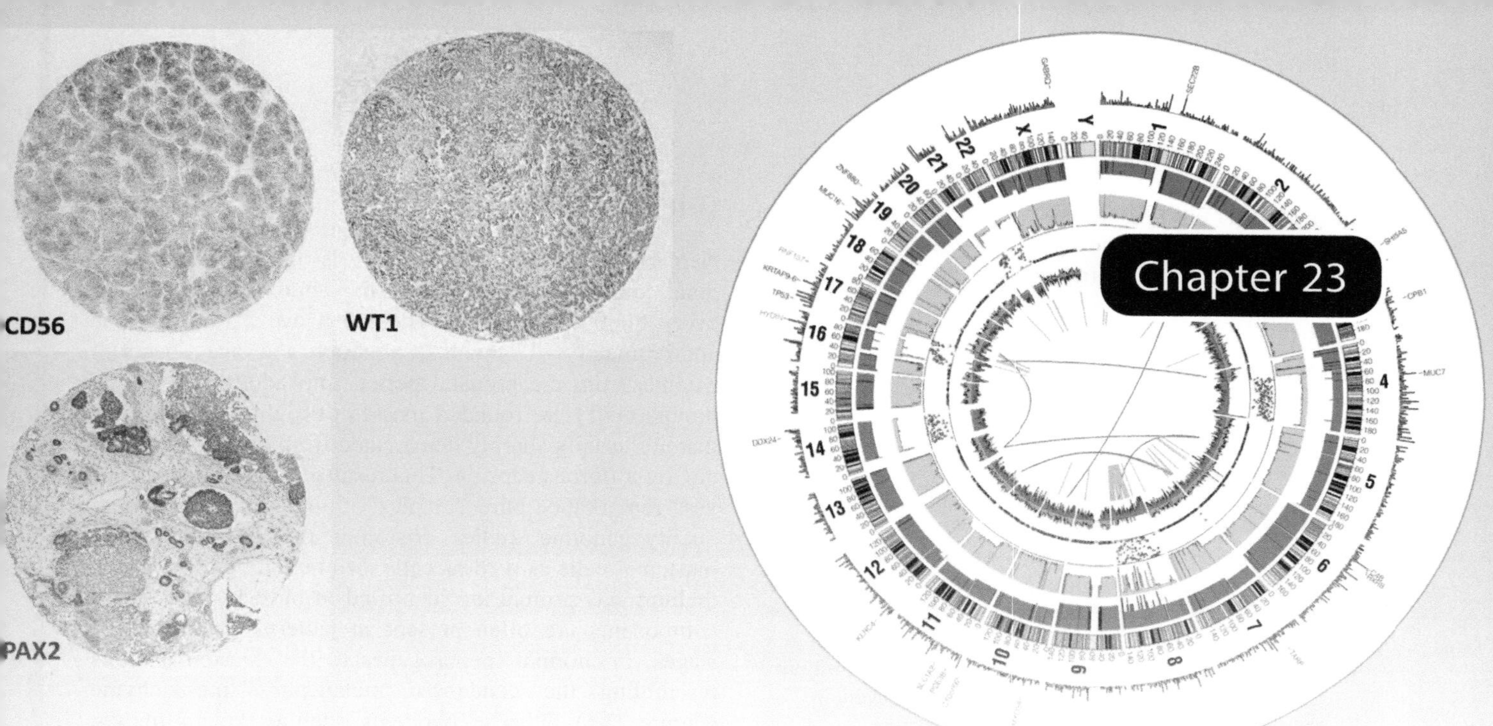

Cancer Genomics

Wilms Tumors

Elizabeth J. Perlman[1]
Peter Hohenstein[2]

[1]*Ann & Robert H. Lurie Children's Hospital of Chicago, Northwestern University Feinberg School of Medicine, Chicago, IL, USA*
[2]*The Roslin Institute, University of Edinburgh, Easter Bush Campus, Midlothian, UK*

Contents

Cancer Genomics. DOI: http://dx.doi.org/10.1016/B978-0-12-396967-5.00023-2

Key Concepts

- Wilms tumors are pediatric kidney tumors found in 1:8000 children with a mean age of 40 months
- Wilms tumors often develop within precursor lesions known as nephrogenic rests
- Disturbance of the mesenchymal-to-epithelial transition that drives nephron formation is likely a key event in the formation of most, if not all, Wilms tumors
- Genes known to be mutated in Wilms tumors include *WT1*, *CTNNB1* and *WTX*; all play critical roles during mesenchymal-to-epithelial transition
- Activation of *IGF2* through epigenetic mechanisms is often found
- Transcriptomic analysis of Wilms tumors confirms the different genetic backgrounds, classifies tumors into subsets and suggests an ontology for the origin of different tumor classes
- $NCAM^+$ $ALDH1^+$ cells have been identified as Wilms tumor cancer stem cells

WILMS TUMORS

Nephroblastoma (Wilms tumor) is the most common renal tumor in children, affecting approximately one in every 8000 children [1]. The mean age at diagnosis is approximately 40 months, although the age spectrum extends from the prenatal period into adulthood. Wilms tumors (WT) are rounded masses of highly variable size that are usually sharply demarcated from the adjacent kidney by a fibrous capsule. This results in the ability to harvest remarkably pure samples of tumor, enabling high quality genomic studies. Histologically, undifferentiated blastemal cells as well as cells differentiating toward epithelium and stromal are identified in most lesions. These components are often present in patterns that resemble stages in normal nephrogenesis, the blastemal cells resembling the condensed metanephric mesenchyme (Figure 23.1). Other components, such as skeletal muscle and cartilage, represent the full spectrum of the differentiating potential of these precursor cells and are often referred to as heterologous elements. Despite their seemingly endless histologic diversity, the majority of Wilms tumors are simply classified as "favorable" histology and are often highly responsive to chemotherapy. However, approximately 5% of Wilms tumors show nuclear anaplasia, defined as the presence of both multipolar, polyploid mitoses and marked nuclear enlargement with hyperchromasia (Figure 23.2) [2]. Anaplastic tumors are classified as unfavorable histology and are associated with adverse outcome largely due to decreased responsiveness to therapy.

Among pediatric embryonal neoplasms, a feature unique to Wilms tumors is the presence of precursor

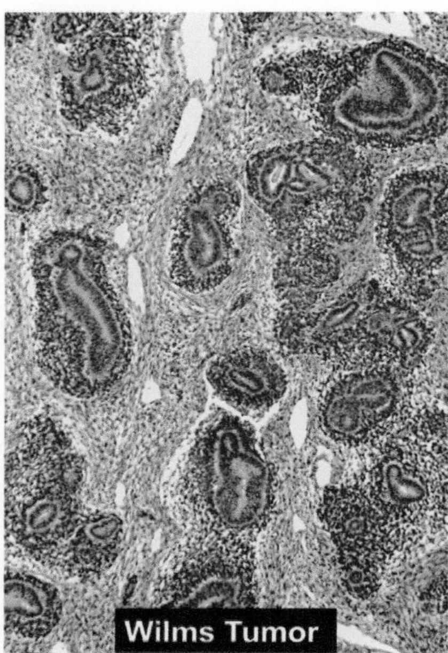

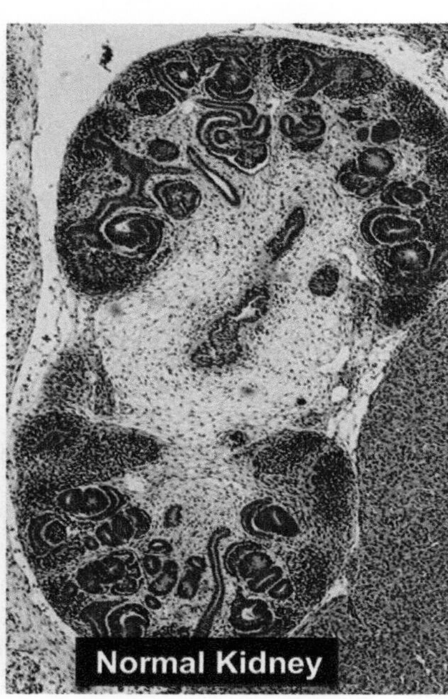

FIGURE 23.1 The histologic resemblance of Wilms tumor to the developing kidney can be striking, with the blastemal cells of Wilms tumor resembling the condensed metanephric mesenchyme.

lesions, known as nephrogenic rests, encountered in 25–40% of patients with nephroblastoma [3–5]. Nephrogenic rests may be subclassified as perilobar (PLNR) or intralobar (ILNR). PLNRs are sharply circumscribed, located at the periphery of the lobule, and contain undifferentiated blastemal cells or evidence of epithelial differentiation (Figure 23.3A). In contrast, ILNRs are typically located in the central areas of the lobule, are poorly circumscribed, and commonly contain stromal elements as well as epithelial tubules (Figure 23.3B). Both PLNRs and ILNRs have the ability either to regress and disappear or, alternatively, to undergo proliferation resulting in hyperplasia.

Hyperplastic nephrogenic rests may be remarkably difficult to distinguish from nephroblastoma by histologic features alone. Finally, clonal expansion within a nephrogenic rest may result in the development of nephroblastoma.

Since the realization by Max Wilms in 1899 that Wilms tumors show many hallmarks of disturbed embryonic renal development, it has been clear that Wilms tumorigenesis cannot be understood without thorough knowledge of normal kidney development. Conversely, better understanding of the events leading to the formation of Wilms tumors will provide valuable information about normal kidney development; the study of Wilms tumors and normal kidney development go hand-in-hand. Indeed, the embryonic kidney can be considered an all-encompassing model for the development of many organs. It involves many fundamental processes also found in the development of other organs and, on the molecular level, the signaling pathways that control proper development of the kidney are found to be essential in many other places in the developing embryo.

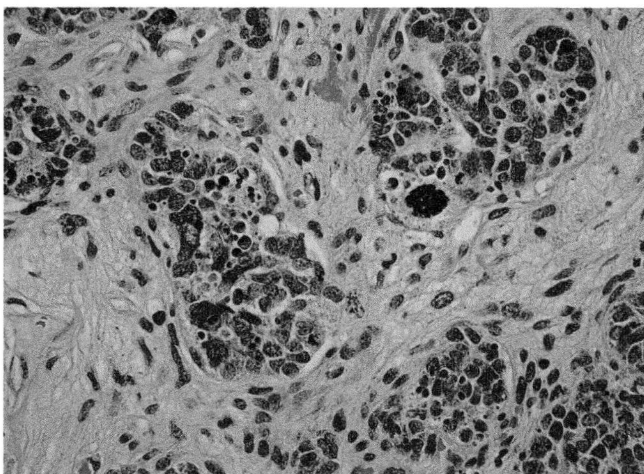

FIGURE 23.2 Approximately 5% of Wilms tumors show both multipolar, polyploid mitoses and marked nuclear enlargement with hyperchromasia and thereby meet the criteria for anaplasia.

Therefore, although excellent up-to-date overviews of kidney development are available (see for instance [6]), a description of the key events in this process here is essential.

RENAL DEVELOPMENT

The description of metanephric development usually starts with the reciprocal interaction of mesenchymal cells from the metanephric mesenchyme and the epithelial cells of the ureteric bud. However, from the point of view of Wilms tumors, the description should start at an even earlier stage, the intermediate mesoderm. This tissue, which provides most cell types found in the urogenital system [7], is situated between the paraxial mesoderm, which gives rise to the cartilage, skeletal and muscle lineages, and the lateral plate mesoderm, which will provide, for instance, the body wall and the circulatory system. Kidney development starts when cells in the dorsal part of the intermediate mesenchyme form an epithelial tube called the Wolffian duct, leaving the remaining mesenchymal cells to form the nephrogenic cord. Over the course of mammalian embryonic development this gives rise to three successive kidneys; the pronephros, the mesonephros and the metanephros and, whereas the first two form the functional postnatal kidney in other species, in mammals, these two structures will degenerate leaving only the metanephros, the most caudal one, to consider here.

The first event in metanephric development is the formation of an epithelial outgrowth of the Wolffian duct, termed the ureteric bud, which forms in response to a GDNF signal from the metanephric mesenchyme. Many mutations in human syndromes or mouse models that interfere with the number and/or position of ureteric buds lead to renal agenesis (when no outgrowth is formed), kidney duplications (multiple ureteric buds) or problems with downstream connection of the kidney outflow to the rest of the urinary tract (wrong position of the ureteric bud), indicating this is a central event in renal development. Once it has invaded into the metanephric mesenchyme, the ureteric bud will start to branch and, in response to signals from the branching bud, the mesenchyme will form a condensate of cells around the newly formed tips of the bud (Figure 23.4). By this stage, the mesenchymal compartment has given rise to the *Six2*-positive nephrogenic lineage [8] and the *Foxd1*-positive interstitial lineage [7]. *Six2*-positive cells from the cap mesenchyme subsequently form the pretubular aggregate and, after going through a mesenchymal-to-epithelial transition (MET), the renal vesicle. For this MET, the action of the Wnt4 signaling molecule is necessary [9] and sufficient [10]. At its distal end, the renal vesicle will connect to the ureteric bud. In subsequent

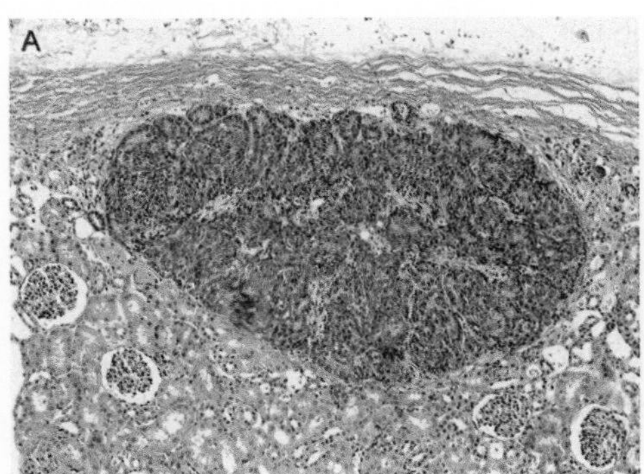

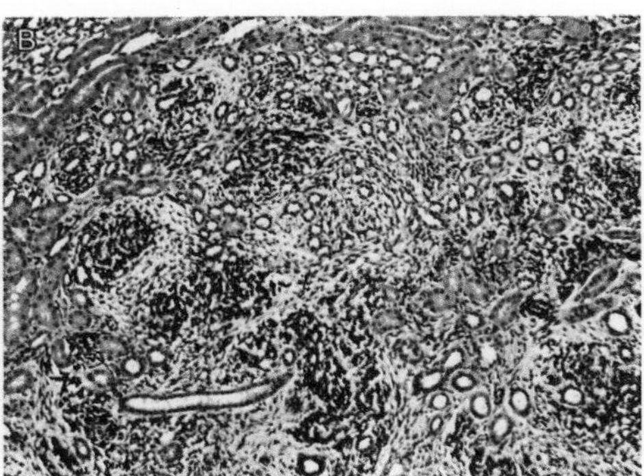

FIGURE 23.3 (A) Perilobar nephrogenic rests (left) are sharply circumscribed, located at the periphery of the lobule, and contain undifferentiated blastemal cells or evidence of epithelial differentiation. (B) Intralobar nephrogenic rests (right) are typically located in the central areas of the lobule, are poorly circumscribed, and commonly contain stromal elements as well as epithelial tubules.

differentiation steps, the renal vesicle will form proximal and distal tubules, as well as the anlage of the loop of Henle. The proximal end of the developing nephron will form the podocyte precursor cells and, eventually, the glomerulus that forms the filtering units of the functional nephron.

WILMS TUMOR DEVELOPMENT

Clues to the potential causes of disruption of normal renal development that may result in the development of Wilms tumor have historically centered on the recognition of syndromes associated with an increased incidence of Wilms tumor, which are identified in approximately 10% of patients with nephroblastoma. The most common syndromes involve genetic or epigenetic changes on the short arm of chromosome 11. Patients with WAGR syndrome (Wilms tumor, aniridia, genitourinary malformation, mental retardation) have a consistent germline deletion of chromosome 11p13. A similar somatic deletion was identified within many Wilms tumors and, in 1990, the gene on 11p13 was cloned and designated *WT1* [11,12]. *WT1* encodes a zinc finger transcription factor that plays a major role in renal and gonadal development [13]. Patients with Denys–Drash syndrome (a syndrome characterized by mesangial sclerosis, pseudohermaphroditism, and a 90% risk of Wilms tumor) have constitutional inactivating point mutations in *WT1* [14]. Wilms tumors in patients with WAGR and Denys–Drash syndromes often arise in ILNRs, and identical *WT1* mutations have been reported in both nephroblastomas and their associated ILNRs [15,16]. Frequently accompanying *WT1* mutation in Wilms tumors are genetic changes resulting in canonical Wnt activation, most commonly due to an activating mutation of β-catenin (*CTNNB1*) [17,18]. Inactivating mutations of *WTX*, a protein that contributes to β-catenin degradation, may also occur in 15–20% of patients with Wilms tumors, regardless of their *WT1* mutation status [19–22].

Another syndrome frequently associated with Wilms tumor development is the Beckwith–Wiedemann syndrome (BWS) which is characterized by hemihypertrophy, macroglossia, omphalocele, and visceromegaly and is linked to chromosome 11p15 [23–26]. The identification of loss of the maternal allele and duplication of the paternal 11p15 allele in sporadic nephroblastoma provided further support for the role of genomic imprinting in the pathogenesis of some Wilms tumors, and resulted in the description of a cluster of imprinted genes at this locus, including *IGF2*, *H19*, and *LIT1*, among others. In particular, *IGF2* loss of imprinting (LOI) occurs in 33–70% of nephroblastomas [27,28]. Patients with BWS often develop multiple PLNRs and *IGF2* loss of imprinting has been identified in perilobar nephrogenic rests. Although 11p15 clearly plays a critical role in the pathogenesis of WT, the observation of 11p15 LOH in normal tissue from some WT patients [29] and the lack of tumors arising in mutant mice with ICR1 LOI [30], the (non-coding) "imprinting center 1" that controls the imprinting of the locus, imply that biallelic expression of *IGF2* alone is insufficient for tumor development.

Familial nephroblastoma has rarely been identified in patients lacking abnormalities on chromosome 11. Additional chromosomal loci implicated in familial nephroblastoma include 17q12-21 (*FWT1*), and 19q13.3-13.4 (*FWT2*), although the genes responsible have not been identified [31,32]. It is therefore apparent that nephroblastoma can be initiated by multiple factors.

The above syndromes associated with Wilms tumors provide the best starting point for gaining a better understanding of Wilms tumors. They might lead us to identification of the cell of origin of the tumors and, together with knowledge of the normal behavior of these cells and the signals that control them, they will provide clues for the design of better treatments.

WT1

The identification of the *WT1* gene quickly led to a first description of its expression pattern in human [33] and mouse [34] embryonic development. Within the

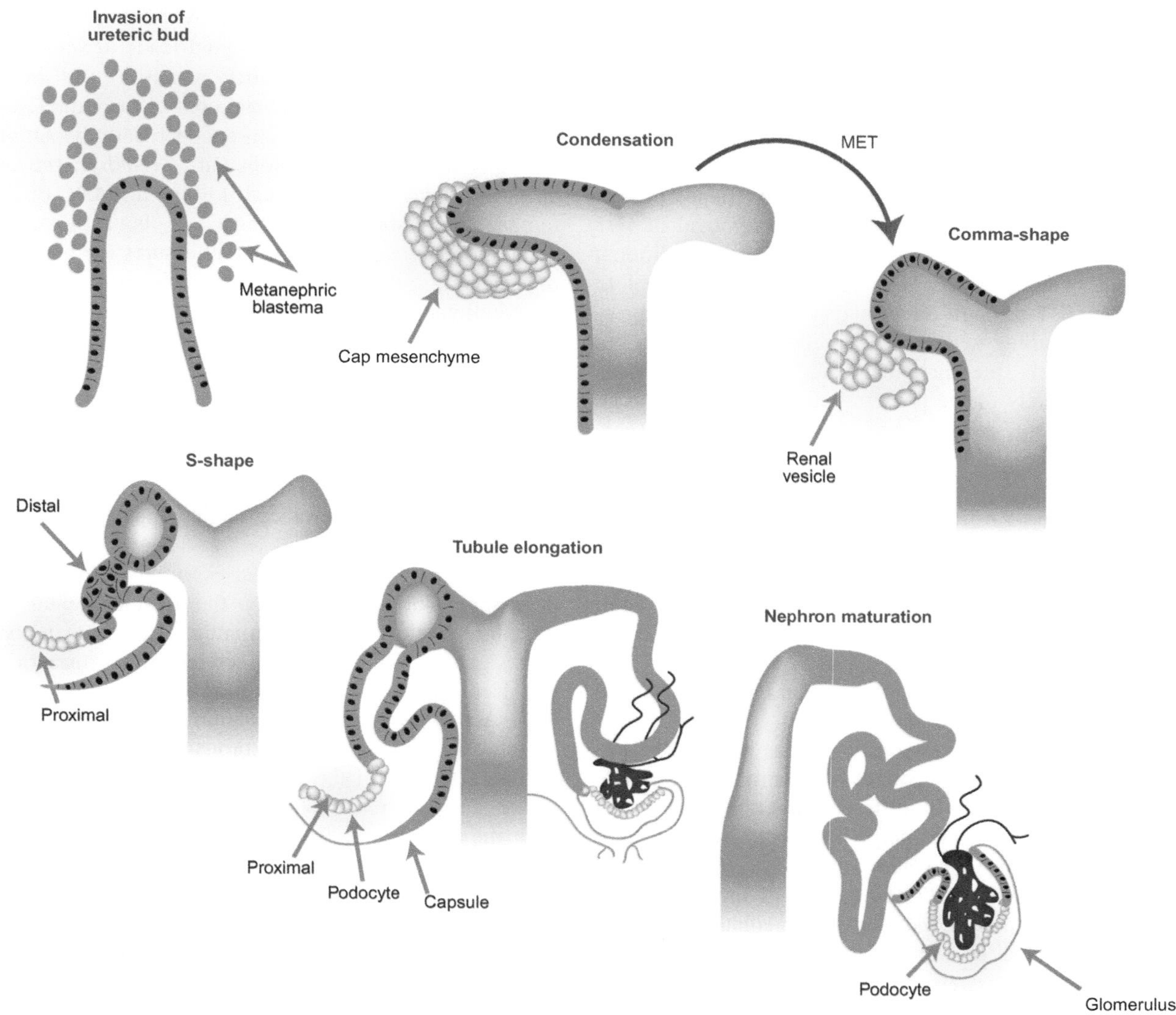

FIGURE 23.4 Schematic overview of nephron development.

developing kidney, the earliest signs of *Wt1* expression are found in the intermediate mesoderm. As the metanephric mesenchyme becomes induced by the invasion of the ureteric bud, *Wt1* expression begins to increase. A further rise in expression levels is seen in the cap-mesenchyme prior to the MET. After this step, the protein is found throughout the renal vesicle, but its expression rapidly becomes restricted to the proximal half of the nephron where it keeps increasing. Eventually, the podocytes are the only cells in the postnatal kidney that express high levels of *Wt1*. This expression pattern was confirmed using a reporter mouse model that expresses a lacZ reporter from a large human YAC construct [35] and is consistent with the hypothesis that disturbance of normal kidney development, in particular at the MET stage is the cause of Wilms tumors [36]. Outside the kidney, *Wt1* expression is found in a variety of tissues and cells, in particular cells that are undergoing MET or its opposite, epithelial to mesenchymal transition (EMT) [35], suggesting a role in the control of these processes. It is now believed that this function explains the role of *WT1* in development and diseases [37].

The analysis of different *Wt1*-mutant mouse models has confirmed the importance of the gene in many of the cells where it is expressed. Conventional *Wt1* knockout mice die at mid-gestation with phenotypes in different tissues [38]. The cause of death is likely disturbance of the epicardial EMT that is essential for normal heart development [38,39]. Kidneys, as well as gonads, are completely absent from *Wt1*-deficient embryos due to massive apoptosis of the intermediate mesoderm, showing *Wt1* has an important pro-survival role in these cells [38]. Additional phenotypes have been identified in the diaphragm, spleen, adrenal glands, retina and liver. The pro-survival role of *Wt1* in the intermediate mesoderm has long seemed contradictory with a role as tumor suppressor gene in Wilms tumors, but more recent work has shown that *Wt1* has different roles in later stages of kidney development and function. Through removal of *Wt1* after the intermediate mesoderm stage using conditional knockout mice, *Wt1* has been shown to be essential to drive the MET through direct activation of *Wnt4* expression [40], leading to a lack of nephron formation but expansion of the mesenchymal component instead. This links *Wt1* directly to the process believed to be disturbed in Wilms tumor development. In later stages, *Wt1* is important for proper development of the podocytes, as the *Wt1*$^{+/R394W}$ mouse model, carrying a Denys–Drash type mutation in the Zn-finger region of the gene, develops early onset glomerular sclerosis [41]. *Wt1* is also essential for the maintenance and function of podocytes. Despite *Wt1* being expressed in only a few cell types in adult mice, inducible body-wide loss of *Wt1* was found to cause rapid multiorgan failure; in the kidney within a few days the foot processes had disappeared and proteinuria and glomerular sclerosis were observed [42].

Combined, a picture starts to emerge where *Wt1* has essential roles at different stages of nephron development where it is expressed, with different target genes at different stages. It is likely that only *Wt1* loss at some of these developmental stages would result in Wilms tumor development. Losing *WT1* too early will likely result in an apoptotic response, like the intermediate mesoderm phenotype in the conventional *Wt1* knockouts, unless there would be an extremely rapid selection for additional mutations to counteract this cell death. Alternatively, when *WT1* is lost too late the cells could already have differentiated too far, and *WT1* loss would result in disturbance of the normal function of the cell but not in transformation, as is the case in the podocyte phenotypes in *Wt1* mouse models. The pre-MET loss of *Wt1* that leads to mesenchymal expansion instead of nephron formation [40] would be a good candidate for a first step in tumor development, but it remains to be seen if this is indeed the right stage and sufficient for Wilms tumor initiation.

Wnt Signaling

The second important genetic factor in Wilms tumor development is the Wnt signaling pathway. Wnts consist of a family of 19 secreted signaling molecules in mammals. They bind to heterodimers of the FZD and LRP families of proteins and can activate different intracellular signaling pathways. The best-studied of these is the β-catenin mediated, or canonical, Wnt signaling pathway [43]. In this pathway, a Wnt signal leads to stabilization of the β-catenin protein, after which it will bind to members of the TCF family and control the expression of target genes. The pathway can oncogenically be activated through stabilizing mutations in *CTNNB1*, the gene encoding β-catenin, or through loss of the *APC* or *AXIN* tumor suppressor genes, which are both part of the β-catenin destruction complex that assures the breakdown of β-catenin in the absence of a Wnt signal. As both types of mutations lead to stabilization of the protein, stabilization is generally considered evidence for activity of the pathway. In addition, several alternative, non-canonical Wnt signaling pathways have been described that act independently from β-catenin. These include the planar cell polarity (PCP) pathway and the Wnt-Ca^{2+}-NFAT pathway. Not much is known about these pathways in cancer, but their importance in development is increasingly being recognized.

Wnt signaling is one of the essential pathways in the developing kidney, including the developmental steps linked to Wilms tumor formation. The first important Wnt signal is Wnt9b that is secreted by the ureteric bud that is essential for the induction of nephron formation [44]. Acting through the β-catenin pathway, it is acting on the *Six2*-positive nephron progenitor cells in the mesenchyme to control the balance between cells remaining in the stem/progenitor state and cells forming nephrons; only the *Six2*-positive cells that respond to Wnt9b will go down the nephrogenic route. Accordingly, ectopic increase of β-catenin signaling through genetic or pharmacological means can induce nephron formation; it is, however, incompatible with a full MET [45,46]. This is the moment that normally Wnt4 comes into play [9] and it was long believed that this was also mediated by β-catenin. However, more recent data have suggested this is not the case, but identified a likely role for the Ca^{2+}-NFAT pathway downstream instead [47,48]. In later developmental stages, Wnt9b is again important in the collecting duct system where it controls the polarity of cell division and diameter of the developing tubules through the non-canonical planar cell polarity (PCP) signaling pathway [49]. Wnt7b is essential for the development of the medulla and the extension of the loop of Henle by signaling to the interstitium via the canonical pathway [50]. Finally, Wnt11 is expressed at the tips of the ureteric bud and is involved in control of branching [51]. Not much is known about Wnt signaling in other stages of nephron development.

The potential role of β-catenin mutations in Wilms tumors might be more complex than suggested at first. It is clear that there is a strong selection for β-catenin mutations in *WT1*-mutant Wilms tumors, frequencies of mutations in both genes are about equal (15–20%), and 121/154 Wilms tumors with β-catenin mutations were also found to carry *WT1* mutations [52]. Moreover, immunostaining for β-catenin is mainly found in the muscle compartment of Wilms tumors [53], which would also correspond to the *WT1*-mutant subset [54,55]. This suggests that the reason for the co-occurrence of these mutations might be found in the normal developmental roles of both genes. As discussed above, there are two developmental stages known where *Wt1* expression and Wnt signaling overlap, the mesenchymal progenitor cells where Wnt9b controls the decision to differentiate or remain in a stem cell fate, and the actual MET driven by the Wt1–Wnt4 cascade. However, as discussed, only the former is mediated by β-catenin signaling, pointing to the nephron progenitor cells as an important developmental stage in the origin of these tumors. This does not, however, do justice to the basic cancer genetics aspects of these tumors. Analysis of independent Wilms tumors in the same patient as well as individual nephrogenic rests in the same tumor clearly show that loss of *WT1* is the initiating, rate-limiting step in the development of these tumors, with activating mutations selected for at a later stage [56,57]. As β-catenin mutations are dominant [43], it is very well possible that the reason for activation of this pathway is independent from its endogenous function. In fact, analysis of five *WT1*/β-catenin mutant Wilms tumor cell lines has suggested a paraxial mesodermal instead of intermediate mesodermal origin of at least this subgroup of tumors [58]. *In vitro* culture effects cannot be excluded here, but this could explain the ectopic tissues developing in *WT1*/β-catenin mutant tumors (as these tissues have a paraxial mesodermal origin). Furthermore, *Wt1* was identified, at least in the Sertoli cells in the testis, as a negative regulator of β-catenin signaling [59]. If the same is true for the developing kidney, *Wt1* loss itself would be expected to increase β-catenin activity, making the benefit of selection of an additional mutation in β-catenin unclear. Additionally, one study described a set of 36 Wilms tumors in which 9 samples with β-catenin mutations showed strong nuclear β-catenin staining indicative of activation of the pathway, but of the remaining 27 β-catenin mutation negative samples, another 15 samples showed this same staining pattern [60], suggesting additional players in the canonical Wnt signaling pathway could be mutated in Wilms tumors. *WT1* status was not analyzed here, but the numbers analyzed would suggest these samples must have been largely *WT1* wild type. These examples make clear that the correlation between *WT1* loss and activation of the canonical Wnt signaling pathway might well be more complicated than initially thought, and our present understanding of the role of these two genes in normal kidney development is not sufficient to explain all observations.

The role of canonical Wnt signaling became further complicated by the identification of *WTX* (also known as *AMER1*). As discussed above, it was initially believed that *WTX* and *WT1*/*CTNNB1* mutations were mutually exclusive [20]. Soon after it was shown that *WTX* is a component of the β-catenin degradation complex [19], thereby providing a clear rationale for this finding. Later data showed, again, that the story was more complicated, as mutations in these three genes were shown not to be mutually exclusive after all, but that *CTNNB1* mutations in *WTX*-mutant tumors are usually found outside *CTNNB1* exon 3, which encodes the residues involved in stability control [21] and that the epithelial component of *WTX*-mutant tumors does not show signs of canonical Wnt signaling activity [61]. Germline mutations in *WTX* were identified in families with sclerosing skeletal dysplasia [62] without signs of increased predisposition to any form of cancer, including Wilms tumors. During development, *Wtx* is expressed in a variety of tissues, including regions of the brain, sensory organs, bone, lung, muscle and the kidney [63]. Within the developing kidney, *Wtx* expression is found partially to overlap with *Wt1*, in particular in the early pre-MET stages of kidney development. *Wtx* knockout mice die as neonates with phenotypes in different organs and somatic overgrowth suggestive of a role in mesenchymal progenitor cell fate specification with a clear involvement of aberrantly stabilized β-catenin [64]. Kidneys were either absent or overgrown, so difficult to interpret from a Wilms tumor point of view. In short, functional analysis of *WTX* function supports a role in the control of β-catenin stability but, although mutations in *WTX* can clearly be identified, it remains to be shown whether these are the actual initiating mutations.

IGF2

The last clear genetic link in Wilms tumor development is the activation, through LOI or LOH, of the imprinted 11p15 region, in which the *IGF2* gene is a strong candidate for an active role in Wilms tumor formation. Whereas the role of *WT1*, *CTNNB1* and *WTX* mutations discussed above might eventually help us to understand the ILNR-associated stromal predominant and favorable histology cases, the involvement of the 11p15 region has historically been linked to the PLNR-associated cases [5,65]. The higher frequency of PLNRs (25%) compared with ILNRs (15%) suggests that a better understanding of the role of 11p15 could have a much higher potential clinical impact [5]. As discussed above, LOI of *IGF2* is found in Beckwith–Wiedemann syndrome, an overgrowth syndrome affecting different organs that predisposes to different embryonal tumors, including Wilms tumors [26]. Expression of *IGF2* within the developing kidney is found in the metanephric and cap mesenchyme [66], but it may be more importantly the

IGF1R receptor which acts following IGF2 binding, that is also found in the cap mesenchyme. *WT1* was found to repress the expression of *IGF1R* [67], providing a potential model to explain both groups of Wilms tumors via deregulation of the same pathway. Despite this finding, so far it has been surprisingly difficult to provide a functional rationale for these observations. *Igf2* is essential for growth. Heterozygous *Igf2* knockout mice that inherit an inactivated paternal, but not maternal, allele weigh only 60% of their control littermates [68,69]. Vice versa, hyperactivation of *Igf2* in chimeric mice resulted in an overgrowth phenotype with many characteristics of BWS but no Wilms tumors [70]. Similarly, mice that inherit a maternal mutant allele of *H19*, which normally represses the paternal expression of *Igf2*, display an overgrowth phenotype without Wilms tumor predisposition.

An important factor in the lack of clarity in the functional analysis of the genes genetically linked to Wilms tumor development is the importance of secondary (and further) hits selected for by the tumor. Through their differential association with ILNRs and PLNRs, the *WT1* mutant and *IGF2*-activated Wilms tumors were believed to define separate groups [65]; however, other data have suggested that, in 44% of tumors with *WT1* mutations, paternal isodisomy can be found [71]. Accordingly, combining a conditional *Wt1* knockout model with maternal loss of *H19* resulted in a mouse model for Wilms tumors, whereas each allele separately did not [30]. In contrast, although a reduction of *CDKN1C* (the imprinted gene encoding $p57^{KIP2}$ also found at 11p15) has been found in Wilms tumors [72], combining mice with a maternal mutant copy of this gene and *Igf2* activation through loss of *H19*, resulted in kidney dysplasia but not Wilms tumors, in addition to other symptoms of BWS [73]. The combination of other Wilms tumor genes in mice may provide further insight into this issue. Activation of β-catenin through loss of the *Apc* tumor suppressor gene in the developing kidney leads to renal carcinomas rather than Wilms tumors [74]. It remains to be seen what the effect would be of combining, for instance, β-catenin activation with *Wt1* or *Wtx* loss in mice. By systematically combining different Wilms tumor alleles with each other in mice, it might be possible to generate a series of Wilms tumor models that would not only create preclinical models for different Wilms tumor types, but also provide the means for the functional analysis of these alleles in the developing kidney and therefore their role in the first steps of Wilms tumor development. A second, not mutually exclusive, factor to take into account is the developmental stage in which a mutation occurs. As shown for the different *Wt1* phenotypes in mice [38,40–42], this can have a serious impact on the resulting phenotype. This might very well also be the case for other Wilms tumor genes.

GENOMICS: BENCH TO BEDSIDE

Overview

Historically, the treatment of WT has largely been determined by histology, stage, age, and tumor size. Staging of nephroblastoma relies on the identification of penetration of the renal capsule or involvement of renal sinus (criteria for stage II), identification of positive surgical margins or regional lymph nodes (criteria for stage III), and identification of distant metastasis (criteria for stage IV). Bilateral nephroblastomas occur in approximately 5% of cases and are designated as stage V. Internationally, there are two overall approaches to the therapy of Wilms tumor which differ in the timing of chemotherapy relative to surgery; these approaches result in comparable outcomes, with the overall 4-year survival for patients with favorable histology nephroblastomas of all stages now approaching 90% [75,76]. The most significant unfavorable prognostic factors are age at detection, high stage, and unfavorable histology. Patients with diffuse anaplasia demonstrate a 4-year survival rate of 82% (stage II), 67% (stage III), and 33% (stage IV) [77]. Conversely, in recent years, a small group of patients has been classified as very low risk Wilms tumors (<24 months, stage I tumors <550 g), and these have an overall survival of 98% without adjuvant chemotherapy [78]. Despite these encouraging results, the fact remains that approximately 20% of patients with favorable histology Wilms tumor relapse, and half of these will not survive. There has been little change in this outcome over the last two decades, resulting in broad attempts internationally to utilize the growing armamentarium of genomic technology to identify biologic markers that could further stratify patients, or that could identify therapeutic targets.

DNA Copy Number Analysis

The association of particular chromosomal changes with relapse in patients with favorable histology Wilms tumor (FHWT) was first suggested by a number of publications reporting karyotopic changes, all of which show similar findings. In the largest study, 127/195 tumors had abnormal karyotypes [79]. The most common changes overall were gain of chromosomes 1q (28%), 8 (24%), and 12 (38%), and loss of 1p (13%), 11q (9%), 16q (19%), and 22 (10%). A stepwise Cox proportional hazards regression demonstrated the significant independent predictors of risk to be 1q gain (relative risk (RR) 3.4, $p = 0.005$), stage IV disease (RR 5.0, $p < 0.001$) and monosomy 22 (RR 5.9, $p < 0.001$). Gain of 1q often coincided with loss of 1p and 16q due to recurrent unbalanced chromosomal abnormalities, including translocations between chromosome 1 and a variety of translocation partners, the most frequent being chromosome 16, t(1q;16q). The other finding was the frequent presence of the isochromosome 1q, i(1q), which also resulted in increased copy numbers of 1q. More recently, comparative genomic hybridization (CGH) has been performed to identify additional genetic markers of relapse in FHWT, and to refine the minimal areas of loss on chromosomes 1p and 16q in order to allow for more specific testing and to explain the poor prognosis associated with LOH. Classical CGH of 67 FHWTs again revealed the most frequent alterations to be gain of 1q, 8, and 12, and loss of 1p, 11p, 16q, and 22 [80]. Of these, only gain of 1q was significantly associated with adverse outcome. Gain of 1q was observed in 27/46 relapsed FHWTs (59%) compared with 5/21 (24%) non-relapsed FHWTs (RR 2.5, $p = 0.019$). Most gains involved the entire long arm; however, six tumors demonstrated gain of smaller regions, with the smallest region of common gain spanning 1q21-25. Gain of 1q was also associated with advanced disease – 70% of tumors with 1q gain were stage III or IV whereas 43% of tumors lacking 1q gain were stage III or IV ($p = 0.05$). In eight cases (25%), 1p loss coexisted with 1q gain. In tumors with 1q gain that relapsed, 1p or 16q loss was also identified in 26 and 37%, respectively. These observations support a strong association between 1q gain and poor outcome, and suggest that unbalanced chromosomal abnormalities result in 1q gain. This was followed by a genome-wide BAC array study of 76 FHWTs in which gains of clones on 1q were found in 55% of relapsing and 20% of non-relapsing cases (RR 2.75) [81]. Similar to findings by classic cytogenetics and classical CGH, gain of 1q correlated with loss of 1p and/or 16q in some tumors. A smaller recurrent region of gain was narrowed to 1q22-q23 in a handful of patients.

In North America, the National Wilms Tumor Study Group (NWTSG), later included within the Children's Oncology Group (COG), has long focused on loss of heterozygosity as a marker for relapse using microsatellite analysis. The initial study surveyed LOH for 33 autosomal arms (all except 8p, 11q, 13p, 14p, 15p, and 18p) and reported 11p LOH in 40%, 1p LOH in 12%, and 16q LOH in 20% of tumors examined [82]. These discoveries led to the analysis of 250 children registered on NWTS4, which demonstrated LOH for markers on 1p and 16q in 12 and 17% of patients, respectively, and each was associated with a poorer relapse-free and overall survival when adjusted for stage and histology [83]. This association was then prospectively confirmed in 1727 informative FHWTs registered on NWTS5 [75]. Among the 970 patients with low-stage disease and LOH for both 1p and 16q, the RR for relapse was 2.9 and for overall survival was 4.25. Among the 686 patients with advanced disease (stages III, IV) and LOH for both 1p and 16q, the RR for relapse was 2.41 and for survival 2.66. Variables including histologic pattern and site of relapse did not correlate with LOH for 1p or 16q. On the basis of these studies, children registered on the current COG protocols are now stratified according to their combined 1p and 16q LOH status. Children with tumors containing both 1p and 16q LOH are treated more aggressively for each stage. However, 1p and 16q LOH is present in only 5% of FHWTs, and predicts only ~9% of relapses. It is important to note that the above measurements of LOH rely on the ability of polymorphic loci to distinguish the two alleles of a particular locus in normal samples, and documenting loss of one of the alleles in the tumor sample. While determining LOH has important strengths as a biomarker, it does not measure copy number, and copy-neutral LOH (two identical copies of the same chromosomal segment) is common in cancer.

In summary, multiple large studies based on cytogenetic, classic CGH, array CGH, single nucleotide polymorphism (SNP), and gene expression analysis all indicate a consistent frequency of 1p and 16q loss (and LOH), and LOH for 1p and 16q is now included in the stratification for therapy in Children's Oncology Group Studies. However, these studies also consistently show 1q gain in ~25% of FHWTs, and a strong association between 1q gain and relapse, as well as between 1q gain and 1p and 16q loss. The combination of high prevalence and high relative risk indicates 1q gain is a potentially strong biomarker for relapse. The immediate clinical impact of analysis of 1q gain only (assuming a conservative projected RR of at least 2.0 and a prevalence of 25%), is the accurate prediction of ~40% of relapses. Preliminary studies of patients registered on NWTS4 confirm the association between 1q gain and relapse [84]. Studies are ongoing to validate these findings in over 1000 NWTS5 patients, including stratification by stage and correlation with 1p and 16q LOH. If validated, 1q gain will be available for use in future protocols. Patients with 1q gain in their tumor could receive more aggressive therapy, similar to the strategy currently used with 1p and 16q LOH. Another potential long-term opportunity would be to *decrease* therapy for low-stage FHWTs that *lack* 1q gain. This would have a far greater impact by reducing toxicity to larger numbers of patients [85].

In addition to identifying genetic changes associated with the pathogenesis and outcome of Wilms tumors, recently a genome-wide association study of blood samples from over 1500 patients with Wilms tumors has reported 10 SNPs to be significantly associated with Wilms tumor development. The most significant of these were identified at 2p24 (rs3755132, $p = 1.03 \times 10^{-14}$; rs807624, $p = 1.32 \times 10^{-14}$) and 11q14 (rs790356, $p = 4.25 \times 10^{-15}$. Both regions contain genes with functions that may be related to Wilms tumorigenesis, including *DLG2* (discs, large homolog 2), involved in the planar cell polarity pathway, and *DDX1* (DEAD box helicase 1), involved in the initiation of translation and in RNA splicing and modification and in DNA double-strand break repair [86].

Transcriptomic Analysis

Several studies have performed gene expression analysis in Wilms tumors, correlating the differentially expressed genes with those involved in renal development. Li et al. analyzed a series of Wilms tumors and compared their gene expression with that of fetal kidney at 16–22 weeks' gestation. Following a rigorous exclusion process, 27 genes were noted to be highly expressed in WT over fetal kidney, including *SIX1*, *PAX2*, *EYA1*, *HOXA11*, *HOXA9*, *MEOX1*, *SALL2*, *PRAME*, *NNAT*, *CRAPB2*, *FZD7*, and *HMGA2*. As the majority of these are highly expressed in early renal development, the authors concluded that Wilms tumors are arrested at an early stage of fetal kidney development [87]. Other studies have likewise provided gene expression evidence connecting early renal development with Wilms tumor development, particularly within those tumors showing blastemal histology [88,89]. Huang et al. compared the expression of FHWT with other pediatric renal tumors (clear cell sarcoma of kidney, rhabdoid tumor, mesoblastic nephroma) and also demonstrated strong upregulation and differential expression of *EYA1*, *PAX2*, *GPR64*, and *WASF3* [90]. Additional studies compared groups of Wilms tumors that lacked *WT1* and/or *CTNNB1* mutations with those with these mutations [91,92]. These studies provided important patterns of gene expression associated with *WT1* and *CTNNB1* mutations in Wilms tumors.

The largest gene expression study of Wilms tumors comprehensively evaluated a case-cohort of 224 FHWTs registered on NWTS5 for global gene expression patterns, *WT1*, *CTNNB1*, *WTX* mutation status, and 11p15 copy number and methylation pattern [93]. Five subsets were identified and the gene expression pattern of each subset was then compared with published gene expression profiles during normal renal development [94,95] and with published gene expression profiles of Wilms tumors with *WT1* and *CTNNB1* mutations [91,92]. These subsets expand on those previously described [96].

Subset 1 tumors (S1) present in infancy and have an epithelial differentiated histology without evidence of associated nephrogenic rests. They have an excellent prognosis, with no relapses reported to date. Genes expressed in the renal epithelium following MET are enriched in S1 tumors, including high *WT1* expression, decreased *IGF2* expression, and decreased expression of Wnt targets. They lack *WT1*, *CTNNB1*, and *WTX* mutations, retain heterozygosity for 1p, and 16q, and retain the normal imprinting pattern at 11p15. Given the previous discussion, these findings are consistent with perturbation in the late post-induction metanephric mesenchyme (MM) which is characterized by high expression of *WT1*, low expression of *IGF2*, and repression of canonical Wnt signaling. Of note, additional epithelial predominant tumors occurred in the group of 224 tumors, but did not show this gene expression pattern, were older at diagnosis, presented at higher stages, demonstrated LOI at 11p15, and half relapsed.

Three subsets (subsets 2, 3, 4) are characterized by low *WT1* expression and an association with intralobar nephrogenic rests. They differ in their clinical and genetic features. *Subset 2 tumors* present at a young age, have an excellent prognosis, and show a high degree of skeletal muscle differentiation histologicially. Their gene expression pattern is characterized by enrichment of genes expressed in the intermediate mesoderm and early MM [94,95], including loss of *WT1* expression due to a *WT1* mutation in approximately 50% of tumors; canonical Wnt activation accompanied by a high *CTNNB1*-exon 3 mutation frequency; and divergent mesenchymal differentiation, a feature previously attributed to *WT1* loss and/or canonical Wnt activation [53–55]. Genes previously reported to be differentially expressed between Wilms tumors with and without a *WT1* mutation and in tumors with a *CTNNB1* mutation [91,92] were concordantly differentially expressed in S2 tumors, including striking upregulation of *WIF1* and downregulation of *HAS2* and *KLK6*. Noteworthy is the enrichment of NFAT and TGFB signaling in S2 tumors, supporting the hypothesis that *WT1* loss may lead to TGFB activation, resulting in NFAT induction which, in turn, mediates the switch of TGFB from growth inhibitor to growth promoter [97,98]. These findings suggest that S2 tumors arise within cells of the intermediate mesoderm or early MM following inactivation of *WT1*. Loss of *WT1* perturbs normal nephron development with preferential expansion of mesenchymal elements (which do not require *WT1*), resulting in an ILNR. Within the ILNR, an additional genetic event (such as *CTNNB1* or *WTX* mutation) results in Wnt activation accompanied by continued proliferation of primitive mesenchymal cells and increased diversion into alternate pathways, including muscle differentiation. The mixed histology observed in S2 tumors supports an origin before the delineation of nephron- or stroma-fated compartments occurs.

Subset 3 tumors show similarities with S2 tumors due to their common primary pathogenetic features (low *WT1* expression, high incidence of *WT1* mutation). However, they differ by their decreased evidence of canonical Wnt activation and decreased divergent mesenchymal differentiation (including skeletal muscle differentiation), accompanied by a lower incidence of *CTNNB1* mutation. S3 tumors present at an older median age and have a worse prognosis when compared to S2 tumors. S3 tumors lack enrichment of genes expressed in the intermediate mesoderm and instead show the same expression of MM genes seen in S5 tumors. These findings suggest that S3 tumors show a similar sequence of genetic events to S2, although these events occur in cells of the MM later in development, at a time when *WT1* expression is increasing and *IGF2* expression is decreasing. In addition, the different timing is associated with a different sensitivity to (or requirement for) Wnt activating mutations and therefore to differences in divergent mesenchymal differentiation.

S4 tumors have a gene expression pattern similar to S2, although they have a lower incidence of *WT1*, *CTNNB1*, and *WTX* mutations (and corresponding expression patterns) and a high incidence of biallelic methylation of *ICR1* (80%), similar to S5. Nonetheless, they are defined by their *WT1*-associated gene expression pattern and by enrichment of genes expressed in the intermediate mesoderm. These tumors may therefore have alternative genetic/epigenetic abnormalities up- or downstream to *WT1*. The small number of tumors within this subset suggests caution is needed with regard to subset 4. Intriguing clinical and genetic differences, particularly a high relapse rate and an older age at presentation, support the continued retention of this subset pending further knowledge. In summary, the S4 gene expression pattern supports an origin very early in renal development, with less evidence of Wnt activation and *WT1* loss.

The 11p15 methylation pattern in S2–4 tumors is complex and revealing. Approximately 50% of S2–4 tumors show evidence of copy-neutral 11p15 LOH, known as uniparental disomy (UPD). In cells that sustain a *WT1* mutation in one allele, UPD for chromosome 11 represents a common mechanism for losing the remaining normal *WT1* allele. UPD (LOH) at 11p15 will also occur in this situation as an epiphenomenon. In contrast, the mechanisms that cause 11p15 LOI are confined to 11p15 and can be independent of the status of 11p13 where the *WT1* gene is localized. S2, S3, and S4 tumors show an increasing prevalence of 11p15 LOI (4.5, 19 and 30%, respectively). LOI was identified in 6/16 S2–4 tumors lacking *CTNNB1* mutations, but in none of the 21 S2–4 tumors with *CTNNB1* exon 3 mutations. These observations suggest that 11p15-biallelic methylation (either LOH or LOI) may be a second genetic event in the pathogenesis of at least some S2–4 tumors and may functionally substitute for *CTNNB1* mutations with regard to tumorigenesis. These findings are in keeping with a previous study reporting the 11p15 copy number and methylation analysis, *WT1* mutation parameters, and the *CTNNB1* status of 36 WT with *WT1* alterations [71]. Interestingly, a mouse model was recently reported that required the introduction of *both* a *WT1* mutation and *IGF2* overexpression to produce murine WT [30].

Subset 5 is the largest subset and displays the full range of histologic patterns. They show the gene expression pattern of the MM previously reported, including expression of *SIX1*, *PAX2*, *EYA1*, and *SALL2*, among others, and over 80% demonstrate biallelic methylation of ICR1 at 11p15. This supports the previous assertions that approximately 70% of WT arise, in part, due to abnormal expression of normally imprinted 11p15 genes [27,28]. It also corroborates an association between an older age at presentation, histologic pattern, and 11p15 LOI in WT [65]. S5 is notable for being the only subset in which PLNRs were observed, although ILNRs were also seen and some tumors contained both features. Patients with BWS are known to develop both ILNRs and PLNRs [5]. This suggests that factors such as the cell type or the timing of the initial genetic event may determine the type of nephrogenic rest and the WT histology that results. In this model, S5 tumors arise within cells of the MM at a time when *IGF2* expression is diminishing. Biallelic methylation of *ICR1* during this developmental window results in increased *IGF2* expression and the development of a nephrogenic rest. If biallelic methylation of *ICR1* occurs prior to induction, the persistent elevation of *IGF2* causes preferential mesenchymal proliferation and prevents nephron development, resulting in an ILNR. If manifested after induction within early nephronic cells, persistent *IGF2* elevation prevents terminal epithelial differentiation, resulting in the development of a PLNR. This is supported by the presence of high *WT1* and *IGF2* expression in primitive epithelial structures of WT, whereas *IGF2* expression is low in normal terminally differentiated epithelial renal elements [66]. Within the resulting nephrogenic rest (be it ILNR or PLNR), a second genetic event likely occurs resulting in tumor development. The absence of tumor development in a mutant mouse strain with biallelic *Igf2* expression, but no other genetic alteration, would suggest that a second event is required [30], although the nature of this putative second event in S5 tumors is still unknown.

In summary, the transcriptomic patterns of Wilms tumor have resulted in the delineation of biologically unique subsets with distinctive mutational spectra and clinical outcomes. The key differentiating features for S1, S2, S3, and S5 are provided in Table 23.1. Not only does this provide insight into the pathogenesis of these subsets and an explanation for their heterogeneity but, more importantly, defining subsets driven by different genetic events may allow for both subset-specific and targeted therapeutic strategies. For example, targeting the IGF receptor may be broadly applicable to WT outside of subset 1, whereas the use of therapies involving the Wnt, TGFB, and NFAT pathways, when available, may have activity restricted to particular subsets.

The greatest immediate clinical significance of the transcriptomic analysis will be on the future therapy of

TABLE 23.1 Summary of Features Distinguishing Different Subsets of Favorable Histology Wilms Tumor

Subset	Histology	Skeletal Muscle (%)	11p15 Retention of Imprinting (%)	WT1 Expression	Median Age (months)	Relapse
S1	Epithelial	0	100	High	14	None
S2	Mixed	90	30	Low	13	Low
S3	Mixed	40	50	Low	39	Moderate
S5	Variable	15	20	High	43	Moderate

patients with very low risk Wilms tumor (VLRWT). Currently, patients <24 months of age with stage I FHWT weighing <550 g are defined as VLRWT and are treated with surgery alone. S1 appears to be responsible for approximately 30% of VLRWT who did not receive adjuvant chemotherapy [78,99,100]. S1 tumors also occurred in patients >24 months of age, with nephrectomy weights >550 g, yet who still retained an excellent survival. This may allow for the removal of the arbitrary age and tumor weight restrictions for S1 tumors, thereby expanding the number of patients able to be treated with nephrectomy alone. S2 tumors (likewise common in infants, and therefore in the VLRWT treatment arm) have an excellent outcome when they receive adjuvant chemotherapy. However, VLRWT showing evidence of 11p15 LOH and/or *WT1* mutation show an *increased* risk of relapse ($p < 0.004$) when they do not receive chemotherapy [99,100], suggesting they may benefit from therapy. Therefore, these studies provide the opportunity to define groups of VLRWT with different relapse risks using appropriate biologic markers that are now being validated.

Transcriptomic studies have also been performed in order to define gene expression signatures for relapse. Smaller studies involving fewer than 50 samples have identified potential signatures of relapse that did not overlap with one another [89,101]. In a larger set of 250 tumors registered in NWTS5, Huang et al. demonstrated that the expression pattern of 50 genes could predict relapse, but only in stage III tumors, and at a low sensitivity (47%) and specificity (70%) [102]. Again, neither the genes nor the implicated pathways in these studies overlapped.

WILMS TUMORS WITH ANAPLASIA

Surprisingly little is known about the genetic changes that characterize anaplastic tumors. Early studies demonstrated hyperdiploidy (DNA content of 1.7–3.2 in unfavorable histology WT (UHWT), compared with 1.0–1.4 in FHWT) [103]. Cytogenetic analysis of three anaplastic Wilms tumors demonstrated multiple abnormalities in each; however, all contained abnormalities involving 17p, suggesting the possibility of a *TP53* gene deletion [104]. This finding resulted in the examination of p53 mutation as a marker of anaplasia in 140 Wilms tumors. The mutation was identified in 8/11 anaplastic WT and in no FHWT [105]. However, subsequent studies have demonstrated discrepancies between analysis for p53 nuclear expression by immunohistochemistry and by formal mutation analysis. There have been no recent studies to clarify this important issue. More recently, LOH for 11q, 22q, and 16q has been shown to occur more frequently in anaplastic tumors and to correlate with tumor recurrence and death [106,107]. Array CGH and cDNA microarray expression profiling comparing anaplastic WT with those of favorable histology demonstrated, in addition to *TP53* deletion and 16q loss, recurrent loss and underexpression of 4q and 14q and focal gain of *MYCN*. The data presented are consistent with the overall hypothesis that anaplasia develops within Wilms tumor due to acquisition of additional genomic changes [108].

NEXT-GENERATION SEQUENCING

Currently, a number of studies utilizing whole-genomic or whole-exomic sequencing to provide further knowledge of Wilms tumor have been initiated. Probably the largest to date is a collaboration between the Children's Oncology Group Renal Tumor Committee and the National Cancer Institute Office of Cancer Genomics within the Therapeutically Applicable Research to Generate Effective Treatments (TARGET) initiative (http://ocg.cancer.gov/programs/target). This effort includes the performance of either whole-genome or exome sequencing, global copy number, gene expression, and methylation analysis on 125 high-risk Wilms tumors (those that relapse and those showing diffuse anaplasia). The outcome of these studies is expected within the next 2 years.

WILMS TUMOR CANCER STEM CELLS

An important factor in using functional understanding of kidney development to develop or improve therapies for

Wilms tumor is the cell of origin of the tumors. This concept is closely linked, though not identical, to the cancer stem cell. Increasingly, it has become more accepted that only a minor fraction of a tumor, the cancer stem cells, are responsible for its growth. It is these cells that need to be targeted for therapy, but on account of features of "stemness" they are much more resistant to radio- or chemotherapy. The cancer stem cell theory should provide an excellent model for the origins of embryonal tumors like Wilms tumor. However, to date, it has received surprisingly little attention [109]. Yet identifying the Wilms tumor cancer stem cell will be essential in order to place the genetic origins of tumors into their functional context. Expression, in particular, would be much more relevant when performed on these cells than on the bulk of the tumor or selected (microdissected) populations of cells.

Recently, a Wilms tumor cancer stem cell was identified as neural cell adhesion molecule 1 (NCAM)$^+$ ALDH1$^+$ cells [110]. It was suggested that in this signature NCAM is a marker for the origin of the cells, whereas ALDH1 activity could resemble tumorigenic transformation in them. NCAM expression is found in several stages of nephron development, and it is not yet clear which of these stages is important for the tumor development. More important at this stage, the authors showed that anti-NCAM antibodies conjugated to cytotoxic drugs were capable of efficiently eradicating Wilms tumors in a xenograft model. So although many questions remain open about the nature of Wilms tumor cancer stem cells [111], it is expected that NCAM$^+$ cells in the developing kidney will get much attention in the coming years as targets for therapy.

CONCLUSIONS

To date, the functional analysis of known Wilms tumor genes has revealed the enormous complexity of the origins of the disease. It is difficult to say if at present any functional clues are sufficiently supported to be of any guidance for therapy. Obvious therapeutic target candidates at the moment would be the canonical Wnt signaling pathway and the IGF2/IGFR1 pathway [112,113], but these are more based on Wilms tumor genetics than on functional analysis of the corresponding genes. If anything, the available functional data would suggest that these therapies might only work in certain, maybe not yet well-defined cases. Especially for favorable histology cases, caution in changing therapy based on limited functional data might be warranted. However, before functional analysis could provide useful directions for therapy, the following things would need to be known:

1 The genetic background of all Wilms tumor cases needs to be elucidated and coupled to histological and prognostic parameters. It is not just the initiating mutation that needs to be known, but also additional mutations, different initiating mutations selected for in the subsequent tumorigenic process. Knowing which combinations of mutations are found in particular subsets of tumors will help explain why, in the context of normal kidney development, these combinations are selected for by the tumor. This will discriminate between the real oncogenic events and normal developmental processes that the developing tumor hijacks for its own purpose.
2 More information on the Wilms tumor cancer stem cell is needed, especially in the context of the different genetic backgrounds.

Only with this knowledge will it be possible truly to use our understanding of normal kidney development and the role of Wilms tumor genes in it for the development of better therapies.

GLOSSARY

Aniridia Absence of the iris.

Blastema A mass of undifferentiated cells that has the capability to develop into an organ or an appendage.

Glomerular sclerosis Scarring of the glomeruli.

Hemihypertrophy A condition in which one side of the body or a part of one side is larger than the other.

Hyperchromasia Increased hematoxylin staining of nuclei due to increased DNA content.

Intermediate mesoderm A type of embryological tissue called mesoderm that is located between the paraxial mesoderm and the lateral plate.

Interstitium The tissue surrounding the loop of Henle in the renal medulla.

Isochromosome A chromosome that has lost one of its arms and replaced it with an exact copy of the other arm.

Macroglossia A large tongue.

Nephrogenic rests A fragment of embryonic tissue in the kidney retained after the period of embryonic development.

Omphalocele A type of abdominal wall defect in which the intestines, liver, and occasionally other organs remain outside of the abdomen in a sac because of a defect in the development of the muscles of the abdominal wall.

Paraxial mesoderm The area of mesoderm in the neurulating embryo that flanks and forms simultaneously with the neural tube.

Paternal isodisomy Two copies of a chromosome, or part of a chromosome, inherited from the father.

Sclerosing skeletal dysplasia A rare genetic disorder characterized by the creation of abnormally dense and overgrown bones.

Stroma Connective, supportive framework of a biological cell, tissue, or organ.

Uniparental disomy Two copies of a chromosome, or of part of a chromosome, from one parent and no copies from the other parent.

Ureteric bud A protrusion from the mesonephric duct during the development of the urinary and reproductive organs.

Visceromegaly Enlargement of the internal organs in the abdomen, such as liver, spleen, stomach, kidneys, or pancreas.

Wolffian duct The embryonic duct of the mesonephros which, in the male, becomes the vas deferens and, in both sexes, gives rise to the ureter.

ABBREVIATIONS

ACGH Array comparative genome hybridization
CGH Comparative genome hybridization
COG Children's Oncology Group
EMT Epithelial-to-mesenchymal transition
FHWT Favorable histology Wilms tumor
ICR1 Imprinting control region 1
ILNR Intralobular nephrogenic rests
LOH Loss of heterozygosity
LOI Loss of imprinting
MET Mesenchymal-to-epithelial transition
MM metanephric mesenchyme
NWTSG National Wilms Tumor Study Group
PLNR Perilobular nephrogenic rests
SNP Single nucleotide polymorphism
TARGET Therapeutically Applicable Research to Generate Effective Treatments
UHWT Unfavorable histology Wilms tumor
UPD Uniparental disomy
VLRWT Very low risk Wilms tumor
WAGR Wilms tumor, aniridia, genitourinary malformation, mental retardation
WT Wilms tumor
YAC Yeast artificial chromosome

REFERENCES

[1] Howlader NA, Krapcho M, Neyman N, Aminou R, Waldron W, et al. SEER Cancer Statistics Review, 1975–2008. 2011; Available from: <http://seer.cancer.gov/csr/1975_2008/>.

[2] Zuppan CW, Beckwith JB, Luckey DW. Anaplasia in unilateral Wilms' tumor: a report from the National Wilms' Tumor Study Pathology Center. Hum Pathol 1988;19:1199–209.

[3] Beckwith JB. Precursor lesions of Wilms tumor: clinical and biological implications. Med Pediatr Oncol 1993;21:158–68.

[4] Beckwith JB. Nephrogenic rests and the pathogenesis of Wilms tumor: developmental and clinical considerations. Am J Med Genet 1998;79:268–73.

[5] Beckwith JB, Kiviat NB, Bonadio JF. Nephrogenic rests, nephroblastomatosis, and the pathogenesis of Wilms' tumor. Pediatr Pathol 1990;10:1–36.

[6] Costantini F, Kopan R. Patterning a complex organ: branching morphogenesis and nephron segmentation in kidney development. Dev Cell 2010;18:698–712.

[7] Mugford JW, Sipila P, McMahon JA, McMahon AP. Osr1 expression demarcates a multi-potent population of intermediate mesoderm that undergoes progressive restriction to an Osr1-dependent nephron progenitor compartment within the mammalian kidney. Dev Biol 2008;324:88–98.

[8] Kobayashi A, Valerius MT, Mugford JW, Carroll TJ, Self M, Oliver G, et al. Six2 defines and regulates a multipotent self-renewing nephron progenitor population throughout mammalian kidney development. Cell Stem Cell 2008;3:169–81.

[9] Stark K, Vainio S, Vassileva G, McMahon AP. Epithelial transformation of metanephric mesenchyme in the developing kidney regulated by Wnt-4. Nature 1994;372:679–83.

[10] Kispert A, Vainio S, McMahon AP. Wnt-4 is a mesenchymal signal for epithelial transformation of metanephric mesenchyme in the developing kidney. Development 1998;125:4225–34.

[11] Call KM, Glaser T, Ito CY, Buckler AJ, Pelletier J, Haber DA, et al. Isolation and characterization of a zinc finger polypeptide gene at the human chromosome 11 Wilms' tumor locus. Cell 1990;60:509–20.

[12] Gessler M, Poustka A, Cavenee W, Neve RL, Orkin SH, Bruns GA. Homozygous deletion in Wilms tumours of a zinc-finger gene identified by chromosome jumping. Nature 1990;343:774–8.

[13] Haber DA, Englert C, Maheswaran S. Functional properties of WT1. Med Pediatr Oncol 1996;27:453–5.

[14] Pelletier J, Bruening W, Kashtan CE, Mauer SM, Manivel JC, Striegel JE, et al. Germline mutations in the Wilms' tumor suppressor gene are associated with abnormal urogenital development in Denys-Drash syndrome. Cell 1991;67:437–47.

[15] Charles AK, Brown KW, Berry PJ. Microdissecting the genetic events in nephrogenic rests and Wilms' tumor development. Am J Pathol 1998;153:991–1000.

[16] Park S, Bernard A, Bove KE, Sens DA, Hazen-Martin DJ, Garvin AJ, et al. Inactivation of WT1 in nephrogenic rests, genetic precursors to Wilms' tumour. Nat Genet 1993;5:363–7.

[17] Koesters R, Ridder R, Kopp-Schneider A, Betts D, Adams V, Niggli F, et al. Mutational activation of the β-catenin proto-oncogene is a common event in the development of Wilms' tumors. Cancer Res 1999;59:3880–2.

[18] Maiti S, Alam R, Amos CI, Huff V. Frequent association of β-catenin and WT1 mutations in Wilms tumors. Cancer Res 2000;60:6288–92.

[19] Major MB, Camp ND, Berndt JD, Yi X, Goldenberg SJ, Hubbert C, et al. Wilms tumor suppressor WTX negatively regulates WNT/β-catenin signaling. Science 2007;316:1043–6.

[20] Rivera MN, Kim WJ, Wells J, Driscoll DR, Brannigan BW, Han M, et al. An X chromosome gene, WTX, is commonly inactivated in Wilms tumor. Science 2007;315:642–5.

[21] Ruteshouser EC, Robinson SM, Huff V. Wilms tumor genetics: mutations in WT1, WTX, and CTNNB1 account for only about one-third of tumors. Genes Chromosomes Cancer 2008;47:461–70.

[22] Perotti D, Gamba B, Sardella M, Spreafico F, Terenziani M, Collini P, et al. Functional inactivation of the WTX gene is not a frequent event in Wilms' tumors. Oncogene 2008;27:4625–32.

[23] Feinberg AP. Multiple genetic abnormalities of 11p15 in Wilms tumor. Med Pediatr Oncol 1996;27:484–9.

[24] Ohlsson R, Nystrom A, Pfeifer-Ohlsson S, Töhönen V, Hedborg F, Schofield P, et al. IGF2 is parentally imprinted during human embryogenesis and in the Beckwith-Wiedemann syndrome. Nat Genet 1993;4:94–7.

[25] Ping AJ, Reeve AE, Law DJ, Young MR, Boehnke M, Feinberg AP. Genetic linkage of Beckwith-Wiedemann syndrome to 11p15. Am J Hum Genet 1989;44:720–3.

[26] Weksberg R, Shen DR, Fei YL, Song QL, Squire J. Disruption of insulin-like growth factor 2 imprinting in Beckwith-Wiedemann syndrome. Nat Genet 1993;5:143–50.

[27] Ogawa O, Eccles MR, Szeto J, McNoe LA, Yun K, Maw MA, et al. Relaxation of insulin-like growth factor II gene imprinting implicated in Wilms' tumour. Nature 1993;362:749–51.

[28] Steenman MJ, Rainier S, Dobry CJ, Grundy P, Horon IL, Feinberg AP. Loss of imprinting of IGF2 is linked to reduced expression and abnormal methylation of H19 in Wilms' tumour. Nat Genet 1994;7:433–9.

[29] Chao LY, Huff V, Tomlinson G, Riccardi VM, Strong LC, Saunders GF. Genetic mosaicism in normal tissues of Wilms' tumour patients. Nat Genet 1993;3:127–31.

[30] Hu Q, Gao F, Tian W, Ruteshouser EC, Wang Y, Lazar A, et al. Wt1 ablation and Igf2 upregulation in mice result in Wilms tumors with elevated ERK1/2 phosphorylation. J Clin Invest 2011;121:174–83.

[31] McDonald JM, Douglass EC, Fisher R, Geiser CF, Krill CE, Strong LC, et al. Linkage of familial Wilms' tumor predisposition to chromosome 19 and a two-locus model for the etiology of familial tumors. Cancer Res 1998;58:1387–90.

[32] Rahman N, Abidi F, Ford D, Arbour L, Rapley E, Tonin P, et al. Confirmation of FWT1 as a Wilms' tumour susceptibility gene and phenotypic characteristics of Wilms' tumour attributable to FWT1. Hum Genet 1998;103:547–56.

[33] Pritchard-Jones K, Fleming S, Davidson D, Bickmore W, Porteous D, Gosden C, et al. The candidate Wilms' tumour gene is involved in genitourinary development. Nature 1990;346:194–7.

[34] Armstrong JF, Pritchard-Jones K, Bickmore WA, Hastie ND, Bard JB. The expression of the Wilms' tumour gene, WT1, in the developing mammalian embryo. Mech Dev 1993;40:85–97.

[35] Moore AW, Schedl A, McInnes L, Doyle M, Hecksher-Sorensen J, Hastie ND. YAC transgenic analysis reveals Wilms' tumour 1 gene activity in the proliferating coelomic epithelium, developing diaphragm and limb. Mech Dev 1998;79:169–84.

[36] Hastie ND. The genetics of Wilms' tumor – a case of disrupted development. Annu Rev Genet 1994;28:523–58.

[37] Miller-Hodges E, Hohenstein P. WT1 in disease: shifting the epithelial-mesenchymal balance. J Pathol 2012;226:229–40.

[38] Kreidberg JA, Sariola H, Loring JM, Maeda M, Pelletier J, Housman D, et al. WT-1 is required for early kidney development. Cell 1993;74:679–91.

[39] Martinez-Estrada OM, Lettice LA, Essafi A, Guadix JA, Slight J, Velecela V, et al. Wt1 is required for cardiovascular progenitor cell formation through transcriptional control of Snail and E-cadherin. Nat Genet 2010;42:89–93.

[40] Essafi A, Webb A, Berry RL, Slight J, Burn SF, Spraggon L, et al. A Wt1-controlled chromatin switching mechanism underpins tissue-specific Wnt4 activation and repression. Dev Cell 2011;21:559–74.

[41] Gao F, Maiti S, Sun G, Ordonez NG, Udtha M, Deng JM, et al. The Wt1 + /R394W mouse displays glomerulosclerosis and early-onset renal failure characteristic of human Denys-Drash syndrome. Mol Cell Biol 2004;24:9899–910.

[42] Chau YY, Brownstein D, Mjoseng H, Lee WC, Buza-Vidas N, Nerlov C, et al. Acute multiple organ failure in adult mice deleted for the developmental regulator Wt1. PLoS Genet 2011;7: e1002404.

[43] Clevers H, Nusse R. Wnt/β-catenin signaling and disease. Cell 2012;149:1245–56.

[44] Carroll TJ, Park JS, Hayashi S, Majumdar A, McMahon AP. Wnt9b plays a central role in the regulation of mesenchymal to epithelial transitions underlying organogenesis of the mammalian urogenital system. Dev Cell 2005;9:283–92.

[45] Kuure S, Popsueva A, Jakobson M, Sainio K, Sariola H. Glycogen synthase kinase-3 inactivation and stabilization of β-catenin induce nephron differentiation in isolated mouse and rat kidney mesenchymes. J Am Soc Nephrol 2007;18:1130–9.

[46] Park JS, Valerius MT, McMahon AP. Wnt/β-catenin signaling regulates nephron induction during mouse kidney development. Development 2007;134:2533–9.

[47] Burn SF, Webb A, Berry RL, Davies JA, Ferrer-Vaquer A, Hadjantonakis AK, et al. Calcium/NFAT signalling promotes early nephrogenesis. Dev Biol 2011;352:288–98.

[48] Tanigawa S, Wang H, Yang Y, Sharma N, Tarasova N, Ajima R, et al. Wnt4 induces nephronic tubules in metanephric mesenchyme by a non-canonical mechanism. Dev Biol 2011;352:58–69.

[49] Karner CM, Chirumamilla R, Aoki S, Igarashi P, Wallingford JB, Carroll TJ. Wnt9b signaling regulates planar cell polarity and kidney tubule morphogenesis. Nat Genet 2009;41:793–9.

[50] Yu J, Carroll TJ, Rajagopal J, Kobayashi A, Ren Q, McMahon APA. Wnt7b-dependent pathway regulates the orientation of epithelial cell division and establishes the cortico-medullary axis of the mammalian kidney. Development 2009;136:161–71.

[51] Majumdar A, Vainio S, Kispert A, McMahon J, McMahon AP. Wnt11 and Ret/Gdnf pathways cooperate in regulating ureteric branching during metanephric kidney development. Development 2003;130:3175–85.

[52] Huff V. Wilms' tumours: about tumour suppressor genes, an oncogene and a chameleon gene. Nat Rev Cancer 2011;11:111–21.

[53] Fukuzawa R, Heathcott RW, Sano M, Morison IM, Yun K, Reeve AE. Myogenesis in Wilms' tumors is associated with mutations of the WT1 gene and activation of Bcl-2 and the Wnt signaling pathway. Pediatr Dev Pathol 2004;7:125–37.

[54] Miyagawa K, Kent J, Moore A, Charlieu JP, Little MH, Williamson KA, et al. Loss of WT1 function leads to ectopic myogenesis in Wilms' tumour. Nat Genet 1998;18:15–7.

[55] Schumacher V, Schuhen S, Sonner S, Weirich A, Leuschner I, Harms D, et al. Two molecular subgroups of Wilms' tumors with or without WT1 mutations. Clin Cancer Res 2003;9:2005–14.

[56] Fukuzawa R, Heathcott RW, More HE, Reeve AE. Sequential WT1 and CTNNB1 mutations and alterations of β-catenin localisation in intralobar nephrogenic rests and associated Wilms tumours: two case studies. J Clin Pathol 2007;60:1013–6.

[57] Uschkereit C, Perez N, de Torres C, Kuff M, Mora J, Royer-Pokora B. Different CTNNB1 mutations as molecular genetic proof for the independent origin of four Wilms tumours in a patient with a novel germ line WT1 mutation. J Med Genet 2007;44:393–6.

[58] Royer-Pokora B, Busch M, Beier M, Duhme C, de Torres C, Mora J, et al. Wilms tumor cells with WT1 mutations have characteristic features of mesenchymal stem cells and express molecular markers of paraxial mesoderm. Hum Mol Genet 2010;19:1651–68.

[59] Chang H, Gao F, Guillou F, Taketo MM, Huff V, Behringer RR. Wt1 negatively regulates β-catenin signaling during testis development. Development 2008;135:1875–85.

[60] Koesters R, Niggli F, von Knebel Doeberitz M, Stallmach T. Nuclear accumulation of β-catenin protein in Wilms' tumours. J Pathol 2003;199:68–76.

[61] Fukuzawa R, Anaka MR, Weeks RJ, Morison IM, Reeve AE. Canonical WNT signalling determines lineage specificity in Wilms tumour. Oncogene 2009;28:1063–75.

[62] Jenkins ZA, van Kogelenberg M, Morgan T, Jeffs A, Fukuzawa R, Pearl E, et al. Germline mutations in WTX cause a sclerosing skeletal dysplasia but do not predispose to tumorigenesis. Nat Genet 2009;41:95–100.

[63] Comai G, Boutet A, Neirijnck Y, Schedl A. Expression patterns of the Wtx/Amer gene family during mouse embryonic development. Dev Dyn 2010;239:1867–78.

[64] Moisan A, Rivera MN, Lotinun S, Akhavanfard S, Coffman EJ, Cook EB, et al. The WTX tumor suppressor regulates mesenchymal progenitor cell fate specification. Dev Cell 2011;20:583–96.

[65] Ravenel JD, Broman KW, Perlman EJ, Niemitz EL, Jayawardena TM, Bell DW, et al. Loss of imprinting of insulin-like growth factor-II (IGF2) gene in distinguishing specific biologic subtypes of Wilms tumor. J Natl Cancer Inst 2001;93:1698–703.

[66] Yun K, Fidler AE, Eccles MR, Reeve AE. Insulin-like growth factor II and WT1 transcript localization in human fetal kidney and Wilms' tumor. Cancer Res 1993;53:5166–71.

[67] Werner H, Re GG, Drummond IA, Sukhatme VP, Rauscher FJ, Sens DA, et al. Increased expression of the insulin-like growth factor I receptor gene, IGF1R, in Wilms tumor is correlated with modulation of IGF1R promoter activity by the WT1 Wilms tumor gene product. Proc Natl Acad Sci USA 1993;90:5828–32.

[68] DeChiara TM, Efstratiadis A, Robertson EJ. A growth-deficiency phenotype in heterozygous mice carrying an insulin-like growth factor II gene disrupted by targeting. Nature 1990;345:78–80.

[69] DeChiara TM, Robertson EJ, Efstratiadis A. Parental imprinting of the mouse insulin-like growth factor II gene. Cell 1991;64:849–59.

[70] Sun FL, Dean WL, Kelsey G, Allen ND, Reik W. Transactivation of Igf2 in a mouse model of Beckwith-Wiedemann syndrome. Nature 1997;389:809–15.

[71] Haruta M, Arai Y, Sugawara W, Watanabe N, Honda S, Ohshima J, et al. Duplication of paternal IGF2 or loss of maternal IGF2 imprinting occurs in half of Wilms tumors with various structural WT1 abnormalities. Genes Chromosomes Cancer 2008;47:712–27.

[72] Thompson JS, Reese KJ, DeBaun MR, Perlman EJ, Feinberg AP. Reduced expression of the cyclin-dependent kinase inhibitor gene p57KIP2 in Wilms' tumor. Cancer Res 1996;56:5723–7.

[73] Caspary T, Cleary MA, Perlman EJ, Zhang P, Elledge SJ, Tilghman SM. Oppositely imprinted genes p57(Kip2) and igf2 interact in a mouse model for Beckwith-Wiedemann syndrome. Genes Dev 1999;13:3115–24.

[74] Sansom OJ, Griffiths DF, Reed KR, Winton DJ, Clarke AR. Apc deficiency predisposes to renal carcinoma in the mouse. Oncogene 2005;24:8205–10.

[75] Grundy PE, Breslow NE, Li S, Perlman E, Beckwith JB, Ritchey ML, et al. Loss of heterozygosity for chromosomes 1p and 16q is an adverse prognostic factor in favorable-histology Wilms tumor: a report from the National Wilms Tumor Study Group. J Clin Oncol 2005;23:7312–21.

[76] Reinhard H, Semler O, Burger D, Bode U, Flentje M, Göbel U, et al. Results of the SIOP 93-01/GPOH trial and study for the treatment of patients with unilateral nonmetastatic Wilms Tumor. Klin Padiatr 2004;216:132–40.

[77] Dome JS, Cotton CA, Perlman EJ, Breslow NE, Kalapurakal JA, Ritchey ML, et al. Treatment of anaplastic histology Wilms' tumor: results from the fifth National Wilms' Tumor Study. J Clin Oncol 2006;24:2352–8.

[78] Green DM, Breslow NE, Beckwith JB, Ritchey ML, Shamberger RC, Haase GM, et al. Treatment with nephrectomy only for small, stage I/favorable histology Wilms' tumor: a report from the National Wilms' Tumor Study Group. J Clin Oncol 2001;19:3719–24.

[79] Bown N, Cotterill SJ, Roberts P, Griffiths M, Larkins S, Hibbert S, et al. Cytogenetic abnormalities and clinical outcome in Wilms tumor: a study by the U.K. cancer cytogenetics group and the U.K. Children's Cancer Study Group. Med Pediatr Oncol 2002;38:11–21.

[80] Hing S, Lu YJ, Summersgill B, King-Underwood L, Nicholson J, Grundy P, et al. Gain of 1q is associated with adverse outcome in favorable histology Wilms' tumors. Am J Pathol 2001;158:393–8.

[81] Natrajan R, Williams RD, Hing SN, Mackay A, Reis-Filho JS, Fenwick K, et al. Array CGH profiling of favourable histology Wilms tumours reveals novel gains and losses associated with relapse. J Pathol 2006;210:49–58.

[82] Maw MA, Grundy PE, Millow LJ, Eccles MR, Dunn RS, Smith PJ, et al. A third Wilms' tumor locus on chromosome 16q. Cancer Res 1992;52:3094–8.

[83] Grundy PE, Telzerow PE, Breslow N, Moksness J, Huff V, Paterson MC. Loss of heterozygosity for chromosomes 16q and 1p in Wilms' tumors predicts an adverse outcome. Cancer Res 1994;54:2331–3.

[84] Gratias EJ, Jennings LJ, Anderson JR, Dome JS, Grundy P, Perlman EJ. Gain of 1q is associated with inferior event-free and overall survival in favorable histology Wilms tumor: a report from the Children's Oncology Group. Cancer 2013. Available from: http://dx.doi.org/10.1002/cncr.28239.

[85] Dome JS, Fernandez CV, Mullen EA, Kalapurakal JA, Geller JI, Huff V, et al. Children's Oncology Group's 2013 blueprint for research: Renal tumors. Pediatr Blood Cancer 2013;60:994–1000.

[86] Turnbull C, Perdeaux ER, Pernet D, Naranjo A, Renwick A, Seal S, et al. A genome-wide association study identifies susceptibility loci for Wilms tumor. Nat Genet 2012;44:681–4.

[87] Li CM, Guo M, Borczuk A, Powell CA, Wei M, Thaker HM, et al. Gene expression in Wilms' tumor mimics the earliest committed stage in the metanephric mesenchymal-epithelial transition. Am J Pathol 2002;160:2181–90.

[88] Dekel B, Metsuyanim S, Schmidt-Ott KM, Fridman E, Jacob-Hirsch J, Simon A, et al. Multiple imprinted and stemness genes provide a link between normal and tumor progenitor cells of the developing human kidney. Cancer Res 2006;66:6040–9.

[89] Li W, Kessler P, Williams BR. Transcript profiling of Wilms tumors reveals connections to kidney morphogenesis and expression patterns associated with anaplasia. Oncogene 2005;24:457–68.

[90] Huang CC, Cutcliffe C, Coffin C, Sorensen PH, Beckwith JB, Perlman EJ. Classification of malignant pediatric renal tumors by gene expression. Pediatr Blood Cancer 2006;46:728–38.

[91] Li CM, Kim CE, Margolin AA, Guo M, Zhu J, Mason JM, et al. CTNNB1 mutations and overexpression of Wnt/β-catenin target genes in WT1-mutant Wilms' tumors. Am J Pathol 2004;165:1943–53.

[92] Zirn B, Samans B, Wittmann S, Pietsch T, Leuschner I, Graf N, et al. Target genes of the WNT/β-catenin pathway in Wilms tumors. Genes Chromosomes Cancer 2006;45:565–74.

[93] Gadd S, Huff V, Huang CC, Ruteshouser EC, Dome JS, Grundy PE, et al. Clinically relevant subsets identified by gene expression patterns support a revised ontogenic model of Wilms tumor: a Children's Oncology Group Study. Neoplasia 2012;14:742–56.

[94] Brunskill EW, Aronow BJ, Georgas K, Rumballe B, Valerius MT, Aronow J, et al. Atlas of gene expression in the developing kidney at microanatomic resolution. Dev Cell 2008;15:781–91.

[95] Lin SA, Kolle G, Grimmond SM, Zhou Q, Doust E, Little MH, et al. Subfractionation of differentiating human embryonic stem cell populations allows the isolation of a mesodermal population enriched for intermediate mesoderm and putative renal progenitors. Stem Cells Dev 2010;19:1637–48.

[96] Scott RH, Murray A, Baskcomb L, Turnbull C, Loveday C, Al-Saadi R, et al. Stratification of Wilms tumor by genetic and epigenetic analysis. Oncotarget 2012;3:327–35.

[97] Dey BR, Sukhatme VP, Roberts AB, Sporn MB, Rauscher III FJ, Kim SJ. Repression of the transforming growth factor-beta 1 gene by the Wilms' tumor suppressor WT1 gene product. Mol Endocrinol 1994;8:595–602.

[98] Singh G, Singh SK, Konig A, Reutlinger K, Nye MD, Adhikary T, et al. Sequential activation of NFAT and c-Myc transcription factors mediates the TGF-β switch from a suppressor to a promoter of cancer cell proliferation. J Biol Chem 2010;285:27241–50.

[99] Perlman EJ, Grundy PE, Anderson JR, Jennings LJ, Green DM, Dome JS, et al. WT1 mutation and 11P15 loss of heterozygosity predict relapse in very low-risk wilms tumors treated with surgery alone: a Children's Oncology Group Study. J Clin Oncol 2011;29:698–703.

[100] Sredni ST, Gadd S, Huang CC, Breslow N, Grundy P, Green DM, et al. Subsets of very low risk Wilms tumor show distinctive gene expression, histologic, and clinical features. Clin Cancer Res 2009;15:6800–9.

[101] Williams RD, Hing SN, Greer BT, Whiteford CC, Wei JS, Natrajan R, et al. Prognostic classification of relapsing favorable histology Wilms tumor using cDNA microarray expression profiling and support vector machines. Genes Chromosomes Cancer 2004;41:65–79.

[102] Huang CC, Gadd S, Breslow N, Cutcliffe C, Sredni ST, Helenowski IB, et al. Predicting relapse in favorable histology Wilms tumor using gene expression analysis: a report from the Renal Tumor Committee of the Children's Oncology Group. Clin Cancer Res 2009;15:1770–8.

[103] Douglass EC, Look AT, Webber B, Parham D, Wilimas JA, Green AA, et al. Hyperdiploidy and chromosomal rearrangements define the anaplastic variant of Wilms' tumor. J Clin Oncol 1986;4:975–81.

[104] Peres EM, Savasan S, Cushing B, Abella S, Mohamed AN. Chromosome analyses of 16 cases of Wilms tumor: different pattern in unfavorable histology. Cancer Genet Cytogenet 2004;148:66–70.

[105] Bardeesy N, Falkoff D, Petruzzi MJ, Nowak N, Zabel B, Adam M, et al. Anaplastic Wilms' tumour, a subtype displaying poor prognosis, harbours p53 gene mutations. Nat Genet 1994;7:91–7.

[106] Wittmann S, Zirn B, Alkassar M, Ambros P, Graf N, Gessler M. Loss of 11q and 16q in Wilms tumors is associated with anaplasia, tumor recurrence, and poor prognosis. Genes Chromosomes Cancer 2007;46:163–70.

[107] Klamt B, Schulze M, Thate C, Mares J, Goetz P, Kodet R, et al. Allele loss in Wilms tumors of chromosome arms 11q, 16q, and 22q correlate with clinicopathological parameters. Genes Chromosomes Cancer 1998;22:287–94.

[108] Williams RD, Al-Saadi R, Natrajan R, Mackay A, Chagtai T, Little S, et al. Molecular profiling reveals frequent gain of MYCN and anaplasia-specific loss of 4q and 14q in Wilms tumor. Genes Chromosomes Cancer 2011;50:982–95.

[109] Pode-Shakked N, Dekel B. Wilms tumor – a renal stem cell malignancy? Pediatr Nephrol 2011;26:1535–43.

[110] Pode-Shakked N, Shukrun R, Mark-Danieli M, Tsvetkov P, Bahar S, Pri-Chen S, et al. The isolation and characterization of renal cancer initiating cells from human Wilms' tumour xenografts unveils new therapeutic targets. EMBO Mol Med 2013;5:18–37.

[111] Hohenstein P. The stem and roots of Wilms' tumours. EMBO Mol Med 2013;5:4–6.

[112] Li R, Pourpak A, Morris SW. Inhibition of the insulin-like growth factor-1 receptor (IGF1R) tyrosine kinase as a novel cancer therapy approach. J Med Chem 2009;52:4981–5004.

[113] Polakis P. Drugging Wnt signalling in cancer. EMBO J 2012;31:2737–46.

Part 5

Ethics of Genomics Research

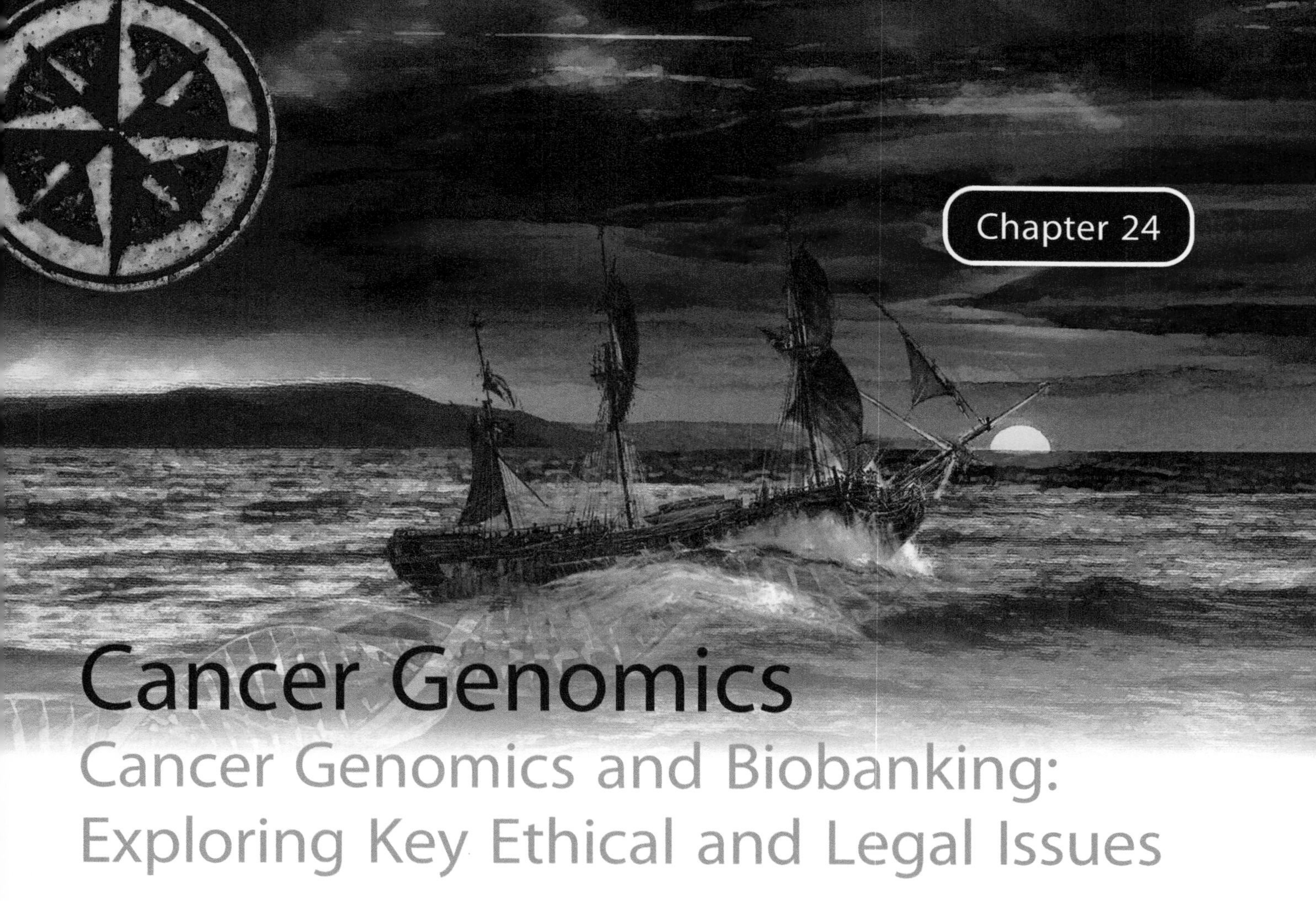

Cancer Genomics and Biobanking: Exploring Key Ethical and Legal Issues

Matthew Herder[1]
Conrad V. Fernandez[2]

[1]Health Law Institute, Faculties of Medicine and Law, Dalhousie University, Halifax, NS, Canada

[2]Departments of Pediatrics and Bioethics, Head, Division of Pediatric Hematology/Oncology, Dalhousie University and IWK Health Centre, Halifax, NS, Canada

Contents

Cancer Genomics. DOI: http://dx.doi.org/10.1016/B978-0-12-396967-5.00024-4

Key Concepts

- The offer to return a summary of research results to participants is increasingly recognized as a moral obligation of researchers
- There has been an explosion of capacity to identify, in the research context, potentially individually relevant results through genomic techniques
- The obligation of researchers to share individual genomic research results is complicated by challenges related to lack of:
  - consent to do so in previously biobanked samples
  - clarity as to if there is any obligation for researchers to share results
  - clarity as to the threshold at which a finding becomes clinically useful
  - clarity as to the duration of responsibility
  - clarity as to the obligations to pediatric participants, deceased participants' survivors or relatives
- The role of biobanks in supporting the offer and sharing of individual genomic research results is in the process of being defined
- The practice of patenting so-called "products of nature" has been the source of controversy and legal dispute for several decades
- The United States' Supreme Court will soon decide whether patents covering DNA sequences associated with breast and ovarian cancers are valid; however, the policy debate about the appropriate scope of patentable subject matter is likely to continue as new technologies and research findings from genomics emerge
- The commercialization of research involving human biological materials has precipitated several court disputes. While courts have not recognized providers of such materials as having property rights, they do have a right to be informed of any efforts to commercialize research
- Guidelines have been developed to ensure that research involving entire communities and/or vulnerable populations adheres to norms of benefit-sharing

INTRODUCTION

Large-scale cancer genomics research projects raise a range of ethical and legal issues. On a general level, many of these issues will arise in the context of any research project. Ethics and law, for example, require that research participants have an opportunity to make an informed choice (also referred to as informed consent) about whether to participate (or not) in any given research project. Reviews of informed consent and other hallmark ethical and legal issues, such as preserving the confidentiality of biomedical data, are thus readily available in the literature [1–7]. Our focus in this chapter will, instead, concern two sets of ethical and legal issues that deserve special attention in the context of cancer genomics and biobanking. Specifically, we will focus on (1) issues surrounding the return of research results to research participants, whether in summary or individualized form, and (2) issues connected to the commercialization of research, including the patenting of biomedical discoveries and biological materials. To make our discussion concrete, we will use two vignettes of a young woman, Olivia, diagnosed with leukemia.

VIGNETTE PART 1: OLIVIA'S LEUKEMIA

Olivia is a 20-year-old university student in a large Canadian city who is Ojibway-Cree in ancestry. She has a 1-month history of fatigue, nose bleeds and low grade fever. She is extremely ill at presentation with a white blood cell count of 175 (normal <12) and requires ICU admission. She is diagnosed with core-binding factor acute myeloid leukemia. Her parents fly in urgently from a northern Canadian reserve.

She is eligible for a cooperative group clinical trial comparing standard chemotherapy with standard chemotherapy and a novel histone deacetylase inhibitor. This trial has a correlative biology component and, in the process of consent for her diagnostic bone marrow, she is approached to have an additional 15 mL of bone marrow collected for biobanking. The 15-page consent form has several standard sections: (A) An explanation of the clinical trial including risks and benefits of the therapeutic agents to be used; (B) An offer to return a summary of the trial results once available; (C) A disclaimer that she has no rights to any commercial findings from her bone marrow; and (D) A claim that she will be told the results of any genetic testing that is completed on her bone marrow that may have clinical relevance to her treatment. The genetic testing discussion is reflected by one paragraph. Although profoundly ill, she consents to both the clinical trial and biology components with the endorsement of her parents. Significant family history includes breast cancer in a maternal aunt at age 42 years but no other cancers in young people. She is randomized to the experimental arm.

PART 1. THE ETHICS OF RETURNING RESEARCH RESULTS TO RESEARCH PARTICIPANTS

The Return of Research Results

Attention to inclusion of participants in the dissemination of research results has really only occurred in the last 10–15 years. Initial discussion focused on the return of a summary of results (aggregate results) and, in more recent years, a shift has occurred in response to the explosion of genomic research capacity and data. This component of the chapter will trace this evolution and situate Olivia's case in current practice and controversies.

Return of a Summary of Results

The provision of a summary of results has until recently primarily occurred in response to specific requests by participants, typically during a follow-up visit to clinic or chance discussion about outcomes. As researchers began to think about this area of ethics, it became apparent that the return of results should be considered from a number of viewpoints – that of the participant, of the researcher and of the research ethics board (REB[1]) and how an obligation to offer research results might be rooted in ethical principles. Historically, when asked, researchers did not offer results primarily and simply because they did not think of it [8,9], although some expressed concerns regarding the potential burden [10].

The offer of return of results has been said to be rooted in respect for the individual. Shalowitz states, "it would be disrespectful to treat research volunteers as conduits for generating scientific data without giving due consideration to their interests in receiving information about themselves derived from participation in research" [11].

This attention to respect thus creates a moral imperative to offer results (not force them on someone who does not wish them) in an accessible and uniform manner to all who participate in research. To do so only in response to participant requests, or to consider scientific publication to be sufficient, fails to respect a principle of equity and fairness to access irrespective of the capacity required to negotiate scientific literature or to request results from researchers who are in relative positions of authority compared to the participant.

We and others [12,13] have argued that the potential benefits outweigh the harms of offering and subsequently providing a summary of results of research but that a formal process should be in place to mitigate these harms. Potential benefits included the concrete, such as providing information that participants could use in terms of anticipating late effects, making choices about lifestyle or prophylaxis that might be beneficial or relieving anxiety about potential harms of a trial arm. Other benefits are less tangible but, theoretically, no less important, such as reducing mistrust in research, affirming the important role that participants play in their volunteering for research and in moving scientific knowledge forward, enhancing accuracy of disseminated information in the lay sphere and raising public awareness of the broad impacts of research [11].

While many agree that the potential benefits are persuasive, other raise concerns such as whether there is truly an ethical obligation inherent in the role of the researcher; distress by the participants in receiving results [14]; harm of misunderstanding or misapplying the results; costs to the research enterprise in diverting scarce research dollars to the process of returning results; and infeasibility in retaining and maintaining contact for the purpose of sharing results [15,16]. Despite these potential barriers, most researchers are supportive of the concept of returning research results, particularly if the potential negative consequences can be mitigated [17,18]. A significant concern raised by some researchers is the difficulty in maintaining contact with participants for the purposes of returning results. In one survey of parents of children with cancer, we found that parents strongly felt it was their personal responsibility to provide up-to-date contact information such as new addresses or e-mails [19]. We have thus incorporated a self-registration mechanism into a return of results website, placing the onus on the individual to retain contact rather than the researcher [20].

It is clear that there is a potential cost to the provision of results related to infrastructure and diversion of researcher efforts to satisfy this obligation [21]. However, an analogy to accepting these costs is that researchers accept the cost of obtaining and documenting consent, as it is rooted in a moral obligation to the participant. Similarly, the costs of offering and returning research results should be considered intrinsic to the appropriate investments of time and effort in any research study; these costs in turn need to be supported by funding agencies.

1 Note: The Research Ethics Board (REB) is an independent, institutionally based body charged with reviewing and approving the ethical conduct of research in Canada and is guided by the Tri-Council Policy version 2. It is the equivalent of similar committees in other jurisdictions that are also charged with ethics review and approval: in the USA – the Institutional Review Board (IRB); in many places in Europe – the independent ethics committee or research ethics committee. For the purpose of simplification, the term REB will be used throughout this chapter but the reader should be aware that the exact scope of responsibility and conduct of these committees vary based on national rules and regulations. All have as a primary aim the protection of human subjects in research.

A number of investigators have examined the attitudes of participants or their guardians with respect to a desire to receive a summary of research results across a variety of settings, although admittedly reported primarily from the cancer settings.

The majority of participants indicate a strong desire to be offered results even if the implications are negative for the individual [22] or demonstrate a poorer outcome for the arm of the trial in which they participated [23]. This strong desire for results is not universal [24] and underlines the importance of recognizing that the moral obligation is to offer results, not to provide them to unwilling participants. When asked, participants will typically favor more direct oral communication of results that have negative or difficult concepts, but will be satisfied with other less costly means of communication, such as newsletters or websites if positive or neutral in impact. In a number of recent studies, participants have shown a reasonably high degree of satisfaction with these indirect methods [25–27]. The Children's Oncology Group has recently established a web portal for the purpose of sharing the results of completed clinical trials and will examine the utilization and impact of this on participants [20].

As a reflection of this emerging demand by participants and recognition of an obligation on the part of researchers, the return of results is increasingly codified in ethics regulations. The Tri-Council Policy of Canada, v2, which regulates the ethical conduct of human research in Canada, mandates the offer of results to participants [28]. Such a requirement does not currently exist in US regulations but is the subject of the Office for Human Research Protections' (OHRP) call for consideration for amendments within the last year. International regulations also speak to the concept of offering a summary of results [29,30]. In the mid-first decade of this century, less than half of research ethics boards in Canada or the USA had a policy addressing the return of results [31,32]. It is expected that this number is by now significantly increased but this has not been formally reported recently. As regulations increasingly require a plan for the offer of a return of a summary of results, in coming years researchers should prospectively incorporate a process by which to do so.

Return of Individual Results

The offer of return of individual results from research stems from the concept that the researcher may discover clinically relevant information in the course of the research. Individual results can be broadly grouped into (1) results directly anticipated as a goal of the research, and (2) results that are in a sense predictable at some (usually low) level but are incidental to the main aims of the study. The former scenario of directly anticipated results would be that of a target measure in conducting a psychometric assessment of the participant such that one could learn their stage of language development [33]. A good example of the latter incidental type of encountering of individual results would occur in the context of functional diagnostic imaging where an incidental finding of a brain tumor might be identified, or in a blood sample taken for a hemoglobin level, where leukemia might be incidentally discovered.

The reflex reaction to such discoveries of individual results, either targeted or incidental, is to plan to offer these back to participants, such that they can act upon them. This makes sense in the usual context of research endeavors, wherein the number of participants and the number of findings that would meet this criterion have typically been manageable, e.g. one MRI in a patient showing a brain tumor [34]. What has been wholly unanticipated is the rapid progress and relative decreasing cost that has been made in genomic techniques, from sequencing to whole genome-wide screening [35]. As a result, researchers are challenged with an enormous wealth of data for each individual participant with weak or non-existent documentation of consent, and consequently many inherent challenges for which little if any regulatory guidance has existed (Table 24.1) [36].

Is There an Ethical Obligation for Researchers to Offer Individual Research Results?

The emerging consensus in the literature is that there is an obligation at some level to offer individual research results, whether the target of the research or incidental to it [37–39]. This obligation is rooted in an inherent respect for the participant; an avoidance of using them solely as an object to the successful conclusion of a study. Recommendations espouse the importance of weighting a net benefit versus the potential harms that may occur in

TABLE 24.1 Challenges in the offer to share individual genomic results

1. Is there an ethical obligation for researchers to offer research results?
2. At what point do the results reach adequate clinical utility to offer to participants?
3. What should researchers do if the genomic results are obtained without prior specific consent?
4. What should researchers do if the results are obtained in a child? Who should be informed?
5. What is the duration of a researcher's obligation to share results?
6. What are the obligations of a biobank from which the specimens are obtained?
7. What are researchers' obligations to next of kin of the participant?

sharing research findings. Of course, the net benefits may be viewed differently from the standpoint of the researcher and the participants and even among individual participants [40]. Despite much writing in this area, it remains unclear who the final arbiter of this equation should be.

The strength of the moral obligation to consider offering individual results to participants is also complicated by the many types of researcher–participant relationships that exist. To list a few that do not nearly capture the full breadth: a researcher may also be a clinician responsible for the ongoing long-term care of the participant (as is frequently the case in pediatric oncology); or she may be a PhD scientist without clinical credentials purely using the biological materials obtained from the participant in a clinic; or he may be a secondary researcher in a different country (with a different set of human research regulations) that has obtained a biobanked specimen of a long-deceased patient. The nature of this pre-existing relationship and its subsequent influence on obligations to participants is thus also very much brought into the equation of how strong an obligation falls to the researcher and, indeed, also how strongly the participant might feel with respect to a researcher fulfilling some duty to them [41].

In general, recommendations very much lean to an obligation if there is ongoing linkage to the participant, the probability of serious harm is high and if the result is actionable in some way [42–44]. Prior expressions of consent to receive or at least be offered such information are highly desirable but frequently not contemplated in long-standing biobanks. Even if consent is expressed, especially in biobanks where the samples may be held for many years, caveats exist. If the participant has said "yes" to the receipt of results, was the consent fully informed (or hidden among the chaos of the consent process for acute treatment as described in our patient scenario)? Were the vast array of possible findings and implications discussed? On the other hand, if the participant said an initial "no" to return of incidental findings, would they still say so now? What if the result is highly deleterious but also highly actionable. . .surely common sense would say most people should say yes. What should the researchers do if the recorded answer does not seem to fit with common sense?

Many would suggest that these questions be taken to the local research ethics board. The objection to this as a solution is that many REBs have no specific policy to address the return of individual results, nor do they necessarily have the clinical expertise to weigh the benefits versus the risks [45–47]. While careful attention to the ethical nuances that may arise may well benefit from experts in research ethics, the risk is that *ad hoc* decisions will not bring the field any closer to a more uniform approach to these challenging areas.

One should also make the distinction as to whether researchers have an obligation actively to look for incidental findings. In a study we conducted, most researchers felt an obligation to share results if found by chance, but limited obligations to go looking for these in a research context [40]. Gliwa and Berkman echo these sentiments [41]. They feel that all three of the following criteria should be met before such an obligation exists and currently argue that, in total, these have not been satisfied: a high proven benefit to the participant; lack of alternative access; and reasonable burden on the researcher. Others have approached this issue by either planned permanent anonymization of the data such that linkage is not possible or placing filters on what will be reported by the bioinformatics software, such that known areas are not examined. This latter strategy has been reported to be successfully employed in a neonatal clinical setting [48].

At What Point Do the Results Reach Adequate Clinical Utility to Offer to Participants?

The application of "adequate clinical utility" to the question of what is offered is complicated by a lack of definition of what meets the threshold for when a result should be offered [39,49]. (1) Does it need to reach a minimum severe clinical significance and by whose definition? (2) Does it need to be actionable and by whose definition? (3) Does the finding need to be an immediate threat to physical health? Or could it be a threat to a family member? Or a threat to mental health or reproductive planning? The Canadian Tri-Council Policy v2 article 3.4 states that "Researchers have an obligation to disclose to the participant any material incidental findings discovered in the course of research" and subsequently defines "material" as including significant implications for health-related, psychological or social well-being of the participant [28]. In the pediatric context, Abdul-Karim et al. [50] have recently proposed three criteria that should be evaluated in defining a known, urgent clinically significant finding including: (1) a clear and direct benefit that would be lost if delayed until the child was able to make their own decision about receiving the information (for example at the age of majority); (2) the known potential benefit outweighs the risk of psychological harm; and (3) genetic variants related to multifactorial disorders that also have an environmental component should be offered only if they indicate a substantial increase in risk over the background risk.

There are currently no recommendations that name specific conditions that meet the threshold of severity and clinical utility in a research context. Green et al. have recently authored a position statement on behalf of the American College of Medical Genetics and Genomics (ACMGG) for the sharing of genomic data ascertained in the clinical setting and define a list of conditions that would be required to be looked for and reported from a clinical laboratory setting [51]. All of these 56 conditions have a severe phenotype and are actionable and most have their onset in childhood. This list would be applied irrespective of the age of the person providing the sample. The recommendations anticipate annual updating based on new knowledge.

While it is tempting to adopt such an inventory for the research setting, many caution that samples collected in a research setting do not undergo the same rigorous high degree of quality assurance and clinical tracking that personal samples in the equivalent of a CLIA-certified laboratory would undergo [52]. Thus replication of the findings in a research laboratory in a clinical laboratory would be essential. At the very least, this proposed ACMGG list would provide some guidance to REBs, when consulted, in deciding if one should consider offering results.

What Should Researchers Do If the Genomic Results Are Obtained without Prior Specific Consent?

As noted earlier, many biological samples in existing biobanks were collected prior to the contemplation and capacity to do such in-depth and rapid genome-wide sequencing. As a result, existing samples may have no genetic-specific consent or have genetic testing allowed but described only in broad terms. Research has demonstrated that participants typically wish to receive results of genomic research, both for themselves and for their child [37,53,54]. When asked, they are frequently accepting of broad consent documents that do not ask for a menu selection of acceptable and unacceptable uses of their DNA [55].

Critics state that participants naively express a broad interest in receiving genomic results. They point to two concerns in cautioning researchers not to accept these statements at face value: first, some comment that participants do not really know what they are asking for and second, when in fact offered results, far fewer go on to accept these results than say they would in theory [56]. Empiric data to guide researchers and REBs are lacking in terms of truly understanding the uptake and impact of returning individual genomic results on participants.

Guidance might also be sought from expert committees with lay representation to determine if a finding merits offering [39]. These could be constructed with representation from ethics, genomics, statistics and counseling.

What Should Researchers Do If the Results Are Obtained in a Child? Who Should Be Informed?

In general, it has been recommended in children to defer genetic testing and sharing of clinical genetic information for adult onset disorders to allow them the autonomy of decision-making for themselves [57] – termed by some as an "open future" [58]. On the other hand, most would support the genetic testing of children if one potentially can identify a condition for which intervention is possible prior to the capacity to consent for themselves. These same principles would seem to apply to the research context, assuming adequate analytic and quality assurance of the samples under study.

A broader acceptance of testing in children may be applied if one considers indirect health benefits for other family members. For example, it makes perfect sense that a child is unlikely to deny the benefit to their own mother or an adult sibling of sharing a BRCA1 gene discovered in the child themselves, even though this is not relevant to childhood healthcare and necessarily discloses genetic information of the child. One could easily argue a direct benefit to the child in saving their mother's life through screening or preventive mastectomy that would fit into the psychosocial category of benefit for the child.

This broadening, of course, is to be constrained at some level by how serious and actionable the condition might be. Similarly, using the child's genetic information for reproductive planning might be limited by the possibility of obtaining personal testing as an adult for carrier status rather than imposing testing on the child.

We (and others) have found that most parents feel a strong right to the incidental and targeted finding related to their child, and that researchers support this if discovered [40]. To balance the right to an open future, some would argue that biobanks and researchers should have a mechanism in place to retain contact with childhood genomic research participants for the purpose of offering genomic results that are of purely adult interest [59]. This leads to a high degree of concern with respect to feasibility of retaining contact. Some have suggested that this could be mitigated by placing the responsibility of retaining and sharing results at the age of maturity upon parents and the child's physician.

In addition to responding to parental interests in receiving research results, Knoppers has described a reciprocal obligation that the parents carry [60]. If the finding is clearly detrimental in childhood and there is proven ameliorative therapy or prevention, then she would argue that the parent must consent to receive this genetic finding based on a parental legal responsibility to act in the best interests of the child.

What Is the Duration of a Researcher's Obligation to Share Results?

Linked to the devolution of responsibility of long-term contact retention for the purpose of researchers providing results to pediatric age participants is the precise duration of responsibility. In studying genomic researchers in the Canadian context, most indicated that they would see the responsibility ending with the operating funding for the research study [40]. This is reasonable from a resource and a contact standpoint. However, some felt strongly that this should be an indefinitely longer responsibility and we speculate that these are researchers that wear a dual hat of researcher and clinician (with ongoing fiduciary responsibilities to their patient). We would suggest therefore that this duration of responsibility be context-specific, and when it is in fact long, that a careful plan be established to track participants and to offer results with appropriate supports (such as genetic counselors).

What Are the Obligations of a Biobank from Which the Specimens Are Obtained?

Biobanks have four central roles including collection and banking of biological specimens, curation and annotation, analysis of samples and facilitation of research [61]. The biobanks may have been created prior to the unanticipated explosion in genomic studies and thus have limited or no consent documentation addressing return of findings. A recent study has demonstrated that only half of biobanks have specific policies addressing this issue [37]. Even when they do, there is considerable variability in what is to be returned, ranging from no results to non-genetic clinical findings (such high blood pressure at ascertainment) to a mix of non-genetic and genetic results, either incidental, targeted or both.

Nonetheless, consensus recommendations would suggest that biobanks have a key responsibility in facilitating the offer of individual results to participants [38,42,61,62]. Wolf et al. suggest four significant areas of responsibility: (1) clarifying the criteria of what is returnable (and thereby, developing a list of returnable results); (2) analyzing a specific finding to determine if it meets the criteria referred to above; (3) identifying the contributing individual or researcher; and (4) re-contacting the contributor to facilitate the sharing of the relevant result(s) [61]. Perhaps as important as defining what should be returned, biobanks play a role in determining what should not be returned – those findings that are unlikely to confer a net benefit. Questions crucial to this decision are: (1) is the result adequately established and linked to a significant phenotype? (2) is the result reliable and accurate? and (3) is the result actionable?

As the biobank itself will often not have a direct relationship with the participant, it is seen by many as a conduit to connecting results from secondary researchers to the clinicians who are best suited to share results with the contributing participant. This then is respectful of knowing the participant's individual context and needs. A high threshold for return of results has been suggested given that biobanks often do not have full consent to return genomic research findings, research conducted under its auspices may be many years from the original sample donation, and there is a lower likelihood of benefit as time elapses from that donation [63,64].

In the absence of biobank policy and, in particular, policy that is applicable across biobanks, a number of recommendations suggest that a partial solution to the current haphazardly applied criteria for returning incidental or individual findings is the creation of a central committee [39,61]. This committee would have the authority and expertise to set guidelines that encompass consent, re-contact, and thresholds for return of results. This is particularly important in biobanks that are pre-existing and in which samples do not have attached explicit consent for return of individual results. Such a committee could also specify a consistent assignment of responsibility to the biobank, secondary researchers and to primary contributors, such that each knows what actions are expected of them. An example is the requirement of secondary researchers to delineate clinically significant findings and propose a plan to report these as a condition of access to the biobank as operative in the Framingham Heart Study [65] and the recommendations from the Canadian Partnership for Tomorrow (Bartha Knoppers, personal communication).

What Are Researchers' Obligations to Next of Kin of the Participant?

Familial communication of clinical genetic results has no specific standard and ranges from no disclosure without the consent of the index patient to allowing exceptions to confidentiality based on risk to the genetic relative [66–68]. It is therefore understandable that disclosure to relatives of discoveries in a research context is even less well defined. In general, the assumption is that if the result does not meet a threshold to share with the index participant than it likely does not meet a threshold for sharing with a relative. One can postulate, however, about exceptions to this rule of thumb; for example, an incidental genomic result found in a person's last few months of anticipated life secondary to smoking-related lung cancer; the patient may have little need for knowing about a breast cancer risk gene but this may be very important for their relatives. This also brings up the issue of whether there is any obligation to communicate genomic results after the death of the index participant [69,70]. This is unresolved and the subject of ongoing research.

Canadian and Australian regulations allow under some circumstances the sharing of research genomic results with relatives without consent, but a number of jurisdictions do not in respecting privacy laws [71]. If one accepts the premise that under some circumstances it may be acceptable to share genomic results with relatives, then a number of questions arise that are unresolved: (1) should the same threshold be utilized as might be applied for the index participant (or are the needs and interests of a relative different)? (2) should the results be shared by the index participant directly, by the researcher with a relative or by some intermediary (such as a genetic counselor) or a combination of these? (3) what are the process/logistic challenges for disclosure – how does one identify all the relevant relatives, who is responsible for their tracking and consent, what is the best method of communication of results, what are the financial ramifications and responsibilities of the researcher, the participant and/or the relative? and (4) if the results are to be shared by the index participant, how do we best prepare them for the task? There is evidence to demonstrate that the sharing of clinically relevant genetic information by the index participant (as "gatekeeper") is not ideal. There is an emotional overlay that clearly may cloud the communication process, participants often feel ill-suited to the task, and recipients of the information may misinterpret what is said [67,72,73]. Although the specific genetic information may be relevant to many kin, sharing is often described as selective in who is chosen and how it is communicated [74]. Alternative mechanisms such as direct communication by the researcher or surrogate, community groups and genetic counselors have all been described to be successful in various ways [74,75].

In summary, definitive obligation standards to relatives are likely to flow from the resolution of guidance and mechanisms that are found to be appropriate and effective in index participants. It would appear most appropriate to resolve if, when and how we communicate as researchers with index participants before branching out to relatives in an organized fashion.

VIGNETTE PART 2: 10 YEARS AFTER OLIVIA'S DONATION

Unfortunately, Olivia's leukemia is refractory to induction chemotherapy and she dies of progressive leukocytosis 4 months after diagnosis. Her parents return to the reserve for local burial respecting traditional customs. A bereavement call occurs from the hematology service 2 months and a year after her death, but no further follow-up beyond that.

Genome-wide testing of her original bone marrow sample in a research lab reveals a novel multidrug resistance (MDR) gene that is highly expressed; in the process a BCR1 mutation is also identified – disease causing but incidental to her leukemia. Over the course of the next 8 years, a small molecule inhibitor is identified to this novel MDR gene, confirmed to be biologically active in myeloid leukemia, and the rights to this drug are purchased by a Virginia biopharmaceuticals company for $200 million dollars. The family is never informed of any of these findings.

PART 2. BIOBANKS AND THE COMMERCIALIZATION OF RESEARCH: LEGAL AND ETHICAL ISSUES

People make tremendous contributions to scientific research such as cancer genomics. Those investments are made in three principal ways. First, the bulk of scientific research carried out in university laboratories is publicly funded. As taxpayers, citizens and residents therefore invest heavily in genomics and other fields. Second, for a variety of reasons, many people become directly involved in scientific research as research participants. By virtue of participating, these people contribute to the generation of knowledge. Third, people contribute to research by providing biological materials such as DNA, cells, hormones, gametes, embryos, tissues, and organs – derived from their own bodies – for research purposes. People often contribute in all of these ways. Olivia, for example, was both a participant in a clinical trial and provided a sample of her bone marrow. Her parents, as taxpayers, meanwhile indirectly support scientific research.

Each of these three contributions is critical to advances in scientific research, which stand to deliver significant health as well as economic benefits to people in general. However, efforts by researchers and organizations to commercialize research, i.e. turn scientific findings into marketable products, that has been funded by, or which depends upon the participation of people or use of biological materials, have, on occasion, threatened to undermine these critical contributions. Providers of biological materials may not agree, for instance, with researchers' plans to patent research findings.

The commercialization of scientific research thus raises a number of legal and ethical questions, including: Should anyone own biological materials like a gene? Who, if anyone, should profit from commercializing research? How should the interests of those who participate in scientific research, or provide biological materials for research purposes, be recognized and protected? These and related questions have been the source of controversy for a long time. Although courts, policy-makers, and bioethicists have repeatedly attempted to address these questions, none has been fully resolved to date. The increasing relevance of large-scale genomics research and biobanking to the development of new treatments of disease, including cancers, means that these issues are likely to be a source of continuing conflict in the years ahead. Below, we describe these issues in depth, explaining how these issues have surfaced in the past, the efforts to address them through laws and policies that have been made to date, the uncertainties that remain, and how they might relate to circumstances captured in Olivia's vignette.

Patenting Human Biological Materials and Related Research Findings

A patent is a legal monopoly, granted by individual countries, for a limited period of time, that gives the owners of the patent the exclusive right to make, use, and sell the invention covered by the patent. All sorts of inventions can be patented if they meet the requirements of a country's patent law. In general, to receive a patent, the invention must be "novel", "non-obvious", and have a certain amount of "utility". These requirements have very specific meanings in law and many patent disputes turn on a court's application of these requirements to a particular invention.

However, in recent years, there has been renewed attention to other patent law requirements, in particular the requirement that an invention be "patentable subject matter" (PSM). Courts have long said that certain things, including "laws of nature, natural phenomena, and abstract ideas," are simply not eligible PSM [76]. Yet, since the early twentieth century, patents have been granted on things that seem to be naturally occurring such as the hormone adrenaline or vitamin B12. Following the invention of recombinant DNA technology in the 1970s, patenting within fields of molecular biology research has increased dramatically [77,78]. According to one controversial estimate, approximately 20% of the human genome has been patented [79–81].

What explains this apparent inconsistency between the rule against patenting "laws of nature, natural phenomena, and abstract ideas" and the fact of widespread patenting of seemingly naturally occurring things? The answer traces back to a court decision from 1911 called *Parke-Davis & Co. v. H.K. Mulford Co.* (Table 24.2) [82].

The invention in dispute in *Parke-Davis* covered a purified form of adrenaline, derived from the suprarenal glands, and methods of making such adrenaline. The inventor named on the patent was Jokichi Takamine, a trained chemist and entrepreneur, who worked in Japan's patent office (most countries have their own patent office) before relocating to the USA. The patent was originally filed in November 1900 and granted in June 1903 after repeated refusals by a patent examiner. Parke-Davis, a pharmaceutical company based in Detroit, began selling the product under the name "Adrenalin" in 1901, and it quickly proved a huge commercial success. This attracted several competitors, culminating in the lawsuit filed by Parke-Davis against H.K. Mulford Co. of Philadelphia for infringement of its patent rights [76].

Curiously, the arguments in court did not focus on whether the invention was PSM. Instead, the lawyer defending H.K. Mulford argued that Parke-Davis' patent was invalid because it was not novel. The technique described by the inventor Takamine was, the lawyer argued, already published in the scientific literature. Yet the decision of Judge Learned Hand, which was later upheld, made sweeping statements about the patentability of adrenaline and, by extension, other products of nature. The following provides an example:

[E]ven if [Adrenalin] were merely an extracted product without change, there is no rule that such products are not patentable. Takamine was the first to make it available for any use by removing it from the other gland-tissue in which it was found, and, while it is of course possible logically to call this a purification of the principle, it became for every practical purpose a new thing commercially and therapeutically. That was a good ground for a patent.

On the strength of this decision, the process of isolating and purifying a natural product has for some time been considered sufficient to overcome the rule against patenting products of nature. *Parke-Davis* was cited with approval in 1958, in a decision upholding a patent on vitamin B12 [*Merck & Co. v. Oline Mathieson Chemical Corp* 1958] [83]. In 2001, the United States Patents and Trademark Office relied on *Parke-Davis* when it issued guidelines with regard to patents on genetic sequences [76].

However, the widespread practice of patenting human genes and related material has been the subject of growing criticism and, with it, questions about the wisdom of the *Parke-Davis* decision are beginning to surface [76]. The focal point for the gene patenting controversy is a company, Myriad Genetics, which, together with the University of Utah, jointly owns a number of patents on genes associated with breast and ovarian cancers. In 2009, a collection of medical societies, researchers, clinicians and patients started a lawsuit against Myriad, arguing that the patents were invalid given that "laws of nature, natural phenomena, and abstract ideas" are not PSM. Myriad's defense depended, in part, on Learned Hand's decision in *Parke-Davis*. At trial, Judge Robert Sweet, ruled in favor of the plaintiffs, noting that "[b]ecause the claimed isolated DNA is not markedly different from native DNA as it exists in nature, it constitutes unpatentable subject matter" [84]. Judge Sweet's ruling caused an incredible stir in the biotechnology industry. The following year, though, the United States' Court of Appeals for the Federal Circuit overturned Judge Sweet on that point [84]. The reasoning of Judge Alan Lourie, who wrote the lead opinion for the Federal Circuit, bears a striking similarity to Judge Learned Hand's logic:

BRCA1 and BRCA2 in their isolated state are not the same molecules as DNA as it exists in the body; human intervention

TABLE 24.2 Summary of Court Decisions

Decision	Year	Ruling
Parke-Davis	1911	Naturally occurring products such as adrenaline are, once purified, patentable subject matter
Merck	1958	Following the precedent set by Parke-Davis, vitamin B12 is patentable subject matter
Prometheus	2012	The process of measuring levels of metabolites in a patient's blood after consuming a drug in order to make a clinical decision about whether a particular dosage of that drug could cause harm or prove ineffective is not patentable subject matter
Myriad	2013	Naturally occurring DNA is not patentable merely because it has been isolated; however, complementary DNA (cDNA) is patentable because it is not naturally occurring
Moore	1990	A physician-researcher is required to obtain a patient's informed consent in order to utilize a patient's biological materials for research purposes; however, patients do not have a property right in biological materials derived from their bodies
Catalona	2006	Following the precedent set in Moore, patients do not have a property interest in biological materials that they previously donated for research purposes; patients cannot require a research institution to transfer biological materials to another research institution
Greenberg	2003	Following the precedent set in Moore, patients do not have a property interest in biological materials that they have previously donated for research purposes. Patients may be able to make a claim of "unjust enrichment" against researchers who commercialize research without their awareness or permission

in cleaving or synthesizing a portion of a native chromosomal DNA imparts on that isolated DNA a distinctive chemical identity from that possessed by native DNA.

The Federal Circuit's decision was subsequently appealed to the United States' Supreme Court, which released its ruling in June 2013. On the question of whether *human* genes are patentable, the Supreme Court answered with a unanimous "no". As commentators have since pointed out, this result should facilitate competition in genetic testing and, over time, lower prices of genetic testing services. At the same time, the Supreme Court drew an important distinction: "naturally occurring DNA segment is a product of nature and not patent eligible merely because it has been isolated, but . . . cDNA [i.e. complementary DNA] is patent eligible because it is not naturally occurring." In other words, even though cDNA is based upon information contained in naturally occurring DNA, the fact that scientists create it through a process of reverse engineering was sufficient to make it patentable. The Supreme Court's decision therefore leaves open the possibility that human intervention may, at times, overcome the rule against patenting naturally occurring things, and "leaves undisturbed most of the intellectual property rights on which the biotechnology industry depends [85]. Whether the *Myriad* decision will facilitate or dampen innovation, much less shield healthcare providers from allegations of patent infringement, remains unclear. There are, in short, conflicting theories and evidence about the social and economic impact of patents on genes and similar discoveries [86–93]. The debate over patenting "products of nature" is thus likely to continue.

In Olivia's case, there was a patent on a novel MDR gene and its use as a diagnostic marker of myeloid leukemia. The validity of the diagnostic marker aspect of the patent is already questionable. Both lower courts in the *Myriad* case determined that the act of comparing two DNA sequences amounts to an "abstract mental process", and thus non-PSM. As well, in another case called *Mayo Collaborative Servs. v. Prometheus Labs., Inc.* [94], the US Supreme Court has already held that a patent predicated on measuring levels of certain metabolites in a patient's blood in order to make a clinical decision about whether a particular dosage of a drug could cause harm or prove ineffective was invalid. To the extent the patent in Olivia's case is worded similarly to the patents in *Myriad* and/or *Prometheus* it is likely invalid. If the Supreme Court decides in *Myriad* that the patents on BRCA1/2 are invalid, it will also call into question whether the other aspect of the patent – on the MDR gene itself – is PSM.

It is important to note, however, such a result in *Myriad* does not automatically invalidate other patents. It simply means that the value of such a patent would be significantly undermined. In other words, the Virginia biopharmaceuticals company may come to regret paying $200 million for the patent rights. Unless and until such a patent is challenged in court, it remains a valid patent (just not a very valuable one). If Olivia's family learns about the patents obtained and the $200 million deal, perhaps a lawsuit would follow. As explained next, similar circumstances have motivated lawsuits and ethical debate in the past.

The Commercialization of Research and Questions of Benefit-Sharing

The use of biological materials in scientific research without the knowledge or permission of those who provided the materials has given rise to several legal disputes. The most famous among these is the case of John Moore, a patient who had undergone clinical treatment for "hairy cell leukemia". He commenced a civil lawsuit against researchers and their parent institution, the University of California, for failing to disclose that they had created a highly profitable cell line based on cells excised from Moore's spleen, and for denying him a share in the profits realized from the cell line. In 1990, the Supreme Court of California found that the defendants had indeed breached their fiduciary duty and violated the principle of informed consent, but it rejected Moore's claim of conversion, refusing to recognize a proprietary interest in his cells or any entitlement to share in the profits that the university and researchers continued to amass [95].

As part of a group of top US research institutions leading the charge into the biotechnology realm during the late 1970s, the University of California was a key player in the lobby for legislation allowing universities to commercialize research [96]. Around that same time, John Moore learned that he suffered from "hairy cell leukemia" and, in October 1976, underwent surgery to have his spleen removed as recommended by his physician, Dr David W. Golde of the University of California at Los Angeles (UCLA) Medical Center. Prior to the surgery, Dr Golde, together with Shirley Quan, a UCLA researcher, began research on spleen tissue that had already been excised from Moore's body. They soon learned that Moore's tissue tended to overproduce a type of protein known as "lymphokines" that was known to possess therapeutic properties.

Over a 7-year span, Dr Golde and Ms Quan requested that Moore return from Seattle to the UCLA Medical Center roughly a dozen times, always under the false pretense of monitoring Moore's continued well-being. Dr Golde took further samples of blood, skin, bone marrow, and sperm but it was not until 1983 – approximately 7 years after the initial surgery and more than 2 years after the university had filed a patent application on his behalf – that Moore was told that his samples were being used for what Golde characterized as purely scientific (as opposed to commercial) research. In actual fact, though, a significant measure of commercial value had already been realized. On the basis of the patent, which encompassed a cell line (dubbed the "Mo" cell line), and a variety of methods of using the cell line to produce lymphokines, the university secured two licensing agreements, one with Genetics Institute, Inc. (for 75 000 company shares plus $330 000) and the other with Sandoz Pharmaceuticals Corporation (for $110 000). For his part, Moore became increasingly suspicious and hired an attorney to investigate the matter. In September 1984, Moore commenced a lawsuit, naming as defendants Dr Golde, Ms Quan, the Regents of the University of California, Genetics Institute, and Sandoz.

The trial Court dismissed the lawsuit for want of a cause of action but the Court of Appeals reversed, recognizing that Moore held a sufficient proprietary interest in his spleen to sustain a conversion claim [97]. However, in the end, while the Supreme Court of California found that the defendants had not met the standard for informed consent and breached their fiduciary duty vis-à-vis Moore, it arrived at a very different conclusion with respect to the claim of conversion than the Court of Appeals.

Following a cautionary prologue in which Justice Panelli took pains to stress that "[n]o court. . .has ever in a reported decision imposed conversion liability for the use of human cells in medical research", and that Moore could not possibly claim ownership over "the patented cell line [because it] is both factually and legally distinct from the cells taken from Moore's body", the Supreme Court cited three reasons for not extending the tort of conversion to the instant case. Two of the three amounted to standard judicial maneuvering: the legislature was the better forum to decide such an issue, Justice Panelli concluded; moreover, the doctrines of informed consent and breach of fiduciary duties adequately protected Moore's interests. However, the third reason reflected a clear policy choice not to "threaten with disabling civil liability *innocent* parties who are engaged in socially useful activities, such as researchers who have no reason to believe that their use of a particular cell sample is, or may be, against a donor's wishes." If it were to uphold Moore's conversion claim, the Supreme Court suggested that the "infant biotechnology industry" would suffer great damage:

> *The extension of conversion law into this area will hinder research by restricting access to the necessary raw materials. Thousands of human cell lines already exist in tissue repositories. . .These repositories respond to tens of thousands of requests for samples annually. Since the patent office requires the holders of patents on cell lines to make samples available to anyone, many patent holders place their cell lines in repositories to avoid the administrative burden of responding to requests. At present, human cell lines are routinely copied and distributed to other researchers for experimental purposes, usually free of charge. This exchange of scientific materials, which is still relatively free and efficient, will surely be compromised if each cell sample becomes the potential subject matter of a lawsuit.*

Moore remains the law in California, and courts elsewhere in the USA have applied it to deny similar claims [98]. For example, in *Greenberg v. Miami Children's Hosp. Research Inst., Inc.* [99], the plaintiffs were largely comprised of a group of parents who had provided a range of biological material in order to facilitate Canavan disease research, a deadly condition afflicting their children. Upon subsequently learning that the scientific researcher had, together with his parent institution (a hospital), patented the gene and corresponding diagnostic test for Canavan disease, the group filed an action, alleging that the defendants had failed to obtain informed consent, breached their fiduciary duties, converted the plaintiffs' property, and were also liable for unjust enrichment, fraudulent concealment, and misappropriation of trade secrets. On a motion to dismiss, the United States District Court for the Southern District of Florida effectively went further than the *Moore* Court, holding that only the claim for unjust enrichment could stand (and this claim was subsequently resolved by a confidential settlement). According to Justice Moreno, the absence of a therapeutic relationship between the plaintiffs and defendants distinguished *Greenberg* from *Moore*. The lack of disclosure about the economic interests and activities did not, in his view, amount to a breach of informed consent:

In declining to extend the duty of informed consent to cover economic interests, the Court takes notes of the practical implications of retroactively imposing a duty of this nature. First, imposing a duty of the character that Plaintiffs seek would be unworkable and would chill medical research as it would mandate that researchers constantly evaluate whether a disclosable event has occurred. Second, this extra duty would give rise to a type of dead-hand control that research subjects could hold because they would be able to dictate how medical research progresses.

This particular finding is troubling. From an ethical perspective, many would argue that the threshold for satisfying the duty of informed consent is higher in the context of research. Indeed, researchers who have failed to disclose the existence of a financial interest in their research have, in some cases, been the subject of significant sanctions [100].

In terms of conversion, the *Moore* decision was again considered persuasive. The Court was again fearful that a finding of conversion would inevitably "cripple medical research as it would bestow a continuing right for donors to possess the results of any research conducted by the hospital." "At the core", in Justice Moreno's view, "these were donations to research without any contemporaneous expectations of return."

While *Moore* and the cases that have followed its rationale are not binding in other states and countries, denying providers of biological materials from having a financial stake in research with those materials dovetails with standards of ethical practice. As part of the process of securing informed consent, researchers typically inform potential research participants that, while the results of the research may have commercial potential, they will not share in any commercial rewards. The primary reason for this appears ethically sound: offering such potential rewards could coerce otherwise unwilling individuals to participate or provide biological materials, contrary to the principle of informed, *voluntary* (i.e. not coerced) consent.

Yet, a number of difficult ethical questions follow from this state of affairs. First, there is a basic unfairness, or at least the perception thereof, in allowing the researchers, institutions, and firms involved – everyone but the providers of biological materials – to potentially profit from research. If any resulting profits can be divided among those actors, how difficult would it be to include providers on the list of potential beneficiaries? This part of the "chilling effect" worry articulated in *Moore* may thus be overblown. In some cases, there is a mechanism to ensure that those who contribute to a large-scale research project are at least given a collective return if the research translates into a blockbuster product [91]. A similar mechanism could be developed in the context of biobanking. Second, recognizing the contribution of those who provide biological materials to research may be less about having a chance to profit, and more about granting providers some level of continuing control. In the *Greenberg* case, the families afflicted with Canavan disease started the lawsuit because they feared the resulting diagnostic tests would not be "affordable and accessible", in turn, hindering "the discovery of more effective prevention techniques and treatments and, eventually,. . .a cure for Canavan disease."

Some research institutions and funding organizations are, in the wake of controversy, beginning to develop alternative approaches that give some decision-making authority to providers of biological materials, particularly when those providers are members of more vulnerable populations [101,102]. For example, following the misuse of biological materials collected from aboriginal communities located in British Columbia, Canada, the Canadian Institutes of Health Research (CIHR), in collaboration with aboriginal communities and researchers, has developed the *CIHR Guidelines for Health Research Involving Aboriginal People* [103]. Any publicly funded researchers engaging in research involving aboriginal people in Canada must adhere to these guidelines which, among other things, require researchers to develop a research agreement with any aboriginal communities involved. In particular, "[c]ommunity and individual concerns over, and claims to, intellectual property should be explicitly acknowledged and addressed in the negotiation with the community prior to starting the research project." These guidelines have become a model for addressing issues of benefit-sharing in the context of research involving aboriginal and other indigenous peoples [102]. Whether these guidelines' underlying principles should be extended to other research settings, such as large-scale genomics and biobanking, which does not necessarily involve aboriginal peoples, and yet where the continuing trust and participation of many people is essential, is an important area of future inquiry [104].

CONCLUSIONS

The power of genomic research promises fantastic opportunity for the advancement and biological understanding of cancer and human health. The same power that allows hitherto unimagined depth of exploration brings with it challenges amplified by the volume and nature of the discoveries. As researchers, ethicists and lawyers in the midst of this revolution in science, we face a rapidly moving field. It is critical to understand the spectrum of ethical challenges related to patenting and commercialization of these findings and to address the emerging quandary of what genomic discoveries do or do not meet a threshold for return to participants. Further empirical research in these ethical and legal areas is essential in appropriately responding to the responsibilities we carry to society and to patients like Olivia.

GLOSSARY

Aggregate research results Refers to the reporting of a summary of research results (as typically would appear in a scientific publication).

Benefit-sharing Refers to the ethical notion of ensuring that any benefits that result from a research project are distributed, not only among researchers and entities funding the research, but also those whose participation allowed the research to take place. Benefit-sharing is based on ethical principles of respect, reciprocity, and equity, and it has been invoked to support the development of best practices and guidelines for research involving particular communities and vulnerable populations that have been exploited by researchers and others in the past.

Biobank Refers to a biorepository that stores an organized collection of biological materials that are typically clinically annotated. The biobank may conduct research itself or be a source of samples and information to be accessed under specific rules by researchers.

Canadian institutes of health research (CIHR) This organization is the Government of Canada's health research investment agency. CIHR's mission is to create new scientific knowledge through grant support, and to enable its translation into improved health, health services and products, and healthcare systems. It is composed of 13 institutes encompassing a range of scientific disciplines ranging from Cancer Research to Population and Public Health.

Clinical utility Refers in the context of research to the concept that the individual finding can be utilized by the participant (and their healthcare provider) to assist in clinical decision making with respect to options of treatment, prevention or predictive counseling. This does not assume that the finding is actionable.

Commercialization Refers broadly to the process of converting scientific research findings into some form of marketable product or service. Commercialization can be achieved in many different ways depending on the type of research, product, or service. In the context of biotechnology, intellectual property, including patents, are commonly relied upon in the commercialization process to attract financing and facilitate research and development.

Incidental research results Refers to results specifically linked to the identity of the research participant but that were not the original aim of the research. Two examples of this concept are: (1) the discovery of a brain tumor when conducting an MRI of the brain looking at functional speech areas; (2) the discovery of a breast cancer risk gene when looking for a gene responsible for urinary tract anomalies.

Individual research results Refers to the results of research specifically linked to the identity of the research participant.

Patent Refers to a legal monopoly over an invention (as defined in the patent itself) that is granted by a country's patent office. In general, the term of a patent is limited to 20 years beginning on the date when the patent application was filed with the patent office. A patent can be obtained in several countries at once (but not consecutively), but the criteria by which a patent application is assessed vary across countries. Once obtained, a patent confers a set of exclusive rights (e.g. to use, make, sell, or import the invention), which only the patent owner is entitled to do unless it gives permission to other(s) to do so.

Research ethics board (REB) The Research Ethics Board is an independent, institutionally based body responsible for the review, approval and ongoing monitoring of the ethical conduct of research. In Canada, this is guided by the Tri-Council Policy version 2 and, in the USA, by the equivalent Institutional Review Board (IRB) governed by the Common Rule.

Tri-Council policy statement (TCPS) A statement that governs the *Ethical Conduct for Research Involving Humans* that is operative in all Canadian institutions that receive funding from the Canadian Institutes of Health Research, the Natural Sciences and Engineering Research Council of Canada, or the Social Sciences and Humanities Research Council of Canada. The second version was published in 2010.

REFERENCES

[1] Read K, Fernandez CV, Gao J, Strahlendorf C, Moghrabi A, Pentz RD, et al. Decision-making by adolescents and parents of children with cancer regarding health research participation. Pediatrics 2009;124: 959–65.

[2] Ferrari A, Montello M, Budd T, Bleyer A. The challenges of clinical trials for adolescents and young adults with cancer. Pediatr Blood Cancer 2008;50(Suppl. 5):1101–4.

[3] Glickman SW, Anstrom KJ, Lin L, Chandra A, Laskowitz DT, Woods CW, et al. Challenges in enrollment of minority, pediatric, and geriatric patients in emergency and acute care clinical research. Ann Emerg Med 2008;51:775–80.e3.

[4] Eder ML, Yamokoski AD, Wittmann PW, Kodish ED. Improving informed consent: suggestions from parents of children with leukemia. Pediatrics 2007;119:e849–59.

[5] Ungar D, Joffe S, Kodish E. Children are not small adults: documentation of assent for research involving children. J Pediatr 2006;149(1):S31–3.

[6] Barfield RC, Church C. Informed consent in pediatric clinical trials. Curr Opin Pediatr 2005;17:20–4.

[7] Simon CM, Siminoff LA, Kodish ED, Burant C. Comparison of the informed consent process for randomized clinical trials in pediatric and adult oncology. J Clin Oncol 2004;22:2708–17.

[8] Di Blasi Z, Kaptchuk TJ, Weinman J, Kleijnen J. Informing participants of allocation to placebo at trial closure: postal survey. Br Med J 2002;325:1329.

[9] Rigby H, Fernandez CV. Providing research results to study participants: support versus practice of researchers presenting at the American Society of Hematology annual meeting. Blood 2005; 106:1199–202.

[10] Fernandez CV, Kodish E, Shurin S, Weijer C. Offering to return results to research participants: attitudes and needs of principal investigators in the Children's Oncology Group. J Pediatr Hematol Oncol 2003;25:704–8.

[11] Shalowitz DI, Miller FG. Communicating the results of clinical research to participants: attitudes, practices, and future directions. PLoS Med 2008;5:e91.

[12] Partridge AH, Winer EP. Informing clinical trial participants about study results. J Am Med Assoc 2002;288:363–5.

[13] Fernandez CV, Kodish E, Weijer C. Informing study participants of research results: an ethical imperative. IRB 2003;25:12–9.

[14] Schulz CJ, Riddle MP, Valdimirsdottir HB, Abramson DH, Sklar CA. Impact on survivors of retinoblastoma when informed of study results on risk of second cancers. Med Pediatr Oncol 2003;41:36–43.

[15] Miller FA, Christensen R, Giacomini M, Robert JS. Duty to disclose what? Querying the putative obligation to return research results to participants. J Med Ethics 2008;34:210–3.

[16] Markman M. What must research subjects be told regarding the results of completed randomized trials? IRB 2004;26:8–10.

[17] Partridge AH, Hackett N, Blood E, Gelman R, Joffe S, Bauer-Wu S, et al. Oncology physician and nurse practices and attitudes regarding offering clinical trial results to study participants. J Natl Cancer Inst 2004;96:629–32.

[18] Rigby H, Fernandez CV. Providing research results to study participants: support versus practice of researchers presenting at the American Society of Hematology annual meeting. Blood 2005;106:1199–202.

[19] Fernandez CV, Santor D, Weijer C, Strahlendorf C, Moghrabi A, Pentz R, et al. The return of research results to participants: pilot questionnaire of adolescents and parents of children with cancer. Pediatr Blood Cancer 2007;48:441–6.

[20] Fernandez CV, Ruccione K, Wells RJ, Long JB, Pelletier W, Hooke MC, et al. Recommendations for the return of research results to study participants and guardians: A report from the Children's Oncology Group. J Clin Oncol 2012;30:4573–9.

[21] Fernandez CV, Skedgel C, Weijer C. Considerations and costs of disclosing study findings to research participants. C maj 2004;170:1417–9.

[22] Partridge AH, Wong JS, Knudsen K, Gelman R, Sampson E, Gadd M, et al. Offering participants results of a clinical trial: sharing results of a negative study. Lancet 2005;365:963–4.

[23] Delany-Moretlwe S, Stadler J, Mayaud P, Rees H. Investing in the future: lessons learnt from communicating the results of HSV/HIV intervention trials in South Africa. Health Res Policy Syst 2011;(Suppl. 1):S8.

[24] Dixon-Woods M, Jackson C, Windridge KC, Kenyon S. Receiving a summary of the results of a trial: qualitative study of participants' views. Br Med J 2006;332:206–10.

[25] Lorimer K, Gray CM, Hunt K, Wyke S, Anderson A, Benzeval M. Response to written feedback of clinical data within a longitudinal study: a qualitative study exploring the ethical implications. BMC Med Res Methodol 2011;11:10.

[26] Mancini J, Genre D, Dalenc F, Ferrero JM, Kerbrat P, Martin AL, et al. Participants' uptake of clinical trial results: a randomised experiment. Br J Cancer 2010;102:1081–4.

[27] Brealey S, Andronis L, Dennis L, Atwell C, Bryan S, Coulton S, et al. Participants' preference for type of leaflet used to feed back the results of a randomised trial: a survey. Trials 2010;11:116.

[28] Canadian Institutes of Health Research, Natural Sciences and Engineering Research Council of Canada, and Social Sciences and Humanities Research Council of Canada. Tri-Council Policy Statement: Ethical Conduct for Research Involving Humans; 2010.

[29] Puri KS, Suresh KR, Gogtay NJ, Thatte UM. Declaration of Helsinki, 2008: implications for stakeholders in research. J Postgrad Med 2009;55:131–4.

[30] Council for International Organizations of Medical Sciences (CIOMS). International Ethical Guidelines for Biomedical Research Involving Human Subjects; 2002.

[31] Kozanczyn C, Collins K, Fernandez CV. Offering results to research subjects: US Institutional Review Board policy. Account Res 2007;14:255–67.

[32] Macneil SD, Fernandez CV. Informing research participants of research results: analysis of Canadian university based research ethics board policies. J Med Ethics 2006;32:49–54.

[33] Lefaivre MJ, Chambers CT, Fernandez CV. Offering parents individualized feedback on the results of psychological testing conducted for research purposes with children: ethical issues and recommendations. J Clin Child Adolesc Psychol 2007;36:242–52.

[34] Borra RJ, Sorensen AG. Incidental findings in brain MRI research: what do we owe our subjects? J Am Coll Radiol 2011;8:848–52.

[35] Speicher MR, Geigl JB, Tomlinson IP. Effect of genome-wide association studies, direct-to-consumer genetic testing, and high-speed sequencing technologies on predictive genetic counselling for cancer risk. Lancet Oncol 2010;11:890–8.

[36] McGuire AL, Basford M, Dressler LG, Fullerton SM, Koenig BA, Li R, et al. Ethical and practical challenges of sharing data from genome-wide association studies: the eMERGE Consortium experience. Genome Res 2011;21:1001–7.

[37] Johnson G, Lawrenz F, Thao M. An empirical examination of the management of return of individual research results and incidental findings in genomic biobanks. Genet Med 2012;14:444–50.

[38] Knoppers BM, Deschenes M, Zawati MH, Tasse AM. Population studies: return of research results and incidental findings policy statement. Eur J Hum Genet 2013;21:245–7.

[39] Fabsitz RR, McGuire A, Sharp RR, Puggal M, Beskow LM, Biesecker LG, et al. Ethical and practical guidelines for reporting genetic research results to study participants: updated guidelines from a National Heart, Lung, and Blood Institute working group. Circ Cardiovasc Genet 2010;3:574–80.

[40] Fernandez CV, Strahlendorf C, Avard D, Knoppers BM, O'Connell C, Bouffet E, et al. Attitudes of Canadian researchers toward the return to participants of incidental and targeted genomic findings obtained in a pediatric research setting. Genet Med 2013;15:558–64.

[41] Gliwa C, Berkman BE. Do researchers have an obligation to actively look for genetic incidental findings? Am J Bioeth 2013;13:32–42.

[42] Zawati MH, Knoppers BM. International normative perspectives on the return of individual research results and incidental findings in genomic biobanks. Genet Med 2012;14:484–9.

[43] Bookman EB, Din-Lovinescu C, Worrall BB, Manolio TA, Bennett SN, Laurie C, et al. Incidental genetic findings in randomized clinical trials: recommendations from the Genomics and Randomized Trials Network (GARNET). Genome Med 2013;5:7.

[44] Wolf SM, Lawrenz FP, Nelson CA, Kahn JP, Cho MK, Clayton EW, et al. Managing incidental findings in human subjects research: analysis and recommendations. J Law Med Ethics 2008;36:219–48.

[45] Williams JK, Daack-Hirsch S, Driessnack M, Downing N, Shinkunas L, Brandt D, et al. Researcher and Institutional Review Board chair perspectives on incidental findings in genomic research. Genet Test Mol Biomarkers 2012;16:508–13.

[46] Simon CM, Williams JK, Shinkunas L, Brandt D, Daack-Hirsch S, Driessnack M. Informed consent and genomic incidental findings: IRB chair perspectives. J Empir Res Hum Res Ethics 2011;6:53–67.

[47] Wolf LE, Catania JA, Dolcini MM, Pollack LM, Lo B. IRB Chairs' perspectives on genomics research involving stored biological materials: ethical concerns and proposed solutions. J Empir Res Hum Res Ethics 2008;3:99–111.

[48] Saunders CJ, Miller NA, Soden SE, Dinwiddie DL, Noll A, Alnadi NA, et al. Rapid whole-genome sequencing for genetic disease diagnosis in neonatal intensive care units. Sci Transl Med 2012;4:154ra135.

[49] Bookman EB, Langehorne AA, Eckfeldt JH, Glass KC, Jarvik GP, Klag M, et al. Reporting genetic results in research studies: summary and recommendations of an NHLBI working group. Am J Med Genet A 2006;140:1033–40.

[50] Abdul-Karim R, Berkman BE, Wendler D, Rid A, Khan J, Badgett T, et al. Disclosure of incidental findings from next-generation sequencing in pediatric genomic research. Pediatrics 2013;131:564–71.

[51] Green RC, Berg JS, Grody WW, Kalia SS, Korf BR, Martin CL, et al. ACMG Recommendations of the reporting of incidental findings in clinical exome and genome sequencing. Genet Med 2013;15:565–74.

[52] Schwartz MK. Genetic testing and the clinical laboratory improvement amendments of 1988: present and future. Clin Chem 1999;45:739–45.

[53] Townsend A, Adam S, Birch PH, Lohn Z, Rousseau F, Friedman JM. "I want to know what's in Pandora's Box": comparing stakeholder perspectives on incidental findings in clinical whole genomic sequencing. Am J Med Genet A 2012;158A:2519–25.

[54] Haga SB, Tindall G, O'Daniel JM. Public perspectives about pharmacogenetic testing and managing ancillary findings. Genet Test Mol Biomarkers 2012;16:193–7.

[55] McMurter B, Parker L, Fraser RB, Magee JF, Kozancyzn C, Fernandez CV. Parental views on tissue banking in pediatric oncology patients. Pediatr Blood Cancer 2011;57:1217–21.

[56] Kohane IS, Masys DR, Altman RB. The incidentalome: a threat to genomic medicine. J Am Med Assoc 2006;296:212–5.

[57] American Society of Clinical Oncology policy statement update: genetic testing for cancer susceptibility. J Clin Oncol 2003 21:2397–406.

[58] Davis DS. Genetic dilemmas and the child's right to an open future. Hastings Cent Rep 1997;27:7–15.

[59] Geller G, Tambor ES, Bernhardt BA, Fraser G, Wissow LS. Informed consent for enrolling minors in genetic susceptibility research: a qualitative study of at-risk children's and parents' views about children's role in decision-making. J Adolesc Health 2003;32:260–71.

[60] Knoppers BM. Paediatric research and the communication of not-so incidental findings. Paediatr Child Health 2012;17:190–2.

[61] Wolf SM, Crock BN, Van Ness B, Lawrenz F, Kahn JP, Beskow LM, et al. Managing incidental findings and research results in genomic research involving biobanks and archived data sets. Genet Med 2012;14:361–84.

[62] Auray-Blais C, Patenaude J. A biobank management model applicable to biomedical research. BMC Med Ethics 2006;7:E4.

[63] Haga SB, Beskow LM. Ethical, legal, and social implications of biobanks for genetics research. Adv Genet 2008;60:505–44.

[64] Cambon-Thomsen A, Rial-Sebbag E, Knoppers BM. Trends in ethical and legal frameworks for the use of human biobanks. Eur Respir J 2007;30:373–82.

[65] Framingham Heart Study. Research application overview, review process and procedures. Clinical implications of findings. <http://www.framinghamheartstudy.org/research/rshapp_add_info.html>.

[66] McClellan KA, Kleiderman E, Black L, Bouchard K, Dorval M, Simard J, et al. Exploring resources for intrafamilial communication of cancer genetic risk: we still need to talk. Eur J Hum Genet 2013;21:903–10.

[67] Nycum G, Avard D, Knoppers BM. Factors influencing intrafamilial communication of hereditary breast and ovarian cancer genetic information. Eur J Hum Genet 2009;17:872–80.

[68] Patenaude AF, Dorval M, DiGianni LS, Schneider KA, Chittenden A, Garber JE. Sharing BRCA1/2 test results with first-degree relatives: factors predicting who women tell. J Clin Oncol 2006;24:700–6.

[69] Tasse AM. Biobanking and deceased persons. Hum Genet 2012;130:415–23.

[70] Sexton AC, Metcalfe SA. Disclosing genetic research results after death of pediatric patients. J Am Med Assoc 2008;300:1693–5.

[71] Alliman S, Veach PM, Bartels DM, Lian F, James C, LeRoy BS. A comparative analysis of ethical and professional challenges experienced by Australian and U.S. genetic counselors. J Genet Couns 2009;18:379–94.

[72] Black L, McClellan KA, Avard D, Knoppers BM. Intrafamilial disclosure of risk for hereditary breast and ovarian cancer: points to consider. J Community Genet 2013;4:203–14.

[73] Black L, McClellan KA. Familial communication of research results: a need to know? J Law Med Ethics 2011;39:605–13.

[74] Vos J, Stiggelbout AM, Oosterwijk J, Gomez-Garcia E, Menko F, Collee JM, et al. A counselee-oriented perspective on risk communication in genetic counseling: explaining the inaccuracy of the counselees' risk perception shortly after BRCA1/2 test result disclosure. Genet Med 2011;13:800–11.

[75] Brandt DS, Shinkunas L, Hillis SL, Daack-Hirsch SE, Driessnack M, Downing NR, et al. A closer look at the recommended criteria for disclosing genetic results: perspectives of medical genetic specialists, genomic researchers, and institutional review board chairs. J Genet Couns 2013;22:544–53.

[76] Harkness J. Dicta on adrenalin(e): Myraid problems with Learned Hand's product-of-nature pronouncements in Parke-Davis v. Muford. J Pat Trad Off Soc 2011;93:363–99.

[77] Hughes SS. Making dollars out of DNA. The first major patent in biotechnology and the commercialization of molecular biology, 1974–1980. Isis 2001;92:541–75.

[78] Mowery D, Nelson R, Sampat B. The growth of patenting and licensing by U.S. universities: an assessment of the effects of the Bayh-Dole act of 1980. Res Policy 2001;30:99–119.

[79] Jensen K, Murray F. Intellectual property landscape of the human genome. Science 2005;310:239–40.

[80] Holman CM. Debunking the myth that whole-genome sequencing infringes thousands of gene patents. Nat Biotechnol 2012;30:240–4.

[81] Rosenfeld J, Mason CE. Pervasive sequence patents cover the entire human genome. Genome Med 2013;5:27.

[82] Parke-Davis & Co. v. H.K. Mulford Co., 189F.95 (C.C.S.D.N.Y. 1911), aff'd in part, rev'd in part, 196 F. 496 (2d Cir. 1912).

[83] Merck & Co. v. Oline Mathieson Chemical Corp., 253 F.2d 156 (4th Cir. 1958).

[84] Ass'n for Molecular Pathology v. U.S. Patent & Trademark Office (Myriad), 702 F.Supp. 2d, 181, 224 (S.D.N.Y 2010), as amended (Apr. 5, 2010), aff'd in part, rev'd in part, 653 F.3d 1329 (Fed. Cir. 2011), aff'd in part, rev'd in part 2013 WL 2631062 (US Sup. Ct., June 13, 2013).

[85] Kesselheim AS, Cook-Deegan RM, Winickoff DE, Mello MM. Gene patenting – The Supreme Court finally speaks. N Eng J Med 2013;369(9):869–75.

[86] Herder M. Choice patents. IDEA 2013;52:309–78.

[87] Walsh J, Cohen W, Cho C. Where excludability matters: Material versus intellectual property in academic biomedical research. Res Policy 2007;36:1184–203.

[88] Murray F, Stern S. Do formal intellectual property rights hinder the free flow of scientfic knowledge?: An empirical test of the anti-commons hypothesis. J Econ Behav Organ 2007;63:648–87.

[89] Cho MK, Illangasekare S, Weaver MA, Leonard DG, Merz JF. Effects of patents and licenses on the provision of clinical genetic testing services. J Mol Diagn 2003;5:3–8.

[90] Heller MA, Eisenberg RS. Can patents deter innovation? The anticommons in biomedical research. Science 1998;280:698–701.

[91] Herder M. Patents & the progress of personalized medicine: biomarkers research as lens. Ann Health Law 2009;18:187–229.

[92] Hopkins MM, Hogarth S. Biomarker patents for diagnostics: problem or solution? Nat Biotechnol 2012;30:498–500.

[93] Huang K, Murray F. Does patent strategy shape the long-run supply of public knowledge? Evidence from human genetics. Acad Manage J 2009;52:1193–221.

[94] Mayo Collaborative Servs. v. Prometheus Labs., Inc., 132 S. Ct. 1289 (2011).

[95] Moore v. Regents of the University of California, 793 P.2d 479 (Cal. 1990) cert. denied 111S.Ct. 1388.

[96] Eisenberg DM, Kessler RC, Foster C, Norlock FE, Calkins DR, Delbanco TL. Unconventional medicine in the United States. Prevalence, costs, and patterns of use. N Engl J Med 1993;328: 246–52.

[97] Moore v. Regents of the University of California, 202 Cal. App. 3d 1230, 249 Cal. Rptr. 494 (Ct. App.), review granted, 252 Cal. Rptr. 816, 763 P.2d 479 (1988).

[98] Catalona v. Wash. Univ., 437 F. Supp. 2d 985 (Dist.Ct. E.D.Mo 2006), aff'd 490F.3d 667 (8th Cir. Mo. 2007).

[99] Greenberg v. Miami Children's Hosp. Research Inst., Inc., 264 F.Supp.2d 1064 (Dist.Ct. S.D. Fla. 2003).

[100] Wilson RF. The death of Jesse Gelsinger: new evidence of the influence of money and prestige in human research. Am J Law Med 2010;36:295–325.

[101] Arbour L, Cook D. DNA on loan: issues to consider when carrying out genetic research with aboriginal families and communities. Community Genet 2006;9:153–60.

[102] Jacobs B, Roffenbender J, Collmann J, Cherry K, Bitsói LL, Bassett K, et al. Bridging the divide between genomic science and indigenous peoples. J Law Med Ethics 2010;38:684–96.

[103] Canadian Institutes of Health Research. CIHR Guidelines for Health Research Involving Aboriginal People; 2005. Available from: <http://www.cihr-irsc.gc.ca/e/29134.html>.

[104] Wolf LE. Advancing research on stored biological materials: Reconciling law, ethics, and practice. Minn J Law, Sci Technol 2010;11:99–156.

Part 6

Future Directions

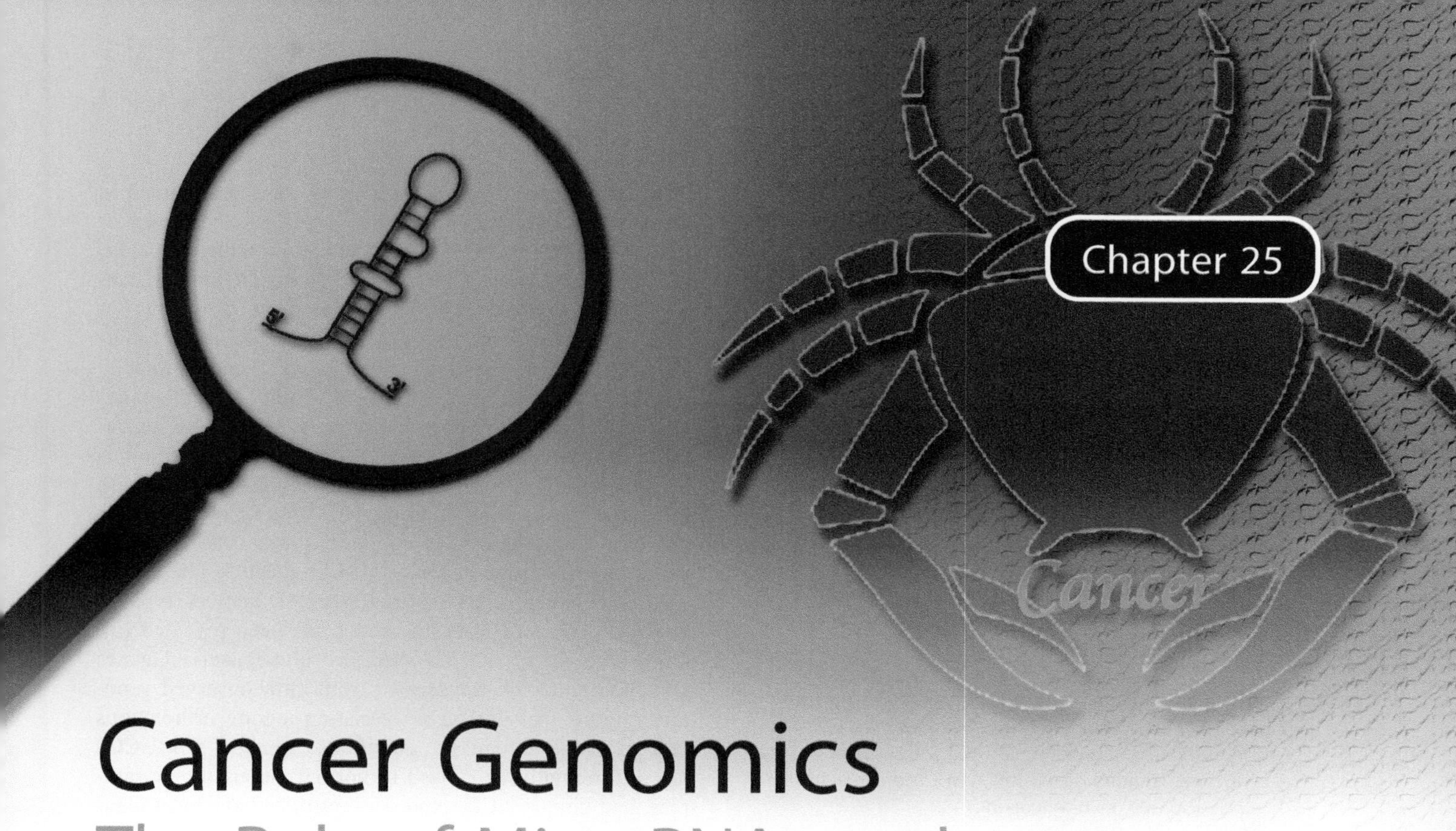

Cancer Genomics

The Role of MicroRNAs and Ultraconserved Non-Coding RNAs in Cancer

Hui Ling
George A. Calin

Department of Experimental Therapeutics and Leukemia, The University of Texas MD Anderson Cancer Center, Houston, TX USA

Contents

Cancer Genomics. DOI: http://dx.doi.org/10.1016/B978-0-12-396967-5.00025-6

Key Concepts

- Protein-coding genes represent less than 2% of the DNA encoded by the genome. The remaining non-coding genes, which are transcribed into non-coding RNAs, once thought to be simply evolutionary relics or junk DNA, are now known to represent a major pathway of gene regulation that is highly sophisticated, impacts virtually every cellular function, and is implicated in a host of pathologies including cancer
- MicroRNAs are a class of non-coding RNAs that are estimated to regulate expression of two thirds of the mammalian genome by binding to promoter, coding and untranslated regions of messenger RNAs, proteins or other non-coding RNAs
- MicroRNAs are frequently located within cancer-associated genomic regions (CAGR) and can act as tumor suppressors or oncogenes
- Aberrant microRNA gene expression signatures characterize cancer cells and microRNA profiling can be applied in diagnosis, prognosis, and treatment of cancer patients
- Circulating microRNAs are potential non-invasive biomarkers in cancer
- Restoring or blocking microRNA function is a potential treatment method for specific types of cancer
- Ultraconserved genes are deregulated in cancer and the unique expression profile of these genes is characterized in chronic lymphocytic leukemia and colorectal cancers

INTRODUCTION

The vast majority (80%) of the human genome participates in gene transcription or regulatory events of gene transcription [1]. Protein-coding genes, the number of which varies little throughout animal phyla, only represent less than 2% of the whole human genome. Therefore, it is unlikely that protein-coding genes are the only molecules regulating physiological activities in complex organisms. By comparison, transcripts with no protein-coding potential, named non-coding RNAs (ncRNAs), comprise the majority of the transcriptome. Moreover, the number of ncRNAs increases correspondingly with the complexity of organisms, suggesting their functional importance in modulating sophisticated regulatory networks [2]. The role of ncRNAs in cancer has been extensively studied in the last decade, with the majority of studies focusing on microRNAs (miRNAs), one type of small conserved ncRNAs. MiRNAs are deregulated in many types of cancer, and their roles as oncogenes or tumor suppressors have been validated by numerous studies [3]. The aberrant miRNA expression signatures in cancer have been actively investigated as a means of developing new diagnostic and prognostic biological markers that could benefit cancer patients. The therapeutic potential of miRNA mimics or miRNA antagonists against cancer has been actively pursued by both academia and the biopharmaceutical industry. The recent clinical trials using miRNA antagonists have achieved success in controlling hepatitis C virus (HCV) infection, and offer cancer patients new hope for effective cancer treatment using miRNA-based agents [4]. More recently, the participation of long ncRNAs (lncRNAs), a distinct type of ncRNAs from miRNAs, in cancer has been revealed [5]. One subgroup of the lncRNAs are ultraconserved genes (UCGs), which are transcribed from ultraconserved genomic regions that exhibit no variation among orthologous regions of the human, rat, and mouse genomes [6]. UCGs are differentially expressed in normal versus tumor samples, and unique UCG expression profiles have been identified in different cancer types [6]. The biological effects of UCGs in cancer have also been experimentally demonstrated. In this chapter, we focus on the involvement of these two types of ncRNAs in cancer. We will first discuss the aberrant miRNA expression in human cancers, followed by a summary of the mechanisms underlying their molecular function in gene regulation. We will then discuss the translational potential of these miRNAs in various clinical applications. Finally, we end the chapter with a brief introduction to UCGs, and a discussion of the recent evidence of their involvement in human cancers.

miRNAs AND CANCER

In 1993, Victor Ambros' and Gary Ruvkun's research groups found a small ncRNA transcript that controls developmental timing of the *Caenorhabditis elegans* through regulation of lin-14 protein [7,8]. This was the first discovery of an miRNA, which represented a new type of regulatory gene that employed a novel mechanism of modulating gene function. Two decades after this initial finding, the recent miRBase database release (version 19) has cataloged over 25 000 mature miRNA sequences in 193 species, of which 2042 mature miRNAs have been identified in human beings [9].

MiRNAs are evolutionarily conserved single-stranded RNAs (19–24 nucleotides in length) without protein-coding function [10]. Initially transcribed from the genome by RNA polymerase II as long, capped and polyadenylated primary transcripts (pri-miRNAs) of various lengths, they are processed by the ribonuclease (RNase)

III Drosha-DGCR8 nuclear complex into hairpin RNAs named precursor miRNAs (pre-miRNAs) which are typically 60–100 nucleotides in length. The pre-miRNAs are exported from the nucleus to the cytoplasm by exportin-5, and further processed by the RNase enzyme Dicer into double-stranded miRNAs. The two strands separate, and the mature strand incorporates into the RNA-induced silencing complex (RISC). By sequence-specific binding to the transcripts of its target genes, an miRNA changes the output of protein-coding genes through post-transcriptional regulation [10]. In Figure 25.1, we illustrate the biogenesis of miRNAs and their functions.

Despite their tiny size, miRNAs play significant roles in human diseases such as cancer [11]. Cancer is a complex disease with a series of genetic and epigenetic changes which alter the balance between oncogenes and tumor suppressors leading to unrestrained cellular proliferation. However, the molecular and cellular complexity of cancer makes it unlikely that any single approach that targets a particular protein-coding gene will be able to reverse oncogenesis or produce a lasting cure. MiRNAs have the advantage of targeting multiple protein-coding genes, within the same signaling pathway, or scattered in different signaling pathways, but with synergistic effect. Therefore, a small change in miRNA expression may cause significant changes in the expression profile of multiple protein-coding genes which, in turn, result in varied and often profound changes in cellular phenotype. The role of miRNAs in cancer has been intensively evaluated and their importance has been demonstrated by *in vitro* and *in vivo* experiments as well as clinical studies [12].

Deregulated miRNA Expression in Cancer

Despite the first discovery of miRNA in 1993, it was nearly a decade later before the association of miRNAs with human cancer was realized. In 2002, in an attempt to find tumor suppressors in the frequently deleted 13q14 region in chronic lymphocytic leukemia (CLL), George Calin and Carlo Croce extensively cloned the genes located in this region, and they made the unexpected finding that two miRNA genes, *miR-15a* and *miR-16-1*, were transcribed from this genomic locus. Subsequently, they found that these two miRNAs are either deleted or

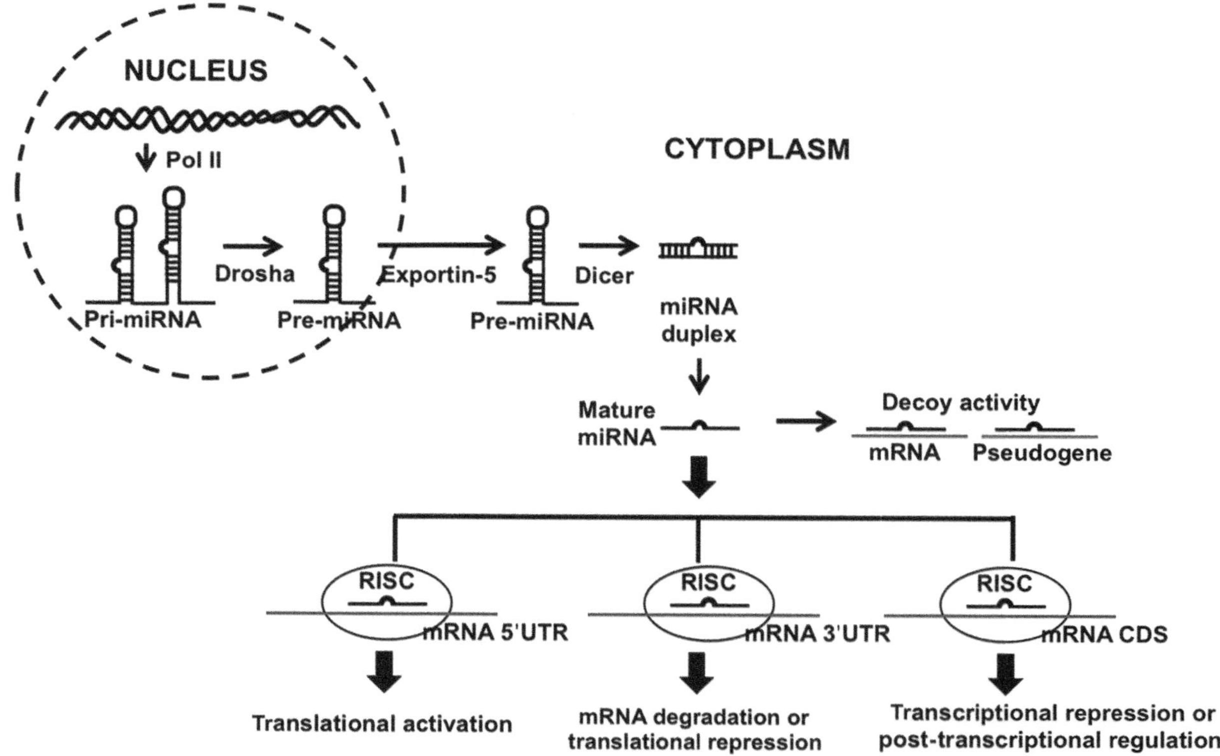

FIGURE 25.1 MicroRNA biogenesis and function. MicroRNAs (miRNAs) are transcribed from the genome by RNA polymerase II (Pol II) into primary miRNA transcripts (pri-miRNAs), which are then processed by Drosha into hairpin structures called precursor miRNAs (pre-miRNAs). The pre-miRNAs are transported by exportin-5 from the nucleus to the cytoplasm, where Dicer further processes them into miRNA duplex. The miRNA duplex unwinds and the mature miRNAs are incorporated onto the RNA-induced silencing complex (RISC). The main mechanism of miRNA action is through its complementary binding with 3-untranslated region (3′-UTR) of target mRNAs. Other mechanisms include the interaction of miRNA with 5′-UTR or coding sequence (CDS), and the RISC-independent decoy mechanism.

TABLE 25.1 Overview of Major miRNAs Involved in Cancer

miRNA	Associated Cancers	Expression (Up or Down)	Oncogene/Tumor Suppressor	CAGR	Major Target Genes
let-7 family	Lung cancer, breast cancer	Down	Tumor suppressor	11q24	*CCND1, CDC25a, CDK6, HOXA9, IMP-1, MYC, RAS, TLR4*
miR-15a/16-1	CLL	Down	Tumor suppressor	13q14	*BCL2, MCL1*
miR-17-92	Lymphoma, lung cancer, breast cancer	Up	Oncogene	13q22	*E2F1, BIM, PTEN*
miR-21	Lymphoma, lung cancer, colorectal cancer	Up	Oncogene	17q23	*PTEN, PDCD4, TPM1*
miR-29	Breast cancer, CLL	Down	Tumor suppressor	7q32 and 1q30	*MCL1, DNMTs*
miR-34a	Colorectal cancer, lung cancer, breast cancer	Down	Tumor suppressor	1p36 and 11q23	*CDK4, CDK6, MYC*
miR-155	CLL, lymphoma, lung cancer	Up	Oncogene	21q21	*SHIP1, CEBPB*

CLL: chronic lymphocytic leukemia; CAGR: cancer-associated genomic region

downregulated in 69% of all CLL cases examined [13]. These findings ushered in a new era of miRNA research by indicating that miRNAs could function as tumor suppressors in CLL development, which provided the first evidence linking changes in miRNA expression to human cancer.

The discovery of *miR-15a/16-1* from the 13q14 region prompted the Croce group to map the chromosomal locations of all known miRNAs in the genome. They observed that miRNA genes are frequently located at fragile sites and cancer-associated genomic regions (CAGRs) – chromosomal loci prone to deletions or amplifications – rather than being randomly distributed in the genome [14]. Afterwards, studies from other groups expanded the initial discovery of miRNA involvement in CLL by identifying a widespread alteration of miRNA expression in many other types of cancers, including lymphomas, breast cancers, lung cancers, papillary thyroid carcinomas, glioblastomas, hepatocellular carcinomas, pancreatic tumors, pituitary adenomas, cervical cancers, brain tumors, prostate cancers, kidney and bladder cancers, and colorectal cancers (Table 25.1) [3]. For instance, the expression levels of *let-7* family, *miR-34* family and *miR-29* family are significantly lower in cancer tissues compared with the corresponding normal tissues; whereas miRNAs such as *miR-21*, *miR-17-92* cluster and *miR-155* are expressed at much higher levels in tumor tissues. The widespread miRNA deregulation in human cancer indicates that miRNAs may play functional roles in cancer pathogenesis.

Besides the localization of the miRNAs at CAGRs, the abnormal expression of miRNAs in cancer can be caused by other mechanisms such as epigenetic regulation, or abnormalities in the miRNA-processing machinery, including mutations in Dicer, TRBP and exportin 5 [15]. In many cases, these different mechanisms can work together to deregulate the expression of a given miRNA. For instance, *miR-34a*, which is located at 1p36, a chromosomal region frequently lost in neuroblastomas, is transcriptionally activated by *TP53*, repressed by *Myc*, and silenced by epigenetic regulation through aberrant CpG methylation [16]. Similarly, the cluster *miR-17-92*, which includes *miR-17*, *miR-18a*, *miR-19a*, *miR-20a*, *miR-19b-1* and *miR-92a-1*, is frequently upregulated in lymphoma through genomic amplification and the miRNA genes in this region are transcriptionally activated by *Myc* [17].

miRNA Function and Its Involvement in Cancer

As demonstrated in 1993, the *lin-4* prototypical miRNA controls lin-14 protein translation by binding to multiple sites in the 3′-untranslated region (UTR) of the *lin-14* messenger RNA (mRNA) [7,8]. This classical mechanism of miRNA function occurs through Watson–Crick base-pairing between the seed sequence of the miRNA (positions 2 to 8) and the 3′-UTR of its target mRNAs [18]. The miRNA–mRNA interaction interferes with the mRNA translation or promotes mRNA degradation which, in turn, leads to a decrease in the production of a specific protein. This canonical mechanism has recently been challenged by several additional observations: (1) the report that human *miR-29b* is localized in the nucleus challenged the previously held view that miRNAs can only localize in the cytoplasm, and thus also challenged the canonical mechanism that miRNAs only function in the cytoplasm [19]; (2) *miR-10a* binds to the 5′-UTR of the targeted protein-coding genes and enhances the translation of these genes [20]; (3) *miR-373* directly binds to the CDH1 promoter and upregulates its transcription [21]; (4) miRNAs can regulate transcription of genes such as *Nanog*, *Pou5f1* and *Sox2* through binding to their protein-coding sequences [22]; (5) *miR-328* directly interacts with poly-C binding protein 2 (PCBP2) as a decoy to release PCBP2-mediated translational inhibition [23]; and (6) the recent evidence of competing endogenous RNA (ceRNA), which has led to the hypothesis that coding and non-coding RNAs can regulate one another through their ability to compete for miRNA binding [24].

Since the main mechanism of miRNA function is based on sequence complementarity, a single miRNA can have hundreds of mRNA targets, and thus play roles in various types of physiological processes. For instance, *miR-181* is involved in B-cell lineage fate [25], *miR-15a* and *miR-16* regulate B-cell survival [26], *miR-125b* and *let-7* are regulators of cell proliferation [27], *miR-430* controls brain patterning [28], *miR-375* functions in pancreatic cell insulin secretion [29], and *miR-143* regulates adipocyte development [30].

According to their roles in cancer, miRNAs can be categorized as either oncogenic miRNAs that activate malignant potential, or tumor suppressor miRNAs that block the malignant progression of a cell. The oncogenic miRNAs include the *miR-21*, *miR-155*, and *miR-17-92* cluster. *MiR-21* has been reported to be elevated in many types of cancers (see Table 25.1). Functional studies showed that ectopic overexpression of *miR-21* leads to pre-B-cell lymphoma in a transgenic mouse model [31], enhances K-Ras-dependent lung carcinogenesis by negatively regulating the Ras/MEK/ERK pathway [32], and promotes metastasis of colorectal cancers by targeting Pdcd4 [33]. These studies demonstrated the oncogenic behaviors of *miR-21* in multiple cancer types. In another example, the *miR-17-92* cluster located at 13q22 is expressed at high level in a large variety of cancers such as lymphoma, lung, breast, stomach, colon, and pancreatic cancers, through gene amplification or transcriptional activation [17]. As direct downstream targets of *Myc*, *miR-17-92* cluster miRNAs cooperate with Myc in promoting carcinogenesis. For instance, transfection of *miR-17-92* facilitates *Myc*-induced B-cell lymphomas in a transgenic mouse model [34]; deletion of this cluster results in excessive apoptosis in *Myc*-induced lymphoma demonstrating an anti-apoptotic role for miRNAs within the *miR-17-92* cluster [35]. Individual miRNA in this cluster has also been demonstrated to suppress negative regulators of PI3 kinase signaling and pro-apoptotic members of the BCL-2 family, thus driving malignancy through increased proliferation and reduced apoptotic events [17]. *MiR-155* is another example of oncogenic miRNAs. The observation that *miR-155* alone is sufficient to cause lymphoblastic leukemia in an animal model clearly suggests that miRNA can be the driving force of carcinogenesis [36]. *MiR-155* was also reported to inactivate mismatch repair and induce microsatellite instability in colorectal cancer [37]. Several miRNAs were reported to participate specifically in tumor metastasis. For instance, *miR-9*, which is transcriptionally activated by *Myc*, was shown to regulate E-cadherin and promote the metastasis of breast cancer cells to the lung in a mouse model [38]. Using the same mouse model, this group also identified *miR-10b* as a pro-metastatic ncRNA. They found that *miR-10b* inhibits *HOXD10* translation, and accordingly leads to upregulation of the pro-metastatic gene *RHOC*[39].

The first identified miRNAs with tumor suppressor function are *miR-15a* and *miR-16-1*[13]. After identification of their downregulation in CLL, the Croce group continued to explore the biological importance of these miRNAs [40]. They found that the expression of these two miRNAs is inversely correlated with BCL-2 protein expression. Ectopic expression of *miR15a/16-1* reduced the BCL-2 protein levels without affecting the gene expression of *BCL-2*, suggesting this effect is through disruption of the translation of the *BCL-2* mRNA. Using microarray analysis, the same group profiled the expression changes in cells transfected with the genomic region containing *miR-15a/16-1* and found that these miRNAs, directly or indirectly, regulated 14% of all the genes in the human genome [41]. They also proved the tumor suppressor function of these miRNAs by showing that overexpression of *miR-15a/16-1* significantly reduced the tumor size and weight in a mouse xenograft model [41]. Further supporting the above findings, studies from another group using a New Zealand black mouse model of spontaneous CLL identified a point mutation in the miR-16-1 genomic region, as well as a decrease of *miR-16-1* expression [42]. Other typical examples of tumor suppressor miRNAs include *miR-34* and *let-7* families. MiRNAs in the *miR-34* family are transcriptionally activated by p53 and mediate in part p53's tumor-suppressive function. Reduced *miR-34* expression confers resistance to p53-induced apoptosis, suggesting that the *miR-34* family is an important downstream executor of the p53 network [43]. The *miR-34* family was also shown to be a potential inhibitor of cancer stem cells in prostate cancer by targeting CD44 [44].

Similarly, the *let-7* family is frequently downregulated in various types of human cancers. The expression levels of *let-7* miRNAs were positively correlated with the length of postoperative survival of lung cancer patients [45]. Ectopic expression of *let-7* miRNA significantly reduced lung cancer formation in mice, suggesting a role of the *let-7* family miRNAs as negative regulators of tumorigenesis [46]. The Weinberg group also identified an miRNA that suppresses tumor metastasis [47]. They found that *miR-31* expression levels are inversely correlated with the metastasis status of breast cancer patients. Experimental evaluation showed that overexpression of *miR-31* caused a significant reduction in tumor metastasis in a mouse model. Consistent with these observations, inhibition of *miR-31* function increases the metastatic capacity of breast cancer cells *in vitro* and *in vivo*. The anti-metastatic effect of *miR-31* was attributed to its negative regulation of many metastasis-promoting genes including *RHOA*.

Although miRNAs may be generally grouped into oncogenes and tumor suppressors, it should be mentioned that the functional roles of miRNA in cancer largely depend on the cellular context, and the same miRNA can work as an oncogene in one case and act as a tumor suppressor in a different situation. For instance, *miR-200* inhibits epithelial–mesenchymal transition (EMT) and reduces tumor invasion by targeting *ZEB1* and *ZEB2*[48]. However, the anti-metastatic effect of *miR-200* was recently challenged by a finding that *miR-200* promotes colonization of metastatic breast cancer cells that entered the circulation [49]. Similarly, *miR-221* inhibits erythroleukemic cell proliferation by targeting the KIT oncogene [50], but promotes carcinogenesis of liver cancer by targeting the DNA damage-inducible transcript 4 (DDIT4) [51]. These findings indicate that care should be taken when determining the function(s) of an miRNA by validating its function in the specific tissue, or physiological circumstance that is relevant for the pathological condition being studied.

The miRNA studies have clearly shown that these ncRNAs are master regulators of tumor biology, and deregulated miRNA expression can affect all of the cellular hallmarks of cancer [52]: (1) self-sufficiency in growth signals (*let-7* family); (2) insensitivity to antigrowth signals (*miR-17-92* cluster); (3) evasion from apoptosis (*miR-34a*); (4) limitless replicative potential (*miR-372/373* cluster); (5) angiogenesis (*miR-210*); (6) genome instability (*miR-155*); and (7) invasion and metastases (*miR-10b*) (Figure 25.2) [53].

miRNAs as Diagnostic Tools

The widespread alteration of miRNA expression levels in cancer offers an excellent opportunity to translate miRNA expression profiles into useful diagnostic and prognostic tools. Various platforms have been developed for high-throughput detection of miRNAs in physiological and pathological conditions: from using custom-made and then commercial miRNA microarrays, bead-based flow cytometric miRNA analysis to the more recent use of the next-generation sequencing (NGS).

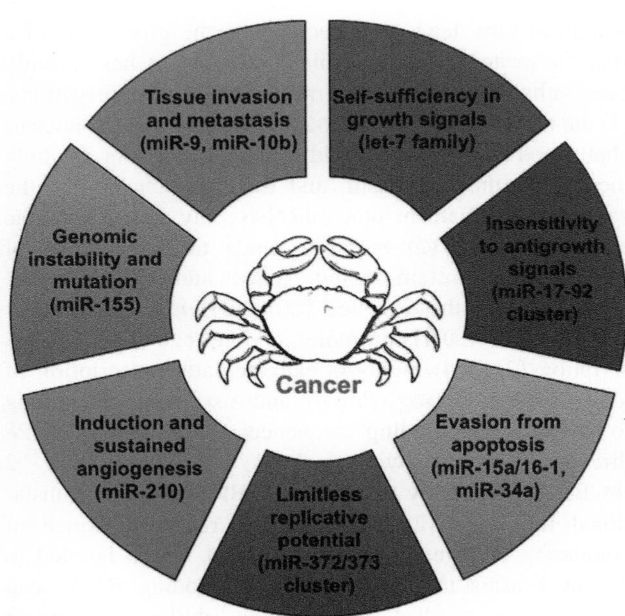

FIGURE 25.2 Relationship between microRNAs and the classic hallmarks of cancer.

The concept of miRNAs as biological markers has been advanced by several miRNA profiling studies that have identified unique miRNA signatures associated with diagnosis, staging, progression, prognosis and treatment response of human cancers. For example by analyzing 17 poorly differentiated metastatic cancer cases from an unknown primary source, Lu et al. showed that the miRNA expression profile was more accurate in establishing the correct diagnosis than the mRNA profile [54]. In another study, the researchers developed a classifier of 48 miRNAs and used this signature accurately to predict the tissue origin in 86% of a blind test set, including 77% of the metastatic tumors [55]. MiRNA biomarkers also have the potential to be used for early cancer diagnosis, as shown by the detection of high levels of *miR-205* and *miR-21* expression in the early stage of ductal adenocarcinoma before overt phenotypic changes in the ducts [56]. The aberrant expression of miRNAs can similarly be used to predict cancer prognosis and response to specific cancer treatment. For instance, an expression signature of 11 miRNAs was found to correlate with the survival rate of patients with acute myelogenous leukemia [57]. In a different study, the expression profile of 13 miRNAs was reported to correlate with disease progression from the time of diagnosis to the time of therapy [58]. In another study, expression levels of *miR-15b*, *miR-34c*, and *miR-361* predicted the risk of liver cancer recurrence following curative resection with an overall accuracy of 90% [59]. We recently reported an miRNA/*TP53*-regulated gene signature, composed of five miRNAs (*miR-15a/miR-16-1*, *miR-34a*, *-34b*, and *-34c*) and four protein-coding genes including the transcription factor *TP53*, the anti-apoptotic oncogenes *BCL2* and *MCL1*, and the 70-kDa-zeta-associated protein (Zap70), that was a powerful predictor of CLL patient survival [60].

MiRNAs are stable under severe conditions that would normally degrade most RNAs, such as boiling, very low or high pH levels, extended storage and up to 10 freeze-thaw cycles [61]. Therefore, it is possible to explore the expression levels of miRNA in body fluids as a useful tool for cancer diagnosis, prognosis and prediction of therapeutic response. As demonstrated by many studies and summarized by Cortez et al., circulating miRNAs in peripheral blood serum can be used for cancer detection and therefore represent a gold mine for non-invasive biomarkers in cancer [62]. For instance, high *miR-21* expression in the serum was found to be associated with relapse-free survival in diffuse large B-cell lymphoma [63]. Mitchell et al. showed that the expression levels of *miR-141* in serum are able to distinguish prostate cancer patients from healthy subjects [64]. A serum miRNA expression signature composed of *miR-486*, *miR-30d*, *miR-1* and *miR-499* was proven to be an independent predictor of patient survival [65]. It was also observed that circulating miRNA levels correlate with therapeutic response to a specific anticancer agent, and this can be applied to predict patient sensitivity to chemotherapeutic drugs. For instance, in hormone-refractory prostate cancer patients, whose disease was resistant to docetaxel-based treatment, the serum *miR-21* levels were higher than those with taxane-sensitive disease [66]. Although the majority of the biomarker studies focused on serum and plasma, several studies assessed the diagnostic potential of miRNA expression in other body fluids. For instance, a study compared *miR-31* expression levels in the saliva between patients with oral squamous-cell carcinoma and healthy individuals [67]. They found significantly higher *miR-31* expression in cancer patients than the controls, and a decrease of *miR-31* expression in cancer patients after tumor resection. Hanke et al. analyzed miRNA expression levels in urine samples, and reported that the ratio of *miR-126* to *miR-152*, and the ratio of *miR-182* to *miR-152*, could predict the presence of bladder cancer with 80% specificity and sensitivity [68]. Future investigations to determine the expression profiles of other circulating miRNAs may provide novel and more precise biomarkers for diagnosis, prognosis, treatment-response prediction and patient selection in clinical trials.

miRNAs as Therapeutic Options

Despite the therapeutic advances in cancer treatment, cancer incidence and cancer-related mortalities have not been dramatically reduced [69]. This highlights the necessity of searching for novel and effective anticancer agents. The deregulation of miRNA expression in cancer offers an excellent opportunity of targeting miRNA for cancer treatment. Two general strategies can be employed: the restoration of tumor suppressor miRNAs and the silencing of oncogenic miRNAs (as summarized in Figure 25.3). The advantage of miRNA-based therapeutics is their multiple-targeting ability, whereby relatively small changes in the miRNA expression or function can exert widespread effects on many protein-coding genes. MiRNAs are able to regulate multiple genes from the same pathways at different levels and prevent compensatory mechanisms such as mutation in the targeted oncogenes, which could cause resistance to cancer treatments. As a result, the therapeutic potential of miRNAs is now under extensive exploration by many research groups. Encouragingly, the effectiveness of miRNA-based therapeutics has been demonstrated by both *in vitro* and *in vivo* experiments, as we will discuss below.

Restoring miRNA Function

The re-establishment of miRNA function can be achieved by miRNA oligonucleotide mimics (miR-mimics), which are double-stranded RNAs designed to mimic the function of the endogenous miRNAs. The mimics can be administered by delivery systems such as nanoparticles or liposomes. Coating the nanoparticles with antibodies allows for precise delivery of miRNA into targeted cancer cells. For instance, nanoparticles coated with a neuroblastoma-specific antibody were shown to deliver the *miR-34a* mimics into the neuroblastoma tumor cells and inhibit the tumor growth in an orthotopic xenograft mouse model [70]. The formulation using neutral lipid emulsions was successful in systematic delivery of *miR-34a* and *let-7* mimics, and reduced the tumor formation in the lung in a K-Ras-activated mouse model [71]. Likewise, delivery of *miR-34a* by a telocollagen effectively inhibited xenografted colon cancer growth in mice. More recently, MRX34, a liposome-formulated*miR-34* mimic developed by Mirna Therapeutics, is scheduled for a phase 1 clinical trial in patients with advanced or metastatic liver cancers [72]. Strikingly, it produced complete tumor regression in two separate orthotopic mouse models of liver cancer and was also effective in mouse models of lung and metastatic colorectal cancer. No immunostimulatory activity and toxicity to normal tissues were observed in the treatments using MRX34 [73].

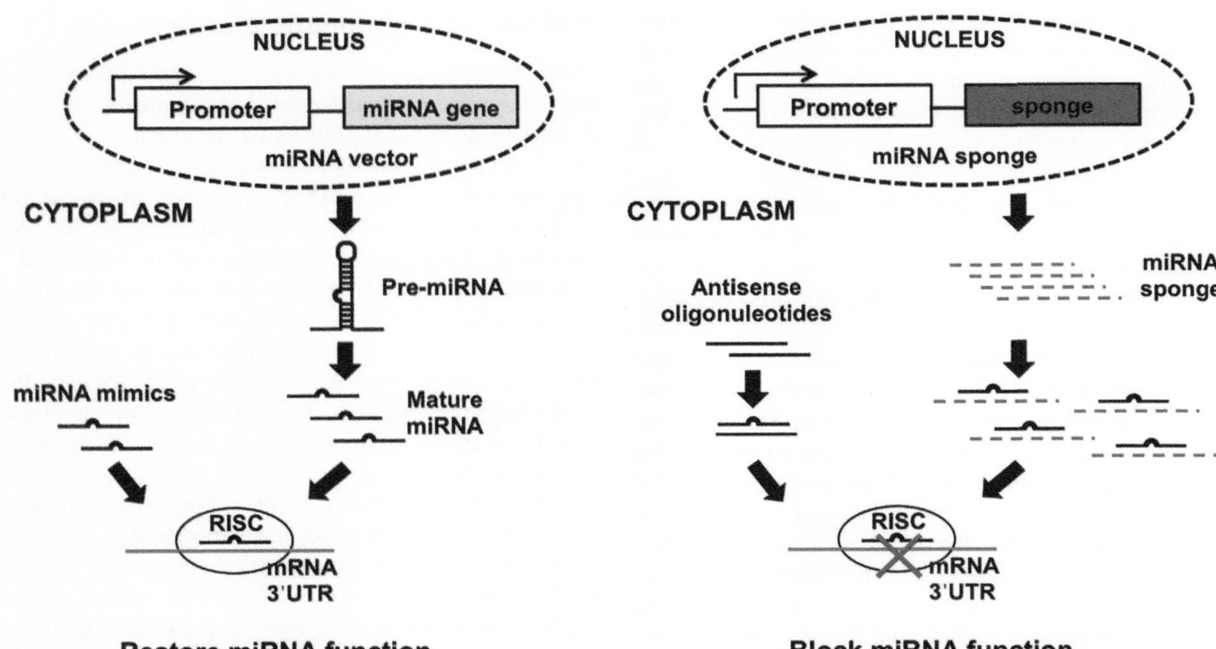

FIGURE 25.3 Strategies of microRNA therapeutics. The function of tumor suppressor miRNAs can be restored by synthetic miRNA mimics, or by viral constructs overexpressing miRNAs. The oncogenic miRNAs can be blocked by using antisense oligonucleotides such as antagomirs, locked nucleic acid (LNA) or by viral constructs overexpressing miRNA sponges, which compete with the miRNA-targeting mRNAs for miRNA binding, and thus block miRNA's function.

The other way to re-supply miRNA function is to use expressing vectors, such as adenovirus-associated virus (AAV) vectors. The benefit of AAV vectors is their high efficiency of transduction and the fact that they are maintained as extrachromosomal episomes or integrate site-specifically into genomes, rather than the random integration associated with retroviruses and lentiviruses which can lead to cellular transformation [74]. Kota et al. used AAV to overexpress the tumor suppressor *miR-26*, an miRNA that is highly expressed in normal liver tissue, but lost in hepatocellular carcinoma [75]. Systemic administration of *miR-26* in a mouse model of liver cancer inhibits tumor formation by reducing proliferation and promoting apoptosis of cancer cells [75]. This study shows for the first time that miRNA replacement may inhibit *in vivo* tumor growth. The AAV system has also been modified for targeted delivery of miRNA in a tumor-specific fashion. For instance, an *miR-26a* expression vector driven by a dual promoter for alpha-fetoprotein (hAFP) and human telomerase reverse transcriptase (hTERT) was shown to induce tumor-specific expression of *miR-26a* and cause reduction in liver tumor growth [76]. Although as noted above, AAV-based gene delivery systems do not run the risk of mutations associated with genetic insertion into the host genome, they can result in the stimulation of an immune response that may be harmful to the host [77].

Blocking miRNA Function

Current strategies for blocking miRNA function use antisense oligonucleotides such as antagomirs, anti-miR constructs, and miRNA sponges to block the oncogenic miRNAs. Due to the functioning mechanism, these strategies may or may not change the expression levels of the target miRNA. In the absence of alteration in miRNA expression, the blocking effects should be monitored by other means such as luciferase reporter fused to miRNA binding sites, or evaluation of the expression of protein-coding genes targeted by the miRNA.

The antisense oligonucleotides are easily degraded by nucleases. Chemical modifications such as 2′-O-methylation (2′-O-Me), 2′-O-methoxyethylation (2′-O-MOE), and 2′-fluoro substitution (2′-F) have been applied to increase their stability and binding affinity to a specific miRNA [78]. Antagomirs are synthetic antisense oligonucleotides characterized by the 2′-O-Me modification on the ribose sugar, the terminal phosphorothioate linkage, and the cholesterol conjugation at the 3′ end. Krutzfeldt et al. developed the first antagomirs, which include antagomir-16, -192 and -194, and delivered them by tail vein injection into mice. They observed a diffused silencing of the respective miRNAs in the liver, lung, kidney, heart, intestine, fat, skin, bone marrow, muscle, ovaries, and adrenal glands [79]. The Weinberg group applied antagomir targeting the pro-metastatic*miR-10b* and successfully decreased the number of tumor metastasis to the lung by 80% in a mouse model of breast cancer [80]. In this study, the treatment with antagomir only caused very minor toxic effects in animals. The effectiveness and safety of antagomirs suggest that they have the potential of translating into clinical applications. The other type of chemical modification is called locked nucleic acid (LNA), which is an RNA analog with the ribose locked in a C3′-endo conformation by the introduction of a 2′-O,4′-C methylene bridge. This modification greatly improved the binding affinity of the oligonucleotide to its mRNA target. Based on the scientific findings that *miR-122* upregulates hepatitis C virus (HCV) RNA replication [81], SantarisPharma A/S developed an LNA-based antisense molecule against *miR-122*, named SPC3649 for the treatment of HCV infection [4]. SPC3649 has been evaluated in phase I and phase 2a clinical trials. As the first miRNA therapeutic agent entering clinical trial, SPC3649 significantly reduced HCV expression and was well tolerated in human subjects [4].

MiRNA sponges are artificial transgenes containing multiple complementary sequences to the targeting miRNA. By binding the miRNA, miRNA sponges block the interaction of the miRNA with its endogenous mRNA targets. The sponges can be stably transfected, and thus the inhibitory effect can be sustained for a long time without repeated administration. This is useful when investigating a biological function of miRNA *in vivo*. Care et al. created the first miRNA sponge by cloning multiple *miR-133* target sequences into an AAV vector, with the aim of studying *miR-133* function in cardiac hypertrophy [82]. The miR-133 sponges effectively blocked *miR-133* function and induced sustained cardiac hypertrophy in mice. Another example is *miR-31*, which was found to be downregulated in aggressive metastatic cancer. After transducing a non-metastatic breast cancer cell line with miR-31 sponge, researchers found that this cell line metastasized to the lung as a result of the silencing of *miR-31* by the sponge [47]. The sponge experiments revealed the biological function of *miR-31* as a suppressor of metastasis.

Challenges

Notwithstanding the encouraging advances in miRNA-based therapeutics, it has been realized that many obstacles have to be overcome before the successful development of miRNA-targeting therapeutics [78]. The main challenge is how to deliver the miRNA therapeutic agents to tumors without compromising their efficacy. For instance, the miRNA-targeting agents are easily degraded by nucleases. Furthermore, renal clearance will also reduce their bioavailability. Chemical modifications, which can stabilize the nucleotides, may yield off-target effects or reduce the efficiency. These limitations of miRNA therapeutics must be considered during drug development in order to achieve a meaningful biological effect.

UCGs AND CANCER

Deregulated UCG Expression in Cancer

In 2004, Bejerano et al. first described the concept of ultraconserved regions, a subset of conserved genomic sequences longer than 200 bp that are absolutely conserved with 100% identity with no insertions or deletions between orthologous regions of the human, rat, and mouse genomes [83]. Their study showed that the ultraconserved regions are often located in the regions involved in RNA processing or regulation of RNA transcription. More than half of the 481 ultraconserved regions described by Bejerano et al. have no evidence of protein-coding potential.

The findings on functional importance of ultraconserved regions prompted Calin et al. to detect transcriptional activity from these genomic regions by microarray analysis. They found that 93% of the ultraconserved genomic regions are transcribed in at least one tissue, and thus represent ultraconserved genes (UCGs) that are expressed both ubiquitously and in a tissue-specific manner in normal human tissues [6]. After identification of UCGs in normal tissues, they performed expression profiling of UCGs in a panel of 133 human leukemia and carcinoma samples and 40 corresponding normal tissues, and identified specific signatures associated with each cancer type. Signatures composed of 19 UCGs (8 up- and 11 downregulated), 61 UCGs (59 up- and 2 downregulated), and 8 UCGs (3 up- and 5 downregulated) were associated with CLL, colorectal cancer and hepatocellular carcinoma, respectively [6]. Table 25.2 lists the most significant of the differentially expressed UCGs involved in CLL and colorectal cancer.

Analysis of UCG genomic location unveiled that UCGs are highly associated with the CAGRs which include minimal regions of amplification, fragile sites, loss of heterozygosity, and common breakpoint regions in or near oncogenes or tumor suppressors [6]. This association of UCGs with CAGRs is comparable to that reported for miRNAs. Most importantly, the cancer-specific UCGs are frequently located in CAGRs that have been linked to that type of cancer. For example, a cluster of seven UCGs is located in the chromosomal 13q21.33-q22.2, a fragile site associated with familial CLL with no aberrant protein-coding genes identified. Two of the UCGs, *uc.349A* and *uc.352*, are differentially expressed in normal versus malignant B-CLL CD5-postitive cells [6]. These findings suggest that UCG expression may

TABLE 25.2 Ultraconserved Genes (UCGs) Involved in Chronic Lymphocytic Leukemia (CLL) and Colorectal Cancer (CRC)

UCG Name	CLL	CRC	CAGR Location
uc.29	Normal	Up	No
uc.73	Down	Up	No
uc.111	Normal	Up	Yes
uc.112	Normal	Up	No
uc.134	Normal	Up	No
uc.135	Down	Normal	Yes
uc.206	Normal	Up	No
uc.230	Normal	Up	No
uc.233	Down	Normal	No
uc.291	Down	Normal	No
uc.292	Normal	Up	No
uc.339	Normal	Up	Yes
uc.341	Normal	Up	Yes
uc.388	Normal	Up	No
uc.399	Normal	Up	No
uc.420	Normal	Up	No

Up: upregulated in cancer versus normal counterpart; Down: downregulated in cancer versus normal counterpart; Normal: no change in cancer versus normal counterpart; CAGR: cancer-associated genomic region

contribute to cancer risk in this genomic region, and thus deregulated expression of UCGs may affect cancer susceptibility or cancer development.

UCGs as Oncogenes or Tumor Suppressors

The deregulated expression of UCGs in cancer indicates their roles as oncogenes or tumor suppressors. For instance, the upregulation of *uc.73A* in colorectal cancer suggests that it may possess oncogenic functions [6]. The researchers screened the colon cancer cell lines for *uc.73A* expression and identified that COLO320 cells have high *uc.73A* expression. Transfection of COLO320 cells with *uc.73A*-targeting small interfering RNAs (siRNAs) effectively reduced *uc.73A* expression levels and, most excitingly, reduced COLO320 cell proliferation and induced apoptosis in this colon cancer cell line. This effect is *uc.73A*-dependent as the regulation of siRNAs on proliferation and apoptosis was not observed in the SW620 cells, which have low *uc.73A* levels comparable to that of normal colon cells [6]. These findings suggest that *uc.73A*may act as an oncogene and promote development of colorectal cancer. The function and regulatory mechanisms of UCGs are still poorly understood. Nevertheless, a negative correlation of five UCGs and miRNA expression signature was observed in CLL samples [6]. The researchers found that three out of five UCGs had significant antisense complementarity with five out of thirteen miRNAs from the signature [6]. They further demonstrated that *miR-155* directly regulates *uc.346A* and *uc.160*, while *miR-24* directly targets *uc.160* [6]. These findings suggest the existence of a complex network of ncRNA interaction, and this interaction may have clinical significance for cancer patients.

Although not the focus of this review, many other types of lncRNAs have been demonstrated to play roles in cancer. For instance, *HOTAIR*, an antisense intergenic lncRNA transcribed from the *HOXC* gene locus, was found to be upregulated by thousands of times in metastatic breast cancer tissue compared with normal breast tissue [84]. The mechanism of *HOTAIR* function in metastasis was demonstrated to be through reprogramming chromatin occupancy of polycomb repressive complex 2 (PRC2) from a pattern of breast epithelial cell to that of an embryonic fibroblast [84].

CONCLUSION

There is no longer any doubt that ncRNAs are important players in cancer initiation, progression, invasion and dissemination. The elucidation of miRNA biogenesis and regulatory mechanisms, the demonstration of their role as tumor suppressors and/or oncogenes, and the discovery that aberrant miRNA expression signatures are associated with cancer, have greatly improved our understanding of cancer biology. In addition, academia and biopharmaceutical companies have enthusiastically pursued the therapeutic application of miRNAs on human diseases. The first miRNA-targeting agent, SPC3649, has successfully launched a phase I and phase II trial, and this is just the beginning. We believe that, in the near future, miRNA profiles identified through high-throughput transcriptome analyses will be widely used in the clinical setting for cancer diagnosis, prognosis, and prediction of drug response, and that miRNAs will provide therapeutic targets for the treatment of patients by combining miRNA-targeting agents with chemotherapy. While miRNAs have been intensively studied, the evidence of UCGs' and other lncRNAs' involvement in cancer is just emerging. Our unpublished work has shown that some UCGs exhibit unique biological functions, and control multiple phenotypes essential for tumor initiation and progression. Once regarded as the "junk" or "dark matter" of the human genome, the power of ncRNAs in human pathology has been gradually recognized and more mysteries are waiting to be uncovered.

GLOSSARY

3′ untranslated region (3′ UTR) Sequences on the 3′ end of mRNA but not translated into protein.

5′ untranslated region (5′ UTR) Sequences on the 5′ end of mRNA but not translated into protein.

Antagomir Synthetic antisense oligonucleotides characterized by the 2′-O-Me modification on the ribose sugar, the terminal phosphorothioate linkage, and the cholesterol conjugation at the 3′ end.

Antisense oligonucleotides Single strands of DNA or RNA that are complementary to a chosen sequence.

Biomarker A term often used to refer to measurable characteristics that reflect the severity or presence of some disease state.

Cancer-associated genomic regions (CAGRs) Fragile chromosomal loci prone to deletions or amplifications in cancer.

Circulating miRNAs miRNAs present in the cell-free component of blood and body fluids.

Chronic lymphocytic leukemia (CLL) Chronic leukemia of the lymphoblastic type, characterized by lymphadenopathy, fatigue, renal involvement, and pulmonary leukemic infiltrates.

Epithelial–mesenchymal transition (EMT) An orchestrated transition from epithelial to mesenchymal phenotype characterized by loss of cell adhesion, repression of E-cadherin expression, and increased cell motility.

Locked-nucleic acid (LNA) RNA analogs where the ribose is locked in a C3′-endo conformation by the introduction of a 2′-O,4′-C methylene bridge.

Long non-coding RNAs (lncRNAs) Transcribed RNA molecules greater than 200 nucleotides in length without protein-coding function.

Messenger RNAs (mRNAs) A subtype of RNA molecules that mediate the transfer of genetic information from the cell nucleus to ribosomes in the cytoplasm, where it serves as a template for protein synthesis.

MicroRNAs (miRNAs) Evolutionarily conserved single-stranded non-coding RNAs of 19–24 nucleotides.

miR-mimics Double-stranded RNAs designed to mimic the function of the endogenous miRNAs.

miRNA profiling The measurement of the relative abundance of a cohort of miRNAs, ranging from a group of several miRNAs of specific biological interest to comprehensive profiling of all miRNAs in a given species.

miRNA sponge Artificial transgenes containing multiple complementary sequences to the targeting miRNA.

Non-coding RNA Transcribed RNA molecules without protein-coding potential.

Primary microRNAs (pri-miRNAs) Initially transcribed microRNAs from genome containing several hundred nucleotides, which are further processed into precursor microRNAs.

Precursor microRNAs (pre-miRNAs) Hairpin RNAs of 60–100 nucleotides that are processed from primary microRNAs by ribonuclease III Drosha-DGCR8 nuclear complex.

Protein-coding genes Same as messenger RNAs.

Ribonuclease III-Drosha-DGCR8 complex An enzyme complex containing the RNase III enzyme Drosha and double-stranded RNA binding protein DGCR8, which imitates the transcription of primary microRNAs.

RNA-induced silencing complex (RISC) A multiprotein complex that incorporates the mature strand of microRNA as a template for recognizing complementary mRNA.

Ultraconserved genes (UCGs) Non-coding RNAs that are transcribed from a subset of genomic sequences which are strictly conserved among orthologous regions of the human, rat, and mouse genomes.

ABBREVIATIONS

3′ UTR 3′ untranslated region
5′ UTR 5′ untranslated region
AAV Adenovirus-associated virus
CAGR Cancer-associated genomic region
ceRNA Competing endogenous RNA
CLL Chronic lymphocytic leukemia
EMT Epithelial–mesenchymal transition
HCV Hepatitis C virus
LNA Locked nucleic acid
LncRNA Long non-coding RNA
miRNA MicroRNA
mRNA Messenger RNA
ncRNA Non-coding RNA
PCG Protein-coding gene
Pre-miRNAs Precursor microRNAs
Pri-miRNAs Primary microRNAs
RISC RNA-induced silencing complex
siRNA Small interfering RNA
UCG Ultraconserved gene

REFERENCES

[1] The ENCODE Project Consortium. An integrated encyclopedia of DNA elements in the human genome. Nature 2012;489:57–74.

[2] Mattick JS. Makunin IV. Non-coding RNA. Hum Molecr Genet 2006;15:R17–29.

[3] Calin GA, Croce CM. MicroRNA signatures in human cancers. Nature Rev Cancer 2006;6:857–66.

[4] Lindow M, Kauppinen S. Discovering the first microRNA-targeted drug. J Cell Biol 2012;199:407–12.

[5] Spizzo R, Almeida MI, Colombatti A, Calin GA. Long non-coding RNAs and cancer: a new frontier of translational research? Oncogene 2012;31:4577–87.

[6] Calin GA, Liu CG, Ferracin M, Hyslop T, Spizzo R, Sevignani C, et al. Ultraconserved regions encoding ncRNAs are altered in human leukemias and carcinomas. Cancer Cell 2007;12:215–29.

[7] Lee RC, Feinbaum RL, Ambros V. The C. elegans heterochronic gene lin-4 encodes small RNAs with antisense complementarity to lin-14. Cell 1993;75:843–54.

[8] Wightman B, Ha I, Ruvkun G. Posttranscriptional regulation of the heterochronic gene lin-14 by lin-4 mediates temporal pattern formation in C. elegans. Cell 1993;75:855–62.

[9] Kozomara A, Griffiths-Jones S. miRBase: integrating microRNA annotation and deep-sequencing data. Nucleic Acids Res 2011;39: D152–7.

[10] Ling H, Zhang W, Calin GA. Principles of microRNA involvement in human cancers. Chin J Cancer 2011;30:739–48.

[11] Calin GA, Croce CM. MicroRNA-cancer connection: the beginning of a new tale. Cancer Res 2006;66:7390–4.

[12] Spizzo R, Nicoloso MS, Croce CM, Calin GA. SnapShot: microRNAs in cancer. Cell 2009;137:586.

[13] Calin GA, Dumitru CD, Shimizu M, et al. Frequent deletions and down-regulation of micro-RNA genes miR15 and miR16 at 13q14 in chronic lymphocytic leukemia. Proc Natl Acad Sci USA 2002;99:15524–9.

[14] Calin GA, Sevignani C, Dumitru CD, Hyslop T, Noch E, Yendamuri S, et al. Human microRNA genes are frequently located at fragile sites and genomic regions involved in cancers. Proc Natl Acad Sci USA 2004;101:2999–3004.

[15] Croce CM. Causes and consequences of microRNA dysregulation in cancer. Nat Rev Genet 2009;10:704–14.

[16] Hermeking H. The miR-34 family in cancer and apoptosis. Cell Death Different 2010;17:193–9.

[17] Olive V, Jiang I, He L. miR-17-92, a cluster of miRNAs in the midst of the cancer network. Internatl J Biochem Cell Biol 2010;42:1348–54.

[18] Bartel DP. MicroRNAs: target recognition and regulatory functions. Cell 2009;136:215–33.

[19] Hwang HW, Wentzel EA, Mendell JT. A hexanucleotide element directs microRNA nuclear import. Science 2007;315:97–100.

[20] Orom UA, Nielsen FC, Lund AH. MicroRNA-10a binds the 5′UTR of ribosomal protein mRNAs and enhances their translation. Molec Cell 2008;30:460–71.

[21] Place RF, Li LC, Pookot D, Noonan EJ, Dahiya R. MicroRNA-373 induces expression of genes with complementary promoter sequences. Proc Natl Acad Sci USA 2008;105:1608–13.

[22] Tay Y, Zhang J, Thomson AM, Lim B, Rigoutsos I. MicroRNAs to Nanog, Oct4 and Sox2 coding regions modulate embryonic stem cell differentiation. Nature 2008;455:1124–8.

[23] Eiring AM, Harb JG, Neviani P, Garton C, Oaks JJ, Spizzo R, et al. miR-328 functions as an RNA decoy to modulate hnRNP E2 regulation of mRNA translation in leukemic blasts. Cell 2010;140:652–65.

[24] Salmena L, Poliseno L, Tay Y, Kats L, Pandolfi PP. A ceRNA hypothesis: the Rosetta Stone of a hidden RNA language? Cell 2011;146:353–8.

[25] Chen CZ, Li L, Lodish HF, Bartel DP. MicroRNAs modulate hematopoietic lineage differentiation. Science 2004;303:83–6.

[26] Klein U, Lia M, Crespo M, Siegel R, Shen Q, Mo T, et al. The DLEU2/miR-15a/16-1 cluster controls B cell proliferation and its deletion leads to chronic lymphocytic leukemia. Cancer Cell 2010;17:28–40.

[27] Lee YS, Kim HK, Chung S, Kim KS, Dutta A. Depletion of human micro-RNA miR-125b reveals that it is critical for the proliferation of differentiated cells but not for the down-regulation of putative targets during differentiation. J Biolog Chem 2005;280:16635–41.

[28] Giraldez AJ, Cinalli RM, Glasner ME, Enright AJ, Thomson JM, Baskerville S, et al. MicroRNAs regulate brain morphogenesis in zebrafish. Science 2005;308:833–8.

[29] Poy MN, Eliasson L, Krutzfeldt J, Kuwajima S, Ma X, Macdonald PE, et al. A pancreatic islet-specific microRNA regulates insulin secretion. Nature 2004;432:226–30.

[30] Esau C, Kang X, Peralta E, Hanson E, Marcusson EG, Ravichandran LV, et al. MicroRNA-143 regulates adipocyte differentiation. J Biolog Chem 2004;279:52361–5.

[31] Medina PP, Nolde M, Slack FJ. OncomiR addiction in an *in vivo* model of microRNA-21-induced pre-B-cell lymphoma. Nature 2010;467:86–90.

[32] Hatley ME, Patrick DM, Garcia MR, Richardson JA, Bassel-Duby R, van Rooij E, et al. Modulation of K-Ras-dependent lung tumorigenesis by MicroRNA-21. Cancer Cell 2010;18:282–93.

[33] Asangani IA, Rasheed SA, Nikolova DA, Leupold JH, Colburn NH, Post S, et al. MicroRNA-21 (miR-21) post-transcriptionally downregulates tumor suppressor Pdcd4 and stimulates invasion, intravasation and metastasis in colorectal cancer. Oncogene 2008;27:2128–36.

[34] He L, Thomson JM, Hemann MT, Hernando-Monge E, Mu D, Goodson S, et al. A microRNA polycistron as a potential human oncogene. Nature 2005;435:828–33.

[35] Mu P, Han YC, Betel D, Yao E, Squatrito M, Ogrodowski P, et al. Genetic dissection of the miR-17∼92 cluster of microRNAs in Myc-induced B-cell lymphomas. Genes Devel 2009;23:2806–11.

[36] Costinean S, Zanesi N, Pekarsky Y, Tili E, Volinia S, Heerema N, et al. Pre-B cell proliferation and lymphoblastic leukemia/high-grade lymphoma in E(mu)-miR155 transgenic mice. Proc Natl Acad Sci USA 2006;103(18):7024–9.

[37] Valeri N, Gasparini P, Fabbri M, Braconi C, Veronese A, Lovat F, et al. Modulation of mismatch repair and genomic stability by miR-155. Proc Natl Acad Sci USA 2010;107:6982–7.

[38] Ma L, Young J, Prabhala H, Pan E, Mestdagh P, Muth D, et al. miR-9, a MYC/MYCN-activated microRNA, regulates E-cadherin and cancer metastasis. Nature Cell Biol 2010;12:247–56.

[39] Ma L, Teruya-Feldstein J, Weinberg RA. Tumour invasion and metastasis initiated by microRNA-10b in breast cancer. Nature 2007;449:682–8.

[40] Cimmino A, Calin GA, Fabbri M, Iorio MV, Ferracin M, Shimizu M, et al. miR-15 and miR-16 induce apoptosis by targeting BCL2. Proc Natl Acad Sci USA 2005;102:13944–9.

[41] Calin GA, Cimmino A, Fabbri M, Ferracin M, Wojcik SE, Shimizu M, et al. MiR-15a and miR-16-1 cluster functions in human leukemia. Proc Natl Acad Sci USA 2008;105:5166–71.

[42] Raveche ES, Salerno E, Scaglione BJ, Manohar V, Abbasi F, Lin YC, et al. Abnormal microRNA-16 locus with synteny to human 13q14 linked to CLL in NZB mice. Blood 2007;109:5079–86.

[43] Chang TC, Wentzel EA, Kent OA, Ramachandran K, Mullendore M, Lee KH, et al. Transactivation of miR-34a by p53 broadly influences gene expression and promotes apoptosis. Molec Cell 2007;26:745–52.

[44] Liu C, Kelnar K, Liu B, Chen X, Calhoun-Davis T, Li H, et al. The microRNA miR-34a inhibits prostate cancer stem cells and metastasis by directly repressing CD44. Nat Med 2011;17:211–5.

[45] Takamizawa J, Konishi H, Yanagisawa K, Tomida S, Osada H, Endoh H, et al. Reduced expression of the let-7 microRNAs in human lung cancers in association with shortened postoperative survival. Cancer Res 2004;64:3753–6.

[46] Esquela-Kerscher A, Trang P, Wiggins JF, Patrawala L, Cheng A, Ford L, et al. The let-7 microRNA reduces tumor growth in mouse models of lung cancer. Cell Cycle 2008;7:759–64.

[47] Valastyan S, Reinhardt F, Benaich N, Calogrias D, Szász AM, Wang ZC, et al. A pleiotropically acting microRNA, miR-31, inhibits breast cancer metastasis. Cell 2009;137:1032–46.

[48] Korpal M, Lee ES, Hu G, Kang Y. The miR-200 family inhibits epithelial-mesenchymal transition and cancer cell migration by direct targeting of E-cadherin transcriptional repressors ZEB1 and ZEB2. J Biol Chem 2008;283:14910–4.

[49] Korpal M, Ell BJ, Buffa FM, Ibrahim T, Blanco MA, Celiá-Terrassa T, et al. Direct targeting of Sec23a by miR-200s influences cancer cell secretome and promotes metastatic colonization. Nat Med 2011;17:1101–8.

[50] Felli N, Fontana L, Pelosi E, Botta R, Bonci D, Facchiano F, et al. MicroRNAs 221 and 222 inhibit normal erythropoiesis and erythroleukemic cell growth via kit receptor down-modulation. Proc Natl Acad Sci USA 2005;102:18081–6.

[51] Pineau P, Volinia S, McJunkin K, Marchio A, Battiston C, Terris B, et al. miR-221 overexpression contributes to liver tumorigenesis. Proc Natl Acad Sci USA 2010;107:264–9.

[52] Hanahan D, Weinberg RA. Hallmarks of cancer: the next generation. Cell 2011;144:646–74.

[53] Santarpia L, Nicoloso M, Calin GA. MicroRNAs: a complex regulatory network drives the acquisition of malignant cell phenotype. Endocrine-Relat Cancer 2010;17:F51–75.

[54] Lu J, Getz G, Miska EA, Alvarez-Saavedra E, Lamb J, Peck D, et al. MicroRNA expression profiles classify human cancers. Nature 2005;435:834–8.

[55] Rosenfeld N, Aharonov R, Meiri E, Rosenwald S, Spector Y, Zepeniuk M, et al. MicroRNAs accurately identify cancer tissue origin. Nat Biotechnol 2008;26:462–9.

[56] du Rieu MC, Torrisani J, Selves J, Al Saati T, Souque A, Dufresne M, et al. MicroRNA-21 is induced early in pancreatic ductal adenocarcinoma precursor lesions. Clin Chem 2010;56:603–12.

[57] Garzon R, Garofalo M, Martelli MP, Briesewitz R, Wang L, Fernandez-Cymering C, et al. Distinctive microRNA signature of acute myeloid leukemia bearing cytoplasmic mutated nucleophosmin. Proc Natl Acad Sci USA 2008;105:3945–50.

[58] Calin GA, Ferracin M, Cimmino A, Di Leva G, Shimizu M, Wojcik SE, et al. A MicroRNA signature associated with prognosis and progression in chronic lymphocytic leukemia. N Engl J Med 2005;353:1793–801.

[59] Chung GE, Yoon JH, Myung SJ, Lee JH, Lee SH, Lee SM, et al. High expression of microRNA-15b predicts a low risk of tumor recurrence following curative resection of hepatocellular carcinoma. Oncol Rep 2010;23:113–9.

[60] Fabbri M, Bottoni A, Shimizu M, Spizzo R, Nicoloso MS, Rossi S, et al. Association of a microRNA/TP53 feedback circuitry with pathogenesis and outcome of B-cell chronic lymphocytic leukemia. J Am Med Assoc 2011;305:59–67.

[61] Chen X, Ba Y, Ma L, Cai X, Yin Y, Wang K, et al. Characterization of microRNAs in serum: a novel class of biomarkers for diagnosis of cancer and other diseases. Cell Res 2008;18:997–1006.

[62] Cortez MA, Bueso-Ramos C, Ferdin J, Lopez-Berestein G, Sood AK, Calin GA. MicroRNAs in body fluids – the mix of hormones and biomarkers. Nat Rev Clin Oncol 2011;8:467–77.

[63] Lawrie CH, Gal S, Dunlop HM, Pushkaran B, Liggins AP, Pulford K, et al. Detection of elevated levels of tumour-associated microRNAs in serum of patients with diffuse large B-cell lymphoma. Br J Haematol 2008;141:672–5.

[64] Michael A, Bajracharya SD, Yuen PS, Zhou H, Star RA, Illei GG, et al. Exosomes from human saliva as a source of microRNA biomarkers. Oral Dis 2010;16:34–8.

[65] Hu Z, Chen X, Zhao Y, Tian T, Jin G, Shu Y, et al. Serum microRNA signatures identified in a genome-wide serum microRNA expression profiling predict survival of non-small-cell lung cancer. J Clin Oncol 2010;28:1721–6.

[66] Zhang HL, Yang LF, Zhu Y, Yao XD, Zhang SL, Dai B, et al. Serum miRNA-21: elevated levels in patients with metastatic hormone-refractory prostate cancer and potential predictive factor for the efficacy of docetaxel-based chemotherapy. Prostate 2011;71:326–31.

[67] Liu CJ, Lin SC, Yang CC, Cheng HW, Chang KW. Exploiting salivary miR-31 as a clinical biomarker of oral squamous cell carcinoma. Head Neck 2012;34:219–24.

[68] Hanke M, Hoefig K, Merz H, Feller AC, Kausch I, Jocham D, et al. A robust methodology to study urine microRNA as tumor marker: microRNA-126 and microRNA-182 are related to urinary bladder cancer. Urol Oncol 2010;28:655–61.

[69] Siegel R, Naishadham D, Jemal A. Cancer statistics, 2013. CA Cancer J Clin 2013;63:11–30.

[70] Tivnan A, Orr WS, Gubala V, Nooney R, Williams DE, McDonagh C, et al. Inhibition of neuroblastoma tumor growth by targeted delivery of microRNA-34a using anti-disialoganglioside GD_2 coated nanoparticles. PloS One 2012;7:e38129.

[71] Trang P, Wiggins JF, Daige CL, Cho C, Omotola M, Brown D, et al. Systemic delivery of tumor suppressor microRNA mimics using a neutral lipid emulsion inhibits lung tumors in mice. Molec Ther 2011;19:1116–22.

[72] Mirna Therapeutics Secures $34.5 Million Series C Financing to Advance Oncology microRNA Pipeline into the Clinic. Business Wire; 2012.

[73] Bader AG. miRNA replacement therapy. Keystone Conference on Noncoding RNAs in Development and Cancer. Vancouver, Canada; 2013.

[74] Michelfelder S, Trepel M. Adeno-associated viral vectors and their redirection to cell-type specific receptors. Adv Genet 2009;67:29–60.

[75] Kota J, Chivukula RR, O'Donnell KA, Wentzel EA, Montgomery CL, Hwang HW, et al. Therapeutic microRNA delivery suppresses tumorigenesis in a murine liver cancer model. Cell 2009;137:1005–17.

[76] Chen L, Zheng J, Zhang Y, Yang L, Wang J, Ni J, et al. Tumor-specific expression of microRNA-26a suppresses human hepatocellular carcinoma growth via cyclin-dependent and -independent pathways. Molec Ther 2011;19:1521–8.

[77] Mingozzi F, High KA. Immune responses to AAV in clinical trials. Curr Gene Ther 2007;7:316–24.

[78] Garzon R, Marcucci G, Croce CM. Targeting microRNAs in cancer: rationale, strategies and challenges. Nat Rev Drug Discov 2010;9:775–89.

[79] Krutzfeldt J, Rajewsky N, Braich R, Rajeev KG, Tuschl T, Manoharan M, et al. Silencing of microRNAs *in vivo* with 'antagomirs'. Nature 2005;438:685–9.

[80] Ma L, Reinhardt F, Pan E, Soutschek J, Bhat B, Marcusson EG, et al. Therapeutic silencing of miR-10b inhibits metastasis in a mouse mammary tumor model. Nat Biotechnol 2010;28:341–7.

[81] Jopling CL, Yi M, Lancaster AM, Lemon SM, Sarnow P. Modulation of hepatitis C virus RNA abundance by a liver-specific MicroRNA. Science 2005;309:1577–81.

[82] Care A, Catalucci D, Felicetti F, Bonci D, Addario A, Gallo P, et al. MicroRNA-133 controls cardiac hypertrophy. Nat Med 2007;13(5):613–8.

[83] Bejerano G, Pheasant M, Makunin I, Stephen S, Kent WJ, Mattick JS, et al. Ultraconserved elements in the human genome. Science 2004;304:1321–5.

[84] Gupta RA, Shah N, Wang KC, Kim J, Horlings HM, Wong DJ, et al. Long non-coding RNA HOTAIR reprograms chromatin state to promote cancer metastasis. Nature 2010;464:1071–6.

Chapter 26

Cancer Genomics

The Impact of the Environment on Cancer Genomics

Louise Parker

Departments of Medicine and Pediatrics, Dalhousie University, Halifax, NS, Canada

Contents

Cancer Genomics. DOI: http://dx.doi.org/10.1016/B978-0-12-396967-5.00026-8

Key Concepts

- Over 90% of cancers are due to exposures to chemicals (exogenous and endogenous) and radiation. The totality of these exposures across a lifetime is the *exposome*. Changing the exposome, i.e. reducing exposures causing or promoting cancer and increasing exposures which prevent cancer will change cancer risk
- Because the world population is aging, the number of new cancer cases diagnosed each year will double to 21 million by 2030. Millions of these cases could be avoided by changing the exposome
- Measuring the exposome is challenging; technological advances in exposome technology are needed on a similar scale to those that have happened in genomic research
- Improved exposomic technologies will support research not only into identifying carcinogenic substances and their interactions, but will also provide insight into their mechanisms of action which, importantly, include epigenomic modifiction through methylation, histone modification and changes to non-coding RNA-mediated signaling pathways. Some of these modifications may be remedial
- Better quantitation of the exposome will, in the long term, reduce the burden of cancer by reducing exposures and mitigating their effects

CANCER: A GLOBAL DISEASE

In the more affluent populations of the world, around half of all men and more than a third of all women will be diagnosed with cancer over the course of their lifetimes. One quarter of all adults die of this disease.

The International Agency for Research in Cancer (IARC) compiles cancer registration and mortality data from all over the world. It estimates that, in 2008, there were 12.6 million new cases of cancer diagnosed worldwide. Cancer is predominantly a disease of older age: in 2008, 55% of newly diagnosed cancers in the USA were in men and women over the age of 65 years [1]. Assuming that the underlying rates of disease remain the same for men and women of different ages, by 2030 there will be over 21 million new cases of cancer diagnosed each year worldwide because of the aging of the world population. In the USA, there were 1.4 million new cases of cancer diagnosed in 2008 and this number will rise to over 2.2 million per year by 2030; 1.5 million (68%) of these cases will be in men and women over the age of 65 years. The cost of this increase in people diagnosed with cancer, in both social and economic terms, is enormous. Using data from the US National Institutes of Health (NIH) on the cost of cancer care in the US for 2010, it is possible to estimate the additional healthcare costs associated with this projected increase in cancer cases due to the demographic shift in the population [2]. Shown in Table 26.1, taking into account only those costs incurred during the first year of cancer treatment and last year of life and assuming that survival and treatment costs remain unchanged, the *additional* financial cost of the increase in the number of cancer cases will be $67 billion per year by 2030. This implies that over the next 20 years care for the additional US cancer patients that will occur as a consequence of the aging population will cost at least a trillion dollars – this is in addition to the 2.3 trillion dollars that the USA would spend on cancer care over the same time period if the number of patients remained the same as in 2008. These estimates are conservative; they do not take into account the fact that healthcare costs have been rising at around 7.5% per year – far higher than the rate of inflation and that new cancer therapies are increasingly more expensive.

This projected vast increase in the number of cancer patients underlines the importance and urgency in better characterizing the avoidable causes of cancer so that public health policy and environmental and workplace protection legislation can support reductions in avoidable exposures to carcinogens and so reduce cancer burden. Further, understanding the mechanisms by which such exposures cause cancer has the potential to contribute to the development of strategies to mitigate the risk of cancer associated with these exposures when it is not possible to eliminate the exposure itself. Understanding the etiological mechanisms in detail also has the potential to uncover new targets for novel therapeutics.

CANCER IS A FAMILY OF DISEASES

Cancer is a collection of more than a hundred different diseases which can be grouped into five broad categories based on their origins (Figure 26.1) [3].

TABLE 26.1 Number of Cancer Cases Diagnosed and Cancer Deaths in 2008 and Projections for 2030 (GLOBOCAN) Additional Direct Costs of Treatment Based on NIH 2010 Costs for US Cancer Care

	Number of New Cases (GLOBOCAN)[a]			Number of Deaths (GLOBOCAN)			
CANCER	2008	2030	Increase 2008 to 2030	2008	2030	Increase 2008 to 2030	Additional Costs[b] by 2030 Due to Increase in Patient Numbers
MALES							
All cases	745 187	1 214 862	469 675	294 115	502 555	208 440	
Brain	11 779	16 551	4772	7419	11 137	3718	$1 049 092 192
Colorectal	79 271	13 1308	52 037	24 509	42 039	17 530	$4 197 953 674
Esophagus	12 970	20 952	7982	11 250	18 515	7265	$1 390 824 834
Lung	114 691	195 399	80 708	90 809	156 869	66 060	$8 688 770 452
Lymphoma	39 853	61 010	21 157	10 491	17 788	7297	$2 135 052 069
Pancreas	18 771	31 245	12 474	17 500	29 580	12 080	$2 540 132 808
Prostate	186 320	313 030	126 710	28 660	53 651	24 991	$4 052 943 922
Stomach	13 189	21 882	8693	6450	10 879	4429	$1 156 470 699
Other	268 343	423 485	155 142	97 027	162 097	65 070	$12 703 390 110
FEMALES							
All cases	692 012	1 005 830	313 818	271 529	424 984	153 455	
Brain	10 029	13 647	3618	5652	8237	2585	$755 555 059
Breast	182 460	251 587	69 127	40 481	59 623	19 142	$2 798 502 458
Colorectal	74 610	116 134	41 524	26 131	41 773	15 642	$3 453 348 546
Esophagus	3500	5531	2031	3030	4849	1819	$351 211 174
Lung	100 330	158 065	57 735	71 032	11 3641	42 609	$7 437 227 87
Lymphoma	34 493	50 918	16 425	10 019	16 052	6033	$1 613 372 211
Ovary	21 652	30 780	9128	15 519	23 866	8347	$1 584 075 069
Pancreas	18 914	30 117	11 203	16 790	27 073	10 283	$2 177 578 089
Stomach	8310	12 888	4578	4429	6973	2544	$589 346 280
Uterus	40 102	56 419	16 317	7469	11 891	4422	$747 201 525
Other	197 612	279 744	82 132	70 977	111 006	40 029	$7 133 546 514
OVERALL							$66 555 595 556

[a]*International Agency for Reseach in Cancer (IARC). GLOBOCAN2008. Retrieved January 12, 2013 from http://globocan.iarc.fr/*
[b]*US National Institutes of Health (NIH) Cancer Prevalence and Cost of Care Projections. Retrieved January 12, 2013 from http://costprojections.cancer.gov/annual.costs.html*

This diversity in disease is an important consideration in the study of cancer: different cancers have different causes, respond differently to treatments and are consequently more or less curable. For example, exposure to tobacco smoke causes more than four out of five incident cases of lung cancer, a disease which responds poorly to current treatment approaches; 5-year survival is low (16% reported for 2005–2009 in North Americans) [4]. In contrast, the majority of breast cancers are related to exposure to endogenous estrogen (i.e. naturally occurring) or exogenous estrogen derived from birth control or hormone replacement therapies. Tobacco smoking plays a relatively minor role in the causation of breast cancer. In addition, both lung cancer and breast cancer have a

Major Cancer Groups (*)

Carcinoma – cancer that begins in the skin or in tissues that line or cover internal organs. There are a number of subtypes of carcinoma, including adenocarcinoma, basal cell carcinoma, squamous cell carcinoma, and transitional cell carcinoma.

Sarcoma – cancer that begins in bone, cartilage, fat, muscle, blood vessels, or other connective or supportive tissue.

Leukemia – cancer that starts in blood-forming tissue such as the bone marrow and causes large numbers of abnormal blood cells to be produced and enter the blood.

Lymphoma and myeloma – cancers that begin in the cells of the immune system.

Central nervous system cancers – cancers that begin in the tissues of the brain and spinal cord

* US National Institutes of Health (NIH) National Cancer Institute. Defining Cancer. Retrieved January 12, 2013 from http://www.cancer.gov/cancertopics/cancerlibrary/what-is-cancer

FIG. 26.1 Summary of major cancer groups.

number of other potential causal agents, such as radiation and diet. It should also be noted that the outcome of therapy is very different for different cancers. Breast cancer is far more curable than lung cancer and current treatment using multimodal approaches (chemotherapy, radiotherapy and surgery) results in a 5-year survival rate approaching 90% in North America [4].

Over 99% of all cancers occur in adults (cancer in childhood is rare and makes up less than 1% of the population burden of this disease) and, as noted above, more than half of all cancers are diagnosed in those over the age of 65 years. In adults, carcinomas are by far the most common form of cancer. In North American men and women, the four most common types of carcinomas account for around 60% of all incident cases of cancer; carcinomas of the prostate, lung, colorectum and bladder in men and of the breast, lung, colorectum and uterus in women. Figure 26.2 illustrates the different distribution of cancers in Canadian men and women [5], which is similar to that of adults in Europe, the USA and the Antipodes.

The pattern of cancer occurrence is not only different between men and women, but also varies strikingly across the world and even within regions of the same country. Table 26.2 demonstrates some of the substantial differences in cancer incidence rates between men and women and between people resident in different parts of the world [1]. These rates are age-adjusted which means that they take into account the different age structures of the populations of different countries and reflect the underlying risk of disease.

In general, rates of cancer tend to be higher in men than in women and adults in Africa and Asia tend to have lower rates of the cancers which are most common in North America, Europe and the Antipodes; prostate, breast, colorectal and lung cancer. Rates of cancer can also vary substantively within the same country. In Canada, for example, the incidence of colon cancer in men in Newfoundland is 60% higher than in men in British Colombia [5] and within the USA, the incidence rate of bladder cancer in Maine is twice that of New Mexico [6].

There are two possible explanations for these major differences in cancer risk between populations: differences in genetic heritage or differences in environmental exposures.

Susceptibility to cancer tends not to be passed down from one generation to another and, for the majority of cases of adult cancer, powerfully acting, inherited cancer genes are not involved [7]. Studies of the incidence of cancer in monzygotic and dizygotic twins have confirmed that inherited genetic factors make a small contribution to individual risk for most types of cancer [8]. The observation that, within a few generations, immigrants tend to exhibit the cancer pattern of their adopted country further supports the view that genetic inheritance is not a major driver of cancer risk – since genes of course remain unchanged by migrant status over this time frame [9]. This leads inevitably to the conclusion that it is exposure to carcinogenic and other substances due to our environment and lifestyle, not our genes, that drives these substantive differences in cancer risk between different populations and the changes in these patterns that have been observed over time, such as the emergence of lung cancer and breast cancer as common cancers in the twentieth century.

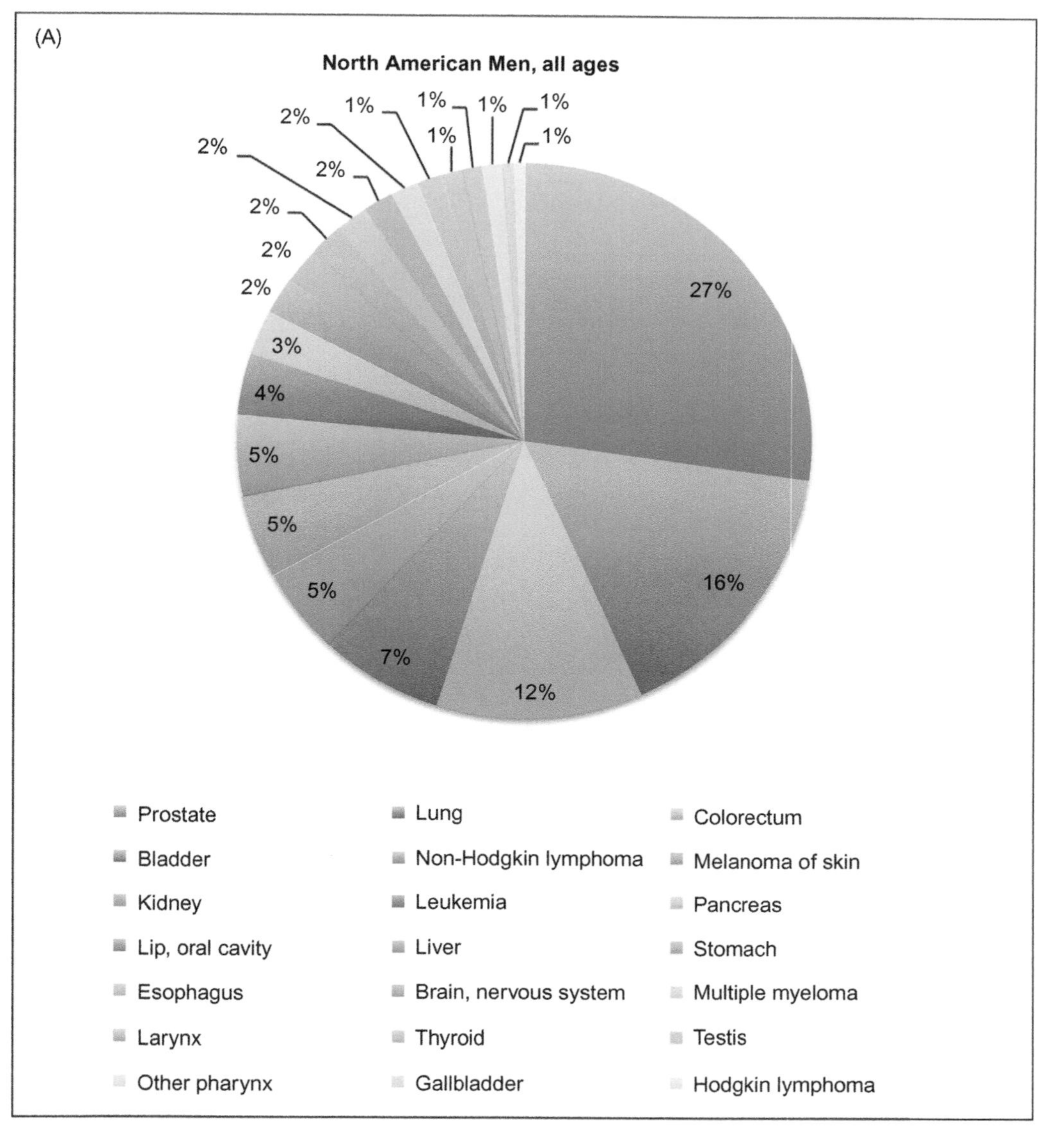

FIG. 26.2 Distribution of common cancers in North American (A) men and (B) women

CANCER IS AN ENVIRONMENTAL DISEASE

Narrow and Broad Definitions of Environmental Exposure

The estimates of the proportion of cancers which are due to "environmental" exposure vary enormously, from 1 to 100%, with the major differences arising largely from variation in the definition of what constitutes an environmental exposure – the "narrow" definition or the "broad" [10]. The "narrow" definition of environmental exposure generally includes exposures derived from air, water and soil. Some of the most common narrow definition environmental exposures affecting cancer risk and classified as "Group 1" carcinogens as defined by the International Agency for Research on Cancer [11] are asbestos, environmental tobacco smoke, household coal emissions, radon in air, arsenic, cadmium, chromium IV, pentachlorobiphenyl, radium in water, and petroleum products, pesticides, heavy metals in soil [10]. The "broader" definition of environment includes not only the contaminants of air, water and soil as captured within the narrow definition, but also the many other exposures that arise exogenously – for example from diet, lifestyle, health behavior, psychosocial environment, and occupational exposures. Clearly, the latter broader definition of environmental exposure will include causal factors for a larger proportion of adult cancers than the former narrow definition.

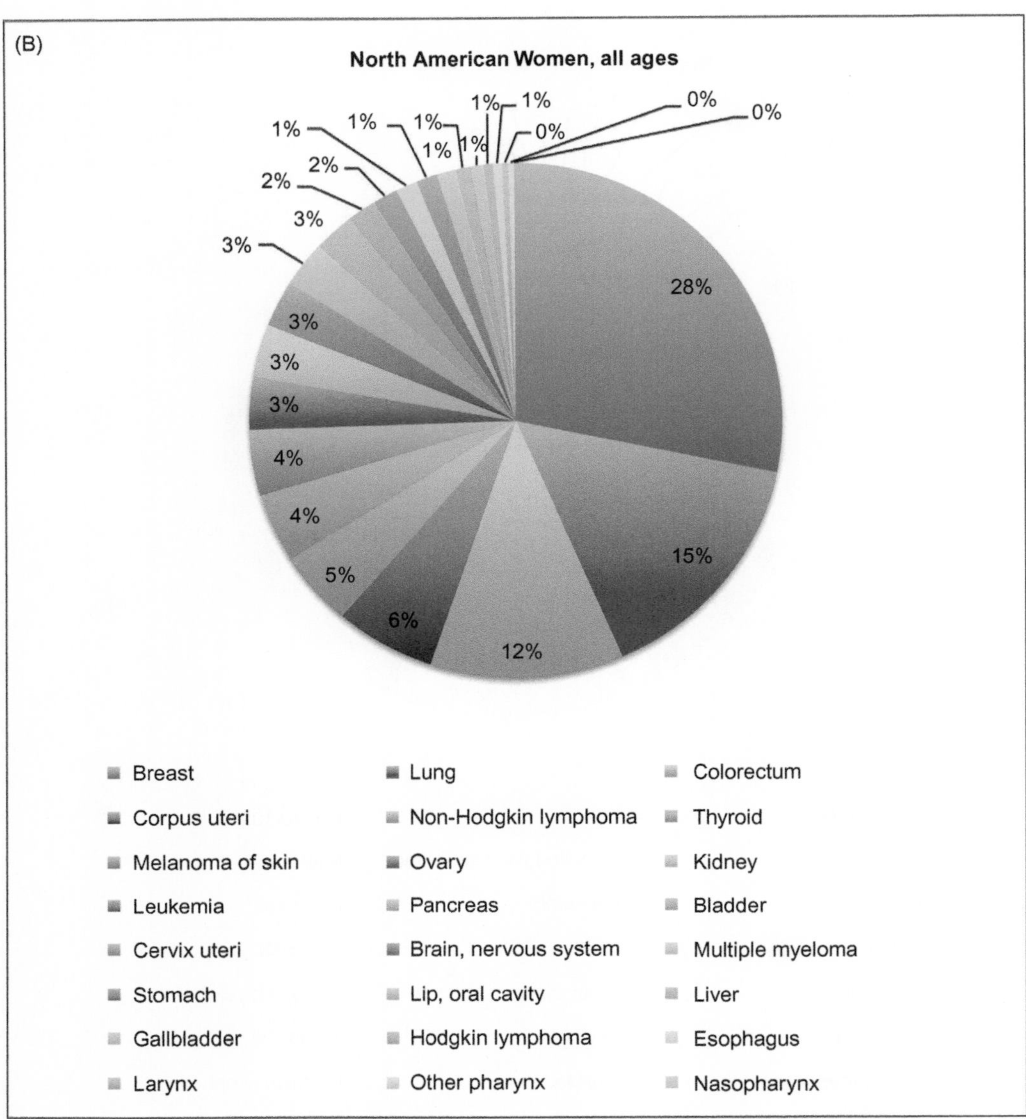

FIG. 26.2 (Continued.)

The Exposome

The average age at which a man or woman develops cancer is over 65 years. Living our long lives in our complex world means that we all experience a myriad of environmental exogenous and endogenous exposures which can act singly, collectively or synergistically to change our cancer risk. We begin to accumulate these exposures from the moment we are conceived and their interaction with our bodies at a nuclear, cellular and organ level are major determinants of our health trajectory. Indeed, some of the most important exposures in the context of long-term health may well occur before we are even born – and the thalidomide tragedy of the 1950s and 1960s which resulted in babies exposed to thalidomide *in utero* being born with phocomelia (the shortening or absence of limbs), demonstrates the specific vulnerability of the developing and rapidly growing fetus to environmental exposures. The exposures which impact on our risk of cancer include not only all of those chemical and radiation exposure captured within the broad definition of the environment, but also endogenous chemical exposures – such as steroid hormones (e.g. estrogen) and chemical changes due to metabolic processes such as chronic inflammation.

This complexity of not only the exogenous, but also the endogenous exposure landscape has been captured by the concept of the "exposome" described by Wild in 2005 as being the totality of environmental exposures affecting the internal chemical exposures from conception

TABLE 26.2 Age Standardized Incidence of Common Cancers in Men and Women Aged 40-74

	FEMALE ASR per 100 000			MALE ASR per 100 000		
	BREAST	COLORECTAL	LUNG	PROSTATE	COLORECTAL	LUNG
AFRICA						
Western Africa	88	12	**4**	52	**15**	9
Middle Africa	**56**	**9**	2	33	**12**	7
Middle-East and Northern Africa (MENA)	88	21	10	25	26	65
Northern Africa	87	15	**6**	18	18	43
Western Sahara	80	17	**0**	27	19	39
Southern Africa	102	18	21	120	49	68
AMERICA						
Northern America	*210*	62	*97*	*240*	88	*126*
Central America	71	16	14	81	18	31
South America	120	29	22	113	33	53
EUROPE						
Central and Eastern Europe	129	57	27	70	86	*162*
Northern Europe	*238*	62	*61*	191	*89*	92
Western Europe	250	*63*	49	*248*	103	122
Southern Europe	189	*61*	28	118	*97*	*132*
OCEANIA						
Melanesia	**60**	**9**	13	34	30	**34**
Micronesia	150	58	46	56	63	125
Micronesia/Polynesia	157	35	44	80	49	125
Australia/New Zealand	*242*	*80*	*53*	*283*	*114*	79
ASIA	70	27	33	**15**	38	83
South-Central Asia	**63**	**11**	9	**9**	**12**	36
South-Eastern Asia	86	33	31	**18**	39	77
Western Asia	89	26	14	31	33	87
ECONOMIC REGIONS						
Less developed regions	73	24	28	26	31	72
Medium human development	68	23	30	15	30	76
Very high human development	199	63	58	180	99	112
World	105	35	35	67	50	87

Top 3 regions shown in bold, lowest 3 regions shown in italic
GLOBOCAN 2008, IARC-21.1.2013

onwards [12]. The US Centers for Disease Control and Prevention (CDC) [13] extended this concept, defining the exposome as:

the measure of all the exposures of an individual in a lifetime and how those exposures relate to health... Understanding how exposures from our environment, diet, lifestyle, etc., interact with our own unique characteristics such as genes, physiology, and epigenetics impact on our health is how the exposome will be articulated.

There is no doubt that cancer is an environmental disease and it is estimated that while 5–10% of cancers (and other chronic diseases) can be accounted for by genetic defects, around 90–95% are due to the environment with the important components of this including cigarette smoking and diet, alcohol, sun exposure, environmental pollutants, infections, stress, obesity, and physical inactivity [14].

The challenge is to understand the mechanisms by which these environmental exposures impact on the genome, the epigenome and physiological and metabolic functions to change the risk of disease.

The exposome includes all exposures which impact on our cancer risk – not only those toxic chemicals (and radiation) which enter the body from exogenous sources (the traditional concept of environmental exposures) but also, importantly, endogenous exposures – the internal exposures which result from internal processes such as inflammation, metabolic processes, disease, infection and gut flora [15,16].

The exposome is an addition to the "omics" technologies, with the potential to deliver huge amounts of information on individual chemical and radiation exposure landscapes and together with other omic data aid in the unraveling of the determinants of health and disease and the mechanism by which these factors act.

Measuring Exposures and Determining Causation

Environmental exposures act by initiating or promoting the development of malignancy. Accurate and precise measures of these exposures are central to understanding the importance of their role in human health, but remain an extraordinary challenge. The measurement of the complete exposome of an individual may prove to be impossible. However, there is nonetheless the opportunity for great strides to be made in our ability to grasp some of the vastness and complexity of the exposome within the research arena. Historically, exposures have often been considered singly (e.g. exposure to tobacco smoke or asbestos or radiation) and have seldom been measured prospectively. Exposure histories are more commonly constructed retrospectively. There is seldom a contemporaneous measure of exposure and it is often estimated from an individual's recall of their exposure or even, in some cases, by the recall of a family member or friend in studies where potential respondents may have died, for example. Recall can be challenging; in the case of tobacco smoke for example, the average age of taking up cigarette smoking has historically been between 10 and 15 years of age but the average time of diagnosis of smoking-related lung cancer is some 50 years later. Even though tobacco smoking is a discrete activity which incurs significant financial cost and specific procurement activities on behalf of the smoker, it is a challenge to construct a reasonable estimate of lifelong tobacco smoke exposure for an individual. When did they start smoking? How many per day? How many times did they change their consumption – to increase it? To decrease it? Even when a measure of number of cigarettes consumed has been constructed, moving from this to a "dose" of carcinogens remains a further hazard – what type of cigarette: low tar, high tar? Filtered, unfiltered? – and then individual variation in the way cigarettes are smoked can materially affect the ingestion of chemicals – depth of inhalation, rate of draw etc. These factors contribute to the complexity of exposure estimation and inevitably lead to misclassification of exposure.

Exposures relevant to health at age 70 years may not only include a multitude of different complex exposures, but may also include some which occurred many decades earlier – well before any health consequences were apparent or even considered! As with thalidomide, the timing of the exposure may be an integral part of its biological importance.

The general approach to exposure measurement has been "bottom-up", i.e. estimation of exposures by the approaches described above without direct measurement of the internal chemical milieu. Since empirical measures of exposures are challenging to make, research in this area often relies on mathematical modeling of exposure from surrogate data (e.g. environmental monitoring of pollution or food intake reports). Rappaport reported that fewer than 40% of papers published in the *Journal of Exposure Science and Environmental Epidemiology* had empirical measures of exposure [15]. Such reliance on surrogate measures of exposure inevitably leads to misclassification of exposures which generally leads to an underestimate of effect; the importance of the impact of many environmental exposures on risk of chronic disease will have been underestimated or even missed entirely. This is especially true when considering endogenous exposures; the levels of toxic chemicals resulting from internal processes and diet, for example, are generally much greater than the levels of such chemical resulting from exposure to pollutants in water or air, and yet these chemicals, and their complex interactions with physiological processes, such as activity levels and stress, have been very poorly measured [15].

Without effective biological markers (an antibody or DNA adduct for example), it is currently impossible to characterize many past or even current exposures accurately or even at all. However, measurement of the exposome through biomarkers in body tissues (blood, urine, hair, nails, other tissues) is in its early stages. The growing acknowledgement and interest in the importance of the exposome in human health should lead to a focus of research energy and investment in this area; the strides that could be made by such a focus are potentially huge.

Consider the Human Genome Project which moved, in the space of less than two decades, from DNA sequencing technologies that delivered data on 50 kilobases per day to sequencers which delivered data on over a hundred million kilobases per day – a two-million fold improvement [16]! Application of such innovation to the challenges of measuring the exposome would certainly enable far superior characterization than is currently possible.

Many estimates of external environmental exposure use geographical references (such as postal codes or zip codes) to locate individuals in space (and time) to enable an estimation of exposure based on geographically referenced environmental data. Exposures to such factors as air and water pollution are commonly estimated in this way. For occupational exposures, the type and nature of employment may be used to estimate exposure to chemical by the employment of exposure matrices which provide exposure estimates for various occupations. For uranium miners exposed to the radioactive gas radon during their underground activities, radiation exposure was estimated from information on how long they worked underground and at what depth. Such exposure estimates are necessarily approximate and are likely to be associated with considerable misclassification at the individual level, leading again to an underestimate of effect.

The importance of diet in determining health and cancer risk is increasingly apparent and the constituents of diet are an integral part of the exposome. However, measuring food intake in a meaningful way to enable estimation of micronutrient intakes, such as vitamins, minerals, trace elements, is extremely challenging. Many dietary assessments involve recall over a period of a day or week at most – none is able to assess diet over a lifetime.

There are two major epidemiological consequences of the challenge in exposure measurements. First, inaccuracies in measuring exposures results in misclassification of exposure. Some individuals will be considered to have some or high exposure when they had none or very little and, conversely, some who had high exposures will be considered to have had none. The consequence of this misclassification is, in most cases, to reduce the empirical estimate of association or effect which means that, in most cases, the contribution of the exposure to cancer risk will be underestimated. The second challenge is that while it is apparent that there are important interactions between different elements of the exposome, for example arsenic and radon are much more carcinogenic for tobacco smokers than for non-smokers, but with limited ability to measure elements of the exposome either at all or with some degree of accuracy, studying complex interactions between these elements of the exposome is difficult.

While measuring exposures, either singly within either the narrow or broader definition of the environment or the more ambitious consideration of the exposome remains a significant challenge, the understanding of the mechanisms by which these multiple exposures interact to affect cancer risk poses an even greater challenge.

Despite the challenges of exposure measurement, epidemiological studies have correctly identified a number of important environmental carcinogens – soot (known to increase scrotal cancer in chimney sweeps in the 1800s), ionizing radiation (known to be responsible for bone cancer in radium dial painters of the 1930s) and tobacco smoke (known to be a cause of lung cancer in doctors in the 1940s) being among the first environmental carcinogens identified. IARC notes several thousand possible and proven environmental human carcinogens. To be a proven carcinogen, human research findings must be consistent with the Bradford Hill criteria for causation shown in Figure 26.3.

IMPORTANT COMPONENTS OF THE EXPOSOME

Although there are many proven or putative environmental human carcinogens which contribute to the exposome, there are some which stand out as being of major public health importance because they are responsible for a substantial number of cancer cases worldwide and because they are potentially avoidable.

In 2010, Parkin and colleagues estimated the fraction of 18 common cancers attributable to 14 lifestyle and environment risk factors (tobacco, alcohol, four elements of diet (consumption of meat, fruit and vegetables, fiber and salt), overweight/obesity, lack of physical activity, occupation, infections, radiation (ionizing and solar), use of hormones, and breast feeding) known to have a significant impact on cancer risk, and which can potentially be avoided by change in behavior and lifestyle [17]. They did not include exposures, such as natural estrogen for example, which they considered not readily changeable by reasonable lifestyle choices. Their approach was to take reported effect size estimates for these risk factors (such as those reported for diet, physical activity and obesity by the World Cancer Research Fund) [18] and bring these data together with reported data on prevalence of these risk factors in the UK population (e.g. number of smokers, proportion of adults meeting physical activity guidelines) and align this with cancer incidence and mortality data to estimate the impact of the risk factor prevalence on cancer risk in the population.

The key findings of this report were that, overall, around 43% (45% in men and 40% in women) of all newly diagnosed cancer cases and 50% of cancer deaths

Bradford Hill Criteria for Causal Association (*)

Strength of association (relative risk, odds ratio): Stronger associations are generally considered to be more likely to reflect a causal relationship than weaker associations. However, this is not always the case in cancer epidemiology: increased risk associated with low-level exposures may be small, (e.g. low levels of radiation), but the association is nonetheless causal.

Consistency: Results should be consistent between different studies involving different people and different places. This is the basis of the meta-analysis approach.

Specificity: The association should generally be specific (e.g. cause a specific cancer rather than all cancers or cancer mortality rather than all cause mortality). However, this is not always the case in cancer epidemiology where different cancers share a common etiology (e.g. in utero X-rays increase the risk of all childhood cancers) and cancer shares a common etiology with other diseases (e.g. smoking causes heart disease, stroke and lung disease as well as cancer).

Temporal relationship (temporality): The exposure must precede the onset of disease.

Biological gradient (dose–response relationship): In general the greater the exposure the greater the effect.

Plausibility: There must be a biologically plausible explanation by which the exposure under consideration could cause the disease identified. This is more complicated than it would first appear since the biological plausibility of an association depends on the extent of knowledge of biological systems; prior to the appreciation of the potential role of epigenetics in carcinogenesis, for example, it would have not been considered plausible for an exposure that did not cause direct change to the genome to be carcinogenic.

Coherence: The finding needs to be consistent with what else is known about the disease and the exposure

Experiment: For example, does an intervention to reduce the exposure also reduce the disease?

Analogy (consideration of alternate explanations)

* Bradford Hill A. The environment and disease: Association or causation? Proc R Soc Med 1965;58:295-300

FIG. 26.3 Bradford Hill criteria.

were accounted for by the 14 exposures they considered and were theoretically preventable. The implication of this is that 134 000 cases of avoidable cancer are diagnosed each year in the UK [19]. The equivalent number for the USA is in excess of 600 000.

Tobacco

Despite clear evidence from Sir Richard Doll's study of lung cancer in doctors, first published 60 years ago [20], exposure to tobacco smoke remains far and away the single most important environmental carcinogen on Earth. Tobacco smoking remains common – even though rates of tobacco smoking in North America have halved in the past 20 years, around one in five North American adults still smokes tobacco contributing to the one billion smokers in the world today. Despite the fact that tobacco kills up to half of its users, world consumption of tobacco products is still increasing, though it is falling in some high-income and upper middle-income countries [21]. Some seven trillion cigarettes are currently manufactured every year.

Tobacco is the single biggest avoidable cause of cancer. Parkin et al. report that tobacco smoking accounted for 19% of all new cancer cases in the UK in 2010 [22].

Men who smoke are more than 20 times more likely to develop lung cancer than non-smokers and over 85% of all lung cancer in men and 80% of that in women is caused by tobacco smoking. The risk of many cancers, including those of the oral cavity and pharynx, esophagus, stomach, liver, pancreas, colorectal, larynx, cervix, ovary, urinary tract and acute myeloid leukemia, is increased by between 20 and 1460% by tobacco smoking [22]. The proportion of new cases of these cancers considered to be due to tobacco smoking in the UK in 2010 is summarized in Table 26.3.

TABLE 26.3 Proportion of Selected Cancers Attributable to Smoking, UK 2010 [22]

Cancer	Number of Cases	Number Due to Smoking	Population-Attributable Fraction (%)
Oral Cavity	6928	4474	65
Esophagus	8528	5583	65
Stomach	7043	1565	22
Liver	3568	819	23
Pancreas	8362	2396	29
Colorectal	39 911	3235	8
Larynx	2185	1727	79
Cervix	2693	195	7
Ovary	6820	177	3
Bladder	9282	3403	37
Kidney and Ureter	9061	2181	24
Myeloid Leukemia	3642	486	13
Lung	40 405	33 586	83
OVERALL	148 428	59 827	40

Overall Parkin et al. considered that 40% of these selected cancers were directly due to cigarette smoking [22]. It is worth noting also that even though the population-attributable fraction of colorectal cancer due to smoking is low (around 7%), because colorectal cancer is one of the most common cancers in men and women, the number of cases of colorectal cancer due to smoking is still in the order of several thousands per year.

Parkin et al. also considered environmental tobacco smoke (ETS) exposure and concluded that 7% of lung cancers in men and 10% in women were attributable to exposure to ETS while cohabiting with a smoking partner and that, overall, 14% of lung cancer is due to exposure of non-smokers to ETS [22]. Eighty-seven percent of lung cancers are attributable either to smoking or to exposure to ETS in non-smokers.

Dietary, Physical Activity and Overweight

The 2007 report Food, Nutrition, Physical Activity and the Prevention of Cancer: a Global Perspective [18], produced by the World Cancer Fund and chaired by Sir Michael Marmot weighed the evidence for cancer risk for diet, overweight/obesity and physical activity. Concluding that poor diet, overweight and under activity are major, avoidable causes of cancer, their main recommendations are listed in Figure 26.4.

Parkin et al. in their 2010 study of environmental causes of cancer in the UK [17] identified that, after tobacco smoking for men, the next most important risk factors were insufficient intake of fruit and vegetables (accounting for 6% of all cancers); for women, overweight and obesity are responsible for 7% of cancers.

The developed, and increasingly the developing world, is in the midst of an epidemic of obesity. In North America, more than two thirds of the adult population is overweight and one third is obese. Of particular concern is that the onset of obesity is occurring earlier: childhood obesity even 50 years ago was rare, now it is commonplace and it is estimated that nearly 20% of North American children are obese. Obesity increases the risk of a number of important diseases including osteoarthritis, diabetes, cardiovascular disease and, importantly, cancer. In the Million Women Study, Reeves et al. reported that obesity increases the risk of cancer by around 12% overall, but the increase in risk associated with obesity was much higher for some cancers, such as endometrial cancer and adenocarcinoma of the esophagus where the increased risk was more than 100% [23].

Over and above the substantial increase in cancer cases due to the general aging of the population, as discussed earlier, increases in obesity in North America and elsewhere will further add to the world cancer burden. The majority of older adults who are obese now and on whom these estimates were based were not obese as children. The risk of adult cancer in those who have been obese since childhood and hence experienced lifelong obesity is unknown but will certainly become apparent in the decades to come.

Recommendations of the World Cancer Research Fund [18]

Body fatness: Be as lean as possible within the normal range of body weight

Physical activity: Be physically active as part of everyday life

Foods and drinks that promote weight gain: Limit consumption of energy-dense foods; Avoid sugary drinks

Plant foods: Eat mostly foods of plant origin

Animal foods: Limit intake of red meat and avoid processed meat

Alcoholic drinks: Limit alcoholic drinks

Preservation, processing, and preparation: Limit consumption of salt; avoid mouldy cereals (grains) or pulses (legumes)

Dietary supplements: Aim to meet nutritional needs through diet alone

Breastfeeding: Mothers to breastfeed; children to be breastfed

FIG. 26.4 Lifestyle recommendations from the World Cancer Research Fund.

Parkin et al. [19] estimated that 17 294 excess in cancer cases occurring in the UK 2010 were due to overweight and obesity (5.5% of all cancers).

Parkin et al. further identified that, in the UK, low physical activity levels resulted in an estimated 3.4% of breast cancer cases, 3.8% of endometrial cancer cases and 5.3% of colon cancer cases during 2010 [24]. These estimates correspond to 1.0% of all cancer cases, 0.4% in men and 1.7% in women.

The World Cancer Research Fund recommends that all should remain as thin as possible within a normal healthy weight, i.e. body mass index (weight (kg)/height $(m)^2$) 19–25 kg/m^2 [18].

Alcohol

IARC considers that alcohol is a carcinogen causally implicated in cancer of the oral cavity, pharynx, larynx, esophagus, liver and breast [11]. The increased risk of cancer is dose related such that the excess risk of cancer is proportional to the alcohol intake.

Overall Parkin et al. concluded that alcohol is the third most important avoidable carcinogen in men, accounting for 4.5% of all male cancers [25]. They report that alcohol is responsible for 30% of oral and pharynx cancers, 21% of esophagus, 12% of colorectal, 9% of liver, 24% of larynx and 6% of breast cancer.

Historically, the relationship between alcohol and cancer has been difficult to study because of confounding with tobacco smoking. The important causal relationship between alcohol intake and breast cancer was reported definitively in research performed on the UK Million Women Study, which was of sufficient size to demonstrate convincingly that it is alcohol rather than tobacco smoking that plays the major role in breast cancer carcinogenesis [26].

Occupational Exposures

Cancer-causing occupational exposures include radiation, particulates (such as asbestos and wood dust), metals such as cadmium and solvents such as aromatic hydrocarbons and dyes.

Parkin concluded that around 5% of all cancers in men in the UK in 2010 were caused by occupational exposure [27].

Radiation

Ionizing radiation has been recognized as a cause of cancer for over 100 years. Both Marie Curie, who discovered radium and coined the term radioactivity, and her daughter Irene died of leukemia after a lifetime of radiation research. Exposure to ionizing radiation is one of the very few established causes of childhood cancer; X-ray exposure *in utero* was found to increase the risk of both solid tumors and leukemia in children by around 30% [28]. Low-dose radiation exposure during childhood is also implicated in childhood leukemia [29]. Occupational exposure to radiation was first confirmed as carcinogenic in studies of women occupationally exposed to radium paint while painting luminous clock dials during the early years of the twentieth century [30]. Studies of cancer rates in the atomic bomb survivors from Hiroshima and Nagasaki, in men and women exposed to radiation for cancer and other therapies (such as for tinea skin infections) and in occupationally exposed groups, such as nuclear workers and radiographers, are regularly reviewed by UNSCEAR (United Nations Standing Committee on the Effects of Atomic Radiation) which publishes reports of the main findings of these studies with regard to dose response for specific cancers and effect modifiers, such as age (younger people are more radiosensitive, presumably because they have more dividing cells) and cigarette smoking (uranium miners exposed to the radioactive gas radon during the course of their work are much more at risk of consequent radiation-induced cancer if they also smoke) [31].

Cancers are not equally inducible by radiation, for example breast cancer is highly radiogenic while prostate cancer is much less so. Melanoma is highly UV-radiogenic and excess exposures to this through tanning beds and unprotected sunbathing have translated directly into higher rates of this skin cancer in young women in particular. Recent studies have been able to measure the excess in leukemia and brain tumors in young people due to their exposure to radiation through diagnostic CT scans [32].

The Social Environment: Impact on the Exposome

The way we live can have dramatic unforeseen consequences on our exposome and have major impact on our cancer risk. Changes in patterns of breast cancer are an example of this.

Many factors increase risk of breast cancer, for example alcohol intake, smoking, obesity, inactivity, diet and the presence of *BRCA1* or *BRCA2* genes. However, breast cancer risk is highly dependent on life-time exposure to estrogen, both endogenous (i.e. produced by the ovaries) and exogenous (taken as a component of birth control or hormone replacement therapy). Around 70% of breast tumors have estrogen receptors which enables estrogen to act as a growth factor. Over time, dramatic societal and physiological changes have contributed substantially to the rise in the risk of breast cancer. As described by Greaves, earlier menarche, later menopause, a dramatic reduction in the number of pregnancies and in the duration of breast feeding mean that a Western woman of the twenty-first century has perhaps three times as many menstrual cycles than her historical compatriot [7]. This leads her to be highly exposed to estrogen which brings with it the elevated risk of breast cancer. Within and between contemporary communities, variations in these characteristics account for substantive differences in the risk of breast cancer.

MECHANISMS OF ENVIRONMENTAL CARCINOGENESIS

Cancer is a complex multifactorial disease in which environmental chemicals and radiation interact with normal cells in a way that supports the initiation of cancer – the transformation of a normal cell into a malignant one, or the progression of a cancerous clone. Cancer has historically been regarded as a genetic disease, with inherited or spontaneous mutation leading to abnormal cell function. However, it is now accepted that changes in the epigenome – even in the absence of genetic change – plays important roles in the initiation and progression of cancer [33].

Not everyone who is exposed to a carcinogen will develop cancer – it is a stochastic process rather than a deterministic one – not every life-long smoker will develop a smoking-related cancer. Hence, the relationship between environmental exposure and cancer is multifactorial; individuals have different susceptibilities to different exposures, which will depend on their inherent characteristics including their genes and gene expression for example, but also on the totality of their exposome and the interactions between its components.

Genetic and Epigenetic Change

Despite the excitement and eager anticipation surrounding the unraveling of the mysteries of the human genome in the Human Genome Project, genetics has been found to account for only about 5–10% of disease, and the remaining causes appear to be environmental. It is apparent that genetic variation alone is not sufficient to explain the phenotypic differences observed between individuals. Monozygotic twins, for example, despite sharing an identical genome often do not experience the same diseases or share the same traits. Additional mechanisms over and above genetic variation must be involved, and this is where epigenetic phenomena play a crucial role in determining and explaining diversity. Epigenetic alteration leads to heritable changes in gene regulation (by DNA methylation, post-translational histone modification and non-coding RNA-mediated signaling pathways), which do not involve change in the DNA sequences [34,35]. The importance of epigenetics – the variation in gene expression resulting from variable modifications in DNA methylation and chromatin structure in chronic disease – and, importantly, in the relationship between environmental exposures and disease risk and in cancer etiology is becoming increasingly apparent. For example, a commonly reported epigenetic change in cancer cells is hypermethylation of cytosine-phosphate-guanine (CpG)-rich regions of gene promoters which blocks the initiation of gene transcription. Transcription of a number of tumor suppressor genes, e.g. $p16^{INK4a}$, *BRCA1*, *p53* and *hMLH-1* is inhibited by this mechanism in several tumors. On the other hand, widespread demethylation of the cancer genome is an early event in carcinogenesis and can result in activation of oncogenes such as *H-RAS* [34,35]. These changes in epigenetic patterning have been observed in pre-cancerous cells suggesting that they are early events in the process of carcinogenesis and may initiate changes in oncogenic pathways, i.e. they are likely to be the cause rather than the *consequence* of malignant transformations. Indeed, global hypomethylation of peripheral blood mononuclear cells has been proposed as a biomarker of cancer risk since it has been found to discriminate between those with and without cancer [36]. Epigenetic gene regulation is also likely to play a role in cancer progression, supporting tumor growth and metastasis [37]. Cancer is a disease of abnormal epigenetics as much as a disease of abnormal genetics.

Modification of chromatin structure by modification of histone proteins, such as deacetylation, also results in either gene silencing or activation. Epigenetic changes resulting in reversible changes in gene expression in response to intracellular or extracellular factors are increasingly implicated in carcinogenesis. Further, it is becoming apparent that the exposome can act at a cellular level to induce epigenomic change leading to changes in gene expression despite an absence of genotoxicity; a chemical does not need to be mutagenic to cause cancer. The changes in gene expression resulting from epigenetic modification consequent to environmental exposure, can be profound since epigenetic modification can occur throughout the genome rather than being confined to a small number of sites. Epigenetic modification can be acquired, but equally and importantly, it has the potential to be reversed which has important implications for the development of novel therapeutics [35,37].

Epigenetics and Environmental Carcinogenesis: Three Examples

Radiation

It was originally considered that radiation caused cancer as a result of the misrepair of DNA strand breaks. However, there is increasing evidence that other mechanisms, including epigenetic modification, are also involved. For example, irradiated rodent thymus tissue has been shown to have DNA strand breaks which are repaired within a month or so, but also to have persistent widespread DNA hypomethylation – opening the possibility of a role of epigenetic regulation (radiation-induced DNA hypomethylation) in radiation-induced genome instability and aberrant gene expression [38].

The mechanisms of non-targeted or bystander effects of radiation exposure, whereby the function of non-irradiated cells is altered by the irradiation of other cells remain to be explained and brings the likelihood that both extranuclear and extracellular events contribute to the biological consequences of exposure to low doses of radiation [39] and, like the role of epigenetic modification in the response to radiation, challenges the early paradigm of direct disruption to the DNA helix being the source of radiogenic disease.

Arsenic

Arsenic is a major, established human carcinogen and a significant contaminant of drinking water in many parts of the world including the northeastern part of North America. Arsenic occurs naturally in the Earth's crust and can be dissolved in deep water acquifers which are the source of water for drilled wells. Airborne arsenic can also be a risk in an occupational setting. Arsenic is a known cause of lung, bladder, kidney, liver and skin cancers. Arsenic is genotoxic. It induces deletion mutations and chromosomal alterations. Arsenic-related tumors have been found to have specific genetic characteristics. For example, bladder tumors in those exposed to arsenic, and where arsenic is presumed to have played an etiologic role, are associated with higher levels of chromosomal instability than bladder tumors not associated with arsenic exposure. These arsenic-associated changes are also considered to result in more aggressive, less treatment-responsive disease [40]. This led to the proposed mechanism of arsenic tumorigenesis as being the induction of numerous structural chromosomal abnormalities through indirect effects on DNA. However, arsenic has also been shown to induce change in the epigenome.

Ren et al. conducted a review of the role of epigenetic dysregulation in arsenic-induced cancer [41]. They reported that arsenic has been shown to alter the epigenome in a number of different ways. *In vitro* studies looking especially at epigenomic change have demonstrated that arsenic can induce changes in:

- methylation levels of both global DNA and gene promoters
- histone acetylation, methylation and phosphorylation.

The extent to which these changes reflect the mechanisms by which arsenic induces cancer in humans is as yet unknown; human studies have yet to be performed [41].

Folate

Folate (vitamin B9) is present naturally in many foods including vegetables (especially dark green leafy vegetables), fruits and fruit juices, nuts, beans, peas, dairy products, poultry and meat, eggs, seafood, and grains [42]. Folate has an essential role in a number of cellular processes including DNA repair and of relevance here is that it acts as a methyl donor supporting epigenetic regulation through methylation. Folate supplementation during pregnancy dramatically decreases the risk of fetal neural tube defects and this is the reason that many countries, including the USA and Canada, fortify breads, cereals, flour and grain products with folate.

Folate supplementation during pregnancy changes DNA methylation patterns in newborn infants and folate deficiency can result in global DNA hypomethylation which is a biomarker of cancer risk [43].

Folate availability therefore plays a crucial role in dynamic, reversible alteration of the epigenome in ways that have been shown to be related to cancer risk [44,45].

The fortification of flours and grains with folate can be considered to be national experiments in altering the exposome to deliver epigenomic programming at a population level. While there is no doubt that this has reduced the rate of infants born with congential malformation (which is the reason it was introduced), there is a question as to whether folate fortification has also changed cancer risk. Two systematic reviews of folate supplementation (i.e. taking dietary folate supplements in addition to any food fortification) and cancer risk have been published recently. Qin et al. reported a significant decrease in melanoma, but not other cancers in patients enrolled into clinical trials involving folate supplementation [46]; while Wien et al. reported a small increase in the risk of prostate cancer following folate supplementation [47]. Few studies have investigated the changes in population cancer risk due to folate food fortification (distinct from individual supplementation). A study from Australia [48] reported a decline in childhood leukemia, which the authors attributed to increased maternal folate consumption but studies from Canada [49], Germany [50], and the USA [51] have been largely negative, with the exception of a folate-related decline in Wilm's tumor.

Moving Knowledge on Environmental Causes of Cancers Forward: The Role of Omics

Understanding the causes of cancer and the plethora of interactions between the individual and his or her environment, including endogenous and exogenous exposures, and the impact of lifestyle and societal choices will provide insights that will reduce the burden of cancer by the reduction of harmful exposures, the promotion of beneficial exposures, the reduction in the impact of exposures, and screening tools better to identify those at risk of disease.

The emergent technologies of metabolomics, epigenomics and exposomics, complementing genomics, transcriptomics and proteomics are poised greatly to increase the power of scientific studies to address the challenging issues of the diversity of the human response to the plethora of chemical and radiation exposures with which we are constantly bombarded.

These technologies will need to be applied in large population studies where exposure history and phenotype are characterized as well as possible and a range of biological specimens are collected. Several countries have initiated such studies including the UK (UK BIOBANK, recruiting 500 000 individuals) [52] and Canada (Canadian Partnership for Tomorrow Project, recruiting 300 000 individuals) [53].

Metabolomics

Metabolomics is the measurement, commonly by NMR or other high-throughput spectrometry technologies, of all endogenous metabolites either within the cell or in the body fluids – plasma, serum or urine, describing the individual metabolotype [54]. Human metabolism can be considered as the intersection of upstream (genome, transcriptome, proteome) and downstream (environment) systems [54]. Hence the metabolotype describes the end result of the interactions between the genome, epigenetic gene regulation and the exposome. The metabolotype will describe the individual response to the exposome and metabolomics techniques are sensitive enough to distinguish different phenotypes associated with specific gene mutations [55]. The metabolome provides four levels of information on the metabolotype: presence or absence of metabolites, the concentration levels, relative levels or ratios of different metabolites, and metabolic profiles [55]. This information will better support the accrual of knowledge on variation in response to carcinogens, the interaction between the multiple components of the exposome, and the mechanisms by which these exposites affect health. It is likely that clusters of metabolites within the metabolome will provide exposite exposure assessments, indicate individual disease risk profiles and offer insight into the interaction of metabolic pathways. Insight into the metabolome, especially in multi-omic approaches, e.g. genetic and exposomic studies, will identify new targets for prophylactic and therapeutic interventions and will support the premise of personalized medicine [56,57].

CONCLUSION

Over 90% of cancers are due to endogenous or exogenous exposure to chemicals and radiation; many of these carcinogenic exposures, to tobacco smoke and poor diet, for example, are avoidable and thus the consequent disease is preventable. Many of these environmental carcinogens act not on the genome but on the epigenome; a carcinogen does not need to have genotoxic properties. The "exposome" is the concept which captures all carcinogenic exposures across a lifetime, quantified by direct biomarker assessment. Exposomics is in its infancy but, if progress such as has occurred in the Human Genome Project occurs in this field, it has the potential greatly to enhance our understanding of the complex interactions and mechanisms of action of environmental exposures. In particular, evidence is increasing that cancer is as much a disease of the epigenome as the genome and that many of these environmental factors act to alter gene expression through changes in the epigenome. Exposomics and emergent technologies such as metabolomics will enable us better to avoid carcinogenic exposures, to limit the carcinogenic effect of these exposures and to identify potential new targets for developmental therapeutics.

GLOSSARY

Exposome The totality of external and internal exposures from birth onwards which influence health.

Exposite An individual exposure, part of the exposome.

Metabolome The totality of all molecules resulting from the process of metabolism.

Metabolite An individual molecule resulting from metabolism, part of the metabolome.

Metabolotype The characterization of individuals based on the specifics of their metabolome, which reflects differences in metabolism.

ABBREVIATIONS

CDC US Centers for Disease Control and Prevention

IARC International Agency for Research in Cancer

NIH US National Institutes of Health

UNSCEAR United Nations Standing Committee on the Effects of Atomic Radiation

REFERENCES

[1] International Agency for Reseach in Cancer (IARC). GLOBOCAN2008. Retrieved January 12, 2013 from <http://globocan.iarc.fr/>.

[2] US National Institutes of Health (NIH) Cancer Prevalence and Cost of Care Projections. Retrieved January 12, 2013 from <http://costprojections.cancer.gov/annual.costs.html>.

[3] US National Institutes of Health (NIH) National Cancer Institute. Defining Cancer. Retrieved January 12, 2013 from <http://www.cancer.gov/cancertopics/cancerlibrary/what-is-cancer>.

[4] National Cancer Institute. (NCI) Surveillance Epidemiology and End Results. Retrieved January 2013, from <http://seer.cancer.gov/statfacts/html/lungb.html#incidence-mortality>.

[5] Canadian Cancer Society's Steering Committee. (2009). Canadian Cancer Statistics, 2009. Toronto, Ontario, Canada: Canadian Cancer Society.

[6] US National Institutes of Health (NIH) State Cancer Profiles. Retrieved January 17, 2013, from <http://statecancerprofiles.cancer.gov/incidencerates/>.

[7] Greaves M. Cancer, the evolutionary legacy. Oxford: Oxford University Press; 2000.

[8] Lichenstein P, Holm NV, Verkasalo PK, Iliadou A, Kaprio J, Koskenvuo M, et al. Environmental and heritable factors in the causation of cancer. Analyses of cohorts of twins from Sweden, Denmark, and Finland. N Engl J Med 2000;347:78–85.

[9] Hemminki K, Li X. Cancer risks in second-generation immigrants to Sweden. Internatl J Cancer 2002;99:229–37.

[10] McGuinn LA, Ghazarian AA, Ellison GL, Harvey CC, Kaefer CM, Reid BC. Cancer and environment: Definitions and misconceptions. Environ Res 2012;112:230–4.

[11] International Agency for Research on Cancer (IARC). (2013, January 15). Agents Classified by the IARC Monographs vols 1–102. Retrieved from <http://monographs.iarc.fr/ENG/Classification/>.

[12] Wild CP. Complementing the genome with an "exposome": the outstanding challenge of environmental exposure measurement in molecular epidemiology. Cancer Epidemiol Biomarkers Prevent 2005;14:1847–50.

[13] Centers for Disease Control and Prevention. (n.d.). Exposome and exposomics. Retrieved January, 2013, from CDC-Exposome and Exposomics- NIOSH Workplace Safety and Health Topics: <www.cdc.gov/niosh/topics/exposome/>.

[14] Anand P, Kunnumakara AB, Sundaram C, Harikumar KB, Tharakan ST, Lai OS, et al. Cancer is a preventable disease that requires major lifestyle changes. Pharm Res 2008;25:2097–116.

[15] Rappaport SM. Implications of the exposome for exposure science. J Expos Sci Environ Epidemiol 2011;21:5–9.

[16] Rappaport SM, Smith MT. Epidemiology. Environment and disease risks. Science 2010;330:460–1.

[17] Parkin DM, Boyd L, Darby SC, Mesher D, Sasieni P, Walker LC. The fraction of cancer attributable to lifestyle and environmental factors in the UK in 2010. Br J Cancer 2011;105:S1–81.

[18] World Cancer Research Fund/American Institute for Cancer Research. Food, nutrition, physical activity, and the prevention of cancer: a global perspective.Washington, DC: AICR; 2007.

[19] Parkin DM, Boyd L, Walker LC. The fraction of cancer attributable to lifestyle and environmental factors in the UK in 2010. Chapter 16. Summary and Conclusions. Br J Cancer 2011;105:S77–81.

[20] Doll R, Hill AB. The mortality of doctors in relation to their smoking habits; a preliminary report. Br Med J 1954;1:1451–5.

[21] World Health Organization. (n.d.). Tobacco. Retrieved 2013, from <http://www.who.int/mediacentre/factsheets/fs339/en/index.html>.

[22] Parkin DM. Tobacco-attributable cancer burden in the UK in 2010. Br J Cancer 2011;105:s6–13.

[23] Reeves GK, Pirie K, Beral V, Green J, Spencer E, Bull D. Million Women Study Collaboration. Cancer incidence and mortality in relation to body mass index in the Million Women Study: cohort study. Br Med J 2007;335:1134.

[24] Parkin DM. Cancers attributable to inadequate physical exercise in the UK in 2010. Br J Cancer 2011;105:S38–41.

[25] Parkin DM. Cancers attributable to consumption of alcohol in the UK in 2010. Br J Cancer 2011;105:S14–8.

[26] Allen NE, Beral V, Casabonne D, Wan Kan S, Reeves GK, Brown AJ, et al. Moderate alcohol intake and cancer incidence in women. J Natl Cancer Inst 2009;101:296–305.

[27] Parkin DM. Cancers attributable to occupational exposures in the UK in 2010. Br J Cancer 2011;105:S70–2.

[28] Doll R, Wakeford R. Risk of childhood cancer from fetal irradiation. Br J Radiol 1997;70:130–9.

[29] Wakeford R. The risk of childhood leukaemia following exposure to ionising radiation – a review. J Radiol Prot 2013;33:1–25.

[30] Clark C. Physicians, reformers and occupational disease: the discovery of radium poisoning. Women Health 1987;12:147–67.

[31] United Nations Scientific Committee on the Effects of Atomic Radiation. Effects of ionizing radiation, Vol. I. New York: United Nations; 2006.

[32] Pearce MS, Salotti JA, Little MP, McHugh K, Lee C, Kim KP, et al. Radiation exposure from CT scans in childhood and subsequent risk of leukaemia and brain tumours: a retrospective cohort study. Lancet 2012;380:499–505.

[33] Chik F, Szyf M, Rabbani SA, Chik F, Szyf M, Rabbani SA. Role of epigenetics in cancer initiation and progression. In: Rhim JS, Kremer R, editors. Human cell transformation: role of stem cells and the microenvironment. New York: Springer Science + Business Media LLC; 2011.

[34] Feinberg AP, Tycko B. The history of cancer epigenetics. Nat Rev Cancer 2004;4:143–53.

[35] Stein RA. Epigenetics and environmental exposure. J Epidemiol Commun Hlth 2012;66:8–13.

[36] Friso S, Udali FS, Guarini P, Pellegrini C, Pattini P, Moruzzi S, et al. Global DNA hypomethylation in peripheral blood mononuclear cells as a biomarker of cancer risk. Cancer Epidemiol Biomarkers Prev 2013;22:348–55.

[37] Weidman JR, Dolinoy DC, Murphy SK, Jirtle RL. Cancer susceptibility: epigenetic manifestations of environmental exposures. Cancer J 2007;13:9–16.

[38] Koturbash I, Pogribny I, Kovalchuk O. Stable loss of global DNA methylation in the radiation-target tissue – A possible mechanism contributing to radiation carcinognesis. Biochem Biophys Res Commun 2005;337:526–33.

[39] Hei TK, Zhou H, Ivanov V, Hong M, Lieberman HB, Brenner DJ, et al. Mechanism of radiation-induced bystander effects: a unifying model. J Pharm Pharmacol 2008;60:943–50.

[40] Moore LE, Smith AH, Eng C, Kalman D, DeVries S, Bhargava V, et al. Arsenic-related chromosomal alterations in bladder cancer. J Natl Cancer Inst 2002;94:1688–96.

[41] Ren X, McHale CM, Skibola CF, Smith AH, Smith MT, Zhang L. An emerging role for epigenetic dysregulation in arsenic toxicity and carcinogenesis. Environ Health Perspect 2011;11:11–9.

[42] US National Institute of Health (NIH) Office of Dietary Supplementation. Folate. Retrieved January 12, 2013 at <http://ods.od.nih.gov/factsheets/Folate-HealthProfessional/>.

[43] Haggarty P, Hoad G, Campbell DM, Horgan GW, Piyathilake C, McNeill G. Folate in pregnancy and imprinted gene and repeat element methylation in the offspring. Am J Clin Nutr 2013;97:94–9.

[44] Davis CD, Uthus EO. DNA methylation, cancer susceptibility, and nutrient interactions. Exp Biol Med (Maywood) 2004;229:988–95.

[45] Gagliardi G, Biricotti M, Failli A, Orsini G, Consolin R, Migheli F, et al. Colorectal carcinoma and folate. Ann Ital Chir 2013;84:123–31.

[46] Qin X, Cui Y, Shen L, Sun N, Zhang Y, Li J, et al. Folic acid supplementation and cancer risk: A meta-analysis of randomized controlled trials. Int J Cancer 2013;133:1033–41.

[47] Wien TN, Pike E, Wisløff T, Staff A, Smeland S, Klemp M. Cancer risk with folic acid supplements: a systematic review and meta-analysis. Br Med J 2012;2:e000653.

[48] Milne E, Laurvick CL, de Klerk N, Robertson L, Thompson JR, Bower C. Trends in childhood acute lymphoblastic leukemia in Western Australia, 1960–2006. Int J Cancer 2008;122:1130–4.

[49] Grupp SG, Greenberg ML, Ray JG, Busto U, Lanctôt KL, Nulman I, et al. Pediatric cancer rates after universal folic acid flour fortification in Ontario. J Clin Pharmacol 2011;51:60–5.

[50] Schüz J, Weihkopf T, Kaatsch P. Medication use during pregnancy and the risk of childhood cancer in the offspring. Eur J Pediatr 2007;166:433–41.

[51] Linabery AM, Johnson KJ, Ross JA. Childhood cancer incidence trends in association with US folic acid fortification (1986–2008). Pediatrics 2012;129:1125–33.

[52] The UK Biobank. About the UK Biobank. Retrieved January 12, 2013 from <http://www.ukbiobank.ac.uk/about-biobank-uk/>.

[53] Borugian MJ, Robson P, Fortier I, Parker L, McLaughlin J, Knoppers BM, et al. The Canadian Partnership for Tomorrow Project: building a pan-Canadian research platform for disease prevention. Can Med Assoc J 2010, April 26.

[54] Geiger C, Geistlinger L, Altmaier E, Hrabé de Angelis M, Kronenberg F, Meitinger T, et al. Genetics meets metabolomics: a genome-wide assocaition study of metabolite profiles in human serum. PLoS Genet 2008;4:e1000282.

[55] Semmar N. Computational metabolomics. Waltham, MA: Nova Biomedical; 2011.

[56] Griffin JL. Understanding mouse models of disease through metabolomics. Curr Opin Chem Biol 2008;10:309–15.

[57] Pontoizeau C, Fearnside JF, Navratil V, Domange C, Cazier JB, Fernández-Santamaría C, et al. Broad-ranging natural metabolyte variation drives physiological plasticity in healthy control inbred rat strains. J Proteome Res 2011;10:1675–89.

Abbreviations

α-KG	α-Ketoglutarate
2-HG	2-Hydroxyglutarate
3′ UTR	3′ untranslated region
5′ UTR	5′ untranslated region
6 MP	6-Mercaptopurine
AAV	Adenovirus-associated virus
ABC	Activated B-cell
ACC	Adrenocortical carcinoma
aCGH	Array comparative genomic hybridization
ADHD	Attention deficit hyperactivity disorder
ADR	Adverse drug reaction
AF1q	Gene fusion partner with MLL also called MLLT11 for myeloid/lymphoid or mixed-lineage leukemia (trithorax homolog, Drosophila); translocated to 11
AFAP	Attenuated familial adenomatous polyposis
AHRQ	Agency for Healthcare Research and Quality
ALCL	Anaplastic large cell lymphoma
ALK	Anaplastic lymphoma kinase
ALL	Acute lymphoblastic leukemia
AMKL	Acute megakaryoblastic leukemia
AML	Acute myeloblastic leukemia
AML	Acute myeloid leukemia
AMPK	Adenine monophosphate-activated protein kinase
APA	Alternative cleavage and polyadenylation
APC	Adenomatous polyposis coli
API	Application programming interfaces
APL	Acute promyelocytic leukemia
AR	Androgen receptor
ARID1A/1B	AT-rich interactive domain-containing protein 1A/1B
ARMS	Alveolar rhabdomyosarcoma
ASCT	Autologous stem cell transplant
ASE	Allele-specific expression
ASK1	Apoptosis signaling kinase 1
ASPS	Alveolar soft part sarcoma
ASXL1	Additional sex combs-like 1
ATC	Anaplastic thyroid carcinoma
ATF2	Activating transcription factor 2
ATM	Ataxia telangiectasia mutated
ATP	Adenosine triphosphate
ATRX	Alpha thalassemia/mental retardation syndrome X-linked
BAALC	Brain and acute leukemia, cytoplasmic
BAC	Bacterial artificial chromosome
BARD1	BRCA-1 associated RING domain-1
BCBS TEC	Blue Shield Technology Evaluation Center
BCR	B-cell receptor
BFM	Berlin–Frankfurt–Munich
BL	Burkitt lymphoma
BM	Bone marrow
BMFS	Bone marrow failure syndrome
BNX	beige/nude/xid
BRCA1/2	Breast cancer type 1/2 susceptibility protein
BRFR	Biochemical relapse-free rate
BTIC	Brain tumor-initiating cell
BWA	Burrows–Wheeler aligner
BWS	Beckwith–Wiedemann syndrome
CA 27.29	Cancer antigen 27.29
CAGR	Cancer-associated genomic regions
CAM	Chorioallantoic membrane
CAP	College of American Pathologists (laboratory accreditation program)
CaP	Prostate cancer
CBF	Core binding factor
CBL	Proto-oncogene derived from Cas-Br-M (murine) ecotropic retroviral transforming sequence
CCHS	Congenital central hypoventilation syndrome
CCLE	Cancer Cell Line Encyclopedia
CD11	Cluster of differentiation
CDC	Centers for Disease Control and Prevention
CDK	Cyclin-dependent kinase
cDNA	Complementary DNA
CEA	Carcinoembryonic antigen
CEBPA	CCAAT/enhancer binding protein (C/EBP), alpha
ceRNA	Competing endogenous RNA
CFFA2	Gene mutated in congenital amegakaryocytic thrombocytopenia
CFS	Congenital fibrosarcoma
CGAP	Cancer Genome Anatomy Project
CGCI	Cancer Genome Characterization Initiative
CGH	Comparative genomic hybridization (arrays)
CGHub	Cancer Genomic Hub
CGP	Cancer Genome Project
CHK2	Checkpoint kinase 2
CIMP	CpG island methylator phenotype
CIN	Chromosomal instability
CIS	Carcinoma *in situ*
CLIA	Clinical Laboratory Improvement Amendments (certification)
CLL	Chronic lymphocytic leukemia
CML	Chronic myelogenous leukemia
CML	Chronic myeloid leukemia
CMPT	Center for Medical Technology Policy
CNA	Copy number aberration
CNS	Central nervous system

CNV	Copy number variation
COG	Children's Oncology Group
COG-STS	Children's Oncology Group–Soft Tissue Sarcoma (STS) Committee
COL1A1	Collagen, type I, alpha 1
COSMIC	Catalogue of Somatic Mutations in Cancer
CPC-GENE	Canadian Prostate Cancer Genome Network
CpG	Cytosine-phosphodiester-guanine dinucleotides
CPNDS	Canadian Pharmacogenomics Network for Drug Safety
CPS	Cancer predisposition syndrome
CR	Complete response
CRC	Colorectal carcinoma
CSC	Cancer stem cell
CTC	Circulating tumor cells
CTD^2	Cancer Target Discovery and Development program
CtDNA	Circulating tumor DNA
dbGaP	Database of Genes and Phenotypes
dbSNP	Database of Single Nucleotide Polymorphisms
DCC	Data Coordination Center
DCIS	Ductal carcinoma *in situ*
DDLPS	De-differentiated liposarcoma
DDS	Denys–Drash syndrome
DFSP	Dermatofibrosarcoma protuberans
DHFR	Dihydrofolate reductase
DIPG	Diffuse intrinsic pontine glioma
DLBCL	Diffuse large B-cell lymphoma
DNA	Deoxyribonucleic acid
DNMT	DNA methyl transferase
DTC	Differentiated thyroid cancer
Dvl	Dishevelled
EBI	European Bioinformatics Institute
EC2	Elastic Compute Cloud
ECS	Exome capture sequencing
EES	Extraosseous Ewing sarcoma
EFS	Event-free survival
EGA	European Genome-Phenome Archive
EGAPP	Evaluation of Genomic Applications in Practice and Prevention
EGFR	Epidermal growth factor receptor
ELA2	Elastase gene
EMT	Epithelial–mesenchymal transition
ENCODE	Encyclopedia of DNA Elements
eQTL	Expression quantitative trait loci
ER	Estrogen receptor
ERG	ETS-related gene
ES	Embryonic stem
ES	Ewing sarcoma
EST	Expressed sequence tag
ETS	Erythroblast transformation-specific
ETV6	ETS variant gene 6
EVS	Exome variant server
EWSR1	Ewing sarcoma breakpoint region 1
FA	Follicular adenoma
FAP	Familial adenomatous polyposis
FDA	Food and Drug Administration
FFPE	Formalin-fixed, paraffin-embedded (tissue)
FHWT	Favorable histology Wilms tumor
FISH	Fluorescence *in situ* hybridization
FKHR	Forkhead homolog 1 rhabdomyosarcoma
FL	Follicular lymphoma
FLCN	Folliculin
FLI1	Friend leukemia virus integration 1
FLT	FMS-like tyrosine kinase
FMS	Feline sarcoma virus
FMTC	Familial medullary thyroid carcinoma
FOB	Fecal occult blood
FOXO1	Forkhead box protein O1
FPKM	Fragments per kilobase per million reads mapped
FTC	Follicular thyroid carcinoma
FUS	Fused in sarcoma
GBM	Glioblastoma
GBM	Glioblastoma multiforme
Gbp	Giga base pair
GCB	Germinal center B-cell
G-CIMP	CpG island methylator phenotype
GCP	Good Clinical Practice
GEMCap	Genomic evaluators of metastatic prostate cancer
GEMM	Genetically engineered mouse model
GFP	Green fluorescent protein
GIST	Gastrointestinal stromal tumor
Gli	Glioma-associated oncogene homolog
GLP	Good Laboratory Practice
GO	Gene ontology
GPU	Graphics processing unit
GSEA	Gene set enrichment analysis
GSK-3β	Glycogen synthase kinase 3
GST	Glutathione S-transferase
GWA	Genome-wide association
GWAS	Genome-wide association study
Gy	Gray
HACE1	HECT domain and ankyrin repeat containing E3 ubiquitin protein ligase 1
HBV	Hepatitis B virus
HCC	Hepatocellular carcionoma
HCC	Hurthle cell carcinoma
HCV	Hepatitis C virus
HDAC	Histone deacetylase
HDAC1	Histone deacetylase 1
HDACI	Histone deacetylase inhibitor
HER2	Epidermal growth factor receptor 2
HGP	Human Genome Project
hh	Hedgehog
HLA	Human leukocyte antigen
HPC	High performance computing
HPF	Hours post-fertilization
HPV	Human papilloma virus
IARC	International Agency for Research in Cancer
IC	Invasive carcinoma
ICGC	International Cancer Genome Consortium
ICH	International Conference on Harmonization
ICR1	Imprinting control region 1
IDC	Invasive ductal carcinoma
IDC-P	Intra-ductal carcinoma of the prostate
IDE	Investigational Device Exemption
IDH	Isocitrate dehydrogenase

IDH1	Isocitrate dehydrogenease 1
IGF	Insulin-like growth factor
IgH	Immunoglobulin heavy chain
IGV	Integrative Genomics Viewer
IHC	Immunohistochemistry
ILNR	Intralobular nephrogenic rests
IM	Immunomodulatory
Indels	Small insertions or deletions
INRG	International Neuroblastoma Risk Group
IPA	Ingenuity Pathway Analysis
JMML	Juvenile myelomonocytic leukemia
KEGG	Kyoto Encyclopedia of Genes and Genomes
KIR	Killer cell immunoglobulin-like receptor
KIT	From the v-kit Hardy-Zuckerman 4 feline sarcoma viral oncogene
LAR	Luminal androgen receptor
LBL	Lymphoblastic lymphoma
LCIS	Lobular carcinoma *in situ*
LD	Linkage disequilibrium
LFS	Li–Fraumeni syndrome
LIF	Leukemia inhibitory factor
lincRNA	Long non-coding intergenic RNA
LKB1	Liver kinase B1
LMO1	LIM-domain-only 1
LNA	Locked nucleic acid
lncRNA	Long non-coding RNA
LOF	Loss of function
LOH	Loss of heterozygosity
LOI	Loss of imprinting
LS	Liposarcoma
M	Mesenchymal
MALT	Mucosa-associated lymphoid tissue
MAPK	Mitogen activated protein kinase
MAQ	Mapping and assembly with quality
MB	Medulloblastoma
MDM2	Murine double minute 2
MDR	Minimum deleted region
MDR	Multidrug resistance
MEN	Multiple endocrine neoplasia
MET	Mesenchymal-to-epithelial transition
MFH	Malignant fibrous histiocytoma
MGI	Molecular grade index
MGMT	O6-methylguanine-DNA methyltransferase
MGUS	Monoclonal gammopathy of undetermined significance
MIBG	*meta*-Iodobenzylguanadine
miR	MicroRNA
miRNA	MicroRNA
MLL	Mixed-lineage leukemia
MLPA	Multiplex ligation probe amplification
MLPA	Multiplex ligation probe analysis
MLPS	Myxoid liposarcoma
MM	Multiple myeloma
MMR	Mismatch repair
MMT	Malignant mesenchymal tumor
MMTV	Murine mammary tumor virus
MNA	MYCN amplification
MPL	Myeloproliferative leukemia virus oncogene
MPSS	Massively parallel signature sequencing
MRI	Magnetic resonance imaging
mRNA	Messenger RNA
MSI	Microsatellite instability
MSKCC	Memorial Sloan-Kettering Cancer Center
MSL	Mesenchymal stem-like
MSS	Microsatellite stable
MTC	Medullary thyroid carcinoma
MTHFR	Methylene-tetrahydrofolate reductase
mTOR	Mammalian target of rapamycin
mTOR	Murine target of rapamycin
MYCN	V-myc myelocytomatosis viral-related oncogene, neuroblastoma derived
MYH9	Myosin heavy chain gene involved in autosomal dominant macrothrombocytopenia
NB	Neuroblastoma
NBPF23	Neuroblastoma breakpoint family 23
NBS1	Nijmegen breakage syndrome 1
NCA	Numerical chromosomal abnormality
NCBI	National Center for Biotechnology Information
NCCN	National Cancer Comprehensive Network
NCI	National Cancer Institute
ncRNA	Non-coding RNA
NEPC	Neuroendocrine prostate cancer
NF1	Neurofibromin 1
nGEMM	Non-germline genetically engineered mouse model
NGS	Next-generation sequencing
NHGRI	National Human Genome Research Institute
NHL	Non-Hodgkin lymphoma
NHLBI	National Heart, Lung, and Blood Institute
NICE	National Institute for Health and Clinical Excellence
NIH	National Institutes of Health
NK cell	Natural killer cell
NK	Natural killer
NOD	Non-obese diabetic
NPM	Nucleophosmin
NTRK3	Neutrophilic tyrosine kinase receptor, type3
NWTSG	National Wilms Tumor Study Group
OCT	Optimum cutting temperature
OMIM	Online Mendelian Inheritance in Man
OR	Odds ratio
OregAnno	Open regulatory annotation
OS	Overall survival
PAM50	Prediction analysis of microarrays 50
PARP	Poly(ADP)-ribose polymerase
PAX3	Paired box 3
PAX7	Paired box 7
PCG	Protein-coding gene
PCGP	Pediatric Cancer Genome Project
PCL	Plasma cell leukemia
PCR	Polymerase chain reaction
PCs	Plasma cells
PDB	Protein Data Bank
PDGFβ	Platelet-derived growth factor beta
PDGFRα	Platelet-derived growth factor receptor alpha
PDGFRA	Platelet-derived growth factor receptor alpha
PDTC	Poorly differentiated thyroid carcinoma
PFS	Progression-free survival

PgP P-glycoprotein
PharmGKB Pharmacogenomics Knowledgebase
PHF6 Plant homeodomain finger 6
PHOX2B Paired-like homeobox 2b
PI3K Phosphatidylinositol 3-kinase
PI3K Phosphoinositide 3-kinase
PLNR Perilobular nephrogenic rests
PLPS Pleomorphic liposarcoma
PolyPhen Polymorphism phenotyping
PPB Pleuropulmonary blastoma
PR Partial response
PR Progesterone receptor
Pre-miRNAs Precursor microRNAs
pre-mRNA Precursor messenger RNA
Pri-miRNAs Primary microRNAs
PSA prostate-specific antigen
PTC Papillary thyroid carcinoma
PTEN Phosphatase and tensin homolog
PTPN11 Protein tyrosine phosphatase, non-receptor type 11 often mutated in JMML; also called SHP2
QC Quality control
qPCR Quantitative polymerase chain reaction
qRT-PCR Reverse transcriptase quantitative polymerase chain reaction
RAS Rat sarcoma
RB Retinoblastoma
RB1 Retinoblastoma 1
RFLP Restriction fragment length polymorphism
RIN RNA integrity number
RISC RNA-induced silencing complex
RMS Rhabdomyosarcoma
RNA Ribonucleic acid
RNAi RNA interference
RNA-seq RNA sequencing
ROC Receiver operating control
ROI Region of interest
RPKM Reads per kilobase per million reads mapped
RPPA Reverse phase protein array
RR Response rate
rRNA Ribosomal RNA
RT Radiotherapy
RT–PCR Reverse transcription polymerase chain reaction
RUNX1 Runt-related transcription factor 1 gene, also called aml1 for acute myeloid leukemia 1 gene
SAGE Sequential analysis of gene expression
SBDS Shwachman–Bodian–Diamond syndrome
SCA Segmental chromosomal abnormality
SCID Severe combined immunodeficiency
SD Stable disease
SDH Succinyldehydrogenase
Shh Sonic hedgehog
SHP2 Src homology phosphatase 2
SHRiMP Short read mapping package
SIFT Sorting Intolerant From Tolerant
SIOP Société Internationale d'Oncologie Pediatrique
SIOPEN International Society of Paediatric Oncology Europe Neuroblastoma
siRNA Small interfering RNA
SLC19A1 Solute carrier family 19
snoRNA Small nucleolar RNA
SNP Single nucleotide polymorphism
SNV Single nucleotide variant
SNV Single nucleotide variation
SOAP Short Oligonucleotide Alignment Program
SOP Standard operating procedures
SPOP Speckle-type POZ protein
SRA Sequence read archive
STSs Soft tissue sarcomas
SV Structural variation
TAILORx Trial assigning individualized options for treatment
TALEN Transcription activator-like effector nuclease
T-ALL T-cell precursor ALL
TAM Tumor-associated macrophage
TAML Therapy-associated acute myeloid leukemia
TARGET Therapeutically Applicable Research to Generate Effective Treatments
TC Thyroid cancer
TCA Tricarboxylic acid
TCGA The Cancer Genome Atlas
TDLU Terminal ductal-lobular units
TET Gene name derived from ten-eleven translocation-1
TH Tyrosine hydroxylase
TK Tyrosine kinase
TKIs Tyrosine kinase inhibitors
TLS Translocated in liposarcoma
TMA Tissue microarray
TMD Transient myeloproliferative syndrome seen in infants with Down syndrome
TMPRSS2 Transmembrane protease, serine 2
TNBC Triple-negative breast cancer
TNM Tumor size, lymph node, and metastasis
TP53 p53 tumor suppressor gene
TP53 Tumor protein 53
TPMT Thiopurine-methyltransferase
TRANSFAC Transcription Factor Database
Trk Tyrosine receptor kinsase
tRNA Transfer RNA
TSC1 Tuberous sclerosis 1
TTD Therapeutic Target Database
UCG Ultraconserved gene
UCSC University of California Santa Cruz
UHWT Unfavorable histology Wilms tumor
UNSCEAR United Nations Standing Committee on the Effects of Atomic Radiation
UPD Uniparental disomy
UPR Unfolded protein response
UPS Undifferentiated pleomorphic sarcoma
UTR Untranslated region
VHL von Hippel–Lindau disease
VLRWT Very low risk Wilms tumor
WAGR Wilms tumor, aniridia, genitourinary malformation, mental retardation
WDLPS Well-differentiated liposarcoma
WES Whole-exome sequencing
WGS Whole-genome sequencing

WGSS Whole-genome shotgun sequencing
WHO World Health Organization
Wnt Wingless type MMTV integration site
WT Wilms tumor
WTSS Whole-transcriptome shotgun sequencing
XID X-linked immunodeficiency
YAC Yeast artificial chromosome
ZFN Zinc finger nucleases

Glossary

3′ untranslated region (3′ UTR) Sequences on the 3′ end of mRNA but not translated into protein.

5′ untranslated region (5′ UTR) Sequences on the 5′ end of mRNA but not translated into protein.

Accuracy The closeness of agreement between the value that is accepted either as a conventional true value or an accepted reference value and the value found experimentally.

Acetyltransferase An enzyme that moves an acetyl group from an acetyl donor to an acceptor. Acetylation is important in regulating the function of proteins and can affect gene expression through histone modification.

Actionable target A mutated protein which is shown to be causally involved in disease initiation, progression and/or maintenance and which can be directly targeted by drugs.

Acute lymphoblastic leukemia A neoplasm of immature B or T cells that exhibits extensive blood/bone marrow involvement.

Adenocarcinoma Epithelial cell cancer, occurring in a gland.

Adenoma A benign tumor of glandular origin. Adenomas are a possible precursor of malignancy. In individuals with heritable colorectal cancer syndromes, monitoring for adenoma formation is an important strategy for early diagnosis.

Adjuvant A pharmaceutical delivered in conjunction with another drug whose action it modifies. 5-Fluorouracil is a common adjuvant administered with other chemotherapy agents during treatment of some colorectal cancers

Adjuvant chemotherapy Chemotherapy administered post-primary treatment (surgery, anti-hormone therapy, radiation) to reduce the risk of recurrence or to offer palliative care in patients with advanced metastatic disease.

Adverse drug event Harm associated with the use of a medication at any dose (normal dose, overdosing, or underdosing) or any harm associated with a drug such as events associated with discontinuation of the drug.

Adverse drug reaction Harm associated with the use of a medication at normal dosage during normal use. This is one form of an adverse drug event.

Aggregate research results Refers to the reporting of a summary of research results (as typically would appear in a scientific publication).

AKT A protein-serine-threonine kinase that is activated by phosphorylation in response to growth factors or insulin, playing a major role in cell metabolism, growth, and survival.

Alignment The process of mapping short sequence reads to the reference genome.

Allele dropout Failure to amplify one of the two alleles in a diploid genome.

Allele frequency The proportion of the allele of interest observed in a population; a measure of the incidence of gene variants; allele frequency is calculated by dividing the number of copies of the allele of interest by the number of copies of all alleles at the same locus in a population.

Allele One of variant forms of a gene that can exist at the same locus (location on the chromosome).

Allele-specific expression (ASE) Gene expression distinguished from a particular allele. In diploid organisms, this is the relative expression of homologous genes. A gene that exhibits allele-specific expression is considered to have imbalanced expression between alleles.

Allelic imbalance See allele-specific expression.

Alternative splicing A process by which different mRNA transcripts are produced from the same precursor transcript through alternate ways to remove introns and connect exons.

Amplicon A piece of DNA or RNA that is the source and/or product of natural or artificial amplification or replication events.

Amplification bias Bias introduced from preferential amplification of nucleotide molecules due to extreme base composition (e.g. GC-rich), poly A slippage etc., resulting in a different coverage distribution in the amplified sample compared to the original unamplified sample.

Analytical validation Traditionally, the assessment of a diagnostic test and measurement of its performance characteristics, for example, the range of conditions under which the assay will give reproducible and accurate data. With respect to omics, this term refers to the assessment of a test's ability to measure accurately and reliably the analyte(s) of interest in the clinical laboratory, as well as in specimens representative of the population of interest.

Anaplastic large cell lymphoma T-cell lymphoma typically characterized by large cell size, CD30 expression, and *ALK* rearrangement.

Anaplastic thyroid carcinoma (ATC) A highly malignant thyroid tumor composed in part or wholly of undifferentiated cells.

Ancestry Inherited properties associated with the descendants of one individual.

Androgen deprivation A therapy for prostate cancer involving the use of drugs that block either the synthesis or function of androgen hormones, which are often required for prostate cancer cell survival.

Aneuploidy Aneuploidy is a condition where the number of chromosomes in the cells is not the typical one for the species. An increase in the normal number of chromosomes is called hyperdiplody, while a reduction is named hypodiplody.

Angiogenesis The formation and differentiation of blood vessels.

Animal model An animal that is used to study disease processes, such as cancer.

Aniridia Absence of the iris.

Antagomir Synthetic antisense oligonucleotides characterized by the 2′-O-Me modification on the ribose sugar, the terminal phosphorothioate linkage, and the cholesterol conjugation at the 3′ end.

Antagonist A chemical that acts within the body to reduce the physiological activity of another chemical substance.

Anthracyclines DNA intercalating class of antibiotics which are derived from the bacterium *Streptomyces* and are used as potent anticancer agents. The most common anthracycline is doxorubicin.

Antisense oligonucleotides Single strands of DNA or RNA that are complementary to a chosen sequence.

APC The adenomatous polyposis coli gene. A tumor suppressor that acts as a regulator of β-catenin. Truncating germline mutations in APC cause both familial adenomatous polyposis and attenuated familial adenomatous polyposis.

Apoptosis A genetically determined process of cell self-destruction (programmed cell death) that is activated either by the presence of a stimulus or by the removal of a stimulus or suppressing agent and, when halted (as by genetic mutation), may result in uncontrolled cell growth and tumor formation.

Array comparative genomic hybridization (aCGH) A technique based on the hybridization of differentially labeled DNA (usually from normal and pathological sample DNAs) aimed to detect DNA gains and losses based on changes in signal ratios.

Assembly Assembly in genomics refers to the merging of short sequence reads into longer pieces of DNA contigs.

Autologous stem cell transplant (ASCT) A medical procedure in which hematopoietic stem cells from an individual are isolated, stored, and infused in the same individual after high-dose chemotherapy or radiation therapy to speed up and improve immune system recovering.

Autophagy Autophagy is a catabolic process involving the degradation of intracellular components in the lysosomal compartment, in order to reallocate nutrients from unnecessary processes to more essential ones. It maintains the balance between the synthesis, degradation, and subsequent recycling of cellular products and is exploited by MM cells to survive in critical starving conditions, although when maximally activated, autophagy could lead to apoptosis.

Autosome Any human chromosome other than the X and Y sex chromosomes.

Autozygosity A genomic region of homozygosity where both strands of DNA were inherited from a common ancestor.

Bayesian framework A statistical framework based on Bayes' theorem.

Bayesian hierarchical model A probabilistic graphical model or network that depicts the relationships between random variables and their conditional dependencies.

Benefit-sharing Refers to the ethical notion of ensuring that any benefits that result from a research project are distributed, not only among researchers and entities funding the research, but also those whose participation allowed the research to take place. Benefit-sharing is based on ethical principles of respect, reciprocity, and equity, and it has been invoked to support the development of best practices and guidelines for research involving particular communities and vulnerable populations that have been exploited by researchers and others in the past.

Beta-binomial distribution An extension of the binomial probability distribution where the prior distribution is modeled as a beta-distribution. The beta-binomial distribution is used to handle overdispersion in the binomial distribution (where the variance is more than expected).

Biobank Refers to a biorepository that stores an organized collection of biological materials that are typically clinically annotated. The biobank may conduct research itself or be a source of samples and information to be accessed under specific rules by researchers.

Bioinformatics The application of computer technology and programming to the field of biology.

Biomarker Biological marker; also called molecular marker. A characteristic that can be objectively measured as a sign of a normal, abnormal or pathogenic process. A biomarker can also be used to indicate pharmacologic responses to a therapy.

Blastema A mass of undifferentiated cells that has the capability to develop into an organ or an appendage.

Bone marrow Bone marrow is a soft, gelatinous tissue which fills the cavities of the bones and is the site where blood cell biogenesis takes place.

Bortezomib A drug used in multiple myeloma treatment regimens, which blocks proteasome activity leading to accumulation of misfolded proteins and thereby multiple myeloma growth inhibition and death.

Brachytherapy A form of radiotherapy in which a radioactive source is placed adjacent to or within the area requiring treatment.

B-RAF Cytoplasmic protein with serine/threonine activity, involved in the regulation of the MAP kinase/ERKs signaling pathway, which affects cell division, differentiation, and secretion. The *BRAF* gene was shown to be mutated in human cancers.

Bridge PCR A PCR process in which the DNA molecule is clonally amplified from an adapter-ligated DNA template molecule that arches over and hybridizes to complementary adapter on a flow cell, serving as the substrate for amplification.

Burkitt lymphoma Highly proliferative mature B-cell lymphoma, typically characterized by *MYC* translocations.

C cell (parafollicular cell) A cell type of the thyroid gland, situated between or within the walls of follicles, producing and secreting calcitonin.

Calibration curve A functional relationship between the analyte concentration in the standards (calibrators) and the measured response. The calibration curve is used to estimate the analyte concentration in test samples by dose interpolation.

Calibration standards Samples having a known concentration of analyte that are used in an assay to gauge the relationship between system responses (for example, absorbance units) and concentrations of an analyte.

Canadian Institutes of Health Research (CIHR) This organization is the Government of Canada's health research investment agency. CIHR's mission is to create new scientific knowledge through grant support, and to enable its translation into improved health, health services and products, and healthcare systems. It is composed of 13 institutes encompassing a range of scientific disciplines ranging from Cancer Research to Population and Public Health.

Cancer genomics The field of study which examines cancers on the molecular level in the hope of identifying cancer-causing mutations and pathways.

Cancer predisposition syndrome A constellation of cancers in a family that is associated with hereditable alterations in cancer susceptibility genes.

Cancer-associated genomic regions (CAGRs) Fragile chromosomal loci prone to deletions or amplifications in cancer.

Carcinoma A cancerous tumor derived from epithelial cells.

Carcinoma *in situ* Carcinoma which remains confined within the lobule or duct from which it arises.

Cardiomyopathy The measurable deterioration of the function of the myocardium usually leading to heart failure.

Castrate-resistant prostate cancer (CRPC) An advanced, incurable form of prostate cancer which has ceased to respond to androgen deprivation therapy.

Chemotherapy Administration of chemical agents that generally target proliferating cancer cells.

Chimera A term used in genetics to describe an animal such as a mouse that is composed of a mixture of wild-type and genetically engineered cells that are derived from different animals.

Chimeric transcript A fusion gene transcript that arise from chromosomal rearrangement.

ChIP-seq This technique uses next-generation sequencing to discover the DNA bound by proteins of interest. This is an extension of chromatin immunoprecipitation (ChIP) where specific interactions between proteins and DNA are investigated.

Chorioallantoic membrane A double-layered mesoderm that develops over 4–10 days post-fertilization in the chick embryo and contains an extensive network of blood vessels that connects to the developing embryo's circulation.

Chromatin A complex of DNA and proteins that forms chromosomes. Two chromatin forms exist; euchromatin is the less condensed form that is often involved in active transcription. Heterochromatin is the condensed form that is typically not involved with transcription.

Chromatin remodeling A process facilitated by specific protein complexes (known as chromatin remodeling complexes), which changes chromatin structure by restructuring, mobilizing or ejecting nucleosomes to regulate access of various regulatory factors to the DNA.

Chromoplexy From the Greek *plexy*, meaning weave or braid, this term refers to large chains of rearrangements of chromosomes than can affect multiple chromosomes.

Chromosomal instability One of many genomic instabilities observed in cancer. It is characterized by large-scale rearrangements of regions of a chromosome. Includes duplications and deletions as well as translocations.

Chromosomal rearrangements Structural rearrangements caused by chromosome breakage followed by reconstitution in an abnormal combination, resulting in inversion, translocation or deletion.

Chromosomal translocation A chromosome abnormality caused by rearrangement of parts between non-homologous chromosomes. A gene fusion may be created when the translocation joins two otherwise separated genes and it is frequent in cancers.

Chromosome conformation capture (3C) This technique allows researchers to examine the higher order structure of the genome (the chromatin) by cross-linking the physical interactions of chromatin with formaldehyde followed by enzymatic digestion and ligation with the frequency of ligation of two restriction fragments a measure of the frequency of interaction within the nucleus.

Chromosome A thread-like structure of nucleic acids and protein found in the nucleus of most living cells, carrying genetic information in the form of genes.

Chromosome conformation capture carbon-copy (5C) 5C is an extension of 3C using next-generation sequencing (NGS) technology to identify the ligation products, using ligation-mediated amplification to copy and amplify a subset of the 3C library followed by detection using NGS. 5C is allowing three-dimensional interaction maps of the genome to be generated uncovering long-range interactions of promoters and distal elements that can potentially affect gene regulation.

Chromothripsis The occurrence of thousands of DNA rearrangements during a single, massive event during the development of cancer.

Chronic lymphocytic leukemia (CLL) Chronic leukemia of the lymphoblastic type, characterized by lymphadenopathy, fatigue, renal involvement, and pulmonary leukemic infiltrates.

Circulating miRNAs miRNAs present in the cell-free component of blood and body fluids.

Circulating tumor cells (CTCs) Tumors cells that disseminate away from the primary tumor into the circulation and are often indicative of distant-site metastasis.

Circulating tumor DNA (ctDNA) Tumor DNA which sheds from the primary tumor upon cell lysis and can be detected by highly sensitive molecular techniques.

C-KIT A tyrosine kinase growth factor receptor, also known as CD117. Gain-of-function mutation is a major contributor to GIST pathogenesis.

Clinical Laboratory Improvement Amendments (CLIA) certification Regulatory standards set up by the federal government of the USA to ensure accuracy, reliability and timeliness of all clinical laboratory testing on humans.

Clinical Laboratory Improvements Amendments (CLIA) United States federal regulatory standards that apply to all clinical laboratory testing.

Clinical utility Refers in the context of research to the concept that the individual finding can be utilized by the participant (and their healthcare provider) to assist in clinical decision making with respect to options of treatment, prevention or predictive counseling. This does not assume that the finding is actionable.

Clinical/biological validation The assessment of a test's ability to predict accurately and reliably the clinically defined pathological state or phenotype of interest.

Clonal frequency of mutation The percentage of cells within a tumor which have a specific mutation relative to the total number of cells.

Cloud computing The use of computational resources that are delivered over a network, which is often the Internet. Third parties provide software or computing services that can be used on an "as needed" basis, providing the user with instant scalability if required. Examples include Amazon's EC2 cloud computing service and Microsoft Azure, both of which allow one to rent servers or clusters of servers to carry out complex calculations.

Codon A specific sequence of three consecutive nucleotides that is part of the genetic code and that specifies a particular amino acid in a protein or starts or stops protein synthesis – called also triplet.

Cohort A group of subjects who have shared a particular event together during a particular time span.

College of American Pathologists (CAP) accreditation program Internationally recognized laboratory accreditation program from the College of American Pathologists. The accreditation is based on rigorous inspection of the quality of pathology and laboratory services.

Commercialization Refers broadly to the process of converting scientific research findings into some form of marketable product or service. Commercialization can be achieved in many different ways depending on the type of research, product, or service. In the context of biotechnology, intellectual property, including patents, are commonly relied upon in the commercialization process to attract financing and facilitate research and development.

Contig A contiguous sequence of nucleotides that is the result of assembling and joining overlapping reads; contigs provide the consensus sequence of the source DNA/RNA.

Copy number aberration (CNA) See Copy number alteration.

Copy number alteration (CNA) A class of genomic structural variation characterized by gain or loss of one or more copies of a section of DNA, distinguished from normal copy number variation (CNV) by its association with a pathological state (e.g. cancer).

Copy number variation (CNV) Two copies of every gene are present in human cells, one inherited from each parent. Gain or loss of large regions of the genome leads to the gain or loss of gene copies. These are referred to as copy number variations. Usually common in a population and non-pathogenic.

Coverage Sequence coverage is the number of reads that span a specific position in the genome. The higher the coverage, the more confident the genotype calls.

Cross-validation A statistical method for preliminary confirmation of a computational model's performance using a single dataset, by dividing the data into multiple segments, and iteratively fitting the model to all but one segment and then evaluating its performance on the remaining segment.

Cytogenetics Branch of genetics aimed at studying number and structure abnormalities of chromosomes.

Dark matter transcripts RNA molecules that are not associated with annotated genes.

Data access control/policy Certain data/categories of data are often subject to controlled access. This is determined by the data producer and usually data access is granted after an application and verification process.

Data access portal A user-friendly web-based resource for accessing mainly category 3 and 4 data from large-scale genome sequencing projects.

Data integration Data integration takes many types of sequencing data, for example whole genome and transcriptome and attempts to combine them in a meaningful way. If a structural variation is found through whole genome sequencing of the DNA then integrating this with transcriptome sequencing could provide further evidence for the variation. Genomic data categories (the level of analysis usually dictates the category of data): (1) raw; (2) processed/normalized; (3) interpreted; and (4) summarized within and across diseases.

Data warehouse/archive A third-party provider of large-scale storage of category 1 and 2 data generated by genome sequencing projects. Data warehouses ensure universal accessibility to raw data under "open" or "controlled" access policies.

de Bruijn graph A graph representation of the overlap between short sequences. Nodes represent individual sequences of length k and edges represent the overlap between two sequences.

***De novo* transcriptome assembly** Transcriptome sequencing reads assembly without reference genome guidance.

Diagnostic biomarker A biomarker that identifies the presence or absence of pathological state or phenotype of interest.

Differential expression A significant difference in the abundance of total mRNA from a gene or a specific isoform between two samples or biological states.

Differentiation The sum of the developmental processes whereby apparently unspecialized cells, tissues, and structures attain their adult form and function.

Diffuse large B-cell lymphoma Genetically heterogeneous group of mature B-cell lymphomas with large cell size and aggressive clinical behavior.

DNA methylation An epigenetic modification that results in the addition of a methyl group (-CH_3) to the C5 position of cytosine pyrimidine ring. DNA methylation can directly or indirectly repress transcription of a gene.

DNA microarrays High-throughput gene expression quantification technology based on DNA probe-target hybridization (of RNA or DNA) and subsequent fluorescence detection. Also referred to as gene chips.

DNA sequence A succession of letters that indicate the order of nucleotides within a DNA (using GACT) molecule.

DNA sequencing A process of identifying the precise order of nucleotides in DNA molecules.

Domain Any of the three-dimensional subunits of a protein that together make up its tertiary structure, that are formed by folding its linear peptide chain.

Dominant oncogenes Genes that when mutated result in an activation of a key protein or pathway, usually directly, such as a mutation in a tyrosine kinase receptor that results in the constitutive activation of that receptor.

Driver event An event that is responsible for cancer initiation and/or progression.

Driver mutations These are the somatic mutations in cancers which are responsible for providing the cell with growth advantage. As a result, they contribute to disease initiation, progression and maintenance.

Driving gene A gene as the driver of tumorigenesis. (see driver mutation).

Ductal carcinoma Cancer that arises from the cells which line the milk ducts.

Dysplasia An abnormality of tissue development characterized by an overabundance of immature cells versus mature and properly differentiated cells.

Effect modifier A measure that identifies patients most likely to be sensitive or resistant to a specific treatment regimen or agent. An effect modifier is particularly useful when that measure can be used to identify the subgroup of patients for whom treatment will have a clinically meaningfully favorable benefit-to-risk profile.

EGFR Epidermal growth factor receptor is the cell-surface receptor for members of the epidermal growth factor family (EGF-family) of extracellular protein ligands.

Encode To specify the genetic code for.

Epidermal growth factor receptor (HER2) A cell surface receptor that binds to external growth stimuli (e.g. EGF; epidermal growth factors) and signals cell proliferation and inhibits cell death.

Epigenetic A term used to describe the regulation of gene expression by mechanisms that modify DNA but do not change the sequence of the gene. These mechanisms include the methylation of DNA as well as the post-translational modification of histones by a host of events including phosphorylation, methylation, acetylation and ubiquitination. Although epigenetic changes are dynamic they are also heritable and can persist during cell division and be transmitted to offspring.

Epigenetic alterations Changes in cellular behavior or the activity state of chromatins without altering DNA nucleotide sequence. Epigenetic alterations include DNA methylation, chromatin remodeling and histone modification.

Epigenetic modification Non-genetic changes that affect gene expression and may be inherited.

Epigenomics The genome-wide analysis of epigenetic features. Most often involves the description of modified DNA or DNA associated proteins occupying promoter regions which affect gene expression, turning them off or on. Epigenetic alterations do not result in changes to a DNA sequence but are chemical modifications that affect the interaction of DNA with chromatin, histones, and transcription factors.

Epithelial–mesenchymal transition (EMT) An orchestrated transition from epithelial to mesenchymal phenotype often characterized by loss of cell adhesion, repression of E-cadherin expression, and increased cell motility.

Estrogen The primary steroid female hormone that is important for development and function of several organs primarily the human breast.

Estrogen receptor An intracellular receptor that activates cell growth and proliferation when it is bound by its ligand estrogen.

Event-free survival The proportion of people who are expected to remain free of a disease (the event) after treatment.

Exome The portion of the human genome that is made up of the exons of protein-coding genes. Includes untranslated regions but not introns (although partial sequencing of intronic regions occurs due to the nature of exon capture). Is estimated to only make up approximately 1–2% of the human genome.

Exome capture sequencing (ECS) This method involves capture of just the coding exons of the human genome using specially designed probes which are complementary to these exons. The DNA of interest is hybridized to these probes and all other DNA washed away.

Exome sequencing (Exome-seq) Sequencing only the protein-coding regions of the genome, the set of complete exons.

Exon A nucleotide sequence in DNA that carries the code for the final mRNA molecule and thus defines a protein's amino acid sequence.

Exposite An individual exposure, part of the exposome.

Exposome The totality of external and internal exposures from birth onwards which influence health.

Expression analysis Refers to the analysis of the expression level of all or a subset of genes. Expression level can be quantified using the number of transcribed copies of each gene.

Expression quantitative trait loci (eQTL) A genomic loci where genetic variation correlates with expression level.

External beam radiotherapy Delivery of ionizing radiation to the prostate gland using an external radiation source.

External RNA Control Consortium (ERCC) A consortium with the aims to develop industry-wide standard controls for gene expression assays.

Extra-Poisson variation A Poisson-like distribution where the variance is proportional rather than equal to the mean.

Familial adenomatous polyposis A type of inherited colorectal cancer syndrome caused by defects in the *APC* gene. Features chromosomal instability and the development of polyps within the colon beginning at a young age. An attenuated form (AFAP) is also caused by defects in *APC* but features fewer polyps.

Fecal occult blood Blood in the feces that is not readily apparent. FOB tests are aimed at detecting globin, DNA, or other factors to indicate the presence of blood.

Flow cell Solid glass surface onto which DNA templates are tethered for clonal amplification and sequencing using second-generation sequencing technologies such as Illumina sequencing method.

Fluorescence *in situ* hybridization (FISH) A cytogenetic technique that is used to detect and localize DNA sequences on chromosomes.

Follicular thyroid carcinoma (FTC) Thyroid neoplasm, not belonging to papillary thyroid carcinoma, with evidence of capsular and/or vascular invasion.

Formalin-fixed, paraffin-embedded samples A process used for archiving tissues and conducting histopathological diagnosis of diseases. Formalin fixation prevents decay by creating covalent chemical bonds between proteins in tissue. Paraffin embedding involves dehydrating tissues and then infiltration with wax. This process provides sample stability for multiple years.

Functional validation Refers to proving the functional role of a driver mutation at the bench.

Fusion gene A chimerical gene produced by abnormal fusion of two separate genes usually as a result of chromosomal deletion or translocation. A fusion gene can be oncogenic in a cancer as a "driver" event.

Gene expression profile Measurement of the expression of a well-defined (or genome-wide) set of genes that differentially classifies individual tumors.

Gene mutation A permanent change in the DNA sequence that makes up a gene. Mutations range in size from a single DNA building block (DNA base) to a large segment of a chromosome.

Gene set enrichment analysis (GSEA) Method of analysis of gene expression profiling data which compares two sample groups to identify predetermined gene subsets concomitantly dysregulated in one group versus the other.

Genome The entire genic and intragenic sequences of DNA in an organism.

Genome-wide association study (GWAS) A protocol in which single nucleotide variants (SNVs) are identified in the genome of individuals that either do or do not have a particular cancer (or other disease). SNVs that occur more frequently in those that have the disease than those that do not may be important for predicting disease onset

Genomic biomarkers Biomarkers based on the analysis of DNA (deoxyribonucleic acid) profiles, especially the analysis of SNPs (single nucleotide polymorphisms).

Genomic "driver" event Genomic alterations that cause a selective growth or survival advantage to tumor cells (see driver mutation).

Genomic "passenger" event Genomic alterations that have no effect on the selective growth or survival advantage of tumor cells.

Genomics Study of structure, evolution, function and mapping of genetic information conducted at the genome level.

Genotype The genetic constitution of an organism; the specific set of alleles inherited at a particular locus. The expression of the genotype constitutes to an observable trait of an organism.

Germline A term used in genetics to describe the cells within an animal that contribute to gamete production and therefore can pass on genetic information to the offspring.

Germline mutations Mutations which are present in the gametes and are passed from one generation the next.

Gleason score A description of the morphological differentiation of prostate cancer, as determined by a pathologist under the microscope. The Gleason score represents the sum of two Gleason grades, ranging from 1 to 5 (where 1 is well-differentiated and 5 is very poorly differentiated), representing the most common and the second most common grade within an area of tumor.

Glomerular sclerosis Scarring of the glomeruli of the kidney.

Good Clinical Practice (GCP) An international quality standard that is provided by International Conference on Harmonization (ICH), an international body that defines standards, which governments can transpose into regulations for clinical trials involving human subjects.

Good Laboratory Practice (GLP) In the experimental (non-clinical) research arena, good laboratory practice specifically refers to a quality system of management controls for research laboratories and organizations to try to ensure the uniformity, consistency, reliability, reproducibility, quality, and integrity of chemical (including pharmaceuticals) non-clinical safety tests; from physio-chemical properties through acute to chronic toxicity tests.

Haploinsufficient A term in genetics used to describe a condition in diploid organisms where a single functional copy of a gene allele is incapable of producing enough gene product to allow a normal growth or development of the organism resulting often in a disease phenotype.

Haplotype phase Determining the arrangements of alleles that are co-located on the same chromosome using correlated single nucleotide polymorphisms (i.e. haplotype).

Hazard ratio An expression of the hazard or chance of events occurring in the treatment arm as a ratio of the hazard of the events occurring in the control arm.

Hemihypertrophy A condition in which one side of the body or a part of one side is larger than the other.

Hemiparesis A condition of muscular weakness causing inability to move one side of the body.

Heterodimer A macromolecular complex formed by two different macromolecules.

Heterotopic A term used to describe the engraftment or transplantation of cells to an abnormal or atypical anatomical location (e.g. lung cells transplanted under the epidermis).

Heterozygosity Having different alleles at one or more corresponding chromosomal loci.

High-dimensional data Large data sets characterized by the presence of many more predictor variables than observations, such as data sets that result from measurements of hundreds to thousands of molecules in a relatively small number of biological samples. The analysis of such data sets requires appropriate computing power and statistical methods.

High-throughput screening Type of techniques allowing a researcher to conduct quickly millions of chemical, genetic or pharmacological tests.

High-throughout sequencing DNA sequencing technologies characterized by their ability to sequence thousands to millions of DNA fragments in parallel.

Histology Study of the microscopic structure of tissues.

Histone Small chromosomal proteins that form a structural unit, around which DNA is wrapped in a nucleosome.

Histone demethylases Enzymes which remove methyl groups from histones and other proteins.

Histone modification A process in which chemical groups are added to or removed from histones to alter gene expression.

Homodimers A macromolecular complex formed by two identical molecules.

Homologs Also known as homologues indicating chromosome pairs of approximately the same length, centromere position, and staining pattern, with genes for the same characteristics at corresponding loci.

Homozygous Identical alleles of a gene are found on both homologous chromosomes.

Hormone replacement therapy Supplement of hormones due to the lack or insufficiency of endogenous hormones as part of a medical treatment.

Hybridization A process of complementary DNA strands binding together. Typically, a probe is used to detect a specific DNA molecule by hybridization of a probe and the target DNA strands.

Hyperchromasia Increased hematoxylin staining of nuclei due to increased DNA content.

Hyperplasia An abnormal increase in cell number within an organ causing it to become enlarged.

IGF1R pathway A crucial cellular signaling pathway involving the IGF1 receptor (IGF1R) and IGF1 and 2 ligands that provides a potent survival and proliferative signals to stimulate growth in many different cell types.

Immune-privileged A term usually referring to an anatomical site in the body that is not subjected to immune surveillance and thus cells transplanted to that site or tissue are not rejected by the immune system.

Immunocompromised An animal or individual that is incapable of mounting an effective immune response.

Immunoglobulin A protein produced by plasma cells and B lymphocytes, which plays a role in the immune response to identify and neutralize pathogens. Antibodies are composed of two large heavy chains and two small light chains and present a constant region and a variable region, which recognizes pathogen antigens.

Immunohistochemistry (IHC) A technique allowing the colorimetric detection of antigens in cells or tissue sections with antibodies.

Incidental research results Refers to results specifically linked to the identity of the research participant but that were not the original aim of the research. Two examples of this concept are: (1) the discovery of a brain tumor when conducting an MRI of the brain looking at functional speech areas; (2) the discovery of a breast cancer risk gene when looking for a gene responsible for urinary tract anomalies.

Indels Insertions or deletions of a single nucleotide or whole sequences.

Individual research results Refers to the results of research specifically linked to the identity of the research participant.

Integument General term describing the outer covering of an organism, e.g. human skin.

Intermediate mesoderm A type of embryological tissue called mesoderm that is located between the paraxial mesoderm and the lateral plate.

Interstitium The tissue surrounding the loop of Henle in the renal medulla.

Intra-ductal carcinoma of the prostate (IDC-P) A phenomenon observed in approximately 25% of prostate cancers in which cancer cells invade the prostatic ducts. IDC-P is strongly associated with high-grade cancers, and may be an independent prognostic factor for relapse and metastasis.

Intra-tumoral heterogeneity Represents the pathological, molecular and genomic variability that exists among cells within the same tumor.

Isochromosome A chromosome that has lost one of its arms and replaced it with an exact copy of the other arm.

Isoform Any of two or more functionally similar proteins that have a similar but not identical amino acid sequence and are either encoded by different genes or by RNA transcripts from the same gene which have had different exons removed.

Karyotype Karyotype is the number and appearance of chromosomes in the nucleus of cells. A normal human karyotype includes 22 pairs of autosomal chromosomes and one pair of sex chromosomes (46,XX in females or 46,XY in males). Abnormal karyotypes can be due to abnormalities in the number of chromosomes (aneuploidy) or in the structure of chromosomes.

Kataegis A pattern of clustered hypermutation found in some cancers including breast. The term kataegis is derived from the Greek word for "thunder". The mutations are predominately cytosine to thymine in the context of a TpC dinucleotide and the AID/APOBEC family of cytosine deaminases is responsible for the process of kataegis.

Ki-67 A nuclear antigen used commonly to determine proliferating cells within a tumor biopsy; it is a nucleolar protein that is highly expressed in S-phase of the cell cycle.

Kinase Enzyme that covalently attaches phosphate groups to substrate molecules, typically proteins.

Knudson's two hit hypothesis A theory established by Dr Alfred Knudson to explain the pattern of tumor development in hereditary and sporadic retinoblastoma, predicting the existence of tumor suppressor genes.

KRAS A protein encoded by the *KRAS* gene which performs an essential function in normal tissue signaling, and the mutation of a *KRAS* gene is an essential step in the development of many cancers.

Leucoencephalopathy Disease of the white matter of the brain, often due to localized areas of cell death.

Leukemia stem cell Also referred to as a leukemia initiating cell, refers to a cell that is self-renewing and can give rise to leukemia when transplanted into an immunodeficient mouse as a xenograft.

Library construction The process for building a sequencing library. Follows extraction of RNA or DNA and precedes sequencing. In many cases this involves fragmentation of the sample and ligation of sequencing-specific primers.

Li–Fraumeni syndrome A cancer predisposition syndrome, originally described in 1969, caused by germline mutations in the *TP53* tumor suppressor gene and associated with a wide spectrum of early-onset child- and adult-onset cancers.

Ligand A group, ion, or molecule coordinated to a central atom or molecule in a complex.

Limit of detection The lowest concentration of analyte for which the response can be reliably distinguished from background noise.

Linkage disequilibrium A term in genetics that refers to the non-random association of two or more genetic loci within individuals in a population, which may or may not be on the same chromosome.

Linkage Genetic markers that are physically close to one another on a chromosome tend not to be separated from one another, or broken up, during recombination. These markers (often SNPs) are said to be in linkage. Linkage analysis can be performed in suitable pedigrees to identify regions of a chromosome with a greater degree of linkage than expected and that is preferentially shared among disease carriers than unaffected members of the family. It is a powerful tool for reducing the search space from a whole genome to part of a chromosome in disease genetics research.

Lobe Gross anatomical extension of the breast which consists of lobule units.

Lobectomy Excision of a whole lobe of the lung.

Lobular carcinoma Cancer that arises from the milk-producing glandular cells in the lobules.

Lobule Glandular part of the lobe of the breast and is responsible for milk production.

Locked-nucleic acid (LNA) RNA analogs where the ribose is locked in a C3′-endo conformation by the introduction of a 2′-O,4′-C methylene bridge.

Locus A location on a chromosome.

Long non-coding RNAs (lncRNAs) Transcribed RNA molecules greater than 200 nucleotides in length without protein-coding function.

Long-range PCR A type of polymerase chain reaction technology that allows amplification of long DNA fragments (typically greater than 5000 bases).

Loss of heterozygosity Loss of heterozygosity results in the loss of a functional tumor suppressor in a two hit manner. The first deleterious mutation in a single allele results in a heterozygous state. In individuals with a hereditary syndrome, the first hit is a germline mutation. Most often the second hit is a deletion of the remaining functional allele although often LOH is used as a more general term to also describe the case where the second hit is a second deleterious mutation. The end result of both cases is the lack of a functional gene product.

Lower limit of quantitation The lowest concentration (amount) of analyte in a test sample that can be determined quantitatively with suitable accuracy (mean bias) and precision.

Lung cancer Malignancy tumors originally arising from lung tissue and the vast majority are bronchogenic in origin.

Lymphatic system Part of the circulatory system consisting of a network of connecting vessels that transfer a clear fluid called lymph throughout the body. The lymphatic system drains lymph from tissues back into the blood.

Lymphoblastic lymphoma A neoplasm of immature B or T cells that exhibits extramedullary involvement (does not extensively involve the bone marrow or blood).

Lynch syndrome A hereditary colorectal syndrome caused by defects in mismatch repair enzymes, resulting in microsatellite instability.

Macroglossia A large tongue.

Mammography Imaging technique that delivers a low-dose X-ray for the early detection and diagnosis of breast cancer.

MAPK (mitogen-activated protein kinase) A serine/threonine-specific protein kinase involved in directing cellular responses to a diverse array of stimuli, such as mitogens, osmotic stress, heat shock and proinflammatory cytokines. They regulate cell proliferation, gene expression, differentiation, mitosis, cell survival, and apoptosis.

Mapping (RNA-Seq) The computational approach to assigning sequencing reads to their origin within a genome. Mapping strategies for RNA need to be aware that sequencing reads may contain individual-specific variants and also cross splice-junctions.

Mass spectrometry An instrumental method for identifying the chemical constitution of a substance by means of the separation of gaseous ions according to their differing mass and charge.

Massively parallel sequencing A term used to refer collectively to high-throughput DNA sequencing approaches that employ miniaturized and highly parallel platforms to sequence millions of short sequence reads of usually 50–400 nucleotides. These techniques are also referred to as next-generation sequencing (NGS) or second-generation sequencing approaches.

Medullary thyroid carcinoma (MTC) Thyroid cancer of aggressive nature originating from C cells and in close association with multiple endocrine neoplasia syndromes.

Menarche age Age at which the first menstrual cycle (or first menstrual bleeding) occurs in females.

Menopause Age, often 40 to 50 years, at which the monthly menstrual cycles cease and signifies the end of the fertile phase of a woman's life.

Messenger RNAs (mRNAs) A subtype of RNA molecules that mediate the transfer of genetic information from the cell nucleus to ribosomes in the cytoplasm, where it serves as a template for protein synthesis.

Meta-analysis An experimental design where the results or data of multiple experiments are combined together and analyzed.

Metabolite An individual molecule resulting from metabolism, part of the metabolome.

Metabolome The totality of all molecules resulting from the process of metabolism.

Metabolomic biomarkers A biomarker based on the analysis of metabolites (metabolites are the intermediates and products of metabolism).

Metabolotype The characterization of individuals based on the specifics of their metabolome, which reflects differences in metabolism.

Metagenomics A field of genomics that studies the collective microbial genomes obtained from an environmental sample such as soil or human gut.

Metastasis The spread of a disease-producing agency (as cancer cells or bacteria) from the initial or primary site of disease to another part of the body.

Metastatic tumor Tumor that spreads from the original place of malignancy to another location in a body.

Methylation The addition of a methyl group to a substrate or the substitution of an atom or group by a methyl group. It can be involved in modification of heavy metals, regulation of gene expression, regulation of protein function, and RNA metabolism.

Methyltransferase An enzyme that transfers a methyl group from a methyl donor to an acceptor molecule. Methylation of DNA and/or histones is important in epigenetic regulation of gene expression.

Microarrays Microarrays contain small sequences of thousands of genes or other genomic regions embedded on a solid surface such as a glass slide, which is subsequently hybridized with DNA or RNA. While less expensive and faster than NGS technologies, they offer only one or two data types per array type.

MicroRNA Small double-stranded, non-protein coding RNAs, 21–25 nucleotides in length involved in repressing translation of target RNA by binding to homologous 3′ UTR region as an imperfect match.

Microsatellite instability Found in a subset of colorectal cancer, MSI is caused by defects in the mismatch repair pathway. Strand slippage in microsatellite repeats during DNA replication causes expansion/contraction of microsatellite repeat regions which can cause frameshifts and loss of function in genes with such repeats in their coding regions.

Minimal deleted region A region of the genome that is concomitantly lost in several samples.

miR-mimics Double-stranded RNAs designed to mimic the function of the endogenous miRNAs.

miRNA See MicroRNA.

miRNA expression miRNA microarrays have been developed to measure the expression of these small non-coding RNAs.

miRNA profiling The measurement of the relative abundance of a cohort of miRNAs, ranging from a group of several miRNAs of specific biological interest to comprehensive profiling of all miRNAs in a given species.

miRNA-seq Like RNA sequencing small non-coding RNAs can be prepared from cells and sequenced. miRNA-seq sequencing data provide the nucleotide sequence in addition to the expression levels.

miRNA sponge Artificial transgenes containing multiple complementary sequences to the targeting miRNA.

Mismatch repair The molecular means of recognizing and repairing the inappropriate insertion, deletion, or misincorporation of incorrect base pairs during DNA replication. Mismatch repair is also used for repairing certain kinds of DNA damage. Defects in certain mismatch repair enzymes are a major cause of certain types of colorectal cancer, in particular Lynch syndrome.

Mitotic index Measure of the proliferative capacity of cells and is defined as the ratio of the number of cells undergoing mitosis over the total number of cells.

Molecular profiling A set of data exhibiting the significant features of something in molecular level such as gene mutations.

Monoclonal gammopathy of undetermined significance (MGUS) A condition characterized by the presence of a monoclonal paraprotein in the blood, less than 10% of plasma cells in the bone marrow and no clinical symptoms. It can evolve to active multiple myeloma at a rate of 1% per year.

Morbidity Physical state of being ill, diseased or unhealthy.

Mortality The state of being susceptible to death.

Mosaic A term used in genetics to refer to an animal composed of two or more cytogenetically distinct subpopulations of cells that were derived from the same zygote. This term is often used to describe the results of transgene experiments involving injection of genetic material into mouse and zebrafish single-cell embryos (i.e. the zygote).

mRNA See messenger RNA.

mTOR A serine/threonine kinase downstream of PI3K.

Mucositis Inflammation and ulceration of the mucous membranes lining the digestive tract.

Multiple endocrine neoplasia type 2 (MEN2) Group of medical disorders associated with tumors of the endocrine system and transmitted in an autosomal dominant pattern, in which gain-of-function mutations of the *RET* gene frequently play a causative role.

Multiple myeloma Multiple myeloma is a clonal proliferation of neoplastic plasma cells in the bone marrow, which synthesize abnormal amounts of immunoglobulin or immunoglobulin fragments.

Multiplex PCR A type of polymerase chain reaction technology that allows simultaneous amplification of multiple target DNA fragments.

Multivariate analyses Based on the statistical principle of multivariate statistics, which involves observation and analysis of more than one statistical outcome variable at a time.

Mutation A change in a gene or a chromosome that differs from the wild type. mutations are associated with cancer.

MYCN (N-MYC) A member of the MYC family of transcription factors which encodes a protein with a basic helix-loop-helix (bHLH) domain.

ncRNA Non-coding RNA. Non-Coding RNA refers to RNA genes that do not encode for protein. A broad class of genes with several subsets including the more common tRNA. New classes with presumed regulatory roles include microRNA (small, approximately 20 bp) and long non-coding RNA.

Negative predictive value The proportion of those truly without the condition that are correctly identified with a negative test result (ratio of true negatives to combined true and false negatives).

Neoadjuvant chemotherapy Chemotherapy that is given before primary treatment to shrink the primary tumor and reduce the need for aggressive surgical resection.

Neoplastic Relating to, or constituting, a tumor or neoplasia.

Nephrogenic rests A fragment of embryonic tissue in the kidney retained after the period of embryonic development.

Neuritogenesis The process of neural development in which neurons sprout and extend neurites to form axons and dendrites and create a functional neuronal network.

Neurocristopathy Developmental abnormality involving neural crest-derived tissues.

Neuropathy A functional deterioration or pathological change in one or more of the peripheral nerves.

Neurotoxicity Damage to the central and/or peripheral nervous system.

Next-generation sequencing (NGS) Modern high-throughput sequencing technologies that parallelize sequencing allowing thousands to millions of DNA molecules to be sequenced simultaneously, lowering the cost and making the sequencing of large genomes possible in weeks. Whole-genome, -exome, -transcriptome and ChIP-Seq all use next-generation sequencing technology.

Nomogram A graphical calculator representing the relationship between three or more variables. In the context of prostate cancer, nomograms such as the Kattan nomogram are prognostic tools that can be used to decide which treatment approaches will result in the greatest benefit for men at various stages of prostate cancer.

Non-coding RNA RNA that does not encode a protein but has a separate biochemical function. This RNA includes multiple species such as transfer RNA, ribosomal RNA, microRNA small nuclear RNA and linc-RNA (also lnc-RNA).

Non-germline A term used in genetics to describe the cells within an animal that do not contribute to gamete production and therefore do not pass on genetic information to the offspring.

Non-small cell lung cancer A group of subtypes of lung cancer which excludes small cell lung cancer.

Non-synonymous mutation A single nucleotide substitution mutation that results in an amino acid change.

***NTRK1* (TRKA)** Proto-oncogene encoding a tyrosine protein kinase that serves as a receptor for nerve growth factor (NGF), which primarily regulates growth, differentiation, and programmed cell death of neurons in both the peripheral and central nervous system.

Nucleosome A basic unit of eukaryotic chromosomal structure that consists of 146 bp of DNA that wraps around eight histone proteins (an octamer consisting of duplicates of H2A, H2B, H3, H4). Nucleosomes are linearly arranged like "beads-on-a-string".

Nulliparity The state of not having offspring.

Oligomer A molecule that consists of a few monomers.

Oligonucleotides A relatively short single-stranded nucleic-acid chain (as an oligodeoxynucleotide or oligoribonucleotide) usually consisting of up to approximately 20 nucleotides.

Omics Scientific disciplines comprising study of related sets of biological molecules. Examples of omics disciplines include genomics, transcriptomics, proteomics, metabolomics, and epigenomics.

Omics-based test An assay composed of or derived from many molecular measurements and interpreted by a fully specified computational model to produce a clinically actionable result.

Omphalocele A type of abdominal wall defect in which the intestines, liver, and occasionally other organs remain outside of the abdomen in a sac because of a defect in the development of the muscles of the abdominal wall.

Oncogene Gene, usually playing a role in cell growth and proliferation, whose gain-of-function alteration leads to neoplastic cell transformation.

Orthotopic A term used to describe the engraftment or transplantation of cells to the same anatomical location from which they were derived (e.g. mammary cells transplanted to the mammary fat pad).

Overall survival The proportion of people who are expected to live after a certain amount of time (e.g. 5 or 10 years). Overall survival takes into account death from any cause which may be unrelated to the disease.

Overdispersion A property of a statistical distribution where more variance is observed than predicted.

Overfitting Occurs when the model-fitting process unintentionally exploits characteristics of the data that are due to noise, experimental artifacts, or other chance effects that are not shared between datasets, rather than to the underlying biology that is shared between datasets. Overfitting leads to a statistical or computational model that exhibits very good performance on the particular dataset on which it is fit, but poor performance on other datasets. Although not unique to omics research, the chance of overfitting increases when the model has a large number of measurements relative to the number of samples.

Paired-end sequencing The sequencing technique where the two ends of a DNA molecule of a known size are sequenced.

Paired-sample analysis Somatic mutation detection techniques which consider the tumor and matched normal samples simultaneously when identifying the most probable somatic aberrations.

Palliative chemotherapy To use chemical agents to extend life and alleviate symptoms but not used as an attempt to cure cancer completely.

Papillary thyroid carcinoma (PTC) Thyroid neoplasm deriving from the thyroid follicular epithelial cells, displaying papillary architecture and characteristic nuclear features.

Paraxial mesoderm The area of mesoderm in the neurulating embryo that flanks and forms simultaneously with the neural tube.

Passenger event An event which results from increased genomic instability and does not contribute to the progression of cancer.

Passenger mutations Any cancer cell could have hundreds of mutated genes, not all are contributing to the disease however. The majority are the result of the unstable cancer genome and do not provide the cell with any growth advantage. These are referred to as passenger mutations.

Patent Refers to a legal monopoly over an invention (as defined in the patent itself) that is granted by a country's patent office. In general, the term of a patent is limited to 20 years beginning on the date when the patent application was filed with the patent office. A patent can be obtained in several countries at once (but not consecutively), but the criteria by which a patent application is assessed vary across countries. Once obtained, a patent confers a set of exclusive rights (e.g. to use, make, sell, or import the invention), which only the patent owner is entitled to do unless it gives permission to other(s) to do so.

Paternal isodisomy Two copies of a chromosome, or part of a chromosome, inherited from the father.

Pathogenesis The mechanism that causes the disease and can also describe the origin and development of the disease.

Pathway analysis Refers to identifying a pathway, a common cellular process governed by a complex of proteins and other molecules, which is affected by the accumulated mutations in the cell.

PDGFRA The platelet-derived growth factor receptor alpha is a gene that encodes a cell surface tyrosine kinase receptor, related to c-KIT.

PEER framework The probabilistic estimation of expression residuals (PEER) framework assesses global hidden factors which influence expression variability. Such factors may be hidden biological or technical variables which influence the expression of multiple genes.

Penetrance of a mutation Magnitude of impact on patient outcome.

Personalized medicine The adaption of therapeutic modalities to individual patients based on the use of their (mainly) genetic information. An example of this approach would be the use of mutation information to deliver a drug that is known to have activity against this mutation (e.g. patients with a *BRAF* V600E mutation and respond more favorably to the BRAF inhibitor Vemurafenib).

Personalized targeted therapy A type of medication that blocks the growth of cancer cells by interfering with specific targeted molecules needed for carcinogenesis and tumor growth that occurs in individual patients.

Pharmacogenetics Genetic differences that can influence an individual's response to a particular drug.

Pharmacogenomics The branch of pharmacology that deals with influence of genetic variation on drug response in individuals. It is sometimes referred to being the whole genome application of pharmacogenetics.

Philadelphia chromosome A chromosomal abnormality created by the translocation of human chromosomes 9 and 22 (i.e. t(9;22)(q34;q11)) that is associated with chronic myelogenous leukemia (CML). The translocation fuses the BCR and ABL kinase genes resulting in the expression of an oncogenic fusion protein that can be targeted by the tyrosine kinase inhibitor imatinib mesylate.

Phosphorylation The addition of phosphate to an organic compound through the action of a phosphorylase or kinase.

Physical coverage A number of DNA fragments that cover a particular nucleotide in the reconstructed sequence such as the reference genome.

PI3K Phosphoinositide 3-kinase is an enzyme involved in cellular functions such as cell growth, proliferation, differentiation, motility, survival and intracellular trafficking.

PI3K/AKT/mTOR pathway An intracellular signaling pathway involved in cellular proliferation, growth, survival and mobility and often exhibits alterations of its components in a wide variety of tumors, therefore representing an attractive target for anticancer therapy.

Plasma cell A plasma cell is a fully differentiated, mature lymphocyte of the B-cell lineage, which produces and secretes antibodies.

Plasma cell leukemia Presence of more than 20% of malignant plasma cells in the bloodstream.

Pneumonectomy Surgical treatment to excise one side of the lung.

Point mutation Alteration in DNA sequence caused by a single nucleotide base change, insertion, or deletion resulting in silent, missense or nonsense mutations.

Poisson distribution A statistical distribution where the variance is equal to the mean and which assesses the probability that a number of events occur in a fixed interval.

Polymorphism Natural variations in a gene, DNA sequence, or chromosome that have no adverse effects on the individual and occur with fairly high frequency in the general population.

Polyp An abnormal growth of tissue that protrudes from a mucous membrane.

Positive predictive value The proportion of true positives with the condition that are correctly identified with a positive test result (ratio of true positives to combined true and false positives).

Preanalytical variables Aspects of sample collection and handling that need to be standardized and documented prior to test development and use.

Precision The closeness of agreement (degree of scatter) between a series of measurements obtained from multiple sampling of the same homogeneous sample under the prescribed conditions. Precision may be considered at three levels: repeatability, intermediate precision and reproducibility.

Preclinical A term that usually refers to experimentation or medical treatment in animals, with the aim of collecting data in support of the safety and efficacy of a new treatment prior to the initiation of clinical trials in humans.

Precursor microRNAs (pre-miRNAs) Hairpin RNAs of 60–100 nucleotides that are processed from primary microRNAs by ribonuclease III Drosha-DGCR8 nuclear complex.

Predictive biomarker A biomarker that correlates with the effectiveness of treatment, which is sometimes also referred to as an "effect modifier" of treatment.

Predictive value Overall capacity of a method to predict an outcome (e.g. survival, recurrence risk, treatment response etc.).

Primary microRNAs (pri-miRNAs) Initially transcribed microRNAs from genome containing several hundred nucleotides, which are further processed into precursor microRNAs.

Primary tumor The original tumor.

Probe A nucleotide sequence that can be used to identify a specific DNA molecule whose sequence is complementary to the probe sequence.

Progesterone The primary steroid female hormone required for the female menstrual cycle as well as different stages of embryogenesis.

Progesterone receptor (PR) A nuclear receptor that can act as a transcription factor when it is activated by its ligand progesterone.

Prognostic biomarker A biomarker correlated with a clinical outcome in the setting of natural history or a standard of care regimen; it is a variable used to estimate the risk of or time to clinical outcomes.

Progression-free survival The length of time during and after treatment during which the disease being treated does not get worse.

Prostate specific antigen (PSA) A protein secreted by the prostate gland in response to androgen stimulation. The presence of PSA in the blood can be an indicator of the presence of prostate cancer.

Proteasome Protein complexes, which degrade unneeded or misfolded proteins by proteolysis.

Protein expressions A subcomponent of gene expression consisting of the stages after DNA has been transcribed to messenger RNA (mRNA), then translated into polypeptide chains, which are ultimately folded into proteins.

Protein-coding genes See messenger RNAs.

Proteomic biomarkers A biomarker based on the analysis of the protein profiles.

Proteosome An intracellular complex that degrades damaged or unnecessary proteins that have been targeted for elimination.

PTEN Phosphatase and tensin homolog, a tumor suppressor gene through the action of its phosphatase protein product to dephosphorylate phosphatidylinositol (3,4,5)-trisphosphate.

Quality control (QC) samples Pre-study validation and in-study samples having a known concentration (nominal) of analyte that are treated as unknowns in an assay. During pre-study validation, QC samples are used to generate information to demonstrate the method is suitable for its intended purpose. During in-study runs, QC values are used as the basis for accepting and rejecting bioanalytical method batches.

Radiation therapy The medical use of ionizing radiation to control or kill malignant cells.

Radical prostatectomy Surgical removal of the entire prostate gland.

RAS Family of guanine nucleotide-binding proteins involved in cell signaling pathways, cell growth and apoptosis that have been frequently found mutated in human cancers.

Read Read-out of the nucleotide sequence of a DNA fragment during sequencing.

Receptor tyrosine kinases Class of transmembrane proteins that have intrinsic protein–tyrosine kinase activity and act as receptors for growth factors, neurotrophic factors, and other extracellular signaling molecules.

Reference material Defined by the National Institute of Standards and Technology as a material that is homogeneous and stable enough to be fit for use in a measurement process. The USP and the National Formulary list certified reference materials that meet the standards of the International Organization for Standardization (ISO) for reference materials in their ISO 17025 and ISO Guide 34 for reference material producers.

Repeatability The precision under the same operating conditions over a short interval of time. Repeatability is also termed as intra-batch or intra-run precision.

Reproducibility Precision of repeated measurements between laboratories and is termed interlaboratory precision. Usually applies to collaborative studies that involve the standardization of a bioanalytical method across multiple laboratories.

Research Ethics Board (REB) The Research Ethics Board is an independent, institutionally-based body responsible for the review, approval and ongoing monitoring of the ethical conduct of research. In Canada, this is guided by the Tri-Council Policy version 2 and, in the USA, by the equivalent Institutional Review Board (IRB) governed by the Common Rule.

Resection The partial surgical removal of an organ.

Response rate (RR) Percentage of patients who respond to a specific treatment or fulfill the specific criteria of response established for a particular clinical trial.

RET Proto-oncogene encoding a tyrosine kinase receptor for members of the glial cell line-derived neurotrophic factor (GDNF) family.

RET/PTC Activated version of the *RET* proto-oncogene, found in papillary thyroid carcinoma.

Retrovirus A single-stranded RNA virus that produces reverse transcriptase by means of which DNA is produced using its RNA as a template and replicates in a host cell that is often tumorigenic.

Reverse phase protein array (RPPA) Antibody-based, reverse phase protein arrays for protein expression levels and phosphorylation state. Measures levels of phosphorylated isoforms.

Ribonuclease III-Drosha-DGCR8 complex An enzyme complex containing the RNase III enzyme Drosha and double-stranded RNA binding protein DGCR8, which imitates the transcription of primary microRNAs.

Risk of recurrence Chance of a cancer patient to present with recurring disease post-treatment.

RNA expression array RNA expression arrays ("microarrays") measure the differences in expression of genes from two different populations of samples. RNA "probes" are, like DNA microarrays, spotted onto glass slides or chips.

RNA processing A biological process aimed at generating mature mRNA, or a functional tRNA or rRNA from the primary transcript. Processing of pre-mRNA involves the steps of capping the 5′ end, polyadenylation of the 3′ end, splicing and sometimes RNA editing.

RNA spike-ins A positive control that is introduced into a sequencing reaction or library. Spike-ins may be at known quantities to assess the dynamic range of quantification in an experiment.

RNA-induced silencing complex (RISC) A multiprotein complex that incorporates the mature strand of microRNA as a template for recognizing complementary mRNA.

RNA-sequencing (RNA-seq) Also referred to as whole transcriptome shotgun sequencing (WTSS). It is the technique for finding the sequence of the transcribed regions of the genome. cDNA is made using the cell's RNA as template and is then sequenced.

rRNA ribosomal RNA; a class of RNA molecules that form part of ribosome, the protein-synthesizing organelle in a cell.

Sample turnover time Time required for processing of patient sample from time of receiving to the availability of the results.

Sanger sequencing A protocol involving the use of dideoxynucleotide-based chain termination for sequencing short strands of DNA. Sanger sequencing is serial, in that only a single DNA sequence can be determined in any given experimental reaction.

Sclerosing skeletal dysplasia A rare genetic disorder characterized by the creation of abnormally dense and overgrown bones.

Segmental chromosomal abnormality (SCA) Genomic abnormality involving loss or gain of one or more individual parts of one or multiple chromosomes; in contrast to numerical chromosomal abnormality that involves loss or gain of whole chromosomes.

Selectivity The extent to which a bioanalytical method can measure particular analyte(s) in a complex mixture without interference from other components of the mixture.

Sensitivity Analytical sensitivity is the ability of an analytical method to detect small quantities of the measured component. Diagnostic sensitivity is the probability that the individual will test positive when an individual has the disease.

Sequence coverage A number of sequenced reads that cover a particular nucleotide in the reconstructed sequence such as the reference genome.

Sequencing library A population of nucleic acid fragments that are flanked by sequencing adapters.

Serial analysis of gene expression (SAGE) A technique based on sequencing short concatenated cDNA tags to identify and quantify transcripts. cDNA tags are created through application of special restriction enzymes with cut sites several nucleotides from the recognition site.

Serrated A pathological term that, in this case, is used to describe the appearance of certain types of polyps. Serrated applies to the sawtooth-like appearance of this type of polyp.

Signal transduction It occurs when an extracellular signaling molecule activates a cell surface receptor. In turn, this receptor alters intracellular molecules creating a response.

Single nucleotide polymorphism (SNP) A single base pair substitution which occurs at a high frequency in the human population, often contributes to inter-individual DNA sequence variability and does not contribute to disease.

Single nucleotide variant or variation (SNV) A single base DNA variation that may or may not be common (includes mutations).

Smoldering myeloma A condition characterized by more than 10% of plasma cells in the bone without evidence of end-organ damage

SNP array This is a DNA microarray (DNA probes immobilized on a glass slide/chip) used to determine single nucleotide changes as well as copy number variation in the genome. It can also be used to estimate copy number variations.

Soft tissue sarcomas (STSs) A heterogeneous group of malignancies of mesenchymal origin arising from anywhere in the body, usually aggressive.

Somatic mutations DNA mutations that are acquired during a person's life time which can be due to endogenous metabolism and/or exogenous exposure to carcinogens or radiation.

Somatic point mutations A change of the nucleotide sequence of the genome of a cell.

Specificity Analytical specificity is the ability unequivocally to measure the analyte in the presence of other components that may be expected to be present in the biological specimen, including impurities, metabolites and endogenous matrix components. Diagnostic specificity is the probability that the individual will test negative when that individual does not have the disease.

Staging The process of assigning a tumor to its correct pathological stage using various tests and diagnostic indicators. More advanced stages of cancer have worse prognoses and may have already involved lymph nodes or metastasis to other organs.

STAT Signal transducer and activator of transcription, a protein which regulates many aspects of cell growth, survival and differentiation.

Statistics and bioinformatics validation Verifying that the omics-based test can perform its intended task. Ideally, this involves assuring that the test can accurately predict the clinical outcome of interest in an independent set of samples that were not used in developing the test. Such validation is particularly important as omics tests typically involve computational models whose parameters can be overfit in any single dataset, leading to an overly optimistic sense of the test's accuracy.

Stem cells The biological cells able to divide (through mitosis) and differentiate into diverse specialized cell types and can self-renew to produce more stem cells.

Stroma A structural network of cells, blood vessels, and extracellular material that forms the connective tissue underlying tissue and organ formation.

Structural variation These refer to variations that encompass large areas of the genome and could include large insertions and deletions, inversions, duplications, translocations and gene fusions.

Subclonal heterogeneity Presence of different clonal populations (with variable percentage) with diverse genetic background within the malignant cells of the same individual.

Succinyl dehydrogenase Enzyme complex responsible for converting succinate to fumarate as part of the Krebs cycle.

Suppressor gene A gene that protects a cell from one step on the path to cancer. When this gene is mutated to cause a loss or reduction in its function, the cell can progress to cancer, usually in combination with other genetic changes.

Supratentorium A region of the brain located above tentorium cerebelli.

Synonymous coding variant A genetic variation in the open reading frame of a gene that does not alter the amino acid composition of the protein produced from the gene.

Synthetic lethality A genetic interaction in which a combination of mutations in more than one gene leads to cell death, while a mutation in one of these genes does not lead to cell death.

Targeted sequencing Like exon capture, targeted sequencing uses probes, but in this technique they are designed to specific regions of the genome that are of interest, e.g. mutated genes. Usually targeted sequencing is used for the purpose of validation and high-throughput screening of recurring mutations.

Taxanes Microtubule-inhibiting class of drugs which are widely used as anticancer agents. They are derived from yew plants (genus *Taxus*). The most common taxanes are docetaxel and paclitaxel.

Terminal ductal-lobular units (TDLUs) The basic functional units of the breast which are composed of extralobular and intralobular ducts. The TDLUs provide the conduit from which the milk drains from the lobules to allow secretion from the nipple.

Thyrocytes Epithelial cells (also called follicular cells) of the thyroid gland that are responsible for the production and secretion of thyroid hormones, that is, thyroxine (T4) and triiodothyronine (T3).

Tissue microarray (TMA) A technology allowing the transposition of punches of tissue cores from tissues donor blocks, arranged in an array fashion.

TMPRSS2–ERG fusion A fusion of two genes on chromosome 21 (*TMPRSS2* and *ERG*), which is observed in approximately 40–50% of prostate cancers.

TNM staging A system that describes the extent of a cancer based on the size of the tumor (T), the presence of tumor cells in the regional lymph nodes (N), and the presence or absence of distant metastases (M).

TP53 The most frequently altered gene in human cancer.

Transcript quantification Computational methods for assessing the abundance of transcripts. These methods usually involve deconvoluting the transcripts present for a given gene and then resolving their individual quantities.

Transcription factor A protein that binds to a DNA sequence to regulate its transcription to RNA.

Transcriptional profile A higher-order view of gene expression in a disease or tissues. A profile may be generated from any of a number of technologies (microarray, RNA-seq) and describes the active transcriptional state of a cell, tissue, or tumor.

Transcriptome The collection of RNA molecules that are expressed in a cell or tissue. This collection includes mRNA, rRNA, tRNA and various types of non-coding RNAs.

Transcriptomics biomarkers A biomarker based on the analysis of RNA expression profiles.

Transcriptomics The sequencing and analysis of gene expression. While this can apply to the use of microarray technology, most often it is used when referring to the use of sequencing-based methods. Used to study messenger RNA as well as non-coding RNA.

Transgenic A term that is used to describe an animal that has been genetically engineered to express other genetic material, referred to as a transgene, that is not normally found in that animal or in that chromosomal position if the gene is derived from an endogenous sequence.

Tri-Council Policy Statement (TCPS) A statement that governs the *Ethical Conduct for Research Involving Humans* that is operative in all Canadian institutions that receive funding from the Canadian Institutes of Health Research, the Natural Sciences and Engineering Research Council of Canada, or the Social Sciences and Humanities Research Council of Canada. The second version was published in 2010.

Triple-negative breast cancer A subtype of breast cancer that lacks the expression of all three receptors (ER, PR, and HER2) and predicts a poor prognosis as it is not responsive to anti-hormone therapies.

Tumor grade Histological assessment of the differentiation state of a tumor tissue, i.e. how much it resembles the normal tissue it arose from. Well-differentiated tumor: tumor closely resembles normal tissue, poorly-differentiated: tumor does not represent the original normal tissue.

Tumor niche The specialized microenvironment in which a tumor grows, that includes stromal fibroblasts, blood vessels and immune cells.

Tumor stage A numerical assessment that reflects tumor size and degree of spread (i.e. local invasion and metastasis). It is considered as the best predictor of prognosis.

Tumor suppressor genes Genes whose loss (or inactivation) can contribute to cell proliferation and survival and can predispose a cell to cancerous transformation.

Tumorigenesis A process by which tumor is formed.

Type I mutation Genes mutated in AML that involve proliferation of leukemia cells.

Type II mutation Genes mutated in AML that restrict the differentiation potential of leukemia cells.

Tyrosine kinase An enzyme that transfers a phosphate group from ATP to tyrosine residues, typically contributing to activation of a signal transduction pathway.

Tyrosine kinase inhibitors (TKIs) These agents bind to the adenosine triphosphate (ATP)-binding pocket, crucial for phosphorylation and activation of a tyrosine kinase enzyme, inhibiting tumor growth.

Ultraconserved genes (UCGs) Non-coding RNAs that are transcribed from a subset of genomic sequences which are strictly conserved among orthologous regions of the human, rat, and mouse genomes.

Unfolded protein response (UPR) A cellular conserved stress response which is activated by the accumulation of unfolded or misfolded proteins in the lumen of the endoplasmic reticulum. UPR initially restores normal cell function by halting protein translation and activating signaling pathways which lead to the synthesis of chaperones to help protein folding. However, if the cellular unbalance is too prolonged, the UPR induces apoptosis.

Uniparental disomy Two copies of a chromosome, or of part of a chromosome, from one parent and no copies from the other parent.

Univariate analysis The simplest form of quantitative (statistical) analysis carried out with the description of a single variable and its attributes of the applicable unit of analysis.

Ureteric bud A protrusion from the mesonephric duct during the development of the urinary and reproductive organs.

Variant detection Computational methods for detecting genetic or somatic variants.

Verification Mutations identified through computational approaches need to be confirmed in the corresponding samples using techniques such as targeted capture and re-sequencing. This process is referred to as verification.

Visceromegaly Enlargement of the internal organs in the abdomen, such as liver, spleen, stomach, kidneys, or pancreas.

Wedge resection A type of surgery to remove a tumor and surrounding normal lung tissue.

Whole exome sequencing Selective sequencing of the gene coding regions of the genome which offers a cheaper and faster alternative to whole-genome sequencing. This technique sequences cDNA that is derived from total RNA and hence lack intragenic sequences.

Whole genome sequencing A protocol involving sequencing of short fragments of the entire genome. Modern whole-genome sequencing involves massively parallel sequencing of millions of DNA fragments in a single reaction.

Whole transcriptome sequencing Massively parallel sequencing of cDNA (reverse transcribed mRNA) to gain information about the RNA content of a tumor.

Wild type A phenotype, genotype, or gene that predominates in a natural population of organisms in contrast to that of mutant forms.

Wolffian duct The embryonic duct of the mesonephros which, in the male, becomes the vas deferens and, in both sexes, gives rise to the ureter.

Xenotransplantation Cross-species transplantation of cells from one organism to another (e.g. transplantation of human leukemia cells in the mouse).

Index

Note: Page numbers followed by "*f*", "*t*" and "*b*" refer to figures, tables and boxes, respectively.

T

U

V

W

X

Z

Printed and bound by CPI Group (UK) Ltd, Croydon, CR0 4YY
08/05/2025
01865024-0003